FRONTIER ORBITALS AND REACTION PATHS

WORLD SCIENTIFIC SERIES IN 20TH CENTURY CHEMISTRY

Published:

Vol. 1: Molecular Structure and Statistical Thermodynamics
— Selected Papers of Kenneth S. Pitzer
by Kenneth S. Pitzer

Vol. 2: Modern Alchemy
— Selected Papers of Glenn T. Seaborg
by Glenn T. Seaborg

Vol. 3: Femtochemistry: Ultrafast Dynamics of the Chemical Bond
by Ahmed H. Zewail

Vol. 4: Solid State Chemistry
— Selected Papers of C. N. R. Rao
by S. K. Joshi and R. A. Mashelkar

Vol. 5: NMR in Structural Biology
— A Collection of Papers by Kurt Wüthrich
by Kurt Wüthrich

Vol. 6: Reason and Imagination: Reflections on Research in Organic Chemistry
— Selected Papers of Derek H. R. Barton
by Derek H. R. Barton

World Scientific Series in 20th Century Chemistry – Vol. 7

FRONTIER ORBITALS AND REACTION PATHS

Selected Papers of Kenichi Fukui

Editors

Kenichi Fukui
Institute for Fundamental Chemistry

Hiroshi Fujimoto
Division of Molecular Engineering,
Kyoto University

World Scientific
Singapore • New Jersey • London • Hong Kong

Published by

World Scientific Publishing Co. Pte. Ltd.
5 Toh Tuck Link, Singapore 596224
USA office: 27 Warren Street, Suite 401-402, Hackensack, NJ 07601
UK office: 57 Shelton Street, Covent Garden, London WC2H 9HE

Library of Congress Cataloging-in-Publication Data
Fukui, Kenichi, 1918–
Frontier orbitals and reaction paths : selected papers of Kenichi Fukui / editors, Kenichi Fukui, Hiroshi Fujimoto.
p. cm. -- (World Scientific series in 20th century chemistry; vol. 7)
ISBN-13 978-981-02-2241-3 -- ISBN-10 981-02-2241-6
1. Molecular orbitals. 2. Chemical reactions. 3. Chemistry, Physical organic. I. Fujimoto, Hiroshi. II. Title. III. Series.
QD461.F934 1997
541.2'8--dc21 97-36333
CIP

British Library Cataloguing-in-Publication Data
A catalogue record for this book is available from the British Library.

Kenichi Fukui

Kenichi Fukui was co-recipient (together with Roald Hoffmann) of the 1981 Nobel Prize in Chemistry for his "frontier orbital theory" concerning the course of chemical reactions. While he began his career as an experimentalist, in the course of his investigation of organic reactions he built up a subgroup of theoreticians. In 1952, Dr. Fukui discovered the correlation between the frontier orbital density and the chemical reactivity in aromatic hydrocarbons, leading to the development of the more general chemical reactivity theory by his group. Dr. Fukui was a professor at Kyoto University from 1945 until 1982, when he assumed the presidency of the Kyoto Institute of Technology for six years. In 1988, he became the founding director of the Institute for Fundamental Chemistry, also in Kyoto. He is a foreign associate of the National Academy of Sciences, a foreign member of the Royal Society, and has been honoured with many other memberships and awards.

PREFACE

It is my great pleasure to accept an invitation to the "World Scientific Series in 20th Century Science" to publish a collection of some of the papers I have written. This volume comprises of some 50 papers, chosen somewhat arbitrarily but representative of my works covering a period of about 50 years.

The papers included in this volume are divided into six groups, though they are related intimately to each other. Part I includes papers in which the frontier orbital theory has been formulated and developed. One will find how I have come to the idea and how the theory has been developed in early days. Part II covers the applications of the theory to the important problems of stereoselection in chemical reactions. The utility of the theory has been expanded remarkably in dealing with those problems. Papers in Part III discuss some basic aspects of chemical reactions. The factors that should govern the selectivities of reactions are examined. Part IV covers another subject, reaction coordinate. It is shown how the ideal reaction path is defined on the potential energy surface. Part V includes papers in which the frontier electron theory is extended to large molecules. Local characteristics of reactions are clarified and the frontier orbitals are connected with useful notions in chemistry, such as reaction sites, functional groups, substituent effects and so on. The concept of frontier orbital interactions is combined also with the theory of reaction coordinate. Part VI covers other topics I have studied.

At present, many theoreticians do sophisticated quantum chemical calculations to find the detailed structure and energy of reacting systems. No doubt, some of experiments will be replaced by MO calculations in near future, depending heavily on the developments of computers. But, I still believe that simple theories deserve well of their presence in chemistry. Simple theories can stimulate intuition of chemists. Simple theories can provide new insights into chemistry and help chemists to understand their experimental results and design new experiments. It is an urgent problem to save natural resources and lift poverty and illness from the world. I believe it is an important role of chemists to find solutions of this problem.

Finally, I would like to express my sincere gratitude to my old collaborators for their efforts in developing the theory with me and to Dr. Grace Tan of World Scientific Publishing Co. for providing me this pleasant opportunity.

Kyoto, 1995

Kenichi Fukui

The frontier orbital theory seems to be accepted by a number of chemists as an interpreting tool of their findings. It is a fan to me to find very often in scientific journals that the concept of HOMO-LUMO interaction is nicely applied to a wide variety of chemical phenomena. But, I feel the frontier orbital theory is a little more penetrating and foreseeing. I sincerely hope that undergraduate and graduate students and young research chemists, in both theoretical and in experimental fields, would find in this book some principles underlying chemistry and develop their own ideas.

I would like to express my heartfelt thanks to Dr. Grace Tan for giving me such an enjoyable opportunity.

Hiroshi Fujimoto

CONTENTS

Acknowledgements

The editor and publisher would like to thank the authors and the publishers of various journals and books for their assistance and permission to reproduce the selected reprints found in this volume:

1. Academic Press, Inc.
2. American Association for Cancer Research
3. American Chemical Society
4. American Institute of Physics
5. Chemical Society of Japan
6. Institute of Physics Publishing
7. International Union of Pure and Applied Chemistry
8. John Wiley & Sons, Inc.
9. Kluwer Academic Publishers
10. The Nobel Foundation

While every effort has been made to contact the publishers of reprinted papers prior to publication, we have not been successful in some cases. Where we could not contact the publishers, we have acknowledged the source of the material. Proper credit will be accorded to these publishers in future editions of this work after permission is granted.

Introductory Remarks to the Frontier Orbital Theory

Chemical reactions are essentially a series of formations and breakings of chemical bonds between atoms. Chemical reactivity theories have to explain appropriately these processes. The frontier orbital theory is a very practical way of thinking in this respect, currently applied by chemists in various experimental fields. This volume collects papers that have been published in the term over thirty years. There are some variations in the way of representing interactions between molecules. We feel that some introductory remarks to the theory will be of use for the readers who are not accustomed to use the MO theory. To begin with, we look at some aspects of molecular orbitals (MO's).

1. Ways of Presenting Molecular Orbitals

1-1. A Simple Example – π MO's of Ethylene

Let us look at here the π electronic part of an ethylene molecule as one of the simplest applications of the MO theory. We start from the 2p atomic orbitals (AO's) of two carbons C_1 and C_2 that have the direction of extension perpendicular to the molecular plane. We denote them by χ_1 and χ_2. The interaction between these two AO's gives rise to two MO's of the ethylene molecule, ϕ_1 and ϕ_2. They are given conventionally by the linear combination of the AO's (LCAO);

$$\phi_i = c_{1,i}\chi_1 + c_{2,i}\chi_2 \quad (i = 1, 2) \tag{1}$$

The coefficients $c_{1,i}$ and $c_{2,i}$ are obtained by solving the secular equations. In the case of ethylene molecule, C_1 and C_2 are equivalent and, therefore, $c_{1,i} = \pm c_{2,i}$ The π-type MO's of ethylene are;

$$\phi_1 = \frac{1}{\sqrt{2+2s_{1,2}}}(\chi_1 + \chi_2) \tag{2}$$

$$\phi_2 = \frac{1}{\sqrt{2-2s_{1,2}}}(\chi_1 - \chi_2) \tag{3}$$

in which $s_{1,2}$ is the overlap integral:

$$s_{1,2}=\int \chi_1 \chi_2 \, dv \qquad (4)$$

The absolute value of the overlap integral $s_{1,2}$ is about 0.24 for the AO's χ_1 and χ_2 centered on the carbons C_1 and C_2 that are separated from each other by a normal C–C double bond length, 0.134 nm.

The orbital energy ε_1 of ϕ_1 and ε_2 of ϕ_2 are given respectively by;

$$\varepsilon_1 = (h_{1,1} + h_{2,2} + 2h_{1,2}) / (2 + 2s_{1,2}) \qquad (5)$$

$$\varepsilon_2 = (h_{1,1} + h_{2,2} - 2h_{1,2}) / (2 - 2s_{1,2}) \qquad (6)$$

where $h_{1,1}$ and $h_{2,2}$ are the Coulomb integrals of the carbon 2p AO's, χ_1 and χ_2, and $h_{1,2}$ is the bond (or resonance) integral between the two AO's.

$$h_{1,1} = \int \chi_1 h \chi_1 \, dv \; (\equiv \alpha) \qquad (7)$$

$$h_{1,2} = \int \chi_1 h \chi_2 \, dv \qquad (8)$$

The Coulomb integral is related to the electronegativity of an atom, i.e., the power of an atom to attract electrons in a molecule, while the bond integral represents the strength of interaction between the two AO's. The MO's ϕ_1 and ϕ_2 are normalized so that:

$$\int \phi_i^{\,2} \, dv = 1 \qquad (i = 1, 2) \qquad (9)$$

Each 2p AO of a carbon atom has a lobe with the positive sign and a lobe with the negative sign. Then, we have two ways of choosing the AO's χ_1 and χ_2 to construct the π-type MO's of ethylene, as illustrated schematically in Figure 1. The overlap integral $s_{1,2}$ is positive in Figure 1(a), while it is negative in Figure 1(b). It is conventional to take the sign of the 2p AO's as shown in Figure 1(a) without explanation, but one can take the sign of χ_1 and of χ_2 as shown in Figure 1(b) without loss of generality.

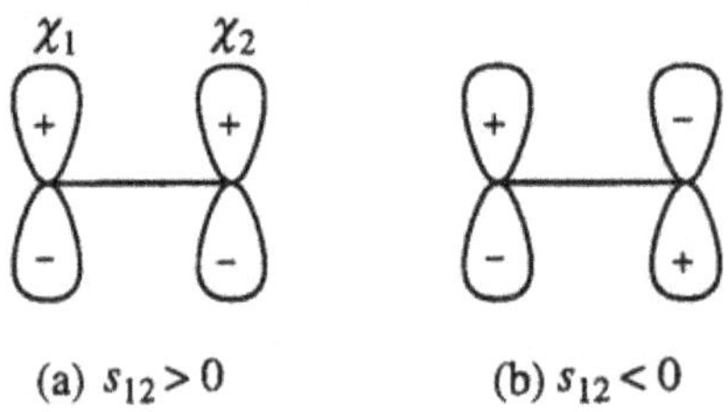

Figure 1. Ways of taking the sings of AO's in the ethylene π conjugation

The resonance integral $h_{1,2}$ is assumed to be roughly proportional and opposite in sign to $s_{1,2}$. In the extended Hückel MO scheme, for instance, $h_{1,2}$ is evaluated by the equation, $h_{1,2} = Ks_{1,2}(h_{1,1} + h_{2,2})/2$ with $K > 1$. The Coulomb integrals $h_{1,1}$ and $h_{2,2}$ are negative and, therefore, $h_{1,2}$ will be negative for positive $s_{1,2}$ and positive for negative $s_{1,2}$.

When one takes the signs of χ_1 and χ_2 as shown in Figure 1(a), one has $\varepsilon_1 < \varepsilon_2$. The AO's χ_1 and χ_2 have the coefficients $(2+2s_{1,2})^{-1/2}$ in the MO ϕ_1 of the lower orbital energy and, therefore, the MO is sketched as given in the bottom of Figure 2. Here the shaded and unshaded lobes represent respectively the positive and negative regions of the real MO. An important point to be noted is that the MO ϕ_1 of a lower orbital energy has no node between C_1 and C_2. This MO is called π. In contrast, the AO's χ_1 and χ_2 have the coefficients opposite in sign in the MO ϕ_2. The MO ϕ_2 of a higher orbital energy is sketched on the top of Figure 2. This MO has a node between C_1 and C_2 and is called π^*.

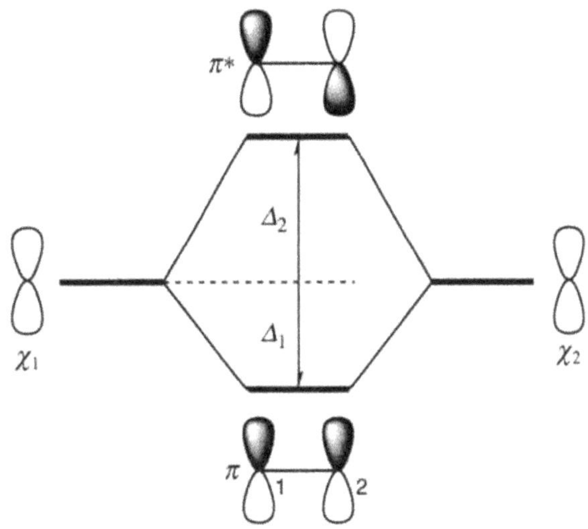

Figure 2. Construction of ethylene π and π^* MO's from the AO's

In contrast, if one takes the signs of χ_1 and χ_2 as illustrated in Figure 1(b), the bond integral $h_{1,2}$ is positive and, consequently, ϕ_2 having the coefficients with different signs on χ_1 and χ_2 is the MO that has the lower orbital energy. One sees, however, that the product of the coefficient $(2-2s_{1,2})^{-1/2}$ and the AO χ_1 and the product of the coefficient $-(2-2s_{1,2})^{-1/2}$ and the AO χ_2 result in the shaded and unshaded lobes that are characteristic to the π MO given above. On the other hand, the MO ϕ_1 of the higher orbital energy has the same coefficient on χ_1 and on χ_2. But, since the sign of the AO's has been taken as shown in Figure 1(b), the MO ϕ_1 gives the shaded and unshaded regions that are characteristic to the π^* MO in this case. In discussing stereoselection in reactions, there appear the cases in which two reaction paths are differentiated from each other by the sign of overlap between the AO's of the reaction sites. The method employed in Figure 2 provides a convenient way of specifying MO's in a chemical sense. It is independent of the way of taking signs of the constituent AO's, except for an arbitrary factor of ±1 that changes the sign of whole MO.

1-2. *In-Phase* and *Out-of-Phase* Overlap between AO's

The example given above demonstrates that neither the coefficient of each AO nor a product of the coefficients of two AO's does not specify an MO. It is the product of coefficients of two AO's and the overlap integral (or the bond integral) between those AO's that is of physical and chemical significance. For example, when we take $s_{1,2}$ positive in the ethylene π part, the product is positive in ϕ_1 and is negative in ϕ_2. But, if we take $s_{1,2}$ negative, the product is positive in ϕ_2 and negative in ϕ_1. In any case, the product is positive in the MO of a lower orbital energy and negative in the MO of a higher orbital energy. Then, it is convenient to specify MO's in the following manner. An MO ϕ_i has an "*in-phase overlap*" with respect to the AO's χ_r and χ_s when the product of the coefficients $c_{r,i}$ and $c_{s,i}$ and the overlap integral $s_{r,s}$ is positive. On the other hand, an MO ϕ_i has an "*out-of-phase overlap*" with respect to the AO's χ_r and χ_s when $c_{r,i}c_{s,i}s_{r,s}$ is negative. The *in-phase* and *out-of-phase* interferences of two AO functions are illustrated compactly by the shaded and unshaded lobes of AO's as presented above. Incidentally, the sum of the products $c_{r,i}c_{s,i}s_{r,s}$ over all the occupied MO's ϕ_i multiplied by 4 is called the overlap population between the AO's χ_r and χ_s.

An illustration of the π and π^* MO's of an ethylene molecule by the shaded and unshaded lobes is now compared with the calculated result of the MO's presented in Figure 3. The contours are plotted in a plane which is parallel to and 0.04 nm above the molecular plane. The positive and negative regions of the MO's are represented by solid lines and broken lines, respectively. The outcome of the *in-phase* and *out-of-phase* overlaps of the two AO's, χ_1 and χ_2, is recognized clearly. The π MO having a lower orbital energy does not change its sign along the C_1–C_2 bond, whereas the π^* MO having a higher orbital energy changes its sign on going from C_1 to C_2. The π MO of ethylene is *symmetric* for reflection through the plane which intersects the midpoint of the C–C bond and is perpendicular to the molecular plane. The π^* MO is *antisymmetric* for reflection through that plane.

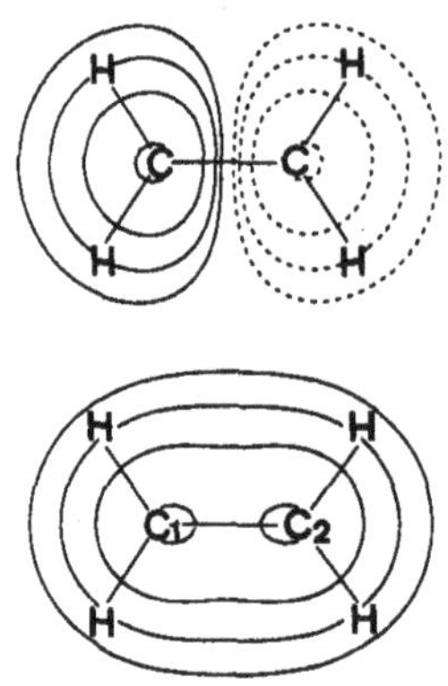

Figure 3. The π and π^* MO's of ethylene

1-3. Bonding and Antibonding MO's

The π MO is located lower in orbital energy than the AO's χ_1 and χ_2. In the ground-state of ethylene, a pair of electrons occupy this MO. Then, the π-electron density of an ethylene molecule is defined by:

$$\rho_\pi = 2\phi_1^2 = (\chi_1^2 + \chi_2^2 + 2\chi_1\chi_2)/(1 + s_{1,2}) \qquad (10)$$

The MO function ϕ_1 is normalized and, therefore, we obtain:

$$\int \rho_\pi \, dv = 2 \tag{11}$$

By subtracting from ρ_π the electron density of two p AO's, χ_1^2 and χ_2^2, of the carbon atoms in an isolated state, one obtains:

$$\Delta\rho_\pi = \{-(\chi_1^2 + \chi_2^2)s_{1,2} + 2\chi_1\chi_2\} / (1 + s_{1,2}) \tag{12}$$

The difference density $\Delta\rho_\pi$ shows how the distribution of π electrons changes upon the interaction between χ_1 and χ_2. It is sketched in Figure 4 for the points 0.04 nm above the line connecting C_1 and C_2. The electron density is reduced on the carbon centers, but shows an increase in the region between the two carbons. The electrons housed in χ_1 and χ_2 in separated carbon atoms are now *delocalized* over the C–C bond region by means of the overlap between the two AO's. It is natural to assume that the two carbon nuclei are attracted by the π electron density accumulated in the interatomic region to strengthen the C–C bond. Hence, the π MO is called a *"bonding MO"*.

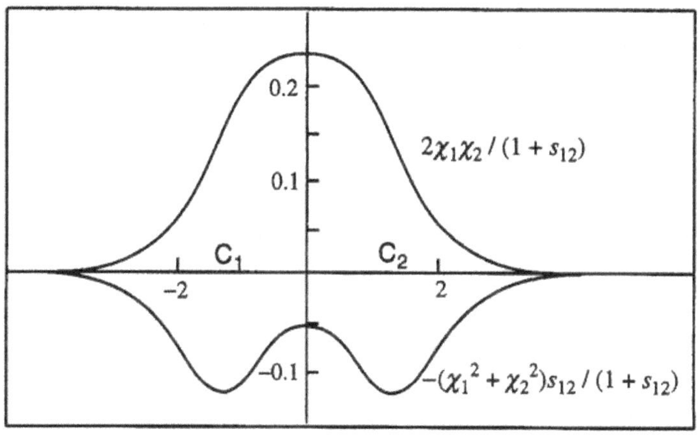

Figure 4. Difference density of the ethylene π bond (in au)

The π* MO is located higher in orbital energy than the AO's χ_1 and χ_2. One will find easily that the electron density is shifted out of the C–C bond region in this MO. Then, if a pair of electrons occupy this MO, two carbons are repelled away from each other and the C–C bond will be weakened. Hence, the π* MO is an *"antibonding MO"*.

The terms *"bonding MO"* and *"antibonding MO"* seem to have another meaning in their usage. The ethylene molecule is stabilized by $2\Delta_1$ when a pair of electrons occupy the π MO, i.e., $\varepsilon_1 = \alpha + \Delta_1$ ($\Delta_1 < 0$), whereas the molecule is destabilized by $2\Delta_2$ when a pair of electrons occupy the π* MO, i.e., $\varepsilon_2 = \alpha + \Delta_2$ ($\Delta_2 > 0$). The MO's which stabilize a molecule by housing electrons are called usually the *bonding* MO's, while the MO's which destabilize a molecule by accepting electrons are called the *antibonding* MO's. In the molecules larger in size, both *in-phase* and *out-of-phase* overlaps appear in an MO. As one sees in Figure 5 given later, an MO lower in orbital energy has a greater number of bonds with *in-phase* overlaps between two AO's neighbored to each other, relative to an MO of a higher orbital energy. The terms *"bonding MO"* and *"antibonding MO"* have been used conventionally to indicate a global feature of MO's. However, in studying mechanisms of chemical reactions, we have to focus our attention on changes in some specific bonds in the reagent and reactant parts. It becomes more important, therefore, to examine whether an MO is *bonding* or *antibonding* with respect to a certain pair of atoms in a molecule than to see whether the acceptance of electrons in that MO will stabilize or destabilize the molecule. The terms *bonding MO* and *antibonding MO* are regarded as indicating the local characteristics of MO's with respect to a bond.

In summary, when the two AO functions χ_r and χ_s interfere *in-phase* in an MO, the MO is *bonding* with respect to the two atoms to which the AO's χ_r and χ_s belong. Transfer of a part of electron population from this MO to an unoccupied MO of another molecule or promotion of an electron within the molecule from this MO to an unoccupied MO leads to a weakening or breaking of that bond. When the two AO's χ_r and χ_s overlap *out-of-phase* in an MO, the MO is *antibonding* with respect to those atoms. Electronic charge transferred to this MO from an occupied MO of another molecule or from an occupied MO of the same molecule weakens that bond. The breaking of chemical bonds in reactions is initiated by an acceptance of electron charge into antibonding MO's and/or a donation of electron charge from bonding MO's.

1-4. Another Example – π MO's of Butadiene

By taking the positive lobes of the 2p AO's χ_1, χ_2, χ_3, and χ_4 of butadiene on the same side of the molecular plane, the four π-type MO's, π_1, π_2, π_3, and π_4, calculated with the minimal basis set at the STO-3G level are given by:

$$\pi_1 = 0.3714\chi_1 + 0.4750\chi_2 + 0.4750\chi_3 + 0.3714\chi_4$$

$$\pi_2 = -0.5208\chi_1 - 0.4040\chi_2 + 0.4040\chi_3 + 0.5208\chi_4$$

$$\pi_3 = 0.6283\chi_1 - 0.4753\chi_2 - 0.4753\chi_3 + 0.6283\chi_4$$

$$\pi_4 = -0.5125\chi_1 + 0.6985\chi_2 - 0.6985\chi_3 + 0.5125\chi_4 \tag{13}$$

The occupied MO π_1 of the lowest orbital energy has larger coefficients on the AO's χ_2 and χ_3 of the inner carbons than on the AO's χ_1 and χ_4 of the terminal carbons. The MO π_1 is *bonding* in all of the C_1–C_2, C_2–C_3 and C_3–C_4 bond regions. The MO π_2 has a larger coefficient on the terminal carbons than on the central carbons. This MO is *bonding* with respect to the C_1–C_2 and C_3–C_4 bonds, but is *antibonding* with respect to the C_2–C_3 bond. It shows a larger amplitude on C_1 and C_4, as illustrated in Figure 5. In the ground-state of a butadiene molecule, two pairs of electrons occupy the π_1 and π_2 MO's. Thus, the C_1–C_2 and C_3–C_4 bonds have a strong double-bond character with a shorter bond length. In contrast, the C_2–C_3 bond is of a weak double-bond nature with a relatively longer bond length, the *in-phase* overlap between χ_2 and χ_3 in the π_1 MO is counterbalanced by the *out-of-phase* overlap between χ_2 and χ_3 in the π_2 MO. The molecule has two unoccupied π-type MO's, π_3 and π_4, in this simple description. The π_3 MO has two nodes one in the C_1–C_2

bond and one in the C_3–C_4 bond, while the π_4 MO has three nodes. The former is bonding with respect to the C_2–C_3 bond, but the latter is antibonding with respect to all the three C–C bonds.

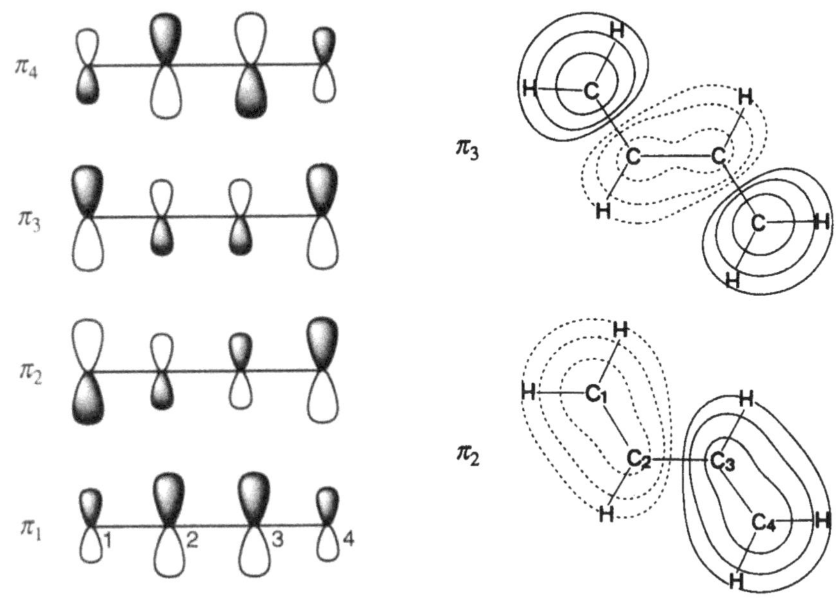

Figure 5. The π-type MO's of butadiene

1-5. A Molecule and Fragments

A molecule contains several structural units, e.g., functional groups. In discussing unimolecular reactions, it is sometimes convenient to think of a molecule as consisting of fragments. For instance, the π-electronic part of butadiene may be constructed from two ethylene-like π conjugations as follows. Each fragment has a π MO and a π^* MO. First, the π MO's of the two fragments overlap at C_2 and C_3 to split into two orbitals ϕ_1' and ϕ_2', as illustrated in Figure 6. The situation is much the same as that in constructing the π and π^* MO's of an ethylene molecule from two 2p AO's. The lowest orbital, ϕ_1' is an *in-phase* combination of two π MO's. It has no node along the C_1–C_2–C_3–C_4 linkage. The *out-of-phase* combination of the π MO's of two ethylene-like fragments at C_2 and C_3 gives rise to the orbital ϕ_2'. In a similar manner, the *in-phase* and *out-of-phase* combinations of the π^* MO's of the two ethylene-like fragments yield ϕ_3' and ϕ_4'. These orbitals have the same nodal properties as the π-type MO's of butadiene given above. Note here that ϕ_1' and ϕ_3' are not orthogonal to each other and ϕ_2' and ϕ_4' are not orthogonal to each other.

The diene conjugation should be destabilized or not stabilized as will be seen later, if the orbitals ϕ_1' and ϕ_2' are occupied by pairs of electrons. There must be additional orbital interactions that have not been involved in the splitting of the MO's presented in Figure 6. One sees that the π_1 and π_2 MO's of butadiene look considerably different from the orbitals ϕ'_1 and ϕ'_2. This difference comes from the fact that the interactions between the π MO of an ethylene-like fragment and the π^* MO of the other fragment are not taken

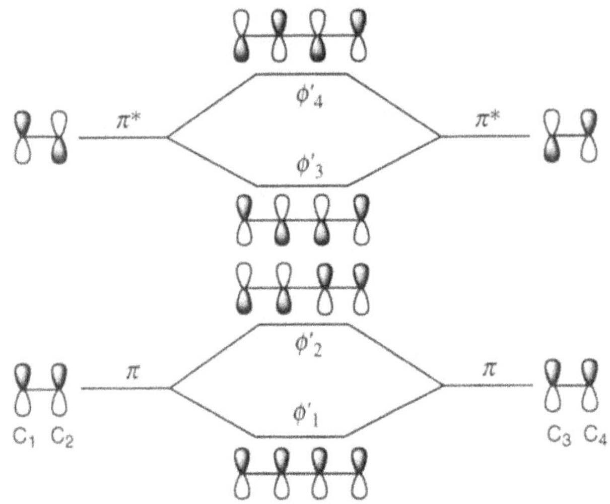

Figure 6. Splitting of the ethylene-type π and π^* MO's first-order in energy

into account in Figure 6. The π^* MO's of the two fragments participate in ϕ_1' and ϕ_2', though smaller in extent relative to the π MO's, to give the π_1 and π_2 MO's of butadiene. The π and π^* MO's intermix in each fragment in such a manner as to strengthen the C_2–C_3 *in-phase* overlap in ϕ_1' and weaken the C_2–C_3 *out-of-phase* overlap in ϕ_2', as illustrated schematically in Figure 7. Details of *orbital mixings* will be found in the papers included in this volume.

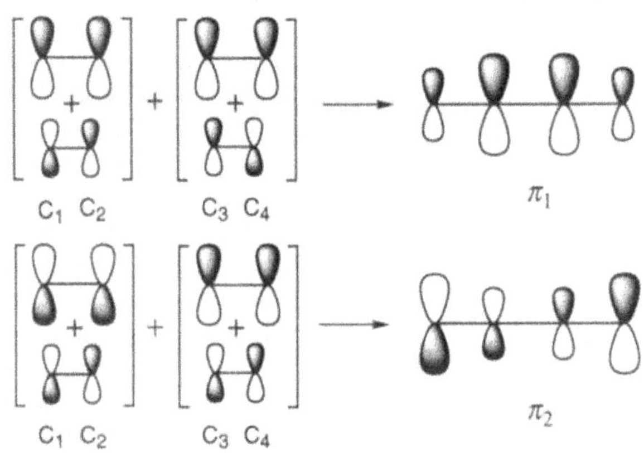

Figure 7. An illustration of mixings between occupied and unoccupied fragment MO's

This example shows the basic feature of MO's do not depend much on the method of calculations. The interaction between the occupied MO's of one part and the unoccupied MO's of the other part is the principal driving force of chemical reactions. The orbital interaction stabilizes the interacting system and brings about the formation of new bonds and breaking of old bonds along the reaction path.

2. Frontier Orbital Interactions

As theoretical calculations become more familiar, it will be more important to have simple methods for analyzing the results. It is helpful to know some of the factors that will control the stability of bonds and the selectivity of reactions. We have already a very useful way of clarifying the nature of chemical interactions and chemical bonds that is known currently as *"orbital interactions"*. The MO's of the reactant and reagent are used to represent the interactions on the basis of perturbation theory.

2-1. Orbital Interactions

The interaction between two molecular species is represented by the combination of the MO's of two fragments. We employ here the simplest case, an interaction between two systems A and B, A having only an occupied MO ϕ_i and B having only an unoccupied MO φ_l, as schematically illustrated in Figure 8. We assume here that the occupied MO ϕ_i of A is located lower in orbital energy than the unoccupied MO φ_l of B.

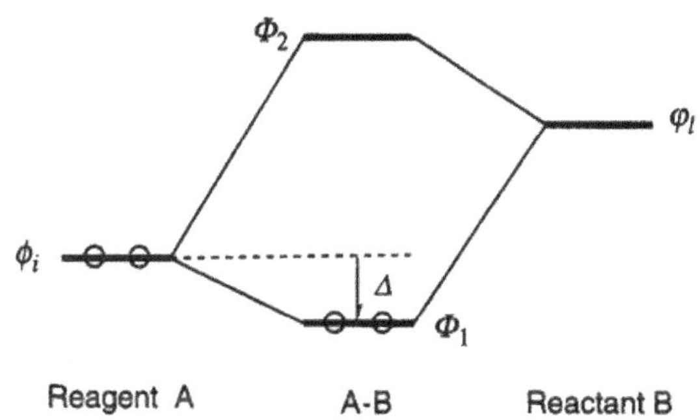

Figure 8. Illustration of an interaction between an occupied MO of A and an unoccupied MO of B

When these two MO's begin to overlap, they split by the interaction into the two MO's of the composite system, A-B. One of these MO's denoted here by Φ_1 should have a lower orbital energy than ϕ_i and the other MO denoted by Φ_2 should be higher in orbital energy than φ_l. The electron pair will occupy the MO Φ_1 of A-B to stabilize electronically the composite system. The MO's Φ_1 and Φ_2 of the composite system are given by linear combinations of the fragment MO's ϕ_i and φ_l. They are calculated by solving the secular equations;

$$c_1(h_{i,i} - \varepsilon) + c_2(h_{i,l} - s_{i,l}\varepsilon) = 0 \tag{14}$$

$$c_1(h_{i,l} - s_{i,l}\varepsilon) + c_2(h_{l,l} - \varepsilon) = 0 \tag{15}$$

in which;

$$h_{i,i} = \int \phi_i h \phi_i \mathrm{d}v \tag{16}$$

$$h_{l,l} = \int \varphi_l h \varphi_l \mathrm{d}v \tag{17}$$

$$h_{i,l} = \int \phi_i h(1)\varphi_l \mathrm{d}v \tag{18}$$

$$s_{i,l} = \int \phi_i \varphi_l(1)\mathrm{d}v \tag{19}$$

where h represents the averaged *one-electron* Hamiltonian of the interacting system. In order that nontrivial solutions exist, the following determinant must vanish.

$$\begin{vmatrix} h_{i,i} - \varepsilon & h_{i,l} - s_{i,l}\varepsilon \\ h_{i,l} - s_{i,l}\varepsilon & h_{l,l} - \varepsilon \end{vmatrix} = 0 \tag{20}$$

Equation 20 can be solved easily, since the integrals $h_{i,i}$, $h_{l,l}$, $h_{i,l}$ and $s_{i,l}$ are given for the interacting system, being evaluated by applying some approximations, e.g., the extended Hückel MO scheme.

The strength and mode of mixings of ϕ_i with φ_l in the MO's Φ_1 and Φ_2 are elucidated numerically by substituting the solution of eq. 20 into eqs. 14 and 15 and finding a set of coefficients c_1 and c_2 under the condition that Φ_1 and Φ_2 are normalized. However, in order to get insight into basic aspects of interactions between molecules, we employ here another way of obtaining approximately the solutions of eqs. 14 and 15. It will tell us how the MO's of two fragments are recombined in the interaction.

If the interaction is not yet strong, one of the solutions ε_1 will be close to $h_{i,i}$. Then, we may write $\varepsilon_1 = h_{i,i} + \Delta$, $c_1 = 1$ and $c_2 = \delta$. The assumption that the interaction is not strong signifies that $|h_{i,l} - s_{i,l}h_{i,i}|$ is small in comparison with $|h_{l,l} - h_{i,i}|$. Then, the secular equations are rewritten as:

$$-\Delta + (h_{i,l} - s_{i,l}h_{i,i} - s_{i,l}\Delta)\delta = 0 \tag{21}$$

$$h_{i,l} - s_{i,l}h_{i,i} - s_{i,l}\Delta + (h_{l,l} - h_{i,i} - \Delta)\delta = 0 \tag{22}$$

By neglecting the terms that are second order or higher with respect to $h_{i,l}$, $s_{i,l}$, Δ and δ, one gets;

$$\delta = \frac{h_{i,l} - s_{i,l}h_{i,i}}{h_{i,i} - h_{l,l}} \tag{23}$$

and:

$$\Delta = (h_{i,l} - s_{i,l}h_{i,i}) \times \delta = \frac{(h_{i,l} - s_{i,l}h_{i,i})^2}{h_{i,i} - h_{l,l}} \tag{24}$$

The approximate orbital energy and the MO Φ_1 of A-B are given then by:

$$\varepsilon_1 \approx h_{i,i} + \frac{(h_{i,l} - s_{i,l}h_{i,i})^2}{h_{i,i} - h_{l,l}} \tag{25}$$

$$\Phi_1 \approx \phi_i + \frac{(h_{i,l} - s_{i,l}h_{i,i})}{h_{i,i} - h_{l,l}}\varphi_l \tag{26}$$

The other solution and the associated MO function Φ_2 are derived similarly:

$$\varepsilon_2 \approx h_{l,l} + \frac{(h_{i,l} - s_{i,l} h_{l,l})^2}{h_{l,l} - h_{i,i}} \tag{27}$$

$$\Phi_2 \approx \varphi_l - \frac{(h_{i,l} - s_{i,l} h_{l,l})}{h_{i,i} - h_{l,l}} \phi_i \tag{28}$$

Note here that the MO's Φ_1 and Φ_2 are not yet normalized in eqs. 26 and 28. The integrals $h_{i,i}$ and $h_{l,l}$ may be replaced by the orbital energy ε_i of the MO ϕ_i and ε_l of the MO φ_l, respectively. This is a treatment based on conventional perturbation theory. In some of the papers in this volume, a method of "*perturbed secular determinant*" is presented. One will find there a beauty of the simple Hückel MO theory from a mathematical point of view.

2-2. Electron Delocalization

A pair of electrons which are housed originally in the occupied MO ϕ_i of A now occupy the MO Φ_1 having the lower orbital energy in the composite system A-B. This provides the composite system with a stabilization of 2Δ.

$$\Delta E = 2\Delta \approx \frac{2(h_{i,l} - s_{i,l}\varepsilon_i)^2}{\varepsilon_i - \varepsilon_l} \tag{29}$$

The MO Φ_1 is shown above to be given by a combination of the MO ϕ_i of A and the MO φ_l of B. There arises a possibility of finding electrons on the atoms in B. That is, the orbital interaction between an occupied MO of one molecule with an unoccupied MO of the other molecule brings about "*delocalization of electrons*".

Reagent and reactant molecules have usually many MO's. The MO's of A may be denoted by ϕ_1, ϕ_2, .., ϕ_m, ϕ_{m+1}, .., ϕ_M from the lowest in orbital energy and the MO's of B by φ_1, φ_2, .., φ_n, φ_{n+1}, .., φ_N. When A and B are separated at the infinite distance, the $2m$ electrons of A occupy the MO's $\phi_1 \sim \phi_m$ and the $2n$ electrons of B occupy the MO's $\varphi_1 \sim \varphi_n$, as represented schematically by the electron configuration illustrated in Figure 9. An MO which has a pair of electrons is called an "*occupied MO*". In MO calculations, one obtains also empty MO's which are located above the occupied levels. An MO without electrons is called an "*unoccupied MO*". In the earlier papers collected in this volume, the term "*vacant MO*" is used in place of unoccupied MO. In a radical or a in molecule which is in an excited electronic state, there appear an MO or MO's which have unpaired electrons. An MO which has an unpaired electron is called a "*singly occupied MO*". The occupation number of each MO is either 0 or 1 or 2 in the simplest description of the electronic structure of molecules. In a description of electronic structures of molecules including the electron correlation effects, the occupation number is not necessarily integer, but close to 0 or 1 or 2. In this book, the simplest method of describing electron configurations and electron orbitals is adopted.

If the reagent molecule A has the occupied MO's ϕ_i (i = 1, 2, ..., m) and the unoccupied MO's ϕ_j (j = m+1, m+2, ..., M) and the reactant molecule B has the occupied MO's φ_k (k = 1, 2, ..., n) and the unoccupied

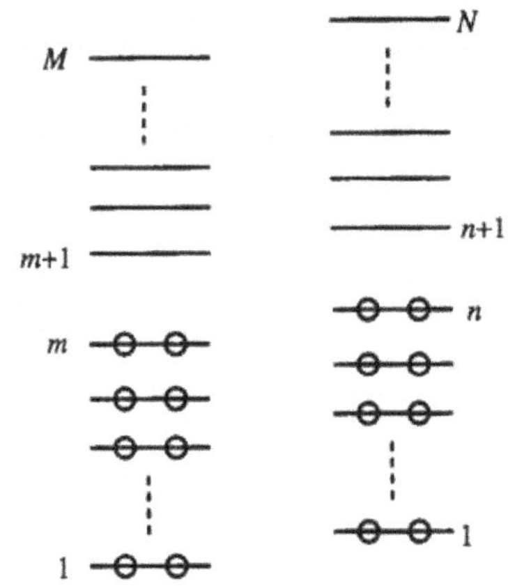

Figure 9. A sketch of electron configuration

MO's ϕ_l ($l = n+1, n+2, ..., N$), the delocalization stabilization may be given by taking the sum over all the combinations of the occupied MO's of one fragment and the unoccupied MO's of the other fragment:

$$\Delta E_D = 2\sum_{i=1}^{m}\sum_{l=n+1}^{N}\frac{(h_{i,l} - s_{i,l}\varepsilon_i)^2}{\varepsilon_i - \varepsilon_l} + 2\sum_{k=1}^{n}\sum_{j=m+1}^{M}\frac{(h_{k,j} - s_{k,j}\varepsilon_k)^2}{\varepsilon_k - \varepsilon_j} \tag{30}$$

The stabilization originating from electron delocalization will be the central subject in the papers collected in this volume.

2-3. Overlap Repulsion

We return here to the simplest system treated above. When both the reagent A and the reactant B have the occupied MO's, ϕ_i and φ_k, two pairs of electrons will occupy both of the two MO's, Φ_1 and Φ_2, of the composite system as illustrated in Figure 10.

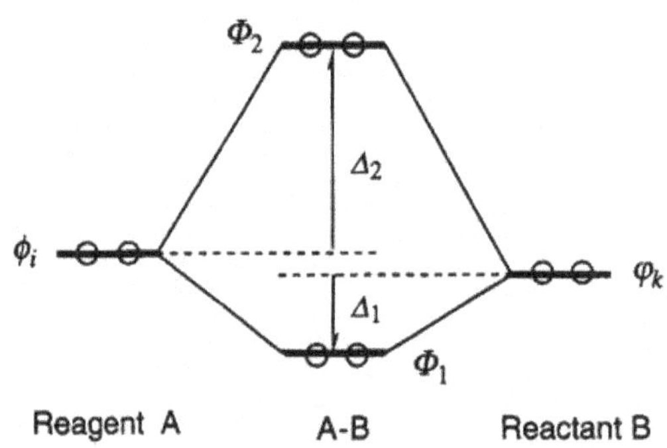

Figure 10. Interactions between the occupied MO's of A and B

The change in energy is given in this case by:

$$\Delta E = 2(\varepsilon_1 + \varepsilon_2) - 2(\varepsilon_i + \varepsilon_k) \tag{31}$$

By introducing eqs. 25 and 27 into eq. 31 and by replacing the suffix l by k to be consistent with the notations of the occupied MO's given above, we obtain;

$$\begin{aligned}\Delta E &\approx 2\frac{(h_{i,k} - s_{i,k}h_{i,i})^2}{h_{i,i} - h_{k,k}} + 2\frac{(h_{i,k} - s_{i,k}h_{k,k})^2}{h_{k,k} - h_{i,i}} \\ &= -2\{2h_{i,k} - s_{i,k}(h_{i,i} + h_{k,k})\}s_{i,k}\end{aligned} \tag{32}$$

Though it is not easy to prove in a strictly mathematical sense, ΔE given above is positive in most cases. That is, interaction between the occupied MO's of two fragment species destabilizes the interacting system. However, if we employ the zero-differential overlap approximation that makes all the overlap integral $s_{i,k}$ to be zero, this repulsive interaction disappears. Hence, this repulsive interaction between the occupied MO's is called *"overlap repulsion"*. A more general formula of overlap repulsion is derived by taking the sum over all the combinations of the occupied MO's of two fragment species:

$$\Delta E_K = -2\sum_{i=1}^{m}\sum_{k=1}^{n}\{2h_{i,k} - s_{i,k}(\varepsilon_i + \varepsilon_k)\}\, s_{i,k}/(1 - s_{i,k}{}^2) \tag{33}$$

Unlike the electrostatic interactions between the electronic charges, overlap repulsion appears even in the interactions between neutral and nonpolar molecules and brings about so-called steric repulsion. To facilitate an interaction between two molecular species, electron delocalization should occur effectively to make ΔE_D overcome ΔE_K.

2-4. Frontier Orbitals

The contribution of an occupied MO ϕ_i of A and an unoccupied MO φ_l of B to electron delocalization depends on their orbital energies. Equation 29 tells us that the extent of electron delocalization is inversely proportional to the energy gap between these MO's, (ε_i - ε_l). Among the combinations of the occupied MO's of A and the unoccupied MO's of B, it is the interaction between the *"highest occupied (HO) MO"* of A and the *"lowest unoccupied (LU) MO"* of B that has the smallest energy gap. Similarly, the combination of the HOMO of B and the LUMO of A has the smallest energy gap among all the combinations of the occupied MO's of B and the unoccupied MO's of A.

It is not easy to prove strictly, but the numerator of eq. 29, ($h_{i,l}$ - $s_{i,l}\varepsilon_i$), is roughly proportional and is opposite in sign to the overlap integral $s_{i,l}$, for major orbital interactions. Actually, in the case of electron delocalization from a low-lying occupied MO of a fragment to a high-lying unoccupied MO of the other fragment, the interaction integral ($h_{i,l}$ - $s_{i,l}\varepsilon_i$) happens to give a value that has the same sign as the overlap integral. The integral is small in absolute value in such a case and, therefore, the contribution of such an orbital pair to electron delocalization is not significant. For an important combination of the MO's like the HOMO of one part and the LUMO of the other part, the numerator will be negative if one takes the signs of MO's ϕ_i and φ_l as to give a positive overlap integral. The sign of the interaction integral is important to see the way of mixings of the MO's ϕ_i and φ_l in the MO's of the composite system and, therefore, to examine the bonds to be formed between A and B.

The important conclusion derived from this simple treatment of orbital interactions is that *chemical reactions should take place preferentially at the position and in the direction in the space where the overlap between the HOMO of A and the LUMO of B and the overlap between the HOMO of B and the LUMO of A are most effective. In the interaction of an electron donor with an electron acceptor, the interaction between the HOMO of the former and the LUMO of the latter governs the interaction.* The HOMO and LUMO are called the "*frontier orbitals*". In the systems which have singly occupied (SO) MO's, radicals or molecules in excited electronic states, *the SOMO's play the role of the HOMO and/or the LUMO.*

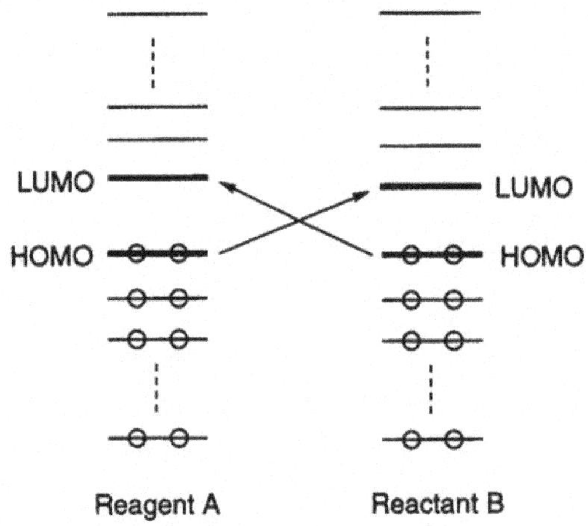

Figure 11. Frontier orbital interactions

To symbolize *electron delocalization* from the HOMO of A to the LUMO of B or from the HOMO of B to the LUMO of A, an arrow directing from the HOMO to the LUMO is used as shown in Figure 11. This has essentially the same chemical meaning as the orbital interaction diagram or orbital splitting diagram, presented in Figure 8.

2-5. Single- and Multi-Centered Reactions

The numerator of eq. 29 may be expanded in terms of the AO's of the fragment species:

$$h_{i,l} - s_{i,l}\varepsilon_i = \sum_r \sum_t c_{r,i} c_{t,l}\, \gamma^i_{r,t} \tag{34}$$

in which;

$$\gamma^i_{r,t} = h_{r,t} - s_{r,t}\varepsilon_i \tag{35}$$

and:

$$h_{r,t} = \int \chi_r h \chi_t \mathrm{d}v \tag{36}$$

$$s_{r,t} = \int \chi_r \chi_t \mathrm{d}v \tag{37}$$

The integral $\gamma_{r,t}$ represents the strength and mode of interaction between the AO χ_r of A and the AO χ_t of B. Note here that it depends on the energy of the MO ϕ_i which donates electron population in the interaction. The integral $\gamma_{r,t}$ will be greater in absolute value for high-lying occupied MO's, e.g., the HOMO.

Electrophilic substitution reaction of aromatic compounds is a typical example of single-centered reaction. The orbital of electrophile is regarded as common and, therefore, the reaction will take place at the site with the largest AO coefficient in the HOMO of aromatic compounds. For example, anthracene gives the largest amplitude on the 9 and 10 positions in the HOMO. The frontier orbital theory suggests that aromatic substitution should occur preferentially at these positions in agreement with experimental results.

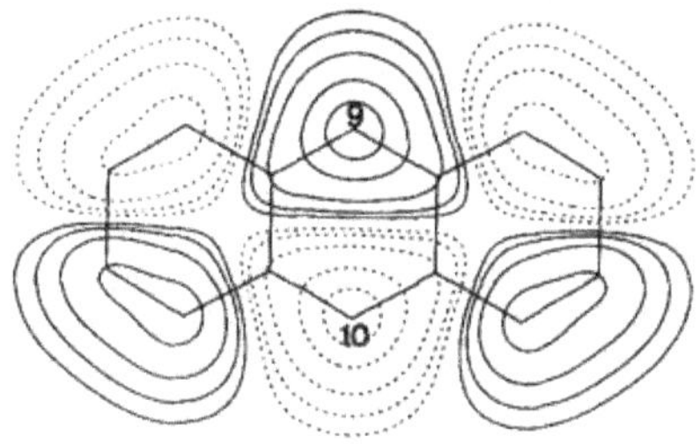

In a two-centered interaction in which the AO χ_r of A overlaps with the AO χ_t of B and the AO χ_s of A overlaps with the AO χ_u of B, eq. 34 is written as:

$$h_{i,l} - s_{i,l}\varepsilon_i = c_{r,i}c_{t,l}\gamma^i_{r,t} + c_{s,i}c_{u,l}\gamma^i_{s,u} \tag{38}$$

The Diels-Alder cycloaddition between a diene and a dienophile is one of the typical two-centered reactions. As a model, we examine the interaction between the frontier MO's of butadiene and ethylene. If we take the positive lobes of the 2p AO's of C_1, C_2, C_3, and C_4 on the same side of the molecular plane of a butadiene molecule in *cis* conformation and if we take the positive lobes of the 2p AO's of $C_{1'}$ and $C_{2'}$ of an ethylene molecule on the same side of the molecular plane, the interaction integrals $\gamma_{1,1'}$ and $\gamma_{4,2'}$ at the two reaction sites are the same in sign, both positive or both negative. By inspecting the coefficients of the AO's χ_1 and χ_4 in the HOMO and in the LUMO of butadiene given in eq. 13 or in Figure 7 and the coefficients of the AO's $\chi_{1'}$ and $\chi_{2'}$ in the MO's of ethylene given in eqs. 2 and 3, one will find that the products of the coefficients and the integral γ have the same sign at the two reaction sites. That is, if the addition reaction takes place by a synchronous mechanism, we have;

$$c_{1,\mathrm{HOMO}} \times c_{1',\mathrm{LUMO}} \times \gamma_{1,1'} = c_{4,\mathrm{HOMO}} \times c_{2',\mathrm{LUMO}} \times \gamma_{4,2'}$$

and:

$$c_{1,\mathrm{LUMO}} \times c_{1',\mathrm{HOMO}} \times \gamma'_{1,1'} = c_{4,\mathrm{LUMO}} \times c_{2',\mathrm{HOMO}} \times \gamma'_{4,2'}$$

The interaction integral defined by eq. 34 will be larger in magnitude for the concerted process than for a two-step process in which a butadiene molecule and an ethylene molecule interacts first at a single site, between C_1 and $C_{1'}$ or between C_4 and $C_{2'}$. One will find that the concerted addition between a diene and a dienophile is favorable from an orbital interaction point of view.

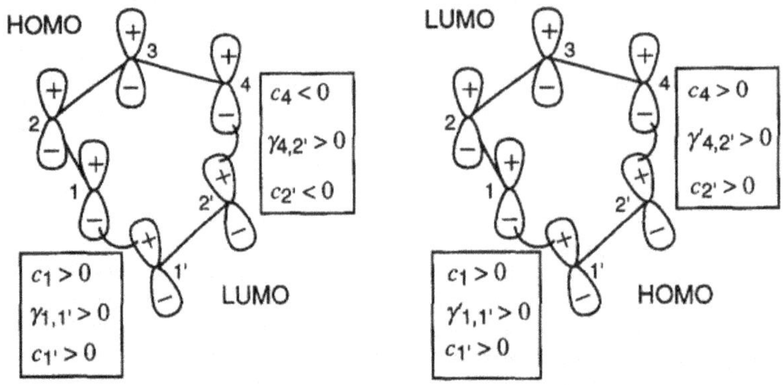

Figure 12. Frontier orbital interactions in Diels-Alder type cycloaddition
(The signs + and − indicate simply the way of taking the positive and negative lobes of 2p AO's.)

If we characterize the frontier MO's by the shaded and unshaded lobes of the constituent p AO's as stated above, the interactions are shown as follows:

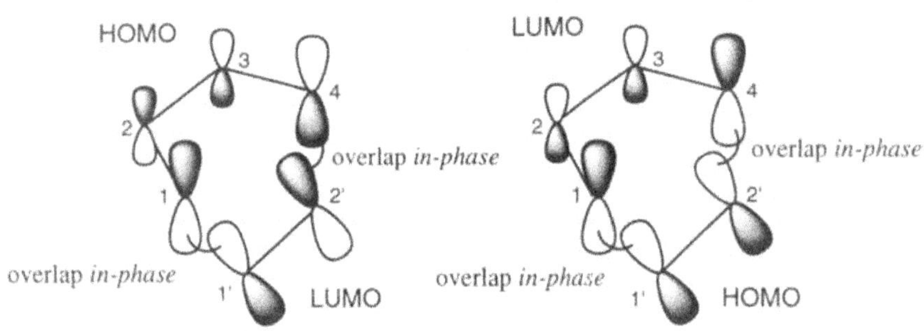

Figure 13. Another way of showing orbital interactions in Diels-Alder type cycloaddition

It is seen that the HOMO of butadiene and the LUMO of butadiene overlap *in-phase* at the two reaction centers (or *out-of-phase* at the two reaction centers). The HOMO of ethylene and the LUMO of butadiene interfere also *in-phase* at the two centers (or *out-of-phase* at two centers). Incidentally, the *out-of-phase* overlaps at the two reaction centers are converted to the *in-phase* overlap at the two centers by multiplying one of the MO's by a factor −1. Electron delocalization from the HOMO of butadiene to the LUMO of ethylene weakens the π conjugation of the C_1-C_2 and C_3-C_4 bonds in butadiene and the $C_{1'}-C_{2'}$ π bond in ethylene and strengthens the π conjugation of the C_2-C_3 bond in butadiene. Electron delocalization from the HOMO of ethylene to the LUMO of butadiene gives the same result. This bond interchange promotes the formation of cyclohexene as the addition product.

In contrast, one finds immediately in Figure 14 that the thermal dimerization of alkenes does not favor a concerted process. Interaction integral vanishes in this case.

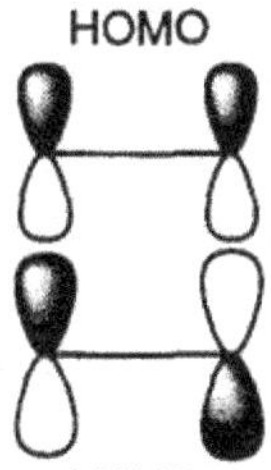

Figure 14. Frontier orbital interaction in thermal dimerization of ethylene

2-6. Numerical Analyses of Interactions

A way of studying chemical reaction mechanisms theoretically is to apply the MO theory directly to a reacting system by regarding the composite system of the reagent and reactant, as if it was a molecule. The composite system A-B is often called a supermolecule, while A and B are called fragments. At an equilibrium structure of A-B, the first derivatives of potential energy vanish for all the coordinates. In addition, the second derivatives are all positive, since the stable structure is located at a minimum of the potential energy surface. At a saddle point, the second derivative of potential energy has one and only one negative value for an internal coordinate. The normal coordinate corresponding to this negative value or the imaginary vibration frequency indicates the reaction coordinate. The *"intrinsic reaction coordinate"* describes how the geometry of a reacting system changes along the ideal reaction path in which reaction proceeds slowly by releasing the kinetic energy at every moment.

The changes in the shape of the frontier orbitals along the intrinsic reaction coordinate are sometimes very informative. For example, oxirane has two low-lying unoccupied MO's (LUMO's), one being symmetric and the other antisymmetric. These MO's mix with each other in the interaction with a nucleophile attacking from the back-side of a C–O bond to give the LUMO that has a large amplitude on that C–O bond. Electron delocalization from the nucleophile to this LUMO causes a deformation of the oxirane ring. This deformation results in a further localization of the LUMO, leading finally to the breaking of that bond.

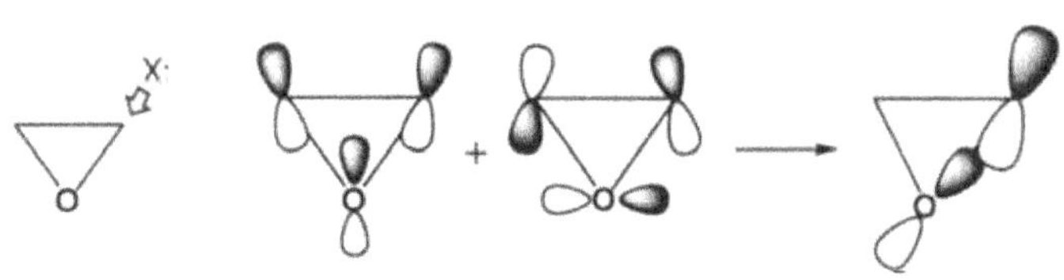

The MO's of the composite system are obtained by solving the secular equations that are similar to eqs. 14 and 15 or approximately by applying perturbation theory. However, if we could have the MO's of a composite system from the beginning, they are expanded numerically in terms of the MO's of the two fragment species, provided that the AO's are fixed. Then, we may be able to discuss chemical interactions on the basis of the electronic structures of the reagent and reactant in an isolated state.

The MO's of the two fragment species, A and B, are given conventionally by the linear combinations of the AO's of each species:

$$\phi_i = \sum_{r=1}^{M} c_{r,i}\chi_r \quad \text{(for A)} \tag{39}$$

$$\varphi_k = \sum_{t=1}^{N} c_{t,k}\chi_t \quad \text{(for B)} \tag{40}$$

An MO of the composite system is given by a linear combination of the AO's of A and the AO's of B.

$$\Phi_f = \sum_{v=1}^{M+N} c_{v,f}\chi_v \tag{41}$$

To see the strength and mode of interactions between the MO's of two fragments, we may expand the MO's of the composite interacting system in terms of the fragment MO's:

$$\Phi_f = \sum_{i=1}^{m} d_{i,f}\phi_i + \sum_{j=m+1}^{M} d_{j,f}\phi_j + \sum_{k=1}^{n} d_{k,f}\varphi_k + \sum_{l=n+1}^{N} d_{l,f}\varphi_l \quad (f = 1 \sim M+N) \tag{42}$$

in which ϕ_i and ϕ_j represent the occupied and unoccupied MO's of A, respectively, and φ_k and φ_l represent the occupied and unoccupied MO's of B, respectively. The coefficients, $d_{i,f}$, $d_{j,f}$, $d_{k,f}$ and $d_{l,f}$ are calculated easily by solving the following equations:

$$c_{r,f} = \sum_{i=1}^{m} d_{i,f}c_{r,i} + \sum_{j=m+1}^{M} d_{j,f}c_{r,j} \quad (r = 1, 2, \ldots, m, m+1, \ldots, M) \tag{43}$$

$$c_{t,f} = \sum_{k=1}^{n} d_{k,f}c_{t,k} + \sum_{l=n+1}^{N} d_{l,f}c_{t,l} \quad (t = 1, 2, \ldots, n, n+1, \ldots, N) \tag{44}$$

The interaction between A and B may now be characterized by the intermixing of MO's of the two fragment species. For example, the π-type MO's of butadiene can be expanded numerically in terms of the π and π^* MO's of two ethylene-like fragments.

2-7. Electron Configurations

In some of the papers collected in this volume, chemical interactions between molecular species are illustrated in terms of *"electron configurations"*. It may provide some imagination of interactions. For example, in the interaction between two closed-shell molecules A and B in the ground state, the electron configuration denoted by Ψ_0 in Figure 15 should be the dominant term, provided that A and B are not a pair of very strong electron-donor and very strong electron-acceptor. The change in the electronic structure brought about by the interaction is represented by the mixing in of other electron configurations. The configurations $\Psi_{i\to l}$ and $\Psi_{k\to j}$ represent one-electron transfer from A to B and from B to A, respectively. On the other hand, $\Psi_{i\to j}$ stands for local promotion of an electron within A. Note that each electron configuration does not signify by itself the existence of a charge-transfer complex or a locally-exited state. These configurations represent in all a state and the electronic structure of the composite interacting system, A-B.

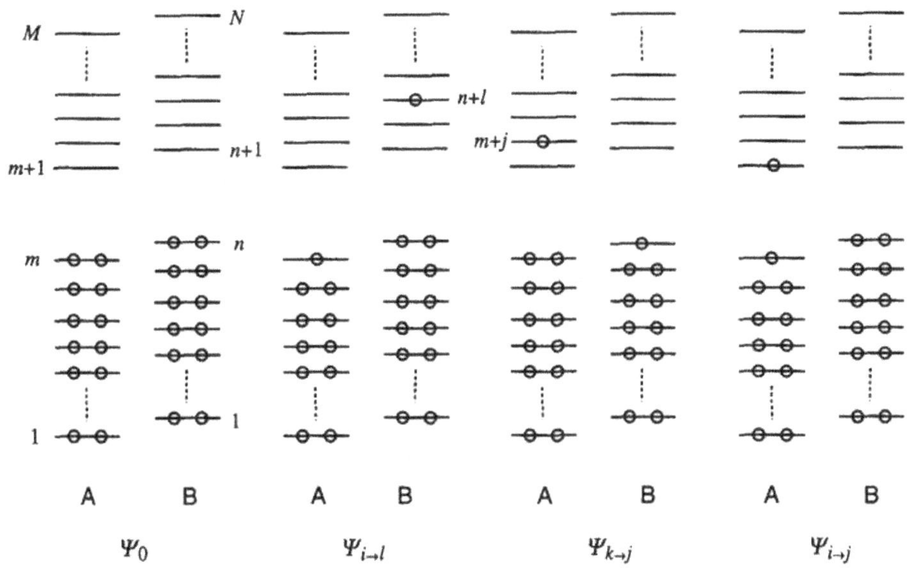

Figure 15. An illustration of electron configurations

Now, the wavefunction of the composite system, A-B, is written as:

$$\Psi = C_0\Psi_0 + C_{i\to l}\Psi_{i\to l} + C_{k\to j}\Psi_{k\to j} + C_{i\to j}\Psi_{i\to j} + C_{k\to l}\Psi_{k\to l} + \ldots\ldots \tag{45}$$

The coefficients of electron configurations, C_0, $C_{i\to l}$, $C_{i\to j}$, ..., are obtained in principle by solving the secular equations that are similar to eqs. 14 and 15 or eq. 20.

$$\begin{vmatrix} H_{0,0} - E & H_{0,p} - S_{0,p}E & H_{0,q} - S_{0,p}E & \ldots\ldots \\ H_{0,p} - S_{0,p}E & H_{p,p} - E & H_{p,q} - S_{p,q}E & \ldots\ldots \\ H_{0,q} - S_{0,q}E & H_{p,q} - S_{p,q}E & H_{q,q} - E & \ldots\ldots \\ & \ldots\ldots\ldots\ldots & & \\ & \ldots\ldots\ldots\ldots & & \end{vmatrix} = 0 \tag{46}$$

where p, q = $i\to l$, $k\to j$, $i\to j$, $k\to l$ and so on. The matrix elements are defined by;

$$H_{0,0} = \int \Psi_0{}^* H \Psi_0 \, d\tau \tag{47}$$

$$H_{p,q} = \int \Psi_p{}^* H \Psi_q \, d\tau \tag{48}$$

$$S_{0,p} = \int \Psi_0{}^* \Psi_p \, d\tau \tag{49}$$

$$S_{p,q} = \int \Psi_p{}^* \Psi_q \, d\tau \tag{50}$$

where H is the *total* Hamiltonian for $2m+2n$ electrons and $d\tau$ stands for integration over space and spin coordinates of those electrons. There are many other electron configurations that are not shown in Figure 15. It is not practical then to evaluate integrals given above for many electron configurations. Alternatively, the coefficients, C_0, $C_{i\to l}$, ..., can be calculated by using the coefficients $d_{i,f}$'s, $d_{j,f}$'s, $d_{k,f}$'s and $d_{l,f}$'s obtained by solving eqs. 43 and 44. For example, the π conjugation of butadiene is represented in terms of the electron configurations of two ethylene-like fragments by applying the procedure of *"configuration analysis"*.

The wavefunction must be antisymmetric for interchange of electronic space and spin coordinates. There is a simple way to write wavefunctions guaranteeing that they will be antisymmetric: one writes the wavefunction as a Slater determinant or as a combination of determinants. The original electron configuration Ψ_0 is given by;

$$\Psi_0 = NA \left| \phi_1(1)\bar{\phi}_1(2)\phi_2(3) \ldots \phi_m(2m-1)\bar{\phi}_m(2m)\varphi_1(2m+1) \ldots\ldots \bar{\varphi}_n(2m+2n) \right| \tag{51}$$

where $\phi_1(1)$ and $\bar{\phi}_1(2)$ are the abbreviated forms of the one-electron orbitals with the α and β spin functions, $\phi_1(1)\alpha(1)$ and $\phi_1(2)\beta(2)$, respectively. The symbol NA in front of the determinant is the normalization-antisymmetrization operator

In the case of A and B, each having one occupied MO, ϕ_i and φ_k, the wavefunction Ψ_0 is given by:

$$\Psi_0 = \frac{1}{\sqrt{4!}} \frac{1}{(1-s_{i,k}^2)} \begin{vmatrix} \phi_i(1) & \bar{\phi}_i(2) & \phi_i(3) & \bar{\phi}_i(4) \\ \phi_i(1) & \bar{\phi}_i(2) & \phi_i(3) & \bar{\phi}_i(4) \\ \varphi_k(1) & \bar{\varphi}_k(2) & \varphi_k(3) & \bar{\varphi}_k(4) \\ \varphi_k(1) & \bar{\varphi}_k(2) & \varphi_k(3) & \bar{\varphi}_k(4) \end{vmatrix} \tag{52}$$

The normalizing factor of wavefunction involves $1/\{(2m+2n)!\}^{1/2}$ ($m=1$ and $n=1$ in this case) for interchange of electronic space spin coordinates and $1/(1 - s_{ik}^2)$ for the overlap between the MO's, ϕ_i of A and φ_k of B. Expanding this according to the usual rules governing determinants gives 4! ways of housing electrons 1, 2, 3 and 4 in the four different spinorbitals, each having the space and spin parts.

When B has an unoccupied MO φ_l, there appears an electron configuration ($i \rightarrow l$) in which an electron is shifted from the MO ϕ_i to the MO φ_l.

$$\Psi_{i\rightarrow l} = (1/\sqrt{2})[1/\sqrt{4!}\begin{vmatrix} \phi_i(1) & \bar{\phi}_i(2) & \phi_i(3) & \bar{\phi}_i(4) \\ \varphi_l(1) & \bar{\varphi}_l(2) & \varphi_l(3) & \bar{\varphi}_l(4) \\ \varphi_k(1) & \bar{\varphi}_k(2) & \varphi_k(3) & \bar{\varphi}_k(4) \\ \varphi_k(1) & \bar{\varphi}_k(2) & \varphi_k(3) & \bar{\varphi}_k(4) \end{vmatrix} - 1/\sqrt{4!}\begin{vmatrix} \phi_i(1) & \bar{\phi}_i(2) & \phi_i(3) & \bar{\phi}_i(4) \\ \varphi_l(1) & \bar{\varphi}_l(2) & \varphi_l(3) & \bar{\varphi}_l(4) \\ \varphi_k(1) & \bar{\varphi}_k(2) & \varphi_k(3) & \bar{\varphi}_k(4) \\ \varphi_k(1) & \bar{\varphi}_k(2) & \varphi_k(3) & \bar{\varphi}_k(4) \end{vmatrix}] \quad (53)$$

The normalizing factor arising from nonorthogonality of the fragment MO's has been omitted in eq. 53.

The original configuration Ψ_0 is not a simple product of the wavefunction of A and the wavefunction of B in an isolated state, but is subjected to an interchange of electronic space and spin coordinates between the MO's of A and B. Two kinds of interaction terms are involved in $H_{0,0}$, the Coulombic interaction and exchange repulsion. The former corresponds to the interaction that arises from the electron density ρ_A and ρ_B of two isolated molecules, A and B. The difference in electron density between the electron configuration Ψ_0 and two isolated systems is given by:

$$\Delta\rho_{0,0} = \rho_{0,0} - (\rho_A + \rho_B) = [2\{\phi_i^2 + \varphi_k^2\}s_{i,k}^2 - 4\phi_i\varphi_k s_{i,k}] / (1 - s_{i,k}^2) \quad (54)$$

It is seen that the interaction between the occupied MO's ϕ_i and φ_k increases the population of these MO's to the amount of $2s_{i,k}^2$. This amount of electron population comes from the second term on the right-hand side of eq. 54 which indicates that electron density is reduced in the region between the two molecules A and B. Thus, the orbital interaction between occupied MO's is antibonding and destabilizes the reacting system. This interaction is called *"exchange repulsion"* or *"closed-shell repulsion"* or, in some cases, *"four-electrons in two-orbitals problem"*. If the two species are separated at a long distance or if one employs the zero-differential overlap approximation, this repulsion vanishes. The situation is much the same as the overlap repulsion in one-electron orbital description of chemical interactions mentioned above. One finds the simplest example of this type of interaction in He_2. One will also find that two pairs of electrons occupying both of the ethylene π and π^* MO's gives rise to the same result as eq. 54.

2-8. Formation and Breaking of Chemical Bonds

The frontier orbital interaction in this expression is interpreted in terms of the mixing of the one-electron-transferred configuration $\Psi_{i\rightarrow l}$ into the original configuration Ψ_0. For the purpose of clarifying the significance of $\Psi_{i\rightarrow l}$, we assume here the simplest case the wavefunction of which is given by:

$$\Psi = C_0\Psi_0 + C_{i\rightarrow l}\Psi_{i\rightarrow l} \quad (55)$$

The coefficients C_0 and $C_{i\to l}$ should satisfy the following normalization condition in this wavefunction:

$$1 = C_0^{\,2} + 2C_0 C_{i\to l} S_{0,i\to l} + C_{i\to l}^{\,2} \tag{56}$$

By neglecting small terms, the change in electron density brought about by the mixing in of $\Psi_{i\to l}$ into Ψ_0 is given by:

$$\Delta\rho = \{\rho_{0,0} - (\rho_A + \rho_B)\} + C_{i\to l}^{\,2}\{\rho_{i\to l,i\to l} - \rho_{0,0}\} + 2C_0 C_{i\to l}\{\rho_{0,i\to l} - S_{0,i\to l}\rho_{0,0}\}$$

$$\approx \Delta\rho_{0,0} + C_{i\to l}^{\,2}\{\varphi_l^{\,2} - \phi_i^{\,2}\} + 2\sqrt{2}\,C_0 C_{i\to l}\{\phi_i\varphi_l - s_{i,l}\phi_i^{\,2}\} \tag{57}$$

The first term on the right-hand side of eq. 57 includes exchange repulsion as mentioned above.

The second term stands for electron shift from the occupied MO ϕ_i of A to the unoccupied MO φ_l of B. The difference in electron density between the configuration $\Psi_{i\to l}$ and the configuration Ψ_0 is given by:

$$\rho_{i\to l,i\to l} - \rho_{0,0} \approx \varphi_l^{\,2} - \phi_i^{\,2} \tag{58}$$

An electron which has occupied originally the MO ϕ_i of A has been shifted to the unoccupied MO φ_l of B. Equation 57 tells us that the occupied MO ϕ_i of A donates a part of its electron population to the amount of $C_{i\to l}{}^{2}$ to the unoccupied MO φ_l of B.

The third term is of particular significance as the principal origin of bond formation in chemical reactions. We have:

$$\int \Psi_0^{\,*}\Psi_{i\to l}\,d\tau = S_{0,i\to l} \approx \sqrt{2}\,s_{i,l} \tag{59}$$

Then, the third term indicates that the occupied MO ϕ_i of A donates another part of its electron population of an amount of $2\sqrt{2}C_0C_{i\to l}s_{i,l}$. This portion of electron population is retained in the intermolecular region between A and B through the overlap between the MO's ϕ_i and φ_l. It is obvious that this portion is utilized for the formation of new bonds between the reaction sites of A and B.

The mixing in of one-electron-transferred electron configurations is called usually *"charge-transfer"*. This must originate from the changes in charge distribution represented by the second term on the right-hand side of eq. 57. In the interaction between two neutral molecules, electron delocalization takes place both from A to B and from B to A. Apparent changes in electronic charge distribution may not be significant. Even in that case, mutual electron delocalization plays a crucial role in binding the two molecules.

In chemical reactions, bonds are broken between the reaction centers and the adjacent atoms in the reagent and reactant molecules. Assume here that an ethylene molecule reacts with a reagent. The HOMO is bonding between the two carbons and, therefore, donation of an electronic charge from this MO weakens the carbon-carbon π bond. In contrast, the LUMO is antibonding and, therefore, acceptance of an electronic charge in this MO weakens the π bond. The 2pπ AO's χ_1 and χ_2 have the coefficients, $c_1 = c_2 = (2+2s_{1,2})^{-1/2} = 0.635$ with $s_{12} = 0.24$ in the HOMO. Then, the carbon-carbon overlap population in the HOMO is calculated to be 0.387 ($= 0.635 \times 0.635 \times 0.24 \times 4$). On the other hand, the overlap population of the carbon-carbon bond in the LUMO is -0.632. It is suggested, therefore, that acceptance of electrons into the LUMO weakens the bond more effectively than donation of electrons from the HOMO does so. A similar argument applies also to other molecules, e.g., butadiene given above.

In order to bring a synchronized breaking of bonds in the reagent and reactant parts, the occupied MO's of the electron-donor part should lose a greater amount of electron population than that the unoccupied MO's of the electron-acceptor part receive. This proposition is connected smoothly with eq. 57. It has been shown above that an occupied MO of the electron-donor part transfers a part of its electron population to the unoccupied MO's of the electron-acceptor part, but the former donates another part to the intermolecular region to form new bonds. This signifies that an occupied MO loses a greater amount of electron population than an unoccupied MO accepts in orbital interactions. Thus, the formation of new bonds and breakings of old bonds take place in a harmonized manner in chemical reactions both in the reagent part and in the reactant part.

Chemical reactions are essentially a series of formations and breakings of bonds between atoms. One may recognize clearly that the basic feature of chemical reactions is represented by electron delocalization between molecules. We believe that the readers will find in the applications included in this volume that the HOMO and LUMO have the properties that are suitable for formations of new bonds and breakings of old bonds along reaction paths.

Part I

The Concept and Formulation

The first paper on the *"frontier orbital theory"* came out in 1952. At that time there were several other theories of chemical reactivities for aromatic compounds. They were the static approaches by Ri, by Coulson and Longuet-Higgins and by Pullman and the localization approaches by Wheland and by Dewar. It was attempted at that time to replace the old "electronic theory" by the theory based on quantum mechanics. Our new reactivity theory was classified as delocalization approach, together with the theory by Brown.

It was the π electron systems of conjugated molecules to which the frontier orbital (electron) theory was applied for the first time. We had neither computers nor electronic calculators at that time and, therefore, we could obtain only the π MO's by using the simple Hückel MO method. In our first paper on the frontier orbital theory (paper no. 1), we found that the electron density of the highest occupied MO had a good correlation with the most reactive positions against electrophiles and oxydizing reagent for fifteen condensed aromatic compounds. Theoretical considerations were given there to reason why the electrons in the highest occupied MO were of particular significance, but the way was quite different from those developed later. The theory was introduced at the beginning based mainly on our chemical intuition. In the succeeding paper (paper no. 2), the frontier orbital concept was extended to the reactions of aromatic compounds with nucleophiles and with radicals. Then, the parallelism found in the reactivity indices existed at that time and our scale, *"superdelocalizability"*, was proved (paper no. 3).

In 1959 we proposed a novel perturbation theory for the simple Hückel MO method called *"method of perturbed secular determinant"* (paper no. 5). Useful equations giving the orbital energy, electron density and bond orders of perturbed systems were derived there. This method was extended further to more complicated systems (papers no. 6 and 7) and was applied to the problem of aromaticity (paper no. 9). The application of the method to the orbital interactions in the complexes of aromatic hydrocarbons with a silver cation reached a conclusion that the reactivity was proportional to the sum of the coefficients of the reaction sites in the frontier MO, $|c_r^f + c_s^f|$, of the aromatic compounds (paper no. 11). This finding turned out later to be very important in applying the frontier orbital theory to the problems of stereoselection in chemical reactions. The frontier orbital theory was utilized also to explain carcinogenic activity of condensed aromatic systems (paper no. 13) and reactivities of saturated hydrocarbons (paper no. 11). It was in 1960 that we could start calculations of the electronic structure of saturated hydrocarbons in our laboratory.

The developments in the theory and applications to many types of reactions are found in other papers (no. 14, 15, 16 and 17).

Reprinted from THE JOURNAL OF CHEMICAL PHYSICS, Vol. 20, No. 4, 722–725, April, 1952
Printed in U. S. A.

A Molecular Orbital Theory of Reactivity in Aromatic Hydrocarbons

KENICHI FUKUI, TEIJIRO YONEZAWA, AND HARUO SHINGU
Faculty of Engineering, Kyoto University, Kyoto, Japan
(Received October 29, 1951)

In the search for a quantitative correlation between reactivity and electronic configuration of aromatic hydrocarbons, the electron density, at each carbon atom, of the highest occupied π-orbital in the ground state of the molecule is calculated by means of the LCAO method. Comparing the result of such a calculation on fifteen condensed aromatic hydrocarbons with their chemical reactivities, we find that the position at which the electron density is largest is most readily attacked by electrophilic or oxidizing reagents.

It is, therefore, concluded that distinct from other π-electrons the pair of π-electrons occupying the highest orbital, which is referred to as *frontier* electrons, plays a decisive role in chemical activation of these hydrocarbon molecules. The theoretical significance of this discrimination of the *frontier* electrons in relation to the chemical activation is discussed.

INTRODUCTION

CONSIDERABLE progress has been made of late in the quantum-mechanical treatments for explaining the chemical reactivity of organic compounds. Especially, the problem of predicting the reactivity of an organic molecule in a particular position has been solved to some extent by means of the LCAO molecular orbital method in the case of the well-known "orienting effect" of substituents in aromatic nuclei. On treating such a problem, two methods have been applied by several authors: the one may be called the "transition state method"[1,2] and the other the "π-electron density method.[3,4]

In the former method the configuration of transition complex is assumed *a priori* on the basis of some other theoretical or empirical arguments, and this method is usually confined to tedious calculations, as it is necessary to compute the energies of all the possible configurations.*

In the latter method, on the other hand, the π-electron density was computed as a resultant of the contributions of all π-electrons in the molecule, and no distinction between individual π-electrons has ever been considered. The defect of the method arising from this indiscrimination appears most evidently in the treatment of nonsubstituted aromatic hydrocarbons, in which the π-electron density has always the same value at each position in the molecule according to the Coulson-Rushbrooke theorem,[2] so that this method is of no use for predicting the reactivity of different positions in such molecules.†

In the present paper it is shown that on the ground of the latter method, if we distinguish the pair of π-electrons in the highest occupied orbital in the ground state from the others and assume that this pair of π-electrons plays a decisive role in the reaction in question, we can obtain an illuminating explanation of the difference of reactivity at each position in a molecule.

In the first place we will report on the treatment of nonsubstituted aromatic hydrocarbons. The orienting effect of substituents in aromatic nuclei will be treated in the next publication.

CALCULATION OF ELECTRON DENSITY

In the present paper our treatment is restricted to condensed aromatic hydrocarbons which are "alternant."[5] Namely, they consist of several fused six-membered rings and have the Kékulé-type structure in their ground state.

The energy spectrum can be obtained by the LCAO method, using one-electron approximation function

$$\Psi=\sum_{i=1}^{2N} c_i\varphi_i, \tag{1}$$

where φ_i is the $2p\pi$-atomic orbital of the ith carbon atom.‡

The secular equations for a series of aromatic hydrocarbons are reduced according to the modes shown in Appendix, where the symmetry operations σ_h, σ_v, and i are not considered.

The density of each electron occupying the nth electronic level at the ith carbon atom is represented by $|c_i|^2$, where c_i is the coefficient in Eq. (1) corresponding to the nth eigenvalue.

[1] G. W. Wheland, J. Am. Chem. Soc. 64, 900 (1942).
[2] H. C. Longuet-Higgins, J. Chem. Phys. 18, 283 (1950).
[3] G. W. Wheland and L. Pauling, J. Am. Chem. Soc. 57, 2086 (1935).
[4] H. C. Longuet-Higgins and C. A. Coulson, Trans. Faraday Soc. 43, 87 (1947).
* This method has never been applied to unsubstituted hydrocarbons except to naphthalene (see reference 1). Whether such a calculation is successfully applicable to the condensed aromatic hydrocarbons which are treated in the present paper, is not known.
† In relation to this problem two other quantum-mechanical methods have been proposed in which the distribution of π-electron density is discussed without using LCAO method. The one is called by O. Schmidt (Z. physik. Chem. 39, 59 (1938); 42, 83 (1939); 44, 185, 194 (1939)) the "Kastenmodel" method in which certain ambiguities are involved as to the model used, resulting in an insufficient accuracy of the method. The other is the "molecular diagram" method after A. and B. Pullman (Rev. sci. 84, 145 (1946); 86, 219 (1948); Chem. Abst. 43, 6905 (1949)) which is almost inapplicable to complex molecules, owing to the laboriousness in calculating the contributions of all possible bond structures, including excited ones, to the normal state of the molecule.
[5] C. A. Coulson and H. C. Longuet-Higgins, Proc. Roy. Soc. (London) A192, 16 (1947).
‡ All σ-electrons which form a closed shell in the ground state are left out of account.

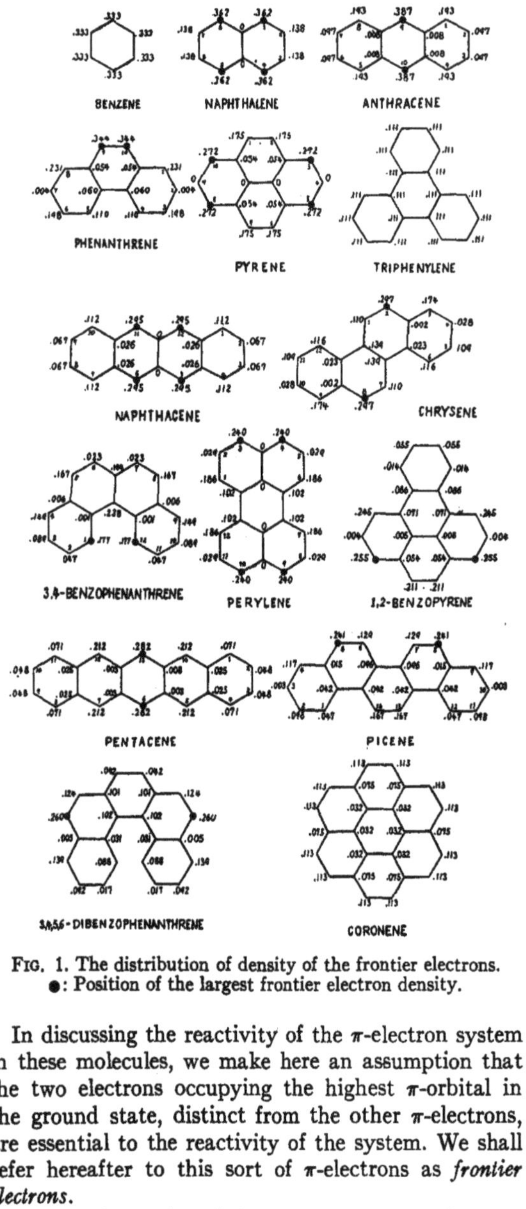

FIG. 1. The distribution of density of the frontier electrons. ●: Position of the largest frontier electron density.

In discussing the reactivity of the π-electron system in these molecules, we make here an assumption that the two electrons occupying the highest π-orbital in the ground state, distinct from the other π-electrons, are essential to the reactivity of the system. We shall refer hereafter to this sort of π-electrons as *frontier electrons*.

The discrimination of the pair of *frontier* π-electrons in relation to the reactivity is considered reasonable so far as, in photochemical processes in which a conjugated double bond system takes part, the triplet state has often been demonstrated as an intermediate step essential to the reaction.[6]

It is our opinion, furthermore, that for ordinary, nonphotochemical, organic reactions—not only for oxidation and other reactions of radical nature, but also, with certain reservations, for reactions of ionic character such as the electrophilic substitutions in aromatic nuclei—a transition of the *frontier* electrons into the triplet state under the influence of the attacking reagent may be essential to the process of chemical activation. Thus, this will be described as a rate process in which the *frontier* electrons solely play an orienting or guiding role, so that the density distribution of the *frontier* electrons will determine the configuration of the transition complex for the reaction.

In view of these considerations we have numerically calculated the density distribution of the *frontier* electrons at each carbon atom in the molecule. The result for 15 hydrocarbons is shown in Fig. 1. The points marked in each diagram of molecules indicate the position of the largest density of the *frontier* electrons.

In this calculation the exchange integrals, except those between two adjacent carbon atoms, and all the overlap integrals, are neglected and the Coulomb integral, as well as the exchange integral, is assumed to be constant.

It is a simplifying circumstance for the present treatment that the distribution of the π-electron density in the triplet state is always the same as that for the ground state, so far as the hydrocarbon systems in the present consideration are concerned. We can, therefore, proceed with our discussion without settling the question as to which of these two states is essential to the reaction. This problem is discussed in some detail in the last section.

COMPARISON WITH EXPERIMENT

In Table I the structures of the reaction products of oxidation, halogenation, nitration, and sulfonation of ten condensed aromatic hydrocarbons are tabulated along with, in the second column, the positions of the highest density of the *frontier* electrons in each hydrocarbon indicated in Fig. 1. No experimental results could be found in the literature for 1,2-benzopyrene and 3,4,5,6-dibenzophenanthrene.

As can be seen from the Table I, there is a perfect agreement between calculation and the result of experiment in the cases of naphthalene, anthracene, phenanthrene, chrysene, naphthacene, perylene, pentacene, and picene.

In the case of 3,4-benzophenanthrene our calculation seems to disagree with the experiment. But this is explicable from the consideration that the position of the largest electron density, 1, is sterically hindered, so that the position 5 where the density is second to the position 1 may be attacked, in conformity with the experiment.

The apparent exceptions in the examples of sulfonation, that of anthracene in particular, may be attributed to the instability of primary sulfonation products, as is widely recognized by preparative organic chemists.

As to picenequinone, the 13,14-structure has been reported by Burg,[7] but later on Cook[8] proved experi-

[6] G. N. Lewis and M. Kasha, J. Am. Chem. Soc. 67, 994 (1945).

[7] O. Burg, Ber. deut. chem. Ges. 13, 1836 (1880).

[8] J. W. Cook, J. Chem. Soc. 1941, 685.

TABLE I. Comparison of the predicted positions of the highest reactivity with experimental results. "1-Cl" means, e.g., "1-Chloronaphthalene," "3,9-di-SO_3H" means, e.g., "Perylene-3,9-disulfonic acid," etc.

Compounds	Position of the highest density of the frontier electrons	Oxidation	Structure of the reaction products Halogenation	Nitration	Sulfonation
Naphthalene	1,4,5,8	1,4-quinone[a] (5,8-quinone)	1-F[l] 1-Cl[m] 1-Br[n] 1,4-di-Cl[o] 1,5-di-Cl[o]	1-NO_2[cc] 1,5-di-NO_2[dd] 1,8-di-NO_2[dd]	1-SO_3H[ii] (2-SO_3H at higher temp.)
Anthracene	9,10	9,10-quinone[b]	9-Cl[p] 9,10-di-Cl[q] 9-Br[p] 9,10-di-Br[r]	9-NO_2[ee] 9,10-di-NO_2[ee]	(9-SO_3H unstable?) 1,5-, 1,8-di-SO_3H[jj]
Phenanthrene	9,10	9,10-quinone[c]	9,10-di-Cl[s] 9-Br[t]	9-NO_2[ff]	9-SO_3H[kk] (2 or 3-SO_3H at higher temp.)
Pyrene	3,5,8,10	3,8-quinone[d] 3,10-quinone[e]	3,5,8,10-tetra-Cl[u] 3,5,8-tri-Br[v]	...	3,5,8,10- tetra-SO_3H[ll]
Naphthacene	5,6,11,12	5,12-quinone[f] (6,11-quinone)	5,12-di-Cl[w] 5,12-di-Br[w]	...	...
3,4-Benzophenanthrene	1,12	5,6-quinone[g]	...	...	...
Chrysene	2,8	1,2-quinone[h]	2-Cl[x] 2-Br[x]	2,8-di-NO_2[gg]	2-SO_3H[mm]
Perylene	3,4,9,10	3,9-quinone[i] 3,10-quinone(?)[i]	3,9-di-Cl[y] 3,4,9,10-tetra-Cl[z] 3,9-di-Br[aa] 3,10-di-Br[aa]	3,10-di-NO_2[hh] 3,4,9,10-tetra-NO_2[hh]	3,9-di-SO_3H[nn] 3,10-di-SO_3H[nn]
Pentacene	6,13	6,13-quinone[j]	...	...	...
Picene	5,8	5,6-quinone[k]	13,14-di-Br(?)[bb]	...	...

[a] C. E. Groves, Ann. Chem. **167**, 357 (1873).
[b] J. Fritzsche, J. prakt. Chem. (1) **106**, 287 (1869).
[c] R. Fittig and E. Ostermayer, Ann. Chem. **166**, 365 (1873).
[d] R. Scholl and C. Seer, Ann. Chem. **394**, 125 (1912).
[e] Rütgerswerke, A.-G. Germ. Pat. 613, 914 (May 31, 1935).
[f] S. Gabriel and E. Leupold, Ber. deut. chem. Ges. **31**, 1277 (1898).
[g] J. W. Cook, J. Chem. Soc., 3524 (1931).
[h] C. Liebermann, Ann. Chem. **158**, 309 (1871).
[i] A. Zinke and E. Unterkreuter, Monatshefte Chemie **40**, 408 (1919).
[j] E. Phillipi, Monatshefte Chemie **53/54**, 639 (1929).
[k] J. W. Cook, J. Chem. Soc., 685 (1941).
[l] Schiemann, Gueffroy, and Wikelmüller, Ann. Chem. **487**, 270 (1931).
[m] A. Töhl and O. Eberhard, Ber. deut. chem. Ges. **26**, 2945 (1893).
[n] C. Glaser, Ann. Chem. **135**, 41 (1865).
[o] J. S. Salkind and S. B. Faerman, J. Russ. Phys.-Chem. Ges. (U.R.S.S.) **62**, 1021 (1930).
[p] W. H. Perkin, Chem. News **34**, 145 (1876).
[q] Ed. Lippmann and I. Pollak, Ber. deut. chem. Ges. **34**, 2768 (1901).
[r] C. Graebe and C. Liebermann, Ber. deut. chem. Ges. **1**, 186 (1868).
[s] J. Schmidt and R. Schall, Ber. deut. chem. Ges. **39**, 3893 (1907).
[t] L. F. Fieser and C. C. Price, J. Am. Chem. Soc. **58**, 1838 (1936).
[u] I. G. Farbenind, A.-G. Fr., 261, 627 (March 23, 1934).
[v] E. Clar, Ber. deut. chem. Ges. **69**, 1671 (1936).
[w] I. J. Postovskii and R. G. Beyles, Compt. rend. Acad. Sci. U.R.S.S. **39**, 102 (1943).
[x] I. G. Farbenind, A.-G. Fr., 793, 893 (February 3, 1936).
[y] Zinke, Funke, and Pongratz, Ber. deut. chem. Ges. **58**, 330 (1925).
[z] A. Zinke and K. Funke, Ber. deut. chem. Ges. **60**, 580 (1927).
[aa] Zinke, Linner, and Wolfbauer, Ber. deut. chem. Ges. **58**, 328 (1925).
[bb] O. Burg, Ber. deut. chem. Ges. **13**, 1837 (1880).
[cc] R. Piria, Ann. Chem. **78**, 32 (1851).
[dd] Ch. Gassman, Ber. deut. chem. Ges. **29**, 1243, 1521 (1896).
[ee] J. Meisenheimer and E. Connerade, Ann. Chem. **330**, 164 (1903).
[ff] J. Schmidt and M. Strobel, Ber. deut. chem. Ges. **36**, 2511 (1903).
[gg] I. G. Farbenind, A.-G. Fr., 794, 534 (February 19, 1936).
[hh] Zinke, Hirsch, and Brozek, Monatshefte Chemie **51**, 205 (1929).
[ii] O. N. Witt, Ber. deut. chem. Ges. **48**, 743 (1915).
[jj] J. Houben, *Das Anthracen und die Anthrachinone* (1929), p. 109.
[kk] A. Werner, Ann. Chem. **321**, 270 (1902).
[ll] I. G. Farbenind, A.-G. Germ. Pat., 620, 635 (October 26, 1935).
[mm] I. G. Farbenind, A.-G. Fr., 789, 461 (October 29, 1935).
[nn] C. Marschalk, Chem. Zentr. **I**, 1833 (1927).

mentally that the quinone obtained from picene by oxidation must have the 5,6-structure which coincides with our results.

In view of these excellent agreements of our theory with experiment we believe that some essential relations must exist between the reactivity and the density of the *frontier* electrons.

As for the significance of the absolute value of the *frontier* electron density, two examples may be of value to be mentioned here: triphenylene is more stable than naphthacene in relation to oxidation;[9] and anthracene is more readily oxidized than phenanthrene.[10] Comparing the largest values of the density in those hydrocarbons having the same number of carbon atoms and π-electrons, we can observe a parallelism between the reactivity and the absolute value of the *frontier* electron density.

[9] Ward, Kirner, and Howard, J. Am. Chem. Soc. **67**, 246 (1945).
[10] R. Anschütz and G. Schultz, Ann. Chem. **196**, 38 (1879).

THEORETICAL DISCUSSION

In our treatments so far disclosed the physical ground of discriminating the highest occupied π-orbital in the ground state from the other π-orbitals is left untouched. We shall try here to add some theoretical considerations.

Anyhow, it can be concluded from the successful results obtained above that certain differences must exist between the *frontier* electrons and the other π-

electrons in respect to their behaviors in the course of electronic interaction during the chemical activation. This fact may be understood in two different ways: the one way of understanding consists in the possibility that the frontier electrons, on account of their valence-electron-like nature, can pass into the next higher orbital, forming a triplet state, presumably under the influence of the field arising from the approaching reagent. From the symmetry consideration of the wave function of the whole system, it seems to be quite reasonable that, in order to retain the singlet spin symmetry of the whole system, the spin state of two electrons in the C$-$Hσ bond to be substituted by the electrophilic reagent is simultaneously changed into a triplet in the course of the reaction which is considered to take place adiabatically.

To understand that the triplet formation of the *frontier* electrons can be effective for bond formation between the molecule and the electrophilic reagent, we can make use of the following analogy. Considering that, as shown by Bartlett,[11] a strong bond of a $^3\Sigma_u^+$ state is formed between two $p\sigma$ orbitals, we may assume that between the hydrocarbon molecule and the electrophilic reagent in a transition complex a $\{(sp^3)\sigma, (sp^3)\sigma\}$-like bond is formed in the same way by a participation of the *frontier* electrons with parallel spins.

The reagent is thus attracted to the position of the largest *frontier* electron density, while the hydrogen ion, formerly in the C$-$Hσ-bond, may be repelled from the molecule because of the triplet property of the C$-$Hσ-part of the electronic system in the transition complex.

On the other hand, the contributions of all the π-electrons in the molecule must, in general, be taken into account on the occasion of chemical reaction. From this point of view, we come to the second way of understanding. When it is assumed that in the electronic interaction with an electrophilic attacking reagent the highest occupied π-orbital would always act as *attractive* and the other π-orbitals as *repulsive*, it can be expected that the formation of a transition complex at the position of the largest *frontier* electron density might be preferred in consequence of the resultant effect of all π-electrons, in either attractive or repulsive orbitals, upon the potential energy of the whole system in the process of formation of the transition complex. In another expression, this way of understanding consists in attributing the distinction of the highest occupied orbital from the others to the difference in the nature of each orbital in reference to the interaction with the attacking reagent. Of course, the repulsive effect represents a resultant contribution of all π-electrons other than *frontiers*. Some of them may give attractive effects but in this case the contribution must be smaller than those of *frontier* electrons. The possibility of a fact that a certain orbital in an atom attracts and that another orbital in the same atom repels a certain orbital of another atom has been shown by Araki and Watari[12] in the case of C_2-molecule.

The two ways of theoretical approach as discussed above may be considered as corresponding to the two different interpretations in the description of the process of electronic interaction. A correlation between them needs to be confirmed by further investigations.

The concept of the frontier electron, as well as the principle involved in the above discussion, is hoped to find applications more widely to problems on the reactivity of π-electron systems.

ACKNOWLEDGMENTS

The authors are grateful to Professor S. Kodama and Professor G. Araki for valuable suggestions and encouragements. Especially are the authors indebted to Professor Araki for enlightening them on the arguments developed in the part of discussion.

APPENDIX. SYMMETRY GROUP AND MODES OF REDUCTION OF SECULAR EQUATIONS.[a]

Compounds	Symmetry group	Modes of reduction of secular equations
Benzene	D_6	$\Gamma(\Psi) = \Gamma A_1 + \Gamma B_1 + \Gamma E_1 + \Gamma E_2$
Coronene	D_6	$\Gamma(\Psi) = 3\cdot\Gamma A_1 + \Gamma A_2 + \Gamma B_1 + 3\cdot\Gamma B_2 + 4\cdot\Gamma E_1 + 4\cdot\Gamma E_2$
Triphenylene	D_3	$\Gamma(\Psi) = 3\cdot\Gamma A_1 + 3\cdot\Gamma A_2 + 6\cdot\Gamma E$
Anthracene	D_2	$\Gamma(\Psi) = 4\cdot\Gamma A_1 + 3\cdot\Gamma B_1 + 3\cdot\Gamma B_2 + 4\cdot\Gamma B_3$
Naphthacene	D_2	$\Gamma(\Psi) = 5\cdot\Gamma A_1 + 4\cdot\Gamma B_1 + 4\cdot\Gamma B_2 + 5\cdot\Gamma B_3$
Naphthalene	D_2	$\Gamma(\Psi) = 3\cdot\Gamma A_1 + 2\cdot\Gamma B_1 + 2\cdot\Gamma B_2 + 3\cdot\Gamma B_3$
Pentacene	D_2	$\Gamma(\Psi) = 6\cdot\Gamma A_1 + 5\cdot\Gamma B_1 + 5\cdot\Gamma B_2 + 6\cdot\Gamma B_3$
Perylene	D_2	$\Gamma(\Psi) = 6\cdot\Gamma A_1 + 4\cdot\Gamma B_1 + 6\cdot\Gamma B_2 + 4\cdot\Gamma B_3$
Pyrene	D_2	$\Gamma(\Psi) = 5\cdot\Gamma A_1 + 3\cdot\Gamma B_1 + 3\cdot\Gamma B_2 + 5\cdot\Gamma B_3$
3,4-Benzo-phenanthrene	C_2	$\Gamma(\Psi) = 10\cdot\Gamma A + 8\cdot\Gamma B$
1,2-Benzopyrene	C_2	$\Gamma(\Psi) = 10\cdot\Gamma A + 10\cdot\Gamma B$
3,4,5,6-Dibenzo-phenanthrene	C_2	$\Gamma(\Psi) = 11\cdot\Gamma A + 11\cdot\Gamma B$
Phenanthrene	C_2	$\Gamma(\Psi) = 7\cdot\Gamma A + 7\cdot\Gamma B$
Picene	C_2	$\Gamma(\Psi) = 11\cdot\Gamma A + 11\cdot\Gamma B$
Chrysene	C_2	$\Gamma(\Psi) = 9\cdot\Gamma A + 9\cdot\Gamma B$

[a] $\Gamma(\Psi)$: reducible representation. ΓA, ΓB, $\cdots$: irreducible representations of character A, B, $\cdots$.

[11] J. H. Bartlett, Phys. Rev. 37, 505 (1931).

[12] G. Araki and W. Watari, Prog. Theoret. Phys. 6, 945 (1951).

Reprinted from The Journal of Chemical Physics, Vol. 22, No. 8, 1433–1442, August, 1954
Printed in U. S. A.

Molecular Orbital Theory of Orientation in Aromatic, Heteroaromatic, and Other Conjugated Molecules

Kenichi Fukui, Teijiro Yonezawa, Chikayoshi Nagata, and Haruo Shingu
Faculty of Engineering, Kyoto University, Kyoto, Japan

(Received March 31, 1953)

As to the LCAO MO treatment of the orientation problem in chemical reactions of π-electron systems, the *frontier* electron concept which has been previously introduced by the authors for explaining the reactivities in aromatic hydrocarbons is subjected to an extension in the sense that the *frontier orbitals* are specified according to the type of reaction. Thus, *fundamental postulates* relating to the reactivities of π-electron systems are set up, which are believed to include general principles involved in the mechanism of both substitution and addition of electrophilic, nucleophilic, or radical type. On the basis of these postulates it is possible to predict the position of attack in conjugated molecules in all the three types of substitution as well as addition in a consistent manner. There is a nearly perfect agreement between the theoretical conclusions and experimental results hitherto reported.

1. INTRODUCTION

ONE of the recent successes achieved by quantum mechanics in the field of physical organic chemistry is the development of the theory in explaining the orienting effect of substituents in aromatic molecules. The quantitative treatments of this problem so far reported may be roughly classified into two groups. The one,[1] which may be called the "static method," is based on the hypothesis that the position of higher (or lower) calculated density of total π electrons is more easily attacked by electrophilic (or nucleophilic) reagents. The other,[2] which may be called the "localization method," is related to the calculation of the difference in unsaturation energy of the hypothetical transition complexes.

Prominence of the latter method should be admitted in that it is capable of explaining the reactivity in all the three types, i.e., electrophilic, nucleophilic, and radical type, of substitution from a unified standpoint, whereas by means of the former method it is difficult to account for the point of attack in the radical substitution, and in this respect the "free valency method"[5] serves as filling up that defect of the static method.

In this paper it is shown that another unifying method can also be established on the basis of the "*frontier electron*" concept which has been previously introduced by some of the present authors in the case of electrophilic substitution in condensed aromatic hydrocarbons.[3] On treating the orientation problem in substituted atomatic and other related molecules, it is found that the concept should naturally be extended in the sense that the *frontier orbitals* are specified according to the type of reaction. Thus, we are led to set up the *fundamental postulates*, in the light of which the position of attack by electrophilic, nucleophilic, and radical reagents can be predicted not only for substitution but also for addition in an excellent agreement with experiment.

[1] E. Hückel, Z. Physik 72, 312 (1931); Z. physik. Chem. B35, 163 (1937); Z. Elektrochem. 43, 827 (1937); G. W. Wheland and L. Pauling, J. Am. Chem. Soc. 57, 2086 (1935), T. Ri, Rev. Phys. Chem. Japan 17, 1, 16 (1943) (in Japanese); M. J. S. Dewar, Trans. Faraday Soc. 42, 764 (1946); H. C. Longuet-Higgins and C. A. Coulson, Trans. Faraday Soc. 43, 87 (1947); C. A. Coulson and H. C. Longuet-Higgins, Proc. Roy. Soc. (London) A191, 39; A192, 16 (1947); A. Pullman, Rev. sci. 86, 219 (1948); A. Pullman and J. Metzger, Bull. soc. chim. France 1948, 1021 (1948); H. C. Longuet-Higgins and C. A. Coulson, J. Chem. Soc. (London) 1949, 971 (1949); C. Sandorfy, Bull. soc. chim. France 1949, 615 (1949); H. C. Longuet-Higgins, J. Chem. Phys. 18, 283 (1950); Orgel, Cottrel, Dick, and Sutton, Trans. Faraday Soc. 47, 113 (1951); J.-I. F. Alonso, Compt. rend. 233, 403 (1951); L. Szabó, Compt. rend. 233, 625 (1951).

[2] G. W. Wheland, J. Am. Chem. Soc. 64, 900 (1942); H. C. Longuet-Higgins, J. Chem. Phys. 18, 283 (1950); M. J. S. Dewar, J. Chem. Soc. (London) 1949, 463 (1949); C. A. Coulson, Research 4, 307 (1951); M. J. S. Dewar, J. Am. Chem. Soc. 74, 3355, 3357 (1952).

[3] Fukui, Yonezawa, and Shingu, J. Chem. Phys. 20, 722 (1952).

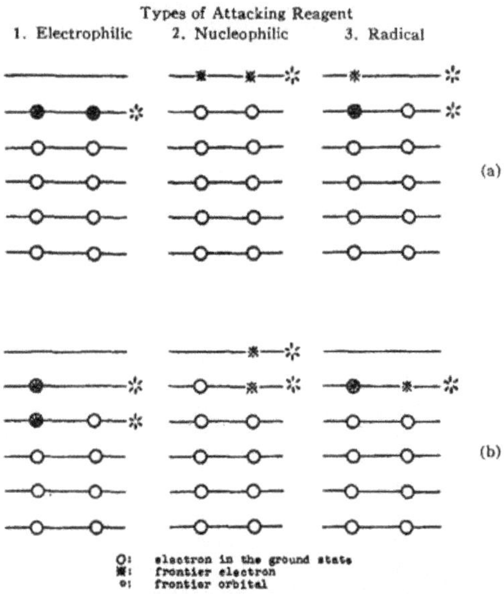

FIG. 1. Frontier electrons and frontier orbitals. (a) Even π-electron system. (b) Odd π-electron system.

2. FUNDAMENTAL POSTULATES

As a general characteristic of chemical activation it may be assumed that in the vicinity to the transition state the electrons in the reagent as well as those in the reactant molecule will be delocalized and a transfer of electrons will occur between the reagent and the reactant. From the standpoint of the *frontier* electron concept we suppose that the *frontiers* are most susceptible to the electron transfer and that this transfer of electrons, which may serve to lower the energy of the activated complex, will form the essential part of the electronic interaction in question. *Two* electrons should be characterized in this connection as *frontiers* in that they are most closely related to the formation of a covalent σ bond between the π-electron system and the reagent, and these two are provided either from the highest occupied orbital in the π-electron system to the reagent in an electrophilic reaction, or from the reagent to the lowest empty orbital in the π-electron system in a nucleophilic reaction, and in a radical reaction one electron is provided from the reagent and the other from the π-electron system.

From this point of view, in order to make reference to the reactivity in terms of the electron distribution in an *isolated* π-electron system, it is convenient to define a new concept of *frontier orbital* in relation to the isolated π-electron system as characterized by partial occupation by the *frontier* electron in case of the electron transfer at the transition state, according to the type of reaction.

On the basis of these considerations, the following *fundamental postulates* are set up, which enable us to predict the reactivity of π-electron system in a most general and consistent manner:

1. *In case of the reaction with an electrophilic reagent, that position is more susceptible to attack which has the higher density of the two electrons occupying the highest molecular orbital in the ground state.*

2. *In case of the reaction with a nucleophilic reagent, that position is more susceptible to attack which would have the higher density of the two electrons assumed to occupy the lowest unoccupied orbital of the ground state.*

3. *In case of the reaction with a radical reagent, that position is more susceptible to attack which would have the higher density of the two electrons, one occupying the highest orbital and the other occupying the lowest unoccupied orbital of the ground state.*

As a consequence of these postulates we can classify the *frontier* electrons and orbitals for various types of reactants and reagents, which is illustrated in Fig. 1, including the cases of even and odd π-electron systems. In the latter case where the π-electron system is a radical, the above postulates should be somewhat modified as shown in Fig. 1(b). Also in the case of an excited state of the π-electron system, e.g., in the case of a diradical state, the principle involved in the postulates may be correspondingly applicable.

3. ELECTROPHILIC SUBSTITUTION

Substitution in Heterocyclic Compounds

For the quantum-mechanical treatment of reactivity in heterocyclic compounds several parameters representing the influence of heteroatom as to the Coulomb and exchange integrals have been introduced by several authors[1,2] from some empirical or theoretical arguments.

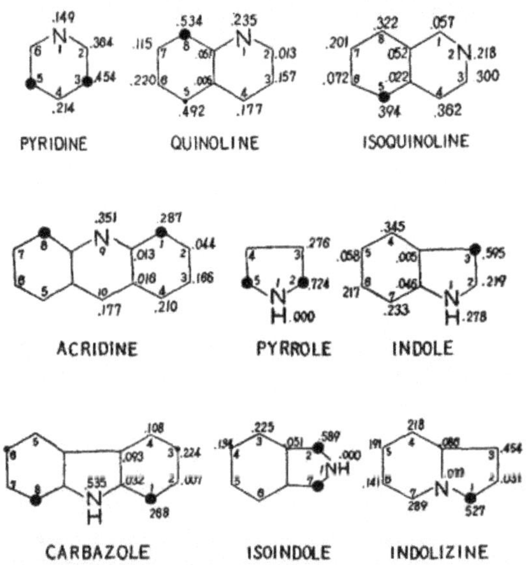

FIG. 2. Frontier electron density for electrophilic substitution in heterocyclic compounds. As to pyridine, the calculation is made with the help of Eq. (1).

TABLE I. Coulomb and exchange integrals for heteromolecules containing nitrogen atom.

Type of nitrogen atom	Pyridine type					Pyrrole type				
	Coulomb[a,b,c]			Exchange[d]		Coulomb			Exchange	
Integrals	α_1	α_2	α_r	β_{12}	β_{rs}	α_1	α_2	α_r	β_{12}	β_{rs}
Wheland-Pauling[e] Longuet-Higgins-Coulson (1947)[e] Pullman[e]	$\alpha+2\beta$	$\alpha+\frac{1}{4}\beta$	α	β	β	$\alpha+2\beta$	$\alpha+\frac{1}{4}\beta$	α	β	β
Orgel *et al.*[e]	$\alpha+\beta$	$\alpha+\frac{1}{10}\beta$	α	β	β	$\alpha+2\beta$	$\alpha+\frac{1}{3}\beta$	α	β	β
Dewar[e]	...	...	...	...	...	α	α	α	0.55β	β
Present authors	$\alpha+\beta$	α	α	β	β	$\alpha+\beta$	α	α	β	β

[a] Subscript 1: nitrogen atom.
[b] Subscript 2: carbon atom adjacent to nitrogen.
[c] Subscript r: other carbon atoms.
[d] Subscript rs: any pair of mutually adjacent atoms except 12 (exchange integrals between nonadjacent atoms and all overlap integrals neglected entirely).
[e] See reference 1.

These parameters are tabulated in Table I together with those which are adopted by the present authors.*

The calculated densities of *frontier* electrons in pyridine, quinoline, isoquinoline, acridine, pyrrole, indole, carbazole, isoindole, and indolizine are indicated in Fig. 2, and the reaction products of halogenation, nitration, and sulfonation are tabulated in Table II. The agreement between calculated and experimental results is almost satisfactory.

Nuclear Substitution in Benzene Derivatives

In this paragraph the object of calculation is restricted to mono-substituted benzenes.† The twofold degeneracy of *frontier* orbitals in nonsubstituted benzene is removed by the influence of the substituent. If the perturbation is small, two separate orbitals, the corresponding energies of which are in close proximity, are produced as the result of the removal of degeneracy. In such a case, the contribution of the density of electrons in the lower orbital must be taken into account in the calculation of the *frontier* electron density.

In order to evaluate the degree of contribution of the lower orbital the following Eq. (1) is assumed as giving the *frontier* electron density for the electrophilic reaction at the rth atom in the molecule $f_r^{(E)}$:

$$f_r^{(E)}=2\cdot\frac{|C_r^{(1)}|^2+|C_r^{(2)}|^2 e^{-D\Delta\lambda}}{1+e^{-D\Delta\lambda}}, \tag{1}$$

where $\Delta\lambda$ is the energy difference between the two orbitals in units of β and $C_r^{(1)}$, $C_r^{(2)}$ are the coefficients of LCAO MO at the rth atom corresponding to the highest and the next orbital respectively, and D is a constant which determines the degree of contribution of the lower MO to the *frontier* electron density. Throughout the present paper‡ the value of D is taken, tentatively, to be 3 in consideration of the complete agreement between calculation and experiment which has been obtained[3] in aromatic hydrocarbons without considering the contribution of the next orbital.

In regard to LCAO MO treatment, mono-substituted benzenes are classified into several types of *isoaromatic*§ hydrocarbon, that is, benzene type, benzyl anion type, styrene type, and 2-phenylpropenyl anion type (Fig. 3), and other types.

Benzene Type

To this type belong, for example, *t*-butyl benzene and benzene sulfonic acid in which the effect of substituent is only inductive. This effect can be represented after Sandorfy[1] and Dewar[2] (1949) by two parameters a and ϵ in Eq. (2).

$$\alpha_r=\alpha+\epsilon^{r-1}a\beta, \tag{2}$$

TABLE II. Products of electrophilic substitution in heterocyclic compounds.

Compounds	Predicted position	Structure of reaction products: Halogenation	Nitration	Sulfonation
Pyridine	3,5	3-Cl[a] 3-Br[a]	3-NO_2[i]	3-SO_3H[p]
Quinoline	8, (5)	5-I[b]	5-NO_2, 8-NO_2[j]	5-SO_3H, 8-SO_3H[q]
Isoquinoline	5	...	5-NO_2, 8-NO_2[k]	...
Acridine	1	...	1-NO_2, 3-NO_2[l] 1,3-di-NO_2[m]	...
Pyrrole	2,5	2-Cl[c] 2,5-di-Cl[d]	...	...
Indole	3	2,3-di-Cl[e] 3-I[f]	3-NO_2[n]	...
Carbazole	1, (3)	3-Cl[g] 3-Br[h]	1-NO_2, 3-NO_2[o]	...

[a] H. J. Den Hertog and J. P. Wibaut, Rec. trav. chim. **51**, 382, 940 (1932)
[b] W. La Coste, Ber. deut. chem. Ges. **18**, 781 (1885).
[c] G. Mazzara and A. Borgo, Gaz. chim. ital. **35** II, 20 (1905).
[d] G. Mazzara and A. Borgo, Gaz. chim. ital. **35** I, 477 (1905).
[e] G. Mazzara and A. Borgo, Gaz. chim. ital. **35** II, 536 (1905).
[f] H. Pauly and K. Gunderman, Ber. deut. chem. Ges. **41**, 4006 (1908).
[g] Mazzara, Lamberti, and Zanardi, Gaz. chim. ital. **26** II, 238 (1896).
[h] W. Vaubel, Z. angew. Chem. **14**, 784 (1901).
[i] P. Schorigin and A. Toptschiew, Ber. deut. chem. Ges. **69**, 1874 (1936).
[j] W. Koenigs, Ber. deut. chem. Ges. **12**, 449 (1879).
[k] Ad. Claus and K. Hoffmann, J. prakt. Chem. (2) **47**, 253 (1893).
[l] C. Graebe and H. Caro, Ann. Chem. **158**, 275 (1871).
[m] K. Lehmsteat, Ber. deut. chem. Ges. **65**, 834 (1924).
[n] F. Angelico and G. Velardi, Gaz. chim. ital. **34** II, 60 (1904).
[o] H. Lindemann, Ber. deut. chem. Ges. **57**, 557 (1924).
[p] O. Fischer, Ber. deut. chem. Ges. **15**, 62 (1882).
[q] A. Kaufmann and H. Hüssig, Ber. deut. chem. Ges. **41**, 1735 (1908).

* As it is intended in the present paper to examine the general validity of the frontier electron theory using the smallest number of uncertain parameters, the effect of the heteroatom on the Coulomb integral of the adjacent carbon atom is neglected, as is shown in Table I.

† For di- or polysubstituted aromatic systems the present method can be also useful, but especially in benzene derivatives it becomes important to consider the steric circumstances.

‡ Also in nucleophilic or radical reaction, the equation similar to Eq. (1) is necessary, if any orbital lies in close proximity to the frontier orbitals.

§ See H. C. Longuet-Higgins, J. Chem. Phys. 18, 283 (1950).

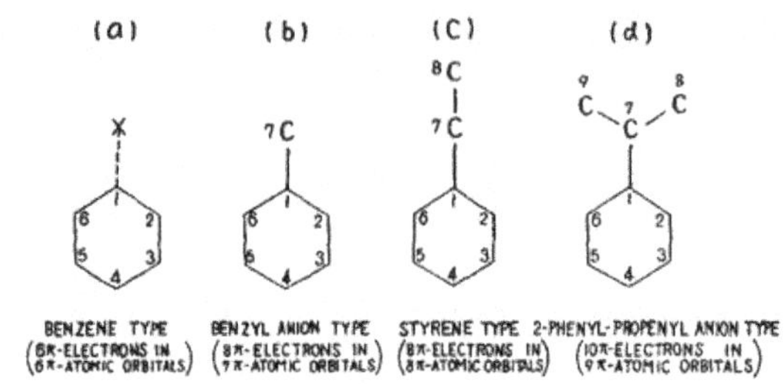

FIG. 3. Characteristic types of π-electron system in mono-substituted benzenes.

in which α_r is the Coulomb integral at the rth carbon atom (in Fig. 3). The parameter a is negative for the substituent which repells the σ electrons in the X$-$C σ bond, and positive for the substituent which attracts the σ electrons. The parameter ϵ, which represents the diminishing factor of the inductive effect, is taken here to be zero.|| In Fig. 4, the *frontier* electron densities at *ortho-*, *meta-*, and *para-*positions are given for various values of a ranging from -1 to $+1$. Qualitatively the results are as satisfactory as that obtained by other methods.[1,2]¶

Benzyl Anion Type

Halogenobenzenes, phenol, aniline, etc., belong to this type. Two parameters a and b** in Eq. (3) are adopted here as being characteristic of each substituent.

$$\alpha_1=\alpha+a\beta,$$
$$\alpha_7=\alpha+b\beta. \tag{3}$$

In Fig. 5, the calculated *frontier* electron densities are plotted for various values of a and b. In halogenobenzenes, phenol, and aniline, the values of a are assumed to be positive and small, so that, as is shown in Fig. 5 (b), the substituents are *ortho-para* directing.

The reactivity of toluene has been hitherto understood by considering the hyperconjugation†† or, otherwise, the purely inductive effect.‡‡ In consideration of the possible contribution of the structure

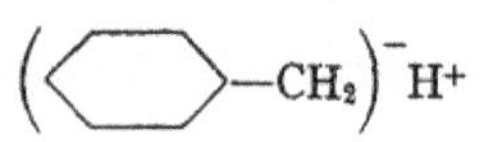

to the electronic state, it seems reasonable that toluene may be put into the class of benzyl anion type, where b is taken to be a large positive value and a a small negative value.§§

Styrene Type

Molecules which have the following substituents belong to this type:

$$-\mathrm{CH{=}CH_2},\quad -\mathrm{C{\equiv}N},\quad -\overset{+}{\mathrm{N}}{\equiv}\mathrm{N},\quad -\overset{\overset{\displaystyle \mathrm{R}}{|}}{\mathrm{C}}{=}\mathrm{O},\quad \text{etc.}$$

In this type another similar parameter c in addition to a and b in Eq. (3), is considered concerning the side-chain atom C_8. The calculated results are shown in Fig. 6. If $a\simeq 0$ and b, $c\ll 1$, the substituent is *ortho-para* directing, and when $a\simeq 0$ and $c>1$, it is *meta*-directing. In Fig. 6(a) the *frontier* electron density at C_8 atom in styrene is seen to be very large in conformity with the high reactivity of ω position of that compound.

2-Phenyl-propenyl Anion Type

Nitrobenzene, benzoate anion, benzoic esters, etc., belong to this type, and parameters a, b, d, and d' are adopted relating to the C_1, C_7, C_8, and C_9 atom, respectively. Calculated result for nitrobenzene is shown in Fig. 6(c).

Other Types

Diphenyl, stilbene, azobenzene, etc., belong to none of the above four types. These conjugated systems may be treated separately without classification. Calculated *frontier* electron densities for some of these molecules are shown in Fig. 6(d), (e), and (f).

Substitution in Substituted Condensed Aromatics

It is, in principle, possible to apply here the same method as used in the case of substituted benzenes,|| ||

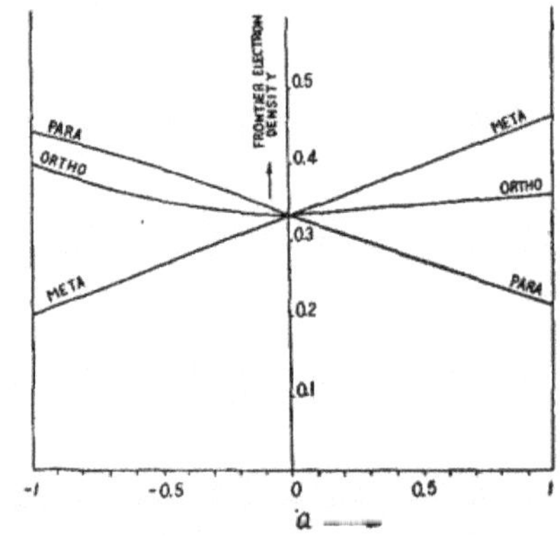

FIG. 4. Frontier electron density *vs* a in benzene type molecules.

|| For some appropriate values of ϵ which are not equal to zero the calculation is also carried out, but the results remain unchanged as to whether *ortho-para* directing or *meta*-directing effect is concluded.

¶ We reserve here the detailed discussions on the problem of $o:p$ ratio, as it seems hopeless to expect any valuable result, having neither sufficient knowledge for determining the exact values of a and ϵ nor a quantitative method of taking into account the influence of steric factors.

** According to Sandorfy (reference 1), Dewar (reference 2), and Jaffé [J. Chem. Phys. **20**, 279 (1952)], the relation between a and b is represented by $a=\epsilon b$, but it may not always be necessary to adhere to this relation when, as in this case, phenolate ion and other anions are also included in the benzyl anion type.

†† For example, Wheland (reference 2) considered

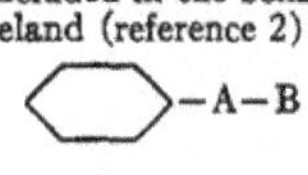

type.

‡‡ E.g., see Wheland and Pauling (reference 1) and Longuet-Higgins (reference 2).

§§ As an example, if $b=5$ and $a=-0.1$, following frontier electron densities are obtained: *ortho*$=0.3442$, *para*$=0.3697$, and *meta*$=0.2820$.

|| || For example, in 1-vinyl naphthalene, the positions ω and 4 are predicted to be most susceptible to attack, and in 2-vinyl

but this requires rather tedious calculations. Here we make use of a simplification similar to that adopted by Longuet-Higgins as to the treatment of substituted condensed aromatics.[4] Namely, the electronic effect of a substituent is represented by only one parameter, either a or b which are given by $\alpha_C=\alpha+a\beta$, and $\alpha_X=\alpha+b\beta$ (X: substituent; C: adjacent carbon). For the electron-repelling¶¶ substituents, we take $a<0$ or $b\geq 0$, and for the electron-attracting, $a>0$ or $b<0$.*** Thus the calculation is carried out as to the following three cases:

(1) The effect is purely inductive and small (the parameter a ($|a|\ll 1$) is adopted),
(2) The effect is mesomeric and small ($|b|\gg 1$),
(3) The effect is mesomeric and large ($b=0$).

The perturbation method is applied in the cases (1) and (2).

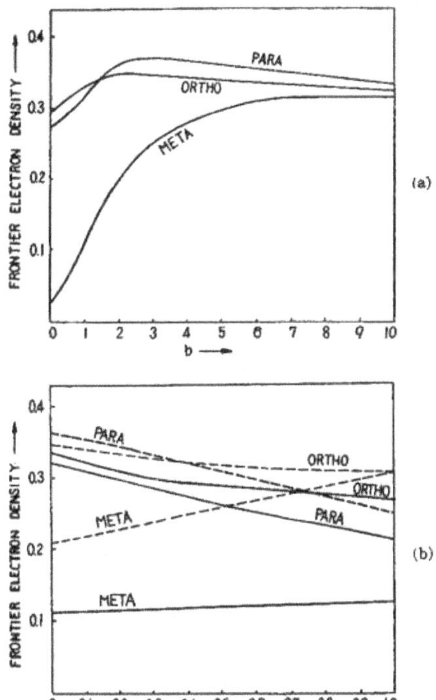

FIG. 5. Frontier electron density in benzyl anion types. (a) Frontier electron density *vs* b when $a=0$. (b) Frontier electron density *vs* a (>0) when $b=1$ (full line curves) and $b=2$ (dotted curves).

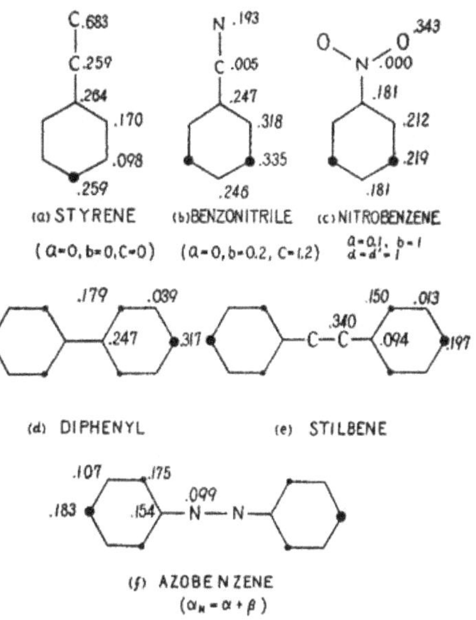

FIG. 6. Frontier electron density in some styrene type and 2-phenyl-propenyl anion type molecules, and in some complex benzene derivatives. Experimental results:

Compound	Reaction	Products
styrene	nitration	ω-NO$_2$[a]
benzonitrile	nitration	3-NO$_2$[b]
diphenyl	nitration	2-NO$_2$, 4-NO$_2$[c]
stilbene	...	...
azobenzene	nitration	p-NO$_2$, p,p'-di-NO$_2$[d]

[a] E. Simon, Ann. Chem. **31**, 269 (1839).
[b] M. Schöpff, Ber. deut. chem. Ges. **18**, 1063 (1885).
[c] Schultz, Schmidt, and Strasser, Ann. Chem. **207**, 352 (1881).
[d] A. Werner and E. Stiasny, Ber. deut. chem. Ges. **32**, 3268 (1899).

As to 1- and 2-substituted naphthalenes the results of calculation are shown in Fig. 7. As can be seen in the figure the *frontier* electron distributions in the case $|b|\gg 1$ coincides exactly with that in the case $|a|\ll 1$ if a is taken to be $-1/b$. The coincidence can be easily proved mathematically for any aromatic hydrocarbon. Thus *the two effects, mesomeric and inductive, may be discussed quite indiscriminately if they are small and in the same direction.*

In Table III experimental results as to the orientation in substituted naphthalenes are compared with the predictions obtained from Fig. 7. The agreement is good, when, as for electron-repelling substituents, b is taken to be zero for amino and hydroxyl and to be large for halogen and methyl and when, as for electron-attracting substituents, a as well as b is assumed to be small. The results of similar treatments of other substituted condensed aromatics are tabulated in Table IV, showing a good agreement of the theory with experiment.

4. NUCLEOPHILIC SUBSTITUTION

The nucleophilic substitution treated here is restricted to the direct replacement of a hydrogen atom with

naphthalene, the positions 1 and ω. These predicted positions are identical for electrophilic, nucleophilic, and radical reactions.

[4] H. C. Longuet-Higgins (see reference 2); A. and B. Pullman, Rev. sci. **84**, 145 (1946). Longuet-Higgins attempted to explain the orientation in 2-naphthylamine by calculating the densities of total π electrons, taking $\alpha_N=\alpha$. He considered that the effect of amino group was approximately the same as that of $-CH_2$ group in the corresponding isoaromatic system.

¶¶ Here, the substituents which in *benzene* derivatives are *ortho-para*-directing are referred to as electron-repelling substituents, and those which are *meta*-directing, as electron-attracting substituents.

*** The case $b=0$ is omitted for the convenience of calculation.

TABLE III. Products of electrophilic substitution in substituted naphthalenes.

(a) Electron-repelling substituents

Compounds	Predicted position	Structure of the reaction products: Chlorination	Bromination	Nitration	Sulfonation
1-F-naphthalene	4	...	4-Br[b]	...	4-SO_3H[h]
1-Cl-naphthalene	4	4-Cl[a]	4-Br[i]	4-NO_2[a]	4-SO_3H, 5-SO_3H[cc]
1-Br-naphthalene	4	4-Cl[b]	4-Br[j] 5-Br[k]	...	4-SO_3H[dd]
1-I-naphthalene	4	4-Cl[c]	4-Br[l]	...	...
1-OH-naphthalene	2,4	2-Cl[d] 4-Cl[e] 2,4-di-Cl[e]	...	2-NO_2[t] 2,4-di-NO_2[u]	2-SO_3H[ee] 4-SO_3H[ff] 2,4-di-SO_3H[ee]
1-NH_2-naphthalene	2,4	4-Cl[f]	2,4-di-Br[m]	(4-NO_2, 5-NO_2, 8-NO_2[v])	4-SO_3H[gg] 2,4-di-SO_3H[gg]
1-CH_3-naphthalene	4	...	4-Br[n]	4-NO_2[w]	4-SO_3H[hh]
2-F-naphthalene	1,8	...	...	1-NO_2[h]	...
2-Cl-naphthalene	1,8	...	...	...	8-SO_3H, 6-SO_3H[ii]
2-Br-naphthalene	1,8	...	...	1-NO_2, 3-NO_2[x]	8-SO_3H[jj]
2-I-naphthalene	1,8	...	1-Br[o]	...	8-SO_3H[jj]
2-OH-naphthalene	1	1-Cl[g]	1-Br[p] 8-Br[q] 1,6-di-Br[p]	1-NO_2[y] 1,6-di-NO_2[z]	1-SO_3H[kk] 6-SO_3H[ll]
2-NH_2-naphthalene	1	...	1,6-di-Br[r]	(8-NO_2, 5-NO_2[aa])	(5-SO_3H, 8-SO_3H[mm])
2-CH_3-naphthalene	1,8	...	...	1-NO_2[bb]	6-SO_3H[hh]

(b) Electron-attracting substituents

Compounds	Predicted position	Structure of the reaction products: Halogenation	Nitration	Sulfonation
1-COOH-naphthalene	8,5	5-Cl[nn] 8-Cl[nn] 5-8-di-Cl[oo] 5-Br[pp]	5-NO_2[aaa] 8-NO_2[aaa]	5-SO_3H[ggg] 6-SO_3H[ggg] 7-SO_3H[ggg]
1-NO_2-naphthalene	8,5	5-Cl[qq] 8-Cl[qq] 5,8-di-Cl[rr] 5-Br[ss]	5-NO_2[bbb] 8-NO_2[bbb]	5-SO_3H[hhh]
1-SO_3H-naphthalene	8,5	5-Br[tt]	4-NO_2[ccc] 5-NO_2[ccc] 8-NO_2[ccc]	5-SO_3H[iii]
1-CHO-naphthalene	8,5	5-Br[uu]	...	...
1-CN-naphthalene	8,5	5-Cl[vv] 5-Br[ww]	5-NO_2[ddd]	...
2-COOH-naphthalene	5,4	5,8-di-Cl[xx] 5-Br[yy]	5-NO_2[eee] 8-NO_2[eee]	5-SO_3H[jjj] 8-SO_3H[kkk]
2-NO_2-naphthalene	5,4	...	...	5-SO_3H[lll] 8-SO_3H[lll]
2-SO_3H-naphthalene	5,4	5-Cl[zz] 8-Cl[zz]	5-NO_2[fff] 8-NO_2[fff]	5-SO_3H[lll] (6 and 7-SO_3H[mmm])

[a] H. E. Armstrong, Chem. News **66**, 189 (1892).
[b] I. Guareschi and P. Biginelli, Chem. Zentr. **1877**, 518 (1877).
[c] C. Willgerodt and P. Schlöser, Ber. deut. chem. Ges. **33**, 693 (1900).
[d] R. Lesser and G. Gad, Ber. deut. chem. Ges. **56**, 972 (1923).
[e] P. T. Cleve, Ber. deut. chem. Ges. **21**, 894 (1886).
[f] P. Seidler, Ber. deut. chem. Ges. **11**, 1201 (1878).
[g] Ioffe, Kuznetzov, and Litovskii, J. Gen. Chem. (U.S.S.R.) **5**, 1685 (1935).
[h] Schieman, Gueffroy, and Wikelmüller, Ann. Chem. **487**, 270 (1931).
[i] I. Guareschi and P. Biginelli, Gaz. chim. ital. **16**, 152 (1886).
[j] C. Glaser, Ann. Chem. **135**, 42 (1865).
[k] Yu. S. Zalkind and S. B. Faermann, J. Russ. Phys.-Chem. Soc. **62**, 1021 (1930).
[l] H. Hirtz, Ber. deut. chem. Ges. **29**, 1409 (1896).
[m] R. Condon and J. Kenyon, J. Chem. Soc. (London) **1935**, 1591.
[n] L. F. Fieser and M. Fieser, *Organic Chemistry* (D. C. Heath and Company, Boston, 1950), second edition, p. 792.
[o] H. Hirtz, Ber. deut. chem. Ges. **29**, 1409 (1896).
[p] Verma, Mozumdar, and Rajah, J. Indian. Chem. Soc. **10**, 591 (1933).
[q] K. Fries and K. Schimmelschmidt, Ber. deut. chem. Ges. **58**, 2840 (1925).
[r] L. Michaellis, Ber. deut. chem. Ges. **26**, 2196 (1893).
[s] A. Atterberg, Ber. deut. chem. Ges. **9**, 927 (1876).
[t] H. H. Hodgson and E. W. Smith, J. Chem. Soc. (London) **1935**, 671 (1935).
[u] J. Schmidt, Ber. deut. chem. Ges. **33**, 3245 (1900).
[v] R. Meloda and J. Streatfeld, J. Chem. Soc. (London) **63**, 1055 (1893).
[w] R. Lesser, Ann. Chem. **402**, 11 (1914).
[x] Yu. Zalkind and F. Filinov, J. Gen. Chem. (U.S.S.R.) **4**, 979 (1943).
[y] A. Pictet de Krijanowski, Chem. Zentr. **1903**, II, 1109 (1903).
[z] J. Schmidt, Ber. deut. chem. Ges. **33**, 3246 (1900).
[aa] P. Friedländer and St. Szymanski, Ber. deut. chem. Ges. **25**, 2076 (1892).
[bb] H. E. Fierz-David and E. Mannhart, Helv. Chim. Acta **20**, 1024 (1937).
[cc] H. E. Armstrong and W. P. Wynne, Chem. News **61**, 285 (1890).
[dd] H. E. Armstrong and S. Williamson, Chem. News **54**, 256 (1886).
[ee] M. Conrad and W. Fischer, Ann. Chem. **273**, 105 (1893).
[ff] K. Holdermann, Ber. deut. chem. Ges. **39**, 1225 (1906).
[gg] W. H. Hunter and M. M. Sprung, J. Am. Chem. Soc. **53**, 1432 (1931).
[hh] L. F. Fieser and M. Fieser, *Organic Chemistry* (D. C. Heath and Company, Boston, 1950), second edition, p. 793.
[ii] H. E. Armstrong and W. P. Wynne, Chem. News **55**, 91 (1887).
[jj] H. E. Armstrong and W. P. Wynne, Chem. News **60**, 58 (1889).
[kk] H. E. Armstrong, Ber. deut. chem. Ges. **15**, 202, 207 (1882).
[ll] L. Schaeffer, Ann. Chem. **152**, 296 (1869).
[mm] H. Erdmann, Ann. Chem. **275**, 280 (1893).
[nn] A. G. Ekstrand, J. prakt. Chem. **38**, 148 (1888).
[oo] A. G. Ekstrand, J. prakt. Chem. **38**, 151, 153 (1888).
[pp] O. Gausmann, Ber. deut. chem. Ges. **9**, 1520 (1875).
[qq] A. G. F. Anilinf, Germ. Pat. 99, 758; Chem. Zentr. **1899** I, 463 (1899).
[rr] Bayer & Co. Germ. Pat. 293, 318; Chem. Zentr. **1916** II, 359.
[ss] I. Guareschi, Ann. Chem. **222**, 291 (1884).
[tt] L. Darmstaedter and H. Wichelhaus, Ann. Chem. **152**, 303 (1869).
[uu] P. Ruggli and R. Preus, Helv. Chim. Acta **24**, 1345 (1941).
[vv] A. G. Ekstrand, J. prakt. Chem. **38**, 147 (1888).
[ww] O. Hausmann, Ber. deut. chem. Ges. **9**, 1516 (1876).
[xx] A. G. Ekstrand, Ber. deut. chem. Ges. **17**, 1605 (1884).
[yy] H. Goldstein and R. Matthey, Helv. Chim. Acta **21**, 62 (1938).
[zz] Chr. Rudolf, Germ. Pat. 101, 348; Chem. Zentr. **1899** I, 960 (1899).
[aaa] A. G. Ekstrand, J. prakt. Chem. **38**, 156, 241 (1888).
[bbb] P. Friedländer, Ber. deut. chem. Ges. **32**, 3531 (1899).
[ccc] H. Erdmann and C. Süvern, Ann. Chem. **275**, 231 (1893).
[ddd] Fr. Graeff, Ber. deut. chem. Ges. **16**, 2246 (1833).
[eee] A. G. Ekstrand, J. prakt. Chem. **42**, 273 (1890).
[fff] H. E. Armstrong and W. P. Wynne, Chem. News **59**, 94 (1889).
[ggg] M. Stumpf, Ann. Chem. **188**, 10 (1877).
[hhh] H. Erdmann, Ann. Chem. **247**, 311 (1888).
[iii] A. E. Armstrong and W. P. Wynne, Chem. News **55**, 136 (1887).
[jjj] M. Stumpf, Ann. Chem. **188**, 10 (1877).
[kkk] M. Stumpf, Ann. Chem. **188**, 12 (1877).
[lll] H. Kappeler, Ber. deut. chem. Ges. **45**, 633 (1912).
[mmm] H. E. Armstrong, Ber. deut. chem. Ges. **15**, 204 (1882).

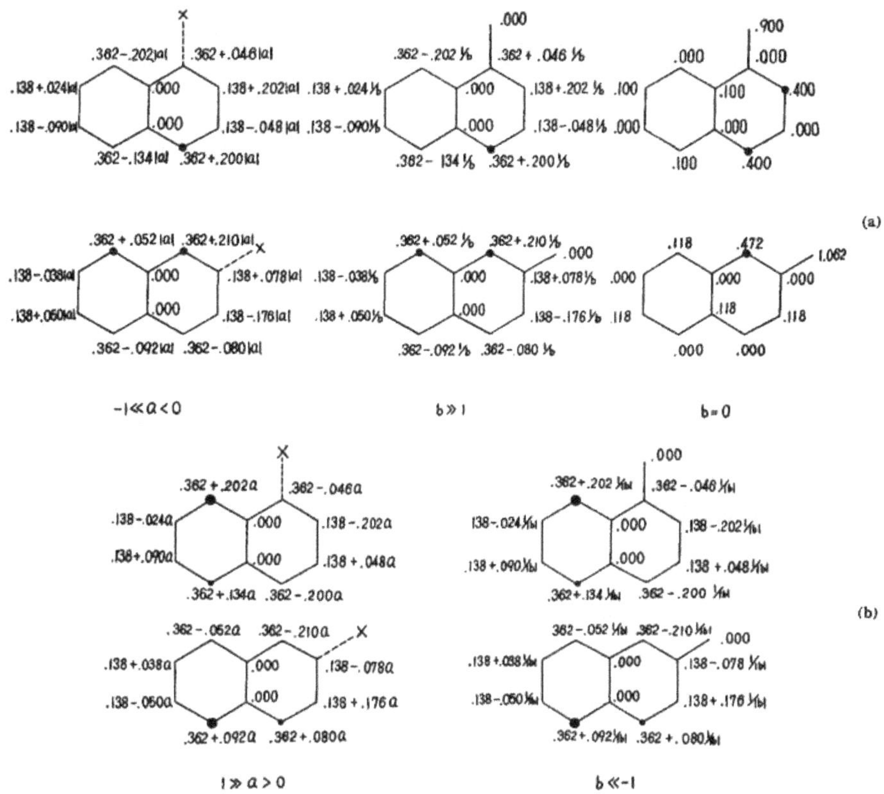

FIG. 7. The frontier electron densities in 1- and 2-substituted naphthalenes. (a) Electron-repelling substituent. (b) Electron-attracting substituent.

nucleophilic reagents such as OH^- and NH_2^-.††† The results of the calculation as to the five known cases, are shown in Fig. 8 together with the experimental results.

5. RADICAL SUBSTITUTION

The calculated *frontier* electron densities‡‡‡ as to the reactivity in radical substitution in several aromatic and

††† A typical example of the nucleophilic substitution in conjugated systems may be the hydrolysis of aryl halides, which, however, cannot be the object of the present treatment, because in such a reaction the point of attack is originally fixed. Of course it is possible, by means of the frontier electron method, to predict which one of all the possible mono-halogeno-derivatives is most liable to hydrolysis.

‡‡‡ In this calculation the contributions of two unpaired electrons in different two orbitals are assumed to be in equal weight.

TABLE IV. Products of electrophilic substitution in substituted condensed aromatic hydrocarbons.

(a) Electron-repelling substituents				
		Structure of the reaction products		
Compounds	Predicted position	Halogenation	Nitrosation or nitration	Sulfonation
9-Cl-anthracene	10	10-Cl[a]	...	...
9-Br-anthracene	10	10-Br[b]	...	...
9-OH-anthracene	10	10-Br[c]	...	...
1-OH-anthracene	2,4	...	2-NO, 4-NO[m]	...
2-OH-anthracene	1	...	1-NO[m]	...
9-Cl-phenanthrene	10	10-Cl[d]	...	...
9-Br-phenanthrene	10	...	10-NO_2[n]	(2-SO_3H[p])
9-OH-phenanthrene	10	10,3-di-Br[e]	...	...
2-Cl-chrysene	8	8-Cl[f]	8-NO_2[o]	...
2-Br-chrysene	8	...	8-NO_2[o]	...
3-Br-pyrene	5,8	5,8-di-Br[g]	...	...
3-NH_2-pyrene	5,8	5-Cl[h]	...	5-SO_3H, 8-SO_3H[q]
12-Cl-naphthacene	5	5-Cl[i], 6-Cl[j]	...	...
12-Br-naphthacene	5	5-Br[i], 6-Br[j]	...	...
3-Cl-perylene	10	9-Cl[k]	...	...
3-Br-perylene	9,10	9-Br, 10-Br[l]	...	...

(TABLE IV. *Continued on next page with footnotes.*)

TABLE IV—*Continued.*

(b) Electron-attracting substituents

Compounds	Predicted position	Structure of the reaction products: Halogenation	Nitration	Sulfonation
9-NO_2-anthracene	10	...	10-NO_2[s]	...
9-COOH-anthracene	10	10-Br,10-Cl[r]	...	...
2-NO_2-chrysene	8	...	8-NO_2[t]	...
2-CN-chrysene	8	...	8-NO_2[t]	...
2-COOH-chrysene	8	...	8-NO_2[t]	...
3-SO_3H-pyrene	5	...	...	5-SO_3H[v]
3-NO_2-perylene	4,(9)	...	(10-NO_2[u])	...
3-SO_3H-perylene	4,(9)	...	...	(9-SO_3H, 10-SO_3H[w])
12-substituted naphthacene	11	...	...	...
13-substituted pentacene	6	...	...	...

[a] Ed. Lippmann and I. Pollack, Ber. deut. chem. Ges. **34**, 2768 (1901).
[b] C. Graebe and C. Liebermann, Ber. deut. chem. Ges. 1, 186 (1868).
[c] Fr. Goldmann, Ber. deut. chem. Ges. **20**, 2437 (1887).
[d] H. Sandqvist and A. Hagelin, Ber. deut. chem. Ges. **51**, 1521 (1919).
[e] J. Schimel and O. Spoun, Ber. deut. chem. Ges. **43**, 1802 (1910).
[f] Soc. Pour. L'ind. Chim. à Bâle, Swiss Pat. 204, 239 (July 17, 1939).
[g] E. Clar, Ber. deut. chem. Ges. **69**, 1671 (1936).
[h] Corell, Vollman, and Becker, U.S.P. 2,185,661 (Jan. 2, 1939).
[i] I. J. Postovskii and R. G. Beyles, Compt. Rend. Acad. Sci. U.R.S.S. **39**, 102 (1943).
[j] C. Marschalk and C. Stumm, Bull. soc. chim. France **1948**, 418 (1948).
[k] Zinke, Funke, and Pongratz, Ber. deut. chem. Ges. **58**, 330 (1925).
[l] Zinke, Kinner, and Wolfbauer, Ber. deut. chem. Ges. **58**, 328 (1925).
[m] R. Anschütz, *Chemie der Kohlenstoffverbindungen* (1935), p. 702.
[n] R. K. Callow and J. M. Gulland, J. Chem. Soc. (London) **1929**, 2424.
[o] I. G. Farbenind. A.-G. Fr. Pat., 794, 534 (February 19, 1936).
[p] L. E. May and E. Mosettig, J. Org. Chem. **11**, 15 (1946).
[q] M. Corell, U.S.P. 2,046,249 (June 30, 1936).
[r] G. Behla, Ber. deut. chem. Ges. **20**, 704 (1887).
[s] J. Meisenheimer, and E. Connerade, Ann. Chem. **330**, 167 (1904).
[t] I. G. Farbenind. A.-G. Fr. Pat., 794,534 (Feb. 19, 1936).
[u] Zinke, Hirsch, and Brozek, Monatshefte Chem. **51**, 205 (1929).
[v] E. Tietze and O. Bayer, Ann. Chem. **540**, 189 (1939).
[w] C. Marschalk, Chem. Zentr. **1927** I, 1833 (1927).

heteroaromatic molecules are shown in Fig. 9 together with experimental results. The agreements of theory with experiment are satisfactory. (Similar results are obtainable by means of the localization method[2] and also by computing the free valency number.[5] The latter method has been developed by Daudel, Vroelant, and Coulson *et al.*[5] and most recently examined experimentally by Kooyman and Farenhorst,[6] and Roitt and Waters.[7] It can, however, be pointed out that the free valency method contains an ambiguity in evaluating N_{max} and requires a most laborious calculation of bond order in larger organic molecules.)

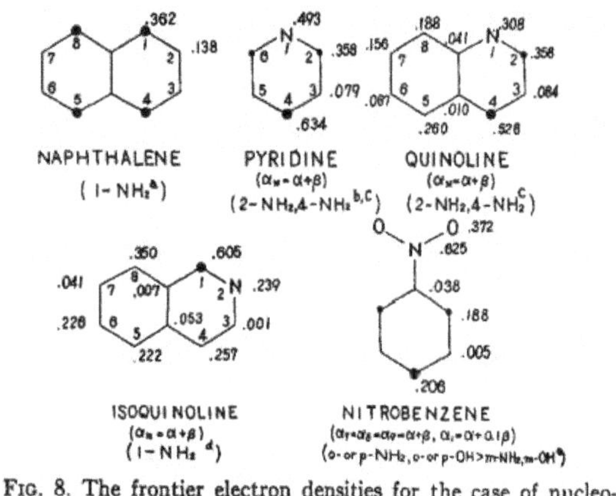

FIG. 8. The frontier electron densities for the case of nucleophilic substitution. (Experimental results are shown in brackets. The parameters used are the same as before.)

Literatures cited in Fig. 8:

[a] F. Sachs, Ber. deut. chem. Ges. **39**, 3006 (1906); C. Graebe, *ibid.* **34**, 1778 (1901).
[b] A. E. Chichibabin, Ber. deut. chem. Ges. **56**, 1879 (1923).
[c] A. E. Chichibabin, J. Russ. Phys.-Chem. Soc. **46**, 1216 (1914).
[d] M. T. Leffler, *Organic Reaction* (1942), p. 91.
[e] A. Wohl, Ber. deut. chem. Ges. **32**, 3486 (1899); Bergstrom, Granare, and Erickson, J. Organ. Chem. **7**, 98 (1942).

[5] C. Vroeland and R. Daudel, Bull. soc. chim. France **16**, 217 (1949); Buu-Hoi, Daudel, and Vroelant, Bull. soc. chim. France **16**, 211 (1949); Daudel, Sandorfy, Vroelant, Yvan, and Chalvet, Bull. soc. chim. France **19**, 66 (1950); Burkitt, Coulson, and Longuet-Higgins, Trans. Faraday Soc. **47**, 553 (1951).

[6] E. C. Kooyman and E. Farenhorst, Nature **169**, 153 (1952).

[7] I. M. Roitt and W. A. Waters, J. Chem. Soc. (London) **1952**, 2695 (1952).

In *alternant* hydrocarbons,[8] the electron distribution in the lowest unoccupied orbital is identically the same

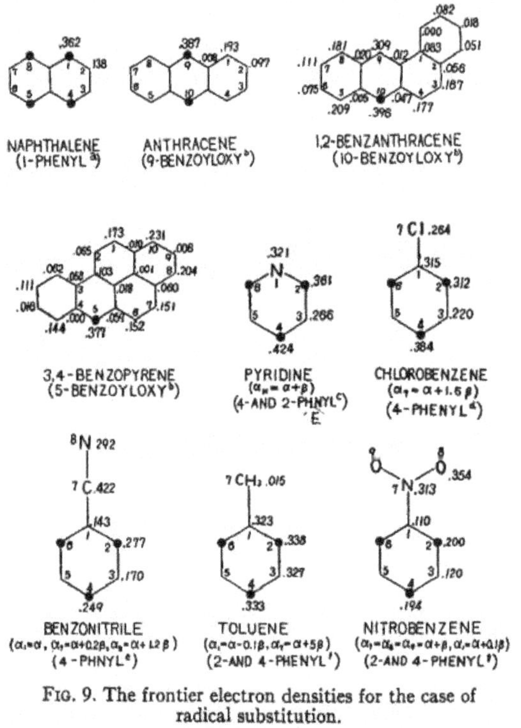

FIG. 9. The frontier electron densities for the case of radical substitution.

Literatures cited in Fig. 9:

[a] W. A. Waters, J. Chem. Soc. (London) **1939**, 864 (1939).
[b] See reference 7.
[c] Haworth, Heilbron, and Hey, J. Chem. Soc. (London) **1940**, 349 (1940).
[d] D. H. Hey, J. Chem. Soc. (London) **1934**, 1966.
[e] M. Gomberg and W. E. Bachmann, J. Am. Chem. Soc. **46**, 2389 (1924).
[f] M. Gomberg and J. C. Pernert, J. Am. Chem. Soc. **48**, 1372 (1926).
[g] D. Detar and H. Scheifele, Jr., J. Am. Chem. Soc. **73**, 1442 (1951).

[8] See C. A. Coulson and H. C. Longuet-Higgins, Proc. Roy. Soc. (London) **A192**, 16 (1947).

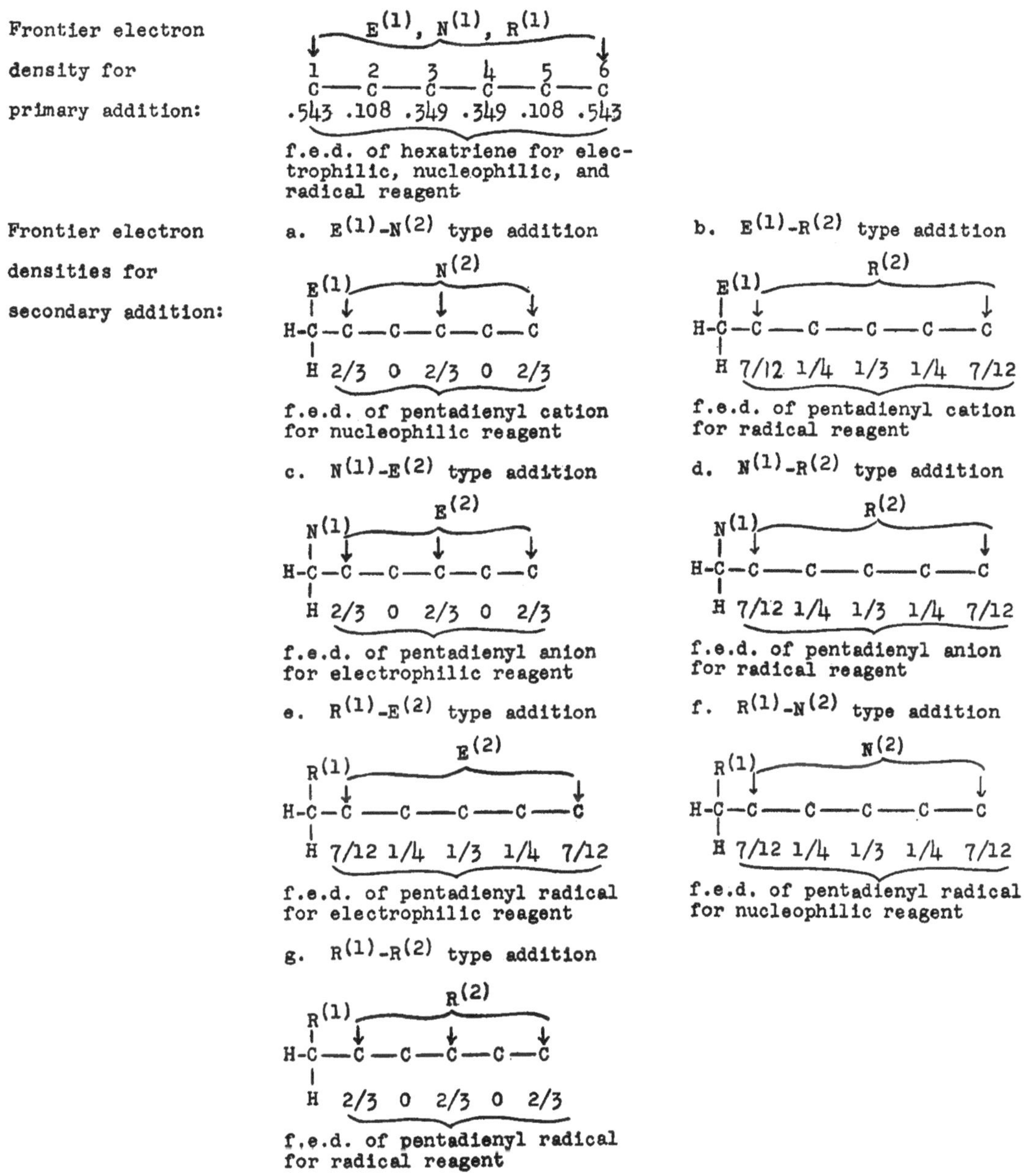

FIG. 10. The frontier electron densities for addition to nexatriene (f.e.d. = frontier electron density).

as in the highest occupied orbital,[9] so that the predicted points of attack are identical in any of electrophilic, nucleophilic, and radical substitution. In most of hydrocarbons treated there predicted positions agree very well with the points of the maximum free valency number, the values of which are found by Kooyman and Farenhorst[6] to be parallel with the rates of radical substitution.

[9] See our previous paper (reference 3).

6. ADDITION

Addition reactions to a conjugated system may be considered to proceed successively through the transition state similar to that of substitution. (Brown[10] and Dewar[11] assumed a simultaneous addition in their treatment of the Diels-Alder diene synthesis. But, from the reaction-kinetical standpoint, successive mechanism

[10] R. D. Brown, J. Chem. Soc. (London) **1950**, 691, 2720, 3249 (1950).

[11] M. J. S. Dewar, J. Am. Chem. Soc. **74**, 3357 (1952).

TABLE V. Comparison of predicted structures of addition products with experimental results.

Conjugated molecules	Predicted positions	Experimental results
butadiene	1:2, 1:4	1:2, 1:4
hexatriene	1:2, 1:4, 1:6	1:2, 1:6[a]
styrene	$\alpha:\omega$	$\alpha:\omega$
stilbene	$\alpha:\alpha'$	$\alpha:\alpha'$
anthracene	9:10	9:10
phenanthrene	9:10	9:10

[a] Farmer, Laroria, Switz, and Thorpe, reference 12.

is naturally to be preferred.) Then we can apply here the *frontier* electron method analogously, using the same classification into three types, *viz.*, electrophilic (E), nucleophilic (N), and radical (R).

In discussing the reactivity in polyene and some aromatic molecules, seven cases§§§ of successive addition are to be considered according to the type of reagent, which are indicated in Fig. 10.

Taking hexatriene as an example, the primary attack is predicted to occur at the terminal carbon atom, in any case of $E^{(1)}$, $N^{(1)}$, and $R^{(1)}$. In the secondary attack, as one atomic π orbital at the terminal carbon has disappeared as a result of the primary addition, the point of attack is now controlled by the *frontier* electron density of a conjugated system consisting of five carbon atoms. From the results shown in Fig. 10, it can be concluded that the structures of the addition products are predicted to be 1:2, 1:4, or 1:6, taking all the possible cases of addition into consideration. Experimentally, addition product of bromine to hexatriene is reported to be a mixture of 1:2- and 1:6-dibromide.[12]

Similar calculations for all the possible modes of addition are carried out as to butadiene, stilbene, styrene, anthracene, and phenanthrene. The results are in a complete agreement with experiment, as is shown in Table V.‖‖‖

It is possibly of importance in obtaining a knowledge of the true feature of activated complexes to consider the theoretical foundation of the fundamental postulates, which, however, is omitted in the present paper and will be published elsewhere.

The authors are grateful to the Education Ministry of the Japanese Government for a grant-in-aid.

§§§ Combining the three types of primary attack, *viz.*, $N^{(1)}$, $E^{(1)}$, and $R^{(1)}$, with three types of secondary attack, *viz.*, $E^{(2)}$, $N^{(2)}$, and $R^{(2)}$, we have nine varieties of mode of addition. But the two among these, $E^{(1)}-E^{(2)}$ and $N^{(1)}-N^{(2)}$ type additions, which are not likely to happen, are left out of consideration.

[12] Farmer, Laroria, Switz, and Thorpe, J. Chem. Soc. (London) **1927**, 2937 (1927).

‖‖‖ The frontier electron method is also useful in the treatment of cationoid, anionoid, and radical polymerization, which will be published elsewhere.

Reprinted from THE JOURNAL OF CHEMICAL PHYSICS, Vol. 26, No. 4, 831–841, April, 1957
Printed in U. S. A.

Interrelations of Quantum-Mechanical Quantities Concerning Chemical Reactivity of Conjugated Molecules

KENICHI FUKUI, TEIJIRO YONEZAWA, AND CHIKAYOSHI NAGATA
Faculty of Engineering, Kyoto University, Kyoto, Japan
(Received June 25, 1956)

It is attempted here to disclose the interrelations underlying the parallelism found in several quantum-mechanical methods previously proposed to explain chemical reactivity in conjugated molecules. This is carried out mathematically in alternant hydrocarbons and hetero-alternant molecules mainly by the use of Coulson's integral formulas. In an alternant hydrocarbon, provided a condition is satisfied between minors of the secular determinant, the orders of reactivity of each position predicted by the self-polarizability, free valence, superdelocalizability, Wheland's localization energy, and Dewar's approximate localization energy by NBMO are shown to be in complete agreement with one another. The relation of the frontier electron density and superdelocalizability is discussed. It is also demonstrated that a nonperturbation treatment for the so-called static method leads us to the same conclusion as the perturbation treatment. The parallelism of various quantities is discussed in their relation to the effect of heteroatom in hetero-alternant molecules.

INTRODUCTION

SEVERAL quantum-mechanical quantities which relate to accounting for chemical reactivity in conjugated molecules have been proposed in LCAO MO treatments. These are such as π-electron density,[1] self-polarizability,[2] free valence,[3] localization energy,[4] frontier electron density,[5] superdelocalizability of π electrons,[6] ion-free valence,[7] atom stabilization energy,[8] etc.

Most of these quantities may be said to have their specific physical grounds in the sense that they can be derived from some considerations relating to the process of reaction. The physical standpoints on which these quantities are based, however, differ considerably from each other; nevertheless, each of these quantities has proved to explain well the experimental results to an almost equal extent. This is not always self-explanatory, and therefore it is to be expected that any correlation of these quantities might exist with each other. Several previous authors have partly attempted to reveal these relationships.[3,9–13]

In the present paper, by using mainly the Coulson integral formulas of these quantities, it is shown that they are closely correlated mathematically with each other in alternant hydrocarbons and also in hetero-alternant molecules.

[1] G. W. Wheland and L. Pauling, J. Chem. Phys. **1**, 606 (1933), etc.
[2] C. A. Coulson and H. C. Longuet-Higgins, Proc. Roy. Soc. (London) **A191**, 39 (1947); **A192**, 16 (1947).
[3] Burkitt, Coulson, and Longuet-Higgins, Trans. Faraday Soc. **47**, 553 (1951).
[4] G. W. Wheland, J. Am. Chem. Soc. **64**, 900 (1940).
[5] Fukui, Yonezawa, and Shingu, J. Chem. Phys. **20**, 722 (1952); Fukui, Yonezawa, Nagata, and Shingu, J. Chem. Phys. **22**, 1433 (1954).
[6] Fukui, Yonezawa, and Nagata, Bull. Chem. Soc. Japan **27**, 423 (1954).
[7] S. Nagakura and J. Tanaka, J. Chem. Soc. Japan. Pure Chem. Sec. **75**, 993 (1954).
[8] T. Nakajima, J. Chem. Phys. **25**, 587 (1955).
[9] H. H. Greenwood, Trans. Faraday Soc. **48**, 585 (1952); J. Chem. Phys. **20**, 1333 (1952).
[10] H. Baba, *Summary of the Symposium on π Electrons* (Chemical Society of Japan, 1954), p. 60.
[11] M. J. S. Dewar, J. Am. Chem. Soc. **74**, 3355 (1952).
[12] B. Pullman, Cahier phys. **48**, 42 (1954).
[13] R. D. Brown, Quart. Rev. Chem. Soc. **6**, No. 1, 63 (1952).

NOTATION

n—the number of atomic π orbitals in an isolated molecule.

$2m$—the number of π electrons in the isolated molecule.

C_r^j—the coefficient of the atomic π orbital for the rth atom in the jth molecular orbital in the isolated molecule.

$\alpha+\epsilon_j$—the energy of the jth molecular orbital ($\epsilon_1<\epsilon_2<\cdots<\epsilon_n$) in the isolated molecule.

$\alpha+\alpha_r$—the Coulomb integral for the rth atom.

α—the Coulomb integral for a carbon atom in benzene.

β—the resonance integral between two adjacent carbon atoms in benzene

ν—a number representing 0 for an electrophilic substitution, 1 for a radical substitution, and 2 for a nucleophilic substitution.[6]

$\Delta(\epsilon)$—the secular determinant for the isolated molecule.

$\Delta_{rs}(\epsilon)$—the minor of $\Delta(\epsilon)$ obtained by striking off its rth row and sth column

$$\Delta'(\epsilon)=\frac{\partial\Delta(\epsilon)}{\partial\epsilon},\quad \Delta_{rr}'(\epsilon)=\frac{\partial\Delta_{rr}(\epsilon)}{\partial\epsilon}.$$

$\alpha+\eta_{r,j}$—the jth root of $\Delta_{rr}(\epsilon)=0$ ($\eta_{r1}<\eta_{r2}<\cdots<\eta_{r,n-1}$).

$D_r(\epsilon)$—the secular determinant for the total reacting system including the quasi-π orbital for the pseudoatom attached to the rth atom in the molecule[6]; i.e.

$$D_r(\epsilon)=(\alpha_h-\epsilon)\Delta(\epsilon)-\gamma^2\Delta_{rr}(\epsilon). \tag{1}$$

$\alpha+\alpha_h$—the Coulomb integral for the pseudoatom

γ—the resonance integral between the quasi-π orbital and the atomic π orbital for the rth atom.

$\alpha+\mu_{rj}$—the jth root of $D_r(\epsilon)=0$

$(\mu_{r1}<\mu_{r2}<\cdots<\mu_{r,n+1})$.

q_r—the π-electron density at the rth atom.[1,2]

P_{rs}—the mobile order of bond rs.[2]

π_{rr}—the self-polarizability[2] of the rth atom in units of $1/\beta$.

F_r—the free valence of the rth atom.[3]

$$N_r=N_{\max}-F_r=\sum_s{}' P_{rs}. \qquad (2)^3$$

(As to the following symbols, see the equation indicated at the right of each)

f_r Eq. (4) $\mathfrak{H}_r(\gamma)$ Eq. (8) $g_r^{(l)}$ Eq. (20)

S_r Eq. (5) $G_r(y)$ Eq. (10) $U^{(l)}$ Eq. (21)

F_r^+, F_r^- Eq. (6) a_i, $b_{r,i}$ Eq. (15) $\mathfrak{A}_r(\Delta\alpha)$ Eq. (25)

E_r, E_r^+, E_r^- Eq. (7) $T_r^{(l)}$ Eq. (19) $\mathfrak{B}_r(\Delta\beta)$ Eq. (26)

l_r Eq. (28)

DEFINITIONS OF REACTIVITY QUANTITIES*

Localization energy at the rth atom[4]:

$$L_r=2\sum_{j=1}^{m}(\eta_{rj}-\epsilon_j)+(2-\nu)(\alpha_r-\eta_{rm}). \qquad (3)$$

Frontier electron density at the rth atom[5]:

$$f_r=(2-\nu)C_r^{m^2}+\nu C_r^{(m+1)^2}. \qquad (4)$$

Superdelocalizability of π electrons at the rth atom[6]:

$$S_r=(2-\nu)\sum_{j=1}^{m}\frac{C_r^{j^2}}{\alpha_h-\epsilon_j}(-\beta)+\nu\sum_{j=m+1}^{n}\frac{C_r^{j^2}}{\epsilon_j-\alpha_h}(-\beta). \qquad (5)$$

Ion free valence[7]

for an electrophilic substitution:

$$F_r^+=F_r+\sum_s{}' C_r^m C_s^m;$$

for a nucleophilic substitution:

$$F_r^-=F_r-\sum_s{}' C_r^{m+1}C_s^{m+1}. \qquad (6)$$

Atom stabilization energy[8]

for a radical substitution:

$$E_r=2\sum_{j=1}^{m}C_r^{j^2}\epsilon_j/(-\beta);$$

for an electrophilic substitution:

$$E_r^+=E_r+C_r^{m^2}\epsilon_m/(-\beta); \qquad (7)$$

for a nucleophilic substitution:

$$E_r^-=E_r-C_r^{(m+1)^2}\epsilon_{m+1}/(-\beta).$$

* As to the π-electron density, self-polarizability, and free valence, see references 2 and 3.

Hyperconjugation Energy

The present authors have previously proposed a theory of reactivity concerning the mechanism of substitution in conjugated molecules and have pointed out the possible importance of the role of hyperconjugation which might occur near the transition state between the quasi-π orbital and the π orbital of the point of attack.[6] The occurrence of this hyperconjugation has been derived reasonably from a simple quantum-theoretical treatment of the total reacting system.[6] A quantity, superdelocalizability of π electrons, has been defined, and it has been proved experimentally that this quantity can be a good measure of the chemical reactivity of conjugated molecules.[14]

The superdelocalizability is a quantity obtained by the perturbation treatment, and, in discussing the correlation with other quantities in the present paper, another quantity which directly represents the stabilization attributable to the hyperconjugation is also to be defined as follows:

$$\mathfrak{H}_r(\gamma)=2\sum_{j=1}^{m}(\epsilon_j-\mu_{rj})+\nu(\alpha_h-\mu_{r,m+1}). \qquad (8)$$

This quantity, $\mathfrak{H}_r(\gamma)$, may be called "hyperconjugation energy" of the molecule and the pseudoatom.†

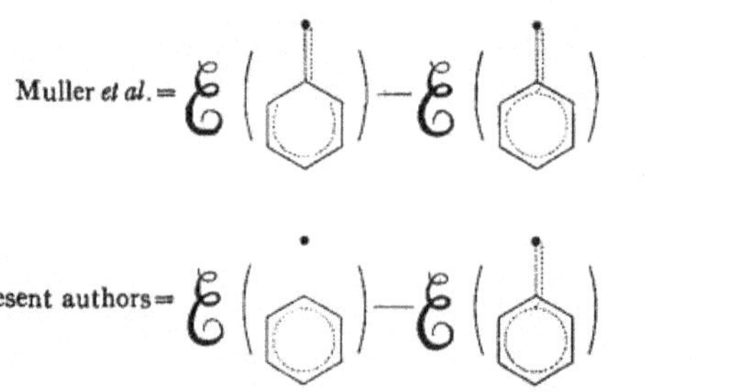

($\mathcal{E}$: π-electron energy)

ALTERNATANT HYDROCARBON

Integral Formulas of Reactivity Quantities

The LCAO MO theory of conjugated molecules has been established particularly by Coulson and Longuet-Higgins,[2] who have given so-called integral formulas for q_r, π_{rr}, P_{rs}, and other polarizabilities. Self-polarizability has been expressed as

$$\pi_{r,r}=\frac{(-\beta)}{\pi}\int_{-\infty}^{\infty}y^2\left|\frac{\Delta_{rr}(iy)}{iy\Delta(iy)}\right|^2dy$$

$$=\frac{(-\beta)}{\pi}\int_{-\infty}^{\infty}y^2\{G_r(y)\}^2dy, \qquad (9)$$

[14] K. Fukui, J. Japan Chem. 10, 289 (1956) in Japanese (to be published elsewhere in English).

† The definition of hyperconjugation energy by Muller, Pickett, and Mulliken [J. Am. Chem. Soc. 76, 4770 (1954)] differs a little from that of present authors. The relation of the two is shown schematically.

where the function of y in the integrand, $\Delta_{rr}(iy)/\{iy\Delta(iy)\}$, is designated as $G_r(y)$ for the sake of convenience in further discussions, namely

$$G_r(y)\equiv\frac{\Delta_{rr}(iy)}{iy\Delta(iy)}. \tag{10}$$

This function is also the integrand of S_r.

In an alternant hydrocarbon,[2] N_r, L_r, S_r, and $\mathfrak{H}_r(\gamma)$ are also easily represented by the following integral formulas:

$$N_r=\frac{1}{\pi(-\beta)}\int_{-\infty}^{\infty}\{1-y^2G_r(y)\}dy, \tag{11}$$

$$L_r=\frac{1}{\pi}\int_{-\infty}^{\infty}\left\{1+iy\left(\frac{\Delta_{rr}'(iy)}{\Delta_{rr}(iy)}-\frac{\Delta'(iy)}{\Delta(iy)}\right)\right\}dy \tag{12}$$

$$=\frac{1}{\pi}\int_{-\infty}^{\infty}\ln\frac{1}{y^2G_r(y)}dy, \tag{12'}$$

$$S_r=\frac{(-\beta)}{\pi}\int_{-\infty}^{\infty}G_r(y)dy, \tag{13}$$

$$\mathfrak{H}_r(\gamma)=\frac{1}{\pi}\int_{-\infty}^{\infty}\frac{\gamma^2G_r(y)}{1+\gamma^2G_r(y)}\times\left\{1-iy\left(\frac{\Delta_{rr}'(iy)}{\Delta_{rr}(iy)}-\frac{\Delta'(iy)}{\Delta(iy)}\right)\right\}dy, \tag{14}$$

$$=\frac{1}{\pi}\int_{-\infty}^{\infty}\ln\{1+\gamma^2G_r(y)\}dy. \tag{14'}$$

In the expression for S_r and $\mathfrak{H}_r(\gamma)$, the Coulomb integral for the pseudoatom, $\alpha+\alpha_h$, is taken as equal to α in the transition state.[6]

Among the quantities concerning reactivity mentioned above, the ion-free valence and the atom stabilization energy are transformed into the following forms in the alternant hydrocarbon:

$$F_r^{\pm}=F_r+\frac{\epsilon_m}{2\beta}f_r,$$

$$E_r=N_{\max}-F_r,$$

$$E_r^{\pm}=N_{\max}-F_r^{\pm}.$$

Hence, by these equations F_r^+, F_r^-, E_r, E_r^+, and E_r^- are discussed in terms of F_r and f_r and, therefore, subsequent discussions on these quantities are tentatively omitted.

Correlation of π_{rr}, L_r, F_r, S_r, and $\mathfrak{H}_r(\gamma)$

From Eqs. (9), (11), (12′), (13), and (14′), it is observed that the relative magnitude of these quantities for each position in a molecule is determined by the magnitude of $G_r(y)$, and hence the property of this function is to be investigated. In the alternant hydrocarbon, $\Delta(iy)$ and $\Delta_{rr}(iy)$ are expanded in the following forms:

$$\begin{aligned}(-1)^m\Delta(iy)&=y^{2m}+a_1\beta^2y^{2m-2}+a_2\beta^4y^{2m-4}+\cdots+a_m\beta^{2m}\\(-1)^m\frac{\Delta_{rr}(iy)}{iy}&=y^{2m-2}+b_{r,1}\beta^2y^{2m-4}+b_{r,2}\beta^4y^{2m-6}+\cdots+b_{r,m-1}\beta^{2m-2}.\end{aligned} \tag{15}$$

It is easily proved algebraically that

$$\begin{aligned}a_i&>0,\quad(i=1,2,\cdots m)\\b_{r,i}&>0,\quad(i=1,2,\cdots m-1)\end{aligned} \tag{16}$$

and that

$$\frac{b_{r,j}}{a_j}\geqq\frac{b_{r,j+1}}{a_{j+1}},\quad(j=0,1,2,\cdots m-1)$$

$$\frac{b_{r,j-1}}{a_j}\leqq\frac{b_{r,j}}{a_{j+1}}.\quad(j=0,1,2,\cdots m-1) \tag{17}$$

$$(b_{r,-1}=0,\ a_0=b_{r,0}=1,\ b_{r,m}=0).$$

By using Eqs. (15), (16), and (17) it can be shown that all the integrands in Eqs. (9), (11), (12), (12′), (13), (14), and (14′) as well as $G_r(y)$ are positive with respect to all the real values of y.

In discussing the correlation of π_{rr} and L_r, an attempt which may be very worthy of attention has been made by Baba.[10] He has pointed out the importance of the coefficients in the expanded formula of $\Delta_{rr}(iy)$ and has shown by using a quantity which he termed "partial polarization energy" that, if the inequalities

$$b_{r,i}\geqq b_{s,i}\quad(i=1,2,\cdots m-1)$$

hold with regard to any pair of Δ_{rr} and Δ_{ss} of the secular determinant, π_{rr} and L_r tell us the same order of reactivity of each position in a molecule.

The relation stated above can more simply be recognized by the integral formula (12′), that is, if the integrands of L_r, etc., are designated as integrand (L_r), etc., respectively, it is easily seen from Eqs. (9), (12′), and (15) that, provided the foregoing inequalities hold between the positions r and s,

$$G_r(y)\geqq G_s(y) \tag{18}$$

with regard to any real value of y and hence

$$\text{int. }(\pi_{rr})\geqq\text{int. }(\pi_{ss}),$$

$$\text{int. }(L_r)\geqq\text{int. }(L_s);$$

and, moreover, it is evident from Eqs. (11), (13), and (14′) that

$$\text{int. }(N_r)\leqq\text{int. }(N_s),$$
$$\text{int. }(S_r)\geqq\text{int. }(S_s),$$
$$\text{int. }(\mathfrak{H}_r(\gamma))\geqq\text{int. }(\mathfrak{H}_s(\gamma)).$$

In this connection, therefore, it is convenient for future discussions to resolve the foregoing relation in the following form:

If the relation $G_r(y)\geqq G_s(y)$ holds with respect to all the pairs of carbon atoms attached to at least one hydrogen atom (this condition may tentatively be referred to as Condition A), *the orders of easiness of substitution, irrespective of electrophilic, radical, or nucleophilic, at each position in an alternant hydrocarbon, which are deduced from π_{rr}, L_r, F_r, S_r and $\mathfrak{H}_r(\gamma)$, are utterly coincident with one another.*‡§

The Relation of the Frontier Electron Density and the Other Reactivity Quantities

For the treatment of this problem, a series of new quantities, $T_r^{(l)}$'s, are defined as

$$T_r^{(l)}=2\beta^{2l}\sum_{j=1}^{m}\frac{C_{rj}^{2}}{\epsilon_j^{2l}}.\quad(l=1,2,3,\cdots).\tag{19}$$

These quantities can be transformed into another form, which is

$$T_r^{(l)}=-2\beta^{2l}\sum_{j=1}^{m}\frac{\Delta_{rr}(\epsilon_j)}{\epsilon_j^{2l}\Delta'(\epsilon_j)}=\frac{\beta^{2l}}{(2l-1)\,!}\left[\frac{d^{2l-1}}{d\epsilon^{2l-1}}\frac{\Delta_{rr}(\epsilon)}{\Delta(\epsilon)}\right]_{\epsilon=0}$$
$$=\frac{(-)^{l-1}\beta^{2l}}{(2l-2)\,!}\left[\frac{d^{2l-2}}{dy^{2l-2}}G_r(y)\right]_{y=0}\tag{19'}$$

by the use of Cauchy's integral representation, and hence are easily evaluated numerically in sequence by the following formulas:

$$T_r^{(1)}=\frac{b_{r,m-1}}{a_m},$$

$$T_r^{(2)}=\frac{a_{m-1}}{a_m}T_r^{(1)}-\frac{b_{r,m-2}}{a_m},$$

$$T_r^{(3)}=\frac{a_{m-1}}{a_m}T_r^{(2)}-\frac{a_{m-2}}{a_m}T_r^{(1)}+\frac{b_{r,m-3}}{a_m},$$

$$\cdots\cdots\cdots\cdots\cdots$$

$$T_r^{(m-1)}=\frac{a_{m-1}}{a_m}T_r^{(m-2)}-\frac{a_{m-2}}{a_m}T_r^{(m-3)}+\cdots+(-1)^{m-1}\frac{a_2}{a_m}T_r^{(1)}-(-1)^{m-1}\frac{b_{r,1}}{a_m},$$

$$T_r^{(m)}=\frac{a_{m-1}}{a_m}T_r^{(m-1)}-\frac{a_{m-2}}{a_m}T_r^{(m-2)}+\cdots+(-1)^{m}\frac{a_1}{a_m}T_r^{(1)}-(-1)^{m}\frac{1}{a_m},$$

$$T_r^{(m+1)}=\frac{a_{m-1}}{a_m}T_r^{(m)}-\frac{a_{m-2}}{a_m}T_r^{(m-1)}+\cdots+(-1)^{m+1}\frac{1}{a_m}T_r^{(1)},$$

$$\cdots\cdots\cdots\cdots\cdots$$

$$T_r^{(m+k)}=\frac{a_{m-1}}{a_m}T_r^{(m+k-1)}-\frac{a_{m-2}}{a_m}T_r^{(m+k-2)}+\cdots+(-1)^{m+1}\frac{1}{a_m}T_r^{(k)}.\quad(k=1,2,3,\cdots)$$

Now, if we put

$$\frac{2T_r^{(l)}}{\sum_{s=1}^{n}T_s^{(l)}}=g_r^{(l)},\tag{20}$$

it is expected from the definition of $T_r^{(l)}$ that

$$\lim_{l\to\infty}g_r^{(l)}=2C_{rm}^{2}=f_r.$$

The results of calculation are listed in Table I, taking butadiene and naphthalene as examples. The convergence of the sequence $g_r^{(1)}$, $g_r^{(2)}$, $\cdots$ is rapid enough to enable us to apply this procedure to numerical computation of the frontier electron density without solving the secular equation. In the calculation of $g_r^{(l)}$ for a few particular positions only, it is convenient to use the

‡ On this point the present authors are very grateful to Dr. Baba for his private discussion on Condition A.

§ The relations $b_{r,i}\geqq b_{s,i}$ $(i=1,2,\cdots m-1)$ and $G_r(y)\geqq G_s(y)$ are not equivalent. For instance, with respect to the pair of positions 4 and 6 in 3,4-benzphenanthrene, $b_{4,i}\geqq b_{6,i}$ $(i=1,2,3,4,5,6,8)$ and $b_{4,7}<b_{6,7}$, but $G_4(y)>G_6(y)$ $(-\infty<y<+\infty)$. The proportion to the total of the alternant hydrocarbons which satisfy Condition A is seen in the following examples:

14 in 19 theoretically possible straight and branched chain alternant hydrocarbons of $\leqq C_{10}$ with no ring;
12 in 15 theoretically possible alternant hydrocarbons of $\leqq C_{10}$ with one 6-membered ring;
4 in 9 theoretically possible condensed aromatic hydrocarbons of 2∼4 rings with no side chain.

In polyenes, $G_r(y)$ can be written in the form

$$G_r(y)=\frac{\sinh(2m+1)x+\sinh(2m-2r+1)}{4\sinh x\cosh x\cosh(2m+1)x}\;(r=1,2,\cdots 2m-1)$$

where $y=2\sinh x$. Therefore, it is known that

$$G_1(y)>G_3(y)>\cdots>G_{2m-1}(y),$$

and hence the reactivity of each position is in the following order:

$$1>3>\cdots>(2m-1)$$

In all the straight polyacenes, Condition A seems to hold. In bent-ring condensed aromatic hydrocarbons, it seems in general not to hold. For instance, in Δ_{11} and Δ_{99} in phenanthrene, it does not hold. In this case $\pi_{99}>\pi_{11}$, $L_9<L_1$, and $S_9>S_1$, but $F_9<F_1$.

following formulas. From Eq. (19′):

$$\sum_{s=1}^{n} T_s^{(l)} = -\frac{\beta^{2l}}{(2l-1)!}\left[\frac{d^{2l-1}}{d\epsilon^{2l-1}}\frac{\Delta'(\epsilon)}{\Delta(\epsilon)}\right]_{\epsilon=0},$$

and hence

$$U^{(l)} \equiv \tfrac{1}{2}\sum_{s=1}^{n} T_s^{(l)} \quad (l=1, 2, 3, \cdots) \qquad (21)$$

are obtained numerically in sequence by the following formulas:

$$U^{(1)} = \frac{a_{m-1}}{a_m},$$

$$U^{(2)} = \frac{a_{m-1}}{a_m}U^{(1)} - 2\frac{a_{m-2}}{a_m},$$

$$U^{(3)} = \frac{a_{m-1}}{a_m}U^{(2)} - \frac{a_{m-2}}{a_m}U^{(1)} + 3\frac{a_{m-3}}{a_m},$$

$$U^{(m-1)} = \frac{a_{m-1}}{a_m}U^{(m-2)} - \frac{a_{m-2}}{a_m}U^{(m-2)} + \cdots + (-1)^{m-1}\frac{a_2}{a_m}U^{(1)} - (-1)^{m-1}(m-1)\frac{a_1}{a_m},$$

$$U^{(m)} = \frac{a_{m-1}}{a_m}U^{(m-1)} - \frac{a_{m-2}}{a_m}U^{(m-2)} + \cdots + (-1)^{m}\frac{a_1}{a_m}U^{(1)} - (-1)^{m}m\frac{1}{a_m},$$

$$U^{(m+1)} = \frac{a_{m-1}}{a_m}U^{(m)} - \frac{a_{m-2}}{a_m}U^{(m-1)} + \cdots + (-1)^{m+1}\frac{1}{a_m}U^{(1)},$$

$$U^{(m+k)} = \frac{a_{m-1}}{a_m}U^{(m+k-1)} - \frac{a_{m-2}}{a_m}U^{(m+k-2)} + \cdots + (-1)^{m+1}\frac{1}{a_m}U^{(k)}. \quad (k=1, 2, 3, \cdots)$$

From Eq. (19′), it follows that

$$T_r^{(l)} - T_s^{(l)} = \frac{(-1)^{l-1}\beta^{2l}}{(2l-2)!}\left[\frac{d^{2l-2}}{dy^{2l-2}}\{G_r(y) - G_s(y)\}\right]_{y=0}.$$

If, therefore, the right-hand side of this equation is positive for $l=1, 2, 3, \cdots$ then it follows that

$$T_r^{(l)} \geqq T_s^{(l)}, \quad (l=1, 2, 3, \cdots)$$

$$g_r^{(l)} \geqq g_s^{(l)}, \quad (l=1, 2, 3, \cdots)$$

TABLE I. The values of $g_r^{(l)}$ and f_r in butadiene and naphthalene.

		Butadiene 		Naphthalene 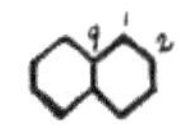		
r		1	2	1	2	9
$g_r^{(l)}$	$l=1$	0.6667	0.3333	0.2558	0.1860	0.1163
	$l=2$	0.7143	0.2857	0.3122	0.1620	0.0516
	$l=3$	0.7222	0.2778	0.3418	0.1482	0.0201
	$l=4$	0.7234	0.2766	0.3540	0.1422	0.0076
	$l=5$	0.7236	0.2764	0.3588	0.1397	0.0029
	$l=6$			0.3606	0.1388	0.0011
	$l=7$			0.3613	0.1384	0.0004
	$l=8$			0.3616	0.1383	0.0002
	$l=9$			0.3617	0.1382	0.0001
f_r		0.7236	0.2764	0.3618	0.1382	0.0000

and from Eq. (20), also

$$f_r \geqq f_s.$$

Therefore if the inequalities

$$(-1)^k\left[\frac{d^{2k}}{dy^{2k}}\{G_r(y) - G_s(y)\}\right]_{y=0} \geqq 0 \qquad (k=0, 1, 2, \cdots) \quad (22)$$

hold with regard to any pair of r and s in an alternant hydrocarbon (this condition may be termed Condition B)‖ the order of magnitude of f_r coincides with that of $b_{r,m-1}$. But the order of magnitude of $b_{r,m-1}$ is the same as that of $G_r(y)$ provided Condition A is satisfied, and thus the order of f_r agrees with that of S_r or other reactivity quantities.¶

In polyacenes, in which Condition A seems to be satisfied, the order of magnitude of f_r coincides with that of S_r. The results of calculation are shown in Table II. The relation (22) seems to hold between almost all the pairs of r and s in other alternant hydrocarbons in which Condition A is satisfied. An exception is the case of styrene, in which Condition A is satisfied, but the relation (22) does not hold between positions *ortho* and *para*. The values of $g_{ortho}^{(l)}$ and $g_{para}^{(l)}$ in this molecule are indicated in Table III.

Thus, we have been led to a conclusion: Namely, *in an alternant hydrocarbon in which both Conditions A and B are satisfied, the order of reactivity at each position in the molecule which is predicted by the frontier electron density agrees exactly with those which are predicted by* π_{rr}, L_r, F_r, S_r, *and* $\mathfrak{H}_r(\gamma)$. For discussing the reactivity of such a molecule, therefore, it may be recommended

‖ Whether this condition is satisfied or not is confirmed by expanding $\{G_r(y) - G_s(y)\}$ in terms of y^2, i.e., in the form of $c_0 + c_1y^2 + c_2y^4 + \cdots$, and subsequently by examining the relation $c_0 \geqq 0$, $c_1 \leqq 0$, $c_2 \geqq 0$, $c_3 \leqq 0$, $\cdots$.

¶ In polyenes, it is evident that the order of magnitude of f_r is always

$$f_1 > f_3 > \cdots > f_{2m-1}$$

and is the same as that of $G_r(y)$, S_r, or other quantities.

TABLE II. The relation of f_r and S_r in polyacenes.

Polyacenes	r	1	2	3	4	5	6	7
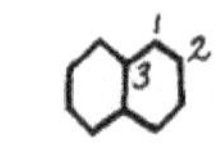	f_r	0.3618	0.1382	0.0000				
Naphthalene	S_r	0.9944	0.8729	0.7032				
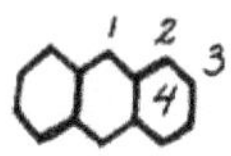	f_r	0.3868	0.1934	0.0966	0.0166			
Anthracene	S_r	1.3137	1.0734	0.9215	0.7198			
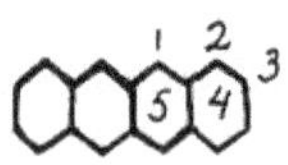	f_r	0.2947	0.1126	0.0671	0.0256	0.0000		
Naphthacene	S_r	1.5049	1.1215	0.9611	0.7503	0.7165		
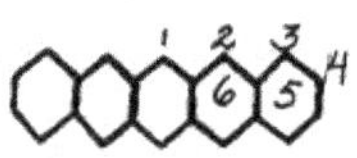	f_r	0.2825	0.2118	0.0706	0.0475	0.0254	0.0034	
Pentacene	S_r	1.9170	1.6344	1.1539	1.9909	0.7775	0.7344	
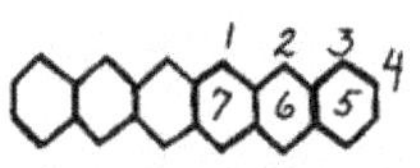	f_r	0.2372	0.1526	0.0470	0.0343	0.0221	0.0068	0.0000
Hexacene	S_r	2.0054	1.7295	1.1777	1.0135	0.7995	0.7533	0.7440

to use the frontier electron density, since it is the simplest to calculate and leads us to the same result as that obtained by using other quantities.

The Mathematical Relation between Perturbation and Nonperturbation Treatments

The method of presenting the reactivity through q_r, π_{rr}, and F_r is based on the magnitude of change in total π-electron energy caused by the perturbation of the reagent. These circumstances[2,13] are shown by

$$\delta\mathcal{E}=q_r\delta\alpha+\frac{1}{2\beta}\pi_{rr}\delta\alpha^2+\cdots, \tag{23}$$

$$\delta\mathcal{E}=2N_r\delta\beta+\cdots, \tag{24}$$

and in this sense these quantities have been said to tell us the relative height of the energy curve at an early stage of the reaction path, and the method in which they are used has been called the "static method"[9] or the "isolated molecule approximation."[13]

But it can be shown that it is also possible to discuss the energy change caused by the electrostatic influence of the reagent by means of a nonperturbation treatment. Thus, if we denote by $\mathfrak{A}_r(\Delta\alpha)$ the change in total π-electron energy of a molecule caused by the change in Coulomb integral for the rth atom from α to $\alpha+\Delta\alpha$, it is easy to derive the integral expression

$$\mathfrak{A}_r(\Delta\alpha)=\Delta\alpha-\frac{1}{2\pi}\int_{-\infty}^{\infty}\ln[1+(\Delta\alpha)^2y^2\{G_r(y)\}^2]dy. \tag{25}$$

As is easily understood by Eq. (25), if

$$\pi_{rr}\geqq\pi_{ss},$$

it follows in an alternant hydrocarbon satisfying Condition A that

$$\mathfrak{A}_r(\Delta\alpha)\leqq\mathfrak{A}_s(\Delta\alpha).$$

TABLE III. The relation of S_r, $g_r^{(l)}$, and f_r in *ortho*- and *para*-positions in styrene.

$$(s_r=2S_r/\sum_{s=1}^{n}S_s)$$

r	s_r	$g_r^{(1)}$	$g_r^{(2)}$	$g_r^{(3)}$	$g_r^{(4)}$	$g_r^{(5)}$	$g_r^{(6)}$	$g_r^{(7)}$	$g_r^{(8)}$	f_r
Ortho	0.2528	0.2500	0.2308	0.2115	0.2000	0.1942	0.1915	0.1903	0.1898	0.1894
Para	0.2472	0.2500	0.2692	0.2885	0.3000	0.3058	0.3085	0.3097	0.3102	0.3106

TABLE IV. A classification of various quantities concerned with chemical reactivity in conjugated molecules.

Methods		Quantities: Perturbation treatments	Quantities: Nonperturbation treatments, General case	Quantities: Nonperturbation treatments, Extreme case
(1) The method in which the change in Coulomb and resonance integrals are considered in the course of reaction	heterolytic	q_r, π_{rr}	$\mathfrak{A}_r(\Delta\alpha)$	$L_r=\mathfrak{A}_r(+\infty)$
	homolytic	N_r (or F_r), l_r	$\mathfrak{B}_r(\Delta\beta)$	$L_r=\mathfrak{B}_r(-\beta)$
(2) The method in which is considered the hyperconjugation between the molecule and the pseudoatom in the transition state		S_r	$\mathfrak{H}_r(\gamma)$	

Similarly, if we designate as $\mathfrak{B}_r(\Delta\beta)$ the change in total π-electron energy caused by the increases in resonance integrals between the rth and the adjacent atoms by the equal amount of $\Delta\beta$, an equation

$$\mathfrak{B}_r(\Delta\beta)=-\frac{1}{\pi}\int_{-\infty}^{\infty}\ln\left[1+\frac{2\beta\Delta\beta+(\Delta\beta)^2}{\beta^2}\times\{1-y^2G_r(y)\}\right]dy\quad(0\leqq\Delta\beta\leqq-\beta)\qquad(26)$$

is obtained and if

$$N_r\leqq N_s\quad(\text{or}\quad F_r\geqq F_s)$$

it follows that

$$\mathfrak{B}_r(\Delta\beta)\leqq\mathfrak{B}_s(\Delta\beta)$$

in an alternant hydrocarbon satisfying Condition A.

Equations (25) and (26) stand for the nonperturbation treatments of Eqs. (23) and (24), respectively, and $\mathfrak{A}_r(\Delta\alpha)$ and $\mathfrak{B}_r(\Delta\beta)$ serve as reactivity quantities in place of q_r, π_{rr}, and F_r. That is, when $\Delta\alpha$ and $\Delta\beta$ are small, Eqs. (25) and (26) become

$$\mathfrak{A}_r(\Delta\alpha)\simeq\Delta\alpha+\frac{1}{2\beta}\pi_{rr}\Delta\alpha^2,$$

$$\mathfrak{B}_r(\Delta\beta)\simeq 2N_r\Delta\beta.$$

On the other hand, it is also not difficult to show that, when $\Delta\alpha$ in Eq (25) tends to $+\infty$, $\mathfrak{A}_r(\Delta\alpha)$ approaches L_r, and when $\Delta\beta$ in Eq. (26) becomes $-\beta$, $\mathfrak{B}_r(\Delta\beta)$ also coincides with L_r. These facts may be of significance in clarifying the relation of the static and the localization methods. Since there is no restriction on the magnitude of $\Delta\alpha$ and $\Delta\beta$ (provided that $0\leqq\Delta\beta<-\beta$) in Eqs. (25) and (26), the method of discussing reactivity by $\mathfrak{A}_r(\Delta\alpha)$ and $\mathfrak{B}_r(\Delta\beta)$ includes the static method, which uses π_{rr} and F_r, and the localization method, to which L_r relates, as its special or extreme cases.

The relation between S_r and $\mathfrak{H}_r(\gamma)$, in the method concerning the frontier electron theory, is mathematically analogous to that of π_{rr} and $\mathfrak{A}_r(\Delta\alpha)$ or of N_r and $\mathfrak{B}_r(\Delta\beta)$. Thus the equation

$$\delta\mathcal{E}=\frac{1}{(-\beta)}S_r\gamma^2+\cdots\qquad(27)$$

corresponds to Eq. (23) or (24), and the expression (14′) corresponds to Eq. (25) or (26). Since $\mathfrak{H}_r(\gamma)$ with small value of γ can be written in the form

$$\mathfrak{H}_r(\gamma)\simeq\frac{1}{(-\beta)}S_r\gamma^2,$$

the method using $\mathfrak{H}_r(\gamma)$ includes the method in which S_r is used as a reactivity quantity.**

The Relation between Dewar's Localization Energy and Other Quantities

An approximate method for calculating L_r has been developed by Dewar.[15] This is based on his NBMO (nonbonding molecular orbital) method and may be of

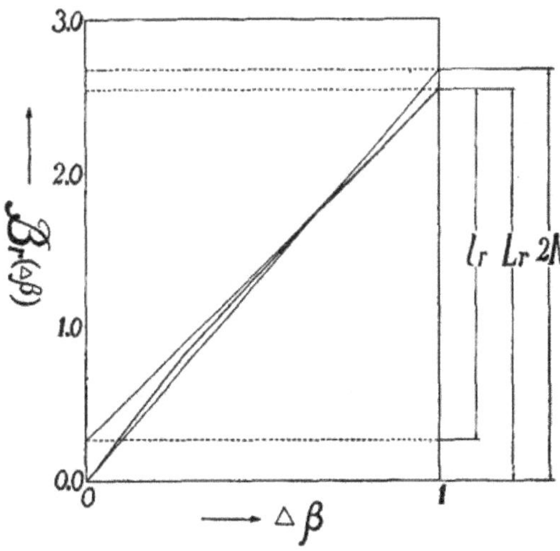

FIG. 1. Relationship between l_r, N_r, L_r, and $\mathfrak{B}_r(\Delta\beta)$. In this figure $\Delta\beta$ and $\mathfrak{B}_r(\Delta\beta)$ are given in units of $(-\beta)$; $l_r=2.306$, $L_r=2.536$, $N_r=1.334$.

** It is also possible to connect $\mathfrak{H}_r(\gamma)$ with L_r. A quantity, $P_{hr}(\gamma)$, which may be called the "mobile order of quasi-bond in the activated complex," is expressed in the form

$$P_{hr}(\gamma)=\frac{1}{\pi(-\gamma)}\int_{-\infty}^{\infty}\frac{\gamma^2G_r(y)}{1+\gamma^2G_r(y)}dy,$$

and another new quantity, $\mathfrak{M}_r(\gamma)$, which is defined as

$$\mathfrak{M}_r(\gamma)=2P_{hr}(\gamma)\cdot(-\gamma)-\mathfrak{H}_r(\gamma),$$

can be derived. Then, it can be shown that

$$\lim_{\gamma\to-\infty}\mathfrak{M}_r(\gamma)=L_r.$$

[15] M. J. S. Dewar, J. Am. Chem. Soc. 74, 3357 (1952).

TABLE V. Reactivity of a hetero-alternant molecule in reference to that of the corresponding parent alternant hydrocarbon. Case (I): rth and sth atoms are in different sets. Case (II): rth and sth atoms are in the same set.

Cases	Quantities	Signs of derivatives Electrophilic	Radical	Nucleophilic	Changes in reactivity (+: increase; −: decrease) Electrophilic $\delta\alpha_s>0$	Electrophilic $\delta\alpha_s<0$	Radical $\delta\alpha_s>0$	Radical $\delta\alpha_s<0$	Nucleophilic $\delta\alpha_s>0$	Nucleophilic $\delta\alpha_s<0$
	q_r and F_r	$\frac{\partial q_r}{\partial \alpha_s}>0$	$\frac{\partial F_r}{\partial \alpha_s}=0$, $\frac{\partial^2 F_r}{\partial \alpha_s^2}>0$	$\frac{\partial q_r}{\partial \alpha_s}>0$	+	−	+	+	−	+
	$\mathfrak{A}_r(\Delta\alpha)$ and $\mathfrak{B}_r(\Delta\beta)$	$\frac{\partial \mathfrak{A}_r(\Delta\alpha)}{\partial \alpha_s}<0$	$\frac{\partial \mathfrak{B}_r(\Delta\beta)}{\partial \alpha_s}=0$, $\frac{\partial^2 \mathfrak{B}_r(\Delta\beta)}{\partial \alpha_s^2}<0$	$\frac{\partial \mathfrak{A}_r(\Delta\alpha)}{\partial \alpha_s}>0$	+	−	+	+	−	+
[I]	S_r	$\frac{\partial S_r}{\partial \alpha_s}>0$	$\frac{\partial S_r}{\partial \alpha_s}=0$, $\frac{\partial^2 S_r}{\partial \alpha_s^2}>0$	$\frac{\partial S_r}{\partial \alpha_s}<0$	+	−	+	+	−	+
	$\mathfrak{H}_r(\gamma)$	$\frac{\partial \mathfrak{H}_r(\gamma)}{\partial \alpha_s}>0$	$\frac{\partial \mathfrak{H}_r(\gamma)}{\partial \alpha_s}=0$, $\frac{\partial^2 \mathfrak{H}_r(\gamma)}{\partial \alpha_s^2}>0$	$\frac{\partial \mathfrak{H}_r(\gamma)}{\partial \alpha_s}<0$	+	−	+	+	−	+
	L_r	$\frac{\partial L_r}{\partial \alpha_s}<0$	$\frac{\partial L_r}{\partial \alpha_s}=0$, $\frac{\partial^2 L_r}{\partial \alpha_s^2}<0$	$\frac{\partial L_r}{\partial \alpha_s}>0$	+	−	+	+	−	+

particular interest on account of its extreme simplicity in calculation.

In his theory, the reactivity of a position is measured by a quantity, l_r [$(\Delta E)_\pi/\beta$ in his notation], which can by his definition be expressed as

$$l_r=\lim_{\Delta\to 0}\frac{\mathfrak{B}_r(-\beta+\Delta)-\mathfrak{B}_r(-\beta)}{\Delta}, \quad (28)$$

and, by using Eq. (26), can easily be rewritten as

$$l_r=\lim_{\Delta\to 0}\left[-\frac{1}{\pi\Delta}\int_{-\infty}^{+\infty}\ln\left\{1+\frac{\Delta^2}{\beta^2}\frac{1-y^2G_r(y)}{y^2G_r(y)}\right\}dy\right]$$
$$=\frac{2}{(-\beta)[G_r(0)]^{\frac{1}{2}}}=\frac{2}{\left(\frac{b_{r,m-1}}{a_m}\right)^{\frac{1}{2}}}=\frac{2}{(T_r^{(1)})^{\frac{1}{2}}}. \quad (29)\dagger\dagger$$

†† The final result of Eq. (29) can also be derived directly from Dewar's original expression.

Thus, Dewar's localization energy is inversely proportional to the square root of $G_r(0)$ or $T_r^{(1)}$. Therefore, in a molecule satisfying Condition A, the order of reactivity at each position predicted by l_r agrees exactly with those predicted by π_{rr}, L_r, F_r, S_r, $\mathfrak{H}_r(\gamma)$, $\mathfrak{A}_r(\Delta\alpha)$, and $\mathfrak{B}_r(\Delta\beta)$.‡‡

The relationship of l_r, N_r, L_r, and $\mathfrak{B}_r(\Delta\beta)$ is portrayed in Fig. 1, using benzene as the example. Namely,

‡‡ The quantity l_r is analogous to S_r in the sense that is shown below schematically (e.g., benzene):

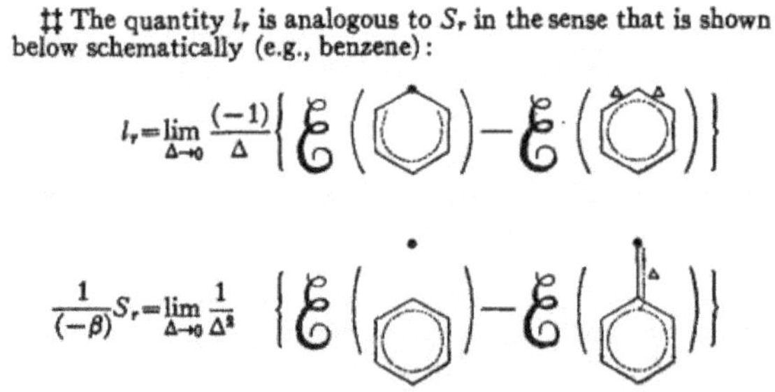

(Δ=a small resonance integral; $\mathcal{E}$=π-electron energy)

TABLE V—*Continued.*

Cases	Quantities	Signs of derivatives: Electrophilic	Signs of derivatives: Radical	Signs of derivatives: Nucleophilic	Changes in reactivity (+: increase; −: decrease) Electrophilic $\delta\alpha_s>0$	Electrophilic $\delta\alpha_s<0$	Radical $\delta\alpha_s>0$	Radical $\delta\alpha_s<0$	Nucleophilic $\delta\alpha_s>0$	Nucleophilic $\delta\alpha_s<0$
	q_r and F_r	$\frac{\partial q_r}{\partial\alpha_s}<0$	$\frac{\partial F_r}{\partial\alpha_s}=0$, $\frac{\partial^2 F_r}{\partial\alpha_s^2}<0$	$\frac{\partial q_r}{\partial\alpha_s}<0$	−	+	−	−	+	−
	$\mathcal{A}_r(\Delta\alpha)$ and $\mathcal{B}_r(\Delta\beta)$	$\frac{\partial\mathcal{A}_r(\Delta\alpha)}{\partial\alpha_s}>0$	$\frac{\partial\mathcal{B}_r(\Delta\beta)}{\partial\alpha_s}=0$, $\frac{\partial^2\mathcal{B}_r(\Delta\beta)}{\partial\alpha_s^2}>0$	$\frac{\partial\mathcal{A}_r(\Delta\alpha)}{\partial\alpha_s}<0$	−	+	−	−	+	−
[II]	S_r	$\frac{\partial S_r}{\partial\alpha_s}=0$, $\frac{\partial^2 S_r}{\partial\alpha_s^2}<0$	$\frac{\partial S_r}{\partial\alpha_s}=0$, $\frac{\partial^2 S_r}{\partial\alpha_s^2}<0$	$\frac{\partial S_r}{\partial\alpha_s}=0$, $\frac{\partial^2 S_r}{\partial\alpha_s^2}<0$	−	−	−	−	−	−
	$\mathfrak{H}_r(\gamma)$	$\frac{\partial\mathfrak{H}_r(\gamma)}{\partial\alpha_s}=0$, $\frac{\partial^2\mathfrak{H}_r(\gamma)}{\partial\alpha_s^2}<0$	$\frac{\partial\mathfrak{H}_r(\gamma)}{\partial\alpha_s}=0$, $\frac{\partial^2\mathfrak{H}_r(\gamma)}{\partial\alpha_s^2}<0$	$\frac{\partial\mathfrak{H}_r(\gamma)}{\partial\alpha_s}=0$, $\frac{\partial^2\mathfrak{H}_r(\gamma)}{\partial\alpha_s^2}<0$	−	−	−	−	−	−
	L_r	$\frac{\partial L_r}{\partial\alpha_s}=0$, $\frac{\partial^2 L_r}{\partial\alpha_s^2}>0$	$\frac{\partial L_r}{\partial\alpha_s}=0$, $\frac{\partial^2 L_r}{\partial\alpha_s^2}>0$	$\frac{\partial L_r}{\partial\alpha_s}=0$, $\frac{\partial^2 L_r}{\partial\alpha_s^2}>0$	−	−	−	−	−	−

l_r is the slope of the curve of $\mathcal{B}_r(\Delta\beta)$ *vs* $\Delta\beta$ at the point $\Delta\beta=-\beta$, while N_r is that of $\Delta\beta=0$.

The Classification of Reactivity Quantities

From the results thus far obtained, the quantities which concern the LCAO MO treatment of the chemical reactivity in conjugated molecules may be classified according to the schema indicated in Table IV.

HETERO-ALTERNANT MOLECULE

The elegant mathematical relations obtained in the alternant hydrocarbon are not to be expected in other molecules, since in this case the secular determinant cannot in general be written in the form of Eq. (15) and involves certain arbitrary parameters in its elements. In the hetero-alternant molecule, however, some qualitative discussions on the interrelation of reactivity quantities being possible, several authors have attacked this problem, in individual cases or in regard to some of the quantities.[2,9,13,16]

In the present paper the mathematical interrelation of the methods in which q_r, F_r, S_r, L_r, $\mathcal{A}_r(\Delta\alpha)$, $\mathcal{B}_r(\Delta\beta)$, and $\mathfrak{H}_r(\gamma)$ are used as the scale of reactivity is discussed more generally. An alternant hydrocarbon (even molecule) is considered in which the Coulomb integral for the sth carbon atom, $\alpha+\alpha_s$, is changed to $\alpha+\alpha_s+\delta\alpha_s$. This is at least approximately realized, for instance, in the cases of s-aza-aromatic hydrocarbons and s-substituted hydrocarbons in which the substituent is purely or almost purely inductive; e.g., $-SO_2R$, $-CCl_3$, etc.

In discussing the reactivity of the rth atom ($r\neq s$) in such molecules, the problem divides into two cases: (1), the case in which the rth and the sth atoms belong to different sets and (2), to the same set.[2]

The derivatives with respect to α_s of q_r, F_r, S_r, L_r, $\mathcal{A}_r(\Delta\alpha)$, $\mathcal{B}_r(\Delta\beta)$, and $\mathfrak{H}_r(\gamma)$ are represented by the

[16] H. C. Longuet-Higgins, J. Chem. Phys. 18, 283 (1950); Nature 166, 139 (1950).

following formulas§§:

$$\frac{\partial q_r}{\partial \alpha_s}=\frac{1}{\pi}\int_{-\infty}^{+\infty}\left\{\frac{\Delta_{rs}(iy)}{\Delta(iy)}\right\}^2 dy, \tag{30}$$

$$\left.\begin{aligned}&\frac{\partial F_r}{\partial \alpha_s}=0,\\ &\frac{\partial^2 F_r}{\partial \alpha_s^2}=\frac{2}{\pi(-\beta)}\int_{-\infty}^{+\infty} y^2 G_s(y)\left\{\frac{\Delta_{rs}(iy)}{\Delta(iy)}\right\}^2 dy,\end{aligned}\right\} \tag{31}$$

$$\left.\begin{aligned}&\frac{\partial S_r}{\partial \alpha_s}=(1-\nu)(-\beta)\left\{\frac{\Delta_{rs}(0)}{\Delta(0)}\right\}^2,\\ &\frac{\partial^2 S_r}{\partial \alpha_s^2}=\frac{2(-\beta)}{\pi}\int_{-\infty}^{+\infty} G_s(y)\left\{\frac{\Delta_{rs}(iy)}{\Delta(iy)}\right\}^2 dy,\end{aligned}\right\} \tag{32}$$

$$\left.\begin{aligned}&\frac{\partial L_r}{\partial \alpha_s}=-(1-\nu)\frac{\{\Delta_{rs}(0)\}^2}{\Delta(0)\Delta_{rr}'(0)},\\ &\frac{\partial^2 L_r}{\partial \alpha_s^2}=-\frac{1}{\pi}\int_{-\infty}^{+\infty}\left\{\frac{\Delta_{rs}(iy)}{\Delta(iy)}\right\}^2\cdot\frac{2y^2G_r(y)G_s(y)+\left\{\dfrac{\Delta_{rs}(iy)}{\Delta(iy)}\right\}^2}{y^2\{G_r(y)\}^2}dy,\end{aligned}\right\} \tag{33}$$

$$\frac{\partial \mathfrak{A}_r(\Delta\alpha)}{\partial \alpha_s}=\frac{\Delta\alpha}{\pi}\int_{-\infty}^{+\infty}\left\{\frac{\Delta_{rs}(iy)}{\Delta(iy)}\right\}^2\cdot\frac{dy}{1+(\Delta\alpha)^2y^2\{G_r(y)\}^2}, \tag{34}$$

$$\left.\begin{aligned}&\frac{\partial \mathfrak{B}_r(\Delta\beta)}{\partial \alpha_s}=0,\\ &\frac{\partial^2 \mathfrak{B}_r(\Delta\beta)}{\partial \alpha_s^2}=\frac{2\beta\Delta\beta+(\Delta\beta)^2}{\pi\beta^2}\int_{-\infty}^{+\infty} y^2\left\{\frac{\Delta_{rs}(iy)}{\Delta(iy)}\right\}^2\cdot\frac{2\left(\dfrac{\beta+\Delta\beta}{\beta}\right)^2G_s(y)-\dfrac{2\beta\Delta\beta+(\Delta\beta)^2}{\beta^2}\left[2y^2G_r(y)G_s(y)+\left\{\dfrac{\Delta_{rs}(iy)}{\Delta(iy)}\right\}^2\right]}{\left\{\left(\dfrac{\beta+\Delta\beta}{\beta}\right)^2-\dfrac{2\beta\Delta\beta+(\Delta\beta)^2}{\beta^2}y^2G_r(y)\right\}^2}dy,\end{aligned}\right\} \tag{35}$$

$$\left.\begin{aligned}&\frac{\partial \mathfrak{H}_r(\gamma)}{\partial \alpha_s}=(1-\nu)\frac{\gamma^2\left\{\dfrac{\Delta_{rs}(0)}{\Delta(0)}\right\}^2}{1+\gamma^2\dfrac{\Delta_{rr}'(0)}{\Delta(0)}},\\ &\frac{\partial^2 \mathfrak{H}_r(\gamma)}{\partial \alpha_s^2}=\frac{\gamma^2}{\pi}\int_{-\infty}^{+\infty}\left\{\frac{\Delta_{rs}(iy)}{\Delta(iy)}\right\}^2\cdot\frac{2G_s(y)+\dfrac{\gamma^2}{y^2}\left[2y^2G_r(y)G_s(y)+\left\{\dfrac{\Delta_{rs}(iy)}{\Delta(iy)}\right\}^2\right]}{\{1+\gamma^2G_r(y)\}^2}dy.\end{aligned}\right\} \tag{36}$$

By using the properties of $\Delta_{rs}(iy)$ and $\Delta_{rs}(0)$,[2] it is easy to show that the signs of these derivatives are as indicated in Table V. The influence of hetero-atom or substituent upon the change in reactivity is also given in the same table.‖‖ In Table V, an almost complete

§§ In obtaining these formulas care should be observed to avoid direct differentiation of Eqs. (11), (12), (13), etc. This might lead us to erroneous results, because they hold only in alternant hydrocarbons. The general expressions for N_r, L_r, S_r, etc. must be used instead.

‖‖ In using Table V for the prediction of reactivity, attention must be paid to the following point. The change of a quantity at the rth atom, arising from the second derivative with respect to the Coulomb integral of the sth atom belonging to the same set as the rth atom, may be small and be canceled or overcome by the change of the quantity which might come from the first derivative

parallelism is seen in the tendency toward change in reactivity.

Moreover, it is seen from Eqs. (32), (33), (34), and (36) that the relations

$$\frac{\partial S_r^{(E)}}{\partial \alpha_s}=\left|\frac{\partial S_r^{(N)}}{\partial \alpha_s}\right|, \tag{37}$$

$$\left|\frac{\partial L_r^{(E)}}{\partial \alpha_s}\right|=\frac{\partial L_r^{(N)}}{\partial \alpha_s}, \tag{38}$$

$$\left|\frac{\partial \mathfrak{A}_r^{(E)}(\Delta\alpha)}{\partial \alpha_s}\right|=\frac{\partial \mathfrak{A}_r^{(N)}(\Delta\alpha)}{\partial \alpha_s}, \tag{39}$$

$$\frac{\partial \mathfrak{H}_r^{(E)}(\gamma)}{\partial \alpha_s}=\left|\frac{\partial \mathfrak{H}_r^{(N)}(\gamma)}{\partial \alpha_s}\right|, \tag{40}$$

exist between the derivatives of S_r, L_r, $\mathfrak{A}_r(\Delta\alpha)$, and $\mathfrak{H}_r(\gamma)$ for the electrophilic and nucleophilic reactions. In these formulas, superscripts E and N designate the electrophilic and nucleophilic substitutions, respectively. Equation (39) is a generalization of the result which has been previously obtained by Greenwood.[9]

DISCUSSION

As is stated in previous sections, it is obvious that both the static and localization methods can be discussed inclusively by the quantities $\mathfrak{A}_r(\Delta\alpha)$ and $\mathfrak{B}_r(\Delta\beta)$, and the order of reactivity of each position predicted by $\mathfrak{A}_r(\Delta\alpha)$ or $\mathfrak{B}_r(\Delta\beta)$ is not changed by the magnitude of $\Delta\alpha$ or $\Delta\beta$ if Condition A is satisfied. That is, in this case, the larger the value of π_{rr} or F_r of a position, the smaller the value of L_r of that position. Brown has made reference to the so-called "noncrossing rule" in the process of reaction with regard to the relation between π_{rr} and L_r or F_r and L_r.[13] However, in a molecule where Condition A is satisfied, the parallelism between π_{rr} and L_r or F_r and L_r is always complete and mathematically obvious. For instance, when we plot $\mathfrak{B}_r(\Delta\beta)$ *vs* $\Delta\beta$ with respect to each r in a molecule satisfying Condition A, curves do not cross with each other. The larger the slope of a curve at $\Delta\beta=0$, which is $2N_r$, the larger the height of the curve at $\Delta\beta=-\beta$, which is nothing but L_r. It is to be investigated further, therefore, whether the coincidence of predictions by π_{rr} and L_r or F_r and L_r should be attributed to the existence of the "noncrossing rule" as affecting the "true" reaction path, or to the existence of some mathematical relation such as Condition A.

Moreover, in an alternant hydrocarbon satisfying Condition A, the results of prediction by π_{rr}, F_r, L_r, S_r, l_r, $\mathfrak{A}_r(\Delta\alpha)$, $\mathfrak{B}_r(\Delta\beta)$, and $\mathfrak{H}_r(\gamma)$ have proved to agree with one another. Also in hetero-alternant molecules or mono-substituted alternant hydrocarbons, predictions on the effect of hetero-atom or substituent are almost parallel with each other. On the other hand, these quantities have, more or less, their own physical grounds which are considerably different by nature from each other. It is of significance from the organic chemical point of view to decide as to which one of these quantities is closer to the "true" mechanism of reaction than the others. This must be carried out, of course, by comparison of results of prediction with experimental data. Each of these quantities has so far proved to predict the reactivity as equally well as the others. However, this might possibly be due to the fact that previous experimental data have principally been limited to alternant hydrocarbons and their derivatives. But, so far as in these molecules the close mathematical interrelations stated above exist between these quantities, it is not adequate to carry out the comparison of reliabilities of the quantities with respect to these molecules. Therefore, it is desirable that the comparison will be made regarding e.g. nonalternant molecules. A discussion on this point will be made elsewhere.

ACKNOWLEDGMENTS

The authors wish to express thanks to Dr. Baba, Hokkaido University, Japan, for his discussion. They are also grateful to the Ministry of Education, Japanese Government, for a grant in aid.

with respect to the Coulomb integral of the atom adjacent to the sth; this, in the present treatment, is assumed unchanged by the influence of the hetero-atom or the substituent. For example, the case of the nucleophilic substitution of β chloropyridine belongs to Case (II), and Table V (II) tells us that β chloropyridine is less reactive than chlorobenzene, which is inconsistent with experience. But if the effect of the change in Coulomb integral of the carbon atom adjacent to nitrogen is taken into account, and if the latter effect overcomes the former, a second-order perturbation term, β chloropyridine, will be more reactive than chlorobenzene [see Table V(I)].

Reprinted from THE JOURNAL OF CHEMICAL PHYSICS, Vol. 27, No. 6, 1247–1259, December, 1957
Printed in U. S. A.

MO-Theoretical Approach to the Mechanism of Charge Transfer in the Process of Aromatic Substitutions

KENICHI FUKUI, TEIJIRO YONEZAWA, AND CHIKAYOSHI NAGATA
Faculty of Engineering, Kyoto University, Kyoto, Japan

(Received March 6, 1957)

An MO-theoretical investigation on the mechanism of aromatic substitution is made. Supposing that the approaching reagent and the atom to be replaced can be treated as an atom (pseudoatom) which has a π-type orbital in the course of reaction, an important role of hyperconjugation taking place between the pseudoatom and the aromatic compound is remarked. Since the electron density at the pseudoatom relates to the amount of charge transfer through this hyperconjugation and varies as the reaction proceeds, the proceeding of the reaction can be measured by this electron density. Then, the electron density is expressed, in the form of a contour diagram, in terms of the Coulomb integral of the pseudoatom and the resonance integral between the pseudoatom and carbon atom to be attacked. Thus, a reaction path can be represented by a locus plotted on this diagram. The discriminating property of frontier orbitals can be observed in this diagram. It has been confirmed that an electrophilic, or a nucleophilic, substitution takes place only when a certain condition is satisfied with respect to the energy of the reagent. By determining the transition state which corresponds to the end point of the locus, several reactivity indices, i.e., frontier electron density, superdelocalizability, its one-term approximation and a generalized reactivity index are derived from the hyperconjugation energy at the transition state, which is also available as a generalized reactivity index. Correlation between these and the existing indices is examined.

INTRODUCTION

RECENT developments in MO-theoretical treatment of the chemical reactivity in conjugated molecules have contributed much to the problem of orientation in aromatic substitution.[1,2] These treatments remain, however, in a stage which provides only a small amount of information on the essential feature of the reaction process.

A few attempts have been made to throw light on the mechanism of the process of aromatic substitutions. One of these presented by Mulliken and his co-workers,[3] was a theoretical treatment of hyperconjugation which occurs with CH_2 group on $C_6H_7^+$ ion

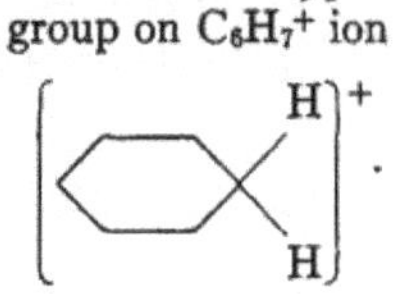

On this ion, the H_2 group is considered to be an atom (pseudoatom) that has a $(1s-1s)$ π-type orbital (quasi-π orbital) according to its C_{2v} symmetry. In addition to this, the aforementioned authors have suggested that when an aromatic compound is substituted by an electrophilic reagent S^+, a similar theoretical treatment of hyperconjugation with CSH group on transition complex

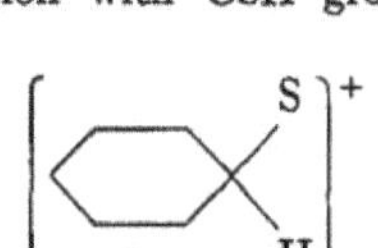

can be made.

Another theoretical treatment on the mechanism of substitution was made by Nagakura and Tanaka.[4] Referring to the energy levels of benzene and attacking reagents estimated from empirical data, they concluded that an electrophilic substitution occurs only when the energy of the vacant orbital of the reagent is less than that of the highest occupied level of benzene. They believe that this condition must be satisfied before electron transfer can take place.

A further investigation of the problem has been presented by these authors,[5] which also gives some physical grounds of frontier electron theory[6] proposed by some of us. In that paper, a theory of aromatic substitution (B: reagent, X: atom to be replaced)

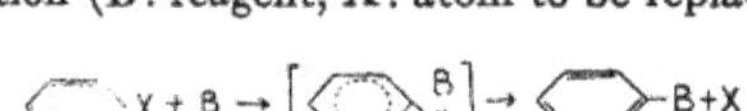

has been constructed in a simple manner taking into considerations the σ electrons in the CB and CX bonds to be formed and to be broken, respectively. In the transition state, the possible importance of hyperconjugation which occurs with CBX group on

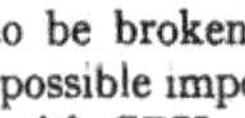

is pointed out as was done by Mulliken *et al.*[3] and a reactivity index, superdelocalizability, is derived from the stabilization energy due to this hyperconjugation. Although the theory of substitution in conjugated molecules presented in that paper[5] involves many points

[1] G. W. Wheland and L. Pauling, J. Chem. Phys. **1**, 606 (1933); C. A. Coulson and H. C. Longuet-Higgins, Proc. Roy. Soc. (London) **A191**, 39 (1947); **A192**, 16 (1947).
[2] G. W. Wheland, J. Am. Chem. Soc. **64**, 900 (1940); M. J. S. Dewar, J. Am. Chem. Soc. **74**, 3431, 3357 (1952).
[3] Pickett, Muller, and Mulliken, J. Chem. Phys. **19**, 1400 (1953); Muller, Pickett, and Mulliken, J. Am. Chem. Soc. **76**, 4770 (1954).
[4] S. Nagakura and J. Tanaka, J. Chem. Phys. **22**, 563 (1954); J. Chem. Soc. Japan, Pure Chem. Sect. **75**, 933 (1954).
[5] Fukui, Yonezawa, and Nagata, Bull. Chem. Soc. Japan **27**, 423 (1954).
[6] Fukui, Yonezawa, and Shingu, J. Chem. Phys. **20**, 722 (1952); Fukui, Yonezawa, Nagata, and Shingu, J. Chem. Phys. **22**, 1433 (1954).

which need further investigation as to some assumptions made, that treatment may provide a key to a more reasonable explanation of the transition state, since the theory has, at least to some extent, enabled us to follow the process of reaction.* In that treatment, it is expected that charge transfer is governed by the type of reaction and may be influenced by the electronic state of the substrate molecule. Owing to this charge transfer, the electron density at the pseudoatom varies as the reaction proceeds, so that the requirement of "self-consistency"† in the system may be satisfied.

Under these considerations, in the present paper, a quantum-mechanical approach to the mechanism of aromatic substitution that enables us to follow the process of substitution is made by laying stress on the change in electron density at the pseudoatom in the course of reaction.

SECULAR EQUATION AND ITS ROOTS

The secular equation of the π part of the total system which includes the quasi-π orbital and substrate π-electron system is, in a simple LCAO MO treatment (overlap neglected), written as[5]

$$D(\lambda)=\begin{vmatrix} h-\lambda & 0 & 0 & 0 & \cdots & \gamma & \cdots & 0\;0 & 0 \\ 0 & a_{11}-\lambda & a_{12} & a_{13} & \cdots & a_{1r} & & \cdots & a_{1n} \\ 0 & a_{21} & a_{22}-\lambda & a_{23} & \cdots & a_{2r} & & \cdots & a_{2n} \\ \cdot & \cdot & \cdot & \cdot & \cdot & \cdot & \cdot & \cdot & \cdot \\ \gamma & a_{r1} & & & & & & \cdots & a_{rn} \\ \cdot & \cdot & \cdot & \cdot & \cdot & \cdot & \cdot & \cdot & \cdot \\ 0 & a_{n1} & & & & & & \cdots & a_{nn}-\lambda \end{vmatrix}$$
$$=(h-\lambda)\Delta(\lambda)-\gamma^2\Delta_{rr}(\lambda)=0, \quad (1)$$

where

$$\Delta(\lambda)=\begin{vmatrix} a_{11}-\lambda & a_{12} & \cdots & a_{1n} \\ a_{21} & a_{22}-\lambda & \cdots & a_{2n} \\ \cdot & \cdot & \cdot & \cdot \\ \cdot & \cdot & \cdot & \cdot \\ \cdot & \cdot & \cdot & \cdot \\ a_{n1} & a_{n2} & \cdots & a_{nn}-\lambda \end{vmatrix} \quad (2)$$

is the secular determinant for the substrate π-electron system, in which the number of atomic π orbital is n and $\Delta_{rr}(\lambda)$ is the minor of $\Delta(\lambda)$ corresponding to the element $a_{rr}-\lambda$, and a_{ii} and $a_{ij}(i\neq j)$ are the Coulomb and the resonance integrals, respectively, the values of which are, for simplicity, assumed to be the same as those in an isolated molecule,‡ and h and γ are the Coulomb integral of the pseudoatom and the resonance integral between the pseudoatom and the rth atom of the substrate molecule at which the reaction takes place, respectively. All the quantities which represent energy are expressed in units of $(-\beta)$, and the orbital energies as well as Coulomb integrals are measured with reference to the value of Coulomb integral in benzene, and the resonance integrals of nonadjacent orbitals and all the overlap integrals are neglected.

Before entering into the calculation of the electron density at the pseudoatom, it is necessary to examine some fundamental relations of the roots of equations $D(\lambda)=0$, $\Delta(\lambda)=0$, and $\Delta_{rr}(\lambda)=0$. If we designate the roots of $\Delta(\lambda)=0$, $\Delta_{rr}(\lambda)=0$, and $D(\lambda)=0$ in decreasing order as

$$\lambda_1, \lambda_2, \cdots, \lambda_n;\ \nu_1, \nu_2, \cdots, \nu_{n-1};\ \mu_1, \mu_2, \cdots, \mu_{n+1},$$

respectively, by Sylvester's theorem the following relations hold, since $\Delta(\lambda)$ is a minor of $D(\lambda)$ and $\Delta_{rr}(\lambda)$ is of $\Delta(\lambda)$.

Case A: When $\Delta(\lambda)=0$ has no multiple root:

$$\begin{gathered}\lambda_1\geqq\nu_1\geqq\lambda_2\geqq\nu_2\geqq\cdots\geqq\nu_{n-1}\geqq\lambda_n,\\ \mu_1\geqq\lambda_1\geqq\mu_2\geqq\lambda_2\geqq\cdots\geqq\mu_n\geqq\lambda_n\geqq\mu_{n+1}.\end{gathered} \quad (3)$$

Case B: When $\Delta(\lambda)=0$ has some double roots, i.e., $\lambda_p=\lambda_{p+1}$, $\lambda_q=\lambda_{q+1}$, $\cdots$ and the other simple roots:

$$\begin{aligned}&\lambda_1\geqq\nu_1\geqq\cdots\geqq\nu_{p-1}\geqq\lambda_p=\nu_p=\lambda_{p+1}\geqq\nu_{p+1}\geqq\cdots\nu_{q-1}\\ &\qquad\geqq\lambda_q=\nu_q=\lambda_{q+1}\geqq\nu_{q+1}\geqq\cdots\geqq\nu_{n-1}\geqq\lambda_n,\\ &\mu_1\geqq\lambda_1\geqq\mu_2\geqq\cdots\geqq\mu_p\geqq\lambda_p=\mu_{p+1}=\lambda_{p+1}\geqq\mu_{p+2}\geqq\cdots\\ &\qquad\geqq\mu_q\geqq\lambda_q=\mu_{q+1}=\lambda_{q+1}\geqq\mu_{q+2}\geqq\cdots\geqq\mu_n\geqq\lambda_n\geqq\mu_{n+1}.\end{aligned} \quad (3')$$

Other cases may be theoretically possible also, but these may be neglected since they are of little practical significance.

ELECTRON DENSITY AT THE PSEUDOATOM AS A FUNCTION OF h AND γ

The kth molecular orbital for the total system corresponding to the kth root of the secular determinant (1), is

$$\Psi^{(k)}=\sum_{s=1}^{n}C_s^{(k)}\phi_s+C_h^{(k)}\phi_h, \quad (4)$$

where ϕ_s is the atomic π orbital at the sth atom in the isolated substrate molecule and ϕ_h is the quasi-π orbital at the pseudoatom. The electron density at the pseudoatom is given[5] by

$$q_h^{(E)}=2\sum_{k=1}^{m}(C_h^{(k)})^2 \text{ for an electrophilic reaction,}$$

$$q_h^{(R)}=2\sum_{k=1}^{m}(C_h^{(k)})^2+(C_h^{(m+1)})^2 \text{ for a radical reaction,} \quad (5)$$

$$q_h^{(N)}=2\sum_{k=1}^{m+1}(C_h^{(k)})^2 \text{ for a nucleophilic reaction.}$$

* In that theory, the progress of the reaction can be represented by a reaction parameter ρ which is introduced with respect to resonance integrals of the bonds to be formed and to be broken in substitution.

† This kind of "self-consistency" is the concept realized by the following semiempirical procedure in the simple LCAO MO treatment (cf., e.g., reference 3). The total π-electron density of an atom is considered to effect the potential field around it. This is accompanied with a change of Coulomb integral of the atom. Therefore, a correlation is set up between the total density and Coulomb integral of that atom in order that the total density and Coulomb integral determined from the relation become consistent.

‡ In regard to the ground of this assumption, see the discussion stated in the last section of the present paper.

where $2m$ is the number of π electrons in the isolated π-electron system, and it is easily shown that

$$(C_h^{(k)})^2 = -\frac{\Delta(\mu_k)}{D'(\mu_k)} \text{ (reference 1)}$$
$$= \left\{1+\gamma^2\frac{\Delta(\mu_k)\Delta_{rr}'(\mu_k)-\Delta'(\mu_k)\Delta_{rr}(\mu_k)}{\Delta(\mu_k)^2}\right\}^{-1}, \quad (6)$$

where $D'(\lambda)\equiv(d/d\lambda)D(\lambda)$. Thus, the π-electron density at the pseudoatom will be determined by the values of h and γ. It may be calculated numerically by using Eqs. (5) and (6), and it can be plotted on a contour diagram with h and γ axes. When the conjugation between the pseudoatom and the substrate is very small or very large, a simple algebraic manipulation leads us to some general results which will be useful in drawing the (q_h-h, γ) diagram and is given in the following two sections.

ELECTRON DENSITY AT THE PSEUDOATOM WITH VERY SMALL CONJUGATION

When the conjugation between the substrate molecule and the pseudoatom is very small, i.e., in the case of $0<\gamma\leqq 1$, one of μ_k's will be very close to the value of h and the other μ_k's will be very close to λ_i's $(i=1, 2, \cdots, n)$. The former is denoted by $\lambda^{(h)}$ and the latter μ_k's by $\lambda^{(1)}, \lambda^{(2)}, \cdots\lambda^{(n)}$, corresponding to $\lambda_1, \lambda_2, \cdots, \lambda_n$, respectively. Namely $\lambda^{(i)}$'s and $\lambda^{(h)}$ are written as $\lambda^{(i)}=\lambda_i+\delta_i$, $\lambda^{(h)}=h+\delta_h$ in which $|\delta_i|$ and $|\delta_h|$ are very small. Here, in the interests of brevity, we assumed that $\Delta(\lambda)=0$ has no multiple root. Even in the case in which $\Delta(\lambda)$ has double roots, the final result is not altered. This latter case is treated briefly in Appendix I.

(i) When h is not equal to any λ_i:

$$\delta_i = \frac{\Delta_{rr}(\lambda_i)}{(h-\lambda_i)\Delta'(\lambda_i)}\gamma^2+[\gamma^4], \quad (i=1, 2, \cdots n)$$
$$\delta_h = -\frac{\Delta_{rr}(h)}{\Delta(h)}\gamma^2+[\gamma^4],$$

where $[\gamma^4]$ denotes the higher terms, and hence it follows that

$$(C_h^{(i)})^2 = -\frac{\Delta_{rr}(\lambda_i)}{(h-\lambda_i)^2\Delta'(\lambda_i)}\gamma^2+[\gamma^4],$$
$$(C_h^{(h)})^2 = 1+\sum_{i=1}^{n}\frac{\Delta_{rr}(\lambda_i)}{(h-\lambda_i)^2\Delta'(\lambda_i)}\gamma^2+[\gamma^4]. \quad (7)$$

(ii) When h is equal to one of λ_i's, say λ_l, we obtain:

$$(C_h^{(i)})^2 = -\frac{\Delta_{rr}(\lambda_i)}{(h-\lambda_i)^2\Delta'(\lambda_i)}\gamma^2+[\gamma^4], \quad (i=1, 2, \cdots, n; i\neq l)$$

$$\left.\begin{matrix}(C_h^{(l)})^2\\(C_h^{(h)})^2\end{matrix}\right\} = \tfrac{1}{2}\mp\tfrac{1}{2}\left(-\frac{\Delta'(\lambda_l)}{\Delta_{rr}(\lambda_l)}\right)^{\frac{1}{2}}\sum_{\substack{i=1\\(i\neq l)}}^{n}\frac{\Delta_{rr}(\lambda_i)}{(\lambda_l-\lambda_i)\Delta'(\lambda_i)}\gamma + \tfrac{1}{2}\sum_{\substack{i=1\\(i\neq l)}}^{n}\frac{\Delta_{rr}(\lambda_i)}{(\lambda_l-\lambda_i)^2\Delta'(\lambda_i)}\gamma^2+[\gamma^3]. \quad (8)$$

Hence, the electron density at the pseudoatom with very small conjugation may be expressed by the following formulas.

I. Electrophilic Reaction

Since in this type of reaction the pseudoatom has initially no electron which will enter into hyperconjugation with the substrate molecule, the occupied orbitals will correspond to $\mu_1, \mu_2, \cdots, \mu_m$, and hence it follows when a small conjugation occurs between the pseudoatom and the substrate molecule that

(a) When $h>\lambda_m$,

$$q_h^{(E)} = 2+2\sum_{i=m}^{n}\frac{\Delta_{rr}(\lambda_i)}{(h-\lambda_i)^2\Delta'(\lambda_i)}\gamma^2+[\gamma^4],$$

(b) When $h=\lambda_m$,

$$q_h^{(E)} = 1-\tfrac{1}{2}\left(-\frac{\Delta'(\lambda_m)}{\Delta_{rr}(\lambda_m)}\right)^{\frac{1}{2}}\sum_{\substack{i=1\\(i\neq m)}}^{n}\frac{\Delta_{rr}(\lambda_i)}{(\lambda_m-\lambda_i)\Delta'(\lambda_i)}\gamma - \left\{\left(\sum_{i=1}^{m-1}-\sum_{i=m+1}^{n}\right)\frac{\Delta_{rr}(\lambda_i)}{(\lambda_m-\lambda_i)^2\Delta'(\lambda_i)}\right\}\gamma^2+[\gamma^3], \quad (9)$$

(c) When $h<\lambda_m$,

$$q_h^{(E)} = -2\sum_{i=1}^{m}\frac{\Delta_{rr}(\lambda_i)}{(h-\lambda_i)^2\Delta'(\lambda_i)}\gamma^2+[\gamma^4].$$

II. Radical Reaction

Since in this case the pseudoatom has one mobile electron which will conjugate with the substrate molecule, the orbitals corresponding to $\mu_1, \mu_2, \cdots\mu_m$, will be fully occupied and that corresponding to μ_{m+1} will be half-occupied, and hence it follows that

(a) When $h>\lambda_m$,

$$q_h^{(R)} = 2+\left\{2\sum_{i=m+1}^{n}\frac{\Delta_{rr}(\lambda_i)}{(h-\lambda_i)^2\Delta'(\lambda_i)} + \frac{\Delta_{rr}(\lambda_m)}{(h-\lambda_m)^2\Delta'(\lambda_m)}\right\}\gamma^2+[\gamma^4],$$

(b) When $h=\lambda_m$,

$$q_h^{(R)} = \tfrac{3}{2}-\tfrac{1}{4}\left(-\frac{\Delta'(\lambda_m)}{\Delta_{rr}(\lambda_m)}\right)^{\frac{1}{2}}\sum_{\substack{i=1\\(i\neq m)}}^{n}\frac{\Delta_{rr}(\lambda_i)}{(\lambda_m-\lambda_i)\Delta'(\lambda_i)}\gamma - \left\{\left(\tfrac{1}{2}\sum_{i=1}^{m-1}-\tfrac{3}{2}\sum_{i=m+1}^{n}\right)\times\frac{\Delta_{rr}(\lambda_i)}{(\lambda_m-\lambda_i)^2\Delta'(\lambda_i)}\right\}\gamma^2+[\gamma^3],$$

TABLE I. $\lim_{\gamma\to 0} q_h(h,\gamma)$ as a function of h.

	Value of h	$h<\lambda_{m+1}$	$h=\lambda_{m+1}$	$\lambda_{m+1}<h<\lambda_m$	$h=\lambda_m$	$h>\lambda_m$
Type of reaction	Electro-philic	0	0	0	1	2
	Radical	0	$\frac{1}{2}$	1	$\frac{3}{2}$	2
	Nucleo-philic	0	1	2	2	2

(c) When $\lambda_m>h>\lambda_{m+1}$,

$$q_h^{(R)}=1-\left\{(\sum_{i=1}^{m}-\sum_{i=m+1}^{n})\frac{\Delta_{rr}(\lambda_i)}{(h-\lambda_i)^2\Delta'(\lambda_i)}\right\}\gamma^2+[\gamma^3], \quad (10)$$

(d) When $h=\lambda_{m+1}$,

$$q_h^{(R)}=\tfrac{1}{2}-\tfrac{1}{4}\left(-\frac{\Delta'(\lambda_{m+1})}{\Delta_{rr}(\lambda_{m+1})}\right)^{\frac{1}{2}}\sum_{\substack{i=1\\(i\neq m+1)}}^{n}\frac{\Delta_{rr}(\lambda_i)}{(\lambda_{m+1}-\lambda_i)\Delta'(\lambda_i)}\gamma$$

$$-\left\{(\tfrac{3}{2}\sum_{i=1}^{m}-\tfrac{1}{2}\sum_{i=m+2}^{n})\times\frac{\Delta_{rr}(\lambda_i)}{(\lambda_{m+1}-\lambda_i)^2\Delta'(\lambda_i)}\right\}\gamma^2+[\gamma^3],$$

(e) When $h<\lambda_{m+1}$,

$$q_h^{(R)}=-\left\{2\sum_{i=1}^{m}\frac{\Delta_{rr}(\lambda_i)}{(h-\lambda_i)^2\Delta'(\lambda_i)}+\frac{\Delta_{rr}(\lambda_{m+1})}{(h-\lambda_{m+1})^2\Delta'(\lambda_{m+1})}\right\}\gamma^2+[\gamma^4].$$

It is seen from the third formula of (10) that a branch of the contour line corresponding to $q_h^{(R)}=1$, on the $(q_h^{(R)}-h,\gamma)$ diagram, starts from a point on the h axis. This point corresponds to the value of h which satisfies the conditions

$$\sum_{i=1}^{m}\frac{\Delta_{rr}(\lambda_i)}{(h-\lambda_i)^2\Delta'(\lambda_i)}=\sum_{i=m+1}^{n}\frac{\Delta_{rr}(\lambda_i)}{(h-\lambda_i)^2\Delta'(\lambda_i)} \quad (11)$$

and $\lambda_m>h>\lambda_{m+1}$. In the interval of h from λ_{m+1} to λ_m, there necessarily exists such a value except the very rare case when $\Delta_{rr}(\lambda_m)=0$ or $\Delta_{rr}(\lambda_{m+1})=0$. This value of h is denoted by h_R. In an alternant hydrocarbon, h_R is always equal to zero.

III. Nucleophilic Reaction

In this case the pseudoatom has two mobile electrons. The occupied orbitals will correspond to the roots μ_1, μ_2, $\cdots\mu_{m+1}$, and hence it follows that

(a) When $h>\lambda_{m+1}$,

$$q_h^{(N)}=2+2\sum_{i=m+1}^{n}\frac{\Delta_{rr}(\lambda_i)}{(h-\lambda_i)^2\Delta'(\lambda_i)}\gamma^2+[\gamma^4],$$

(b) When $h=\lambda_{m+1}$,

$$q_h^{(N)}=1-\tfrac{1}{2}\left(-\frac{\Delta'(\lambda_{m+1})}{\Delta_{rr}(\lambda_{m+1})}\right)^{\frac{1}{2}}\times\sum_{\substack{i=1\\(i\neq m+1)}}^{n}\frac{\Delta_{rr}(\lambda_i)}{(\lambda_{m+1}-\lambda_i)\Delta'(\lambda_i)}\gamma \quad (12)$$

$$-\left\{(\sum_{i=1}^{m}-\sum_{i=m+2}^{n})\frac{\Delta_{rr}(\lambda_i)}{(\lambda_{m+1}-\lambda_i)^2\Delta'(\lambda_i)}\right\}\gamma^2+[\gamma^3],$$

(c) When $h<\lambda_{m+1}$,

$$q_h^{(N)}=-2\sum_{i=1}^{m+1}\frac{\Delta_{rr}(\lambda_i)}{(h-\lambda_i)^2\Delta'(\lambda_i)}\gamma^2+[\gamma^4].$$

Thus, from Eqs. (9), (10), and (12), we obtain the final result indicated in Table I which will be required in constructing a contour diagram of q_h in terms of h and γ. Thus, it is found that the function $q_h(h,\gamma)$ is discontinuous at the points

$(h=\lambda_m,\ \gamma=0)$ for an electrophilic substitution,
$(h=\lambda_m,\ \gamma=0)$ and $(h=\lambda_{m+1},\ \gamma=0)$ for a radical substitution,
$(h=\lambda_{m+1},\ \gamma=0)$ for a nucleophilic substitution.

As to an approximate expression of the function $q_h(h,\gamma)$ in the vicinity of these points, see Appendix II.

ELECTRON DENSITY AT THE PSEUDOATOM WITH VERY LARGE CONJUGATION

The discussion of this section is of no physica' significance, but it is useful for the purpose of obtaining a knowledge about a mathematical property of q_h diagram. When γ is very large, it is seen from Eq. (1) that μ_k's become very close to ν_j's ($j=1, 2, \cdots, n-1$), and $+\gamma(\to\infty)$ and $-\gamma(\to-\infty)$. Thus $\mu_1\simeq\gamma$, $\mu_2\simeq\nu_1$, $\mu_3\simeq\nu_2$, $\cdots$, $\mu_n\simeq\nu_{n-1}$, $\mu_{n+1}\simeq-\gamma$. And, therefore, from Eq. (6) the values of $\lim_{\gamma\to\infty}(C_h^{(k)})^2$ are obtained easily as follows:

$$\lim_{\gamma\to\infty}(C_h^{(1)})^2=\tfrac{1}{2},\quad \lim_{\gamma\to\infty}(C_h^{(n+1)})^2=\tfrac{1}{2},$$

and

$$\lim_{\gamma\to\infty}(C_h^{(2)})^2=\cdots=\lim_{\gamma\to\infty}(C_h^{(n)})^2=0.$$

Hence, it follows that

$$\lim_{\gamma\to\infty}q_h^{(E)}=\lim_{\gamma\to\infty}q_h^{(R)}=\lim_{\gamma\to\infty}q_h^{(N)}=1. \quad (13)$$

Moreover, it can be proved that the contour line $q_h=1$

has its branch which is asymptotic to the straight line $h=0$ at very large conjugation (see Appendix III).

SOME GENERAL PROPERTIES OF q_h CONTOUR DIAGRAM IN AN ALTERNANT HYDROCARBON

In alternant hydrocarbons, such as benzene, butadiene, styrene, naphthalene, and anthracene, the following relation holds since in these molecules $\Delta(\lambda)$ and $\Delta_{rr}(\lambda)$ are an even and an odd function of λ, respectively:

$$D(\lambda;h,\gamma)=-D(-\lambda;-h,\gamma).$$

Therefore, if μ_k is the kth root of $D(\lambda;h,\gamma)=0$, $-\mu_k$ is shown to be the $(n+2-k)$th root of $D(\lambda;-h,\gamma)=0$, and hence

$$[C_h^{(k)}(-h,\gamma)]^2=[C_h^{(n+2-k)}(h,\gamma)]^2 \tag{14}$$

ince $(C_h^{(k)})^2$ can be expressed in terms of μ_k^2 and γ^2 only, as is seen in Eq. (6). Therefore, from Eqs. (5) and (6),

$$\begin{aligned} q_h^{(E)}(h,\gamma)+q_h^{(N)}(-h,\gamma) &=2\{\sum_{k=1}^{m}[C_h^{(k)}(h,\gamma)]^2+\sum_{k=1}^{m+1}[C_h^{(k)}(-h,\gamma)]^2\} \\ &=2\{\sum_{k=1}^{m}[C_h^{(k)}(h,\gamma)]^2+\sum_{k=m+1}^{n+1}[C_h^{(k)}(h,\gamma)]^2\} \\ &=2 \end{aligned} \tag{15}$$

and also

$$\begin{aligned} q_h^{(R)}(h,\gamma)+q_h^{(R)}(-h,\gamma) &=2\sum_{k=1}^{m}[C_h^{(k)}(h,\gamma)]^2+[C_h^{(m+1)}(h,\gamma)]^2 \\ &\quad+2\sum_{k=1}^{m}[C_h^{(k)}(-h,\gamma)]^2+[C_h^{(m+1)}(-h,\gamma)]^2 \\ &=2\sum_{k=1}^{m}[C_h^{(k)}(h,\gamma)]^2+2[C_h^{(m+1)}(h,\gamma)]^2 \\ &\quad+2\sum_{k=m+2}^{n+1}[C_h^{(k)}(h,\gamma)]^2 \\ &=2\sum_{k=1}^{n+1}[C_h^{(k)}(h,\gamma)]^2 \\ &=2, \end{aligned} \tag{16}$$

since

$$\sum_{k=1}^{n+1}(C_h^{(k)})^2=1.$$

Using Eqs. (15) and (16), we can obtain the contour diagram for $q_h^{(N)}(h,\gamma)$ from that of $q_h^{(E)}(h,\gamma)$ and also the diagram for $q_h^{(R)}(h,\gamma)$ by calculating only half of it.

Also, when $h=0$, it is obvious that

$$\begin{aligned} q_h^{(E)} &=2\sum_{k=1}^{m}(C_h^{(k)})^2=1-(C_h^{(m+1)})^2 \\ &=1-\frac{1}{1+\gamma^2(\Delta_{rr}'(0)/\Delta(0))}, \end{aligned} \tag{17-a}$$

since

$$2\sum_{k=1}^{m}(C_h^{(k)})^2=\sum_{k=1}^{n+1}(C_h^{(k)})^2-(C_h^{(m+1)})^2$$

from the relation (14). And, from Eq. (16), it is easily shown that

$$q_h^{(R)}(0,\gamma)=1, \tag{17-b}$$

and from Eq. (15), it follows that

$$q_h^{(N)}(0,\gamma)=1+\frac{1}{1+\gamma^2(\Delta_{rr}'(0)/\Delta(0))}. \tag{17-c}$$

These equations are useful in drawing the q_h diagram.

THE q_h CONTOUR DIAGRAM FOR BENZENE AS AN EXAMPLE

The electron density at the pseudoatom is calculated numerically as a function of h and γ using benzene as an example.

From Eqs. (5) and (6), we obtain

$$\begin{aligned} q_h &=2\sum_{k=1}^{3}(C_h^{(k)})^2+n_h(C_h^{(4)})^2 \\ &=2\sum_{k=1}^{3}\left[1+\gamma^2\frac{\mu_k^6-4\mu_k^4+3\mu_k^2+12}{(\mu_k^2-1)^2(\mu_k^2-4)^2}\right]^{-1} \\ &\quad+n_h\left[1+\gamma^2\frac{\mu_4^6-4\mu_4^4+3\mu_4^2+12}{(\mu_4^2-1)^2(\mu_4^2-4)^2}\right]^{-1}, \end{aligned} \tag{18}$$

where $n_h=0$, 1 or 2 according to whether the reaction is electrophilic, radical, or nucleophilic, respectively. So far as $\gamma\neq 0$ in this molecule, $\mu_1>2$, $2>\mu_2>1$, $\mu_3=1$, $1>\mu_4>-1$, and the occupied orbitals are (i) μ_1, μ_2 and μ_3 for an electrophilic reaction, (ii) μ_1, μ_2, μ_3 and μ_4 (half-occupied) for a radical reaction, and (iii) μ_1, μ_2, μ_3 and μ_4 for a nucleophilic reaction. From Eq. (6), it is also seen that the orbital corresponding to $\mu_3=1$ does not contribute to q_h. Also it can be shown from Eqs. (17-a, b, c,) that at $h=0$,

$$q_h^{(E)}=1-\frac{1}{1+\frac{3}{4}\gamma^2},\quad q_h^{(R)}=1,\quad q_h^{(N)}=1+\frac{1}{1+\frac{3}{4}\gamma^2}.$$

With the help of these relations, the calculation can be carried out numerically. The results are indicated in Figs. 1, 2, and 3 for the electrophilic, radical, and nucleophilic reactions, respectively.

As is seen from these figures, each diagram has one or

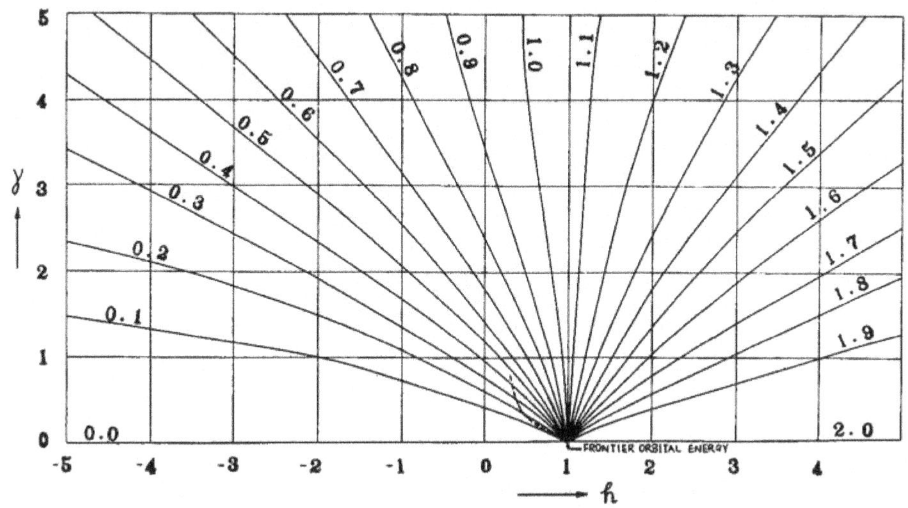

FIG. 1. The $(q_\lambda - h, \gamma)$ diagram of benzene for electrophilic attack.

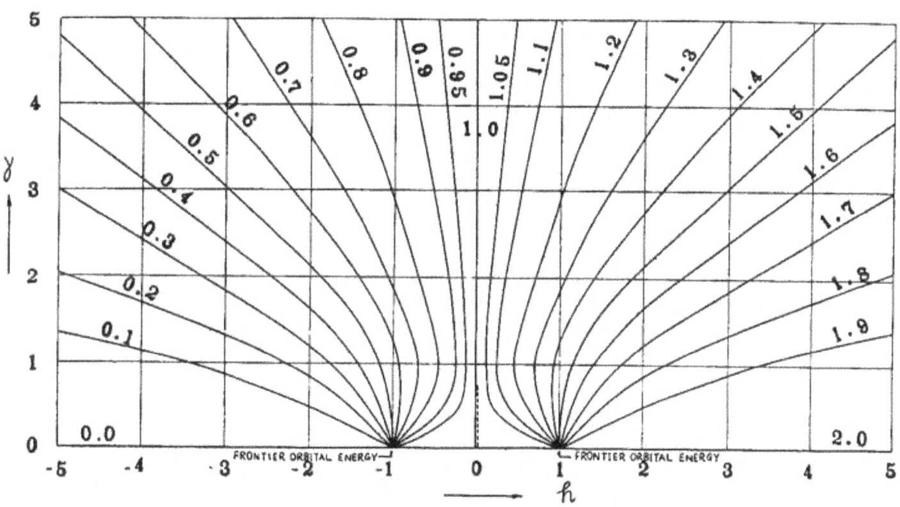

FIG. 2. The $(q_\lambda - h, \gamma)$ diagram of benzene for radical attack.

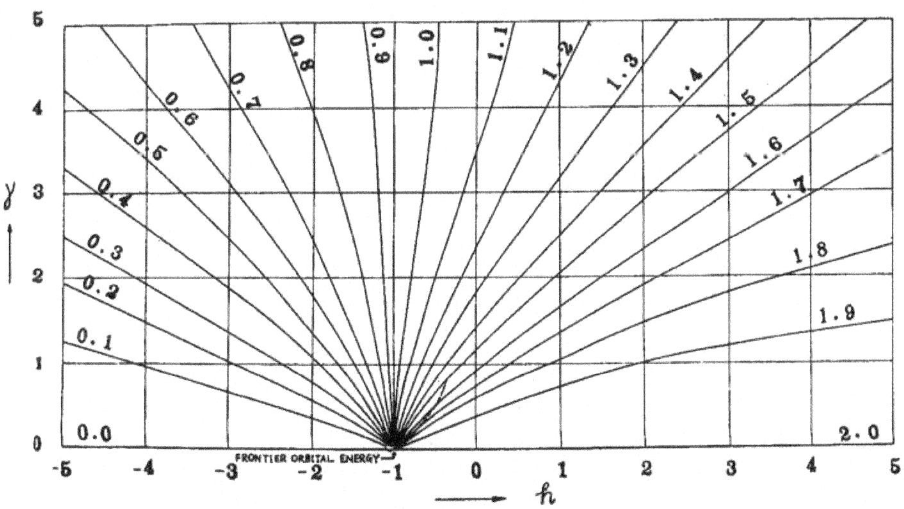

FIG. 3. The $(q_\lambda - h, \gamma)$ diagram of benzene for nucleophilic attack.

two singular points. The shape of assembly of contour lines near these singular points resembles closely a Japanese folding fan in a spread form. These singularities appear at the points where $\gamma=0$ and h is equal to the value corresponding to the highest occupied orbital in the diagram for an electrophilic reaction, the lowest vacant orbital for a nucleophilic reaction, and both the orbitals for a radical reaction. These particular orbitals are nothing but those which we had called the frontier orbitals.[6]

THE INTERPRETATION OF THE MECHANISM OF AROMATIC SUBSTITUTION BY (q_h-h, γ) DIAGRAM

On the diagram obtained in the preceding section, we may be able to follow the process of aromatic substitution if we use the following two relations which might hold between q_h, h, and γ in the course of reaction.

The first of these relations is given by the equation stated in the preceding section, being of the form

$$q_h=f(h,\gamma). \tag{19}$$

In the case of benzene, this corresponds to Eq. (18).

The second relation comes from the requirement that the energy of quasi-π orbital must vary accordingly as q_h increases or decreases. Namely, the electron density at the pseudoatom depends upon the change in its Coulomb integral and the value of this integral cannot be independent upon the change in electron density. The second relation may thus be considered to prescribe what should be satisfied between the Coulomb integral and the π-electron density in order that the total system be self-consistent.§ For expressing this relation, we may introduce the following equation

$$h=g(q_h). \qquad (20)\|$$

By using Eqs. (19) and (20), we can draw on the diagram the locus of the reaction path by eliminating q_h from them.

It is attempted to discuss the process of reaction by considering the case of electrophilic substitution in benzene as an example. We consider first the case when the energy of the electrophilic reagent¶ lies below that of the highest occupied orbital in an isolated benzene molecule. The reagent will approach the position of attack of the substrate molecule with no appreciable amount of charge transfer until the Coulomb integral of the quasi-π orbital becomes close to the energy of the frontier orbital in benzene, since if any small conjugation between the reagent and the substrate should occur, the electron density must immediately become 2, as is seen from Fig. 1, so that according to the relation (20), the Coulomb integral at the pseudoatom become very large (the value of h becomes small). Thus, the reaction proceeds in a first stage according to a line along the h axis on the diagram in Fig. 1. The locus may start from a certain point on the h axis ($h>1$) which corresponds to the energy of the isolated electrophilic reagent, and then, accordingly as the reagent is thrown into conjugation with benzene molecule, the value of h may gradually decrease along the h axis to the point $h=1$ which in this case corresponds to the energy of the frontier orbital for an electrophilic substitution.

After the point corresponding to the frontier orbital has been reached, q_h varies along the locus which is determined by the relations (19) and (20). Thus, the process of reaction will be represented by a certain dotted line as is given in that figure schematically.

It is to be noted that, as is easily seen from the foregoing discussion, when the energy of the isolated reagent is higher than that of the highest occupied orbital of the substrate molecule, the charge transfer resulting from hyperconjugation cannot occur (cf. Fig. 1) and accordingly benzene will never be subjected to an electrophilic substitution by such a reagent.**

Also with regard to the nucleophilic substitution, an entirely analogous discussion can be made.

According to these considerations, two important conclusions may be derived. The first of these is as follows: in order that an electrophilic substitution can occur, the energy of the lowest unoccupied orbital of the isolated electrophilic reagent must be lower than that of the highest occupied orbital of the substrate molecule, and in order that a nucleophilic reaction can occur, the energy of the highest occupied orbital of the isolated nucleophilic reagent must be higher than that of the lowest vacant orbital of the substrate molecule. There is no such special limitation as to the energy of the half-occupied orbital of a radical reagent in radical reaction. This conclusion is in accordance with the proposition made by Nagakura and Tanaka in regard to electrophilic substitution,[4] and may serve as a justification of it.

The second of the conclusions is that frontier orbitals are, in the present theory, distinctly discriminated from the other orbitals in relation to the charge transfer, as is understood at a glance upon the (q_h-h, γ) diagram. It may safely be said that the frontier orbital plays a significant role in the following sense. The energy of that orbital is closely related to the condition which determines whether an electrophilic, as well as a nucleophilic, substitution takes place or not.

In order to derive, from the mechanism stated above, any quantity that serves as a reactivity index of conjugated molecules, it is necessary to determine the end of locus, namely, the values of h and γ corresponding to the transition states.

The first method to determine the end point of the

§ Cf., for instance, Mulliken *et al.* (reference 3) and also M. Kurita and M. Kubo, Bull. Chem. Soc. Japan **24**, 13 (1951).

‖ It is no doubt possible to set up similar relations with regard to the other positions, but these are tentatively ignored for simplicity. In fact, the effect of considering them all is shown to be small.

¶ It corresponds to the lowest vacant orbital of electrophilic reagent in MO-theoretical sense.

** A molecular complex may be formed in this case (cf. reference 4), this fact, however, has no concern with the subject of the present paper.

locus may be possible by energetic consideration of the total system. Namely, we may, in principle, determine such values of h and γ that satisfy Eqs. (19) and (20) and make the value of total energy, including that of the σ part, maximum. Unfortunately, however, we have no sufficient experimental data available for such a quantitative, numerical treatment. Therefore, further discussion is inevitably reserved, and this problem is treated below in a simple manner.

The second method may be as follows: We assume that one of the values of q_h, h, and γ at the transition state (say, $q_h(t)$, h_t, and γ_t, respectively) might be a known constant with regard to a given reagent and a given reaction, being independent of individual substrate molecules. Then we may be able to calculate the other two, by using the relations (19) and (20). Among these three values at the transition state, $q_h(t)$ and h_t are considered to be influenced mainly by the property of the substrate, since $q_h(t)$ may depend obviously upon the size of substrate molecule and h_t may change accordingly as the frontier orbital energy changes.†† On the contrary, the quantity, γ_t, seems to concern itself almost exclusively with the neighborhood of the reaction center in the substrate molecule and is considered to be almost independent of the size or structure of the substrate molecule. From these considerations, we may proceed further, assuming that γ_t is a known constant‡‡ which is determined by the reagent and the type of reaction and is independent of individual substrate conjugated system.

A brief discussion is made here about the reaction path after the transition state has been reached. The following procedure, upon which this discussion is based, seems to be a useful third method of determining the end point of the locus on the (q_h-h, γ) diagram. For this purpose, it is necessary to construct another q_h diagram, substituting the product molecule for the substrate molecule. For instance, in the case of the electrophilic substitution of benzene by NO_2^+, the second diagram to be drawn is for nitrobenzene, which is a different π-electron system from benzene. On the diagram for nitrobenzene, the locus of the reaction path also may be drawn for the hypothetical reaction in which the nitro group in nitrobenzene is replaced by the reagent, H^+. If this diagram is superimposed on the diagram for the original reaction in which the H atom in benzene is replaced by the reagent, NO_2^+, the two loci will cross each other at the transition state. This procedure seems applicable in determining the value of h_t and γ_t. But in order to carry it out, it is necessary to take into account the changes in various integrals in the secular determinant during the course of the reaction. In view of the cumbersomeness of this procedure, we recommend adopting the second treatment introduced in the foregoing, on which the following discussion is based.

†† In some nonalternant hydrocarbons, e.g., in fulvalene, the frontier orbital for a nucleophilic reaction lies below the zero orbital. In such cases, h_t must be considerably lower than in the case of benzene.

‡‡ If the distance of the pseudoatom and π orbital at the attacked position of the substrate molecule may be given, the value of γ at the transition state can be estimated. As the value of γ_t, Mulliken *et al.* (reference 3) took the value 2 (overlap included) in the case of $C_6H_7^+$ complex. In the case of larger reagent, it might be reasonable to take γ_t less than 2.

FRONTIER ELECTRON DENSITY AND SUPERDELOCALIZABILITY AS REACTIVITY INDEXES

From the general considerations made in the preceding section, we may be able to obtain the values of h_t and γ_t. If we solve the equation

$$D_t(\lambda)=(h_t-\lambda)\Delta(\lambda)-\gamma_t^2\Delta_{rr}(\lambda)=0, \tag{21}$$

and obtain its roots, the hyperconjugation energy[6,7] at the transition state is calculated according to the following formula:

$$\mathcal{H}_r=2\sum_{i=1}^{m}(\mu_i-\lambda_i)+n_h(\mu_{m+1}-h_t). \tag{22}$$

This quantity relates to the activation energy of substitutions in such a manner that

$$\Delta E=A-\mathcal{H}_r-n_hh_t, \tag{23}$$

where ΔE is the activation energy, and A is a constant in substitution by a given reagent.[5] Thus, $(\mathcal{H}_r+n_hh_t)$ and accordingly $\mathcal{H}_r$, when $h_t=0$, may be considered as the most general type of reactivity index in our frontier electron theory, since reactivity indices, the frontier electron density, and the superdelocalizability can be derived from this quantity as will be discussed later.

In calculating the value of $\mathcal{H}_r$, for simplicity, we may adopt the perturbation theory, by considering γ_t a small perturbation term. Thus, the hyperconjugation energy at the transition state is obtained from Eqs. (21) and (22) as a second-order perturbation term.

$$\mathcal{H}_r\cong\left\{2\sum_{i=1}^{m}\frac{\Delta_{rr}(\lambda_i)}{(h_t-\lambda_i)\Delta'(\lambda_i)}-n_h\sum_{i=1}^{n}\frac{\Delta_{rr}(\lambda_i)}{(h_t-\lambda_i)\Delta'(\lambda_i)}\right\}\gamma_t^2. \tag{24}$$

The right side of this equation contains h_t explicitly. Since we have assumed that γ_t is a known constant, we have to express h_t in terms of γ_t. For this purpose, we have to assume the form of the function g in Eq. (20). For simplicity, the following linear form may be adopted:

$$h=(h_0-h_1)(1-q_h)+h_1 \tag{25}$$

where h_0 and h_1 represent the values of h when $q_h=0$ and 1, respectively. In this point, another assumption may be added: when $h=0$, q_h is equal to unity, thereby satisfying the self-consistency requirement concerning

[7] Fukui, Yonezawa, and Nagata, J. Chem. Phys. **26**, 831 (1957).

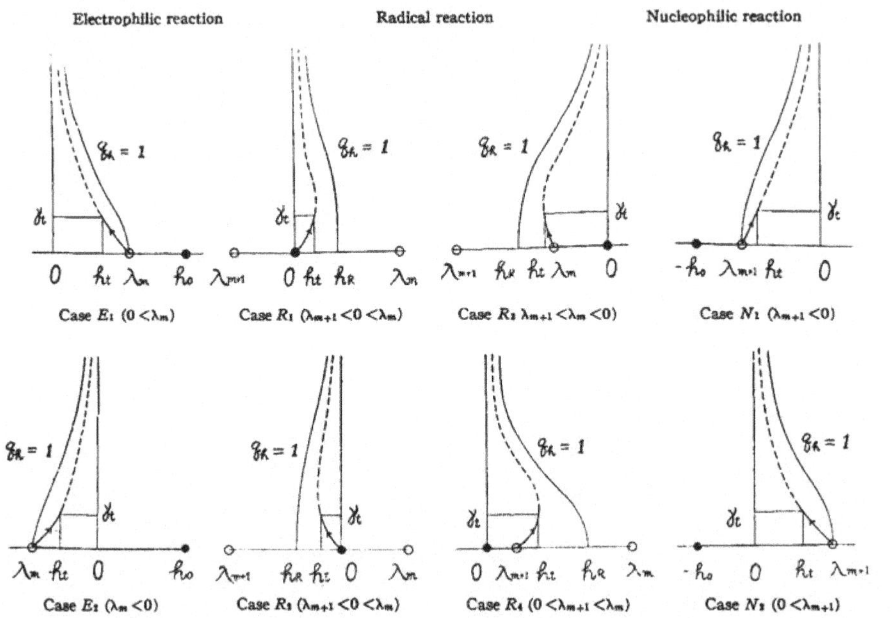

FIG. 4. Loci of reaction path on the $(q_\lambda-h, \gamma)$ diagram.

the relation of π-electron density and Coulomb integral at any π or quasi-π orbital. Hence, Eq. (25) becomes

$$h=h_0(1-q_\lambda). \qquad (25\text{-a})$$

Now one may draw the locus of the reaction path on the $(q_\lambda-h, \gamma)$ diagram. Several cases occur according to the position of λ_m or λ_{m+1} in relation to the origin, as illustrated in Fig. 4. Cases E_1, N_1, R_1, and R_2 are normal cases and are realized in almost all of the conjugated molecules, some of the polyphenols belong to Cases E_2 and R_3, and Cases N_2 and R_4 are seen in fulvalene and various quinones.

I. Cases E_1 and N_1

In these cases, h_t lies always between 0 and λ_f, where λ_f means the frontier orbitals, i.e., λ_m for the electrophilic, and λ_{m+1} for the nucleophilic reaction. As one extreme case, when $h_t=0$, the hyperconjugation energy is calculated as

$$\mathcal{H}_r\cong\left\{-(2-n_h)\sum_{i=1}^{m}\frac{\Delta_{rr}(\lambda_i)}{\lambda_i\Delta'(\lambda_i)}+n_h\sum_{i=m+1}^{n}\frac{\Delta_{rr}(\lambda_i)}{\lambda_i\Delta'(\lambda_i)}\right\}\gamma_t^2, \qquad (26)$$

and

$$\Delta E=A-\mathcal{H}_r \qquad (26\text{-a})$$

by the use of Eq. (24). As the other extreme case, when $h_t=\lambda_f$, the $\mathcal{H}_r$ is represented by a first-order perturbation term

$$\mathcal{H}_r\cong\sqrt{2}\left(2\frac{-\Delta_{rr}(\lambda_f)}{\Delta'(\lambda_f)}\right)^{\frac{1}{2}}\gamma_t, \qquad (27)$$

and

$$\Delta E\cong A-\mathcal{H}_r-n_h\lambda_f. \qquad (27\text{-a})$$

The term in parentheses on the right side of Eq. (26) is the quantity which we have previously termed the "superdelocalizability,"[5] S_r, and the term $2[-\Delta_{rr}(\lambda_f)/\Delta'(\lambda_f)]$ in Eq. (27) the frontier electron density, since $-\Delta_{rr}(\lambda_i)/\Delta'(\lambda_i)$ is equal to the density of π electron at the rth atom in the ith molecular orbital for the isolated π-electron system.

II. Cases R_1 and R_2

In these cases, h_t always lies between 0 and h_R. If we take $h_t=0$, $\mathcal{H}_r$ is expressed by Eq. (26) (putting $n_h=1$), and in this case the superdelocalizability becomes the reactivity index. If we take $h_t=h_R$,

$$\mathcal{H}_r\cong\left\{\sum_{i=1}^{m}\frac{\Delta_{rr}(\lambda_i)}{(h_R-\lambda_i)\Delta'(\lambda_i)}-\sum_{i=m+1}^{n}\frac{\Delta_{rr}(\lambda_i)}{(h_R-\lambda_i)\Delta'(\lambda_i)}\right\}\gamma_t^2, \qquad (28)$$

and

$$\Delta E=A-\mathcal{H}_r. \qquad (28\text{-a})$$

Remembering Eq. (11) and taking only the largest terms in both sides of Eqs. (11) and (28) for approximation, $\mathcal{H}_r$ may be written approximately as

$$\mathcal{H}_r\sim\frac{\left\{\left(-\frac{\Delta_{rr}(\lambda_m)}{\Delta'(\lambda_m)}\right)^{\frac{1}{2}}+\left(-\frac{\Delta_{rr}(\lambda_{m+1})}{\Delta'(\lambda_{m+1})}\right)^{\frac{1}{2}}\right\}^2}{\lambda_m-\lambda_{m+1}}, \qquad (29)§§$$

§§ In the very special case when $\Delta_{rr}(\lambda_m)$ or $\Delta_{rr}(\lambda_{m+1})=0$, h_R does not necessarily lie between λ_m and λ_{m+1}, and λ_{m-1} in place of λ_m, or λ_{m+2} in place of λ_{m+1} may appear in Eq. (29). In such a case, the corresponding position is in general to have a relatively small $\mathcal{H}_r$.

TABLE II. Reactivity indexes and experimental results for radical reaction of aromatic compounds.

Compound	Position, r[a]	S_r for radical attack	S_r' for radical attack	f_r for radical attack	Methyl affinity[b]	Rate constant[c] of radical reaction by CCl_3	Relative reactivity,[d] towards CCl_3 radical
Naphthacene	5	1.5049 (1)[e]	0.9992 (1)[e]	0.2947 (10)[e]	9250	4.4	1.50
Anthracene	9	1.3137 (2)	0.9340 (2)	0.3868 (3)	820	0.9	1.10
1,2-Benzanthracene	7	1.2502 (3)	0.8757 (3)	0.3961 (2)	468	1.25	1.50
1,2,5,6-Dibenzanthracene	7	1.1460 (4)	0.5738 (8)	0.2717 (11)	···	0.16	0.20
Stilbene	7	1.1380 (5)	0.7617 (4)	0.3842 (4)	183	···	−0.25
Quinoline	4	1.1207 (6)	0.7139 (5)	0.3540 (6)	29	···	···
Pyrene	1	1.1150 (7)	0.6103 (6)	0.2716 (12)	125	0.04	−0.60
Chrysene	12	1.0442 (8)	0.5716 (9)	0.2973 (9)	57.5	···	−1.30
Phenanthrene	9	0.9975 (9)	0.5692 (10)	0.3445 (7)	27	0.01	−2.30
Naphthalene	1	0.9944 (10)	0.5854 (7)	0.3618 (5)	22	···	−1.95
Biphenyl	2	0.9770 (11)	0.2540 (12)	0.1790 (13)	5	···	−3.10
Pyridine	2	0.9100 (12)	0.5041 (11)	0.4291 (1)	3	···	···
Benzene		0.8333 (13)	0.3333 (13)	0.3333 (8)	1	···	−4.0

[a] On the numbering of positions in each compound, reference is made to Chemical Abstracts, the American Chemical Society. Concerning each compound, the three reactivity indices S_r, S_r', and f_r predict the same position to be most reactive.
[b] M. Szwarc, J. Polymer Sci. 16, 367 (1955).
[c] E. C. Kooyman and J. W. Heringua, Nature 170, 663 (1952); figures indicate the specific rate constant (reference substance: cetene).
[d] E. C. Kooyman and E. Fahrenhorst, Nature 169, 153 (1952); figures indicate the logarithm of specific rate constant (reference substance: styrene).
[e] The figures in parentheses represent the order of reactivity predicted by each reactivity index.

and hence

$$\Delta E \cong A-\mathcal{H}_r-\frac{\lambda_{m+1}|C_r^{(m)}|+\lambda_m|C_r^{(m+1)}|}{|C_r^{(m)}|+|C_r^{(m+1)}|}$$

$$\sim A-\mathcal{H}_r-\frac{\lambda_{m+1}+\lambda_m}{2}. \quad \text{(29-a)}$$

In this approximation, the reactivity index is expressed in terms of frontier electron densities.

The rare cases, i.e., Cases E_2, N_2, R_3, and R_4 are treated in Appendix IV.

Thus, it is shown that the superdelocalizability and the frontier electron density are to become the reactivity indices for the extreme value of h_t. The experimental support for the availability of the frontier electron density has already been presented.[6] The result of comparison of the value of superdelocalizability with experimental data will be reported later in full details.‖| Generally, h_t may depend upon the position of a substrate molecule as well as the nature of the reagent. But the reactivity index corresponding to any intermediate value of h_t should always be of an intermediate nature between the superdelocalizability and the frontier electron density.

In this connection, it may be possible to adopt the following quantities as the reactivity indices:

‖| The superdelocalizability serves as the reactivity index in comparing the reactivities of different molecules, while the frontier electron density becomes the reactivity index in comparing the reactivities of the different positions of a given molecule, as is seen in Eqs. (26-a), (27-a), (28-a), and (29-a). In our previous paper (reference 6), we have adopted the expression $(-\Delta_{rr}(\lambda_m)/\Delta'(\lambda_m))+(-\Delta_{rr}(\lambda_{m+1})/\Delta'(\lambda_{m+1}))$ as the frontier electron density for the radical attack. From the result of Eq. (29) of the present paper, a modification may be needed. But these two expressions lead us almost always to the same conclusion in predicting the reactive positions.

(1) One-Term Approximation of Superdelocalizability, S_r'

In Eq. (26), it is seen that the terms which contribute the most to the superdelocalizability are those which correspond to the frontier orbitals. Hence, as an approximate formula of $\mathcal{H}_r$, we obtain

$$\mathcal{H}_r\sim\gamma_t^2\left\{(2-n_h)\frac{-\Delta_{rr}(\lambda_m)}{\lambda_m\Delta'(\lambda_m)}-n_h\frac{-\Delta_{rr}(\lambda_{m+1})}{\lambda_{m+1}\Delta'(\lambda_{m+1})}\right\}$$

$$=S_r'\gamma_t^2, \quad (30)$$

where S_r' is "the one term approximation of superdelocalizability" and is given by

$$S_r'=(2-n_h)\frac{-\Delta_{rr}(\lambda_m)}{\lambda_m\Delta'(\lambda_m)}-n_h\frac{-\Delta_{rr}(\lambda_{m+1})}{\lambda_{m+1}\Delta'(\lambda_{m+1})}, \quad (31)$$

which is a better approximation of the degree of contribution of the frontier electron density to the superdelocalizability. These circumstances are illustrated clearly in Table II concerning several aromatic compounds.

(2) A Quantity $V_r^{(s)}$

$$V_r^{(s)}=(2-n_h)\sum_{i=1}^{m}\frac{-\Delta_{rr}(\lambda_i)}{(\lambda_i)^s\Delta'(\lambda_i)}$$

$$+(-1)^s n_h\sum_{i=m+1}^{n}\frac{-\Delta_{rr}(\lambda_i)}{(\lambda_i)^s\Delta'(\lambda_i)}. \quad (32)$$

In this quantity, if we let $s=1$, $V_r^{(1)}$ becomes equal to the superdelocalizability, and if we let $s=\infty$, $\lim_{s\to\infty}(2V_r^{(s)}/\sum_r V_r^{(s)})$ is the frontier electron density provided that $\lambda_{m+1}<0<\lambda_m$. Since the superdelocalizability $V_r^{(1)}$ and the frontier electron density are reactivity indices which correspond to extreme values of h_t, $V_r^{(s)}(1<s<\infty)$ may be expected to become a "good" reactivity index for intermediate values of h_t. For instance, if we let $s=2$, and consider the case of alternant hydrocarbons, we obtain

$$V_r^{(2)}=2\sum_{i=1}^{m}\frac{-\Delta_{rr}(\lambda_i)}{\lambda_i^2\Delta'(\lambda_i)}. \qquad (33)$$

It has been proved by the present authors[7] that this quantity, $V_r^{(2)}$, relates to Dewar's approximate localization energy,[2] l_r, in such a way that

$$V_r^{(2)}=4/l_r^2. \qquad (34)$$

Since the quantity, l_r, is actually a good reactivity index for alternant hydrocarbons,[2] $V_r^{(2)}$ can be so equally as well as l_r.

The most suitable reactivity index is chosen according to the nature of the reaction. For instance, in Table II, it is seen that the best reactivity index for the trichloromethyl radical reaction is the superdelocalizability one. It has been proved[8] recently, that for discussing the carcinogenic activity of aromatic molecules S_r' produces a better result than S_r.

It should be noted that the conclusion mentioned above relates only to problems which are concerned with the comparison of the reactivity of *different molecules*. For discussing the reactivity of different positions within *a molecule*, the frontier electron density is the best reactivity index since it is easy to calculate.

CONCLUSION

The present MO-theoretical approach to the mechanism of charge transfer in aromatic substitution may contain some points which are open to criticism and therefore have to be improved in the future. For example, in the present treatment, the change in energy levels of the substrate molecule, as a whole, is utterly neglected due to the charge transfer.¶¶ Also, the value of various integrals in the secular determinant (1) are assumed to be constant throughout the course of the reaction. This assumption is incompatible with the so-called static method which considers principally the changes in Coulomb and exchange integrals caused by the approach, of the attacking reagent.[1] However, the frontier electron theory of the authors does hold that the electrostatic interaction between the reagent and the substrate molecule may actually be small in the vicinity of the transition state where a considerable amount of charge transfer may occur and a large portion of the charge of the reagent will be neutralized.

The authors are grateful to the Education Ministry of the Japanese Government for financial support.

[8] Nagata, Fukui, Yonezawa, and Tagashira, Cancer Research 15, 233 (1955).

¶¶ It may be true, however, that this effect is small if the substrate molecule is considerably large.

APPENDIX I. VALUE OF $\lim_{\gamma\to 0} q_h^{(B)}$ WHEN $\Delta(\lambda)$ HAS SOME DOUBLE ROOTS

When $\Delta(\lambda)=0$ has some double roots $\lambda_p=\lambda_{p+1}$, $\lambda_q=\lambda_{q+1}$, $\cdots$ and other simple roots, the following formulas are obtained in place of Eqs. (7) and (8).

(i) When h is not equal to any λ_i it is obvious that

$$\lim_{\gamma\to 0}(C_h^{(i)})^2=0, \quad (i=1,2,\cdots n;\, i\neq p, p+1, q, q+1,\cdots)$$

and

$$\lim_{\gamma\to 0}(C_h^{(h)})^2=1,$$

which are obtained as in the case of no multiple root. Furthermore, when i is equal to one of p, $p+1$, q, $q+1$, $\cdots$, it follows that

$$\left.\begin{matrix}\delta_p\\ \delta_{p+1}\end{matrix}\right\}=\begin{cases}0\\ 2\dfrac{\Delta_{rr}'(\lambda_p)}{(h-\lambda_p)\Delta''(\lambda_p)}\gamma^2+[\gamma^4].\end{cases}$$

Hence,

$$\left.\begin{matrix}\lim_{\gamma\to 0}(C_h^{(p)})^2\\ \lim_{\gamma\to 0}(C_h^{(p+1)})^2\end{matrix}\right\}=\begin{cases}0\\ \lim_{\gamma\to 0}\left(-2\dfrac{\Delta_{rr}'(\lambda_p)}{(h-\lambda_p)^2\Delta''(\lambda_p)}\gamma^2\right)\end{cases}=\begin{cases}0\\ 0\end{cases}$$

because both $(h-\lambda_p)$ and $\Delta''(\lambda_p)$ are not equal to zero. Similarly

$$\lim_{\gamma\to 0}(C_h^{(q)})^2=\lim_{\gamma\to 0}(C_h^{(q+1)})^2=\cdots=0.$$

(ii) When h is equal to one of the simple roots, say λ_l it is clear that

$$\lim_{\gamma\to 0}(C_h^{(i)})^2=0, \quad (i=1,2,\cdots n;\, i\neq l)$$

and also

$$\lim_{\gamma\to 0}(C_h^{(l)})^2=\lim_{\gamma\to 0}(C_h^{(h)})^2=\tfrac{1}{2}.$$

(iii) When h is equal to certain double roots, such as $\lambda_p(=\lambda_{p+1})$

$$\lim_{\gamma\to 0}(C_h^{(i)})^2=0, \quad (i=1,2,\cdots n;\, i\neq p, p+1),$$

when i is equal to one of h, p, and $p+1$, by introducing

$$\lambda^{(p)}=\lambda_p+\delta_p=h+\delta_p,$$
$$\lambda^{(h)}=h+\delta_h,$$

and

$$\lambda^{(p+1)}=\lambda_{p+1}+\delta_{p+1}=h+\delta_{p+1}$$

into Eq. (1), we obtain

$$\left.\begin{matrix}\delta_p\\ \delta_h\\ \delta_{p+1}\end{matrix}\right\}=\begin{cases}\left(-\dfrac{2\Delta_{rr}'(h)}{\Delta''(h)}\right)^{\frac{1}{2}}\gamma+[\gamma^2]\\ 0\\ -\left(-\dfrac{2\Delta_{rr}'(h)}{\Delta''(h)}\right)^{\frac{1}{2}}\gamma+[\gamma^2].\end{cases}$$

Incorporating these equations into (6) gives

$$\left.\begin{array}{l}\lim_{\gamma\to0}(C_h{}^{(p)})^2\\ \lim_{\gamma\to0}(C_h{}^{(k)})^2\\ \lim_{\gamma\to0}(C_h{}^{(p+1)})^2\end{array}\right\}=\left\{\begin{array}{l}\lim_{\gamma\to0}\frac{1}{2}-\dfrac{2\Delta_{rr}'(h)\Delta'''(h)-3\Delta''(h)\Delta_{rr}''(h)}{12\Delta_{rr}'(h)\Delta''(h)}\left(-\dfrac{2\Delta_{rr}'(h)}{\Delta''(h)}\right)^{\frac{1}{2}}\gamma\\ 0\\ \lim_{\gamma\to0}\frac{1}{2}+\dfrac{2\Delta_{rr}'(h)\Delta'''(h)-3\Delta''(h)\Delta_{rr}''(h)}{12\Delta_{rr}'(h)\Delta''(h)}\left(-\dfrac{2\Delta_{rr}'(h)}{\Delta''(h)}\right)^{\frac{1}{2}}\gamma.\end{array}\right.$$

From the limiting values obtained above, it is easy to arrive at the same result as is obtained in the case of no double roots and is shown in Table I.

It should be noted that, in the special case in which λ_{m+1} equals λ_m, a different result will be obtained. This occurs e.g., in cyclobutadiene. In this case, only in a radical reaction, Table I must be modified in the following manner.

(i) When $h>\lambda_m$, $\lim_{\gamma\to0} q_h{}^{(R)}=2$,

(ii) When $h=\lambda_m=\lambda_{m+1}$, $\lim_{\gamma\to0} q_h{}^{(R)}=1$,

(iii) When $h<\lambda_{m+1}$, $\lim_{\gamma\to0} q_h{}^{(R)}=0$.

APPENDIX II. APPROXIMATE EXPRESSION OF $q_h(h,\gamma)$ NEAR THE SINGULARITIES

In order to obtain an approximate formula of $q_h(h,\gamma)$ in the vicinity of the singularities which appear in the diagram at the two points, $(\lambda_m,0)$ and $(\lambda_{m+1},0)$. In the diagram for an electrophilic reaction, this point is $(\lambda_m,0)$. We let

$$h=\lambda_m+b\gamma,\quad(\gamma:\text{small}),$$

introduced this into Eq. (1), and solved. Following the same procedure used to obtain Eq. (9) we derive the equation

$$q_h{}^{(E)}=1+\frac{b}{(b^2+2f_r{}^{(E)})^{\frac{1}{2}}}+[\gamma],$$

where $f_r{}^{(E)}$ is equal to $(-2\Delta_{rr}(\lambda_m)/\Delta'(\lambda_m))$ which is simply the frontier electron density for an electrophilic attack.[2] Introducing $b=-(\lambda_m-h)/\gamma$ into this equation, we have

$$q_h{}^{(E)}\cong1-\frac{\lambda_m-h/\gamma}{\{(\lambda_m-h/\gamma)^2+2f_r{}^{(E)}\}^{\frac{1}{2}}}$$

as an approximate formula of $q_h{}^{(E)}(h,\gamma)$ near the point $(\lambda_m,0)$. Similarly, for a radical reaction, the following two equations are obtained near the points $(\lambda_m,0)$ and $(\lambda_{m+1},0)$, respectively:

$$q_h{}^{(R)}\cong\frac{3}{2}-\frac{1}{2}\frac{\lambda_m-h/\gamma}{\{(\lambda_m-h/\gamma)^2+2f_r{}^{(E)}\}^{\frac{1}{2}}},$$

$$q_h{}^{(R)}\cong\frac{1}{2}+\frac{1}{2}\frac{h-\lambda_{m+1}/\gamma}{\{(h-\lambda_{m+1}/\gamma)^2+2f_r{}^{(N)}\}^{\frac{1}{2}}},$$

where $f_r{}^{(N)}$ represents $(-2\Delta_{rr}(\lambda_{m+1})/\Delta'(\lambda_{m+1}))$ and is the same as the frontier electron density for a nucleophilic attack.[2] In the diagram for a nucleophilic reaction, the equation

$$q_h{}^{(N)}\cong1+\frac{h-\lambda_{m+1}/\gamma}{\{(h-\lambda_{m+1}/\gamma)^2+2f_r{}^{(N)}\}^{\frac{1}{2}}}$$

holds near the point $(\lambda_{m+1},0)$.

APPENDIX III. ASYMPTOTE WHICH THE CONTOUR LINE $q_h=1$ APPROACHES AT VERY LARGE CONJUGATION

At very large conjugation, by expanding μ_k in powers of γ^{-1}, the following equations are obtained:

$$\mu_k=\nu_{k-1}+\frac{(h-\nu_{k-1})\Delta(\nu_{k-1})}{\Delta_{rr}'(\nu_{k-1})}\gamma^{-2}+[\gamma^{-4}],\qquad(k=2,3,\cdots n)$$

$$\mu_1=\gamma+\tfrac{1}{2}(h-s_1-t_1)+[\gamma^{-1}],$$

$$(C_h{}^{(k)})^2=\frac{\Delta(\nu_{k-1})}{\Delta_{rr}'(\nu_{k-1})}\gamma^{-2}+[\gamma^{-4}],\quad(k=2,3,\cdots n)$$

$$(C_h{}^{(1)})^2=\tfrac{1}{2}\{1+\tfrac{1}{2}(s_1+h+t_1)\gamma^{-1}\}+[\gamma^{-2}]$$

and hence

$$q_h=1+\tfrac{1}{2}(s_1+h+t_1)\gamma^{-1}+[\gamma^{-2}],$$

where s_1 and t_1 are the coefficients in the expansions

$$\Delta(\lambda)=\lambda^n+s_1\lambda^{n-1}+\cdots$$

and

$$\Delta_{rr}(\lambda)=-\lambda^{n-1}+t_1\lambda^{n-2}+\cdots,$$

respectively. In an alternant hydrocarbon, $s_1=t_1=0$. In nonalternant hydrocarbons and substituted aromatic hydrocarbons, s_1 and t_1 can be different from zero. But in these cases t_1 is equal to $-s_1$, as is easily understood by considering the secular determinant (2), since we are interested only in substitutions in which the atom to be replaced is a hydrogen atom, and therefore the Coulomb term, a_{rr}, of the position of reaction is always equal to zero.

Therefore, even in the molecules which are different from alternant hydrocarbons, it holds that

$$q_h=1+\tfrac{1}{2}h\gamma^{-1}+[\gamma^{-2}].$$

Hence the contour line, $q_\lambda=1$, has an asymptote which is expressed by $h=0$, a straight line parallel to the γ axis, on the $(q_\lambda-h, \gamma)$ diagram.

APPENDIX IV. THE REACTIVITY INDICES IN CASES E_2, N_2, R_3, AND R_4

1. Cases E_2 and N_2

In these cases, h_t lies always between 0 and the frontier orbital, λ_f. If we let $h_t=0$, $\mathcal{H}_r$ is represented by***

Case E_2

$$\mathcal{H}_r=-2\lambda_m+2\sum_{i=m}^{n}\frac{\Delta_{rr}(\lambda_i)}{\lambda_i\Delta'(\lambda_i)}\gamma_i^2$$

$$=-2\lambda_m+S_r^{(f)}\gamma_i^2+S_r^{(N)}\gamma_i^2, \qquad \text{(IV-1)}$$

Case N_2

$$\mathcal{H}_r=2\lambda_{m+1}-2\sum_{i=1}^{m+1}\frac{\Delta_{rr}(\lambda_i)}{\lambda_i\Delta'(\lambda_i)}\gamma_i^2$$

$$=2\lambda_{m+1}+S_r^{(f)}\gamma_i^2+S_r^{(E)}\gamma_i^2, \qquad \text{(IV-2)}$$

where $S_r^{(N)}$ and $S_r^{(E)}$ are the quantities which are obtained by putting $n_\lambda=2$ and 0, respectively, into the terms in the parentheses of the right side of Eq. (26). And $S_r^{(f)}$ means $(-2\Delta_{rr}(\lambda_f)/|\lambda_f|\Delta'(\lambda_f))$.

The first zero-order terms are constant in a given substrate molecule, and the reactivity of each position is determined by the second and third terms. The term in each equation which contributes most corresponds to the frontier orbital, λ_f, and the reaction will take place at the position of the largest frontier electron density.

If we take $h_t=\lambda_f$, $\mathcal{H}_r$ becomes

$$\mathcal{H}_r\cong2\left(-\frac{\Delta_{rr}(\lambda_f)}{\Delta'(\lambda_f)}\right)^{\frac{1}{2}}\gamma_t,$$

expressed in terms of the frontier electron density.

*** Here it is assumed that $\lambda_m<0<\lambda_{m-1}$ in Cases E_2 and R_3, and $\lambda_{m+2}<0<\lambda_{m+1}$ in Cases N_2 and R_4, since the other cases are of no practical significance.

2. Cases R_3 and R_4

These cases occur in the molecules corresponding to Cases E_2 and N_2. Of the two frontier levels, the one nearer to the γ axis is denoted by λ_f. Two theoretically possible subcases occur according to whether $2|\lambda_f|$ is smaller or larger than h_0. In the former subcase, the tangent of the locus at the point $(h=\lambda_f, \gamma=0)$ and λ_f have the same sign, and in the latter subcase, the opposite sign.

The value of h_t lies always between 0 and h_R. If we take $h_t=0$, $\mathcal{H}_r$ becomes

Case R_3

$$\mathcal{H}_r\cong-\lambda_m+\frac{\Delta_{rr}(\lambda_m)}{\lambda_m\Delta'(\lambda_m)}\gamma_i^2+2\sum_{i=m+1}^{n}\frac{\Delta_{rr}(\lambda_i)}{\lambda_i\Delta'(\lambda_i)}\gamma_i^2$$

$$=|\lambda_f|+\tfrac{1}{2}S_r^{(f)}\gamma_i^2+S_r^{(N)}\gamma_i^2, \qquad \text{(IV-3)}$$

Case R_4

$$\mathcal{H}_r\cong\lambda_{m+1}-\frac{\Delta_{rr}(\lambda_{m+1})}{\lambda_{m+1}\Delta'(\lambda_{m+1})}\gamma_i^2-2\sum_{i=1}^{m}\frac{\Delta_{rr}(\lambda_i)}{\lambda_i\Delta'(\lambda_i)}\gamma_i^2$$

$$=|\lambda_f|+\tfrac{1}{2}S_r^{(f)}\gamma_i^2+S_r^{(E)}\gamma_i^2, \qquad \text{(IV-4)}$$

The second term in the right side of each equation, which is most important in determining the position of attack, corresponds to λ_f.

If we take $h_t=h_R$, $\mathcal{H}_r$ is represented by Eq. (28) and approximately, by (29).

The sum of the second and third terms of the right sides of Eqs. (IV-1), (IV-2), (IV-3), and (IV-4) can be said to be the "modified superdelocalizability" for the present cases. Namely, in Case E_2, they are the superdelocalizability of (substrate)$^{++}$ for the nucleophilic attack; in Case N_2, that of (substrate)$^{--}$ for the electrophilic attack; in Case R_3, that of (substrate)$^{+}$ for the nucleophilic attack; and in Case R_4, that of (substrate)$^{-}$ for the electrophilic attack, respectively. Thus, these superdelocalizabilities correspond to those for the substrate molecule which have been subjected to electron transfer that can occur adiabatically through the paths illustrated by the thick lines on the diagrams in Fig. 4.

Reprinted from the JOURNAL OF CHEMICAL PHYSICS, Vol. 31, No. 2, 287–293, August, 1959
Printed in U. S. A.

Novel Perturbation Theory in Simple LCAO Treatment of Conjugated Molecules—Method of Perturbed Secular Determinant

KENICHI FUKUI, CHIKAYOSHI NAGATA, TEIJIRO YONEZAWA, HIROSHI KATO, AND KEIJI MOROKUMA

Faculty of Engineering, Kyoto University, Kyoto, Japan

(Received November 5, 1958)

A new method of perturbation theory is developed in the simple LCAO MO treatment of conjugated molecules. Several useful equations are derived, giving orbital energy, electron density, and bond order of the perturbed system.

INTRODUCTION

THE perturbation theory in the simple LCAO MO method has been developed extensively by many researchers, *viz.*, by Coulson and Longuet-Higgins,[1] Longuet-Higgins,[2] Matsen,[3] Jaffé,[4] Masse,[5] and Dewar,[6] and has been applied successfully to the problems of absorption spectra[6,7] and chemical reactivity[8–10] of the molecules containing π electrons, especially of large molecules.

Coulson and Longuet-Higgins[1] have derived the equations which give the changes in electron density and bond order caused by a perturbation within a molecule; that is, the effects of an alteration in electronegativity of the constituent atom. A later work of Dewar[6] has made possible the perturbation treatment of such a system as is composed of two conjugated molecules or groups connected with each other by a bridgelike bond. In his paper, however, the equation which gives the electron density at a certain atom in a conjugated molecule when a mesomeric group is attached to it has not been correctly derived.

In the present paper, a new mathematical method of perturbation theory is proposed and the equations are derived which give the orbital energy, the electron density, and the bond order of the perturbed system.

PRELIMINARIES

The secular equation of a conjugated system which has n atomic orbitals may be converted into the following form:

$$\left.\begin{array}{l}(\alpha_1-\epsilon)c_1^j+\beta_{12}c_2^j+\beta_{13}c_3^j+\cdots+\beta_{1n}c_n^j=(\epsilon_j-\epsilon)c_1^j\\ \beta_{21}c_1^j+(\alpha_2-\epsilon)c_2^j+\beta_{23}c_3^j+\cdots+\beta_{2n}c_n^j=(\epsilon_j-\epsilon)c_2^j\\ \cdots\cdots\cdots\cdots\cdots\cdots\cdots\cdots\cdots\cdots\cdots\cdots\\ \beta_{n1}c_1^j+\beta_{n2}c_2^j+\beta_{n3}c_3^j+\cdots+(\alpha_n-\epsilon)c_n^j=(\epsilon_j-\epsilon)c_n^j\end{array}\right\}, \quad (1)$$

where α_r and β_{rs} are the well-known Coulomb and resonance integrals, respectively; ϵ_j is the jth orbital energy, and c_s^j is the coefficient of the sth atomic orbital in the jth molecular orbital, respectively. From Eq. (1), it follows that

$$c_s^j=\frac{\epsilon_j-\epsilon}{\Delta(\epsilon)}\sum_{r=1}^{n}(-1)^{r+s}c_r^j\Delta_{rs}(\epsilon), \quad (2)$$

where $\Delta(\epsilon)$ and $\Delta_{rs}(\epsilon)$ are the secular determinant of the conjugated system and the minor of its rsth ele-

[1] C. A. Coulson and H. C. Longuet-Higgins, Proc. Roy. Soc. (London) **A191**, 39 (1947); **A192**, 16 (1947); **A193**, 447, 456 (1948); **A195**, 188 (1948).
[2] H. C. Longuet-Higgins, J. Chem. Phys. **18**, 265, 275, 283 (1950).
[3] F. A. Matsen, J. Am. Chem. Soc. **72**, 5243 (1950).
[4] H. H. Jaffé, J. Am. Chem. Soc. **77**, 274 (1955); **76**, 5843 (1954).
[5] J-L. Masse, Cahiers phys. **70**, 31 (1956).
[6] M. J. S. Dewar, J. Am. Chem. Soc. **74**, 3341 (1952).
[7] M. J. S. Dewar, J. Chem. Soc. **1950**, 2329.
[8] Yonezawa, Hayashi, Nagata, Okamura, and Fukui, J. Polymer Sci. **14**, 312 (1954).
[9] Hayashi, Yonezawa, Nagata, Okamura, and Fukui, J. Polymer Sci. **20**, 537 (1956).
[10] Yonezawa, Higashimura, Katagiri, Hayashi, Okamura, and Fukui, J. Polymer Sci. **26**, 311 (1958).

ment, respectively. It is easy to obtain the following equations from Eq. (2).

$$\frac{c_s^j c_t^j}{\epsilon_j-\epsilon}=\frac{1}{\Delta(\epsilon)}\sum_{r=1}^{n}(-1)^{r+s}c_r^j c_t^j\Delta_{rs}(\epsilon), \tag{3}$$

$$\sum_{j}^{\text{all}}\frac{c_s^j c_t^j}{\epsilon_j-\epsilon}=\frac{(-1)^{s+t}\Delta_{st}(\epsilon)}{\Delta(\epsilon)}, \tag{4}$$

$$(-1)^{r+s+1}c_r^j c_s^j=\frac{\Delta_{rs}(\epsilon_j)}{\Delta'(\epsilon_j)}, \tag{5}$$

where $\Delta'(\epsilon_j)\equiv\partial\Delta(\epsilon_j)/\partial\epsilon_j$. Here it is assumed that the equation $\Delta(\epsilon)=0$ has no double root. In the case of a degenerate system, corresponding formulas may also easily be obtained.

By using Eq. (5), we obtain the following formulas from Eqs. (A8), (A18), and (A20) of Appendix I.

$$(-1)^{r+s+1}\frac{\Delta'_{rs}(\epsilon_j)}{\Delta'(\epsilon_j)}=\sum_{k(\neq j)}^{\text{all}}\frac{c_r^j c_s^j+c_r^k c_s^k}{\epsilon_j-\epsilon_k}, \tag{6}$$

$$(-1)^{r+s+1}\frac{\Delta''_{rs}(\epsilon_j)}{\Delta'(\epsilon_j)}=\sum_{k(\neq j)}^{\text{all}}\sum_{l(\neq j,k)}^{\text{all}}\frac{c_r^j c_s^j+2c_r^k c_s^k}{(\epsilon_j-\epsilon_k)(\epsilon_j-\epsilon_l)}, \tag{7}$$

$$(-1)^{r+s+1}\frac{\Delta_{rs}'''(\epsilon_j)}{\Delta'(\epsilon_j)}=\sum_{k(\neq j)}^{\text{all}}\sum_{l(\neq j,k)}^{\text{all}}\sum_{m(\neq j,k,l)}^{\text{all}}\times\frac{c_r^j c_s^j+3c_r^k c_s^k}{(\epsilon_j-\epsilon_k)(\epsilon_j-\epsilon_l)(\epsilon_j-\epsilon_m)}. \tag{8}$$

METHOD OF THE PERTURBED SECULAR DETERMINANT

Perturbed Secular Determinant

We consider a conjugated molecule whose secular determinant is $\Delta(\epsilon)$. When this molecule is subjected to some sort of perturbation, its secular determinant will be changed. For example, when carbon atom, say the rth atom in the molecule, is replaced by a hetero-atom, the Coulomb integral of that atom, α_r, will become $\alpha_r+\alpha$. We may regard this change as a perturbation and write the secular determinant of the perturbed system as

$$D(\epsilon)=\Delta(\epsilon)+\alpha\Delta_{rr}(\epsilon). \tag{9}$$

Similarly, when the resonance integral of bond (rs), β_{rs}, is increased by β, we can write

$$D(\epsilon)=\Delta(\epsilon)+2(-1)^{r+s}\beta\Delta_{rs}(\epsilon)-\beta^2\Delta_{rrss}(\epsilon). \tag{10}$$

Furthermore, a conjugated molecule, whose secular determinant is $\Delta_1(\epsilon)$, will be able to conjugate with another conjugated system, whose secular determinant is $\Delta_2(\epsilon)$, through a weak bond whose resonance integral is γ. The perturbed determinant can be written as

$$D(\epsilon)=\Delta_1(\epsilon)\Delta_2(\epsilon)-\gamma^2\Delta_{1rr}(\epsilon)\Delta_{2ss}(\epsilon), \tag{11}$$

where r and s are the positions of bonding in the two systems, respectively.

In the following, we will treat the three types of perturbation stated above, and derive the equations by which the mth perturbation terms can be evaluated. Now, for the sake of convenience, the first, the second, and the last types of perturbation, corresponding to Eqs. (9), (10), and (11) are referred to as Case A, Case B, and Case C, respectively.

Orbital Energy

In Case A, each orbital energy of the system is varied continuously according to the change of Coulomb integral at the rth atom; therefore, the perturbed orbital energy, ϵ_j', can be expanded in powers of α, namely,

$$\epsilon_j'=\sum_{m=0}^{\infty}a_r^{j(m)}\alpha^m. \tag{12}$$

By using the Laurent theorem, $a_r^{j(m)}$ is expressed by the following formula

$$a_r^{j(m)}=\frac{1}{2\pi i}\int_C\frac{\epsilon_j'(\alpha)}{\alpha^{m+1}}d\alpha, \tag{13}$$

where C is a small circular contour which moves counterclockwise around the origin in the α plane. From Eq. (13) and the relation of ϵ_j' and α, that is $\Delta(\epsilon_j')+\alpha\Delta_{rr}(\epsilon_j')=0$, $a_r^{j(m)}$ is written as the function of ϵ_j', and by converting the variable ϵ_j' to Z, it follows that

$$a_r^{j(m)}=\frac{(-1)^m}{2\pi i}\int_{C'}\times\frac{Z\{\Delta_{rr}(Z)\}^{m-1}\{\Delta'(Z)\Delta_{rr}(Z)-\Delta(Z)\Delta_{rr}'(Z)\}}{\{\Delta(Z)\}^{m+1}}dZ \tag{14}$$

$$=\frac{(-1)^m}{2\pi i}\int_{C'}f(Z)dZ, \tag{14'}$$

where C' is a small circular contour which moves counterclockwise around the point $Z=\epsilon_j$ in the Z plane. The function $f(Z)$ has a pole of order $(m+1)$ at the point $Z=\epsilon_j$. Hence, $a_r^{j(m)}$ is easily given by the following formula.

$$a_r^{j(m)}=\frac{(-1)^m}{m!}\cdot\left[\frac{d^m}{dZ^m}\frac{Z\{\Delta_{rr}(Z)\}^{m-1}\{\Delta'(Z)\Delta_{rr}(Z)-\Delta(Z)\Delta'_{rr}(Z)\}}{\{\Delta(Z)/(Z-\epsilon_j)\}^{m+1}}\right]_{Z=\epsilon_j}$$

$$(m=0,1,2,\cdots). \tag{15}$$

This is the general formula by which the mth perturbed term in orbital energies can be evaluated directly.

That is,

$$a_r^{j(0)}=\epsilon_j, \tag{16}$$

$$a_r^{j(1)}=\frac{-\Delta_{rr}(\epsilon_j)}{\Delta'(\epsilon_j)}=(c_r^j)^2, \tag{17}$$

$$a_r^{j(2)}=\frac{\Delta_{rr}(\epsilon_j)\Delta_{rr}'(\epsilon_j)}{\{\Delta'(\epsilon_j)\}^2}-\frac{1}{2}\frac{\{\Delta_{rr}(\epsilon_j)\}^2\Delta''(\epsilon_j)}{\{\Delta'(\epsilon_j)\}^3}=\sum_{k(\neq j)}^{\text{all}}\frac{(c_r^jc_r^k)^2}{\epsilon_j-\epsilon_k}, \tag{18}$$

......................................,

are obtained. The furthest right sides of Eqs. (17) and (18) are derived by using Eqs. (6) to (8) and Appendix I. The results coincide exactly with those obtained by the existing method of perturbation.

The perturbed orbital energy for Case *B* is obtained in like manner as in Case *A*. The perturbed orbital energies can be expanded in powers of β; namely,

$$\epsilon_j'=\sum_{m=0}^{\infty}a_{rs}^{j(m)}\beta^m, \tag{19}$$

then $a_{rs}^{j(m)}$ is expressed by the following formula:

$$a_{rs}^{j(m)}=\frac{1}{2\pi i}\int_C\frac{\epsilon_j'(\beta)}{\beta^{m+1}}d\beta, \qquad (m=0,1,2,\cdots) \tag{20}$$

where

$$\beta=(-1)^{r+s}\frac{\Delta_{rs}(\epsilon_j')}{\Delta_{rrss}(\epsilon_j')}\left\{1-\left(1+\frac{\Delta(\epsilon_j')\Delta_{rrss}(\epsilon_j')}{\{\Delta_{rs}(\epsilon_j')\}^2}\right)^{\frac{1}{2}}\right\}. \tag{20'}$$

Thus, are obtained

$$a_{rs}^{j(0)}=\epsilon_j, \tag{21}$$

$$a_{rs}^{j(1)}=2c_r^jc_s^j, \tag{22}$$

$$a_{rs}^{j(2)}=\sum_{k(\neq j)}^{\text{all}}\frac{(c_r^jc_s^k+c_s^jc_r^k)^2}{\epsilon_j-\epsilon_k}, \tag{23}$$

....

In deriving these formulas, the relations shown in Appendix II are put to use.

For Case *C*, if we put the perturbed orbital energies as ϵ_{1j}' and ϵ_{2k}' corresponding to the orbital energies of $\Delta_1(\epsilon)$ and $\Delta_2(\epsilon)$, respectively, they can be expanded in powers of γ^2 as

$$\left.\begin{aligned}\epsilon_{1j}'&=\sum_{m=0}^{\infty}b_{1rs}^{j(m)}\gamma^{2m}\\ \epsilon_{2k}'&=\sum_{m=0}^{\infty}b_{2rs}^{k(m)}\gamma^{2m}\end{aligned}\right\} \tag{24}$$

In just the same way as in Cases *A* and *B*, $b_{1rs}^{j(m)}$ is given by the following formula:

$$b_{1rs}^{j(m)}=\frac{1}{m!}\left[\frac{d^m}{dZ^m}\frac{Z\{\Delta_{1rr}(Z)\Delta_{2ss}(Z)\}^{m-1}[A]}{\{\Delta_1(Z)\Delta_2(Z)/(Z-\epsilon_{1j})\}^{m+1}}\right]_{Z=\epsilon_{1j}}, \qquad (m=0,1,2,\cdots) \tag{25}$$

where

$$[A]=\{\Delta_1'(Z)\Delta_2(Z)+\Delta_1(Z)\Delta_2'(Z)\}\Delta_{1rr}(Z)\Delta_{2ss}(Z)-\{\Delta_{1rr}'(Z)\Delta_{2ss}(Z)+\Delta_{1rr}(Z)\Delta_{2ss}'(Z)\}\Delta_1(Z)\Delta_2(Z). \tag{25'}$$

Thus, are obtained

$$b_{1rs}^{j(0)}=\epsilon_{1j}, \tag{26}$$

$$b_{1rs}^{j(1)}=\sum_{k}^{\text{all}}\frac{(c_{1r}^jc_{2s}^k)^2}{\epsilon_{1j}-\epsilon_{2k}}, \tag{27}$$

$$b_{1rs}^{j(2)}=\left\{\sum_{k}^{\text{all}}\frac{(c_{1r}^jc_{2s}^k)^2}{\epsilon_{1j}-\epsilon_{2k}}\right\}\times\left[\sum_{k}^{\text{all}}\sum_{m(\neq j)}^{\text{all}}\frac{(c_{1r}^mc_{2s}^k)^2}{(\epsilon_{1j}-\epsilon_{2k})(\epsilon_{1j}-\epsilon_{1m})}-\sum_{k}^{\text{all}}\frac{(c_{1r}^jc_{2s}^k)^2}{(\epsilon_{1j}-\epsilon_{2k})^2}\right], \tag{28}$$

....

The formula of $b_{2rs}^{k(m)}$ is obtained if the suffixes *j* and 1 are replaced by *k* and 2, respectively, in Eq. (25).

A special case of importance involved in Case *C* is that in which $\Delta_2(\epsilon)$ is a single atom or a mesomeric group, X. In this case the coefficient of the perturbed orbital energy is given by

$$b_{r\text{X}}^{j(m)}=\frac{1}{m!}\left[\frac{d^m}{dZ^m}\frac{Z[\Delta_{rr}(Z)\}^{m-1}[A']}{(\alpha_\text{X}-Z)^{m+1}\{\Delta(Z)/(Z-\epsilon_j)\}^{m+1}}\right]_{Z=\epsilon_j}, \qquad (m=0,1,2,\cdots), \tag{29}$$

where

$$[A']=(\alpha_\text{X}-Z)\{\Delta_{rr}(Z)\Delta'(Z)-\Delta_{rr}'(Z)\Delta(Z)\}-\Delta(Z)\Delta_{rr}(Z), \tag{29'}$$

and α_X is the Coulomb integral of the atom X. Hence, the coefficients of the perturbed energies are given by

$$b_{r\text{X}}^{j(0)}=\epsilon_j, \tag{30}$$

$$b_{r\text{X}}^{j(1)}=-\frac{(c_r^j)^2}{(\alpha_\text{X}-\epsilon_j)}, \tag{31}$$

$$b_{r\text{X}}^{j(2)}=\frac{(c_r^j)^2}{\epsilon_j-\alpha_\text{X}}\left[\sum_{k(\neq j)}^{\text{all}}\frac{(c_r^k)^2}{(\epsilon_j-\alpha_\text{X})(\epsilon_j-\epsilon_k)}-\frac{(c_r^j)^2}{(\epsilon_j-\alpha_\text{X})^2}\right]. \tag{32}$$

On the other hand, the coefficient of the *m*th order perturbed orbital energy corresponding to the orbital energy of the substituent, X, is given by

$$b_{r\text{X}}^{\text{X}(m)}=\frac{(-1)^{m+1}}{m!}\left[\frac{d^m}{dZ^m}\frac{Z\{\Delta_{rr}(Z)\}^{m-1}[A']}{\{\Delta(Z)\}^{m+1}}\right]_{Z=\alpha_\text{X}}, \qquad (m=0,1,2,\cdots) \tag{33}$$

and thus,

$$b_{r\mathrm{X}}{}^{(0)}=\alpha_{\mathrm{X}}, \tag{34}$$

$$b_{r\mathrm{X}}{}^{(1)}=\sum_{j}^{\mathrm{all}}\frac{(c_r{}^j)^2}{\alpha_{\mathrm{X}}-\epsilon_j}, \tag{35}$$

$$b_{r\mathrm{X}}{}^{(2)}=\left\{\sum_{j}^{\mathrm{all}}\frac{(c_r{}^j)^2}{\alpha_{\mathrm{X}}-\epsilon_j}\right\}\left\{-\sum_{j}^{\mathrm{all}}\frac{(c_r{}^j)^2}{(\alpha_{\mathrm{X}}-\epsilon_j)^2}\right\}. \tag{36}$$

Bond Order and Electron Density

The bond order, p_{tu}, and the electron density, q_t of the nonperturbed system, are given as

$$p_{tu}=\sum_{j}^{\mathrm{all}}\nu_j p_{tu}{}^j, \qquad q_t=\sum_{j}^{\mathrm{all}}\nu_j q_t{}^j, \tag{37}$$

where ν_j is the number of electron in the jth orbital, and $p_{tu}{}^j$ and $q_t{}^j$, following Löwdin,[11] will be interpreted as the bond order of the bond (tu) of the jth orbital, and the electron density of the tth atom of the jth orbital, and are designated as "separate bond order" and "separate electron density," respectively.[11] Similarly, the bond order and the electron density for the perturbed system, P_{tu} and Q_t, are obtained by summing the perturbed separate bond orders $P_{tu}{}^j$'s and perturbed separate electron densities $Q_t{}^j$'s over all the occupied orbitals; namely,

$$P_{tu}=\sum_{j}^{\mathrm{all}}\nu_j P_{tu}{}^j, \qquad Q_t=\sum_{j}^{\mathrm{all}}\nu_j Q_t{}^j. \tag{38}$$

First, for Case A, $P_{tu}{}^j$ can be expanded in powers of α, that is,

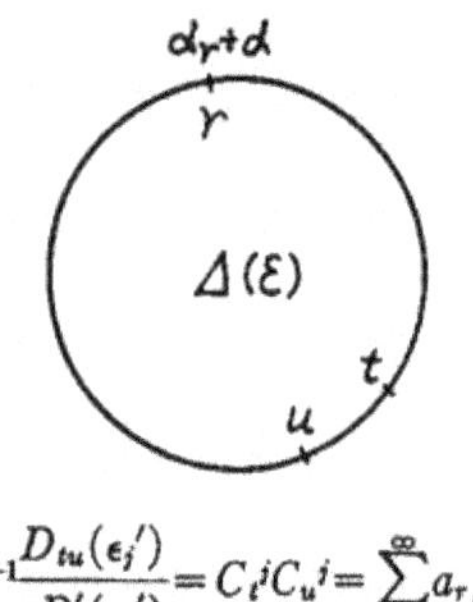

$$P_{tu}{}^j=(-1)^{t+u+1}\frac{D_{tu}(\epsilon_j{}')}{D'(\epsilon_j{}')}=C_t{}^jC_u{}^j=\sum_{m=0}^{\infty}a_{r,tu}{}^{j(m)}\alpha^m, \tag{39}$$

where $C_r{}^j$ is the coefficient of the rth atomic orbital in the jth perturbed MO. Then, the formula of $a_{r,tu}{}^{j(m)}$, is given by

$$a_{r,tu}{}^{j(m)}=\frac{(-1)^{m+1}}{m!}\times\left[\frac{d^m}{dZ^m}\frac{\{\Delta_{rr}(Z)\}^{m-1}\Delta_{tr}(Z)\Delta_{ru}(Z)}{\{\Delta(Z)/(Z-\epsilon_j)\}^{m+1}}\right]_{Z=\epsilon_j}$$

$$(m=0,1,2,\cdots). \tag{40}$$

[11] P.-O. Löwdin, J. Chem. Phys. **21**, 496 (1953).

Hence are obtained

$$P_{tu}=p_{tu}+2\sum_{j}^{\mathrm{occ}}\sum_{k}^{\mathrm{unocc}}\frac{c_r{}^jc_r{}^k(c_t{}^jc_u{}^k+c_u{}^jc_t{}^k)}{\epsilon_j-\epsilon_k}\alpha+\cdots \tag{41}$$

$$Q_t=q_t+4\sum_{j}^{\mathrm{occ}}\sum_{k}^{\mathrm{unocc}}\frac{c_r{}^jc_t{}^jc_r{}^kc_t{}^k}{\epsilon_j-\epsilon_k}\alpha+\cdots. \tag{42}$$

The coefficients of the second terms in the right sides of Eqs. (41) and (42) are the bond-atom and the atom-atom polarizability, respectively. Similarly, for Case B, P_{tu} and Q_t are

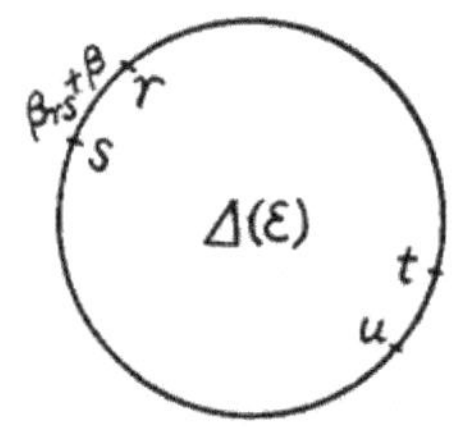

$$P_{tu}=p_{tu}+2\sum_{j}^{\mathrm{occ}}\sum_{k}^{\mathrm{unocc}}\frac{(c_r{}^jc_s{}^k+c_s{}^jc_r{}^k)(c_t{}^jc_u{}^k+c_u{}^jc_t{}^k)}{\epsilon_j-\epsilon_k}\beta+\cdots \tag{43}$$

$$Q_t=q_t+4\sum_{j}^{\mathrm{occ}}\sum_{k}^{\mathrm{unocc}}\frac{c_t{}^jc_t{}^k(c_r{}^jc_s{}^k+c_s{}^jc_r{}^k)}{\epsilon_j-\epsilon_k}\beta+\cdots \tag{44}$$

The coefficients of the second terms in the right sides of Eqs. (43) and (44) are the bond-bond and the atom-bond polarizability, respectively. For Case C, are obtained

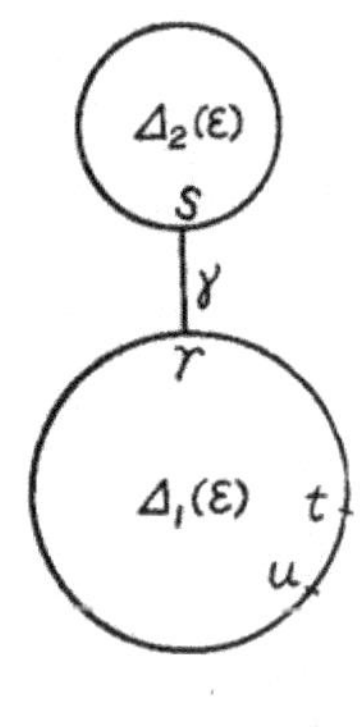

$$P_{tu}=p_{tu}+2\left(\sum_{k}^{\mathrm{occ}}\sum_{j}^{\mathrm{unocc}}-\sum_{k}^{\mathrm{unocc}}\sum_{j}^{\mathrm{occ}}\right)\left[\frac{(c_{2s}{}^k)^2}{\epsilon_{1j}-\epsilon_{2k}}\left\{\frac{(c_{1r}{}^j)^2c_{1t}{}^jc_{1u}{}^j}{\epsilon_{1j}-\epsilon_{2k}}-\sum_{l(\neq j)}^{\mathrm{all}}\frac{c_{1r}{}^jc_{1r}{}^l(c_{1t}{}^jc_{1u}{}^l+c_{1u}{}^jc_{1t}{}^l)}{\epsilon_{1j}-\epsilon_{1l}}\right\}\right]\gamma^2+\cdots, \tag{45}$$

$$Q_t=q_t+2\left(\sum_{k}^{\mathrm{occ}}\sum_{j}^{\mathrm{unocc}}-\sum_{k}^{\mathrm{unocc}}\sum_{j}^{\mathrm{occ}}\right)\left[\frac{c_{1r}{}^jc_{1t}{}^j(c_{2s}{}^k)^2}{\epsilon_{1j}-\epsilon_{2k}}\left\{\frac{c_{1r}{}^jc_{1t}{}^j}{\epsilon_{1j}-\epsilon_{2k}}-2\sum_{l(\neq j)}^{\mathrm{all}}\frac{c_{1r}{}^lc_{1t}{}^l}{\epsilon_{1j}-\epsilon_{1l}}\right\}\right]\gamma^2+\cdots. \tag{46}$$

When an atom X, is attached to a conjugated system,

P_{tu} etc. are given by

$$P_{tu}=p_{tu}+\left(\nu\sum_j^{\text{all}}-\sum_j^{\text{occ}}\right)\left\{\frac{(c_r^j)^2c_t^jc_u^j}{(\alpha_X-\epsilon_j)}\right.$$
$$\left.+\frac{c_r^j}{(\alpha_X-\epsilon_j)}\sum_{k(\neq j)}^{\text{all}}\frac{c_r^k(c_t^jc_u^k+c_u^jc_t^k)}{\epsilon_j-\epsilon_k}\right\}\gamma^2+\cdots, \qquad (47)$$

$$Q_t=q_t+\left\{\nu\left(\sum_j^{\text{all}}\frac{c_t^jc_r^j}{\alpha_X-\epsilon_j}\right)^2-2\sum_j^{\text{occ}}\frac{(c_t^jc_r^j)^2}{(\alpha_X-\epsilon_j)^2}\right.$$
$$\left.+4\sum_j^{\text{occ}}\frac{c_t^jc_r^j}{\alpha_X-\epsilon_j}\sum_{k(\neq j)}^{\text{all}}\frac{c_t^kc_r^k}{\epsilon_k-\epsilon_j}\right\}\gamma^2+\cdots, \qquad (48)$$

$$Q_X=\nu-\left(\nu\sum_j^{\text{all}}-2\sum_j^{\text{occ}}\right)\frac{(c_r^j)^2}{(\alpha_X-\epsilon_j)^2}\gamma^2+\cdots, \qquad (49)$$

where Q_X and ν are the electron density and the number of mobile electrons in the substituent, X. Here, we assumed the following restriction. When $\nu=2$, the energy level of the substituent atom should not be higher than the lowest unoccupied level of the conjugated system, and when $\nu=0$, it should not be lower than the highest occupied level.

In deriving Eqs. (47), (48), and (49), we assumed that the conjugated system is not degenerate. But it is also easy to treat a degenerate system. If the conjugated system $\Delta(\epsilon)$ has two levels of equal energy, say lth and $(l+1)$th levels which are both occupied, Eq. (48) becomes

$$Q_t=q_t+\left\{\nu\left(\sum_j^{\text{all}}\frac{c_t^jc_r^j}{\alpha_X-\epsilon_j}\right)^2-2\sum_j^{\text{occ}}\frac{(c_t^jc_r^j)^2}{(\alpha_X-\epsilon_j)^2}\right.$$
$$+4\sum_{j(\neq l,l+1)}^{\text{occ}}\frac{c_t^jc_r^j}{\alpha_X-\epsilon_j}\sum_{k(\neq j)}^{\text{all}}\frac{c_t^kc_r^k}{\epsilon_k-\epsilon_j}$$
$$+4\frac{c_t^lc_r^l+c_t^{l+1}c_r^{l+1}}{\alpha_X-\epsilon_l}\sum_{k(\neq l,l+1)}^{\text{all}}\frac{c_t^kc_r^k}{\epsilon_k-\epsilon_l}$$
$$\left.-4\frac{c_t^lc_r^lc_t^{l+1}c_r^{l+1}}{(\alpha_X-\epsilon_l)^2}\right\}\gamma^2+\cdots. \qquad (50)$$

Furthermore, when $\nu=2$ and the level of the substituent atom happens to coincide with the degenerate levels of the conjugated system, Q_t is given by

$$Q_t=q_t+\left\{2\left(\sum_j^{\text{unocc}}\frac{c_t^jc_r^j}{\epsilon_l-\epsilon_j}\right)^2+4\sum_j^{\text{unocc}}\sum_k^{\text{occ}}\frac{c_t^jc_r^jc_t^kc_r^k}{(\epsilon_l-\epsilon_j)(\epsilon_j-\epsilon_k)}\right\}\gamma^2$$
$$+\cdots. \qquad (51)$$

The above-derived equations are very useful. In fact, the present authors have calculated the charge at the oxygen in the carboxyl group of the plant growth compounds such as benzoic acid derivatives by using Eq. (50).[12] Also the electron density at the "carcinogenophore" (or *K* and *L* region) of carcinogenic compounds such as polycondensed aromatic hydrocarbons and dimethylaminozaobenzene derivatives has been calculated by Eq. (48).[13] We are sure that these equations will also be applicable to other biological problems.

TABLE I. The π-electron density of a substituted benzene calculated by the usual variation method, by Dewar's equation, and by the present authors' equation. For simplicity, the Coulomb integral, at the atom X, that is, the value of α_X is taken to be equal to α, the Coulomb integral for a carbon atom in benzene, and δ is a parameter appearing in each method, representing the conjugation between the substituent and the benzene nucleus.

The number of electrons at the substituent, ν	2	0
Variation method	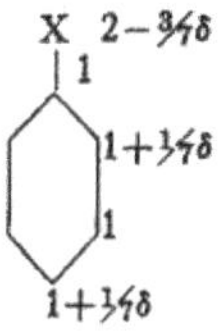 X $2-\frac{3}{4}\delta$; 1; $1+\frac{1}{4}\delta$; 1; $1+\frac{1}{4}\delta$	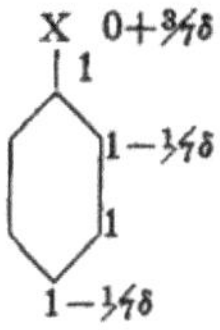 X $0+\frac{3}{4}\delta$; 1; $1-\frac{1}{4}\delta$; 1; $1-\frac{1}{4}\delta$
Dewar's equation	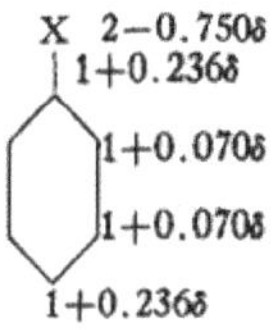 X $2-0.750\delta$; $1+0.236\delta$; $1+0.070\delta$; $1+0.070\delta$; $1+0.236\delta$	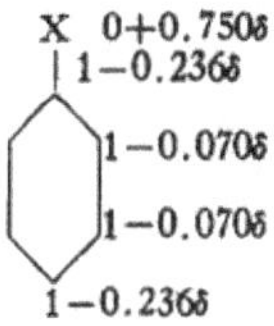 X $0+0.750\delta$; $1-0.236\delta$; $1-0.070\delta$; $1-0.070\delta$; $1-0.236\delta$
The present authors' equation	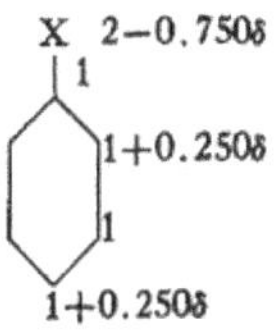 X $2-0.750\delta$; 1; $1+0.250\delta$; 1; $1+0.250\delta$	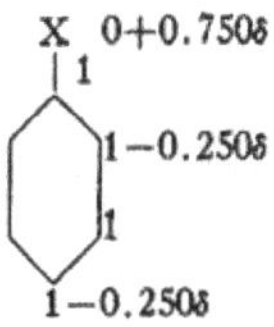 X $0+0.750\delta$; 1; $1-0.250\delta$; 1; $1-0.250\delta$

Correction of Dewar's Formula for Perturbed Electron Density

Dewar[6] has derived by using the perturbation method an equation which gives the electron density at any position in a conjugated system when a mesomeric group is attached to it, namely

$$Q_t=q_t+\left\{\nu\sum_j^{\text{all}}\left(\frac{c_r^jc_t^j}{\epsilon_j-\alpha_X}\right)^2-2\sum_j^{\text{occ}}\left(\frac{c_r^jc_t^j}{\epsilon_j-\alpha_X}\right)^2\right\}\gamma^2+\cdots. \qquad (52)$$

This equation does not agree with Eq. (48) derived by the present authors. In order to be correct, Eq. (52) should be amended so that it coincides with Eq. (48). Namely, the term

$$\nu\sum_j^{\text{all}}\left(\frac{c_t^jc_r^j}{\alpha_X-\epsilon_j}\right)^2$$

should be replaced by

$$\nu\left(\sum_j^{\text{all}}\frac{c_t^jc_r^j}{\alpha_X-\epsilon_j}\right)^2.$$

[12] Fukui, Nagata, and Yonezawa, J. Am. Chem. Soc. **80**, 2267 (1958).

[13] K. Fukui and C. Nagata (unpublished papers).

and the term appearing in the last in Eq. (48) should be added. The contribution of the latter term to the value of electron density is so serious that no equation lacking in this term can give any good result. In fact, we can realize the circumstance in Table I. We will take substituted benzene molecule as an example. The results of calculation by Dewar's and by the present authors' equation are compared with the direct variation calculation. It is clearly seen that the result obtained by the present authors' equation coincides exactly with that of variation calculation and explains well the experimentally well-known ortho-para directing or meta directing effect of the mesomeric substituent. On the other hand, the result by Dewar's equation is utterly different from that of variation calculation and corresponds to no experimental fact. This is true for other basic conjugated compounds such as butadiene and for other more complicated molecules. The reason for this poor result of Dewar's equation is mainly attributed to the lack of the above-mentioned important term

$$4\sum_j^{\rm occ}\frac{c_i^jc_r^j}{\alpha x-\epsilon_j}\sum_{k(\neq j)}^{\rm all}\frac{c_i^kc_r^k}{\epsilon_k-\epsilon_j}.$$

It is not difficult to disclose the cause of error of Dewar's equation. He has derived his equation by using the following wave functions:

$$\Xi_m=\Phi_m-\frac{a_{mr}\beta_{rs}}{E_m-\alpha}\psi, \tag{53}$$

$$\Xi_s=\psi-\sum_m\frac{a_{mr}\beta_{rs}}{\alpha-E_m}\Phi_m. \tag{54}$$

In order to get a correct result, one should take the second-order perturbation term into account. That is,

$$\Xi_m=\Phi_m-\frac{a_{mr}\beta_{rs}}{E_m-\alpha}\psi+\sum_{k(\neq m)}\frac{a_{mr}a_{kr}(\beta_{rs})^2}{(E_m-E_k)(E_m-\alpha)}\Phi_k. \tag{55}$$

For Ξ_s, Eq. (54) is correctly used because the second-order perturbation term of Ξ_s becomes zero. When Eqs. (54) and (55) are used, just the same formula as the present authors' is obtained, giving the correct perturbed electron density.

APPENDIX I. LAGRANGE'S INTERPOLATION FORMULA

As well known, Lagrange's interpolation formula is

$$\frac{g(\chi)}{f(\chi)}=\sum_{j=1}^n\frac{g(\chi_j)}{f'(\chi_j)(\chi-\chi_j)}, \tag{A1}$$

where $f(\chi)$ and $g(\chi)$ are polynomials of χ, and $g(\chi)$ is of lower degree than $f(\chi)$ which is assumed to have no multiple root. From (A1) it follows that

$$\frac{\Delta'(\epsilon)}{\Delta(\epsilon)}=\sum_{j=1}^n\frac{1}{\epsilon-\epsilon_j}, \tag{A2}$$

$$\frac{\Delta_{rs}(\epsilon)}{\Delta(\epsilon)}=\sum_{j=1}^n\frac{\Delta_{rs}(\epsilon_j)}{\Delta'(\epsilon_j)(\epsilon-\epsilon_j)}. \tag{A3}$$

Lagrange's formula is easily modified as follows;

$$\frac{\{g(\chi)\}^2}{\{f(\chi)\}^2}=\sum_{j=1}^n\left[\frac{\{g(\chi_j)\}^2}{\{f'(\chi_j)\}^2(\chi-\chi_j)^2}+\frac{2f'(\chi_j)g'(\chi_j)g(\chi_j)-f''(\chi_j)\{g(\chi_j)\}^2}{\{f'(\chi_j)\}^3(\chi-\chi_j)}\right], \tag{A4}$$

from which we obtain at once the following formula:

$$\frac{\{\Delta_{rs}(\epsilon)\}^2}{\{\Delta(\epsilon)\}^2}=\sum_{j=1}^n\left[\frac{\{\Delta_{rs}(\epsilon_j)\}^2}{\{\Delta'(\epsilon_j)\}^2(\epsilon-\epsilon_j)^2}+\frac{2\Delta'(\epsilon_j)\Delta'_{rs}(\epsilon_j)\Delta_{rs}(\epsilon_j)-\Delta''(\epsilon_j)\{\Delta_{rs}(\epsilon_j)\}^2}{\{\Delta'(\epsilon_j)\}^3(\epsilon-\epsilon_j)}\right]. \tag{A5}$$

From Eqs. (A2) and (A3), we obtain

$$\frac{1}{2}\frac{\Delta''(\epsilon_k)}{\Delta'(\epsilon_k)}=\sum_{j=1;j\neq k}^n\frac{1}{\epsilon_k-\epsilon_j}, \tag{A6}$$

$$\frac{\Delta'_{rs}(\epsilon_k)}{\Delta'(\epsilon_k)}-\frac{1}{2}\frac{\Delta_{rs}(\epsilon_k)\Delta''(\epsilon_k)}{\{\Delta'(\epsilon_k)\}^2}=\sum_{j=1;j\neq k}^n\frac{\Delta_{rs}(\epsilon_j)}{\Delta'(\epsilon_j)(\epsilon_k-\epsilon_j)}, \tag{A7}$$

and hence

$$\frac{\Delta'_{rs}(\epsilon_k)}{\Delta'(\epsilon_k)}=\sum_{j=1;j\neq k}^n\frac{[\Delta_{rs}(\epsilon_j)/\Delta'(\epsilon_j)]+[\Delta_{rs}(\epsilon_k)/\Delta'(\epsilon_k)]}{\epsilon_k-\epsilon_j}. \tag{A8}$$

Also, differentiating Eqs. (A2) and (A3) with respect to ϵ, it follows that

$$\frac{-\Delta''(\epsilon)\Delta(\epsilon)+\{\Delta'(\epsilon)\}^2}{\{\Delta(\epsilon)\}^2}=\sum_{j=1}^n\frac{1}{(\epsilon-\epsilon_j)^2}, \tag{A9}$$

$$\frac{-\Delta'_{rs}(\epsilon)\Delta(\epsilon)+\Delta_{rs}(\epsilon)\Delta'(\epsilon)}{\{\Delta(\epsilon)\}^2}=\sum_{j=1}^n\frac{\Delta_{rs}(\epsilon_j)}{\Delta'(\epsilon_j)(\epsilon-\epsilon_j)^2}. \tag{A10}$$

From Eqs. (A2), (A9), (A3), and (A10), we obtain

$$\frac{\Delta''(\epsilon)}{\Delta(\epsilon)}=\sum_{j=1}^n\sum_{k=1,k\neq j}^n\frac{1}{(\epsilon-\epsilon_j)(\epsilon-\epsilon_k)}, \tag{A11}$$

$$\frac{\Delta'_{rs}(\epsilon)}{\Delta(\epsilon)}=\sum_{j=1}^n\sum_{k=1,k\neq j}^n\frac{\Delta_{rs}(\epsilon_j)}{\Delta'(\epsilon_j)(\epsilon-\epsilon_j)(\epsilon-\epsilon_k)}. \tag{A12}$$

From Eqs. (A9) and (A10), in the way in which Eqs. (A6) and (A7) are derived from Eqs. (A2) and (A3), we obtain

$$-\frac{1}{3}\frac{\Delta'''(\epsilon_k)}{\Delta'(\epsilon_k)}+\frac{1}{4}\left\{\frac{\Delta''(\epsilon_k)}{\Delta'(\epsilon_k)}\right\}^2=\sum_{j=1;j\neq k}^n\frac{1}{(\epsilon_k-\epsilon_j)^2}, \tag{A13}$$

$$\frac{1}{6}\frac{\Delta'''(\epsilon_k)\Delta_{rs}(\epsilon_k)}{\{\Delta'(\epsilon_k)\}^2}-\frac{1}{4}\left\{\frac{\Delta''(\epsilon_k)}{\Delta'(\epsilon_k)}\right\}^2\cdot\frac{\Delta_{rs}(\epsilon_k)}{\Delta'(\epsilon_k)}+\frac{1}{2}\frac{\Delta''(\epsilon_k)\Delta'_{rs}(\epsilon_k)}{\{\Delta'(\epsilon_k)\}^2}-\frac{1}{2}\frac{\Delta''_{rs}(\epsilon_k)}{\Delta'(\epsilon_k)}=\sum_{j=1;j\neq k}^n\frac{\Delta_{rs}(\epsilon_j)}{\Delta'(\epsilon_j)(\epsilon_k-\epsilon_j)^2}. \tag{A14}$$

Similarly, by differentiating Eqs. (A9) and (A10) with

respect to ϵ.

$$\frac{1}{8}\frac{\Delta'''(\epsilon_k)}{\Delta'(\epsilon_k)}-\frac{1}{4}\frac{\Delta'''(\epsilon_k)\Delta''(\epsilon_k)}{\{\Delta'(\epsilon_k)\}^2}+\frac{1}{8}\left\{\frac{\Delta''(\epsilon_k)}{\Delta'(\epsilon_k)}\right\}^3 = \sum_{j=1;\,j\neq k}^{n}\frac{1}{(\epsilon_k-\epsilon_j)^3}, \quad \text{(A15)}$$

$$\frac{1}{24}\frac{\Delta'''(\epsilon_k)}{\Delta'(\epsilon_k)}\cdot\frac{\Delta_{rs}(\epsilon_k)}{\Delta'(\epsilon_k)}+\frac{1}{6}\frac{\Delta''(\epsilon_k)\Delta'''(\epsilon_k)}{\{\Delta'(\epsilon_k)\}^2}\cdot\frac{\Delta_{rs}(\epsilon_k)}{\Delta'(\epsilon_k)} -\frac{1}{8}\left\{\frac{\Delta''(\epsilon_k)}{\Delta'(\epsilon_k)}\right\}^3\cdot\frac{\Delta_{rs}(\epsilon_k)}{\Delta'(\epsilon_k)}+\frac{1}{4}\frac{\{\Delta''(\epsilon_k)\}^2\Delta'_{rs}(\epsilon_k)}{\{\Delta'(\epsilon_k)\}^3} -\frac{1}{6}\frac{\Delta'''(\epsilon_k)\Delta'_{rs}(\epsilon_k)}{\{\Delta'(\epsilon_k)\}^2}-\frac{1}{4}\frac{\Delta''(\epsilon_k)\Delta''_{rs}(\epsilon_k)}{\{\Delta'(\epsilon_k)\}^2}+\frac{1}{6}\frac{\Delta'''_{rs}(\epsilon_k)}{\Delta'(\epsilon_k)} = \sum_{j=1;\,j\neq k}^{n}\frac{\Delta_{rs}(\epsilon_j)}{\Delta'(\epsilon_j)(\epsilon_k-\epsilon_j)^3}. \quad \text{(A16)}$$

Also from Eqs. (A6) and (A13),

$$\frac{1}{3}\frac{\Delta'''(\epsilon_k)}{\Delta'(\epsilon_k)}=\sum_{j=1;\,j\neq k}^{n}\sum_{l=1;\,l\neq j,k}^{n}\frac{1}{(\epsilon_k-\epsilon_j)(\epsilon_k-\epsilon_l)}, \quad \text{(A17)}$$

and from Eqs. (A6), (A8), (A14), and (A17), we obtain

$$\frac{\Delta_{rs}''(\epsilon_k)}{\Delta'(\epsilon_k)}=\sum_{j=1;\,j\neq k}^{n}\sum_{l=1;\,l\neq j,k}^{n} \times\frac{[\Delta_{rs}(\epsilon_k)/\Delta'(\epsilon_k)]+2[\Delta_{rs}(\epsilon_j)/\Delta'(\epsilon_j)]}{(\epsilon_k-\epsilon_j)(\epsilon_k-\epsilon_l)}. \quad \text{(A18)}$$

From Eqs. (A6), (A15), and (A17)

$$\frac{1}{4}\frac{\Delta''''(\epsilon_k)}{\Delta'(\epsilon_k)}=\sum_{j=1;\,j\neq k}^{n}\sum_{l=1;\,l\neq j,k}^{n}\sum_{m=1;\,m\neq j,k,l}^{n} \times\frac{1}{(\epsilon_k-\epsilon_j)(\epsilon_k-\epsilon_l)(\epsilon_k-\epsilon_m)}, \quad \text{(A19)}$$

and from Eqs. (A6), (A8), (A17), (A18), and (A19), we obtain

$$\frac{\Delta_{rs}'''(\epsilon_k)}{\Delta'(\epsilon_k)}=\sum_{j=1;\,j\neq k}^{n}\sum_{l=1;\,l\neq j,k}^{n}\sum_{m=1;\,m\neq j,k,l}^{n} \times\frac{[\Delta_{rs}(\epsilon_k)/\Delta'(\epsilon_k)]+3[\Delta_{rs}(\epsilon_j)/\Delta'(\epsilon_j)]}{(\epsilon_k-\epsilon_j)(\epsilon_k-\epsilon_l)(\epsilon_k-\epsilon_m)}. \quad \text{(A20)}$$

APPENDIX II. EXTENSION OF JACOBI'S THEOREM

From Jacobi's theorem,

$$\Delta(\epsilon)\Delta_{rrst}(\epsilon)=\Delta_{rr}(\epsilon)\Delta_{st}(\epsilon)-\Delta_{rs}(\epsilon)\Delta_{rt}(\epsilon),$$

it follows that

$$\Delta_{rrst}(\epsilon_j)=\frac{1}{\Delta'(\epsilon_j)}[\Delta_{rr}'(\epsilon_j)\Delta_{st}(\epsilon_j)+\Delta_{rr}(\epsilon_j)\Delta_{st}'(\epsilon_j) -\Delta_{rt}'(\epsilon_j)\Delta_{rs}(\epsilon_j)-\Delta_{rt}(\epsilon_j)\Delta_{rs}'(\epsilon_j)],$$

$$\Delta'_{rrst}(\epsilon_j)=\frac{1}{2\Delta'(\epsilon_j)}[\Delta''(\epsilon_j)\Delta_{st}(\epsilon_j)+2\Delta'_{rr}(\epsilon_j)\Delta'_{st}(\epsilon_j) +\Delta_{rr}(\epsilon_j)\Delta_{st}''(\epsilon_j)-\Delta_{rs}''(\epsilon_j)\Delta_{rt}(\epsilon_j)-\Delta_{rs}(\epsilon_j)\Delta_{rt}''(\epsilon_j) -2\Delta_{rs}'(\epsilon_j)\Delta_{rt}'(\epsilon_j)]-\frac{\Delta''(\epsilon_j)}{2\{\Delta'(\epsilon_j)\}^2}[\Delta_{rr}'(\epsilon_j)\Delta_{st}(\epsilon_j) +\Delta_{rr}(\epsilon_j)\Delta_{st}'(\epsilon_j)-\Delta_{rs}'(\epsilon_j)\Delta_{rt}(\epsilon_j)-\Delta_{rs}(\epsilon_j)\Delta_{rt}'(\epsilon_j)]$$

[Reprinted from the Bulletin of the Chemical Society of Japan, Vol. 32, No. 5 (1959)]

450 Kenichi FUKUI, Chikayoshi NAGATA, Teijiro YONEZAWA and Akira IMAMURA [Vol. 32, No. 5

Expansion Formulae of Secular Determinant in Simple LCAO MO Treatment of Aromatic Hydrocarbons

By Kenichi FUKUI, Chikayoshi NAGATA, Teijiro YONEZAWA and Akira IMAMURA

(Received November 17, 1958)

In the LCAO MO treatment of conjugated molecules, it goes without saying that the secular determinant must be known in its expanded form in order to solve it numerically. The use of secular determinant is not necessarily restricted to finding its roots. Coulson and Longuet-Higgins[1] obtained the integral formulae for electronic energy, bond order and various kinds of polarizability in which the integrands were expressed in terms of secular determinants. Also Baba[2] and the present authors[3] pointed out the importance of the coefficients in the expansion of minors of secular determinant for predicting the reactivity of alternant hydrocarbons.

In this paper, we derive several formulae which are useful for expanding the secular determinant of aromatic hydrocarbons. By using these formulae, the secular determinant can be expressed in terms of that of a smaller molecule and the laborious calculation for large molecules can be considerably simplified.

The Expansion Formulae of Secular Determinant for the Aromatic Hydrocarbon.—In the first, the types of "growth" of aromatic hydrocarbons are classified into five as shown in Fig. 1. The first, the third, and the fifth are tentatively named "even growth", and the second, and the fourth "odd growth". Then, an even molecule is formed either by an even growth of an even molecule, or otherwise by an odd growth of an odd molecule, and the formation of an odd molecule results from an odd growth of an even molecule or from an even growth of an odd molecule.

If we put the secular determinants of a molecule before and after the growth as $\Delta(\lambda)$ and $D(\lambda)$, respectively, the expansion formulae for each type are obtained

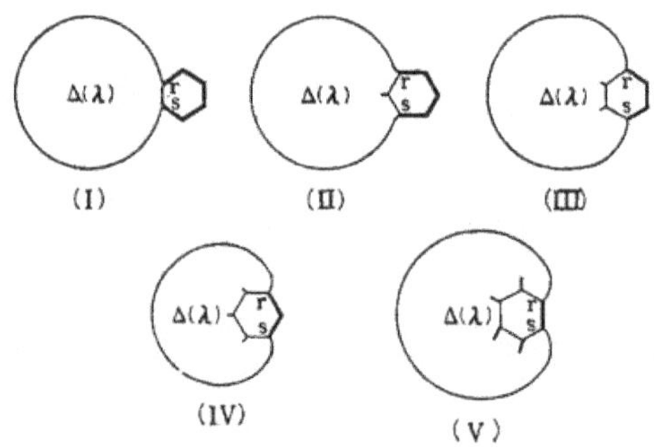

Fig. 1. The types of "growth" of aromatic hydrocarbons are shown. (I) is the first type of growth and (II) is the second type of growth and so on. The thick lines indicate the bonds formed as a result of "growth".

as follows:

For the first type of growth,

$$D(\lambda)=(\lambda^4-3\lambda^2+1)\Delta(\lambda)+\lambda(\lambda^2-2)\{\Delta_{rr}(\lambda)+\Delta_{ss}(\lambda)\}+(\lambda^2-1)\Delta_{rrss}(\lambda)-2(-1)^{r+s+1}\Delta_{rs}(\lambda) \quad (1)$$

For the second type of growth,

$$D(\lambda)=-\lambda(\lambda^2-2)\Delta(\lambda)-(\lambda^2-1)\{\Delta_{rr}(\lambda)+\Delta_{ss}(\lambda)\}-\lambda\Delta_{rrss}(\lambda)+2(-1)^{r+s+1}\Delta_{rs}(\lambda) \quad (2)$$

For the third type of growth,

$$D(\lambda)=(\lambda^2-1)\Delta(\lambda)+\lambda\{\Delta_{rr}(\lambda)+\Delta_{ss}(\lambda)\}+\Delta_{rrss}(\lambda)-2(-1)^{r+s+1}\Delta_{rs}(\lambda) \quad (3)$$

For the fourth type of growth,

$$D(\lambda)=-\lambda\Delta(\lambda)-\Delta_{rr}(\lambda)-\Delta_{ss}(\lambda)+2(-1)^{r+s+1}\Delta_{rs}(\lambda) \quad (4)$$

For the fifth type of growth,

$$D(\lambda)=\Delta(\lambda)-\Delta_{rrss}(\lambda)-2(-1)^{r+s+1}\Delta_{rs}(\lambda) \quad (5)$$

Besides the above types of nuclear growth, the types of branching and bridging also exist in conjugated systems.

1) C. A. Coulson and H. C. Longuet-Higgins, *Proc. Roy. Soc.*, **A191**, 39 (1947).
2) H. Baba, This Bulletin, **30**, 147 (1957).
3) K. Fukui, T. Yonezawa, and C. Nagata, *J. Chem. Phys.*, **26**, 831 (1957).

They are shown in Fig. 2 and are tentatively designated as type 1, 2, ······ , 9. The expansion formulae for them are given as:

For the type 1,

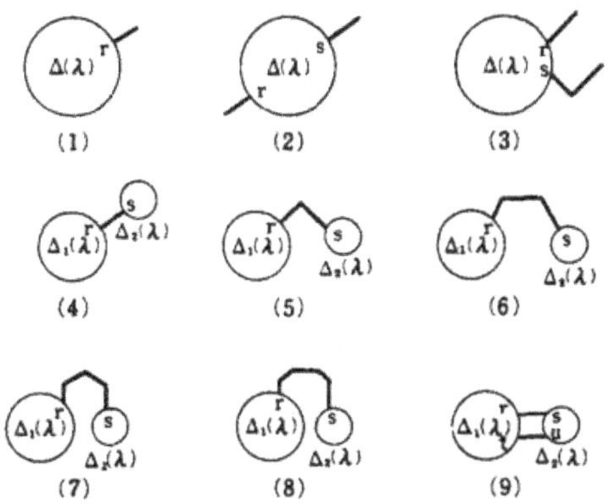

Fig. 2. The types of branching and bridging in conjugated molecules are shown. The thick lines indicate the branches {(1), (2) and (3)} and bridges {(4), (5), (6)······, (9)} respectively.

$$D(\lambda)=-\lambda\Delta(\lambda)-\Delta_{rr}(\lambda) \tag{6}$$

For the type 2,

$$D(\lambda)=\lambda^2\Delta(\lambda)+\lambda\{\Delta_{rr}(\lambda)+\Delta_{ss}(\lambda)\}+\Delta_{rrss}(\lambda) \tag{7}$$

For the type 3,

$$D(\lambda)=(\lambda^2-1)\{-\lambda\Delta(\lambda)-\Delta_{rr}(\lambda)\} -\lambda^2\Delta_{ss}(\lambda)-\lambda\Delta_{rrss}(\lambda) \tag{8}$$

For the type 4,

$$D(\lambda)=\Delta_1(\lambda)\Delta_2(\lambda)-\Delta_{1rr}(\lambda)\Delta_{2ss}(\lambda) \tag{9}$$

For the type 5,

$$D(\lambda)=-\lambda\Delta_1(\lambda)\Delta_2(\lambda)-\Delta_{1rr}(\lambda)\Delta_2(\lambda) -\Delta_{2ss}(\lambda)\Delta_1(\lambda) \tag{10}$$

For the type 6,

$$D(\lambda)=(\lambda^2-1)\Delta_1(\lambda)\Delta_2(\lambda) +\lambda\{\Delta_{1rr}(\lambda)\Delta_2(\lambda)+\Delta_{2ss}(\lambda)\Delta_1(\lambda)\} +\Delta_{1rr}(\lambda)\Delta_{2ss}(\lambda) \tag{11}$$

For the type 7,

$$D(\lambda)=-\lambda(\lambda^2-2)\Delta_1(\lambda)\Delta_2(\lambda) -(\lambda^2-1)\{\Delta_{1rr}(\lambda)\Delta_2(\lambda)+\Delta_{2ss}(\lambda)\Delta_1(\lambda)\} -\lambda\Delta_{1rr}(\lambda)\Delta_{2ss}(\lambda) \tag{12}$$

For the type 8,

$$D(\lambda)=(\lambda^4-3\lambda^2+1)\Delta_1(\lambda)\Delta_2(\lambda)+\lambda(\lambda^2-2) \times\{\Delta_{1rr}(\lambda)\Delta_2(\lambda)+\Delta_{2ss}(\lambda)\Delta_1(\lambda)\} +(\lambda^2-1)\Delta_{1rr}(\lambda)\Delta_{2ss}(\lambda) \tag{13}$$

For the type 9,

$$D(\lambda)=\Delta_1(\lambda)\Delta_2(\lambda)-\Delta_{1rr}(\lambda)\Delta_{2ss}(\lambda) -\Delta_{1tt}(\lambda)\Delta_{2uu}(\lambda)+\Delta_{1rrtt}(\lambda)\Delta_{2ssuu}(\lambda) -2(-1)^{r+t+1}\Delta_{1rt}(\lambda)(-1)^{s+u+1}\Delta_{2su}(\lambda) \tag{14}$$

where $\Delta(\lambda)$, $\Delta_1(\lambda)$ and $\Delta_2(\lambda)$ may be secular determinants of any conjugated system. $\Delta_{rr}(\lambda)$, or $\Delta_{ss}(\lambda)$ is the (r, r), or the (s, s) minor and $\Delta_{rrss}(\lambda)$ is the (rs, rs) minor of $\Delta(\lambda)$, and these are obtained by striking off the indicated rows and columns of $\Delta(\lambda)$. The equation of $\Delta_{rs}(\lambda)$ can be obtained according to the theorem proposed by Coulson and Longuet-Higgins[1].

The expansion formulae derived above, are very useful for obtaining secular equations of conjugated molecules, especially of large molecules.

Faculty of Engineering
Kyoto University
Sakyo-ku, Kyoto

[Reprinted from the Bulletin of the Chemical Society of Japan, Vol. 33, No. 7 (1960)]

A Further Study of Simple LCAO MO Perturbation Treatment of Conjugated Molecules

By Kenichi FUKUI, Keiji MOROKUMA, Teijiro YONEZAWA and Chikayoshi NAGATA

(Received December 19, 1959)

The perturbation theory in the simple LCAO MO method has been studied by many authors, for instance, by Coulson and Longuet-Higgins[1] and by Dewar[2]. Recently, by the present authors[3], a new mathematical method of perturbation theory, named the method of perturbed secular determinant, has been developed and several useful equations have been derived which give the orbital energy, the electron density and the bond order of the perturbed system.

The only system involved in these works, however, was the one which suffered only one perturbation, that is, a small change of a Coulomb integral or of a resonance integral, or a weak bonding to another system. There could be seen no systematic perturbation treatment of the system which was affected by two or more perturbations, for instance, changes of two Coulomb integrals, changes of a Coulomb integral and of a resonance integral, and a weak bonding with a change of a Coulomb integral. To deal with such a system would be of practical interest in discussing the chemical reactivity, absorption spectrum, dipole moment and other electronic properties of large conjugated molecules.

In the present paper are derived several equations or relations which give the molecular orbital, its orbital energy, the electron density and the bond order of systems affected by two perturbations, and lastly they are extended to a more general case of many perturbations.

Non-degenerate System

1. Summary of Results for Single Perturbation.—In order to simplify the expression of various equations appearing in the treatment of multiple-perturbed systems, it would be convenient to summarize the results for single perturbation. The following formulae, obtained by the method of perturbed secular determinant and also by the existing method of perturbation, were given in the previous paper, though some of them not explicitly.

In this section it is assumed that the system to be treated is not degenerate in itself, nor has it any degenerate level between another system to conjugate with.

First, if the Coulomb integral, α_r, of the rth atom in a conjugated system, $\Delta(\varepsilon)$,* is increased by α_r' (Case A of Ref. 3), then, as its consequence, the jth orbital energy, ε_j, will increase by $(\delta\varepsilon_j)_r$, the separate bond order, $p_{uv}{}^j$, by $(\delta p_{uv}{}^j)_r$, the separate electron density, $p_{uu}{}^j$, by $(\delta p_{uu}{}^j)_r$, the total π electronic energy, ε, by $(\delta\varepsilon)_r$, the bond order, p_{uv}, by $(\delta p_{uv})_r$, the electron density, p_{uu}, by $(\delta p_{uu})_r$ and the jth molecular orbital, Ψ_j, by $(\delta\Psi_j)_r$, in such a way that**

$$(\delta\varepsilon_j)_r=(c_r{}^j)^2\alpha_r'+\sum_{k(\neq j)}^{\text{all}}\frac{(c_r{}^jc_r{}^k)^2}{\varepsilon_j-\varepsilon_k}(\alpha_r')^2+\cdots\cdots \tag{1}$$

$$\begin{aligned}(\delta p_{uv}{}^j)_r=&\sum_{k(\neq j)}^{\text{all}}\frac{c_r{}^jc_r{}^k(c_u{}^jc_v{}^k+c_v{}^jc_u{}^k)}{\varepsilon_j-\varepsilon_k}\alpha_r'\\&+\sum_{k(\neq j)}^{\text{all}}\frac{c_r{}^jc_r{}^k}{\varepsilon_j-\varepsilon_k}\left[\sum_{m(\neq j)}^{\text{all}}\frac{c_r{}^m\{c_r{}^jc_u{}^kc_v{}^m+c_r{}^m(c_u{}^jc_v{}^k+c_v{}^jc_u{}^k)\}}{\varepsilon_j-\varepsilon_m}\right.\\&\left.-\frac{c_r{}^j\{c_r{}^kc_u{}^jc_v{}^j+c_r{}^j(c_u{}^jc_v{}^k+c_v{}^jc_u{}^k)\}}{\varepsilon_j-\varepsilon_k}\right](\alpha_r')^2+\cdots\cdots\qquad(u=v\text{ or }u\neq v)\end{aligned} \tag{2}$$

1) C. A. Coulson and H. C. Longuet-Higgins, *Proc. Roy. Soc.* (*London*), **A191**, 39 (1947); **A192**, 16 (1947): **A193**, 447, 456 (1948); **A195**, 188 (1948).
2) M. J. S. Dewar, *J. Am. Chem. Soc.*, **74**, 3341 (1952).
3) K. Fukui, C. Nagata, T. Yonezawa, H. Kato and K. Morokuma, *J. Chem. Phys.*, **31**, 287 (1959).

* $\Delta(\varepsilon)$ is the secular determinant of the system and will be used also to characterize a conjugated system.

** As for the nomenclature, see Ref. 3. Electron density will be looked upon as a special case ($u=v$) of bond order in the further discussion.

$$(\delta\varepsilon)_r = p_{rr}\alpha_r' + 2\sum_j^{\text{occ}}\sum_k^{\text{unocc}}\frac{(c_r^j c_r^k)^2}{\varepsilon_j-\varepsilon_k}(\alpha_r')^2+\cdots\cdots \tag{3}$$

$$(\delta p_{uv})_r = 2\sum_j^{\text{occ}}\sum_k^{\text{unocc}}\frac{c_r^j c_r^k(c_u^j c_v^k + c_v^j c_u^k)}{\varepsilon_j-\varepsilon_k}\alpha_r'$$
$$+2\sum_j^{\text{occ}}\{\text{the second order term in}(\delta p_{uv}^j)_r\}+\cdots\cdots \qquad (u=v \text{ or } u\neq v) \tag{4}$$

$$(\delta\Psi_j)_r = \sum_{k(\neq j)}^{\text{all}}\frac{c_r^j c_r^k}{\varepsilon_j-\varepsilon_k}\alpha_r'\Psi_k - \frac{1}{2}\left\{\sum_{k(\neq j)}^{\text{all}}\frac{(c_r^j c_r^k)^2}{(\varepsilon_j-\varepsilon_k)^2}\right\}(\alpha_r')^2\Psi_j$$
$$+\sum_{k(\neq j)}^{\text{all}}\frac{c_r^j c_r^k}{\varepsilon_j-\varepsilon_m}\left\{\sum_{m(\neq j)}^{\text{all}}\frac{(c_r^m)^2}{\varepsilon_j-\varepsilon_m}-\frac{(c_r^j)^2}{\varepsilon_j-\varepsilon_k}\right\}(\alpha_r')^2\Psi_k+\cdots\cdots \tag{5}$$

Secondly, when the resonance integral, β_{st}, of the bond (st) in the system $\Delta(\varepsilon)$ is increased by β_{st}' (Case B of Ref. 3), increases of those quantities are written as follows:

$$(\delta\varepsilon_j)_{st} = 2c_s^j c_t^j\beta_{st}' + \sum_{k(\neq j)}^{\text{all}}\frac{(c_s^j c_t^k + c_t^j c_s^k)^2}{\varepsilon_j-\varepsilon_k}(\beta_{st}')^2+\cdots\cdots \tag{6}$$

$$(\delta p_{uv}^j)_{st} = \sum_{k(\neq j)}^{\text{all}}\frac{(c_s^j c_t^k + c_t^j c_s^k)(c_u^j c_v^k + c_v^j c_u^k)}{\varepsilon_j-\varepsilon_k}\beta_{st}'$$
$$+\sum_{k(\neq j)}^{\text{all}}\frac{(c_s^j c_t^k + c_t^j c_s^k)}{\varepsilon_j-\varepsilon_k}\left\{\sum_{m(\neq j)}^{\text{all}}\frac{(c_s^j c_t^m + c_t^j c_s^m)c_u^k c_v^m + (c_s^k c_t^m + c_t^k c_s^m)(c_u^j c_v^m + c_v^j c_u^m)}{\varepsilon_j-\varepsilon_m}\right.$$
$$\left.-\frac{(c_s^j c_t^k + c_t^j c_s^k)c_u^j c_v^j + 2c_s^j c_t^j(c_u^j c_v^k + c_v^j c_u^k)}{\varepsilon_j-\varepsilon_k}\right\}(\beta_{st}')^2+\cdots\cdots \qquad (u=v \text{ or } u\neq v) \tag{7}$$

$$(\delta\varepsilon)_{st} = 2p_{st}\beta_{st}' + 2\sum_j^{\text{occ}}\sum_k^{\text{unocc}}\frac{(c_s^j c_t^k + c_t^j c_s^k)^2}{\varepsilon_j-\varepsilon_k}(\beta_{st}')^2+\cdots\cdots \tag{8}$$

$$(\delta p_{uv})_{st} = 2\sum_j^{\text{occ}}\sum_k^{\text{unocc}}\frac{(c_s^j c_t^k + c_t^j c_s^k)(c_u^j c_v^k + c_v^j c_u^k)}{\varepsilon_j-\varepsilon_k}\beta_{st}'$$
$$+2\sum_j^{\text{occ}}\{\text{the second order term in}(\delta p_{uv}^j)_{st}\}+\cdots\cdots \qquad (u=v \text{ or } u\neq v) \tag{9}$$

$$(\delta\Psi_j)_{st} = \sum_{k(\neq j)}^{\text{all}}\frac{(c_s^j c_t^k + c_t^j c_s^k)}{\varepsilon_j-\varepsilon_k}\beta_{st}'\Psi_k - \frac{1}{2}\sum_{k(\neq j)}^{\text{all}}\frac{(c_s^j c_t^k + c_t^j c_s^k)^2}{(\varepsilon_j-\varepsilon_k)^2}(\beta_{st}')^2\Psi_j$$
$$+\sum_{k(\neq j)}^{\text{all}}\frac{c_s^j c_t^k + c_t^j c_s^k}{\varepsilon_j-\varepsilon_k}\left\{\sum_{m(\neq j)}^{\text{all}}\frac{c_s^k c_t^m + c_t^k c_s^m}{\varepsilon_j-\varepsilon_m}-\frac{2c_s^j c_t^j}{\varepsilon_j-\varepsilon_k}\right\}\Big](\beta_{st}')^2\Psi_k+\cdots\cdots \tag{10}$$

Thirdly, when a conjugated system, $\Delta(\varepsilon)$, conjugates at the rth atom through a weak bonding, whose resonance integral is γ_r, with another conjugated system, whose secular determinant is $\Delta_1(\varepsilon)$, (Case C of Ref. 3), then the increases of molecular quantities by this perturbation are given by the following formulae. The orbital energy of the kth orbital ψ_{1k} of the system $\Delta_1(\varepsilon)$ is η_{1k} and the coefficient of the ξth atomic orbital in that molecular orbital is written as d_ξ^{1k}.

$$(\delta\varepsilon_j)_{r\xi} = \sum_k^{\text{all}}\frac{(c_r^j d_\xi^{1k})^2}{\varepsilon_j-\eta_{1k}}(\gamma_r)^2 + \left\{\sum_k^{\text{all}}\frac{(c_r^j d_\xi^{1k})^2}{\varepsilon_j-\eta_{1k}}\right\}\left[\sum_k^{\text{all}}\frac{(d_\xi^{1k})^2}{\varepsilon_j-\eta_{1k}}\left\{\sum_{m(\neq j)}^{\text{all}}\frac{(c_r^m)^2}{\varepsilon_j-\varepsilon_m}-\frac{(c_r^j)^2}{\varepsilon_j-\eta_{1k}}\right\}\right](\gamma_r)^4+\cdots\cdots \tag{11}$$

$$(\delta p_{uv}^j)_{r\xi} = \sum_k^{\text{all}}\frac{c_r^j d_\xi^{1k}}{\varepsilon_j-\eta_{1k}}\left\{\sum_{m(\neq j)}^{\text{all}}\frac{c_r^m d_\xi^{1k}(c_u^j c_v^m + c_v^j c_u^m)}{\varepsilon_j-\varepsilon_m}-\frac{c_r^j d_\xi^{1k} c_u^j c_v^j}{\varepsilon_j-\eta_{1k}}\right\}(\gamma_r)^2+\cdots\cdots$$
$$(u=v \text{ or } u\neq v) \tag{12}$$

$$(\delta p_{r\xi}^j)_{r\xi} = \sum_k^{\text{all}}\frac{(c_r^j d_\xi^{1k})^2}{\varepsilon_j-\eta_{1k}}\gamma_r + 2\sum_k^{\text{all}}\sum_n^{\text{all}}\frac{(c_r^j d_\xi^{1k} d_\xi^{1n})^2}{(\varepsilon_j-\eta_{1k})(\varepsilon_j-\eta_{1n})}\left\{\sum_{m(\neq j)}^{\text{all}}\frac{(c_r^m)^2}{\varepsilon_j-\varepsilon_m}-\frac{(c_r^j)^2}{\varepsilon_j-\eta_{1k}}\right\}(\gamma_r)^3+\cdots \tag{12'}$$

$$(\delta p_{uv}^{1k})_{r\xi} = \sum_j^{\text{all}}\sum_m^{\text{all}}\frac{c_r^j c_r^m (d_\xi^{1k})^2 c_u^j c_v^m}{(\eta_{1k}-\varepsilon_j)(\eta_{1k}-\varepsilon_m)}(\gamma_r)^2+\cdots\cdots \qquad (u=v \text{ or } u\neq v) \tag{13}$$

$$(\delta\Psi_j)_{r\xi}=\sum_k^{\text{all}}\frac{c_r^j d_\xi^{1k}}{\varepsilon_j-\eta_{1k}}\gamma_r\phi_k-\frac{1}{2}\sum_k^{\text{all}}\frac{(c_r^j d_\xi^{1k})^2}{(\varepsilon_j-\eta_{1k})^2}(\gamma_r)^2\Psi_j+\sum_{m(\neq j)}^{\text{all}}\frac{c_r^j c_r^m}{\varepsilon_j-\varepsilon_m}\sum_k^{\text{all}}\frac{(d_\xi^{1k})^2}{\varepsilon_j-\eta_{1k}}(\gamma_r)^2\Psi_m+\cdots\cdots \quad (14)$$

where the atom u and the bond (uv) are in the system $\Delta(\varepsilon)$, and ξ is the position of weak conjugation in the system $\Delta_1(\varepsilon)$. Similar formulae would be obtained for the system $\Delta_1(\varepsilon)$. Thus the change of bond order and that of electron density and total π electronic energy change of the combined system are obtained as follows:

$$(\delta\varepsilon)_{r\xi}=2\sum_j^{\text{occ}}(\delta\varepsilon_j)_{r\xi}+2\sum_k^{\text{occ}}(\delta\eta_{1k})_{r\xi}$$

$$=2\Big(\sum_j^{\text{occ}}\sum_k^{\text{unocc}}-\sum_j^{\text{unocc}}\sum_k^{\text{occ}}\Big)\frac{(c_r^j d_\xi^{1k})^2}{\varepsilon_j-\eta_{1k}}(\gamma_r)^2+2\sum_j^{\text{occ}}\{\text{the fourth order term in}(\delta\varepsilon_j)_{r\xi}\}$$

$$+2\sum_k^{\text{occ}}\{\text{the fourth order term in}(\delta\eta_{1k})_{r\xi}\}+\cdots\cdots \quad (15)$$

$$(\delta p_{uv})_{r\xi}=2\sum_j^{\text{occ}}(\delta p_{uv}^j)_{r\xi}+2\sum_k^{\text{occ}}(\delta p_{uv}^{1k})_{r\xi}$$

$$=2\Big(\sum_k^{\text{occ}}\sum_j^{\text{unocc}}-\sum_k^{\text{unocc}}\sum_j^{\text{occ}}\Big)\left[\frac{(d_\xi^{1k})^2}{\varepsilon_j-\eta_{1k}}\left\{\frac{(c_r^j)^2c_u^jc_v^j}{\varepsilon_j-\eta_{1k}}-\sum_{m(\neq j)}^{\text{all}}\frac{c_r^jc_r^m(c_u^jc_v^m+c_v^jc_u^m)}{\varepsilon_j-\varepsilon_m}\right\}\right](\gamma_r)^2+\cdots\cdots \qquad (u=v \text{ or } u\neq v) \quad (16)$$

$$(\delta p_{r\xi})_{r\xi}=2\Big(\sum_j^{\text{occ}}\sum_k^{\text{unocc}}-\sum_j^{\text{unocc}}\sum_k^{\text{occ}}\Big)\frac{(c_r^j d_\xi^{1k})^2}{\varepsilon_j-\eta_{1k}}\gamma_r+2\sum_j^{\text{occ}}\{\text{the third order term in}(\delta p_{r\xi}^j)_{r\xi}\}$$

$$+2\sum_k^{\text{occ}}\{\text{the third order term in}(\delta p_{r\xi}^{1k})_{r\xi}\}+\cdots\cdots \quad (16')$$

Here it was assumed that the highest occupied orbitals of the systems $\Delta(\varepsilon)$ and $\Delta_1(\varepsilon)$ are lower than the lowest unoccupied orbitals of the systems $\Delta_1(\varepsilon)$ and $\Delta(\varepsilon)$, respectively. The total charge transfer, δQ, from the system $\Delta_1(\varepsilon)$ to the system $\Delta(\varepsilon)$ is defined by

$$\delta Q=\sum_{u=1}^{n}\delta p_{uu} \quad (17)$$

where n is the number of atomic orbitals in the system $\Delta(\varepsilon)$.

For this case δQ is given by the following equation.

$$(\delta Q)_{r\xi}=2\Big(\sum_k^{\text{occ}}\sum_j^{\text{unocc}}-\sum_k^{\text{unocc}}\sum_j^{\text{occ}}\Big)\frac{(c_r^j d_\xi^{1k})^2}{(\varepsilon_j-\eta_{1k})^2}(\gamma_r)^2+\cdots\cdots \quad (18)$$

As a special case involved in the Case C, when the system $\Delta_1(\varepsilon)$ is an atom or a mesomeric group, X, Eqs. 11 to 16 and 18 become

$$(\delta\varepsilon_j)_{rX}=\frac{(c_r^j)^2}{\varepsilon_j-\alpha_X}(\gamma_r)^2+\frac{(c_r^j)^2}{\varepsilon_j-\alpha_X}\left\{-\frac{(c_r^j)^2}{(\varepsilon_j-\alpha_X)^2}+\sum_{m(\neq j)}^{\text{all}}\frac{(c_r^m)^2}{(\varepsilon_j-\alpha_X)(\varepsilon_j-\varepsilon_m)}\right\}(\gamma_r)^4+\cdots\cdots \quad (11\text{-a})$$

$$(\delta\alpha_X)_{rX}=\sum_j^{\text{all}}\frac{(c_r^j)^2}{\alpha_X-\varepsilon_j}(\gamma_r)^2+\left\{\sum_j^{\text{all}}\frac{(c_r^j)^2}{\alpha_X-\varepsilon_j}\right\}\left\{-\sum_j^{\text{all}}\frac{(c_r^j)^2}{(\alpha_X-\varepsilon_j)^2}\right\}(\gamma_r)^4+\cdots\cdots \quad (11\text{-b})$$

$$(\delta p_{uv}^j)_{rX}=\left\{\sum_{m(\neq j)}^{\text{all}}\frac{c_r^jc_r^m(c_u^jc_v^m+c_v^jc_u^m)}{(\varepsilon_j-\varepsilon_m)(\varepsilon_j-\alpha_X)}-\frac{(c_r^j)^2c_u^jc_v^j}{(\varepsilon_j-\alpha_X)^2}\right\}(\gamma_r)^2+\cdots\cdots \quad (12\text{-a})$$

$$(\delta p_{rX}^j)_{rX}=\frac{(c_r^j)^2}{\varepsilon_j-\alpha_X}\gamma_r+\frac{2(c_r^j)^2}{(\varepsilon_j-\alpha_X)^2}\left\{\sum_{m(\neq j)}^{\text{all}}\frac{(c_r^m)^2}{\varepsilon_j-\varepsilon_m}-\frac{(c_r^j)^2}{\varepsilon_j-\alpha_X}\right\}(\gamma_r)^3+\cdots\cdots \quad (12'\text{-a})$$

$$(\delta p_{uv}^X)_{rX}=\left\{\sum_j^{\text{all}}\sum_m^{\text{all}}\frac{c_r^jc_r^mc_u^jc_v^m}{(\varepsilon_j-\alpha_X)(\varepsilon_m-\alpha_X)}\right\}(\gamma_r)^2+\cdots \quad (13\text{-a})$$

$$(\delta p_{rX}^X)_{rX}=\sum_j^{\text{all}}\frac{(c_r^j)^2}{\alpha_X-\varepsilon_j}\gamma_r-2\left\{\sum_j^{\text{all}}\sum_m^{\text{all}}\frac{(c_r^jc_r^m)^2}{(\alpha_X-\varepsilon_j)^2(\alpha_X-\varepsilon_m)}\right\}(\gamma_r)^3+\cdots\cdots \quad (13'\text{-a})$$

$$(\delta p_{XX}{}^{j})_{rX}=\frac{(c_r{}^j)^2}{(\varepsilon_j-\alpha_X)^2}(\gamma_r)^2+\cdots\cdots \tag{12-b}$$

$$(\delta p_{XX}{}^{X})_{rX}=-\sum_j^{\text{all}}\frac{(c_r{}^j)^2}{(\alpha_X-\varepsilon_j)^2}(\gamma_r)^2+\cdots\cdots \tag{13-b}$$

$$(\delta p_{uv})_{rX}=(\nu\sum_j^{\text{all}}-2\sum_j^{\text{occ}})\left\{\frac{(c_r{}^j)^2c_u{}^jc_v{}^j}{(\varepsilon_j-\alpha_X)^2}+\sum_{m(\neq j)}^{\text{all}}\frac{c_r{}^jc_r{}^m(c_u{}^jc_v{}^m+c_v{}^jc_u{}^m)}{(\alpha_X-\varepsilon_j)(\varepsilon_j-\varepsilon_m)}\right\}(\gamma_r)^2+\cdots\cdots \tag{16-a}$$

$$(\delta p_{XX})_{rX}=(2\sum_j^{\text{occ}}-\nu\sum_j^{\text{all}})\frac{(c_v{}^j)^2}{(\alpha_X-\varepsilon_j)^2}(\gamma_r)^2+\cdots\cdots \tag{16-b}$$

$$(\delta p_{rX})_{rX}=(2\sum_j^{\text{occ}}-\nu\sum_j^{\text{all}})\frac{(c_r{}^j)^2}{\varepsilon_j-\alpha_X}\gamma_r+2\sum_j^{\text{occ}}\{\text{the third order term in}\,(\delta p_{rX}{}^j)_{rX}\}$$

$$+\nu\{\text{the third order term in}\,(\delta p_{rX}{}^X)_{rX}\}+\cdots\cdots \tag{16'-a}$$

$$(\delta Q)_{rX}=(\nu\sum_j^{\text{all}}-2\sum_j^{\text{occ}})\frac{(c_r{}^j)^2}{(\alpha_X-\varepsilon_j)^2}(\gamma_r)^2+\cdots\cdots \tag{18-a}$$

where ν is the number of electrons in the atom X. The equations corresponding to Eqs. 14 and 15 are omitted for simplicity.

In the following paragraphs some important cases of doubly perturbed systems will be treated.

2. Case I. Changes of Two Coulomb Integrals.—If two atoms, say the rth and the sth atoms, in a conjugated molecule, $\Delta(\varepsilon)$, are replaced by two heteroatoms, the Coulomb integrals of these atoms will become $\alpha_r+\alpha_r'$ and $\alpha_s+\alpha_s'$, respectively. Then the perturbed secular determinant of the system will be written

$$D(\varepsilon)=\Delta(\varepsilon)+\alpha_r'\Delta_{rr}(\varepsilon)+\alpha_s'\Delta_{ss}(\varepsilon)+\alpha_r'\alpha_s'\Delta_{rrss}(\varepsilon) \tag{19}$$

where $\Delta_{rr}(\varepsilon)$ and $\Delta_{rrss}(\varepsilon)$ are the (rr) minor and the $(rr,\,ss)$ minor of $\Delta(\varepsilon)$, respectively. Regarding these changes, α_r' and α_s', as perturbations, one can easily obtain the perturbed orbital energy, of the jth molecular orbital, the perturbed bond order and the perturbed electron density by the use of the method of perturbed secular determinant, as will be briefly explained in the following.

First if one intermediately considers a system where α_s' is zero, the intermediate perturbed secular determinant is given as:

$$\square(\varepsilon)=\Delta(\varepsilon)+\alpha_r'\Delta_{rr}(\varepsilon) \tag{20}$$

According to the procedure utilized in the previous paper, the intermediate perturbed orbital energy, $\tilde{\varepsilon}_j$, and the intermediate perturbed separate bond order, $\tilde{p}_{uv}{}^j$, are expanded in powers of α_r':

$$\left.\begin{aligned}\tilde{\varepsilon}_j&=\sum_{m=0}^{\infty}a_r{}^{j(m)}(\alpha_r')^m\\ \tilde{p}_{uv}{}^j&=\sum_{m=0}^{\infty}a_{r,uv}{}^{j(m)}(\alpha_r')^m\end{aligned}\right\} \tag{21}$$

and the coefficients, $a_r{}^{j(m)}$ and $a_{r,uv}{}^{j(m)}$, are given by the following formulae.

$$\left.\begin{aligned}a_r{}^{j(m)}&=\frac{(-1)^m}{m!}\left[\frac{d^m}{dz^m}\frac{z\{\Delta_{rr}(z)\}^{m-1}\{\Delta'(z)\Delta_{rr}(z)-\Delta(z)\Delta_{rr}'(z)\}}{\{\Delta(z)/(z-\varepsilon_j)\}^{m+1}}\right]_{z=\varepsilon_j}\\ &\qquad(m=0,1,2,\cdots\cdots)\\ a_{r,uv}{}^{j(m)}&=\frac{(-1)^{m+1}}{m!}\left[\frac{d^m}{dz^m}\frac{\{\Delta_{rr}(z)\}^m\Delta_{ur}(z)\Delta_{rv}(z)}{\{\Delta(z)/(z-\varepsilon_j)\}^{m+1}}\right]_{z=\varepsilon_j}\\ &\qquad(m=0,1,2,\cdots\cdots)\end{aligned}\right\} \tag{22}$$

Directly from these formulae the following expressions are obtained

$$\left.\begin{aligned}\tilde{\varepsilon}_j &= \varepsilon_j + (\delta\varepsilon_j)_r \\ \tilde{p}_{uv}{}^j &= p_{uv}{}^j + (\delta p_{uv}{}^j)_r\end{aligned}\right\} \tag{23}$$

where $(\delta\varepsilon_j)_r$ and $(\delta p_{uv}{}^j)_r$ are given by Eqs. 1 and 2.

Then the second perturbation, α_s', is to be taken into account. The perturbed secular determinant of the system in question is expressed with the aid of the secular determinant of the intermediate system,

$$D(\varepsilon) = \square(\varepsilon) + \alpha_s' \square_{ss}(\varepsilon) \tag{24}$$

A similar procedure leads to the expansion formulae

$$\left.\begin{aligned}\varepsilon_j' &= \sum_{m=0}^{\infty} b_s{}^{j(m)} (\alpha_s')^m \\ P_{uv}{}^j &= \sum_{m=0}^{\infty} b_{s,uv}{}^{j(m)} (\alpha_s')^m\end{aligned}\right\} \tag{25}$$

where the coefficients, $b_s{}^{j(m)}$ and $b_{s,uv}{}^{j(m)}$, are the functions of α_r' and are given by

$$\left.\begin{aligned}b_s{}^{j(m)} &= \frac{(-1)^m}{m!}\left[\frac{d^m}{dz^m}\frac{z\{\square_{ss}(z)\}^{m-1}\{\square'(z)\square_{ss}(z) - \square(z)\square_{ss}'(z)\}}{\{\square(z)/(z-\tilde{\varepsilon}_j)\}^{m+1}}\right]_{z=\tilde{\varepsilon}_j} \\ &\qquad (m = 0, 1, 2, \cdots\cdots) \\ b_{s,uv}{}^{j(m)} &= \frac{(-1)^{m+1}}{m!}\left[\frac{d^m}{dz^m}\frac{\{\square_{ss}(z)\}^m \square_{us}(z)\square_{sv}(z)}{\{\square(z)/(z-\tilde{\varepsilon}_j)\}^{m+1}}\right]_{z=\tilde{\varepsilon}_j} \\ &\qquad (m = 0, 1, 2, \cdots\cdots)\end{aligned}\right\} \tag{26}$$

Thus we finally reach the following results for this perturbed system.

$$\varepsilon_j' = \varepsilon_j + (\delta\varepsilon_j)_r + (\delta\varepsilon_j)_s + 2\sum_{k(\neq j)}^{\text{all}} \frac{c_r{}^j c_r{}^k c_s{}^j c_s{}^k}{\varepsilon_j - \varepsilon_k}\alpha_r'\alpha_s' + \cdots\cdots \tag{27}$$

$$\begin{aligned}P_{uv}{}^j &= p_{uv}{}^j + (\delta p_{uv}{}^j)_r + (\delta p_{uv}{}^j)_s \\ &+ \left[\sum_{k(\neq j)}^{\text{all}}\sum_{m(\neq j)}^{\text{all}} \frac{c_r{}^j c_s{}^j (c_r{}^k c_s{}^m + c_s{}^k c_r{}^m) c_u{}^k c_v{}^m + c_r{}^m c_s{}^m (c_r{}^j c_s{}^k + c_s{}^j c_r{}^k)(c_u{}^j c_v{}^k + c_v{}^j c_u{}^k)}{(\varepsilon_j - \varepsilon_k)(\varepsilon_j - \varepsilon_m)}\right. \\ &\left. - \sum_{k(\neq j)}^{\text{all}} \frac{c_r{}^j c_s{}^j \{(c_r{}^j c_s{}^k + c_s{}^j c_r{}^k)(c_u{}^j c_v{}^k + c_v{}^j c_u{}^k) + 2c_r{}^k c_s{}^k c_u{}^j c_v{}^j\}}{(\varepsilon_j - \varepsilon_k)^2}\right]\alpha_r'\alpha_s' + \cdots\cdots\end{aligned} \tag{28}$$

where $(\delta\varepsilon_j)_r$ and $(\delta p_{uv}{}^j)_r$ are the change of the orbital energy and the separate bond order (or the separate electron density) respectively, of the *j*th molecular orbital caused by the perturbation α_r', as were given in Eqs, 1 and 2.

On inspection of Eqs. 27 and 28 it is clearly seen that, as concerns the first order terms, two perturbations have no effect upon each other and, therefore, the so-called "additivity rule" comes into being. That is to say, the change of the orbital energy, the bond order or the electron density caused by two perturbations is expressed by the mathematical sum of its changes caused by each perturbation. In the second order a cross term appears, which may become comparable in magnitude with the non-cross term and cannot be neglected. Also as for the total π electronic energy, the bond order and the electron density, which are expressed by the sum of corresponding separate quantity, the circumstances are the same, as follows:

$$\varepsilon' = \varepsilon + (\delta\varepsilon)_r + (\delta\varepsilon)_s + 2\sum_j^{\text{occ}}\sum_k^{\text{unocc}} \frac{c_r{}^j c_r{}^k c_s{}^j c_s{}^k}{\varepsilon_j - \varepsilon_k}\alpha_r'\alpha_s' + \cdots\cdots \tag{29}$$

$$P_{uv} = p_{uv} + (\delta p_{uv})_r + (\delta p_{uv})_s + \sum_j^{\text{occ}} \{\text{cross term of the order } \alpha_r'\alpha_s' \text{ in } P_{uv}{}^j\} + \cdots\cdots \tag{30}$$

where $(\delta\varepsilon)_r$ and $(\delta p_{uv})_r$, given by Eqs. 3 and 4, are the changes of total π electronic energy and the bond order (or the electron density) respectively, caused by the perturbation α_r'. Eqs. 27 to 30 can also be derived by the existing method of perturbation.

The explicit expression of perturbed molecular orbital would be useful for some purposes. It is

$$\Psi_j'=\Psi_j+(\delta\Psi_j)_r+(\delta\Psi_j)_s-\sum_{k(\neq j)}^{\text{all}}\frac{c_r^jc_r^kc_s^jc_s^k}{(\varepsilon_j-\varepsilon_k)^2}\alpha_r'\alpha_s'\Psi_j$$
$$+\sum_{k(\neq j)}^{\text{all}}\frac{(c_r^jc_s^k+c_s^jc_r^k)}{\varepsilon_j-\varepsilon_k}\left\{\sum_{m(\neq j)}^{\text{all}}\frac{c_r^mc_s^m}{\varepsilon_j-\varepsilon_m}-\frac{c_r^jc_s^j}{\varepsilon_j-\varepsilon_k}\right\}\alpha_r'\alpha_s'\Psi_k+\cdots\cdots \tag{31}$$

where $(\delta\Psi_j)_r$ is given by Eq. 5.

3. Case II. Changes of a Coulomb Integral and a Resonance Integral.—When at the same time the Coulomb integral of the rth atom is increased by α_r' and the resonance integral of the bond (st), β_{st}, by β_{st}', then the perturbed secular determinant of the system is expressed* by

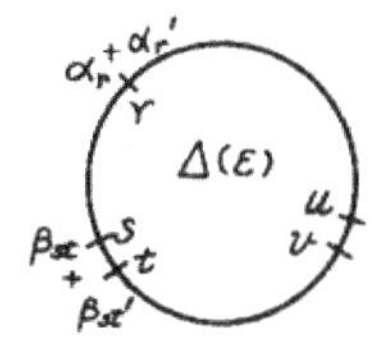

$$D(\varepsilon)=\Delta(\varepsilon)+\alpha_r'\Delta_{rr}(\varepsilon)+(-1)^{s+t}2\beta_{st}'\Delta_{st}(\varepsilon)-(\beta_{st}')^2\Delta_{sstt}(\varepsilon)$$
$$+(-1)^{s+t}\alpha_r'\beta_{st}'\Delta_{rrst}(\varepsilon)-\alpha_r'(\beta_{st}')^2\Delta_{rrsstt}(\varepsilon) \tag{32}$$

As well as in Case I, the following expressions of the orbital energy, the separate bond order and the separate electron density are obtained.

$$\varepsilon_j'=\varepsilon_j+(\delta\varepsilon_j)_r+(\delta\varepsilon_j)_{st}+2\sum_{k(\neq j)}^{\text{all}}\frac{c_r^jc_r^k(c_s^jc_t^k+c_t^jc_s^k)}{\varepsilon_j-\varepsilon_k}\alpha_r'\beta_{st}'+\cdots\cdots \tag{33}$$

$$P_{uv}^j=p_{uv}^j+(\delta p_{uv}^j)_r+(\delta p_{uv}^j)_{st}$$
$$+\left[\sum_{k(\neq j)}^{\text{all}}\sum_{m(\neq j)}^{\text{all}}\frac{c_r^jc_r^k(c_s^jc_t^m+c_t^jc_s^m)(c_u^kc_v^m+c_v^kc_u^m)}{(\varepsilon_j-\varepsilon_k)}\right.$$
$$\times\frac{\{c_r^kc_r^m(c_s^jc_t^m+c_t^jc_s^m)+c_r^mc_r^j(c_s^kc_t^m+c_t^kc_s^m)\}(c_u^jc_v^k+c_v^jc_u^k)}{(\varepsilon_j-\varepsilon_m)}$$
$$-\left.\sum_{k(\neq j)}^{\text{all}}\frac{2\{2c_r^jc_r^kc_s^jc_t^j+(c_r^j)^2(c_s^jc_t^k+c_t^jc_s^k)\}(c_u^jc_v^k+c_v^jc_u^k)c_r^jc_r^k(c_s^jc_t^k+c_t^jc_s^k)c_u^jc_v^j}{(\varepsilon_j-\varepsilon_k)^2}\right]\alpha_r'\beta_{st}'$$
$$+\cdots\cdots \tag{34}$$

where $(\delta\varepsilon_j)_r$, $(\delta\varepsilon_j)_{st}$, $(\delta p_{uv}^j)_r$ and $(\delta p_{uv}^j)_{st}$ are given by Eqs. 1, 6, 2 and 7, respectively. Also in this case a cross term between the two perturbations, both in the orbital energy and in the bond order (and the density), first arises in the second order, and therefore the additivity rule is correct in the first order. The circumtances are the same for the total energy, the bond order and the electron density.

$$\left.\begin{aligned}&\varepsilon'=\varepsilon+(\delta\varepsilon)_r+(\delta\varepsilon)_{st}+2\sum_j^{\text{occ}}\{\text{cross term of the order }\alpha_r'\beta_{st}'\text{ in }\varepsilon_j'\}+\cdots\cdots\\&P_{uv}=p_{uv}+(\delta p_{uv})_r+(\delta p_{uv})_{st}+2\sum_j^{\text{occ}}\{\text{cross term of the order }\alpha_r'\beta_{st}'\text{ in }P_{uv}^j\}\\&\quad+\cdots\cdots\end{aligned}\right\} \tag{35}$$

where $(\delta\varepsilon)_{st}$ and $(\delta p_{uv})_{st}$ are given by Eqs. 8 and 9. The perturbed molecular orbital is obtained as follows:

$$\Psi_j'=\Psi_j+(\delta\Psi_j)_r+(\delta\Psi_j)_{st}+\sum_{k(\neq j)}^{\text{all}}\left\{\sum_{m(\neq j)}^{\text{all}}\frac{c_r^kc_r^m(c_s^jc_t^m+c_t^jc_s^m)+c_r^jc_r^m(c_s^kc_t^m+c_t^kc_s^m)}{(\varepsilon_j-\varepsilon_k)(\varepsilon_j-\varepsilon_m)}\right.$$
$$\left.-\frac{2c_r^jc_r^kc_s^jc_t^j+(c_r^j)^2(c_s^jc_t^k+c_t^jc_s^k)}{(\varepsilon_j-\varepsilon_k)^2}\right\}\alpha_r'\beta_{st}'\Psi_k$$
$$-\sum_{m(\neq j)}^{\text{all}}\frac{c_r^jc_r^k(c_s^jc_t^k+c_t^jc_s^k)}{(\varepsilon_j-\varepsilon_k)^2}\alpha_r'\beta_{sr}'\Psi_k+\cdots\cdots \tag{36}$$

where $(\delta\Psi_j)_{st}$ is given by Eq. 10.

4. Case III. A Weak Bonding with a Change of a Coulomb Integral.—When a conjugated system, $\Delta(\varepsilon)$, conjugates at the sth atom through a weak bonding, whose

* When $r=t$, though the perturbed secular determinant becomes $D(\varepsilon)=\Delta(\varepsilon)+\alpha_r'\Delta_{rr}(\varepsilon)+(-1)^{s+r}2\beta_{sr}'\Delta_{sr}(\varepsilon)-(\beta_{sr}')^2\Delta_{ssrr}(\varepsilon)$, Eqs. 33 to 36 are valid.

resonance integral is γ_s, with another conjugated system $\Delta_1(\varepsilon)$, and at the same time the Coulomb integral of the rth atom is increased by α_r', as illustrated in the figure, then the following formulae are obtained.

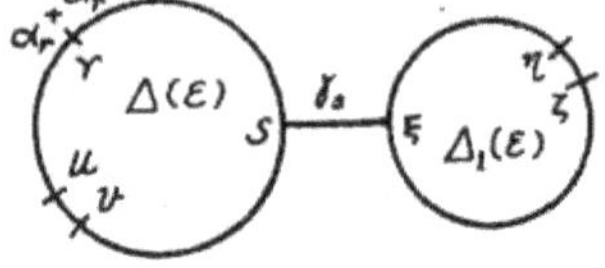

$$D(\varepsilon)=\Delta(\varepsilon)\Delta_1(\varepsilon)+\alpha_r'\Delta_{rr}(\varepsilon)\Delta_1(\varepsilon)-(\gamma_s)^2\Delta_{ss}(\varepsilon)\Delta_{1\xi\xi}(\varepsilon)-\alpha_r'(\gamma_s)^2\Delta_{rrss}(\varepsilon)\Delta_{1\xi\xi}(\varepsilon) \tag{37}$$

$$\varepsilon_j'=\varepsilon_j+(\delta\varepsilon_j)_r+(\delta\varepsilon_j)_{s\xi}+\sum_k^{\text{all}}\frac{c_r^jc_s^j(d_\xi^{1k})^2}{\varepsilon_j-\eta_{1k}}\left\{2\sum_{m(\neq j)}^{\text{all}}\frac{c_r^mc_s^m}{\varepsilon_j-\varepsilon_m}-\frac{c_r^jc_s^j}{\varepsilon_j-\eta_{1k}}\right\}\alpha_r'(\gamma_s)^2+\cdots \tag{38}$$

$$\eta_{1k}'=\eta_{1k}+(\delta\eta_{1k})_{s\xi}+\sum_j^{\text{all}}\frac{c_r^jc_s^j(d_\xi^{1k})^2}{\eta_{1k}-\varepsilon_j}\left\{2\sum_{m(\neq j)}^{\text{all}}\frac{c_r^mc_s^m}{\varepsilon_j-\varepsilon_m}-\frac{c_r^jc_s^j}{\varepsilon_j-\eta_{1k}}\right\}\alpha_r'(\gamma_s)^2+\cdots \tag{39}$$

$$P_{uv}^j=p_{uv}^j+(\delta p_{uv}^j)_r+(\delta p_{uv}^j)_{s\xi}+\{\text{cross term of the order }\alpha_r'(\gamma_s)^2\}+\cdots \tag{40}$$

$$P_{s\xi}^j=(\delta p_{s\xi}^j)_{s\xi}+\sum_k^{\text{all}}\frac{c_r^jc_s^j(d_\xi^{1k})^2}{\varepsilon_j-\eta_{1k}}\left\{2\sum_{m(\neq j)}^{\text{all}}\frac{c_r^mc_s^m}{\varepsilon_j-\varepsilon_m}-\frac{c_r^jc_s^j}{\varepsilon_j-\eta_{1k}}\right\}\alpha_r'\gamma_s+\cdots \tag{40'}$$

$$P_{uv}^{1k}=(\delta p_{uv}^{1k})_{s\xi}+\{\text{cross term of the order }\alpha_r'(\gamma_s)^2\}+\cdots \tag{41}$$

$$P_{\eta\zeta}^{1k}=p_{\eta\zeta}^{1k}+(\delta p_{\eta\zeta}^{1k})_{s\xi}+\{\text{cross term of the order }\alpha_r'(\gamma_s)^2\}+\cdots \tag{42}$$

$$P_{s\xi}^{1k}=(\delta p_{s\xi}^{1k})_{s\xi}+\sum_j^{\text{all}}\frac{c_r^jc_s^j(d_\xi^{1k})^2}{\eta_{1k}-\varepsilon_j}\left\{2\sum_{m(\neq j)}^{\text{all}}\frac{c_r^mc_s^m}{\varepsilon_j-\varepsilon_m}-\frac{c_r^jc_s^j}{\varepsilon_j-\eta_{1k}}\right\}\alpha_r'\gamma_s+\cdots \tag{42'}$$

$$P_{\eta\zeta}^j=(\delta p_{\eta\zeta}^j)_{s\xi}+\{\text{cross term of the order }\alpha_r'(\gamma_s)^2\}+\cdots \tag{43}$$

where ξ is the position of weak conjugation in the system $\Delta_1(\varepsilon)$, the bond $(\eta\zeta)$ is in the system $\Delta_1(\varepsilon)$; and $(\delta\varepsilon_j)_{s\xi}$, $(\delta p_{uv}^j)_{s\xi}$, $(\delta p_{s\xi}^j)_{s\xi}$ and $(\delta p_{uv}^{1k})_{s\xi}$ are given by Eqs. 11, 12, 12′ and 13, from which $(\delta\eta_{1k})_{s\xi}$, $(\delta p_{\eta\zeta}^j)_{s\xi}$, $(\delta p_{\eta\zeta}^{1k})_{s\xi}$ and $(\delta p_{s\xi}^{1k})_{s\xi}$ will easily be formulated. Therefore the total π electronic energy of the combined system is

$$\varepsilon'=\varepsilon+(\delta\varepsilon)_r+(\delta\varepsilon)_{s\xi}+2\left(\sum_j^{\text{occ}}\sum_k^{\text{unocc}}-\sum_j^{\text{unocc}}\sum_k^{\text{occ}}\right)\times\left[\frac{c_r^jc_s^j(d_\xi^{1k})^2}{\varepsilon_j-\eta_{1k}}\left\{2\sum_{m(\neq j)}^{\text{all}}\frac{c_r^mc_s^m}{\varepsilon_j-\varepsilon_m}-\frac{c_r^jc_s^j}{\varepsilon_j-\eta_{1k}}\right\}\right]\alpha_r'(\gamma_s)^2+\cdots \tag{44}$$

And

$$P_{uv}=p_{uv}+(\delta p_{uv})_r+(\delta p_{uv})_{s\xi}+\{\text{cross term of the order }\alpha_r'(\gamma_s)^2\}+\cdots \tag{45}$$

$$P_{s\xi}=(\delta p_{s\xi})_{s\xi}+2\left(\sum_j^{\text{occ}}\sum_k^{\text{unocc}}-\sum_j^{\text{unocc}}\sum_k^{\text{occ}}\right)\left[\frac{c_r^jc_s^j(d_\xi^{1k})^2}{\varepsilon_j-\eta_{1k}}\left\{2\sum_{m(\neq j)}^{\text{all}}\frac{c_r^mc_s^m}{\varepsilon_j-\varepsilon_m}-\frac{c_r^jc_s^j}{\varepsilon_j-\eta_{1k}}\right\}\right]\alpha_r'\gamma_s+\cdots \tag{45'}$$

$$P_{\eta\zeta}=p_{\eta\zeta}+(\delta p_{\eta\zeta})_{s\xi}+\{\text{cross term of the order }\alpha_r'(\gamma_s)^2\}+\cdots \tag{46}$$

$$\delta Q=(\delta Q)_{s\xi}+\{\text{cross term of the order }\alpha_r'(\gamma_s)^2\}+\cdots \tag{47}$$

where $(\delta\varepsilon)_{s\xi}$, $(\delta p_{uv})_{s\xi}$, $(\delta p_{s\xi})_{s\xi}$ and $(\delta Q)_{s\xi}$ are given by Eqs. 15, 16, 16′ and 18. $(\delta p_{\eta\zeta})_{s\xi}$ is easily written on inspection of Eq. 16. As is seen in Eqs. 45 to 47, a cross term first appears in the third order, that is, the order of $\alpha_r'(\gamma_s)^2$, and the additivity rule is valid up to the second order perturbation.

The wave functions of this perturbed system are represented by

$$\Psi_j'=\Psi_j+(\delta\Psi_j)_r+(\delta\Psi_j)_{s\xi}+\sum_k^{\text{all}}\frac{c_r^jd_\xi^{1k}}{\varepsilon_j-\eta_{1k}}\left\{\sum_{m(\neq j)}^{\text{all}}\frac{c_r^mc_s^m}{\varepsilon_j-\varepsilon_m}-\frac{c_r^jc_s^j}{\varepsilon_j-\eta_{1k}}\right\}\alpha_r'\gamma_s\phi_{1k}+\cdots \tag{48}$$

$$\phi_{1k}'=\phi_{1k}+(\delta\phi_{1k})_{s\xi}+\sum_j^{\text{all}}\frac{c_r^jd_\xi^{1k}}{\eta_{1k}-\varepsilon_j}\sum_{m(\neq j)}^{\text{all}}\frac{c_r^mc_s^m}{\eta_{1k}-\varepsilon_m}\alpha_r'\gamma_s\Psi_j+\cdots \tag{49}$$

where $(\delta\Psi_j)_{s\xi}$ is given by Eq. 14, from which also $(\delta\phi_{1k})_{s\xi}$ may easily be written.

5. Case IV. Weak Bondings to Two Conjugated Systems.—Here we consider a case in which a conjugated system, $\Delta(\varepsilon)$, conjugates at the rth and the sth atoms with two other conjugated systems, $\Delta_1(\varepsilon)$ and $\Delta_2(\varepsilon)$, respectively, making weak bondings whose resonance integrals are γ_r and γ_s, respectively. The molecular orbital, its orbital energy and the coefficient of atomic orbital in the non-perturbed $\Delta_1(\varepsilon)$ system are designated by ψ_{1k}, η_{1k} and d_μ^{1k}, respectively, and in the system $\Delta_2(\varepsilon)$ by ψ_{2k}, η_{2k} and d_ν^{2k}. Then the following expressions are easily obtained for the perturbed combined system.

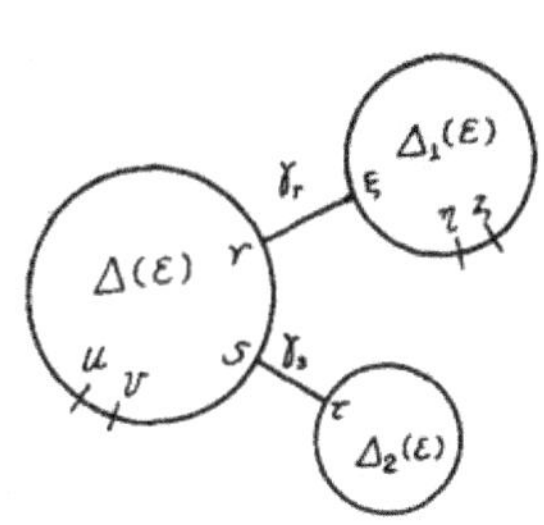

$$D(\varepsilon)=\Delta(\varepsilon)\Delta_1(\varepsilon)\Delta_2(\varepsilon)-(\gamma_r)^2\Delta_{rr}(\varepsilon)\Delta_{1\xi\xi}(\varepsilon)\Delta_2(\varepsilon)-(\gamma_s)^2\Delta_{ss}(\varepsilon)\Delta_1(\varepsilon)\Delta_{2\tau\tau}(\varepsilon)+(\gamma_r\gamma_s)^2\Delta_{rrss}(\varepsilon)\Delta_{1\xi\xi}(\varepsilon)\Delta_{2\tau\tau}(\varepsilon) \tag{50}$$

$$\varepsilon_j'=\varepsilon_j+(\delta\varepsilon_j)_{r\xi}+(\delta\varepsilon_j)_{s\tau}+\sum_k^{\text{all}}\frac{c_r^j(d_\xi^{1k})^2}{\varepsilon_j-\eta_{1k}}U_{jk}(\gamma_r\gamma_s)^2+\cdots \tag{51}$$

$$U_{jk}=\sum_l^{\text{all}}\frac{c_s^j(d_\tau^{2l})^2}{\varepsilon_j-\eta_{2l}}\left\{2\sum_{m(\neq j)}^{\text{all}}\frac{c_r^m c_s^m}{\varepsilon_j-\varepsilon_m}-\frac{c_r^j c_s^j}{\varepsilon_j-\eta_{1k}}-\frac{c_r^j c_s^j}{\varepsilon_j-\eta_{2l}}\right\} \tag{51-a}$$

$$\eta_{1k}'=\eta_{1k}+(\delta\eta_{1k})_{r\xi}+\sum_j^{\text{all}}\frac{c_r^j(d_\xi^{1k})^2}{\eta_{1k}-\varepsilon_j}U_{jk}(\gamma_r\gamma_s)^2+\cdots \tag{52}$$

$$P_{uv}^j=p_{uv}^j+(\delta p_{uv}^j)_{r\xi}+(\delta p_{uv}^j)_{s\tau}+\{\text{cross term of the order }(\gamma_r\gamma_s)^2\}+\cdots \tag{53}$$

$$P_{r\xi}^j=(\delta p_{r\xi}^j)_{r\xi}+\sum_k^{\text{all}}\frac{c_r^j(d_\xi^{1k})^2}{\varepsilon_j-\eta_{1k}}U_{jk}\gamma_r(\gamma_s)^2+\cdots \tag{53'}$$

$$P_{uv}^{1k}=(\delta p_{uv}^{1k})_{r\xi}+\{\text{cross term of the order }(\gamma_r\gamma_s)^2\}+\cdots \tag{54}$$

$$P_{r\xi}^{1k}=(\delta p_{r\xi}^{1k})_{r\xi}+\sum_j^{\text{all}}\frac{c_r^j(d_\xi^{1k})^2}{\eta_{1k}-\varepsilon_j}U_{jk}\gamma_r(\gamma_s)^2+\cdots \tag{54'}$$

$$P_{uv}^{2l}=(\delta p_{uv}^{2l})_{s\tau}+\{\text{cross term of the order }(\gamma_r\gamma_s)^2\}+\cdots \tag{55}$$

$$P_{r\xi}^{2k}=\sum_k^{\text{all}}\frac{(d_\xi^{1k})^2}{\eta_{2l}-\eta_{1k}}\sum_j^{\text{all}}\sum_m^{\text{all}}\frac{c_r^j c_r^m c_s^j c_s^m(d_\tau^{2l})^2}{(\eta_{2l}-\varepsilon_j)(\eta_{2l}-\varepsilon_m)}\gamma_r(\gamma_s)^2+\cdots \tag{55'}$$

$$P_{\eta\zeta}^{1k}=p_{\eta\zeta}^{1k}+(\delta p_{\eta\zeta}^{1k})_{r\xi}+\{\text{cross term of the order }(\gamma_r\gamma_s)^2\}+\cdots \tag{56}$$

$$P_{\eta\zeta}^j=(\delta p_{\eta\zeta}^j)_{r\xi}+\{\text{cross term of the order}(\gamma_r\gamma_s)^2\}+\cdots \tag{57}$$

$$P_{\eta\zeta}^{2l}=\{\text{cross term of the order}(\gamma_r\gamma_s)^2\}+\cdots \tag{58}$$

where ξ and τ are the positions of the weak conjugation in the systems $\Delta_1(\varepsilon)$ and $\Delta_2(\varepsilon)$, respectively, and the bond (uv) or the atom u is assumed to be in the system $\Delta(\varepsilon)$, and the bond $(\eta\zeta)$ or the atom η in the system $\Delta_1(\varepsilon)$. Similar expressions as above may be written for the system $\Delta_2(\varepsilon)$. Thus the total π electronic energy and the bond order of this perturbed system are given as follows:

$$\begin{aligned}\varepsilon'&=2\sum_j^{\text{occ}}\varepsilon_j'+2\sum_k^{\text{occ}}\eta_{1k}'+2\sum_l^{\text{occ}}\eta_{2l}'\\&=\varepsilon+(\delta\varepsilon)_{r\xi}+(\delta\varepsilon)_{st}\\&\quad+\left\{\sum_j^{\text{occ}}\Big(\sum_k^{\text{unocc}}\sum_l^{\text{unocc}}-\sum_k^{\text{occ}}\sum_l^{\text{occ}}\Big)-\sum_j^{\text{unocc}}\Big(\sum_k^{\text{occ}}\sum_l^{\text{all}}+\sum_k^{\text{all}}\sum_l^{\text{occ}}\Big)\right\}\frac{c_r^j c_s^j(d_\xi^{1k})^2(d_\tau^{2l})^2}{(\varepsilon_j-\eta_{1k})(\varepsilon_j-\eta_{2l})}\\&\quad\times\left(2\sum_{m(\neq j)}^{\text{all}}\frac{c_r^m c_s^m}{\varepsilon_j-\varepsilon_m}-\frac{c_r^j c_s^j}{\varepsilon_j-\eta_{1k}}-\frac{c_r^j c_s^j}{\varepsilon_j-\eta_{2l}}\right)(\gamma_r\gamma_s)^2+\cdots\end{aligned} \tag{59}$$

$$P_{uv}=p_{uv}+(\delta p_{uv})_{r\xi}+(\delta p_{uv})_{s\tau}+\{\text{cross term of the order }(\gamma_r\gamma_s)^2\}+\cdots \tag{60}$$

$$\begin{aligned}P_{r\xi}=(\delta p_{r\xi})_{r\xi}&+\Bigg[2\Big(\sum_j^{\text{occ}}\sum_k^{\text{unocc}}-\sum_j^{\text{unocc}}\sum_k^{\text{occ}}\Big)\frac{c_r^j(d_\xi^{1k})^2}{\varepsilon_j-\eta_{1k}}U_{jk}\\&+2\sum_l^{\text{occ}}\sum_k^{\text{all}}\frac{(d_\xi^{1k})^2}{\eta_{2l}-\eta_{1k}}\sum_j^{\text{all}}\sum_m^{\text{all}}\frac{c_r^j c_r^m c_s^j c_s^m(d_\tau^{2l})^2}{(\eta_{2l}-\varepsilon_j)(\eta_{2l}-\varepsilon_m)}\Bigg]\gamma_r(\gamma_s)^2+\cdots\end{aligned} \tag{60'}$$

$$P_{\eta\zeta}=p_{\eta\zeta}+(\delta p_{\eta\zeta})_{r\xi}+\{\text{cross term of the order }(\gamma_r\gamma_s)^2\}+\cdots\cdots \tag{61}$$

The changes of total charge involved in the systems $\Delta(\varepsilon)$ and $\Delta_1(\varepsilon)$ are designated as δQ and δQ_1, respectively, which are written as follows:

$$\delta Q=(\delta Q)_{r\xi}+(\delta Q)_{s\tau}+\{\text{cross term of the order }(\gamma_r\gamma_s)^2\}+\cdots\cdots \tag{62}$$

$$\delta Q_1=(\delta Q)_{r\xi}+\{\text{cross term of the order }(\gamma_r\gamma_s)^2\}+\cdots\cdots \tag{63}$$

A similar equation may be written also for the system $\Delta_2(\varepsilon)$.

The molecular orbitals of this perturbed system are

$$\Psi_j'=\Psi_j+(\delta\Psi_j)_{r\xi}+(\delta\Psi_j)_{s\tau}+\{\text{cross term of the orders }\gamma_r(\gamma_s)^2\text{ and }(\gamma_r)^2\gamma_s\}+\cdots\cdots \tag{64}$$

$$\phi_{1k}'=\phi_{1k}+(\delta\Psi_{1k})_{r\xi}+\sum_j^{\text{all}}\sum_l^{\text{all}}\frac{c_r{}^jc_s{}^jd_\xi{}^{1k}d_\eta{}^{2l}}{(\eta_{1k}-\eta_{2l})(\eta_{1k}-\varepsilon_j)}\gamma_r\gamma_s\phi_{2k}+\cdots\cdots \tag{65}$$

As to special cases in which either or both of the systems $\Delta_1(\varepsilon)$ and $\Delta_2(\varepsilon)$ are single atoms, one may simplify the results of Cases III and IV by using Eqs. 11-a to 18-a instead of Eqs. 11 to 18 for non-cross terms and by abridging cross terms.

6. More General Case of Many Perturbations. — Perturbation treatment of systems affected by two perturbations stated above may be extended to a more general case of many perturbations. As illustrated in the figure, we consider a conjugated system, $\Delta(\varepsilon)$, and increase the Coulomb integrals of the rth atoms in it ($r=1', 2', \cdots\cdots, f'$) by α_r' and the resonance integrals of the bonds (st) ($st=1'', 2'', \cdots\cdots, g''$) by β_{st}', and make weak bondings, whose resonance integral is γ_q, between the qth atoms ($q=1, 2, \cdots\cdots, h$) in the system and the ξ_qth atoms in other conjugated systems $\Delta_q(\varepsilon)$. Here some of the r's, s's and q's may be the identical atoms.

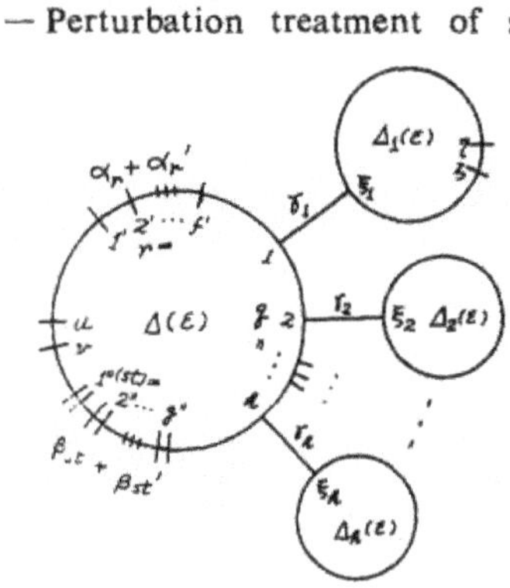

Then the perturbed jth orbital energy of the system $\Delta(\varepsilon)$ is

$$\begin{aligned}\varepsilon_j'=\varepsilon_j&+\sum_{r=1'}^{f'}(\delta\varepsilon_j)_r+\sum_{(st)=1''}^{g''}(\delta\varepsilon_j)_{st}+\sum_{q=1}^{h}(\delta\varepsilon_j)_{q\xi}\\&+C(\alpha^2,\alpha\beta,\beta^2)+C(\alpha^3,\alpha^2\beta,\alpha\beta^2,\beta^3,\alpha\gamma^2,\beta\gamma^2)\\&+C(\alpha^4,\alpha^3\beta,\alpha^2\beta^2,\alpha\beta^3,\beta^4,\alpha^2\gamma^2,\alpha\beta\gamma^2,\beta^2\gamma^2,\gamma^4)+\cdots\cdots\end{aligned} \tag{66}$$

where $(\delta\varepsilon_j)_r$, $(\delta\varepsilon_j)_{st}$ and $(\delta\varepsilon_j)_{q\xi}$ are energy changes caused by perturbations α_r', β_{st}' and γ_q, respectively, and are given by Eqs. 1, 6 and 11. and C means the cross term among perturbations. For instance $C(\alpha^2,\alpha\beta,\beta^2)$ consists of cross terms of orders $\alpha_1'\alpha_2', \alpha_1'\alpha_3', \cdots\cdots, \alpha_2'\alpha_3', \cdots\cdots, \alpha_1'\beta_1', \alpha_1'\beta_2', \cdots\cdots, \alpha_2'\beta_1', \cdots\cdots, \beta_1'\beta_2', \beta_1'\beta_3', \cdots\cdots, \beta_2'\beta_3', \cdots\cdots$; and $C(\alpha^3,\alpha^2\beta, \alpha\beta^2, \beta^3, \alpha\gamma^2, \beta\gamma^2)$ of orders $(\alpha_1')^2\alpha_2', \cdots\cdots, \alpha_1'\alpha_2'\alpha_3', \cdots\cdots, (\alpha_1')^2\beta_1', \cdots\cdots, \alpha_1'\alpha_2'\beta_1', \cdots\cdots, \alpha_1'(\beta_1')^2, \cdots\cdots, \alpha_1'\beta_1'\beta_2', \cdots\cdots, (\beta_1')^2\beta_2', \cdots\cdots, \beta_1'\beta_2'\beta_3', \cdots\cdots, \alpha_1'(\gamma_1)^2, \cdots\cdots, \beta_1'(\gamma_1)^2, \cdots\cdots$. The perturbed separate bond order by this orbital of the bond (uv) in the system $\Delta(\varepsilon)$ is

$$\begin{aligned}P_{uv}{}^j=p_{uv}{}^j&+\sum_{r=1'}^{f'}(\delta p_{uv}{}^j)_r+\sum_{(st)=1''}^{g''}(\delta p_{uv}{}^j)_{st}+\sum_{q=1}^{h}(\delta p_{uv}{}^j)_{q\xi}\\&+C(\alpha^2,\alpha\beta,\beta^2)+C(\alpha^3,\alpha^2\beta,\alpha\beta^2,\beta^3,\alpha\gamma^2,\beta\gamma^2)\\&+C(\alpha^4,\alpha^3\beta,\alpha^2\beta^2,\alpha\beta^3,\beta^4,\alpha^2\gamma^2,\alpha\beta\gamma^2,\beta^2\gamma^2,\gamma^4)+\cdots\cdots\end{aligned} \tag{67}$$

where $(\delta p_{uv}{}^j)_r$, $(\delta p_{uv}{}^j)_{st}$ and $(\delta p_{uv}{}^j)_{q\xi}$ are the changes caused by α_r', β_{st}' and γ_q, respectively, and are given by Eqs. 2, 7 and 12. The perturbed separate bond order by this orbital of the bond $(\eta\zeta)$ in another system, say, the system $\Delta_1(\varepsilon)$, is

$$\begin{aligned}P_{\eta\zeta}{}^j=(\delta p_{\eta\zeta}{}^j)_{1\xi}&+C\{\alpha(\gamma_1)^2,\beta(\gamma_1)^2\}\\&+C\{\alpha^2(\gamma_1)^2,\alpha\beta(\gamma_1)^2,\beta^2(\gamma_1)^2,\gamma^2(\gamma_1)^2\}+\cdots\cdots\end{aligned} \tag{68}$$

The perturbed kth orbital energy of a system, say, the system $\Delta_1(\varepsilon)$, is

$$\eta_{1k}'=\eta_{1k}+(\delta\eta_{1k})_{1\xi}+C\{\alpha(\gamma_1)^2,\ \beta(\gamma)_1)^2\}$$
$$+C\{\alpha^2(\gamma_1)^2,\ \alpha\beta(\gamma_1)^2,\ \beta^2(\gamma_1)^2,\ \gamma^2(\gamma_1)^2\}+\cdots\cdots \tag{69}$$

The perturbed separate bond orders by this orbital of bonds $(\eta\zeta)$ and (uv) are

$$P_{\eta\zeta}^{1k}=p_{\eta\zeta}^{1k}+(\delta p_{\eta\zeta}^{1k})_{1\xi}+C\{\alpha(\gamma_1)^2,\ \beta(\gamma_1)^2\}$$
$$+C\{\alpha^2(\gamma_1)^2,\ \alpha\beta(\gamma_1)^2,\ \beta^2(\gamma_1)^2,\ \gamma^2(\gamma_1)^2\}+\cdots\cdots \tag{70}$$

$$P_{uv}^{1k}=(\delta p_{uv}^{1k})_{1\xi}+C\{\alpha(\gamma_1)^2,\ \beta(\gamma_1)^2\}$$
$$+C\{\alpha^2(\gamma_1)^2,\ \alpha\beta(\gamma_1)^2,\ \beta^2(\gamma_1)^2,\ \gamma^2(\gamma_1)^2\}+\cdots\cdots \tag{71}$$

Also this bond order by this orbital of the bond $(\mu\nu)$, which is located in another system, say, the system $\Delta_k(\varepsilon)$, is given by

$$P_{\mu\nu}^{1k}=C\{(\gamma_1)^2(\gamma_k)^2\}+\cdots\cdots \tag{72}$$

The perturbed separate bond order of the bond which is made anew by perturbation between conjugated systems, say, the bond $(1\xi_1)$ between the systems $\Delta(\varepsilon)$ and $\Delta_1(\varepsilon)$, may be written as

$$P_{1\xi}^{j}=(\delta p_{1\xi}^{j})_{1\xi}+C\{\alpha\gamma_1,\ \beta\gamma_1\}+C\{\alpha^2\gamma_1,\ \alpha\beta\gamma_1,\ \beta^2\gamma_1,\ \gamma^2\gamma_1\}$$
$$+C\{\alpha^3\gamma_1,\ \alpha^2\beta\gamma_1,\ \alpha\beta^2\gamma_1,\ \beta^3\gamma_1,\ \alpha\gamma^2\gamma_1,\ \beta\gamma^2\gamma_1,\ \alpha(\gamma_1)^3,\ \beta(\gamma_1)^3\}+\cdots\cdots \tag{73}$$

$$P_{1\xi}^{1k}=(\delta p_{1\xi}^{1k})_{1\xi}+C\{\alpha\gamma_1,\ \beta\gamma_1\}+C\{\alpha^2\gamma_1,\ a\beta\gamma_1,\ \beta^2\gamma_1,\ \gamma^2\gamma_1\}$$
$$+C\{\alpha^3\gamma_1,\ \alpha^2\beta\gamma_1,\ \alpha\beta^2\gamma_1,\ \beta^3\gamma_1,\ \alpha\gamma^2\gamma_1,\ \beta\gamma^2\gamma_1,\ \alpha(\gamma_1)^3,\ \beta(\gamma_1)^3\}+\cdots\cdots \tag{74}$$

$$P_{1\xi}^{ql}=C\{(\gamma_q)^2\gamma_1\}+C\{\alpha(\gamma_q)^2\gamma_1,\ \beta(\gamma_q)^2\gamma_1\}+\cdots\cdots \qquad (q=2,3,\cdots\cdots,h) \tag{75}$$

As for the total π electronic energy an equation similar to Eq. 66, as for the total bond order (and the electron density) one similar to Eq. 67, and as for the total bond order of a newly formed bond one similar to Eq. 75 may be obtained.

Thus it has become clear to what order of perturbation terms the additivity rule applies, and what kinds of terms appear in several molecular quantities. If it is necessary, one could formulate cross terms in reference to the results of two perturbations.

Degenerate System

In the previous section it was assumed that no molecular orbital is degenerate with any other molecular orbital in the same molecule, not with molecular orbitals in any other system to be subjected to conjugation.

Even if some levels in the non-perturbed systems are degenerate, so far as they are chosen mutually orthogonal, the perturbed orbital energy and the perturbed separate bond order (and electron density) corresponding to non-degenerate orbitals are not affected at all, that is, equations derived in the preceding section, giving separate quantities, are valid also in this case.

As for degenerate levels, on the other hand, the sum of the perturbed orbital energies or of the perturbed bond orders over all the degenerate levels is easily obtained by taking a limit of the corresponding sum which is derived as if they were not degenerate. For instance, in Case I, when the lth and the $(l+1)$th levels are degenerate,

$$\varepsilon_l'+\varepsilon_{l+1}'=\lim_{\varepsilon_{l+1}\to\varepsilon_l}[\varepsilon_l'{}_{,\,\text{nondeg}}+\varepsilon_{l+1}'{}_{,\,\text{nondeg}}]$$
$$=2\varepsilon_l+[\delta(\varepsilon_l+\varepsilon_{l+1})]_r+[\delta(\varepsilon_l+\varepsilon_{l+1})]_s$$
$$+2\sum_{k\,(\neq l,l+1)}^{\text{all}}\frac{c_r^k c_s^k(c_r^l c_s^l+c_r^{l+1}c_s^{l+1})}{\varepsilon_l-\varepsilon_k}(\alpha_r'\alpha_s')+\cdots\cdots \tag{76}$$

where

$$[\delta(\varepsilon_l+\varepsilon_{l+1})]_r=[(c_r^l)^2+(c_r^{l+1})^2]\,\alpha_r'$$
$$+\sum_{k\,(\neq l,l+1)}^{\text{all}}\frac{(c_r^k)^2[(c_r^l)^2+(c_r^{l+1})^2]}{\varepsilon_l-\varepsilon_k}(\alpha_r')^2+\cdots\cdots \tag{77}$$

and $\varepsilon_l'{}_{,\,\text{nondeg}}$ is given by Eq. 27, which is derived for non-degenerate system. Explicit formulae for other various degenerate cases are omitted here for the sake of brevity, but one might write them down without difficulty according to the above principle. The total π electronic energy and the bond order (or the electron density) being represented as the sums of orbital energies and separate orders (or densities) over the occupied orbitals, it is concluded that, even in the degenerate cases, if all the degenerate levels are occupied, the degeneracy does not affect the additivity rule, though concrete formulae involved may differ from the case of no degeneracy.

Discussion

By the results obtained in the preceding sections the additivity rule concerning multiple perturbations and its limits of use have been clarified.

When a hetero-substitution of an atom in a conjugated system causes changes of the Coulomb integral of the atom and of the resonance integrals between neighboring atoms, their effects as a whole may be additively estimated in the first order perturbation treatment. When introduction of a conjugable group into a conjugated system at its certain atom gives rise simultaneously to a change of the Coulomb integral of the atom, the additivity rule applies to the first order terms.

Further, for the introduction of many substituents and hetero-substitutions, the additivity of these effects are materialized, and various examples are seen in experiments. The additivity of shift of absorption sprctra would be explained by the additivity in the orbital energy. The vector additivity of dipole moment would be understood as the additivity in the electron density or in the amount of charge transfer. Aono and Higuchi[4] interpreted the additivity of hyperfine splitting in the electron spin resonance spectrum of substituted semiquinones in terms of the additivity in the unpaired electron density.

Speaking of reactivity indices, i. g. the frontier electron density, $P_{uu}{}^f$, localization energy, ε'(localized state) $-\varepsilon'$(non-localized state), total π electron density, P_{uu}, free valence, $\sqrt{3}-\sum_v{}' P_{uv}$, and most other indices, the additivity rule is applicable to a certain order of perturbation terms. Superdelocalizability is also proved to fulfil the additivity rule.

If one practically evaluates the cross terms appearing in the formulae, one would be able to discuss the deviation from the additivity rule, which is also often realized in experiments.

The authors are much indebted to Mr. Shigeyuki Aono for his helpful suggestions.

Faculty of Engineering
Kyoto University
Sakyo-ku, Kyoto

4) S. Aono, J. Higuchi, *Busseiron Kenkyu*, 7, 197 (1960).

THE JOURNAL OF CHEMICAL PHYSICS VOLUME 32, NUMBER 6 JUNE, 1960

Charge Transfer Mechanism of Reaction of Conjugated Molecules

KENICHI FUKUI, KEIJI MOROKUMA, TEIJIRO YONEZAWA, AND CHIKAYOSHI NAGATA
Faculty of Engineering, Kyoto University, Kyoto, Japan

(Received January 22, 1960)

Mathematical interrelations are found between the superdelocalizability and the Z value, i.e., two of the reactivity indexes of conjugated molecules. MO theoretical discussions on the mechanism of aromatic substitution are also made.

FOR all the MO (molecular-orbital) theoretical contributions which have been made on the chemical reactivity of conjugated molecules, there remains much to be revealed on the essential feature of aromatic substitution or of other reactions of conjugated molecules.

A theory of chemical reactivity of conjugated molecules, called the "frontier electron theory," has been presented by the present authors.[1] In that theory the highest occupied and the lowest unoccupied MO's in the ground state played a dominant role. In other words, the intramolecular orientation could well be predicted by the density distribution of the electrons occupying these particular orbitals. The discovery of the peculiar behavior of these electrons (frontier electrons) may thus be said to have afforded materials which might be a clue to elucidating the mechanism of aromatic substitution. Nagakura and Tanaka[2] concluded from experimental data that an electrophilic substitution takes place only when the lowest unoccupied level of the attacking reagent is of lower energy than the highest occupied one of the substrate and that a nucleophilic reaction occurs only when the lowest unoccupied level of the substrate lies energetically lower than the highest occupied level of the nucleophilic reagent. (This condition will hereafter be called condition A.) On the other hand the present authors proposed a mechanism of aromatic substitution which involves a charge transfer through a hyperconjugation between the substrate and the reagent. In this mechanism the approaching reagent and the leaving atom were treated as the constituents of a pseudoatom having a π-type orbital (quasi-π orbital). By adopting this mechanism the characteristic behavior of frontier electrons was successfully explained, and a reactivity index, superdelocalizability, which is applicable to predicting the intermolecular relative reactivity, was simultaneously derived.[3] Further, in the latest paper,[4] the mechanism of charge transfer, as previously mentioned, was investigated more minutely. The theoretical conclusions are summarized as follows:

1. In order that an electrophilic, or a nucleophilic, substitution can occur, the energy of reagents must satisfy condition A. But it is to be noted that the meaning of the orbital energies is somewhat different from that of Nagakura and Tanaka. They settled condition A on the basis of the ionization potential and the electron affinity, in principle, of the reagent having been submitted to the transfer of one electron, whereas we meant the Coulomb integral of the orbital, which will participate in the bond formation, of the reagent without charge transfer.

2. The energy of the quasi-π orbital in an electrophilic, or a nucleophilic, substitution rises or falls swiftly to the height of a frontier orbital instantaneously with hyperconjugation and then increases or decreases continuously with increasing extent of hyperconjugation to the transition state, where it will be near the Coulomb integral of a carbon atom.

3. Frontier electron density, superdelocalizability, and a generalized index are derived as reactivity indexes from the hyperconjugation energy at the transition state.

Recently, another theoretical contribution to the mechanism of electrophilic substitution has been presented by Brown.[5] At the transition state of the rate-determining step in some electrophilic substitutions of aromatic hydrocarbons, he thought, a complex is formed between the substrate and the reagent because of a charge transfer through the position of substitution, and that thus the stabilization energy of the complex is the most important factor affecting the reactivity. After condition A concerning energy levels of the substrate and the reagent, he treated this complex quantum-chemically as an intermediate between the configuration without charge transfer and the configurations corresponding to the transfer of one electron from occupied orbitals of the substrate to the unoccupied orbital of the electrophile, that is, in a simple form of configuration interaction treatment of the charge-transfer complex discussed by Mulliken.[6] Thus he concluded that a quantity, what he called the "Z

[1] K. Fukui, T. Yonezawa, and H. Shingu, J. Chem. Phys. **20**, 722 (1952); K. Fukui, T. Yonezawa, C. Nagata, and H. Shingu, J. Chem. Phys. **22**, 1433 (1954).

[2] S. Nagakura, and J. Tanaka, J. Chem. Phys. **22**, 563 (1954); J. Chem. Soc. Japan, Pure Chem. Sec. **75**, 933 (1954); Bull. Chem. Soc. Japan **32**, 734 (1959).

[3] K. Fukui, T. Yonezawa, and C. Nagata, Bull. Chem. Soc. Japan **27**, 423 (1954).

[4] (a) K. Fukui, T. Yonezawa, and C. Nagata, J. Chem. Phys. **27**, 1247 (1957); (b) *ibid.*, Eq. (24).

[5] R. D. Brown, J. Chem. Soc. **1959**, 2224, 2232.

[6] R. S. Mulliken, J. Am. Chem. Soc. **74**, 811 (1952); J. Phys. Chem. **56**, 801 (1952).

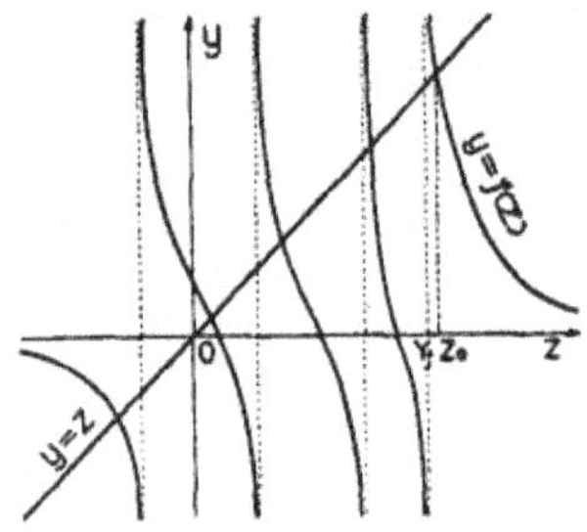

FIG. 1. Functional relations $y=Z$ and $y=f(Z)$ for $Y_f>0$.

value," which refers only to the frontier orbital of the substrate would be an index of reactivity.

In this paper, mathematical interrelations between the superdelocalizability and the Z value are pointed out both in heterolytic and homolytic reactions. A few comparative discussions on the theories of the mechanism of aromatic substitution are also made.

MATHEMATICAL INTERRELATIONS BETWEEN SUPERDELOCALIZABILITY AND Z VALUE

1. Heterolytic Reaction

First of all the electrophilic reaction will be considered. Suppose $\epsilon_E(=\alpha+h\beta)$ and $\epsilon_f(=\alpha+\lambda_f\beta)$ are the energy of the unoccupied orbital of the electrophile and that of the frontier orbital for an electrophilic attack (the highest occupied orbital) of the substrate, respectively, where α and β are the Coulomb integral of a carbon atom and the resonance integral of a benzene bond, then condition A is written as

$$\epsilon_f-\epsilon_E=(-\beta)(h-\lambda_f)=(-\beta)Y_f>0. \qquad (1)$$

Since $(-\beta)$ is positive, Y_f must be positive.

It would now be desirable to summarize the procedures and results given by Brown.[5] It was proved that the stabilization energy due to the contribution of all possible charge-transfer configurations is expressed in units of $(-\beta)$ by the largest root (Z_0) of the following equation:

$$F(Z)=Z-\sum_i^{\text{occ}}[2(C_r^i)^2\gamma^2/(Z-Y_i)]=0, \qquad (2)$$

where C_r^i is the coefficient of the rth atomic orbital (χ_r) in the ith MO whose energy is $\alpha+\lambda_i\beta$, Y_i is equal to $(h-\lambda_i)$, γ is a positive quantity which represents the extent of interaction between the rth atomic orbital and the electrophile at the transition state, and $\sum_{\text{occ}}$ should cover all occupied MO's. For small γ, after utilizing a perturbation theory and neglecting higher orders of γ, Z_0 is directly obtained from Eq. (2), as follows:

$$Z_0=Y_f+[2(C_r^f)^2\gamma^2/Y_f]. \qquad (3)$$

(the larger its magnitude, the more reactive that position).

Brown adopted Z_0, calling it the Z value, as a reactivity index for the electrophilic substitution, and by adjusting the values of h and γ succeeded in interpreting the rate of nitration of conjugated hydrocarbons. It is very interesting that in his theory the frontier orbital plays an exclusive role in deciding not only the intramolecular orientation but the intermolecular relative reactivity. In this sense his theory may serve as an alternate physical basis of the frontier electron theory. But on this point some discussions are to be given in the last section.

We might add a few remarks on his procedure with respect to the following points:

1. Figure 1 illustrates two relations $y=f(Z)$ and $y=Z$, where

$$f(Z)=\sum_i^{\text{occ}}[2(C_r^i)^2\gamma^2/(Z-Y_i)]. \qquad (4)$$

Of the values of Z at intersecting points of the two relations the largest one corresponds to Z_0. This figure will be useful later for comparing with the case where Y_f is negative.

2. In Brown's paper[5] it seems as if Eq. (3) resulted from deliberately taking only two configurations (one without charge transfer and one with charge transfer from the frontier orbital to the electrophile) into account from all possible configurations and then regarding γ small. It is of importance to notice that the more general Eq. (2), which includes all configurations, leads directly to Eq. (3), provided γ is small enough.

Now from the point of view of the quasi-π orbitals used in our previous theoretical studies, the configuration interaction treatment of the mechanism of an electrophilic substitution will be given.

According to our theory the energy of the quasi-π orbital gets and stays higher than that of the highest occupied orbital of the substrate at the transition state. If we again put the energy of the quasi-π orbital $\alpha+h\beta$, Y_f is negative in this case. These mathematical circumstances are essentially different from Brown's theory. Equation (2) is valid also in this case; but, as illustrated in Fig. 2, the largest root is never given by Eq. (3). It is easily expanded with respect to γ and, neglecting higher terms of γ, is written:

$$Z_0=\sum_i^{\text{occ}}[2(C_r^i)^2/(-Y_i)]\gamma^2=\sum_i^{\text{occ}}[2(C_r^i)^2/(\lambda_i-h)]\gamma^2. \qquad (5)$$

This is the same as that previously obtained by the present authors,[4(b)] a generalized form of the superdelocalizability. As an extreme case, when the adjustable parameter h is taken as zero, Eq. (5) becomes simply

$$Z_0=\sum_i^{\text{occ}}[2(C_r^i)^2/\lambda_i]\gamma^2\equiv S_r^{(E)}\gamma^2, \qquad (6)$$

where $S_r^{(E)}$ is the superdelocalizability for an electrophilic attack.

Thus the superdelocalizability, which was theoretically derived in the hyperconjugation consideration,[3,4] is now acquired also in this configuration interaction treatment. Apart from the difference of the models of substitution mechanism, especially of the meanings of h involved in this interrelation, the superdelocalizability and the Z value were both proved to be given as the largest positive root of Eq. (2) corresponding to the positive and the negative values of Y_f, respectively.

If λ_f is equal to h, i.e., Y_f is equal to zero, Z_0 is written in a modified form as follows:

$$Z_0=\sqrt{2}\,|C_r^f|\,\gamma+\sum_i^{\mathrm{occ}-f}[2(C_r^i)^2/Y_i]\gamma^2. \tag{7}$$

Here the frontier electron density comes to determine a greater part of the reactivity.

For a nucleophilic reaction, though Brown did not treat this case, if one similarly defines the Z value according to condition A, one will easily find that the Z value is mathematically related to the superdelocalizability in the same way as for an electrophilic reaction.

2. Homolytic Reaction

In order to clarify the interrelations between the indexes derived in various methods, an analogous treatment would be significant also for a radical reaction. Possible doublet configurations are classified into three classes: the configuration without charge transfer ($R\cdot\ \cdots$ ArH, wave function Ψ_0), those corresponding to one-electron transfer from the ith occupied orbital of the attacked molecule to the half-occupied one of the attacking radical ($R^-\cdots ArH^+$, wave function Ψ_i), and those from the half-occupied orbital of the radical to the jth unoccupied one of the molecule ($R^+\cdots ArH^-$, wave function Ψ_j). Interconfigurational integrals have the following values:

$$\left.\begin{aligned}&\int\Psi_0\mathbf{H}\Psi_k dv=(C_r^k)\int\chi_R H\chi_r dv=C_r^k\gamma\beta,\\&\qquad\qquad(k\text{: every MO}),\\&\int\Psi_k\mathbf{H}\Psi_l dv=0,\\&\qquad\qquad(k,l\text{: every MO};\ k\neq l),\end{aligned}\right\} \tag{8}$$

where $\mathbf{H}$ is as usual the total Hamiltonian, expressed by the sum of effective one-electron Hamiltonians H, and χ_R is the half-occupied orbital of the attacking radical or the half-occupied quasi-π orbital at the transition state, which has the energy of $\alpha+h\beta$. Thus $F(Z)$ function is represented by

$$F(Z)=Z-\sum_i^{\mathrm{occ}}[(C_r^i)^2\gamma^2/(Z-Y_i)]-\sum_j^{\mathrm{unocc}}[(C_r^j)^2\gamma^2/(Z+Y_j)]. \tag{9}$$

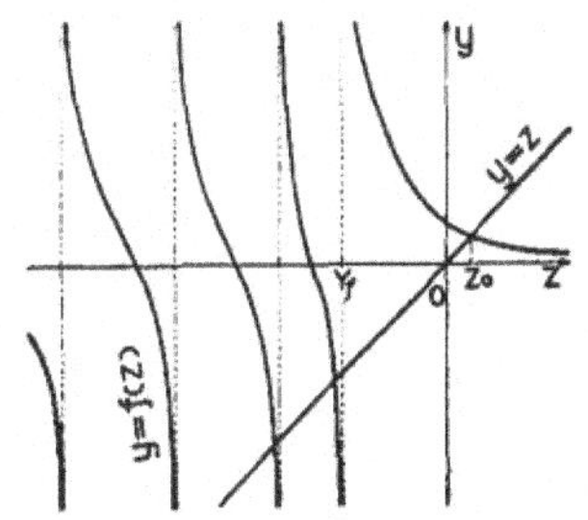

FIG. 2. Functional relations $y=Z$ and $y=f(Z)$ for $Y_f<0$.

According to our theory the half-occupied quasi-π orbital lies between the highest occupied and the lowest unoccupied orbitals.[4] Then the stabilization energy in units of $(-\beta)$ at the transition state is

$$Z_0=\{\sum_i^{\mathrm{occ}}[(C_r^i)^2/(\lambda_i-h)]+\sum_j^{\mathrm{unocc}}[(C_r^j)^2/(h-\lambda_j)]\}\gamma^2. \tag{10}$$

If we put h equal to zero, this reduces to

$$Z_0=\{\sum_i^{\mathrm{occ}}[(C_r^i)^2/\lambda_i]+\sum_j^{\mathrm{unocc}}[(C_r^j)^2/-\lambda_j]\}\gamma^2, \tag{11}$$

which is, namely, the superdelocalizability for a radical attack, also acquired in our hyperconjugation treatment of this problem.[4,7] In this connection Nagakura and Tanaka[2] state that the energy of the half-occupied level of most radicals is lower than that of the lowest unoccupied level of the attacked molecule and higher than that of the highest occupied level. Then in the configuration interaction treatment, it follows that the stabilization energy at the transition state, that is, the reactivity index, is expressed by Eq. (10).

Thus, in a radical reaction, although not in heterolytic reactions, the configuration interaction treatment, no matter to which of the quasi-π orbital or the half-occupied orbital of the radical our attention may be paid, may be said to present the identical index to the hyperconjugation treatment.

DISCUSSION

In the preceding section we derived some mathematical correlations between the Z value and the superdelocalizability. As to heterolytic reactions the former corresponds to the most positive root of the equation $F(Z)$ for a negative value of Y_f, while the latter does so for its positive value.

[7] In more scarce cases, when the half-occupied level lies lower than the highest occupied one of the molecule, that is, $Y_f>0$, Z_0 is obtained as

$$Z_0=Y_f+(C_r^f)^2\gamma^2/Y_f.$$

And when the half-occupied level stands higher than the lowest unoccupied one, a similar equation would be given. To these representations, only one or the other of the frontier orbitals is contributing, rather resembling the case of the Z value for heterolytic reaction.

TABLE I. Frontier electron density and extent of charge transfer calculated by the present authors' equation and by Brown's equation.

Compound	Position	Frontier electron density	Δq_λ by Eq. (13) in units of γ^2	Δq_E by Eq. (12)[a]
Benzene		0.333	0.750	0.760[b]
Naphthalene	1	0.362	1.222	0.932
	2	0.138	0.889	0.974
Anthracene	9	0.387	2.500	0.945
	1	0.193	1.625	0.972
	2	0.097	1.125	0.986
Naphthacene	5	0.295	3.800	0.964
	1	0.112	2.000	0.986
	2	0.067	1.400	0.992
Phenanthrene	9	0.344	1.240	0.936
	1	0.231	1.160	0.957
	3	0.198	1.140	0.963
	4	0.110	0.960	0.979
	2	0.004	0.840	0.999

[a] Following values of parameters are assumed in accord with Brown: $h_E=2.0$ and $\gamma=0.6$.

[b] For the system which has doubly degenerate frontier levels Eq. (12) should be modified as follows:

$$\Delta q_E=1-2[(c_r^f)^2+(c_r^{f-1})^2]\gamma^2/(h_E-\lambda_f)^2.$$

This value was calculated with this equation.

In spite of the mathematical interrelations and the similar viewpoints of two theories, both of which attach great importance to the frontier orbital, Brown's theory[5] of aromatic substitution seems very different from the present authors' theory.[1,3,4] In this section we will make a comparative discussion of the two theories and point out some problems in the mechanism of aromatic substitution.

1. Extent of Charge Transfer

First, the discussion will be confined for simplicity to electrophilic substitution. According to Brown's theory the extent of the charge transfer, Δq_E, from the substrate to the electrophile at the transition state, is given, regarding γ small, as

$$\Delta q_E=1-[2(C_r^f)^2\gamma^2/(h_E-\lambda_f)^2], \tag{12}$$

where h_E is the coefficient in the energy of the lowest unoccupied orbital of the electrophile at the transition state, which is expressed as $\alpha+h_E\beta$. Inspection of the third and the fifth columns of Table I and comparison of Eq. (12) with the corresponding reactivity index, Eq. (3), make it obvious that the larger the frontier electron density, which is the intramolecular reactivity index as seen in Eq. (3), on an atom is, the smaller the quantity of charge transferred through the atom. That is to say, a molecule suffers substitution reaction at the position through which the least amount of charge is transferred to the electrophile. This prediction would seem to conflict with chemical intuition.

On the contrary, the extent of charge transfer in the present authors' theory is expressed for a small conjugation by

$$\Delta q_\lambda=\sum_j^{\text{occ}}[2(C_r^j)^2\gamma^2/(\lambda_j-h_\pi)^2], \tag{13}$$

where h_π is the Coulomb integral of the quasi-π orbital, being variable depending on Δq_λ. With the progress of reaction, that is, with the increase of γ, the charge transfers continuously to the quasi-π orbital oozing out of the substrate according to Eq. (13). This situation was explained in detail with the $(\Delta q_\lambda-h, \gamma)$ diagram (see footnote 4). At the transition state, where h_π was taken to be zero, as clearly seen in the third and the fourth columns of Table I, the extent of charge transfer is completely parallel in a molecule to the frontier electron density, that is, the most reactive position is also the one through which the largest amount of charge transfers.

For a nucleophilic reaction the circumstances are quite the same.

Thus we have encountered two mutually contradictory conclusions as to the relation of transferred charge and reactivity of one position. Further experiments will unravel this entanglement.

2. Energy Levels of Reagents

Condition A was set up by Nagakura and Tanaka in regard to the energy level of reagents in their isolated state, which has almost no concern with the transition state.[2] Brown, however, applied this rule also to the transition state. Indeed he formally distinguished the transition state by marking ‡ from the separated state; however, by adopting condition A, Y_f for the transition state was made a positive value. It might be dangerous to discuss the relative height of the unoccupied level of the electrophile at the transition state to the height of the highest occupied one of the substrate on the basis of the values in the separate state.

According to our theory, as it was stated in the earlier part of this paper, in order for an electrophilic substitution to take place, the highest occupied level of the substrate must be higher than the quasi-π orbital energy which corresponds to the unoccupied level of the reagent. And, since the Coulomb integral of the quasi-π orbital is decided by its charge density in order to satisfy self-consistency, it is natural that it lies lower than the highest occupied level of the substrate at the stage where no charge is yet transferred.

3. Variation of Level Energy during Reaction

It is well expected that the energy of the reagent may vary considerably as the reaction proceeds. By introducing the self-consistent relation between the energy of the quasi-π orbital and its electron density that the former gets higher with the increase of the latter, the present authors indicated without difficulty that the energy of the quasi-π orbital rises during the progress of reaction in an electrophilic substitution and it falls in a nucleophilic substitution, and consequently that it approaches the Coulomb energy of a carbon atom. Since mathematical expressions were given to these relations, one can easily trace the proceedings of the substitution reaction. (See also footnote 4.)

On the other hand, Brown did not take into account such a change in the energy level due to the transference of charge but rather that due to the solvation of the electrophile. But his discussion on the effects of solvation seems to be too complicated to formulate, and to make it difficult to follow after the completion of the reaction, including the feature of the activated complex, according to his theory.

4. Transition State

It does not seem clear in Brown's model what the activated complex, or the transition state, of the first step of his mechanism is. The first step is

$$\mathrm{Ar{-}H + E^+ \rightleftharpoons \begin{array}{c} \mathrm{E} \\ \vdots \\ {}^{+}\mathrm{Ar{-}H} \end{array}} \qquad \text{(II)}.$$

Though in his formalism he distinguished the activated complex of this step from the charge-transfer complex (II), the state to which his conclusion actually pertains seems to correspond rather to the complex (II). To make a distinction between the two complexes may be difficult.

The definition of the transition state is clear in our theory.[3] If one regards the three-center system consisting of the attacked carbon atom, the hydrogen atom, and the electrophile as the reaction center, the lowest occupied orbital of the system will be found to be a σ orbital in nature. The transition state was defined as the state at which this orbital has the maximum energy during the course of the reaction. Near the state the next orbital was proved to have a π character, and the stabilization energy due to the conjugation of this quasi-π orbital with the substrate was calculated as reactivity index. The superdelocalizability is a scale of this stabilization energy. This treatment would be rational in an aromatic substitution which implies dissociation and formation of a σ bond.

[Reprinted from the Bulletin of the Chemical Society of Japan, Vol. 33, No. 11 (1960)]

A Quantum-mechanical Approach to the Theory of Aromaticity

By Kenichi FUKUI, Akira IMAMURA, Teijiro YONEZAWA and Chikayoshi NAGATA

(Received May 28, 1960)

The term "aromatic" originated from a characteristic odor which was possessed by certain substances isolated from various natural products such as benzoic acid (e. g. from gum benzoin), benzaldehyde (bitter-almond oil), cymene (carraway oil), toluene (tolu balsam) and the like. Hence this terminology essentially involved much ambiguousness since originally it was not based upon any scientific concept of classification. Recently the term "aromaticity" has attracted anew the attention of many organic chemists[1] since the discovery and the synthesis of multifarious non-benzenoid "aromatic" compounds such as tropolone, sydnone, ferrocene and numerous derivatives of these compounds[2].

A quantum-mechanical attack on the problem of aromaticity was for the first time delivered by Hueckel in regard to benzene, who presented the so-called $4n+2$ rule[3]. According to his theory, a conjugated compound has $4n+2$ π electrons on its ring should exhibit an aromatic character. Although long years have passed since his report, only a very few attempts have been made to approach this problem theoretically. Recent conspicuous developments in the synthetic organic chemistry of non-benzenoid compounds, however, forced pressingly a re-examination of the significance of aromaticity. Craig investigated the compounds containing dπ electrons by simple LCAO MO treatment and defined two types of aromaticity according to the type of $d\pi$ orbital involved[4]. One of them obeys the $4n+2$ rule while the other is of a novel type in which the rule does not hold.

The researchers mentioned above, however, considered only the energy of π electrons and neglected the internuclear repulsion energy. A more dexterous device should accordingly be desired for the elucidation of aromaticity.

The present authors attempted to investigate the aromaticity more minutely from the quantum-chemical point of view. Prior to entering into the main subject it is desirable to define the term "aromaticity". What is meant by aromaticity is of no clear-cut nature and is usually represented by the following characteristics of aromatic compounds.

Aromatic compounds have a plane structure.

Aromatic compounds show a remarkable stability.

Aromatic compounds have a characteristic reactivity.

Aromatic compounds show a peculiarity in such physical properties as ultraviolet absorption spectra[5], molecular refraction[6], magnetic anisotropy[7], and so forth.

Among these properties some have been already solved and explained satisfactorily, while others have yet been left untouched. The distinguished stability of aromatic compounds aroused the present authors' interest above all other things and is mainly discussed in the present paper.

Energetical Consideration

Classical explanation of the stability of aromatic compounds had been based upon the strain theory, referring to the proposition that a six-membered ring is stable since it undergoes no strain force. However, by means of this theory we can not understand the reason why some five-membered rings such as furan are stable while others such as cyclopentadiene

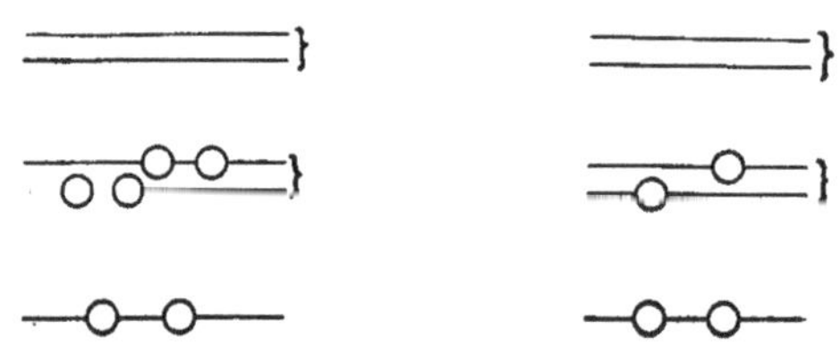

$4n+2$ electrons $\qquad$ $4n$ electrons

Fig. 1. π-electron configuration of $(4n+2)$- and $(4n)$-membered rings.

The brace } indicates that two levels are degenerate.

1) R. Robinson, *Tetrahedron*, 3, 323 (1958); H. C. Longuet-Higgins, *Proc. Chem. Soc.*, 1957, 157.
2) W. Baker and W. D. Ollis, *Quart. Revs.* (*London*), 11, 15 (1957); A. W. Johnsor, *J. Chem. Soc.*, 1954, 1331; T. Nozoe, "Gendai Kagaku", Vol. 10, Iwanami, Tokyo (1957).
3) E. Hueckel, *Z. Physik*, 70, 204 (1931).
4) D. P. Craig, *J. Chem. Soc.*, 1959, 997.

5) R. A. Friedel and M. Orchin, "Ultraviolet Spectra of Aromatic Compounds", John Wiley & Sons, Inc., New York (1951); R. Pariser and P. G. Parr, *J. Chem. Phys.*, 21, 466, 767 (1953).
6) M. Kotake, "Daiyûkikagaku", Vol. 1, Asakura, Tokyo (1958) p. 204..
7) L. Pauling, *J. Chem. Phys.*, 4, 673 (1936); F. London, *J. de Physique*, 8, 397 (1937).

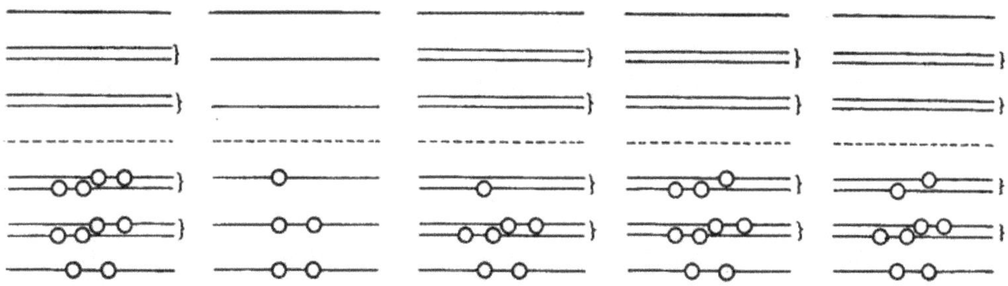

Case A Case B Case C Case C Case D

Fig. 2. Electronic configurations which are taken into account in the calculation. The brace } indicates that two levels are degenerate.

are not. Moreover, recent experiments elucidated that some seven-membered conjugated ring compounds such as tropone and tropolone can exist in a stable form. These facts suggest that, in order to answer the question of aromatic stability, other factors than the strain force, e. g. the magnitude of total π electronic energy, should be taken into consideration. For the purpose of attacking this problem, quantum mechanics must obviously be a powerful weapon. Hueckel's $4n+2$ rule was derived by simple LCAO MO treatment[3]. He showed that the electronic configuration of the compound having $4n+2$ electrons on its ring forms a set of closed shells in MO sense analogous to the atomic structure. In the case of $4n$ electrons the complete closed shell structure is impossible as is seen in Fig. 1.

In Hueckel's $4n+2$ rule, however, the effect of electron-electron repulsion and internuclear repulsion had not been taken into account. Accordingly, this rule is not capable of predicting the number of π electrons which can exist and be stable on a given membered carbon ring. To overcome this difficulty, the π electronic energy is obtained by the formulas utilized by Pople[8]. In the present method the internuclear and electron-electron repulsion energy are considered explicitly and the antisymmetrized product of Hueckel MO were used for the sake of simplicity. Therefore MO is written as

$$\psi_i=\sum_r C_{ir}\phi_r \qquad (1)$$

where ψ_i is the ith MO and ϕ_r represents the AO on the rth carbon atom. Electronic configurations which are taken into account are classified into four cases A, B, C and D and these cases are schematically shown in Fig. 2.

For case A Pople gave the total π electronic energy including nuclear-nuclear repulsion as follows:

$$\varepsilon_\pi=\sum_r P_{rr}\left(U_{rr}+\frac{1}{4}P_{rr}\gamma_{rr}\right)+2\beta\sum_{r<s}{}^{*}P_{rs} + \sum_{r<s}\left\{(P_{rr}-1)(P_{ss}-1)-\frac{1}{2}(P_{rs})^2\right\}\gamma_{rs} \qquad (2)$$

where U_{rr} is a core matrix element, β_{rs} is the resonance integral.

$$\gamma_{rs}=\int\phi_r^*(1)\phi_s^*(2)\frac{e^2}{r_{12}}\phi_r(1)\phi_s(2)\mathrm{d}v_1\mathrm{d}v_2$$

and

$$P_{rs}=2\sum_{i=1}^{n}C_{ir}C_{is} \qquad (3)$$

Summation $\sum$ and $\sum^*$ should cover all the atoms and the nearest neighboring atoms, respectively. For the case of regular n-membered ring molecules, U_{rr} and β_{rs} are constant values represented by U and β, respectively. In Eq. 2 the internuclear repulsion energy is included in the following form

$$\varepsilon'=\sum_{r<s}Z_rZ_s\gamma_{rs} \qquad (4)$$

and Z_r and Z_s are assumed to be unity. For case B the total π-electronic energy of the non-degenerate system with odd electrons was obtained as follows[8]:

$$\varepsilon_\pi=\sum_r(P^0_{rr}U+P^\alpha_{rr}P^\beta_{rr}\gamma_{rr})+2\beta\sum_{r<s}{}^{*}P^0_{rs} + \sum_{r<s}\{(P^0_{rr}-1)(P^0_{ss}-1) - [(P^\alpha_{rs})^2+(P^\beta_{rs})^2]\}\gamma_{rs} \qquad (5)$$

where

$$P^\alpha_{rs}=\sum_{i=1}^{n}C_{ir}C_{is},\quad P^\beta_{rs}=\sum_{i=1}^{n-1}C_{ir}C_{is}$$

and $$P^0_{rs}=P^\alpha_{rs}+P^\beta_{rs} \qquad (6)$$

For case C the total π electronic energy is obtained by only one electronic configuration for regular n-membered ring system, so that the formula for this case is equal to that of case B. For case D the wave functions are chosen as follows.

8) J. A. Pople, *Trans. Faraday Soc.*, **49**, 1375 (1953); A. Brickstock and J. A. Pople, ibid., **50**, 901 (1954).

TABLE I. THE DELOCALIZATION ENERGY OF FIVE-, SIX- AND SEVEN-MEMBERED RING SYSTEMS

Number of carbon atoms on ring	Number of π electrons	Energy of non-conjugated system, e.V.	Energy of conjugated system, e.V.	Difference e.V.
5	5	5U	5U− 9.80	9.80
	6	6U+10.53	6U− 0.48	11.01
	7	7U+26.78	7U+17.23	9.55
6	4	4U+ 4.90	4U− 6.30	11.20
	5	5U	5U−13.50	13.50
	6	6U	6U−13.88	13.88
	7	7U+10.53	7U− 2.97	13.50
	8	8U+25.96	8U+14.76	11.20
7	5	5U+ 4.38	5U−10.17	14.55
	6	6U	6U−17.32	17.32
	7	7U	7U−14.17	14.17

$$\chi=\frac{1}{\sqrt{2}}\frac{1}{\sqrt{2n!}}\{[\psi_1\bar{\psi}_1\cdots\cdots\bar{\psi}_n\bar{\psi}_{-n}]$$

$$-[\psi_1\bar{\psi}_1\cdots\cdots\bar{\psi}_n\psi_{-n}]\} \quad (7)$$

where ψ_n and ψ_{-n} are the molecular orbitals for degenerate levels. In this case we get the total π electronic energy by the same method that Pople used as follows:

$$\varepsilon_\pi=\sum_r P_{rr}U+\frac{1}{4}\sum_r\{(P_{rr})^2-(C_{nr})^4-(C_{-nr})^4$$
$$+6(C_{nr})^2(C_{-nr})^2\}\gamma_{rr}+2\beta\sum_{r<s}{}^{*}P_{rs}$$
$$+\sum_{r<s}\{(P_{rr}-1)(P_{ss}-1)-\frac{1}{2}(P_{rs})^2$$
$$-\frac{1}{2}(C_{nr}C_{ns})^2-\frac{1}{2}(C_{-nr}C_{-ns})^2$$
$$+3(C_{nr}C_{ns})(C_{-nr}C_{-ns})\}\gamma_{rs} \quad (8)$$

where

$$P_{rs}=2\sum_{j=1}^{n-1}C_{jr}C_{js}+C_{nr}C_{ns}+C_{-nr}C_{-ns} \quad (9)$$

By using these formulas the total π electronic energy including internuclear repulsion energy is obtained.

Before performance of calculation it is necessary to consider what kind of quantity can be the measure for aromaticity. In many cases the resonance energy was adopted as the measure for aromaticity[9]. Recently, Craig tried to specify the aromatic character by its symmetry of molecular structures and spin functions[9]. This method, however, is inapplicable to the molecule which lacks sufficient symmetry. In this paper, we intend to compare the stability of the systems of regular n-membered ring having m π electrons. Since the calculation is restricted to π electronic energy and the effect of σ bond is neglected, the comparison of stability should be confined to the ring system with the same number of members and variable number of π electrons. We, therefore, adopt the energy difference between conjugated regular n-membered ring molecule and non-conjugated system that has electrons on each atom as the measure for aromaticity as is shown in Eq. 10.

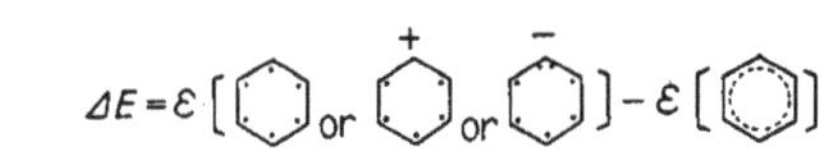

(10)

This measure is nothing but the delocalization energy of the system. The energy of the neutral or ionic non-conjugated system can be given as follows where electron-electron and internuclear repulsions and electron-nuclear attraction are taken into account.

$$\varepsilon_{\text{non-conjugated}}=\sum_{\mu>\nu}(Z_\mu-2)^2\gamma_{\mu\nu}+\sum_{r>s}(Z_r-1)^2\gamma_{rs}$$
$$+\sum_{r\mu}(Z_\mu-2)(Z_r-1)\gamma_{r\mu}+\sum_{\alpha>\beta}Z_\alpha{}^2\gamma_{\alpha\beta}$$
$$+\sum_{\alpha,\gamma}Z_\alpha(Z_r-1)\gamma_{\alpha r}+\sum_\alpha\gamma_{\alpha\alpha}+nU \quad (11)$$

where r and s denote the carbon atom which has only one electron on it, α and β two electrons, and μ and ν none, respectively, and n is the number of π electrons. For the sake of simplicity the values of Z_r, Z_α and Z_μ are assumed to be unity. Thus the calculated value is the measure of the stability by delocalization of non-conjugated electrons. Therefore the greater the magnitude of delocalization energy is, the more stable the molecule is. Accordingly this measure is expected to be a good index for aromaticity, since the stability of the molecule is the essential character for aromatic compounds. The atomic integral used in calculation is gained by the semiempirical formulas adopted by Pariser and Parr[5]. For brevity, all the C–C bond distances are assumed to be 1.39 Å referring to that of benzene. Thus

9) D. Ginsburg, "Non-benzenoid Aromatic Compounds", Interscience Publishers Inc., New York (1959).

the obtained results of the delocalization energy are listed in Table I.

As is shown in Table I, infive-, six- and seven-membered ring systems the molecule is most stable when it has six π electrons on its ring. In a five-membered ring system this result corresponds to the fact that the five-membered ring molecule with six π electrons such as furan, sydnone exists in stable state. Moreover in the seven-membered ring system the stability of tropone and tropolone corresponds to this result. Accordingly, we can explain and extend Hueckel's $4n+2$ rule on the different physical grounds from Hueckel's and combine the rule with the fact that some five- and seven-membered ring molecules with six π electrons are stable.

In this connection, it may be worth while to add that another approximate method will contribute to the illustration of $4n+2$ rule. This is the method originally presented by Peter[10] in which the resonance energy is approximately obtained by the perturbation method. The wave function of linear polyenes is utilized as a zero order wave function. The intramolecular bonding between the atoms r and s, as is indicated in Fig. 3, is regarded as the perturbation and the stabilization energy due to this bonding is obtained. Result of simple LCAO MO calculation is

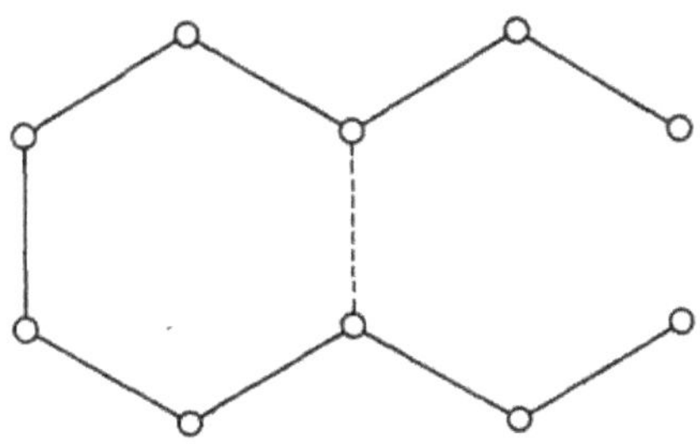

Fig. 3. Scheme for the perturbation method. Dotted line indicates the new bonding engendered.
Circle and line represent the carbon atom and the C-C bond constructing the linear polyene.

$$\delta\varepsilon=2P_{rs}\delta\beta_{rs}+\Pi_{rs,rs}(\delta\beta_{rs})^2+\cdots\cdots \qquad (12)$$

where $\delta\varepsilon$ is the stabilization energy, P_{rs} and $\Pi_{rs,rs}$ are the order of imaginary bond between the atoms r and s and the selfpolarizability of the imaginary bond $r-s$, respectively, each being of the original polyene, and $\delta\beta_{rs}$ denotes the resonance integral of the bond $r-s$.

A glance at Eq. 4 makes it comprehensible that the magnitude of the stabilization energy predominantly depends on the bond order P_{rs}. Peter obtained the resonance energy of various

10) D. Peter, *J. Chem. Soc.*, **1958**, 1023.

TABLE II. THE VALUE OF BOND ORDER OF 4-, 6-, 8-, 10-, 12- AND 14-MEMBERED POLYENES

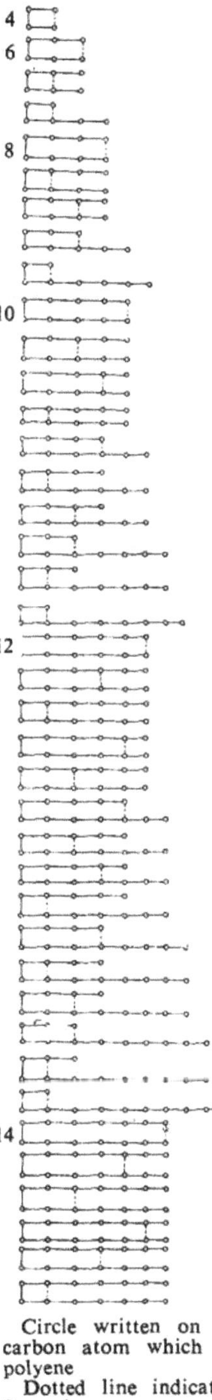

n	Type of bond formation	Bond order P_{rs}	Number of canonical structure
4		$P_{1.4}=-0.4473$	2
6		$P_{1.6}=0.3019$	2
		$P_{2.5}=-0.0866$	1
		$P_{1.4}=-0.3874$	2
8		$P_{1.8}=-0.2293$	2
		$P_{2.7}=-0.3333$	2
		$P_{3.6}=0.0340$	1
		$P_{1.6}=0.2632$	2
		$P_{1.4}=-0.3673$	2
10		$P_{1.10}=0.1855$	2
		$P_{3.8}=0.2297$	2
		$P_{2.9}=-0.0170$	1
		$P_{4.7}=-0.1278$	1
		$P_{1.8}=-0.2028$	2
		$P_{3.6}=-0.2293$	2
		$P_{2.7}=0.0440$	1
		$P_{1.6}=0.2469$	2
		$P_{2.5}=-0.1107$	1
		$P_{1.4}=-0.3576$	2
12		$P_{1.12}=-0.1562$	2
		$P_{3.10}=-0.1798$	2
		$P_{5.8}=-0.2938$	2
		$P_{2.11}=0.0100$	1
		$P_{4.9}=0.0586$	1
		$P_{1.10}=0.1660$	2
		$P_{3.8}=0.2146$	2
		$P_{2.9}=-0.0234$	1
		$P_{4.7}=-0.1377$	1
		$P_{1.8}=-0.1896$	2
		$P_{3.6}=-0.3038$	2
		$P_{2.7}=0.0485$	2
		$P_{1.6}=0.2382$	2
		$P_{2.5}=-0.1142$	1
		$P_{1.4}=-0.3523$	2
14		$P_{1.14}=0.1348$	2
		$P_{3.12}=0.1491$	2
		$P_{5.10}=0.2000$	2
		$P_{2.13}=-0.0063$	1
		$P_{4.11}=-0.0330$	1
		$P_{6.9}=-0.1491$	1

Circle written on the line represents the carbon atom which is a member of linear polyene
Dotted line indicates the new bond to be formed.

compounds by applying this method and the $4n+2$ rule generally, that is, the molecule in the case that the bond formation makes $4n+2$ ring has the largest resonance energy and therefore more stable than that in the case of $4n$ ring.

If we assume that internuclear repulsion energy stated above does not vary so much according to its geometrical structure, so far as the number of the atoms constituting a molecule does not change, the change of total energy of a molecule may approximately be discussed with an aid of Eq. 4. In many cases that the number of carbon atoms constituting the linear polyene is altered, the value of the bond order was calculated between several carbon-carbon atoms in simple LCAO MO treatment.

The result of calculation is collected in Table II for the cases of four, six, eight, ten, twelve and fourteen carbon atoms. Investigation of this table leads to the conclusion that in a polyene molecule the addition of a bond accompanied with the formation of a $4n+2$ ring stabilizes the molecule, while with the formation of a $4n$ ring it becomes unstable. It may be of interest to examine Table II more

TABLE III. THE PARALLELISM BETWEEN BOND ORDER AND CONJUGATION ENERGY

Molecule	Conjugation energy per electron, β	Bond order P_{rs}
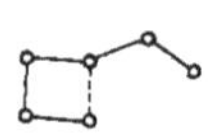	1.100	−0.3874
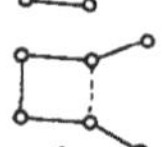	1.148	−0.0866
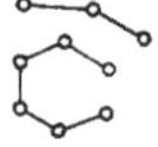	1.165	0.0000
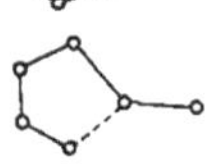	1.244	0.0000
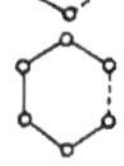	1.333	0.3019

Dotted line indicates the bond formed between atom r and s. β in parenthesis is the resonance integral of carbon-carbon bond for hydrocarbon.

carefully to find that in the case of the bond formation with a $4n+2$ ring the corresponding value of bond order increases with the number of describable canonical structures which have alternate single and double bonds, while in $4n$ ring formation the relation is reversed. Hence the number of canonical structures can be a measure for the stability of a $4n+2$ ring molecule but no longer so for $4n$ ring molecules.

The way of estimating the stabilization energy mentioned above can be replaced by the calculation of conjugation energy per electron in perturbed molecule with simple LCAO MO treatment. The relation between these two methods is shown in Table III, in which a good parallelism is seen.

Electron Distribution

According to the $4n+2$ rule, it is seen that π electron distribution should be correlated closely with the stability of aromatic compounds and molecules with the ring containing $4n+2$ π electrons may be more stable than those with $4n$ π electrons. Therefore the aromatic compound should have the ring which has $4n+2$ π electrons on it. With regard to this rule it is non-benzenoid aromatic compounds with substituents such as sydnone, tropone and so on that attract our interest more strongly.

The sum of π electron density on the whole ring was obtained in Fig. 4 in the cases of sydnone, tropone and tropolone. These results suggest that in any cases the aromatic compounds tend to form a ring involving $4n+2$ π electrons. When the sum of π electron density exceeds 6, the 6-membered ring repels electrons to the substituent, and when it is less than 6, the ring attracts electrons to the ring. Further calculation of the electron distribution in some hydrocarbons containing odd-membered ring is done. The result is shown in Table IV, which illustrates that the $(4n+3)$-membered ring repels electrons and $(4n+1)$-membered ring attracts them independently of the ring size. Thus in azulene the seven-membered ring is positively

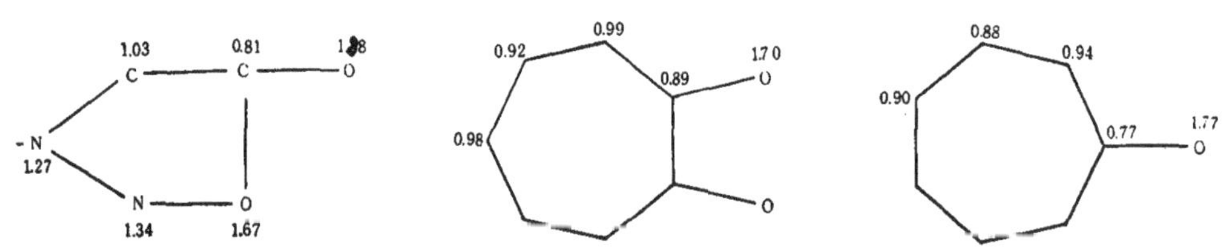

Sydnone 6.12 Tropolone 6.58 Tropone 6.21

Fig. 4. Total π electron distribution of non-benzenoid aromatic compounds. The value written under the figure is the sum of the π electron density of all member atoms constituting the ring.

TALBE IV. π ELECTRON DISTRIBUTION OF ODD-MEMBERED RING

	Sum of the total π electron density for $(4n+1)$-membered ring	Formal charge of $(4n+1)$-membered ring	Sum of the total π electron density for $(4n+3)$-membered ring	Formal charge of $(4n+3)$-membered ring
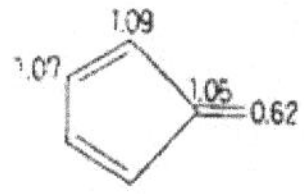	5.37	−0.37	—	—
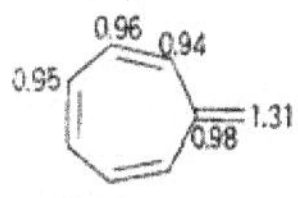	—	—	6.68	+0.32
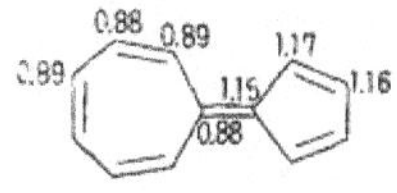	5.81	−0.81	6.19	+0.81
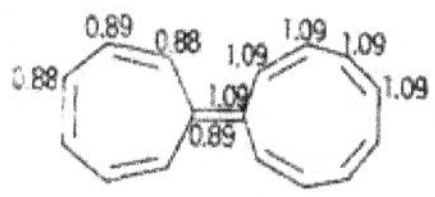	9.81	−0.81	6.19	+0.81
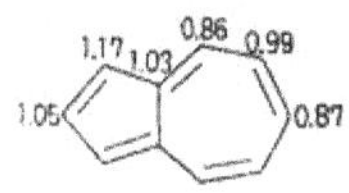	5.45	−0.45	6.63	+0.37

The value written in figures is the total π electron density at each position.

and the five-membered ring negatively charged representing the Fries' rule[11].

The same tendency can also be ascertained by such a perturbation treatmeat as is described above. A $(4n+1)$- or $(4n+3)$-membered ring can be formed as a result of the bond formation between appropriate two atoms in corresponding polyene molecule, as is shown in Fig. 5. This bond formation is considered as

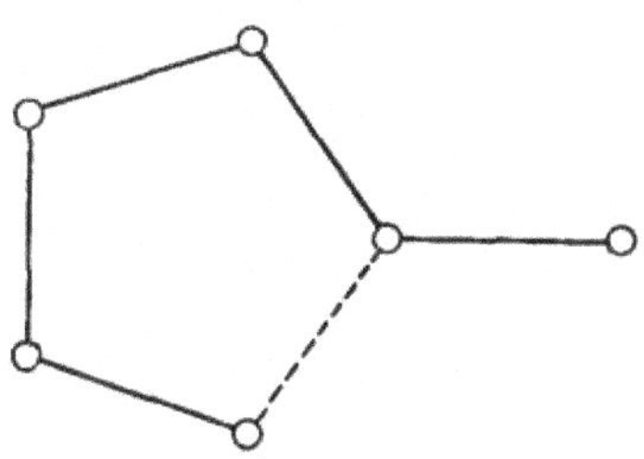

Fig. 5. Adopted model in perturbation method. Circle and line represent the carbon atom and the C-C bond constructing the linear polyene. Dotted line indicates the new bonding engendered.

a perturbation. The wave function for linear polyene is available as a zero order wave function. The total π electron density at the rth carbon atom, q_r, is represented by

$$q_r = 1 + \Pi_{r,st}\delta\beta_{st} + \cdots \tag{13}$$

where $\delta\beta_{st}$ is the resonance integral of the new bond $s-t$ and $\Pi_{r,st}$ is the atom-bond polarizability of the atom r and the bond $s\cdots\cdots t$ of the original polyene given by

$$\Pi_{r,st} = \frac{\partial q_r}{\partial \beta_{st}} = \sum_{j=1}^{m}\sum_{k=m+1}^{n} \frac{C_{rj}C_{rk}(C_{sj}C_{tk}+C_{tj}C_{sk})}{\varepsilon_j - \varepsilon_k} \tag{14}$$

TABLE V. THE VALUE OF ATOM-BOND POLARIZABILITY AND THE ENERGY DIFFERENCE BETWEEN THE HIGHEST OCCUPIED ORBITAL OF $(4n+1)$- AND $(4n+3)$-MEMBERED RINGS AND U

$(4n+1)$-membered ring

n	$\Pi_{1,2,4n+2}$ $(1/\beta)$	$\varepsilon_f - U(\beta)$
1	$\Pi_{1,2,6} = -0.4675$	0.6180
2	$\Pi_{1,2,10} = -0.3148$	0.3474
3	$\Pi_{1,2,14} = -0.2383$	0.2408
4	$\Pi_{1,2,18} = -0.1920$	0.1848
5	$\Pi_{1,2,22} = -0.1242$	0.1474

$(4n+3)$-membered ring

n	$\Pi_{1,2,4n+4}$ $(1/\beta)$	$\varepsilon_f - U(\beta)$
1	$\Pi_{1,2,8} = 0.2964$	−1.2468
2	$\Pi_{1,2,12} = 0.2709$	−0.8310
3	$\Pi_{1,2,16} = 0.2127$	−0.6180
4	$\Pi_{1,2,20} = 0.1750$	−0.4914
5	$\Pi_{1,2,24} = 0.1146$	−0.4068

$\Pi_{r,st}$ represents atom-bond polarizability.
ε_f is energy level of highest occupied orbitals of $(4n+1)$- and $(4n+3)$-membered ring.
α and β are Coulomb integral of carbon atom and resonance integral of carbon-carbon bond in polyene, respectively.

11) K. Fries, *Ann. Chem.*, **516**, 248 (1935).

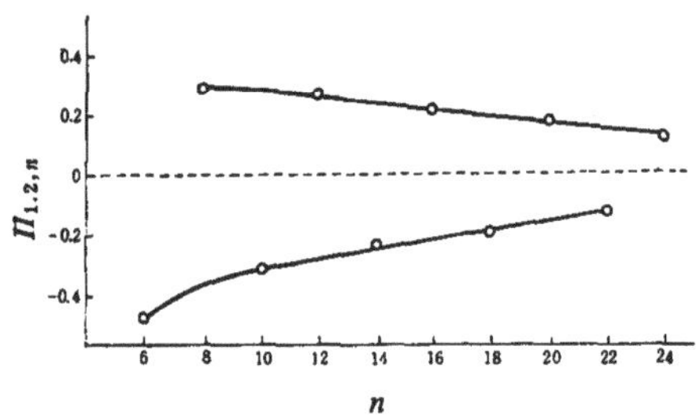

Fig. 6. Atom-bond polarizability and the number of members of linear polyene.

where C_{rj} is the coefficient of the atomic orbital of the rth atom in the jth molecular orbital in the simple LCAO MO treatment, ε_j is the energy of the jth molecular orbital, and subscript 1, 2, 3,……, m and subscript $m+1, m+2$ ……, n denote the occupied and unoccupied molecular orbitals, respectively. The calculated result of $\Pi_{r,st}$ is tabulated in Table V and is also plotted against the number of atoms, n, in Fig. 6. In the series of $(4n+3)$-membered rings the value of $\Pi_{r,st}$ is decreased with n and the sum of π electron density of the whole ring tends to approach to the limiting value, $4n+2$. A quite reverse tendency is seen in the series of $(4n+1)$-membered rings. That is, $\Pi_{r,st}$ is increased with n and the sum of π electron density of the whole ring approaches to the value of $4n+2$. Moreover, it is worth noticing that the absolute value of electronic charge transferred from side chain to ring or in the inverse direction decreases with n in regard to both the $(4n+1)$- and $(4n+3)$-membered rings.

These tendencies may qualitatively be understood from an energetical point of view. The energy levels of $(4n+1)$- and $(4n+3)$-membered rings are given schematically in Fig. 7. In the case of five- and nine-membered rings the charge transfer may occur from side chain to ring and the molecule forms a "closed shell" in a MO sense in order to stabilize itself. On the other hand, in the case of seven-membered ring the direction of the charge transfer is converse.

These circumstances can also be elucidated analytically by using the integral formula for atom-bond polarizability of Coulson[12]. Integral formula for atom-bond polarizability is written as

$$\Pi_{r,st} = (-1)^{s+t} \frac{2}{\pi} \int_{-\infty}^{+\infty} \frac{\Delta_{rt}(iy)\Delta_{st}(iy)}{\{\Delta(iy)\}^2} dy \tag{15}$$

where Δ is the secular determinant for the corresponding linear polyene and Δ_{rt} denotes the secular determinant with row r and column t struck out. Consider a $(4n+1)$-membered ring with a side chain of one carbon atom formed from a $(4n+2)$-membered polyene by a bridging between the 2nd and the $(4n+2)$th atoms of the polyene. The atom-bond polarizability $\Pi_{1,2.4n+2}$ is given by

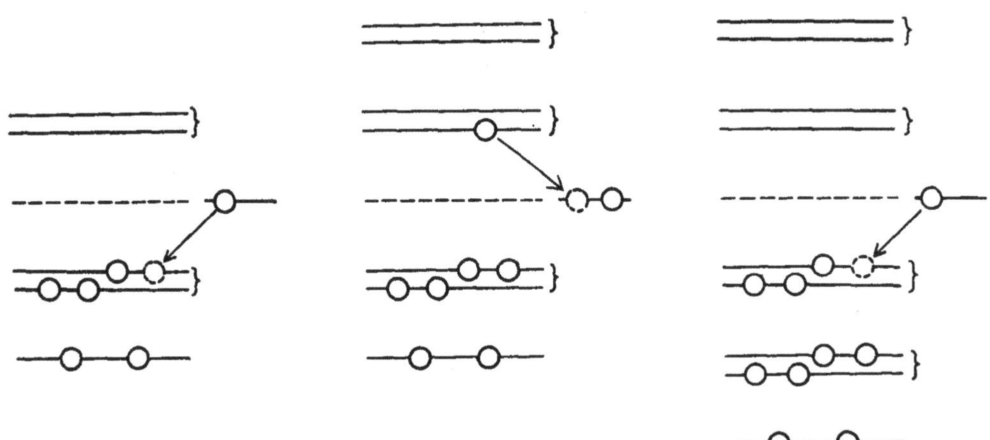

Five-membered ring with five π electrons | Seven-membered ring with seven π electrons | Nine-membered ring with nine π electrons

Fig. 7. Energy level of odd-membered ring.
The energy written aside represents that of the side-chain carbon.
The dotted line indicates the π electrons transferred and an arrow shows the direction of the electron transfer.

12) C. A. Coulson and H. C. Longuet-Higgins, *Proc. Roy. Soc.*, **A191**, 39 (1947); **A192**, 16 (1947).

TABLE VI. REACTIVITY INDEXES FOR $(4n+1)$- AND $(4n+3)$-MEMBERED RINGS

Compound	Position	f_r(N)	S_r(N)	q_r
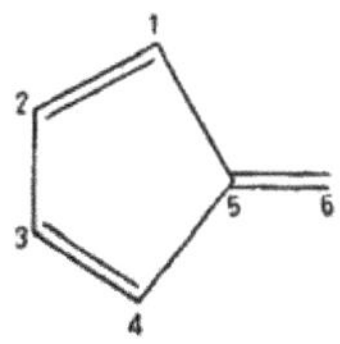	1	0.246	0.906	1.092
	2	0.156	0.784	1.073
	5	0.073	0.598	1.047
	6	1.124	2.701	0.623

Compound	Position	f_r(E)	S_r(E)	q_r
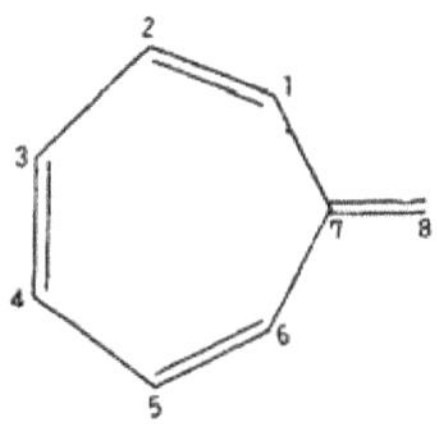	1	0.223	2.049	0.942
	2	0.100	1.071	0.962
	3	0.163	1.502	0.953
	7	0.046	0.756	0.976
	8	0.982	4.744	1.311

f_r, S_r and q_r are frontier electron density, superdelocalizability and total π-electron density at position r, respectively. (N) and (E) denote the reactivity indexes for nucleophilic and electrophilic reaction, respectively.

$$\Pi_{1,2\cdot 4n+2}=(-1)^{2+4n+2}\frac{2}{\pi}\times\int_{-\infty}^{+\infty}\frac{\Delta_{1,2}(iy)\Delta_{1,4n+2}(iy)}{\{\Delta^{(4n+2)}(iy)\}^2}dy \tag{16}$$

Following relations are easily proved[12)]

$$\Delta_{1,4n+2}(iy)=1,\quad \Delta_{1,2}(iy)=\Delta^{(4n)}(iy)>0$$

and
$$\Delta^{(4n+2)}(iy)<0 \tag{17}$$

where $\Delta^{(4n)}$ and $\Delta^{(4n+2)}$ denote the secular determinants for $(4n)$- and $(4n+2)$-membered polyene, respectively. From Eqs. 13 and 16 with the aid of Eq. 17 it follows that

$$\Pi_{1,2\cdot 4n+2}>0 \quad\text{and}\quad q_1=1+\Pi_{1,2\cdot 4n+2}\delta\beta<1. \tag{18}$$

Hence the total density at the atom of side chain is less than unity, since $\delta\beta$ is negative. In the case of $(4n+3)$-membered ring originated from a $(4n+4)$-membered polyene the following inequalities are obtained in a similar manner.

$$\Pi_{1,2\cdot 4n}<0 \quad\text{and}\quad q_1=1+\Pi_{1,2\cdot 4n}\delta\beta>1 \tag{19}$$

Thus the total density becomes larger than unity.

Moreover, a close relation is observed between the atom-bond polarizability of a polyene and the highest occupied levels of $(4n+1)$- and $(4n+3)$-membered rings formed from the polyene. Table V illustrates this relation with regard to the cases of $(4n+1)$-membered ring formed from $(4n+2)$-membered polyene and $(4n+3)$-membered ring formed from a $(4n+4)$-membered polyene. In the table it is clearly observed that the lower the highest occupied level of $(4n+1)$-membered ring is, …… and also the higher the highest occupied level of $(4n+3)$-membered ring is, the larger the amount of charge transferred from ring to side chain is. This result also coincides with the fact that the absolute value of the charge transferred decreases with n.

Thus a ring in the molecule has a tendency to approach to a $4n+2$ electron ring by the charge transfer to stabilize the molecule. The position to be attacked by a reagent was predicted for azulene in the paper symposium on aromatic character[13)]. The attack of reagent was supposed to occur so as to make the ring possess $4n+2$ electrons by the process of reaction. Similarly, from the consideration stated above, it is predicted that the molecule containing a $(4n+1)$-membered ring and a side chain attached to it such as fulvene may be apt to react with a nucleophilic reagent at the side chain carbon. Inversely, in the case of $(4n+3)$-membered ring, an electrophilic reagent is inclined to attack the side chain carbon. Therefore, $4n+2$ rule can be said to govern the orientation of chemical reaction. Further, this rule also suggests a guiding principle in the syntheses of new aromatic compounds. Namely, we should choose an appropriate substituent attached to the ring that makes the ring possess $4n+2$ electrons. If a molecule

13) D. H. Reid, *Tetrahedron*, 3, 339 (1958).

has excessive electrons on its ring, the substituent to be chosen is electron-attracting, and if it has insufficient electrons, an electron-repelling substituent should be introduced to stabilize it.

This has also been supported by several quantum-mechanical theories of chemical reactivity such as the static method[12] and the frontier electron theory[14]. The reactivity indexes used in these theories have been calculated and tabulated in Table VI, showing that the above-mentioned prediction is correct.

Faculty of Engineering
Kyoto University
Sakyo-ku, Kyoto

14) K. Fukui, T. Yonezawa and C. Nagata, *J. Chem. Phys.*, **26**, 831 (1957); **27**, 1247 (1957).

[Reprinted from the Bulletin of the Chemical Society of Japan, Vol. 34, No. 8 (1961)]

A Theoretical Treatment of Molecular Complexes. I. Silver-Aromatic Hydrocarbon Complexes

By Kenichi FUKUI, Akira IMAMURA, Teijiro YONEZAWA and Chikayoshi NAGATA

(Received January 11, 1961)

Theoretical researches of molecular complexes have become active since Mulliken combined the old concepts of molecular compound and established a new theory[1]. Prior to this research many qualitative discussions on molecular complexes had been presented. These are the covalent bond theory by Bennett and Willis[2], the polarization aggregate theory by Briegleb[3], the ionic complex theory by

1) R. S. Mulliken, *J. Am. Chem. Soc.*, **74**, 811 (1952).

2) G. M. Bennett and G. H. Willis, *J. Chem. Soc.*, **1929**, 256.

3) G. Briegleb, *Z. physik. Chem.*, **B16**, 249 (1932).

Weiss[4], and the complex resonance by Brackmann[5]. Weiss had pointed out that molecular complexes have an ionic structure B^+A^-, so that the lower the ionization potential for an electron donor is, and the higher the electron affinity for an electron accceptor is, the more easily the molecular complex is formed. Brackmann regarded the structure of complex as a resonance hybrid to which no-bond and dative structures contribute.

$$D:A \leftrightarrow D^+:A^-$$

Based upon these concepts, Mulliken presented a quantum mechanical theory on the charge transfer spectra of molecular complex[1]. Further, Mulliken predicted the configuration of benzene-iodine complex, benzene-silver complex and BF_3-$N(CH_3)_3$ complex, and his assumption was proved to be correct except in the case of benzene-iodine complex. With regard to benzene-iodine complex, Aono discussed Mulliken's model and presented a corrected one[6]. It may be true that in spite of these successes, Mulliken's procedure is not very useful to discuss the relative ease of complex formation among different molecules. There are often found some trials of discussing this problem by comparing the magnitude of ionization potentials of electron donors. For example, with respect to the silver-aromatic molecular complex, a correlation between the magnitude of ionization potential of methyl derivatives of 1, 2-benzanthracene and their equilibrium constants in complex formation was pointed out by Nakajima[7]. However, this treatment has only a limited application since the ionization potential is not the sole factor relating to the ease of complex formation, and is hence not suitable in discussing the configuration of complex.

In the present paper we make use of the delocalization method in order to approach the conformation of molecular compound, finding a clear correlation between the equilibrium constant and the calculated value of the delocalization energy for silver-aromatic hydrocarbon complexes.

Model and Formulation

In this paper the delocalization method was applied to silver-aromatic hydrocarbon complexes. Since it has been confirmed by X-ray diffraction that in silver-benzene complex silver cation is located above the middle of a carbon-carbon bond of benzene plane[8], we assume that silver cation approaches from above the molecular plane to the middle of a carbon-carbon bond of a benzene ring in aromatic hydrocarbons to form the molecular complex. The π electronic stabilization energy due to delocalization from the occupied orbitals of the aromatic hydrocarbon to the lowest vacant orbital of the silver cation is calculated by the perturbation theory. This procedure is quite similar to the one which has been used in the frontier electron theory presented by some of the present authors[9]. When the aromatic molecule is confined to alternant hydrocarbons and their derivatives, and the silver cation is combined with the rth and the sth carbon atoms of the aromatic hydrocarbon by a weak bond, the secular equation of the π part of the total system which includes the silver cation and the aromatic molecule is written in the simple LCAO MO treatment (overlap neglected) as follows:

$$D(\lambda)=\begin{vmatrix} h-\lambda & 0 & 0\cdots\cdots 0 & \gamma & \gamma & 0\cdots\cdots 0 \\ 0 & a_{11}-\lambda & a_{12}\cdots\cdots & a_{1r} & a_{1s} & \cdots\cdots a_{1n} \\ 0 & a_{21} & a_{22}-\lambda\cdots & a_{2r} & a_{2s} & \cdots\cdots a_{2n} \\ \vdots & & & & & \\ 0 & & & & & \\ \gamma & a_{r1} & a_{r2}\cdots\cdots & a_{rr}-\lambda & a_{rs} & \cdots\cdots a_{rn} \\ \gamma & a_{s1} & a_{s2}\cdots\cdots & a_{sr} & a_{ss}-\lambda & \cdots\cdots a_{sn} \\ 0 & & & & & \\ \vdots & & & & & \\ 0 & a_{n1} & a_{n2}\cdots\cdots & a_{nr} & a_{ns} & \cdots a_{nn}-\lambda \end{vmatrix}$$

$$=(h-\lambda)\Delta(\lambda)$$
$$-\gamma^2\{\Delta_{rr}(\lambda)+\Delta_{ss}(\lambda)-2\Delta_{rs}(\lambda)\} \quad (1)$$

where

$$\Delta(\lambda)=\begin{vmatrix} a_{11}-\lambda & a_{12} & \cdots\cdots & a_{1n} \\ a_{21} & a_{22}-\lambda\cdots\cdots & & a_{2n} \\ \vdots & \vdots & & \vdots \\ a_{n1} & a_{n2} & \cdots\cdots & a_{nn}-\lambda \end{vmatrix} \quad (2)$$

is the secular determinant for the isolated aromatic molecule, and $\Delta_{rr}(\lambda)$, $\Delta_{ss}(\lambda)$, and $\Delta_{rs}(\lambda)$ are the minors of $\Delta(\lambda)$ corresponding to (rr), (ss) and (rs) elements, respectively; a_{ii} and $a_{ij}(i\neq j)$ are the Coulomb and the resonance integrals, respectively; and h and γ are the Coulomb integral of the silver cation and the resonance integral between the silver cation and the rth and the sth atoms of the aromatic molecule at which the delocalization of π electron takes place, respectively. In presenting the value of energies, we make the zero point equal to the Coulomb integral of

4) J. Weiss, *J. Chem. Soc.*, 1942, 245; 1943, 462.
5) W. Brackmann, *Rec. trav. chim.*, 68, 147 (1949).
6) S. Aono, *Progr. Theoret. Phys. Japan*, 22, 313 (1959).
7) T. Nakajima, *The Science Reports of the Tôhoku Univ. Ser. I*, 41, 171 (1957).

8) R. E. Rundle and J. H. Goring, *J. Am. Chem. Soc.*, 72, 5337 (1950).
9) K. Fukui, T. Yonezawa and C. Nagata, This Bulletin, 27, 423 (1954).

benzene carbon and express them in units of $(-\beta)$ which is the resonance integral of the benzene C-C bond. If we designate the roots of $\Delta(\lambda)=0$ and $D(\lambda)=0$ as

$$\lambda_1, \lambda_2, \cdots\cdots \lambda_n ; \lambda^{(1)}, \lambda^{(2)} \cdots\cdots {}^{(n)}, \lambda^{(h)}$$
$$(\lambda_1 > \lambda_2 > \cdots\cdots > \lambda_n)$$

respectively, among which $\lambda^{(h)}$ will be very close to the value of h and $\lambda^{(j)}$ to λ_j ($j=1, 2 \cdots\cdots n$) since γ is small. Namely $\lambda^{(j)}$ and $\lambda^{(h)}$ can be written as $\lambda^{(j)}=\lambda_j+\delta_j$, $\lambda^{(h)}=h+\delta_h$ in which $|\delta_j|$ and $|\delta_h|$ are small. For simplicity, we exclusively treat the following two cases in which $\Delta(\lambda)$ has no multiple root.

Case 1: When h is not equal to any λ_j, we can obtain δ_j as

$$\left.\begin{aligned} \delta_j &= 2\left\{\frac{\Delta_{rr}(\lambda_j)}{(h-\lambda_j)\Delta'(\lambda_j)}+\frac{\Delta_{ss}(\lambda_j)}{(h-\lambda_j)\Delta'(\lambda_j)} - \frac{2\Delta_{rs}(\lambda_j)}{(h-\lambda_j)\Delta'(\lambda_j)}\right\}\gamma^2+[\gamma^4] \quad (j=1, 2\cdots\cdots n) \\ \delta_h &= -\sum_{j=1}^{n}\left\{\frac{\Delta_{rr}(\lambda_j)}{(h-\lambda_j)\Delta'(\lambda_j)} + \frac{\Delta_{ss}(\lambda_j)}{(h-\lambda_j)\Delta'(\lambda_j)}-\frac{2\Delta_{rs}(\lambda_j)}{(h-\lambda_j)\Delta'(\lambda_j)}\right\}\gamma^2 + [\gamma^4] \end{aligned}\right\} \quad (3)$$

where $[\gamma^4]$ denotes the higher terms. The delocalization energy, ΔE, is defined as follows:

$$\Delta E=2\sum_{j=1}^{m}(\lambda^{(j)}-\lambda_j) \quad (4)$$
$$=2\sum_{j=1}^{m}\left\{\frac{\Delta_{rr}(\lambda_j)+\Delta_{ss}(\lambda_j)-2\Delta_{rs}(\lambda_j)}{(h-\lambda_j)\Delta'(\lambda_j)}\right\}\gamma^2+[\gamma^4]$$
$$=-2\sum_{j=1}^{m}\frac{(C_r^j+C_s^j)^2}{h-\lambda_j}\gamma^2+[\gamma^4] \quad (5)$$

where $j=1, 2\cdots\cdots m$ denote the occupied orbitals, and the rth and the sth carbon atoms are adjacent to each other. And C_r^j is the coefficient of the rth atomic orbital in the jth molecular orbital of the isolated aromatic molecule. When h is equal to zero, then the total delocalization energy becomes

$$\Delta E=\{S_r^{(E)}+S_s^{(E)}+2S_{rs}^{(E)}\}\gamma^2+[\gamma^4] \quad (6)$$

where $S_r^{(E)}$, $S_s^{(E)}$ are superdelocalizabilities at the rth and the sth carbon atoms for the electrophilic reaction[9], and $S_{rs}^{(E)}$ is given in Eq. 7 and may be named bond-delocalizability of the bond rs.

$$S_{rs}=\sum_{j=1}^{n}\frac{(\nu_j-\nu)C_r^jC_s^j}{\lambda_j} \quad (7)$$

where ν_j is the number of π electrons in the jth molecular orbital and ν is the number of π electrons in the approaching orbital to or from which the delocalization of electrons occurs. $S_{rs}^{(E)}$ designates S_{rs} in which $\nu=0$.

The formula obtained by Eq. 5 indicates that the magnitude of delocalization energy depends sensitively upon the local symmetry of the molecular orbital; that is to say, when C_r^j and C_s^j have same signs, the magnitude of delocalization energy is large, while it is small in case of different signs. It may be stressed that a conclusion similar to that of Mulliken has thus been derived by the present treatment.

Case 2: When h is equal to the energy of the highest occupied orbital of the isolated aromatic molecule λ_f,

$$\left.\begin{aligned} \delta_j &= 2\left\{\frac{\Delta_{rr}(\lambda_j)}{(h-\lambda_j)\Delta'(\lambda_j)}+\frac{\Delta_{ss}(\lambda_j)}{(h-\lambda_j)\Delta'(\lambda_j)} - \frac{2\Delta_{rs}(\lambda_j)}{(h-\lambda_j)\Delta'(\lambda_j)}\right\}\gamma^2+[\gamma^4] \quad (j=1, 2, \cdots\cdots n) \\ \left.\begin{matrix}\delta_h\\ \delta_f\end{matrix}\right\} &= \pm\sqrt{-\frac{\Delta_{rr}(h)+\Delta_{ss}(h)-2\Delta_{rs}(h)}{\Delta'(h)}} \times\gamma+[\gamma^2] \end{aligned}\right\} \quad (8)$$

and the delocalization energy depends dominantly upon the term of first order. Hence, we obtain

$$\Delta E=2\sqrt{-\frac{\Delta_{rr}(h)+\Delta_{ss}(h)-2\Delta_{rs}(h)}{\Delta'(h)}}\gamma+[\gamma^2]$$
$$=2|C_r^f+C_s^f|\gamma+[\gamma^2] \quad (9)$$

It may be remarked that in case 2, the magnitude of ΔE depends upon the local symmetry of the highest occupied orbital, the frontier orbital at the bond to be attacked.

Then the following conclusion can be derived. When h is not equal to any λ_j, the larger the value of $-\sum_j \frac{(C_r^j+C_s^j)^2}{h-\lambda_j}$ is, the easier the molecular complex formation is, and when h is equal to λ_f, the reactivity is proportional to $|C_r^f+C_s^f|$.

Results and Discussion

The easiness of the molecular complex formation depends upon many factors such as delocalization energy, dispersion energy, exchange repulsion energy and electrostatic energy. Therefore, the reactivity should be discussed under consideration of all these factors. In discussing the relative easiness of complex formation of a series of similar compounds, however, it is expected that the energies, other than delocalization energy, contributing to the activation energy, may be approximately constant or negligible. In such a case, the delocalization energy is considered to play a dominant role.

In the present paper, an attempt is made to

TABLE I. THE VALUES OF THE INDEX FOR COMPLEX FORMATION AND THE EQUILIBRIUM CONSTANT FOR THE COMPLEX BETWEEN SILVER CATION AND AROMATIC COMPOUNDS

Aromatic compound	K_1 (mol./l.)$^{-1}$	$S_r^{(E)}+S_s^{(E)}+2S_{r,s}^{(E)}$	Energy of the highest occupied orbital
	2.41	1.333	−1.000
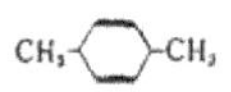	2.63	1.402	−0.669
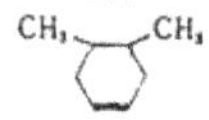	2.89	1.553	−0.707
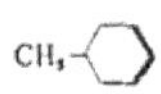	2.95	1.418	−0.823
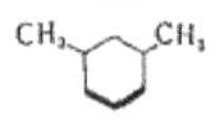	3.03	1.466	−0.763
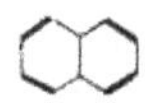	3.08	1.600	−0.618
	3.67	1.798	−0.605
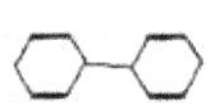	3.94	1.370	−0.705
	6.3	2.139	−0.504
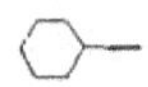	18.2	2.089	−0.662

Thick lines in the figure of aromatic compounds indicate the bonds with maximum value of the index, which are assumed to be the positions of attack by silver cations.

With regard to toluene and xylene, and carbon-carbon bond adjacent to the methyl group is not taken into account for the sake of steric circumstances.

K_1 is the equilibrium constant for the following reaction

$$Ag^+ + Ar = AgAr^+ \qquad K_1 = (AgAr^+)/(Ag^+)(Ar)$$

explain the reactivity of complex formation of a series of similar compounds by the consideration of the delocalization energy only. The delocalization energies calculated by the formulas stated above are tabulated in Table I together with the equilibrium constants of the first step of the molecular complex formation between silver cation and aromatic compound[10]. A clear parallelism is seen between the index and the equilibrium constant. The parameters used in calculation for methyl group are as follows.

α_{CH_3} (Coulomb integral for methyl group in Eq. 1) = 3

$\beta_{C\cdot CH_3}$ (resonance integral for CH_3-C bond in Eq. 1) = 1

The energies of the highest occupied levels thus obtained are compared with the observed ionization potentials in Table II. At a glance at this table, a good parallelism is observed between the ionization potential and the energy of the highest occupied level. This implies that the adopted parameters should be appropriate.

10) L. J. Andrews and R. R. M. Keefer, *J. Am. Chem. Soc.*, **71**, 3644 (1949); **72**, 5034 (1950).

TABLE II. RELATION BETWEEN THE IONIZATION POTENTIAL AND THE HIGHEST OCCUPIED LEVEL IN AROMATIC COMPOUNDS

Aromatic compound	Ionization potential eV.	Highest occupied level
Benzene	9.24[a]	−1.000
Toluene	8.82[b]	−0.823
m-Xylene	8.58[a]	−0.763
o-Xylene	8.58[a]	−0.707
p-Xylene	8.48[a]	−0.669
Naphthalene	8.12[a]	−0.618

a; From F. H. Field and J. L. Franklin, "Electron Impact Phenomena", Academic Press, Inc., New York (1957).

b; K. Watanabe, *J. Chem. Phys.*, **26**, 542 (1957).

The value of index for *o*-xylene seems too large. This discrepancy may partly be ascribed to the neglect of the effect of the steric hindrance between two methyl groups. With regard to biphenyl, the calculation is carried out with

TABLE III. THE VALUES OF THE INDEX FOR COMPLEX FORMATION AND THE EQUILIBRIUM CONSTANT FOR THE COMPLEX BETWEEN SILVER CATION AND POLYCONDENSED AROMATIC HYDROCARBONS

Aromatic compound	K_1	$S_r^{(E)}+S_s^{(E)}+2S_{rs}^{(E)}$	Energy of the highest occupied orbital
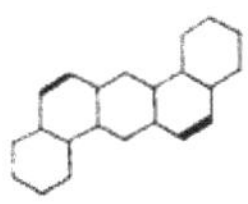	1.65	1.864	−0.474
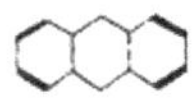	1.35	1.747	−0.414
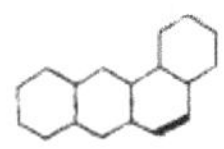	1.33	1.900	−0.452
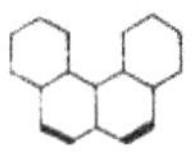	1.30	1.737	−0.568
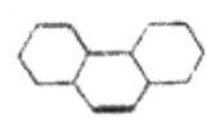	1.09	1.798	−0.605

Thick lines in the figure of aromatic compounds indicate the bonds with maximum value of the index, which are assumed to be the positions of attack by silver cations.

K_1 is the equilibrium constant for the following reaction.

$$Ag^+ + Ar = AgAr^+ \qquad K_1 = (AgAr^+)/(Ag^+)(Ar)$$

the assumption that this compound has a plane structure. Experimental data show that this compound is not planar. If two benzene nuclei are but little conjugated with each other, the value of the equilibrium constant for this compound may reasonably be expected to be approximately twice as large as that of benzene as was pointed out by Andrews et al.[10] On the other hand, the index is scarcely affected by the degree of conjugation between two benzene nuclei. On the occasion of comparison with the value of the index, therefore, the observed value of equilibrium constant must be divided by a factor which is larger than unity and smaller than two. This is the reason why the observed value of the equilibrium constant seems too large in comparison with the magnitude of the index calculated.

It seems that also the value of energy of the highest occupied orbital can be a good measure for complex formation. This fact shows that the ionization potential is also useful as a rough index to discuss the easiness of complex formation, though it is impossible to predict the position of attack as stated above.

Furthermore, the complex formation in polycondensed aromatic hydrocarbons is treated in a similar manner and the results are shown in Table III. In this series the parallelism between the value of index and equilibrium constant[11] is not so good. In these polycondensed aromatic hydrocarbons their molecular size is not uniform. The coordination with silver cations may take place at more than one positions in a molecule which has more than one positions of equal or nearly equal reactivity. One silver cation which has already added to a position in a hydrocarbon molecule will not exert a substantial influence upon the later addition of another silver cation to another position in the same molecule, provided these two positions are sufficiently apart from each other. In such a case, in order to be compared with the calculated value of the theoretical index, the observed value of the equilibrium constant must be modified by considering the number of sites in one molecule as is stated above in the case of biphenyl.

Another factor influencing the easiness of the complex formation is the polarizability which may be different in various cases. This means that ion-induced dipole interactions are not uniform from molecule to molecule, and it is indispensable for the prediction of reactivity to take the ion-dipole interaction into account. This point will be discussed in future.

Faculty of Engineering
Kyoto University
Sakyo-ku, Kyoto

11) R. E. Kofahl and H. J. Lucas, *J. Am. Chem. Soc.*, 76, 3931 (1954).

[Reprinted from the Bulletin of the Chemical Society of Japan, Vol. 34, No. 8 (1961)]

A New Quantum-mechanical Reactivity Index for Saturated Compounds

By Kenichi FUKUI, Hiroshi KATO and Teijiro YONEZAWA

(Received January 24, 1961)

In our previous paper[1], the frontier electron density[2] was shown to become a good intramolecular index of the reactivity in a metathetical reaction of saturated compounds.

In the present paper, we treat several lower paraffins and halomethanes by using the same method as that used in the previous treatment[1], and attempt to derive a new intermolecular index of the reactivity, which is hereafter referred to as the delocalizability, D_r. This index is similar to the superdelocalizability[3] of conjugated molecules. It is found that the magnitude of D_r in these compounds has an intimate correlation with the value of activation energy observed in metathetical reactions.

Electronic Structure of *n*-Paraffins and Methyl Fluoride

The calculated values of orbital energies and electron distributions in methane, ethane, propane, *n*-butane and methyl fluoride are listed in Table I.

With regard to *n*-paraffins, we can point out the following characteristics:

(1) The bonding electrons tend to localize in the particular bonds in each molecular orbital. For example, the electrons in the highest occupied and the lowest vacant orbitals in these compounds except methane, have a general trend to localize in the central C-C bonds as shown in Fig. 1*.

(2) The electron distributions of occupied and unoccupied orbitals are almost symmetrical with respect to the energy equal to α (see below) as is seen in Fig. 1.

(3) The distances of levels of molecular orbital energy in these compounds are much larger than those of conjugated molecules. The excitation energy of the N-V transition in a simple LCAO MO sense is, therefore, estimated to be 9~10 eV. referring the presumed value

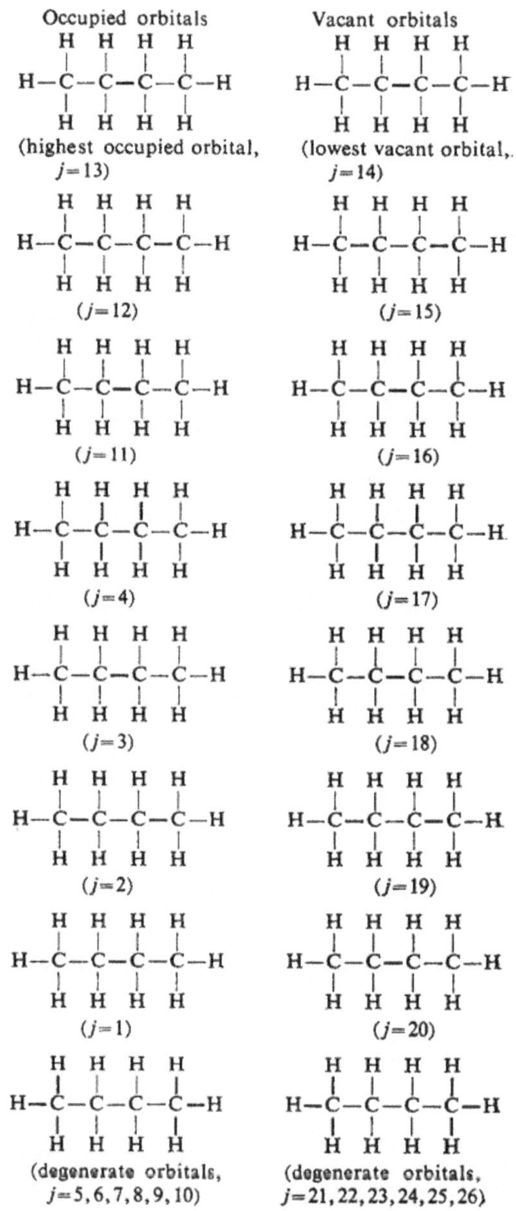

Fig. 1. The electron distributions of *n*-butane.
(Thick lines denote the most densely populated bonds of bonding electrons.)

---

1) K. Fukui, H. Kato and T. Yonezawa, This Bulletin, 34, 442 (1961).
2) See e. g. K. Fukui, T. Yonezawa, C. Nagata and H. Shingu, *J. Chem. Phys.*, 22, 1432 (1954).
3) K. Fukui, T. Yonezawa and C. Nagata, This Bulletin, 27, 423 (1954).
* This result suggests that in lower *n*-paraffins their first ionization can be attributed to the removal of one electron localized in C-C bonds.

TABLE I. THE ORBITAL ENERGIES AND ELECTRON DISTRIBUTIONS IN SEVERAL SATURATED COMPOUNDS

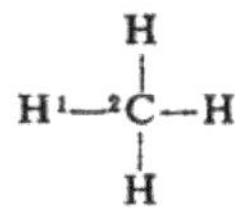

Orbital energy j	ε_j	Electron distribution $(C^j_{H1})^2$	$(C^j_{C2})^2$
6,7,8	$\alpha-1.3722\beta$*	0.1171	0.1329
5	$\alpha-0.8478\beta$	0.1856	0.0644
2,3,4	$\alpha+0.8322\beta$*	0.1329	0.1171
1	$\alpha+1.6678\beta$	0.0644	0.1856

* threefold degenerate

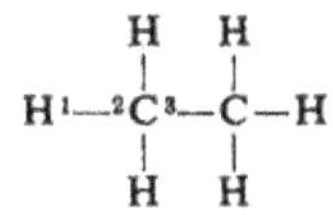

Orbital energy j	ε_j	Electron distribution $(C^j_{H1})^2$	$(C^j_{C2})^2$	$(C^j_{C3})^2$
11,12,13,14	$\alpha-1.3722\beta$*	0.0781	0.0886	0
10	$\alpha-1.3527\beta$	0.0323	0.0355	0.2966
9	$\alpha-1.0019\beta$	0.1057	0.0561	0.0145
8	$\alpha-0.6881\beta$	0.0878	0.0173	0.1847
7	$\alpha+0.7254\beta$	0.0336	0.0238	0.3278
3,4,5,6	$\alpha+0.8322\beta$*	0.0886	0.0781	0
2	$\alpha+1.5208\beta$	0.0465	0.1139	0.0187
1	$\alpha+1.7566\beta$	0.0274	0.0867	0.1575

* fourfold degenerate

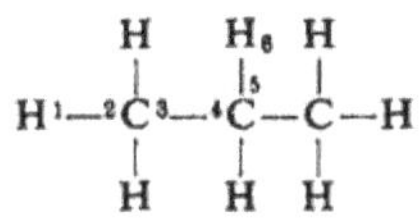

Orbital energy j	ε_j	Electron distribution $(C^j_{H1})^2$	$(C^j_{C2})^2$	$(C^j_{C3})^2$	$(C^j_{C4})^2$	$(C^j_{C5})^2$	$(C^j_{H6})^2$
16,17,18,19,20	$\alpha-1.3722\beta$*	0.062	0.071	0	0	0.053	0.047
15	$\alpha-1.3576\beta$	0.011	0.012	0.105	0.106	0.115	0.104
14	$\alpha-1.3482\beta$	0.021	0.023	0.183	0.186	0	0
13	$\alpha-1.0614\beta$	0.058	0.036	0.035	0	0.070	0.113
12	$\alpha-0.8535\beta$	0.090	0.032	0.028	0.105	0	0
11	$\alpha-0.6169\beta$	0.045	0.006	0.144	0.100	0.013	0.089
10	$\alpha+0.6997\beta$	0.020	0.014	0.207	0.191	0	0
9	$\alpha+0.7548\beta$	0.013	0.009	0.117	0.128	0.081	0.108
4,5,6,7,8	$\alpha+0.8322\beta$*	0.071	0.062	0	0	0.047	0.053
3	$\alpha+1.4502\beta$	0.026	0.058	0.001	0.043	0.142	0.063
2	$\alpha+1.6419\beta$	0.035	0.099	0.079	0.020	0	0
1	$\alpha+1.7919\beta$	0.014	0.045	0.100	0.121	0.078	0.024

* fivefold degenerate

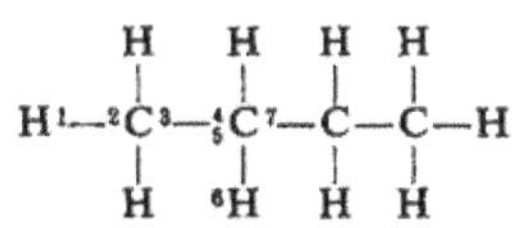

Orbital energy j	ε_j	Electron distribution $(C^j_{H1})^2$	$(C^j_{C2})^2$	$(C^j_{C3})^2$	$(C^j_{C4})^2$	$(C^j_{C5})^2$	$(C^j_{H6})^2$	$(C^j_{C7})^2$
21,22,23,24,25,26	$\alpha-1.3722\beta$*	0.052	0.059	0	0	0.044	0.039	0
20	$\alpha-1.3612\beta$	0.005	0.006	0.047	0.048	0.074	0.066	0.092
19	$\alpha-1.3546\beta$	0.014	0.016	0.133	0.134	0.037	0.034	0
18	$\alpha-1.3436\beta$	0.012	0.013	0.104	0.101	0.005	0.005	0.200
17	$\alpha-1.0870\beta$	0.035	0.022	0.032	0.002	0.057	0.087	0.007
16	$\alpha-0.9473\beta$	0.070	0.032	0	0.032	0.012	0.027	0.083
15	$\alpha-0.7567\beta$	0.064	0.016	0.077	0.116	0.006	0.023	0.009
14	$\alpha-0.5822\beta$	0.026	0.003	0.099	0.058	0.008	0.063	0.114
13	$\alpha+0.6971\beta$	0.012	0.008	0.128	0.113	0.001	0.002	0.192
12	$\alpha+0.7339\beta$	0.016	0.011	0.151	0.155	0.024	0.033	0
11	$\alpha+0.7627\beta$	0.005	0.004	0.047	0.053	0.055	0.072	0.119
5,6,7,8,9,10	$\alpha+0.8322\beta$*	0.059	0.052	0	0	0.039	0.044	0
4	$\alpha+1.4135\beta$	0.014	0.031	0	0.037	0.107	0.050	0.014
3	$\alpha+1.5585\beta$	0.029	0.074	0.024	0.002	0.036	0.014	0.065
2	$\alpha+1.7073\beta$	0.023	0.070	0.094	0.064	0.022	0.007	0.004
1	$\alpha+1.8098\beta$	0.008	0.026	0.063	0.084	0.057	0.017	0.103

* sixfold degenerate

TABLE I (Continued)

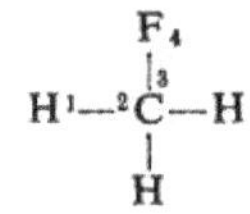

Orbital energy		Electron distribution			
j	ε_j	$(C^j_{H1})^2$	$(C^j_{C2})^2$	$(C^j_{C3})^2$	$(C^j_{F4})^2$
7,8	$\alpha-1.3722\beta$*	0.1561	0.1773	0	0
6	$\alpha-1.0729\beta$	0.1756	0.1112	0.1291	0.0083
5	$\alpha-0.2183\beta$	0.0656	0	0.6691	0.1338
3,4	$\alpha+0.8322\beta$*	0.1773	0.1561	0	0
2	$\alpha+1.0527\beta$	0.0188	0.0244	0.0724	0.7962
1	$\alpha+1.6184\beta$	0.0723	0.1978	0.1285	0.0618

* doubly degenerate

of $-\beta$ (6~7 eV.). This excitation energy is consistent with the spectroscopic data[4].

The characteristics of orbital energies and of the electron distribution stated above agree with the properties observed experimentally in saturated compounds and also with calculated results obtained by Lennard-Jones and Hall's equivalent orbital method[5].

In the doubly degenerate highest occupied orbitals of methyl fluoride, the bonding electrons localize dominantly in the methyl group. In the next highest occupied orbital, an exceedingly large density is observed on the fluorine atom in accordance with the facts pointed out by Frost and McDowell[6], who measured the ionization potentials of methyl halides.

The Coulomb and resonance integrals used in the present paper are summarized in the following table.

$\alpha_H = \alpha-0.2\beta$	$\beta_{cc} = 0.34\beta$*
$\alpha_F = \alpha+0.9\beta$	$\beta_{CH} = 1.1\beta$
$\alpha_{Cl} = \alpha+0.5\beta$	$\beta_{CF} = 0.5\beta$
$\alpha_{Br} = \alpha+0.45\beta$	$\beta_{CCl} = 0.65\beta$
$\alpha_I = \alpha+0.40\alpha$	$\beta_{CBr} = 0.58\beta$
$\alpha_C = \alpha$	$\beta_{CI} = 0.53\beta$

* β_{cc} refers to the resonance integral between two sp^3 orbitals attached to one and the same carbon atom.

A New Reactivity Index for Radical Reactions of Saturated Compounds, $D_r^{(R)}$

The radical reactions of alkyl compounds are covered by the general equation

$$\text{R-X}+\text{Y}\cdot \rightarrow [\text{R}\cdots\text{X}\cdots\text{Y}] \rightarrow \text{R}\cdot+\text{X-Y} \qquad (1)$$

where R is an alkyl group, X is the group to be extracted and Y is an attacking radical or atom. The bracket in the equation refers to the transition state.

The secular determinant for the activated complex, [R-X···Y], is written

$$\Delta_\sigma = \begin{vmatrix} \alpha_Y-\varepsilon & \gamma\beta & 0 & 0 & 0 & 0 & 0 \\ \gamma\beta & \alpha_X-\varepsilon & \beta_{CX} & 0 & 0 & 0 & 0 \\ 0 & \beta_{CX} & \alpha_C-\varepsilon & \beta_{CH} & \beta_{cc} & & \\ 0 & 0 & \beta_{CH} & \alpha_H-\varepsilon & 0 & & \\ 0 & 0 & \beta_{cc} & 0 & \alpha_C-\varepsilon & & \\ 0 & \multicolumn{6}{l}{\text{(The secular equation for the R-X molecule)}} \end{vmatrix}$$

where α_A refers to the Coulomb integrals of atom A, β_{AB} is the resonance integral between bonding orbitals of atom A and B, and γ is a parameter for the resonance integral of the X-Y incipient bond in the transition state.

By using the perturbation theory, we can derive a new reactivity index, named D_r, for metathetical radical reactions in saturated compounds, in place of the superdelocalizability (S_r) in our previous paper[3].

$$D_r^{(R)} = \sum_j^{\text{occ}} \frac{(C_r^j)^2}{\varepsilon_j-\alpha}\beta + \sum_j^{\text{unocc}} \frac{(C^j_r)^2}{\alpha-\varepsilon_j}\beta \qquad (3)$$

where C_r^j is the coefficient of the rth atomic orbital in the jth molecular orbital, ε_j is the jth molecular orbital energy, α is the Coulomb integral of a sp^3 orbital in a carbon atom, and β is the resonance integral between two sp^3 orbitals in a C-C bond.

In the derivation of Eq. 3, we assume that α_x is equal to α, and the values of α and β in the activated complex are the same as those of the isolated molecule.

It may be worthy of notice that $D_r^{(R)}$ represents the stabilization energy due to σ type interaction in a radical reaction, while $S_r^{(R)}$ denotes the stabilization energy which

4) P. G. Wilkinson and H. L. Jhonston, *J. Chem. Phys.*, **18**, 190 (1950); W. P. Potts, ibid., **20**, 809 (1952).

5) J. Lennard-Jones and G. G. Hall, *Disc. Faraday Soc.*, **10**, 18 (1951); *Trans. Faraday Soc.*, **47**, 581 (1952); K. Fueki and K. Hirota, *J. Chem. Soc. Japan, Pure Chem. Sec.* (*Nippon Kagaku Zasshi*), **81**, 212 (1960).

6) D. C. Frost and C. A. McDowell, *Proc. Roy. Soc.*, **A241**, 194 (1957).

TABLE II (a). THE CALCULATED VALUES OF $D_r^{(R)}$ AND OBSERVED ACTIVATION ENERGIES IN SEVERAL *n*-PARAFFINS

$R\text{-}H + R'\cdot \rightarrow R\cdot + R'\text{-}H$

Compound R-H		$D_r^{(R)}$(H)	E(kcal./mol.)[8] $CH_3\cdot$	$CH_3CH_2\cdot$	$CH_3CH_2CH_2\cdot$
CH_3-H		0.9926	11.5	15.0	16.5
CH_3CH_2-H	(prim-H)	1.0029	10.0	14.8	16.2
$CH_3CH_2CH_2$-H	(prim-H)	1.0032	9.5	14.2	15.5
$CH_3CH_2CH_2CH_2$-H	(prim-H)	1.0043	9.3	13.9	15.3
$(CH_3)_2CHCH_2$-H	(prim-H)	1.0049	—	—	—
$(CH_3)_2CH$-H	(sec-H)	1.015	8.7	13.3	14.5
$CH_3CH_2CH(CH_3)$-H	(sec-H)	1.016	8.2	12.7	14.1
$(CH_3)_3C$-H	(tert-H)	1.027	7.7	11.6	13.4

TABLE II (b). THE AVERAGE VALUES OF $D_r^{(R)}$ (H) AND RATE CONSTANTS CHARACTERISTICS OF VARIOUS TYPES OF HYDROGEN ATOM IN ALKANES

Type of hydrogen atom	$D_r^{(R)}$(H)	$k_C \times 10^{-6}$ at 182°C (mol^{-1} cc. sec^{-1}*[7])
H in methane	0.993	—
Primary H	1.003	0.3
Secondary H	1.016	2.1
Tertiary H	1.027	15

* k_C means the rate constant divided by the number of H atoms of the same kind.

arises from the hyperconjugation taking place between a reagent and a conjugated molecule.

Results and Discussions

In our previous paper[1], the magnitudes of the frontier electron densities at hydrogen atoms in several saturated hydrocarbons ran parallel with the experimental reactivities[7] of hydrogen atoms in metathetical reactions. The frontier electron density for radical attack was shown to be a good intramolecular index of reactivity. This index, however, does not indicate the correct order of reactivity in different molecules. Therefore, we adopt $D_r^{(R)}$ which is useful not only as an intramolecular index but also as an intermolecular index. The calculated values of $D_r^{(R)}$ in several saturated compounds are listed together with the observed values of activation energy[8,9] in various metathetical reactions in Tables II and III.

Hydrogen Extraction Reaction.—In the case of reaction $RH + R'\cdot \rightarrow R\cdot + R'H$, results of calculation and experiments are cited in Table II (a) and (b), where R and R′ are two different alkyl groups. The values of $D_r^{(R)}$ at hydrogen atoms in several lower paraffins indicate a tendency that the larger the molecule RH is, the smaller is the magnitude of activation energy, this being in accord with experiments[8]. Furthermore, these values of $D_r^{(R)}$ agree with the activities of various kinds of hydrogen atoms estimated by Trotman-Dickenson[7] as shown in Table II (b).

In Table III (a), (b) and (c), the calculated values of $D_r^{(R)}$ at hydrogen atoms in various halomethanes are listed together with the observed activation energies[9]. In Table III (a), the radical reactions between methyl radical and chlorine-substituted methanes are indicated. The greater the number of chlorine atoms in methane is, the smaller is the activation energy, and also the magnitude of $D_r^{(R)}$ at hydrogen atoms increases with the number of chlorine atoms. The results on fluorine-substituted methanes are shown in Table III (b), and

TABLE III. THE VALUES OF $D_r^{(R)}$(H) AND ACTIVATION ENERGIES OF HALOMETHANES IN METATHETICAL REACTIONS

(a) $CH_nCl_{4-n} + CH_3\cdot \rightarrow CH_{n-1}Cl_{4-n}\cdot + CH_4$

Compound	$D^{(R)}$(H)	E(kcal./mol.)[9]
$CHCl_3$	2.181	5.8
CH_2Cl_2	1.237	7.2
CH_3Cl	1.058	9.4

(b) $CH_nF_{4-n} + CH_3\cdot \rightarrow CH_{n-1}F_{4-n}\cdot + CH_4$

Compound	$D_r^{(R)}$(H)	E(kcal./mol.)[8]
CH_3F	1.181	8.7
CH_2F_2	1.235	6.2

(c) $CH_nX_{4-n} + Br\cdot \rightarrow CH_{n-1}X_{4-n}\cdot + HBr$

Compound	$D_r^{(R)}$(H)	E(kcal./mol.)[9]
CH_4	0.993	17.8
CH_3Br	1.071	15.6
$CHCl_3$	2.181	10.0

7) A. F. Trotman-Dickenson, *Quart. Revs.*, 7, 198 (1953).

8) N. N. Tikhomirova and V. V. Voevodski, *Chem. Abstr.*, 45, 9940 (1951).

9) E. W. R. Steacie, "Atomic and Free Radical Reactions", Reinhold Co., New York. (1946); ibid., Second Ed. Vol. II (1954).

the case where a bromine atom attacks methane derivatives is indicated in Table III (c). Also in the last two cases, although experimental data are few, a parallelism between the values of $D_r^{(R)}$ and the activation energies[9] is clearly observed.

Halogen Abstraction Reaction.—In Table IV (a), (b) and (c) are illustrated both the calculated values of $D_r^{(R)}$ at the halogen atom in various halomethanes and the observed values of activation energy in the halogen extraction reactions[9]. In Table IV (a), a good agreement is observed between the values of $D_r^{(R)}$ and the activation energies of various monosubstituted halomethanes. In Table IV (b), the values of $D_r^{(R)}$ at chlorine atoms in various chlorine-substituted methanes and the activation energies in the hydrogen-atom reactions are indicated. In Table IV (c), the values of $D_r^{(R)}$ at bromine atoms in several bromomethanes are compared with the activation energies in the sodium-atom reaction.

TABLE IV. THE VALUES OF $D_r^{(R)}(X)$ AND ACTIVATION ENERGIES IN HALOMETHANES IN METATHETICAL REACTIONS

(a) $CH_3X+H\cdot \rightarrow CH_3\cdot+HX$

CH_3X	$D_r^{(R)}(X)$	E(kcal./mol.)[9]
CH_3I	1.851	5
CH_3Br	1.669	—
CH_3Cl	1.496	7~9
CH_3F	1.415	9

(b) $CH_nCl_{4-n}+H\cdot \rightarrow CH_nCl_{3-n}\cdot+HCl$

Compound	$D_r^{(R)}(Cl)$	E(kcal./mol.)[9]
CH_3Cl	1.496	7~9
CH_2Cl_2	1.756	6
$CHCl_3$	3.200	4.3
CCl_4	3.829	3.5

(c) $CH_nBr_{4-n}+Na\cdot \rightarrow CH_nBr_{3-n}\cdot+NaBr$

Compound	$D_r^{(R)}(Br)$	E(kcal./mol.)[9]
CH_3Br	1.669	5.9
CH_2Br_2	2.045	2.6
$CHBr_3$	4.889	0.3

In spite of the fact that this is the first attempt of the quantum-mechanical investigation of the reactivity of saturated compounds, the agreement of the theory with experience is almost satisfactory, and hence the new quantum-mechanical quantity $D_r^{(R)}$ can be a useful index for predicting the reactivity of saturated compounds, equally with S_r in conjugated molecules. There remain, however, some points to be checked in future. These are summarized as follows:

(1) We assume that α_x is equal to α to derive D_r. Meanwhile, a slight change in activation energy is experimentally observed according to the species of attacking reagents. Hence, if the values of α_x would be modified according to the change of the attacking reagent, this effect can be interpreted in the present procedure.

(2) In the present treatment, we do not take into account the stabilization energy due to the resonance in an attacking alkyl radical or the radical formed. It has been made clear that this resonance energy has a remarkable influence on the magnitude of the activation energy[10]. In this connection, the present treatment should be modified so as to involve the contribution of this kind of resonance to the activation energy.

(3) A disagreement between the values of $D_r^{(R)}$ in methyl halides and the experimental activation energies is observed. This is probaly because the energies of the vacant orbitals of some methyl halides are nearly equal to α. Thus, the values of the $D_r^{(R)}$ become extremely large. If carefully selected values of parameters for halogen atoms should be adopted, these disagreements may be avoidable.

Faculty of Engineering
Kyoto University
Sakyo-ku, Kyoto

10) M. G. Evans, *Disc. Faraday Soc.*, **10**, 1 (1951); C. A. Coulson, ibid., **2**, 9 (1947).

[Reprinted from the Bulletin of the Chemical Society of Japan, Vol. 34, No. 8 (1961)]

An Attempt to Discuss Reactivities of Excited Molecules by the Molecular Orbital Method

By Kenichi FUKUI, Keiji MOROKUMA and Teijiro YONEZAWA

(Received December 29, 1960)

In order to elucidate the mechanisms of reactions of conjugated or non-conjugated organic compounds which are in their ground state, a number of quantum mechanical, especially molecular orbital, studies have been performed and have achieved remarkable success in predicting theoretically their chemical reactivities[1-4]. In connection with the recent advance of techniques of photochemical investigation, chemical reactivities of electronically excited molecules have aroused our interest.

Because of the complicated characters and the multifariousness of reactions of excited molecules, only a few theoretical discussions have ever been made. As regards the problem of their self-dissociation, some qualitative discussions in terms of the potential energy curve have prevailed[5]. For discussing the energy transfer reaction between an excited molecule and an unexcited one, the collision theory has prevailed as a powerful tool[6]. On account of the lack in systematic experimental data and, in addition, because of the difficulties in the theoretical treatment, only a few quantum chemical studies have been presented on excited molecules involved in the addition reactions of large organic molecules. Photopolymerization of hydrocarbons has been discussed in terms of free valence of the molecules in their lowest excited state. The intramolecular orientation in the reaction of anthracene or naphthacene has been explained well[7]. But that of acenaphthylene has not been[8]; free valence of its lowest excited states is greatest at positions 5 and 6, and, on the contrary, the photodimer of the componud is produced through positions 1 and 2.

In the present paper, as was previously done with reference to the molecules in their ground state by the present authors[9,10] and by Brown[11], the reactivites of excited molecules are assumed to be parallel with the π stabilization energy at the transition state due to the interaction of the molecule with the reagent (molecule, ion or radical) through the position of attack. Under this assumption the reactivity indexes are derived by the use of the molecular orbital method.

1) G. W. Wheland, *J. Am. Chem. Soc.*, **64**, 900 (1942).
2) C. A. Coulson and H. C. Longuet-Higgins, *Proc. Roy. Soc.*, **A191**, 39 (1947).
3) F. H. Burkitt, C. A. Coulson and H. C. Longuet-Higgins, *Trans. Faraday Soc.*, **47**, 553 (1951).
4) K. Fukui, T. Yonezawa and C. Nagata, This Bulletin, 27, 423 (1954); *J. Chem. Phys.*, **27**, 1247 (1957),
5) K. Yagi, Z. Yoshida and T. Tabata, "Fluorescence", Nanko-do, Tokyo (1958); and references cited there.
6) M. Shimizu, "Symposium on Molecular Structure and Chemical Reaction", held in Nov. 1960, Tokyo, p. 149; and references cited there.
7) C. Sandorfy, *Compt. rend.*, **230**, 961 (1950).
8) V. A. Crawford and C. A. Coulson, *J. Chem. Soc.*, **1948**, 1990.
9) K. Fukui, K. Morokuma, T. Yonezawa and C. Nagata, *J. Chem. Phys.*, **32**, 1743 (1960).
10) T. Yonezawa, K. Hayashi, C. Nagata, S. Okamura and K. Fukui, *J. Polymer Sci.*, **14**, 312 (1954).
11) R. D. Brown, *J. Chem. Soc.*, **1959**, 2224.

Formulation and Comparison with Experiments

Reactions of an excited molecule with an unexcited molecule are classified into several cases from the point of view of the spin multiplicity of reactants. The assumptions, the formulations and general considerations of the present molecular orbital treatments leading to the theoretical measures of reactivity will be described rather circumstantially in the first few cases, and in other cases only the indexes of reactivity derived will be given.

Reaction of an Excited Triplet Molecule with an Unexcited Radical.—First of all, attention is paid to the reaction of an excited triplet molecule (^{3}A) with an unexcited radical (^{2}B), where the numbers on the left shoulder refer to the multiplicity of the molecules. Here it would be rational for practical purposes in treating photochemical reactions to assume that the molecule ^{3}A has only one excited electron. Then the wave functions of these molecules are written as follows:

$$\Psi(^3\mathrm{A}) = A\left[\cdots(\psi_1\alpha)^{m-3}(\psi_1\beta)^{m-2}(\psi_{01})^{m-1}(\psi_{02})^{m} \times \begin{Bmatrix} \alpha(m-1)\alpha(m) \\ \beta(m-1)\beta(m) \\ (1/\sqrt{2})\{\alpha(m-1)\beta(m)+\beta(m-1)\alpha(m)\} \end{Bmatrix}\right] \tag{1}$$

$$\Psi(^2\mathrm{B}) = A\left[\cdots(\phi_1\alpha)^{n-2}(\phi_1\beta)^{n-1}\begin{Bmatrix}(\phi_0\alpha)^n \\ (\phi_0\beta)^n\end{Bmatrix}\right]$$

where ψ_i and ϕ_j are the molecular orbitals of molecules A and B, respectively, whose numbering is schematically shown in Fig. 1, 01 and 02 denoting the lower and the higher half-occupied levels, of A, respectively, and (0) denoting the half-occupied level of the radical B; m, an even integer, is the number of electrons of the molecule A and n, an odd number, is that of the radical B; $\boldsymbol{A}$ is the antisymmetrizing operator including the normalization constant; and each line in braces may constitute one wave function.

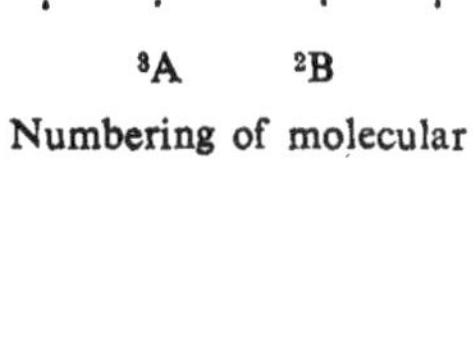

Fig. 1. Numbering of molecular orbitals.

The states of the total system consisting of the triplet A and the doublet B would be a quartet and a doublet. But the quartet, which would be proved later to be less significant in this discussion, is abandoned. The doublet wave function for $M_S=1/2$* is

$$^2\Psi_{1/2}[^2(\mathrm{AB})] \equiv \Psi[0] = \boldsymbol{A}[\cdots(\psi_1\alpha)^{m-3}(\psi_1\beta)^{m-2} \cdots(\phi_1\alpha)^{m+n-2}(\phi_1\beta)^{m+n-1}(\psi_{01})^{m-1}(\psi_{02})^{m} \times(\phi_0)^{m+n}\{(\alpha\beta+\beta\alpha)\alpha-2(\alpha\alpha)\beta\}/\sqrt{6}] \tag{2}$$

where, for example, $\alpha\beta\alpha$ means $\alpha(m-1)\beta(m)\alpha(m+n)$. That for $M_S=-1/2$ is not written because the energetical argument is similar. Thus the state $\Psi[0]$ is assumed to correspond to the initial system of the reaction in question. If A and B are far separated from each other, the energy of the state is the same as that of the product function $\Psi(^3\mathrm{A})\cdot\Psi(^2\mathrm{B})$. The molecular orbitals may include all π ones and σ ones, but in this paper we consider the reactions in company with only the π electron excitation, and hence σ orbitals will rationally be omitted as making an invariant framework.

At the transition state of the reaction the two molecules, ^{3}A and ^{2}B, will stay at a distance, making a weak interaction through the atoms at which the reaction takes place. This interaction is assumed to be a kind of $\pi-\pi$ conjugation and in this simple LCAO MO approach it will be represented by a small resonance integral $\gamma\beta$ between the atomic π orbitals r and s of the reactants, A and B, respectively, where β is the resonance integral of the C-C π bond in benzene. This perturbation makes the initial state $\Psi[0]$ come to the direct or indirect interaction with various electronic configurations of the combined system. Of these configurations only those, which have energy matrix elements with the initial state in scope of the simple LCAO MO approximation, are taken into account. This means that the configurations (of course, doublet and $M_S=1/2$) in which one electron transfers from ^{3}A to ^{2}B and vice versa have concern with the transition state of the reaction, whereas those which correspond to intramolecular one-electron transitions do not.

Under these assumptions the electron configurations to be taken into account are those corresponding to the following one-electron transfers, represented by $l\rightarrow m$, from an occupied or half-occupied orbital, l, to an unoccupied or half-occupied orbital, m:

$$\begin{aligned} &02\rightarrow(0),\ 02\rightarrow(j'),\ 01\rightarrow(0),\ 01\rightarrow(j') \\ &i\rightarrow(0),\ i\rightarrow(j'),\ (0)\rightarrow01,\ (0)\rightarrow02 \\ &(0)\rightarrow i',\ (j)\rightarrow01,\ (j)\rightarrow02,\ (j)\rightarrow i' \end{aligned} \tag{3}$$

* M_S denotes the eigenvalue of S_z, the z component of the spin momentum operator.

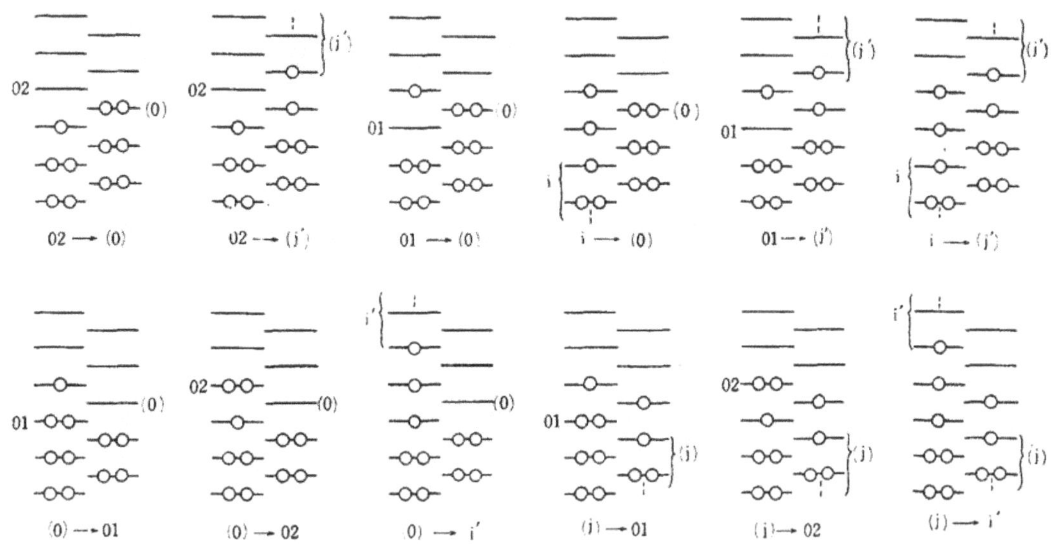

Fig. 2. Electronic configurations taken into account.

where

$i=1, 2, \cdots$	(occupied orbitals of ^{3}A)
$i'=1', 2' \cdots$	(unoccupied orbitals of ^{3}A)
$(j)=(1), (2), \cdots$	(occupied orbitals of ^{2}B)
$(j')=(1'), (2'), \cdots$	(unoccupied orbitals of ^{2}B)

They are schematically illustrated in Fig. 2. For an illustration, the configuration associated with the transfer 02→(0), will have the following wave function:

$$\Psi[02\to(0)] = \boldsymbol{A}[\cdots(\phi_0\alpha)^{m+n-2}(\phi_0\beta)^{m+n-1}(\phi'_0\alpha)^{m+n}] \tag{4}$$

The energy matrix element between these configurations and the initial state can be expressed in terms of the integral $\gamma\beta$. For example,

$$\left|\int \Psi^*[02\to(0)]H\Psi(0)d\tau\right| = |c_r^{02}d_s^0\gamma\beta| \tag{5}$$

where c_r^k and d_s^l are the coefficient of the atomic orbitals r and s in the molecular orbital k and l of the molecules A and B, respectively.

The transition state of the reaction is the lowest energy state resulting from the interaction mentioned above. The difference in π electronic energy, i. e. the stabilization energy, between the initial state and the transition state, was assumed in the preceding section to parallel the activation energy, of the reaction, and accordingly the reactivity of the excited molecule. This stabilization energy would be expanded in powers of γ by regarding it to be small. There arise three main subcases according to the mutual relation of the energy levels 01 and 02 of the molecule ^{3}A and (0) of the molecule ^{2}B*. In the following $Y_{k,l}$ denotes the energy in units of $(-\beta)$ turned out in one electron transfer from the molecular orbital k of the one molecule to the molecular orbital l of the other one. That is, $Y_{k,l}$ is positive if the level k is higher than the level l and vice veras. Thus we reach the final reactivity indexes of excited molecules.

(a) If $Y_{02,0}>Y_{0,01}>0$, that is, if the half-occupied level of the molecule ^{2}B lies between the levels 01 and 02 of ^{3}A and is nearer to 01 than to 02, as is shown in Fig. 3 (a), the stabilization energy to the second order of γ is written in the following expression**:

$$\Delta E = \left[Y_{02,0} + \frac{(3/2)(c_r^{02})^2(d_s^0)^2}{Y_{02,0}}\gamma^2\right](-\beta) \tag{6}$$

From this formula some general conclusions on reactivities of excited molecules may be

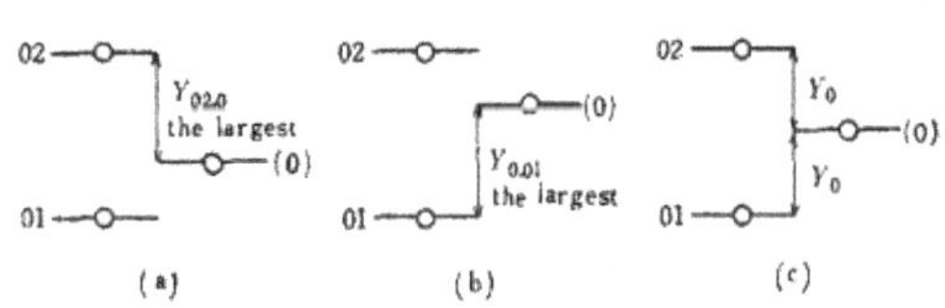

Fig. 3. Mutual relation of molecular levels.

* Some rare cases, for instance, $Y_{0,02}>0$ or $Y_{01,0}>0$, are not included for the sake of brevity.

** In general the stabilization energies for each M_S component may differ from each other. In the surroundings of reaction no magnetic field being applied, no specialization of the component is possible and the mean of all components would be significant as a real stabilization energy. In this case the energy for $M_S=1/2$ is the same as that for $M_S=1/2$; also the mean energy becomes the same.

TABLE I. RATE CONSTANT AND π STABILIZATION ENERGY OF POLYMERIZATION OF STYRENE

Reaction[a]	Rate constant[b] k (l. mol^{-1} sec^{-1})	π Stabilization energy (in units of $-\beta$)
$R\cdot + {}^3A \rightarrow RA\cdot$	$\sim 10^{11}$	$0.414 + 0.801\,\gamma^2$ (position 9)
		$0.414 + 0.400\,\gamma^2$ (position 1)
		$0.414 + 0.200\,\gamma^2$ (position 2)
$R\cdot + A \rightarrow RA\cdot$	3.4×10^2	$0.926\,\gamma^2$
$R\cdot + M \rightarrow R\cdot$	1.2×10^2	$0.932\,\gamma^2$
$RA\cdot + M \rightarrow R\cdot$	0.8	$0.739\,\gamma^2$

a) Abbreviations, R·: polymer radical of styrene; ^{3}A: triplet anthracene; A: anthracene; M: styrene.

b) Taken from Ref. 10.

derived. In most reactions between unexcited molecules, terms first appear in the second order of γ, whereas in Eq. 6 the term in zero order exists. The appearance of the zero order term is universal in all the following cases and subcases, making clear the reason why the reactions including excited molecules proceed so fast. The difference in reactivity of several excited molecules would thus dominantly be decided by this term. The intramolecular orientation, on the other hand, depends on the term in the second order. More minutely, so far as attention is paid to the lowest excitation of the molecule A as is expected in real photochemical reactions of conjugated molecules, the atom where $(c_r^{02})^2$, the frontier electron density for a nucleophilic attack in the ground state, is the greatest is the most reactive in an excited molecule. The coefficient (3/2) is characteristic to the triplet A molecule. If the molecule A is excited in the singlet state it amounts to (1/2), as will be mentioned in the next section. The difference would not necessarily mean the difference in reactivity of these two states, since the approximation is too simple to discuss at this point, on which there will be some discussion in a later section.

To the transition state the electron configuration corresponding to the one-electron transfer $02 \rightarrow (0)$ has the greatest contribution, since the other configurations have the first contribution to ΔE in the fourth order. Therefore, the direction of charge transfer is from molecule A to molecule B.

(b) If $Y_{0,01} > Y_{02,0} > 0$, i.e., if the half-occupied level of ^{2}B lies between the levels 02 and 01, and nearer to 02 than to 01, as shown in Fig. 3 (b), then

$$\Delta E = \left[Y_{0,01} + \frac{(3/2)(c_r^{01})^2 (d_s^{0})^2}{Y_{0,01}} \gamma^2 \right] (-\beta) \quad (7)$$

In this subcase also the rate of reaction is great and the intramolecular orientation of the excited molecule is controlled by $(c_r^{01})^2$, the frontier electron density for an electrophilic attack in the ground state. The most contributive configuration in this subcase corresponds to the transition $(0) \rightarrow 01$, resulting in a charge transfer from B to A.

(c) If $Y_0 \equiv Y_{02,0} = Y_{0,01} > 0$, i.e., if the level (0) is just in the midst of the levels 01 and 02, as shown in Fig. 3 (c), then

$$\Delta E = \left[Y_0 + \frac{(3/2)\left[(c_r^{01})^2 + (c_r^{02})^2\right](d_s^{0})^2}{Y_0} \gamma^2 \right] \times (-\beta) \quad (8)$$

A large rate constant is confirmed also in this subcase, and $[(c_r^{01})^2 + (c_r^{02})^2]$, the frontier electron density for a radical attack in the ground state, comes to the point.

So far as alternant hydrocarbons are concerned, however, the equality $(c_r^{01})^2 = (c_r^{02})^2$ holds in the simple LCAO MO approximation, so that in all subcases and the following cases the most reactive position in a molecule in its lowest excited state is the same as that in the ground state. This explains the great reactivity of meso position of anthracene. Further experimental studies are anticipated with respect to the intramolecular orientation in nonalternant hydrocarbons and heterocyclic compounds where the equality mentioned above no more exists.

Here a comparison with experimental data may be given. Excited triplet anthracene was reported by Norrish et al.[12] to enter into the chain of polymerization of styrene in the following way:

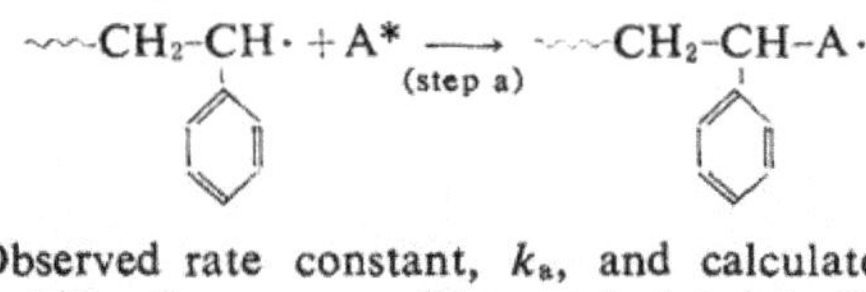

(9)

Observed rate constant, k_a, and calculated π stabilization energy ΔE are tabulated in Table I. The stabilization in the reaction a is calculated by using Eq. 8. Those of other reactions in which only molecules in their ground state participate are given by the usual formula for stabilization energy[10]:

12) V. S. Anderson and R. G. W. Norrish, *Proc. Roy. Soc.*, **A251**, 1 (1959); R. G. W. Norrish and J. P. Simons, ibid., **A251**, 4 (1959).

$$\Delta E=\left[\sum_{j}^{\text{unocc}}\sum_{i}^{\text{all}}\nu_i-\sum_{j}^{\text{occ}}\sum_{i}^{\text{all}}(2-\nu_i)\right]\times\frac{(c_r{}^i)^2(d_s{}^j)^2}{Y_{ji}}\gamma^2(-\beta) \tag{10}$$

where ν_i is the number of electrons in the ith molecular orbital of A. The extraordinary great magnitude of rate constant of reaction a is well understood in connection with the presence of the zero order term. Further, the position 9 is the most active, at which anthracene would enter into the polymer chain. An appropriate choice of the value of γ ($=0.7$) makes the $\log k_a-\Delta E$ plots practically linear.

The wave functions (Eq. 1) of ^{3}A and ^{2}B may be combined so as to construct a quartet instead of the doublet (Eq. 2). The quartet state has matrix elements with some configurations in Eq. 3, only to give the stabilization energy in the second order. Because of the lack of the zero order terms and, accordingly, because of the instability of the transition state, reactions would not be considered to proceed by way of the quartet transition state.

Reaction of an Excited Singlet Molecule with an Unexcited Radical.—The reaction of an excited singlet molecule (^{1}A) with an unexcited radical (^{2}B) can be discussed in a way similar to that of a triplet molecule (^{3}A) mentioned in the preceding reaction. The doublet wave function for $M_S=1/2$ of the combined system is, in place of Eq. 2 in the triplet case,

$$\Psi[0]=\boldsymbol{A}[\cdots(\psi_{01})^{m-1}(\psi_{02})^m(\psi_0)^{m+n}\times\{(\alpha\beta-\beta\alpha)\alpha\}/\sqrt{2}] \tag{11}$$

Also, we are led to the same equations as Eqs. 3, 4 and 5, excepting that the numerical coefficient in the second term should be read (1/2) in this case, instead of (3/2). The meanings of the frontier electron density and the direction of charge transfer in each subcase are entirely the same as in the last case; and therefore discussions may be omitted for brevity.

Reaction of an Excited Tripled Molecule with an Unexcited Molecule.—The reaction of an excited triplet molecule (^{3}A) with an unexcited singlet molecule (^{1}B)* now comes to our attention. Discussions are almost alike, so that the only important formulas would be summarized in this case and in the following case. The numbering of molecular orbitals is illustrated in Fig. 4. The states of the combined system at the initial stage of reaction are represented by a set of three triplet wave functions, one with $M_S=1$, one with $M_S=0$ and the other with $M_S=-1$, as follows:

* The singlet molecule ^{1}B may be an ion whose spin is singlet.

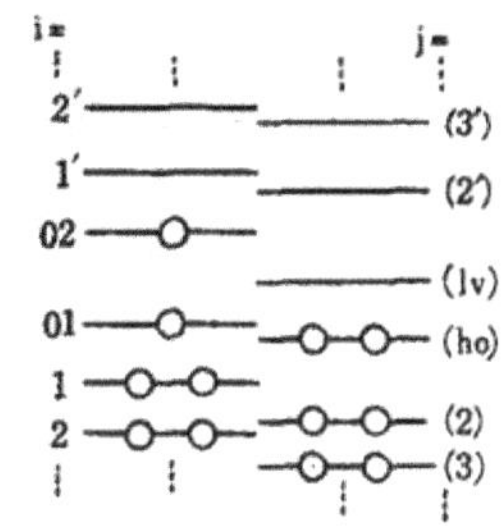

Fig. 4. Numbering of molecular orbitals. lv and ho mean the lowest unoccupied and the highest occupied orbitals, respectively.

$$\begin{aligned}
{}^3\Psi_1&=\boldsymbol{A}[\cdots(\psi_{01})^{m+n-1}(\psi_{02})^{m+n}\\
&\quad\times\alpha(m+n-1)\alpha(m+n)]\\
{}^3\Psi_0&=\boldsymbol{A}[\cdots(\psi_{01})^{m+n-1}(\psi_{02})^{m+n}\\
&\quad\times(\alpha(m+n-1)\beta(m+n)\\
&\quad+\beta(m+n-1)\alpha(m+n))/\sqrt{2}]\\
{}^3\Psi_{-1}&=\boldsymbol{A}[\cdots(\psi_{01})^{m+n-1}(\psi_{02})^{m+n}\\
&\quad\times\beta(m+n-1)\beta(m+n)]
\end{aligned} \tag{12}$$

Electronic configurations which are taken into account in the perturbation calculation are summarized in Fig. 5 with illustrations. Each M_S component of the functions 12 will have matrix elements with the same M_S component of the wave functions for the configurations

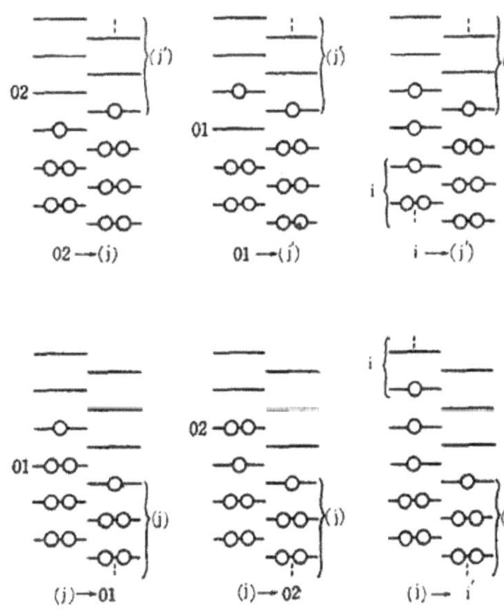

Fig. 5. Electronic configurations taken into account.
$i=1,2,\cdots$ (occupied orbitals of ^{3}A); $i'=1',2',\cdots$ (unoccupied orbitals of ^{3}A); $j=$ ho, 2, $\cdots$ (occupied orbitals of ^{1}B); and $j'=$ lv, $2',\cdots$ (unoccupied orbitals of ^{1}B).

TABLE II. π STABILIZATION ENERGY IN PHOTODIMERIZATION OF POLYACENES AND ACENAPHTHYLENE

Compound	Pairs of positions to react with	π Stabilization energy (in units of $-\beta$)
Anthracene	9-9, 10-10 (central)	0.547 γ
	1-1, 4-4 (end)	0.273 γ
Naphthacene	5-5, 12-12 (central)	0.417 γ
	1-1, 4-4 (end)	0.159 γ
Pentacene	6-6, 13-13 (central)	0.400 γ
	5-5, 14-14	0.299 γ
	1-1, 4-4 (end)	0.100 γ
Acenaphthylene	1-1, 2-2	0.568 γ
	5-5, 6-6	0.375 γ

in Fig. 5. Then the stabilization energy* is obtained as follows, according to the mutual relation of energy levels:

(a) If $Y_{02,\mathrm{lv}}$ is positive and the most positive of all Y's, then

$$\varDelta E=\left[Y_{02,\mathrm{lv}}+\frac{(c_r^{02})^2(d_s^{\mathrm{lv}})^2\gamma^2}{Y_{02,\mathrm{lv}}}\right](-\beta) \qquad (13)$$

This is the case for most electrophilic reagents (cations) serving as ^{1}B.

(b) If $Y_{\mathrm{ho},01}$ is positive and the most positive of all Y's, then

$$\varDelta E=\left[Y_{\mathrm{ho},01}+\frac{(c_r^{01})^2(d_s^{\mathrm{ho}})^2\gamma^2}{Y_{\mathrm{ho},01}}\right](-\beta) \qquad (14)$$

Most nucleophilic reagents (anions) as ^{1}B are included in this subcase.

(c) If $Y_{02,\mathrm{lv}}=Y_{\mathrm{ho},01}$, and they are positive (then of course the most positive), then, putting them equal to Y_0,

$$\varDelta E=\left[Y_0+\frac{\{(c_r^{02})^2(d_s^{\mathrm{lv}})^2+(c_r^{01})^2(d_s^{\mathrm{ho}})^2\}\gamma^2}{Y_0}\right]\times(-\beta) \qquad (15)$$

Such a conjugated molecule as has its lowest excitation energy smaller than the excitation energy to the triplet of the molecule A would be classified in this subcase.

(d) If all Y's are negative, then

$$\varDelta E=\left[\sum_j^{\mathrm{unocc}}\sum_i^{\mathrm{all}}\nu_i-\sum_j^{\mathrm{occ}}\sum_i^{\mathrm{all}}(2-\nu_i)\right]\frac{(c_r^i)^2(d_s^j)^2\gamma^2}{Y_{ji}}\times(-\beta) \qquad (16)$$

In this subcase the stabilization energy begins in the second order terms, predicting a slow rate in the process of reaction.

(e) If $Y_{02,\mathrm{lv}}=0$ and the others are negative, then

$$\varDelta E=\left[\,|\,(c_r^{02})(d_s^{\mathrm{lv}})\,|\,\gamma+\left\{\sum_j^{\mathrm{unocc}}\sum_i^{\mathrm{all}}\nu_i-\sum_j^{\mathrm{occ}}\sum_i^{\mathrm{all}}(2-\nu_i)\right\}'\frac{(c_r^i)^2(d_s^j)^2\gamma^2}{2Y_{ji}}\right](-\beta) \qquad (17)$$

where the mark $'$ on the summation tells to omit the set ($j=\mathrm{lv}$, $i=02$).

(f) If $Y_{\mathrm{ho},01}=0$ and others are negative, then, similarly

$$\varDelta E=\left[\,|\,(c_r^{01})(d_s^{\mathrm{ho}})\,|\,\gamma+\left\{\sum_j^{\mathrm{unocc}}\sum_i^{\mathrm{all}}\nu_i-\sum_j^{\mathrm{occ}}\sum_i^{\mathrm{all}}(2-\nu_i)\right\}'\frac{(c_r^i)^2(d_s^j)^2\gamma^2}{2Y_{ji}}\right] \qquad (18)$$

$'$ indicating the neglect of the set ($j=\mathrm{ho}$, $i=01$).

(g) If $Y_{02,\mathrm{lv}}=Y_{\mathrm{ho},01}=0$ and others are negative, we obtain

$$\varDelta E=\left[\sqrt{(c_r^{02})^2(d_s^{\mathrm{lv}})^2+(c_r^{01})^2(d_s^{\mathrm{ho}})^2}\,\gamma+\left\{\sum_j^{\mathrm{unocc}}\sum_i^{\mathrm{all}}\nu_i-\sum_j^{\mathrm{occ}}\sum_i^{\mathrm{all}}(2-\nu_i)\right\}'\times\frac{(c_r^i)^2(d_s^j)^2\gamma^2}{2Y_{ji}}\right](-\beta) \qquad (19)$$

neglecting in the summation the sets ($j=\mathrm{ho}$, $i=01$) and ($j=\mathrm{lv}$, $i=02$) In the latter three subcases the stabilization is of the first order, and for the lowest excited molecule the frontier electron density controls a main part of the reactivity indexes. The subcase g includes reactions of an excited molecule with an unexcited molecule of the same kind.

The photodimerization of aromatic hydrocarbons has been considered to pass through the following mechanism:

$$\mathrm{A}^*+\mathrm{A}\xrightarrow[\text{(step b)}]{}(\mathrm{AA}^*)\xrightarrow[\text{(step c)}]{}\mathrm{AA} \qquad (20)$$

Though the detail of the intermediate (AA*) is not clear, assuming the present type of complex for the transition state of step b, one can discuss the reactivity of this reaction as belonging to the subcase g. In Table II calculated values of the π stabilization energy are collected. In that table, reagents are assumed to make two weak bondings through two meso positions, as illustrated in Fig. 6; the assumption of one bonding would only need to divide the values by the factor 2.

* Each M_S component gives the same stabilization energy, and the mean is, therefore, the same.

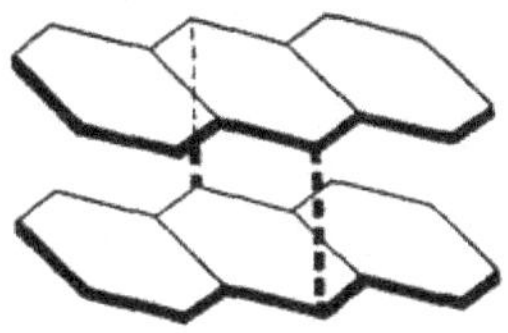

Fig. 6. Model of transition state of photodimerization of anthracene.

Anthracene is well known to make a photodimer at 9 and 10 positions, in accord with our calculation. Also in other polyacenes the most central meso-positions have the largest stabilization energy. This result is compatible with the experiment[13]. Bowen and Marsh[14] studied the photodimerization of acenaphthylene, finding that the structure of the photodimer was

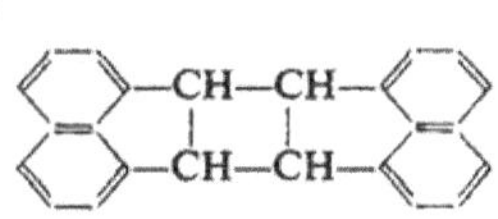

This finding was first theoretically discussed by Crawford and Coulson[8]. They calculated the free valence of this molecule in its lowest excited state and found that the values of positions 5 and 6 (0.62) are greater than those of positions 1 and 2 (0.57), which conflicted with the above experimental results. Thus they stated that the free valence is not a sufficient guide to reactivity. Recently this experiment has been interpreted successfully by the present authors in terms of the frontier electron density for a radical reaction[15]. The frontier electron density of positions 1 and 2 is greater than that of positions 5 and 6. In that paper the excited acenaphthylene was regarded as an ordinary radical reagent attacking an acenaphthylene molecule in the ground state. At this time it would be interesting to apply the present theory which involves an explicit consideration of the excitation. This belongs to the subcase g. As is seen in Table II the stabilization energy of the reaction through pairs of positions 1-1 and 2-2 is greater than that through 5-5 and 6-6, in accordance with the experiment.

In intermolecular comparison it is said that the ease of photodimerization is greater in pentacene than in anthracene[13]. But this is not a direct comparison of the rate constant k_b of the step b but a rough estimation from the amount of dimers produced. The total rate of dimer formation would be affected not only by the rate constant k_b but also by various competitive reactions such as the collisional or spontaneous degradation of the excited A molecule, dissociation of (AA*), and so forth. To measure and compare the rate constant with calculated stabilization energy would be interesting.

Reaction of an Excited Singlet Molecule with an Unexcited Molecule.—The method and the results of the treatment for this kind of reaction is exactly the same as that in the last case if only the wave function, Eq. 12, is replaced by the following equation:

$$ {}^1\Psi_0=\boldsymbol{A}[\cdots(\psi_{01})^{m+n-1}(\psi_{02})^{m+n}\{\alpha(m+n-1)\times\beta(m+n)-\beta(m+n-1)\alpha(m+n)\}/\sqrt{2}] \quad (21) $$

Eqs. 13 to 20 are also valid in this case.

Reaction of an Excited Molecule with Another Excited Molecule.—We derived also the equations giving the stabilization energy in a reaction of an excited molecule (singlet or triplet) with another excited molecule (singlet or triplet) in a similar way. But as they are less meaningful for practical purposes and rather intricate, we do not give their explicit forms. It would be noteworthy that in this case also the frontier electron densities decide the intramolecular orientation.

Discussion

We have derived formulas giving the π stabilization energy and the reactivity index, discussing the rate of reactions of excited molecules. But we do not, at all, regard this theory as complete but merely as a simple pioneering attempt in the molecular orbital method; the assumptions included may have to be subjected to some improvements in future. For discussing the difference in rate constant caused by the difference in multiplicity of states, interactions of electrons should be explicitly taken into account. For discussing various dissociation reactions, attention would have to be paid to σ electrons in the bond to be broken. Furthermore, in order to understand all the figures of reactions of an excited molecule, discussions on relations between various elementary steps involved in the reaction would be expected.

Faculty of Engineering
Kyoto University
Sakyo-ku, Kyoto

13) For example, M. Nakagawa, "Gendai Kagaku", III-D, Iwanami, Tokyo (1956); M. Koizumi and M. Shiba, ibid., I-E, (1956).

14) E. L. Bowen and J. D. F. Marsh, *J. Chem. Soc.*, **1947**, 109.

15) K. Fukui, T. Yonezawa and C. Nagata, This Bulletin, **34**, 37 (1961).

Reprinted from CANCER RESEARCH Vol. 15, No. 4, pp. 233–39, May 1955

Electronic Structure and Carcinogenic Activity of Aromatic Compounds

I. Condensed Aromatic Hydrocarbons*

CHIKAYOSHI NAGATA, KENICHI FUKUI, TEIJIRO YONEZAWA, AND YUSAKU TAGASHIRA

(Departments of Fuel Chemistry and of Pathology, Kyoto University, Kyoto, Japan)

Numerous attempts have been made from many points of view, both chemical and physical, to account for the differences in carcinogenic activity of various aromatic compounds (6, 7) and, thereby to explore the riddle of carcinogenesis due to these substances. The experimental approaches included investigations of chemical constitution (26), fluorescence (5), phosphorescence (20), absorption spectra (16, 17), etc., but no satisfactory explanation has ever been given for the exact nature of carcinogenic compounds.

A large majority of the important carcinogens belong to the aromatic hydrocarbons, aromatic heterocyclics, and aromatic amines, which are characterized by having a certain number of mobile electrons, so-called π-electrons. O. Schmidt (27) first attempted to relate the carcinogenicity of aromatic hydrocarbons with the distribution of π-electrons in these compounds, and thereafter many physicists and theoretical chemists such as A. and B. Pullman (22, 25), Daudel (9), Berthier and Coulson (2), Greenwood (14), and Dewar (10) have endeavored to explain carcinogenicity from the nature of π-electrons. Among these works the Pullmans' "K-region theory" (8, 25) appears to be a basis for further study.

The Pullmans carried out the calculations of the π-electron distribution in molecules and developed their theory by means of a "molecular diagram method." However, this method was not applicable to large molecules on account of the complexities of the calculations. The calculation is too laborious to make in the case of polycondensed aromatics, i.e., for the molecules with five or more condensed benzene rings, to which group many potent carcinogens belong. Thus, by the molecular diagram method they could give only qualitative information as to these compounds.[1] Moreover, as Boyland (3) and Kooyman and Heringua (18) pointed out, by considering the K-region alone it seems difficult to explain the mechanism of carcinogenesis. In spite of all the quantum-mechanical approaches attempted, therefore, it has not yet been possible to construct a complete theory of carcinogenicity by a consideration of the π-electron system.

Some of the present authors (12, 13) have previously discovered the significance of what we call *frontier electrons* in chemical reactions of π-electron systems and have succeeded in explaining the experimental results satisfactorily by means of calculating the frontier electron density. Furthermore, it has been found that a distinct parallelism exists between carcinogenicity and the frontier electron distributions. In the present paper the relation between the calculated frontier electron distributions and carcinogenicities of nonsubstituted aromatic hydrocarbons is reported, and at the same time some discussions are given about the activity of carcinogens. The results on substituted aromatic hydrocarbons, aromatic heterocyclics, and aromatic amines will be reported in the near future.

OUTLINE OF THEORETICAL TREATMENTS

All the compounds which the present paper treats belong to the π-electron system. It is well known that π-electrons are very mobile and would be considerably influenced by outer disturbances. Consequently, absorption spectra, magnetic anisotropy, chemical reactivity, and many other physical and chemical properties which are charac-

* This work was aided by a grant from the Ministry of Education, Japanese Government.

Received for publication August 11, 1954.

[1] Under these circumstances the Pullmans used in their recent papers (23, 24) the localization method, of which mention is made in the next section.

teristic of the π-electron system are ascribed to π-electrons.

Two methods exist in the quantum-mechanical treatment of π-electron systems. One is the "valence bond (V.B.) method" and the other is the "molecular orbital (M.O.) method." Because of the excessive tediousness of the exact V.B. treatment for large molecules, several approximate methods have been developed. They involve, however, a considerable amount of arbitrariness which makes the results doubtful. On the other hand, the M.O. method has proved to be more readily applicable to large aromatic molecules such as the carcinogenic compounds. Under these circumstances most of the chemical and physical treatments of π-electron systems are based on the M.O. method.

The discussion of chemical reactivity of the π-electron system has been made through two different ways of approach in the M.O. theory. The first is called "localization method" of Wheland (31). In this method, it is assumed that the rate of reaction at a certain position in a molecule is determined by the "localization energy" which is required in localizing a requisite number of π-electrons at that position. The second is called the "static method."[2] In this method the chemical reactivity at a position in a molecule is correlated with the density of all the π-electrons at that position in the isolated molecule.

The fundamental supposition in the localization method is that a parallelism would exist between the localization energy and the chemical reactivity. On the other hand, the basis of the static method may be valid when the interaction between the molecule and the reagent is very small. In both of the two methods mentioned above, the chemical feature of the process of reaction is not directly considered. Namely, in these methods the mechanism of formation and dissociation of σ-bonds is not taken into consideration. Accordingly, these two methods are not considered to be sufficiently satisfactory to describe the true process of chemical reaction.

Frontier electron method.—Some of the present authors introduced the frontier electron method as a third one of molecular orbital treatments of chemical reactivity (11–13). In that theory the electronic interaction between the molecule and the reagent has been divided into σ- and π-parts. When certain relations hold between some integrals of σ-electrons, the σ-part of the energy of the system has a maximum in the process of reaction, by which the transition state is defined. In the transition state a hyperconjugation may take place between the initial conjugated system and the quasi-π-orbital which may appear near the reaction center. According to the theoretical treatment which has been given in detail in the literature (11), the activation energy, ΔE, of the reaction is given by

$$\Delta E = (\Delta E)_\sigma + (\Delta E)_\pi , \qquad (1)$$

where

$$(\Delta E)_\pi = \sum_{j=1}^{N} \frac{(\nu_j - \nu)\, C_r^{j2}}{\epsilon_j - \alpha_h} \gamma^2 + \nu\,(\alpha_h - \alpha_R), \qquad (2)$$

in which N is the total number of π-orbitals in the isolated conjugated molecule, ν_j and C_r^j are the number of electrons (0, 1, or 2) and the coefficient of the rth atomic π-orbital in the jth molecular orbital in the conjugated molecule, respectively, the energy of which is ϵ_j. The value of ν is determined by the type of reaction in the sense of organic electronic theory, and is taken as 0, 1, or 2, according to whether the reagent is electrophilic, radical, or nucleophilic, respectively. α_h and α_R are the Coulomb integrals of the quasi-π-orbital and of the orbital of the reaction center in the reagent, respectively. By the reason given in the original paper, α_h may in the present case be put equal to α, which is the Coulomb integral at a carbon atom in benzene.

The terms ΔE_σ and $\nu(\alpha_h - \alpha_R)$ as well as γ are constant and may be disregarded in discussing reactivity if only the same type of reactions are considered, as in the present case. The relative ease of reaction, therefore, can be measured by the amount of the coefficient of γ^2 in Equation (2). This quantity is referred to as "super-delocalizability" and denoted by S_r, that is

$$S_r = \sum_{j=1}^{N} \frac{(\nu_j - \nu)\, C_r^{j2}}{\lambda_j}, \qquad (3)$$

where λ_j is given by $\epsilon_j = \alpha + \lambda_j\beta$ in which β is the exchange integral between two adjacent π-orbitals in benzene.

In the right side of Equation (3) one or two terms (denoted by suffix f) exist in which $|\lambda_f|$ is small and $\nu_f - \nu \neq 0$. Hence, the magnitude of S_r is determined predominantly by these terms, especially in a large molecule such as condensed aromatic hydrocarbons. We call the molecular orbitals corresponding to such terms the *frontier orbitals*. Thus, reactivity in these molecules can be discussed by calculating only the contribution of

[2] See, for example, E. Hückel, *Z. Physik*, **72**:312, 1931; G. W. Wheland and L. Pauling, *J. Am. Chem. Soc.*, **57**:2086, 1935; H. C. Longuet-Higgins and C. A. Coulson, *Trans. Far. Soc.*, **43**:87, 1947; C. A. Coulson and H. C. Longuet-Higgins, *Proc. Roy. Soc. London*, *s.A*, **191**:39; *s.A*, **192**:16, 1947.

the frontier orbitals.[3] In almost all the cases, the frontier orbitals coincide with the highest occupied orbital in the electrophilic reaction, with the lowest vacant orbital in the nucleophilic reaction, and with both the above orbitals in the radical reaction.

APPLICATION TO THE CARCINOGENIC PROBLEM

It has become clear that an intimate correlation exists between the carcinogenic activities and the frontier electron distributions of nonsubstituted aromatic hydrocarbons.

We designated the position corresponding to the phenanthrene double bond and the position corresponding to *meso* of anthracene the "principal carcinogenophore"[4] and the "subsidiary carcinogenophore," respectively. The following relations have been found, and, according to these relations, nonsubstituted aromatic hydrocarbons can be classified into three groups, i.e., (A), (B), and (C).

A) These compounds have both the principal and the subsidiary carcinogenophores and at the same time have sufficiently large values of super-delocalizability at the principal carcinogenophore (S_p). They are carcinogenic.

B) These compounds have the principal carcinogenophore (p.c.) only. They are slightly carcinogenic or noncarcinogenic, even if the values of S_p are above the critical value.

C) These compounds have the subsidiary carcinogenophore (s.c.) only, are devoid of both the principal and the subsidiary carcinogenophore, or have both carcinogenophores but small values of S_p. They are all noncarcinogenic.

The frontier electron densities calculated at every carbon atom ($f_r \equiv 2C_r^{f2}$) in each compound and the positions of principal and subsidiary carcinogenophores are indicated in Charts 1 and 2. The values for the compounds which belong to Class (C) are omitted.

The calculated values of super-delocalizability at the principal and subsidiary carcinogenophores are shown in Table 1 and compared with the experimental carcinogenic activity. The value of S_p is the sum of the values of super-delocalizability of two carbon atoms at the principal carcinogenophore.[5]

[3] However, in the case when the next orbital is in close proximity to the frontier orbital, the corresponding term must be taken into account.

[4] This corresponds to the so-called K-region of the Pullmans.

[5] The theoretical ground for this summation is easily given by the above-mentioned treatment of the frontier electron theory extended to the case of simultaneous 1,2-addition or substitution, to which, we assume, the carcinogenic reaction should belong.

As generally used, the signs + and − indicate the degree of carcinogenicity. The greater the number of + signs, the greater is the activity. The sign ± indicates that the compound is very feebly carcinogenic; i.e., in some cases it has been reported to be carcinogenic, but in other experiments was not carcinogenic. The − sign indicates that the compound has not been reported to induce tumors.

It can be seen in Table 1 that a close relation exists between the value of super-delocalizability at p.c. and carcinogenic activity. It is rather striking from the physical point of view that such a relation, connecting a biological phenomenon with physics, has been obtained by a calculation in which no arbitrary parameter was introduced.

The critical value of S_p, below which a compound is not carcinogenic, is found to be about 0.65 (Table 1). This numerical value has a definite meaning, because S_p is a dimensionless quantity.

The compounds of Class (A) have both the p.c. and s.c.; the values of S_p are larger than the threshold value, and almost all these compounds are carcinogenic. The order of the values of S_p roughly parallels the degree of activity of these compounds. However, in this connection, it should be noted that an attempt to explain very small differences in activity by means of S_p alone may be rather meaningless, since the carcinogenicity varies with the experimental procedure, with the species of test animal, and with differences in the diffusion of the carcinogen into the cell.

Hitherto it has been considered that 1,2-benzanthracene is noncarcinogenic (1, 4, 15, 19, 21, 28), but, according to some recent experiments by Steiner *et al.* (29, 30), it has become clear that this compound is carcinogenic. As is seen from the calculations in Table 1, this compound would be expected to carcinogenic. The existing theories based on the supposition that this compound was noncarcinogenic, therefore, must be subjected to a modification, as has been pointed out by Steiner and Falk (30).

As exceptional cases we must refer to 2,3,7,8-dibenzphenanthrene and 2′3′-naphtho-3,4-pyrene. In spite of the reported inactivity of these compounds, the calculated values of S_p are larger than the threshold value. Perhaps they should be retested by experiments of longer duration as in the case of 1,2-benzanthracene.

In general, the value of S_s is not so important as that of S_p, though the existence of s.c. relates to the activity essentially, as will be stated below. Pullman's opinion (23, 24) that the larger the electron density of the meso position (corresponding to s.c.) the lower the carcinogenicity has not been confirmed by our results. Thus, the values of

S_s were rather large for 3,4-benzpyrene, which is one of the most potent carcinogens, as well as for 3,4,8,9-dibenzpyrene, which is a relatively potent carcinogen.

The compounds of Class (B) have only p.c. and not s.c., and the carcinogenicity of these compounds is relatively weak, even though the value of S_p is comparatively large. From this fact, it may be possible to conclude that the existence of s.c. is a necessary, though not a sufficient, condition for carcinogenicity. This is interesting in relation to the opinions of Boyland (3) and Kooyman and Heringua (18) who have pointed out the importance of the anthracene meso position.[6]

Although 3,4-benzphenanthrene and 1,2,3,4-dibenzphenanthrene are definitely carcinogenic, the values of S_p for these compounds are below the threshold value. However, we feel this might be due to some other factors, for instance, the special steric circumstances in these compounds.

The compounds of Class (C) are devoid of the p.c. or otherwise have a p.c. whose S_p is below the threshold value. Experimentally, these compounds have been found to be noncarcinogenic. Therefore the agreement of the calculated results with experiments is very satisfactory.

Thus, the present authors conclude that the existence of a p.c. whose S_p is larger than the threshold value should be necessary for carcinogenic activity; further, the value of S_p is the most important factor determining the degree of activity. From this point of view it can be said that the main center of reaction is p.c. With the existence of s.c. alone, there appears to be no activity, even if the value of S_s is large. But, as is seen from

[6] However, Pullman (23, 24) has reported, against them, that the existence of K-region alone is sufficient for the carcinogenicity.

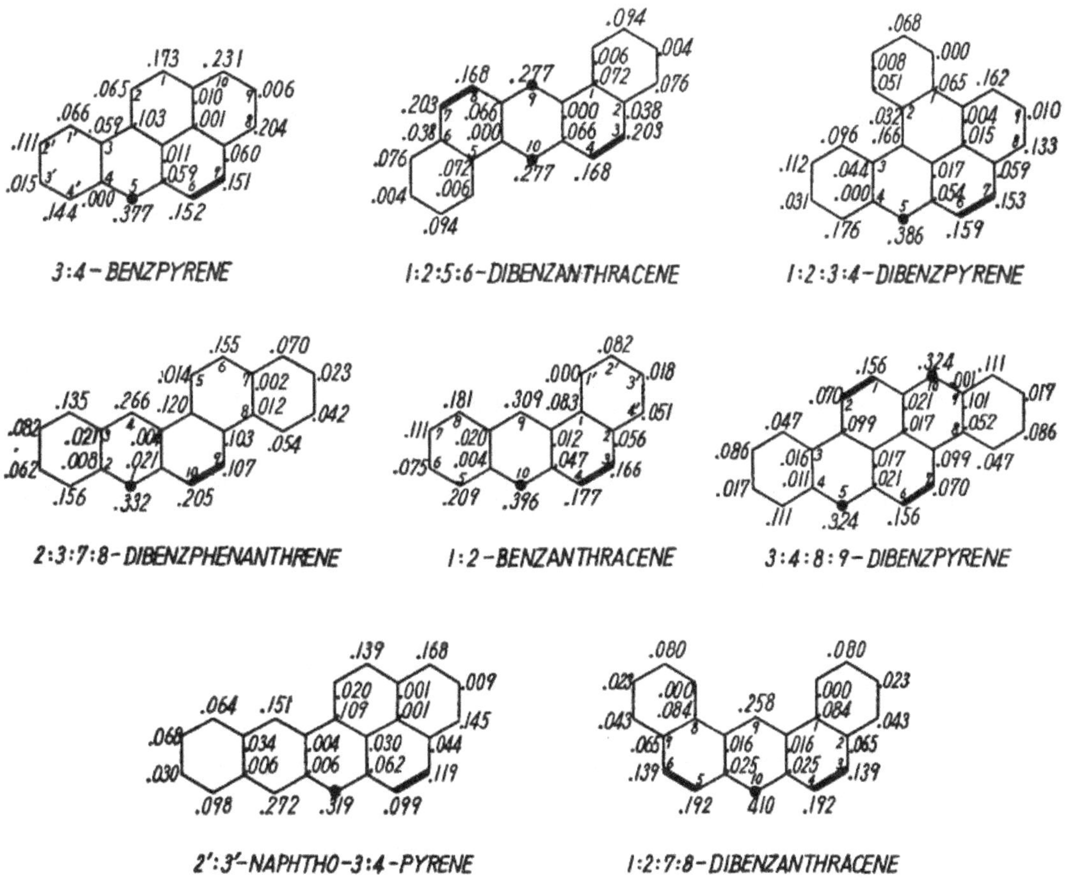

CHART 1.—Frontier electron distributions and the positions of principal and subsidiary carcinogenophores in the compounds of Class (A). (Numerical values in the Chart indicate the frontier electron density at each position in the molecule f_r; the principal carcinogenophore is shown by a thick line; the subsidiary carcinogenophore is shown by a black spot.)

Table 1, the activity is remarkably decreased when the carcinogen lacks the s.c. even though the value of S_p is sufficiently large. Therefore, it seems certain that, in a sense, the s.c. is connected with the process of carcinogenesis. Consequently, it is assumed that the s.c., though it is not the main center of reaction, plays a sort of subsidiary role in making more easy the chemical combination between the p.c. and protein. It is concluded, therefore, that the existence of s.c. is also necessary for the occurrence of activity.

The frontier electron method also appears applicable for explaining the structures of metabolites of carcinogens. The results will be reported in the near future.

SUMMARY

1. The frontier electron method, one of the quantum mechanical methods for explaining chemical reactivity, which has been proposed by some of the present authors, was applied to the problem of the carcinogenic activity of nonsubstituted aromatic hydrocarbons. Frontier electron distributions and the super-delocalizabilities, which are the quantities determining the chemical reactivity of these compounds, were calculated, and the results were reported.

2. By comparing the calculated results with experimental data, the present authors concluded that two different positions were necessary for the

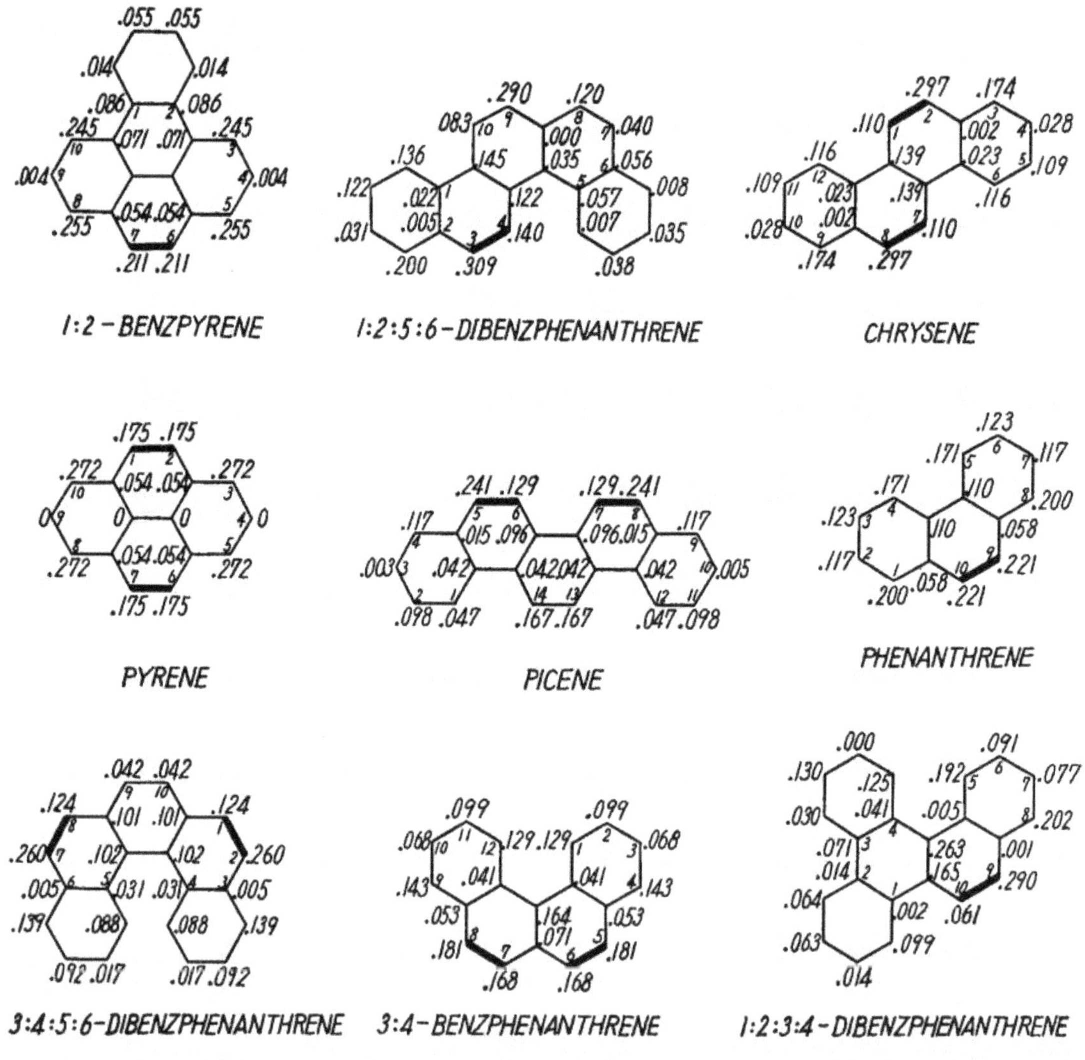

CHART 2.—Frontier electron distributions and the positions of principal and subsidiary carcinogenophores in the compounds of Class (B). (Cf. notes in Chart 1.)

carcinogenic activity. They were designated the "principal" and the "subsidiary" carcinogenophores, respectively. An intimate correlation existed between the frontier electron densities or the super-delocalizabilities at the two kinds of carcinogenophores and the magnitudes of carcinogenic activity of these compounds. The value of super-delocalizability at the principal carcinogenophore was the most important factor determining the magnitude of carcinogenic activity.

3. The center of reaction in carcinogenesis might be the principal carcinogenophore, and the subsidiary carcinogenophore might play some subsidiary role in carcinogenesis.

TABLE 1

COMPARISON OF THE VALUES OF SUPER-DELOCALIZABILITIES AT CARCINOGENOPHORES WITH CARCINOGENIC ACTIVITIES

Class	Compounds	Super-delocalizability at principal carcinogenophore (S_p)	Super-delocalizability at subsidiary carcinogenophore (S_s)	Carcinogenic activity
	3,4-Benzpyrene	0.8151	1.0148	++++
	1,2,5,6-Dibenzanthracene	0.7844	0.5725	+++
	1,2,3,4-Dibenzpyrene	0.7818	0.9696	+
A	2,3,7,8-Dibenzanthracene	0.7722	0.8189	−
	1,2-Benzanthracene	0.7594	0.8757	+
	3,4,8,9-Dibenzpyrene	0.7440	1.0704	+
	2′,3′-Naphtho-3,4-pyrene	0.7209	1.0542	−
	1,2,7,8-Dibenzanthracene	0.6719	0.8930	+
	1,2-Benzpyrene	0.8491		−
	1,2,5,6-Dibenzphenanthrene	0.8172		+
	Pyrene	0.7865		−
	Chrysene	0.7811		±
B	Picene	0.7371		−
	Phenanthrene	0.7306		±
	3,4,5,6-Dibenzphenanthrene	0.7020		−
	3,4-Benzphenanthrene	0.6144		+
	1,2,3,4-Dibenzphenanthrene	0.5753		+
	2′,1′-Anthra-1,2-anthracene	0.6295	0.5970	−
	2,3,5,6-Dibenzphenanthrene	0.6034	0.8089	−
	Anthanthrene	0.5945	0.9725	−
	1,2,7,8-Dibenznaphthacene	0.5703	0.7899	−
	1,2,3,4,5,6-Tribenzanthracene	0.5676	0.3989	−
	1,2-Benznaphthacene	0.5530	0.9741	−
	1,2,9,10-Dibenznaphthacene	0.5332	0.8649	−
C	Pentaphene	0.4841	0.4237	−
	1′,2′-Anthra-1,2-anthracene	0.3632	0.5253	−
	Pentacene		1.2818	−
	Anthracene		0.9348	−
	1,2,3,4-Dibenzanthracene		0.7159	−
	Naphthacene		0.4876	−
	Triphenylene			−
	Perylene			−
	1,2,6,7-Dibenzpyrene			−

REFERENCES

1. ANDERVONT, H. B., and SHIMKIN, M. B. Biologic Testing of Carcinogens. II. Pulmonary-Tumor-Induction Technique. J. Nat. Cancer Inst., **1:**225–39, 1940.
2. BERTHIER, G.; COULSON, C. A.; GREENWOOD, H. H.; and PULLMAN, A. Structure électronique des hydrocarbures aromatiques à 4 noyaux benzéniques accolés. Étude par la méthode des orbitales moléculaires. Compt. rend Acad. Sc., Paris, **226:**1906–8, 1948.
3. BOYLAND, E. Different Types of Carcinogens and Their Possible Mode of Action: *A Review*. Cancer Research, **12:**77–84, 1952.
4. BOYLAND, E., and BURROWS, H. The Experimental Production of Sarcoma in Rats and Mice by a Colloidal Aqueous Solution of 1:2:5:6-Dibenzanthracene. J. Path. & Bact., **41:**231–38, 1935.
5. BRUCE, W. F. Carcinogenic Hydrocarbons. V. A Comparison of the Fluorescence Intensity of Cholanthrene and Certain of Its Homologs. J. Am. Chem. Soc., **63:**304–5, 1941.
6. COOK, J. W.; HASLEWOOD, G. A. D.; HEWETT, C. C; HIEGER, I.; KENNAWAY, E. L.; and MAYNEORD, W. V. Chemical Compounds as Carcinogenic Agents. Am. J. Cancer, **29:**219–59, 1937.
7. COOK, J. W., and KENNAWAY, E. L. Chemical Compounds as Carcinogenic Agents. First Supplementary Report: Literature of 1937. Am. J. Cancer, **33:**50–97, 1938.
8. COULSON, C. A. Electronic Configuration and Carcinogenesis. Adv. Cancer Research, **1:**1–96, 1953.
9. DAUDEL, R. Sur les applications de la mécanique ondulatoire à l'étude des phénomènes de cancérisation. Bull. Cancer, **35:**110–24, 1948.
10. DEWAR, M. J. S. A Molecular Orbital Theory of Organic Chemistry. VI. Aromatic Substitution and Addition. J. Am. Chem. Soc., **74:**3357–63, 1952.
11. FUKUI, K.; YONEZAWA, T.; and NAGATA, C. Theory of

Substitution in Conjugated Molecules. Bull. Chem. Soc. Japan, **27:**423–27, 1954.

12. FUKUI, K.; YONEZAWA, T.; NAGATA, C.; and SHINGU, H. Molecular Orbital Theory of Orientation in Aromatic, Heteroaromatic, and Other Conjugated Molecules. J. Chem. Phys., **22:**1433–42, 1954.
13. FUKUI, K.; YONEZAWA, T.; and SHINGU, H. A Molecular Orbital Theory of Reactivity in Aromatic Hydrocarbons. J. Chem. Phys., **20:**722–25, 1952.
14. GREENWOOD, H. H. The Reactivity of Aromatic Bonds, with Reference to Carcinogenic Compounds of 1:2-Benzanthracene. Brit. J. Cancer, **5:**441–57, 1951.
15. HIEGER, I. On the Mechanism of Carcinogenesis by Chemical Compounds. Am. J. Cancer, **28:**522–29, 1936.
16. IVERSON, S. A Possible Correlation between Absorption Spectra and Carcinogenicity. Copenhagen: Einar Munksgaard, 1949.
17. JONES R. N. The Ultraviolet Absorption Spectra of Some Derivatives of 1:2-Benzanthracene. J. Am. Chem. Soc., **62:** 148–52, 1940.
18. KOOYMAN, E. C., and HERINGUA, J. W. Extension of the K-Region Hypothesis of Carcinogenic Chemical Compounds. Nature, **170:**663, 1952.
19. LEES, J. C. The Inhibition of Growth by 1:2:5:6-Dibenzanthracene and Other Agents. Quart. J. Exper. Physiol., **27:**161–70, 1937.
20. MOODIE, M. M., and REID, C. Inter- and Intramolecular Energy Transfer Process. 3. Phosphorescence Bands of Some Carcinogenic Aromatic Hydrocarbons. J. Chem. Phys., **22:**252–54, 1954.
21. ORR, J. W. The Changes Antecedent to Tumor Formation during the Treatment of Mouse Skin with Carcinogenic Hydrocarbons. J. Path. & Bact., **46:**495–515, 1938.
22. PULLMAN, A. Structure électronique et pouvoir cancérigène des hydrocarbures aromatiques condensés. Bull. Assoc. franç. l'étude du cancer, **33:**120–30, 1946.
23. ———. Sur les facteurs conditionnant l'apparition de l'activité cancérogène dans les hydrocarbures aromatiques. Compt. rend., **236:**2318, 1953.
24. ———. Compléments sur les facteurs déterminant l'existence de l'activité cancérogène dans les hydrocarbures aromatiques. *Ibid.*, pp. 2508.
25. PULLMAN, A., and PULLMAN, B. Structure électronique et pouvoir cancérigène des molécules organiques. Rev. Sci., **84:**145–58, 1946.
26. ROBINSON, R. Some Aspects of Chemistry Related to Medicine. Brit. M.J., **1:**943–46, 1946.
27. SCHMIDT, O. Die Charakterisierung der einfachen und Krebs erzeugenden arometischen Kohlenwasserstoffe durch die Dichteverteilung bestimmter Valenzelektronen (B-Elektronen). Ztschr. f. Physikal. Chem., **42:**88–110, 1939.
28. SHEAR, M. J., and LEITER, J. Studies in Carcinogenesis. XVI. Production of Subcutaneous Tumors in Mice by Miscellaneous Polycyclic Compounds. J. Nat. Cancer Inst., **2:**241–58, 1941.
29. STEINER, P. E., and EDGCOMB, J. H. Carcinogenicity of 1,2-Benzanthracene. Cancer Research, **12:**657–59, 1952.
30. STEINER, P. E., and FALK, H. L. Summation and Inhibition Effects of Weak and Strong Carcinogenic Hydrocarbons: 1,2-Benzanthracene, Chrysene, 1,2,5,6-Dibenzanthracene and 20-Methylcholanthrene. Cancer Research, **11:**56–63, 1951.
31. WHELAND, G. W. A Quantum Mechanical Investigation of the Orientation of Substituents in Aromatic Molecules. J. Am. Chem. Soc., **64:**900–908, 1942.

Reprinted from MOLECULAR ORBITALS IN CHEMISTRY, PHYSICS, AND BIOLOGY

A Simple Quantum-Theoretical Interpretation of the Chemical Reactivity of Organic Compounds

KENICHI FUKUI

Faculty of Engineering, Kyoto University, Kyoto, Japan

I. Nonempirical and Empirical Approaches

The recent remarkable progress in physical measuring technique has revealed the structure of molecules in more and more detail. In order to understand these results on a physical basis, radically and fundamentally, more and more accuracy will obviously be required with respect to the theoretical treatment of molecules. But to fulfill this requirement actually is by no means an easy problem.

Namely, from the point of view of the nonempirical interpretation of chemical phenomena, it may not be merely rhetorical to say that even the hydrogen molecule is not at all a simple molecule: it can provide a series of, so far, unsolved problems of quantum-mechanical interest, including the purely wave-mechanical calculation of the rate of one of the simplest chemical reactions

$$H + H_2 \rightarrow H_2 + H$$

On the other hand, a large majority of experimental chemists are concerned daily with the reactions between molecules having several nuclei other than protons as well as scores of electrons. It may be needless to say that extreme difficulty would inevitably be encountered by the theoretical chemists who might try to treat these phenomena quantum mechanically.

Accordingly, the appearance of some powerful approximative methods of treating other than small molecules had been anticipated, so as to obtain insight which might be useful in chemistry. The combination of statistical mechanics with quantum theory was attempted for this purpose (Tolman, 1927; Pelzer and Wigner, 1932; La Mer, 1933). One of the useful approximative theories was what we called "the theory of absolute reaction rates" (Eyring, 1935; Wynne-Jones and Eyring, 1935; Polanyi and Evans, 1935; Polanyi, 1937; Evans and Polanyi, 1937). As is well known, this theory could widen the route of the theoretical approach to the true chemical process. But the applicability of this theory did not satisfy the chemists who tried to employ it for the purpose of discussing the reactivity of, for instance, large organic molecules.

On the side of organic chemistry, on the other hand, a large pile of experimental facts already had been accumulated before the birth of quantum mechanics. In order to understand these experiences from a unified point of view, a great many efforts had been made by Robinson, Ingold, and many other organic chemists, until the so-called "electronic theory" was established (Ingold, 1934). In the electronic theory of organic chemistry, the attempt was to understand the chemical and physical behavior of each atom and each group of atoms in organic molecules through the electrostatic effect caused by the migration of electrons, in both the statical and dynamical sense.

Organic electronic theory had served to arrange and put in order a huge accumulation of experience in organic chemistry so as to make it far more comprehensible than without the contribution of this theory. In this sense much work carried out on this line had great significance in the field of practical chemistry. Since this work, however, leaned too much on experience and lacked a solid theoretical foundation, many inconsistencies came to light with respect to the explanation of various electronic effects. The essentially qualitative nature of the organic electronic theory was also a serious weak point.

In the aromatic substitution, in particular, the problem of the so-called "orienting effect" of the substituent had attracted the attention of many organic chemists, since there existed in such a molecule more than one similar, but nonequivalent, position which was more or less capable of reaction. To estimate and compare with experimental results the relative rate of substitution in different positions in a molecule was of considerable interest, but, owing to its qualitative nature, the electronic theory was not adequately qualified to give an illuminating explanation of this problem.

II. Static Approach

The first plausible quantum-mechanical interpretation of the orientation in aromatic substitutions was given in 1931 by Hückel (1931, 1937a,b), whose approximation, i.e., the well-known HMO (the Hückel molecular orbital) approximation, has been widely used to date as one suitable for treating a large conjugated molecule. Hückel's work started the subsequent progress which immediately followed it in the field of the theoretical interpretation of chemical phenomena of large conjugated molecules. Concerning the problem of chemical reactivity, Wheland and Pauling (1933), Ri and Eyring (1940), Ri (1943), Pullman and Pullman (1946), Dewar (1946), Coulson (1947, 1948), Coulson and Longuet-Higgins (1947a,b), Longuet-Higgins and Coulson (1947), Daudel *et al* (1949), Sandorfy (1949), Brown (1948, 1949), Pullman (1954), and many other theoretical chemists (Pullman and Metzger, 1948; Berthier *et al.*, 1948; Pullman and Berthier, 1949; Burkitt *et al.*, 1951; Fernández-Alonso, 1951; Kurita and Kubo, 1951; Szabo, 1951) developed the theory. In these treatments, the electron migration in a conjugated molecule due to the approach of reagents was calculated quantum mechanically. The energy of interaction between the disturbed substrate molecule and the reagent was calculated, and the resulting electrostatic effects seemed to make it possible to explain the orientation in conjugated molecules. Besides the reactivity problems, these treatments also succeeded in interpreting several "static" problems, such as dipole moment, acidity and basicity, equilibrium constant, color, and so forth.

In heterolytic aromatic substitution, the electric charge on the reagent will bring about a disturbance of the electron distribution in the substrate molecule at their approach to each other. Thus, for instance, the larger the electron density of a position is, the more easily a cationic reagent will be able to approach to the position. Such a polarization effect and the subsequent Coulombic interaction are important only at a longer distance, because in a very short distance the electron transfer will occur according to the difference in electronegativity between the substrate molecule and the ionic reagents, so that "neutralization" might erase the electrostatic effect. Thus, in this model, the energy change required for reaction is calculated with reference to a state corresponding to a certain stage much earlier than the transition state. In this sense the model used in these treatments was called "static."

In the homolytic reaction the attacking radical is considered to weaken the bonds adjacent to the position of attack without any serious alteration of the electronic structure of the substrate molecule. The energy required to weaken the bonds is calculated and used as a measure of reactivity (Bown, 1952). This is the way of thinking of the process of a homolytic reaction in the "static" model.

In the quantum-chemical theory of reactivity, in general, a theoretical index representing the measure of ease of reaction is used. This, we may call the "reactivity index." The reactivity index is usually defined in a concise form on the basis of the absolute reaction rate theory. The reactivity indices employed in the "static" model are indicated in Table I. Total π-electron density and self-polarizability (Coulson and Longuet-Higgins, 1947a,b) are used for ionic reactions, and the free valence (Coulson, 1947, 1948) is for a

TABLE I
REACTIVITY INDICES IN THE "STATIC" MODEL

Reactivity index	Symbol	Model
Total π-electron density[a]	q_r	$-\delta$, $+\delta$ --- R^+
Self-polarizability[a]	π_{rr}	
Free valence[b]	F_r	····•R

[a] Coulson and Longuet-Higgins (1947a,b).
[b] Coulson (1947, 1948).

radical attack. The total π-electron density was successfully used to explain the reactivity of polar molecules. For the simple treatment of netural hydrocarbons self-polarizability is used instead. Also free valence proved to be a good index for the radical attack of unsaturated molecules. But this theory predicted some inconsistent results. For instance, the following few, but important examples can be pointed out. (Fig. 1). The theoretical position of

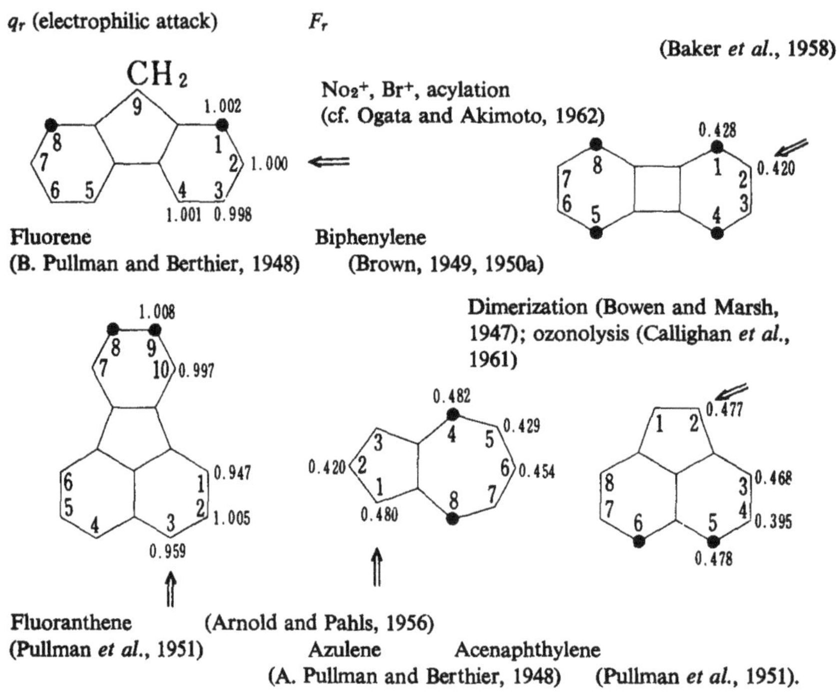

FIG. 1. Some exceptional examples of the static approach.

attack is denoted by a black circle, and the experimental position of attack by an arrow. In fluorene, for instance, electrophilic reagents attack position 2 (for example, see Ogata and Akimoto, 1962), while the position of the largest q_r is 1 (B. Pullman and Berthier, 1948). Also, electrophilic reagents attack most dominantly position 3 of fluoranthene (Streitwieser, Fahey, 1962), while q_r predicts position 8 as the most reactive [see Pullman *et al.* (1951), who also accounted for this discrepancy by calculating localization energies].

Moreover, there is a more serious difficulty in this approach. That is, more exact calculations of total π-electron density lead to a totally unsatisfactory result. For instance, the result of self-consistent field (SCF) calculation is indicated in Fig. 2. As is seen in Fig. 2, the electronic interaction causes a

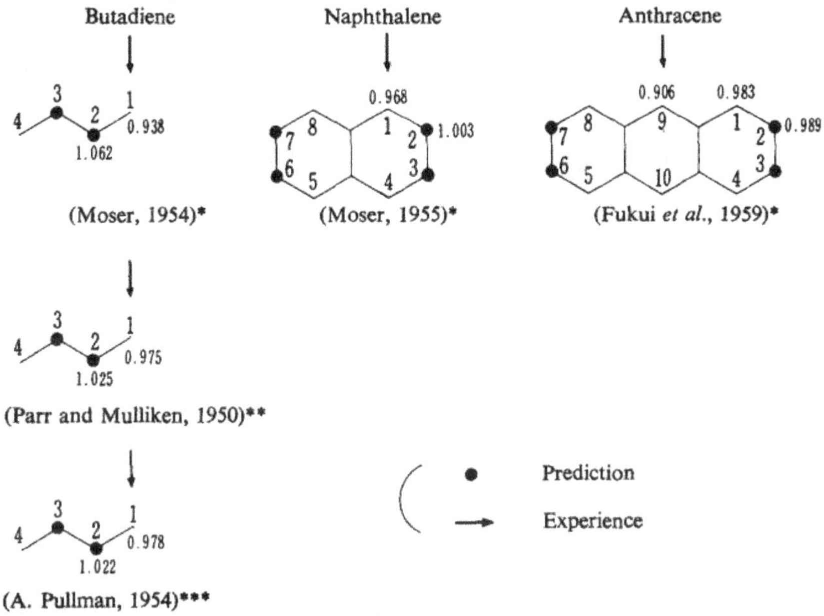

FIG. 2. SCF q_r (electrophilic attack).

* Semiempirical SCF calculation; ** nonempirical SCF calculation; *** nonempirical SCF-CI (configurational interaction) calculation.

concentration of electron density at the positions at which the electrons of lower energies stay dominantly. This brings about a disagreement between theory and experiment [for similar observations, see Kolboe and Pullman (1958)].

It is very difficult to attribute this discrepancy to the inadequacy of the quantum mechanical method of calculation, because such a discrepancy is not seen in other reactivity indices.

A more elaborate calculation of free valence revealed that this reactivity index was extremely sensitive to the degree of approximation adopted. In some approximations it succeeded but in some other cases it failed. A reactivity index should be stable in order to be reliable.

III. Localization Approach

As stated above, the Hückel method first enabled us to discuss the reactivity of large conjugated molecules and provided for us a foundation of the quantum chemistry of complicated chemical reactions. Coming after the "static" approach, there appeared in 1942 a theory of reactivity which was based upon a model closely connected with the "activated complex" concept in the absolute rate theory. This may be called the "localization" model, since the transition state model of this approach corresponds to an extremely localized π-electron model. Wheland (1942), Dewar (1952), and others (Mackor *et al.*, 1958; Streitwieser, 1961) have contributed to the development of this theory. They calculated quantum mechanically the measure of activation energy based upon the model of the activated complex indicated in

TABLE II
REACTIVITY INDICES IN THE "LOCALIZATION" MODEL

Reactivity index	Symbol	Model
Wheland's localization energy (1942)	L_r	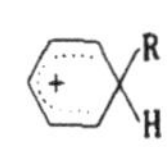
Dewar's approximate localization energy (1952)	l_r	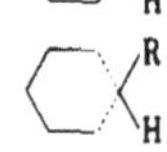
Omega localization energy (cf. Streitwieser, 1961)	L_r^{ω}	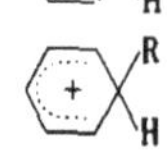

Table II. The reactivity index, called the "localization energy" in this approach, explained very well the relative rate of substitution in conjugated molecules with very few exceptions, indicated, for instance, in Fig. 3.

In Fig. 4, the relative rate of nitration of various aromatic hydrocarbons is plotted against the value of localization energy at the corresponding position of the molecule. The inclination is found from the figure to be ca. 8. If the entropy of activation is assumed to be essentially constant for these reactions, it follows that

$$\log_{10} k = \frac{(-\beta)}{2.303RT}(-L_r) + \text{const.}$$

We can estimate the value of resonance integral β from this equation. The value of inclination 8 corresponds to the value of resonance integral $|\beta|$ of approximately 10 kcal/mole. This value of $|\beta|$ is only about one-fourth of that obtained by experimental determination, such as absorption spectra,

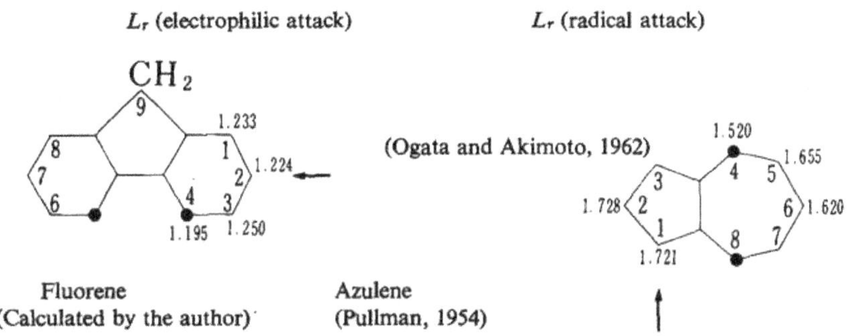

FIG. 3. Some exceptional examples of the localization approach.

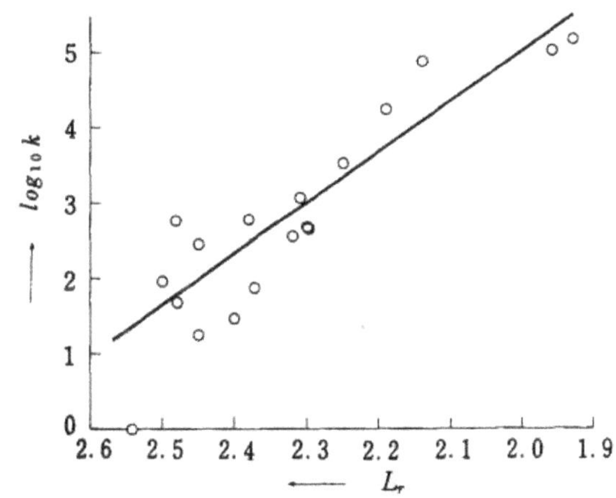

FIG. 4. Relative rate of nitration and the localization energy. k = Rate constant of nitration; β = Resonance integral of benzene C–C bond; RT = 0.6 kcal/mole; $(-\beta) \cong 10$ kcal/mole. (Numerical values of L_r were taken from Streitwieser, 1961, p. 336; the values of k were recited from Brown, 1959, p. 2237, Table I.)

ionization potential, and many other physico-chemical measurements. Similar plots with respect to other reactivity indices give more reasonable values for β.

This result seems to indicate what a high-energy state the model of activated complex used in the localization approach corresponds to. That is to say, the localization energy serves as a megaphone announcing the reactivity of conjugated molecules. Of course, the "compensation" relation (Leffler, 1955) holding between the enthalpy of activation and the entropy of activation should be taken into account. But the value of $|\beta|$ obtained above is too small. A hyperconjugation was also considered (Daudel and Chalvet, 1956; Mackor *et al.*, 1958) between the localized π-electron system and the reagent. The resultant model clearly approaches close to the delocalization model which will be mentioned below.

IV. Delocalization Approach

A. Reactivity Indices

It is quite natural that, after the "static" and "localization" approaches, there should appear what is called the "delocalization" approach (Fueno, 1961). In this method, a model of the transition state is used in which the delocalization of π-electrons in reactant molecules takes place. There are several varieties of methods which belong to this category. They differ slightly in the activated complex model from one another. The reactivity indices derived from this approach are listed in Table III.

Frontier electron density, superdelocalizability, and hyperconjugation energy constitute the system of reactivity indices in frontier electron theory. In this approach, the transition state is assumed to correspond to a model in which a hyperconjugation (Muller *et al.*, 1954) occurs between the π-electron system and the "pseudo-atom" at the position of reaction. The pseudo-atom is considered to have a quasi π orbital which is composed of the reagent orbital and one orbital of the atom to be expelled. The hyperconjugation will then take place through the bridge between the quasi π orbital and the π orbital at the position of attack. Because of this hyperconjugation the total system may be stabilized by a charge migration. The stabilization energy was referred to as "hyperconjugation energy." Calculation of hyperconjugation energy is rather impractical: it is troublesome because we have to determine the quasi π orbital energy in order to obtain its numerical value. A reactivity index should be of a practical nature considering its origin, and more desirably it could be represented by a theoretical quantity belonging to the isolated substrate molecule. By the perturbation treatment, the hyperconjugation energy may be expressed in terms of the second-order perturbation term which we called the "superdelocalizability." This index is therefore only applicable for comparing the reactivity of different positions of different molecules with a definite reagent. The frontier electron density serves as a reactivity index which can very simply tell the orientation or position of attack in a molecule.

In most cases, the superdelocalizability and the hyperconjugation energy give almost the same results with respect to the prediction of the relative reactivity. Also, in most hydrocarbons with no odd-membered ring, the order of frontier electron density of different positions parallels that of superdelocalizability in a molecule. Even in large odd-membered ring compounds, the experimentally most reactive positions in a molecule are sufficiently well given by both the frontier electron density and the superdelocalizability.

TABLE III
REACTIVITY INDICES IN THE "DELOCALIZATION" MODEL

Reactivity index	Symbol	Model
Frontier electron density (Fukui *et al.*, 1952, 1954a)	f_r	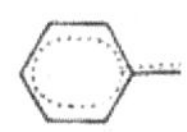
Superdelocalizability (Fukui *et al.*, 1954b)	S_r	e.g., 6 electrons for electrophilic attack
Hyperconjugation energy (Fukui *et al.*, 1957a,b)	H_r	
Ion free valence (Nagakura and Tanaka, 1954)	$F_r^{\pm}$	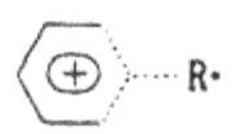 (for electrophilic attack)
Z value (Brown, 1959)	Z_r	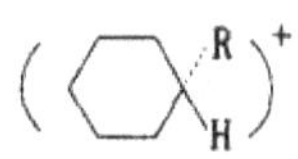 (for electrophilic attack)
Interconjugation energy of Yonezawa *et al.*, perturbation treatment (1954, 1957; also see Hayashi *et al.*, 1956)	E_{rs}	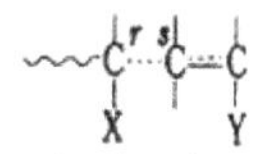 (for vinyl polymerization)

Streitwieser's development of π-extension concept for various complicated organic reactions (1961)

The Z value was recently proposed by Brown (1959), and is very closely related to the indices of the frontier electron theory (Fukui *et al.*, 1960c). Brown stated in his paper (1959) that his theory of electrophilic aromatic substitution provided the theoretical foundation of the frontier electron theory through this reactivity index, the Z value. In one and the same molecule, the order of Z value in a molecule is entirely the same as that of the frontier electron density, as is evident from its definition. The mathematical relationship between Z value and superdelocalizability was also made clear.

Other indices on this line were also proposed in several specific problems of chemical reaction. The ion free valence was introduced by Nagakura and

Tanaka (1954, 1959) for heterolytic reactions of conjugated molecules. For instance, in cationic reagents, one electron of a conjugated molecule is transferred to the reagent and the resultant radical attacks the position of largest free valence of the conjugated molecule cation. Thus, the reactivity index is called the "ion free valence." In all the conjugated hydrocarbons which do not contain odd-membered rings, the ion free valence, $F_r^{\pm}$, is represented by

$$F_r^{\pm} = F_r \pm \text{const.} f_r$$

where F_r is Coulson's free valence of a neutral hydrocarbon and f_r is the frontier electron density. Therefore, we may discuss this index in terms of the free valence and the frontier electron density. Recently Nagakura (1961) proposed a new theoretical approach to the charge transfer mechanism of heterolytic aromatic substitution.

Yonezawa and his co-workers' perturbation calculation (1954) and its modification by Levinson (1962) treating the vinyl polymerization problem also may belong to this class in the present classification of approaches.

Streitwieser (1961) has developed, in his excellent book "Molecular Orbital Theory for Organic Chemists," the concept of the π-extension model and applied this to various organic reactions such as rearrangements, cycloaddition of two olefins and diene additions, and carbonium cation reactions.

B. Simple Applications

Useful comments regarding the result of the application of the frontier electron theory to some specific problems in organic chemistry have already been published (Fernández-Alonso *et al.*, 1957a,b; Fernández-Alonso and Domingo, 1958; Pilar, 1958, 1959; Sato and Cvetanović, 1959; Pauncz and Cohen, 1960; Mestechkin, 1960; Zollinger, 1960; Kobayashi, 1960, 1961; Koutecký *et al.*, 1961; Bhattacharya and Basu, 1962; Rasch, 1962; Zahradník *et al.*, 1962a,b), and some applications to the problems of biochemical interest were also made (Muir and Hansch, 1961; Koshimizu *et al.*, 1960; Nagata *et al.*, 1955; Fukui *et al.*, 1958a,b; 1960a,c,d,f; 1961d,e,f, 1962c).

An accumulation of numerical values (Fukui *et al.*, 1952, 1954a) of reactivity indices in frontier electron theory has not been reproduced in the the present review. Only a very few examples will be mentioned here. In Fig. 5, theoretical prediction and the experimental results are indicated with respect to a few examples of larger conjugated molecules. The reactivity of compound (a) has also been predicted by other methods (Chalvet and Peltier, 1956).

Another example concerns substitution in dibenzopyrene. De La Mare and Ridd indicated in their book (1959) a diagram of 3,4,8,9-dibenzopyrene with arrows showing the experimental positions of attack of cationic reagents.

(a)

(i) f_r for the electrophilic attack

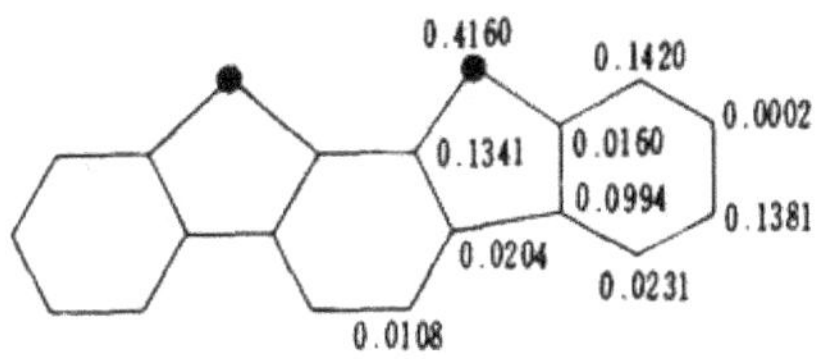

(ii) f_r for the nucleophilic attack

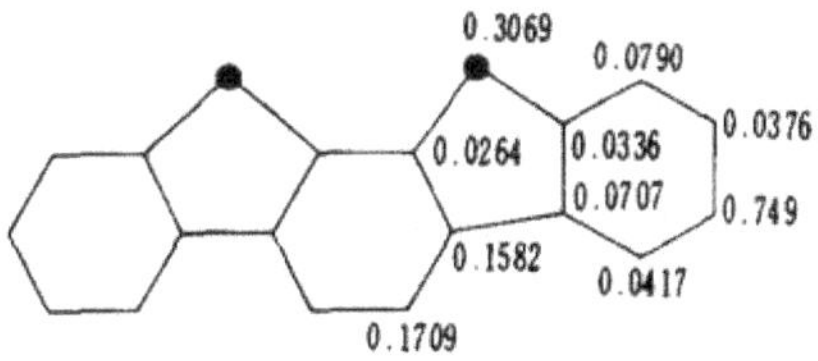

(iii) f_r for the radical attack

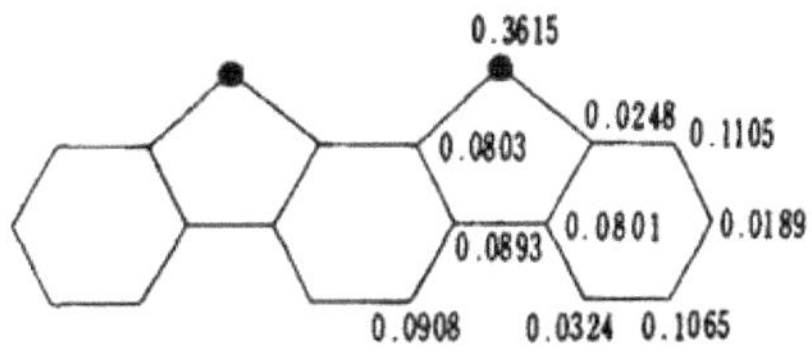

(b)

f_r for the electrophilic attack
NO_2^+, dehydrogenation; addition of
dienophile (Aitken and Reid, 1956).

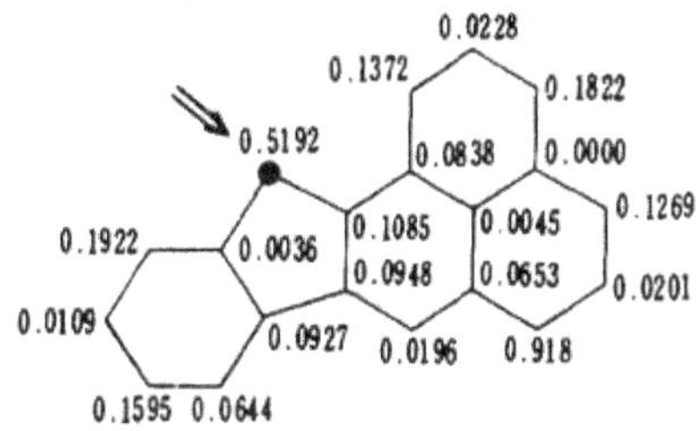

FIG. 5. Frontier electron density of complicated molecules.

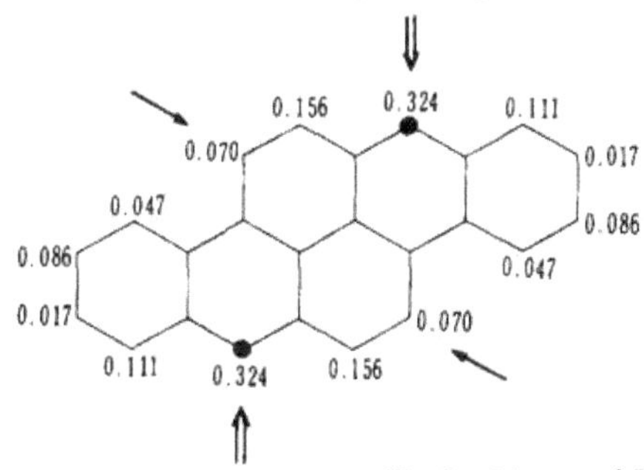

(De La Mare and Ridd, 1959)

(Ioffe and Efros, 1946)

FIG. 6. Substitution in 3,4,8,9-dibenzopyrene (numerical values are the frontier electron density).

(a)

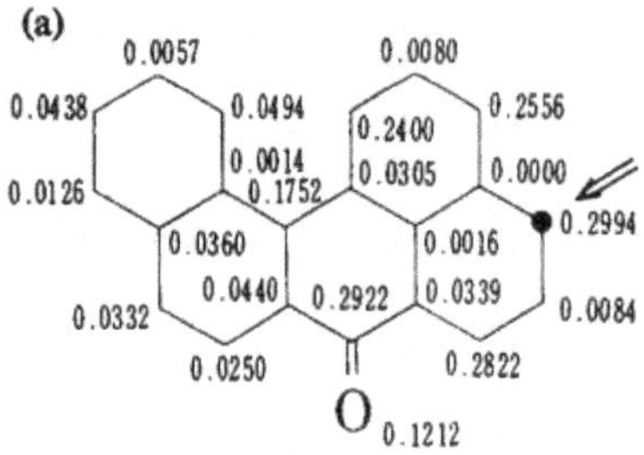

($\varepsilon^N = \alpha - 0.1731\beta$)
Benzo[10,11]mesobenzanthrone

OH (240°C, fused KOH with MnO2)
(Stephenson and Sutcliffe, 1962)

(b)

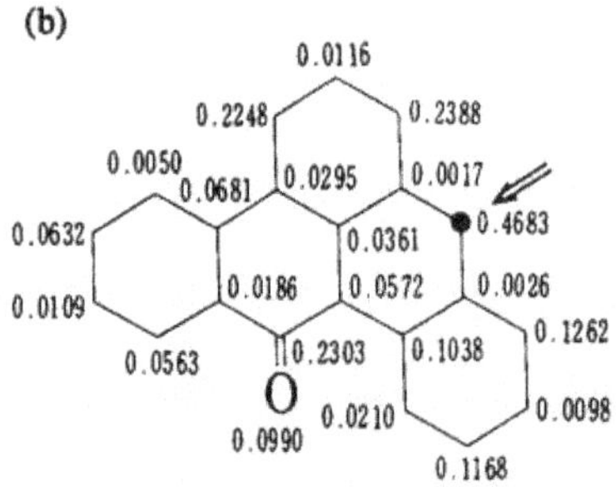

($\varepsilon^N = \alpha - 0.1351\beta$)
Benzo[5,6]mesobenzanthrone

OH (110°C, alcoholic KOH)
(Bradley and Sutcliffe, 1952)

(c)

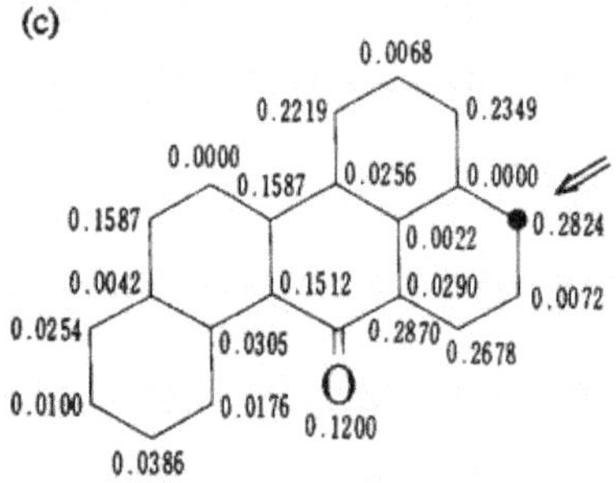

($\varepsilon^N = \alpha - 0.1648\beta$)
Benzo[8,9]mesobenzanthrone

OH (220°C, fused KOH)
(Bradley and Sutcliffe, 1952)

FIG. 7. Hydroxylation of benzomesobenzanthrones (numerical values are the frontier electron density for the nucleophilic attack, f_r^N). ε^N = the frontier orbital energy for the nucleophilic attack; α and β are the Coulomb and resonance integrals in benzene.

As is seen in Fig. 6, the arrows are directed to phenanthrene *meso*-type positions of this molecule. These positions of attack are, however, doubtful in the light of the numerical value of theoretical indices available. Reexamination of the original literature hopefully will indicate that this comes from an erroneous citation by the author of the above book.

Figure 7 illustrates an interesting example which is provided by the case of the hydroxylation of benzobenzanthrones. The extraordinarily large reactivity of the compound (b) in comparison with (a) and (c) may be understood by the numerical values of the frontier electron density and the frontier orbital energy.

C. Polycentric Reactions

The delocalization approach is employed also in various polycentric reactions. The reactivity index, in this case the stabilization by interconjugation at the transition state, is derived from a model indicated in Fig. 8. Of course, the two-step addition mechanism was also proposed in very sterically constrained cases (see, e.g., Newman, 1961):

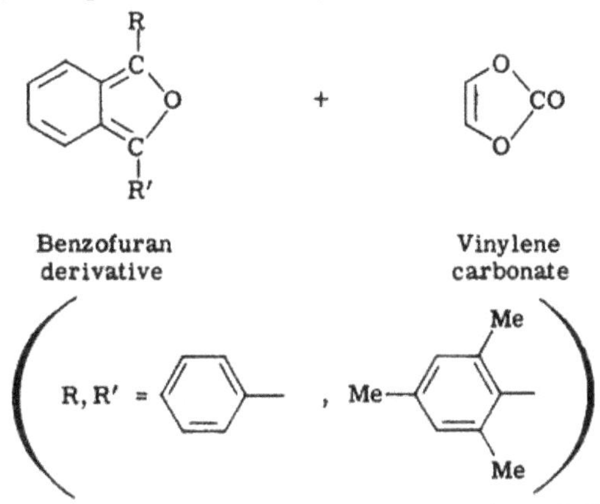

But in the present stage I could say that the interaction of dienes and dienophiles might take place simultaneously at two centers each, as is depicted in Fig. 8. [see, for instance, Fig. 10 (a). Were it a two-step addition, the first attack would most likely occur at the position of the highest frontier electron density; experimentally, the addition takes place at positions 1,4 to each other at which the sum of $(f_r)^{1/2}$ is largest; we can also give other such examples.]

If we take for granted the simultaneous addition mechanism for the Diels–Alder addition, the transition state must be such as indicated in Fig. 8 (a). In most cases dienophiles have larger ionization potential than dienes. The relationship of energy levels may be like Fig. 8(b). At the transition state, the electron in the highest occupied orbital of the diene will conjugate most dominantly with the lowest unoccupied orbital of the dienophile. The stabilization through this interconjugation at the transition state may be represented by Eq. (1) (second-order perturbation):

$$\Delta E = 2\left(\sum_{i}^{\text{occ}} \sum_{i'}^{\text{unocc}} - \sum_{i}^{\text{unocc}} \sum_{i'}^{\text{occ}}\right) \frac{(C^{i}_{1r}C^{i'}_{2r'} + C^{i}_{1s}C^{i'}_{2s'})^2}{\varepsilon_{1i} - \varepsilon_{2i'}} \gamma^2 \qquad (1)$$

(a)

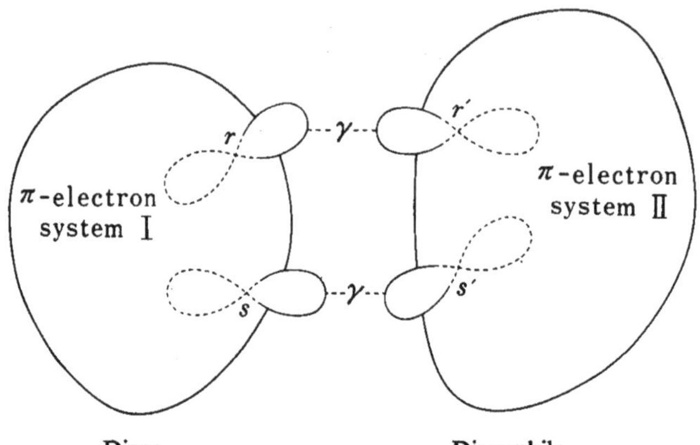

(b)

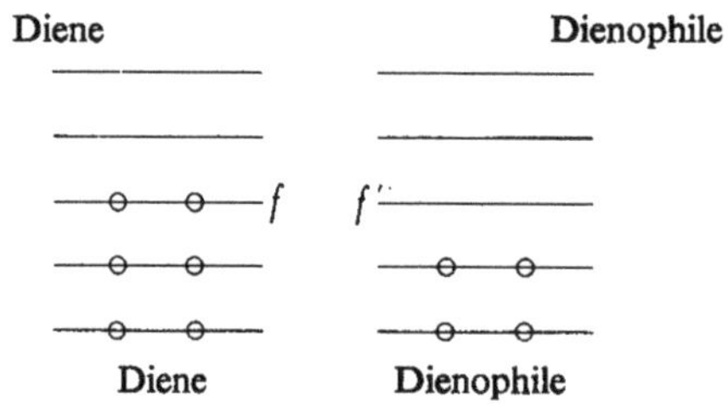

(c)

An example:

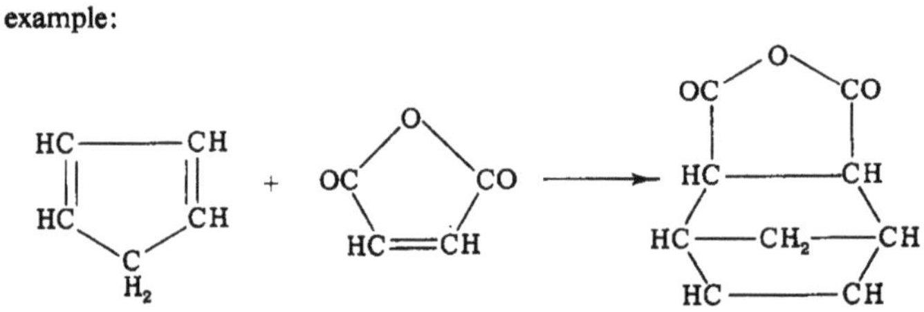

(*endo* addition)

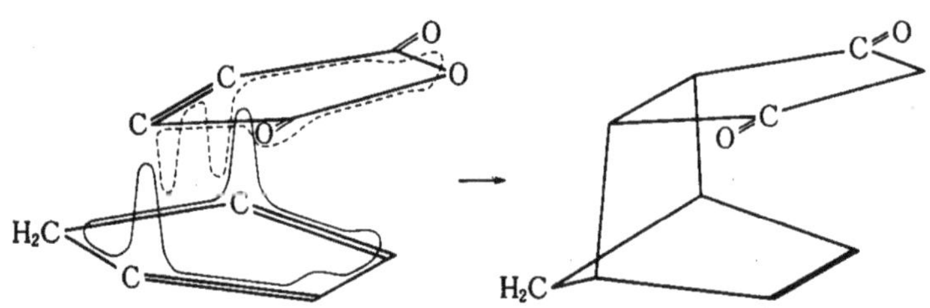

FIG. 8. Interaction of dienes and dienophiles in the Diels–Alder addition. ———, Distribution of frontier electrons for electrophilic attack; ...., distribution of frontier electrons for nucleophilic attack.

where the C_{1r}^{i} denotes the coefficient of the rth atomic orbital of the ith molecular orbital in the compound 1, ε_{1i} is the energy of the ith molecular orbital of the compound 1, and γ represents the resonance integral between two interacting atomic orbitals of each compound.

If the highest occupied orbital of the diene, which we may call the frontier orbital of the diene (denoted by f), lies very near to the lowest unoccupied orbital of the dienophile, which we may call the frontier orbital of the dienophile (denoted by f'), the stabilization energy may be approximated by Eq. (2) (first-order perturbation):

$$\Delta E \cong 2|C_{1r}^{f}C_{2r'}^{f'} + C_{1s}^{f}C_{2s'}^{f'}|\gamma \tag{2}$$

In Eq. (2), if the sign, plus or minus, of $C_{1r}^{f}C_{2r'}^{f'}$ is equal to the sign of $C_{1s}^{f}C_{2s'}^{f'}$, the stabilization must be large, while if they are opposite, the stabilization energy will become small. It is accordingly very interesting to investigate this symmetry relation of wave functions of both dienes and dienophiles which actually take part in the diene addition. In Table IV, we can see that the signs are favorable for making the stabilization energy large in all of the common dienes and dienophiles indicated in the table. Such a discussion is hardly possible in the bond localization approach (see, e.g., Brown 1950b, 1951). The possible electronic overlapping near the transition state is depicted in Fig. 8 (c) in regard to an example.

(10 hr 80°C) (Boekelheide and Vick, 1956) *peri*-Cycloheptanaphthalene

(≫8½ days, 80°C) (Boekelheide and Vick, 1956) *peri*-Cycloheptaacenaphthylene

FIG. 9. S_r and the Diels–Alder reactivity.

The Diels–Alder reactivity can also be interpreted in terms of the frontier electron density, provided we content ourselves with the so-called "isolated molecule approximation." Namely, if we assume that in Eq. (2) we take the ethylene molecule as the dienophile, where

$$C_{2r'}^{f'} = -C_{2s'}^{f'} = \frac{1}{\sqrt{2}}$$

then we obtain

$$\Delta E \cong (f_{1r})^{1/2} + (f_{1s})^{1/2} \tag{3}$$

TABLE IV
SYMMETRY OF FRONTIER ORBITAL WAVE FUNCTIONS

Frontier orbital of diene for electrophilic reagents	Frontier orbital of dienophile for nucleophilic reagents
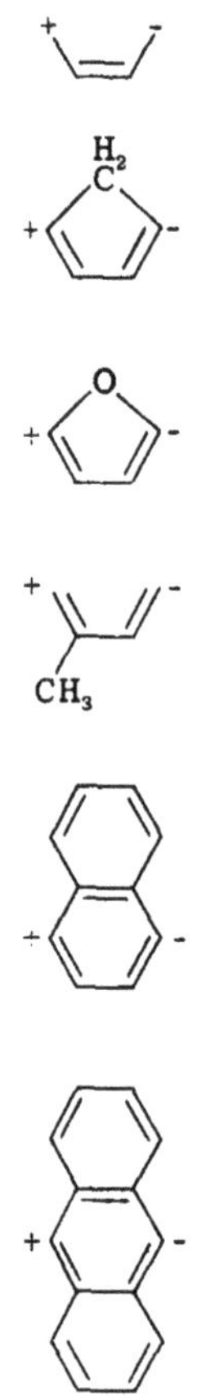	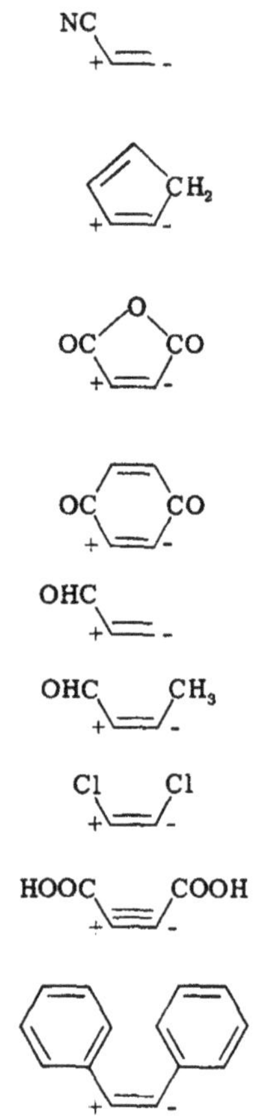

for the stabilization energy by the proposition that the above-mentioned sign relations hold between C^{f}_{1r} and $C^{f'}_{2r'}$. In this way we can find out the two positions in a molecule which are located mutually on 1,4 positions where the sum of $(f_r)^{1/2}$ is the largest. First, the Diels–Alder reactivity of *peri*-cycloheptanaphthalene[1](I) and *peri*-cycloheptaacenaphthylene[2](II) is compared

[1] This name follows IUPAC nomenclature. Boekelheide and Vick (1956) called this hydrocarbon "pleiadiene."

[2] Similarly, Boekelheide and Vick (1956) called this compound "acepleidylene."

with the value of superdelocalizability in Fig. 9. Experimentally, compound (I) is found to be more reactive than (II) toward Diels–Alder reagents. Namely, at 80°C (I) completely reacts with the reagent within 10 hours, while (II) requires more than 8 days to accomplish the reaction. In most conjugated molecules, the reactivity increases with the increase of the number of fused rings, and accordingly this result is not self-evident. Calculation of superdelocalizability indicates, however, that the experimental result is quite reasonable. Similar results were obtained by B. Pullman and Berthier (1956).

Next, we can see in Fig. 10 three complicated condensed aromatic hydro-

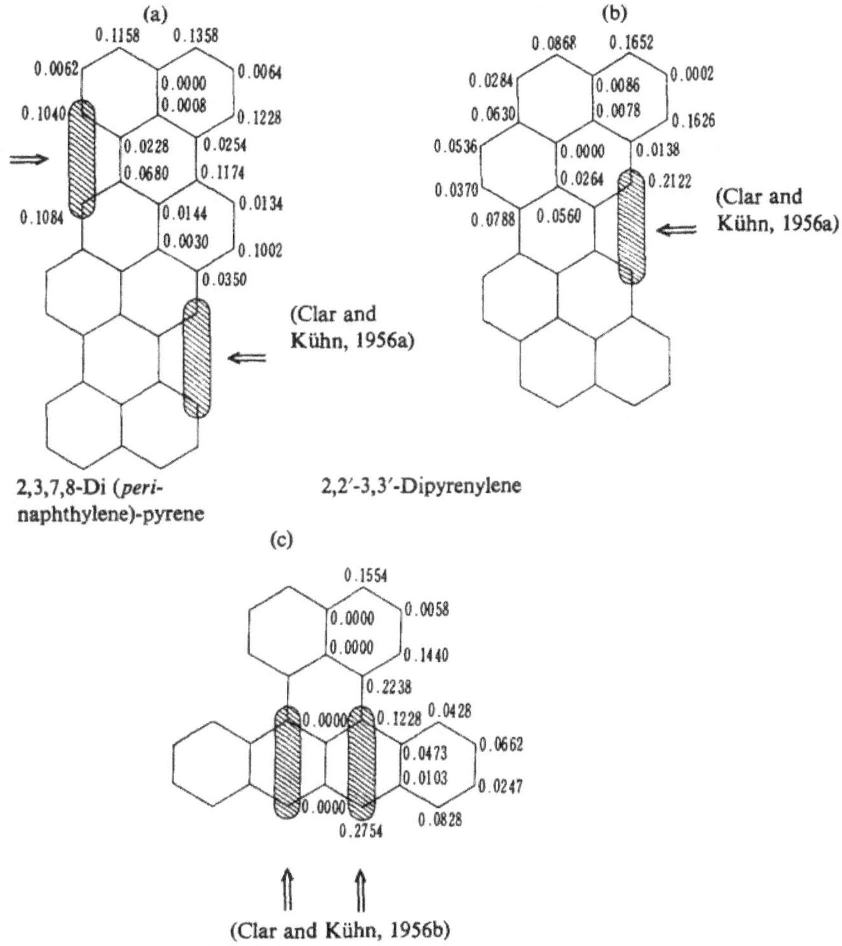

2,3,7,8-Di (*peri*-naphthylene)-pyrene

2,2′-3,3′-Dipyrenylene

1,2,5,6-Dibenzoperylene

FIG. 10. f_r and the Diels–Alder reactivity.

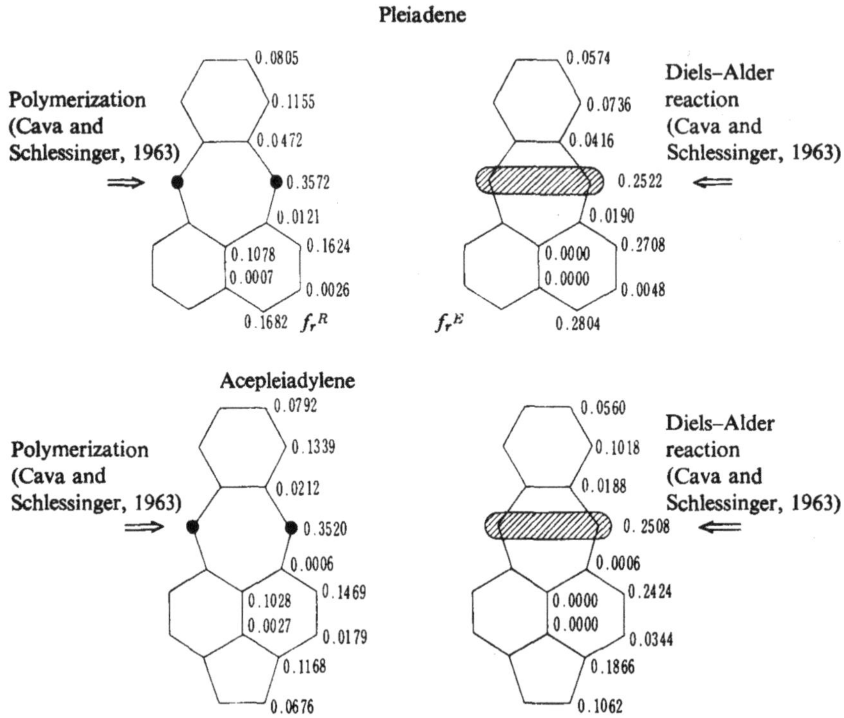

FIG. 11. Reactivity of pleiadene hydrocarbons. f_r^R = Frontier electron density for radical attack; f_r^E = frontier electron density for electrophilic attack.

carbon molecules which can be subjected to the Diels–Alder reaction. The actual position of attack is indicated by an arrow. Can any experimental organic chemists tell this position of attack exactly by their intuitive insight? We may never venture to say "Never." But one can most safely say that the frontier electron density can more dependably tell that position within a few minutes of calculation on an electronic computer.

Another example of the problem concerning the reactivity of large conjugated molecules is shown in Fig. 11, and other examples are provided by intramolecular rearrangements, for instance the Claisen rearrangement, Cope rearrangement, benzidine rearrangement, etc. (Streitwieser, 1961).

V. Interrelation of Various Approaches

The theoretical interpretation of the chemical reactivity of conjugated compounds has been outlined above, where a little stress has been put on the delocalization approach. The three approaches mentioned above, that is, static approach, localization approach, and delocalization approach, were all to some extent successful in interpreting the chemical reactivity of common aromatic hydrocarbons. But these three approaches are considerably different from one another in regard to the model from which these indices are derived. This fact may imply the possibility of finding some mathematical interrelation between any two of these indices.

In the common condensed aromatic hydrocarbons, in particular those having no odd-membered rings, we found some very beautiful mathematical relationships among various reactivity indices. This relation is shown in Table V by means of mathematical formulas. How intimately these are connected with one another may be quite clear at a glance at the Table. The explanation of each formula may be unnecessary. All of these indices, when they are expressed in integral representation, are represented only in terms of a function G_r. If $G_r > G_s$ with respect to a couple of positions in a molecule, then we have

$$\begin{array}{ll} S_r > S_s & L_r < L_s \\ F_r > F_s & H_r > H_s \\ \pi_{rr} > \pi_{ss} & l_r < l_s \\ & \bar{S}_r > \bar{S}_s \end{array}$$

Accordingly, we have the same predicted order of reactivity within one molecule with any one of these indices. In Eq. (4) (Table V), a surprisingly simple relation is given between Dewar's approximate localization energy and the approximate superdelocalizability (Fukui *et al.*, 1961b). Koutecký *et al.*, (1961) pointed out that the applicability of the Dewar index might be based on the relation (4). Equation (5) indicates how Brown's Z value relates to the frontier electron density (Fukui *et al.*, 1960b). Within one molecule, the order of Z value is utterly the same as that of the frontier electron density, as is immediately understood from this equation.

So far as these intimate connections exist in reactivity indices, the results of prediction by these indices with respect to common aromatic hydrocarbons and also with respect to their derivatives must be almost compatible with one another. But, accordingly, it is very difficult to pick up the correct approach to the true mechanism of aromatic substitution from a variety of theoretical treatments.

For this purpose, we suggest testing the applicability of various indices to the reaction of odd-membered hydrocarbons in which the mathematical correlation does not always hold, and also raising the accuracy of calculation of indices for comparison of their applicability, because most calculations of these indices have been based on the roughest approximate methods. It is hoped that investigations are to be carried out on this line.

TABLE V

MATHEMATICAL INTERRELATIONS OF CHEMICAL REACTIVITY INDICES[a]

For alternant aromatic hydrocarbons[b]:

$$\left\langle G_r(y) = \frac{\Delta_{rr}(iy)}{iy\Delta(iy)} \right\rangle$$

$$S_r = \frac{(-\beta)}{\pi} \int_{-\infty}^{+\infty} G_r(y)\, dy$$

$$L_r = \frac{1}{\pi} \int_{-\infty}^{+\infty} \ln \frac{1}{y^2 G_r(y)}\, dy$$

$$F_r = N_{\max} - \frac{1}{\pi(-\beta)} \int_{-\infty}^{+\infty} \{1 - y^2 G_r(y)\}\, dy$$

$$H_r = \frac{1}{\pi} \int_{-\infty}^{+\infty} \ln\,\{1 + \gamma^2\, G_r(y)\}\, dy$$

$$\pi_{rr} = \frac{(-\beta)}{\pi} \int_{-\infty}^{+\infty} y^2\, \{G_r(y)\}^2\, dy$$

$$l_r = \frac{2}{(-\beta)} G_r(0)^{-1/2} = 2(\bar{S}_r)^{-1/2} \tag{4}$$

where S_r = superdelocalizability; L_r = Wheland's localization energy; F_r = free valence; H_r = hyperconjugation energy; π_{rr} = self-atom-polarizability; l_r = Dewar's approximate localization energy; $\bar{S}_r$ = approximate superdelocalizability[c]; $\Delta(\varepsilon)$ = secular determinant; $\Delta_{rr}(\varepsilon)$ = (rr)-minor determinant.

For all conjugated systems:

$$\left\langle Z_r = \lambda_R - \lambda_f + \frac{f_r}{\lambda_R - \lambda_f}\gamma^2 \right\rangle \tag{5}$$

where f_r = frontier electron density; Z_r = Z value; γ = the resonance energy between the reagent and the reaction site of the π-electron system; λ_R = the energy of the reagent orbital and λ_f = the energy of the highest occupied π orbital (both in units of β with reference to α, where α and β are the Coulomb and the resonance integral of benzene).

[a] Fukui *et al.* (1957b).
[b] Coulson and Longuet-Higgins (1947a,b).
[c] Fukui *et al.* (1954a).

VI. Approach to the Reactivity of Saturated Compounds

Thus, the quantum-mechanical theory of reactivity has successfully been applied to unsaturated compounds. But it was no earlier than several years ago that the quantum-mechanical theory proved itself a powerful means of treating the reactivity of saturated compounds. Many papers have already been published (Lennard-Jones and Hall, 1952; Hall, 1953; Brown, 1953; Dewar and Pettit, 1954; Del Re, 1958; Lester, 1959; Fueki and Hirota, 1960a,b;

TABLE VI

THE REACTIVITY OF PARAFFIN HYDROCARBONS TOWARD HYDROGEN ABSTRACTION AND THE FRONTIER ELECTRON DENSITY AT HYDROGEN ATOMS

(i) Chlorination of 2-methyl butane:

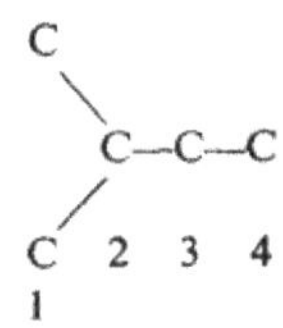

Position of attack	Frontier electron density	Relative rate[a]
2	0.0745	22
3	0.0481	14
1	0.0286	5.6
4	0.0278	5.5

(ii) Oxidation of 2-methyl pentane (reaction rate of initial stage):

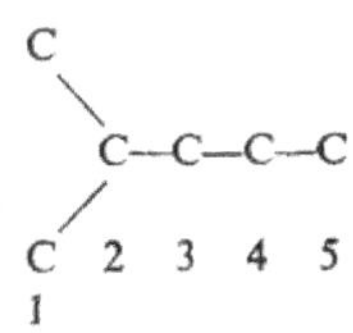

Position of attack	Frontier electron density	Relative activity[b]
2	0.0638	23.6
3	0.0479	5.1
4	0.0295	2.6
1	0.0225	2.4
5	0.0155	1

[a] Walling (1957).
[b] Cullis *et al.* (1958, 1959).

Yamazaki and Shida, 1960; Fukui *et al.*, 1960e, 1961a,c, 1962a,b; Yonezawa *et al.*, 1962; Morokuma *et al.*, 1963). Review of these results is omitted here. Only two results of such a calculation are instanced in Table VI with respect to the reactivity of 2-methylbutane and 2-methylpentane. We can find a very close relation between the relative rate of abstraction and the frontier electron density of each hydrogen atom. Also, Fig. 12 shows the probability of C–C

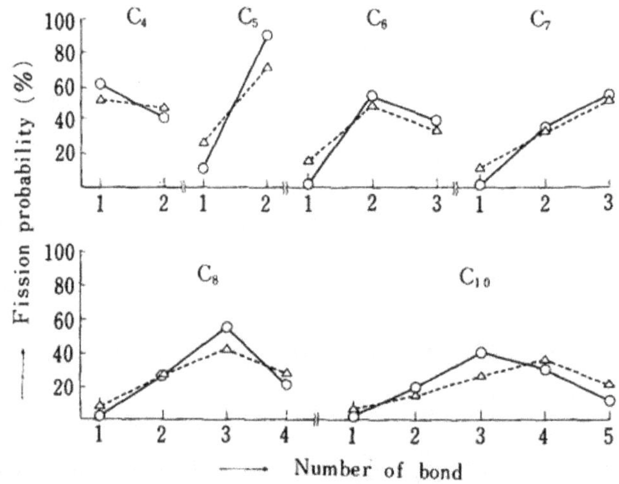

FIG. 12. The probability of C–C bond fission by electron impact (Fueki and Hirota, 1960b) and the bond frontier electron density (Fukui *et al.*, 1960e) of normal paraffins.

bond fission by electron impact and the bond frontier electron density (the bond density of the highest occupied molecular orbital (cf. Fueki and Hirota, 1960a; Fukui *et al.*, 1960e) of normal paraffins. In this case, too, a beautiful correlation can be observed between them.

REFERENCES

Aitken, I. M., and Reid, D. H. (1956). *J. Chem. Soc.* p.3487.
Arnold, H., and Pahls, K. (1956). *Chem. Ber.* **89**, 121.
Baker, W., Barton, J. W., and McOmie, J. F. W. (1958). *J. Chem. Soc.* p.2666.
Berthier, G., Coulson, C. A., Greenwood, H. H., and Pullman, A. (1948). *Compt. Rend. Acad. Sci.* **227**, 1906.
Bhattacharya, R., and Basu, S. (1962). *J. Chim. Phys.* **59**, 125.
Boekelheide, V., and Vick, G. K. (1956). *J. Am. Chem. Soc.* **78**, 653.
Bowen, E. L., and Marsh, L. D. F. (1947). *J. Chem. Soc.* p.109.
Bradley, W., and Sutcliffe, F. K. (1952). *J. Chem. Soc.* p.1247.
Brown, R. D. (1948). *Trans. Faraday Soc.* **44**, 984.
Brown, R. D. (1949). *Trans. Faraday Soc.* **45**, 296.

Brown, R. D. (1950a). *Trans. Faraday Soc.* **46**, 146.
Brown, R. D. (1950b). *J. Chem. Soc.* p.3249.
Brown, R. D. (1951). *J. Chem. Soc.* p.1950.
Brown, R. D. (1952). *Quart. Rev.* (*London*) **6**, 63.
Brown, R. D. (1953). *J. Chem. Soc.* p.2615.
Brown, R. D. (1959). *J. Chem. Soc.* p.2232.
Burkitt, F. H., Coulson, C. A., and Longuet-Higgins, H. C. (1951). *Trans. Faraday Soc.* **47**, 553.
Callighan, R. H., Tarker, M. F., Jr., and Wilt, M. H. (1961). *J. Org. Chem.* **26**, 1379.
Cava, M. P., and Schlessinger, R. H. (1963). *J. Am. Chem. Soc.* **85**, 835.
Chalvet, O., and Peltier, J. (1956). *Bull. Soc. Chim. France.* p.1667.
Clar, E., and Kühn, O. (1956a). *Ann. Chem.* **601**, 181.
Clar, E., and Kühn, O. (1956b). *Ann. Chem.* **601**, 193.
Coulson, C. A. (1947). *Discussions Faraday Soc.* **2**, 9.
Coulson, C. A. (1948). *J. Chim. Phys.* **45**, 243.
Coulson, C. A., and Longuet-Higgins, H. C. (1947a). *Proc. Roy. Soc.* (*London*) **A191**, 39.
Coulson, C. A., and Longuet-Higgins, H. C. (1947b). *Proc. Roy. Soc.* (*London*) **A192**, 16.
Cullis, C. F., Harday, F. R. F., and Turner, D. W. (1958). *Proc. Roy. Soc.* (*London*) **A244**, 57.
Cullis, C. F., Harday, F. R. F., and Turner, D. W. (1959). *Proc. Roy. Soc.* (*London*) **A251**, 265.
Daudel, R., and Chalvet, O. (1956). *J. Chim. Phys.* **53**, 943.
Daudel, R., Jacques, R., Jean, M., Sandorfy, C., and Vroelant, C. (1949). *J. Chim. Phys.* **46**, 249.
De La Mare, P. B., and Ridd, J. H. (1959). "Aromatic Substitutions," p.182. Butterworths, London.
Del Re, G. (1958). *J. Chem. Soc.* p.4031.
Dewar, M. J. S. (1946). *Trans. Faraday Soc.* **42**, 764.
Dewar, M. J. S. (1952). *J. Am. Chem. Soc.* **74**, 3357.
Dewar, M. J. S., and Pettit, R. (1954). *J. Chem. Soc.* p.1625.
Evans, M. G., and Polanyi, M. (1937). *Trans. Faraday Soc.* **33**, 448.
Eyring, H. (1935). *J. Chem. Phys.* **3**, 107.
Fernández-Alonso, J. I. (1951). *Compt. Rend. Acad. Sci.* **233**, 403.
Fernández-Alonso, J. I., and Domingo, R. (1958). *Selecta Chim.* No. 17, 13.
Fernández-Alonso, J. I., Vila, L. C., and Domingo, R. (1957a). *J. Am. Chem. Soc.* **79**, 5839.
Fernández-Alonso, J. I., Domingo, R., Vila, L. C., and Peradejordi, F. (1957b). *Anales Real Soc. Espan. Fis. Quim.* (*Madrid*) **B53**, 109.
Fueki, K., and Hirota, K. (1960a). *Nippon Kagaku Zasshi* **81**, 209 (in Japanese).
Fueki, K., and Hirota, K. (1960b). *Nippon Kagaku Zasshi* **81**, 212 (in Japanese).
Fueno, T. (1961). *Ann. Rev. Phys. Chem.* **12**, 303.
Fukui, K., Yonezawa, T., and Shingu, H. (1952). *J. Chem. Phys.* **20**, 722.
Fukui, K., Yonezawa, T., and Nagata, C. (1954a). *Bull. Chem. Soc. Japan* **27**, 423.
Fukui, K., Yonezawa, T., Nagata, C., and Shingu, H. (1954b). *J. Chem. Phys.* **22**, 1433.
Fukui, K., Yonezawa, T., and Nagata, C. (1957a). *J. Chem. Phys.* **26**, 831.
Fukui, K., Yonezawa, T., and Nagata, C. (1957b). *J. Chem. Phys.* **27**, 1247.
Fukui, K., Nagata, C., and Yonezawa, T. (1958a). *J. Chem. Am. Soc.* **80**, 2267.
Fukui, K., Nagata, C., and Yonezawa, T. (1958b). *Gann.* **49**, 43.
Fukui, K., Morokuma, K., and Yonezawa, T. (1959). *Bull. Chem. Soc. Japan*, **32**, 853.
Fukui, K., Nagata, C., and Imamura, A. (1960a). *Science.* **132**, 87.
Fukui, K., Morokuma, K., Yonezawa, T., and Nagata, C. (1960b). *J. Chem. Phys.* **32**, 1743.
Fukui, K., Nagata, C., Yonezawa, T., Inamoto, Y., and Imamura, A. (1960c). *Gann.* **51**, 67.

Fukui, K., Imamura, A., and Nagata, C. (1960d). *Bull. Chem. Soc. Japan*, **33**, 122.
Fukui, K., Kato, H., and Yonezawa, T. (1960e) *Bull. Chem. Soc. Japan*, **33**, 1201.
Fukui, K., Morokuma, K., and Nagata, C. (1960f). *Bull. Chem. Soc. Japan* **33**, 1214.
Fukui, K., Kato, H., and Yonezawa, T. (1961a). *Bull. Chem. Soc. Japan* **34**, 442.
Fukui, K., Kato, H., Nagata, C., and Yonezawa, T. (1961b). *Bull. Chem. Soc. Japan* **34**, 933.
Fukui, K., Kato, H., and Yonezawa, T. (1961c). *Bull. Chem. Soc. Japan* **34**, 1111.
Fukui, K., Morokuma, K., Nagata, C., and Imamura, A. (1961d). *Bull. Chem. Soc. Japan* **34**, 1224.
Fukui, K., Nagata, C., Yonezawa, T., Morokuma, K., and Imamura, A. (1961e). *Gann.* **52**, 1.
Fukui, K., Nagata, C., Imamura, A., and Tagashira, Y. (1961f). *Gann.* **52**, 127.
Fukui, K., Kato, H., Yonezawa, T., Morokuma, K., Imamura, A., and Nagata, C. (1962a). *Bull. Chem. Soc. Japan* **35**, 38.
Fukui, K., Kato, H., and Yonezawa, T. (1962b). *Bull. Chem. Soc. Japan* **35**, 1475.
Fukui, K., Nagata, C., Imamura, A., Yonezawa, T., and Tagashira, Y. (1962c). *Gann.* **53**, 25.
Hall, G. G., (1953). *Trans. Faraday Soc.* **49**, 217.
Hayashi, K., Yonezawa, T., Nagata, C., Okamura, S., and Fukui, K. (1956). *J. Polymer Sci.* **20**, 537.
Hückel, E. (1931). *Z. Physik*, **72**, 312.
Hückel, E. (1937a). *Z. Physik. Chem.* **B35**, 163.
Hückel, E. (1937b). *Z. Elektrochem.* **43**, 827.
Ingold, C. K. (1934). *Chem. Rev.* **15**, 225.
Ioffe, I. S., and Efros, L. S. (1946). *Zh. Obshch. Khim.* **16**, 111; (1947).
Kobayashi, H. (1960). *Nippon Kagaku Zasshi* **81**, 519 (in Japanese).
Kobayashi, H. (1961). *Nippon Kagaku Zasshi* **82**, 272 (in Japanese).
Kolboe, S, and Pullman, A. (1958). "*Calcul des Functions d'Onde Moléculaire*," p. 213. C.N.R.S.
Koshimizu, K., Fujita, T., and Mitsui, T. (1960). *J. Am. Chem. Soc.* **82**, 4041.
Koutecký, J., Zahradník, R., and Čížek, J. (1961). *Trans. Faraday Soc.* **57**, 169.
Kurita, Y., and Kubo, M. (1951). *Bull. Chem. Soc. Japan* **24**, 13.
La Mer, V. K. (1933). *J. Chem. Phys.* **1**, 289.
Leffler, J. E. (1955). *J. Org. Chem.* **20**, 1202.
Lennard-Jones, J., and Hall, G. G. (1952). *Trans. Faraday Soc.* **47**, 581.
Lester, G. R. (1959). "Advances in Mass Spectrometry," p.287. Pergamon Press, New York.
Levinson, G. S. (1962). *J. Polymer Sci.* **60**, 43.
Longuet-Higgins, H. C., and Coulson, C. A. (1947). *Trans. Faraday Soc.* **43**, 87.
Mackor, E. L., Hofstra, A., and van der Waals, J. H. (1958). *Trans. Faraday Soc.* **54**, 66.
Mestechkin, M. M. (1960). *Vestn. Leningr. Univ.* No. 22, 98.
Morokuma, K., Fukui, K., Yonezawa, T., and Kato, H. (1963). *Bull. Chem. Soc. Japan* **36**, 47.
Moser, C. M. (1954). *Compt. Rend. Acad. Sci.* **238**, 1585.
Moser, C. M. (1955) *J. Chim. Phys.* **52**, 24.
Muir, R. M., and Hansch, C. (1961). "Plant Growth Regulations," p.249. Iowa State Univ. Press, Ames, Iowa.
Muller, N., Picket, L. W., and Mulliken, R. S. (1954). *J. Am. Chem. Soc.* **76**, 4770.
Nagakura, S. (1961). *Technical Report of Institute for Solid State Physics, Tokyo Univ. Ser. A.* No. 35.
Nagakura, S., and Tanaka, J. (1954). *Nippon Kagaku Zasshi* **75**, 933 (in Japanese).
Nagakura, S., and Tanaka, J. (1959). *Bull. Chem. Soc. Japan* **32**, 734.

Nagata, C., Fukui, K., Yonezawa, T., and Tagashira, Y. (1955). *Cancer Res.* **15**, 233.
Newman, M. S. (1961). *J. Org. Chem.* **26**, 2630.
Ogata, Y., and Akimoto, H. (1962). *J. Org. Chem.* **27**, 294.
Parr, R. G., and Mulliken, R. S. (1950). *J. Chem. Phys.* **18**, 1338.
Pauncz, R., and Cohen, A. (1960). *J. Chem. Soc.* p.3288.
Pilar, F. L. (1958). *J. Chem. Phys.* **29**, 1119.
Pilar, F. L. (1959). *J. Chem. Phys.* **30**, 375.
Pelzer, H., and Wigner, E. (1932). *Z. Physik. Chem.* **B15**, 445.
Polanyi, M. (1937). *J. Chem. Soc.* p.629.
Polanyi, M., and Evans, M. G., (1935). *Trans. Faraday Soc.* **31**, 875.
Pullman, A. (1954). *J. Chim. Phys.* **51**, 188.
Pullman, A., and Berthier, G. (1948). *Compt. Rend. Acad. Sci.* **227**, 677.
Pullman, A., and Berthier, G., (1949). *Compt. Rend. Acad. Sci.* **229**, 567.
Pullman, A., and Metzger, J. (1948). *Bull. Soc. Chim. France* p.1021.
Pullman, A., and Pullman, B. (1946). *Experientia* **2**, 364.
Pullman, A., Pullman, B., Bergmann, E. O., Berthier, G., Fischer, E., Hirshberg, Y., and Pouks, Y. (1951). *J. Chim. Phys.* **48**, 359.
Pullman, B. (1954). *Cahiers Phys.* **48**, 42.
Pullman, B., and Berthier, G. (1948). *Bull. Soc. Chim. France*, p.551.
Pullman, B., and Berthier, G. (1956). *Compt. Rend. Acad. Sci.* **241**, 2563.
Rasch, G. (1962). *Z. Physik. Chem.* **219**, 180.
Ri, T. (1943). *Rev. Phys. Chem. Japan* **17**, 1, 16.
Ri, T., and Eyring, H. (1940). *J. Chem. Phys.* **8**, 433.
Sandorfy, C. (1949). *Bull. Soc. Chim. France*, p.615.
Sato, S., and Cvetanović, R. J. (1959). *J. Am. Chem. Soc.* **81**, 3223.
Stephenson, M., and Sutcliffe, F. K. (1962). *J. Chem. Soc.* p.3516.
Streitwieser, A., Jr. (1961). "Molecular Orbital Theory for Organic Chemists," pp. 307–448. Wiley, New York.
Streitwieser, A., Jr., and Fahey, R. C. (1962). *J. Org. Chem.* **27**, 2352.
Szabo, L. (1951). *Compt. Rend. Acad. Sci.* **233**, 625.
Tolman, R. C. (1927). "Statistical Mechanics," p.257. Chem. Catalog Co., New York.
Walling, C. (1957) "Free Radicals in Solution," p.359. Wiley, New York.
Wheland, G. W. (1942). *J. Am. Chem. Soc.* **64**, 900.
Wheland, G. W., and Pauling, L. (1933). *J. Chem. Phys.* **1**, 606.
Wynne-Jones, W. F. K., and Eyring, H. (1935). *J. Chem. Phys.* **3**, 492.
Yamazaki, H., and Shida, S. (1960). *J. Chem. Phys.* **32**, 950.
Yonezawa, T., Hayashi, K., Nagata, C., Okamura, S., and Fukui, K. (1954). *J. Polymer Sci.* **14**, 312.
Yonezawa, T., Higashimura, T., Katagiri, K., Hayashi, K., Okamura, S., and Fukui, K. (1957). *J. Polymer Sci.* **26**, 311.
Yonezawa, T., Kato, H., Saito, H., and Fukui, K. (1962). *Bull. Chem. Soc. Japan* **35**, 1814.
Zahradník, R., Párkányi, C., and Koutecký. J. (1962a). *Collection Czech. Chem. Commun.* **27**, 1242.
Zahradník, R., Vlachová, D., and Koutecký, J. (1962b). *Collection Czech. Chem. Commun.* **27**, 2336.
Zollinger, H. (1960). "Diazo and Azo Chemistry," pp.293, 374. Interscience, New York.

Reprinted from LES PRIX NOBEL EN 1981

KENICHI FUKUI

I was born the eldest of three sons of Ryokichi Fukui, a foreign trade merchant and factory manager, and Chie Fukui, in Nara, Japan, on October 4, 1918. In my high school years, chemistry was not my favourite subject, but the most decisive occurrence in my educational career came when my father asked the advice of Professor Gen-itsu Kita of Kyoto Imperial University concerning the course I should take. Prof. Kita suggested that Ryokichi, one of his juniors from the same native province, should send me to the Department of Industrial Chemistry with which he was then affiliated.

For a few years after my graduation from Kyoto Imperial University in 1941, I was engaged in experimental research on synthetic fuel chemistry in the Army Fuel Laboratory. The result brought me a prize in 1944. I became lecturer in the Fuel Chemistry Department of Kyoto Imperial University in 1943, assistant professor in 1945, and professor in 1951. In 1947 I married Tomoe Horie. I have two children, Tetsuya (son) and Miyako (daughter).

While I started originally as an experimentalist, I had built up a subgroup of theoreticians in my group before 1956. My work on experimental organic chemistry continued along with this, and the results were mostly published in Japanese papers, the number of which amounted to 137 during the period 1944–1972, together with my papers on reaction engineering and catalytic engineering.

But the nature of my main work in chemistry can be better represented by more than 280 English publications, of which roughly 200 concern the theory of chemical reactions and related subjects. Other English papers relate to statistical theory of gellation, organic synthesis by inorganic salts, and polymerization kinetics and catalysts.

My first scientific delight came in 1952 when I found a correlation between the frontier electron density and the chemical reactivity in aromatic hydrocarbons. This success led my theoretical group to the chemical reactivity theory, extending more and more widely the range of compound and reactions that were discussed.

The year in which my 1952 paper was published was the same as the Professor Mulliken's publication of the important paper on the charge-transfer force in donor-acceptor complexes. Influenced by this paper, I gave a theoretical foundation for the findings mentioned above. The basic idea was essentially the consideration of the importance of the electron delocalization between the frontier orbitals of reactant species. The frontier orbital approach was further developed in various directions by my own group and many other scientists, both theoretical and experimental.

I was also interested in formulating the path of chemical reactions. The first paper appeared in 1970. This simple idea served to provide information on the geometrical shape of reacting molecules, and was able to make the role of the frontier orbitals in chemical reactions more distinct through visualization, by drawing their diagrams.

I must confess that, when I was writing the 1952 paper, I never imagined I would be coming to Stockholm to receive the Nobel Prize 30 years later. But I have to add that already at that time Professor Gen-itsu Kita encouraged me by suggesting the possibility of the growth of my theory leading me one day to this supreme prize. The possibility became a reality through the good circumstances in which I found myself: with my teachers, my colleagues and students, and, of course, my parents and family.

Some other details which perhaps should be mentioned are:
The Japan Academy Medal, May 1962.
National Science Foundation Senior Foreign Scientist Fellow, Feb.–July, 1970.
Member, International Academy of Quantum Molecular Science (France), from July 1970 to the present.
Councillor, Kyoto University, Nov. 1970–March 1973.
Dean, Faculty of Engineering, Kyoto University, April 1971–March 1973.
US–Japan Eminent Scientist Exchange Program Chemist, Sept. 1973.
Counsellor, Institute for Molecular Science, from Jan. 1976 to the present.
Vice-President, Chemical Society of Japan, March 1978–Feb. 1979.
Chairman, Executive Committee, 3rd International Congress of Quantum Chemistry, Kyoto, Oct.–Nov. 1979.
Foreign Associate, National Academy of Sciences, April 1981.
Order of Culture, Nov. 1981.
Person of Cultural Merits, Nov. 1981.

THE ROLE OF FRONTIER ORBITALS IN CHEMICAL REACTIONS

Nobel lecture, 8 December, 1981

by

KENICHI FUKUI

Department of Hydrocarbon Chemistry, Kyoto University, Sakyo-ku, Kyoto 606, Japan

Since the 3rd century for more than a thousand years chemistry has been thought of as a complicated, hard-to-predict science. Efforts to improve even a part of its unpredictable character are said to have born fruit first of all in the success of the "electronic theory". This was founded mainly by organic chemists, such as Fry, Stieglitz, Lucas, Lapworth and Sidgwick, brought to a completed form by Robinson and Ingold, and developed later by many other chemists.[1] In the electronic theory, the mode of migration of electrons in molecules is noted and is considered under various judgements. For that purpose, a criterion is necessary with respect to the number of electrons which should originally exist in an atom or a bond in a molecule. Therefore, it can be said to be the concept by Lewis of the sharing of electrons that has given a firm basis to the electronic theory.[2]

In the organic electronic theory, the chemical concepts such as acid and base, oxidation and reduction and so on, have been conveniently utilized from a long time ago. Furthermore, there are terms centring closer around the electron concept, such as electrophilicity and nucleophilicity, and electron donor and acceptor both being pairs of relative concepts.

One may be aware that these concepts can be connected qualitatively to the scale of electron density or electric charge. In the electronic theory, the static and dynamic behaviours of molecules are explained by the electronic effects which are based on nothing but the distribution of electrons in a molecule.

The mode of charge distribution in a molecule can be sketched to some extent by the use of the electronegativity concept of atoms through organic chemical experience. At the same time, it is given foundation, made quantitative, and supported by physical measurements of electron distribution and theoretical calculations based on quantum theory.

The distribution of electrons or electric charge—with either use the result is unchanged—in a molecule is usually represented by the total numbers (generally not integer) of electrons in each atom and each bond, and it was a concept easily acceptable even to empirical chemists as having a tolerably realistic meaning. Therefore, chemists employed the electron density as a fundamental concept to explain or to comprehend various phenomena. In particular, for the purpose of promoting chemical investigations, researchers usually rely upon the analogy through experience, and the electron density was very effectively and widely used as the basic concept in that analogy.

When the magnitude of electron density is adopted as the criterion the electrostatic attraction and repulsion caused by the electron density are taken into account. Therefore, it is reasonable to infer that an electrophilic reagent will attack the position of large electron density in a molecule while a nucleophilic reaction will occur at the site of small electron density. In fact, Wheland and Pauling[3] explained the orientation of aromatic substitutions in substituted benzenes along these lines, and theoretical interpretations of the mode of many other chemical reactions followed in the same fashion.

However, the question why one of the simple reactions known from long before, the electrophilic substitution in naphthalene, for instance, such as nitration, yields α-substituted derivatives predominantly was not so easy to answer. That was because, in many of such unsubstituted aromatic hydrocarbons, both the electrophile and the nucleophile react at the same location. This point threw some doubt on the theory of organic reactivity, where the electron density was thought to do everything.

THE CONCEPT OF FRONTIER ORBITAL INTERACTIONS

The interpretation of this problem was attempted by many people from variously different angles. Above all, Coulson and Longuet-Higgins[4] took up the change of electron density distribution under the influence of approaching reagent. The explanation by Wheland[5] was based on the calculation of the energy required to localize electrons forcibly to the site of reaction. But I myself tried to attack this problem in a way which was at that time slightly unusual. Taking notice of the principal role played by the valence electrons in the case of the molecule formation from atoms, only the distribution of the electrons occupying the highest energy π orbital of aromatic hydrocarbons was calculated. The attempt resulted in a better success than expected, obtaining an almost perfect agreement between the actual position of electrophilic attack and the site of large density of these specified electrons as exemplified in Fig. 1.[6]

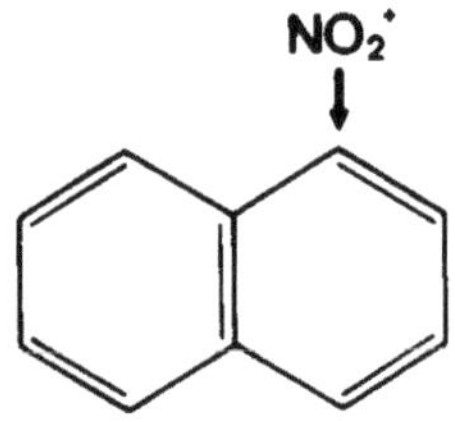

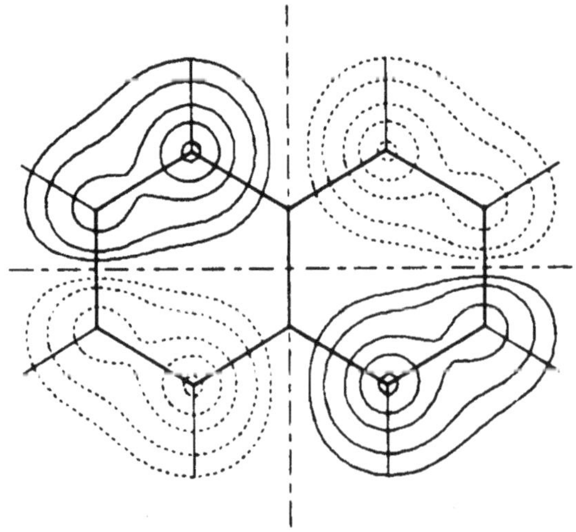

Fig. 1. Nitration of naphthalene.

The "orbital" concept, which was established and developed by many scientists, such as Pauling, Slater, Mulliken, Roothaan, Löwdin, Hückel, Parr and so on, had till then been employed to construct the wave function of a molecule, through which molecular properties were usually interpreted.[7] It seemed that the electron distribution in an orbital was directly connected to chemical observations and this fact was certainly felt to be interesting by many chemists.

But the results of such a rather "extravagant" attempt was by no means smoothly accepted by the general public of chemists. That paper received a number of controversial comments. This was in a sense understandable, because, for lack of my experiential ability, the theoretical foundation for this conspicuous result was obscure or rather improperly given. However, it was fortunate for me that the paper on the charge-transfer complex of Mulliken[8] was published in the same year as ours.

The model of Mulliken et al. for protonated benzene was a good help.[9] Our work in collaboration with Yonezawa, Nagata and Kato provided a simple and pointed picture of theoretical interpretation of reactions,[10] as well as the "overlap and orientation" principle proposed by Mulliken with regard to the orientation in molecular complexes.[11] Subsequent to the electrophilic substitution, the nucleophilic substition was discussed and it was found that in this case the lowest energy vacant orbital played this particular part.[12] In reactions with radicals, both of the two orbitals mentioned above, became the particular orbitals.

There was no essential reason to limit these particular orbials to π orbitals, so that this method was properly applied not only to unsaturated compounds but also to saturated compounds. The applicability to saturated compounds was a substantial advantage in comparison with many theories of reactivity which were then available only for π electron compounds. The method displayed its particular usefulness in the hydrogen abstraction by radicals from paraffinic hydrocarbons, the S_N2 and E2 reactions in halogenated hydrocarbons, the nucleophilic abstraction of α-hydrogen of olefins, and so forth.[13]

These two particular orbitals, which act as the essential part in a wide range of chemical reactions of various compounds, saturated or unsaturated, were referred to under the general term of "frontier orbitals", and abbreviated frequently by HOMO (highest occupied molecular orbital) and LUMO (lowest unoccupied molecular orbital).

In this way, the validity of the theory became gradually clearer. The vein of ore discovered by chance was found to be hopefully more extensive than expected. But it was attributed to the role of the symmetry of particular orbitals pointed out in 1964 with regard to Diels-Alder reactions[14] that the utility of our studies was further broadened. It was remarked that as is seen in Fig. 2, the symmetries of HOMO and LUMO of dienes and those of LUMO and HOMO of dienophiles, respectively, were found to be in a situation extremely favourable for a concerted cyclic interaction between them.

This signified the following important aspects: First, it pointed out a possible correlation between the orbital symmetry and the rule determining the sub-

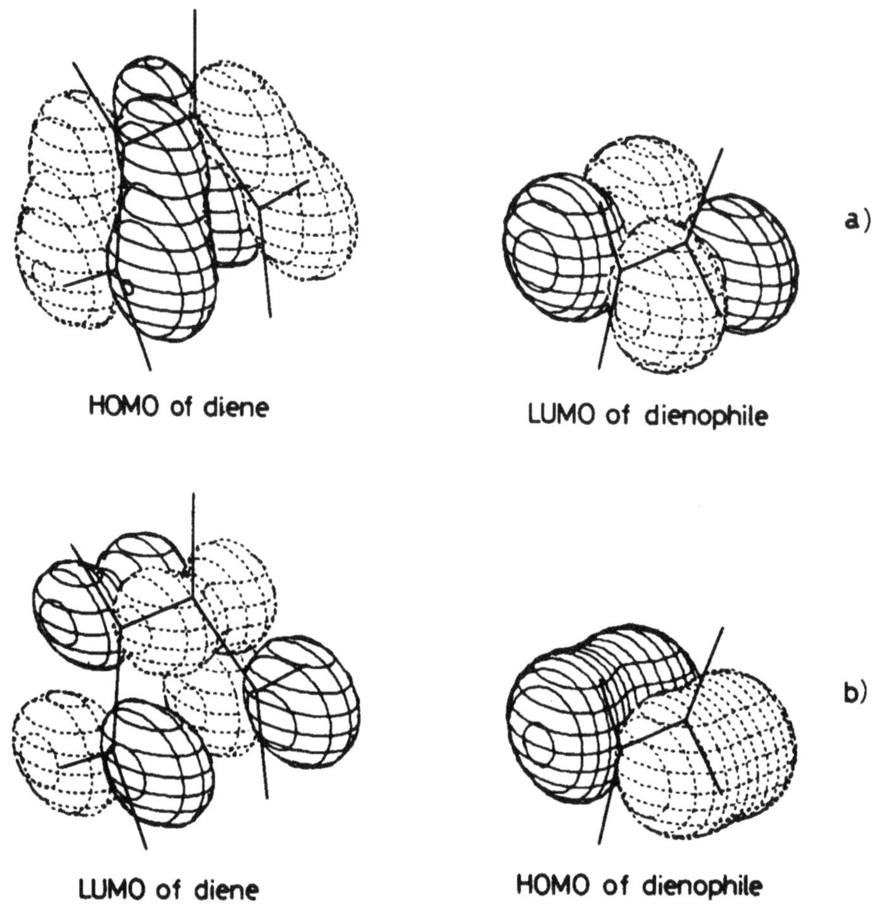

Fig. 2. The significance of orbital symmetry in the HOMO-LUMO overlapping in Diels-Alder reactions.

stantial occurrence or non-occurrence of a chemical reaction, which may be called the "selection rule", in common with the selection rule in molecular spectroscopy. Second, it provided a clue to discuss the question concerning what was the "concertedness" in a reaction which forms a cycle of electrons in conjugation along the way.

In 1965 Woodward and Hoffmann proposed the stereoselection rules which are established today as the "Woodward-Hoffmann" rules.[15, 16] An experimental result developed in Havinga's important paper[17] was extended immensely. It is only after the remarkable appearance of the brilliant work by Woodward and Hoffmann that I have become fully aware that not only the density distribution but also the nodal property of the particular orbitals have significance in such a wide variety of chemical reactions. In fact, we studied previously the noted (4n+2) rule proposed by Hückel[18] and noticed that the sign of the bond order in the highest energy electron orbital of an open-chain conjugation should be closely related to the stabilization of the corresponding conjugated rings.[19] We did not imagine, however, on that occasion that the discussion might be extended to the so-called Möbius-type ring-closure![20]

By considering the HOMO-LUMO interactions between the fragments of a conjugated chain divided into parts,[21] the frontier orbital theory can yield selection rules which are absolutely equivalent to those obtained from the principle called "the conservation of orbital symmetry" by Woodward and Hoffmann. One point that I may stress here is, as was pointed out by Fujimoto, Inagaki and myself,[22] that the electron delocalization between the particular orbitals interprets definitely in terms of orbital symmetries the formation and breaking of chemical bonds which, I believe, should be a key for perceiving chemical reaction processes.

In the cycloaddition of butadiene and ethylene shown in Fig. 2, both the interaction between the HOMO of diene and the LUMO of dienophile and that between the LUMO of diene and the HOMO of dienophile stabilize the interacting system. If one is interested in the local property of interaction, however, one may recognize the clear distinction between the roles of the two types of orbital interactions. The HOMO of ethylene and the LUMO of butadiene are both symmetric with regards to the symmetry plane retained throughout the course of cycloaddition. This signifies that each of the carbon atoms of ethylene are bound to both of the terminal carbons of butadiene. The chemical bonding between the diene and dienophile thus generated may be something like the one in a loosely bound complex, e.g., protonation to an olefinic double bond. On the contrary, the HOMO of butadiene and the LUMO of ethylene are antisymmetric. The interaction between these orbitals leads, therefore, to two separated chemical bonds, each of which combines a carbon atom of ethylene and a terminal carbon atom of butadiene. Needless to say it is the interaction between the HOMO of diene and the LUMO of dienophile that is of importance for the occurrence of concerted cycloaddition.[22]

In this way, it turned out in the course of time that the electron delocalization between HOMO and LUMO generally became the principal factor determining the easiness of a chemical reaction and the stereoselective path, irrespective of intra- and intermolecular processes, as illustrated in Fig. 3. Besides our own school, a number of other chemists made contributions. I want to refer to several names which are worthy of special mention.

First of all, the general perturbation theory of the HOMO-LUMO interaction between two molecules was built up by Salem.[23–25] One of Salem's papers[25] was in line with the important theory of Bader,[26] which specified the mode of decomposition of a molecule or a transition complex by means of the symmetry of the normal vibration. Furthermore Pearson[27] investigated the relation between the symmetry of reaction coordinates in general and that of HOMO and LUMO.

The discussion so far may seem to be an overestimation of these selected orbitals, HOMO and LUMO. This point was ingeniously modified by Klopman.[28] He carefully took into account the factors to be considered in the perturbation theory of reacting systems and classified reactions into two cases: the one was "frontier-controlled" case in which the reaction was controlled by the particular orbital interaction, and the other was the "charge-controlled"

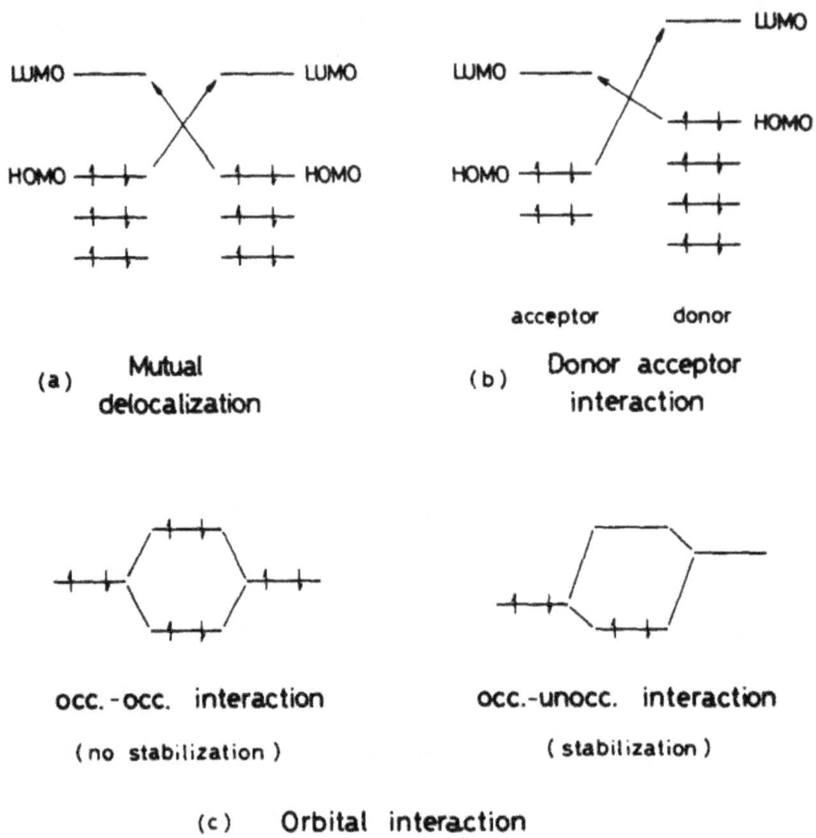

Fig. 3. The mode of interaction between orbitals of two molecules.

case, where it was controlled by the electrostatic interaction of charges. This classification was conveniently used by many people. In this context the review articles of Herndon[29] and of Hudson[30] appeared to be very useful. The names of Coulson[4] and Dewar[31] should also be noted here as those who contributed to the development of reactivity theories.

Returning to the subject again, let us assume that two molecules approach each other and orbital overlapping takes place. The perturbation theory[32] of this sort of interaction indicates that, the larger the orbital overlapping is and the smaller the level separation of two overlapping orbitals is, the larger is the contribution of the orbital pair to the stabilization of an interacting system. Accordingly, at least at the beginning, a reaction will proceed with a mutual nuclear configuration which is most favourable for the HOMO-LUMO overlapping.

Now let us suppose an electron flow from the HOMO of molecule I to the LUMO of molecule II. In each molecule the bonds between the reaction centre—the place at which the orbitals overlap with those of the other molecule—and the remaining part of the molecule are weakened. On this occasion, in molecule I the bonds which are bonding in HOMO are weakened and those antibonding in HOMO are strengthened, while in molecule II the bonds which are antibonding in LUMO are weakened and those bonding in LUMO are strengthened. Consequently, the HOMO of molecule I particularly destabilizes as compared with the other occupied orbitals, and the LUMO of molecule II discriminatively stabilizes among unoccupied orbitals, so that the HOMO-LUMO level separation between the two molecules is decreased. Such a circumstance is clearly understandable in Fig. 2.

The following tendency is further stressed. When the bond weakenings specified above have arisen, the HOMO and the LUMO tend to become more localized at these weakened bonds in each molecule. Besides, the weakening of the bonds between the reaction centre and the remaining part causes an increase of the amplitudes of HOMO and LUMO at the reaction centres, resulting in a larger overlapping of HOMO and LUMO.[33] Such a trend of the characteristic change in the orbital pattern is made numerically certain by actual calculations. The role of interaction between HOMO and LUMO turns out in this way to become more and more important as the reaction proceeds.

A series of studies on chemical interactions were attempted in which the interaction of reactants was divided into the Coulomb, the exchange, the polarization, and the delocalization interactions, and their magnitude of contribution to the interaction energy was quantitatively discussed.[32, 34] The interactions discussed by this method were the dimerization[35] and the addition to ethylene[36] of methylene and the dimerization of BH_3,[37] and also several donor-acceptor interactions—BH_3-NH_3,[38] BH_3-CO,[39] NH_3-HF,[40] etc. The method was applied also to reactions of radicals, such as the abstraction of a methane hydrogen by methyl radical, the addition of methyl radical to ethylene,[41] and recombinations, disproportionations, and self-reactions of two radicals.[42] In these calculations, the configuration analysis proposed originally by Baba[43] was also utilized conveniently. We could show numerically the mode of increase of the electron delocalization from HOMO to LUMO along with the proceeding of reaction, the increasing weight of contribution of such a delocalization to the stabilization of the reacting system, the driving force of the reaction in terms of orbital interactions, and so on.

The question "Why HOMO and LUMO solely determine the reaction path?" was one which I very frequently received from the audiences in my lectures given in the past in different places. The discussion so far made here is thought to correspond, at least partly, to that answer. But one may not adhere so strictly to the HOMO and LUMO. In one-centre reactions like substitutions, which the orbital symmetry has nothing to do with, any occupied orbitals which are very close to HOMO should properly be taken into account.[12] In large paraffin molecules a number of HOMO's (high-lying occupied MO's), and furthermore as will shortly be referred to later, in metal crystals, even "HOMO-band" must be taken along the line of reactivity argument. If HOMO or LUMO happens to be inadequate owing to its extension, the symmetry, or the nodal property, the next orbital should be sought for. One of the simplest examples of such an instance is the protonation of pyridine. In this case, the nitrogen lone-pair orbital is not HOMO, but the addition of proton to the nitrogen lone-pair so as not to disturb the π conjugation will evidently be more advantageous than the addition to higher occupied π orbitals which may intercept the π conjugation. Thus, the reason why proton dare not add to the positions of large amplitude of π HOMO in this case will easily be understood. It is not completely satisfactory to dispose of a disagreement between the HOMO-LUMO argument and the experimental fact formally as an *exception* to the theory. A so-called exception does possess its own reason. To investigate what the reason is will possibly yield a novel finding.

The HOMO-LUMO interaction argument was recently pointed out[44] to be in an auxiliary sense useful for the interpretation of the sign of a reaction constant and the scale of a substituent constant in the Hammett rule[45] which has made an immeasurable contribution to the study of the substituent effect in chemical reactivity. In the cyclic addition, like Diels-Alder reactions and 1,3-dipolar additions, the relative easiness of occurrence of reactions, various subsidiary effects, and interesting phenomena like regioselection and periselection were interpreted with considerable success simply by the knowledge of the height of the energy level of HOMO and LUMO, the mode of their extension, their nodal structure, etc.[46]—I defined these in a mass: the *"orbital pattern"*.

Other topics that have been discussed in terms of HOMO-LUMO interactions are thermal formation of excited states,[47] singlet-triplet selectivity,[48] the chemical property of biradicals and excited molecules,[49] the interaction of the central atom and ligands in transition metal complexes,[50] the interaction of three or more orbitals[51] and so forth. Inagaki et al. included in the theory the polarization effect in HOMO and LUMO due to the mixing in of other orbitals and gave an elucidation for a number of organic chemical problems which were not always easy to explain. The unique stereoselection in the transannular cross-bond formation, the lone-pair effect, the d orbital effect, and the orbital polarization effect due to substituents were the cases.[52]

As was partly discussed above, the method of orbital interaction was applied not only to the ground electronic state but to the excited states, giving an explanation of the path of even complicated photochemical isomerizations.[13,21] In a majority of cases the HOMO and the LUMO of the ground-state molecule were also found to be the essential orbitals. Even the ground-state reaction of a strong electron acceptor (or donor) causes a mixing in of an ionized electron configuration or an excited electron configuration in another molecule. In consequence, a partial HOMO-HOMO or LUMO-LUMO interaction, which would be trivial if there were no influence of the acceptor (or donor), becomes important in stabilizing the interacting system.[47]

The problems so far discussed have been limited to chemical reactions. However, the HOMO-LUMO interaction must come into relation also with other chemical phenomena in almost the same mechanism—with the exception of one different point that they usually do not bring about so remarkable a change in the nuclear configuration as in the case of chemical reactions. Now let us examine the possibility of applying the theory to so-called "aromaticity"—one of the simplest, but the hardest-to-interpret problems. There seem to be few problems so annoying to theoreticians as the explanation of this chemically classical concept. I greatly appreciate the contribution of Dewar's theory[53, 54] based on a quantitative energy values argument. Here, however, I want to give a quality comment through a totally different way of consideration.

It is easily ascertained[55] in Fig. 4 that in benzene, naphthalene, phenanthrene, etc., any virtual division of the molecule into two always produces the parts in which their HOMO and LUMO overlap in-phase at the two junctions.

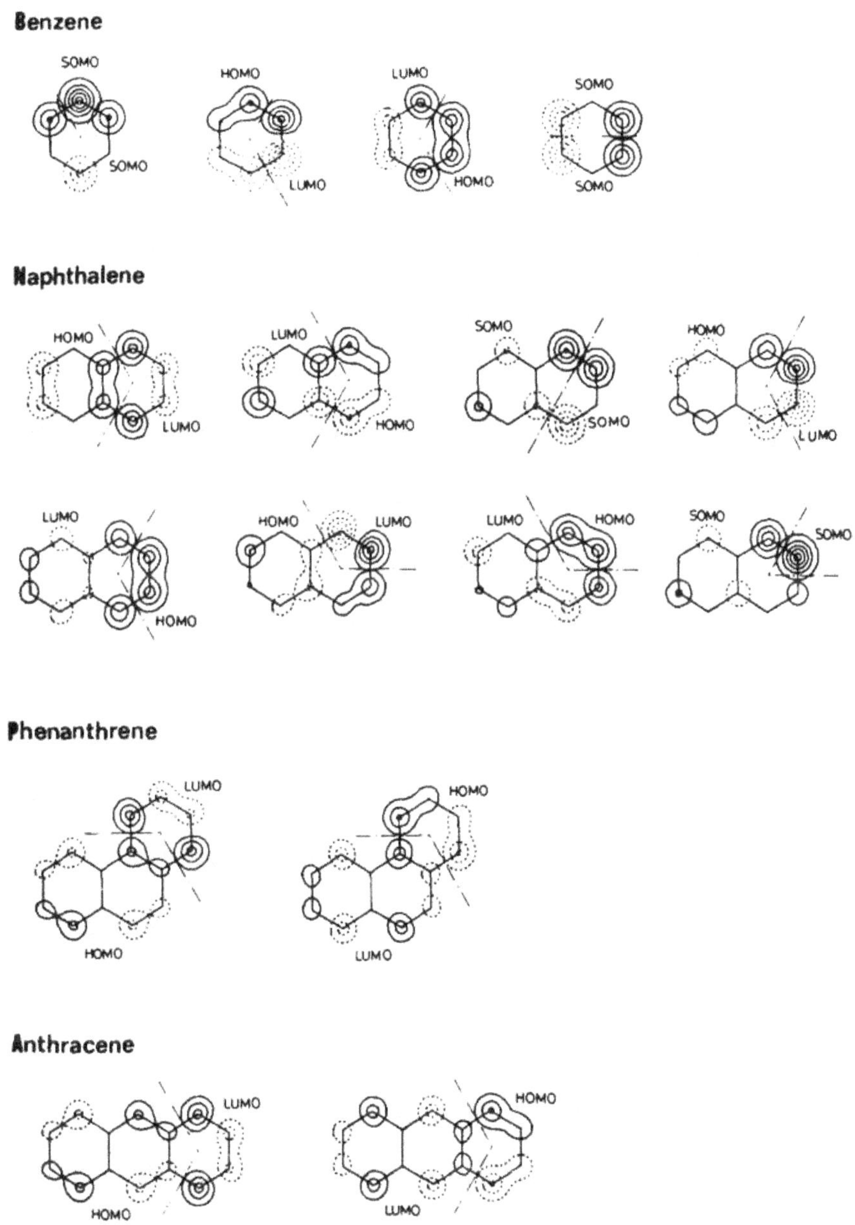

Fig. 4. The HOMO-LUMO phase relationship in virtual division of aromatic hydrocarbons. (SOMO: a singly occupied MO of a radical)

But these circumstances are not seen in anthracene which is usually looked upon as one of the typical representatives of aromatic compounds. Hosoya[55] pointed out from the comparison with phenanthrene indicated in Fig. 4, that the ring growth of type (II) was less stable than that of (I).

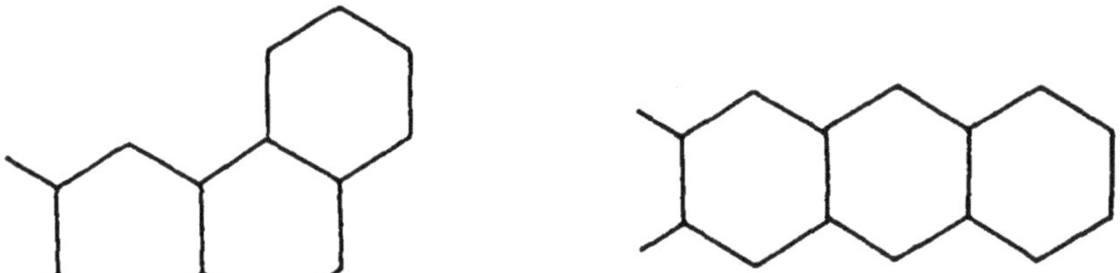

(I) (II)

It is well known that anthracene occasionally exhibits a reactivity of olefin-like additions.

In view of so-called Hückel's (4n+2)-rule mentioned above, an anthracene molecule has 14π electrons and fulfils the stability condition for "aromaticity." Actually, if one considers a molecule of anthracene with the two inside bonds deleted,

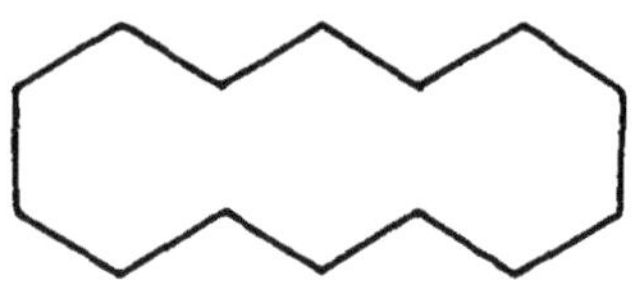

it is really seen that the HOMO and the LUMO of the two parts overlap in an in-phase manner at both of the junctions:

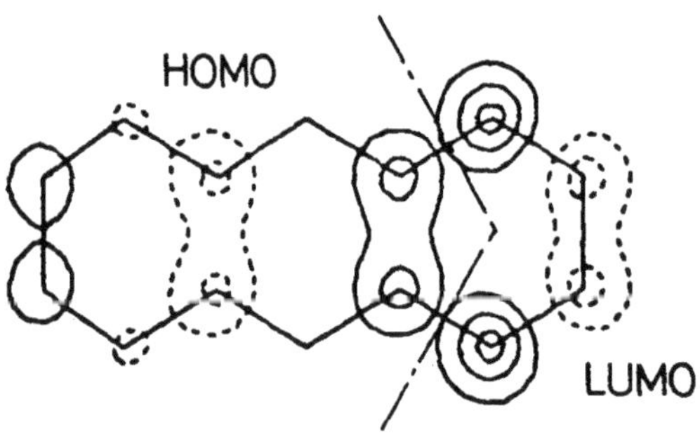

In this way, it is understood that the two bonds which were deleted above exerted a certain unfavourable influence for aromaticity. Such an influence bears a close resemblance to that of impurity scattering in the wave of a free electron moving in a metal crystal.

This discussion seems to be a digression but, as a matter of fact, it relates to the essential question as to how an electron in a molecule can delocalize. As will be mentioned later, Anderson[56] solved the question how an electron in a random system can localize. In a molecule, there are potential barriers between atoms which should be got over by the aid of a certain condition to be satisfied, in order for an electron to move around it freely. Although the question how valence electrons can delocalize in a molecule may have not yet been solved satisfactorily under the condition of unfixed nuclear configuration, the in-phase relation of HOMO and LUMO at the junctions of the two parts of the molecule seems to be at least one of the conditions of intramolecular delocalization of electrons.

Generally speaking, the electron delocalization gives rise to a stabilization due to "conjugation" which is one of the old chemical concepts. If so, similar stabilization mechanisms must be chemically detected in other systems than aromatic compounds. The discussion of this delocalization stabilization at the transition state or on the reaction path was nothing but the reactivity theory hitherto mentioned. The term "delocalizability" was attached to the reactivity indices we derived,[10] and our reactivity theory itself was sometimes called "delocalization approach."[14] The "hyperconjugation" of various sorts is explained in the same manner. The stabilization due to homoaromaticity or bicycloaromaticity of Goldstein,[57] the stability in spirocycles, pericycles,[58] "laticycles" and "longicycles" of Hoffmann and Goldstein,[59] that of spirarenes of Hoffmann and Imamura,[60] and so on, are all comprehended as examples of the stabilization due to the delocalization between HOMO and LUMO, although other explanations may also be possible.

You may be doubtful to what extent such a qualitative consideration is reliable. In many cases, however, a considerably accurate nonempirical determination of the stable conformation of hydrocarbon molecules[61, 62] results in a conclusion qualitatively not much different from the expectation based on the simple orbital interaction argument mentioned above.

CHEMICAL REACTION PATHWAYS

It has already been pointed out that the detailed mechanism of a chemical reaction was discussed along the reaction path on the basis of the orbital interaction argument. For that purpose, however, it is required that the problem as to how the chemical reaction path is determined should have been solved. The method in which the route of a chemical reaction was supposed on the potential energy surface and the rate of the reaction was evaluated by the aid of a statistical-mechanical formulation was established by Eyring.[63] Many people wrote papers where the rate expression was derived wave-mechanically with the use of the potential energy function. Besides, the problem of obtaining the trajectory of a given chemical reaction with a given initial condition was treated by Karplus.[64]

The centre line of the reaction path, so to speak, the idealized reaction coordinate—which I called "intrinsic reaction coordinate" (IRC)[65]—seemed to have been, rather strangely, not distinctly defined till then. For that reason, I began with the general equation which determines the line of force mathematically. [34, 66, 67] Although my papers themselves were possibly not very original, they turned out later to develop in a very interesting direction.[68-74] These papers opened the route to calculate the quasistatic change of nuclear configuration of the reacting system which starts from the transition state proceeding to a stable equilibrium point.[66] I termed the method of automatic determination of the molecular deformation accompanying a chemical reaction as "reaction ergodography." [34 67] This method was applied to a few definite examples by Kato and myself[67] and by Morokuma.[72, 73] Those examples were: abstraction and substitution of methane hydrogen by hydrogen atom,[67] nucleophilic replacement in methane by hydride anion,[72] and isomerization of methylcarbylamine to acetonitrile.[73] All of these reactions thus far treated are limited to the simplest cases, but there seems to be no principal difficulty in extending the applicability to larger systems. Once IRC was determined in this way, the driving force of a chemical reaction was analyzed on the basis of the orbital interaction argument.[66]

In the reacting system with no angular momentum it is possible to obtain the IRC by the use of the space-fixed Cartesian coordinate system. All of the calculated examples mentioned above belong to this case. However, in the reaction in which rotational motion exists, it is required to discuss the IRC after separating the nuclear configuration space from the rotational motion.[74-77] For that purpose, it is essential to derive the general classical Hamiltonian of the reacting systems and then to separate the internal motion which is determined only by the internal coordinates. The nuclear configuration space thus separated out is in general a Riemannian space. The classical Lagrangian form to be obtained in that process of constructing the Hamiltonian is used to derive the IRC equation in the presence of rotational motion. It is thus understood that the rotational motion of the reacting system generally causes a deviation of IRC.[74]

Once the method of determining unique reaction pathways is obtained, the next problem we are concerned with is to see if the calculated pathways are interpreted in terms of the frontier orbital interactions. A method referred to as the "interaction frontier orbitals" or "hybrid molecular orbitals" has been developed very recently by Fujimoto and myself in order to furnish a lucid scheme of frontier orbital interactions with the accuracy of nonempirical calculations now and in the future.[78-80] By including properly contributions of other MOs than the HOMO and the LUMO, we realized in terms of orbital diagrams how ingenious the empirically established chemical concepts—"reaction sites" and "functional groups"—and the empirical concept of reaction pathways could be. Fig. 5 compares the HOMO of styrene and its interaction frontier orbital for protonation to the olefinic double bond. The latter is seen to be localized very efficiently in the frontier of chemical interaction. The double bond is evidently the functional unit in this case. Innovation of the frontier orbital concept will hopefully be continued by young people to make it useful for one of our ultimate targets: theroretical design of molecules and chemical reactions.

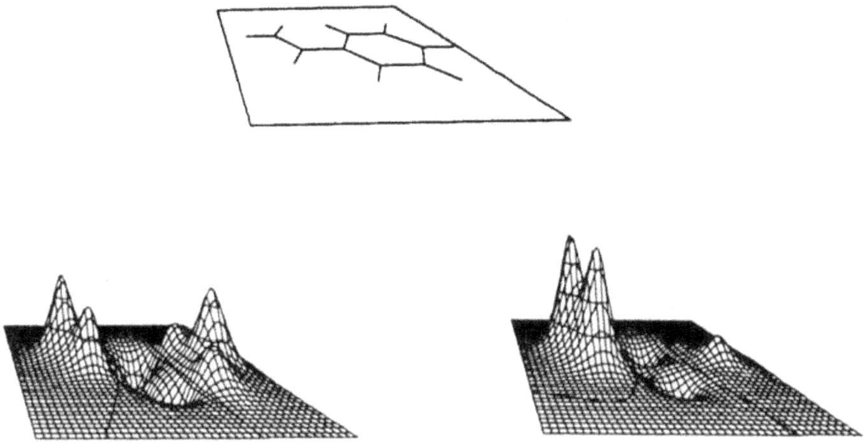

Fig. 5. A comparison of HOMO and the interaction frontier orbital for protonation in styrene.

FRONTIER ORBITALS IN RELATED FIELDS

Theoretical treatments of the property of solid crystals, or chemisorption on a solid surface, appear to have hitherto been almost monopolistically treated by the methodology of physics. But the orbital pattern technique has also advanced gradually in this field.

The "cluster approach," [81, 82] in which a portion of the metal crystal is drawn out as the form of a cluster of atoms and its catalytic actions or other properties are investigated, has contributed to the development of the orbital pattern approach, because the physical methods mentioned above can hardly be applied to such sizable systems. It is expected that, if clusters of various sizes and various shapes are studied to look into the characteristic feature of their HOMO's (high-lying occupied MO's) and LUMO's (low-lying unoccupied MO's), the nature of chemisorption and catalytic action, the mode of surface chemical reaction, and several related subjects of interest can be investigated theoretically.

As is the case of molecular interactions in usual chemical reactions, only the HOMO- and LUMO-bands lying in the range of several electron volts near the Fermi level can participate in the adsorption of molecules and surface reactions on solid crystals. You may recollect here that, in the BCS (Bardeen-Cooper-Schrieffer) theory of superconductivity, too, only the HOMO's and LUMO's in close proximity to the Fermi surface can be concerned in forming electron pairs as the result of interaction with lattice vibration. In the case of solid catalysts mentioned above, the discrimination of particular orbitals and electrons from the others have made the situation much easier.

Consider a system composed of a regular repetition of a molecular unit, for instance, a one-dimensional high polymer chain or a one-dimensional lattice, in which a certain perturbation is imposed at a definite location. Sometimes it is convenient to discuss the influence of this perturbation by transforming the orbitals belonging to the HOMO-band to construct the orbitals localized at that place. One such technique was proposed by Tanaka, Yamabe and myself.[83] This method is expected to be in principle applied to a local discussion of such problems as the adsorption of a molecule on the two-dimensional surface of catalysts, surface reactions, and related matters. This approach may be called a little more chemical than the method using the function of local density of states[84] or similar ones, in that the former can be used for the argument of the reactivity of molecules on a catalyst surface in terms of the phase relationship of localized orbitals.

What is called low-dimensional semiconductors and some superconductors have also been the objects of application of the orbital argument. In these studies, the dimerization of S_2N_2 to S_4N_4[85] and the high-polymerization to $(SN)_x$[86] were discussed, and the energy band structure of $(SN)_x$ polymer chain was analyzed to investigate the stable nuclear arrangement and the mode of inter-chain interactions.[87]

The modern technique employed in solid state physics to interpret the interesting characteristic behaviours of noncrystalline materials, in particular of amorphous materials, in which the nuclear arrangements were not regular, was certainly striking. Anderson showed generally that in a system of random lattice the electron localization should take place.[56] Mott, stating in his 1977 Nobel lecture that he thought it the first prize awarded for the study of amorphous materials, answered the question, "How can a localized electron be conducted?" with the use of the idea of hopping. Here, too, the HOMO-LUMO interaction—in this case the consideration of spin is essential—would play an important part.

Here in a few words, I want to refer to the meaning and the role of virtual orbitals. The LUMO, which has been one of the stars in orbital arguments hitherto discussed, is the virtual orbital which an external electron is considered to occupy to be captured by a molecule to form an anion. Virtual orbitals always play an essential part in producing metastable states of molecules by electron capture.[88] To discuss such problems generally, Tachibana et al.[89] systematized the theory of resonant states from the standpoint of complex eigenvalue problem. The idea of resonant states will take a principal part in chemical reactions, particularly in high-energy reactions which will be developed more in the future.

PROSPECT

In introducing above a series of recent results of the studies carried out mainly by our group, I have ventured to make those things the object of my talk which are no more than my prospective insight and are not yet completely established. This is just to stimulate, by specifying what are the fields I believe promising in the future, the intentional efforts of many younger chemists in order to develop them further.

In my opinion, quantum mechanics has two different ways of making participations in chemistry. One is the contribution to the nonempirical comprehension of empirical chemical results just mentioned. However, we should not overlook another important aspect of quantum mechanics in chemistry. That is the promotion of empirical chemistry from the theoretical side. But, also for this second purpose, as a matter of course, reliable theoretical foundations and computational methods are required. The conclusions of theories should be little affected by the degree of sophistication in approximations adopted.

On the other hand, for theoreticians to make the second contribution, the cases where predictions surpassing the experimental accuracy are possible by very accurate calculations are for the present limited to those of a very few, extremely simple molecules. In order to accomplish this object in regard to ordinary chemical problems, it becomes sometimes necessary to provide qualitative theories which can be used even by experimental chemists. If one can contribute nothing to chemistry without carrying out accurate calculations with respect to each problem, one can not be said to be making the most of quantum mechanics for the development of chemistry. It is certainly best that the underlying concepts are as close to experience as possible, but the sphere of chemical experience is steadily expanding. Quantum chemistry has then to perform its duty by furnishing those concepts with the theoretical basis in order to make them chemically available and serviceable for the aim of promoting empirical chemistry.

Even the same atoms of the same element, when they exist in different molecules, exhibit different behaviours. The chemical symbol H even seems to signify atoms of a completely different nature. In chemistry, this terrible individuality should never be avoided by "averaging," and, moreover, innumerable combinations of such atoms form the subject of chemical research, where it is not the "whole assembly of compounds of different kinds but each individual kind of compounds" that is of chemical interest. On account of this formidable complexity, chemistry possesses inevitably one aspect of depending on the analogy through experience. This is in a sense said to be the fate alloted to chemistry, and the source of a great difference in character from physics. Quantum chemistry, too, so far as it is chemistry, is required to be useful in promoting empirical chemistry as mentioned before.

ACKNOWLEDGEMENTS

Lastly, I want to mention at this opportunity out of a sense of gratefulness the names of many people in our group who have been walking on the same road as mine since my first paper (1952) on quantum chemistry, particularly Drs. T. Yonezawa, C. Nagata, H. Kato, A. Imamura, K. Morokuma, T. Yamabe and H. Fujimoto, and also I can not forget the names of younger doctors mentioned in the text who made a contribution in opening new circumstances in each field. Among them, Prof. T. Yonezawa was helpful in performing calculations in our 1952 paper, and also, it is to be mentioned with appreciation that the attractive title "frontier orbitals" of my lecture originated from the terminology I adopted in that paper by the suggestion of Prof. H. Shingu, who kindly participated in that paper as an organic chemist to classify the relevant experimental results. Furthermore, many other collaborators are now distinguishing themselves in other important fields of chemistry, which, however, have not been the object of the present lecture.

It was the late Prof. Yoshio Tanaka of the University of Tokyo and Prof. Masao Horio of Kyoto University who recognized the existence and significance of my early work in advance to others. I owe such a theoretical work, which I was able to carry out in the Faculty of Engineering, Kyoto University, and moreover in the Department of *Fuel Chemistry,* to the encouragement and kind regard of Prof. Shinjiro Kodama, who fostered the Department. What is more, it was the late Prof. Gen-itsu Kita, my life-teacher, and the founder of the Department, who made me enter into chemistry, one of the most attractive and promising fields of science, and led me to devote my whole life to it. For all of these people no words of gratitude can by any means be sufficient.

REFERENCES

1. For instance, see Ingold, C. K., Structure and Mechanism in Organic Chemistry, Cornell University Press (1953), Ithaca, N.Y.
2. For instance, see Lewis, G. N., Valence and the Structure of Atoms and Molecules, Chemical Catalog Co. (1923), New York, N.Y.
3. Wheland, G. W. and Pauling, L., J. Am. Chem. Soc. *57,* 2086 (1935).
4. Coulson, C. A. and Longuet-Higgins, H. C., Proc. Roy. Soc. (London) *A191,* 39; *A192,* 16 (1947).
5. Wheland, G. W., J. Am. Chem. Soc. *64,* 900 (1942).
6. Fukui, K., Yonezawa, T. and Shingu, H., J. Chem. Phys. *20,* 722 (1952).
7. For instance, see Parr, R. G., The Quantum Theory of Molecular Electronic Structure, Benjamin (1963), New York, N.Y. and references cited therein.
8. Mulliken, R. S., J. Am. Chem. Soc. *74,* 811 (1952).
9. Picket, L. W., Muller, N. and Mulliken, R. S., J. Chem. Phys. *21,* 1400 (1953).
10. Fukui, K., Yonezawa, T. and Nagata, C., Bull. Chem. Soc. Jpn. *27,* 423 (1954); Fukui, K., Kato, H. and Yonezawa, T., ibid. *34,* 1112 (1961).
11. Mulliken, R. S., Rec. Trav. Chim. *75,* 845 (1956).
12. Fukui, K., Yonezawa, T., Nagata, C. and Shingu, H., J. Chem. Phys. *22,* 1433 (1954).
13. For instance, see Fukui, K., Theory of Orientation and Stereoselection, Springer (1970), Berlin.
14. Fukui, K. In: Molecular Orbitals in Chemistry, Physics and Biology, Löwdin, P.-O. and Pullman, B. eds., Academic (1964), New York, N.Y., p. 513.
15. Woodward, R. B. and Hoffmann, R., Angew. Chem. *81,* 797 (1969); The Conservation of Orbital Symmetry, Academic Press (1969), New York, N.Y.; see papers In: Orbital Symmetry Papers, Simmons, H. E. and Bunnett, J. F., eds., ACS (1974), Washington D.C.
16. Woodward, R. B. and Hoffmann, R., J. Am. Chem. Soc. *87,* 395 (1965).
17. Havinga, E., de Kock, R. J. and Rappoldt, M. P., Tetrahedron *11,* 276 (1960); Havinga, E. and Schlatmann, J. L. M. A., ibid. *15,* 146 (1961).
18. Hückel, E., Z. Phys. *70,* 204 (1931); *76,* 628 (1932).
19. Fukui, K., Imamura, A., Yonezawa, T. and Nagata, C., Bull. Chem. Soc. Japan *33,* 1501 (1960).
20. Heilbronner, E., Tetrahedron Lett., *1964,* 1923.
21. Fukui, K., Acc. Chem. Res. *4,* 57, (1971).
22. Fujimoto, H., Inagaki, S. and Fukui, K., J. Am. Chem. Soc. *98,* 2670 (1976).
23. Salem, L., J. Am. Chem. Soc. *90,* 543, 553 (1968).
24. Devaquet, A. and Salem, L., J. Am. Chem. Soc. *91,* 3743 (1969).

25. Salem, L., Chem. Brit. *5,* 449 (1969).
26. Bader, R. F. W., Can., J. Chem. *40,* 1164 (1962).
27. Pearson, R. G., Symmetry Rules for Chemical Reactions, John Wiley (1976), New York, N.Y. and references cited therein.
28. Klopman, G., J. Am. Chem. Soc. *90,* 223 (1968). Also see Klopman, G., Chemical Reactivity and Reaction Paths, John Wiley (1974), New York, N.Y.
29. Herndon, W. C., Chem. Rev. *72,* 157 (1972).
30. Hudson, R. F., Angew. Chem. Int. Ed. Eng. *12,* 36 (1973).
31. For instance, see Dewar, M. J. S. and Dougherty, R. C., The PMO Theory of Organic Chemistry, Plenum (1975), New York, N.Y. and many papers cited therein; see also Dewar, M. J. S., Thetrahedron *S8,* Part I, p. 85 (1966).
32. Fukui, K. and Fujimoto, H., Bull. Chem. Soc. Japan *41,* 1989 (1968).
33. Fukui, K. and Fujimoto, H., Bull. Chem. Soc. Japan *42,* 3392 (1969).
34. Fukui, K. In: The World of Quantum Chemistry, Daudel, R. and Pullman, B. eds., Reidel (1974), Dordrecht, p. 113.
35. Fujimoto, H., Yamabe, S. and Fukui, K., Bull. Chem. Soc. Japan *45,* 1566 (1972).
36. Fujimoto, H., Yamabe, S. and Fukui, K., Bull. Chem. Soc. Japan *45,* 2424 (1972).
37. Yamabe, S., Minato, T., Fujimoto, H. and Fukui, K., Theoret. Chim. Acta (Berl.) *32,* 187 (1974).
38. Fujimoto, H., Kato, S., Yamabe, S. and Fukui, K., J. Chem. Phys. *60,* 572 (1974).
39. Kato, S., Fujimoto, H., Yamabe, S., Fukui, K., J. Am. Chem. Soc. *96,* 2024 (1974).
40. Yamabe, S., Kato, S., Fujimoto, H. and Fukui, K., Bull. Chem. Soc. Japan *46,* 3619 (1973); Yamabe, S., Kato, S., Fujimoto, H. and Fukui, K., Theoret. Chim. Acta (Berl.) *30,* 327 (1973).
41. Fujimoto, H., Yamabe, S., Minato, T. and Fukui, K., J. Am. Chem. Soc. *94,* 9205 (1972).
42. Minato, T., Yamabe, S., Fujimoto, H. and Fukui, K., Bull. Chem. Soc. Japan *51,* 1 (1978); Minato, T., Yamabe, S., Fujimoto, H. and Fukui, K., Bull. Chem. Soc. Japan *51,* 682 (1978).
43. Baba, H., Suzuki, S. and Takemura, T., J. Chem. Phys. *50,* 2078 (1969).
44. Henri-Rousseau, O. and Texier, F., J. Chem. Education *55,* 437 (1978).
45. For instance, see Hammett, L. P., Physical Organic Chemistry, McGraw-Hill (1940), New York, N.Y.
46. Houk, K. N., Acc. Chem. Res. *8,* 361 (1975) and references cited therein.
47. Inagaki, S., Fujimoto, H. and Fukui, K., J. Am. Chem. Soc. *97,* 6108 (1975).
48. See Fleming, I., Frontier Orbitals and Organic Chemical Reactions, John Wiley (1976), New York, N.Y.; Gilchrist, T. L. and Storr, R. C., Organic Reactions and Orbital Symmetry, Cambridge Univ. Press, 2nd ed. (1979), London.
49. Fukui, K. and Tanaka, K., Bull. Chem. Soc. Japan *50,* 1391 (1977).
50. Fukui, K. and Inagaki, S., J. Am. Chem. Soc. *97,* 4445 (1975) and many papers cited therein.
51. Inagaki, S., Fujimoto, H. and Fukui, K., J. Am. Chem. Soc. *98,* 4693 (1976).
52. Inagaki, S. and Fukui, K., Chem. Lett. *1974,* 509; Inagaki, S., Fujimoto, H. and Fukui, K., J. Am. Chem. Soc. *98,* 4054 (1976).
53. Dewar, M. J. S., The Molecular Orbital Theory of Organic Chemistry, McGraw-Hill (1969), New York, N.Y.
54. Dewar, M. J. S., Angew. Chem. *83,* 859 (1971).
55. Fukui, K., Kagaku to Kogyo *29,* 556 (1976); Hosoya, H., Symposium on Electron Correlation in Molecules, Res. Inst. for Fund. Phys., Dec. 18, 1976.
56. Anderson, P. W., Phys. Rev. *109,* 1492 (1958); cf. 1977 Nobel Lectures in Physics by Anderson, P. W., and by Mott, N. F.
57. Goldstein, M. J., J. Am. Chem. Soc. *89,* 6357 (1967).
58. Simmons, H. E. and Fukunaga, T., J. Am. Chem. Soc. *89,* 5208 (1967).
59. Goldstein, M. J. and Hoffmann, R., J. Am. Chem. Soc. *93,* 6193 (1971).
60. Hoffmann, R., Imamura, A. and Zeiss, G., J. Am. Chem. Soc. *89,* 5215 (1967).
61. For instance, see Hehre, W. J. and Pople, J. A., J. Am. Chem. Soc. *97,* 6941 (1975).
62. Hehre, W. J., Acc. Chem. Res. *8,* 369 (1975).
63. For instance, see Glasstone, S., Laider, K. J. and Eyring, H., The Theory of Rate Processes, McGraw-Hill (1941), New York, N.Y. and references cited therein.

64. Wang, I. S. Y. and Karplus, M., J. Am. Chem. Soc. *95,* 8060 (1973) and references cited therein.
65. Fukui, K., J. Phys. Chem. *74,* 4161 (1970).
66. Fukui, K., Kato, S. and Fujimoto, H., J. Am. Chem. Soc. *97,* 1 (1975).
67. Kato, S. and Fukui, K., J. Am. Chem. Soc. *98,* 6395 (1976).
68. Tachibana, A. and Fukui, K., Theoret. Chim. Acta (Berl.) *49,* 321 (1978).
69. Tachibana, A. and Fukui, K., Theoret. Chim. Acta (Berl.) *51,* 189, 275 (1979).
70. Tachibana, A. and Fukui, K., Theoret. Chim. Acta (Berl.) *5,* (1979).
71. Fukui, K., Rec. Trav. Chim., Pays-Bas *98,* 75 (1979).
72. Joshi, B. D. and Morokuma, K., J. Chem. Phys. *67,* 4880 (1977).
73. Ishida, K., Morokuma, K. and Komornicki, A., J. Chem. Phys. *66,* 2153 (1977).
74. Fukui, K., Tachibana, A. and Yamashita, K., Intern. J. Quant. Chem., Quant. Chem. Symp. *15,* 621 (1981).
75. Miller, W. H., Handy, N. C. and Adams, J. E., J. Chem. Phys. *72,* 99 (1980); Gray, S. K., Miller, W. H., Yamaguchi, Y. and Schaefer, H. F., III., ibid. *73,* 2732 (1980).
76. Fukui, K., Intern. J. Quant. Chem., Quant. Chem. Symp. *15,* 633 (1981).
77. Fukui, K., Acc. Chem. Res. *14,* 363 (1981).
78. Fukui, K., Koga, N. and Fujimoto, H., J. Am. Chem. Soc. *103,* 196 (1981).
79. Fujimoto, H., Koga, N., Endo, M. and Fukui, K., Tetrahedron Letters *22,* 1263; 3427 (1981).
80. Fujimoto, H., Koga, N. and Fukui, K., J. Am. Chem. Soc. *103,* 7452 (1981).
81. For instance, see Johnson, K. H. In: The New World of Quantum Chemistry, Pullman, B. and Parr, R. G. eds., Reidel (1976), Dordrecht, p. 317 and references cited therein.
82. Kobayashi, H., Kato, H., Tarama, K. and Fukui, K., J. Catalysis *49,* 294 (1977); Kobayashi, H., Yoshida, S., Kato, H., Fukui, K. and Tarama, K., Surface Science *79,* 189 (1979).
83. Tanaka, K., Yamabe, T. and Fukui, K., Chem. Phys. Letters *48,* 141 (1977).
84. For instance, Schrieffer, J. R. In: The New World of Quantum Chemistry, Pullman, B. and Parr, R. G. eds., Reidel, D., Dordrecht (1976), p. 305; Danese, J. B. and Schrieffer, J. R., Intern. J. Quant. Chem., Quant. Chem. Symp. *10,* 289 (1976).
85. Tanaka, K., Yamabe, T., Noda, A., Fukui, K. and Kato, H., J. Phys. Chem. *82,* 1453 (1978).
86. Yamabe, T., Tanaka, K., Fukui, K. and Kato, H., J. Phys. Chem. *81,* 727 (1977).
87. Tanaka, K., Yamabe, T. and Fukui, K., Chem. Phys. Letters *53,* 453 (1978).
88. For instance, see Ishimaru, S., Yamabe, T., Fukui, K. and Kato, H., J. Phys. Chem. *78,* 148 (1974); Ishimaru, S., Fukui, K. and Kato, H., Theoret. Chim. Acta (Berl.) *39,* 103 (1975).
89. Tachibana, A., Yamabe, T. and Fukui, K., J. Phys. B. Atom. Molec. Phys. *10,* 3175 (1977); Adv. Quant. Chem. *11,* 195 (1978).

Reprinted from CHEMICAL REACTIVITY AND REACTION PATHS
Edited by Gilles Klopman.

CHAPTER 3

Intermolecular Interactions and Chemical Reactivity

H. Fujimoto and K. Fukui
Faculty of Engineering
Kyoto University
Kyoto, Japan

I. INTRODUCTION

Chemical reactions have long been an attractive subject of quantum chemistry. Especially, molecular orbital (MO) theory has experienced a brilliant success in interpreting and predicting the orientation and the stereoselection of a large number of chemical reactions. The fundamental character of

24 INTERMOLECULAR INTERACTIONS AND CHEMICAL REACTIVITY

the various sorts of chemical reactions, intermolecular and intramolecular, have been discussed in terms of MOs. Chemical reactivity indices for π and σ electronic systems which have been developed based upon several reaction models seem now to be in common use by organic chemists.[1-5] The Woodward-Hoffmann selection rules for pericyclic reactions have disclosed the basic principles underlying apparently different kinds of reactions and have stimulated new experiments on the prediction of the rules.[6-9] Recent progress in the high-speed computer has supplied us with several semiempirical MO methods, which can conveniently be applied to complicated chemically interacting systems with tolerable accuracy in a chemical sense.[10-18] The paths of some reactions have been studied using semiempirical MO methods.[19-26] *Ab initio* MO calculations have also been attempted on some chemical interactions between sizable molecules.[27-31] On the other hand, some extensive studies on the modes of chemical interaction have been developed by the use of MOs which are obtained with respect to the reactant and reagent in their isolated states.[32-39] Application of these methods of representing chemically interacting systems is in general limited to the early stage of reaction, since the energy of interacting systems is usually expanded in power series of overlap integrals, interaction integrals, and so on. However, these approaches are expected to be able to give detailed information on the mechanisms of chemical reactions. In the following sections, we will make a brief survey of the chemical reactivity indices and the stereoselection rules, which have been established on the basis of simplified reaction models, in relation to more comprehensive studies of chemical reactions. We will not discuss developments in the theories of long-range forces the main interest of which are not directed toward chemical reactions. The variational calculations on reaction intermediates or activated complexes will also mostly be disregarded here.

A. Chemical Reactivity Indices

The chemical reactivity indices that have been proposed by the use of MOs are conveniently divided into three groups: the static, the localization, and the delocalization approaches.[2]

In the early stage of calculations on π electrons of planar conjugated systems, the π electron densities were regarded as a measure of chemical reactivity toward ionic reactions.[40-42] In the case of alternant hydrocarbons, the π electron densities of every carbon atom calculated by the simple Hückel MO method are unity.[43] The chemical reactivity index, self-atompolarizability, was proposed on the basis of the perturbation theory.[44] For the purpose of discussing the chemical reactivity toward radical reagents, the chemical reactivity index, free valence, was put forth.[45] These indices are classified as belonging to the static approaches.

The localization energy is a measure of the change in the π electron energy of a system by deleting the atom under attack by a reagent from the remainder of the π conjugated part.[46] The position of the smallest unstablization, obtained by putting 0, 1, and 2 electrons in the deleted atom, is the most reactive toward an attacking nucleophile, radical, and electrophile, respectively. Dewar's reactivity number is an approximation of the localization energy for alternant hydrocarbons, which can be obtained by much simpler calculations.[47] These indices belong to the localization approaches.

In 1952, the new reactivity index, frontier electron density, was proposed.[48] It was first pointed out that electrophilic aromatic substitutions took place most dominantly at the position where the partial electron density in the highest occupied (HO) MO was the largest. Afterwards, it was stated that the position of the largest partial electron density in the lowest unoccupied (LU) MO was usually the most favorable position of nucleophilic aromatic substitutions.[49] These particular MOs were termed frontier orbitals and the partial densities in these MOs were called frontier electron densities. In homolytic reactions both HOMO and LUMO serve as frontier orbitals. Assuming hyperconjugation between the aromatic π electron system of a molecule and the pseudo π orbital composed of the atomic orbital (AO) of the attacking reagent and the AO of hydrogen to be replaced, at the transition state of aromatic substitution, the chemical reactivity index, superdelocalizability, was derived.[50,51] While frontier electron density is employed for an intramolecular comparison of reactive positions of a given molecule, superdelocalizability can be used for the purpose of comparing the relative reactivities of different molecules for a given reaction. The chemical reactivity indices, frontier electron density, and superdelocalizability are regarded as the delocalization model of chemical interaction, together with Brown's Z value[52] and with the ion free valence by Nagakura and Tanaka.[53]

With the progress of MO methods for σ electrons, some theoretical treatments of the chemical reactivity of saturated compounds have been developed. Delocalizability was defined in order to discuss the reactivities of hydrogens in paraffin hydrocarbons and their derivatives.[3,54] Frontier electron density was also used as a reactivity measure of hydrogens in some halogenated hydrocarbons.[55–57] The relative ease of thermal cleavage of carbon–carbon bonds of alkanes has been elucidated by means of σ MO perturbation calculations.[2,58]

B. Stereoselection Rules

Woodward and Hoffmann derived significant selection rules of electrocyclic reactions, sigmatropic reactions, and concerted cycloaddition reactions, by an ingenious use of correlation diagrams.[6–9] The correlation of the state functions in electrocyclic reactions was discussed by Longuet-Higgins and

26 INTERMOLECULAR INTERACTIONS AND CHEMICAL REACTIVITY

Abrahamson.[59] The discovery of Woodward-Hoffmann selection rules have attracted much attention from organic chemists and have quickly been applied to numerous cases.[60–62]

The chemical reactivity indices, free valence, and localization energy were applied to Diels-Alder reactions.[63,64] In 1964, the importance of nodal properties of the MOs of reactant and reagent in concerted multicentric processes was first pointed out by Fukui.[2] Diels-Alder additions of dienes and dienophiles were discussed by the use of perturbation MO method. The conjugation stabilization energy due to the concerted interaction between the two termini of diene and dienophile is given by[65–67]

$$\Delta E = 2\sum_{i}^{\text{occ}}\sum_{l}^{\text{uno}} \frac{(c_t^{(i)}c_u^{(l)}\gamma_{tu} + c_{t'}^{(i)}c_{u'}^{(l)}\gamma_{t'u'})^2}{\varepsilon_{ai} - \varepsilon_{bl}} + 2\sum_{k}^{\text{occ}}\sum_{j}^{\text{uno}} \frac{(c_t^{(j)}c_u^{(k)}\gamma_{tu} + c_{t'}^{(j)}c_{u'}^{(k)}\gamma_{t'u'})^2}{\varepsilon_{bk} - \varepsilon_{aj}} \quad (1)$$

where $c_t^{(i)}$, $c_{t'}^{(i)}$ are the coefficients of the AOs t, t' in the LCAO MO a_i of diene and $c_u^{(l)}$, $c_{u'}^{(l)}$ are those of the AOs u, u' in the MO b_l of dienophile, ε_{ai} and ε_{bl} are the MO energies and γ_{tu} and $\gamma_{t'u'}$ stand for the interaction integrals, as is illustrated in Figure 3-1. The first term on the right-hand side of Eq. (1) means the summation over all the combinations of the occupied MOs a_i of diene and the unoccupied MOs b_l of dienophile and the second term covers all the combinations of the unoccupied MOs a_j of diene and the occupied MOs b_k of dienophile. Here it should be noted that the interaction is still not too strong, thus still satisfying the condition that the numerators of Eq. (1) are small. Of the terms appearing in Eq. (1), the interaction between HOMO of diene and

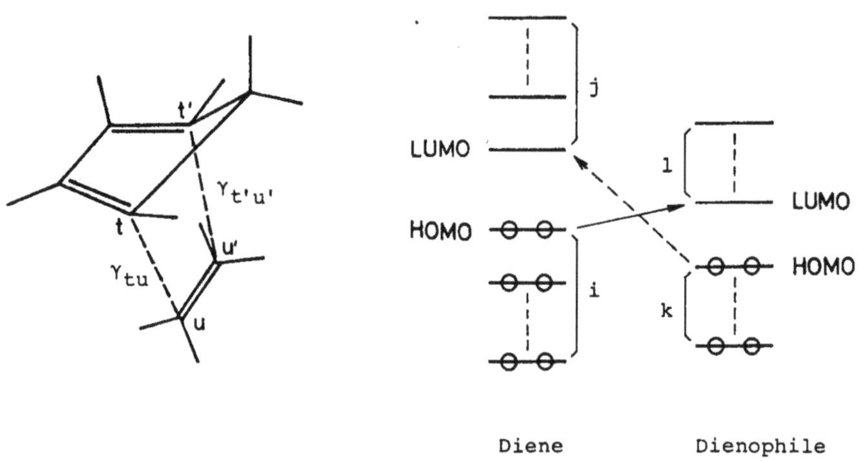

Fig. 3-1. Mode of orbital interaction in Diels-Alder reactions.

LUMO of dienophile usually plays a discriminative role. Then Eq. (1) can be approximated by the following one-term expression:

$$\Delta E \cong 2\frac{(c_t^{(\mathrm{HO})}c_u^{(\mathrm{LU})}\gamma_{tu} + c_{t'}^{(\mathrm{HO})}c_{u'}^{(\mathrm{LU})}\gamma_{t'u'})^2}{\varepsilon_{a\mathrm{HO}} - \varepsilon_{b\mathrm{LU}}} \tag{2}$$

It was shown that the products of the AO coefficients of both reaction centers, $c_t^{(\mathrm{HO})}c_u^{(\mathrm{LU})}$ and $c_{t'}^{(\mathrm{HO})}c_{u'}^{(\mathrm{LU})}$, did have the same sign with respect to a number of combinations of diene and dienophile, making the concerted process favorable.[2]

The stereoselection rules for thermal pericyclic reactions were found to be governed by a single basic principle in which the number of electrons participating in the formation of the cycle had a key importance.[6–9,67–74] The selection rules for photochemical interactions seem to be not necessarily defined in a unique manner, although they are usually contrary to thermal ones.[6–9,67–74] Existence of singly occupied (SO) MOs makes it difficult to derive general selection rules.

The concept of stereoselections, represented by the Woodward-Hoffmann selection rules, has been applied to other interesting systems. Hoffmann and Woodward extended the basic concept of Eq. (2) to discuss the exo-endo selectivity in Diels-Alder reactions, regarding the interaction as three-centric with respect to both diene and dienophile.[7,8] Herndon and Hall pointed out the possible difference in the interaction integrals in the exo-endo processes.[75]

The importance of the secondary attractive force in Diels-Alder addition of butadiene, which was different from that proposed by Hoffmann and Woodward, was pointed out by Dewar.[71] Salem also disclosed from numerical calculations that the same kind of secondary interaction as was proposed by Dewar was operative in the dimerization of butadiene and of acrolein.[35] Sustmann and Binsch investigated the exo-endo selectivity in the interaction of dienes and dienophiles which had no extra π electrons available for the secondary interaction.[39] Goldstein obtained a selection rule for bicycloaromaticity.[76] Spironconjugation was discussed by Simmons and Fukunaga and by Hoffmann, Imamura, and Zeiss.[77,78] Mango and Schachtschneider showed how the unstabilization due to the thermal interaction between two olefinic double bonds was removed through the electron-donating and the electron-accepting interactions with the d orbitals of transition-metal catalysts.[79–81] While they took only the π electrons of olefins into account, the potential significance of the σ electrons of the carbon-carbon bonds was also pointed out.[5,82] These interactions can also be discussed by means of Eqs. (1) and (2).[5]

Although the selection rules are of profound significance in elucidating the basic principle of chemical interactions, they are qualitative in their nature. Some semiempirical and *ab initio* MO calculations have been carried out on electrocyclic reactions of conjugated systems.[83–88] Valence bond (VB) calculation of the photochemical ring-closing reaction of butadiene was also carried out by van der Lugt and Oosterhoff.[89]

II. DEVELOPMENT OF MO THEORIES OF CHEMICAL REACTIONS

As has been discussed, several chemical reactivity indices have been proposed using MOs, each depending upon an assumed reaction model. In some fortunate cases, like the substitution reactions of alternant hydrocarbons, any of the reactivity indices mentioned before was found to give satisfactory results. In some other unfortunate cases, some of them failed to give correct prediction. In recent years, several theories were presented which tried to treat chemical reactions more comprehensively.[33–39] In these, the chemical reactivity is not expressed by a single term, but by a combined sum of more than two different terms. Klopman discussed the interaction between an electrophile and a nucleophile, taking the two effects into account, neighboring effect and electron-transfer effect.[34] The interaction energy was represented by the sum of the three energy terms

$$\Delta E = -\Delta q_t \Delta q_u \frac{\Gamma}{\varepsilon} + \Delta\mathrm{solv} + \sum_{i}^{\mathrm{occ}} \sum_{l}^{\mathrm{uno}} \frac{2(c_t^{(i)})^2 (c_u^{(l)})^2 \gamma_{tu}{}^2}{\varepsilon_{ai}^* - \varepsilon_{bl}^*} \tag{3}$$

where Δq_t is the total initial charge of the atom t, Γ is the Coulomb repulsion term between atoms t and u, ε is the local dielectric constant of the solvent, ε_{ai}^* is the energy of the ith occupied MO of the nucleophile under the influence of the electrophile and the possible effect of solvent, and ε_{bl}^* is associated with the lth unoccupied MO of the electrophile. The first term on the right-hand side of Eq. (3) stands for the Coulombic interaction between the charges of nucleophile and electrophile, the second term is the change in solvation energy, and the last represents the stabilization due to electron-transfer interaction. In the case of interactions between a hard base (with low ε_{ai}^* value) and a hard acid (with high ε_{bl}^* value), the first term dominates, while the third term plays the most dominant role in the interaction between a soft base (with high ε_{ai}^* value) and a soft acid (with low ε_{bl}^* value). Klopman called the former case "charge controlled" and the latter case "frontier controlled." He elegantly interpreted the changes in the position of additions and substitutions according to the differences in the softness and/or the hardness of attacking reagents.

In deriving Eq. (1), the overlap integrals between two interacting AOs were neglected. Including the overlap integrals explicitly, Salem obtained a new equation for the ground-state interaction of two closed-shell systems, A and B:

$$\begin{aligned}\Delta E = &-\sum_{t}^{A}\sum_{u}^{B}(q_t + q_u)\gamma_{tu}s_{tu}\\ &-2\sum_{i}^{\text{occ}}\sum_{l}^{\text{uno}}\left[\frac{\left(\sum_{t,u} c_t^{(i)}c_u^{(l)}\gamma_{tu}\right)^2}{\varepsilon_{bl}-\varepsilon_{ai}} + \tfrac{1}{4}(\varepsilon_{bl}-\varepsilon_{ai})\left(\sum_{t,u} c_t^{(i)}c_u^{(l)}s_{tu}\right)^2\right]\\ &-2\sum_{k}^{\text{occ}}\sum_{j}^{\text{uno}}\left[\frac{\left(\sum_{t,u} c_t^{(j)}c_u^{(k)}\gamma_{tu}\right)^2}{\varepsilon_{aj}-\varepsilon_{bk}} + \tfrac{1}{4}(\varepsilon_{aj}-\varepsilon_{bk})\left(\sum_{t,u} c_t^{(j)}c_u^{(k)}s_{tu}\right)^2\right]\end{aligned} \quad (4)$$

where q_t is the charge density of the AO t and s_{tu} is the overlap integral between the two AOs t of A and u of B. Salem estimated the second term in the brackets. This term was smaller than the first term in the brackets, having the same form as Eq. (1). The first term on the right-hand side of Eq. (4) behaves repulsively in the interaction and Salem ascribed this to the penetration of electrons into the "exclusion core." Applying the theory to Diels-Alder reactions of several dienes and dienophiles, Salem proved the important role of the HOMO-LUMO interaction.[35] An equation similar to Eq. (4) was also derived for the photochemical interactions of two conjugated systems. Devaquet and Salem tried to represent the energy of the combined system of two reactants applying the perturbation theory in the Fock-Roothaan SCF MO formalism[90] and introduced an additional energy term, $\sum_{t,u} \Delta q_t \Delta q_u / R_{tu}$, representing the Coulombic interaction between the net charges of two reactants.[37,38] They calculated the interaction energy of the dimerization of acrolein and found that the Coulombic interaction energy was important in determining the favorable reaction path. Although Devaquet and Salem put an approximation in constructing the matrix elements of the combined system of two reactants, Sustmann and Binsch got the interaction energy between two closed-shell systems by an iterative determination of the Fock perturbed matrix elements of the combined system.[39] Murrell, Randić, and Williams expanded the interaction energy between two systems in the power series of interaction potential and overlap integral, dividing it into Coulomb, exchange, polarization, exchange polarization and charge-transfer interaction energies.[32] The theory has been applied to rather static systems like hydrogen bonds, charge-transfer complexes, and so on.[91–93]

In order to clarify the basic principles governing chemical reactions, the interaction energy between two reactants was represented by the combined sum of four energies, Coulomb, exchange, delocalization, and polarization interactions, of which chemically graspable formulas were derived using LCAO SCF MOs of reactant and reagent.[36] When the theory is applied to π electron systems, the Coulomb interaction energy is related to the reactivity measure "total π electron density," the delocalization interaction energy to "superdelocalizability" and the polarization interaction energy to "self-atompolarizability." The exchange interaction energy emerges from the antisymmetrization of the wave functions of two reactants obtained for each of them in an isolated state. The examination of the magnitudes of these energy terms suggested the increasing importance of the delocalization interaction in stabilizing chemically interacting systems with the progress of the reactions. The delocalization interaction is also important as the major origin of bond interchange, that is, the formation of new bonds and the disappearance of old bonds, in chemical reactions.[94,95] In the following, we will discuss how the chemical interaction is represented by the MOs of two reactants in some details, with the aid of some illustrative examples.

A. Interaction Between Two Systems

Let us represent the wave function of the system composed of two independent systems A and B, interacting with each other, by a linear combination of the wave functions of adiabatic interaction, charge-transfer interaction and polarization interaction:

$$\Psi = C_0\Psi_0 + \sum_p C_p\Psi_p \tag{5}$$

We assume that both of the two reactants A and B have closed-shell electronic structures. Although it is desirable to include as many electron con-

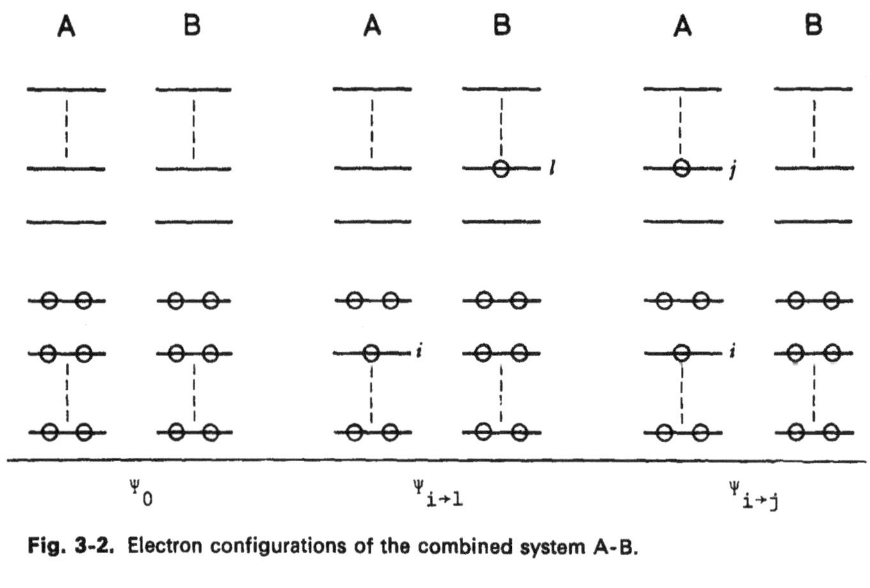

Fig. 3-2. Electron configurations of the combined system A-B.

figurations as possible, this is rather a tedious chore and, hence, we may here take only "one-electron transferred" and "one-electron excited" electron configurations with respect to the original state. These are schematically illustrated in Figure 3-2. The contributions of other electron configurations are small in the early stage of chemical interaction.[36] The wave functions of these states are represented by the Slater determinants composed of spin orbitals of A and B. The state of the adiabatic interaction is given simply by a single determinantal wave function

$$\Psi_0 = \mathcal{N}\mathcal{A}|a_1\bar{a}_1a_2\bar{a}_2\cdots a_m\bar{a}_mb_1\bar{b}_1b_2\bar{b}_2\cdots b_n\bar{b}_n| \tag{6}$$

where $\mathcal{N}\mathcal{A}$ is the normalization-antisymmetrization operator, barred and nonbarred MOs are the spin orbitals with spin functions α and β, respectively. Similarly, the wave function of a singlet one-electron transferred state is given by

$$\Psi_{i\to l} = (1/\sqrt{2})\{\mathcal{N}\mathcal{A}|a_1\bar{a}_1\cdots a_i\cdots a_m\bar{a}_mb_1\bar{b}_1\cdots b_n\bar{b}_n\bar{b}_l| - \mathcal{N}\mathcal{A}|a_1\bar{a}_1\cdots \bar{a}_i\cdots a_m\bar{a}_mb_1\bar{b}_1\cdots b_n\bar{b}_nb_l|\} \tag{7}$$

Here, the spatial parts of the one electron wave functions, a_is and b_ks are assumed to be represented in LCAO forms:

$$a_i(1) = \sum_t c_t^{(i)}t(1) \quad \text{for system } A$$

and

$$b_k(1) = \sum_u c_u^{(k)}u(1) \quad \text{for system } B$$

The MOs a_i are orthonormalized with respect to the isolated system A. The MOs b_k also form an orthonormal set in an isolated state of B. The MOs a_i and the MOs b_k are not orthogonal in general, except for the case of infinite separation. We can take all the functions as real.

The Hamiltonian operator of the interacting system of A and B in the frame of the Born-Oppenheimer approximation is given by (in atomic units)[96]

$$H = \sum_\lambda\left(-\tfrac{1}{2}\Delta(\lambda) - \sum_\alpha^A\frac{Z_\alpha}{r_{\lambda\alpha}} - \sum_\beta^B\frac{Z_\beta}{r_{\lambda\beta}}\right) + \sum_{\lambda<\lambda'}\frac{1}{r_{\lambda\lambda'}} + \sum_{\gamma<\gamma'}\frac{Z_\gamma Z_{\gamma'}}{R_{\gamma\gamma'}} \tag{8}$$

where Δ is the Laplacian operator, Z_α and Z_β are the positive charges of the nuclei of the atoms α and β, belonging to the systems A and B, respectively, $r_{\lambda\alpha}$ is the distance between the electron λ and the nucleus α, $r_{\lambda\lambda'}$ is the separation between the two electrons λ and λ', and $R_{\gamma\gamma'}$ is the distance between the two nuclei γ and γ'.

32 INTERMOLECULAR INTERACTIONS AND CHEMICAL REACTIVITY

The lowest energy of the interacting system can be obtained by solving the secular equation

$$\begin{vmatrix} H_{0,0} - E & H_{0,p} - S_{0,p}E & \cdots \\ H_{p,0} - S_{p,0}E & H_{p,p} - E & \cdots \\ \cdots & \cdots & \cdots \end{vmatrix} = 0 \tag{9}$$

where

$$H_{p,p'} = \int \Psi_p^* H \Psi_{p'}\, d\tau, \qquad S_{p,p'} = \int \Psi_p^* \Psi_{p'}\, d\tau$$

The interaction energy may be defined as the difference between the ground-state energy of the combined system and the sum of the energies of A and B in their isolated states,

$$\Delta W = W - (W_{A0} + W_{B0}) \tag{10}$$

If the interaction is assumed to be not yet strong, the interaction energy can be expanded in the following form:

$$\Delta W \cong (H_{0,0} - W_{A0} - W_{B0}) - D - \Pi \tag{11}$$

where

$$D \cong \sum_i^{\text{occ}} \sum_l^{\text{uno}} \frac{|H_{0,i\to l} - S_{0,i\to l}H_{0,0}|^2}{H_{i\to l,i\to l} - H_{0,0}} + \sum_k^{\text{occ}} \sum_j^{\text{uno}} \frac{|H_{0,k\to j} - S_{0,k\to j}H_{0,0}|^2}{H_{k\to j,k\to j} - H_{0,0}} \tag{12}$$

and

$$\Pi \cong \sum_i^{\text{occ}} \sum_j^{\text{uno}} \frac{|H_{0,i\to j} - S_{0,i\to j}H_{0,0}|^2}{H_{i\to j,i\to j} - H_{0,0}} + \sum_k^{\text{occ}} \sum_l^{\text{uno}} \frac{|H_{0,k\to l} - S_{0,k\to l}H_{0,0}|^2}{H_{k\to l,k\to l} - H_{0,0}} \tag{13}$$

We call the D and Π energy terms, the "delocalization interaction energy" and the "polarization interaction energy," respectively. The first-order interaction energy $(H_{0,0} - W_{A0} - W_{B0})$ may be divided into the two terms, the Coulomb interaction energy and the exchange interaction energy

$$\varepsilon_Q \cong 2\sum_i^{\text{occ}} V_{Bii} + 2\sum_k^{\text{occ}} V_{Akk} + 4\sum_i^{\text{occ}}\sum_k^{\text{occ}} (ii|kk) + \sum_\alpha^{A}\sum_\beta^{B} \frac{Z_\alpha Z_\beta}{R_{\alpha\beta}} \tag{14}$$

and

$$\varepsilon_K \cong -2\sum_i^{\text{occ}}\sum_k^{\text{occ}} \Big[(ik|ki) + S_{ik}V_{ik} + \sum_{i'}^{\text{occ}} S_{ik}\{2(ik|i'i') - (ii'|i'k)\}$$
$$+ \sum_{k'}^{\text{occ}} S_{ik}\{2(ik|k'k') - (ik'|k'k)\}\Big] \tag{15}$$

where

$$V_{Bii} = \int a_i(1)\left(-\sum_{\beta}^{B} \frac{Z_\beta}{r_{1\beta}}\right) a_i(1)\, dv(1)$$

$$V_{ik} = \int a_i(1)\left(-\sum_{\alpha}^{A} \frac{Z_\alpha}{r_{1\alpha}} - \sum_{\beta}^{B} \frac{Z_\beta}{r_{1\beta}}\right) b_k(1)\, dv(1)$$

$$S_{ik} = \int a_i(1) b_k(1)\, dv(1)$$

$$(ii|kk) = \int\int a_i(1) b_k(2) \frac{1}{r_{12}} a_i(1) b_k(2)\, dv(1)\, dv(2)$$

When we employ the point-charge approximation for nuclear attractions and electron repulsions, then

$$\int t(1) \frac{-Z_\beta}{r_{1\beta}} t(1)\, dv(1) = -\frac{Z_\beta}{R_{\alpha\beta}} \qquad (t \text{ belongs to } \alpha)$$

$$\int\int t(1)u(2) \frac{1}{r_{12}} t(1)u(2)\, dv(1)\, dv(2) = \frac{1}{R_{\alpha\beta}} \qquad (u \text{ belongs to } \beta)$$

the Coulomb interaction energy is given in a chemically graspable form, by the aid of the Mulliken approximation[97]:

$$\varepsilon_Q \cong \sum_{\alpha}^{A} \sum_{\beta}^{B} \frac{(Z_\alpha - N_\alpha)(Z_\beta - N_\beta)}{R_{\alpha\beta}} \tag{16}$$

where N_α is the gross atomic population of the atom α[98]:

$$N_\alpha = 2 \sum_{i}^{\text{occ}} \sum_{t}^{\alpha} \sum_{t'}^{A} c_t^{(i)} c_{t'}^{(i)} s_{tt'}$$

The exchange interaction energy represents a repulsive term in the usual chemical reactions of two closed-shell systems.

Neglecting exchange terms, the numerator of the delocalization interaction energy is given by

$$H_{0,i\to l} - S_{0,i\to l} H_{0,0} \cong \sqrt{2}\left[V_{Bil} + 2\sum_{k}^{\text{occ}} (il|kk) - S_{il}\left\{V_{Bii} + 2\sum_{k}^{\text{occ}} (ii|kk)\right\}\right] \tag{17}$$

Here it was assumed that the MOs satisfy the Fock equation for the isolated systems. When the interaction takes place through the orbital overlapping of paired AOs (t, u) (t', u') and so on, we have

$$H_{0,i\to l} - S_{0,i\to l} H_{0,0} \cong \sqrt{2} \sum_{t,u} c_t^{(i)} c_u^{(l)} \gamma_{tu}^{(t)} \tag{18}$$

34 INTERMOLECULAR INTERACTIONS AND CHEMICAL REACTIVITY

where

$$\gamma_{tu}^{(t)} \cong -\int t(1)\left(\sum_{\beta}^{B}\frac{Z_\beta - N_\beta}{r_{1\beta}}\right)u(1)\,dv(1) + S_{tu}\sum_{\alpha}^{A}\sum_{\beta}^{B}\frac{n_\alpha^{(ii)}(Z_\beta - N_\beta)}{R_{\alpha\beta}} \quad (19)$$

$$n_\alpha^{(ii)} = \sum_{t}^{\alpha}\sum_{t'}^{A} c_t^{(i)}c_{t'}^{(i)}s_{tt'}$$

Equation (18) has a form similar to the numerator of Eqs. (1), (3), and (4) derived from simpler calculations. We see that the interaction integral γ depends on the net charges of the atoms of the electron accepting molecule.

The denominator of the energy term D is given by

$$H_{i\to l,i\to l} - H_{0,0} \cong -(E_{Bl} + \Delta E_{Bl}) + (I_{Ai} + \Delta I_{Ai}) + a_{il} \quad (20)$$

where

$$E_{Bl} = -\left[H_{Bll} + \sum_{k}^{occ}\{2(ll|kk) - (lk|kl)\}\right]$$

$$\Delta E_{Bl} = -\left\{V_{All} + 2\sum_{i'}^{occ}(ll|i'i')\right\} + \sum_{i'}^{occ}\{V_{li'}S_{li'} + (li'|i'l)\}$$

$$I_{Ai} = -\left[H_{Aii} + \sum_{i'}^{occ}\{2(ii|i'i') - (ii'|i'i)\}\right]$$

$$\Delta I_{Ai} = -\left\{V_{Bii} + 2\sum_{k}^{occ}(ii|kk)\right\} + \sum_{k}^{occ}\{V_{ik}S_{ik} + (ik|ki)\}$$

$$a_{il} = -(ii|ll) + 2\{V_{il}S_{il} + (il|li)\}$$

$$H_{Aii} = \int a_i(1)\left(-\tfrac{1}{2}\Delta(1) - \sum_{\alpha}^{A}\frac{Z_\alpha}{r_{1\alpha}}\right)a_i(1)\,dv(1)$$

Therefore, the denominator is not simply given by the difference between the energy of the MO a_i and that of the MO b_l of B as in the case of a simple Hückel calculation. Klopman took the effect of opponent reactant upon the MO energies in the interaction, still neglecting differential overlaps.[34] I_{Ai} is the vertical ionization potential of the electron in the MO a_i and ΔI_{Ai} is the change in the I_{Ai} value due to the interaction with B.[99] Similarly, E_{Bl} is the vertical electron affinity of the MO b_l of B and ΔE_{Bl} is the change in the E_{Bl} value due to the interaction with A, When both A and B are neutral nonpolar molecules, the absolute values of ΔI_{Ai} and ΔE_{Bl} may be relatively small in comparison with the absolute value of a_{il}, making $(H_{i\to l,i\to l} - H_{0,0})$ smaller than $(I_{Ai} - E_{Bl})$. By taking explicitly the electron repulsions, nuclear attractions, and overlap integrals into account, we see that the charge-transfer interaction between the HOMO of A and the LUMO of B is relatively more important over other terms than expected simply from $(I_{Ai} - E_{Bl})$. The same can be said for the charge-transfer from B to A. When one of the two, say A, is an anion, then $-\Delta E_{Bl}$ will have a large positive value which can barely be compensated by adding a_{kl}. However, in this case I_{Ak} is small in comparison with the cases of neutral A, making the interaction between the HOMO of A and the LUMO of B dominant over all other charge-transfer terms. In the case of the interaction between a neutral system A and a cationic system B, E_{Bl} will have a large positive value and the interaction between the HOMO of A and the LUMO of B will be important. In some cases, the energy difference between the electron transferred state from the HOMO of one reactant into the LUMO of another, $\Psi_{\mathrm{HO}\to\mathrm{LU}}$, and the original state, Ψ_0, is so small that the second-order perturbation expression like Eq. (12) is no more valid. Then it may be more appropriate to write the energy term D as

$$D \cong \sqrt{2}|H_{0,\mathrm{HO}\to\mathrm{LU}} - S_{0,\mathrm{HO}\to\mathrm{LU}}H_{0,0}| \tag{21}$$

When one of the charge-transferred states, say c, happens to have a lower energy than the original state, Eq. (11) should be replaced by[36,52,53]

$$\Delta W = H_{c,c} - (W_{A0} + W_{B0}) - \frac{|H_{0,c} - S_{0,c}H_{c,c}|^2}{H_{0,0} - H_{c,c}} + \cdots \tag{22}$$

The numerator of the polarization energy Π is given by

$$H_{0,i\to j} - S_{0,i\to j}H_{0,0} \cong \sqrt{2}\left\{V_{Bij} + 2\sum_{k}^{\mathrm{occ}}(ij|kk)\right\} \tag{23}$$

where

$$V_{Bij} = \int a_i(1)\left(-\sum_{\beta}^{B}\frac{Z_\beta}{r_{1\beta}}\right)a_j(1)\,dv(1)$$

The denominator of Π is written as

$$H_{i\to j,i\to j} - H_{0,0} \cong -(E_{Aj} + \Delta E_{Aj}) + (I_{Ai} + \Delta I_{Ai}) + a_{ij} \tag{24}$$

where

$$a_{ij} = -(ii|jj) + 2(ij|ji)$$

The overlap integral $S_{0,i\to j}$ is the second order of the overlap integrals between the MOs of A and B. The denominators of Π cannot usually be so small as those of D in the majority of chemical reactions. Therefore, the contribution of Π to the stabilization of the interacting system may become less important than that of D as chemical interaction gets strong through the orbital overlappings of the MOs of A and B. In addition to D and Π, the dispersion energy may be important in the interaction between two completely nonpolar systems.[32,91] However, in most chemical reactions, one or both of the two reactants have local dipoles. Thus the dispersion energy may tentatively be neglected here.

B. Application to Chemically Reacting Systems

As has been discussed above, the interaction energy between two interacting systems can be divided into several terms, in an approximate sense. In some cases, one or two terms are dominant and others can be neglected. It is one of the characteristic features of ionic reactions that the Coulomb interaction energy plays a dominant role in the early stage of interaction. On the other hand, the contribution of the delocalization interaction energy to the stabilization becomes increasingly important with the progress of the reaction, both in polar and in nonpolar cases. It may be suitable for a better understanding of the important role of the delocalization interaction to look at some numerical results. In Table 3-1 are shown the results of a simple Hückel MO perturbation calculation for some two-centric interactions, using Eq. (1) with $|\gamma_{tu}| = |\gamma_{t'u'}|$.[67] Although the contributions of the HOMO-LUMO interactions are underestimated in Eq. (1) in comparison with the estimate of Eq. (12), the values may still serve as measures for the purpose of discussing the favorable reaction paths. In two centric cases we may assume, in principle, two modes of interactions. One is the case in which γ_{tu} and $\gamma_{t'u'}$ have the same sign, and another is the case in which γ_{tu} and $\gamma_{t'u'}$ have different signs. We may call the

Table 3-1. The Favorable Mode of Cyclic Interactions

		$\Delta E(\beta/\gamma^2)$ [a]	
Reagent (u, u')	Reactant (t, t')	syn	anti
Ethylene 1,2	Ethylene 1,2	0	2.000
	Allyl anion 1,3	2.414	0.414
	Butadiene 1,2	0.106	1.895
	Butadiene 1,4	1.789	0.422
	Hexatriene 1,6	0.622	1.657
	Benzene 1,2	0.333	1.444
	Benzene 1,4	1.333	0.444
	Anthracene 1,2	0.250	1.604
	Anthracene 1,4	1.475	0.424
	Anthracene 9,10	1.616	0.404
Butadiene 1,4	Butadiene 1,2	1.744	0.492
	Butadiene 1,4	0.716	1.789
	Naphthalene 1,2	1.454	0.573
	Naphthalene 1,4	0.658	1.427
Allyl cation 1,4	Ethylene 1,2	0.414	2.414
	Butadiene 1,4	2.789	0.789
Allyl anion 1,3	Butadiene 1,4	0.789	2.789

[a] $\gamma = |\gamma_{tu}| = |\gamma_{t'u'}|$, β: resonance integral of C=C double bond.

former case "syn" interaction and the latter "anti" interaction.[67,70,74] Here it is assumed that all the π AOs have the same sign on the same side of the molecular plane of each conjugated system. The modes of syn and anti interactions are schematically illustrated in Figure 3-3. The interaction integral γ is negative when the AOs overlap mainly with the lobes of the same

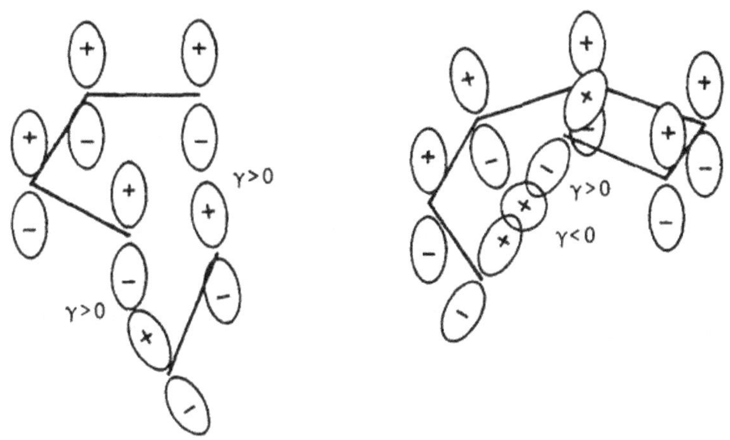

syn interaction (Diels-Alder addition) *anti* interaction (1, 7-hydrogen migration)

Fig. 3-3. A schematic illustration of syn and anti interactions.

sign, while it is positive when the two AOs interact in opposite phase. Dewar classified the two modes of interactions as Hückel and anti-Hückel interactions in his PMO theory.[71,100] Zimmerman used the terms Hückel and Möbius in order to distinguish the two types of interactions.[72,73] Table 3-1 shows that the thermal interaction between ethylene and butadiene 1,4 prefers to take place through syn interaction while interaction between two ethylenes favours anti interaction. These examples correspond to the $[4s + 2s]$ and $[2s + 2a]$ cases of Woodward and Hoffmann.[8] In general, thermal syn interaction is favored when the number of electrons forming a cycle is $4n + 2$, and thermal anti interaction is facilitated when it is $4n$. The conclusion is consistent with those of Woodward and Hoffmann,[6–9] Dewar,[71,100] Zimmerman,[72,73] Salem,[35] and Fukui and Fujimoto.[70] Woodward and Hoffmann derived the selection rules, considering the correlations of the MOs by the symmetries conserved throughout reactions.[8] The reason why the symmetries of MOs of a reacting system are retained was discussed by Fukui in the formulation of reaction coordinate.[101] We can also regard a

38 INTERMOLECULAR INTERACTIONS AND CHEMICAL REACTIVITY

two-centric unimolecular reaction as if it were a cyclic intermolecular interaction between two parts of a molecule appropriately partitioned.[67–69,74,102]

Next let us consider single-centric reactions. When the attacking reagents is a single orbital species with the energy α, we have the reactivity index, superdelocalizability, from Eq. (1) by multiplying the energy ΔE by β/γ^2 to have dimensionless quantity. When the reagent orbital is unoccupied,

$$S_t^{(E)} = 2\sum_i^{\text{occ}} \frac{(c_t^{(i)})^2}{\lambda_i} \qquad (\varepsilon_i = \alpha + \lambda_i\beta)$$

When the reagent orbital is occupied,

$$S_t^{(N)} = 2\sum_j^{\text{uno}} \frac{(c_t^{(j)})^2}{-\lambda_j} \qquad (\varepsilon_j = \alpha + \lambda_j\beta)$$

Since the SOMO of the reagent can interact with both the occupied and unoccupied MOs of the reactant, superdelocalizability for radical attack is defined by

$$S_t^{(R)} = \sum_i^{\text{occ}} \frac{(c_t^{(i)})^2}{\lambda_i} + \sum_j^{\text{uno}} \frac{(c_t^{(j)})^2}{-\lambda_j}$$

When the one-term approximation like Eq. (2) is valid, the chemical reactivity index, frontier electron density, can be used as a measure of the relative reactivity of the various positions of the molecule

$$f_t^{(E)} = 2(c_t^{(\mathrm{HO})})^2$$
$$f_t^{(N)} = 2(c_t^{(\mathrm{LU})})^2$$
$$f_t^{(R)} = (c_t^{(\mathrm{HO})})^2 + (c_t^{(\mathrm{LU})})^2$$

These indices were first defined in the framework of the simple Hückel MO method.[48–51] When we use the MOs which are normalized including overlap integrals, the partial population of the AO t in the ith MO consists of two parts,

$$n_t^{(i)} = p_t^{(i)} + v_t^{(i)}$$
$$p_t^{(i)} = 2(c_t^{(i)})^2$$
$$v_t^{(i)} = 2\sum_{\substack{t' \\ (\neq t)}} c_t^{(i)} c_{t'}^{(i)} s_{tt'}$$

where $p_t^{(i)}$ is the valence-inactive part and $v_t^{(i)}$ is the valence-active part.[103] Equations (12) and (18) indicate that it is appropriate to take the valence-inactive part which feels free from intramolecular bonding as a measure of chemical reactivity.[104]

The rate-determining step of the base catalyzed rearrangement of olefins has been shown to be the abstraction of hydrogen by base.[105]

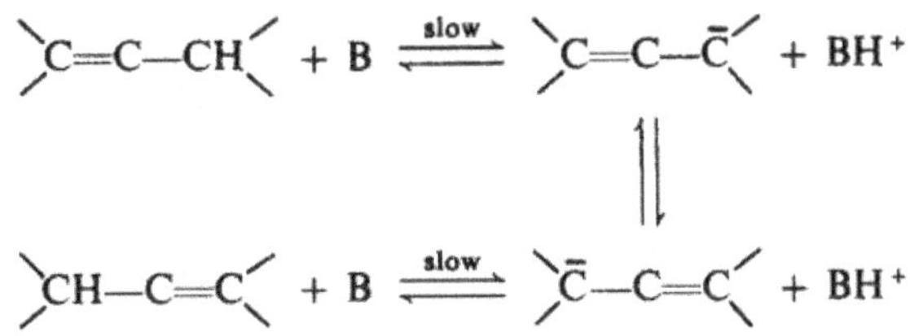

Table 3-2 gives the frontier electron densities of allylic hydrogens of olefins toward a nucleophilic attack.[106,107] The extended Hückel MO calculation gives almost the same energy value for LUMOs of a series of olefins. While the atomic population of the hydrogens fail to give the correct order, we find a parallel between the frontier electron densities and the observed reaction rates.

The nucleophilic bimolecular elimination reactions of substituted hydrocarbons are known to occur stereoselectively with the preferential abstraction

Table 3-2. Frontier Electron Densities of Hydrogens and Rate Constants of Isomerization Reaction.

	$CH_2{=}CHCH_2R$				
R	CH_3	C_2H_5	*i*-C_3H_7	*t*-C_4H_9	$CH_2{=}CH$
k (sec^{-1})[105]	8.4×10^{-4}	4.8×10^{-4}	1.5×10^{-5}	6.0×10^{-6}	3.6×10^{-3}
$p_t^{(LU)}$	0.0371	0.0365	0.0361	0.0353	0.0765
$v_t^{(LU)}$	−0.0135	−0.0133	−0.0133	−0.0130	−0.0282
N_t	0.873	0.872	0.870	0.869	0.854

	$CH_2{=}CHCH(CH_3)R$			
R	*H*	CH_3	C_2H_5	*i*-C_3H_7
k (sec^{-1})[105]	8.4×10^{-4}	2.0×10^{-4}	1.1×10^{-4}	3.0×10^{-5}
$p_t^{(LU)}$	0.0371	0.0324	0.0321	0.0312
$v_t^{(LU)}$	−0.0135	−0.0120	−0.0119	−0.0115
N_t	0.873	0.890	0.889	0.887

$CH_2{=}CHCH_2R$ $CH_2{=}CHCH(CH_3)R$

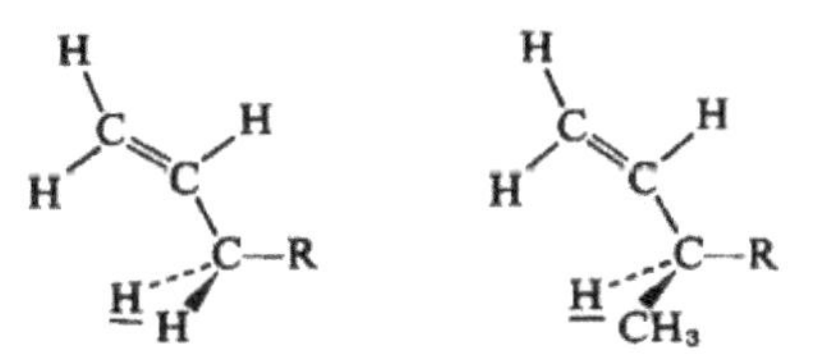

of *trans*-β-hydrogen.[108] In Fig. 3-4 are given the frontier electron densities of hydrogens for nucleophilic attack.[55] We see that the β-hydrogens which are located trans to chlorine are predicted to be the most reactive in both ethyl chloride and 2-chlorobutane. In 2-*exo*-chloronorbornane, the *exo*-hydrogen attached to the carbon 3 has the largest frontier electron density value. The E2 reaction of 2-*exo*-bromonorbornane was shown to take place predominantly in cis fashion to yield norbornene.[109] In these single-centric reactions, the stereoselection or spatial direction of reagent attack is governed by the inequality of the chemical reactivities of the same sort of atoms at different

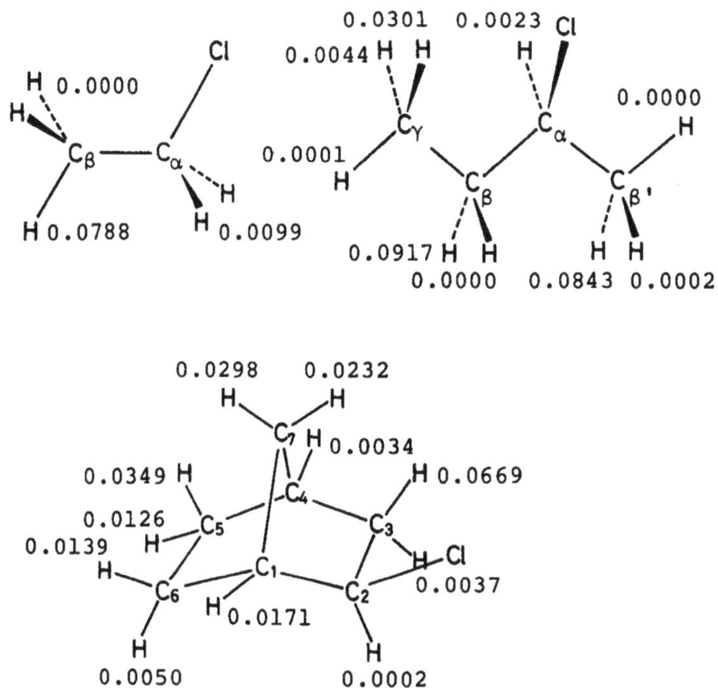

Fig. 3-4. Frontier electron densities of hydrogens toward nucleophilic reagents.

spatial positions. Another kind of inequality of spatial direction of reagent approach will play an important role in the homolytic halogenation of norbornane. The chlorination of norbornane was reported to yield 2-*exo*-chloronorbornane as the major product.[110] The first step of this reaction may be hydrogen abstraction by a radical. The extended Hückel MO calculation on 2-norbornyl radical showed that the trivalent carbon 2 did not have the sp^2 planar structure, having the hydrogen attached to the carbon 2 bent toward the endo direction.[111] Therefore, as is illustrated in Figure 3-5, the

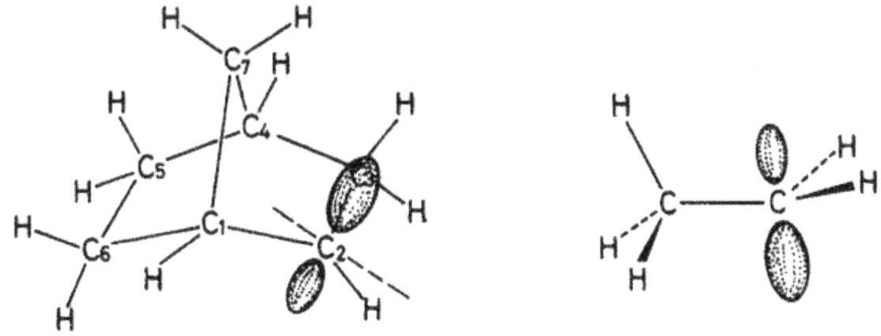

Fig. 3-5. Orbital extension in 2-norbornyl radical and in ethyl radical.

unpaired electron orbital would have larger extension to the exo direction than to the endo direction, by the mixing of the $2s$ and the $2p$ AOs of that carbon atom. Although the coefficient of this hybrid in SOMO has a definite value, the absolute value of the interaction integral γ may be larger for the exo approach of chlorine than for the endo approach. Such a deformation of the radical center of 2-norbornyl from planar structure was recently observed.[112] Inequality in the spatial direction of orbital extension may also have a significance in determining the stereoselectivity of addition of hydrogen halides, halogens, and so on, to olefins. *Ab initio* calculation on the ethyl radical suggested strongly the potentiality of such a directing effect.[113]

C. Isolated Molecule Approximation

In the above discussion, the interaction energy ΔW was discussed in the framework of no configuration change. That is, the interaction was considered, neglecting the destabilization due to the geometrical changes of both reagent and reactant from the most stable equilibrium nuclear configurations. Destabilization due to the molecular deformations of reagent and reactant as well as solvation and desolvation energy in liquid-phase reactions may unquestionably account for a part of the activation energy. Moreover, the use of the method discussed here is limited to the cases of weak interaction, owing to the approximations employed. We have assumed that the reaction path which yields the largest stabilization of the interaction in the early stage of a reaction is the path of the lowest activation energy. In other words, our discussion presented here is based on the assumption of the "noncrossing" of potential curves. Hence, it must be examined whether such an approximation is really sound or not.

In the simple Hückel MO theory, an atom is usually bonding with the neighboring atoms in occupied MOs and antibonding in unoccupied MOs.[114] That is, we have (for hydrocarbons)

$$\sum_{t'}^{\text{nei}} c_{t'}^{(i)} c_t^{(i)} = \frac{\alpha - \varepsilon_i}{-\beta} (c_t^{(i)})^2 = \lambda_i (c_t^{(i)})^2 \qquad (25)$$

42 INTERMOLECULAR INTERACTIONS AND CHEMICAL REACTIVITY

where $\sum_{t'}^{\text{nei}}$ means the summation over all the neighboring atoms t' of the atom t. As usual, the energies of occupied MOs are lower than α and $\sum_{t'}^{\text{nei}} c_t^{(i)} c_{t'}^{(i)}$ is positive for any t in bonding levels. On the contrary, the energies of the unoccupied MOs are higher than α, in general, making $\sum_{t'}^{\text{nei}} c_t^{(j)} c_{t'}^{(j)}$ negative for any t. Equation (25) indicates that the sum of the partial bond order(s) with the neighboring atom(s) is parallel to the partial electron density of the central atom. The following electron configurations contribute to the interactions. In electrophilic reactions, an electron is removed from occupied MOs (most dominantly from HOMO) of the reactant. In nucleophilic reactions, an electron is transferred into unoccupied MOs (most dominantly into LUMO). Both the charge-transfer from occupied MOs and that into unoccupied MOs bring about the decrease in the sum of bond order(s) of an atom with the neighboring atom(s). This will cause the isolation of the atom under the attack of the reagent from the remaining part of the reactant. The decrease in bond order reflects the stretching of the bond, which will result in the decrease in the absolute value of the resonance integral. The loosening of a bond will cause the unstabilization of MOs which are bonding in that bond region, and the stabilization of MOs which are antibonding in that bond region. When the reagent is not a single atom, the same may be said for the reagent part. The charge-transfer interaction between HOMO of the reagent and LUMO of the reactant and that between LUMO of the reagent and HOMO of the reactant will bring about the elevation of HOMOs and the lowering of LUMOs, resulting in the narrowing of the interfrontier level separations. By this, the HOMO-LUMO interactions become more and more dominant over other terms. The change in nuclear framework in the case of charge-transfer interaction which is expected from the nodal properties of the frontier orbitals usually conforms to the molecular deformation which really takes place in the reaction. For instance, we may consider the addition of a reagent to butadiene. The most reactive position of butadiene is 1 or 4, both for electrophilic and for nucleophilic reagents as shown in Fig. 3-6. HOMO of butadiene is bonding in 1,2- and 3,4-π bonds and antibonding in 2,3-π bond. LUMO is antibonding in 1,2- and 3,4-π bonds and bonding in 2,3-π bond. Charge transfer from the reagent into LUMO of butadiene and

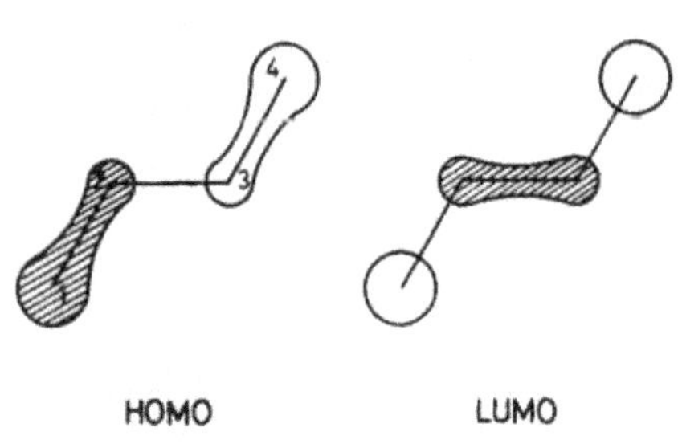

Fig. 3-6. Nodal properties of frontier orbitals of butadiene (hatched and nonhatched areas indicate the plus and minus parts of real MOs).

that into the reagent from HOMO of butadiene through the overlap between the reagent orbital and the π AO of a terminal carbon of butadiene, say 1, tend to isolate the carbon atom from the remainder of the molecule. With this, the HOMO is elevated and the LUMO is lowered as shown in Fig. 3-7. At the same time, the 2,3-π bond is strengthened, while the 3,4-π bond is weakened. This implies that the remainder part of butadiene by deleting the carbon 1 gets allylic properties. In many cases, the frontier electron density at the reaction center becomes larger as it is isolated from the remainder of the

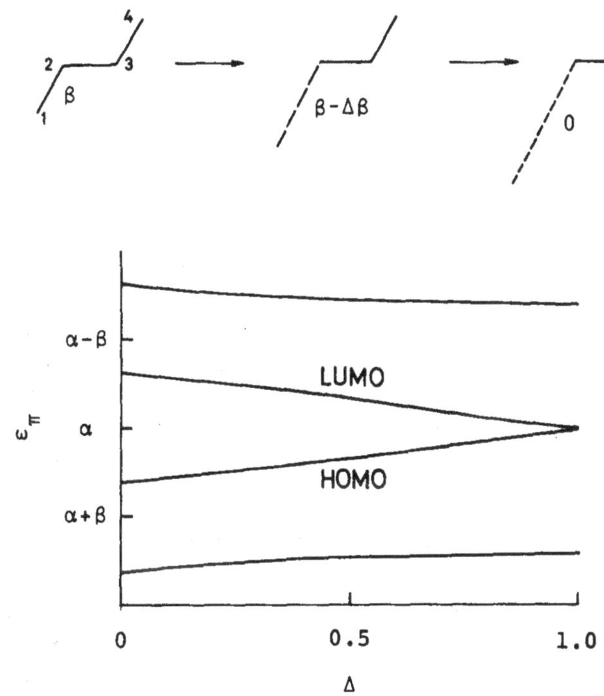

Fig. 3-7. Changes in MO energies of butadiene due to the stretching of C_1—C_2 bond.

molecule. In the example mentioned above, frontier electron density of the carbon 1 increases from 0.724 of butadiene to 1.000 of an isolated π AO, as shown in Fig. 3-8, since both the single π AO of the carbon 1 and the nonbonding MO of the allyl have the same energy $\varepsilon = \alpha$. Even in the cases in which the remainder part of a reactant by deleting an AO has no nonbonding level, the growing of the frontier electron density with the progress of the reaction seems to be a general tendency of chemical reactions. Thus, the cooperation of three principles of chemical reactions mentioned above:

44 INTERMOLECULAR INTERACTIONS AND CHEMICAL REACTIVITY

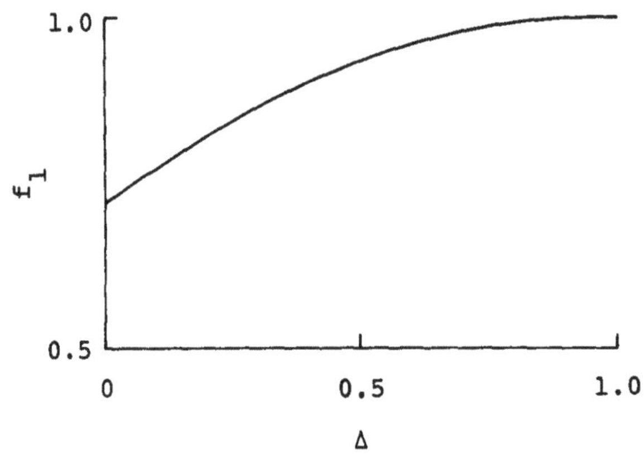

Fig. 3-8. Change in frontier electron density of the carbon 1 of butadiene due to the stretching of C_1—C_2 bond.

a positional parallelism between the charge-transfer and bond interchange, the narrowing of interfrontier level separation with bond interchange, and the growing frontier electron density along the reaction path, may not only supply the reason why HOMO and LUMO can predict the most reactive positions but also give a reasonable basis why we could extend the comparison of reaction paths in the early stage of chemical reaction to the neighborhood of the transition state. The self-acceleration of chemical interaction in this manner might be one of the basic natures of chemical reactions.

D. Charge-Transfer Interactions and Intermolecular Bond Formation

Charge-transfer interactions are responsible for the formation of new bonds and disappearance of old bonds in the reactant and the reagent. The nodal properties of HOMO and LUMO are important in predicting what change in molecular shape will take place in chemical reactions. Such a concept was also applied to the metal-catalyzed disproportionation of olefins.[79] Another problem to be considered here concerns the origin of the intermolecular bond formation.

The Coulomb interaction may be responsible for purely ionic bonds between reactant and reagent. The increase in electron densities in intermolecular region originates from the overlapping of the occupied MOs (particularly HOMO) of reactant and the unoccupied MOs (particularly LUMO) of reagent and the converse.[94,95] We consider here, as an example, the interaction between two two-electron systems A and B, A having a doubly occupied MO a_i and B having a doubly occupied MO b_k and an unoccupied

MO b_l. Neglecting the polarization interaction, the wave function is approximated by

$$\Psi = C_0\Psi_0 + C_{i\to l}\Psi_{i\to l} \tag{26}$$

The electron density is given by

$$\rho = C_0^2\rho_{0,0} + C_0C_{i\to l}(\rho_{0,i\to l} + \rho_{i\to l,0}) + C_{i\to l}^2\rho_{i\to l,i\to l} \tag{27}$$

where

$$\rho_{p,p'} = 4\int\int\int\int \Psi_p^*(1, 2, 3, 4)\Psi_{p'}(1, 2, 3, 4)\, d\tau_1\, d\tau_2\, d\tau_3\, d\tau_4$$

The electron density in the Ψ_0 state is given by

$$\rho_{0,0} = (\rho_{A0} + \rho_{B0}) + \frac{2(a_i^2 + b_k^2)S_{ik}^2 - 4a_ib_kS_{ik}}{(1 - S_{ik}^2)} \tag{28}$$

where ρ_{A0} and ρ_{B0} give the electron densities of A and B, respectively, in their isolated states. The second term on the right-hand side of Eq. (28) implies the distortion of the electron cloud due to the interference of the wave functions of A and B. Here we have

$$-\int a_i(1)b_k(1)S_{ik}\, dv(1) = -S_{ik}^2$$

This means that the adiabatic interaction between A and B cannot be the origin of intermolecular bond formation, because the changes in electron densities in intermolecular region due to electron exchange sum up to $\sim -4S_{ik}^2$.[94,95,115] The same can be said from MO population analysis.[116,117]

Introducing the charge-transfer state $\Psi_{i\to l}$, we have

$$\rho_{0,i\to l} \cong \sqrt{2}(\rho_{A0} + \rho_{B0})S_{il} + \sqrt{2}(a_ib_l - a_i^2S_{il} + b_kb_lS_{ik}) \tag{29}$$

and

$$\begin{aligned}\rho_{i\to l,i\to l} \cong \rho_{A0} + \rho_{B0} - a_i^2 + b_l^2 \\ + \{a_i^2(S_{ik}^2 - S_{il}^2) + b_k^2S_{ik}^2 - b_l^2S_{il}^2 \\ - 2a_ib_kS_{ik} + 2a_ib_lS_{il} - b_kb_lS_{ik}S_{il} - b_lb_kS_{ik}S_{il}\}\end{aligned} \tag{30}$$

The integral of the second term on the right-hand side of Eq. (29) over all the space vanishes. The first term in the parentheses clearly serves as an origin of increase in electron densities in the intermolecular region between A and B, because we have

$$\int a_i(1)b_l(1)\, dv(1) = \sum_t^A \sum_u^B c_t^{(i)}c_u^{(l)}s_{tu} = S_{il}$$

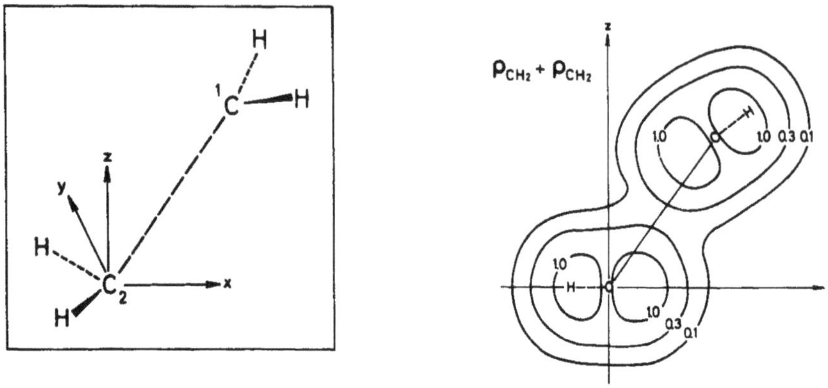

Fig. 3-9. Electron densities of two singlet methylenes without interaction in a non-least motion approach. The figures show the distribution of valence electrons in x–z cross section at $y = 0$ ($e^-/\text{Å}^3$).

and the product of the coefficients C_0 and $C_{t\to l}$ have the same sign as that of S_{tl}, making $C_0C_{t\to l}S_{tl}$ positive for the lowest-energy state. The second term in the parentheses indicates that the electrons which participate in the intermolecular bond formation are supplied from the occupied MO a_t of electron donor A. The last term represents the intramolecular electron rearrangement in B due to the interaction with A.

We show the results of a semiempirical MO calculation on the changes in electron densities in the interaction of two singlet methylenes.[95] Dimerization of methylenes was first discussed by Hoffmann, Gleiter, and Mallory[21] and later by Kollmar,[118] by Sustmann and Binsch,[39] and by Basch[119] mainly from an energetic viewpoint. In Fig. 3-9 is shown the simple sum of electron densities of two singlet methylenes located 2.5 Å apart from each other in a non-least-motion approach. Figure 3-10 gives the electron densities in the Ψ_0

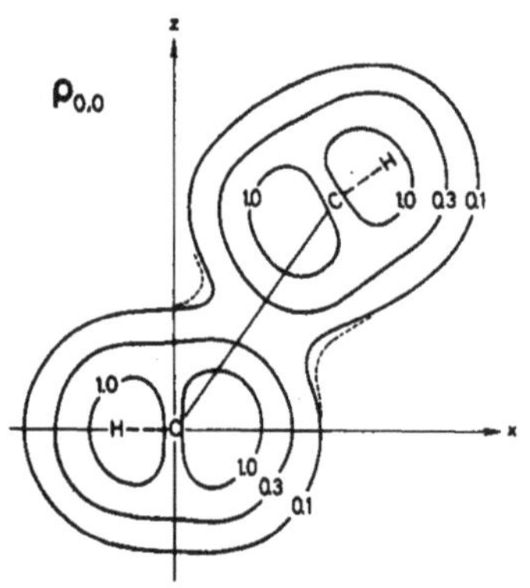

Fig. 3-10. Electron densities of two singlet methylenes without charge-transfer interactions in a non-least-motion approach.

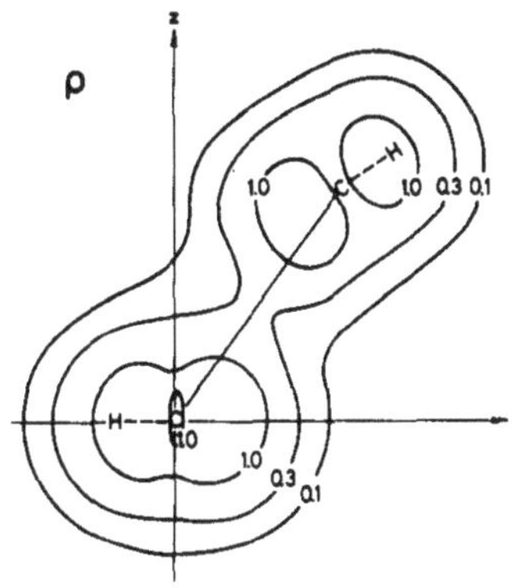

Fig. 3-11. Electron densities of two singlet methylenes with charge-transfer interactions in a non-least-motion approach.

state. We see that the electron densities in the region between the carbon atoms of two methylenes decrease by the electron exchange interaction in that state. By including charge-transferred states, we have the electron densities shown in Fig. 3-11. Considerable increase in electron densities in the region between the two carbons is observed. Figure 3-12 gives the electron densities of the points on the line connecting two carbon atoms. The electron densities of the least-motion approach of two singlet methylenes are shown in Fig. 3-13. No effective bond formation is achieved, since the HOMO-LUMO interaction is symmetry unfavorable in this process.

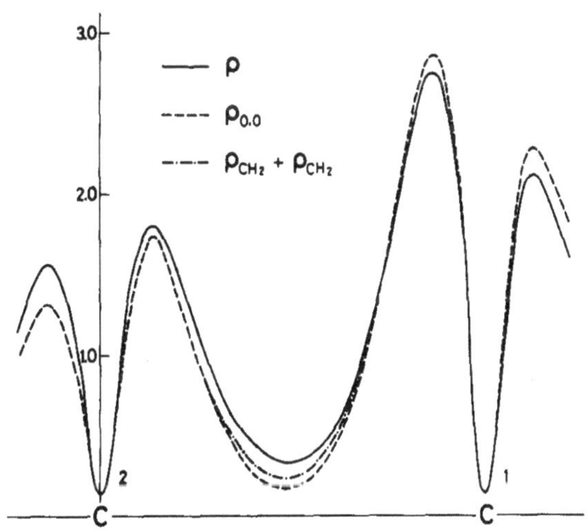

Fig. 3-12. Electron densities of two singlet methylenes at the points on the line connecting two carbons in a non-least-motion approach.

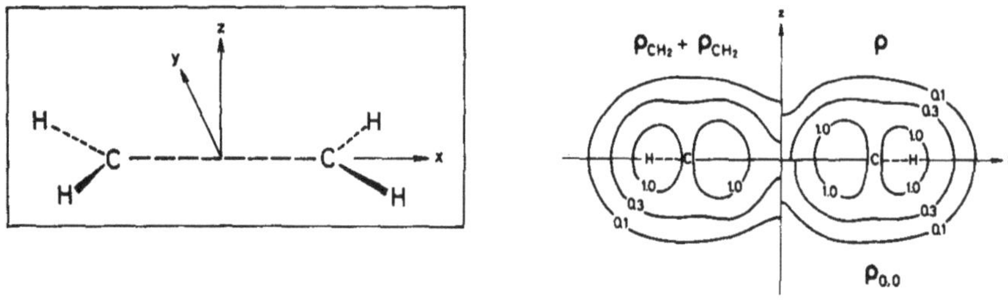

Fig. 3-13. Electron densities of two singlet methylenes in a least-motion approach.

E. Application to Other Systems

In the above discussion, we have limited ourselves to the interactions between two closed-shell systems. Some modifications are needed for the systems in which radicals and excited molecules take part. Figure 3-14 shows the mode of orbital interaction between a radical and a closed-shell molecule. SOMO of radical interacts both with HOMO and with LUMO of the closed-shell partner. The discussion mentioned in the previous sections for the interaction between two closed-shell systems almost holds for the doublet interaction between a radical and a closed-shell system.[120] In the case of photochemical reactions, there appear two SOMOs. The mode of interaction is illustrated in Fig. 3-15.[68,69,74,121] Comparing Figs. 3-1 and 3-15, we may understand why the stereoselectivity in photochemical reactions is opposite to that of thermal reactions. Photochemical additions of carbonyls to olefins were studied by Herndon and Giles, using the perturbation method in the framework of the simple Hückel MO theory.[122] Salem and Devaquet studied photochemical

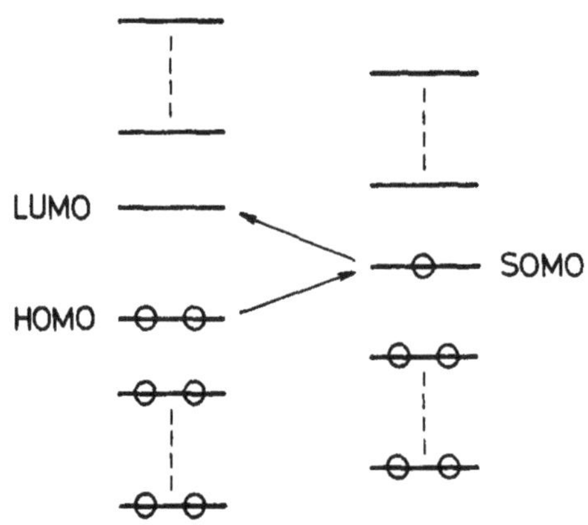

Fig. 3-14. Mode of orbital interaction between a closed-shell system and a radical.

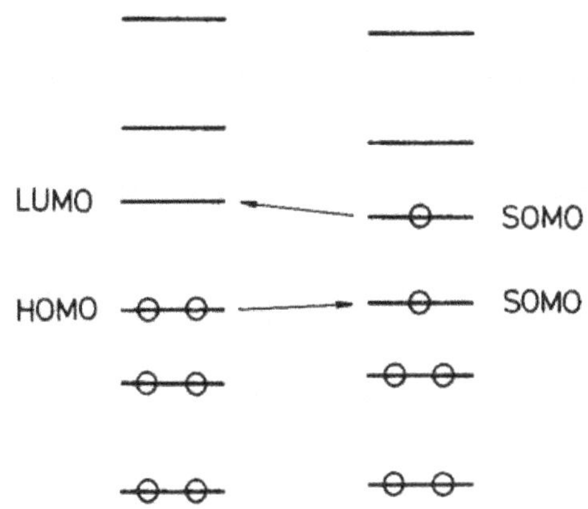

Fig. 3-15. Mode of orbital interaction between a closed-shell system and an excited open-shell system.

dimerization of dienes and keto steroids.[35,37] Detailed treatments of photo-induced reactions now seem to be somewhat difficult, because the nuclear configurations and the electronic structures of photoexcited molecules are not necessarily easy to find and the excited singlet MO wave functions cannot be obtained as simply as those of the ground state and excited triplet state. In the interaction between two radicals, the exchange interaction is responsible for the multiplet separation and may contribute to the stabilization of singlet interaction.[36] Charge-transfer between SOMOs may still be important in this case.

III. CONCLUSION

The methods of discussing chemical reactivity of molecular systems in terms of MOs in isolated states have been presented. In order to make the comparison of chemical reactivities distinct and quantitative, the interaction energy between two systems has been given by the combined sum of several energy terms. In some reactions, one of them will be important and in some other reactions, another term will dominate. In the early stage of ionic reactions, the Coulomb interaction will be the major source of stabilization. In nonionic reactions of nonpolar molecules, the reaction path may be governed by the exchange interaction energy and the delocalization interaction energy. Even in ionic cases, the role of the delocalization stabilization will be enhanced with the progress of reactions, as the net charges will be neutralized through charge-transfer interactions.[123,124] Charge-transfer interaction is also important as the major origin of bond interchange accompanying chemical reactions. Orientation in single-centric reactions and stereo-

50 INTERMOLECULAR INTERACTIONS AND CHEMICAL REACTIVITY

selection in multicentric reactions have been discussed in a unified manner, using the second-order perturbation equation of the delocalization interaction energy.

A chemical reactivity index is a theoretical measure for chemical reactivities of molecules which is obtained by the employment of the most simple quantum chemical tool applicable for chemical reactions, namely, MO methods. The chemical reactivity index, calculated from the MOs of isolated reactants, is correlated with reaction path, under the assumption of the noncrossing of potential energy curves for more than two different modes of interactions. Although it is not easy to verify the correctness of such an assumption, we have discussed about one possible reason why we could recognize the pertinency of the isolated molecule approximation.

Before finishing our discussion, we may point out the following rule for the "frontier controlled" reactions:

The majority of chemical reactions are liable to take place at the position and in the direction where the overlapping of HOMO and LUMO of respective reactants is at its maximum; in an electron-donating species, HOMO predominates in the overlapping interaction, while LUMO does so in an electron-accepting reactant; in the reacting species which possess SOMOs, these play the part of HOMO or of LUMO or of both.

These particular MOs may be termed the generalized frontier orbitals.

References

1. A. Streitwieser, Jr., *Molecular Orbital Theory for Organic Chemists*, Wiley, New York, 1961, pp. 307–356.
2. K. Fukui, in *Molecular Orbitals in Chemistry, Physics, and Biology*, P.-O. Löwdin and B. Pullman, Eds., Academic, New York, 1964.
3. K. Fukui, in *Modern Quantum Chemistry, Istanbul Lectures, Part I*, O. Sinanoğlu, Ed., Academic, New York, 1965.
4. K. Fukui, *Sigma Molecular Orbital Theory*, O. Sinanŏglu and K. B. Wiberg, Eds., Yale Univ. Press, New Haven, Conn., 1970.
5. K. Fukui, Fortschr. *Chem. Forsch.* **15**, 1 (1970).
6. R. B. Woodward and R. Hoffmann, *J. Am. Chem. Soc.* **87**, 395, 2511 (1965).
7. R. Hoffmann and R. B. Woodward, *J. Am. Chem. Soc.* **87**, 2046, 4388, 4389 (1965).
8. R. B. Woodward and R. Hoffmann, *Angew. Chem. Int. Ed. Engl.* **8**, 781 (1969); *The Conservation of Orbital Symmetry*, Academic, New York, 1969.
9. R. Hoffmann and R. B. Woodward, *Acct. Chem. Res.* **1**, 17 (1969).
10. R. Hoffmann, *J. Chem. Phys.* **39**, 1397 (1963); **40**, 2047, 2474, 2480 (1964).
11. J. A. Pople, D. P. Santry, and G. A. Segal, *J. Chem. Phys.* **43**, s129 (1965).

12. J. A. Pople and D. L. Beveridge, *Approximate Molecular Orbital Theory*, McGraw-Hill, New York, 1970.
13. J. A. Pople, D. L. Beveridge, and P. A. Dobosh, *J. Chem. Phys.* **47**, 2026 (1967).
14. M. J. S. Dewar and G. Klopman, *J. Am. Chem. Soc.* **89**, 3089 (1967).
15. N. C. Baird and M. J. S. Dewar, *J. Chem. Phys.* **50**, 1262 (1969).
16. N. C. Baird, M. J. S. Dewar, and R. Sustmann, *J. Chem. Phys.* **50**, 1275 (1969).
17. T. Yonezawa, K. Yamaguchi, and H. Kato, *Bull. Chem. Soc. Jap.* **40**, 536 (1967).
18. H. Kato, H. Konishi, H. Yamabe, and T. Yonezawa, *Bull. Chem. Soc. Jap.* **40**, 2761 (1967).
19. H. Kato, K. Morokuma, T. Yonezawa, and K. Fukui, *Bull. Chem. Soc. Jap.* **38**, 1749 (1965).
20. R. Hoffmann, *J. Am. Chem. Soc.* **90**, 1475 (1968).
21. R. Hoffmann, R. Gleiter, and F. B. Mallory, *J. Am. Chem. Soc.* **92**, 1460 (1970).
22. R. Hoffmann, S. Swaminathan, B. G. Odell, and R. Gleiter, *J. Am. Chem. Soc.* **92**, 7091 (1970).
23. R. Hoffmann, C. C. Wan, and V. Neagu, *Mol. Phys.* **19**, 113 (1970).
24. M. J. S. Dewar, E. Haselbach, and M. Shanshal, *J. Am. Chem. Soc.* **92**, 3505 (1970).
25. J. R. Hoyland, *Theor. Chim. Acta* **22**, 229 (1971).
26. M. J. S. Dewar and W. W. Schoeller, *J. Am. Chem. Soc.* **93**, 1481 (1971).
27. E. Clementi, *J. Chem. Phys.* **46**, 3851 (1967).
28. W. Th. A. M. van der Lugt and P. Ros, *Chem. Phys. Lett.* **4**, 389 (1969).
29. G. Berthier, D.-J. David, and A. Veillard, *Theor. Chim. Acta* **14**, 329 (1969).
30. A. Dedieu and A. Veillard, *Chem. Phys. Lett.* **5**, 328 (1970).
31. A. J. Duke and R. F. W. Bader, *Chem. Phys. Lett.* **10**, 631 (1971).
32. J. N. Murrell, M. Randić, and D. R. Williams, *Proc. Roy. Soc. Lond.* Ser. A **284**, 566 (1965).
33. G. Klopman and R. F. Hudson, *Theor. Chim. Acta* **8**, 165 (1967).
34. G. Klopman, *J. Am. Chem. Soc.* **90**, 223 (1968).
35. L. Salem, *J. Am. Chem. Soc.* **90**, 543, 553 (1968).
36. K. Fukui and H. Fujimoto, *Bull. Chem. Soc. Jap.* **41**, 1989 (1968).
37. A. Devaquet and L. Salem, *J. Am. Chem. Soc.* **91**, 3793 (1969).
38. A. Devaquet, *Mol. Phys.* **18**, 233 (1970).
39. R. Sustmann and G. Binsch, *Mol. Phys.* **20**, 1, 9 (1971).
40. T. Ri and H. Eyring, *J. Chem. Phys.* **8**, 433 (1940).
41. A. Pullman and B. Pullman, *Experientia* **2**, 364 (1946).
42. M. J. S. Dewar, *Trans. Faraday Soc.* **42**, 764 (1946).
43. E. Hückel, *Z. Physik* **70**, 204 (1931).
44. C. A. Coulson and H. C. Longuet-Higgins, *Proc. Roy. Soc. Lond.*, Ser. A **191**, 39 (1947); **192**, 16 (1947).
45. C. A. Coulson, *Discussions Faraday Soc.* **2**, 9 (1947).

46. G. W. Wheland, *J. Am. Chem. Soc.* **64**, 900 (1942).
47. M. J. S. Dewar, *J. Am. Chem. Soc.* **74**, 3357 (1952).
48. K. Fukui, T. Yonezawa, and H. Shingu, *J. Chem. Phys.* **20**, 722 (1952).
49. K. Fukui, T. Yonezawa, C. Nagata, and H. Shingu, *J. Chem. Phys.* **22**, 1433 (1954).
50. K. Fukui, T. Yonezawa, and C. Nagata, *Bull. Chem. Soc. Jap.* **27**, 423 (1954).
51. K. Fukui, T. Yonezawa, and C. Nagata, *J. Chem. Phys.* **26**, 831 (1957).
52. R. D. Brown, *J. Chem. Soc.* 2232 (1959).
53. S. Nagakura and J. Tanaka, *J. Chem. Soc. Jap., Pure Chem. Sec.* **75**, 993 (1954).
54. K. Fukui, H. Kato, and T. Yonezawa, *Bull. Chem. Soc. Jap.*, **34**, 1111 (1961).
55. K. Fukui and H. Fujimoto, *Tetrahedron Lett.* 4303 (1965).
56. K. Fukui, H. Hao, and H. Fujimoto, *Bull. Chem. Soc. Jap.* **42**, 348 (1969).
57. H. Fujimoto, S. Yambe, and K. Fukui, *Bull. Chem. Soc. Jap.* **44**, 971 (1971).
58. K. Fueki and K. Hirota, *J. Chem. Soc. Jap., Pure Chem. Sec.* **81**, 209 (1960).
59. H. C. Longuet-Higgins and E. W. Abrahamson, *J. Am. Chem. Soc.* **87**, 2045 (1965).
60. S. I. Miller, in *Advances in Physical Organic Chemistry*, V. Gold, Ed., Academic, New York, 1968, Vol. 6, pp. 185–332.
61. N. T. Anh, *Les Regles de Woodward-Hoffmann*, Ediscience, Paris, 1970.
62. G. B. Gill, *Quart. Rev.* **22**, 338 (1968).
63. R. D. Brown, *J. Chem. Soc.* 691, 2730 (1950).
64. R. D. Brown, *J. Chem. Soc.* 691, 2730 (1950).
65. K. Fukui, C. Nagata, T. Yonezawa, H. Kato, and K. Morokuma, *J. Chem. Phys.* **31**, 287 (1959).
66. K. Fukui, K. Morokuma, T. Yonezawa, and C. Nagata, *Bull. Chem. Soc. Jap.* **33**, 963 (1960).
67. K. Fukui and H. Fujimoto, *Bull. Chem. Soc. Jap.* **39**, 2116 (1966).
68. K. Fukui, *Tetrahedron Lett.* 2009 (1965).
69. K. Fukui, *Bull. Chem. Soc. Jap.* **39**, 498 (1966).
70. K. Fukui and H. Fujimoto, *Bull. Chem. Soc. Jap.* **40**, 2018 (1967).
71. M. J. S. Dewar, *Tetrahedron Suppl.* **8**, Part I, 75 (1966).
72. H. E. Zimmerman, *J. Am. Chem. Soc.* **88**, 1564, 1566 (1966).
73. H. E. Zimmerman, *Acct. Chem. Res.* **4**, 272 (1971).
74. K. Fukui and H. Fujimoto, *Mechanisms of Molecular Migrations*, B. S. Thyagarajan, Ed., Interscience, New York, 1969, Vol. 2, pp. 117–190.
75. W. C. Herndon and L. H. Hall, *Tetrahedron Lett.* 3095 (1967).
76. M. J. Goldstein, *J. Am. Chem. Soc.* **89**, 6357 (1967).
77. H. E. Simmons and T. Fukunaga, *J. Am. Chem. Soc.* **89**, 5208 (1967).
78. R. Hoffmann, A. Imamura, and G. D. Zeiss, *J. Am. Chem. Soc.* **89**, 5215 (1967).
79. F. D. Mango and J. H. Schachtschneider, *J. Am. Chem. Soc.* **89**, 2484 (1967); **91**, 1030 (1969); **93**, 1123 (1971).
80. F. D. Mango, *Tetrahedron Lett.* 505 (1971).

81. W. Th. A. M. van der Lugt, *Tetrahedron Lett.* 2281 (1970).
82. G. S. Lewandos and R. Pettit, *Tetrahedron Lett.* 789 (1971).
83. W. Kutzelnigg, *Tetrahedron Lett.* 4965 (1967).
84. D. T. Clark and G. Smale, *Tetrahedron Lett.* 3673 (1968).
85. D. T. Clark and G. Smale, *Tetrahedron* **25**, 13 (1969).
86. D. T. Clark and D. R. Armstrong, *Theor. Chim. Acta* **13**, 365 (1969); **14**, 370 (1969).
87. K. Hsu, R. J. Buenker, and S. D. Peyerimhoff, *J. Am. Chem. Soc.* **93**, 2117 (1971).
88. M. J. S. Dewar and S. Kirschner, *J. Am. Chem. Soc.* **93**, 4290, 4291, 4292 (1971).
89. W. Th. A. M. van der Lugt and L. J. Oosterhoff, *J. Am. Chem. Soc.* **91**, 6042 (1969).
90. C. C. J. Roothaan, *Rev. Mod. Phys.* **23**, 69 (1951).
91. F. B. van Duijneveldt and J. N. Murrell, *J. Chem. Phys.* **46**, 1759 (1967).
92. J. L. Lippert, M. W. Hanna, and P. J. Trotter, *J. Am. Chem. Soc.* **91**, 4035 (1969).
93. E. G. Cook, Jr. and J. C. Schug, *J. Chem. Phys.* **53**, 723 (1970).
94. H. Fujimoto, S. Yamabe, and K. Fukui, *Tetrahedron Lett.* 443 (1971).
95. H. Fujimoto, S. Yamabe, and K. Fukui, *Bull. Chem. Soc. Jap.* **44**, 2936 (1971).
96. M. Born and J. R. Oppenheimer, *Ann. Physik* **84**, 457 (1927).
97. R. S. Mulliken, *J. Chim. Phys.* **46**, 497 (1949).
98. R. S. Mulliken, *J. Chem. Phys.* **23**, 1833 (1955).
99. T. Koopmans, *Physica* **1**, 104 (1933).
100. M. J. S. Dewar, *The Molecular Orbital Theory of Organic Chemistry*, McGraw-Hill, New York, 1969.
101. K. Fukui, *J. Phys. Chem.* **74**, 4161 (1970).
102. K. Fukui, *Acct. Chem. Res.* **4**, 57 (1971).
103. K. Ruedenberg, *Rev. Mod. Phys.* **34**, 326 (1962).
104. K. Fukui and H. Fujimoto, *Bull. Chem. Soc. Jap.* **40**, 2787 (1967).
105. A. Schriesheim, J. E. Hofmann, and C. A. Rowe, Jr., *J. Am. Chem. Soc.* **83**, 3731 (1961), and their subsequent papers.
106. H. Fujimoto, H. Oba, and K. Fukui, *J. Chem. Soc. Jap., Pure Chem. Sec.* **90**, 1005 (1969).
107. H. Fujimoto, S. Sugihara, S. Yamabe, and K. Fukui, *Bull. Chem. Soc. Jap.* **44**, 2565 (1971).
108. E. S. Gould, *Mechanism and Structure in Organic Chemistry*, Holt and Rinehart, New York, 1959, pp. 472–513.
109. H. Kwart, T. Takeshita, and J. L. Nyce, *J. Am. Chem. Soc.* **86**, 2606 (1964).
110. E. C. Kooyman and G. C. Vegter, *Tetrahedron* **4**, 382 (1958).
111. H. Fujimoto and K. Fukui, *Tetrahedron Lett.* 5551 (1966).
112. J. Gloux, M. Guglielmi, and H. Lemaire, *Mol. Phys.* **19**, 833 (1970).
113. W. A. Lathan, W. J. Hehre, and J. A. Pople, *J. Am. Chem. Soc.* **93**, 808 (1971).
114. K. Fukui and H. Fujimoto, *Bull. Chem. Soc. Jap.* **42**, 3399 (1969).

115. V. Magnasco, *Theor. Chim. Acta* **21**, 267 (1971).
116. R. S. Mulliken, *J. Chem. Phys.* **23**, 1841 (1955).
117. L. Salem, *Proc. Roy. Soc. Lond.*, Ser. A **264**, 379 (1961).
118. H. Kollmar, *Tetrahedron Lett.* 3337 (1970).
119. H. Basch, *J. Chem. Phys.* **55**, 1700 (1971).
120. H. Fujimoto, S. Yamabe, and K. Fukui, unpublished results.
121. K. Fukui, K. Morokuma, and T. Yonezawa, *Bull. Chem. Soc. Jap.* **34**, 1178 (1961).
122. W. C. Herndon and W. B. Giles, *Mol. Photochem.* **2**, 277 (1970).
123. H. Fujimoto, S. Yamabe, and K. Fukui, *Tetrahedron Lett.* 439 (1971).
124. K. Fukui, H. Fujimoto, and S. Yamabe, *J. Phys. Chem.*, in press.

Pure & Appl.Chem., Vol.54, No.10, pp.1825-1836, 1982.
Printed in Great Britain.

0033-4545/82/101825-12$03.00/0
Pergamon Press Ltd.

CHEMICAL REACTIVITY THEORY—ITS PRAGMATISM AND BEYOND

Kenichi Fukui

Presidential Office
Kyoto University of Industrial Arts and Textile Fibres
Matsugasaki, Sakyo-ku, Kyoto 606, Japan

Abstract - The development of the theory of chemical reactions is sketched. The use of a perturbation molecular orbital theory, the primordial frontier orbital theory, is stressed in view of the reaction design. Examples of the frontier orbital polarization approach are shown. A theory of chemical reaction path, called the IRC (intrinsic reaction coordinate) approach, is mentioned. It is shown that this approach is useful for making the chemical reactivity theory go beyond the perturbation method. Several applications of the IRC approach are discussed.

INTRODUCTION

Since G. N. Lewis introduced the concept of "sharing of electrons" in 1916, the electronic theory had been developed by a number of chemists. Particularly R. Robinson and C. K. Ingold contributed much to establish the electronic theory so as to be suitable for discussing the chemical reactivity of organic compounds(Ref. 1 & 2). In that theory, the electron density distribution and its change in a compound played a principal role. Various chemical concepts were understood in terms of the electronic charge. The position of attack in chemical reactions was discussed by the electrostatic principle. In this fashion, the electronic theory was able to explain a majority of complicated experimental results in a unified manner.

The success of the electron density criterion was great. However, a weak point also existed. It was involved in the substitution reaction of non-substituted aromatic hydrocarbons. For instance, naphthalene reacts with both electrophiles and nucleophiles at the same position 1. Evidently, the simple electronic theory was not well qualified for explaining this result. So, Coulson and Longuet-Higgins considered the change of electron density under the influence of the attacking reagent(Ref. 3). Wheland calculated the energy required to localize electrons to the reaction site(Ref. 4). These two papers successfully interpreted the experimental results.

A PERTURBATION THEORY—FRONTIER ORBITAL APPROACH

In this connection, another attempt was made to approach this problem. The principal role of valence electrons in the formation of a molecule from constituent atoms was noticed. By analogy, the density distribution of the highest occupied molecular orbital(HOMO) was calculated for the attack of an electrophilic reagent. An agreement was obtained between the position of attack and the site of the largest density of this particular orbital. For a nucleophilic attack, the distribution of the lowest unoccupied MO(LUMO) was calculated and a correlation with reactivity was found. In the radical substitution the distribution of (HOMO + LUMO) possessed a correlation with reactivity(Ref. 5 & 6). The reasoning for these results was simply made by a perturbation theory(Ref. 7).

These selected orbitals, HOMO and LUMO, were called "frontier orbitals," and the reactivity criteria using the frontier orbitals were soon extended to other compounds than conjugated molecules and to other reactions than aromatic substitutions. Such an extension was made with no essential difficulty, since the method was based upon the electron delocalization between the frontier orbitals of the substrate and the reagent and there was no reason to put any limitation with respect to the molecule and the reaction(Ref. 8 & 9). Some reactivity indices were derived from this delocalization picture. The method showed its usefulness in the field of the hydrogen abstraction from hydrocarbons by radicals, the S_N2 and the E2 reactions in halogenated hydrocarbons, the nucleophilic abstraction of hydrogen in olefins, and so forth. These results were collected in a single volume(Ref. 9). All of other reactivity theories then available were applicable only to conjugated molecules.

FRONTIER ORBITAL SYMMETRY

The usefulness of the theory was much more broadened when the symmetry of the frontier orbitals was employed in the discussion of the reactivity in Diels-Alder reaction(Ref. 10). By this work, the orbital symmetry was connected to the selective occurrence of chemical reactions. It was shown that the symmetries of HOMO and LUMO of dienes and dienophiles were in a situation favourable for a concerted cyclic interaction between them. The electron delocalization from the diene HOMO to the dienophile LUMO causes the weakening of the double bonds of both diene and dienophile and the formation of new bonds between the two molecules and also inside the diene. The nodal structure of diene HOMO and the dienophile LUMO is satisfactorily adapted for the formation of two new independent bonds. These circumstances simply explain why diene normally behaves as an electron-donor and dienophile as an electron-acceptor(Ref. 11). The bond exchange takes place smoothly by the aid of such a favourable situation of orbital symmetry. The orbital symmetry relationship in actually occurring Diels-Alder reaction is always favourable in the sense mentioned above.

The "selection rule" for the occurrence of concerted two-centre reactions seemed to be the conformity of the symmetries of HOMO and LUMO at the reaction centres. But one should not be too hasty in concluding so with only one example. In fact, it was after the proposal of a more general rule by Woodward and Hoffmann(Ref. 12) that the selection rule was systematized in the foregoing style(Ref. 13).

After the discovery of the Woodward-Hoffmann rule, the recognition of the frontier orbital theory became more rapid than before. It was therefore particularly fortunate for the development of the reactivity theory based mainly on perturbation methods that the Woodward-Hoffmann rule independently appeared just in time.

In process of time, the nature of chemical reactions had gradually been disclosed. The electron delocalization between HOMO and LUMO was found to be the principal factor. So, the easiness of a chemical reaction was controlled by the favourableness of the HOMO - LUMO interaction. It also determined the stereoselective path. Intramolecular processes were discussed similarly to intermolecular processes. A number of chemists contributed much to develop the HOMO - LUMO interaction approach(Ref. 14-21).

It is often useful to discuss the interaction energy divided into its constituents——the coulomb, the exchange, the polarization, and the delocalization energies. Such a partitioning technique is based on a perturbation or non-perturbation evaluation of the interaction energy(Ref. 22 & 23). Several examples are given(Ref. 24-31).

ORBITAL POLARIZATION APPROACH

The usefulness of the HOMO - LUMO interaction criterion in chemical reactivity problems has been shown. An effective discussion is in some cases possible even without quantitative calculation of molecular energy. A qualitative theory of orbitals for that purpose is most nicely mentioned in a lovely and unique book of Jorgensen and Salem(Ref. 32).

In relation to such a qualitative reactivity theory the frontier orbital polarization approach is frequently convenient for the reaction design. The orbital interaction causes orbital polarization in the course of reaction. Such as regioselectivity in cycloadditions, various subsidiary effects in cycloadditions, substituent effects, d-orbital effects, lone-pair orbital effects, plane-asymmetric stereoselective reactions, and other selective reactions were interpreted(Ref. 9, 13, 19-21, 33-37).

An interesting application of orbital polarization approach is the explanation of transannular cross- and noncross-bond formations. The following four cases are to be explained:

Case I.

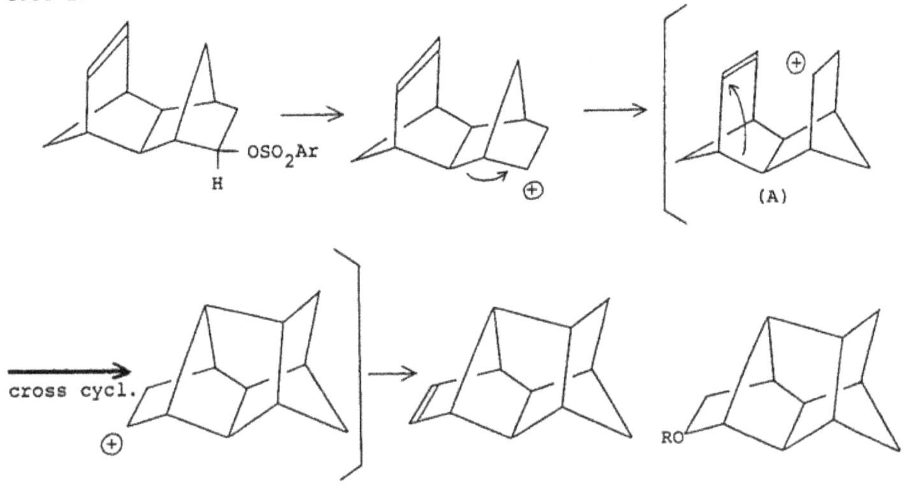

(Ref. 38)

Case II.

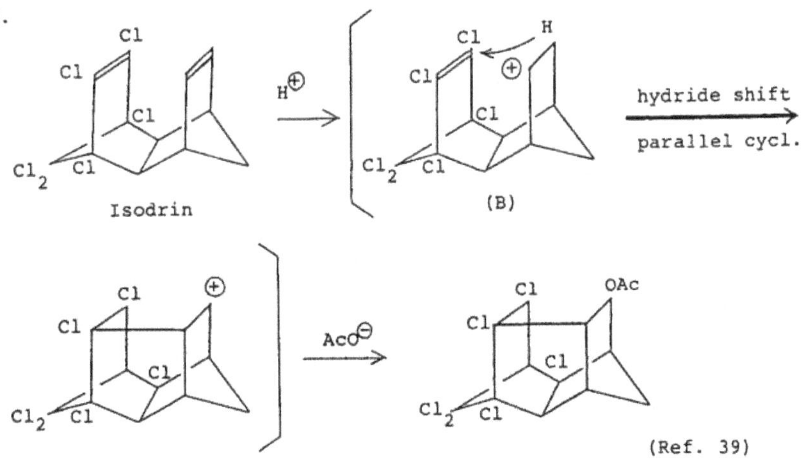

(Ref. 39)

Case III.

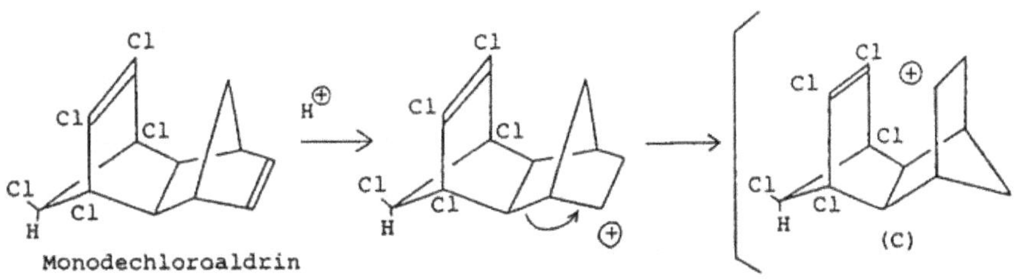

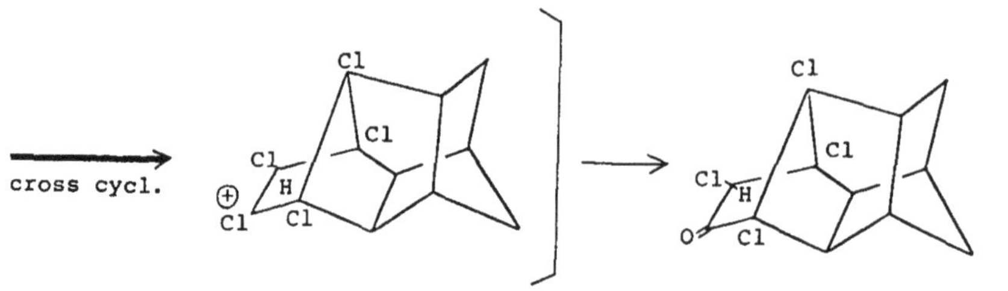

and epimer about Cl-C-H

(Ref. 40)

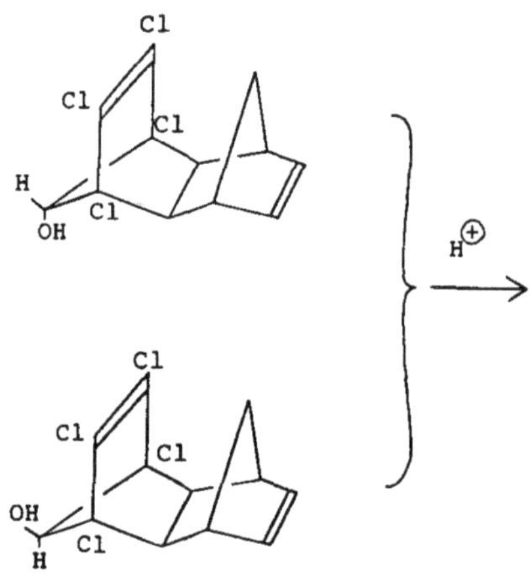

The same two products, which are considered to result from cross and parallel cyclizations.

(Ref. 41)

The direction of polarization in the π^* orbital of the etheno-bridge due to the cationic charge in (A) can easily be given(Ref. 37).

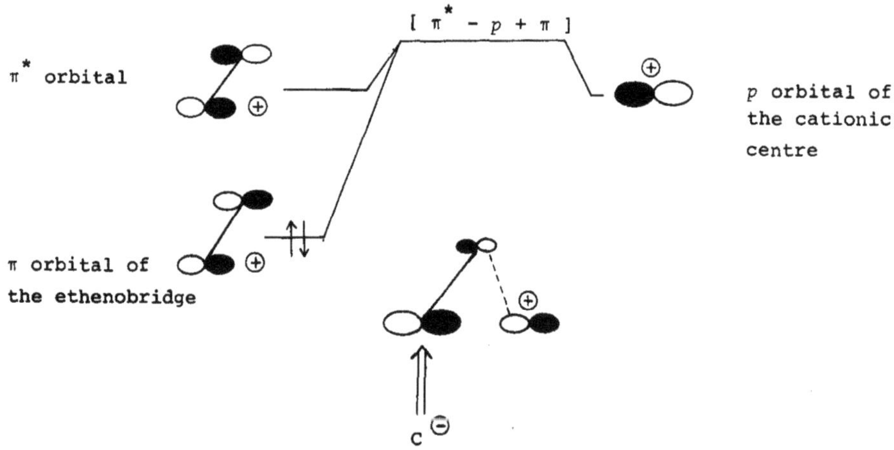

The path of the Wagner-Meerwein rearrangement is thus understood.

In contrast, the direction of orbital polarization is opposite in (B), in which the effect of polyhalogenation brings about a lowering of π orbital energies, so as to cause the shift of the hydride.

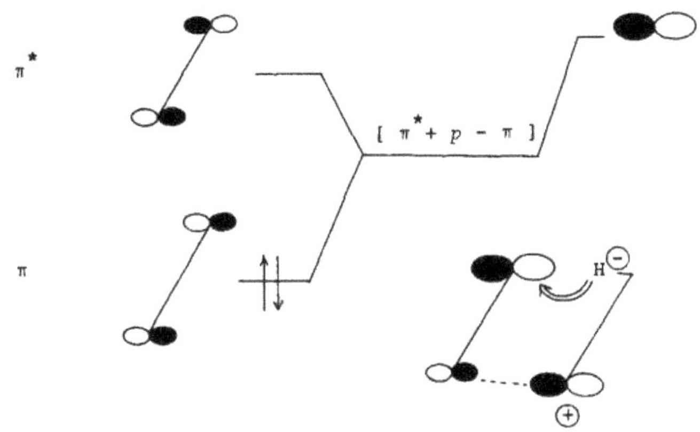

In the hypothetical ion, (C), the lowering of π levels could be insufficient to cause the hydride shift, since the strong participation of the axial C-Cl bond to absorb electrons disappears in this case, so that the cross cyclization occurs again like in (A). The result of Case IV indicates that the effect of OH substituent instead of axial-Cl at the methylene bridge might make the cross and parallel cyclizations compete.

Next example is the stereospecificity in the Diels-Alder reaction.

Case I.

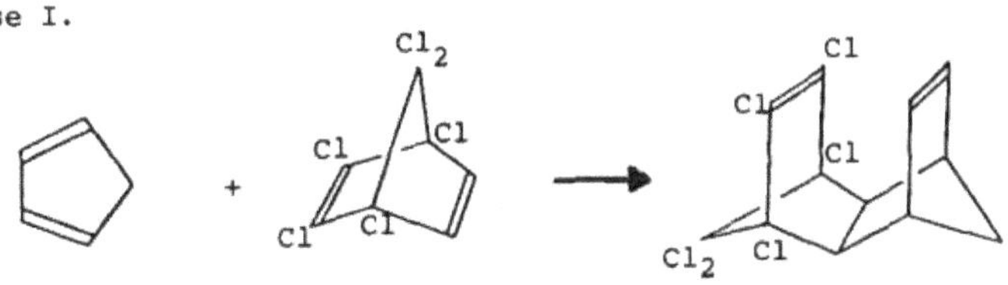

(Ref. 42)

Isodrin

(endo-endo adduct)

Case II.

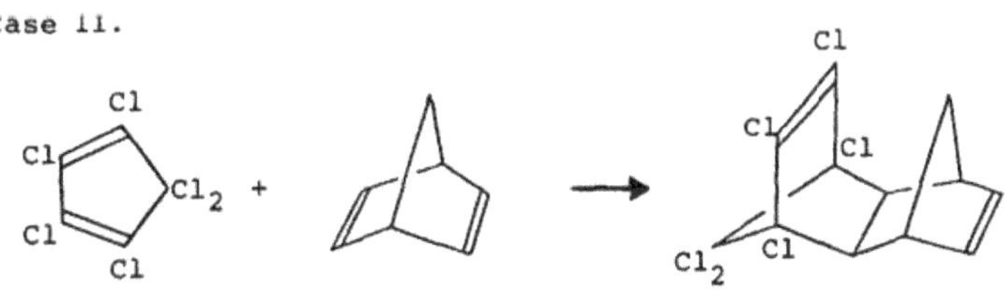

(Ref. 43)

Aldrin

(endo-exo adduct)

Case III.

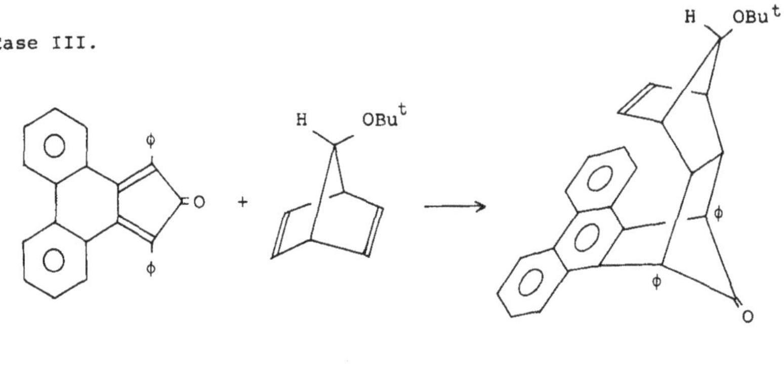

(endo-endo-anti-Bu^{t}O adduct)

(Ref. 44)

Case I is an example for the case occurring in normal electron demand. The regioselection in hexachloronorbornadiene could be attributed to the interference of the antibonding p lobes at the adjacent chlorine atoms in the LUMO of the dichlorinated ethylenic bond. The endo-endo terminology of the product implies that the first "endo" criterion relates to the two remaining double bonds and the second "endo" corresponds to the norbornadiene double bond. The dominance of endo-adduct in all these cases with respect to remaining double bonds is easily understood by considering the auxiliary orbital effect which is favourable for the "endo" configuration.

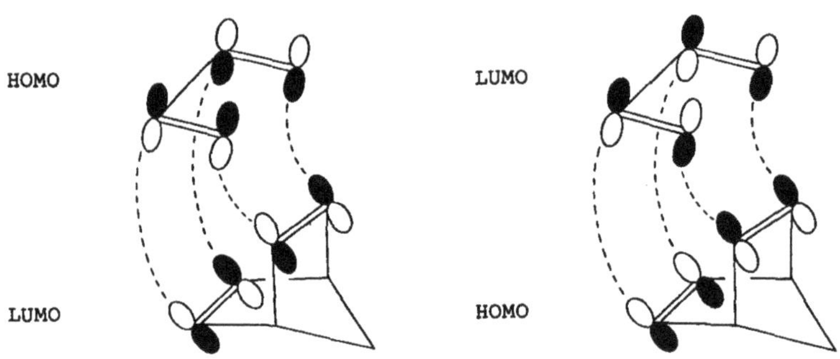

In the formation of exo-adduct there is no favourableness like this.

The exo-endo selectivity with respect to the norbornene double bond is the next problem. Usually the exo addition is predominant as in Case II. Case I and Case III must involve particular reasons to produce 'endo' adducts. Such a reason for Case I could be the steric circumstances at the methylene chlorine substituents. In Case III the regioselection in norbornadiene is the first problem to be solved. The influence of the tert.-butoxy substituent at the methylene bridge could be interpreted as an orbital polarization which follows:

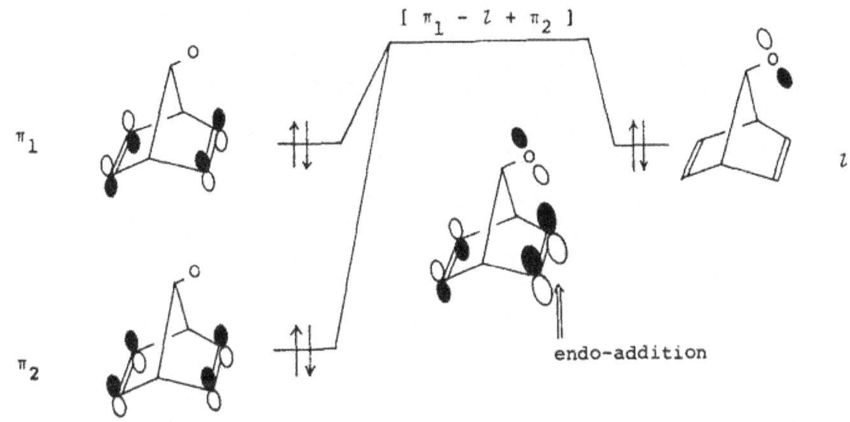

The direction of polarization of two double bonds is in conformity with endo-syn(with respect to Bu^tO) adduct, resulting in anti-Bu^tO(with respect to the remaining double bond) isomer. Such an explanation is only possible in regard to the HOMO of the parallel ethylenic bonds. The repulsive effect of oxygen atom would raise the energy of the dienophile, which in consequence comes to act as a donor. In this way the Diels-Alder addition will take place in inverse electron demand, as is pointed out by Mackenzie(Ref. 44).

THE PATH OF A CHEMICAL REACTION

In the papers already referred to(Ref.24-31),the change of geometry of reacting molecules is treated. The range of the change discussed there is sufficiently wide, so that the theory may be called as going beyond the perturbation method. But in this series of works the determination of the geometry of reacting molecules is not yet theoretically complete. It was in pace with the development of the theoretical calculation of the potential energy function that the fundamental idea to decide the change of geometry of reacting molecules proceeded.

The adiabatic potential of a chemically reacting system is now theoretically available, in principle, as a function of the variables which can determine the atomic arrangement. Such a potential function, say V, possesses several equilibrium points where all of their derivatives vanish. Among these equilibrium points, those of interest are the initial stable point, the transition-state point, and the final stable point of a reaction.

The reaction path is a chemical concept involving an ambiguity or flexibility. It was attempted by many authors to formulate the classical reaction path mathematically(Ref. 45-47). Two different approaches are introduced here. One is the trajectory approach and the other is the reaction coordinate approach. The trajectory of a reaction varies with the initial condition of the reacting system while the reaction coordinate is not affected by the initial condition. The trajectory approach is essentially compared to solving the classical Newtonian equation of motion directly, and was successfully developed in particular by Karplus and other chemists(Ref. 48). But here the focus is put on the reaction coordinate in view of the connection with the perturbation theory of reactivity.

The reaction coordinate is determined by integrating Newton's equation of motion with respect to the time considering an infinitely slow change of the atomic configuration. The following relation holds every moment:

$$M_i(dX_i / d\tau) = \partial V / \partial X_i \quad (i=1,2,\cdots 3N)$$

where X_i are the space-fixed cartesian coordinates of atom i of mass M_i, and N is the number of atoms of the reacting system. The parameter τ plays the part of "time" in the quasistatic motion. The solution which is determined by this differential equation and passes the transition-state point and a stable point is referred to as an <u>intrinsic reaction coordinate</u>(IRC) (Ref. 49). The IRC is the locus of infinitely slow motion of the reacting system along a chemical reaction, keeping the significance to represent the "centre-line" of the reaction path which is of empirical origin and never a simple concept. This differential equation may be called the IRC-equation.

The first application of the IRC-equation is to obtain the continuous change of the geometry of reacting molecules by a successive plot starting from the transition-state point to a stable equilibrium point—"reaction ergodography" (Ref. 52).

The method was successfully applied to the following reactions:

i) The hydrogen migration in enol-form malonaldehyde (Ref. 50)

ii) $T + CH_4 \longrightarrow CH_3 + H$

$T + CH_4 \longrightarrow CH_3 + HT$

$T + CD_4 \longrightarrow CD_3T + D$ (Ref. 51, 52)

$T + CD_4 \longrightarrow CD_3 + DT$

iii) $HNC \longrightarrow HCN$ (Ref. 53)

iv) $H^- + CH_4 \longrightarrow CH_4 + H^-$ (Ref. 53, 54)

v) $HFCO \longrightarrow HF + CO$ (Ref. 55)

vi) Isomerization of triplet butadiene (Ref. 56)

vii) $CH_3CH_2F \longrightarrow CH_2CH_2 + HF$ (Ref. 57)

viii) $H_2CO \longrightarrow HCOH$

$H_2CO \longrightarrow H_2 + CO$ (Ref. 58)

The sixth example refers to an excited state. The IRC equation can be equally applied to the geometry change from any non-equilibrium point.

The determination of the geometry of reacting molecules along the IRC supplied information on various properties of a chemical reaction. First of all, one can obtain the energy of each point on the IRC, which gives the shape and height of potential barrier. The activation energy of the Eyring-Polanyi theory is obtained together with the breadth of the barrier. In some favourable cases one can discuss or guess the effect of isotope exchange simply by the broadening of the barrier breadth(Ref. 52). Similarly, one can calculate the force acting on each atom at any point on the IRC to promote the change of geometry directly from the IRC equation.

The IRC plot makes it possible to correlate any molecular constant which is calculable along the reaction path. For instance, one can obtain a <u>quantitative</u> orbital correlation diagram, which makes the Woodward-Hoffmann <u>correlation</u> diagram quantitative if it is so desired. Also one can obtain a <u>vibrational</u> correlation diagram in which all normal vibrations and vibrational frequencies are correlated to the geometry along the IRC(Ref. 55, 58, 59). This diagram can be correlated with the curvature of the IRC path so that one can discuss the composition of the reaction coordinate with regard to normal vibrations and the contribution of each normal vibration to the vibrational energy transformation. Such a discussion is no doubt useful in relation to laser chemistry, in particular to mode-selective chemical reactions. Yamashita et al.(Ref. 58) treated the asymptotic behaviour of the IRC in the vicinity of the stable equilibrium points of the reacting system. The vibrational correlation diagrams for the isomerization and decomposition of formaldehyde were obtained to discuss the strategy to give rise to a mode-selective reaction.

The energy(Ref. 59, 60) and the density partitioning techniques(Ref. 61) along the reaction path are applied to discuss the detailed reaction mechanism. One can apply various quantum-chemical methods to analyze submolecular or electronic processes involved in the reaction based on the calculated geometry.

An attempt to use the IRC approach was made to treat the role of solvent molecules in a chemical reaction(Ref. 62). A powerful effect of one water molecule hypothetically relaying the 1, 3-hydrogen shift in formamidine on the reduction of the potential barrier was theoretically disclosed(Ref. 63). Such a model will be useful in discussing biomolecular phenomena.

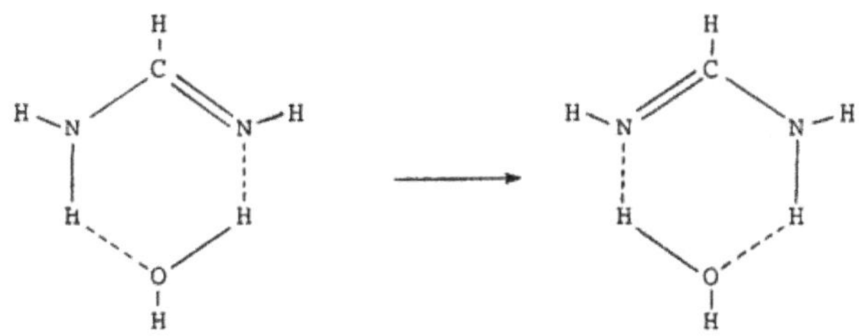

The IRC approach is useful for the purpose of visualization of the mode of proceeding of a chemical reaction. This can be done by constructing a sort of localized orbitals which represent the intermolecular interaction most effectively. Such localized orbitals are mathematically obtained by diagonalizing any quantity which corresponds to that intermolecular interaction. If one uses the magnitude of delocalization energy as the quantity to be diagonalized, one gets interaction frontier orbitals(IFO)(Ref. 64). If one takes the intermolecular overlap population instead of the delocalization energy one obtains interaction hybrid molecular orbitals(IHMO)(Ref. 65).

On account of a mathematical property of such localized orbitals, the mode of interaction is described by a few pairs of localized orbitals each belonging to one of the two reacting molecules. If one of the reactants possesses only one orbital, like proton or hydride anion, the mode of interaction is represented by a single localized orbital of the other reactant. One can trace the change of the pattern of such a single "frontier orbital" to visualize a given chemical reaction. An animated film for chemical reactions could be made in this way for the purpose of instruction.

Two examples of IHMO are indicated below:

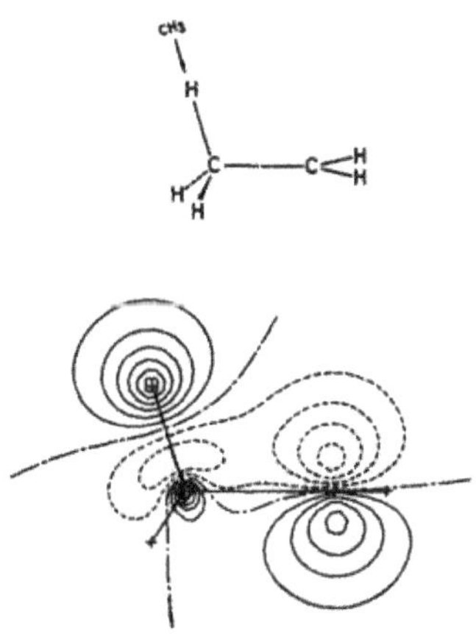

The IHMO of ethyl radical for the abstraction of a hydrogen atom by methyl radical.

Full lines indicate the lobes with + sign, and dotted lines with - sign. The line —·—·— indicates the nodal plane.

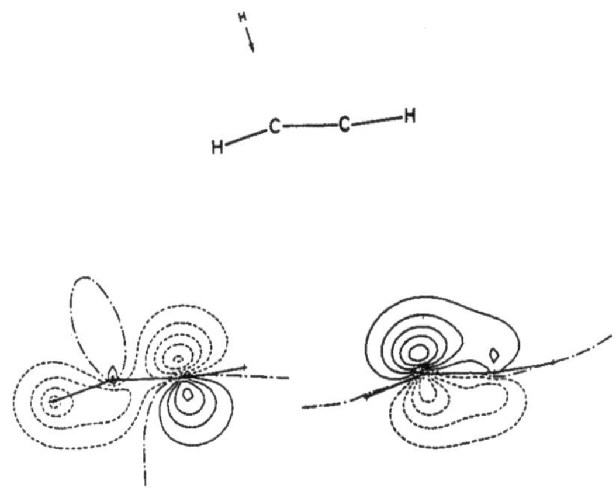

The IHMO of acetylene for the addition of hydrogen atom.

REFERENCES

1. G. N. Lewis, Valence and the Structure of Atoms and Molecules, Chemical Catalog Co., New York, N.Y. (1923).
2. C. K. Ingold, Structure and Mechanism in Organic Chemistry, Cornell University Press, Ithaca, N.Y. (1953).
3. C. A. Coulson and H. C. Longuet-Higgins, Proc. Roy. Soc. (London) A191, 39; A192, 16 (1947).
4. G. W. Wheland, J. Am. Chem. Soc. 64, 900 (1942).
5. K. Fukui, T. Yonezawa, and H. Shingu, J. Chem. Phys. 20, 722 (1952).
6. K. Fukui, T. Yonezawa, C. Nagata, and H. Shingu, J. Chem. Phys. 22, 1433 (1954).
7. K. Fukui, T. Yonezawa, and C. Nagata, Bull. Chem. Soc. Japan 27, 423 (1954).
8. K. Fukui, H. Kato, and T. Yonezawa, Bull. Chem. Soc. Japan 34, 1112 (1961).
9. K. Fukui, Theory of Orientation and Stereoselection, Springer-Verlag, Berlin (1970).
10. K. Fukui, in: Molecular Orbitals in Chemistry, Physics, and Biology, P.-O. Löwdin and B. Pullmann, eds., Academic Press, New York, N.Y. (1964), p.513.
11. H. Fujimoto, S. Inagaki, and K. Fukui, J. Chem. Soc. 98, 2670 (1976).
12. R. B. Woodward and R. Hoffmann, Angew. Chem. 81, 797 (1969); The Conservation of Orbital Symmetry, Academic Press, New York, N.Y. (1969).
13. K. Fukui, Acc. Chem. Res. 4, 57 (1971).
14. L. Salem, Chem. Brit. 5, 449 (1969).
15. G. Klopman, Chemical Reactivity and Reaction Paths, John Wiley, New York, N.Y. (1974).
16. R. G. Pearson, Symmetry Rules for Chemical Reactions, John Wiley, New York, N.Y. (1976).
17. W. C. Herndon, Chem. Rev. 72, 157 (1972).
18. R. F. Hudson, Angew. Chem. Int. Ed. Eng. 12, 36 (1973).
19. K. N. Houk, Acc. Chem. Res. 8, 361 (1975).

20. I. Fleming, Frontier Orbitals and Organic Chemical Reactions, John Wiley, New York, N.Y. (1976).

21. T. L. Gilchrist and R. C. Storr, Organic Chemical Reactions and Orbital Symmetry, Cambridge University Press, Cambridge, 2nd Ed. (1979).

22. K. Fukui and H. Fujimoto, Bull. Chem. Soc. Japan 41, 1989 (1968).

23. K. Fukui, in: The World of Quantum Chemistry, R. Daudel and B. Pullmann, Eds., Reidel, Dordrecht (1974), p.113.

24. H. Fujimoto, S. Yamabe, and K. Fukui, Bull. Chem. Soc. Japan 45, 1566 (1972).

25. H. Fujimoto, S. Yamabe, and K. Fukui, Bull. Chem. Soc. Japan 45, 2424 (1972).

26. S. Yamabe, T. Minato, H. Fujimoto, and K. Fukui, Theoret. Chim. Acta (Berl.) 32, 187 (1974).

27. H. Fujimoto, S. Kato, S. Yamabe, and K. Fukui, J. Chem. Phys. 60, 572 (1974).

28. S. Kato, H. Fujimoto, S. Yamabe, and K. Fukui, J. Am. Chem. Soc. 96, 2024 (1974).

29. S. Yamabe, S. Kato, H. Fujimoto, and K. Fukui, Bull. Chem. Soc. Japan 46, 3619 (1973); S. Yamabe, S. Kato, H. Fujimoto, and K. Fukui, Theoret. Chim. Acta (Berl.) 30, 327 (1973).

30. H. Fujimoto, S. Yamabe, T. Minato, and K. Fukui, J. Am. Chem. Soc. 94, 9205 (1972).

31. T. Minato, S. Yamabe, H. Fujimoto, and K. Fukui, Bull. Chem. Soc. Japan 51, 1 (1978); T. Minato, T. Yamabe, M. Miyake, K. Tanaka, H. Kato, and K. Fukui, ibid., 51, 682 (1978).

32. W. L. Jorgensen and L. Salem, The Organic Chemist's Book of Orbitals, Academic Press, New York, N.Y. (1973).

33. S. Inagaki, H. Fujimoto, and K. Fukui, J. Am. Chem. Soc. 97, 6108 (1975).

34. K. Fukui and K. Tanaka, Bull. Chem. Soc. Japan 50, 1391 (1977).

35. K. Fukui and S. Inagaki, J. Am. Chem. Soc. 97, 4445 (1975).

36. S. Inagaki, H. Fujimoto, and K. Fukui, J. Am. Chem. Soc. 98, 4693 (1976).

37. S. Inagaki and K. Fukui, Chem. Lett. 1974, 509; S. Inagaki, H. Fujimoto, and K. Fukui, J. Am. Chem. Soc. 98, 4054 (1976).

38. L. de Vris and S. Winstein, J. Am. Chem. Soc. 82, 5363 (1960).

39. S. B. Soloway, A. M. Damiana, J. W. Sims, H. Bluestone, and R. E. Lidov, J. Am. Chem. Soc. 82, 5377 (1960).

40. C. H. M. Adams, D. J. Cawley, and K. Mackenzie, J. Chem. Soc. Perkin Trans. II, 1973, 909.

41. A private communication from Prof. K. Mackenzie.

42. S. B. Soloway, A. M. Damiana, J. W. Sims, H. Bluestone, and R. E. Ridov, J. Am. Chem. Soc. 82, 5377 (1960).

43. S. B. Soloway, J. Am. Chem. Soc. 74, 1027 (1952).

44. K. B. Astin and K. Mackenzie, J. Am. Chem. Soc. Perkin Trans. II, 1975, 1004.

45. L. Hofacker, Z. Naturforsch. 18a, 607 (1963); L. Hofacker and N. Rosch, Ber. Bunsen-Ges. Phys. Chem. 77, 661 (1973).

46. R. A. Marcus, J. Chem. Phys. 45, 4493, 4500 (1966); 49, 2610 (1968).

47. J. Brickmann, Z. Naturforsch. 28a, 1757 (1973); P. Russegger and J. Brickmann, J. Chem. Phys. 62, 1086 (1975).

48. I. S. Y. Wang and M. Karplus, J. Am. Chem. Soc. 95, 8160 (1973); A. Warshel and M. Karplus, Chem. Phys. Lett. 32, 11 (1975); D. L. Thompson, Acc. Chem. Res. 9, 338 (1976); R. B. Walker, E. B. Steel, and J. C. Light, J. Chem. Phys. 69, 2922 (1978).

49. K. Fukui, J. Phys. Chem. 74. 4161 (1970); The World of Quantum Chemistry, D. Reidel, Dordrecht, Holland, 1974, p.113.

50. S. Kato, H. Kato, and K. Fukui, J. Am. Chem. Soc. 99, 684 (1977).

51. K. Fukui, S. Kato, and H. Fujimoto, J. Am. Chem. Soc. 97, 1 (1975).

52. S. Kato and K. Fukui, J. Am. Chem. Soc. 98, 6395 (1976).

53. K. Ishida, K. Morokuma, and K. Komornicki, J. Chem. Phys. 66, 2153 (1977). Also see S. K. Gray, W. H. Miller, Y. Yamaguchi, and H. F. Schaefer, III., J. Chem. Phys. 73, 2372 (1980).

54. B. D. Joshi and K. Morokuma, J. Chem. Phys. 67, 4880 (1977).

55. S. Kato and K. Morokuma, Ann. Rev. Inst. Mol. Sci. 1980, 10.

56. I. Ohmine and K. Morokuma, Ann. Rev. Inst. Mol. Sci. 1978, 11.

57. S. Kato and K. Morokuma, J. Chem. Phys. 73, 3900 (1980).

58. K. Yamashita, T. Yamabe, and K. Fukui, Chem. Phys. Lett. 84, 123 (1981).

59. S. Kato and K. Morokuma, J. Chem. Phys. 73, 3900 (1980).

60. S. Kato and K. Morokuma, J. Chem. Phys. 72, 206 (1980).

61. H. Fujimoto and K. Fukui, Israel J. Chem. 19, 317 (1980).

62. K. Yamashita, M. Kaminoyama, T. Yamabe, and K. Fukui, Theoret. Chim. Acta (Berl.) 60, 303 (1981).

63. K. Yamashita, M. Kaminoyama, and T. Yamabe, unpublished.

64. K. Fukui, N. Koga, and H. Fujimoto, J. Am. Chem. Soc. 103, 196 (1981).

65. H. Fujimoto, N. Koga, M. Endo, and K. Fukui, Tetrahedron Lett. 22, 1263 (1981); 22, 3427 (1981); H. Fujimoto, N. Koga, and K. Fukui, J. Am. Chem. Soc. 103, 7452 (1981).

Part II

Stereoselection

In 1965 and in 1966, Woodward and Hoffmann published a series of communications in which they stressed importance of orbital symmetry in determining stereoselectivity in the reactions of conjugated systems. The arguments were closely related to the frontier orbital concept, though slightly different in some respects. We had already studied frontier orbital interactions in the complex between a silver cation and aromatic compounds (paper no. 10 in Part 1) and those in Diels-Alder reactions (paper 14 in Part 1). The importance of the signs of the AO coefficients at the two reaction sites in the frontier MO's was disclosed there. Thus, we began the study of these problems by applying the frontier orbital theory. The first paper in this part (paper no. 18) provided the ways of discussing multi-centered reactions. Included there were ring-closure of conjugated polyenes, ring-opening of cyclic olefins, 1, 2-*cis* additions, 1, 3-dipolar additions and others. In the second paper (paper no. 19), we studied the stereoselection that would be controlled by the hybridization at two reaction centers. The favorable and unfavorable structural changes due to the mixing of σ MO's and π MO's within a molecule under the perturbation of a reagent were discriminated. In addition, the delocalization stabilization was evaluated for many types of interactions between two conjugated molecules. A slightly more rigorous theory of stereoselection in electrocyclic reactions was presented in the third paper of this part (paper no. 20). There, the same conclusion as obtained from perturbation treatments was derived. Stereoselection in a variety of organic reactions was shown later to be interpreted very easily by dividing a molecule into fragments (paper no. 21). The frontier orbital theory found many problems for its application as presented also in the next part.

BULLETIN OF THE CHEMICAL SOCIETY OF JAPAN VOL. 39 498—503 (1966)

An MO-Theoretical Illumination for the Principle of Stereoselection

By Kenichi FUKUI

Faculty of Engineering, Kyoto University, Sakyo-ku, Kyoto

(Received June 23, 1965)

The principle of the stereoselection in organic chemistry has been discussed from the point of view of the molecular orbital theory. The stereoselectivity in the stereospecific ring-closure of conjugated polyenes and the stereospecific ring-cleavage of unsaturated carbon cycles, *trans*-1,2-noncycloadditions, *cis*-1,2-additions and *cis*-1,3-dipolar additions have been accounted for with a qualitative generalization. In many cases the symmetry relationship of certain particular molecular orbitals ("frontier orbitals") plays a conspicuously significant role in the steric control of stereochemical processes. This may be put in contrast with the role of the frontier electron density in the orientation process in organic reactions.

The purpose of the present paper is to make an attempt to furnish organic stereochemistry with a theoretical strategy. In view of the multifariousness noticed in the accumulation of recent stereochemical studies, it is hoped that a guiding principle to unravel the entanglements involved will appear. In order to be of practical applicability, such a guiding principle must first of all possess generality and simplicity. Therefore, the molecular orbital (MO) theory seems suitable.

The recent experimental information on the stereoselectivity in several cyclic reactions of unsaturated compounds[1] has been an incentive to start an MO-theoretical attack.[2-4] All these papers pointed out the importance of the symmetry of particular MO's in discussing the problem in question. These particular MO's were the highest-occupied MO (HO) and the lowest vacant MO (LV) of the ground state molecule. Oosterhoff[2] has suggested that the difference in stereochemical behavior between thermal and photo-induced reactions may be attributed to the symmetry characteristics of HO and LV. The important role of HO and LV (the frontier orbitals) in the course of the chemical interaction of organic compounds was early well established in the frontier electron theory,[5,6] and the symmetry of the frontier orbitals was also duly employed for the elucidation of Diels-Alder reactivity.[7]

However, it should be noted that the discrimination of the frontier orbitals from the other MO's should be based upon a proper comprehension of the theoretical grounds. Woodward and Hoffmann[3] discussed, in their elegant generalization of the problem of stereoselectivity, the ring-closure of conjugated olefins, and suggested that the steric course should be determined by the symmetry of HO only. However, this is evidently an "over-discrimination" of HO, since the overlap stabilization between two π atomic orbitals in a single conjugated system should be evaluated as the sum of the contributions from all occupied MO's.[4] The mere coincidence of the results does not justify such improper or less authentic neglect of the other MO's.

The basic idea, used throughout the present paper, for discussing the steric course in a consistent manner is to determine, from the MO perturbation calculation of the energy change of interaction, the direction of the overlap stabilization between interacting atomic orbitals; this will lead to the most likely correct steric course. The details will be illustrated by examples.

The Ring-Closure of Conjugated Polyenes.—Consider a linear conjugated polyene with k π electrons. So-called Hückel MO's of the simplest type are given by[8]:

$$\phi_j=\sum_{r=1}^{k} C_r{}^j\phi_r=\left(\frac{2}{k+1}\right)^{1/2}\sum_{r=1}^{k}\phi_r\sin\frac{jr\pi}{k+1}\qquad (j=1,\ 2,\cdots k)$$

where ϕ_r is rth carbon $2p\pi$ atomic orbital (AO), and $C_r{}^j$ is the coefficient of ϕ_r in jth MO. These MO's are obtained by solving the usual secular equations, which yield the secular determinant:

1) a) E. Havinga, R. J. de Kock and M. P. Rappoldt, *Tetrahedron*, **11**, 276 (1960); b) E. Havinga and J. L. M. A. Schlatmann, ibid., **16**, 146 (1961); c) G. J. Fonken, *Tetrahedron Letters*, **1962**, 549; d) R. E. K. Winter, ibid., No. 17, 1207 (1965).

2) L. J. Oosterhoff, private suggestion to the authors of the paper cited in Ref. 1(b), a suggestion quoted on p. 151 of that paper.

3) R. B. Woodward and R. Hoffmann, *J. Am. Chem. Soc.*, **87**, 395 (1965).

4) K. Fukui, *Tetrahedron Letters*, No. 24, 2009 (1965).

5) a) K. Fukui, T. Yonezawa and H. Shingu, *J. Chem. Phys.*, **20**, 722 (1952); b) K. Fukui, T. Yonezawa, C. Nagata and H. Shingu, ibid., **22**, 1433 (1954).

6) a) K. Fukui, H. Kato and T. Yonezawa, This Bulletin, **33**, 1201 (1960); b) K. Fukui, H. Kato and T. Yonezawa, ibid., **34**, 442 (1961); c) K. Fukui, H. Kato and T. Yonezawz, ibid., **34**, 1111 (1961).

7) K. Fukui, "Molecular Orbitals in Chemistry, Physics, and Biology," Ed. by P.-O. Löwdin and B. Pullman, Academic Press, New York (1964), p. 513.

$$\Delta(\varepsilon)=\begin{vmatrix} -\varepsilon & \beta & \cdots & 0 & 0 \\ \beta & -\varepsilon & \cdots & 0 & 0 \\ \cdots & \cdots & \cdots & \cdots & \cdots \\ 0 & 0 & \cdots & -\varepsilon & \beta \\ 0 & 0 & \cdots & \beta & -\varepsilon \end{vmatrix} = \beta^k \frac{\sin(k+1)\theta}{\sin\theta} \quad (1)$$

$$(\varepsilon = -2\beta\cos\theta)$$

(k rows and k columns)

where the zero point of the energy, ε, is taken as equal to the Coulomb integral of carbon $2p\pi$ AO, β being the resonance integral between two adjacent $2p\pi$ AO's.

The change in the π-electronic energy in the case of the overlapping of the first and kth AO's is clearly approximated by:

$$\Delta E = 2\sum_{j=1}^{k} \nu_j C_1^j C_k^j \gamma = 2P_{1k}\gamma \quad (2)$$

by the first-order perturbation calculation, where ν_j is the number of π electrons occupying the jth MO, while γ is the resonance integral between ϕ_1 and ϕ_k in the case of the σ-like overlapping of these two orbitals in the distorted π electron system. The sign of γ is obvious, as is indicated by the scheme of Fig. 1.

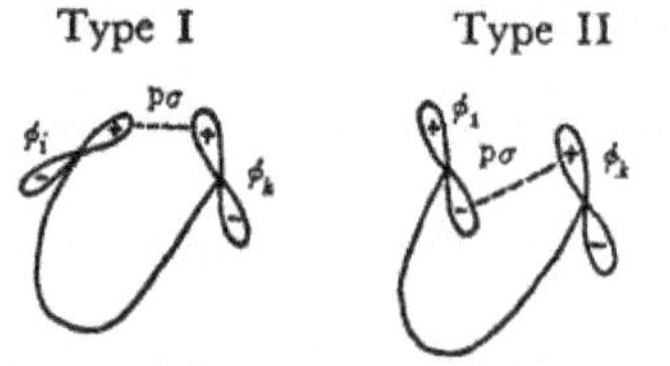

Fig. 1. Two types of overlap stabilization.

The quantity, P_{1k}, has the same form as that defined as the well-known "bond order," but in the present case it serves to determine the stability due to the overlapping of two π orbitals; it may thus appropriately be tentatively named the "overlap stabilization." If this quantity is positive, the overlapping of Type I is preferred, while if it is negative, Type II is more preferable, in the case of cyclization.

For the ground state of linear conjugated polyenes, P_{1k} is known to be easily expressed in an integral representation[9] as:

$$P_{1k} = \frac{\beta^{k-1}}{\pi}\int_{-\infty}^{+\infty}\frac{dy}{\Delta(iy)} = 4(-1)^{k/2+1}\frac{1}{\pi} \times \int_0^{\infty}\frac{\cosh^2\varphi\, d\varphi}{\cosh(k+1)\varphi} \quad (3)$$

$$(y = -2\beta\sinh\varphi)$$

which is immediately convertible into the form:

$$P_{1k} = \frac{(-1)^{k/2+1}}{k+1}\left(1+\frac{1}{\cos(\pi/(k+1))}\right) \quad (4)$$

Therefore, it follows clearly that, when $k=4m+2$, P_{1k} is positive, so that the cyclization of Type I occurs, while when $k=4m$, P_{1k} is negative, so that a Type II cyclization results; this is consistent with experimental facts in the case of $k=6$.[1]

This relationship is not limited to the cyclization due to the bonding between two termini. It is likewise very easy to prove mathematically that P_{rs} is positive in the cyclization forming a $(4m+2)$-membered ring, leading to a Type I course, whereas the cyclization forming a $4m$-membered ring would follow a Type II course. The discussion of this sort has already been made by numerical calculations more comprehensively and also more elaborately in connection with the theory of aromaticity by Fukui et al.[10]

A reverse relation holds in the first excited state, in which we assume the jumping of one π electron of HO into LV. In the present simple treatment, both the triplet and the singlet states will be discussed in terms of the same theoretical quantity. The overlap stabilization, which is in this case denoted by P_{1k}^*, is easily represented by:

$$P_{1k}^* = \frac{(-1)^{k/2}}{k+1}\frac{\cos(\pi/(k+1))+1}{\cos(\pi/(k+1))} \times \left(2\cos\frac{\pi}{k+1}-1\right) \quad (5)$$

Therefore, it is positive when $k=4m$, and negative when $k=4m+2$, provided $k\geqq 4$. Photo-induced cyclizations are known experimentally to take place by the Type II course when $k=6$.[1]

The steroselection in polyene cyclization, whether Type I or Type II, may thus be understood to be governed by the following rule:

	Thermal reaction	Photo-induced reaction
$k = 4m$	Type II	Type I
$k = 4m+2$	Type I	Type II

The steric courses, Type I and Type II, were named by Woodward and Hoffmann[3] "disrotatory" and "conrotatory" processes respectively, in their first generalization of this interesting cyclic conversion.

The Ring-Opening of Cyclic Olefins.— The reverse process to the above-mentioned cyclization have also been studied and found to follow the same stereoselection rule;[1] these reactions in two opposite directions were explained *en bloc* by Woodward and Hoffmann by emphasizing the symmetry of HO and LV.[3] This sort of ring-opening reaction is, however, supposed to

8) See, for instance, A. Streitwieser, Jr., "Molecular Orbital Theory for Organic Chemists," John Wiley & Sons, New York (1961), p. 50.

9) C. A. Coulson and H. C. Longuet-Higgins, *Proc. Roy. Soc.*, **A191**, 39 (1947); **A192**, 16 (1947).

10) K. Fukui, A. Imamura, T. Yonezawa and C. Nagata, This Bulletin, **33**, 1591 (1960).

be governed essentially by a sigma-pi interaction between a part of the π-electron system and the σ bond to be broken. A more plausible, separate explanation must be given on the basis of the direction of stabilization in this sigma-pi interaction.

The electrons to be taken into account are two σ-bonding electrons in the a—b bond to be broken and (k-2) π electrons in the conjugated part of the ring, r-t-s in Fig. 2.

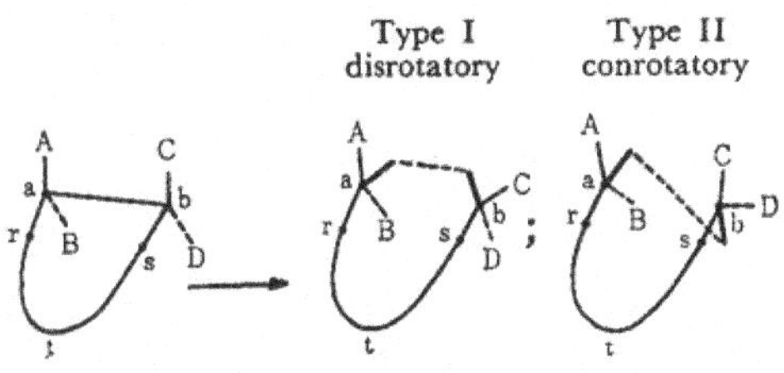

Fig. 2. Two types of interaction in the ring-cleavage.

The open-chain polyene to be produced has k π electrons. At the initial stage, the sigma and pi parts evidently have no appreciable interconjugation. As soon as the ring-cleavage begins to proceed, however, a π-like conjugation will take place between each of the two "incipient" orbitals appearing at the carbon atoms, a and b, and each end of the conjugated part, r-t-s.

The components of the two incipient orbitals perpendicular to the plane of the conjugated part are parallel and antiparallel to each other in the interactions of Type I and Type II respectively.[11]

The change in the energy of the system, ΔE, may be evaluated most simply by a perturbation treatment as a second-order term, if we regard the above-mentioned π-like conjugation as a small perturbation. The result is[12]:

$$\Delta E = \text{const.} + \left\{\sum_j \nu_j \frac{(C_r{}^j C_a{}'\gamma_a + C_s{}^j C_b{}'\gamma_b)^2}{\varepsilon_j - \varepsilon_0{}'} - \sum_j (2-\nu_j) \frac{(C_r{}^j C_a \gamma_a + C_s{}^j C_b \gamma_b)^2}{\varepsilon_j - \varepsilon_0}\right\} + 0(\gamma^4) \quad (6)$$

where (1) ε_0 and $\varepsilon_0{}'$ are the energies of a—b bonding and antibonding orbitals, which are denoted by $\chi = C_a\varphi_a + C_b\varphi_b$ and $\chi = C_a{}'\varphi_a + C_b{}'\varphi_b$ respectively, in which φ_a and φ_b are the initially-existing, carbon sp^3 hybridized orbitals of the a—b bond: (2) ε_j is the energy of the jth MO of the ($k-2$) π electron system, which is written as $\varphi_j = \sum_t C_t{}^j \phi_t$, where ϕ_t is tth atomic orbital; (3) ν_j is the number of electrons occupying the jth MO; (4) γ_a and γ_b are the "resonance" integrals (although this terminology may be inadequate for their positive value) between each of the two incipient orbitals appearing at the carbon atoms, a and b, with the adjacent 2pπ orbital at the carbon atoms, r or s; and (5) the term const. signifies the term which is involved almost always in both Type I and Type II interactions.

As may clearly be understood from the illustration, the sign, γ_a and γ_b which favors the minimization of ΔE will control the direction of the steric course. The same sign of γ_a and γ_b means the interaction of Type I, whereas Type II corresponds to γ_a and γ_b with opposite signs.

In the light of our knowledge about the parameters used in the MO treatment of σ electron systems developed by Fukui et al.,[13] the chief contributing terms on the right side of Eq. 6 are demonstrated to be those associated with HO and LV. Figure 3 indicates an example for the case of $k=6$. The charge transfer interaction of an HO electron with the a—b antibonding orbital will drive the σ bond into cleavage so as to cause a narrowing of the separation of $\varepsilon_0{}'$ and ε_0, resulting in a stronger and stronger interaction.

If we take into account, for the sake of simplicity, only the most dominant term in each summation of the right side, Eq. 6 may be reduced to the following form:

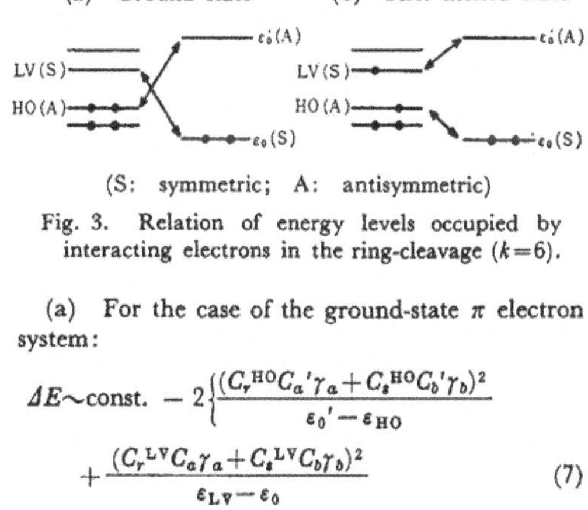

Fig. 3. Relation of energy levels occupied by interacting electrons in the ring-cleavage ($k=6$).

(a) For the case of the ground-state π electron system:

$$\Delta E \sim \text{const.} - 2\left\{\frac{(C_r{}^{\text{HO}} C_a{}'\gamma_a + C_s{}^{\text{HO}} C_b{}'\gamma_b)^2}{\varepsilon_0{}' - \varepsilon_{\text{HO}}} + \frac{(C_r{}^{\text{LV}} C_a \gamma_a + C_s{}^{\text{LV}} C_b \gamma_b)^2}{\varepsilon_{\text{LV}} - \varepsilon_0}\right. \quad (7)$$

(b) For the case of the first-excited-state π electron system:

$$\Delta E \sim \text{const.} - \left\{\frac{(C_r{}^{\text{LV}} C_a{}'\gamma_a + C_s{}^{\text{LV}} C_b{}'\gamma_b)^2}{\varepsilon_0{}' - \varepsilon_{\text{LV}}} + \frac{(C_r{}^{\text{HO}} C_a \gamma_a + C_s{}^{\text{HO}} C_b \gamma_b)^2}{\varepsilon_{\text{HO}} - \varepsilon_0}\right\} \quad (8)$$

where the super- and subscript HO's and LV's signify the highest occupied and the lowest vacant MO's in the ground state.

11) The description of line 9, page 3 of Ref. 4 should be reversed with regard to "conrotatory" and "disrotatory" processes.

12) K. Fukui, C. Nagata, T. Yonezawa, H. Kato and K. Morokuma, *J. Chem. Phys.*, **31**, 287 (1959); also see Ref. 7.

13) K. Fukui, H. Kato and T. Yonezawa, This Bulletin, **33**, 1198 (1960); also see Ref. 6a).

The orbital symmetry[14] of HO is antisymmetric (A) when $k-2=4m$, and symmetric (S) when $k-2=4m-2$, while LV is S when $k-2=4m$ and A when $k-2=4m-2$. The symmetry of the ε_0 level is, of course, S, and ε_0' is A. Therefore, in order for ΔE to be minimized,

(a) For the ground state: γ_a and γ_b should have opposite signs when $k=4m$, and the same sign when $k=4m+2$;

(b) For the first excited state: γ_a and γ_b should have the same sign when $k=4m$, and opposite signs when $k=4m+2$.

Hence we are led to the following stereoselection rule, which has already been pointed out by Woodward and Hoffmann,[3] in accordance with the experimental results[1,3]:

	Thermal reaction	Photo-induced reaction
$k=4m$	Type II, conrotatory	Type I, disrotatory
$k=4m+2$	Type I, disrotatory	Type II, conrotatory

It should be noted that the orbital symmetry of HO and LV plays an essential role in determining the steric course. The discrimination of the "frontier orbitals" is thus justified in this case. This may be put in contrast with the fact that the "frontier electron density" plays an important role in determining the intramolecular orientation.

The *trans*-1, 2-Noncycloadditions. — The predominance of noncyclic *trans*-1, 2-additions in olefins has been well established in many experimental results in both ionic and radical additions.[15] It therefore appears appropriate to discuss the problem by dividing it into two items:

i) In the case in which the attacking reagent, XY, is used in highly dissociative conditions, the *trans*-selectivity may be attributed to the well-discussed symmetrical adduct of X to the ethylenic bond, which is supposed to be formed in the first stage.[16] No further, new way of explanation seems to be necessary in such a case, except a consideration of the orbital symmetry relation of X and the ethylenic bond which creates the possibility of forming such a symmetrical adduct. This point will be discussed in detail in the next section, on *cis*-1, 2-additions.

ii) For the case of almost simultaneous noncyclic additions, however, there seems to be a more plausible explanation than the existing ones if we consider the relationship of the orbital symmetry[14] between the reagent and the olefin. The direction of the approach of the reagent is supposed to be perpendicular to the plane of the ethylenic molecule.[17] The characteristic feature of the noncyclic addition of XY may be specified as the unfavorable square-form interaction of XY with the C-C double bond in the course of approach. For instance, a non-dissociated Br_2 molecule does not have any MO with a symmetry suitable for entering into square-form conjugation with ethylenic π orbitals. Its lowest vacant MO is antisymmetric, so that the interaction with ethylenic π electrons could be facilitated only by orbital overlapping in opposite directions at two terminals, as is illustrated in Fig. 4.[18] The formulation of this relation is in essence identical to the treatment of Eqs. 1—4 (Fig. 4(a)); it is also possible to apply Eq. 7 for the same purpose (Fig. 4(b)). The electron transfer from the π bond to the X-Y antibonding MO will give rise to the dissociation of XY, leading to the stage

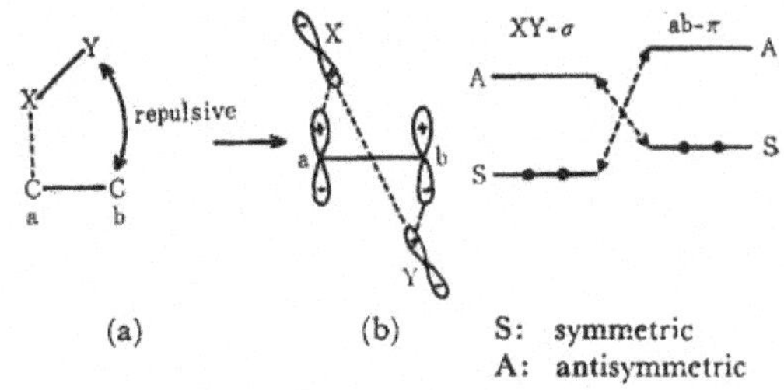

Fig. 4. The mode of initial interaction of the attacking reagent with an ethylenic bond.

indicated by (b) in Fig. 4. The explanation mentioned above may lead to a presumption of 1, 2-*cis*-noncycloaddition in the photo-excited state under suitable conditions.

No experimental results seem to be available for the stereoselectivity in noncyclic additions more distant than to 1, 2-positions of conjugated

14) More generally, it is, of course, not required to be exactly symmetric or antisymmetric. Only the relation between the signs of the coefficients in Eq. 7 is essential.

15) See, for instance, a) E. S. Gould, "Mechanism and Structure in Organic Chemistry," Henry Holt & Co., New York (1959), p. 514, 733 ff.; b) B. A. Bohm and P. I. Abell, *Chem. Revs.*, 62, 599 (1962).

16) Evidence has been given early for the two-step additions (A. W. Francis, *J. Am. Chem. Soc.*, 47, 2340 (1925)).

17) The opinion of Bader et al. (A. R. Bader, R. P. Buckley, F. Leavitt and M. Szwarc, *J. Am. Chem. Soc.*, 79, 5621 (1957)) that the approach of the methyl radical in the additions to olefins was along the double bond axis was later corrected (M. Matsuoka and H. Szwarc, ibid., 83, (1961)).

18) The author does not intend to reject the assumption of a bridged intermediate like the cyclic bromonium ion. Depending on conditions, we might have to take into account the interaction of X and C_b in addition to that of X and C_a. The essential point of the present discussion lies in pointing out the possibility of a mechanism with which we need no further, too-far-fetched views.

19) It should be noted that the terms "1, 4-" or "1, 6-addition" are used here in a purely mechanistic sense. They do not include a reaction composed of a series of successive processes. Hence, it is desirable that the starting conjugated polyene have a specific favorable structure: i. e., the *cis*-butadiene form for 1, 4-addition and the *cis-cis*-hexatriene form for 1, 6-addition. Also, we have to avoid confusion due to the term "*trans*-1, 4-addition," by which, e. g., the following reaction is meant:

Butadiene + Cl_2 → *trans*-1, 2-Bischloromethyl ethylene

The term "*cis*-1, 4-addition" employed above signifies an addition of XY by the attack of X and Y from the same side of the molecular plane of conjugation.

olefins. Discussions similar to those for 1, 2-additions are possible if we make a consideration analogous to that of Fig. 4. The conclusion is that 1, 4-noncycloaddition should be *cis*, 1, 6-, *trans*, and so forth.[19]

The *cis*-1,2-Additions.—The *cis*-1,2-addition to an ethylenic bond can take place when the reagent has a vacant MO (of course, preferably LV, for the reason understood through Eq. 6), of a symmetry that is advantageous for interaction with the occupied MO of the ethylenic π bond. For this reason the reagents which have the LV orbital symmetrical with respect to the middle of their two reaction centers are qualified as having a possibility of *cis*-addition. The formation of the intermediates considered in the two-step mechanism of stereospecific *trans*-additions, such as the bridged cation of protonated ethylene, may clearly be understood by the fact that the reagent ion has a symmetrical LV orbital. The other reagents which can satisfy this symmetry requirement are the phenyl cation in the phenonium ion,[20] the oxygen atom in epoxidation, carbenes in their singlet-state adducts to olefins,[21] the permanganate ion,[22] osmium tetroxide,[23] dienes in the Diels-Alder addition, and so on. Among these the symmetry of LV in various dienes has already been investigated by numerical calculations.[7] With respect to the other species, it is also easy to show by an appropriate MO treatment that their LV satisfies the symmetry conditions mentioned above.

The *cis*-1, 2-addition in hydroboration is well known. The result of an extended Hückel calculation by Kato et al. for boron compounds[24] may be useful in comprehending how borane can add to the *cis* positions. The LV of the BH_3 molecule is completely localized at the central boron $2p\pi$ AO, so that a strong charge-transfer interaction is possible with bonding π electrons of the C-C double bond (Fig. 5). As the $p\sigma$-overlapping proceeds, the bond populations of B-H sigma as well as of C_a-C_b pi are seriously reduced, so as to repel the hydrogen atom and force it to combine with the C_b carbon. Thus, a square-form interaction[25] is performed to give rise to *cis*-addition. From this explanation it is easy to realize that the hydroboration is a reaction of an electrophilic nature.

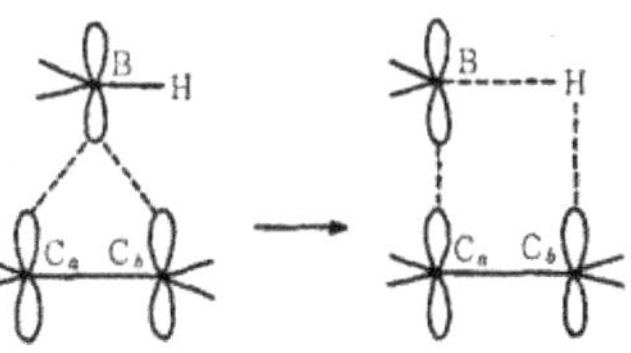

Fig. 5. The square-form interaction of BH_3 with C-C double bond.

Under these circumstances it is interesting that the easiness of the photo-induced cyclodimerization of olefins[26] in comparison with thermal reactions is also intimately connected with the symmetry relationship of HO and LV orbitals. In such cases, the *cis*-addition is geometrically desirable in order for the cyclodimerization to take place. Consequently, the orbital symmetry must be favorable for the square form interaction. As may easily be understood from the illustration in Fig. 6, in which acenaphthylene is chosen as an example, Eqs. 7 and 8 show that the square form interaction is thus facilitated in *excited-state* reactions.[27] The relation of the opposite symmetries of HO and LV in the ground state also holds as a general rule in alternant hydrocarbons, such as anthracene and pentacene, whose photodimerization has been well studied.

We have to remember that these circumstances must be similar in the case of the strong charge-transfer interaction in which one electron jumps from the olefin into the reagent orbital. The orbital symmetry requirement for a square-form

20) S. Winstein, E. Grunwald and L. L. Ingraham, *J. Am. Chem. Soc.*, **70**, 821 (1948); D. J. Cram, ibid., **71**, 3863 (1949); as regards the MO-theory, see Ref. 8, p. 382; the LV orbital of the phenyl cation is localized at the sp^2-like sigma orbital on the attached carbon atom (K. Fukui, H. Kato and T. Yonezawa, This Bulletin, **35**, 1475 (1962); K. Morokuma, S. Ohnishi, T. Masuda and K. Fukui, ibid., **36**, 1228 (1963)), which is, therefore, symmetrical with respect to the bisecting plane of the C-C double bond in the phenonium cation.

21) P. S. Skell and A. Y. Garner, *J. Am. Chem. Soc.*, **78**, 3409, 5430 (1956).

22) J. Boeseken, *Rec. Trav. Chim.*, **40**, 553 (1921); **47**, 683 (1928); K. B. Wiberg and K. A. Saegebarth, *J. Am. Chem. Soc.*, **79**, 2822 (1957); concerning the MnO_4^{2-} cis-addition, see J. S. F. Pode and W. A. Waters, *J. Chem. Soc.*, **1956**, 717.

23) R. Criegee, B. Marchand and H. Wannowius, *Ann.*, **522**, 75 (1936); **550**, 99 (1938).

24) H. Kato, T. Yonezawa and K. Fukui, Summary, 18th Annual Meeting of the Chemical Society of Japan (1965) (in Japanese), p. 440; H. Kato, K. Yamaguchi, T. Yonezawa and K. Fukui, This Bulletin, (in press).

25) Some alkyl borane compounds are known to react as a dimeric form. In such cases, the mode of interaction would resemble that of the S_N2 reaction of the halogen anion with alkyl halide.

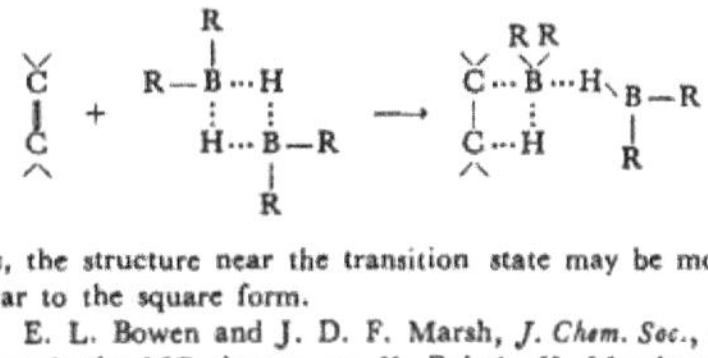

Thus, the structure near the transition state may be more or less similar to the square form.

26) E. L. Bowen and J. D. F. Marsh, *J. Chem. Soc.*, **1947**, 109; as regards the MO theory, see K. Fukui, K. Morokuma and T. Yonezawa, This Bulletin, **34**, 1178 (1960); A Schönberg and A. Mustafa, *J. Chem. Soc.*, **1948**, 2126; A. Mustafa, *Chem. Revs.*, **51**, 1 (1952); O_2 *cis*-addition is also described in G. O. Schenck and K. Ziegler, *Naturwissenschaften*, **32**, 157 (1944).

27) The perturbation theory is modified in case b where the second-order term in Eq. 8 should be replaced by the first-order term,

$$\{|C_r^{LV}C_{r'}^{LV}\gamma_a+C_s^{LV}C_{s'}^{LV}\gamma_b|^2 + |C_r^{HO}C_{r'}^{HO}\gamma_a+C_s^{HO}C_{s'}^{HO}\gamma_b|^2\}^{1/2}$$

where rs and $r's'$ are the two pairs of the reaction centers of the M and M′ molecules respectively.

(a) Thermal dimerization (b) Photodimerization
Molecule M Molecule M′ Molecule M Molecule M′

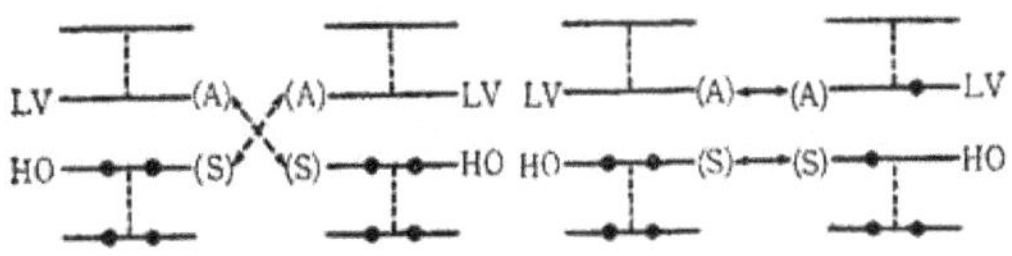

(S and A mean respectively symmetric and anti-symmetric with respect to the symmetry plane.)

Fig. 6. The symmetry relationship of OH and LV of acenaphthylene in connection with dimerization reactivity.

Isolated system Charge-transfer system
Reagent Olefin Reagent Olefin

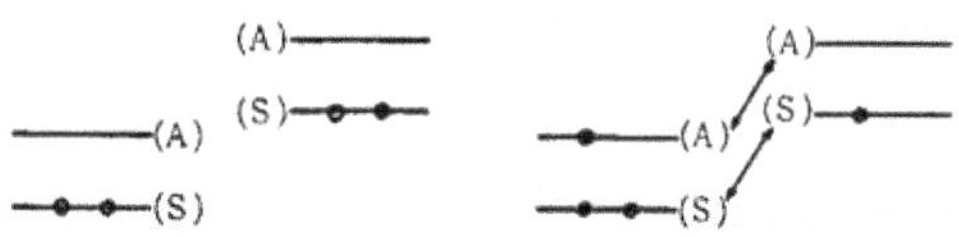

Fig. 7. The relation of orbital symmetry with charge-transfer interaction.

interaction, which the isolated systems are unable to fulfil, will then be satisfied. An instance for such a case is indicated in Fig. 7.

Here we should make reference to the recent finding in the field of polymerization chemistry that several stereospecific ionic polymerizations are thought to occur through successive *cis*-openings of the monomer double bond.[28] An interpretation of this fact has been given from a purely sterical point of view.[29] However, in view of the reasons above set forth it seems likely that the orbital symmetry of the system, including the growing chain end and the counter ion or catalyst part, should be examined after reliable information on the catalyst structure has been obtained. Also, whether or not circumstances like those in Fig. 7 might be realized in these cases is a matter of deep interest. We hope to re-examine the working mechanism of stereomodifiers in stereospecific vinyl polymerizations from the point of view of orbital symmetry. In addition, we may have to consider the participation of the neighboring group in the activated complex influencing the orbital symmetry when the monomer has any conspicuously influential group in the molecule. On this point, we shall present a report in the near future.

The 1, 3-Dipolar Additions.—The addition of 1, 3-dipoles to olefins occurs at positions *cis* with respect to both reactants.[30] Although various species of 1, 3-dipoles are known, the essential structure under reactive conditions may be characterized by the conjugated system which is indicated in Fig. 8.

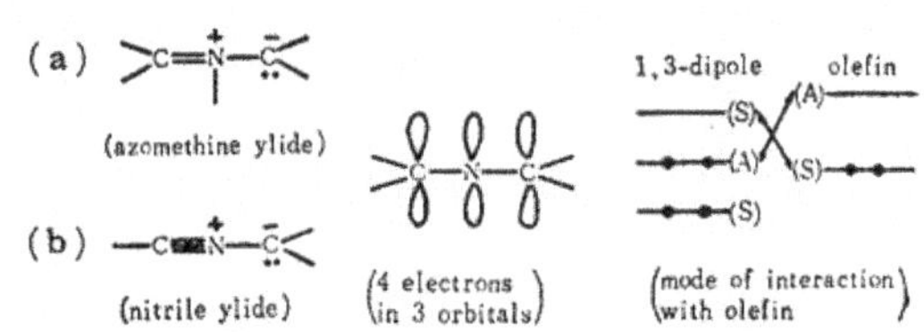

Fig. 8. The schematic representation of 1, 3-dipoles.

The LV orbital of 1, 3-dipoles is symmetric with respect to the two side AO's, so *cis*-interaction with olefin may be made favorable.

Other interesting problems, like the *trans*-1, 2-additions and the *endo*- and *exo*-additions in Diels-Alder reactions, as well as the stable conformations of organic molecular skeletons and the problem of *trans*-β-elimination, are left untouched in the present paper; they will be considered in several separate papers later.

28) G. Natta, M. Farina, M. Peraldo, P. Corradini, G. Bressan and P. Ganis, *Rend. Accad. naz. Lincei* (8), **28**, 442 (1960); G. Natta, M. Peraldo, M. Farina and G. Bressan, *Makromolekulare Chem.*, **55**, 139 (1962); M. Peraldo and M. Farina, *Chem. e Ind.* (*Milano*), **42**, 1349 (1960); T. Miyazawa and Y. Ideguchi, *Polymer Letters*, **1**, 389 (1963); H. Tadokoro, M. Ukita, M. Kobayashi and S. Murahashi, ibid., **1**, 405 (1963); M. Tasumi, T. Shimanouchi, H. Tanaka and S. Ikeda, *J. Polymer Sci.*, **A2**, 1607 (1964); G. Natta, G. Dall'Asta and G. Mazzanti, *Angew. Chem.*, **76**, 765 (1964).

29) G. Dall'Asta and G. Mazzanti, *Makromolekulare Chem.*, **61**, 178 (1963).

30) R. Huisgen, *Proc. Chem. Soc.*, **1961**, 357.

BULLETIN OF THE CHEMICAL SOCIETY OF JAPAN VOL. 39 2116—2126 (1966)

Sigma-Pi Interaction Accompanied by Stereoselection

By Kenichi Fukui and Hiroshi Fujimoto

Faculty of Engineering, Kyoto University, Sakyo-ku, Kyoto

(Received January 7, 1966)

A simple molecular orbital theoretical treatment of the sigma-pi interaction caused by chemical reactions of planar-conjugated systems has been developed. The problem of the streo-selection of the two alternative directions of configuration altering due to the hybridization change at the centers of interaction has been considered by means of a simple perturbation treatment. Two theoretical indices, which represent the conjugation stabilization by way of the sigma-pi and reagent-pi interactions, have been introduced for this purpose. The theoretical results are consistent with the results of experience so far accumulated in various organic reactions, such as additions, eliminations, substitutions, and rearrangements, of conjugated molecules.

The electronic wave function for planar-conjugated molecules is divided into the so called sigma and pi parts. The former is composed of the molecular orbitals (MO) which are symmetric with respect to the molecular plane and have the direction of extension on that plane, while the latter consists of the MO's antisymmetric with respect to that plane, to which their direction of maximum extension is perpendicular. The condition of this separability has been thoroughly discussed by Lykos and Parr.[1] When such a planar system is exposed to a certain perturbation, say, for instance, to a chemical interaction due to the approach of atoms, molecules, radicals, or ions from definite sides of the plane toward definite positions of the molecule, both the sigma and pi MO's will be affected. The change in MO's as well as in energies is obtainable from the usual perturbation calculation on an assumption of an unchanged molecular configuration.[2] However, the actual effect will cause a change in the nuclear distances

1) P. G. Lykos and R. G. Parr, *J. Chem. Phys.*, **24**, 1166 (1956); **25**, 1301 (1956).

2) Hosoya and Nagakura (H. Hosoya and S. Nagakura, Preprint of Symposium on Molecular Structure, Japan (1962), p. 48 (in Japanese)) calculated the deformation of MO's with the mixing of the 3s orbital with the $2p\pi$ orbitals due to the approach of a point charge to an ethylenic bond.

and valence angles near the center of approach, giving rise to changes in the hybridization of carbon atoms. Such an altering in the molecular configuration is not calculable by a mere perturbation treatment. The most favorable configuration would be obtained as the one which gives the minimum adiabatic potential due to the Born-Oppenheimer approximation adopted for the perturbed system. This procedure is not, however, simple enough to be practical for organic chemists.

This paper will present a simple MO treatment of organic chemical processes in planar-conjugated molecules, thus providing a method by which we can discuss very simply the direction of the configuration-altering of molecules associated with the hybridization change in carbon atoms. This configuration's frequent alterations arise stereospecifically; it is the purpose of the present paper to elucidate the rule which governs this stereoselection.

Simple Perturbation Treatment

Let us consider a planar-conjugated hydrocarbon molecule having several carbon $2p\pi$ atomic orbitals (AO). The carbon atom, t, has one π orbital, ϕ_t, and three sp^2 hybridized σ orbitals, χ_t, $\chi_{t'}$, and $\chi_{t''}$, to combine with neighboring carbon sp^2 orbitals or otherwise with hydrogen 1s orbitals, ζ_p's. The direction of the extension of these real hybridized orbitals conforms to the direction of bonding, so that they have always positive values in the region of their maximum extension (Fig. 1). All $2p\pi$ orbitals are taken as

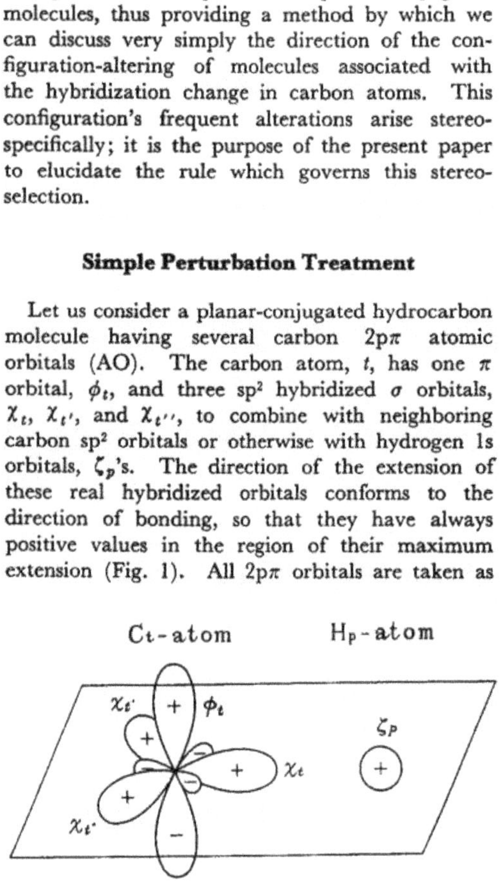

Fig. 1. The atomic orbitals.

having the same sign on the same side of the molecular plane. The total unperturbed wave function is assumed to be a product of the π part and the σ part, both of which are represented by a mere product of the usual simple Hückel MO's, which are written as:

$$\psi_i = \sum_t c_t^i \phi_t \quad \text{for the } \pi \text{ part, and} \quad (1a)$$

$$\varphi_k = \sum_t b_t^k \chi_t + \sum_p d_p^k \zeta_p \quad \text{for the } \sigma \text{ part,}^{3)} \quad (1b)$$

where ψ_i and φ_k are ith π MO and kth σ MO respectively. The first term on the right side of the latter equation signifies the summation with respect to each of three σ hybrids, namely:

$$\sum_t (b_t^k \chi_t + b_{t'}^k \chi_{t'} + b_{t''}^k \chi_{t''})$$

Instead of proceeding with the usual perturbation concept, we may here adopt the "method of perturbed secular determinant."[4),*] This method is suitable for calculating the energy difference between two systems for which the secular determinants differ slightly. The characteristic feature of this method is that it makes possible the calculation of the changes in energy, electron density, bond order, and other MO-theoretical quantities without an explicit knowledge of the change in the Hamiltonian operator or in that of wave functions.

The nonperturbed MO's of the π part are orthogonal to those of the σ part. The resonance integral between ϕ_t and χ_t's for the nonperturbed field is, of course, zero. Let us consider the configuration alteration which is indicated in Fig. 2, corresponding to the hybridization change due to chemical interaction with conjugated molecules. Figure 2 illustrates the case of a two-center interaction. Since, in all cases, our interest is limited to the direction of configuration alteration, each carbon atom is assumed to change its hybridized state from sp^2 to sp^x, where x is a definite number between 2 and 3. With this change the Coulomb integrals of a π orbital and a σ hybrid vary by $(\Delta\alpha)_\pi$ and $(\Delta\alpha)_\sigma$ respectively. The changes in the resonance integrals between a varied and the neighboring π AO's, and between a varied and the neighboring σ hybrids, are $(\Delta\beta)_\pi$ and $(\Delta\beta)_\sigma$ respectively, while the change in the resonance integral between the two σ hybrids of the same carbon atom[3)] is $(\Delta\beta)_\sigma'$. The resonance integral between the π orbital and a σ hybrid of the same carbon atom, say t, is designated by γ_t^σ. Since the way of taking the sign of basis orbitals is as illustrated in Fig. 1, we have two cases with respect to the value of γ^σ: namely, the one is the case of the carbons t and u of Fig. 2(i) (or the carbon t of ii), and the other is the case of the carbon u of Fig. 2(ii).

3) a) K. Fukui, H. Kato, T. Yonezawa, K. Morokuma, A. Imamura and C. Nagata, This Bulletin, **35**, 38 (1962); b) see also K. Fukui, H. Kato and T. Yonezawa, ibid., **34**, 442, 1111 (1961); **35**, 1475 (1962); T. Yonezawa, H. Kato, H. Saito and K. Fukui, ibid., **35**, 1814 (1962); K. Morokuma, K. Fukui, T. Yonezawa and H. Kato, ibid., **36**, 47 (1963); H. Kato, K. Fukui and T. Yonezawa, ibid., **38**, 189 (1965).

4) K. Fukui, C. Nagata, T. Yonezawa, H. Kato and K. Morokuma, *J. Chem. Phys.*, **31**, 287 (1959); K. Fukui, K. Morokuma, T. Yonezawa and C. Nagata, This Bulletin, **33**, 963 (1960).

* The term "perturbation" is here employed to imply the origin of the change in particular parameters in the secular determinant. The change in any quantum-mechanical quantity derived by solving it can be obtained by a procedure similar to the usual perturbation calculations.

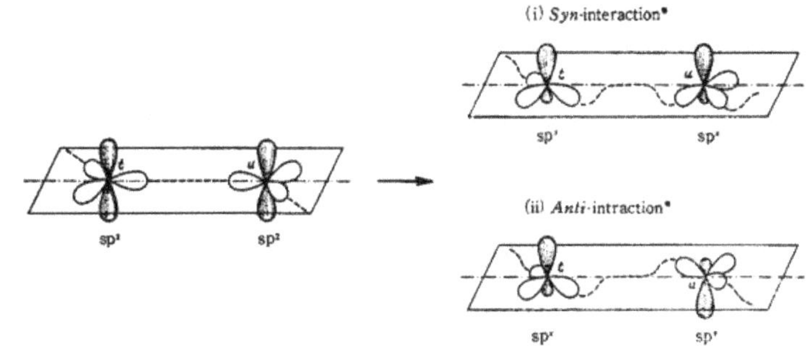

(x: a number between 2 and 3)

Fig. 2. A schematic representation of the mode of two-center σ-π interaction with a conjugated molecule.
* The terms *syn*- and *anti*- are used in order to evade confusion with *cis*- and *trans*-configurations of the remaining double bonds.

In the former case γ^σ is clearly negative, while in the latter it is positive. In addition, we have to consider the interaction with the reagent system. The resonance integral for the overlapping at the rth π orbital of the conjugated molecule with the reagent system (mostly one or two independent systems) is denoted by γ_r^{R}.

The increase in the total energy, ΔE, is thus represented by the following equations:[4]

$$\Delta E = (\Delta E)_\sigma + (\Delta E)_\pi + (\Delta E)_{\mathrm{R}} + (\Delta E)_{\sigma\pi} + (\Delta E)_{\mathrm{R}\pi} \tag{4}$$

$$\begin{aligned}(\Delta E)_\sigma &= 2\sum_k^{\mathrm{occ}}\sum_t (b_t^k)^2(\Delta\alpha)_\sigma \\ &+ 4\sum_k^{\mathrm{occ}}\sum_t \left\{\sum_u^{\mathrm{nei}} b_t^k b_u^k(\Delta\beta)_\sigma + \sum_{t'}^{\mathrm{sam}} b_t^k b_{t'}^k(\Delta\beta)_\sigma'\right\} \\ &+ 2\sum_k^{\mathrm{occ}}\sum_l^{\mathrm{uno}} \frac{1}{\eta_k-\eta_l}\left\{\sum_t b_t^k b_t^l(\Delta\alpha)_\sigma\right. \\ &+ \sum_t\sum_u^{\mathrm{nei}}(b_t^k b_u^l + b_u^k b_t^l)(\Delta\beta)_\sigma \\ &+ \left.\sum_t\sum_{t'}^{\mathrm{sam}}(b_t^k b_{t'}^l + b_{t'}^k b_t^l)(\Delta\beta)_\sigma'\right\}^2\end{aligned} \tag{5}$$

$$\begin{aligned}(\Delta E)_\pi &= 2\sum_i^{\mathrm{occ}}\sum_t (c_t^i)^2(\Delta\alpha)_\pi \\ &+ 4\sum_i^{\mathrm{occ}}\sum_t\sum_u^{\mathrm{nei}} c_t^i c_u^i(\Delta\beta)_\pi \\ &+ 2\sum_i^{\mathrm{occ}}\sum_j^{\mathrm{uno}}\frac{1}{\varepsilon_i-\varepsilon_j}\left\{\sum_t c_t^i c_t^j(\Delta\alpha)_\pi\right. \\ &+ \left.\sum_t\sum_u^{\mathrm{nei}}(c_t^i c_u^j + c_u^i c_t^j)(\Delta\beta)_\pi\right\}^2\end{aligned} \tag{6}$$

$$\begin{aligned}(\Delta E)_{\sigma\pi} &= 2\left(\sum_i^{\mathrm{occ}}\sum_k^{\mathrm{uno}} - \sum_k^{\mathrm{occ}}\sum_i^{\mathrm{uno}}\right)\frac{1}{\varepsilon_i-\eta_k} \\ &\times\left(\sum_t c_t^i b_t^k \gamma_t^\sigma\right)^2\end{aligned} \tag{7}$$

$$\begin{aligned}(\Delta E)_{\mathrm{R}\pi} &= 2\left(\sum_i^{\mathrm{occ}}\sum_l^{\mathrm{uno}} - \sum_l^{\mathrm{occ}}\sum_i^{\mathrm{uno}}\right)\frac{1}{\varepsilon_i-e_l} \\ &\times\left(\sum_r c_r^i d_{r'}^l \gamma_r^{\mathrm{R}}\right)^2 + 2\left(\sum_i^{\mathrm{occ}}\sum_{l'}^{\mathrm{uno}} - \sum_{l'}^{\mathrm{occ}}\sum_i^{\mathrm{uno}}\right) \\ &\times\frac{1}{\varepsilon_i-e_{l'}}\left(\sum_s c_s^i d_{s'}^{l'}\gamma_s^{\mathrm{R}'}\right)^2 + \cdots\end{aligned} \tag{8}$$

In addition to the right-side terms in Eq. 4, there may happen to exist zero-order terms which have no connection with stereochemical problems.

We may proceed further without mentioning the modification well-known to be necessary for the case of any type of degeneracy with respect to MO energies which may yield first-order terms in these equations in lieu of the corresponding second-order terms. The terms $(\Delta E)_\sigma$, $(\Delta E)_\pi$ and $(\Delta E)_{\mathrm{R}}$ signify the change in energy due to the interaction in the sigma, pi, and reagent parts respectively, while $(\Delta E)_{\sigma\pi}$ and $(\Delta E)_{\mathrm{R}\pi}$ are the conjugation stabilizations through sigma-pi and reagent-pi interactions respectively. In these equations occ and uno mean all the occupied and unoccupied MO's respectively, while nei and sam imply all neighboring orbitals and all hybrids of the same carbon atom respectively. When a radical plays a part, the terms due to the half-occupied orbital must not be doubled. The energy of ith π-MO and kth σ-MO are denoted by ε_i and η_k respectively. The summation with respect to t on the right sides of Eqs. 5 and 7 has the same meaning as in Eq. 1b. For the reagent part, e_l and $d_{r'}^l$ are the energy and the coefficient of r'th AO (at which the overlapping takes place with the rth π orbital of the reactant) of the Reagent System 1 (which is, of course, not limited to a conjugated system). Similarly, $e_{l'}$ and $d_{s'}^{l'}$ are for Reagent System 2, and so forth. The term $(\Delta E)_{\mathrm{R}}$ represents the energy change in the reagent, which may have a form similar to $(\Delta E)_\sigma$ and $(\Delta E)_\pi$.

The stereoselectivity of this type of reaction is thus governed by the terms of $(\Delta E)_{\sigma\pi}$ and $(\Delta E)_{R\pi}$ since, in such a problem, $(\Delta E)_{\sigma}$, $(\Delta E)_{\pi}$ and $(\Delta E)_{R}$ are deemed as common in the two opposite configurations in Fig. 2. In addition, the geometrical circumstances in the interaction of the reagent system with the reactant system, or the interesting effect of neighboring-group participation, should never be left out of account; however, this is obviously a problem to be discussed in a separate paper.

The Parametrization

In order to calculate the value of ΔE, we need to evaluate the integrals of both σ and π parts on a unified basis. These we have already made available in several of our previous papers,[3] and also summarized in a recent paper in which we adopted a slightly different estimation.[5] In the present paper we will use the latter values, with slight modifications, plus the values of newly-evaluated integrals for aromatic skeletons, as are listed in Table I. These numerical values have been determined with reference to the analytical form of the Slater orbitals and the theoretical evaluation of the overlap integrals in order to be consistent with various empirical data.

TABLE I. THE PARAMETERS EMPLOYED FOR HYDROCARBONS

Coulomb integrals:

- for carbon p orbital: $\alpha-0.15\beta$
- sp^3 hybrid: $\alpha-0.10\beta$
- sp^2 hybrid: α
- sp hybrid: $\alpha+0.10\beta$
- for all hydrogens: $\alpha-0.20\beta$

Resonance integrals:

- for C-C single bonds:
  - sp^3-sp^3: 0.84β sp^3-sp^2: 0.88β sp^3-sp: 0.96β
  - sp^2-sp^2: 0.92β sp^2-sp: 0.98β sp-sp: 1.04β
- for C-C double bonds:
  - p-p (π): 0.30β
  - sp^2-sp^2 (σ): 1.00β sp^2-sp (σ): 1.04β
  - sp-sp (σ): 1.08β
- for C-C triple bonds:
  - p-p (π): 0.35β
  - sp-sp (σ): 1.13β
- for aromatic C-C bonds:
  - sp^2-sp^2: 0.96β
- for C-C bonds of polyenyl radicals:
  - sp^2-sp^2: 0.96β
- for two σ hybrids belonging to the same carbon:
  - sp^3-sp^3: 0.30β sp^2-sp^2: 0.38β sp-sp: 0.60β
- for C-H* bond:
  - sp^3 carbon: 0.90β
  - sp^2 carbon: 0.94β
  - sp carbon: 0.96β

(α: the Coulomb integral of a carbon sp^2 hybrid
β: the resonance integral for two bonding carbon sp^2 hybrids in the ethylenic double bond.)

(The resonance integrals for C-C single bond other than sp^3-sp^3 are given for the single bonds of such types as C-C= (sp^3-sp^2), C-C≡ (sp^3-sp); =C-C= (sp^2-sp^2), and so on. Such a clasification obviously does not possess any essential meaning. The slight difference in these resonance integrals does not affect seriously the result.)

The Sigma-pi Interaction

The quantity, $(\Delta E)_{\sigma\pi}$, which has been introduced above is useful in various problems in which the hybridization change participates. We assume, in this connection, that the arbitrary value of r_t^σ in Eq. 7 is to be regarded as equal to $g_t\gamma^\sigma$, in which g_t is $+1$ (as in the case of the carbons t and u of Fig. 2(i) and the carbon t of ii) and -1 (in the case of the carbon u of Fig. 2 (ii) and γ^σ is a negative constant; thus we may difine a new index, which we tentatively denote by $h_{tu\cdots}$, as:

$$h_{tu\cdots} = (\Delta E)_{\sigma\pi}\beta/(\gamma^\sigma)^2$$
$$= 2\left(\sum_i^{\text{occ}}\sum_k^{\text{uno}} - \sum_k^{\text{occ}}\sum_i^{\text{uno}}\right)\frac{(-\beta)}{\eta_k-\varepsilon_i}$$
$$\times\ (c_t^i c'^k_t g_t + c_u^i c'^k_u g_u + \cdots\cdots)^2 \qquad (9)$$

where $c'^k_t = b^k_t + b^k_{t'} + b^k_{t''}$, where β is the resonace integral between two bonding carbon sp^2 hybrids in the ethylene double bond, and where t, u,$\cdots$ imply the carbon atoms at which the hybridization change takes place.* Several of the simplest cases are specified by:

$$h_t = 2\left(\sum_k^{\text{occ}}\sum_i^{\text{uno}} - \sum_i^{\text{occ}}\sum_k^{\text{uno}}\right)\frac{(-\beta)}{\varepsilon_i-\eta_k}(c_t^i)^2(c'^k_t)^2 \qquad (10a)$$

$$h_{tu}^{(syn)} = 2\left(\sum_k^{\text{occ}}\sum_i^{\text{uno}} - \sum_i^{\text{occ}}\sum_k^{\text{uno}}\right)\frac{(-\beta)}{\varepsilon_i-\eta_k}$$
$$\times\ (c_t^i c'^k_t + c_u^i c'^k_u)^2 \qquad (10b)$$

$$h_{tu}^{(anti)} = 2\left(\sum_k^{\text{occ}}\sum_i^{\text{uno}} - \sum_i^{\text{occ}}\sum_k^{\text{uno}}\right)\frac{(-\beta)}{\varepsilon_i-\eta_k}$$
$$\times\ (c_t^i c'^k_t - c_u^i c'^k_u)^2 \qquad (10c)$$

For the systems having a half-occupied level, i, the terms due to that level must not be doubled but must be summed over all the k's.

The index h_t represents the ease of hybridization of the tth π orbital, namely, the ease of mixing the s-nature into the p AO; thus we may refer to

---

5) K. Fukui in "Modern Quantum Chemistry," Vol. 1, Ed. by O. Sinanoglu, Academic Press, New York (1965), p. 49.

* In the two-center interaction at the equivalent positions (t, u) of a molecule with C_{2h}-symmetry, in which we assume that $|g_u|=a|g_t|$ ($a>0$), it is easy to show that:

$$h_{tu}^{(syn)} - h_{tu}^{(anti)} = a\,[h_{tu}^{(syn)} - h_{tu}^{(anti)}]_{a=1}$$

it as the "hybridizability" of the tth $p\pi$ atomic orbital. This quantity is closely related to the problem of olefinicity (and, therefore, also to that of aromaticity), which will be discussed elsewhere.

The index h_{tu} is connected with the stability of the activated complex of the two-center reactions of conjugated molecules, a discussion of which will immediately follow.

Acyclic Two-center Interactions

Let us proceed further by dividing the problem into two cases, those of cyclic and of acyclic interactions. Also, let us limit the discussion to the two-center interaction only.

By an "acyclic two-center interaction" we mean a simultaneous or quasi-simultaneous interac-

TABLE II. THE COMPARISON OF $h_{tu}^{(syn)}$ AND $h_{tu}^{(anti)}$ FOR NONCYCLIC TWO-CENTER INTERACTIONS IN SEVERAL CONJUGATED HYDROCARBONS

Conjugated hydrocarbons	Positions of interaction, tu	$h_{tu}^{(syn)}$	$h_{tu}^{(anti)}$	Favorable configurations
(I) α,β-Interaction				
Ethylene	1,2	3.1194	5.5244	*anti*
(excited)		(8.9912)	(5.5180)	(*syn*)
Butadiene	1,2	3.2486	5.4947	*anti*
Hexatriene	1,2	3.2994	5.4671	*anti*
Benzene	1,2	3.4428	5.0468	*anti*
Styrene	α,β	3.2301	5.5457	*anti*
Naphthalene	1,2	3.3883	5.2540	*anti*
Anthracene	1,2	3.3728	5.3021	*anti*
Acenaphthylene*1	1,2	3.4096	5.4107	*anti*
Fulvene*1	1,2	3.4124	5.3581	*anti*
(II) α,γ-Interaction				
Allyl cation	1,3	5.0844	4.7323	*syn*
Pentadienyl cation	1,3	4.7313	4.5092	*syn*
Allyl	1,3	4.7528	5.1758	*anti*
Pentadienyl	1,3	4.5738	4.9189	*anti*
Allyl anion	1,3	4.4212	5.6192	*anti*
Pentadienyl anion	1,3	4.4163	5.3286	*anti*
(III) α,δ-Interaction				
Butadiene	1,4	4.6130	4.4241	*syn*
Hexatriene	1,4	4.5679	4.3888	*syn*
Benzene	1,4	4.3817	4.1079	*syn*
Naphthalene	1,4	4.5559	4.2012	*syn*
Anthracene	1,4	4.5840	4.2128	*syn*
Anthracene	9,10	4.8058	4.3194	*syn*
Fulvene	1,4	4.2373	4.7009	*anti**2
(IV) α,ε-Interaction				
Pentadienyl cation	1,5	4.7367	4.7737	*anti*
Pentadienyl	1,5	4.8185	4.7737	*syn*
Pentadienyl anion	1,5	4.9002	4.7737	*syn*
(V) α,ζ-Interaction				
Hexatriene	1,6	4.5570	4.5775	*anti*

*1 Numbering for acenaphthylene and fulvene:

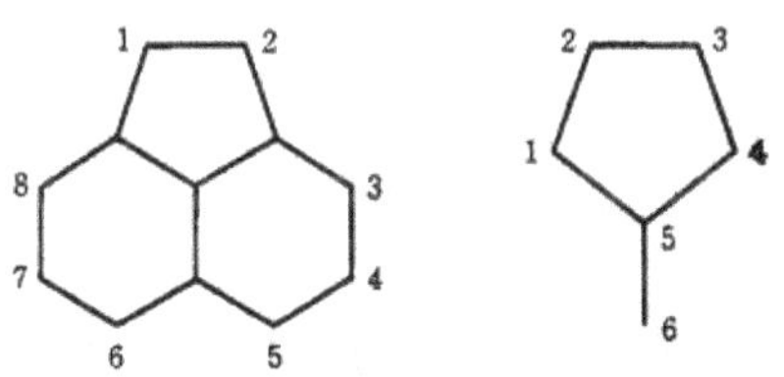

*2 The prediction of *anti*-predominance of 1,4-interaction in fulvene may be attributed to the odd-membered ring, in which this interaction is regarded also as α,γ-.

TABLE III. THE FAVORABLE MODE OF CYCLIC INTERACTIONS AS BASED UPON $(\Delta E)_{R\pi}$ (1)

Conjugated systems	Positions of interaction *rs*	$(\Delta E)_{R\pi}(\beta'/(\gamma^R)^2)$*1 *syn*	*anti*	Favorable configuration
Case Ia. The reagent is a vacant single orbital.				
Ethylene	1,2	3.4286	0.0000	*syn*
Butadiene	1,2	3.4757	0.1937	*syn*
	1,4	0.6195	3.6886	*anti*
Hexatriene	1,2	3.5307	0.3237	*syn*
	1,4	0.6599	3.5654	*anti*
	1,6	3.7701	0.9883	*syn*
Benzene*2	1,2	2.3297	0.5714	*syn*
	1,4	0.6154	2.2857	*anti*
Styrene	α,β	3.4606	0.1480	*syn*
Naphthalene	1,2	2.6992	0.4561	*syn*
	1,4	0.6024	2.7136	*anti*
Anthracene	1,2	2.8563	0.4232	*syn*
	1,4	0.5941	2.8733	*anti*
	9,10	0.5729	3.4608	*anti*
Acenaphthylene	1,2	3.4126	0.4886	*syn*
Fulvene	1,2	3.4243	0.5648	*syn*
Case Ib. The reagent is a half-occupied single orbital.*4				
Allyl	1,3	1.2651	16.9705	*anti*
Pentadienyl	1,3	1.2186	16.5163	*anti*
	1,5	16.3511	1.7143	*syn**3
Heptatrienyl	1,3	1.2134	12.9885	*anti*
	1,5	12.5808	1.6210	*syn*
	1,7	1.9770	12.6326	*anti*
Cyclopropenyl	1,2	1.6000	4.8000	*anti*
Cyclopentadienyl	1,2	3.4075	1.4089	*syn**3
	1,3	1.1279	3.6886	*anti*
Cycloheptatrienyl	1,2	3.2612	6.3476	*anti*
	1,3	1.5412	8.0676	*anti*
	1,4	7.2086	2.4002	*syn*
Indenyl	1,2	3.6098	1.7058	*syn*
	1,3	1.1597	5.6332	*anti*
Case Ic. The reagent is a doubly-occupied single orbital.				
Ethylene	1,2	0.0000	4.8000	*anti*
Butadiene	1,2	0.3066	5.5024	*anti*
	1,4	6.4126	0.7618	*syn*
Hexatriene	1,2	0.6629	6.2231	*anti*
	1,4	7.2193	0.8341	*syn*
	1,6	1.2933	8.0676	*anti*

*1 The value of $(\Delta E)_{R\pi}$ depends on the reagent orbital. The values presented here are tentatively calculated for the case of hydrogen migration.
The quantity β' corresponds to the resonance integral of an ethylenic p-p(π) bond. ($=0.3\beta$)

*2 This case corresponds to the 5,6-proton migration of cyclohexa-1,3-dienyl cation. Other cases are also to be interpreted similarly.

*3 The cyclic and acyclic cases were not distinguished in the theory of Woodward and Hoffmann (Ref. 16); namely e.g. the cases of 1,5-migrations of cyclopentadiene and all-*cis*-pentadiene. The former corresponds to the case of 1,2-interaction of cyclopentadienyl and the latter to that of 1,5-interaction of pentadienyl in this table.

*4 The values in case Ib are given for the systems each of which is composed of a vacant single orbital and the anion corresponding to the original reactant except the cases of cyclopropenyl and cycloheptatrienyl. In these two cases, the values are for a doubly-occupied single orbital with cyclopropenyl cation or cycloheptatrienyl cation. The zero-order terms due to electron transfer interaction are deemed as common both in *syn*- and *anti*-interactions.

TABLE IV. THE FAVORABLE MODE OF CYCLIC INTERACTIONS AS BASED UPON $(\Delta E)_{R\pi}$ (2)

Conjugated systems	Positions of interaction rs	$(\Delta E)_{R\pi}(\beta'/(\gamma^R)^2)$*1 *syn*	*anti*	Favorable configuration
Case II. The reagent is ethylenic.				
Ethylene	1,2	0.0000	2.0000	*anti*
Allyl anion*2	1,3	2.4142	0.4142	*syn*
Butadiene	1,2	0.1056	1.8945	*anti*
	1,4	1.7889	0.4223	*syn*
Hexatriene	1,2	0.1500	1.8711	*anti*
	1,4	1.6560	0.4503	*syn*
	1,6	0.6218	1.6570	*anti*
Benzene	1,2	0.3333	1.4444	*anti*
	1,4	1.3333	0.4444	*syn*
Naphthalene	1,2	0.7668	2.0701	*anti*
	1,4	1.4491	0.4300	*syn*
Anthracene	1,2	0.2500	1.6036	*anti*
	1,4	1.4748	0.4243	*syn*
	9,10	1.6163	0.4041	*syn*
Pleiadiene*3	7,10	1.5269	0.4715	*syn*
Pleiadylene*3	5,8	1.4291	0.5330	*syn*

*1 The values are calculated for the case of reagent of ethylene. $\beta'=0.3\beta$.
*2 This case presents a model for the 1,3-dipolar additions (R. Huisgen, *Proc. Chem. Soc.*, **1961**, 357).
*3 The numbering is:

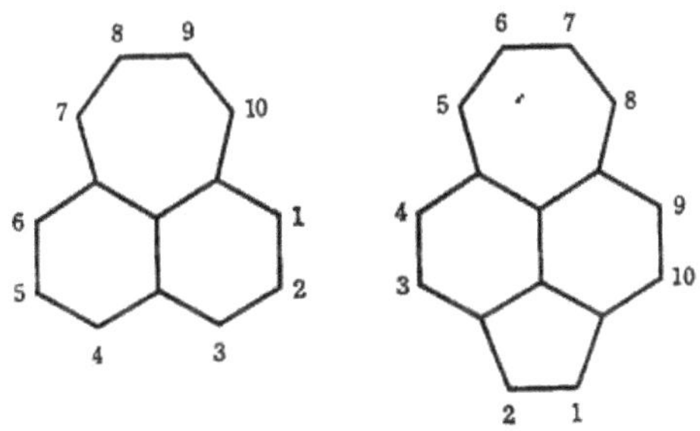

Pleiadiene Pleiadylene

TABLE V. THE FAVORABLE MODE OF CYCLIC INTERACTIONS AS BASED UPON $(\Delta E)_{R\pi}$ (3)

Conjugated systems	Positions of interaction rs	$(\Delta E)_{R\pi}(\beta'/(\gamma^R)^2)$*1 *syn*	*anti*	Favorable configuration
Case IIIa. The reagent is an allyl cation.				
Ethylene	1,2	0.4142	2.4142	*anti*
Butadiene	1,4	2.7889	0.7889	*syn*
Case IIIb. The reagent is an allyl.				
Allyl	1,3	2.0000*3	0.0000*3	*syn**2
Case IIIc. The reagent is an allyl anion.				
Ethylene	1,2	2.4142	0.4142	*syn*
Butadiene	1,4	0.7889	2.7889	*anti*
Naphthalene	1,4	0.7175	2.0508	*anti*
Anthracene	9,10	0.7245	2.7245	*anti*
Acenaphthylene	1,2	2.6915	0.4818	*syn*

*1 $\beta'=0.3\beta$
*2 This case corresponds to Cope rearrangements. The intramolecular rearrangements like that of *cis*-1,2-divinyl cyclobutane to cycloöcta-1,5-diene (cf. E. Vogel, *Ann.*, **615**, 1 (1958)) are also essentially a Cope rearrangement.
*3 These values correspond to the first order terms, $2|c_rc_r'\pm c_sc_s'|$ between two half-occupied orbitals in γ^R.

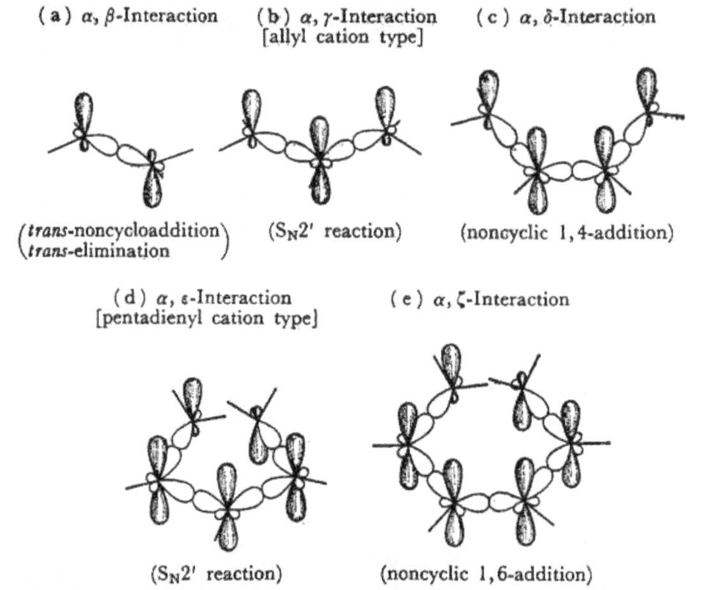

Fig. 3. A schematic representation of favorable configurations in noncyclic two-center interactions of conjugated molecules.

tion of two independent reagents with each of two positions of the reactant conjugated system. To this group may belong additions, eliminations, substitutions, rearrangements, and so forth. In such a case, if we concentrate our attention on the qualitative direction of the configuration alteration, *syn-* or *anti-* in Fig. 2, it is evident that we need to compare the magnitudes of the term $(\Delta E)_{\sigma\pi}$ only, since $(\Delta E)_{R\pi}$ is as an approximation common in these two configurations. A qualitative determination of the preferable configuration may be possible by a comparison of $h_{tu}^{(syn)}$ and $h_{tu}^{(ant)}$ of Eq. 10.

***α, β*-Interaction.**—The two carbon atoms under consideration, neighboring each other in a conjugated system, are designated as α and β. Let us take the ethylene molecule as an example. The values of $h_{12}^{(syn)}$ and $h_{12}^{(ant)}$ are 3.1194 and 5.5244 respectively; thus, they exhibit the predominance of *anti*-interaction.[6] This is consistent with a number of well-known experimental results of *trans*-noncycloadditions[7] and *trans*-eliminations.[8] The results for the α, β-interactions in other conjugated systems are listed in Table II (I); all of them demonstrate the *anti*-interaction to be favorable.[9]

***α, γ*-Interaction.**—Similarly, we can discuss the α, γ-interactions. The essential feature of S_N2' reactions is regarded as 1, 3-noncyclointeraction of the allyl cation type, because the leaving group takes off two electrons.* These reactions are known to proceed by means of a *syn*-configuration;[10] this is in accordance with the theoretical results obtained in Table II (II).

***α, δ*-, *α, ε*-, and *α, ζ*-Interactions.** — These cases seem not yet to have been experimentally established. The theoretical results, however, predict the dominant configurations indicated in Fig. 3, a configuration which corresponds to the calculations of Table II (III, IV, V).

It may be noteworthy that, as is shown in Table II, the differences in the values of $h_{tu}^{(syn)}$ and $h_{tu}^{(ant)}$ become smaller as the two reaction centers become farther apart. Thus, in noncyclic 1, 6-additions, high stereospecificity is not to be expected.

6) With respect to the MO-theory of the stereoselectivity of additions, see also K. Fukui, This Bulletin, **39**, 498 (1966).

7) E. S. Gould, "Mechanism and Structure in Organic Chemistry," Henry Holt & Co., New York (1959), p. 514, 732 ff.; B. A. Bohm and P. I. Abell, *Chem. Revs.*, **62**, 599 (1962).

8) E. Frankland, *J. Chem. Soc.*, **1912**, 654; D. Y. Curtin and D. B. Kellom, *J. Am. Chem. Soc.*, **75**, 6011 (1953); see also many examples indicated by Gould Ref. 7.

9) The predominance of *trans*-addition has also been discussed as an orientation problem. See K. Fukui, *Tetrahedron Letters*, No. 28, 2427 (1965); see also a paper of the present authors to be published in the near future.

10) H. L. Goering, T. D. Nevitt and E. F. Silversmith, *J. Am. Chem. Soc.*, **77**, 4042 (1955); G. Stork and W. N. White, ibid., **78**, 4609 (1956). One of the present authors (K. F.) is grateful to Professor Stork for his private communication.

* The hybridized state of β-carbon is assumed to stay unchanged during the course of this reaction. A similar assumption is made in the discussion of other, more distant two-center interactions, and in cyclic cases.

Cyclic Two-center Interactions

The term "cyclic two-center interaction" signifies a chemical interaction corresponding to the transition state in which a conjugated reactant system interacts simultaneously at two positions with one "connected" reagent system. Thus, for instance, if we regard a diene in the Diels-Alder reaction as a conjugated reactant system, the associated dienophile is a reagent, while if we regard a dienophile as a conjugated reactant system, the associated diene is a reagent. The reagent is, of course, not necessarily a conjugated system; it may be any other system which is capable of interacting simultaneously with two positions of a conjugated system. It may be a proton in a protonated ethylene molecule, a phenyl cation in a phenonium ion, osmium tetroxide in an adduct to conjugated hydrocarbons, and so forth.[11]

MO-theoretical treatments for interactions of such a kind are not new.[12] To such treatments may belong the theoretical studies of the well-known Diels-Alder reaction, the Claisen and the Cope rearrangements, the 1, 3-dipolar cycloadditions, and so on. The significance of the symmetry of the highest-occupied (HO) MO and the lowest-vacant (LV) MO in the problem of selection rule for the Diels-Alder reaction was first pointed out by one of the present authors.[13] Hoffmann and Woodward discussed a similar selection rule for the occurrence of cycloadditions more generally, from the point of view of the MO symmetry conservation.[14]

The problem of the stereoselectivity of these reactions is presented in terms of $(\Delta E)_{R\pi}$ and $(\Delta E)_{\sigma\pi}$ in Eq. 4. However, it seems likely that, in this cyclic case, $(\Delta E)_{R\pi}$ is in general more determinative than $(\Delta E)_{\sigma\pi}$. The term $(\Delta E)_{R\pi}$ in Eq. 8 is reduced to the following form:

$$(\Delta E)_{R\pi} = \sum_i \sum_l \frac{\nu_i - \nu_l}{\varepsilon_i - e_l} \left(\sum_r c_r^i d_{r'}^l \gamma_r^R \right)^2 \qquad (11)$$

where ν_i and ν_l are the numbers of electrons occupying the ith and the lth MO in the reactant and the reagent systems respectively. The sigma-type overlap stabilization of this type has already been discussed in several papers.[13,15] Several important cases will now be treated.

Case I. The reagent orbital is a single s orbital:

(Ia) When the reagent orbital is vacant:

$$(\Delta E)_{R\pi} = \sum_i \frac{\nu_i}{\varepsilon_i - e} (c_r^i \pm c_s^i)^2 (\gamma^R)^2 \qquad (12a)$$

(Ib) When the reagent orbital is half-occupied:

$$(\Delta E)_{R\pi} = \frac{1}{2} \left\{ \sum_i \frac{\nu_i}{\varepsilon_i - e} - \sum_i \frac{(2-\nu_i)}{\varepsilon_i - e} \right\} \times (c_r^i \pm c_s^i)^2 (\gamma^R)^2 \qquad (12b)$$

(Ic) When the reagent orbital is doubly-occupied:

$$(\Delta E)_{R\pi} = - \sum_i \frac{2-\nu_i}{\varepsilon_i - e} (c_r^i \pm c_s^i)^2 (\gamma^R)^2 \qquad (12c)$$

where ν_i is the number of π electrons occupying the ith MO of the reactant system, and where r and s are the centers of interaction. It is assumed that the absolute values of γ_r^R and γ_s^R are equal in the transition state, so that the upper and lower signs in parentheses are for *syn*- and *anti*-interactions respectively.

Case I, in particular the hydrogen migration belonging to Case Ib, was very recently discussed theoretically by Woodward and Hoffmann from a different point of view.[16] By Eqs. 12a—c, the results listed in Table III are obtained for the favorable mode of interaction in the ground-state reactions. The stabilization in some "nonclassical" carbonium ions, and also such reactions as the proton or carbonium cation migration in protonated conjugated hydrocarbons or carbonium cations, are covered by Case Ia; similarly, anion migrations belong to Case Ic. The well-known experimental results so far available are consistent with the theoretical results.

The reagent orbital is frequently a single p orbital. Similar arguments can easily show that such an orbital, when it is vacant, is allowed to enter into interaction simultaneously with two π orbitals of the reactant, for which the most favorable configuration is demonstrated to be "*syn*" in Table III, Case Ia. The well-known

11) For the discussion of such cases, see also the description of *cis*-1, 2-addition in Ref. 6.

12) See the many examples given, for example, in A. Streitwieser, Jr., "Molecular Orbital Theory for Organic Chemists," John Wiley & Sons, New York (1961).

13) K. Fukui, "Molecular Orbitals in Chemistry, Physics, and Biology," Ed. by P.-O. Löwdin and B. Pullman, Academic Press, New York (1964), p. 513.

14) R. Hoffmann and R. B. Woodward, *J. Am. Chem. Soc.*, **87**, 2046 (1965).

15) K. Fukui, *Tetrahedron Letters*, No. 24, 2009 (1965); cf. also Refs. 4 and 6.

16) R. B. Woodward and R. Hoffmann, *J. Am. Chem. Soc.*, **87**, 2511 (1965): These innovators have, no doubt, brought the first streaks of dawn into the field of theoretical stereochemistry through their elegant generalization (see Ref. 14 and also R. B. Woodward and R. Hoffmann, *J. Am. Chem. Soc.*, **87**, 395 (1965)). However, from the quantum-chemical point of view, their papers seem to leave a few points open to criticism. One of the present authors has recently made a comment (Ref. 15). In their discussion of "sigmatropic" reactions (*J. Am. Chem. Soc.*, **87**, 2511 (1965)), to take another instance, they determined the expected steric course by considering only the highest-occupied orbital (HO) symmetry, even in excited states. As was pointed out in Ref. 15, such a discrimination of HO would require proper grounds. Thus, the results of the prediction of excited-state reactions would probably be unreliable. Excited-state radical reactions are mostly too complicated to be treated by simple MO theories. In the ground-state reactions, the prediction by HO symmetry only almost coincides with that which is made by Eq. 12b.

TABLE VI. THE FAVORABLE MODE OF CYCLIC INTERACTIONS AS BASED UPON $(\Delta E)_{R\pi}$ (4)

Conjugated systems	Positions of interaction *rs*	$(\Delta E)_{R\pi}(\beta'/\gamma^R)^2$ *1 *syn*	*anti*	Favorable configuration
Case IV. The reagent is butadienic.				
Ethylene	1,2	1.7889	0.4223	*syn*
Butadiene	1,2	1.7441	0.4919	*syn*
	1,4	0.7155	1.7888	*anti**2
Benzene	1,2	1.3333	0.6111	*syn*
Naphthalene	1,2	1.4540	0.5731	*syn*
	1,4	0.6577	1.4266	*anti*
Anthracene	1,2	1.4913	0.5627	*syn*
	1,4	0.6565	1.4599	*anti*

*1 $\beta' = 0.3\beta$

*2 After E. Vogel (*Ann.*, **615**, 1 (1958)), cycloöcta-1,5-diene from catalytic dimerization of butadiene is not likely to come through the direct 1,4-1′,4′-interaction.

TABLE VII. THE FAVORABLE MODE OF CYCLIC INTERACTIONS AS BASED UPON $(\Delta E)_{R\pi}$ (5)

Conjugated systems	Positions of interaction *rs*	$(\Delta E)_{R\pi}(\beta'/(\gamma^R)^2)$ *1 *syn*	*anti*	Favorable configuration
Case V. The reagent is of the osmium tetroxide*2,*3 type (including permanganate and manganate anions, MnO_4^- *4 and MnO_4^{2-} *5).				
Ethylene	1,2	0.6317	0	*syn*
Butadiene	1,2	0.6095	0.0340	*syn*
	(1,4)	(0.1256)	(0.6025)	(*anti*)
Benzene	1,2	0.4449	0.1053	*syn*
Naphthalene	1,2	0.4934	0.0840	*syn*
Anthracene	1,2	0.5083	0.0784	*syn*
Phenanthrene	9,10	0.5318	0.0642	*syn*

*1 $\beta' = 0.3\beta$

*2 R. Criegee, *Ann.*, **522**, 75 (1936); R. Criegee, B. Marchand and H. Wannowius, idid., **550**, 99 (1938).

*3 These values are for the case of interaction with MnO_4^- (cf. C. J. Ballhausen and H. B. Gray, "Molecular Orbital Theory," W. A. Benjamin, New York (1964), p. 126). In these cases, only the z-components of the two adjacent oxygen atomic orbitals of the degenerate highest occupied orbitals are tentatively taken for the simplicity of calculation. The lowest vacant orbitals which belong to *E* are in favor of *syn* interaction except the case of butadiene 1,4.

*4 J. Boeseken, *Rec. trav. chim.*, **40**, 553 (1921); **47**, 683 (1928); K. B. Wiberg and K. A. Saegebarth, *J. Am. Chem. Soc.*, **79**, 2822 (1957).

*5 J. S. F. Pode and W. A. Waters, *J. Chem. Soc.*, **1956**, 717.

formation of phenonium ions and the singlet-state addition of carbenes, as well as the addition of atomic carbon to olefins,[17] which has recently been found, may involve this type of interaction.

When the reagent p orbital is half-occupied, too, the *syn* addition takes place if the symmetry relation is favorable. The formation of some epoxides, or the reaction with ozone may belong to this case.

Case II. The reagent orbitals are two bonding orbitals, which are equivalent to each other and which can enter into interaction with two orbitals of the reactant. Therefore, we have:

$$(\Delta E)_{R\pi} = \frac{1}{2}\left\{\sum_i \frac{\nu_i}{\varepsilon_i - e'}(c_r^i \mp c_s^i)^2 - \sum_i \frac{2-\nu_i}{\varepsilon_i - e}(c_r^i \pm c_s^i)^2\right\}(\gamma^R)^2 \qquad (13)$$

for the case of a reagent composed of two ethylenic π orbitals, where e and e' are the energies of bonding and antibonding orbitals respectively. The upper and lower signs stand for the *syn*- and *anti*-configurations respectively.

A comparison of the $(\Delta E)_{R\pi}$ values for two configurations of several conjugated systems indicates the dominant direction of change listed in Table IV; this is in agreement with our experience in dienes in Diels-Alder reactions, which belong to this case and which have already been treated by one of the present authors.[13] Excited-state reactions, in which the electron configuration of the reactant is assumed, can also be easily discussed in terms of Eq. 13.

Case III. The reagent orbitals are three

17) P. S. Skell and A. Y. Garner, *J. Am. Chem. Soc.*, **78**, 5430 (1956); P. S. Skell and R. R. Engel, ibid., **87**, 1135, 2493 (1965).

allylic π orbitals, two terminal orbitals of which interact with two reactant π orbitals. The well-known Claisen[18a)] and Cope rearrangements[18b)] are examples of this case. The $(\Delta E)_{R\pi}$ values in Table V demonstrate that, in experimentally-known cases, the theoretically-favorable configuration is *syn*. The 1, 3-dipolar addition occurs at *cis*- with respect dipolarophiles.

Case IV. The reagent orbitals are butadiene-type π orbitals, two terminal orbitals of which interact with two reactant π orbitals. The theoretical results are listed in Table VI. The Diels-Alder reaction is known to take place at *cis* positions in dienophiles.

Case V. The reagent is of another type. The additions of osmium tetroxide, manganate and permanganate anions, and so forth involve transition states in which the reagent system is considered to belong to none of cases, I to IV. The values of $(\Delta E)_{R\pi}$ are calculable separately by Eq. 11, leading to the conclusions indicated in Table VII, which are consistent with experience.

Through all these cases of cyclic interactions, the spatial circumstances of the transition state evidently require *syn* interaction in order for the reaction actually to take place. Therefore, the conclusion obtained in Tables III—VII affords a stereoselection rule: *syn* stands for the "sterically-allowed" case, and *anti*, the "streically-forbidden."

The explanation as to why the dimerization of butadiene yields 1, 2-divinyl cyclobutane compounds only, and not 1, 3-divinyl compounds,[19)] is also an interesting example of Eq. 11, which gives a far larger $(\Delta E)_{R\pi}$ value for the formation of 1, 2- than 1, 3-divinyl compounds.

In all the cases mentioned above, we can easily see that the theoretical conclusion obtained from Eq. 11 for the ground-state reactions is in excellent agreement with that from the consideration of HO and LV only.[6)] These orbitals are nothing but the frontier MO's, which are known to play a significant role in various reactions of organic molecules.

Thus, Eq. 11 has a wide applicability to various other problems of reactivity in conjugated systems, problems which are too manifold to systematize in a single paper. Considerations of them are to be published henceforth in sequence.

18) a) P. de Mayo, "Molecular Rearrangements," Interscience Publishers, New York (1963); b) E. N. Marvell, J. L. Stephenson and J. Ong, *J. Am. Chem. Soc.*, **87**, 1267 (1965).

19) E. Vogel, *Ann.*, **615**, 1 (1958).

BULLETIN OF THE CHEMICAL SOCIETY OF JAPAN VOL. 40 2018—2025 (1967)

The Stereoselection Rule for Electrocyclic Interactions

Kenichi Fukui and Hiroshi Fujimoto

Faculty of Engineering, Kyoto University, Sakyo-ku, Kyoto

(Received February 6, 1967)

A general stereoselection rule governing the electrocyclic interactions has been proposed with a molecular orbital rationale. The reactions discussed from a unified point of view cover the ring-closure of polyenes, the Diels-Alder additions, 1, 4-dipolar additions, the Cope and the Claisen rearrangements, the hydrogen and the proton migrations, 1, 3-dipolar additions, the ring-opening of cyclic olefins, the deamination of cyclic unsaturated imines, 1, 2-noncycloadditions, 1, 2-eliminations, the nucleophilic displacement with allylic rearrangements, and other cyclic intermediate or complex formations. The relation to the molecular orbital symmetry has been also discussed.

The molecular orbital (MO) theory, which has early proved itself suitable for interpreting the chemical reactivity, is exhibiting its usefulness also in the problem of the stereoselectivity. In particular, the symmetry property of MO has been intimately connected to the steric course of chemical reactions. One of the present authors first pointed out the importance of the symmetry of the highest occupied (HO) MO of dienes and the lowest vacant (LV) MO of dienophiles for the occurrence of the Diels-Alder reaction.[1] The MO symmetry was correlated with the stereoselectivity in the ring-closure of linear polyenes and in the ring-opening of cyclic olefins first by Woodward and Hoffmann,[2] and later by Longuet-Higgins and Abrahamson,[3] and by Fukui.[4] Woodward and Hoffmann[2] considered the symmetry of HO only, while Fukui[1,4] HO and LV for discussing the interaction of two systems.

These particular orbitals were coincident with those which played an important role in the MO theory of chemical reactivity of both saturated and unsaturated compounds, called the frontier orbitals by the present authors.[5,6] The physical basis on which these orbitals were distinguished from the others in the course of interaction with a reagent was clearly disclosed.[7] However, the discrimination of HO in the Woodward-Hoffmann interpretation[2] should be given a theoretical basis anew, since they considered the overlap stabilization of two π-atomic orbitals (AO) at both ends of the conjugated chain and it is hence evident that all of the π-MO's must deservedly contribute to such a sort of stabilization energy.

The MO theoretical interpretation was made also for other stereochemical problems. The selection rules in such cyclic additions as Diels-Alder, 1, 3-dipolar, and 1, 2-*cis* types,[1,8,9,12] in hydrogen migration,[10,12] in the Claisen and Cope rearrangements,[8,9,12] in noncycloadditions like 1, 2-*trans*,[9,11,12] and in 1, 2-*trans*-eliminations[12] were discussed theoretically.[13]

1) K. Fukui, "Molecular Orbitals in Chemistry, Physics, and Biology," ed. by P.-O. Löwdin and B. Pullman, Academic Press Inc., New York, N. Y. (1964), p. 513.

2) R. B. Woodward and R. Hoffmann, *J. Am. Chem. Soc.*, **87**, 395 (1965).

3) H. C. Longuet-Higgins and E. W. Abrahamson, *ibid.*, **87**, 2045 (1965).

4) K. Fukui, *Tetrahedron Letters*, **1965**, 2009.

5) a) K. Fukui, T. Yonezawa and H. Shingu, *J. Chem. Phys.*, **20**, 722 (1952); b) K. Fukui, T. Yonezawa, C. Nagata and H. Shingu, *ibid.*, **22**, 1433 (1954); see also many other papers cited in Ref. 1.

6) a) K. Fukui, H. Kato and T. Yonezawa, This Bulletin, **33**, 1201 (1960); b) K. Fukui, H. Kato and T. Yonezawa, *ibid.*, **34**, 442 (1961); c) K. Fukui, H. Kato and T. Yonezawa, *ibid.*, **34**, 1111 (1961); see also many other papers cited in K. Fukui, "Modern Quantum Chemistry," Vol. 1, ed. by O. Sinanoğlu, Academic Press Inc., New York, N. Y. (1965), p. 49.

7) a) K. Fukui, T. Yonezawa and C. Nagata, This Bulletin, **27**, 423 (1954); b) K. Fukui, T. Yonezawa and C. Nagata, *J. Chem. Phys.*, **26**, 831 (1957); c) K. Fukui, T. Yonezawa and C. Nagata, *ibid.*, **27**, 1247 (1957); d) R. D. Brown, *J. Chem. Soc.*, **1959**, 2224, 2232; e) K. Fukui, K. Morokuma, T. Yonezawa and C. Nagata, *J. Chem. Phys.*, **32**, 1743 (1960).

8) R. Hoffmann and R. B. Woodward, *J. Am. Chem. Soc.*, **87**, 2046 (1965).

9) K. Fukui, This Bulletin, **39**, 498 (1966).

10) R. B. Woodward and R. Hoffmann, *J. Am. Chem. Soc.*, **87**, 2511 (1965).

11) K. Fukui, *Tetrahedron Letters*, **1965**, 2427.

12) K. Fukui and H. Fujimoto, This Bulletin, **39**, 2116 (1966).

13) Other problems were discussed in a similar fashion: a) for the stereoselectivity in E2 reactions, see K. Fukui and H. Fujimoto, *Tetrahedron Letters*, **1965**, 4303; b) for *endo-exo* selection in cyclic additions, see R. Hoffmann and R. B. Woodward, *J. Am. Chem. Soc.*, **87**, 4388 (1965); c) for the stereochemistry of the transition state of the Cope rearrangements, see R. Hoffmann and R. B. Woodward, *J. Am. Chem. Soc.*, **87**, 4389 (1965); K. Fukui and H. Fujimoto, *Tetrahedron Letters*, **1966**, 251; and d) for the stereochemistry of excited configuration of conjugated chains, see R. Hoffmann and R. A. Olofson, *J. Am. Chem. Soc.*, **88**, 943 (1966). (Dr. Hoffmann kindly sent a manuscript to us prior to publication.)

The present paper intends to supply a universal stereoselection rule governing these stereochemical phenomena by a unified principle. The rule is given in a surprisingly simple form, making it at the same time possible to understand the reason why only HO does appear to act a discriminative part.[14]

The Electrocyclic Interaction

Woodward and Hoffmann[2] used the term "electrocyclic reactions" in their first elegant generalization of the stereochemical phenomena in the ring-closure of conjugated polyenes and the ring-opening of cyclic olefins. Here we consider a chain-like connected sequence of n AO's containing an even number, let it be $2m$, of electrons,[15] and define the "electrocyclic interaction" of these electrons in case of the approach of both ends of the chain to form a cyclic system. These AO's are not limited to the carbon 2p AO only, but may include nitrogen, oxygen, or other p AO's, 1s AO of hydrogen, or even any type of hybridized orbital to constitute a sigma skeletal chain of which we can adopt the usual approximation that the resonance integrals between non-neighbored AO's are neglected. It is important to choose these AO's as real functions.

The MO's are composed of the linear combination of these AO's which are named $\phi_1, \phi_2, \ldots, \phi_n$ from one end of the chain. The secular determinant in the Hückel MO treatment[16] is given by the following formula:

$$\Delta^{(0)}(\varepsilon) = \begin{vmatrix} \alpha_1-\varepsilon & \beta_{12} & 0 & \cdots & 0 & 0 \\ \beta_{12} & \alpha_2-\varepsilon & \beta_{23} & \cdots & 0 & 0 \\ 0 & \beta_{23} & \alpha_3-\varepsilon & \cdots & 0 & 0 \\ \cdots & \cdots & \cdots & \cdots & \cdots & \cdots \\ \cdots & \cdots & \cdots & \cdots & \cdots & \cdots \\ 0 & 0 & 0 & \cdots & \alpha_{n-1}-\varepsilon & \beta_{n-1n} \\ 0 & 0 & 0 & \cdots & \beta_{n-1n} & \alpha_n-\varepsilon \end{vmatrix} \quad \text{(1-a)}$$

in which $\alpha_1, \alpha_2, \ldots$ are the Coulomb integrals being all negative, and $\beta_{12}, \beta_{23}, \ldots$ are the resonance integrals. The sign (or direction) of AO's is assumed to be chosen so as to make all β_{ii+1}'s ($i=1, 2, \ldots, n-1$) possess negative values. The interaction between the two termini may be characterized by introducing a parameter β_{1n} into the determinant as its $(1, n)$ and $(n, 1)$ elements. The secular determinant becomes

$$\Delta(\varepsilon) = \Delta^{(0)}(\varepsilon) - \beta_{1n}^2 \Delta^{(0)}_{11,nn}(\varepsilon) - 2(-1)^n \beta_{1n}\beta_{12}\beta_{23}\cdots\beta_{n-1n} \quad \text{(1-b)}$$

in which $\Delta^{(0)}_{11,nn}$ is the minor of $\Delta^{(0)}$ obtained by removing the first and the nth columns and rows. The sign of β_{1n}, negative or positive, is made to stand for the direction of interaction of chain ends. Provided such a selection of direction is sterically permitted in this interaction, the reaction will take place through the one of steric paths which gives the lower electronic energy. The comparison of energy is easily made by solving the secular determinant. The secular determinants corresponding to the negative and the positive

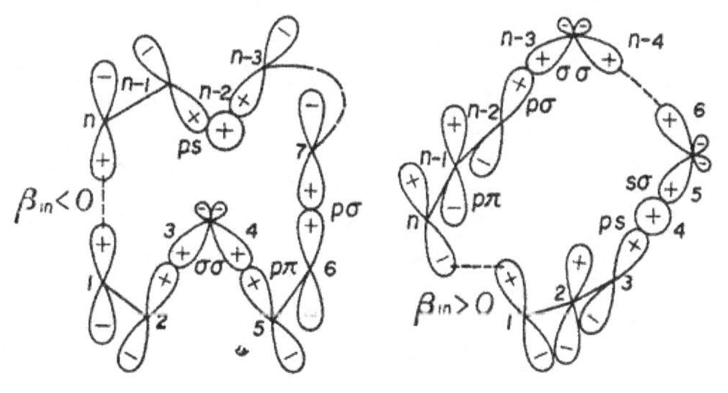

Fig. 1. The electrocyclic interactions with the negative and the positive β_{1n}'s. The signs (+) and (−) imply the positive and the negative parts of the real AO wave functions.

14) Hoffmann and Woodward (Ref. 8) as well as Longuet-Higgins and Abrahamson (Ref. 3) used correlation diagrams in place of the HO MO symmetry criterion employed in Ref. 2.

15) An unequivocal way of taking such a sequence of AO's will be possible with respect to any system if we select the one which is most sensibly affected by the difference in the steric course.

16) More elaborate treatments will be no more than a nuisance in the present case in which only the things essential are necessary. It is well known that the results of prediction of chemical reactivity by relevant MO-theoretical indices are not essentially affected by the roughness of the Hückel MO calculation.

β_{1n}'s are expressed by $\Delta^{(N)}(\varepsilon)$ and $\Delta^{(P)}(\varepsilon)$ respectively.

We replace the argument ε by α_0+z to obtain the functions of z, $D^{(N)}(z)$ and $D^{(P)}(z)$, corresponding respectively to $\Delta^{(N)}(\varepsilon)$ and $\Delta^{(P)}(\varepsilon)$, so that both of the equations, $D^{(N)}(z)=0$ and $D^{(P)}(z)=0$, come to possess m negative and $(n-m)$ positive roots.

We try to compare the sum of the negative roots between the cases of negative and positive β_{1n}'s which are put, for the purpose of comparison, γ and $-\gamma$ $(\gamma<0)$ respectively.[17]

Thus, we write

$$D^{(N)}(z) = D(z) - a$$

$$D^{(P)}(z) = D(z) + a$$

where

$$D(z) = \Delta^{(0)}(\alpha_0+z) - \gamma^2\Delta^{(0)}_{11,nn}(\alpha_0+z)$$

$$a = 2(-1)^n\gamma\beta_{12}\beta_{23}\cdots\beta_{n-1n}>0$$

The equation $D(z)=0$ also has m negative and $(n-m)$ positive roots. Twice the sum of the negative roots of $D(z)\mp a=0$ is denoted by $E^{(N)}$ and $E^{(P)}$. This can be expressed by the Coulson and Longuet-Higgins' integral representation.[18] Thus, we get

$$E^{(N)} - E^{(P)}$$

$$= \frac{1}{\pi}\int_{-\infty}^{+\infty} iy\left\{\frac{D'(iy)}{D^{(N)}(iy)} - \frac{D'(iy)}{D^{(P)}(iy)}\right\}\mathrm{d}y$$

If we put

$$D(iy) = R(y) + iyS(y)$$

where $R(y)$ and $S(y)$ are real even functions of y, an integral by parts leads us to the formula

$$E^{(N)} - E^{(P)}$$

$$= \frac{1}{2\pi}\int_{-\infty}^{+\infty}\log\left\{1+\frac{4aR}{(R-a)^2+y^2S^2}\right\}\mathrm{d}y$$

Now, we regard $D(0)$ as a function of α_0, indicating it by $A_0(\alpha_0)$. The equation $A_0(\alpha_0)=0$ possesses n real roots. In order that $D(z)=0$ possesses m negative and $(n-m)$ positive roots, α_0 must be in the interval ranging from the mth to the $(m+1)$th root of $A_0(\alpha_0)=0$. Such an α_0 obviously satisfies the condition

$$(-1)^m A_0(\alpha_0)>0$$

Further, if we write

$$R(y) = A_0(\alpha_0) - A_2(\alpha_0)y^2 + \cdots \begin{cases}(-1)^{n/2}A_n(\alpha_0)y^n & \text{(for even } n)\\ (-1)^{(n-1)/2}A_{n-1}(\alpha_0)y^{n-1} & \text{(for odd } n)\end{cases}$$

it follows evidently that

$$A_{2q}(\alpha_0) = \frac{1}{(2q)!}\left[\frac{\mathrm{d}^{2q}}{\mathrm{d}z^{2q}}D(z)\right]_{z=0}$$

$$= \frac{1}{(2q)!}\frac{\mathrm{d}^{2q}}{\mathrm{d}\alpha_0^{2q}}A_0(\alpha_0)$$

$$q = 0, 1, 2, \cdots \begin{cases}n/2 & \text{(for even } n)\\ (n-1)/2 & \text{(for odd } n)\end{cases}$$

Now we put

$$R(y) = (-1)^m T(y)$$

where

$$T(y) = f(\alpha_0) - \frac{1}{2!}f''(\alpha_0)y^2 + \frac{1}{4!}f''''(\alpha_0)y^4\cdots \begin{cases}\dfrac{(-1)^{n/2}}{n!}f^{(n)}(\alpha_0)y^n & \text{(for even } n)\\ \dfrac{(-1)^{(n-1)/2}}{(n-1)!}f^{(n-1)}(\alpha_0)y^{n-1} & \text{(for odd } n)\end{cases}$$

and

$$f(\alpha_0) = (-1)^m A_0(\alpha_0)$$

in which α_0 has been taken so as to satisfy

$$f(\alpha_0) > 0$$

Here we limit the discussion, from both mathematical and practical points of view, to the following most common cases:

$$\left.\begin{array}{ll}\text{i.} & n=\text{even},\ m=n/2 \\ & \text{(mostly a neutral system)}, \\ \text{ii.} & n=\text{odd},\ m=(n-1)/2 \\ & \text{(mostly a cationic system)}, \\ \text{iii.} & n=\text{odd},\ m=(n+1)/2 \\ & \text{(mostly an anionic system)}.\end{array}\right\} \quad (2)$$

Moreover, we make an important proposition that it is possible to choose a value of α_0 so as to make $T(y)$ positive for all real values of y. This proposition is strictly right in the cases $n=2$, $n=3$, and $n=4$ for chains with any set of α's and β's and in the cases of any n for chains with equal α's. In the cases $n\geqq5$, where all α's are not necessarily equal, this proposition, in a rigorously mathematical sense, not always holds; but anyone can always ascertain for each given case by a very simple numerical calculation along the line mentioned above, in advance to the employment of the subsequent theoretical result giving the selection rule, whether this proposition is actually true or not. From the results of a number of such calculations,

17) The present paper is connected with those electrocycles in which the comparison of the two cases, $\beta_{1n}=\gamma$ and $-\gamma$, is effectual: hence, the terminal AO's (1 and n) are usually required to be $p\pi$ and the interaction between them is of $p\sigma$ type (Fig. 1). Thus, the stereoselection rule of the present paper has nothing to do, for instance, with the problem of the stability of an $sp^3\sigma$ chain of cycloparaffin like skeletons, in which the interactions of two sp^3 hybrids with positive and negative β_{1n}'s can never be discussed as having an essentially equal efficiency of orbital overlapping.

18) C. A. Coulson and H. C. Longuet-Higgins, *Proc. Roy. Soc. (London)*, **A191**, 39 (1947); **A192**, 16 (1947).

it is known that the exception lies only in the region to which actual chemical systems can hardly belong. Moreover, we can always choose α_0 even for $n \geqq 5$ so as to make

$$T_0(y) = f(\alpha_0) - \frac{1}{2!} f''(\alpha_0) y^2$$

positive for the whole interval of real y. But it is seen from Eq. (3) that only the range of small y contributes to the right-side integral. Therefore, the value of the integral will not appreciably change if we replace $T(y)$ by $T_0(y)$. Such circumstances will not only diminish the error caused by making the general assumption that the proposition applies in almost all practical occasions even in the case of $n \geqq 5$, but also enable us, if necessary, to step outside from the limitation of (2). Numerical calculations can make sure of these circumstances.

In this way, we get, provided the proposition mentioned above holds,

$$E^{(N)} - E^{(P)} = \frac{1}{2\pi} \int_{-\infty}^{+\infty} \log \left\{ 1 + (-1)^m \frac{4aT}{(R-a)^2 + y^2 S^2} \right\} dy \begin{cases} >0 \text{ for even } m, \\ <0 \text{ for odd } m. \end{cases} \quad (3)$$

This result implies that:

For odd m the interaction with negative β_{1n} will dominate, while for even m the interaction with positive β_{1n} will be prevalent. **[*Theorem A*]**

This theorem describes the function of the "electrocyclic effect" which controls the steric course of cyclic interactions.

It is striking that the essential principle of stereoselection originates not from the number of AO's but from the number of electrons, or the number of doubly occupied MO's, although the number of electrons is not independent of the number of AO's through the assumption of (2). The verification of the theorem will follow with respect to a number of examples. Numerical calculations also assured this theorem.[12]

The Ring-Closure of Linear Conjugated Chains

The general rule governing the stereoselectivity of the thermal ring-closing reaction of linear conjugated chains which was proposed by Woodward and Hoffmann with the correspondence to experimental results is as follows:[19]

The ring formation by connecting both ends of a chain containing $2k\pi$ electrons should occur in the "conrotatory" mode for even k and in the "disrotatory" mode for odd k. **[*Rule 1*]**

Theorem A immediately leads one to *Rule 1* if both n and $2m$ are put equal to $2k$, since the disrotatory interaction evidently corresponds to negative β_{1n} and the conrotatory one to positive β_{1n}.

The Cyclic Interaction between Two Systems

The interaction in the Diels-Alder addition, like the reaction between 1, 2-positions of ethylene derivatives and 1, 4-positions of butadiene derivatives, is a special case of the interaction of the present kind in which $n=6$, $m=3$. The 1, 4-dipolar additions which has recently been reported[20] also belong to the case of $n=6$, $m=3$. The transition states in the Cope rearrangement and, in an approximate sense, also in the Claisen rearrangement, correspond also to the case of $n=6$, $m=3$. These are all known to occur in *cis*-mode. On applying *Theorem A* to this problem, we are able to deduce at once the following rule:

The cyclic reaction between two systems forming a cycle composed of 2k electrons in 2k atomic orbitals in the transition state takes place in the cis-trans or trans-cis fashion when k is even, while in the cis-cis or trans-trans fashion when k is odd. **[*Rule 2*]**

The geometrical situation will determine the selection of each of two modes of occurrence; for instance, the Diels-Alder reaction takes place only in the *cis-cis* mode.

The conclusion disclosed above also includes, as a special case, that of Hoffmann and Woodward (for thermal reactions)[8] which gives a theoretical explanation for the stereoselection in cycloadditions by an orbital symmetry consideration. *Rule 2* is applied *in statu quo* also to the cyclic interactions of more than two systems, provided appropriate examples are encountered.

An interesting example of application of *Theorem A* is to the hydrogen migration. The transition state of the $(1, 2k+1)$ hydrogen radical migration of the type

$$H_3C\text{-}(CH{=}CH\text{-})_{k-1}CH{=}CH_2 \rightarrow H_2C{=}CH\text{-}(CH{=}CH\text{-})_{k-1}CH_3$$

belongs to the case of $n=2k+2$, $m=k+1$. So, we obtain the following rule:

The (1, 2k+1) hydrogen radical migration takes place by way of "suprafacial" fashion[21] if k is even, while by way of "antarafacial" fashion[21] if k is odd. **[*Rule 3*]**

This conclusion includes the stereoselection rule for the hydrogen-migration (thermal) given by Woodward and Hoffmann.[10]

The cycle of odd number AO's also supplies an interesting example. The $(1, 2k)$ proton migration of the type

19) Woodward and Hoffmann discussed the excited-state reaction as well as the ground-state reaction. But by the mathematical reason the present authors limit the problem to the ground-state reaction only.

20) R. Huisgen and K. Herbig, *Ann.*, **688**, 98 (1965).

21) This was named by Woodward and Hoffmann (Ref. 10).

$$H_3C\text{-}(CH{=}CH\text{-})_{k-1}CH_2^+ \rightarrow H_2C^+\text{-}(CH{=}CH\text{-})_{k-1}CH_3$$

or the protonation of conjugated olefins belongs to the case of $n=2k+1$, $m=k$. For instance, the 1, 2-migration of α-proton in ethyl cation is known to be allowed ($n=3$, $m=1$). *Theorem A* at once derives the following conclusion:

The (1, 2k) proton migration or protonation can actually take place by way of "suprafacial" fashion if k is odd. **[*Rule 4*]**

The interaction of 1, 3-positions of allyl cation with 1, 4-positions of butadiene ($n=7$, $m=3$) will be in the *cis*-mode. Similarly, allyl anion will interact with ethylene derivatives ($n=5$, $m=3$) in the *cis*-fashion. An example of this type is 1, 3-dipolar additions, which are known to occur in the *cis*-mode. The other examples of an odd-membered cycle with even number electrons are: the phenonium cation, the transition state of the singlet addition of carbenes to olefins, the transition state in hydroboration of olefins, and so on. All of these examples belong essentially to the case $n=3$, $m=1$, and are known to occur actually.

To the contrary, the 1, 4-interaction of two butadiene ($n=8$, $m=4$) will take place in "antarafacial" mode.

The Ring-Opening of Cyclic Olefins and the Deamination of Cyclic Unsaturated Imines.

Here it is thought desirable to add to what has been described above that *Theorem A* is also applicable to the cyclic transition states of the ring-opening of cyclic olefins[2])

$$\left[\begin{array}{l}\text{-}(CH{=}CH\text{-})_k\text{-}\\ \text{-}CH_2 - CH_2\text{-}\end{array}\right] \rightarrow H_2C{=}CH\text{-}(CH{=}CH\text{-})_{k-1}CH{=}CH_2$$

TABLE 1. THE SELECTION RULE FOR SOME RING-CLEAVAGE REACTIONS

	even k	odd k
The ring-opening of cyclic olefins[a]) ($n=2k+2$, $m=k+1$)	Disrotatory opening	Conrotatory opening
The deamination of cyclic imines[b]) ($n=2k+4$, $m=k+2$)	Axisymmetric fragmentation	Sigmasymmetric fragmentation

a) The AO's constituting the electrocycle are $2k$ carbon $p\pi$ AO's and two incipient orbitals appearing in the transition state on each carbon forming the sigma bond to be broken.

b) The AO's of the electrocycle are $2k\pi$ AO's and four incipient orbitals, one each for the two carbons and two for the nitrogen atom of the two N-C bonds to be broken.

and the deamination of unsaturated cyclic imines.[22])

$$\left[\begin{array}{l}-(CH{=}CH\text{-})_k-\\ \text{-}CH_2\text{-}NH\text{-}CH_2\text{-}\end{array}\right] \rightarrow H_2C{=}CH\text{-}(CH{=}CH\text{-})_{k-1}CH{=}CH_2$$

The rule will be what is illustrated in Table 1, in conformity with the previous explanations based on the orbital symmetry[3,22]) consideration.

The Noncyclic Two-Center Interaction in Linear Conjugated Systems

The problem of the stereoselection in the addition to unsaturated bonds was generally and comprehensively discussed already by the present authors by the frontier electron method[11]) and by a perturbation treatment.[12]) It was shown to be the σ-π interaction that might possess an essential meaning for the stereoselection of this sort.[23]) The theoretical results indicated that 1, 2- and 1, 6-additions should take place through the "*anti*"-interaction, whereas 1, 4-addition through the "*syn*"-interaction.[24]) Here, we expect a more general, non-perturbation treatment for demonstrating this theorem.

The energy change is compared with respect to the "*syn*" and "*anti*" two-center σ-π interactions, which are schematically illustrated in Figs. 2 and 3. The change of hybridization is taken into account only at the terminal carbons.

It is recommended[25]) to adopt sp^2 (or sp) hybrids as the basis AO's composing the σ chain since the chain thus made up will most sensibly transmit the effect of the stereoselection. The deformed π AO at the sites of interaction may be represented by a $sp^x (x>3)$ hybrid. The treatment of σ electron

22) D. M. Lemal and S. D. McGregor, *J. Am. Chem. Soc.*, **88**, 1335 (1966).

23) The importance of the σ-π interaction in the stereospecificity of some ring-opening reactions was also pointed out by Fukui (see Ref. 9). Also see the paper of Longuet-Higgins and Abrahamson (Ref. 3).

24) The terminology "*syn*" and "*anti*" is employed, in place of "*cis*" and "*trans*" respectively, in order to avoid confusion in the case of additions to polyenic olefins.

25) Those who are not accustomed to the σ AO treatment are advised to proceed along the following way of consideration. At first, the LCAO MO's based upon the valence AO's taken in the fixed direction (*e. g.* 2s, $2p_x$, $2p_y$, and $2p_z$ of all carbon atoms assigned to the spatially fixed coordinates) are obtained by any appropriate method (*e. g.* R. Hoffmann, *J. Chem. Phys.*, **39**, 1397 (1963) and subsequent papers, or J. A. Pople, D. P. Santry and G. A. Segal, *ibid.*, **43**, S129 (1965); J. A. Pople and G. A. Segal, *ibid.*, **43**, S136 (1965)). The set of MO's thus obtained is then converted by the Lennard-Jones orthogonal transformation to another set of MO's which are substantially localized in one atom making a hybrid like orbital which extends in the direction of bonding (*cf.* J. Lennard-Jones, *Proc. Roy. Soc. (London)*, **A198**, 1, 14 (1949)). The value of β's and γ's may be evaluated by using these newly obtained MO's and Hamiltonian operator.

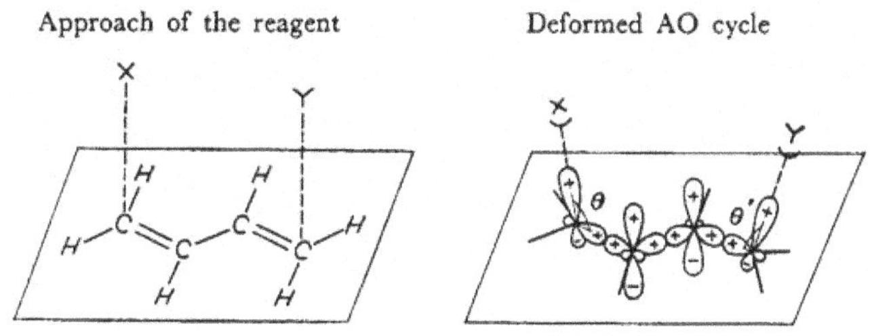

(θ, θ'>90°)

Fig. 2. A schematic representation of the "*syn*"-interaction of σ and π chains of AO's. Butadiene is taken as an example of the conjugated system. The approaching reagent is represented by XY (or X and Y in their "concerted" or "quasi-concerted" approach).

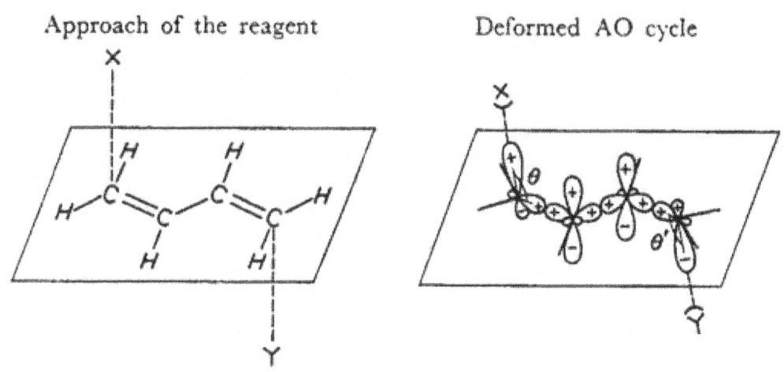

(θ, θ' > 90°)

Fig. 3. A schematic representation of the "*anti*"-interaction of σ and π chains of AO's. Butadiene is taken as an example of the conjugated system. The approaching reagent is represented by XY (or X and Y in their "concerted" or "quasi-concerted" approach).

systems by employing these hybrids as the basis AO's has been developed by many authors.[26] The secular determinant with the σ-π interaction is represented by

$$\Delta^{(\sigma,\pi)}(\varepsilon)=\left|\begin{array}{cccc|cccc}
\alpha_1^{(\pi)}-\varepsilon & \beta_{12}^{(\pi)} & \cdots & & \gamma & \cdots & & \\
\beta_{12}^{(\pi)} & \alpha_2^{(\pi)}-\varepsilon & \cdots & & \cdots & & & \\
\cdots & & & & \cdots & & & \\
\cdots & & & \alpha_k^{(\pi)}-\varepsilon & \cdots & & & \gamma' \\
\hline
\gamma & \cdots & & & \alpha_1^{(\sigma)}-\varepsilon & \beta_{12}^{(\sigma)} & \cdots & \\
\cdots & & & & \beta_{12}^{(\sigma)} & \alpha_2^{(\sigma)}-\varepsilon & \cdots & \\
\cdots & & & & \cdots & & & \\
\cdots & & & \gamma' & \cdots & & & \alpha_{2k-2}^{(\sigma)}-\varepsilon
\end{array}\right| \qquad (4)$$

in which superscript (π) and (σ) denote the originally pi and the sigma parts respectively, γ and γ' being the "resonance" integrals of σ-π interaction at carbon 1 and carbon k respectively. If we take a negative value for γ, the "*syn*"- and "*anti*"-interactions correspond to a negative and a positive γ' respectively. A glance at Eq. (4) makes us to learn that this is nothing but a special case of Eq. (1-b). Accordingly, by putting $k=2\nu$, one is immediately led to a rule which says:

In the (1, 2ν) interaction of a cycle of (6ν−2) electrons the "syn"-mode is predominant for even ν, while for odd ν "anti"-mode prevails, **[*Rule 5*]**

in accordance with experience. Namely, the *trans*-1, 2-addition belongs to the case of ν=1, 1, 4-addition (ν=2) *syn*, and 1, 6-addition (ν=3) *anti*.

The *trans*-mode stereoselection in the elimination to produce an unsaturated bond also seems to have a concern with the relation discussed above (ν=1), although it has been known that for this sort of stereoselection the orienting effect of the nucleophilic group to be repelled is important.[13a]

The nucleophilic replacement with the allylic rearrangement (the S_N2' reaction) is one of

26) a) For the σ MO of conjugated systems, see K. Fukui, H. Kato, T. Yonezawa, K. Morokuma, A. Imamura and C. Nagata, This Bulletin, **35**, 38 (1962); J. A. Pople and D. P. Santry, *Mol. Phys.*, **9**, 301 (1965); b) also refer to the following papers for saturated systems: C. Sandorfy, *Can. J. Chem.*, **33**, 1337 (1955); H. Yoshizumi, *Trans. Faraday Soc.*, **53**, 125 (1957); Ref. 6; K. Fukui, H. Kato and T. Yonezawa, This Bulletin, **35**, 1475 (1962); T. Yonezawa, H. Kato, H. Saito and K. Fukui, *ibid.*, **35**, 1814 (1962); K. Morokuma, K. Fukui, T. Yonezawa and H. Kato, *ibid.*, **36**, 47 (1963); H. Kato, K. Fukui and T. Yonezawa, *ibid.*, **38**, 189 (1965); J. A. Pople and D. P. Santry, *Mol. Phys.*, **7**, 269 (1964); **9**, 311 (1965).

1, 3-noncyclointeractions, known to proceed stereospecifically in the *cis* fashion.[27] This belongs to the case $k=3$ ($n=3k-2=7$, $m=3$) in Eq. (4). If we put $k=2\mu+1$ in Eq. (4), we obtain from *Theorem A* the following general rule:

In the $(1, 2\mu+1)$ interaction of a cycle of 6μ electrons the "syn"-mode is predominant for odd μ and the "anti"-mode is preferable for even μ; and in the cycle containing $(6\mu+2)$ electrons vice versa. [***Rule 6***]

These selection rules will supply means to the prediction of favorable steric courses only under the supposition that any other factors do not play a more determinative part. The cyclic intermediate formation in "noncyclic" addition is no doubt one of such influencing factors, although such cyclic intermediates themselves are the system governed by the selection rules mentioned in the preceding sections. In the cases where more than two electrocycles participate, the most determinative one will control the steric course of the reaction.

Aromaticity and the Hückel's 4n+2 Rule

Aromatic sextet has long been a subject in organic chemistry. A satisfying explanation to this problem was presented by Hückel, leading to a rule, referred to as the $4n+2$ rule.[28] A quantum mechanical approach to this rule was made by applying the perturbation theory.[29] Heilbronner proposed an interesting idea concerning on the conformations of annulenes that the higher members of them with $4n$ electrons might be twisted once to give a Möbius ring.[30] *Theorem A* is intimately connected to the theory of aromaticity and may supply, in rather qualitative sense, a physical basis of the $4n+2$ rule. The rule obtained is as follows:

The monocyclic conjugated system which contains $2m\pi$ electrons prefers the Hückel perimeter (with negative β_{1n}) if m is odd, (while the Möbius perimeter (with positive β_{1n}) prevails if m is even). [***Rule 7***]

The effect of the steric repulsion as well as the change in β's due to the twist[30] should be also taken into consideration in discussing the possibility of Möbius-type conformations, especially in case of the small ring-conjugated polyenes.[13d]

The Relation to the Perturbation Treatment

It is well known that the energy change due to the introduction of a small change of resonance integrals into the secular determinant in the Hückel MO treatment of conjugated systems is represented in terms of the bond order. The essential feature is not altered also in the treatment of σ electron systems. The one-electron energies and the one-electron wave functions for $\Delta^{(0)}(\varepsilon)$ are designated as $\varepsilon_i^{(0)}$ and $\psi_i^{(0)}$ respectively. The MO's are represented by the linear combination of AO's, $\phi_1, \phi_2, \ldots, \phi_n$, such that

$$\psi_i^{(0)} = \sum_r C_{ir}\phi_r$$

The introduction of β_{1n} into $\Delta^{(0)}(\varepsilon)$ makes the energy increase as much as

$$\Delta E = 4\sum_{i=1}^{m} C_{i1}C_{in}\beta_{1n} = 2P_{1n}\beta_{1n}$$

where the quantity P_{1n} is the "bond order" of the virtual bond considered between carbons 1 and n by way of ϕ_1 and ϕ_n. When it is positive, a negative β_{1n} is favorable, while a positive β_{1n} is advantageous for negative P_{1n}. This quantity was called by the present authors the "overlap stabilization" and was applied to the stereoselectivity in the ring-closure of polyenes.[4,9,12] A similar treatment was made to discuss the transition state configuration in some electrocyclic interactions.[13c]

The present paper includes the perturbation treatment as a special case of small β_{1n}'s and therefore falls on an extension of the perturbation treatment to the region of appreciably large interactions, exhibiting a wider availability in discussing similar problems.

The Relation to the Orbital Symmetry

The symmetry property of MO's for linear polyenes is well established. The $(2k+1)$th MO is symmetric (S) and the $2k$th MO is antisymmetric (A) with respect to the center of the molecule both in even and in odd pπ AO systems. In an S-MO the AO coefficients of two termini have the same sign, while in an A-MO they have different signs.

In the more general systems presented above which are composed of a chain of any sort of AO's, similar relations are shown to exist. Namely, it follows from the relation of AO coefficients and the secular determinant that[31]

$$C_{i1}C_{in} = (-1)^n\beta_{12}\beta_{23}\cdots\beta_{n-1n}/\Delta'(\varepsilon_i) \begin{cases} >0 & \text{(for odd } i\text{)} \\ <0 & \text{(for even } i\text{)} \end{cases}$$

since β's are all negative and the sign of $\Delta'(\varepsilon_i)$ agrees with that of $(-1)^i$. Therefore the $(2k+1)$-th MO is quasi-symmetric (qS) and the $2k$th MO is quasi-antisymmetric (qA). The HO ($i=m$) is qS

27) H. L. Goering, T. D. Nevitt and E. F. Silversmith, *J. Am. Chem. Soc.*, **77**, 4042 (1955); G. Stork and W. N. White, *ibid.*, **78**, 4609 (1956). One of the present authors (K. F.) thanks to Professor Stork for his private communication on this reaction.

28) E. Hückel, *Z. Physik*, **70**, 204 (1931); **76**, 628 (1932).

29) K. Fukui, A. Imamura, T. Yonezawa and C. Nagata, This Bulletin, **33**, 1591 (1960).

30) E. Heilbronner, *Tetrahedron Letters*, **1964**, 1923.

31) K. Fukui, C. Nagata, T. Yonezawa, H. Kato and K. Morokuma, *J. Chem. Phys.*, **31**, 287 (1959).

if m is odd, and qA if m is even. Hence, even though the following description might be adopted in place of *Theorem A*, the result of prediction would never change:

"If the highest occupied molecular orbital is symmetric (or quasi-symmetric) the interaction with negative β_{1n} will dominate, while if it is antisymmetric (or quasi-antisymmetric) the interaction with positive β_{1n} will be prevalent."

This might be nothing but the reason why Woodward-Hoffmann's discussion[2] by means of the symmetry consideration of HO MO only was able to succeed. The physical basis underlying their interpretation is thus given by *Theorem A*.

The Odd-Electron Systems and Excited-State Reactions

Mainly by the mathematical reason the discussion disclosed above was limited to the ground-state reaction of even-electron systems. However, the principle to discuss the selection of steric courses in odd-electron systems or for excited-state reactions is completely the same as the former cases, if we employ numerical computations. Namely, we are always able to compare $E^{(N)}$ and $E^{(P)}$ by solving numerically the secular determinant, or to calculate P_{1n} to determine the favorable sign of β_{1n} by way of the perturbation procedure. Numerical computations along the former line was made by Zimmerman on the problem of ring-closure of linear olefins[32] and both numerical calculations and, in some special cases, analytical treatments were made for discussing the stereoselection of excited-state reactions along the latter line by the present authors.[4,9,12] The well-known relation that the virtual bond order between two AO's which belong to the same group (starred or unstarred) of an alternant hydrocarbon system is zero suggests that the formation of an odd-membered electrocycle from a chain composed of AO's of equal Coulomb integrals gives the same values to $E^{(N)}$ and $E^{(P)}$. This can be also shown from the secular equations similar to Eq. (1-b) corresponding to $\Delta^{(N)}(\varepsilon)$ and $\Delta^{(P)}(\varepsilon)$. Although changes in the Coulomb integrals may cause the distinction between the energies of the two steric courses, the difference may be rather small in comparison with the one obtained for even-membered electrocycles.

In general, the prediction of the favorable steric courses in a similar fashion may be possible for any kind of reactions accompanied with stereoselection by solving the secular equations corresponding to the two steric courses, provided the comparison of the two cases (γ and $-\gamma$) is effectual.

$$\Delta(\varepsilon) = \begin{vmatrix} \boxed{\Delta_1(\varepsilon)} & \gamma_{12} & & & \pm\gamma_{1n} \\ \gamma_{12} & \boxed{\Delta_2(\varepsilon)} & \gamma_{23} & & \\ & \gamma_{23} & \ddots & \ddots & \\ & & \ddots & \ddots & \gamma_{n-1n} \\ \pm\gamma_{1n} & & & \gamma_{n-1} & \boxed{\Delta_n(\varepsilon)} \end{vmatrix}$$

The selection rules disclosed above may present the method more simple and with more physical foundation, in comparison with the former ones, to predict the favorable steric course of a given reaction. Anyhow, we do not need the solution of secular equation or any MO properties, *e. g.* HO symmetry to discuss the steric course of a ring-closure reaction, or MO energies of the initial and the final systems to draw a correlation diagram, even when we employ the rules most rigorously ascertaining the proposition leading to the *Theorem A*.

32) H. E. Zimmerman, *J. Am. Chem. Soc.*, **88**, 1564, 1566 (1966).

Recognition of Stereochemical Paths by Orbital Interaction

KENICHI FUKUI[1]

Department of Chemistry, Illinois Institute of Technology, Chicago, Illinois 60616

Received August 3, 1970

The importance of orbital interactions, particularly between the HO (highest occupied) MO (molecular orbital) of one reactant and the LU (lowest unoccupied) MO of the other, in rationalizing chemical reactions is now invincibly evident.[2,3] Several papers have discussed this problem, giving a fair reasoning of the orbital interaction scheme.[4-6] The physical grounds for differentiation of the HOMO and LUMO from the other MO's have also been disclosed.[7]

The scheme of the HOMO–LUMO interaction between reacting species[2-4] has drawn the attention of organic chemists to the interpretation of the steric course of concerted processes, mainly through the elegant generalization of stereoselection rules and the classification of various stereospecific reactions by Woodward and Hoffmann.[5] Although a large part of their interpretation is based on the correlation diagram method, the orbital interaction scheme is also used in some parts, namely, in the explanation of the preferable paths of ring opening for cyclobutenes and for cyclopropyl derivatives, the dimerization of two ethylene molecules by an orthogonal approach, the migration of hydrogen or alkyl across a conjugated chain, the exo–endo selectivity in Diels–Alder-type cycloadditions, and the addition of ketenes to olefins. Obviously the correlation diagram criteria are not employed in some cases, such as interpretation of "sigmatropic" reactions, in which the reaction path has no usable symmetric character.

The purpose of this Account is to show that the HOMO–LUMO interaction scheme can in a unified manner be applied to various processes which have been mentioned and discussed by Woodward and Hoffmann,[5] and also to present, at the same time, an extremely simple and practical method to interpret or predict the favorable steric pathway. While this method is general in character, it is particularly suitable

(1) Address correspondence to: Department of Hydrocarbon Chemistry, Kyoto University, Kyoto, Japan.

(2) For instance, see K. Fukui, "Molecular Orbitals in Chemistry, Physics, and Biology," P.-O. Löwdin and B. Pullman, Ed., Academic Press, New York, N. Y., 1964, p 513, and references cited therein, and subsequent papers on the frontier electron theory.

(3) (a) K. Fukui, *Bull. Chem. Soc. Jap.*, 39, 498 (1966); (b) K. Fukui and H. Fujimoto, *ibid.*, 39, 2116 (1966); (c) K. Fukui and H. Fujimoto, "Mechanisms of Molecular Migrations," Vol. 2, B. S. Thyagarajan, Ed., Interscience, New York, N. Y., 1969, p 118; (d) K. Fukui, "Theory of Orientation and Stereoselection," Springer-Verlag, Heidelberg, 1970.

(4) (a) G. Klopman and R. F. Hudson, *Theor. Chim. Acta*, 8, 165 (1967); (b) R. F. Hudson and G. Klopman, *Tetrahedron Lett.*, 1103 (1967); (c) G. Klopman, *J. Amer. Chem. Soc.*, 90, 223 (1968); (d) L. Salem, *ibid.*, 90, 543 (1968); (e) L. Salem, *ibid.*, 90, 553 (1968); (f) L. Salem, *Chem. Brit.*, 449 (1969); (g) A. Devaguet and L. Salem, *J. Amer. Chem. Soc.*, 91, 3793 (1969).

(5) R. B. Woodward and R. Hoffmann, *Angew. Chem.*, 81, 797 (1969); *Angew. Chem., Int. Ed. Engl.*, 8, 781 (1969); "The Conservation of Orbital Symmetry," Academic Press, New York, N. Y., 1969; and their preceding papers cited therein.

(6) R. G. Pearson, *Theor. Chim. Acta*, 16, 107 (1970).

(7) (a) K. Fukui and H. Fujimoto, *Bull. Chem. Soc. Jap.*, 41, 1989 (1968); (b) K. Fukui and H. Fujimoto, *ibid.*, 42, 3399 (1969).

for the interpretation of rather complicated reactions such as intramolecular rearrangements, to which the correlation diagram method is not always directly applicable. The theoretical basis is evident in light of previous papers.[2,3,7] This method is a practical representation of the fundamental proposition[3,4f,7b] that a *majority of chemical reactions should take place at the position and in the direction of maximum overlapping of the HOMO (or high-lying occupied MO's) and the LUMO (or low-lying unoccupied MO's) of the reacting species; in reacting species possessing a singly occupied (SO) MO, this plays the part of the HOMO or of the LUMO, or of both.*

Procedure

The process is illustrated in Figure 1 for the simple case of the thermal ring opening of cyclobutene. Usually chemical reactions occur in a certain limited region of the molecule.[3d,7b] The region essential to the conversion in this case is composed of one π bond and one σ bond to be broken and two π bonds to be formed. As a first approximation, it may be sufficient to take only these bonds into account. The reaction may be initiated by the mutual charge-transfer interaction between the π bond and the σ bond to be broken. The mode is illustrated in Figure 1. This interaction will take place between the HOMO and the LUMO of each bonding region. The HOMO is bonding and the LUMO is antibonding. The charge transfer to the antibonding LUMO from the bonding HOMO will cause the weakening, and accordingly the stretching, of both bonds. This principle is most easily understood if the diatomic case is thought of. An increase in bond length will cause a lowering of the LUMO level and an elevation of the HOMO energy, so that the charge transfer is further facilitated. Such interactions may be assisted by vibrational motion. As the reaction proceeds, the donor character of the C=C π bond and the acceptor character of the C—C σ bond become distinct, and the contribution of interaction I of Figure 1 becomes more important than that of I′.

The nodal properties of these MO's are characterized by the schematic diagram (Figure 1). The shaded and unshaded regions correspond to the plus and minus lobes of the MO's. Figure 1 indicates that the direction of effective overlapping of HOMO and LUMO in this case is that which is shown by arrows. It is concluded that the conrotatory change[5] would be favorable.

Here, a very elementary but important remark might be necessary. Each molecular orbital may have an indefiniteness by a factor of absolute value unity. Two different ways of shading are possible for each MO, if we are interested in real functions only. Physically, these ways represent an identical MO. Since the interaction energy between two systems, A and B, includes orbital overlapping between MO's belonging to A and B in the form of their absolute value or square,[2,3,7] the mode of change I* in Figure 2 can be expected with the same probability as that of I provided other conditions are the same. Note that these two modes of conrotatory

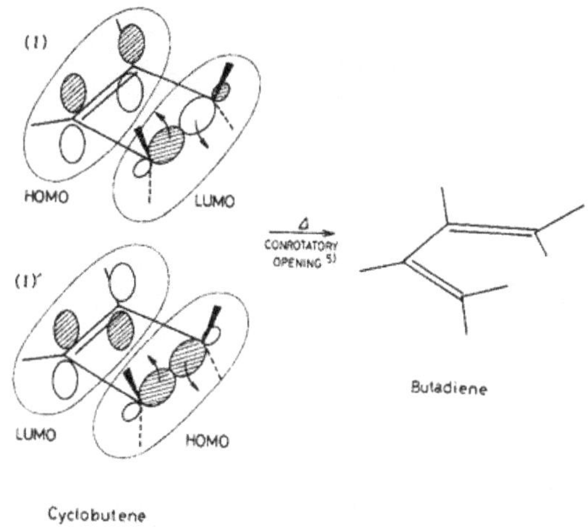

Figure 1. The mode of the HOMO–LUMO interaction in the ring-opening of cyclobutene.

change are not always equivalent if the molecule has appropriate substituents. Similar circumstances apply to the relation of II and II* of Figure 2. However, a useful convention may be adopted here in which we only refer to the "positive" overlapping such as I or II and do not care about the "negative" overlapping such as I* or II* for the purpose of using the maximum overlapping criteria. By this convention, the difference in the way of shading, as that between I and II, which is physically meaningless, turns out to distinguish two *different* modes of conversion. The two ways of shading the LUMO part in the models I and II are thus regarded as independent alternatives. In this way, we are able to consider hereafter only the direction which leads to shaded–shaded and unshaded–unshaded overlappings as the proper one which corresponds to bonding interaction, and at the same time we have to take *all* different ways of shading which may have the possibility of producing a nonequivalent direction. Consideration of Figure 4d and 4f will be helpful in understanding this relation.

In addition to this, a general remark is necessary concerning the choice of the region in the molecule in which we have to consider the HOMO–LUMO interaction. A fundamental guide may be: *the bordering surface of the two parts between which the HOMO–LUMO interaction is considered should be crossed by the newly formed bonds* (lemma A). This is required since the multicenter interaction between these two parts, taking place through the bonds to be formed, is the very origin of the stereoselectivity.

The practical value of such a convention will be clearer if we use a stereomodel like that shown in Figure 3. The HOMO and LUMO parts are specified with colored pegs. Two different colors are understood to represent the shaded and unshaded lobes of atomic orbitals (AO). For the sake of convenience, the σ bond to be broken has pegs just like those of sp^2 carbons and a

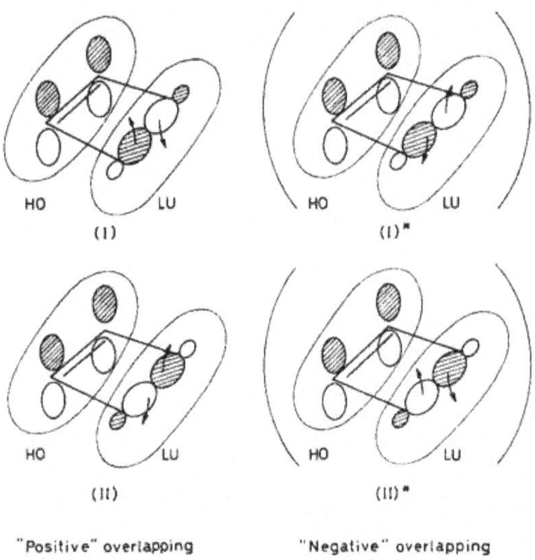

Figure 2. A convention used in shading MO's.

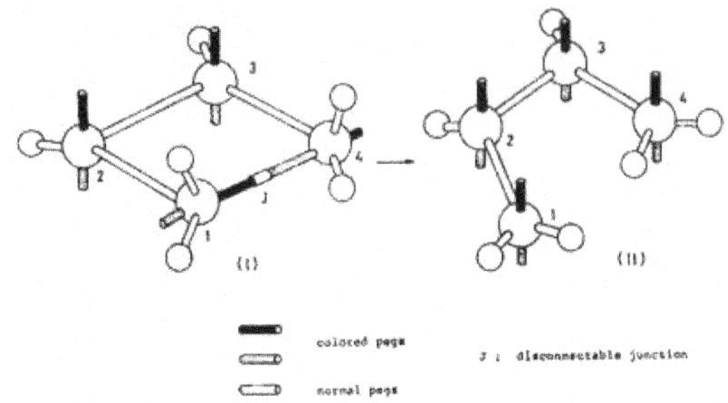

Figure 3. A stereomodel for the explanation by orbital interaction scheme.

disconnectable junction, J. The sp^3 "tail" pegs allow the model to represent the growth of the sp^3 hybrid to a p AO as the reaction proceeds. At first the colored pegs are properly placed so as to represent the nodal properties of the HOMO and LUMO. Then, junction J is removed and carbons 1 and 4 are rotated to form the 1,2 and 3,4 π bonds in such a way that two pegs of the same color become parallel on each side of a π bond that is forming. Of course, as mentioned before, the two nonequivalent ways of coloring lead to two nonequivalent pathways in the case of substituted cyclobutenes. We can simultaneously discuss the favorable or unfavorable steric circumstances with respect to these two paths by the use of such a stereomodel. It is known that the color relation of AO's between the two newly formed π bonds in the resultant molecule in Figure 3II is not physically significant since the color of the pegs only illustrates the direction of favorable change with respect to the partition of the molecule into the two parts illustrated in Figure 1I.

Examples

This simple method is powerful from the practical standpoint since it can reproduce the stereoselective reaction paths discussed in the comprehensive paper of Woodward and Hoffmann.[5] Several examples of intramolecular processes are given in Figure 4 (example o is discussed in ref 8) in which the sketch of a molecular model is replaced by the simpler representation as in Figure 1 for reasons of simplicity. If the reader experiments with a stereomodel, he will find it to be so simple and automatic that no explanation of the figure will be needed. In all of the examples mentioned, the two modes of HOMO–LUMO interactions like I and I′ in Figure 1 lead to the same conclusion. Therefore only the more important one of them is indicated in Figure 4. In excited-state reactions, too, only the SOMO–LUMO interaction is illustrated, since the SO′MO–HOMO interaction leads to the same conclusion. Symbols for the MO's used are defined in Figure 5. At the present stage of development we are forced to proceed with the assumption of reaction *via* the lowest excited state for these photoinduced processes. Literature references are completely omitted since the exemplification follows that of the Woodward–Hoffmann paper.[5]

Also, a few well-known examples of intermolecular processes are indicated in Figure 6 for comparison.[2,3] In addition, this method applies to stereoselective acyclic eliminations and additions.[3] The steric preferences of α,β-trans additions and eliminations, cis-α,γ-SN2′ reactions, α,δ-cis additions, and photoinduced α,β-cis additions are successfully explained.

In several examples of Figure 4, knowledge of a σ LUMO is needed. For such a purpose, the use of hybrid-based MO's[9] is recommended as most convenient. The nodal properties of σ MO's of conjugated molecules can be determined through such calculations.[9e]

Figures 4 and 6 exemplify two-center interactions, and Figure 7 gives examples of three-center reactions. These reactions, on the basis of the assumption of concertedness, can be treated as a combination of two two-center interactions in order to predict the direction of conversion in each bond fission or bond formation. Note that in each two-center interaction lemma A is obeyed.

Such a stepwise recognition of "concerted" reactions may at first sight seem to be self-contradictory, but it may rather be more reasonable than an attempt to understand every concerted multi-center reaction as a strictly "synchronous" change. Sometimes a bond fission may initiate the whole process, or it may happen that the conversion may start with a bond-forming

(8) (a) M. Jones, Jr., and L. T. Scott, *J. Amer. Chem. Soc.*, **89**, 150 (1967). (b) The SOMO in Figure 4o is the same as the LUMO of the ground-state butadiene.

(9) (a) K. Fukui, "Modern Quantum Chemistry. Istanbul Lectures," Part I, O. Sinanoğlu, Ed., Academic Press, New York, N. Y., 1965, p 49, and references cited therein; (b) K. Fukui, "Sigma Molecular Orbital Theory," O. Sinanoğlu and K. B. Wiberg, Ed., The Yale University Press, New Haven, Conn., 1969, p 121, and references cited therein; (c) S. Katagiri and C. Sandorfy, *Theor. Chim. Acta*, **4**, 203 (1966); (d) J. A. Pople and D. P. Santry, *Mol. Phys.*, **7**, 269; **8**, 1 (1964); (e) K. Fukui, H. Kato, T. Yonezawa, K. Morokuma, A. Imamura, and C. Nagata, *Bull. Chem. Soc. Jap.*, **35**, 38 (1962).

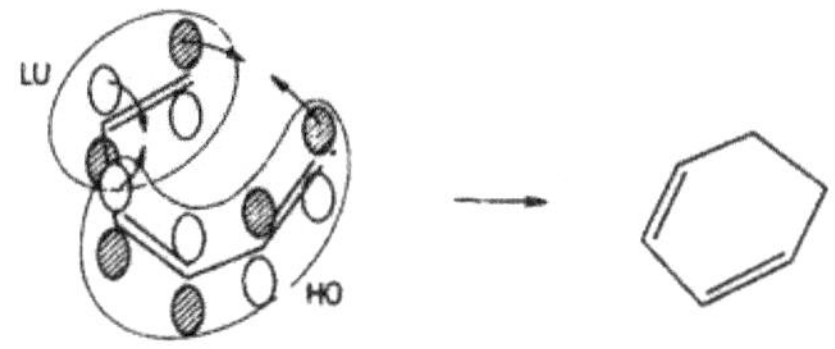

(a) Ring-closure of hexatriene (G).

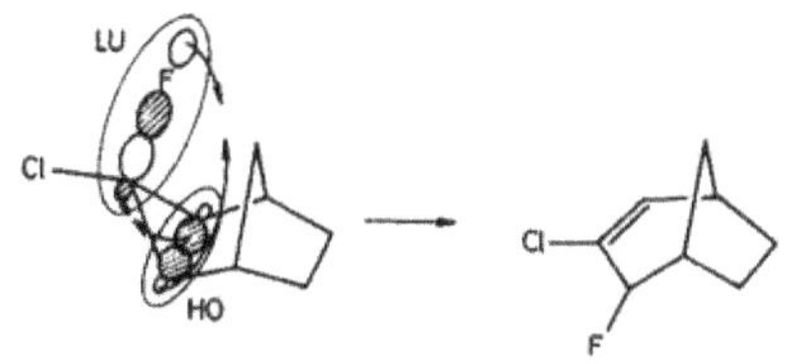

(b) Selective rearrangement of 1,1-dihalocyclopropanes (G).

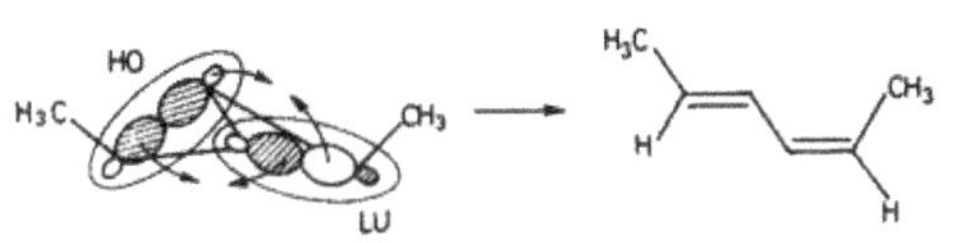

(c) Bicyclobutane rearrangement (G).

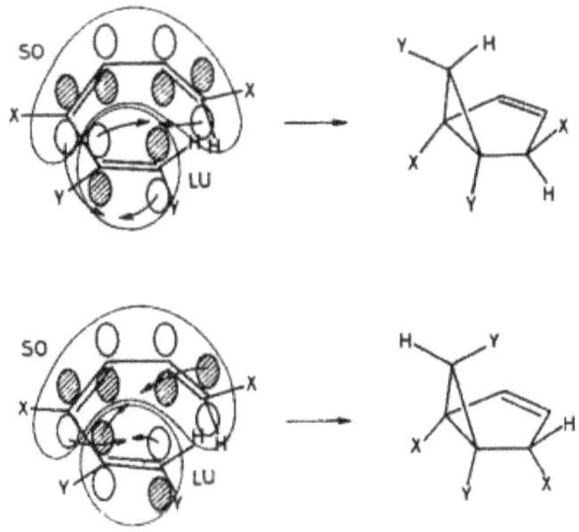

(d) Bicyclization of hexatriene (E).

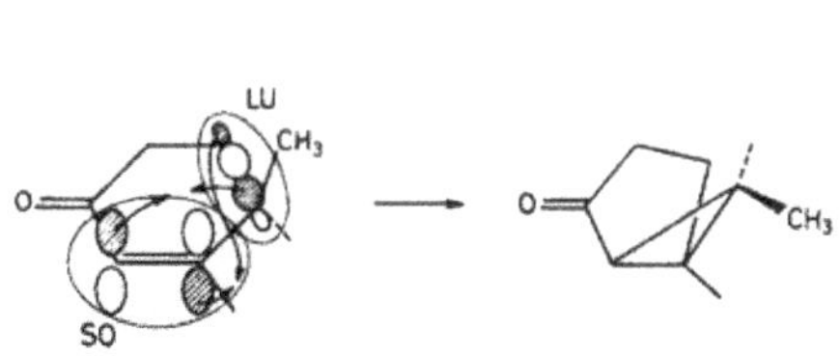

(e) Cyclization of cyclohexenone (E).

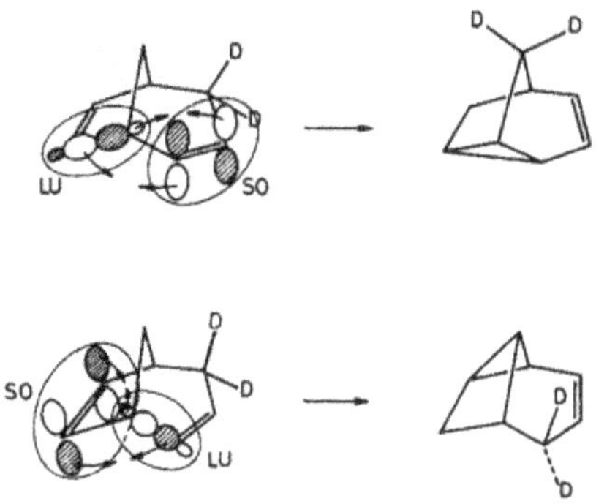

(f) Bicyclooctadiene isomerization (E).

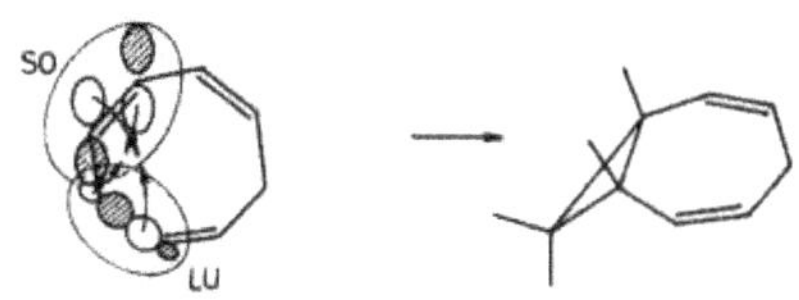

(g) Cyclization of cyclooctatriene (E).

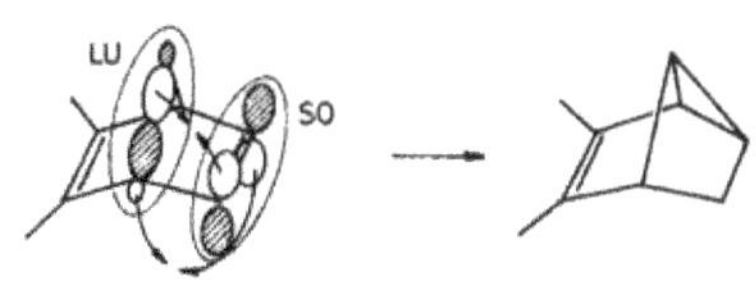

(h) Cyclization of norbornadiene (E).

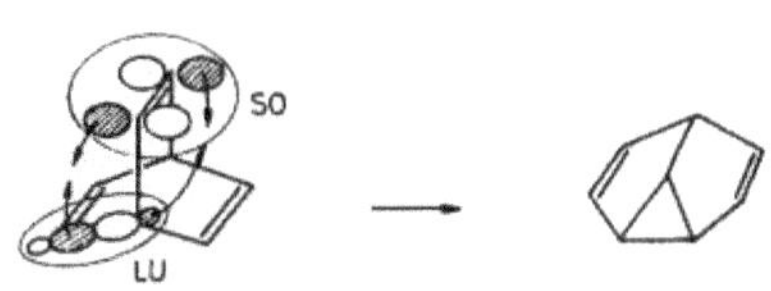

(i) Cyclization of bicyclooctatriene (E).

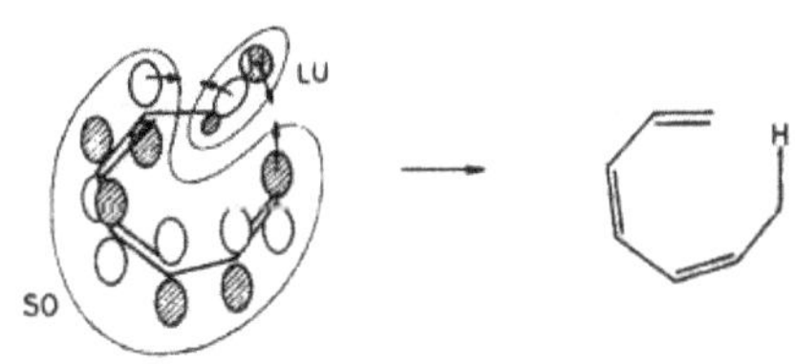

(j) 1,7-Hydrogen migration (E).

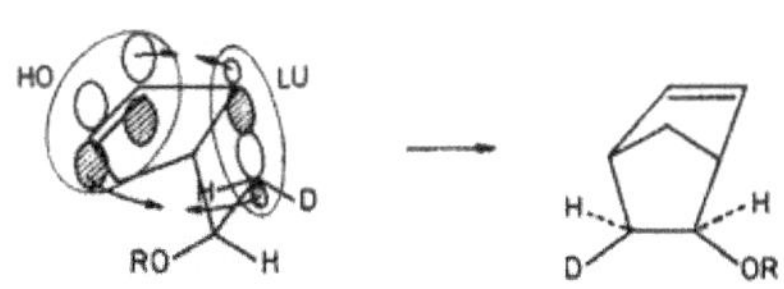

(k) Bicycloheptene rearrangement (G).

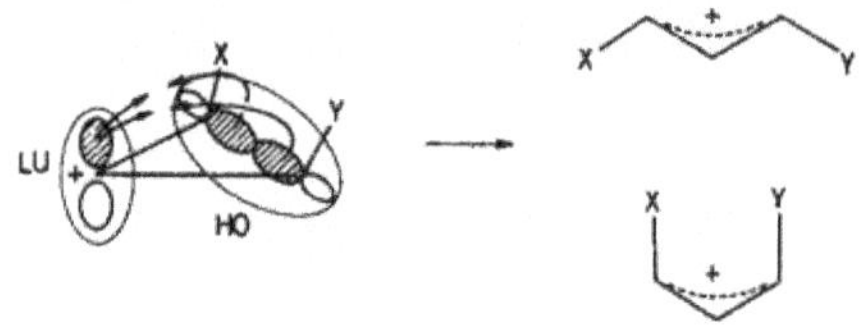

(l) Ring-opening of cyclopropyl cation (G).

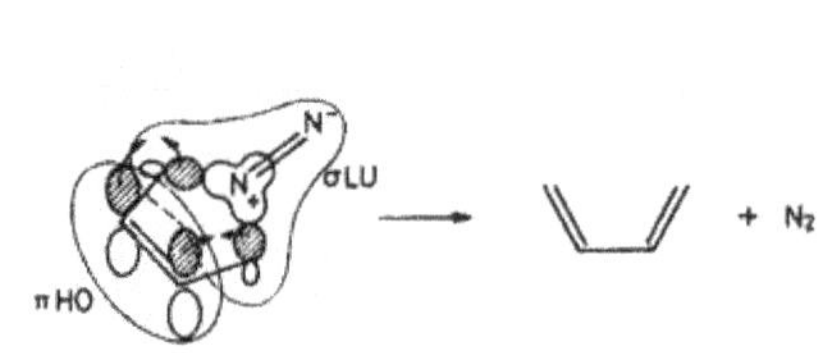

(n) Fragmentation of five-membered diazene ring (G).

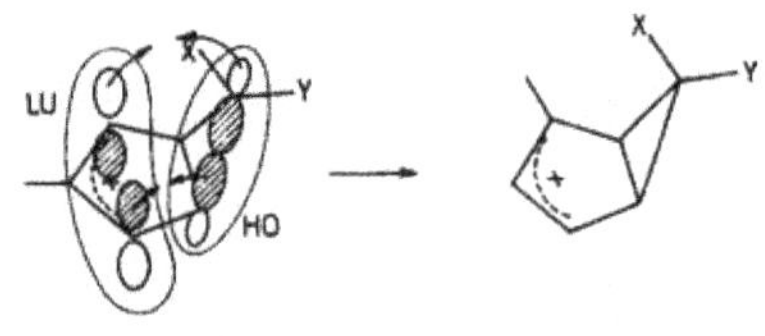

(m) Alkyl migration in bicyclohexenyl cation (G).

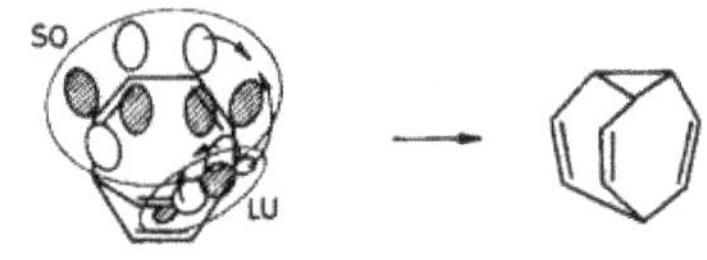

(o) Isomerization of bicyclo[4.2.2]-deca-2,4,7,9-tetraene to bullvalene (E).[8a]

Figure 4. Examples of orbital interaction involved in intramolecular stereospecific processes: (G) ground-state or thermal reaction; (E) excited-state or photoinduced reaction

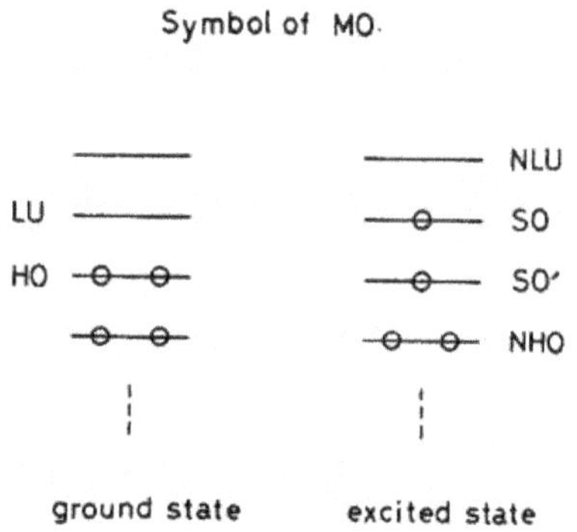

Figure 5. Symbols used to represent particular molecular orbitals: NHO, next highest occupied; NLU, next lowest unoccupied; SO, SO′, singly occupied, high lying and low lying.

interaction. We can connect the result of theoretical investigation directly with the mechanism of reaction and recognize that the driving force in the majority of reactions may be charge-transfer interaction between a certain electron-donating part and a certain electron-accepting part which are temporarily enhanced by molecular deformation through the elevation of the HOMO level and the depression of the LUMO levels of these parts. The particular circumstances of a given case can only be understood by looking into actual examples.

Extension

(a) Subsidiary Effects. As mentioned before, the HOMO–LUMO interaction scheme has been employed for interpretation of the exo–endo selectivity in Diels–Alder-type cycloadditions.[8e,5]

A rather sophisticated way of understanding the boat–chair selectivity in the Cope rearrangement, which was adopted in the explanation given by Woodward and Hoffmann,[5] can be replaced by one based on the orbital interaction scheme. The nonbonding SOMO–

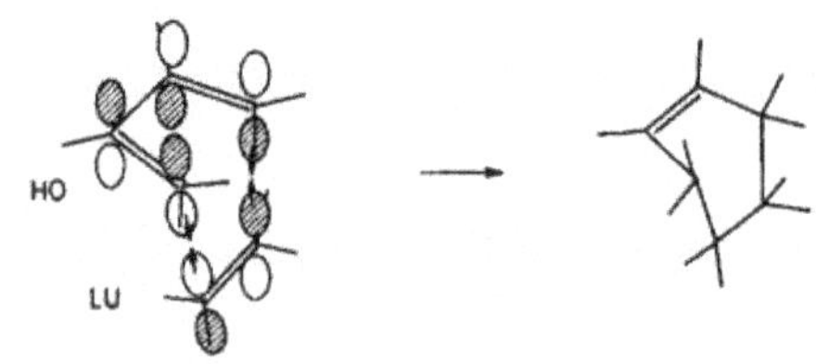

(a) Diels-Alder addition (G).

(b) Dimerization of olefins (E).

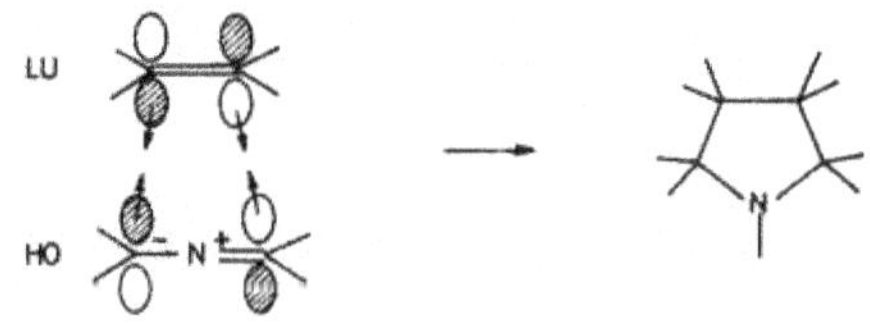

(c) 1,3-Dipolar addition (G).

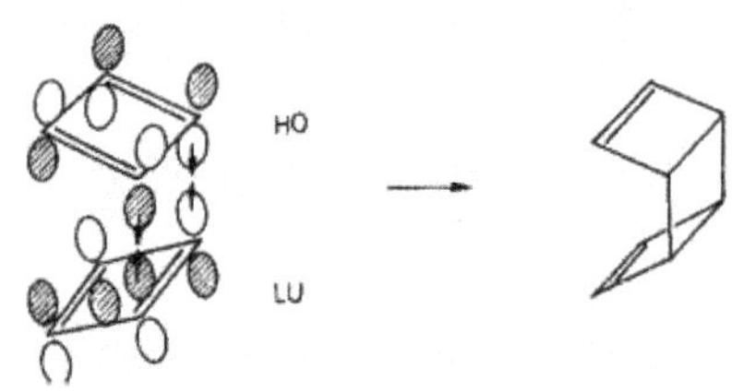

(d) Dimerization of cyclobutadiene (G).

Figure 6. Examples of intermolecular processes.

SOMO interaction has nothing to do with this selectivity since these MO's have a node at the central car-

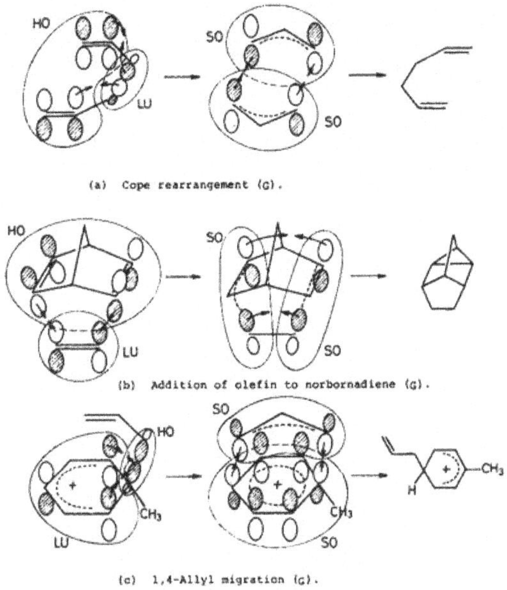

Figure 7. Examples of three-center reactions.

bon, as indicated in Figure 8. Consideration of NHO– and NLU–MO interactions leads to understanding the preference for chair-form transition states.

(b) Multiplicity Selectivity. The behavior of excited intermediates in the triplet state may be specified by their tendency to adopt a configuration in which two unpaired electrons are as far away as possible from each other, as is expected theoretically. For instance, cyclization in an excited triplet state can be specified by stating that cyclization usually takes place so as to pass through an intermediate which does not contain two unpaired electrons in one ring or in one conjugated chain, while an excited singlet-state reaction has no such restriction.[10] In this connection, the following lemma can be combined with the preceding method to predict the difference between triplet and singlet reactivities.[11–13] It is to be noted that concertedness is not necessarily expected in these cases. *In a triplet-state reaction each bond, which is to be newly formed through the overlapping of AO's of the two parts where the orbital interaction is considered, should be formed so as to result in a structure having two unpaired electrons not in conjugation with each other by virtue of that bond* (lemma B). Excited singlet-state reactions are not necessarily subject to such a limitation.

Figure 9 illustrates these circumstances. Figure 9A concerns the excited-state reaction of 1,3-pentadiene.[11,12] In the triplet-state reaction, the main product

(10) S. Kita and K. Fukui, *Bull. Chem. Soc. Jap.*, **42**, 66 (1969).
(11) (a) R. Srinivasan and S. Boué, *Tetrahedron Lett.*, 203 (1970); (b) R. Srinivasan, *J. Amer. Chem. Soc.*, **84**, 4141 (1962).
(12) S. Boué and R. Srinivasan, *ibid.*, **92**, 3226 (1970).
(13) (a) E. Baggiolini, K. Schaffner, and O. Jeger, *Chem. Commun.*, 1103 (1969); (b) J. Ipaktschi, *Tetrahedron Lett.*, 215 (1969); (c) W. G. Dauben, M. S. Kellogg, J. I. Seeman, and W. A. Spitzer, *J. Amer. Chem. Soc.*, **92**, 1786 (1970).

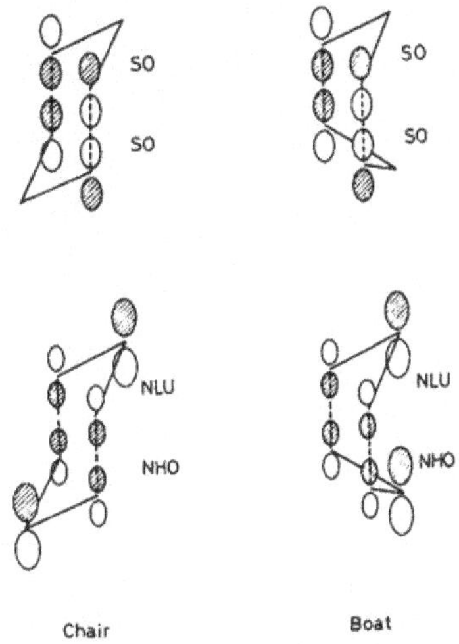

Figure 8. The boat–chair selectivity in the Cope rearrangement.

is 1,3-dimethylcyclopropene, and 4-methylcyclobutene is not found. Diagrams ii-1 and ii-2 show that these processes do not meet the requirement of lemma B.

It is intersting to look into the reason why the biradical

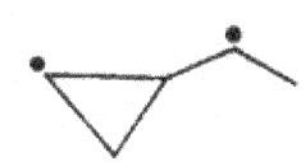

is not formed in the process of i and ii-3. In this connection one should take into account the fact that the LUMO of propylene has a large extension at the carbon atom attached to methyl. An exaggerated sketch for

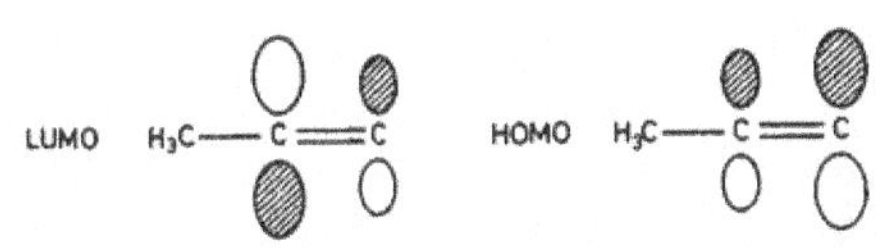

the mode of interaction may be

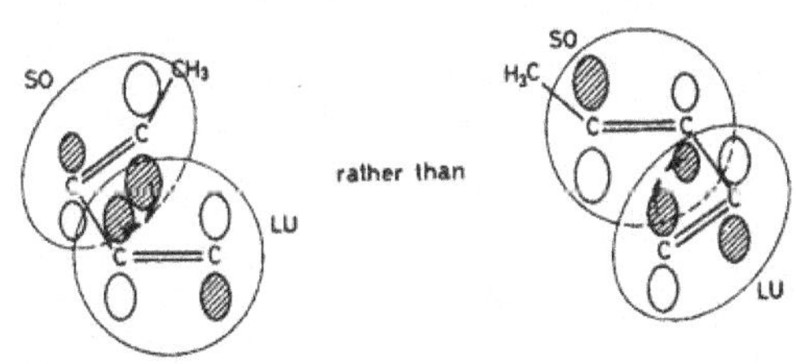

Figure 9B deals with the difference in reactivity in β,γ-unsaturated ketones. All examples[13,14] are considered to take place through the transfer of a n,π^* triplet to a π,π^* triplet. The examples of Figure 4e seem to belong to triplet-state reactions of the same category.[5] It has been established that the reaction of

(14) I am grateful to Dr. A. Kende for supplying me with experimental facts.

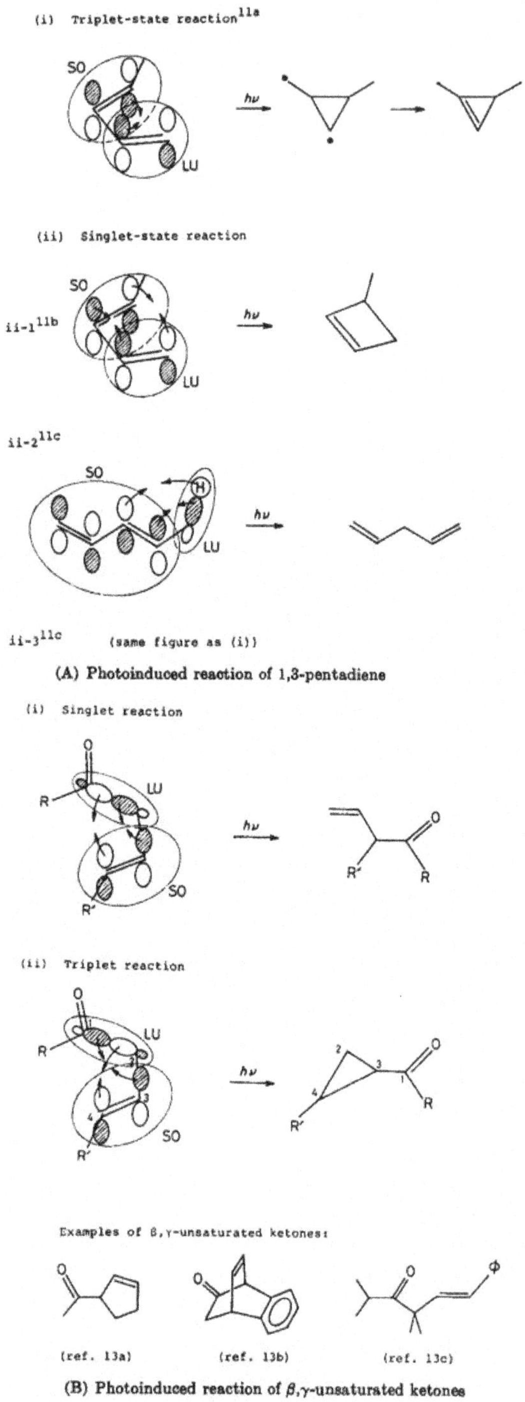

Figure 9. The singlet–triplet selectivity.

Figure 4i proceeds through a triplet state.[15] All of these cases satisfy the requirement of lemma B.

(c) Multicyclization. A useful extension of these ideas might be the prediction of the structure of prod-

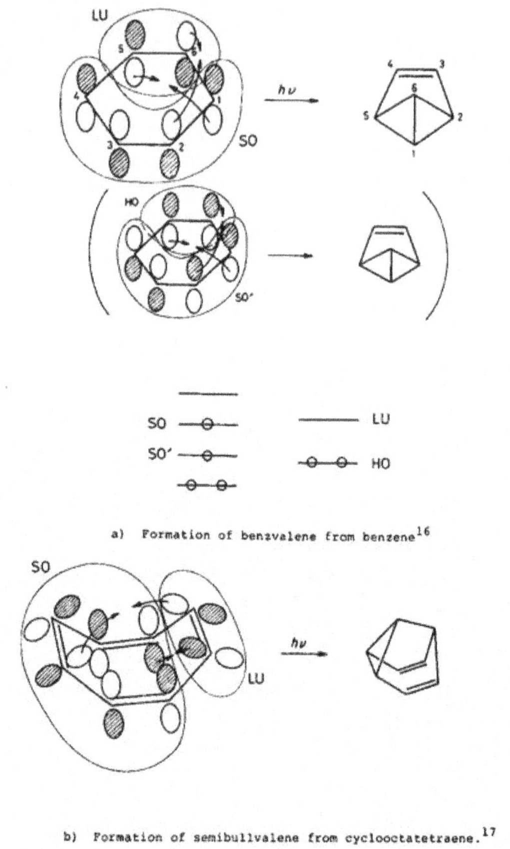

Figure 10. Examples of photoinduced multicyclization.

ucts in complicated photochemical multicyclization processes of unsaturated systems. Although it may not be possible to establish concertedness, this is never an essential point in the present discussion. Examples of some multicyclization processes are illustrated in Figure 10.

Merits

The HOMO–LUMO interaction scheme, in combination with a numerical calculation, is obviously applicable to any system with ambiguous symmetry properties. It may happen that the conjugative influence of neighboring parts exerts an influence on the orbital extension and on the nodal properties of the HOMO or LUMO, and, clearly, this method applies even in such cases and can give information on these effects.

Orbital symmetry is a most important factor in determining the magnitude of delocalization stabilization due to the HOMO–LUMO overlapping.[2,3] However, it

(15) H. E. Zimmerman, R. W. Binkley, R. S. Givens, G. L. Grunewald, and M. A. Sherwin, *J. Amer. Chem. Soc.*, **91**, 3316 (1969).

(16) K. E. Wilzbach, J. S. Ritscher, and L. Kaplan, *ibid.*, **89**, 1031 (1967).

(17) H. E. Zimmerman and H. Iwamura, *ibid.*, **90**, 4763 (1968).

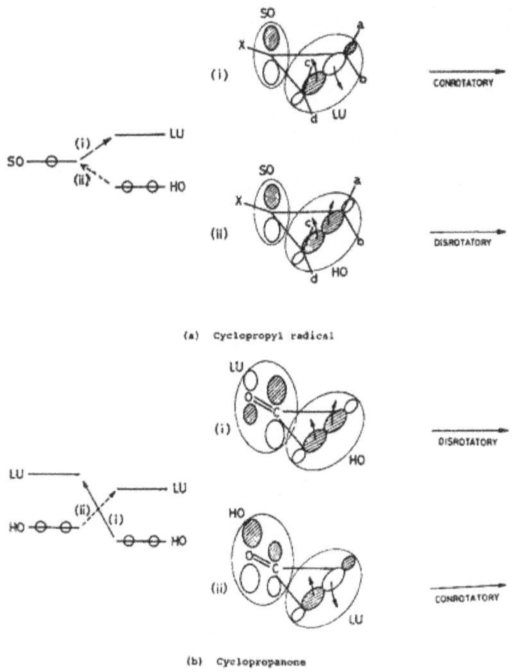

Figure 11. Continuity in the stereoselection in thermal ring-opening reactions.

should be noted that the existence of symmetry elements in orbitals is *not* essential in the recognition of favorable reaction paths. This point should be obvious from the previous discussion.

The examples shown in Figure 11 illustrate that stereospecificity is predicted to be continuous, provided that the related factors are continuous. In the ring-opening reactions of cyclopropyl radical (a), the results of prediction by a LUMO–SOMO interaction (i) and HOMO–SOMO interaction (ii) are opposite. The favorable path may be subject to substituent effects as well as electrocyclic control. Elevation of the SOMO energy will make the LUMO–SOMO interaction dominant, and lowering of the SOMO level may increase the importance of HOMO–SOMO interaction. In this way, the direction of change may vary.[18] A similar situation is expected in the ring opening of cyclopropanone (b). However, in view of the large amplitude of the LUMO at the carbonyl carbon in this instance, case i is expected to prevail.[19]

As one of the merits of the present method, it may be pointed out that the orbital interaction scheme can always give the directional features of the reaction mechanism at the point of initiation and the force which acts along the reaction path and thus drives the system to further conversion. In this connection, the use of stereomodels will facilitate the discarding of difficultly accessible steric paths when it is combined with consideration of *positive* overlap of the particular orbitals previously mentioned.

An interesting application of this method is the explanation of the catalytic action of d-orbital participation which causes an otherwise "symmetry-forbidden" reaction to become "allowed."[3d] Applications are also possible to the prediction of molecular stability and the stability of configurations of molecular species[3d] and to explanations of "spiroconjugation"[20] and "bicycloaromaticity."[21]

This method requires no selection rule and is completely automatic. Usually, only knowledge of the HOMO and LUMO of very simple systems is needed, and this makes the theory quite simple and extendable and gives it a uniform character.

Of course, reactivity in all types of unicentric reactions of saturated and unsaturated compounds can be interpreted by the HOMO–LUMO interaction method.[2,3] Problems of orientation and stereoselection can thus be discussed in a unified fashion by this approach.

So far, few contradictions have been found between predictions based on the correlation diagram method and the orbital interaction criteria, but the latter are able to treat many problems which are beyond the reach of the former method.

The understanding of complicated stereospecific organic reactions and the generalization of stereoselection rules made by Woodward and Hoffmann are unquestionably valuable. A relevant way of reasoning to a result should lead to an understanding of the underlying nature of chemical reactions. The correlation diagram method based on the principle of conservation of orbital symmetry is one way of reasoning; however, the orbital interaction scheme appears to be more unificative and wider in scope, and it gives more direct descriptions of the nature of chemical reactions.

This work was done during my stay at Illinois Institute of Technology as a National Science Foundation Senior Foreign Scientist Fellow. I am also pleased to thank Dr. S. I. Miller of Illinois Institute of Technology and Dr. H. Fujimoto of Kyoto University for helpful comments.

(18) *Cf.* H. C. Longuet-Higgins and E. W. Abrahamson, *J. Amer. Chem. Soc.*, **87**, 2045 (1965).
(19) *Cf.* the prediction made by N. J. Turro, *Accounts Chem. Res.*, **2**, 25 (1969).
(20) H. E. Simmons and T. Fukunaga, *J. Amer. Chem. Soc.*, **89**, 5208 (1967).
(21) M. J. Goldstein, *ibid.*, **89**, 6357 (1967).

Part III

Theoretical Aspects of Chemical Reactions

While the frontier orbital theory was applied to many kinds of reactions, we studied some aspects of chemical interactions between molecules. In the first paper in this part (paper no. 22), we represented the electronic structure of a system consisting of two molecules by a combination of electron configurations and partitioned the interaction energy into several terms. They represented the electrostatic interaction between charges, exchange repulsion, polarization and electron delocalization. Chemical significance of each term was examined in detail by transforming the equations derived there into the forms that could be grasped by intuition. The analysis gave clear insight into the factors that should govern reactivities of molecules and selectivities in reactions. In the second paper in this part (paper no. 23), we could disclose several factors that made the frontier MO's play distinctive role in interactions. It was shown that electron donation from the HOMO and electron acceptance into the LUMO should cause weakening of bonds between the reaction site and the neighboring atoms and, consequently, a distortion of reactant molecules. An important finding was that the distortion of reactant molecules along the reaction path should elevate the HOMO level and lower the LUMO level to reduce the HOMO-LUMO energy gap and should amplify these MO's on the reaction centers to promote further electron delocalization.

Chemical reactions are the process of formations and breakings of chemical bonds. We studied this problem by applying the orbital interaction concept (papers no. 24 and no 28). It was exciting to find that electron delocalization was the origin of the formation of new bonds along the reaction path and, accordingly, it should be the major driving force of chemical reactions. The importance of the frontier orbital interactions was demonstrated also by numerical analyses of the electronic structure of interacting systems (papers no. 25 and no. 33). Thus, together with the analyses mentioned above, we were convinced of the validity of frontier orbital theory.

The basic ideas derived from the analysis of interaction energies (paper no. 22) were extended further to many other problems, including orbital interactions associated with the polarization interaction (paper no. 27) and orbital interactions taking place between three systems (paper no. 26 and no. 30). The orbital mixing rule was derived to interpret how more than three MO's of two fragment species intermix in chemical interactions between them (paper no. 29). We could see qualitatively the way in which the frontier orbitals of a molecule were constructed from the fragment MO's.

The frontier orbital concept was applied to many other problems, e.g., ring-opening polymerization of oxiranes (paper no. 31) and interactions of compounds containing S, Se, and As (papers no. 34~no. 37).

BULLETIN OF THE CHEMICAL SOCIETY OF JAPAN VOL. 41 1989—1997 (1968)

An MO-theoretical Interpretation of the Nature of Chemical Reactions. I. Partitioning Analysis of the Interaction Energy*¹

Kenichi FUKUI and Hiroshi FUJIMOTO

Faculty of Engineering, Kyoto University, Sakyo-ku, Kyoto

(Received February 20, 1968)

The molecular interaction energy, without considering the nuclear configuration change, is discussed by partitioning it into the Coulomb, exchange, delocalization, and polarization terms. A succinct approximate expression for each term is derived, and its magnitude is discussed. The expressions obtained as the second-order perturbation terms for the delocalization and the polarization energies are in accord with those of the well-established reactivity indices, the delocalizability, and the self-atom polarizability respectively. The chemical reactivity can thus be measured by the combined sum of these four terms. In usual ionic reactions the Coulomb and the delocalization terms are more important than the exchange and the polarization terms respectively.

The molecular orbital (MO) approach can be used to interpret the chemical reactivity of organic molecules, both saturated[1] and unsaturated.[2] Most of these MO criteria for the chemical reactivity are based on the use of various reactivity indices defined as the theoretical quantities of isolated reactant molecules, each based on an appropriately-assumed model for the chemical reaction. The classification of these indices has tentatively been made by dividing them into three groups, the static, the localization, and the delocalization approaches.[3] The comparison of these approaches appeared promising for acquiring knowledge of the reaction mechanism.[2] However, the semiempirical character of calculation made an absolute comparison of the various methods difficult. In addition, as a result of the existence of several theoretical correlations among the reactivity indices,[4] no unequivocal conclusion has yet been obtained. Although the formal introduction of

*1 The basic idea of this paper was presented by one of the present authors (K.F.) at the International Summer School of Quantum Chemistry, Istanbul, August, 1964, while a skeletal presentation was made in K. Fukui, *Kogyo Kagaku Zasshi* (*J. Chem. Soc. Japan, Ind. Chem. Sect.*), **69**, 794 (1966).

1) K. Fukui, "Modern Quantum Chemistry," Vol I, ed. by O. Sinanoğlu, Academic Press, New York (1965), p. 49, and the references cited therein.

2) A. Streitwieser, Jr., "Molecular Orbital Theory for Organic Chemists," John Wiley & Sons, New York (1961), Chapter 11, and the references cited therein.

3) K. Fukui, "Molecular Orbitals in Chemistry, Physics and Biology," ed. by P.-O. Löwdin and B. Pullman, Academic Press, New York (1964), p. 513.

4) K. Fukui, T. Yonezawa and C. Nagata, *J. Chem. Phys.*, **26**, 831 (1957), and the references cited therein.

1990 Kenichi FUKUI and Hiroshi FUJIMOTO [Vol. 41, No. 9

the "transition-state" concept in the absolute-rate theory has doubtless been illuminating, the basic principle of the activated complex formation is not yet clear. The nature of chemical reactions, therefore, still remains obscure in many details from the quantum-mechanical point of view.

It is the aim of the present approach to make a contribution to this problem through a formulation of the energy of chemical interaction, to derive at the same time a prototype of several reactivity indices in a unified manner, and also to discuss the characteristic features of the reaction path on the potential energy surface. The method is based chiefly on the Roothaan-type linear combination of atomic orbitals (LCAO) MO approach, with some use of the configuration interaction (CI) procedure.[5]

Electron Delocalization in the Chemical Interaction. Since 1933[6] several theories were proposed one after another to explain the chemical reactivity of conjugated molecules along the static[6-8] and the localization approaches.[9] In 1952 the "frontier electron" concept[10] was put forth; it pointed out that the electrophilic aromatic substitution should take place at the carbon atom of the largest density of the highest energy electrons. Similarly, the nucleophilic replacement was later shown to occur at the position of the largest density of the lowest unoccupied MO. In comparison with the theoretically well-established character of the other reactivity theories then existing, the lack of a solid physical foundation for the infant frontier electron theory suggested instead the possibility of the involvement of a new principle for the mode of chemical interaction.

It was coincidentally in the same year that the elegant theory of the charge-transfer interaction was developed by Mulliken with regard to the molecular complex formation between a donor and an acceptor.[11] In this connection, the "overlap and orientation" principle was proposed;[12] it considered only the overlap interaction between the highest energy occupied MO of the donor and the lowest energy unoccupied MO of the acceptor.

Since the behavior of the frontier electrons was attributed to some type of electron delocalization between the reactant and the reagent,[13] it was natural for the charge-transfer mechanism and the overlap and orientation principle to be connected to the problem of ionic reactions. Brown combined the charge-transfer theory with the frontier electron concept to derive a consistent theory of chemical reactivity.[14] All of these treatments[10-15] together may be termed the "delocalization approach" to chemical reactivity theory.

Applying the donor and acceptor concept to the substitution reactions in conjugated molecules, the electrophilic and the nucleophilic reactions were discussed from a unified point of view. Thus, the highest occupied (HO) MO of the donor molecule and the lowest unoccupied (LU) MO of the acceptor molecule were considered to behave as frontier orbitals in ionic reactions,[10] not only in substitutions but also in additions, and even in multicentric reactions.[3] When the overlap and orientation principle was applied to odd-electron molecules or radicals, the half-occupied nonbonding (NB) MO was responsible for the reactive interaction. A similar behavior of these particular MO's has also been pointed out in nonplanar saturated molecules.[1] In addition, the conspicuous property of HO and LU in determining the steric course of organic reactions has been discovered and has attracted much attention.[8,16]

In this view, it may be of interest to discuss the energy of reacting systems by dividing it into various kinds of interactions. Such a treatment will be useful in helping us to understand what factor is really important in the chemical reaction.

Interaction with No Nuclear Configuration Change

The energy of the interaction between two independent systems, in which, tentatively, no nuclear configuration change is considered, is defined by the following equation:

5) C. C. J. Roothaan, *Revs. Mod. Phys.*, **23**, 69 (1951). Longuet-Higgins and Murrell applied the concept of configuration interaction in discussing the electronic spectra of composite alternant hydrocarbons. They introduced the idea of polarization and charge-transfer interaction between the two fragments of a composite molecule. H. C. Longuet-Higgins and J. N. Murrell, *Proc. Phy. Soc.*, **A68**, 601 (1955); J. N. Murrell, *ibid.*, **A68**, 969 (1955).

6) L. Pauling and G. W. Wheland, *J. Chem. Phys.*, **1**, 362 (1933).

7) T. Ri and H. Eyring, *ibid.*, **8**, 433 (1940); T. Ri, *Rev. Phys. Chem. Japan*, **17**, 1, 16 (1943); A. Pullman and B. Pullman, *Experientia*, **2**, 364 (1946); M. J. S. Dewar, *Trans. Faraday Soc.*, **42**, 764 (1946).

8) a) C. A. Coulson, *Discussions Faraday Soc.*, **2**, 9 (1947); *J. chim. phys.*, **45**, 243 (1948); b) C. A. Coulson and H. C. Longuet-Higgins, *Proc. Roy. Soc.* (*London*) **A191**, 39; **A192**, 16 (1947); other papers cited in Ref. 3.

9) G. W. Wheland, *J. Am. Chem. Soc.*, **64**, 900 (1942); M. J. S. Dewar, *ibid.*, **74**, 3357 (1952).

10) K. Fukui, T. Yonezawa and H. Shingu, *J. Chem. Phys.*, **20**, 722 (1952); see also the subsequent papers: K. Fukui, T. Yonezawa, C. Nagata and H. Shingu, *ibid.*, **22**, 1433 (1954), and the papers cited in Ref. 3.

11) R. S. Mulliken *J. Am. Chem. Soc.*, **74**, 811 (1952).

12) R. S. Mulliken, *Rec. Trav. Chim.*, **75**, 845 (1956).

13) K. Fukui, T. Yonezawa and C. Nagata, This Bulletin, **27**, 423 (1954); K. Fukui, T. Yonezawa and C. Nagata, *J. Chem. Phys.*, **27**, 1247 (1957).

14) R. D. Brown, *J. Chem. Soc.*, **1959**, 2232; *cf.* K. Fukui, K. Morokuma, T. Yonezawa and C. Nagata, *J. Chem. Phys.*, **32**, 1743 (1960).

15) S. Nagakura and J. Tanaka, *Nippon Kagaku Zasshi* (*J. Chem. Soc. Japan, Pure Chem. Sect.*), **75**, 933 (1954); T. Yonezawa, K. Hayashi, C. Nagata, S. Okamura and K. Fukui, *J. Polymer Sci.*, **14**, 312 (1954), and other, related papers cited in Refs. 2 and 3.

16) R. B. Woodward and R. Hoffmann, *J. Am. Chem. Soc.*, **87**, 395, 2511 (1965).

$$\Delta W = W - (W_{A0} + W_{B0}) \tag{1}$$

where ΔW is the interaction energy, W_{A0} and W_{B0} being the energies of the initial stationary states of two isolated systems, A and B, and W, the lowest total energy of the two mutually-interacting systems. The energies of all the systems were calculated in the frame of the Born-Oppenheimer approximation[17] and obtained in the form of the adiabatic potential.

The Hamiltonian operator of the combined system in atomic units is:

$$H = \sum_{\lambda} H(\lambda) + \sum_{\lambda<\lambda'} \frac{1}{r_{\lambda\lambda'}} + \sum_{\gamma<\gamma'} \frac{Z_\gamma Z_{\gamma'}}{R_{\gamma\gamma'}}$$

$$H(\lambda) = -\frac{1}{2}\Delta(\lambda) + V(\lambda)$$

$$V(\lambda) = V_A(\lambda) + V_B(\lambda)$$

$$V_A(\lambda) = -\sum_{\alpha} \frac{Z_\alpha}{r_{\lambda\alpha}}, \quad V_B(\lambda) = -\sum_{\beta} \frac{Z_\beta}{r_{\lambda\beta}} \tag{2}$$

in which:

$H(\lambda)$:	One-electron Hamiltonian operator of the electron λ.
$Z_\alpha, Z_\beta, Z_\gamma$:	The positive charges of the nuclei α, β, and γ, which specify their belonging to the system A, the system B, and the combined system, AB, respectively.
$r_{\lambda\lambda'}$:	The distance between the two electrons λ and λ'.
$r_{\lambda\alpha}$, *etc.*:	The distance of the electron λ from the nucleus α, *etc.*
$R_{\alpha\alpha'}$, *etc.*:	The distance between the nuclei α and α', *etc.*

The unperturbed normalized wave-functions of the A and B systems are assumed to be represented in terms of Slater determinants composed of orthonormal spin-orbitals with LCAO MO spatial functions, which are Roothaan-type SCF (self-consistent-field) MO's with respect to the *ground state* of each isolated system.[18] The wave-functions of an excited or charge-transferred configuration may be made up of the MO's associated with the realistic and non-realistic roots of the Fock equation for the ground state. Therefore, all of the MO's are definite and fixed, with a given nuclear configuration in each isolated system.

The spatial parts of LCAO MO are represented by:

$$a(1) = \sum_t C_t t(1) \quad \text{for the system } A \tag{3a}$$

and

$$b(1) = \sum_u C_u u(1) \quad \text{for the system } B \tag{3b}$$

where t and u are the AO's belonging to the nuclei of A and B respectively. No particular restriction is placed on the type of these AO's. Usually they may be taken to be Slater-type or Gaussian AO's. However, it is often convenient for the later discussion of reactivity criteria to employ the valence atomic orbitals (VAO's).[19] Our present purpose allows us to take all functions as real.

The occupied and unoccupied MO's in the ground state are distinguished in the following way:[20]

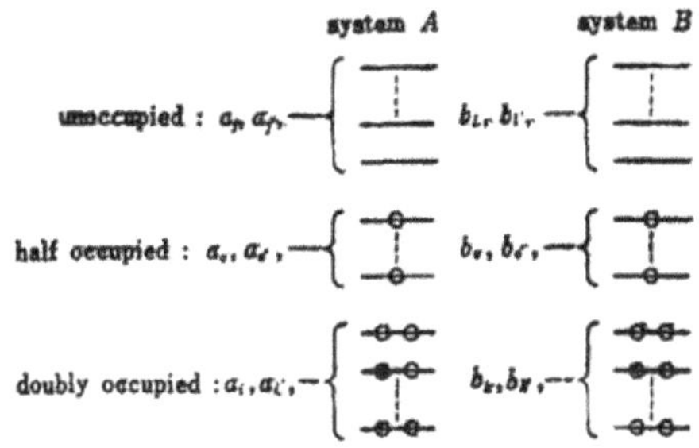

The configurations to be taken into account are as follows:

The electron configuration of the combined system A---B

Zero-configuration	(mono-)excited configuration with regard to zero-configuration	(mono-)transferred configuration with regard to zero-configuration

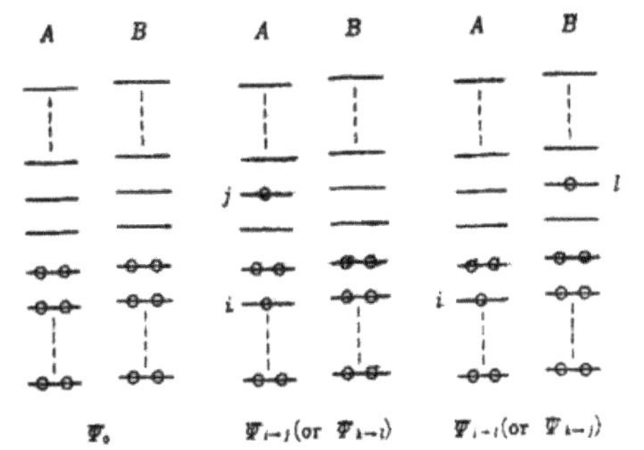

The perturbed normalized wave-functions of the combined-system Ψ_J's are constructed by means of

17) M. Born and J. R. Oppenheimer, *Ann. Physik*, **84**, 457 (1927).

18) Either A and B, or both, may be an odd-electron system. For such a system the "restricted" open-shell SCF MO theory (*e.g.*, see G. Berthier, "Molecular Orbitals in Chemistry, Physics, and Biology," ed. by P.-O. Löwdin and B. Pullman, Academic Press, New York (1964), p. 57) should be applied to compose the wave-function. A calculation similar to the present one was early made by Mori (Y. Mori, This Bulletin, **35**, 1584 (1962)) with respect to the interaction between a molecule and a radical.

19) K. Ruedenberg, *Rev. Mod. Phys.*, **34**, 326 (1962); compare R. McWeeny, *ibid.*, **32**, 335 (1960).

20) For the presentation of an "unoccupied" MO to interact with another molecule, the Roothaan LCAO MO approach, with a basis set of a limited number of AO's chosen so as to be suitable for describing the occupied MO's is not necessarily appropriate.

the Slater determinants, composed of a_i's, a_j's, b_k's, and b_l's according to each electron configuration. The subscript J stands for all configurations, 0, $i\to j$, $k\to l$, $i\to l$, $k\to j$, and other highly-excited or transferred ones, such as $(i\to j,\ i'\to j')$, $(i\to j,\ i'\to l)$, $(i\to l,\ i'\to l')$, and so on.[21] The wave-function Ψ_J may further be specified by the superscript $(2S+1)$ and the subscript M_S as represented by ${}^{2S+1}_{M_S}\Psi_J$, where S and M_S designate the usual total spin quantum numbers. For instance, ${}^{1}_{0}\Psi_{i\to l}$ is represented by:

$$\frac{1}{\sqrt{2}}[NA\,|a_1\bar{a}_1a_2\bar{a}_2\cdots a_{i-1}\bar{a}_{i-1}a_i a_{i+1}\bar{a}_{i+1}\cdots a_n\bar{a}_n b_1\bar{b}_1\cdots b_m\bar{b}_m\bar{b}_l| + NA\,|a_1\bar{a}_1a_2\bar{a}_2\cdots a_{i-1}\bar{a}_{i-1}\bar{a}_i a_{i+1}\bar{a}_{i+1}\cdots a_n\bar{a}_n b_1\bar{b}_1\cdots b_m\bar{b}_m b_l|]$$

where NA signifies the normalization-antisymmetrization operator; the nonbarred and barred MO's represent the spin-orbitals with spin-functions of α and β respectively, and n and m indicate the number of doubly-occupied MO's of A and B respectively in the zero-configuration.

The secular determinant is represented by:

$$\begin{vmatrix} H_{I,I}-E & H_{I,J}-S_{I,J}E\cdots \\ H_{J,I}-S_{J,I}E & H_{J,J}-E\cdots \\ \cdots\cdots & \end{vmatrix} = 0 \quad (4)$$

where the subscript I signifies the initial J, and where:

$$H_{J,J'} = \int \Psi_J^* H \Psi_{J'} d\tau$$

$$S_{J,J'} = \int \Psi_J^* \Psi_{J'} d\tau$$

The orders of magnitude of the overlap integrals are estimated as:

$$S_{0,i\to l} = S_{0,k\to j} = O(s_{ab})$$

$$S_{0,i\to j} = S_{0,k\to l} = S_{i\to l,\,k\to j} = S_{i\to j,\,k\to l} = O(s_{ab}^2)$$

$$S_{i\to j,i'\to l} = S_{i\to j,k\to j'} = S_{i\to l,k\to l'} = S_{k\to j,k'\to l} = O\ (s_{ab}^3)$$

$$S_{i\to j,i'\to j'} = S_{k\to l,k'\to l'} = O\ (s_{ab}^4)$$

$$S_{i\to j,(i\to j,i'\to l)} = S_{i\to j,(i\to j,k\to j')} = S_{i\to j,i\to l} = S_{i\to j,k\to j} = O(s_{ab})$$

$$S_{i\to j,(i\to j,i'\to j')} = S_{i\to j,(i\to j,k\to l)} = S_{i\to j,i'\to j} = O(s_{ab}^2)$$

and so forth, where s_{ab} represents an overlap integral of an MO of the system A and an MO of the system B. The general relation is given by:

$$S_{J,J'} = O\ (s_{ab}^{\nu}) \quad (5)$$

in which ν is the minimum number of electron transfers between A and B required to shift the electron configuration from J to J'.

In order to calculate W in Eq. (1), which is given by the lowest root of the secular equation (4), it is assumed that the lowest $H_{J,J}$ is $H_{I,I}$ and that no others are equal to $H_{I,I}$.[22] On the basis of the second assumption, that $S_{J,J'}$ and $H_{J,J'}$ $(J\neq J')$ are small, W can be expanded in approximately the following form:

$$W = H_{I,I} - \left(\sum_p^{\text{monotr.}} + \sum_p^{\text{monoex.}} + \sum_p^{\text{ditr.}} + \sum_p^{\substack{\text{monotr.-}\\ \text{monoex.}}} + \sum_p^{\text{diex.}} + \cdots\right)\frac{|H_{I,p} - S_{I,p}H_{I,I}|^2}{H_{p,p} - H_{I,I}} + \cdots \quad (6)$$

where monotr., diex., *etc.* imply monotransferred, diexcited, *etc.* configurations with respect to the initial configuration, I.[23] The third term $(+\cdots)$ designates the remaining terms, which cover the difference from the exact root of Eq. (4).

If it is assumed[12] that the interaction between A and B is not yet very strong and that $H_{I,p}$ is almost linear with $S_{I,p}$, it follows that, in light of Eq. (5), the terms from monotransferred configurations will possess magnitudes of the order of s_{ab}^2, and that the monoex. and the ditr. terms are of s_{ab}^4, the monotr.-monoex. term is of s_{ab}^6, and so on. In addition to this, if SCF MO's are employed, the well-known Brillouin theorem[24] makes the $H_{I,p}$ values of monoex. terms small. Moreover, the denominator of the second term, $H_{p,p}-H_{I,I}$, can usually not be small in excited configuration terms, while in transferred configuration terms it can be, and sometimes, as in Eq. (7'), even a first-order term appears in place of the second-order transferred configuration terms of Eq. (6). In these circumstances the monotransferred term will become uniquely important.

21) The wave-functions for these electron configurations should be regarded as those adopted to describe the perturbed wave-function in terms of that of the isolated reactant molecules, and not necessarily as corresponding to any realistic " excited " or " transferred " states.

22) This assumption reflects the adiabaticity requirement for the interaction. There are two other possibilities. In one the lowest $H_{J,J}$ is $H_{I,I}$ and the other $pH_{J,J}$'s $(H_{1,1}, H_{2,2}\cdots, H_{q,q}\cdots, H_{p,p})$ are equal to $H_{I,I}$, where the suffix q stands for one of the monotransferred configurations with reference to I. In this case the equation corresponding to Eq. (7) becomes:

$$W = H_{I,I} - \left\{\sum_{q=1}^{p}|H_{I,q} - S_{I,q}H_{I,I}|^2\right\}^{1/2} + \cdots \quad (7')$$

The other is the case in which the lowest $H_{J,J}$ is not $H_{I,I}$ but $pH_{C,C}$'s $(H_{C1,C1}, H_{C2,C2}\cdots, H_{Cq,Cq}\cdots, H_{Cp,Cp}\cdots)$ of an equal energy, where C_q stands for one of the monotransferred configurations of I. In this case

$$W = H_{C,C} - \frac{\sum_{q=1}^{p}|H_{I,C_q} - S_{I,C_q}H_{C,C}|^2}{H_{I,I} - H_{C,C}} + \cdots \quad (7'')$$

23) When the initial configuration is an excited one, the term " excited " configuration should include "di"-excited configurations, too.

24) L. Brillouin, *Actualités sci. et ind.*, **71** (1933); **159** (1934).

Thus, if we take only the terms not smaller than s_{ab}^2, namely, the monotransferred terms in the second term of Eq. (6), we get:

$$W = H_{I,I} - \sum_{p=1}^{N} \frac{|H_{I,p} - S_{I,p}H_{I,I}|^2}{H_{p,p} - H_{I,I}} \tag{7}$$

where p stands for the pth monotransferred configuration with reference to I, and where N is the number of monotransferred configurations.

Combined Reactivity Index

The interaction energy, ΔW, in Eq. (1), appropriately calculated with respect to a given nuclear configuration, will serve as a reasonable measure of the stability of that configuration. A comparison of ΔW with the various nuclear configurations which are possible in a given chemical interaction will make it possible to discuss the preferred orientation theoretically. Thus, ΔW can be a general scale of reactivity. The singlet interaction energy,[25] ΔW, of Eq. (1), which has reference to the initial state, with a zero-configuration composed of two even-electron ground-state systems, is written by Eq. (6) as:

$$\Delta W = \varepsilon_0 - D - \Pi \tag{8}$$

where:

$$\varepsilon_0 = H_{0,0} - (W_{A0} + W_{B0}) \tag{8a}$$

$$D = \sum_i^{\text{occ}} \sum_l^{\text{uno}} \frac{|H_{0,i\to l} - S_{0,i\to l}H_{0,0}|^2}{H_{i\to l,i\to l} - H_{0,0}} + \sum_k^{\text{occ}} \sum_j^{\text{uno}} \frac{|H_{0,k\to j} - S_{0,k\to j}H_{0,0}|^2}{H_{k\to j,k\to j} - H_{0,0}} \tag{8b}$$

$$\Pi = \sum_i^{\text{occ}} \sum_j^{\text{uno}} \frac{|H_{0,i\to j} - S_{0,i\to j}H_{0,0}|^2}{H_{i\to j,i\to j} - H_{0,0}} + \sum_k^{\text{occ}} \sum_l^{\text{uno}} \frac{|H_{0,k\to l} - S_{0,k\to l}H_{0,0}|^2}{H_{k\to l,k\to l} - H_{0,0}} \tag{8c}$$

Coulomb and Exchange Interaction Terms. Among the three terms on the right side of Eq. (8), ε_0 is a first order term which arises from the Coulomb and the exchange interaction terms between A and B: it can be calculated easily as:

$$\varepsilon_0 = \varepsilon_Q + \varepsilon_K \tag{9}$$

where the Coulomb interaction energy, ε_Q, is:

$$\varepsilon_Q = 2\sum_i V_{Bii} + 2\sum_k V_{Akk} + 4\sum_i \sum_k (ii|kk) + \sum_\alpha \sum_\beta \frac{Z_\alpha Z_\beta}{R_{\alpha\beta}} \tag{10}$$

and where the exchange interaction energy, ε_K, is:

$$\varepsilon_K = -2\sum_i \sum_k [(ik|ki) + s_{ik}[V_{ik} + \sum_{i'} \{2(ik|i'i') - (ii'|i'k)\} + \sum_{k'} \{2(ik|k'k') - (ik'|k'k)\}]] \tag{11}$$

in which:

$$V_{Bii} = \int a_i(1)V_B(1)a_i(1)\mathrm{d}v(1), \ etc.$$

$$V_{ik} = \int a_i(1)V(1)b_k(1)\mathrm{d}v(1), \ etc.$$

$$(ij|kl) = \int a_i(1)b_k(2)\frac{1}{r_{12}}a_j(1)b_l(2)\mathrm{d}v(1)\mathrm{d}v(2), \ etc.$$

$$s_{ik} = \int a_i(1)b_k(1)\mathrm{d}v(1), \ etc.$$

If we here employ the Mulliken approximation:[26]

$$\int\cdots\int t(1)t'(1)f\cdots\mathrm{d}v(1)\cdots \cong \frac{1}{2}s_{tt'}\left\{\int\cdots\int t(1)t(1)f\cdots\mathrm{d}v(1)\cdots + \int\cdots\int t'(1)t'(1)f\cdots\mathrm{d}v(1)\cdots\right\}$$

we can easily obtain:[27]

$$\varepsilon_Q \cong \sum_t N_{At}v_{Btt} + \sum_u N_{Bu}v_{Auu} + \sum_t \sum_u N_{At}N_{Bu}(tt|uu) + \sum_\alpha \sum_\beta \frac{Z_\alpha Z_\beta}{R_{\alpha\beta}} \tag{12}$$

25) The term "singlet interaction" signifies an interaction of two systems, A and B, with a conservation of the total spin zero. Similarly, "doublet interaction" and "triplet interaction" will later be used in the cases of the interaction of a radical with a molecule or with a radical.

26) R. S. Mulliken, *J. chim. phys.*, **46**, 497 (1949).

27) In the doublet interaction between an even-electron system (A) and an odd-electron system (B), a similar equation is obtained:

$$\varepsilon_Q = 2\sum_i V_{Bii} + 2\sum_k V_{Akk} + V_{Ao'o'} + 4\sum_i \sum_k (ii|kk) + 2\sum_i (ii|o'o') + \sum_\alpha \sum_\beta \frac{Z_\alpha Z_\beta}{R_{\alpha\beta}} \cong \sum_\alpha \sum_\beta \frac{(Z_\alpha - N_\alpha)(Z_\beta - N_\beta)}{R_{\alpha\beta}} \tag{12'}$$

The singlet and the triplet interactions between two odd-electron systems give an analogous relation.

where N_{At} is the population of the tth AO of the isolated system, A, and is given by:

$$N_{At} = 2\sum_{i}^{\text{occ}}\sum_{t'} C_t^{(i)} C_{t'}^{(i)} s_{tt'}$$

and:

$$v_{Btt} = \int t(1) V_B(1) t(1) \mathrm{d}v(1) = \sum_{\beta} v_{\beta tt}, \textit{ etc.},$$

in which $C_t^{(i)}$ means the coefficient of Eq. (3a) for MO a_i, and in which

$$s_{tt'} = \int t(1) t'(1) \mathrm{d}v(1)$$

$$v_{\beta tt} = -\int t(1) \frac{Z_\beta}{r_{1\beta}} t(1) \mathrm{d}v(1), \textit{ etc.}$$

Further approximation which puts

$$v_{\beta tt} \cong -\frac{Z_\beta}{R_{\alpha\beta}}$$

etc., where α denotes the nucleus to which the tth AO belongs, may be possible if the distance between A and B is not very short. This approximation at once gives rise to the succinct expression:

$$\varepsilon_Q \cong \sum_{\alpha}\sum_{\beta} \frac{(Z_\alpha - N_\alpha)(Z_\beta - N_\beta)}{R_{\alpha\beta}} \tag{13}$$

where N_α is the population of the atom α, that is, it is equal to $\sum_t^{(\alpha)} N_{At}$.[28] This equation is very convenient for grasping the meaning of the Coulomb interaction term, since $(Z_\alpha - N_\alpha)$ and $(Z_\beta - N_\beta)$ represent the net plus charge of the atom α of A and of the atom β of B respectively. In neutral nonpolar molecules, therefore, ε_Q is not very important.

The exchange interaction term, ε_K, is almost solely responsible for the multiplet separation. In the ground-state interactions it is usually repulsive except for the singlet interaction of two ground-state odd-electron systems.[29]

The case of single-site overlapping, namely, when overlapping between only one AO (say r) of A and one AO (say r') of B is appreciable, may often be important. This is nearly the case in the aromatic substitution by a reagent with essentially one AO. If VAO's are employed as the basis AO's, the MO overlapping with reagent orbitals becomes single-sited in many cases.

In the case of single-site overlapping, Eq. (11) becomes, by the use of the Mulliken approximation:

$$\begin{aligned}\varepsilon_K \cong -\frac{1}{8} s_{rr'}^2 P_{rr} P_{r'r'} [&2v_{rr} + 2v_{r'r'} + (rr|rr) \\ &+ (r'r'|r'r') + 2(rr|r'r') + P_{rr}\{(rr|rr) \\ &+ (rr|r'r')\} + P_{r'r'}\{(r'r'|r'r') + (rr|r'r')\}]\end{aligned} \tag{14}$$

in which:

$$v_{rr} = v_{Arr} + v_{Brr}, \textit{ etc.},$$

$$P_{rr} = 2\sum_{i}^{\text{occ}} C_r^{(i)} C_r^{(i)}, \textit{ etc.}$$

indicating that the exchange interaction varies with the square of the overlap integrals and is, accordingly, less important than the Coulomb interaction at the initial stage of interaction in polar reactions.

Charge-transfer Interaction Term. The second term appearing on the right side of Eq. (8) comes from the electron-delocalization interaction due to the mixing of monotransferred configurations. The calculation of the matrix elements and overlap integrals gives:

$$\begin{aligned}H_{i\to l, i\to l} - H_{0,0} = &-(E_{Bl} + \Delta E_{Bl}) \\ &+ (I_{Ai} + \Delta I_{Ai}) + a_{il}\end{aligned} \tag{15}$$

in which:[30]

$$E_{Bl} = -[H_{Bll} + \sum_k \{2(ll|kk) - (lk|kl)\}]$$

$$I_{Ai} = -[H_{Aii} + \sum_{i'} \{2(ii|i'i') - (ii'|i'i)\}]$$

---

28) If we put $s_{tt'}$'s and $s_{uu'}$'s equal to zero, Eq. (13) becomes identical to the equation obtained by Rein and Pollak (R. Rein and M. Pollak, *J. Chem. Phys.*, **47**, 2039 (1967)). They also assumed $s_{ab}=0$ in obtaining the exchange interaction energy. See also G. Klopman and R. F. Hudson, *Theoret. chim. Acta* (*Berl.*), **8**, 165 (1967).

29) In this case

$$\begin{aligned}\varepsilon_K = &-2\sum_i\sum_k [(ik|ki) + s_{ik}[V_{ik} + \sum_{i'}\{2(ik|i'i') - (ii'|i'k)\} + \sum_{k'}\{2(ik|k'k') - (ik'|k'k)\}]] \\ &- \sum_i\sum_k s_{ik}[\{2(ik|oo) - (io|ok)\} + \{2(ik|o'o') - (io'|o'k)\}] - \sum_i [(io'|o'i) + s_{io'}[V_{io'} \\ &+ \sum_{i'}\{2(io'|i'i') - (ii'|i'o')\} + \sum_k\{2(io'|kk) - (ik|ko')\} + (io'|oo) + (io|oo')]] - \sum_k [(ok|ko) \\ &+ s_{ok}[V_{ok} + \sum_i\{2(ok|ii) - (oi|ik)\} + \sum_{k'}\{2(ok|k'k') - (ok'|k'k)\} + (ok|o'o') + (oo'|o'k)]] + \delta_{oo'}\end{aligned}$$

where:

$$\delta_{oo'} = (oo'|o'o) + s_{oo'}[V_{oo'} + \sum_i\{2(oo'|ii) - (oi|io')\} + \sum_k\{2(oo'|kk) - (ok|ko')\}]$$

Since the magnitude of ε_K is essentially determined by the orbital overlapping between the two systems, the attractive term, $\delta_{oo'}$, is in this case dominant, as is the case of hydrogen molecule formation from two hydrogen atoms.

30) The terms of the s_{ab} $(cd|ef)$ type are omitted in order to avoid intricacy of expression in the subsequent equations, where a, b,⋯ denote MO's. These are obtainable in a manner similar to that of 29). The terms of s_{ab}^2 are all neglected.

$$\Delta E_{Bl} = -\{V_{All} + 2\sum_{i'} (ll|i'i')\} + \sum_{i'} \{V_{li'}s_{li'} + (li'|i'l)\}$$

$$\Delta I_{Ai} = -\{V_{Bii} + 2\sum_{k} (ii|kk)\} + \sum_{k} \{V_{ik}s_{ik} + (ik|ki)\}$$

$$a_{il} = -(ii|ll) + 2\{V_{il}s_{il} + (il|li)\}$$

$$H_{Aii} = \int a_i(1)H_A(1)a_i(1)\mathrm{d}v(1), \text{ etc.}$$

$$H_A(\lambda) = -\frac{1}{2}\Delta(\lambda) + V_A(\lambda), \text{ etc.}$$

and:

$$H_{0,i\to l} - S_{0,i\to l}H_{0,0} = \sqrt{2}[Q_{il} + K_{il}] \qquad (16)^{31)}$$

where

$$Q_{il} = \{V_{Bil} + 2\sum_{k} (il|kk)\} - s_{il}\{V_{Bii} + 2\sum_{k} (ii|kk)\}$$

is the Coulomb interaction part, and where

$$K_{il} = -\sum_{k} \{(ik|kl) - s_{il}(ik|ki)\}$$

is the exchange interaction part.

The Koopmans theorem[32] makes a possible physical interpretation of the quantities in Eq. (15). The quantity I_{Ai} is the vertical ionization potential, with no rearrangement of the nuclear configuration, of the isolated system A with respect to the electron in the ith MO; similarly, E_{Bl} is the vertical "electron affinity"[33] of the isolated system B with respect to the lth MO, while ΔI_{Ai} is the change in I_{Ai} due to the approach of the system B, and ΔE_{Bl} is the change in E_{Bl} due to the approach of the system A. With regard to a_{il}, the following relations hold:

$$a_{il} = I^{(B+l)}_{Ai} - I^{(B)}_{Ai} = E^{(A)}_{Bl} - E^{(A-i)}_{Bl}$$

where:

$$I^{(B)}_{Ai} = I_{Ai} + \Delta I_{Ai}$$

$$E^{(A)}_{Bl} = E_{Bl} + \Delta E_{Bl}$$

where $I^{(B+l)}_{Ai}$ is the value of I_{Ai} in the case of the approach of the system B with an additional electron occupying the lth unoccupied MO, and where $E^{(A-i)}_{Bl}$ is the value of E_{Bl} in the case of the approach of the system A in which one electron occupying the ith MO is subtracted.

Then, we derive the useful equation:

$$H_{i\to l,i\to l} - H_{0,0} = I^{(B)}_{Ai} - E^{(A-i)}_{Bl} = I^{(B+l)}_{Ai} - E^{(A)}_{Bl} \qquad (17)$$

which shows that this quantity is not simply written as $(I_{Ai} - E_{Bl})$. As a matter of fact, Eq. (17) is self-evident, as is directly elucidated by the following diagram:

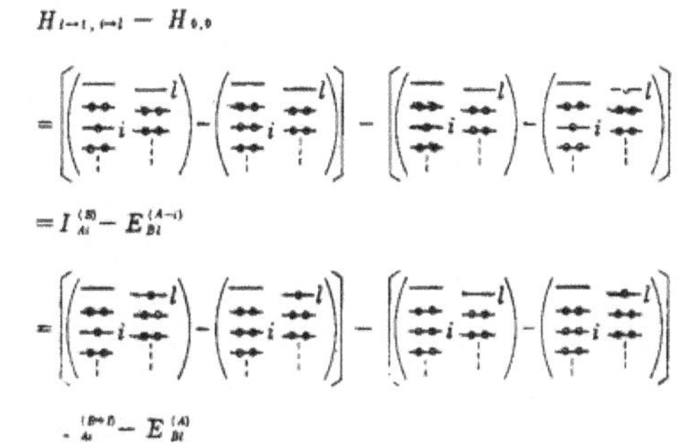

Relations of such a kind are completely general in character (*cf.* Eq. (21)).

Equation (16) indicates that the numerator of each term on the right side of Eq. (8b) consists of two parts, the Coulomb interaction part and the exchange interaction part. Since the latter is small compared with the former except in a perfectly nonpolar case, we will consider only the case in which the exchange part can be neglected.

The approximation of *multiple-site overlapping*, in which the paired AO's are denoted by the r of A and the r' of B, leads to the expression:

$$H_{0,i\to l} - S_{0,i\to l}H_{0,0} \cong \sqrt{2} \sum_{r} C^{(i)}_{r} C^{(l)}_{r'} \gamma^{(i)}_{rr'} \qquad (18)$$

where:

$$\gamma^{(i)}_{rr'} \cong s_{rr'}\left[\frac{1}{2}(v_{Brr} + v_{Br'r'}) + \frac{1}{2}\sum_{u} N_{Bu}\{(rr|uu) + (r'r'|uu)\} + \sum_{\alpha}\sum_{\beta} \frac{n^{(ii)}_{\alpha}(Z_\beta - N_\beta)}{R_{\alpha\beta}}\right] \qquad (19)$$

in which:

$$n^{(ii)}_{\alpha} = \sum_{t}^{(\alpha)} \sum_{t'} C^{(i)}_{t} C^{(i)}_{t'} s_{tt'}$$

Eq. (19) indicates a linear relationship of the right side of Eq. (16) with orbital overlapping. Equation (19) can also be written:

$$\gamma^{(i)}_{rr'} \cong -\int r(1)\left(\sum_{\beta} \frac{Z_\beta - N_\beta}{r_{1\beta}}\right) r'(1)\mathrm{d}v(1) + s_{rr'} \sum_{\alpha}\sum_{\beta} \frac{n^{(ii)}_{\alpha}(Z_\beta - N_\beta)}{R_{\alpha\beta}} \qquad (19')$$

which shows the remarkable characteristic that the charge-transfer interaction is large with the local net

31) To derive this equation, it must be remembered that the MO's are assumed to satisfy the Fock equation for the isolated system.

32) T. Koopmans, *Physica*, **1**, 104 (1933).

33) With respect to the physical significance of the energy level of a Fock-Roothaan unoccupied MO, one should in any event be cautious. To connect E_{Bl} with any experimental value, it is recommended that one adopts an extended-basis set of AO's in the Roothaan LCAO. This point will be discussed in a subsequent paper.

charge near the reaction center of the acceptor molecule.

Thus, we can obtain this approximate expression of D:

$$D \cong 2\left\{\sum_i^{\text{occ}}\sum_l^{\text{uno}} \frac{(\sum_r C_r^{(i)}C_{r'}^{(l)}\gamma_{rr'}^{(i)})^2}{I_{Ai}^{(B)} - E_{Bl}^{(A-i)}} + \sum_k^{\text{occ}}\sum_j^{\text{uno}} \frac{(\sum_r C_r^{(j)}C_{r'}^{(k)}\gamma_{rr'}^{(k)})^2}{I_{Bk}^{(A)} - E_{Aj}^{(B-k)}}\right\} \qquad (20)$$

in which:

$$\gamma_{rr'}^{(k)} \cong -\int r(1)\left(\sum_\alpha \frac{Z_\alpha - N_\alpha}{r_{1\alpha}}\right)r'(1)dv(1) + s_{rr'}\sum_\alpha\sum_\beta \frac{n_\beta^{(kk)}(Z_\alpha - N_\alpha)}{R_{\alpha\beta}}$$

If we consider the special case of a single-site interaction between the rth AO of the reactant, A, and a reagent, B, assuming that the reagent possesses only one AO (designated as r') and, accordingly, only one MO to be taken into account, and that $C_{r'}^{(l)}\gamma_{rr'}^{(i)}(=\gamma_r)$ and $C_{r'}^{(k)}\gamma_{rr'}^{(k)}(=\gamma_r')$ are constant with respect to a given reagent, then it follows that:

$$D \cong \begin{cases} 2\sum_i^{\text{occ}} \dfrac{C_r^{(i)2}}{\varepsilon_B - \varepsilon_{Ai}}\gamma_r^2 & \text{when the reagent orbital is unoccupied} \\ 2\sum_j^{\text{uno}} \dfrac{C_r^{(j)2}}{\varepsilon_{Aj} - \varepsilon_B}\gamma'^2_r & \text{when the reagent orbital is occupied,} \end{cases} \qquad (20')$$

in which:

$$\varepsilon_{Ai} = -I_{Ai}^{(B)}, \quad \varepsilon_{Aj} = -E_{Aj}^{(B-k)}$$

and

$$\varepsilon_B = \begin{cases} -E_B^{(A-i)} & \text{when the reagent orbital is unoccupied} \\ -I_B^{(A)} & \text{when the reagent orbital is occupied} \end{cases}$$

The quantity on the right side of Eq. (20′) is nothing but *superdelocalizability*[10] and *delocalizability*[34] previously defined by one of the present authors in discussing the reactivity of unsaturated and saturated compounds. These relations are useful for estimating the value of the quantities employed in semi-empirical reactivity theories.

The same equations as in Eqs. (17)—(19′) hold in the interaction of an even-electron system and an odd-electron system. In this case, however, Eq. (20) becomes:

$$D \cong 2\left\{\sum_i^{\text{occ}}\sum_l^{\text{uno}} \frac{(\sum_r C_r^{(i)}C_{r'}^{(l)}\gamma_{rr'}^{(i)})^2}{I_{Ai}^{(B)} - E_{Bl}^{(A-i)}} + \sum_k^{\text{occ}}\sum_j^{\text{uno}} \frac{(\sum_r C_r^{(j)}C_{r'}^{(k)}\gamma_{rr'}^{(k)})^2}{I_{Bk}^{(A)} - E_{Aj}^{(B-k)}}\right\}$$
$$+\left\{\sum_i^{\text{occ}} \frac{(\sum_r C_r^{(i)}C_{r'}^{(o')}\gamma_{rr'}^{(i)})^2}{I_{Ai}^{(B)} - E_{Bo'}^{(A-i)}} + \sum_j^{\text{uno}} \frac{(\sum_r C_r^{(j)}C_{r'}^{(o')}\gamma_{rr'}^{(o')})^2}{I_{Bo'}^{(A)} - E_{Aj}^{(B-o')}}\right\} \qquad (21)$$

and, corresponding to Eq. (20′), we obtain:

$$D \cong \sum_i^{\text{occ}} \frac{C_r^{(i)2}}{\varepsilon_B - \varepsilon_{Ai}}\gamma_r^2 + \sum_j^{\text{uno}} \frac{C_r^{(j)2}}{\varepsilon_{Aj} - \varepsilon_B'}\gamma_r'^2 \qquad (21')$$

in which ε_B is equal to $(-E_{Bo'}^{(A-i)})$ and ε_B' is equal to $(-I_{Bo'}^{(A)})$. This equation is similar to the form of superdelocalizability and that of delocalizability for this case of interaction. It is noteworthy that, even in the interaction of a neutral molecule with a neutral radical, the local charge of atoms determines the magnitude of D.[35]

Polarization Term. The third interaction term corresponds to the polarization due to the mutual approach of A and B. The necessary matrix elements to calculate are as follows:

$$H_{i\to j,i\to j} - H_{0,0} = I_{A+j,i}^{(B)} - E_{Aj}^{(B)} = I_{Ai}^{(B)} - E_{A-i,j}^{(B)} \qquad (22)$$

where $I_{A+j,i}^{(B)}$ is the ionization potential of an electron of the ith MO of the system A, with one additional electron occupying the jth MO, in the approach of the system B, and $E_{A-i,j}^{(B)}$ is the electron affinity with respect to the jth unoccupied MO of the system A, with one electron of the ith MO subtracted, in the approach of the system B. Moreover,

$$H_{0,i\to j} - S_{0,i\to j}H_{0,0} = \sqrt{2}[V_{Bij} + 2\sum_k (ij|kk)] + O(s_{ab}^2) \qquad (23)$$

This reduces, by the aid of the Mulliken approximation, to:

$$H_{0,i\to j} - S_{0,i\to j}H_{0,0} \cong -\sqrt{2}\zeta_{AB}^{(ij)} \qquad (23')$$

in which:

$$\zeta_{AB}^{(ij)} = \sum_\alpha\sum_\beta \frac{n_\alpha^{(ij)} + n_\alpha^{(ji)}}{2} \cdot \frac{Z_\beta - N_\beta}{R_{\alpha\beta}} \qquad (24)$$

where:

$$n_\alpha^{(ij)} = \sum_t^{(\alpha)}\sum_{t'} C_t^{(i)}C_{t'}^{(j)}s_{tt'}$$

The case of a single-site interaction between the atom α of A and the atom β of B is taken for the

34) K. Fukui, H. Kato and T. Yonezawa, This Bulletin, **34**, 1111 (1961).

35) This fact does not imply that the term D is not important in the radical interaction of a molecule. One must note that a majority of radicals involve a polar character, so the Coulomb interaction part of Eq. (16) still remains important.

sake of simplicity, in which the overlap integrals are neglected. This is a case of aromatic substitution in which only the 2 pπ AO's are taken into account. Then, we have:

$$\zeta_{AB}^{(ij)} \cong C_r^{(i)} C_r^{(j)} \frac{Z_\beta - N_\beta}{R_{r\beta}} \qquad (24')$$

which is proportional to the net charge of the atom β of the system B. This equation is useful for understanding the physical meaning of Eq. (23). The approximate formula of Π can be obtained by using Eq. (23'):

$$\Pi \cong 2\left\{ \sum_i^{\text{occ}} \sum_j^{\text{uno}} \frac{(\zeta_{AB}^{(ij)})^2}{I_{Ai}^{(B)} - E_{A-i,j}^{(B)}} + \sum_k^{\text{occ}} \sum_l^{\text{uno}} \frac{(\zeta_{BA}^{(kl)})^2}{I_{Bk}^{(A)} - E_{B-k,l}^{(A)}} \right\} \qquad (25)$$

If we consider that the quantity $(Z_\beta - N_\beta)/R_{r\beta}$ in Eq. (24') for a definite distance of approach is consant with regard to a given reagent, B, and denote it by δ_r, we obtain as the first half of Π:

$$\Pi \sim \frac{1}{2} \pi_{Arr} (\delta_r)^2$$

in which:

$$\pi_{Arr} = 4 \sum_i^{\text{occ}} \sum_j^{\text{uno}} \frac{C_r^{(i)2} C_r^{(j)2}}{\varepsilon_{Aj} - \varepsilon_{Ai}} \overline{\beta} \qquad (26)$$

($\varepsilon_{Aj} = -E_{A-i,j}^{(B)}$, $\varepsilon_{Ai} = -I_{Ai}^{(B)}$; $\overline{\beta}$ is an appropriately-chosen energy unit)

is of the same form as the well-known *self-atom polarizability* defined by Coulson and Longuet-Higgins in discussing the property of conjugated molecules.[8a)]

In this way, it has been demonstrated that the generalized scale of reactivity for an interaction between two chemical systems breaks up into the Coulomb interaction term of the atomic charge, the exchange interaction term, the delocalizability term, and the polarizability term. In some chemical reactions, some of these terms may be less important than the others. As has been discussed earlier, the polarizability term is usually less important than the delocalizability term in ionic reactions, and the exchange interaction term is not important except in special cases. In a case in which one of the charge-transfer configurations possesses nearly the same energy as the initial one, the corresponding term becomes particularly important, whereas none of excited configurations can be important since none of them can possess the same energy as the initial one. The Coulomb interaction of the atomic charge is appreciable in a long-range interaction, but it will become less and less important as the reaction proceeds and as an appreciable amount of charge is transferred from the donor to the acceptor, which thus " neutralize " each other.

The nuclear configuration change which is caused by the interaction described above and which is left out of account in the present paper will be discussed in a forth coming paper with reference to the general features of the chemical reaction.

BULLETIN OF THE CHEMICAL SOCIETY OF JAPAN VOL. 42 3399—3409 (1969)

An MO-theoretical Interpretation of the Nature of Chemical Reactions. II. The Governing Principles

Kenichi FUKUI and Hiroshi FUJIMOTO

Department of Hydrocarbon Chemistry, Kyoto University, Kyoto

(Received May 19, 1969)

The importance of the overlapping between the highest occupied (HO) molecular orbital (MO) of an electron donor and the lowest unoccupied (LU) MO of an electron acceptor in determining the favorable position and spatial direction of a chemical reaction is emphasized by setting up the following auxiliary principles: the principle of a positional parallelism between the charge-transfer and the bond-interchange, the principle of a narrowing of the inter-frontier energy-level separation, and the principle of a growing frontier density along the reaction path. These sub-principles work in a cooperative manner to enable us to arrive at this general governing principle: most of the chemical interactions are liable to occur at the position and in the direction where the overlapping of the HO and the LU of the respective reactants is at its maximum; in an electron-donating reactant, HO predominates in the overlapping interaction with the MO's of the other reactant, whereas LU does so in an electron-accepting reactant; in the reactants which possess singly-occupied (SO) MO's, these play the part of HO or of LU, or of both. These particular MO's, HO, LU, and SO are "generalized frontier orbitals" in chemical reactions. This principle is applied to uni-centric reactions to give the orientation rule; if it is applied to multi-centric reactions, the stereoselection rule is obtained.

Our previous paper[1] has treated, on the assumption of no nuclear configuration change, the interac tion energy between two reactant systems; this energy is partitioned into the Coulomb interaction, exchange interaction, delocalization, and polarization terms. It has been shown that the Coulomb interaction and the delocalization terms are important. The former comes from the net atomic charges and dominates the long-range interaction, while the delocalization term is ascribed to the charge-transfer interaction and is dominant in the usual donor-acceptor interaction energy. The quantity D, which represents the delocalization term, plays a dominant role in determining the energy of charge-transfer interaction, serving as a measure of the reactivity.

The expression of D is analogous to the chemical reactivity indices, superdelocalizability[2] and delocalizability,[3] which were originally defined by the Hückel MO formulation. These indices were derived as being closely related to the frontier electron density. The interrelation of theseq uantities is worth discussing.

In the present paper, we will attempt to investigate the essential features of the charge-transfer interaction in connection with the change in nuclear configuration along the reaction path. It may be permitted to discuss such a problem in the framework of the Born-Oppenheimer approximation. The auxiliary principles which characterize the mode of change will be sought in order to derive the fundamental orientation rule and, at the same time, to disclose the relation between the orientation rule and the reactivity indices.

1) K. Fukui and H. Fujimoto, This Bulletin, **41**, 1989 (1968).

2) K. Fukui, T. Yonezawa and C. Nagata, *ibid.*, **27**, 423 (1954).

3) K. Fukui, H. Kato and T. Yonezawa, *ibid.*, **34**, 1111 (1961).

The Delocalization Term in Degenerate Cases

The delocalization term comes from the mixing of the charge-transfer states into the initial state in an adiabatic interaction of two ground-state reactant molecules. The delocalization term, D, is expressed[1] as:

$$D=\sum_{i}^{\text{occ}}\sum_{l}^{\text{uno}}\frac{|H_{0,i\to l}-S_{0,i\to l}H_{0,0}|^2}{H_{i\to l,i\to l}-H_{0,0}}+\sum_{k}^{\text{occ}}\sum_{j}^{\text{uno}}\frac{|H_{0,k\to j}-S_{0,k\to j}H_{0,0}|^2}{H_{k\to j,k\to j}-H_{0,0}} \tag{1}$$

in which the suffix 0 implies the initial configuration and in which $i\to l$ signifies the monotransferred configuration in which an electron in the ith occupied MO of the reactant A is transferred to the lth unoccupied MO of the reactant B. Similarly, k and j represent an occupied MO of B and an unoccupied MO of A respectively. The corresponding wave-functions are designated as Ψ_0, $\Psi_{i\to l}$ and $\Psi_{k\to j}$ and:

$$H_{p,q}=\int\Psi_p{}^*H\Psi_q\mathrm{d}\tau \quad (2)$$

$$S_{p,q}=\int\Psi_p{}^*\Psi_q\mathrm{d}\tau \quad (3)$$

It is assumed, as has been done in our previous paper,[1] that the wave-function for a certain electron configuration, Ψ_p, of the total system composed of interacting reactants, A and B, is represented by an antisymmetrized product of the MO's of the two reactants, constructed so as to correspond to the electron configuration, in which each MO is made SCF (self-consistent-field) of the Roothaan type with respect to each *isolated* ground-state reactant molecule, taken as constructing an orthonormal set. The symbols $\sum^{\text{occ}}$ and $\sum^{\text{uno}}$ imply the summation over occupied and unoccupied MO's. The Hamiltonian operator, H is that of the combined system consisting of mutually-interacting A and B. For a more detailed explanation, see Ref. 1.

Equation (1) is obtained by the usual configuration interaction procedure as a second-order perturbation term, on the assumption that the interaction is still not very strong.[1] If one of the monotransferred configurations happens to have the same energy as the initial configuration, first-order perturbation term will appear. Such a case can arise in reference to the configuration in which one electron is shifted from the highest occupied (HO) molecular orbital (MO) of the donor molecule to the lowest unoccupied (LU) MO of the acceptor molecule in the case of donor-acceptor interaction. In that case the quantity D will be dominated by the first-order term:

$$D\sim|H_{0,\mathrm{HO}\to\mathrm{LU}}-S_{0,\mathrm{HO}\to\mathrm{LU}}H_{0,0}| \quad (4)$$

This equation is obtained from Eq. (7′) of Ref. 1.

Even in a not exactly degenerate case, the monotransferred term corresponding to the one-electron transfer from the HO of the donor to the LU of the acceptor can make an overwhelming contribution to the delocalization term when the denominator of this term in Eq. (2) is sufficiently small in comparison with the other monotransferred terms.

$$D\sim\frac{|H_{0,\mathrm{HO}\to\mathrm{LU}}-S_{0,\mathrm{HO}\to\mathrm{LU}}H_{0,0}|^2}{H_{\mathrm{HO}\to\mathrm{LU},\mathrm{HO}\to\mathrm{LU}}-H_{0,0}} \quad (5)$$

As Mulliken has already pointed out, $H_{0,\mathrm{HO}\to\mathrm{LU}}$ is approximately proportional to $S_{0,\mathrm{HO}\to\mathrm{LU}}$, so that the magnitude of D and, accordingly, the interaction (stabilization) energy can be measured in terms of the extent of orbital overlapping between HO and LU, since:

$$S_{0,\mathrm{HO}\to\mathrm{LU}}\cong\sqrt{2}\,s_{\mathrm{HO,LU}} \quad (6)$$

where:

$$s_{\mathrm{HO,LU}}=\int a_{\mathrm{HO}}(1)b_{\mathrm{LU}}(1)\mathrm{d}v(1) \quad (7)$$

and where a_{HO} and b_{LU} are the HO MO of A and the LU MO of B respectively, both being taken as having a real spatial part.

In this way, Eq. (5) leads to the "overlap and orientation" principle of Mulliken.[4] Equation (4) will lead to the same conclusion with regard to the orientation principle. The favorable orientation in such cases can be determined by the rule of the maximum overlapping of HO and LU. Namely, the interaction takes place at the position and in the direction in which the largest overlapping of these particular MO's is attained.

In addition, there are special circumstances which make these particular orbitals, HO and LU, discriminatively important. The importance of the interaction term between HO and LU prevails over a wider range of combination of two reactants; furthermore, it is enhanced in actually-occurring reactions if the nuclear-configuration change along the reaction path is taken into consideration. This point will be discussed on the basis of auxiliary principles later. In any event, the usefulness of the equations of the type of Eq. (4) or (5) is in fact, far more than could be expected simply from the conditions for the fitness of these equations.

The Principle of a Positional Parallelism between the Charge-transfer and the Bond-interchange

It has already been shown[5-7] that the nodal property of HO and LU MO's plays an essential role in determining the steric course of electrocyclic reactions. The same property is important in promoting the alteration of the molecular shape in the case of orbital interaction.

Usually, an atom is, as a whole, bonding with neighboring atoms in occupied MO's and antibonding in unoccupied MO's. This circumstance can be understood by a very simple example which employs the linear-combination-of-atomic-orbitals (LCAO) MO scheme. If we use the Hückel MO theory (overlap neglected) in the frame of π-electron approximation for a planar conjugated molecule, the following relation holds with respect to the rth π AO:

$$(\alpha_r-\varepsilon_i)(C_r^{(i)})^2+\sum_s^{\text{nei}}C_r^{(i)}C_s^{(i)}\beta_{rs}=0 \quad (8)$$

where α_r is the Coulomb integral of the rth AO; β_{rs}, the resonance integral between the rth and the sth AO's; ε_i, the energy of the ith MO, and $C_r^{(i)}$, the coefficient of the rth AO in the ith MO. The notation $\sum^{\text{nei}}$ implies summation over all the neighboring AO's of the rth AO. In the case of con-

4) R. S. Mulliken, *Rec. Trav. Chim.*, **75**, 845 (1956).
5) K. Fukui, "Molecular Orbitals in Chemistry, Physics, and Biology," ed. by P.-O. Löwdin and B. Pullman, Academic Press, New York (1964), p. 513.
6) K. Fukui, This Bulletin, **39**, 498 (1966).
7) K. Fukui and H. Fujimoto, *ibid.*, **39**, 2116 (1966).

jugated hydrocarbons, all the α_r's and all the β_{rs}'s can be put equal to constant values, α and β respectively. Then, it follows from Eq. (8) that:

$$\sum_{s}^{\text{nei}} C_r^{(i)} C_s^{(i)} = \frac{\alpha-\varepsilon_i}{(-\beta)} (C_r^{(i)})^2 \qquad (9)$$

As usual, the energies of the occupied orbitals are lower than α, while those of the unoccupied orbitals are higher than α. Therefore, the following relations hold for any r: $\sum_{s}^{\text{nei}} C_s^{(i)} C_t^{(i)}$ is positive and proportional to $(C_r^{(i)})^2$ for occupied MO's and is negative and proportional to $(C_r^{(i)})^2$ for unoccupied MO's.

In this way, it is established that the position of the largest HO or LU density is, at the same time, that position which is most liable to weaken the bonds with neighboring atoms in case of electron-releasing or accepting interaction respectively.[8] Namely, the charge-transfer weakens the bonds most greatly at the position of the largest frontier electron density.[5,9]

Some examples of the nodal property of the π HO's and LU's of several conjugated hydrocarbons are shown in Fig. 1, in which the shaded and unshaded areas correspond to the positive and negative regions of the MO coefficients. It can be qualitatively understood from the figure that there is a parallelism between the partial population and the partial sum of the bond-order with neighbors. Even in heteroaromatic systems similar relationships hold. We can see the circumstances in Fig. 2.

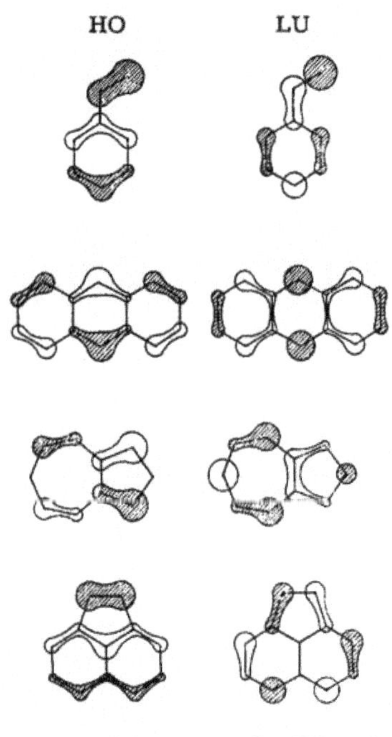

Fig. 1. The nodal property of π HO and LU of some conjugated hydrocarbons.

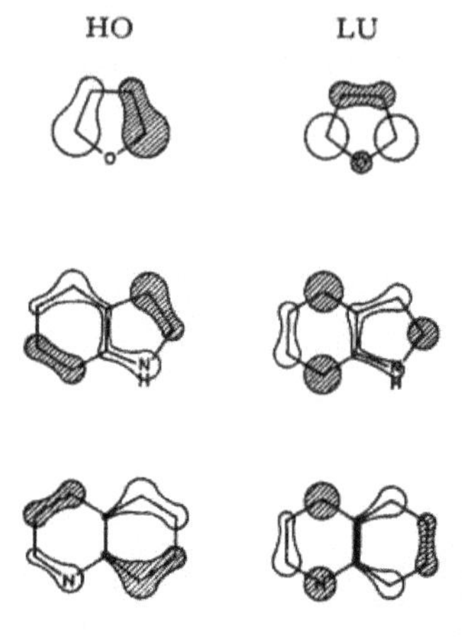

Fig. 2. The nodal property of π HO and LU of some heteroconjugated systems.

Such a character of MO's is to a certain extent perceived in more elaborate calculations, independent of the order of approximation adopted. An example is given in Table 1 using the Pariser-Parr-Pople approximation.[10] The suffix f signifies HO or LU. A distinct parallelism is recognized among the partial population and the partial sum of the bond-order with the neighbors.

Similar circumstances also exist in the extended Hückel treatment[11] of three-dimensional molecules. The orientation in the E2 reaction of halogenated hydrocarbons has previously been discussed.[12,13] If we concentrate our attention on one of the hydrogen 1s AO's, say r, the following equation is

8) When the HO energy happens to be higher than α, the electron-releasing will cause bond-strengthening. The result is the same with the electron-accepting in the LU MO when its energy is lower than α.

9) The extent of this bond-loosening will, therefore, be a measure of the ease of intramolecular electron localization at the reaction center in the case of interaction with another molecule. This may be one of the reasons why the localization energy gives a parallel prediction with the frontier electron density for the chemical reactivity of a majority of conjugated molecules (*cf.* A. Streitwieser, Jr., "Molecular Orbital Theory for Organic Chemists," John Wiley & Sons, New York (1961), p. 305).

10) See, for instance, R. Pariser and R. G. Parr, *J. Chem. Phys.*, **21**, 466, 767 (1953).

11) R. Hoffmann, *ibid.*, **39**, 1397 (1963), and subsequent papers.

12) K. Fukui and H. Fujimoto, *Tetrahedron Letters*, **1965**, 4303.

13) K. Fukui, H. Hao and H. Fujimoto, This Bulletin, **42**, 348 (1969).

TABLE 1. THE POSITIONAL PARALLELISM BETWEEN $(C_r^{(f)})^2$ AND $\sum_s^{\text{nei}} C_r^{(f)}C_s^{(f)}$ IN PARISER-PARR-POPLE CALCULATION

Compound	r	$(C_r^{(f)})^2$	$\sum_s^{\text{nei}} C_r^{(\text{HO})}C_s^{(\text{HO})}$	$\sum_s^{\text{nei}} C_r^{(\text{LU})}C_s^{(\text{LU})}$
Anthracene	1	0.09512	0.03680	−0.03680
	2	0.04770	0.01966	−0.01966
	9	0.19472	0.08745	−0.08745
Phenanthrene	1	0.10550	0.07154	−0.07154
	2	0.00100	0.00034	−0.00034
	3	0.09868	0.06598	−0.05598
	4	0.05843	0.03197	−0.03197
	9	0.16775	0.09051	−0.09051
Chrysene	1	0.07978	0.04403	−0.04403
	2	0.01427	0.00738	−0.00738
	3	0.04871	0.02809	−0.02809
	4	0.06089	0.02985	−0.02985
	5	0.05696	0.02168	−0.02168
	6	0.14446	0.07687	−0.07687

Numberings are as follows:

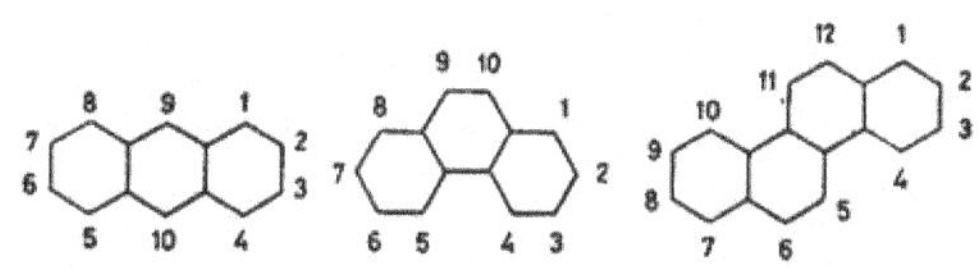

Anthracene Phenanthrene Chrysene

All C-C distances and all C-C-C angles were assumed to be 1.4Å and 120°, respectively, for the sake of simplicity of computations.

obtained in place of Eq. (8):

$$(h_{rr}-\varepsilon_i)(C_r^{(i)})^2+\sum_{s(\neq r)}(h_{rs}-\varepsilon_i s_{rs})C_r^{(i)}C_s^{(i)}=0 \quad (10)$$

where:

$$h_{rs}=\frac{K}{2}(h_{rr}+h_{ss})s_{rs} \quad (11)$$

where h_{rr}, h_{rs} and s_{rs} are the Coulomb, resonance, and overlap integrals respectively, and where K is the Wolfsberg-Helmholtz constant.[11] The other symbols have the same meaning as in the Hückel scheme. Although the value of the Coulomb integral varies, of course, from one AO to another, it may be possible to employ tentatively, in place of the individual h_{ss}, a sort of mean value, h, which makes the second term of the right side of Eq. (10) equal to:

$$\sum_{s(\neq r)}\left\{\frac{K}{2}(h_{rr}+h)-\varepsilon_i\right\}C_r^{(i)}C_s^{(i)}s_{rs} \quad (12)$$

Hence, it follows that:

$$\sum_{s(\neq r)}C_r^{(i)}C_s^{(i)}s_{rs}=-\frac{h_{rr}-\varepsilon_i}{\frac{K}{2}(h_{rr}+h)-\varepsilon_i}(C_r^{(i)})^2 \quad (13)$$

Then, with respect to LU, the relations $h_{rr}-\varepsilon_{\text{LU}}<0$ and $(K/2)(h_{rr}+h)-\varepsilon_{\text{LU}}<0$ usually hold. The left side of Eq. (13) is equal to $(1/2)v_r^{(i)}$, which is the valence-active part[14] of the partial AO population of the rth AO in the ith MO defined by:

$$v_r^{(i)}=2\sum_{(s\neq r)}C_r^{(i)}C_s^{(i)}s_{rs} \quad (14)$$

Therefore, $v_r^{(\text{LU})}$ must be negative and proportional to $(C_r^{(\text{LU})})^2$, indicating that the hydrogen atom having the largest frontier electron density is most liable to a loosening of the bond with the remaining part of the molecule.

This conclusion may be confirmed by actual numerical calculation in which the approximation using the mean h value is, of course, not applied. Examles are shown in Table 2. The table contains the frontier electron density $(C_r^{(\text{LU})})^2$ and the valence-active part of the partial population of the frontier orbital, $v_r^{(\text{LU})}$. Exemplifications by the various calculations mentioned above indicate the existence of a parallelism among actually reactive sites in a molecule; the position of the largest frontier orbital extension is at the same time the position where the bonds with neighboring atoms are, as a whole, most liable to be loosened. Since

14) K. Ruedenberg, *Rev. Mod. Phys.*, **34**, 326 (1962).

TABLE 2. THE RELATION BETWEEN $v_r^{(LU)}$ AND $(C_r^{(LU)})^2$ OF HYDROGENS IN SEVERAL HALOGENATED PARAFFINS

Chlorinated paraffin	r	$(C_r^{(LU)})^2$	$v_r^{(LU)}$
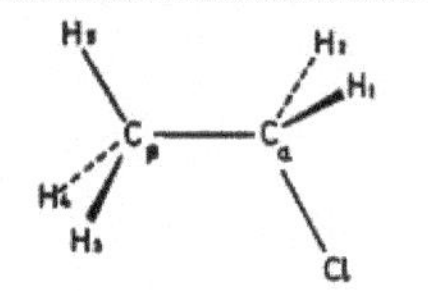	1	0.01791	−0.01192
	3	0.00124	−0.00062
	5	0.05412	−0.03689
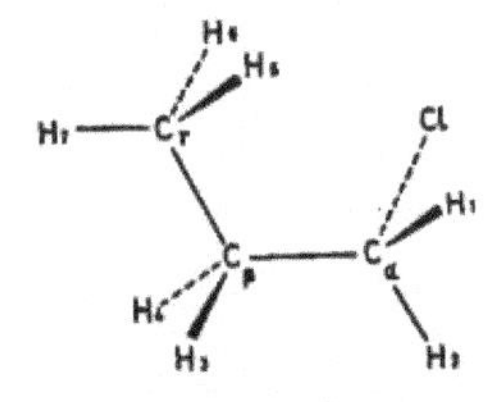	1	0.01658	−0.01163
	2	0.03135	−0.02134
	3	0.07458	−0.05128
	4	0.00067	−0.00018
	5	0.00096	−0.00071
	6	0.01708	−0.01361
	7	0.00063	−0.00027
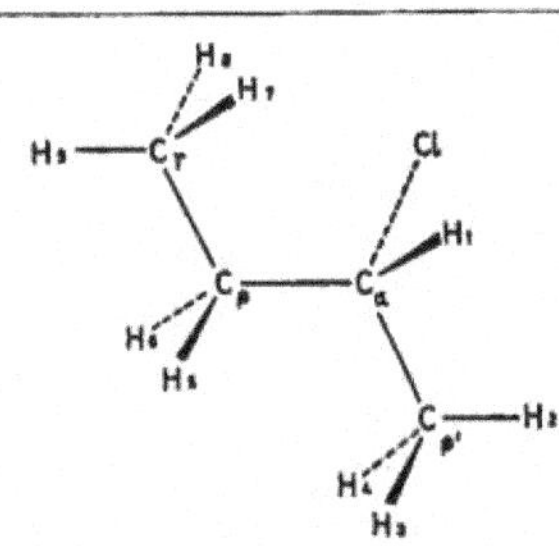	1	0.00978	−0.00661
	2	0.00029	−0.00006
	3	0.06638	−0.04564
	4	0.00044	−0.00011
	5	0.07888	−0.05295
	6	0.00090	−0.00041
	7	0.00158	−0.00131
	8	0.02119	−0.01700
	9	0.00000	−0.00000

the frontier orbital extension is a measure of the ease of charge-transfer interaction, this conclusion represents a parallel relationship between the position of charge-transfer and the position of bond-interchange in a molecule in chemical reactions; it will later be shown to act as one of the two auxiliary principles which lead to the general orientation principle.

The Principle of a Narrowing of the Inter-frontier Level Separation

As has been mentioned above, the charge-transfer interaction takes place at the position of the largest frontier electron density, the position which is simultaneously most susceptible to a weakening of the bonds with the neighboring atoms. This bond-loosening causes the nuclear configuration to change in a definite direction, which should be controlled by the nodal property of the frontier orbitals.

This direction of nuclear-configuration change may be characterized by the mode of change in the energy-level separation between the HO of the donor and the LU of the acceptor. Namely, the HO MO of the donor becomes, in general, higher, while LU of the acceptor becomes lower, in the event of charge-transfer, causing a narrowing of the inter-frontier energy level separation. Sush a conclusion can very easily be understood, since a bonding MO is made unstable by electron-releasing and since an antibonding MO is made unstable by electron-accepting, in both cases through bond-loosening.

Let us consider this problem by quoting several examples. In the first place, let us take the Diels-Alder-type addition of ethylene and butadiene as an instance. The nodal properties of the HO and LU of ethylene and butadiene are shown in Fig. 3, together with the mode of charge-transfer interaction. The HO MO of butadiene is bonding in the region of 1,2- and 3,4-π-bonds and antidonding in the 2,3-region, while the LU of butadiene has the opposite nodal property. The HO of ethylene is bonding, while LU is antibonding. This implies that both

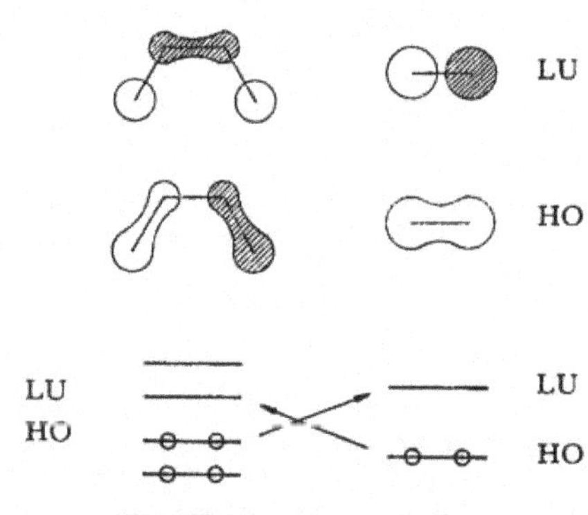

Fig. 3. The nodal property of HO and LU in ethylene and butadiene.

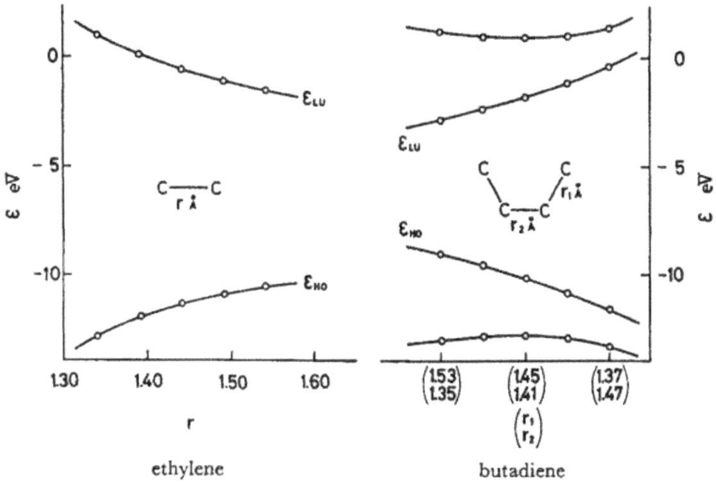

Fig. 4. The change in energy of molecular orbitals in ethylene and butadiene along the reaction path.

the charge-transfer from the HO of ethylene to the LU of butadiene and that from the HO of butadiene to the LU of ethylene will decrease the π bond-order of ethylene and cause a double-bond shift in butadiene.

The results of calculations by the Pariser-Parr-Pople scheme[10] for several altered nuclear coordinates are shown in Fig. 4. The transient nuclear configurations along the reaction path could reasonably be assumed by considering the direction of charge-transfer and the nodal property of the HO and LU. We can see that the changes in the HO and LU can be clearly distinguished from those of the other orbitals. The separation of these MO levels is considerably contracted as the reaction proceeds. Such a property is common in dienes and dienophiles.

The mode of change in the MO energies of this sort is more conspicuous in the sigma electron systems. As a typical example, let us present the results of an extended Hückel calculation on the SN-2 reaction of methyl chloride by the chloride anion. The reaction coordinates were extracted from the calculation of Kato *et al.*[15] The solid line in Fig. 5 indicates the change in the energy of the LU of methyl chloride, while the broken line shows that of the system composed of methyl chloride and chloride anion. The other unoccupied MO's are located much higher than LU,[16] and the changes in these levels were found to be small in comparison with that in LU.

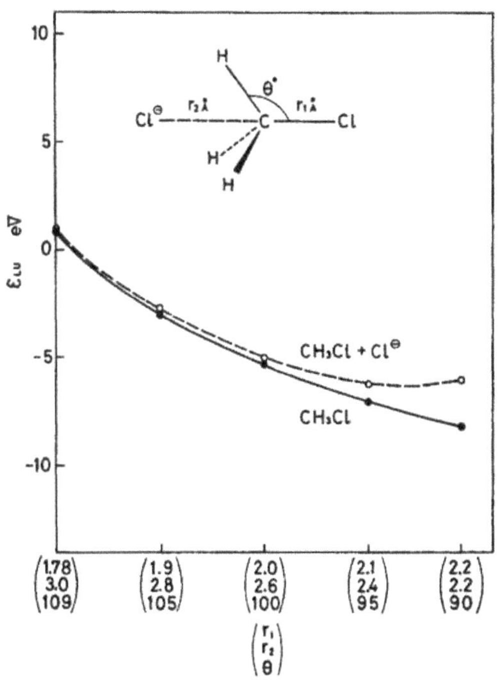

Fig. 5. The change in the energy of LU MO of methyl chloride with and without another approaching chloride anion.

It may be of interest to investigate the content of the activation energy. The stretching of the C–Cl bond in methyl chloride will require enough energy to form an important fraction of the activation energy of the reaction. The LU MO is highly localized at the C–Cl bond and exhibits an antibonding char-

15) H. Kato, K. Morokuma, T. Yonezawa and K. Fukui, This Bulletin, **38**, 1749 (1965).

16) The unoccupied MO's are expressed in terms of a set of limited-basis AO's chosen so as to be suitable in representing the occupied MO's. This is an unsatisfactory point. However, the essential features will never be changed by the employment of any extended-basis set of AO's.

acter. The charge-transfer from the attacking base to LU will, therefore, cause the C–Cl bond stretching. However, this will produce stabilization on account of the lowering of the MO level, as is shown in Fig. 5. In this way, a considerable part of the activation energy is reduced. It can be understood that the charge-transfer and the nuclear configuration change cooperate in driving the chemical reactions.

The circumstances are similar also in nucleophilic elimination reactions. For instance, the LU of ethyl chloride is highly localized at the carbon-chlorine bond, which shows an antibonding nature. Accordingly, the charge-transfer to the LU from the attacking base through *trans*-β-hydrogen may cause a weakening of the C–Cl bond as well as of the β-carbon-hydrogen bond, resulting in a lowering of the LU MO energy. Such a lowering of the MO energy may be of great importance in facilitating the bimolecular elimination of halogenated paraffins by a nucleophile. The results of the numerical calculation have been given in previous papers.[12,13]

Next, let us consider aromatic substitutions. In these reactions the bonds between the reaction center and the neighboring atoms may be lengthened to some extent by the decrease in the π bond-orders of this region. This will cause an elevation of the HO level and a lowering of the LU level. In addition, the leaving hydrogen atom at the reaction center will leave the molecular plane as the reaction proceeds. It may be of interest to examine the energy change in the frontier orbitals in this case. The results of the numerical calculation are shown in Fig. 6.

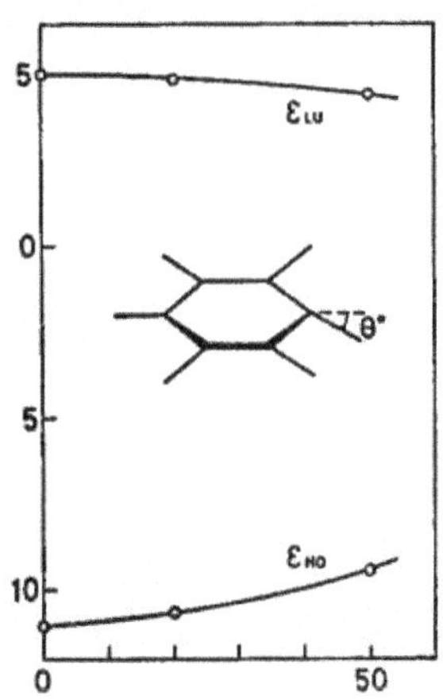

Fig. 6. The change in the energy of HO and LU in benzene due to the C–H bending.

The changes in the energies of HO and LU due to such a change in the bond angle are, in general, small in comparison with those due to the bond-stretching. However, the change is significant in the sense that the effect of valence angles can never be left out of account in discussing the problem.

The ascending HO energy and the descending LU energy of aromatic molecules couple with the descending LU of electrophilic reagents and with the ascending HO of nucleophilic reagents respectively to give rise to a narrowing of the inter-frontier level separation. In the case of liquid-phase ionic reactions, it must frequently be remarked that one should take into account the process of desolvation and solvation in the level change of the reagent orbitals. In homolytic reactions, the ascending HO and descending LU levels both approach the singly-occupied (SO) MO level of reagent radicals, bringing about a simultaneous charge-transfer to and a back-transfer from the SO MO.

In order to confirm these circumstances, the frontier MO's of attacking electrophiles are examined. The hydronium ion is taken as a simple example of solvated electrophiles. The extended Hückel calculation indicates that the LU of hydronium ion lowers as the distance between the proton, which is the bare electrophilic reagent, and the water molecule, that is the solvent, grows with the proceeding of the reaction. The results are shown in Fig. 7.

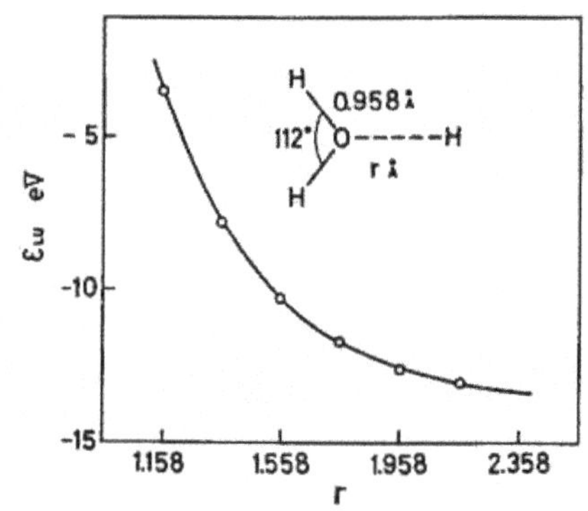

Fig. 7. The change in the energy of LU in hydronium ion due to the O–H stretching.

Lastly, the heterolytic addition to an olefinic double bond will be cited. The reagent is here typified by a chlorine molecule in order to simplify the calculation. The first stage of this reaction is assumed to be the formation of a bridged cation,[17] which may cause the stretching of the carbon-carbon bond.

An extended Hückel calculation is made to obtain a relation between the position of chlorine and the MO energy. The chlorine is shifted in the direction parallel to the carbon-carbon bond. As can be seen in Fig. 8, the total energy, E, increases as the chlorine atom separates from the center plane

17) An energy minimum arises in the event of approaching the C–C bond on the bisecting line. The stability of triangular interaction is discussed in Ref. 6.

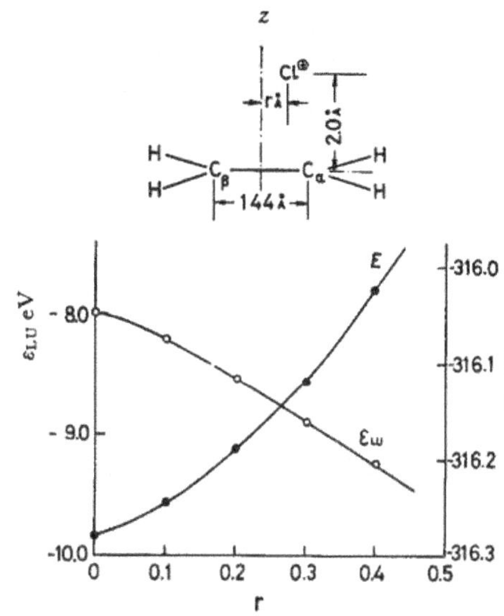

Fig. 8. The change in the energy of LU and in the total energy of ethylene-chorine cation system.

to approach the α-carbon; also, the LU level becomes lower in the event of change in this direction. The progress of the charge-transfer causes the narrowing of the inter-frontier separation, so the contribution of LU of the cation becomes enhanced as the interaction goes on.

The results of numerical calculations with respect to the several examples of chemical reactions cited above suggest the general relationship between the inter-frontier energy level separation and the nuclear configuration change along the reaction path. This implies that the importance of the frontier orbitals is more than would be expected from the case where the separation narrowing is left out of account.

The Principle of a Growing Frontier Electron Density along the Reaction Path

In order to make the third auxiliary principle concrete, it is recommended that we discuss the simplest case first. The well-discussed problem of aromatic substitution fits this purpose. As the reaction proceeds, the sum of the mobile bond-orders of the bonds between the reaction center and the neighboring atoms will decrease according to Eq. (9), causing a loosening of these bonds. The effect of this π-bond weakening may be represented, in the frame of the Hückel MO method, by the decrease in the absolute value of the resonance integrals corresponding to these bonds.

It is important to investigate the change in the frontier electron density at the actual reaction site according to the increase in the resonance integrals (that is, the decrease in the absolute value) with the neighboring atoms. In actual reactions, this change evidently never amounts to as large values as $(-\beta)$, where β is the original value of the resonance integral. However, as an extremely assumptive case, this increment may be taken as $(-\beta)$, so that the general tendency of the effect of such changes might more dinstinctly be realized. In that case the π AO at the reaction center finally becomes isolated (Stage(IV)). Stage (III), immediately before Stage (IV), indicates that a small π conjugation still remains between the reaction center and the neighboring atoms.

The frontier electron densities, $(C_r^{(HO)})^2$ and $(C_r^{(LU)})^2$, usually increase along the (I)→(II) path. This is valid with respect to the positions in alternant hydrocarbons, and also in a majority of positions of nonalternant hydrocarbons.

In order to understand these circumstances qualitatively, although not in a mathematically strict sense, it is convenient to consider Stage (III). In this stage it is very easy to prove that:

$$\lim_{\Delta\beta\to 0} (C_r^{(HO)})^2 \text{ and } (C_r^{(LU)})^2 = \frac{1}{2} \tag{15}$$

provided that the system obtained by deleting the atom r from the original hydrocarbon has a nonbonding level, $\varepsilon=\alpha$.[18] Since the original values of $(C_r^{(HO)})^2$ and $(C_r^{(LU)})^2$ are usually far less than 0.5, this result suggests that the frontier electron density increases along the path (I)—(III). The relation of Eq. (15) holds in alternant hydrocarbons and also in several, actually reactive positions of some nonalternant hydrocarbons, like the $(C_1^{(HO)})^2$ of fulvene, the $(C_1^{(HO)})^2$ of acenaphthylene, and the $(C_7^{(HO)})^2$ of fluoranthene.[19]

When the remaining system has no nonbonding MO, the following two sub-cases are possible. Let

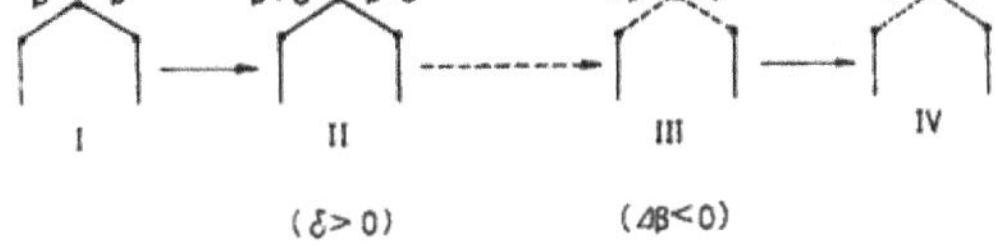

18) If the remaining system possesses n nonbonding MO's, $(C_r^{(HO)})^2$ and $(C_r^{(LU)})^2$ become $1/(n+1)$.

19) The numbering of these nonalternants conforms to: K. Fukui, T. Yonezawa and C. Nagata, This Bulletin, **34**, 37 (1961).

the number of occupied MO's of the original system be n, and let the number of occupied MO's of the remaining system obtained by deleting the atom r from the original system be m. The first sub-case is the one where $m=n$. In this case it follows that:

$$\lim_{\Delta\beta\to 0} (C_r^{(LU)})^2 = 1 \tag{16}$$

Such a relation holds in the position 6 of fulvene, the position 5 of acenaphthylene, the position 6 of azulene, and the position 3 of fluoranthene. The second sub-case is specified by $m=n-1$, so that:

$$\lim_{\Delta\beta\to 0} (C_r^{(HO)})^2 = 1 \tag{17}$$

which is actually satisfied in the position 1 of azulene. All of these cases indicate that the rule of growing frontier electron density might hold in actually reactive positions.

One of the few exceptions is $(C_3^{(HO)})^2$ of fluoranthene. In this case we have $m=n$, so that $\lim_{\Delta\beta\to 0} (C_3^{(HO)})^2=0$. However, it should be noted that in this case the HO MO energy of the remaining system is very close to the nonbonding orbital energy ($\varepsilon=\alpha+0.1317\beta$) whereas the energy of the isolated AO of the position 3 will easily become lower than $\varepsilon=\alpha$ under the influence of an approaching electrophile, so that the isolated AO energy might not become higher than the HO MO of the remaining system. In this way, it is possible even in this case that:

$$\lim_{\Delta\beta\to 0} (C_3^{(HO)})^2 = 0.5 \text{ or } 1 \tag{18}$$

Such a Coulombic effect of the attacking reagent should be considered in the other cases which have already been mentioned. In all of these cases, however, the circumstances are more favorable, in view of this effect, for the validity of the rule of growing frontier electron density along the reaction path.

The heterolytic addition of chlorine to a C=C bond (discussed in preceding section) is also an example. Figure 9 shows that the partial valence-inactive population of the $2p_z$ AO of β-carbon in the LU increases with the progress of the reaction.

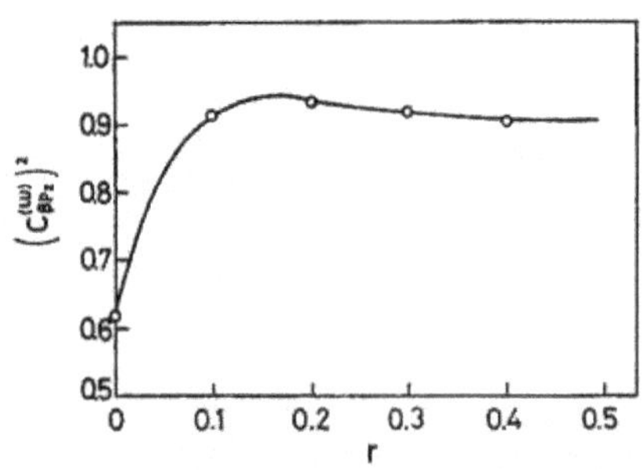

Fig. 9. The change in the partial population of p_z orbital at β-carbon of ethylene-chlorine cation system.

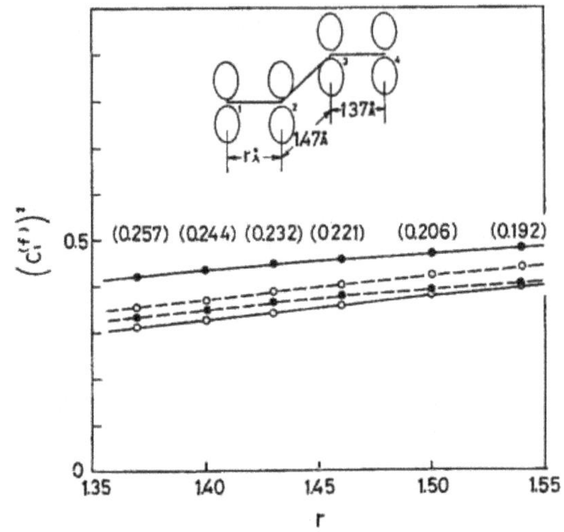

Fig. 10. The change in frontier electron density in the π electronic part of *trans*-butadiene along the departing of the end carbon atom.

—●— LU including overlap
—○— HO including overlap
--●-- LU neglecting overlap
--○-- HO neglecting overlap

(The values in parenthses indicate the overlap integral between two $p\pi$ atomic orbitals at the carbon atoms 1 and 2.)

This rule is not essentially violated by the adoption of more elaborate methods of calculation. With the change in the nuclear configuration, the overlap integrals between the AO's belonging to the departing atom and the AO's belonging to the remainder of the molecule vary. In order to discuss the change in the magnitude of the frontier electron density as a measure of the numerator of Eq. (1), the effect of the change in the overlap integrals upon the frontier electron density through the normalization condition can never be neglected. Figure 10 gives the change in the $(C_r^{(f)})^2$ in the π electronic part of butadiene due to the stretching of the C-C bond. We can see that the essential conclusion mentioned above also holds in the extended Hückel calculation, explicitly including overlap integrals.

The Co-operation of Sub-principles and the General Orientation and Stereoselection Rule

It is obvious that the three sub-principles work co-operatively in promoting the chemical reaction. The charge-transfer interaction takes place predominantly at the position and in the direction in which the overlapping of the HO and LU of the two reacting species is largest, as is to be expected from Eqs. (5) and (6). This interaction brings about a local bond-weakening, which is mainly controlled by the nodal property of the frontier orbitals. The extent of bond-weakening positionally parallels the frontier-orbital extension according to the first sub-principle. The bond-loosening leads to the nuclear configuration change in a definite direction, which causes the narrowing of the inter-frontier energy level separation according to the second sub-principle and, at the same time, the increase in the frontier electron density at the reaction center according to the third sub-principle. These make the denominator on the right side of Eq. (5) smaller and the numerator larger, so the contribution of the frontier orbital interaction term to the delocalization part (Eq. (1)) of the interaction energy becomes larger. Hence, the charge-transfer interaction increases, resulting again in a bond-interchange near the reaction center, a molecular shape deformation, a narrowing of the inter-frontier separation, and an increase in the frontier electron density, in turn. The importance of the term of Eq. (5) thus becomes greater and greater, the form of the delocalization term approaching that of Eq. (4).

It should be remarked in this connection that the MO's which can participate in such a type of co-operation are obviously restricted to the HO and LU. The energies of the other MO's undergo only the minimum change which is absolutely needed for the occurrence of the reaction and is assumed to be almost constant with respect to every possible position of the reaction. This is because a stable molecule originally takes the nuclear configuration which is energetically most favorable, and, in the event of reaction, any change in the nuclear configuration will be accompanied by an unstabilization which resembles the promotion in atoms in the case of molecule formation. Therefore, the change in the nuclear configuration will take place in such a direction that the unstabilization will be most effectively reduced, while a direction of change in which no energetical advantage can be expected will be avoided. As has already been mentioned, the charge-transfer between the HO and LU causes an alteration in the molecualr shape at the site of the greatest HO or LU MO density. In this way, the change in molecular shape is automatically limited to the proximity of the particular position in the reacting species. This self-controlling action seems to underlie the empirical rule which has been known as "the principle of least motion" or "the principle of least molecular deformation."[20]

It is evident from the above discussion that the most effective means to promote a chemical reaction smoothly by reducing the unstabilization attributable to the nuclear configuration change due to the interaction is to give rise to a change by which the charge-transfer from the HO of one species to the LU of the other is effectuated. This charge-transfe rmay be uni-directional or mutual according to the magnitude of the difference between the electron-donating power of one reactant and the electron-accepting power of the other. Other directions of nuclear configuration change are all rejected as bringing about little gain in stabilization energy.

In this way, the reaction pathway becomes dominantly controlled by the frontier-orbital interaction. The position of the reaction will be determined, through Eq. (6), by the rule of the maximum overlapping of the HO and LU (sometimes the SO in place of the HO or LU) of the two reacting species. The general orientation principle is, therefore, as follows: "*A majority of chemical reactions are liable to take place at the position and in the direction where the overlapping of the HO and LU of the respective reactants is at its maximum*; *in an electron-donating species, HO predominates in the overlapping interaction, while the LU does so in an electron-accepting reactant*; *in the reacting species which possess SO MO's, these play the part of the HO or of the LU, or of both.*"

In some intramolecular reactions, like the cyclization of a conjugated olefinic chain, the dominantly controlling factor seems to be the first-order inteaction term,[6,21] in which the HO-LU interaction is concealed; however, the same reaction can be discussed by considering the HO-LU interaction between two parts of a molecule.[22] The HO-LU interaction has also been considered between the sigma- and pi-parts of conjugated molecules in order to derive a stereoselection rule.[7]

In this view, the particular MO's which control the reaction pathway in chemical reactions in general, *i.e.*, HO, LU, and SO, are referred to as "generalized frontier orbitals."

It seems in this way that the principle presented above is a general one which will qualitatively govern almost all sorts of chemical interactions, including intermolecular and intramolecular as well as unicentric and multicentric. The principle can be applied to a unicentric reaction to derive an orientation rule; and if it is employed to discuss multicentric reactions, the result is the stereoselection rule, which has already been discussed.[7,21,22]

The reactivity indices, the frontier electron density, the delocalizability, and the superdelocalizability, are those for unicentric reactions and give predictions consistent with the general orientation principle. In this connection, it is worthwhile to mention the relation between these chemical reactivity indices and the general orientation rule. These indices are derived from Eq. (1), in which the nuclear-configration change along

20) See, for example, J. Hine, *J. Org. Chem.*, **31**, 1236 (1966), and the references cited therein.

21) K. Fukui, *Tetrahedron Letters*, **1965**, 2009.

22) K. Fukui and H. Fujimoto, "Mechanism of Molecular Migrations," Vol. 2, ed. by B. S. Thyagarajan, Wiley-Interscience (1969).

the reaction path is left out. Namely the indices, like most other reactvity indices ever proposed, are calculated with regard to an isolated, undeformed molecule. If the molecular deformation along the reaction path is taken into consideration, the values of these indices will be changed. The change in the value of delocalizability and superdelocalizability will be controlled by the sub-principles discussed above, so the contribution of the terms corresponding to the frontier orbitals will become more dominant than in the original expression.

In the superdelocalizability thus modified, the frontier terms should be weighted more. The index of the type of Eq. (5), the one-term approximation of superdelocalizability,[23] and Brown's *Z*-value[24] are all this line. We prefer the frontier terms and recommend that the frontier electron density be used as a reactivity index.

The calculations have been carried out partly on the HITAC 5020E computer at the Computer Center of the University of Tokyo and partly on the FACOM 230·60 computer at the Computer Center of Kyoto University.

23) K. Fukui, T. Yonezawa and C. Nagata, *J. Chem. Phys.*, **27**, 1247 (1957).

24) R. D. Brown, *J. Chem. Soc.*, **1959**, 2232.

9205

Molecular Orbital Calculation of Chemically Interacting Systems. Interaction between Radical and Closed-Shell Molecules

Hiroshi Fujimoto, Shinichi Yamabe,* Tsutomu Minato, and Kenichi Fukui

Contribution from the Faculty of Engineering, Kyoto University, Sakyo-ku, Kyoto, Japan. *Received June 15, 1972*

Abstract: An approximate molecular orbital (MO) calculation has been carried out on the interaction between methyl radicals and methane and that between methyl radicals and ethylene. The changes in the electron distributions due to the chemical interactions were calculated from the wave functions of the combined systems of radicals and closed-shell molecules which were obtained by the configuration interaction (CI) method of the electronic states of the adiabatic interaction and the charge-transfer interaction. These electronic states were described by the use of the MO's of reactant and reagent in an isolated state. The exchange interaction contributes to the bonds which are newly formed between reactant and reagent in an antibonding fashion, while the charge-transfer interaction is shown to be the main origin of the formation of new bonds as well as the loosening of the bonds to be broken in the reactions. The orbital interactions between the singly occupied (SO) MO of the radical and the highest occupied (HO) MO and the lowest unoccupied (LU) MO of the closed-shell molecule were found to be of importance.

Quantum chemical study of radical reactions may be suitable for an elucidation of the basic nature of chemical reactions, because they are less influenced by the environment of reactions than ionic reactions. The chemical reactivity index, free valence, was derived as a measure of the chemical reactivities of conjugated molecules toward the reagent radical.[1] Szwarc and Binks related the localization energies to the methyl affinities of aromatic and olefinic compounds.[2] In the frontier orbital theory, the important role of the orbital overlap interactions between the SOMO of the radical and the HOMO and the LUMO of the closed-shell system was pointed out.[3-8] The reactivity indices, frontier electron density, superdelocalizability, and delocalizability were defined in order to discuss the reactivities of unsaturated and saturated molecules with radicals in the frame of the isolated-molecule approximation. Recently, some MO calculations of the addition of methyl radical to ethylene and butadiene[9,10] and of the reaction of hydrogen with methane were performed.[11]

Here, it may be interesting to investigate, in detail, how the bond formation between reagent and reactant begins to take place, accompanied by the electron spin transfer, in the chemical reactions between the radical and closed-shell molecule. The aim of the present calculation is not to search for the accurate potential energy surface and to reproduce the experimental activation energy of some specified radical reaction but to discuss, rather qualitatively, the characteristic features of radical reactions by the employment of semiempirical MO's.

Interaction between Radicals and Closed-Shell Molecules

Some theoretical treatments of chemical interactions have been developed.[12-18] Klopman pointed out the importance of the Coulomb interaction and the electron transfer interaction in ionic reactions.[13] He thus used the terms "charge-controlled" and "frontier-controlled" in classifying some kinds of donor–acceptor interactions. Salem discussed thermal and photochemical interactions between two conjugated systems, applying the MO perturbation theory.[14,15] Chemical interaction between the two systems was studied by the present authors by partitioning the interaction energy into several terms in which the Coulomb interaction energy and the charge-transfer interaction energy were included.[18] The important role of the charge-transfer interaction between reagent and reactant in interpreting the formation of new bonds and the weakening of the bonds to be broken in the reaction was pointed out by the use of a CI wave function of a chemically interacting system of two closed shells.[19] In this paper, we present the results of an MO calculation on the chemical interaction of radical with closed-shell molecules, following the line mentioned in our previous papers.[18,19]

The CI method is convenient in order to express the

(1) C. A. Coulson, *Discuss. Faraday Soc.*, **2**, 9 (1947).
(2) J. H. Binks and M. Szwarc, *J. Chem. Phys.*, **30**, 1494 (1959).
(3) K. Fukui, T. Yonezawa, C. Nagata, and H. Shingu, *ibid.*, **22**, 1433 (1954).
(4) K. Fukui, H. Kato, and T. Yonezawa, *Bull. Chem. Soc. Jap.*, **34**, 1111 (1961).
(5) K. Fukui in "Molecular Orbitals in Chemistry, Physics, and Biology," P.-O. Löwdin and B. Pullman, Ed., Academic Press, New York, N. Y., 1964, p 513.
(6) K. Fukui in "Modern Quantum Chemistry. Istanbul Lectures, Part I," O. Sinanoğlu, Ed., Academic Press, New York, N. Y., 1965, p 49.
(7) K. Fukui in "Sigma Molecular Orbital Theory," O. Sinanoğlu and K. B. Wiberg, Ed., Yale University Press, New Haven, Conn., 1970, p 121.
(8) K. Fukui, *Fortschr. Chem. Forsch.*, **15**, 1 (1970).
(9) M. V. Basilevsky and I. E. Chlenov, *Theor. Chim. Acta*, **15**, 174 (1969).
(10) J. R. Hoyland, *ibid.*, **22**, 229 (1971).
(11) K. Morokuma and R. E. Davis, *J. Amer. Chem. Soc.*, **94**, 1060 (1972), and the references cited therein.
(12) J. N. Murrell, M. Randić, and D. R. Williams, *Proc. Roy. Soc., Ser. A*, **284**, 566 (1965).
(13) G. Klopman, *J. Amer. Chem. Soc.*, **90**, 223 (1968).
(14) L. Salem, *ibid.*, **90**, 543, 553 (1968).
(15) A. Devaquet and L. Salem, *ibid.*, **91**, 3793 (1969).
(16) A. Devaquet, *Mol. Phys.*, **18**, 233 (1970).
(17) R. Sustmann and G. Binsch, *ibid.*, **20**, 1, 9 (1971).
(18) K. Fukui and H. Fujimoto, *Bull. Chem. Soc. Jap.*, **41**, 1989 (1968).
(19) H. Fujimoto, S. Yamabe, and K. Fukui, *ibid.*, **44**, 2936 (1971).

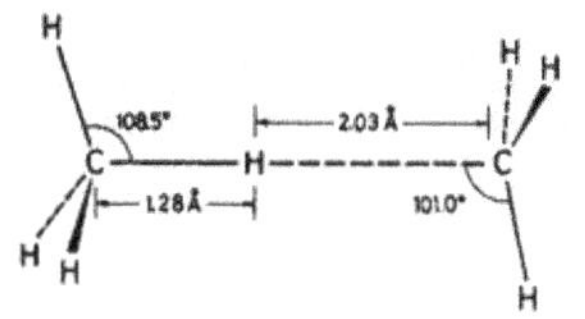

Figure 1. Interaction model of methane and methyl radical.

interacting system by the use of the MO's which are obtained for reactant and reagent in their isolated states. To a first approximation, we may take the electron configurations, representing the adiabatic, one-electron transfer, and local one-electron excitation interactions. Then, the wave function is given by eq 1, where 0 indicates the original state in which

$$\psi = C_0\psi_0 + \sum_i^{\text{occ}}\sum_l^{\text{uno}} C_{i\to l}\psi_{i\to l} + \sum_k^{\text{occ}}\sum_j^{\text{uno}} C_{k\to j}\psi_{k\to j} + \sum_i^{\text{occ}} C_{i\to o}\psi_{i\to o} + \sum_j^{\text{uno}} C_{o\to j}\psi_{o\to j} + \sum_i^{\text{occ}}\sum_j^{\text{uno}} C_{i\to j}\psi_{i\to j} + \sum_k^{\text{occ}}\sum_l^{\text{uno}} C_{k\to l}\psi_{k\to l} + \sum_k^{\text{occ}} C_{k\to o}\psi_{k\to o} + \sum_l^{\text{uno}} C_{o\to l}\psi_{o\to l} \quad (1)$$

neither electron transfer nor electron excitation takes place, i and j are the occupied and unoccupied MO's of the closed-shell molecule, say A, in its isolated state, k, l, and o mean the occupied, unoccupied, and singly occupied MO's of the radical, say B, respectively. The symbols Σ^{occ} and Σ^{uno} imply the summation over occupied and unoccupied MO's, respectively, and $i \to l$, for instance, stands for the one-electron shift from the ith originally occupied MO of A to the lth originally unoccupied MO of B. The coefficients, C_0, $C_{i\to l}$, ..., in eq 1 can be obtained by solving the simultaneous equations

$$\sum_{p'} C_{p'}(H_{p,p'} - S_{p,p'}E) = 0$$
$$(p, p' = 0, i \to l, \ldots) \quad (2)$$

where

$$H_{p,p'} = \int \psi_p{}^* H \psi_{p'} \mathrm{d}\tau$$
$$S_{p,p'} = \int \psi_p{}^* \psi_{p'} \mathrm{d}\tau$$

The interaction energy between the two systems is approximately given by the sum of the four energies: Coulomb (ϵ_Q), exchange (ϵ_K), delocalization (D), and polarization (Π) interaction terms:

$$\Delta W \cong \epsilon_Q + \epsilon_K - D - \Pi \quad (3)$$

The formulas of these energy terms can be obtained rather in a straightforward manner, as in our previous paper.[18] The electron density of the interacting system is calculated by the use of eq 1

$$\rho(1) = M\int \psi^*(1,2,\ldots,M) \times \psi(1,2,\ldots,M)\,\mathrm{d}\xi_1\mathrm{d}\tau_2\ldots\mathrm{d}\tau_M \quad (4)$$

where ψ is assumed to be normalized so as to make the integration of $\rho(1)$ overall space, with respect to the electron referred to as 1, give the total electrons, M, of the interacting system. In the following, we will investigate the change in the electron distribution due to the interaction in detail, since this, we believe, reflects most distinctly what happens in chemical reactions.

Results of the Calculation

The calculation was performed by the use of a semiempirical all-valence electron SCF-MO method including overlap integrals.[20] The method of calculation and the approximations adopted are the same as those in our previous paper.[21] The matrix elements in eq 2 were calculated to the second order of overlap integrals by the use of the Mulliken approximation.[22]

Hydrogen Abstraction of Methane by Methyl Radical. As the first example, we show the result of the calculation on the interaction between methane and methyl radical. The reaction model was assumed to be as shown in Figure 1, making reference to the MINDO calculation by Dewar[23] and our preliminary calculation by the extended Hückel MO method.[24] Although the interaction between a radical and a closed-shell molecule was expressed by the use of restricted open-shell MO's in our previous paper,[18] they cannot necessarily reproduce chemically interesting quantities of radical, such as ionization potential, electron affinity, etc. It may be better to employ the MO's which reflect the properties of independent reactants. We here approximate them by the MO's of the unrestricted open-shell method.[25] The unrestricted MO wave function has a deficiency that it is not the eigenfunction of the operator S^2, but the contamination of higher multiplets to doublet is negligibly small for the present purpose. The vertical ionization potentials of methane (T_d) and methyl radical (assumed to be D_{3h}) obtained by our MO method are 12.79 and 10.41 eV, in fair agreement with the observed values, 12.99 and 9.86 eV, respectively.[26,27] The electron affinity of the methyl radical is calculated to be −1.00 eV, which is about 2 eV smaller than the observed value.[28]

The contributions of local-excited states to the wave function given by eq 1 are usually so small in comparison with those of charge-transferred states that they can be neglected with no serious influence upon the results.[29] We may assume here without loss of generality that the methyl radical has four valence electrons with spin α and three valence electrons with spin β. Then, the electron densities of two spins of the system composed of methane and methyl without any interaction may be given by

$$\rho_A{}^\alpha + \rho_B{}^\alpha = \sum_i^{\text{occ}}(a_i)^2 + \sum_k^{\text{occ}}(b_k)^2 + (b_o)^2 \quad (5)$$

(20) T. Yonezawa, K. Yamaguchi, and H. Kato, *Bull. Chem. Soc. Jap.*, **40**, 536 (1967).
(21) K. Fukui, H. Fujimoto, and S. Yamabe, *J. Phys. Chem.*, **76**, 232 (1972).
(22) R. S. Mulliken, *J. Chim. Phys. Physicochim. Biol.*, **46**, 497 (1949).
(23) M. J. S. Dewar and E. Haselbach, *J. Amer. Chem. Soc.*, **92**, 590 (1970).
(24) R. Hoffmann, *J. Chem. Phys.*, **39**, 1397 (1963); **40**, 2047, 2474, 2480 (1965).
(25) J. A. Pople and R. K. Nesbet, *ibid.*, **22**, 571 (1954).
(26) K. Watanabe, *ibid.*, **26**, 542 (1957).
(27) E. W. C. Clarke and C. A. McDowell, *Proc. Chem. Soc., London*, 69 (1960).
(28) H. O. Pritchard, *Chem. Rev.*, **52**, 529 (1953).
(29) The stabilization due to the energy term Π was estimated to be less than 2% of that due to the energy D in the present calculation.

and

$$\rho_A{}^\beta + \rho_B{}^\beta = \sum_i^{\text{occ}}(a_i)^2 + \sum_k^{\text{occ}}(b_k)^2 \qquad (6)$$

where a_i is the ith occupied MO of A, b_k is the kth occupied MO of B, and b_0 is the MO of the unpaired electron. Here we have

$$\int\{\rho_A{}^\alpha(1) + \rho_B{}^\alpha(1) + \rho_A{}^\beta(1) + \rho_B{}^\beta(1)\}\mathrm{d}v_1 = M \qquad (7)$$

Figure 2 presents the simple sum of the electron densities of methane and methyl in one of the σ_v planes of the C_{3v} symmetry. The difference between the distributions of the two spins is obviously attributable to the unpaired electron which is almost localized to the atomic orbital (AO) of the carbon of methyl, having the extension in the direction along the principal axis.

The wave function ψ_0 may be given by a single determinant which is constructed by a_i's, b_k's, and b_o

$$^1/_2\psi_0 = \mathfrak{N}\mathfrak{A}[a_1\bar{a}_1\ldots a_i\bar{a}_i\ldots a_m\bar{a}_m b_1\bar{b}_1\ldots b_k\bar{b}_k\ldots b_n\bar{b}_n b_o] \qquad (8)$$

where $\mathfrak{N}\mathfrak{A}$ means the normalization-antisymmetrization operator. Taking the terms up to the second order of the overlap integrals of the MO's of A and B (when they are integrated over all space), the electron densities of spin α and spin β in the state of the adiabatic interaction, ψ_0, are given by

$$\rho_o{}^\alpha \cong \sum_i^{\text{occ}}\sum_{i'}^{\text{occ}} a_i a_{i'}\left(\sum_k^{\text{occ}} s_{ik}s_{i'k} + s_{io}s_{i'o}\right) + \left\{\sum_k^{\text{occ}}\sum_{k'}^{\text{occ}} b_k b_{k'}\sum_i^{\text{occ}} s_{ik}s_{ik'} + (b_o)^2\sum_i^{\text{occ}} s_{io}{}^2 + \sum_k^{\text{occ}}(b_k b_o + b_o b_k)\sum_i^{\text{occ}} s_{ik}s_{io}\right\} - 2\left\{\sum_i^{\text{occ}}\sum_k^{\text{occ}} a_i b_k s_{ik} + a_i b_o s_{io}\right\} + \rho_A{}^\alpha + \rho_B{}^\alpha \qquad (9)$$

$$\rho_0{}^\beta \cong \sum_i^{\text{occ}}\sum_{i'}^{\text{occ}} a_i a_{i'}\sum_k^{\text{occ}} s_{ik}s_{i'k} + \sum_k^{\text{occ}}\sum_{k'}^{\text{occ}} b_k b_{k'}\sum_i^{\text{occ}} s_{ik}s_{ik'} - 2\sum_i^{\text{occ}}\sum_k^{\text{occ}} a_i b_k s_{ik} + \rho_A{}^\beta + \rho_B{}^\beta \qquad (10)$$

where

$$s_{ik} = \int a_i(1)b_k(1)\mathrm{d}v_1$$

There appear the terms which represent the distortion of the electron cloud, originating from the electron exchange interaction between the two systems. The exchange interaction between two closed-shell molecules results in the decrease in the electron densities of the intermolecular region and the increase in the electron densities in the neighborhood of the two molecules.[19,30] The same is true for the interaction of a radical with a closed-shell molecule. The third terms in eq 9 and 10 imply the decrease in the electron densities in the region between the two systems, leading to an antibonding contribution to the bonds to be formed between them. The extent of the change in electron distribution is parallel to the sum of the squares of overlap

(30) V. Magnasco, *Theor. Chim. Acta*, **21**, 267 (1971). See, also, M. Dreyfus and A. Pullman, *ibid.*, **19**, 20 (1970); L. Salem, *Proc. Roy. Soc., Ser. A*, **264**, 379 (1961).

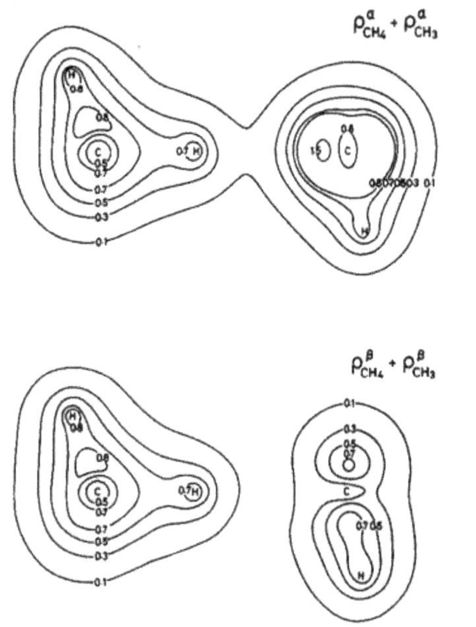

Figure 2. Sum of the electron densities of methane and methyl radical without interaction (in e/Å³). Note that the difference between Figure 2 and Figure 3 or Figure 4 is significant.

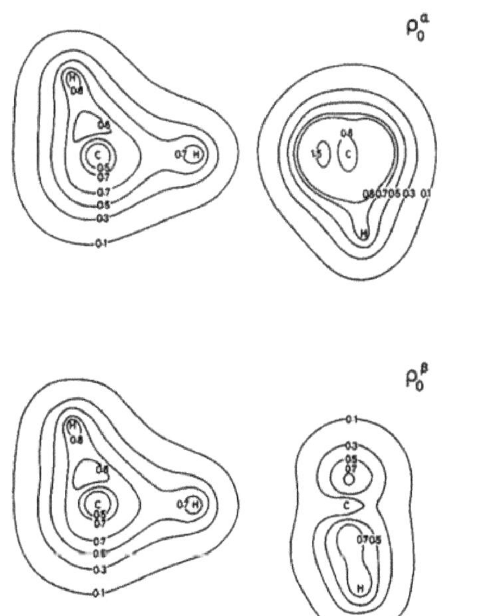

Figure 3. Electron densities of the system composed of methane and methyl radical with adiabatic interaction.

integrals of doubly and singly occupied MO's. In Figure 3 are illustrated $\rho_0{}^\beta$ and $\rho_0{}^\beta$, corresponding to the ψ_0 state. The change in the electron distribution of α spin is conspicuous, while that of spin β is not. The marked change in the electron densities of spin α

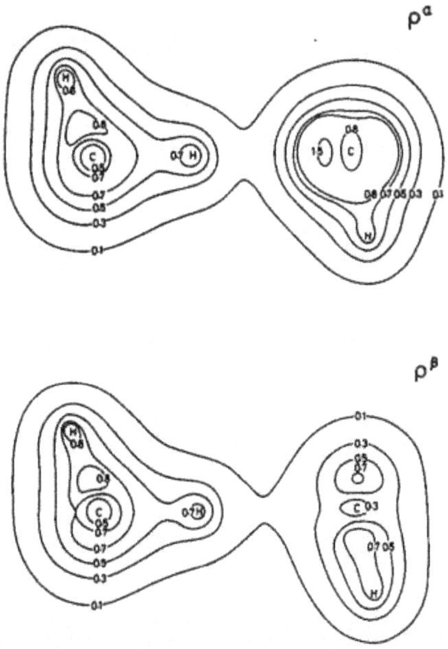

Figure 4. Electron densities of the system composed of methane and methyl radical with adiabatic and charge-transfer interactions.

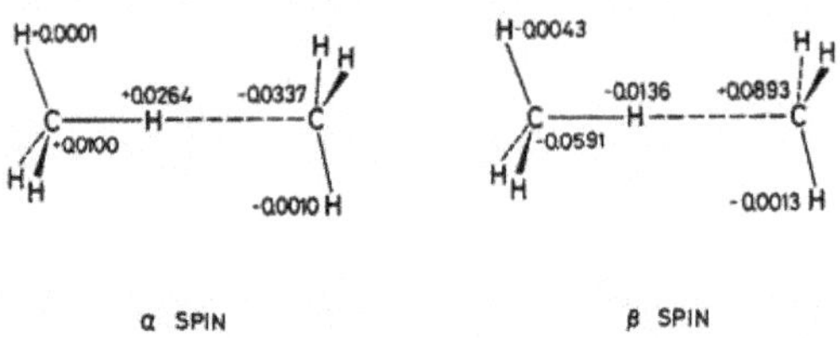

Figure 5. Changes in atomic populations of methane and methyl radical due to interaction.

is clearly owing to the large overlap between a_t^α's and b_o^α.

The electron densities ρ^α and ρ^β of the lowest energy state which is represented by the linear combination of ψ_0 and the charge-transferred electron configurations are shown in Figure 4. The terms to the second order of overlap integrals are taken into consideration. Among the various charge-transferred electron configurations, the one in which an electron is transferred from the HOMO (a_1) of deformed methane (C_{3v}) to b_o^β of methyl was calculated to be the most important and the one from b_o^α to the LUMO (a_1) of methane was found to be the next important. The cross terms of the charge-transferred states and the ψ_0 state in eq 4 are responsible for the increase in electron densities in the intermolecular region which stands for the formation of new bonds between two species.[19,31] However, the increase in the electron densities of spin α in the region between methane and methyl through the charge-transfer interaction is almost cancelled by the decrease caused by the electron exchange interaction. Therefore, the net contribution of α-spin electrons to the bond formation between the hydrogen of methane and the carbon of methyl is of a minor importance in the present model. On the other hand, the growing of the β electron densities in the region between methane and methyl, coming from the introduction of the charge-transferred states, obviously serves as the major origin of the formation of the new carbon–hydrogen bond. Comparison of Figure 4 with Figure 2 indicates that the decrease in the electron densities in the neighborhood of the carbon–hydrogen bond of methane to be broken in the reac-

(31) H. Fujimoto, S. Yamabe, and K. Fukui, *Tetrahedron Lett.*, 443 (1971).

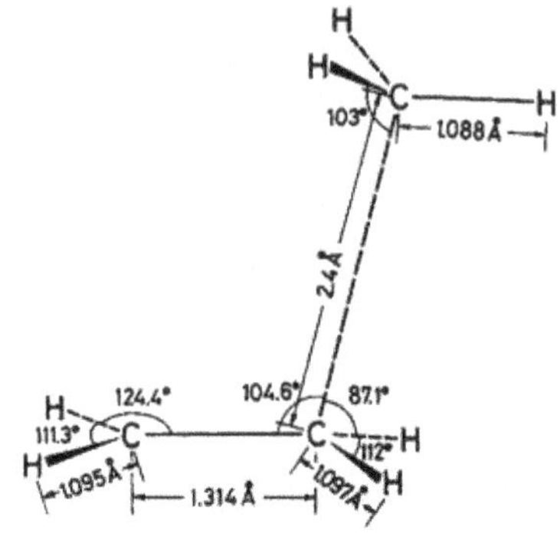

Figure 6. Interaction model of ethylene and methyl radical after Hoyland.[10]

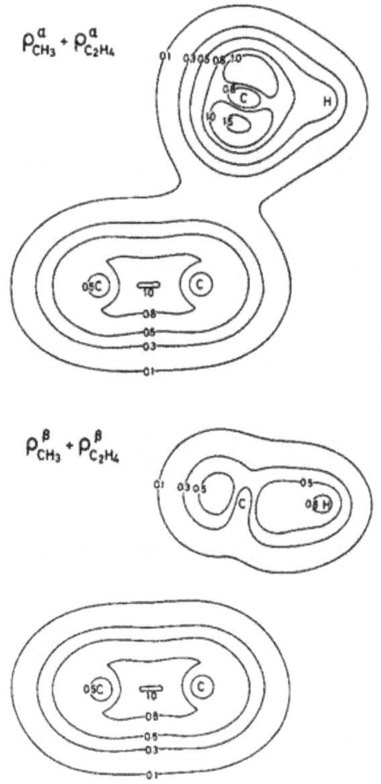

Figure 7. Sum of the electron densities of ethylene and methyl radical without interaction.

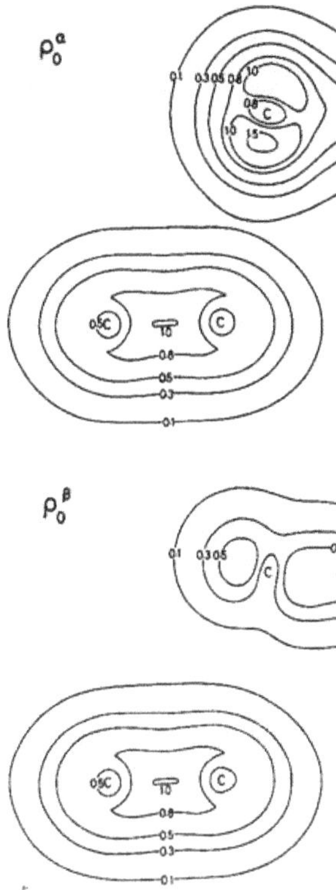

Figure 8. Electron densities of the system composed of ethylene and methyl radical with adiabatic interaction.

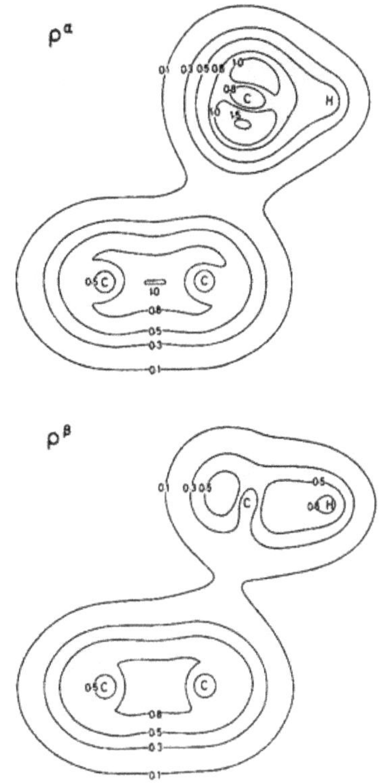

Figure 9. Electron densities of the system composed of ethylene and methyl radical with adiabatic and charge-transfer interactions.

tion is also caused mainly by β-spin electrons. From the above result, we can recognize the interesting behavior of β-spin electrons of methane which tend to form an electron pair with the unpaired electron of methyl.

Methane is the acceptor of the electrons with spin α and the donor of the electrons with spin β. On the contrary, methyl is the donor of the electrons with spin α and the acceptor of the electrons with spin β. The calculated changes in atomic populations, applying the Mulliken population analysis[32] to the AO pairs of methane and methyl, are shown in Figure 5. The amount of electrons with spin α donated from methyl to methane and that with spin β donated from methane to methyl are not the same in the present calculation. This result might be inherent to the approximations used for the calculation of matrix elements of the secular equation. However, it is in agreement with the result of a variational calculation on the whole system composed of methane and a methyl radical that the hydrogen of methane to be abstracted by a methyl radical migrates carrying slightly positive net charge. The essential difference in the behaviors of electrons of opposite spins mentioned above is believed to be independent of the crudeness of approximations, in a qualitative sense.

Addition of Methyl Radical to Ethylene. In order to see if the discussion given above holds for other chemical interactions between a radical and a closed-shell system, we proceed here to another example, the addition of methyl radical to ethylene. The course of the reaction was thoroughly examined by Basilevsky[9] and by Hoyland.[10] They gave good calculated values of the activation energy of the reaction by the use of semiempirical MO methods.[9,10,33]

Figure 6 shows the reaction model employed for the calculation by Hoyland.[10] The simple sums of the electron densities of methyl and ethylene without interaction on the bisecting plane are shown in Figure 7. Here, methyl is assumed to have 4 valence electrons with spin α.

By introducing the electron exchange interaction between the two species, a conspicuous change in the distribution of the electrons with spin α takes place due to the strong repulsive interaction between the unpaired electron of methyl and the π electrons (not in a strict sense) of ethylene as shown in Figure 8. The change in the distribution of the electrons with spin β is relatively small.

Figure 9 demonstrates that the strong antibonding contribution of α-spin electrons to the bond to be newly formed between the carbons of methyl and ethylene, originating from the electron exchange interaction, is barely recovered by the inclusion of charge-transferred states. The important role of β-spin electrons in the formation of the carbon–carbon bond is clearly

(32) R. S. Mulliken, *J. Chem. Phys.*, 23, 1833 (1955).

(33) M. V. Basilevsky, V. A. Tikhomirov, and I. E. Chlenov, *Theor. Chim. Acta*, 23, 75 (1971).

recognized from Figures 7 and 9. Among the various electron configurations except ψ_0, the configuration in which an electron is transferred from the SOMO of methyl to the LUMO of ethylene is calculated to be the most dominant, and the one from the HOMO of ethylene to the SOMO of methyl is the next dominant. The present calculation suggests that methyl is positively charged, while ethylene is negatively charged in the interaction. This tendency seems to be in an agreement with the Szwarc's proposition that a methyl radical adds to olefinic double bonds in a nucleophilic fashion.[34]

Conclusion

A configuration interaction treatment of chemical interactions between methyl and methane and between methyl and ethylene revealed interesting features of the unpaired electron of the radical and the electrons of closed-shell systems with opposite spin. The contribution of an unpaired electron to the formation of new bonds through charge-transfer interaction was found to be cancelled either partly or entirely by the exchange interaction. On the contrary, the charge-transfer interaction from a closed-shell molecule to a radical to form an electron pair with the odd electron contributed significantly to bond formation. Such a behavior of the unpaired electron will have a profound importance in understanding the nature of radical reactions, because the SOMO has the greatest amplitude in the neighborhood of the most reactive position of radical, in general. To estimate quantitatively the ratio of the contribitions of the two spins to the formation of new bonds and the weakening of old bonds is hardly possible at the present time because of the crudeness of the calculation. More elaborate calculations will be needed for such a purpose. Nucleophilicity and electrophilicity of radicals will have a concern with this problem.

Acknowledgment. We would like to express our gratitude to the Data Processing Center, Kyoto University, for its generous permission to use the FACOM 230-60 computer.

(34) A. Rajbenbach and M. Szwarc, *Proc. Roy. Soc., Ser. A*, **251**, 394 (1959).

Molecular orbital calculations of the electronic structure of borazane

Hiroshi Fujimoto, Shigeki Kato, Shinichi Yamabe, and Kenichi Fukui

Faculty of Engineering, Kyoto University, Sakyo-ku, Kyoto, Japan
(Received 2 October 1972)

An *ab initio* molecular orbital calculation of the addition compound of boron hydride and ammonia has been carried out. By expanding the molecular orbitals of the addition compound in terms of the molecular orbitals of boron hydride and of ammonia in a separated state, the origin of charge transfer and bond formation between these two species in the interaction has been studied. A configuration analysis of the wavefunction of the addition compound has been performed.

INTRODUCTION

Organoboranes and boron halides are electron deficient species and seek for electron pairs. They are liable to form addition compounds with Lewis bases.[1] The adducts of trialkylboranes with amines are representative members. Some molecular orbital (MO) calculations have been carried out on the addition compound of boron hydride and ammonia (borazane), discussing its stable configuration and the charge distribution.[2] Some years ago, Klopman pointed out that two factors, viz., the static interaction of net charges and the electron delocalization interaction, should be taken into consideration to study the chemical interaction between electrophile and nucleophile.[3] Thus, he could divide the acid-base interactions into two classes, designating them as "charge controlled" and "frontier controlled." Here it seems interesting to analyze the electronic structure of BH_3-NH_3, with the knowledge of that of BH_3 and NH_3 in a separated state.

Some MO perturbation theories have been proposed to elucidate the detailed mechanism of chemical interaction of two systems.[4-7] A partitioning analysis of interaction energy in a chemical reaction between two systems was made by the present authors, representing the combined system of two reactants by a configuration interaction (CI) wavefunction of charge-transferred and locally excited electron configurations.[8] Succinct expressions of the interaction energy, given by a combined sum of several interaction terms, and the change in the electron distribution due to the interaction, were derived by use of the MO's of reactants in an isolated state. Approximate MO calculations have also been carried out on some chemical reactions.[9] However, such perturbation methods may not necessarily be applied to the BH_3-NH_3 adduct with sufficient accuracy, because the interaction between the two species is too strong to be regarded as a perturbation.

Meanwhile, Lowe suggested a way to construct the perturbed MO's of a reactant placed under the influence of a reagent from the MO's of unperturbed state.[10] The aim of this report is to present the result of a MO calculation of BH_3-NH_3, obtained by expanding the MO's of the adduct in terms of the MO's of BH_3 and the MO's of NH_3 in a separated state.

PROCEDURE FOR CALCULATION

Let us consider an interaction between two systems, e.g., A and B. Their MO's are given by the linear combinations of atomic orbitals (LCAO)

$$a_i(1)=\sum_r^A c_r^{(i)} r(1) \quad \text{for system } A\,(r=1,\ \ldots, m),$$

$$b_k(1)=\sum_s^B c_s^{(k)} s(1) \quad \text{for system } B\,(s=m+1,\ \ldots, m+n),$$

where r and s are AO's. All the MO's are hereafter taken to be real. The MO's a_i $(i=1, \ldots, m)$ form an orthonormal set and the MO's b_k $(k=m+1, \ldots, m+n)$ constitute another orthonormal set. The MO's a_i and b_k are not orthogonal with each other, in general. However, they can easily be transformed into a set of orthonormal one-electron functions ϕ_f $(f=1, \ldots, m+n)$.[11] Then, the MO's of the combined system of two reactants, $A-B$, which are denoted by ψ_g $(g=1, \ldots, m+n)$, are represented by the linear combinations of ϕ,

$$\psi_g=\sum_f d_f^{(g)} \phi_f\,, \tag{1}$$

where

$$d_f^{(g)}=\int \psi_g \phi_f d\tau\,.$$

Carrying out a transformation on $(\phi_1, \phi_2, \ldots, \phi_{m+n})$ into the original basis $(a_1, a_2, \ldots, a_m, b_{m+1}, b_{m+2}, \ldots, b_{m+n})$, we can express the MO's of $A-B$ by the linear combinations of the MO's of A and the MO's of B,

$$\psi_g=\sum_i^{occ} C_i^{(g)} a_i+\sum_j^{uno} C_j^{(g)} a_j+\sum_k^{occ} C_k^{(g)} b_k+\sum_l^{uno} C_l^{(g)} b_l\,, \tag{2}$$

where $\sum^{occ}$ and $\sum^{uno}$ imply taking the sum over all the occupied and over all the unoccupied MO's respectively.[12]

When the combined system $A-B$ has a closed-shell electronic structure and the wavefunction is given by a single Slater determinant, the electron density of the system is defined by

$$
\begin{aligned}
\rho &= 2\sum_{s}^{\mathrm{occ}} \psi_s^2 \\
&= 2\sum_{s}^{\mathrm{occ}}\left[\sum_{i}^{\mathrm{occ}}\{C_i^{(s)}a_i\}^2+\sum_{j}^{\mathrm{uno}}\{C_j^{(s)}a_j\}^2+\sum_{k}^{\mathrm{occ}}\{C_k^{(s)}b_k\}^2+\sum_{l}^{\mathrm{uno}}\{C_l^{(s)}b_l\}^2\right] \\
&+\sum_{s}^{\mathrm{occ}}\left[\sum_{i}^{\mathrm{occ}}\sum_{i'(\neq i)}^{\mathrm{occ}} C_i^{(s)}C_{i'}^{(s)}(a_i a_{i'}+a_{i'}a_i)+2\sum_{i}^{\mathrm{occ}}\sum_{j}^{\mathrm{uno}} C_i^{(s)}C_j^{(s)}(a_i a_j+a_j a_i)+\sum_{j}^{\mathrm{uno}}\sum_{j'(\neq j)}^{\mathrm{uno}} C_j^{(s)}C_{j'}^{(s)}(a_j a_{j'}+a_{j'}a_j)\right] \\
&+\sum_{s}^{\mathrm{occ}}\left[\sum_{k}^{\mathrm{occ}}\sum_{k'(\neq k)}^{\mathrm{occ}} C_k^{(s)}C_{k'}^{(s)}b_k b_{k'}+b_{k'}b_k)+2\sum_{k}^{\mathrm{occ}}\sum_{l}^{\mathrm{uno}} C_k^{(s)}C_l^{(s)}(b_k b_l+b_l b_k)+\sum_{l}^{\mathrm{uno}}\sum_{l'(\neq l)}^{\mathrm{uno}} C_l^{(s)}C_{l'}^{(s)}(b_l b_{l'}+b_{l'}b_l)\right] \\
&+2\sum_{s}^{\mathrm{occ}}\left[\sum_{i}^{\mathrm{occ}}\sum_{k}^{\mathrm{occ}} C_i^{(s)}C_k^{(s)}(a_i b_k+b_k a_i)+\sum_{i}^{\mathrm{occ}}\sum_{l}^{\mathrm{uno}} C_i^{(s)}C_l^{(s)}(a_i b_l+b_l a_i)+\sum_{j}^{\mathrm{uno}}\sum_{k}^{\mathrm{occ}} C_j^{(s)}C_k^{(s)}(a_j b_k+b_k a_j)\right. \\
&\left.+\sum_{j}^{\mathrm{uno}}\sum_{l}^{\mathrm{uno}} C_j^{(s)}C_l^{(s)}(a_j b_l+b_l a_j)\right]. \qquad (3)
\end{aligned}
$$

The terms in the first brackets in Eq. (3) are the major origin of the loosening or strengthening of bonds in A and B in the course of chemical interaction, because they mean partial removing of electrons from occupied MO's and placing a portion of them into unoccupied MO's. The second and third terms have locally nonzero values, in general, although their integration over all the space vanishes. Accordingly, they may be regarded as representing intramolecular electron rearrangement in each A and B. It is the last term that is responsible for the formation of chemical bonds between two reactants. Among the terms appearing in the last brackets, the orbital overlap interaction between the occupied MO's of A and the occupied MO's of B makes an antibonding contribution to the bonds to be formed between A and B, as a whole.[8,13-15] Increase in the electron densities in the intermolecular region between A and B originates mainly from the orbital interaction of the occupied MO's of A with the unoccupied MO's of B and from the interaction of the occupied MO's of B with the unoccupied MO's of A.[8] The contribution of the orbital interaction between the unoccupied MO's of A and the unoccupied MO's of B to the bond formation is supposed to be far less important than other terms, in general.

Equation (3) indicates that a part of electrons removed from the occupied MO's of a reactant is elevated to the unoccupied MO's of the same reactant, another part is transferred to the other reactant, and the remaining part is consumed for the formation of covalent bonds between the two systems. The changes in electron densities which arise in the intermolecular region from the interaction of the MO a_i and A and the MO b_k of B yield the overlap integral between these MO's when they are integrated:

$$
\int a_i(1)b_k(1)dv(1)=S_{ik}=\sum_{r}^{A}\sum_{s}^{B} c_r^{(i)}c_s^{(k)}s_{rs},
$$

$$
s_{rs}=\int r(1)s(1)dv(1).
$$

The sum of the r,s element of the above quantity, multiplied by $4C_i^{(s)}C_k^{(s)}$ over all the doubly occupied MO's of the combined system A–B gives a portion of the overlap population between the AO's r and s. According to the population analysis of Mulliken,[13] the overlap population between the AO's r and s is divided evenly into two parts, one constituting the valence-active part[16] of the gross population of the AO r and the other that of the AO's. Consequently, we may define the following quantity as the change in the occupation number of the initially doubly occupied MO a_i due to the interaction:

$$
\Delta\nu_i=2\sum_{s}^{\mathrm{occ}}\left\{(C_i^{(s)})^2+\sum_{k}^{\mathrm{occ}} C_i^{(s)}C_k^{(s)}S_{ik}+\sum_{l}^{\mathrm{uno}} C_i^{(s)}C_l^{(s)}S_{il}\right\}-2. \qquad (4)
$$

Similarly, the change in the occupation number of the initially unoccupied MO a_j may be given by

$$
\Delta\nu_j=2\sum_{s}^{\mathrm{occ}}\left\{(C_j^{(s)})^2+\sum_{k}^{\mathrm{occ}} C_j^{(s)}C_k^{(s)}S_{jk}+\sum_{l}^{\mathrm{uno}} C_j^{(s)}C_l^{(s)}S_{jl}\right\}. \qquad (5)
$$

Examination of these quantities will supply us with a detailed information on the number of electrons donated and accepted with regard to each MO, as well as the over-all amount of the charge transferred between the two reactants.

CHARGE TRANSFER AND BOND FORMATION

The MO's of BH_3–NH_3 were calculated by an *ab initio* MO method, expanding the Slater AO into three Gaussian functions.[17] The nuclear configuration of the adduct is shown in Fig. 1.[18] The MO's of two components, BH_3 and NH_3, were also calculated by the same method with respect to the same nuclear configuration as in the adduct. They are shown in Fig. 2.

Table I gives the changes in the occupation numbers of MO's, defined in Eqs. (4) and (5).[19] It can be seen that electrons are removed from the highest occupied (HO) MO of NH_3 most distinctly, and that almost the same number of electrons flows into the lowest unoccupied (LU) MO of BH_3. The changes in other MO's are

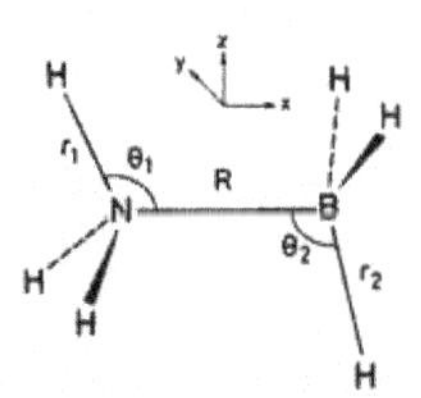

FIG. 1. Nuclear configuration of borazane. $R = 1.561$ Å, $r_1 = 1.011$ Å, $r_2 = 1.222$ Å, $\theta_1 = 116.0°$, and $\theta_2 = 105.4°$.

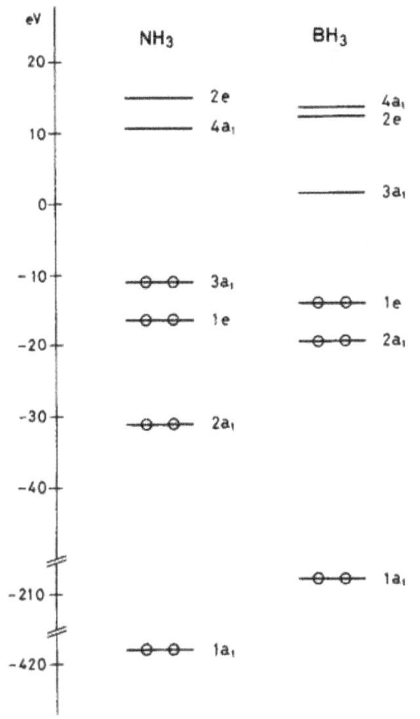

FIG. 2. MO levels of BH_3 part (C_{3v}) and NH_3 part (C_{3v}).

TABLE I. Changes in the occupation number of the MO's of BH_3 and NH_3 due to the interaction.

BH_3 MO	$\Delta\nu$	NH_3 MO	$\Delta\nu$
$1a_1$	+0.0007	$1a_1$	+0.0009
$2a_1$	−0.0453	$2a_1$	+0.0005
$1e$	−0.0060	$1e$	−0.0129
$3a_1$	+0.4759	$3a_1$	−0.4394
$2e$	+0.0057	$4a_1$	+0.0003
$4a_1$	+0.0064	$2e$	+0.0133
total	+0.4370	total	−0.4370

relatively small in comparison with these two MO's. This suggests that the charge-transferred state from the HOMO of NH_3 to the LUMO of BH_3 makes a dominant contribution to the electronic structure of BH_3–NH_3.

The contributions of orbital interactions between the MO's of BH_3 and the MO's of NH_3 to the overlap population of the B–N bond are listed in Table II. It is understood that the bond is formed mainly through the overlappings of the $2s$ and $2p_x$ AO's of boron and nitrogen. Furthermore, Table II demonstrates that the interaction between the occupied MO's of NH_3 and the unoccupied MO's of BH_3 is of substantial importance in explaining the formation of the bond. This may be related to the direction of extension of the HOMO of NH_3 and of the LUMO of BH_3.

TABLE II. Breakdown of the AO overlap populations of the B–N bond.

			B $1s$	B $2s$	B $2p_x$	B $2p_y$	B $2p_z$
N	$1s$	i[a]	0.0000	0.0016	−0.0003	0	0
		ii	0.0000	−0.0027	−0.0060	0	0
		iii	0.0000	0.0000	0.0000	0	0
		iv	0.0000	−0.0001	−0.0002	0	0
	$2s$	i	−0.0008	−0.1091	0.0164	0	0
		ii	−0.0010	0.0993	0.1817	0	0
		iii	0.0002	−0.0011	0.0002	0	0
		iv	−0.0001	0.0046	0.0090	0	0
	$2p_x$	i	−0.0039	−0.0163	0.0021	0	0
		ii	−0.0033	0.1487	0.2529	0	0
		iii	−0.0002	0.0004	−0.0001	0	0
		iv	0.0000	−0.0018	−0.0030	0	0
	$2p_y$	i	0	0	0	−0.0325	0
		ii	0	0	0	0.0085	0
		iii	0	0	0	0.0149	0
		iv	0	0	0	0.0000	0
	$2p_z$	i	0	0	0	0	−0.0325
		ii	0	0	0	0	0.0085
		iii	0	0	0	0	0.0149
		iv	0	0	0	0	0.0000

[a]Overlap populations, arising from the interaction between (i) occ MO's of BH_3 and occ MO's of NH_3, (ii) occ MO's of NH_3 and uno MO's of BH_3, (iii) occ MO's of BH_3 and uno MO's of NH_3, and (iv) uno MO's of BH_3 and uno MO's of NH_3.

WAVE FUNCTION

Each MO of BH_3–NH_3 is expressed by the linear combination of the basis MO's of two separated species. The composition of the highest occupied sigma MO of BH_3–NH_3 is given in Table III. It is shown that the MO consists mainly of the antibonding combination of $3a_1$ MO of NH_3 and $2a_1$ MO of BH_3 and of the bonding combination of $3a_1$ MO of NH_3 and $3a_1$ MO of BH_3.

By using the MO's of the combined system A–B given by the linear combinations of the MO's of two interacting species, the MO wavefunction of the system A–B can be represented in a chemically graspable form. We may write the wavefunction of the system by the combination of various electron configurations, including charge-transferred and polarized states,[8]

$$\Psi = C_0\Psi_0 + \left(\sum_q^{\mathrm{mtr}} + \sum_q^{\mathrm{mex}} + \sum_q^{\mathrm{ditr}} + \sum_q^{\mathrm{diex}} + \cdots\right) C_q \Psi_q \,, \tag{6}$$

where Ψ_0 is the state in which neither electron transfer nor electron excitation takes place, mtr means single electron transferred configuration with respect to Ψ_0, mex denotes locally excited configuration of single electron in one of the two reactants, ditr implies transfer of two electrons, and so on. The electron configurations corresponding to the original state Ψ_0, monotransferred state $\Psi_{i\to l}$, and monoexcited state $\Psi_{i\to j}$ are illustrated in Fig. 3. Following the conventional abbreviation, the wavefunction Ψ_0 is represented as

$$\Psi_0 = N_0 \left| a_1 \bar{a}_1 a_2 \cdots a_M \bar{a}_M b_{m+1} \bar{b}_{m+1} b_{m+2} \cdots b_{m+N} \bar{b}_{m+N} \right| , \tag{7}$$

where N_0 denotes the normalization factor and the number of electrons of A in a separated state is $2M$ and that of B is $2N$. Similarly, the charge-transferred state $\Psi_{i\to l}$ and polarized state $\Psi_{i\to j}$ are written as

$$\begin{aligned} {}^1\Psi_{i\to l} = (2)^{-1/2} \{ & N_{i\to l} \left| a_1 \bar{a}_1 \cdots \bar{a}_{i-1} a_i a_{i+1} \cdots \bar{a}_M b_{m+l} \cdots \right. \\ & \left. \times \bar{b}_{m+N} \bar{b}_{m+l} \right| - N_{i\to l} \left| a_1 \bar{a}_1 \cdots \bar{a}_{i-1} \bar{a}_i a_{i+1} \cdots \right. \\ & \left. \times \bar{a}_M b_{m+1} \cdots \bar{b}_{m+N} b_{m+l} \right| \} \end{aligned} \tag{8}$$

and

TABLE III. Composition of the highest occupied sigma MO of borazane.

	MO	Coefficient[a]
BH_3	$1a_1$	−0.026 14
	$2a_1$	−0.812 52
	$1e$	0
	$3a_1'$	0.29850
	$2e$	0
	$4a_1$	0.037 11
NH_3	$1a_1$	−0.003 31
	$2a_1$	−0.003 30
	$1e$	0
	$3a_1$	0.530 14
	$4a_1$	0.008 65
	$2e$	0

[a]The sign of each component MO was taken to make the 2*s* AO of boron or nitrogen have positive coefficient of LCAO.

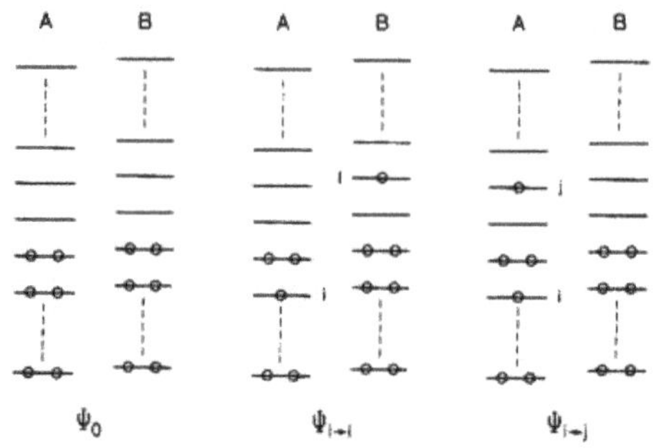

FIG. 3. Schematic representation of the original, monotransferred, and monoexcited electron configuration.

$$\begin{aligned} {}^1\Psi_{i\to j} = (2)^{-1/2} \{ & N_{i\to j} \left| a_1 \bar{a}_1 \cdots \bar{a}_{i-1} a_i a_{i+1} \cdots \right. \\ & \left. \times \bar{a}_M b_{m+1} \cdots \bar{b}_{m+N} \bar{a}_j \right| - N_{i\to j} \left| a_1 \bar{a}_1 \cdots \bar{a}_{i-1} \bar{a}_i a_{i+1} \cdots \right. \\ & \left. \times \bar{a}_M b_{m+1} \cdots \bar{b}_{m+N} a_j \right| \} . \end{aligned} \tag{9}$$

The wavefunctions for other electron configurations can be given in a similar manner by use of the MO's a_i, a_j, b_k, and b_l. Here it is convenient to take the virtual orbitals as unoccupied MO's corresponding to the unrealistic roots of the Fock–Roothaan equations, to obtain the occupied MO's.[20]

By use of the MO's of the combined system A–B expanded in terms of the MO's of A and B, we have

$$\begin{aligned} \Psi &= \{(2M+2N)!\}^{-1/2} \left| \psi_1 \bar{\psi}_1 \psi_2 \bar{\psi}_2 \cdots \psi_{M+N} \bar{\psi}_{M+N} \right| \\ &= \left(\sum_{\gamma_1} \sum_{\gamma_3} \sum_{\gamma_{2(M+N)-1}} C_{\gamma_1}^{(1)} C_{\gamma_3}^{(2)} \cdots C_{\gamma_{2(M+N)-1}}^{(M+N)} \right) \\ &\quad \times \left(\sum_{\gamma_2} \sum_{\gamma_4} \cdots \sum_{\gamma_{2(M+N)}} C_{\gamma_2}^{(1)} C_{\gamma_4}^{(2)} \cdots C_{\gamma_{2(M+N)}}^{(M+N)} \right) \\ &\quad \times \{(2M+2N)!\}^{-1/2} \left| \chi_{\gamma_1} \chi_{\gamma_2} \chi_{\gamma_3} \chi_{\gamma_4} \cdots \chi_{\gamma_{2(M+N)-1}} \chi_{\gamma_{2(M+N)}} \right| , \end{aligned} \tag{10}$$

where χ stands for a and b and each summation is from 1 to $m+n$.

By comparing Eq. (6) with Eq. (10), it is easy to calculate the coefficients C_0, $C_{i\to l}$, $C_{i\to j}$, etc., in the CI wavefunction. For the original state Ψ_0, we have

$$C_0 = \left[\sum_p (-1)^p P\{ C_1^{(1)} C_2^{(2)} \cdots C_M^{(M)} C_{m+1}^{(M+1)} \cdots C_{m+N}^{(M+N)} \} \right]^2 (1/N_0') , \tag{11}$$

where permutation $\sum_p (-1)^p P\{C_1^{(1)} C_2^{(2)}\}$ stands for $(C_1^{(1)} C_2^{(2)} - C_2^{(1)} C_1^{(2)})$, for instance, and $N_0' = N_0 \{(2M+2N)!\}^{1/2}$. The normalization factor is given in the Appendix. With respect to charge-transferred and locally excited electron configurations, we have the following relations:

$$\begin{aligned} C_{i\to l} = (2)^{1/2} & \left[\sum_p (-1)^p P\{ C_1^{(1)} \cdots C_{m+l}^{(i)} \cdots \right. \\ & \left. \times C_M^{(M)} C_{m+1}^{(M+1)} \cdots C_{m+N}^{(M+N)} \} \right] \\ & \times \left[\sum_{p'} (-1)^{p'} P' \{ C_1^{(1)} \cdots C_i^{(i)} \cdots C_M^{(M)} C_{m+1}^{(M+1)} \cdots C_{m+N}^{(M+N)} \} \right] \\ & \times (1/N_{i\to l}') \end{aligned} \tag{12}$$

and

$$C_{i\text{-}j}=(2)^{1/2}\left[\sum_{p}(-1)^{p}P\{C_1^{(1)}\cdots C_j^{(i)}\cdots \times C_M^{(M)}C_{m+1}^{(M+1)}\cdots C_{m+N}^{(M+N)}\}\right]\times\left[\sum_{p'}(-1)^{p'}P'\{C_1^{(1)}\cdots C_i^{(i)}\cdots C_M^{(M)}C_{m+1}^{(M+1)}\cdots C_{m+N}^{(M+N)}\}\right]\times(1/N'_{i\text{-}j})\,. \tag{13}$$

Equations (11)–(13) are the same as the ones derived by Baba *et al.* in their configuration analysis,[21] except for the normalization factors due to the inclusion of the overlap integrals.

In the case of the addition compound of BH_3 and NH_3 mentioned above, the coefficient C_0 was calculated to be 0.607 55. The contributions of various monotransferred electron configurations are shown in Table IV. The expectation that the charge-transferred state from the HOMO of NH_3 to the LUMO of BH_3 is discriminatively more important than other monotransferred terms is really confirmed. The coefficients of monoexcited states are given in Table V. The contribution of polarized states is far smaller than that of charge-transferred states.

Among the ditransferred electron configurations, the most dominant term was found to be the state in which two electrons were shifted from the HOMO of NH_3 to the LUMO of BH_3. The coefficient of this state is 0.087 42. Representing the wavefunction of borazane by the linear combination of the original, monotransferred, and ditransferred electron configurations, Mulliken suggested that $a>b>c$ in the equation[22]

$$\Psi=a\Psi(H_3N,BH_3)+b\Psi(H_3N^{+}-B^{-}H_3)+c\Psi(H_3N^{2+},B^{2-}H_3).$$

The result of the present calculation provides the intuitive suggestion of Mulliken with a quantitative support.

Table VI shows the approximate energy differences between the Ψ_0 state and monotransferred states which have been calculated by use of the equation derived in our previous paper.[8] The charge-transferred state from the HOMO of NH_3 to the LUMO of BH_3 is found to be located about 2.28 eV above the original state in our calculation.

TABLE IV. Coefficients of monotransferred electron configurations.

$\lvert C_{i\text{-}l}\rvert$		l^a: $4a_1$	$2e$		
i^a	$1a_1$	0.002 29	0	0	
	$2a_1$	0.002 50	0	0	
	$1e$	0	0.041 35	0	
		0	0	0.041 35	
$\lvert C_{k\text{-}j}\rvert$		j^a: $3a_1$	$2e$		$4a_1$
k^a	$1a_1$	0.005 33	0	0	0.001 22
	$2a_1$	0.002 72	0	0	0.000 28
	$1e$	0	0.010 84	0	0
		0	0	0.010 84	0
	$3a_1$	0.367 86	0	0	0.022 90

[a]i and j denote the occupied MO and the unoccupied MO of BH_3 part, respectively and k and l indicate the occupied MO and the unoccupied MO of NH_3 part, respectively.

TABLE V. Coefficients of monoexcited electron configurations.

$\lvert C_{i\text{-}j}\rvert$		j^a: $3a_1$	$2e$		$4a_1$
i^a	$1a_1$	0.006 03	0	0	0.001 16
	$2a_1$	0.074 52	0	0	0.025 72
	$1e$	0	0.019 07	0	0
		0	0	0.019 07	0
$\lvert C_{k\text{-}l}\rvert$		l^a: $4a_1$	$2e$		
k^a	$1a_1$	0.001 58	0	0	
	$2a_1$	0.017 01	0	0	
	$1e$	0	0.061 53	0	
		0	0	0.061 53	
	$3a_1$	0.018 08	0	0	

[a]The same as in Table IV.

CONCLUSION

By expanding the MO's of borazane in terms of the MO's of BH_3 and NH_3 in an isolated state, we have investigated the origin of charge transfer and bond formation and have analyzed the MO wavefunction in a chemically graspable form. The particular importance of the charge transfer from the HOMO of NH_3 to the LUMO of BH_3 in understanding the electronic structure of the addition compound has been recognized quantitatively. The method presented here will be helpful in studying the basic features of chemical interactions between two systems to which ordinary perturbation method for chemical interaction can not be applied with sufficient accuracy.

ACKNOWLEDGMENT

We are grateful to the Data Processing Center, Kyoto University for its generous permission to use the FACOM 230·60 computer.

TABLE VI. Energy differences between the original state and monotransferred states.

		l^a: $4a_1$	$2e$		
i^a	$1a_1$	311.14	...	...	
	$2a_1$	19.90	...	...	
	$1e$	...	18.96	...	
		...	...	18.96	
		j^a: $3a_1$	$2e$		$4a_1$
k^a	$1a_1$	414.64	...	...	430.53
	$2a_1$	22.86	...	...	41.55
	$1e$	...	25.24	...	...
		...	...	25.24	...
	$3a_1$	2.28	...	...	23.87

[a]The same as in Table IV.

APPENDIX

The normalization of the wavefunction is made by multiplying the following factor.

The normalization factor of Ψ_0 is given by

$$N_0 = \{(2M+2N)!\}^{-1/2} \begin{vmatrix} S_{11}\, S_{12} \cdots S_{1M}\, S_{1m+1}\, S_{1m+2} \cdots S_{1m+N} \\ S_{21}\, S_{22} \cdots S_{2M}\, S_{2m+1}\, S_{2m+2} \cdots S_{2m+N} \\ \cdots \\ \cdots \\ S_{m+N1} \cdots S_{m+Nm+N} \end{vmatrix}^{-1} .$$

The normalization factor of Ψ_{l-l} can be calculated in the following manner. We transform the set of doubly occupied MO's a_i $(i=1,\cdots,M)$ and b_k $(k=m+1,\cdots,m+N)$ in the original state and the additional MO b_{m+l} into an orthonormal set

$$(a_1, a_2, \cdots, a_M, b_{m+1}, b_{m+2}, \cdots, b_{m+N}, b_{m+l}) = (\phi_1, \phi_2, \cdots, \phi_{M+N}, \phi_{M+N+1})\mathbf{T} ,$$

where we may write

$$\mathbf{T} = \begin{pmatrix} t_1^{(1)}\, t_1^{(2)} \cdots t_1^{(M+N+1)} \\ t_2^{(1)}\, t_2^{(2)} \cdots t_2^{(M+N+1)} \\ \cdots \\ t_{M+N+1}^{(1)} \cdots t_{M+N+1}^{(M+N+1)} \end{pmatrix} .$$

Therefore, we have,

$$\begin{aligned} &|a_1 \bar{a}_1 \cdots \bar{a}_{l-1} b_{m+l} \bar{a}_l \cdots a_M \bar{a}_M b_{m+1} \bar{b}_{m+1} \cdots b_{m+N} \bar{b}_{m+N}| \\ &= \Big(\sum_{\gamma_1}\sum_{\gamma_2} \cdots \sum_{\gamma_{M+N}} t_{\gamma_1}^{(1)} t_{\gamma_2}^{(2)} \cdots t_{\gamma_l}^{(M+N+1)} \cdots t_{\gamma_M}^{(M)} t_{\gamma_{M+1}}^{(M+1)} \cdots t_{\gamma_{M+N}}^{(M+N)}\Big) \\ &\times \Big(\sum_{\gamma'_1}\sum_{\gamma'_2} \cdots \sum_{\gamma'_{M+N}} t_{\gamma'_1}^{(1)} t_{\gamma'_2}^{(2)} t_{\gamma'_l}^{(l)} \cdots t_{\gamma'_M}^{(M)} t_{\gamma'_{M+1}}^{(M+1)} \cdots t_{\gamma'_{M+N}}^{(M+N)}\Big) |\phi_{\gamma_1} \bar{\phi}_{\gamma'_1} \phi_{\gamma_2} \bar{\phi}_{\gamma'_2} \cdots \phi_{\gamma_{M+N}} \bar{\phi}_{\gamma'_{M+N}}| \\ &= \Big(\sum_{\gamma_1<\gamma_2<\cdots<\gamma_{M+N}}^{M+N+1}{}' \sum_{p} (-1)^p P\{t_{\gamma_1}^{(1)} t_{\gamma_2}^{(2)} \cdots t_{\gamma_l}^{(M+N+1)} \cdots t_{\gamma_{M+N}}^{(M+N)}\}\Big) \\ &\times \Big(\sum_{\gamma'_1<\gamma'_2<\cdots<\gamma'_{M+N}}^{M+N+1}{}' \sum_{p'} (-1)^{p'} P' \{t_{\gamma'_1}^{(1)} t_{\gamma'_2}^{(2)} \cdots t_{\gamma'_l}^{(l)} \cdots t_{\gamma'_{M+N}}^{(M+N)}\}\Big) |\phi_{\gamma_1} \bar{\phi}_{\gamma'_1} \phi_{\gamma_2} \bar{\phi}_{\gamma'_2} \cdots \phi_{\gamma_{M+N}} \phi_{\gamma'_{M+N}}| , \end{aligned}$$

where the summation $\sum_{\gamma_w}$ is from 1 to $M+N+1$, and $\sum'_{\gamma_1<\cdots<\gamma_{M+N}}$ means taking the sum over all the ways of separating $(M+N)$ orbitals from $(M+N+1)$ orbitals. Here, we can use the relation

$$\sum_{\gamma_1<\gamma_2<\cdots<\gamma_{M+N}}^{M+N+1} \sum_{w=1}^{M+N} |t_{\gamma_w}^{(u)} t_{\gamma_w}^{(v)}| = \left|\sum_{w=1}^{M+N+1} t_w^{(u)} t_w^{(v)}\right| ,$$

where u and v denote a column and a row. Consequently, the normalization factor is given by

$$N'_{l-1} = (2)^{1/2} \left\{ \begin{vmatrix} S_{11} S_{12} & \cdots & S_{1m+l} & \cdots & S_{1M} S_{1m+1} \cdots S_{1m+N} \\ S_{21} S_{22} & \cdots & S_{2m+l} & \cdots & S_{2M} S_{2m+1} \cdots S_{2m+N} \\ & & \cdots & & \\ S_{m+l1} & \cdots & S_{m+lm+l} & \cdots & S_{m+lm+N} \\ & & \cdots & & \\ S_{m+N1} & \cdots & S_{m+Nm+l} & \cdots & S_{m+Nm+N} \end{vmatrix} \right.$$

$$\times \begin{vmatrix} S_{11} S_{12} \cdots S_{1l} & \cdots & S_{1M} S_{1m+1} & \cdots & S_{1m+N} \\ S_{21} S_{22} \cdots S_{2l} & \cdots & S_{2M} S_{2m+1} & \cdots & S_{2m+N} \\ & & \cdots & & \\ S_{l1} S_{l2} \cdots S_{ll} & \cdots & S_{lM} S_{lm+1} & \cdots & S_{lm+N} \\ & & \cdots & & \\ S_{m+N1} \cdots S_{m+Nl} & & \cdots & & S_{m+Nm+N} \end{vmatrix}$$

$$\left. + \begin{vmatrix} S_{11} S_{12} \cdots & S_{1m+l} & \cdots S_{1M} S_{1m+1} \cdots S_{1m+N} \\ S_{21} S_{22} \cdots & S_{2m+l} & \cdots S_{2M} S_{2m+1} \cdots S_{2m+N} \\ & \cdots & \\ S_{l1} S_{l2} \cdots & S_{lm+l} & \cdots S_{lM} S_{lm+1} \cdots S_{lm+N} \\ & \cdots & \\ S_{m+N1} \cdots & S_{m+Nm+l} & \cdots \quad S_{m+Nm+N} \end{vmatrix}^2 \right\}^{-1/2}$$

The normalization factor of Ψ_{l-j} can be obtained by simply replacing $m+l$ by j in the above equation.

[1]L. Pauling, The Nature of the Chemical Bond (Cornell U. P., Ithaca, N. Y., 1960), p. 124.

[2](a) R. Hoffmann, J. Chem. Phys. **40**, 2474 (1964); (b) A. Veillard, B. Levy, R. Daudel, and F. Gallais, Theor. Chim. Acta **8**, 312 (1967); (c) M.-Cl. Moireau and A. Veillard, *ibid.* **11**, 344 (1968); (d) S. Peyerimhoff and R. J. Buenker, J. Chem. Phys. **49**, 312 (1968); (e) B. Tinland, J. Mol. Struct. **3**, 244 (1968); (f) A. A. Frost, Theor. Chim. Acta **18**, 156 (1970).

[3]G. Klopman, J. Am. Chem. Soc. **90**, 223 (1968).

[4]J. N. Murrell, M. Randić, and D. R. Williams, Proc. R. Soc. A **284**, 566 (1965).

[5]L. Salem, J. Am. Chem. Soc. **90**, 543 (1968); **90**, 553 (1968).

[6](a) A. Devaquet and L. Salem, J. Am. Chem. Soc. **91**, 3793 (1969); (b) A. Devaquet, Mol. Phys. **18**, 233 (1970).

[7]R. Sustmann and G. Binsch, Mol. Phys. **20**, 1 (1971); **20**, 9 (1971).

[8](a) K. Fukui and H. Fujimoto, Bull. Chem. Soc. Jap. **41**, 1989 (1968); (b) K. Fukui and H. Fujimoto, *ibid.* **42**, 3399 (1969);

(c) H. Fujimoto, S. Yamabe, and K. Fukui, *ibid.* **44**, 2936 (1971).

[9]K. Fukui, H. Fujimoto, and S. Yamabe, J. Phys. Chem. **76**, 232 (1972).

[10]J. P. Lowe, J. Am. Chem. Soc. **93**, 301 (1971); **94**, 60 (1972); **94**, 3718 (1972).

[11]P.-O. Löwdin, J. Chem. Phys. 18, 365 (1950).

[12]Alternatively, the coefficients $C_i^{(\ell)}$, $C_j^{(\ell)}$, $C_k^{(\ell)}$, and $C_l^{(\ell)}$ can be obtained as the solutions of m and n diemnsional linear simultaneous equations to adjust the AO coefficients with respect to each MO of the combined system.

[13]R. S. Mulliken, J. Chem. Phys. **23**, 1833, 1841 (1955).

[14]L. Salem, Proc. R. Soc. A **264**, 379 (1961).

[15]V. Magnasco, Theor. Chim. Acta **21**, 267 (1971).

[16]K. Ruedenberg, Rev. Mod. Phys. **34**, 326 (1962).

[17](a) K. O-ohata, H. Taketa, and S. Huzinaga, J. Phys. Soc. Jap. **21**, 2306 (1966); (b) H. Taketa, S. Huzinaga, and K. O-ohata, *ibid.*, **21**, 2313 (1966). The constants α were taken after R. F. Stewart, J. Chem. Phys. **52**, 431 (1970), and the functions F_ν were evaluated by use of the approximation proposed by L. J. Schaad and G. O. Morrel, J. Chem. Phys. **54**, 1965 (1971).

[18](a) S. G. Shore and R. W. Parry, J. Am. Chem. Soc. **77**, 6084 (1955); (b) E. W. Hughes, *ibid.* **78**, 502 (1956); (c) E. L. Lippert and W. N. Lipscomb, *ibid.* **78**, 503 (1956).

[19]The occupation number of some MO's, e.g., $1a_1$ of NH_3, exceeds 2. This may be due to the inappropriateness of the application of the population analysis of Mulliken to such heteropolar bonds. Some other partitioning of electron population will be more reasonable [see, for instance, A. J. Duke and R. F. W. Bader, Chem. Phys. Lett. **10**, 631 (1971). However, the excess over 2 is negligibly small in the present case.

[20]C. C. J. Roothaan, Rev. Mod. Phys. **23**, 69 (1951).

[21]H. Baba, S. Suzuki, and T. Takemura, J. Chem. Phys. **50**, 2078 (1969).

[22]R. S. Mulliken and W. B. Person, *Molecular Complexes* (Wiley, New York, 1969), p. 187.

An Orbital Interaction Rationale for the Role of Catalysts

Kenichi Fukui* and Satoshi Inagaki

Contribution from the Department of Hydrocarbon Chemistry, Kyoto University, Kyoto, Japan. *Received October 15, 1974*

Abstract: The orbital interaction scheme having successfully been applied to interpreting the nature of two-molecule interactions is extended to three-body interactions involving catalytic actions of the third species influencing the two molecules to react with each other. The importance of the interactions among HOMO's and LUMO's of three species is stressed with some perturbation theoretical formulations to explain how the third species can assist an energetically or symmetrically disfavored process to take place. Several illustrative examples are given.

The molecular orbital (MO) perturbation theory based on the orbital interaction of molecules has played a significant role in developing the theory of chemical reactions.[1-4] The reactivity indices derived from such an approach defined in regard to the isolated reactant molecule concisely reflected various chemical data.[1] Recently elucidated, various stereoselective phenomena[5] were also treated in the same fashion.[1c] Theories of organic chemical reactions thus far developed have mostly been engaged in bimolecular interactions but few in three-body interactions.[3c] The marvellous activity of catalysts is essentially based on the role of a third species in two-molecule interactions. An abundance of experimental results has been accumulated and are now ready for our investigation of the detailed mechanism.

Consider a catalytic process in which the interaction between two reactants A and B is promoted in the presence of the third species C. The species C may interact with both A and B simultaneously (Scheme I) and may also interact only with one of them (Scheme II). In the present paper the perturbed energy and wave function are given by extending the theory for bimolecular interactions developed previously by the group of the present authors.[3a]

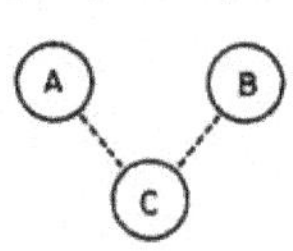

Scheme I

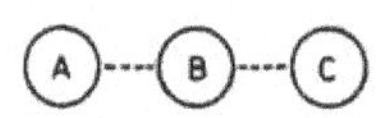

Scheme II

Theory

The ground-state wave function of the total interacting system Ψ_0' is represented by a linear combination of various configuration functions Ψ_K.

$$\Psi_0' = \sum_K C_K \Psi_K$$

The function Ψ_K is constructed by Slater determinants so as to be a spin eigenfunction, each determinant being composed of the MO's of isolated systems A, B, and C of the same geometry as in the interacting system. These MO's may be occupied or unoccupied in the initial states of each isolated molecule.

The interaction energy of the three systems, ΔW, is defined by the following equation

$$\Delta W = W - (W_{AO} + W_{BO} + W_{CO}) \tag{1}$$

where W_{AO}, W_{BO}, and W_{CO} are the ground-state energies of each isolated system, and W is the lowest energy of the whole interacting system.

The secular determinant is represented by

$$\begin{vmatrix} H_{00} - E \ldots & H_{0K} - S_{0K}E \ldots & H_{0L} - S_{0L}E \ldots \\ H_{K0} - S_{K0}E \ldots & H_{KK} - E \ldots & H_{KL} - S_{KL}E \ldots \\ H_{L0} - S_{L0}E \ldots & H_{LK} - S_{LK}E \ldots & H_{LL} - E \ldots \end{vmatrix} = 0 \tag{2}$$

in which the subscripts stand for electronic configurations, H being the Hamiltonian operator of the combined system, and H_{IJ} and S_{IJ} are the integrals represented by

$$H_{IJ} = \int \Psi_I^* H \Psi_J \mathrm{d}\tau$$
$$S_{IJ} = \int \Psi_I^* \Psi_J \mathrm{d}\tau$$

On the assumption that H_{00} is the lowest of all diagonal elements, that no others are equal to H_{00}, and that S_{IJ} and H_{IJ} in the off-diagonal elements are small, the perturbed energy H_{00}' can be expanded in the following form

$$H_{00}' = H_{00} + \sum_K \frac{|H_{0K} - S_{0K}H_{00}|^2}{H_{00} - H_{KK}} + \sum_K \sum_{L(\neq K)} \frac{(H_{0K} - S_{0K}H_{00})(H_{L0} - S_{L0}H_{00})(H_{KL} - S_{KL}H_{00})}{(H_{00} - H_{KK})(H_{00} - H_{LL})} + \text{(small terms of higher than 3rd order)} \tag{3}$$

The subscripts K and L stand for configurations not equal to 0. The perturbed wave function Ψ_0' is given by

$$\Psi_0' = \left\{1 - \sum_K \frac{S_{K0}(H_{0K} - S_{0K}H_{00})}{H_{00} - H_{KK}} - \frac{1}{2}\sum_K \frac{|H_{0K} - S_{0K}H_{00}|^2}{(H_{00} - H_{KK})^2}\right\}\Psi_0 + \sum_K \left\{\frac{H_{K0} - S_{K0}H_{00}}{H_{00} - H_{KK}} + \sum_{L(\neq K)} \frac{(H_{L0} - S_{L0}H_{00})(H_{KL} - S_{KL}H_{00})}{(H_{00} - H_{KK})(H_{00} - H_{LL})}\right\}\Psi_K +$$

$$\text{(small terms of higher than 2nd order)} \quad (4)$$

We will consider the case in which two reactant molecules (A and B) cannot effectively interact with each other until a third species or catalyst (C) intervenes between them (Scheme I). In this way one can assume that the MO overlapping between A and B, s_{ab}, is 0 while those between A and C and between B and C, denoted by s_{ac} and s_{bc}, respectively, have small nonzero values. If it is further assumed that the interaction is not so strong, and that H_{IJ} is nearly proportional to S_{IJ}, one can obtain the general relation of the order of magnitude between the off-diagonal elements and MO overlaps

$$H_{IJ} - S_{IJ}H_{00} \propto S_{IJ} = o(s_{ac}^{\mu}s_{bc}^{\nu})$$

in which μ and ν are the minimum numbers of electron transfer between A and C, and between B and C, respectively, required to shift the electron configuration from I to J.[3a]

We are now concerned with the effect of an intermediary C on the interaction between A and B. For this purpose, one can take out of account those terms in the right-hand sides of eq. 3 and 4 containing the configurations where electron rearrangements take place only either between A and C or between B and C. To be of present interest: (1) the electron configurations of the combined system involved in each term should contain at least one of the transferred or polarized configurations of molecule A and at least one of the transferred or polarized configurations of molecule B; (2) the order of magnitude of the term in Ψ_0' is not higher than the third order of MO overlap ($o(s_{ac}^{\mu}s_{bc}^{\nu})$, $\mu + \nu \leq 3$), and that in H_{00}' is not higher than the fourth order of MO overlap ($o(s_{ac}^{\mu}s_{bc}^{\nu})$, $\mu + \nu \leq 4$).

The terms appearing in Ψ_0' which involve occupied MO's of A and unoccupied MO's of B or the reverse are as follows:

$$\Psi_0': \quad \sum_{i}^{\overset{\text{A}}{\text{occ}}} \sum_{l}^{\overset{\text{B}}{\text{uno}}} \Bigg[\Bigg\{\frac{H_{0,i\to l} - S_{0,i\to l}H_{00}}{H_{00} - H_{i\to l,i\to l}} + \sum_{m}^{\overset{\text{C}}{\text{occ}}} \frac{(H_{0,m\to l} - S_{0,m\to l}H_{00})(H_{m\to l,i\to l} - S_{m\to l,i\to l}H_{00})}{(H_{00} - H_{m\to l,m\to l})(H_{00} - H_{i\to l,i\to l})} +$$

$$\sum_{n}^{\overset{\text{C}}{\text{uno}}} \frac{(H_{0,i\to n} - S_{0,i\to n}H_{00})(H_{i\to n,i\to l} - S_{i\to n,i\to l}H_{00})}{(H_{00} - H_{i\to n,i\to n})(H_{00} - H_{i\to l,i\to l})}\Bigg\}\Psi_{i\to l} +$$

$$\sum_{n}^{\overset{\text{C}}{\text{uno}}} \sum_{m}^{\overset{\text{C}}{\text{occ}}} \Bigg\{\frac{(H_{0,m\to l} - S_{0,m\to l}H_{00})(H_{m\to l,i\to n} - S_{m\to l,i\to n}H_{00})}{(H_{00} - H_{m\to l,m\to l})H_{00} - H_{i\to n,i\to n})}\Psi_{i\to n} + \frac{(H_{0,i\to n} - S_{0,i\to n}H_{00})(H_{i\to n,m\to l} - S_{i\to n,m\to l}H_{00})}{(H_{00} - H_{i\to n,i\to n})(H_{00} - H_{m\to l,m\to l})}\Psi_{m\to l}\Bigg\}$$

$$+ \sum_{n}^{\overset{\text{C}}{\text{uno}}} \sum_{m}^{\overset{\text{C}}{\text{occ}}} \frac{H_{0,(i\to n,m\to l)} - S_{0,(i\to n,m\to l)}H_{00}}{H_{00} - H_{(i\to n,m\to l)(i\to n,m\to l)}} \times \Psi_{i\to n,m\to l}\Bigg]$$

$$+ \sum_{j}^{\overset{\text{A}}{\text{uno}}} \sum_{k}^{\overset{\text{B}}{\text{occ}}} \text{[corresponding terms involving unoccupied MO's of A and occupied MO's of B]} \quad (5)$$

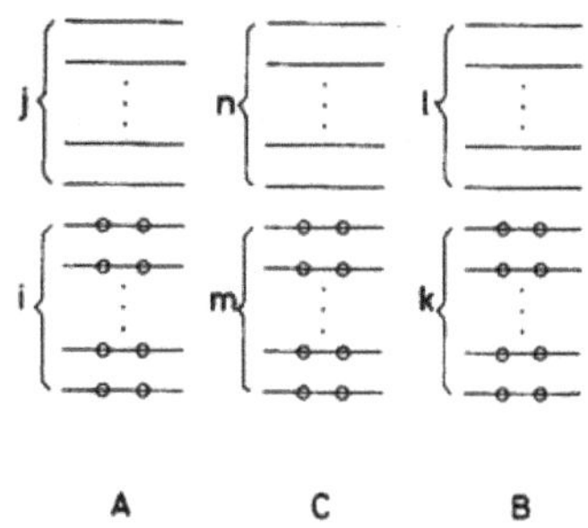

Figure 1.

The letters $i, j, k, l, m,$ and n represent the occupied orbitals and the unoccupied orbitals of A, B, and C molecules (Figure 1). The arrow in the subscripts, e.g., $i \to l$, signifies the electron transfer from the ith occupied MO of A to the lth unoccupied MO of B. The last term of expression 5 includes all the corresponding terms originating from the mathematical equivalency of A and B.

The energy terms in H_{00}' are correspondingly given in the same order of expression 5.

H_{00}':

$$\sum_{i}^{\overset{\text{A}}{\text{occ}}} \sum_{l}^{\overset{\text{B}}{\text{uno}}} \Bigg[\frac{|H_{0,i\to l} - S_{0,i\to l}H_{00}|^2}{H_{00} - H_{i\to l,i\to l}} +$$

$$2\sum_{n}^{\overset{\text{C}}{\text{uno}}} \mathbf{R}\, \frac{(H_{0,i\to n} - S_{0,i\to n}H_{00}((H_{i\to l,0} - S_{i\to l,0}H_{00})(H_{i\to n,i\to l} - S_{i\to n,i\to l}H_{00})}{(H_{00} - H_{i\to n,i\to n})(H_{00} - H_{i\to l,i\to l})} +$$

$$2\sum_{m}^{\overset{\text{C}}{\text{occ}}} \mathbf{R}\, \frac{(H_{0,m\to l} - S_{0,m\to l}H_{00})(H_{i\to l,0} - S_{i\to l,0}H_{00})(H_{m\to l,i\to l} - S_{m\to l,i\to l}H_{00})}{(H_{00} - H_{m\to l,m\to l})(H_{00} - H_{i\to l,i\to l})} +$$

$$2\sum_{n}^{\overset{\text{C}}{\text{uno}}} \sum_{m}^{\overset{\text{C}}{\text{occ}}} \mathbf{R}\, \frac{(H_{0,i\to n} - S_{0,i\to n}H_{00})(H_{m\to l,0} - S_{m\to l,0})(H_{i\to n,m\to l} - S_{i\to n,m\to l}H_{00})}{(H_{00} - H_{i\to n,i\to n,})(H_{00} - H_{m\to l,m\to l})} +$$

$$\sum_{n}^{\overset{\text{C}}{\text{uno}}} \sum_{m}^{\overset{\text{C}}{\text{occ}}} \frac{|H_{0,(i\to n,m\to l)} - S_{0,(i\to n,m\to l)}H_{00}|^2}{H_{00} - H_{(i\to n,m\to l),(i\to n,m\to l)}}\Bigg] +$$

$$\sum_{j}^{\overset{\text{A}}{\text{uno}}} \sum_{k}^{\overset{\text{B}}{\text{occ}}} \text{[corresponding terms involving unoccupied MO's of A and occupied MO's of B]} \quad (6)$$

Here **R** signifies the real part.

We look at the sign of the third-order perturbation energy term in eq 3. One can write one of the third-order perturbation terms as

$$S_{0K}S_{KL}S_{L0} \times \frac{(H_{0K}/S_{0K}) - H_{00})((H_{KL}/S_{KL}) - H_{00})((H_{L0}/S_{L0}) - H_{00})}{(H_{00} - H_{KK})(H_{00} - H_{LL})}$$

The term $((H_{IJ}/S_{IJ}) - H_{00})$ usually is negative if S_{IJ} is small. Accordingly, the condition that this term becomes negative is

$$S_{0K}S_{KL}S_{L0} > 0 \quad (7)$$

since the product of energy gaps in the denominator is of plus sign in the ground-state interaction. Suppose that we can choose, the only one, the most important combination of three configurations 0, K, and L from such third-order terms by considering the donor-acceptor relationship of the three interacting systems A, B, and C. The inequality in (7) plays a conspicuous role in determining the mode of favorable interaction of A, B, and C if one permits the direct interaction between the reactants A and B. This implies the consideration of a cyclic three-molecule intéraction (Scheme III). For instance, consider the case of the following type of interaction:

Scheme III

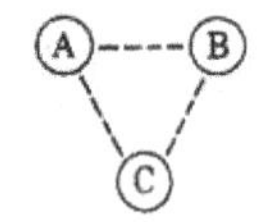

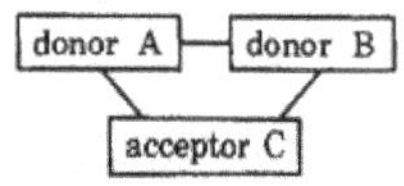

Then the most important combination of electron configurations is

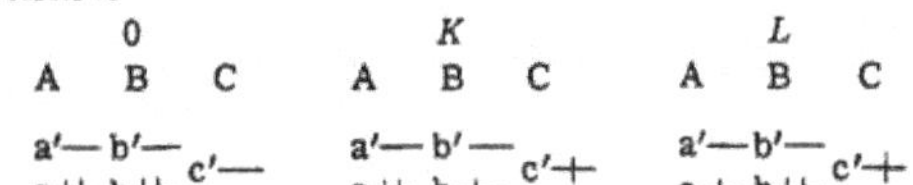

where a, b, and c are HOMO's and a', b', and c' are LUMO's, and obviously

$$S_{0K}S_{KL}S_{L0} \sim -s_{ab}s_{ac'}s_{bc'} > 0$$

It is to be noted that the order of magnitude of these terms is of the third order of MO overlaps. The condition for stabilization is hence schematically represented by

$$\begin{array}{c}(\mathrm{HOMO})_A \overset{-}{—} (\mathrm{HOMO})_B \\ +\diagdown \quad \diagup + \\ (\mathrm{LUMO})_C\end{array} \tag{7a}$$

where + and − denote in-phase and out-of-phase overlapping of MO's.

Similarly, the second important case is

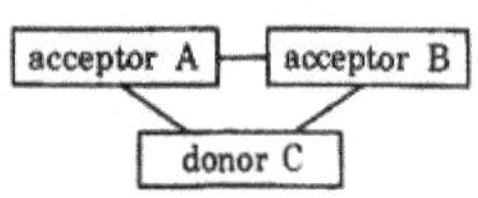

with the important combination of configurations

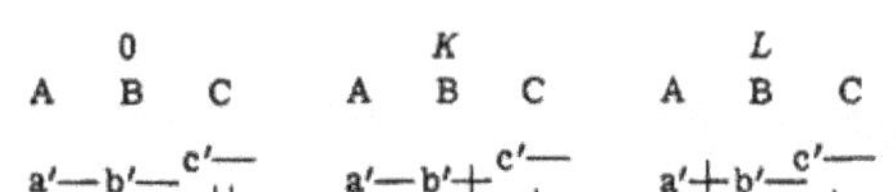

of the relation

$$S_{0K}S_{KL}S_{L0} \sim s_{a'b'}s_{a'c}s_{b'c} > 0$$

The mode of MO overlapping for stabilization is evidently

$$\begin{array}{c}(\mathrm{LUMO})_A \overset{+}{—} (\mathrm{LUMO})_B \\ +\diagdown \quad \diagup + \\ (\mathrm{HOMO})_C\end{array} \tag{7b}$$

The use of these phase relationships is very convenient for discussing the mechanism of chemical reactions in general and will be discussed later with some applications to catalytic phenomena. Thus, it is to be noted that according to Scheme III the third-order terms with respect to MO overlapping come to be important.

Coming back to Scheme I again, it is significant here to investigate the physical reason for the mixing of configuration in the rearrangement of electronic structure of the interacting system. For instance, look at the three kinds of terms of $\Psi_{i\to l}$ in expression 5. The first one comes from the interaction between Ψ_0 and $\Psi_{i\to l}$ due to an indirect MO overlappings between A and B. The remaining two come from the mixing of delocalized configuration $i \to l$ induced by the delocalization $m \to l$, and by the one $i \to n$, respectively. The first three terms in H_{00}' of eq 6 can be understood similarly. The terms of $\Psi_{i\to n}$ and $\Psi_{m\to l}$ in Ψ_0' and the fouth in H_{00}' are interpreted to come from mutual induction between the delocalizations $i \to n$ and $m \to l$. The last terms given explicitly in Ψ_0' and H_{00}' signify a simultaneous delocalization from i to n and from m to l.

In this connection it is desired for practical uses to reduce expressions 5 and 6 into a form represented in terms of the MO's of isolated species A, B, and C. The first step is the reduction of the overlap integral S_{IJ} between two configurations I and J into the MO overlappings s_{ac} and s_{bc}. For instance, $S_{0,i\to l}$ in the first term of the coefficient of $\Psi_{i\to l}$ in eq 5 might have the order of magnitude of $2^{1/2}s_{il}$ ($=o(s_{ab})$) if a direct orbital overlapping between molecules A and B were allowed. Since we have assumed that $s_{ab} = 0$, we have to be concerned with higher order terms, which are of the form $2^{1/2}s_{im}s_{ml} = o(s_{ac}s_{bc})$. It follows therefore that the order of magnitude of Ψ_0' is $o(s_{ac}s_{bc})$ and that of the corresponding term in H_{00}' is $o(s_{ac}^2 s_{bc}^2)$.

Similarly, the order of magnitude of the remaining two terms of the coefficient of $\Psi_{i\to l}$ in eq 5 is also $o(s_{ac}s_{bc})$ and that of the corresponding term in H_{00}' is $o(s_{ac}^2 s_{bc}^2)$.

In this way, one can derive expressions 5 and 6 represented in terms of each MO of the isolated species, A, B, and C. The procedure is similar to the derivation of eq 20 from eq 8b of ref 3a. The corresponding expressions are omitted here for brevity.

The remaining terms satisfying the conditions mentioned before, which appear in Ψ_0', are those which involve the interactions between occupied orbitals of A and B and unoccupied orbitals of C (eq 8 and 9). The first two terms of ex-

$$\Psi_0': \sum_i^{\overset{A}{\mathrm{occ}}} \sum_k^{\overset{B}{\mathrm{occ}}} \sum_n^{\overset{C}{\mathrm{uno}}} \sum_{n'}^{\overset{C}{\mathrm{uno}}} \left[\left\{ \frac{(H_{0,k\to n'} - S_{0,k\to n'}H_{00})(H_{k\to n',i\to n} - S_{k\to n',i\to n}H_{00})}{(H_{00} - H_{i\to n,i\to n})(H_{00} - H_{k\to n',k\to n'})} \Psi_{i\to n} + \frac{(H_{0,i\to n} - S_{0,i\to n}H_{00})(H_{i\to n,k\to n'} - S_{i\to n,k\to n'}H_{00})}{(H_{00} - H_{i\to n,i\to n})(H_{00} - H_{k\to n',k\to n'})} \Psi_{k\to n'} \right\} + \frac{H_{0,(i\to n,k\to n')} - S_{0,(i\to n,k\to n')}H_{00}}{H_{00} - H_{(i\to n,k\to n'),(i\to n,k\to n')}} \Psi_{i\to n,k\to n'} \right] \tag{8}$$

$$H_{00}': \sum_i^{\overset{A}{\mathrm{occ}}} \sum_k^{\overset{B}{\mathrm{occ}}} \sum_n^{\overset{C}{\mathrm{uno}}} \sum_{n'}^{\overset{C}{\mathrm{uno}}} \left[2\mathrm{R} \frac{(H_{0,i\to n} - S_{0,i\to n}H_{00})(H_{k\to n',0} - S_{k\to n',0}H_{00})(H_{i\to n,k\to n'} - S_{i\to n,k\to n'}H_{00})}{(H_{00} - H_{i\to n,i\to n})(H_{00} - H_{k\to n',k\to n'})} + \frac{|H_{0(i\to n,k\to n')} - S_{0,(i\to n,k\to n')}H_{00}|^2}{H_{00} - H_{(i\to n,k\to n'),(i\to n,k\to n')}} \right] \tag{9}$$

pression 8 and the first one in 9 represent the mixing-in of one by way of the other of $i \rightarrow n$ and $k \rightarrow n'$ configurations, the orders of magnitude being ($o(s_{ac}s_{bc}^2)$ or $o(s_{ac}^2s_{bc})$ in Ψ_0', and $o(s_{ac}^2s_{bc}^2)$ in H_{00}'. The last terms in eq 8 and 9 result from the ditransferred configurations in which A and B act as donors and C as an acceptor (the orders of magnitude are $o(s_{ac}s_{bc})$ in Ψ_0' and $o(s_{ac}^2s_{bc}^2)$ in H_{00}').

Similarly, the terms involving the interaction between unoccupied orbitals of A and B and occupied orbitals of C are shown in eq 10 and 11. The criterion for picking up these terms was only the magnitude of the numerators. The second step is the discussion of the magnitude of denominators. The energy difference of two configurations can easily be approximated in terms of the difference between two MO energies. One can see the significance of the highest occupied (HO) MO and the lowest unoccupied (LU) MO of each isolated system in the summations in eq 8-11. Evidently, these particular MO's make the magnitude of corresponding denominators smallest. As well as in two-molecule interaction cases,[3] the conspicuous role of HOMO and LUMO can in this way apply to the present case.

Ψ_0':

$$\sum_{j}^{\mathrm{A\,unocc}}\sum_{l}^{\mathrm{B\,unocc}}\sum_{m}^{\mathrm{C\,occ}}\sum_{m'}^{\mathrm{C\,occ}}\Bigg[\Bigg\{\frac{(H_{0,m'\leftarrow l}-S_{0,m'\leftarrow l}H_{00})(H_{m'\leftarrow l,m\leftarrow j}-S_{m'\leftarrow l,m\leftarrow j}H_{00})}{(H_{00}-H_{m'\leftarrow l,m'\leftarrow l})(H_{00}-H_{m\leftarrow j,m\leftarrow j})}\Psi_{m\leftarrow j}+\frac{(H_{0,m\leftarrow j}-S_{0,m\leftarrow j}H_{00})(H_{m\leftarrow j,m'\leftarrow l}-S_{m\leftarrow j,m'\leftarrow l}H_{00})}{(H_{00}-H_{m\leftarrow j,m\leftarrow j})(H_{00}-H_{m'\leftarrow l,m'\leftarrow l})}\Psi_{m'\leftarrow l}\Bigg\}+\frac{H_{0,(m\leftarrow j,m'\leftarrow l)}-S_{0,(m\leftarrow j,m'\leftarrow l)}H_{00}}{H_{00}-H_{(m\leftarrow j,m'\leftarrow l),(m\leftarrow j,m'\leftarrow l)}}\Psi_{m\leftarrow j,m'\leftarrow l}\Bigg] \quad (10)$$

H_{00}':

$$\sum_{j}^{\mathrm{A\,unocc}}\sum_{l}^{\mathrm{B\,unocc}}\sum_{m}^{\mathrm{C\,occ}}\sum_{m'}^{\mathrm{C\,occ}}\Bigg[2\mathrm{R}\frac{(H_{0,m\leftarrow j}-S_{0,m\leftarrow j}H_{00})(H_{m'\leftarrow l,0}-S_{m'\leftarrow l,0}H_{00})(H_{m\leftarrow j,m'\leftarrow l}-S_{m\leftarrow j,m'\leftarrow l}H_{00})}{(H_{00}-H_{m\leftarrow j,m\leftarrow j})(H_{00}-H_{m'\leftarrow l,m'\leftarrow l})}+\frac{|H_{0,(m\leftarrow j,m'\leftarrow l)}-S_{0,(m\leftarrow j,m'\leftarrow l)}H_{00}|^2}{H_{00}-H_{(m\leftarrow j,m'\leftarrow l),(m\leftarrow j,m'\leftarrow l)}}\Bigg] \quad (11)$$

Hence, the important orbital interactions for Scheme I which are of interest in the theory of catalysis may be

$$\begin{array}{ll} (\mathrm{HOMO})_A\text{–}(\mathrm{LUMO})_C\text{–}(\mathrm{LUMO})_B & \\ (\mathrm{HOMO})_A\text{–}(\mathrm{HOMO})_C\text{–}(\mathrm{LUMO})_B & (12a) \\ (\mathrm{HOMO})_A\text{–}(\mathrm{LUMO})_C\text{–}(\mathrm{HOMO})_B & \\ (\mathrm{LUMO})_A\text{–}(\mathrm{HOMO})_C\text{–}(\mathrm{LUMO})_B & (12b) \end{array}$$

The first two schemes follow from investigating expressions 5 and 6, and the remaining two come from eq 8-11.

It follows similarly that the corresponding ones for Scheme II are represented by

$$\begin{array}{ll} (\mathrm{HOMO})_A\text{–}(\mathrm{HOMO})_B\text{–}(\mathrm{LUMO})_C & \\ (\mathrm{LUMO})_A\text{–}(\mathrm{LUMO})_B\text{–}(\mathrm{HOMO})_C & (12c) \end{array}$$

These particular orbital interaction schemes are derived from eq 5 and 6 by interchanging B and C.

The schemes 12a-c should be taken into account in addition to usual HOMO-LUMO interactions between two components of A, B, and C, when one considers the catalytic action of the third species on the other two.

Such a schematic presentation of particular orbital interactions is useful for a modification of the orbital symmetry rule;[5] that is, even if the overlap integral between $(\mathrm{HOMO})_A$ and $(\mathrm{LUMO})_B$ or $(\mathrm{LUMO})_A$ and $(\mathrm{HOMO})_B$ happens to be zero, an indirect $(\mathrm{HOMO})_A$-$(\mathrm{LUMO})_B$, $(\mathrm{LUMO})_A$-$(\mathrm{HOMO})_B$, $(\mathrm{HOMO})_A$-$(\mathrm{HOMO})_B$, or $(\mathrm{LUMO})_A$-$(\mathrm{LUMO})_B$ interaction through the intermediator C can assist the symmetry-disfavored process to take place.

The present approach is easily extended to the excited state reactions, in the same fashion as in the two-species interaction case.

Discussion

Apparently complicated reactions catalyzed by transition metal complexes are composed of successive elementary processes: coordination of ligands on metal, reaction of ligands, and detachment of product ligands accompanied by regeneration of active catalytic species. Chemical transformations on metal are the present interest, which are classified into a variety of elementary processes: insertion of one ligand between metal and another, recombination of ligands, isomerization or rearrangement of coordinated molecules, and so on. We will discuss and examine the mechanism of such elementary processes in the light of the theoretical framework developed in the previous section.

Insertion Processes. The insertion reaction of coordinated unsaturated bonds into the metal-hydrogen or metal-carbon bond is one of the important elementary processes common to hydrogenation, hydroformylation, polymerization, etc.[6] It should be noted that in most of the insertion processes the ligands H^- or R^- attacking on the unsaturated bonds have occupied orbitals of low ionization potential and at the same time inserting species have unoccupied orbitals at low energy levels. One is tempted by these facts to conclude that the charge transfer from one to the other ligand through metal (scheme 12a) is rather more important than the electron delocalization from both ligands to metal or the reverse (scheme 12b). That is, the electron delocalization from the HOMO of the ligand A of which the bond with metal undergoes fission by insertion to the LUMO of inserting ligand B is in this way expected to play a predominant role. A similar view was proposed for metal-catalyzed polymerization by Furukawa, in which electrons of coordinated alkyl are partly delocalized to an unoccupied orbital of metal and subsequently to the π^* orbital of the monomer olefin.[7]

One can conveniently apply the picture mentioned above to the regioselectivity of insertion, if one considers a direct

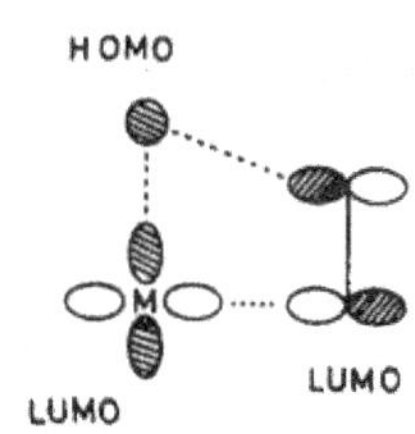

interaction of $(HOMO)_A$ and $(LUMO)_B$, being led to Scheme III. The observed regioselectivity of the following reactions[8]

$$(\pi\text{-}C_5H_5)_2MoH_2 + CH_2{=}CHX \longrightarrow (\pi\text{-}C_5H_5)_2MoH\text{–}CHXCH_3$$
$$X = CN,\ CO_2CH_3$$

is consistent with the fact that LUMO's of the above olefins have the largest amplitude at the terminal carbon.

Metathesis and Dimerization of Olefins

Facile disproportionation reactions of olefins on some transition metals have been observed,[9] i.e.,

$$R_1R_1C{=}CR_1R_1 + R_2R_2C{=}CR_2R_2 \rightarrow 2R_1R_1C{=}CR_2R_2$$

Some intermediates have been proposed: cyclobutane[10] or quasi-cyclobutane[11] intermediates, carbenoid complexes,[12] and five-membered metallocyclic structures.[13] A related reaction, the dimerization of olefins to cyclobutane derivatives, has also been known to take place in the presence of zero valent iron,[14] nickel,[15] and cobalt[16] catalysts.

Notwithstanding the thermal dimerization of ethylenes to form cyclobutane is a symmetry-disfavored process;[5] the metathesis and the dimerization of olefins occur readily in the presence of transition metal catalysts. Several theoretical chemists have paid attention to the orbital and state symmetry aspects and examined the relaxation effect of the metal d orbital on the symmetry restriction.[17]

We can find interesting terms in two pairs of formulas, eq 8 and 9 and 10 and 11, which may be likely to argue for suprafacial-suprafacial approach[5] of two olefins on the catalysts. One pair of these formulas implies the electron delocalization from ligands to metal. The term which involves the HOMO's of coordinated olefins and the LUMO of the catalyst is predicted on the basis of scheme 12b to be the most important. Of the other terms related to the delocalization from metal to ligands (expressions 10 and 11), the one which involves the LUMO's of both ligands and the HOMO of catalyst may be prominent. Here a symmetry restriction arises which does not appear if $n \neq n'$ in eq 8 and 9 or $m \neq m'$ in eq 10 and 11. In the case that n and n' are taken as the $(LUMO)_C$ and m and m' as the $(HOMO)_C$, the symmetry properties of $(HOMO)_A$ and $(HOMO)_B$, and those of $(LUMO)_A$ and $(LUMO)_B$, should be the same in order that A and B can approach favorably with a certain symmetry. The overlapping $S_{l\to n,k\to n}$ or $S_{m\to j,m\to l}$ is zero otherwise. As stated in the last section, this symmetry rule forms a striking contrast to that for the bimolecular system in which the $(HOMO)_A$-$(LUMO)_B$ interaction, or the symmetry property of $(HOMO)_A$ and $(LUMO)_B$ is significant. The $(HOMO)_A$-$(LUMO)_C$-$(HOMO)_B$ and $(LUMO)_A$-$(HOMO)_C$-$(LUMO)_B$ interactions among two olefin molecules and metal are favored by the same symmetries of HOMO's and of LUMO's in contrast to bimolecular interaction, as is illustrated below.

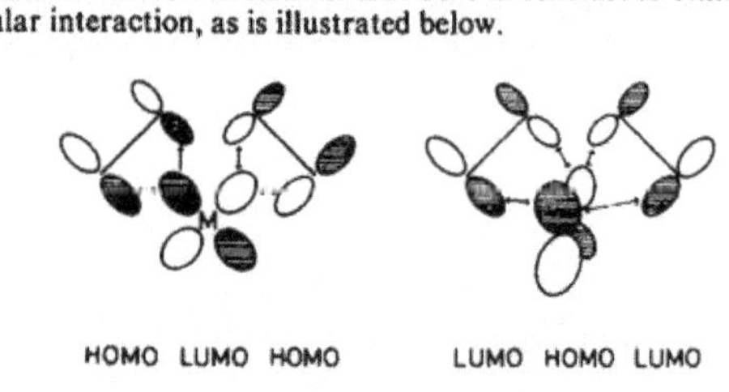

2 + 2 Cycloreversion. Cleavage of the cyclobutane ring to olefins is one of the most investigated reactions catalyzed by transition metal complexes; cubane to *syn*-tricyclooctadiene,[13a] quadricyclane to norbornadiene,[18] prismane to Dewar benzene,[19] and so forth. Recent studies of the conversions of cubane[20] and of quadricyclane[21] by the Rh complex have demonstrated the involvement of oxidative addition of the metal to the hydrocarbons through carbon–carbon bonds. Although these facts seem to show that the catalytic processes proceed by Scheme II (A-B-C), a possibility still remains that the transition state for the cycloreversion may be appropriately represented by Scheme I or by Scheme III, in which a relevant metal orbital interacts with both orbitals simultaneously. It is likely that the metal adduct intermediates may be located in a bypath not leading directly to the products and it may rearrange just prior to the cycloreversion process into such a symmetrical geometry as proposed for olefin dimerization. Interaction according to Scheme I or to Scheme III satisfies the condition $S_{OK}S_{KL}S_{LO} > 0$ in the same way as the interaction among two olefins and metal shown before.

Recombination of Ligands. It would be meaningful to refer to the recombination reactions of alkyl ligands in dialkyl(dipyridyl)nickel complex, promoted by coordination of electron-accepting olefins.[22] The activation energy $\Delta H^\ddagger$ is about 15.6 kcal/mol for R = ethyl in the presence of acrolein while it is 68 kcal/mol in the absence of the olefin. The ligand–ligand recombination reaction can be the object of our present interest when the ligands to be bonded are considered reactants A and B, and the combined system of the rest of the metal complex and the coordinated olefin is considered catalyst C. One can see that the terms of eq 8

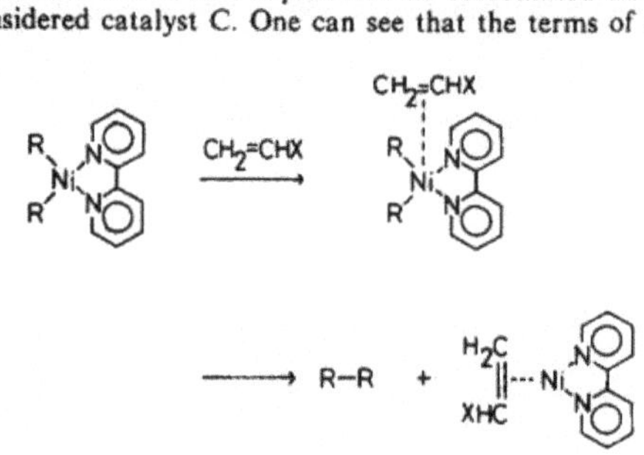

and 9 involving the occupied orbitals of A and B and the unoccupied orbitals of C are more important than the terms of eq 10 and 11, since the reactants A and B are anionic ligands. The contribution of such a delocalization term increases as the ability of the catalyst system C to accept electrons increases, or as the LUMO of coordinating olefins is lowered. As a result the recombination of anionic ligands on metal takes place readily as the interaction between the HOMO's of the reactants by way of the LUMO of the intermediary catalyst is favored (eq 8 and 9). This conclusion is consistent with the order of electron-accepting ability of coordinated olefins.

The formulas derived in the theory section are applicable to another type of catalytic processes, in which catalyst C interacts preferentially with one of two reactants (B). One can obtain the appropriate terms for Scheme II by formally replacing the subscripts k, l, m, and n by m, n, k, and l in each term of Ψ_0' and H_{00}'. The terms originating from the electronic configurations which contain any zero configuration of A or B can be neglected as before. A further restriction may be taken into consideration from the point of view of a chemical aspect. The reactants A and B cannot be in an equivalent chemical situation in contrast to Scheme I. No

$$2\sum_{i}^{\overset{A}{occ}}\sum_{k}^{\overset{B}{occ}}\sum_{n}^{\overset{C}{unoc}} \mathrm{R}\left\{\frac{(H_{0,i\to n} - S_{0,i\to n}H_{00})(H_{k\to n,0} - S_{k\to n,0}H_{00})(H_{i\to n,k\to n} - S_{i\to n,k\to n}H_{00})}{(H_{00} - H_{i\to n,i\to n})(H_{00} - H_{k\to n,k\to n})}\right\} \tag{13}$$

$$2\sum_{j}^{\overset{A}{unoc}}\sum_{l}^{\overset{B}{unoc}}\sum_{m}^{\overset{C}{occ}} \mathrm{R}\left\{\frac{(H_{0,m\to j} - S_{0,m\to j}H_{00})(H_{m\to l,0} - S_{m\to l,0}H_{00})(H_{m\to j,m\to l} - S_{m\to j,m\to l}H_{00})}{(H_{00} - H_{m\to j,m\to j})(H_{00} - H_{m\to l,m\to l})}\right\} \tag{14}$$

mechanisms are conceivable by which electron delocalizations from B to C ($k \to n$) and from C to B ($m \to l$) can promote those from B to A ($k' \to j$) and from A to B ($i \to l'$), respectively. Of the terms surviving such selections, the energy terms involving $i \to n$ and $k \to n$ (eq 13) or $m \to j$ and $m \to l$ (eq 14) are expected to contribute most seriously to catalytic action. The formulas 13 and 14 imply the importance of the interactions between the HOMO's and between the LUMO's of the reactants (12c) in case of electron-donating and -accepting catalysts, respectively.

Electrocyclic Reactions and Valence Isomerizations

Disrotatory ring opening of cyclobutane to butadiene is unfavorable in the ground state reactions.[5]

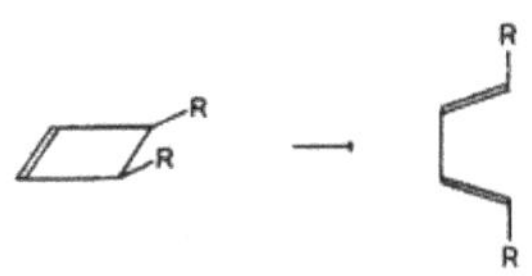

However, such chemical transformations have been observed to occur readily in the presence of certain catalysts. Isomerization of tricyclo[4.2.0.0^{2,5}]octa-3,7-diene to cyclooctatetraene is completed within 40 min in the presence of $AgBF_4$ but does not occur in 24 hr in the absence of catalysts.[23] An analogous phenomenon is observed in the Rh(I)-catalyzed valence isomerization of Dewar benzene to benzene.[19] We believe that the catalyzed electrocyclic reactions are initiated by the coordination of the cationic metal ion or complex on the π bond.

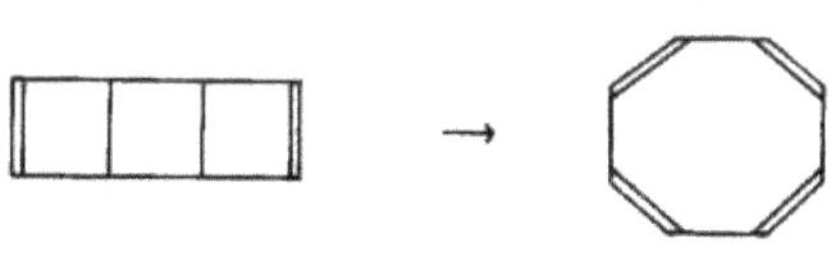

The important orbital interaction, $(\mathrm{HOMO})_A$-$(\mathrm{HOMO})_B$-$(\mathrm{LUMO})_C$, may be represented as follows. $S_{0k}S_{KL}S_{L0} = S_{0,i\to n}S_{i\to n,k\to n}S_{k\to n,0} \simeq s_{ik}{}^2 s_{kn}{}^2 > 0$ (Scheme II). If direct interaction is allowed between A and C, $S_{0K}S_{KL}S_{L0} \simeq -s_{in}s_{kn}s_{ik}$ (Scheme III). In this case $S_{0K}S_{KL}S_{L0}$ cannot be positive.[24] There are two modes of disrotatory ring opening: in the directions syn and anti to the metal.

LUMO LUMO

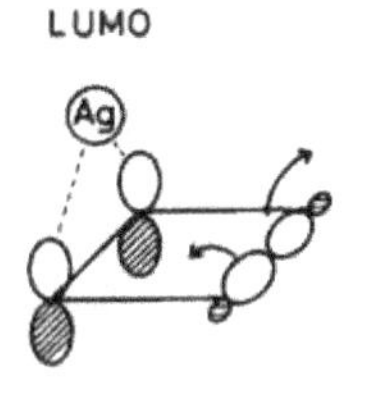

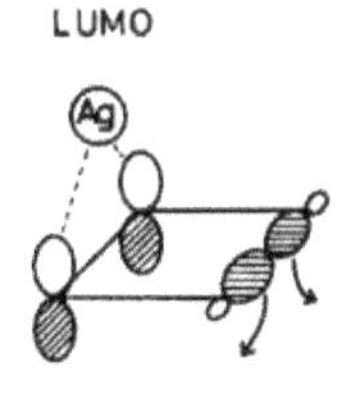

HOMO HOMO HOMO HOMO

The discrimination between syn and anti disrotation is an unprecedented category. This stereochemical problem will be a significant subject of experiments in the near future. The preference of the anti-disrotation is predicted within our present knowledge. Suppose that the whole system is composed of two parts, the electron-accepting combined system of Ag^+ and C=C double bond, and the electron-donating σ bond to be cleaved, in order to appreciate the secondary effects of Ag^+. The preferential anti extension of the π orbital in the LUMO due to anti-bonding overlapping with the metal orbital may favor the interaction between the π orbital and the larger lobes of the original σ bond orbital in the direction anti to the metal. In addition the larger lobes of the σ bond orbital is compelled to be in an opposite sign relation with the metal orbital in case of syn disrotation. This secondary orbital interaction growing as the rotation proceeds may result in less stabilization. In anti disrotation there is no disadvantage of such a phase restriction in the secondary interaction.

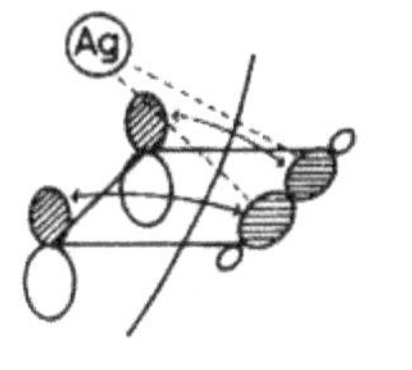

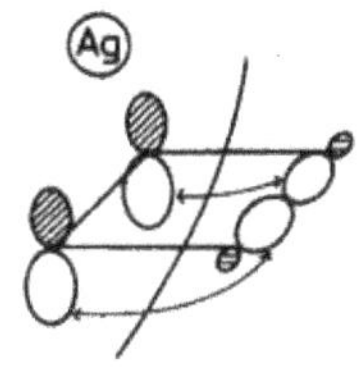

LUMO HOMO LUMO HOMO

Another class of chemical reactions suggesting the importance of the $(\mathrm{HOMO})_A$-$(\mathrm{HOMO})_B$-$(\mathrm{LUMO})_C$ interaction is a valence isomerization of strained saturated compounds. The typical members of the reactions are the skeletal rearrangements of cubane, homocubane, and basketene to cuneane, norsnoutane, and snoutene, respectively,[25] and of *syn*-tricyclo[4.2.0.0^{2,5}]octane to tetrahydrosemibullvalene.[26] The occurrence of thermally "forbidden" $[2\sigma_s + 2\sigma_s]$ transformations[5] is readily understood by the HOMO-HOMO interaction induced by the catalysts. The orbital interactions are illustrated below for cubane and tricyclooctane. The other possible way of assigning the phase relation between two σ orbitals is rejected on account of steric hindrance.

This interaction scheme is supported by the fact that the rate of isomerization of cubane derivatives decreases remarkably as the number of the electron-accepting carbomethoxy substituents increases; the ratio of the reaction rate for the parent structure and mono- and di-substituted cubanes is 17,000:310:1. The considerable changes may be attributed to the lowering of the HOMO energy level by the substitution, or to the increase in the absolute value of the denominator in eq 13. One can see a cause for no change of the *anti*-tricyclo[4.2.0.0^{2,5}]octane under the same condition in the geometrical restriction. Two modes of interaction are conceivable but sterically inaccessible.

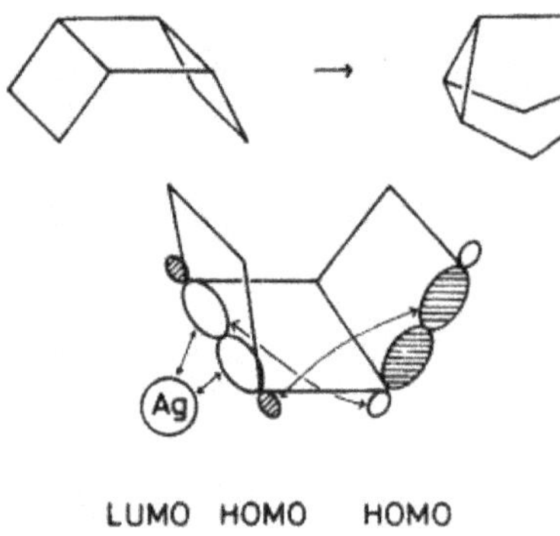

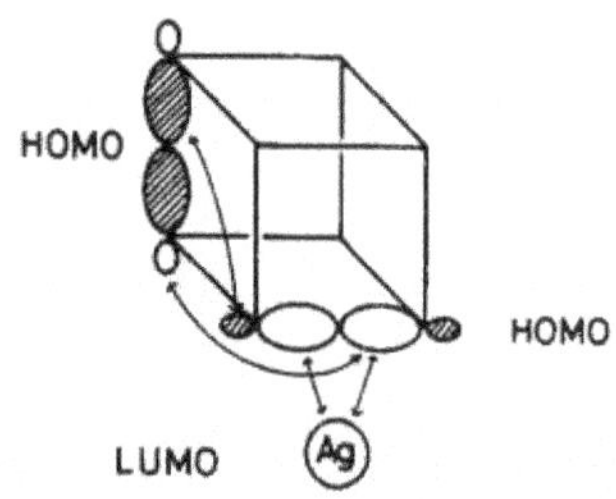

Sigmatropic Shift. A $[_{\pi}2_s + {_\sigma}2_s]$ sigmatropic hydrogen shift[5] was observed to take place in the presence of $Fe(CO)_5$ catalyst, in which the hydrogen transfer occurs in the same side as the coordinated olefin[27] (referred to as "syn" interaction hereafter). The isotope effect on the rate and the de-

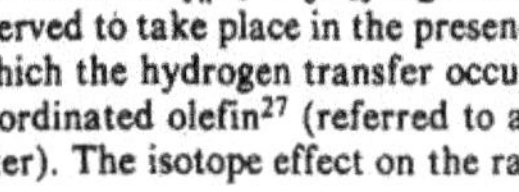

gree of deuterium incorporation in the $DCo(CO)_4$-catalyzed isomerization of 3-phenylpropene to 1-phenylpropene seem to indicate an internal 1,3-hydrogen shift via complex formation.[28] If the metal complexes are electron acceptors, one can draw the $(HOMO)_A$-$(HOMO)_B$-$(LUMO)_C$ interaction among the C—H bond, the C═C bond, and the metal. Although the "anti" interaction is also conceivable in the same sense as in the electrocyclic reactions, the mode of "syn" 1,3-sigmatropic shift is depicted here. The illustra-

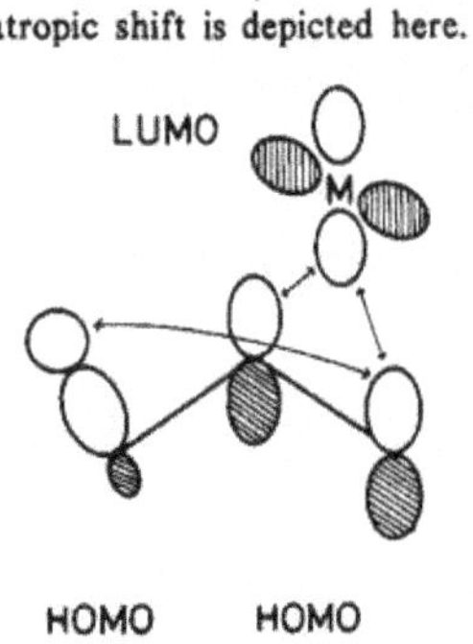

tion clearly shows the possibility of 1,3-sigmatropic rearrangement in a suprafacial-suprafacial manner. The corresponding scheme $(LUMO)_A$-$(LUMO)_B$-$(HOMO)_C$ for electron-donating catalysts suggests the same conclusion on the steric course.[29]

Activation of Molecular Oxygen

The activation and the reactions of the ground state $(^3\Sigma_g{}^-)$ oxygen molecule under the influence of catalysts have been one of the most interesting and important subjects in organic chemistry and also in biochemistry. We consider the following reactions as bimolecular ones of the substrate with activated oxygen which is coordinated to metal surrounded with ligands. The activation mechanism will be understood as a three-body interaction of Scheme I among the ligands (A), metal (C), and oxygen (B).

The frontier orbitals of $^3\Sigma_g{}^-$ O_2 are doubly degenerate singly occupied (SO) MO's. Generally SOMO can donate and accept electrons to unoccupied orbitals and from occupied orbitals of another molecule, respectively. The high electronegativity of oxygen suggests the importance of charge transfer from electron-donating ligands through metal, or of electron acceptance in one of SOMO's (eq 5 and 6). The other SOMO may take part in the reverse delocalization though to a less degree than the former transfer. The electron configuration of the activated oxygen depends on the balance between these two modes of electron delocalization. If electron acceptance of oxygen occurs exclusively, or the ligands are electron-releasing groups, the configuration of O_2 will resemble that of superoxide $O_2{}^-$. When the reverse electron transfer is comparably remarkable the activated oxygen may be endowed with such a property as $^1\Delta_g$ O_2 which is specified by the powerful electron acceptability as pointed out in theoretical investigation of the 2 + 2 cycloaddition mechanism.[30] The electronic structure of coordinated molecular oxygen is expected to be a mixed configuration within the range from $O_2{}^-$ to $O_2(^1\Delta_g)$ according to the nature of metal and ligands.

An addition reaction of carbonyl species with bis(triphenylphosphine)platinum oxide[31] may be classified into the superoxide-like reaction of oxygen activated by a powerful σ donor, triphenylphosphine, although the mechanism has not been clarified by experiment as yet. A parallel approach of an unsaturated bond to the coordinated O_2 is possible since the symmetry of the HOMO of electron-donating superoxide is the same as that of the LUMO of electron-accepting carbonyl species, that is, antisymmetric.

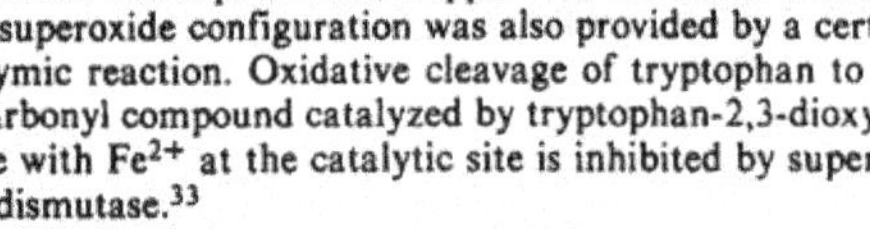

A similar activation mechanism has appeared for onium-catalyzed oxygenation of hydrocarbons.[32] Electron delocalization from the counteranion of sulfonium, phosphonium, and so on, through vacant d orbitals of the cationic center, has been considered to activate the ground state oxygen molecule. An experimental support for the involvement of the superoxide configuration was also provided by a certain enzymic reaction. Oxidative cleavage of tryptophan to the dicarbonyl compound catalyzed by tryptophan-2,3-dioxygenase with Fe^{2+} at the catalytic site is inhibited by superoxide dismutase.[33]

Subsequently another possibility that activated oxygen behaves as $^1\Delta_g$ molecular oxygen is exemplified by the oxygenation of electron-donating tetraphenylcyclopentadienone by diperoxychromium(VI) oxide etherate to 1,2,3,4-tetraphenyl-5,6-dioxabicyclo[2.2.1]hept-2-en-7-one.[34] The HOMO of the substrate and the LUMO of $^1\Delta_g$ molecular oxygen are both antisymmetric. Diels-Alder type reaction of superoxide cannot be favored by the symmetry of the frontier orbital, since the LUMO of the diene is symmetric while the HOMO of superoxide is antisymmetric. Hence, oxygen ligand in diperoxychromium oxide is expected to behave as a $^1\Delta_g$ type species. The difference in electronic structure of the oxygen ligand between bis(triphenylphos-

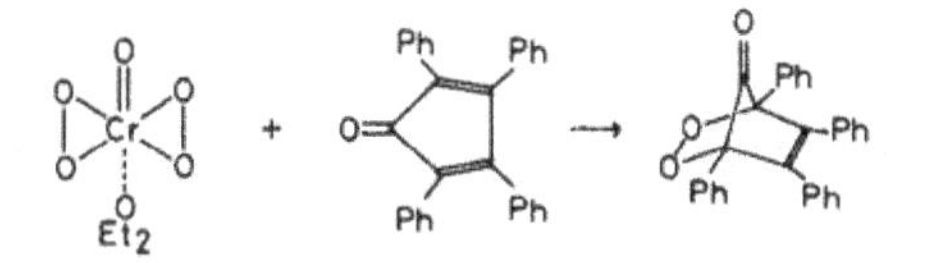

phine)platinum oxide and diperoxychromium oxide may be due to the electron acceptability of nonreacting ligands, i.e., electron-donating phosphine and electron-accepting oxygen.

References and Notes

(1) (a) K. Fukui, "Molecular Orbitals in Chemistry, Physics, and Biology", P.-O. Löwdin and B. Pullman, Ed., Academic Press, New York, N.Y., 1964, p 513; (b) K. Fukui, "Modern Quantum Chemistry", Vol. 1, O. Sinanoglu, Ed., Academic Press, New York, N.Y., 1965, p 49; (c) K. Fukui and H. Fujimoto, "Mechanism of Molecular Migrations", Vol. 2, B. S. Thyagarajan, Ed., Interscience, New York, N.Y., 1969, p 118; K. Fukui, "Theory of Orientation and Stereoselection", Springer-Verlag, Heidelberg, 1970; *Acc. Chem. Res.*, **4**, 57 (1971).

(2) (a) J. N. Murrell, M. Randić, and D. R. Williams, *Proc. R. Soc. London, Ser. A*, **284**, 566 (1965); J. N. Murrell and G. Shaw, *J. Chem. Phys.*, **46**, 1768 (1967); (b) R. Rein and M. Pollak, *ibid.*, **47**, 2039 (1967); (c) G. Klopman, *J. Am. Chem. Soc.*, **90**, 223 (1968); (d) L. Salem, *ibid.*, **90**, 543, 553 (1968).

(3) (a) K. Fukui and H. Fujimoto, *Bull. Chem. Soc. Jpn.*, **41**, 1989 (1968); (b) K. Fukui, *Int. Congr. Pure Appl. Chem. 23rd*, **1**, 65 (1971); (c) H. Fujimoto and K. Fukui, *Adv. Quantum Chem.*, **6**, 177 (1972); (d) K. Fukui and H. Fujimoto, *Bull. Chem. Soc. Jpn.*, **42**, 3399 (1969); H. Fujimoto and K. Fukui, *ibid.*, **44**, 2936 (1971); (e) H. Fujimoto, S. Kato, S. Yamabe, and K. Fukui, *ibid.*, **46**, 1071 (1973).

(4) W. C. Herndon, *Chem. Rev.*, **72**, 157 (1972); R. F. Hudson, *Angew. Chem.*, **85**, 63 (1973); J. D. Bradley and G. C. Gerrans, *J. Chem. Educ.*, **50**, 463 (1973).

(5) R. B. Woodward and R. Hoffmann, *Angew. Chem., Int. Ed. Engl.*, **8**, 781 (1969).

(6) A number of chemical species other than olefins and CO are known as inserting reagents. (a) N_2: M. E. Vol'pin, *Pure Appl. Chem.*, **30**, 607 (1972); M. E. Vol'pin, V. B. Shur, R. V. Kudrjavtsev, and V. B. Prodjako, *Chem. Commun.*, 1038 (1968); J. Chatt, G. A. Heath, and R. L. Richards, *ibid.*, 1010 (1972); J. Chatt, G. A. Heath, and G. J. Leigh, *ibid.*, 444 (1972). (b) CO_2: L. S. Pu, A. Yamamoto, and S. Ikeda, *J. Am. Chem. Soc.*, **90**, 3896 (1968); A. Misono, Y. Uchida, M. Hidai, and T. Kuse, *Chem. Commun.*, 98 (1968); S. Komiya and A. Yamamoto, *J. Organomet. Chem.*, **46**, C58 (1972); V. D. Bianco, S. Doronzo, and M. Rossi, *ibid.*, **35**, 337 (1972); A. Miyashita and A. Yamamoto, *ibid.*, **49**, C57 (1973); M. E. Vol'pin, *Pure Appl. Chem.*, **30**, 607 (1972). (c) CS_2: I. S. Butler and A. E. Fenster, *J. Organomet. Chem.*, **66**, 161 (1974). (d) SO_2: F. A. Hartman and A. Wojcicki, *Inorg. Chem.*, **7**, 1504 (1968).

(7) J. Furukawa, *Shokubai*, **13**, 107 (1971).

(8) A. Nakamura and S. Otsuka, *J. Am. Chem. Soc.*, **94**, 1886 (1972).

(9) For example, see N. Calderon, *Acc. Chem. Res.*, **5**, 127 (1972), and references cited therein.

(10) N. Calderon, E. A. Ofstead, J. P. Ward, W. A. Judy, and K. W. Scott, *J. Am. Chem. Soc.*, **90**, 4133 (1968); E. A. Zuech, W. B. Hughes, D. H. Kubicek, and E. T. Kittleman, *ibid.*, **92**, 528 (1970); W. B. Hughes, *ibid.*, **92**, 532 (1970).

(11) (a) C. P. C. Bradshaw, E. J. Howman, and L. Turner, *J. Catal.*, **7**, 269 (1967); (b) D. L. Crain, *ibid.*, **13**, 110 (1969); (c) J. L. Wang and H. R. Menapace, *J. Org. Chem.*, **33**, 3794 (1968).

(12) (a) G. S. Lewandos and R. Pettit, *J. Am. Chem. Soc.*, **93**, 7067 (1971); (b) G. Dall'asta, *Makromol. Chem.*, **154**, 1 (1972).

(13) (a) J. Halpern, *Acc. Chem. Res.*, **3**, 386 (1970); (b) R. H. Grubbs and T. K. Brunck, *J. Am. Chem. Soc.*, **94**, 2538 (1972). The importance of the nuclear configuration which involves two coordinated olefins or tetramethylene-like structures on metal still remains, even if the metallocyclic species takes part in the disproportionation reactions. The interconversion of the metallocyclic species

is considered to occur through such a symmetrical transition state geometry as is illustrated in the text.

(14) C. W. Bird, D. L. Colinese, R. C. Cookson, J. Hudec, and R. O. Williams, *Tetrahedron Lett.*, 373 (1961); P. W. Jolly, F. G. A. Stone, and K. Mackenzie, *J. Chem. Soc.*, 6416 (1965).

(15) C. W. Bird, R. C. Cookson, and J. Hudec, *Chem. Ind. (London)*, 20 (1960); G. N. Schrauzer and S. Eichler, *Chem. Ber.*, **95**, 2764 (1962).

(16) D. R. Arnold, D. J. Trecker, and E. B. Whipple, *J. Am. Chem. Soc.*, **87**, 2596 (1965).

(17) (a) F. D. Mango and J. H. Schachtschneider, *J. Am. Chem. Soc.*, **89**, 2484 (1967); **91**, 1030 (1969); **93**, 1123 (1971); F. D. Mango, *Tetrahedron Lett.*, 4813 (1969); 505 (1971); 1509 (1973); (b) W. Merk and R. Pettit, *J. Am. Chem. Soc.*, **89**, 4788 (1967); R. Pettit, H. Sugahara, J. Wristers, and W. Merk, *Discuss. Faraday Soc.*, **47**, 71 (1969); (c) W. Th. A. M. van der Lugt, *Tetrahedron Lett.*, 2281 (1970); (d) G. L. Caldow and R. A. MacGregor, *J. Chem. Soc. A*, 1654 (1971); (e) K. Fukui, *Fortschr. Chem. Forsch.*, **15**, 1 (1970); ref 3e.

(18) H. Hogeveen and H. C. Volger, *J. Am. Chem. Soc.*, **89**, 2486 (1967).

(19) H. Hogeveen and H. C. Volger, *Chem. Commun.*, 1133 (1967).

(20) L. Cassar, P. E. Eaton, and J. Halpern, *J. Am. Chem. Soc.*, **92**, 3515 (1970).

(21) L. Cassar and J. Halpern, *Chem. Commun.*, 1082 (1970).

(22) T. Yamamoto, A. Yamamoto, and S. Ikeda, *J. Am. Chem. Soc.*, **93**, 3350 (1971).

(23) W. Merk and R. Pettit, the former paper cited in Ref 17b.

(24) This conclusion originates from the spherical symmetry of the LUMO of A_g^+. If Rh(I) can provide a relevant d orbital for the interaction with substrate, the argument on 2 + 2 cycloreversion may be made on the basis of Scheme III.

(25) L. A. Paquette, *J. Am. Chem. Soc.*, **92**, 5765 (1970); *Acc. Chem. Res.*, **4**, 280 (1971).

(26) J. Wristers, L. Brener, and R. Pettit, *J. Am. Chem. Soc.*, **92**, 7499 (1970).

(27) F. G. Cowheard and J. L. von Rosenberg, *J. Am. Chem. Soc.*, **91**, 2157 (1969).

(28) L. Roos and M. Orchin, *J. Am. Chem. Soc.*, **87**, 5502 (1965).

(29) A cyclic interaction mechanism is also conceivable, which suggests a possibility of involvement of hydride shift to metal as an intermediate step.

(30) S. Inagaki, S. Yamabe, H. Fujimoto, and K. Fukui, *Bull. Chem. Soc. Jpn.*, **45**, 3510 (1972); S. Inagaki and K. Fukui, submitted for publication.

(31) R. Ugo, F. Conti, S. Cenini, R. Mason, and G. B. Robertson, *Chem. Commun.*, 1498 (1968).

(32) K. Fukui, K. Ohkubo, and T. Yamabe, *Bull. Chem. Soc. Jpn.*, **42**, 312 (1969); K. Ohkubo, T. Yamabe, and K. Fukui, *ibid.*, **42**, 2220 (1969); K. Ohkubo, *ibid.*, **47**, 557 (1974).

(33) F. Hirata and O. Hayaishi, *J. Biol. Chem.*, **246**, 7825 (1971).

(34) H. W. S. Chan, *Chem. Commun.*, 1550 (1970).

Chemical Pseudoexcitation and Paradoxical Orbital Interaction Effect

Satoshi Inagaki, Hiroshi Fujimoto, and Kenichi Fukui*

Contribution from the Department of Hydrocarbon Chemistry, Kyoto University, Kyoto, Japan. *Received December 7, 1974*

Abstract: Some third-order configuration interaction (CI) perturbation energy terms for bimolecular interactions have been shown to be interpreted as representing the local excitation within one or both of the interacting molecules, induced by charge transfer (chemical pseudoexcitation). The chemical pseudoexcitation effect was more explicitly expressed by the second-order perturbation energy terms obtained on the assumption of the degeneracy of a particular transferred configuration with the original one. The following rules have been derived. (1) Some thermal reactions between strong donors and strong acceptors are endowed with some characteristics of the photochemical reactions of weaker donor-acceptor cases. (2) Some reactions normally requiring photoinduction may take place on, or may be accelerated by, a complex formation of one of the reactants with catalysts which enhances the donor-acceptor relationship. (3) Paradoxical orbital interaction, in which the HOMO-HOMO and/or the LUMO-LUMO interactions play a dominant role, is expected to govern some significant stages of a variety of reactions in place of the normal HOMO-LUMO interaction. The regioselectivity in two types of thermal 2 + 2 cycloaddition reactions between strong donors and strong acceptors, the mechanisms of thermal and photochemical reactions via charge-transfer complexes, and the mechanisms of thermal and photochemical electrophilic substitution in aromatic systems are generally discussed with some beautiful examples.

Perturbation approaches to chemical reaction mechanisms on the basis of molecular orbital (MO) method have given rise to the concept of orbital interaction. The electron delocalization due to the interaction between the "frontier" orbitals, the highest occupied (HO) MO of donor, and the lowest unoccupied (LU) MO of acceptor contributes dominantly not only to the stabilization of the interacting system but also to the intermolecular bond formation.[1-5] There may be no room for doubt about the role of the interaction between these particular orbitals in determining the stereochemical paths of various reactions which proceed through concerted or least-motion approach of reactants. This implies that the transferred configuration involving an electron shift from the HOMO of donor to the LUMO of acceptor is of a profound importance, when we describe the interaction in terms of the configuration function. Even in nonconcerted transformations composed of several distinguishable elementary processes, a certain electron configuration can be assigned to each step of the reaction. There may, however, be a number of equivocal processes which are located between the two extremes, concerted and step-by-step mechanisms. It would be a poor approximation to discuss such reactions only in terms of HOMO-LUMO interaction, namely, in terms of interaction between the transferred configuration and the original configuration. It is necessary to take the other orbital interactions into consideration, or to perform multiconfiguration interaction analysis.

The possibility of a drastic change in stereochemical path due to an electron transfer was first pointed out by one of the present authors (K.F.) in connection with a "symmetry-forbidden"[6] reaction, an electrophilic cis-1,2 addition to olefinic double bonds.[7] From a series of papers by Epiotis[8] he developed a similar idea and discussed the mechanisms of a variety of thermal and photochemical reactions. The interactions between one transferred configuration and another, between one locally excited configuration and another, and between a transferred configuration and a locally excited configuration should be studied in order to discuss such problems in a strict way. The roles of the above configuration interactions in reaction mechanisms have not been discussed in detail so far[5] while the significance of the interaction of the original electron configuration with the transferred configuration has been fully discussed together with the interaction of the original configuration with the locally excited configuration.[9]

The purpose of the present paper is to present in general a theoretical basis of the roles of the various configuration interactions not necessarily involving the zero configuration, by analyzing the second-order perturbed wave function and the third-order perturbed energy for nondegenerate systems, and the first-order perturbed wave function and the second-order perturbed energy for degenerate systems. It is also aimed to relate the concept of such configuration interactions to the reaction mechanisms, especially to the intermediary processes between the concerted and the stepwise reactions, and between the thermal and the photochemical reactions. The current topics in organic chemistry would be quoted to visualize the theoretical concept in a chemically graspable manner.

Theoretical Background

The background of the present work is found in the perturbation theory based on the concept of configuration interaction as described in our previous paper.[5] Suppose that a ground-state molecule (A) weakly interacts with a ground-state partner (B). The occupied and the unoccupied MO's of A and those of B are denoted by the letters i, j, k, and l, respectively. We use an arrow to signify an electron shift, for instance, $i \rightarrow l$ stands for the electron transfer from the ith occupied orbital of A to the lth unoccupied MO of B, and $i \rightarrow j$ means an electron promotion from the

6109

ith occupied MO to the jth unoccupied MO within A. An electronic state of the interacting system can be represented by a linear combination of various electronic configurations. The zero-configuration function Ψ_0 is described by a single Slater determinant composed of the originally occupied MO's of the isolated molecules. The expectation value of the total Hamiltonian H with respect to Ψ_0 includes the Coulombic and the exchange interaction energies. One can describe the transferred and the locally excited configurations by using the originally unoccupied orbitals of the isolated molecules in addition.

The secular equation for weakly interacting system is given by eq. 1. Then, the approximate solution can be expanded as[10] eq 2, in which $H_{IJ} = \int \Psi_I^* H \Psi_J d\tau$ and $S_{IJ} = \int \Psi_I^* \Psi_J d\tau$, and where the subscripts denote the electronic configurations. The perturbed wave function is also given by eq 3. Although it is desirable to take as many electronic configurations as possible into consideration in order to discuss the details of interaction, our present purpose needs only several interaction terms in eq 2 and 3. The magnitude of each term is, as shown in our previous paper,[5] estimated by the order of MO overlap between A and B, s_{ab}, on the assumption that the off-diagonal elements of the secular eq 1 are proportional to the overlap integrals between the configurations, S_{KL}. The general relation of overlap integrals between S_{KL} and s_{ab} is given by eq 4, in which ν is the minimum number of electron shifts between A and B required to obtain the electronic configuration K from L. The energy terms not higher than $o(s_{ab}{}^4)$ can be classified as follows (the terms derived from the mathematical equivalency between A and B are not presented for brevity).

$$\begin{vmatrix} H_{00} - E & \cdots & H_{0K} - S_{0K}E & \cdots & H_{0L} - S_{0L}E & \cdots \\ \vdots & & \vdots & & \vdots & \\ H_{K0} - S_{K0}E & \cdots & H_{KK} - E & \cdots & H_{KL} - S_{KL}E & \cdots \\ \vdots & & \vdots & & \vdots & \\ H_{L0} - S_{L0}E & \cdots & H_{LK} - S_{LK}E & \cdots & H_{LL} - E & \cdots \\ \vdots & & \vdots & & \vdots & \end{vmatrix} = 0 \tag{1}$$

$$H_{00}' = H_{00} + \sum_{\substack{K \\ (\neq 0)}} \frac{|H_{0K} - S_{0K}H_{00}|^2}{H_{00} - H_{KK}} + \sum_{\substack{K \\ (\neq 0)}} \sum_{\substack{L \\ (\neq 0, K)}} \frac{(H_{0K} - S_{0K}H_{00})(H_{KL} - S_{KL}H_{00})(H_{L0} - S_{L0}H_{00})}{(H_{00} - H_{KK})(H_{00} - H_{LL})}$$
$$+ \text{(small terms of higher than the third order)} \tag{2}$$

$$\Psi_0' = \left\{1 - \sum_{\substack{K \\ (\neq 0)}} \frac{H_{K0} - S_{K0}H_{00}}{H_{00} - H_{KK}} S_{0K} - \frac{1}{2} \sum_{\substack{K \\ (\neq 0)}} \frac{|H_{0K} - S_{0K}H_{00}|^2}{(H_{00} - H_{KK})^2} \right\}\Psi_0 + \sum_{\substack{K \\ (\neq 0)}} \left\{ \frac{H_{K0} - S_{K0}H_{00}}{H_{00} - H_{KK}} + \sum_{\substack{L \\ (\neq 0, K)}} \frac{(H_{L0} - S_{L0}H_{00})(H_{KL} - S_{KL}H_{00})}{(H_{00} - H_{KK})(H_{00} - H_{LL})} \right\} \Psi_K +$$
$$\text{(small terms of higher than the second order)} \tag{3}$$

$$S_{KL} = o(s_{ab}{}^{\nu}) \tag{4}$$

(i) One-electron transfer from A to B

$$\sum_i^{\substack{A \\ occ}} \sum_l^{\substack{B \\ uno}} \frac{|H_{0,i\to l} - S_{0,i\to l}H_{00}|^2}{H_{00} - H_{i\to l,i\to l}} \tag{5a}$$

(ii) One-electron excitation within A

$$\sum_i^{\substack{A \\ occ}} \sum_j^{\substack{A \\ uno}} \frac{|H_{0,i\to j} - S_{0,i\to j}H_{00}|^2}{H_{00} - H_{i\to j,i\to j}} \tag{5b}$$

(iii) One-electron excitation within A induced by one-electron transfer from A to B

$$\sum_i^{\substack{A \\ occ}} \sum_j^{\substack{A \\ uno}} \sum_l^{\substack{B \\ uno}} 2 \times \left\{\frac{(H_{0,i\to l} - S_{0,i\to l}H_{00})(H_{i\to j,0} - S_{i\to j,0}H_{00})}{(H_{00} - H_{i\to l,i\to l})(H_{00} - H_{i\to j,i\to j})} \times (H_{i\to l,i\to j} - S_{i\to l,i\to j}H_{00})\right\} \tag{5c}$$

(iv) One-electron excitation within A induced by one-electron transfer from B to A

$$\sum_i^{\substack{A \\ occ}} \sum_j^{\substack{A \\ uno}} \sum_k^{\substack{B \\ occ}} 2 \times \left\{ (H_{0,k\to j} - S_{0,k\to j}H_{00}) \times \frac{(H_{k\to j,i\to j} - S_{k\to j,i\to j}H_{00})(H_{i\to j,0} - S_{i\to j,0}H_{00})}{(H_{00} - H_{k\to j,k\to j})(H_{00} - H_{i\to j,i\to j})} \right\} \tag{5d}$$

(v) Two-electron transfer from A to B

$$\sum_i^{\substack{A \\ occ}} \sum_{i'}^{\substack{A \\ occ}} \sum_l^{\substack{B \\ uno}} \sum_{l'}^{\substack{B \\ uno}} \frac{|H_{0,(i\to l,i'\to l')} - S_{0,(i\to l,i'\to l')}H_{00}|^2}{H_{00} - H_{(i\to l,i'\to l'),(i\to l,i'\to l')}} \tag{5e}$$

(vi) Two-electron transfer from A to B induced by one-electron transfer from A to B

$$\sum_i^{\substack{A \\ occ}} \sum_{i'}^{\substack{A \\ occ}} \sum_l^{\substack{B \\ uno}} \sum_{l'}^{\substack{B \\ uno}} 2 \left\{ \frac{(H_{0,i\to l} - S_{0,i\to l}H_{00})}{H_{00} - H_{i\to l,i\to l}} \times (H_{i\to l,(i\to l,i'\to l')} - S_{i\to l,(i\to l,i'\to l')}H_{00}) \times \frac{H_{(i\to l,i'\to l'),0} - S_{(i\to l,i'\to l'),0}H_{00}}{H_{00} - H_{(i\to l,i'\to l'),(i\to l,i'\to l')}} \right\} \tag{5f}$$

(vii) Mutual electron transfer between A and B

$$\sum_i^{\substack{A \\ occ}} \sum_j^{\substack{A \\ uno}} \sum_k^{\substack{B \\ occ}} \sum_l^{\substack{B \\ uno}} \frac{|H_{0,(i\to l,k\to j)} - S_{0,(i\to l,k\to j)}H_{00}|^2}{H_{00} - H_{(i\to l,k\to j),(i\to l,k\to j)}} \tag{5g}$$

(viii) Mutual electron transfer induced by one-electron transfer from A to B

$$\sum_i^{\substack{A \\ occ}} \sum_j^{\substack{A \\ uno}} \sum_k^{\substack{B \\ occ}} \sum_l^{\substack{B \\ uno}} 2 \times \left\{ \frac{(H_{0,i\to l} - S_{0,i\to l}H_{00})(H_{i\to l,(i\to l,k\to j)} - S_{i\to l,(i\to l,k\to j)}H_{00})}{H_{00} - H_{i\to l,i\to l}} \times \frac{H_{(i\to l,k\to j),0} - S_{(i\to l,k\to j),0}H_{00}}{H_{00} - H_{(i\to l,k\to j),(i\to l,k\to j)}} \right\} \tag{5h}$$

(ix) Mutual electron transfer induced by one-electron transfer from B to A

$$\sum_i^{\substack{A \\ occ}} \sum_j^{\substack{A \\ uno}} \sum_k^{\substack{B \\ occ}} \sum_l^{\substack{B \\ uno}} 2 \times \left\{ \frac{(H_{0,k\to j} - S_{0,k\to j}H_{00})(H_{k\to j,(i\to l,k\to j)} - S_{k\to j,(i\to l,k\to j)}H_{00})}{H_{00} - H_{k\to j,k\to j}} \times \frac{H_{(i\to l,k\to j),0} - S_{(i\to l,k\to j),0}H_{00}}{H_{00} - H_{(i\to l,k\to j),(i\to l,k\to j)}} \right\} \tag{5i}$$

The configurations which mix into the ground-state function Ψ_0' directly or indirectly are similarly obtained, the order with respect to MO overlap being not higher than second order.

(i) One-electron transfer from A to B

$$\sum_{i}^{\text{A occ}}\sum_{l}^{\text{B uno}} \frac{H_{i\to l,0} - S_{i\to l,0}H_{00}}{H_{00} - H_{i\to l,i\to l}}\Psi_{i\to l} \qquad (6a)$$

(ii) One-electron excitation within A

$$\sum_{i}^{\text{A occ}}\sum_{j}^{\text{B uno}} \frac{H_{i\to j,0} - S_{i\to j,0}H_{00}}{H_{00} - H_{i\to j,i\to j}}\Psi_{i\to j} \qquad (6b)$$

(iii) One-electron excitation within A induced by one-electron transfer from A to B

$$\sum_{i}^{\text{A occ}}\sum_{j}^{\text{B uno}}\sum_{l}^{\text{B uno}} \frac{(H_{i\to l,0} - S_{i\to l,0}H_{00})(H_{i\to l,i\to j} - S_{i\to l,i\to j}H_{00})}{(H_{00} - H_{i\to l,i\to l})(H_{00} - H_{i\to j,i\to j})}\Psi_{i\to j} \qquad (6c)$$

(iv) One-electron excitation within A induced by one-electron transfer from B to A

$$\sum_{i}^{\text{A occ}}\sum_{j}^{\text{B uno}}\sum_{k}^{\text{B occ}} \frac{(H_{k\to j,0} - S_{k\to j,0}H_{00})(H_{i\to j,k\to j} - S_{i\to j,k\to j}H_{00})}{(H_{00} - H_{k\to j,i\to j})(H_{00} - H_{i\to j,i\to j})}\Psi_{i\to j} \qquad (6d)$$

(v) Two-electron transfer from A to B

$$\sum_{i}^{\text{A occ}}\sum_{i'}^{\text{A occ}}\sum_{l}^{\text{B uno}}\sum_{l'}^{\text{B uno}} \frac{H_{(i\to l,i'\to l'),0} - S_{(i\to l,i'\to l'),0}H_{00}}{H_{00} - H_{(i\to l,i'\to l'),(i\to l,i'\to l')}}\Psi_{i\to l,i'\to l'} \qquad (6e)$$

(vi) Two-electron transfer from A to B induced by one-electron transfer from A to B

$$\sum_{i}^{\text{A occ}}\sum_{i'}^{\text{A occ}}\sum_{l}^{\text{B uno}}\sum_{l'}^{\text{B uno}} \frac{H_{i\to l,0} - S_{i\to l,0}H_{00}}{H_{00} - H_{i\to l,i\to l}} \frac{H_{(i\to l,i'\to l'),i\to l} - S_{(i\to l,i'\to l'),i\to l}H_{00}}{H_{00} - H_{(i\to l,i'\to l'),(i\to l,i'\to l')}}\Psi_{i\to l,i'\to l'} \qquad (6f)$$

(vii) Mutual electron transfer between A and B

$$\sum_{i}^{\text{A occ}}\sum_{j}^{\text{A uno}}\sum_{k}^{\text{B occ}}\sum_{l}^{\text{B uno}} \frac{H_{(i\to l,k\to j),0} - S_{(i\to l,k\to j),0}H_{00}}{H_{00} - H_{(i\to l,k\to j),(i\to l,k\to j)}}\Psi_{i\to l,k\to j} \qquad (6g)$$

(viii) Mutual electron transfer induced by one-electron transfer from A to B

$$\sum_{i}^{\text{A occ}}\sum_{j}^{\text{A uno}}\sum_{k}^{\text{B occ}}\sum_{l}^{\text{B uno}} \frac{(H_{i\to l,0} - S_{i\to l,0}H_{00})(H_{(i\to l,k\to j),i\to l} - S_{(i\to l,k\to j),i\to l}H_{00})}{(H_{00} - H_{i\to l,i\to l})(H_{00} - H_{(i\to l,k\to j),(i\to l,k\to j)})} \times \Psi_{i\to l,k\to j} \qquad (6h)$$

(ix) Mutual electron transfer induced by one-electron transfer from B to A

$$\sum_{i}^{\text{A occ}}\sum_{j}^{\text{A occ}}\sum_{k}^{\text{B uno}}\sum_{l}^{\text{B uno}} \frac{(H_{k\to j,0} - S_{k\to j,0}H_{00})(H_{(i\to l,k\to j),k\to j} - S_{(i\to l,k\to j),k\to j}H_{00})}{(H_{00} - H_{k\to j,k\to j})(H_{00} - H_{(i\to l,k\to j),(i\to l,k\to j)})} \times \Psi_{i\to l,k\to j} \qquad (6i)$$

The roles of one-electron transfer (eq 5a and 6a) and one-electron local excitation (eq 5b and 6b) in mechanisms of chemical reactions have been fully examined and discussed already.[5] One can say that the electron delocalization from the HOMO of donor to the LUMO of acceptor, i.e., the interaction between the transferred configuration and the zero configuration, contributes dominantly to the stabilization of the interacting system of neutral or weakly polar species. It is, however, likely that the polarization energy due to the interaction between the locally excited configuration and the zero configuration, and the Coulombic energy contained in H_{00} may play considerable roles in the interaction between charged or highly polarizable species. Let us focus our attention to the other interaction terms. The choice of the terms to be considered of importance here has been made at first by estimating the magnitude of the numerators based upon the MO overlap ($s_{ab}^{\,n}$). It is necessary, moreover, to select relatively significant terms among them by looking at the magnitude of the denominators. On this assumption, the terms involving two-electron transfer may be omitted, since the energy gaps between the zero configuration and the two-electron transferred configurations are usually so large as compared with the ones between the zero configuration and one-electron transferred or locally excited configurations. The terms in eq 5e–5i and 6e–6i are expected to contribute little. It may be reasonable to concentrate our attention on the terms in eq 5c, 5d, 6c, and 6d.

The third-order perturbation energy terms, eq 5c and 5d, contain a transferred configuration and a locally excited configuration. The interpretation of the configuration interaction between the two is not necessarily unequivocal. They may be interpreted as the local excitation induced by the charge transfer or as the delocalization induced by the local excitation. The ambiguity is eliminated by the absence of the terms interpreted as the charge transfer induced by the local excitation and the presence of the terms interpreted as the local excitation induced by the charge transfer in the perturbed wave function (eq 6c and 6d). The nonequivalent contributions from $\Psi_{i\to l}$ and $\Psi_{i\to j}$ (eq 6c) and that from $\Psi_{k\to j}$ and $\Psi_{i\to j}$ (eq 6d) to the second-order perturbed wave function are owing to the difference in the order between the interactions of $\Psi_{i\to l}$ and $\Psi_{i\to j}$ with Ψ_0. The order of the delocalization (former) is $o(s_{ab}^{\,1})$ while the polarization (latter) is $o(s_{ab}^{\,2})$. As a result, the terms in eq 5c and 6c represent the $i \to j$ local excitation within A induced by the $i \to l$ transfer from A to B. The process involves the interactions among the zero (0), the transferred $K(i \to l)$, and the locally excited $L(i \to j)$ configurations, as depicted in Figure 1a. The other mode of the $i \to j$ local excitation within A is induced by the $k \to j$ transfer from B to A, represented by eq 5d and 6d. In this case the $k \to j$ transferred configuration is involved in place of the $i \to l$ configuration (Figure 1b). One can imagine the process in which an electron transfers from i to l and then moves to j, while a hole in k, resulting from the $k \to j$ electron transfer, is filled with an electron from i.[7,8,11] Hereafter, we may call the local excitation induced by the charge transfer simply as "(chemical) pseudoexcitation".[12] A theoretical reasoning should be made for emphasizing the role of the higher-order

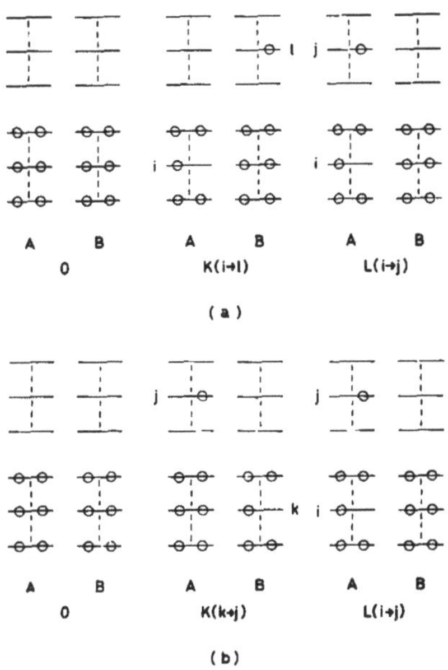

Figure 1. Pseudoexcitation within the A molecule induced by charge transfer from A to B (a) and from B to A (b).

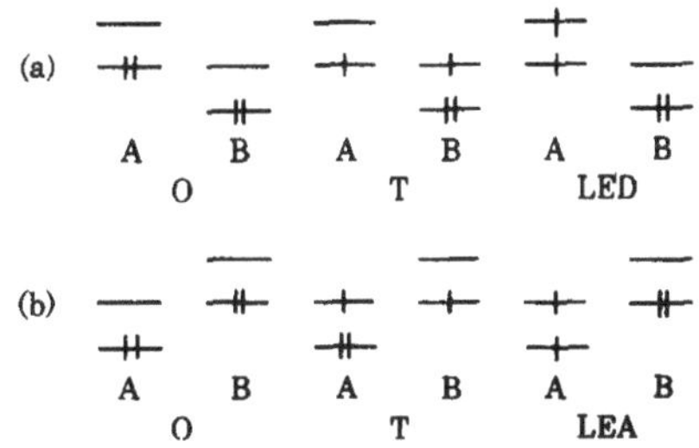

Figure 2. A simplified CI presentation of the pseudoexcitation within donor (a) and within acceptor (b).

local excitation terms (eq 5c, 5d, 6c, and 6d) rather than that of the lower-order terms (eq 5b and 6b). Both perturbation terms are of the same orders with respect to MO overlap: s_{ab}^4 in H_{00}' and s_{ab}^2 in Ψ_0'. However, the higher-order terms are likely to contribute more greatly in the case that the energy separation of the transferred configuration from the zero configuration gets smaller.

The circumstances can be grasped in much more straightforward manner if we apply the perturbation theory for the degenerate system. It can be assumed that a very strong charge transfer is represented better by the degenerate combination of the zero configuration 0 and one of the transferred configurations (D), i.e., $H_{DD} \sim H_{00}$. In such a case, the perturbed energy H_{00}' and the perturbed wave function Ψ_0' are given as follows.

$$H_{00}' = H_{00} - |H_{0D} - S_{0D}H_{00}| + \frac{1}{2}\Bigg[\sum_{\substack{K\\(\neq 0,D)}}\left\{\frac{|H_{0K} - S_{0K}H_{00}|^2}{H_{00} - H_{KK}} + \frac{|H_{DK} - S_{DK}H_{00}|^2}{H_{00} - H_{KK}}\right\} - 2\sum_{\substack{K\\(\neq 0,D)}}\left\{\frac{(H_{0D} - S_{0D}H_{00})(H_{DK} - S_{DK}H_{00})(H_{K0} - S_{K0}H_{00})}{|H_{0D} - S_{0D}H_{00}|(H_{00} - H_{KK})}\right\} - 2\{(H_{D0} - S_{D0}H_{00})S_{0D}\}\Bigg] + \text{(small terms of higher than the second order)} \quad (7)$$

$$\Psi_0' = \frac{1}{\sqrt{2}}\left\{1 - \frac{1}{2}\frac{(H_{D0} - S_{D0}H_{00})}{|H_{0D} - S_{0D}H_{00}|}S_{0D}\right\}\Bigg[\Psi_0 + \frac{(H_{D0} - S_{D0}H_{00})}{|H_{0D} - S_{0D}H_{00}|}\Psi_D + \sum_{\substack{K\\(\neq 0,D)}}\left\{\frac{H_{K0} - S_{K0}H_{00}}{H_{00} - H_{KK}} + \frac{(H_{D0} - S_{D0}H_{00})(H_{KD} - S_{KD}H_{00})}{|H_{0D} - S_{0D}H_{00}|(H_{00} - H_{KK})}\right\}\Psi_K\Bigg] + \text{(small terms of higher than the first order)} \quad (8)$$

The orders of the terms involving the degenerate transferred configuration are lower by one in the degenerate system than in the nondegenerate system. Accordingly, the pseudoexcitation would be conspicuous by way of the interaction with the electron transferred configuration (D), while the mixing-in of the locally excited configurations through the direct interaction with the zero configuration remains intact in the perturbation formulas for the degenerate system. The coefficients of the locally excited configuration Ψ_K, $(H_{D0} - S_{D0}H_{00})(H_{KD} - S_{KD}H_{00})/|H_{0D} - S_{0D}H_{00}|(H_{00} - H_{KK})$ $(o(s_{ab}^1))$, are greater than $(H_{K0} - S_{K0}H_{00})/(H_{00} - H_{KK})$ $(o(s_{ab}^2))$. The same results are also obtained from the perturbed energy (eq 7). The second-order energy $|H_{0K} - S_{0K}H_{00}|^2/(H_{00} - H_{KK})$ is $o(s_{ab}^4)$ while the energy to be compared, $(H_{0D} - S_{0D}H_{00})(H_{DK} - S_{DK}H_{00})(H_{K0} - S_{K0}H_{00})/|H_{0D} - S_{0D}H_{00}|(H_{00} - H_{KK})$, is $o(s_{ab}^3)$. In addition, a new term $|H_{DK} - S_{DK}H_{00}|^2/(H_{00} - H_{KK})$ $(o(s_{ab}^2))$ appears in the perturbed energy for the degenerate system. This term originates from the fourth-order perturbation energy in the nondegenerate system. This can be understood by noting that the transferred configuration and the zero configuration have the same amplitude in the degenerate system.

One can readily see the importance of the configurations where the electron shift involves the HOMO's and the LUMO's, since such configurations are close enough to the zero configuration in energy to make the energy gaps in the denominators the smallest. Let the pseudoexcitation be represented by a simplified CI. The particular MO's of importance are employed in the presentation. The configurations involved in the pseudoexcitation within A induced by the charge transfer from A to B are the zero configuration (0), the transferred configuration (T), where one electron is shifted from the HOMO of A to the LUMO of B, and the locally excited configuration (LED), where one electron is promoted from the HOMO of A to the LUMO of A within the donor, as depicted in Figure 2a. Similarly, the pseudoexcitation within A induced by the charge transfer from B to A is represented by the CI among 0, T and the locally excited configuration (LEA), where the electron promotion occurs from the HOMO to the LUMO within the acceptor A, as depicted in Figure 1b.

Consequently, one can reach the following conclusions.

Rule 1. Some thermal reactions between good donors and good acceptors may tend to exhibit some characteristics which are common with the photochemical reactions of weak donor-acceptor systems.

Rule 2. Some reactions normally requiring photoinduction are expected to be brought about in the absence of light source, or to be promoted, by the catalysts which enhance the donor-acceptor relation through a complex formation with one of the reactants.

An important reservation with rule 1 must be made here. Remark 1: a combination of excessively strong donor and

6112

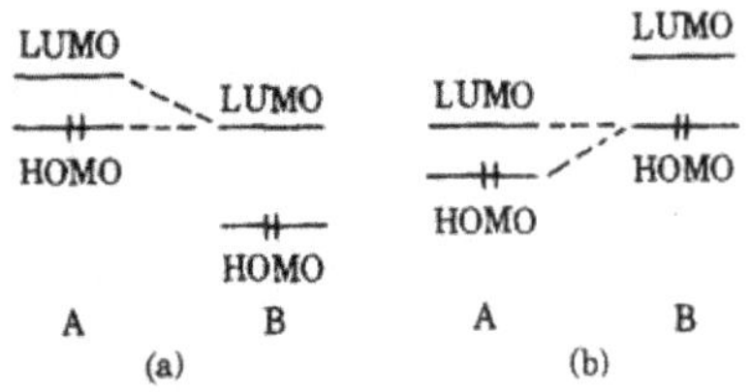

Figure 3. The key orbital interactions for the pseudoexcitation within donor (a) and for that within acceptor (b).

strong acceptor may give rise to a pair of stable cation and anion radicals, may dissociate in solution, may bind loosely, or may stay too stable to undergo the reactions expected from the pseudoexcitation effect.

This is the case with $H_{00} > H_{TT}$. The perturbed ground-state energy and wave function of such a case are given by the expansion at $E = H_{TT}$.

$$H_{TT}' = H_{TT} + \sum_K \frac{|H_{TK} - S_{TK}H_{TT}|^2}{KH_{TT} - H_{KK}} + \ldots \quad (9)$$

$$\Psi_T' = \Psi_T + \sum_K \frac{H_{KT} - S_{KT}H_{TT}}{H_{TT} - H_{KK}} \Psi_K + \ldots \quad (10)$$

The second terms in the right-hand side include those which come from the mixing-in of 0, LED, and LEA configurations. The inequality $|H_{00} - H_{TT}| < |H_{\mathrm{LED,LED}} - H_{TT}| \sim |H_{\mathrm{LEA,LEA}} - H_{TT}|$ can be assumed usually in this case and the difference in the absolute values is expected to be large as the H_{TT} is low. These mean that excessively strong charge transfer may stabilize the system mainly due to the relative increase in the contribution from the zero configuration to the ground-state wave function. The mixing-in of the zero configuration predominates over that of LED and LEA configurations to the extent to which the T configuration predominates over the other transferred configurations in the normal ground-state interaction ($H_{00} < H_{TT}$).

It is desirable and practical to reduce the CI expression at the MO level. The third energy term, eq 5c, or the pseudoexcitation $i \rightarrow j$ induced by electron transfer $i \rightarrow l$ contains the product of the overlap integrals between the configuration, $S_{0,i\rightarrow l}S_{i\rightarrow l,i\rightarrow j}S_{i\rightarrow j,0}$. The product is reduced to the form of the MO overlap product $s_{il}^2 s_{lj}^2$. The product $S_{0,k\rightarrow j}S_{k\rightarrow j,i\rightarrow j}S_{i\rightarrow j,0}$ in the pseudoexcitation term, eq 5d, is reduced to the form $s_{kj}^2 s_{ik}^2$.

Let the HOMO's and the LUMO's be i and k, and j and l, respectively. One can then see that the pseudoexcitation within A induced by the charge transfer from A to B needs the effective $(\mathrm{HOMO})_A$-$(\mathrm{LUMO})_B$-$(\mathrm{LUMO})_A$ interaction (Figure 3a), and that the pseudoexcitation within A induced by the charge transfer from B to A involves the $(\mathrm{HOMO})_A$-$(\mathrm{HOMO})_B$-$(\mathrm{LUMO})_A$ interaction (Figure 3b). The same conclusion can be readily drawn from the second-order terms, eq 6c and 6d, in the perturbed wave function. This is also the case with the perturbation formulas for the degenerate system.

One can reach the third rule accordingly.

Rule 3: the stereochemical course of the reaction, in which the initial HOMO-LUMO interaction directly leads to no stable species and leaves the possibility of selecting one of energetically equivalent or nearly so courses may be determined by the pseudoexcitation, namely, by the HOMO-HOMO interaction or by the LUMO-LUMO interaction.

An important remark should also be made with rule 3. Remember the dynamic aspect of chemical reaction. The pseudoexcitation is induced by the charge transfer. The HOMO-LUMO interaction predominant at the early stage may lead the interacting system to one stereochemical course separated from another by a high energy barrier between them. In such a case, even if the pseudoexcitation occurs in the subsequent process appreciably, the effect is expected to be little on the reaction course, since the pseudoexcitation takes place within the framework determined by the HOMO-LUMO interaction.

Remark 2: the HOMO-HOMO interaction or the LUMO-LUMO interaction does not necessarily determine the stereochemical course of the reactions in which the pseudoexcitation occurs considerably.

A Semiempirical MO Study on Regioselectivity of Thermal 2 + 2 Cycloaddition Reactions

Concerted 2 + 2 cycloaddition reactions are a typical class of excited-state reactions.[6] The addition of electron-donating olefins to strong acceptors, $^1\Delta_g$ molecular oxygen,[13] benzyne,[14] cyano-substituted olefin,[15] azodicarboxylic ester,[16] ketene,[17] keteneimmonium cation,[18] chlorosulfonyl isocyanate,[19] and sulfur trioxide,[20] is, however, known to take place stereoselectively with the retention of the initial configuration of the olefins. These facts support the validity of rule 1. Recent theoretical investigations have suggested that thermal 2 + 2 donor-acceptor condensation reactions may be initiated by the three-centered interaction between the two p orbitals of donor and one of the p orbitals of acceptor,[21-24] followed by the rotation of the acceptor molecule to form the final four-membered ring product.[21,25] The three-centered interaction in the initiation stage was found to be governed by the normal HOMO-LUMO interaction.[23-25] The growing contribution from the transferred configuration to the ground-state wave function, resulting from the foregoing HOMO-LUMO interaction, may facilitate the rotation process through the pseudoexcitation. The interaction between the transferred configuration and the locally excited configuration allows the HOMO-HOMO and the LUMO-LUMO interaction.[25,26] It has been confirmed by a spectroscopic and kinetic study that the thermal 2 + 2 cycloaddition reactions between vinyl ethers and tetracyanoethylene consist of successive charge transfer and cycloaddition processes.[27]

Here we will attempt to investigate the possible mechanisms of the regioselectivity-determining step in the 2 + 2 cycloaddition reactions. The regioselectivity may be considered to be determined by the direction in which the atom of acceptor with the largest LUMO amplitude, constituting the transient three-membered ring at the early stage, moves preferentially. A typical reaction is the addition of allene with acrylonitrile.[28] The terminal carbon of acrylonitrile with the largest LUMO amplitude is known to be combined with the central carbon of allene with the smaller HOMO and the larger LUMO amplitudes. The combination is paradoxical in the sense that the atom of the donor with the *smaller* HOMO amplitude forms a bond with the atom of the acceptor with the largest LUMO amplitude. Similar regioselectivity was observed in the reactions of acrolein and of acrylic esters as acceptors.[28]

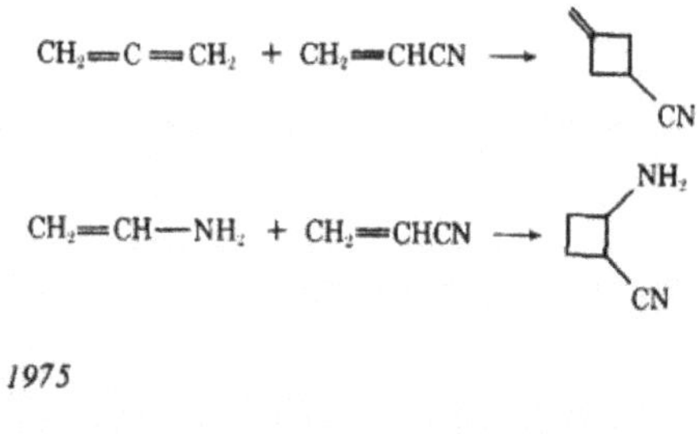

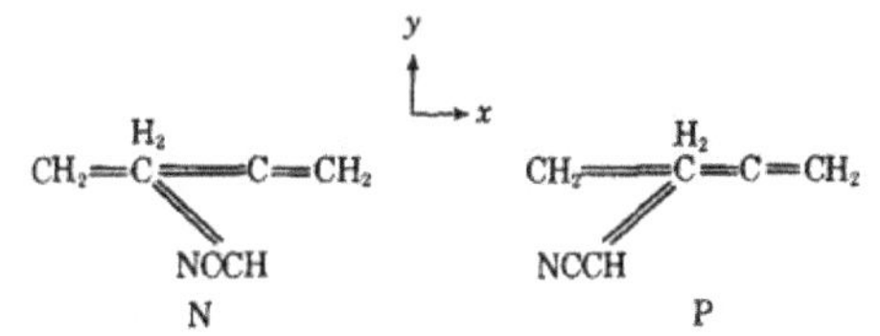

Figure 4. Arrangements of reactants with the normal (N) and the paradoxical (P) orbital interactions.

Another typical reaction is the addition of aminoethylene with acrylonitrile, a simplified model of the reactions of enamines with acrylonitrile,[29] and with propiolate esters.[30] In these cases, the cycloadducts are produced through the normal combination between the β carbon of the donor with the largest HOMO amplitude and the β carbon of the acceptor with the largest LUMO amplitude.

We analyzed atomic orbital (AO) overlap population in order to scrutinize the regioselectivity-determining step. Molecular arrangement at the relevant reaction stage is assumed to be a transient one in which the three-centered interaction geometry is just being transformed to four-membered ring product. This assumption is supposed to be reasonable according to a series of theoretical investigations on the nuclear configuration change along the reaction coordinate as cited before. Now let us compare the two modes of interactions of olefins, for instance, allene and acrylonitrile. The electrophilic center of acrylonitrile was located 1.75 Å above the middle point between the nucleophilic terminal carbon and the center of the C=C bond of allene in N-type interaction, while the carbon of acrylonitrile was placed 1.75 Å above the point halfway from the middle point of a C=C bond to the central carbon of allene in P-type interaction. The planes of the donor and the acceptor were kept parallel and acrylonitrile was rotated by 45° clockwise or counterclockwise to simulate the course of ring closing (Figure 4).

The AO overlap populations, calculated by the extended Hückel MO method,[31] are partitioned into the contributions from each pair of MO's between allene and acrylonitrile.[32] The results of the analysis of the AO overlap population between the p orbital of the electrophilic carbon of acrylonitrile and the terminal carbon of allene in the N model and between the former and the central carbon of allene in the P model are listed in Table I. A similar analysis is also made for the aminoethylene-acrylonitrile system (Table II). The orbitals having the maximum extention in the Z direction are designated by the next (N) HOMO, HOMO, LUMO, and NLUMO. The preference of P to N in the allene-acrylonitrile system, consistent with the experimental observations, is seen in the AO overlap populations, 0.157 in N and 0.175 in P. The difference is attributed mainly to the preferential contribution from the LUMO-LUMO interaction (0.103) in the P model. The dominant contribution in the N model is 0.089 owing to the HOMO-LUMO interaction. The partial AO population coming from the HOMO-HOMO interaction is not as large in both of the models. This calculation suggests that the regioselectivity of the reaction of allene with acrylonitrile is controlled for the most part by the pseudoexcitation within the donor while the pseudoexcitation within the acceptor is of minor effect. In the reaction of aminoethylene the HOMO-LUMO interaction (0.070 in N and 0.032 in P) together with the pseudoexcitation within the acceptor (0.048 in N and 0.022 in P) determines the regioselectivity while the effect of the pseudoexcitation within the donor is comparable in both models, though large (0.072 in N and 0.073 in P).

Table I. Analysis of AO Overlap Populations of Allene and Acrylonitrile[a]

Allene	Acrylonitrile NHOMO	HOMO	LUMO	NLUMO
N (total AO overlap population 0.157)				
NHOMO	0.000	0.000	0.005	−0.001
HOMO	0.001	0.042	0.089	0.008
LUMO	−0.001	−0.024	0.037	0.001
NLUMO	0.000	0.000	0.000	0.000
P (total AO overlap population 0.175)				
NHOMO	0.000	0.006	0.013	0.002
HOMO	0.002	0.042	0.042	0.004
LUMO	−0.005	−0.039	0.103	0.006
NLUMO	0.000	0.000	−0.001	0.000

[a] Analysis of AO overlap population between p_z-AO on the terminal carbon of allene and p_z-AO on the terminal carbon of acrylonitrile (N) and that between p_z-AO on the central carbon of allene and p_z-AO on the terminal carbon of acrylonitrile (P).

Table II. Analysis of AO Overlap Populations of Aminoethylene and Acrylonitrile[a]

Aminoethylene	Acrylonitrile NHOMO	HOMO	LUMO	NLUMO
N (total AO overlap population 0.185)				
NHOMO	0.000	0.007	0.015	0.003
HOMO	0.003	0.048	0.070	0.005
LUMO	−0.004	−0.040	0.072	0.006
P (total AO overlap population 0.147)				
NHOMO	0.001	0.017	0.032	0.004
HOMO	0.001	0.022	0.032	−0.002
LUMO	−0.002	−0.029	0.073	−0.002

[a] Analysis of AO overlap population between p_z-AO's on C_α of aminoethylene and on the terminal carbon of acrylonitrile (P) and that between the p_z-AO's on C_β of aminoethylene and on the terminal carbon of acrylonitrile.

The difference in the regioselectivity mentioned above is supposed to result from the property of the donor HOMO. The perturbation of the ethylenic π orbital due to the neighboring groups, i.e., a lone pair of electrons of the amino group and the C–H σ bond of methylene, may be different appreciably in strength. The lone-pair orbital lies so high as to produce strong polarization leading to the concentration of the HOMO electron density on the terminal carbon of aminoethylene, while the C–H bond orbital of allene is located too low to interact effectively with the ethylenic π orbital. The highly polarized HOMO of aminoethylene may cause the breaking of the three-centered structure at the relatively earlier stage of the reaction since the HOMO-LUMO interaction between the terminal carbon of acrylonitrile and the β carbon of enamine stabilizes the system (remark 2). In contrast, the three-centered structure may continue to exist longer in the approach of acrylonitrile to allene.

At the three-centered geometry, the HOMO-LUMO charge-transfer interaction and the pseudoexcitation within the acceptor involving the HOMO-HOMO interaction can take place sufficiently, while the pseudoexcitation within the donor involving the LUMO-LUMO interaction can hardly be of importance on account of orbital symmetry. Accordingly the pseudoexcitation within the donor is expected to control the orientation of the electrophilic center of acrylonitrile in the second stage. The effect of the orbital interaction on the regioselectivity-determining step of the allene-acrylonitrile condensation reaction is paradoxical in the sense that the LUMO, instead of the HOMO, of the donor play a significant role in governing the stereoselection even on the thermal condition.

Remark 1 must be noted again in connection with the mechanism of the thermal 2 + 2 donor–acceptor condensation reactions. A too strong charge transfer does not necessarily promote the reactions requiring the pseudoexcitation. An experimental evidence is now available. In fact it was recently reported that the rate of the thermal 2 + 2 cycloaddition reaction between vinyl ethers and tetracyanoethylene has an optimum with respect to the changes in the strength of charge-transfer interaction.[27]

It is meaningful to compare briefly the thermal 2 + 2 cycloaddition reactions with the Diels–Alder reactions and with the 1,3-dipolar cycloaddition reaction. The regioselectivity of the concerted reactions has been understood in terms of the normal orbital interaction effect or the interaction of the HOMO of the donor with the LUMO of the acceptor; the chemical combinations usually occur between the atom of the donor with the largest HOMO amplitude and the atom of the acceptor with the largest LUMO amplitude.[33] The concerted cycloaddition reactions can be discussed generally based on the symmetry of the relevant "frontier" orbitals without invoking the significant geometry change on the course of the reaction to the final product. On the contrary, the 2 + 2 thermal cycloaddition reactions should inevitably require the sufficient mixing-in of the locally excited configurations, with the peculiar change in the relative orientation of the reactants.

Applications

The continuity between the photochemical and the thermal reactions is exemplified here. The reactions of electron-donating toluene derivatives with electron-accepting 7,7,8,8-tetracyanoquinodimethane (TCNQ) provide us with an elegant example of rule 1. The stronger donors, mesitylene (R_1 = Me, R_2 = H)[34] and *p*-methoxytoluene (R_1 = H, R_2 = OMe),[35] react only on heating with TCNQ, while in the case of the weaker donors, toluene (R_1 = R_2 = H) and *p*-xylene (R_1 = H, R_2 = Me), the photoirradiation is required. The addition reactions may be initiated by the charge transfer from the aromatics to TCNQ. Strong charge transfer may induce the pseudoexcitation with the electron promotion in the toluenes, presumably, from the HOMO to the σ LUMO of the benzylic C–H to be abstracted, through the LUMO of TCNQ. The LUMO–LUMO interaction (rule 3) may be of key importance in the final stage of the chemical transformation while the normal HOMO–LUMO interaction predominates in the stage of the complex formation.

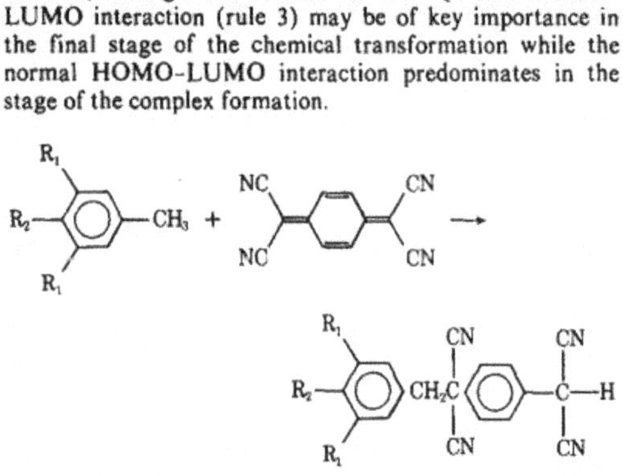

Another example, although not systematically investigated, is found in the reactions of indoles with cyanoethylenes. The reaction of tetracyanoethylene with 2-methylindole gives 2-methyl-3-tetracyanoethylindole in a quantitative yield at room temperature under no irradiation.[36] As indoles are known to form immediately a deep blue-black colored π complex with tetracyanoethylene,[37] this reaction may also take place through the charge transfer accompanied by the pseudoexcitation. A weaker acceptor, acrylonitrile, requires photoinduction to cause a similar kind of reaction, α-cyanoethylation on the 3 position of indoles.[38] On the basis of the fact that the charge-transfer band is absent in the absorption spectra and that the fluorescence of indole is effectively quenched by acrylonitrile, a mechanism of α-cyanoethylation was proposed which involves the direct excitation of indoles followed by the exciplex formation.[38] The donor–acceptor relationship between indoles and acrylonitrile may not be so remarkable as to cause appreciable pseudoexcitation. In the case of the reaction of less distinct donor–acceptor systems, benzene–pyrrole[39] and naphthalene–pyrrole,[40] the mechanism involving the direct excitation was also proposed.

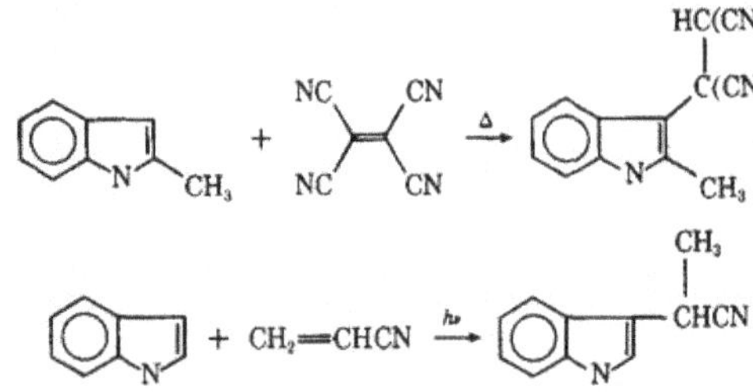

A remarkable effect of $ZnCl_2$ observed in the photochemical reactions of benzene with acrylonitrile[41] should be noted here in connection with the role of catalysts in the pseudoexcitation processes (rule 2). The yield of 1,2 cycloadduct of benzene with acrylonitrile in the photochemical reaction increases more than five times in the presence of $ZnCl_2$. The increase in the yield may be attributed to the pseudoexcitation enhanced or introduced through the complex formation of acrylonitrile with $ZnCl_2$. Another example to show the significance of catalyst is the ground-state 2 + 2 cycloaddition reactions between the acetylenic compounds and the carbonyl compounds, modified by BF_3.[42] The mechanism of the thermal oxetane formation of highly electron-accepting carbonyl compounds has been fully discussed in our previous paper.[25b] The essential feature has been found to be quite similar to the singlet oxygen reactions.[24,25a] It has been recognized that a Lewis acid, BF_3, induces the 2 + 2 cycloaddition even between a weak donor and a weak acceptor.[42] This fact is completely explained by rule 2.

Electrophilic substitution reactions of aromatic compounds provide the preceding theoretical arguments with an experimental support. The substitution reactions are usually caused by strong electrophiles in the thermal conditions. Recently an aromatic substitution with a weak electrophile, boron tribromide, was reported to occur through charge-transfer complex on the photoirradiation, while no detectable reaction took place in the dark.[43] Heating the reaction system without light afforded the different products, presumably derived from the homolysis of BBr_3.[43] The photohaloboronation supports rule 1. It is further interesting that $AlCl_3$ has been reported to enable the haloboronation to take place readily on the thermal condition.[44] This can be explained by means of rule 2.

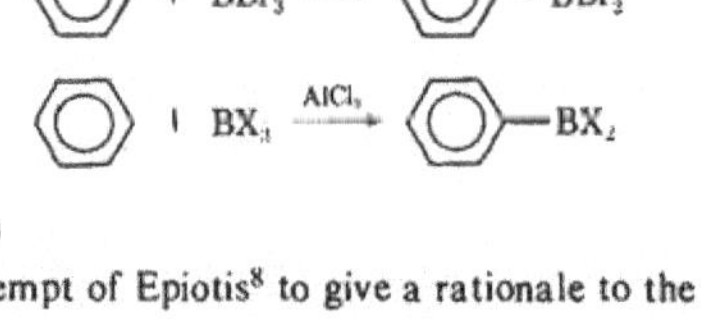

Discussion

The attempt of Epiotis[8] to give a rationale to the facility and the selectivity of the ground-state 2 + 2 cycloaddition reactions was appropriate in the sense that he proposed the significance of the transferred configuration in a strong charge-transfer interaction between the reactants. It may be worth referring to the dynamic aspects of the mechanism of the reactions appreciably involving the pseudoexcitation. The mechanism of the thermal 2 + 2 cycloaddition reaction of the strong donor–acceptor system proposed by the present authors[22] should be distinguished from the s + s and the s + a processes of Woodward and Hoffmann.[6] According to our proposed mechanism, the reaction should begin with the cyclic three-centered interaction among the two nucleophilic atoms of donor and a more electrophilic atom of the acceptor.[21-25] The other atom of the acceptor is far away from the interaction center in the initial stage of the reaction. This supposition came from the consideration of the most favorable interaction between the HOMO of the donor and the LUMO of the acceptor.[22-25,45] At the early stage, the charge transfer from the donor to the acceptor and the accompanied pseudoexcitation within the acceptor are allowed to take place, but the pseudoexcitation within the donor cannot occur because Ψ_{LED} belongs to a symmetry different from that of the other dominant configurations. Then, the wave function of this state Ψ_0' is approximately given by

$$\Psi_0' \sim C_0\Psi_0 + C_T\Psi_T + C_{LEA}\Psi_{LEA}$$
$$(C_0 \gtrsim C_T > C_{LEA}, C_{LED} = 0) \qquad (11)$$

Subsequently, the rotation of the acceptor molecule leading to the four-membered ring product makes possible the pseudoexcitation within the donor. The transient state is similarly approximated by

$$\Psi_0' \sim C_0\Psi_0 + C_T\Psi_T + C_{LEA}\Psi_{LEA} + C_{LED}\Psi_{LED}$$
$$(C_0 \sim C_T \sim C_{LEA} \sim C_{LED}) \qquad (12)$$

In the above case the regioselectivity may be controlled by the pseudoexcitation within the donor, or by the LUMO–LUMO interaction, which newly appears on the breaking of the three-centered structure. In another case the electronic state of the interacting system even at the beginning of the reaction can be described by eq 12, where the coefficients C_0 and C_T dominate over the others. This means that the regioselectivity is determined by the HOMO–LUMO interaction. The selection of the reaction course depends solely on whether the donor HOMO amplitude is localized mainly on one atom or delocalized uniformly between the two, as exemplified.

The significance of the three-centered geometry at the early stage is, though indirectly, supported by some experimental results. An episulfoxide, dibenzoylstilbene episulfoxide, was reported to be photochemically converted to the decomposition products, monothiobenzil and benzil, presumably via a four-membered ring intermediate, oxathietane.[46] It is interesting that the rearrangement of the episulfoxide to oxathietane requires the photoinduction while the reaction of olefins with $^1\Delta_g$ molecular oxygen to dioxetane through a perepoxide quasiintermediate[25a] takes place "thermally". This difference can be explained in terms of the difference in the electron acceptability between O_2 and SO parts (rule 1). Another evidence is provided by the fact that the partial loss of the stereochemical integrity occurs in the acceptor molecule while the stereochemistry of donor is retained.[15c] This may come from our proposal that the stereochemistry of donor is fixed by the three-centered interaction while the tailing atom of the reacting centers in the acceptor is possible to rotate at the early stage.[47]

Acknowledgment. One of the authors (S.I.) expresses his appreciation to Mr. Kazuyuki Yamasaki for valuable information on the experimental results cited in the text, and to Mr. Shigeki Kato for the MO calculation program which S.I. partially supplemented for the present purpose. He also thanks the Data Processing Center of Kyoto University for generous use of the FACOM 230-70 computer.

References and Notes

(1) (a) K. Fukui, "Molecular Orbitals in Chemistry, Physics, and Biology", P. O. Löwdin and B. Pullman, Ed., Academic Press, New York, N.Y., 1964, p 513, and the preceding papers cited therein; (b) K. Fukui and H. Fujimoto, "Mechanisms of Molecular Migration", Vol. 2, B. S. Thygarajan, Ed., Interscience, New York, N.Y., 1969, p 118; (c) K. Fukui, "Theory of Orientation and Stereoselection", Springer-Verlag, Heidelberg, 1970; (d) K. Fukui, *Acc. Chem. Res.*, **4**, 57 (1971).

(2) (a) L. Salem, *J. Am. Chem. Soc.*, **90**, 543, 553 (1968); (b) L. Salem, *Chem. Br.*, 449 (1969).

(3) G. Klopman, *J. Am. Chem. Soc.*, **90**, 223 (1968).

(4) For recent reviews see: (a) W. C. Herndon, *Chem. Rev.*, **72**, 157 (1972); (b) R. F. Hudson, *Angew. Chem.*, **85**, 63 (1973).

(5) K. Fukui and H. Fujimoto, *Bull. Chem. Soc. Jpn.*, **41**, 1989 (1968); (b) K. Fukui, "XXIIIrd International Congress of Pure and Applied Chemistry", Vol. 1, Butterworths, London, 1971, p 65; (c) H. Fujimoto and K. Fukui, *Adv. Quantum Chem.*, **6**, 177 (1972); (d) K. Fukui, "The World of Quantum Chemistry", Proceedings of the First International Congress of Quantum Chemistry, Menton, France, R. Daudel and B. Pullman, Ed., D. Reidel Publishing Co., Boston, Mass., 1974, p 113.

(6) R. B. Woodward and R. Hoffmann, *Angew. Chem., Int. Ed. Engl.*, **8**, 781 (1969).

(7) K. Fukui, *Bull. Chem. Soc. Jpn.*, **39**, 498 (1966).

(8) N. D. Epiotis, *J. Am. Chem. Soc.*, **94**, 1924, 1935, 1941, 1946 (1972). Subsequent to the submission of this manuscript, a review appears [N. D. Epiotis, *Angew. Chem., Int. Ed. Eng.*, **13**, 751 (1974)].

(9) Although the second-order energy and the first-order wave function contain the interaction of the zero configuration with the configurations where more than one electrons are transferred and/or promoted, the contributions from their configurations are estimated to be small due to the small off-diagonal matrix elements and due to the large energy separation from the zero configuration.[5]

(10) (a) The third-order perturbed energy and the second-order perturbed wave function have been discussed in detail in connection with the mechanisms of catalysis.[10b] The relaxation effect of catalysts on the "symmetry forbiddenness" has been presented; (b) K. Fukui and S. Inagaki, *J. Am. Chem. Soc.*, **97**, 4445 (1975).

(11) K. Yamaguchi, T. Fueno, and H. Fukutome, *Chem. Phys. Lett.*, **22**, 461 (1973).

(12) The term "chemical pseudoexcitation" is used to symbolize an appreciable effect of the terms in eq 5c, 5d, 6c, and 6d on a certain condition and to distinguish itself from the well-defined "excitation" standing for a discontinuous transition from an energetically lower state to a higher state. Such a conventional usage may be acceptable since, as is mentioned in the text, through some conditioned interactions a thermal process can give the same type of products as a photoinduced, excited-state reaction.

(13) P. D. Bartlett and A. P. Schaap, *J. Am. Chem. Soc.*, **92**, 3223 (1970).

(14) I. Tabushi, R. Oda, and K. Okazaki, *Tetrahedron Lett.*, 3743 (1968); M. Jones, Jr., and R. H. Levin, *ibid.*, 5593 (1968); L. Friedman, R. J. Oslewicz, and P. W. Rabideau, *ibid.*, 5735 (1968); H. H. Wasserman, A. J. Solodar, and L. S. Keller, *ibid.*, 5597 (1968); P. G. Gassman and H. P. Benecke, *ibid.*, 1089 (1969).

(15) (a) S. Proskow, H. E. Simmons, and T. L. Cairns, *J. Am. Chem. Soc.*, **85**, 2341 (1963); (b) P. D. Bartlett, *Q. Rev., Chem. Soc.*, **24**, 473 (1970); (c) S. Proskow, H. E. Simmons, and T. L. Cairns, *J. Am. Chem. Soc.*, **88**, 5254 (1966).

(16) J. Firl and S. Sommer, *Tetrahedron Lett.*, 4713 (1972).

(17) R. Huisgen, L. Feiler, and G. Binsch, *Angew. Chem.*, **76**, 892 (1964); J. C. Martin, V. W. Goodlett, and R. D. Burpitt, *J. Org. Chem.*, **30**, 4309 (1965); G. Binsch, L. A. Feiler, and R. Huisgen, *Tetrahedron Lett.*, 4497 (1968); R. Montaigne and L. Ghosez, *Angew. Chem., Int. Ed. Engl.*, **7**, 221 (1968); N. S. Isaacs and P. F. Stanbury, *Chem. Commun.*, 1061 (1970).

(18) J. Marchand-Brynaert and L. Ghosez, *J. Am. Chem. Soc.*, **94**, 2870 (1972).

(19) E. J. Moriconi and J. F. Kelley, *Tetrahedron Lett.*, 1435 (1968); H. Bestian, H. Biener, K. Claus and H. Heyn, *Justus Liebigs Ann. Chem.*, **718**, 94 (1968); F. Effenberger and G. Keifer, *Angew. Chem.*, **79**, 936 (1967).

(20) M. Nagayama, O. Okumura, S. Noda, and A. Mori, *J. Chem. Soc., Chem. Commun.*, 841 (1973).

(21) D. M. Hays and R. Hoffmann, *J. Phys. Chem.*, **76**, 656 (1972).

(22) (a) S. Inagaki, S. Yamabe, H. Fujimoto, and K. Fukui, *Bull. Chem. Soc. Jpn.*, **45**, 3510 (1972).

(23) R. Sustmann, A. Ansmann, and F. Vahrenholt, *J. Am. Chem. Soc.*, **94**, 8099 (1972).

(24) S. Inagaki and K. Fukui, *Bull. Chem. Soc. Jpn.*, **46**, 2240 (1973).

(25) (a) S. Inagaki and K. Fukui, *J. Am. Chem. Soc.*, in press; (b) S. Inagaki, T. Minato, S. Yamabe, H. Fujimoto, and K. Fukui, *Tetrahedron*, **30**, 2165 (1974).

(26) The importance of such orbital interaction has been proposed for 2 + 2 cycloaddition reaction mechanism elsewhere without referring to the geometrical change of intermolecular arrangement along the reaction coordinate.[8,11]

(27) T. Arimoto and J. Osugi, *Chem. Lett.*, 271 (1974).

(28) J. D. Roberts and C. M. Sharts, *Org. React.*, **12**, 1 (1962); J. E. Baldwin and U. V. Roy, *Chem. Commun.*, 1225 (1969).

(29) (a) I. Fleming and J. Harley-Mason, *J. Chem. Soc.*, 2165 (1964); (b) K. C. Brannock, A. Bell, R. D. Burpitt, and C. A. Kelley, *J. Org. Chem.*, **29**, 801 (1964).

(30) (a) C. F. Huebner, L. Dorfman, M. M. Robinson, E. Donoghue, W. G. Pierson, and P. Strachan, *J. Org. Chem.*, **28**, 3134 (1963); (b) K. C. Brannock, R. D. Burpitt, V. W. Goodlett, and J. G. Thweatt, *J. Org. Chem.*, **29**, 818 (1964).

(31) R. Hoffmann, *J. Chem. Phys.*, **39**, 1397 (1963).

(32) Consult the following papers: H. Fujimoto, S. Kato, S. Yamabe, and K. Fukui, *J. Chem. Phys.*, **60**, 572 (1971), and references cited therein.

(33) For the Diels–Alder reactions see: O. Eisenstein, J.-M. Lefour, and N. T. Anh, *Chem. Commun.*, 969 (1971). For 1,3-dipolar cycloaddition reactions see: K. N. Houk, *J. Am. Chem. Soc.*, **94**, 8953 (1972); K. N. Houk, J. Sims, C. R. Watts, and L. J. Luskus, *ibid.*, **95**, 7301 (1973); T. Minato, S. Yamabe, S. Inagaki, H. Fujimoto, and K. Fukui, *Bull. Chem. Soc. Jpn.*, **47**, 1619 (1974).

(34) K. Yamasaki, A. Yoshino, T. Yonezawa, and M. Ohashi, *J. Chem. Soc., Chem. Commun.*, 9 (1973); K. Yamasaki, T. Yonezawa, and M. Ohashi, *J. Chem. Soc., Perkin Trans. 1*, 93 (1975).

(35) K. Yamasaki, unpublished result. One of the authors (S.I.) thanks Yamasaki for the personal communication prior to publication.

(36) W. E. Noland, W. C. Kuryla, and R. F. Lange, *J. Am. Chem. Soc.*, **81**, 6010 (1959).

(37) R. E. Merrifield and W. D. Phillips, *J. Am. Chem. Soc.*, **80**, 2778 (1958).

(38) K. Yamasaki, T. Matsuura, and I. Saito, *J. Chem. Soc., Chem. Commun.*, in press.

(39) M. Bellas, D. Bryce-Smith, and A. Gilbert, *Chem. Commun.*, 263 (1967).

(40) J. J. McCullough, W. S. Wu, and C. W. Huang, *J. Chem. Soc., Perkin Trans. 2*, 370 (1972).

(41) M. Ohashi, A. Yoshino, K. Yamasaki, and T. Yonezawa, *Tetrahedron Lett.*, 3395 (1973).

(42) H. Vieregge, H. J. T. Bos, and J. F. Arens, *Recl. Trav. Chim. Pays-Bas*, **78**, 664 (1959).

(43) Y. Ogata, Y. Izawa, H. Tomioka, and T. Ukigai, *Tetrahedron*, **25**, 1817 (1969).

(44) G. A. Olah, "Friedel-Crafts Chemistry", Wiley, New York, N.Y., 1973, p 73.

(45) The wave function representing a state of the whole interacting system can be expanded into a linear combination of various electron configurations of reactants, belonging to the same irreducible representation of the point group. The 2 + 2 cycloaddition via a three-centered interaction followed by the rotation of a reactant, as proposed by the present authors, can be described by a wave function the dominant term of which is the closed-shell configuration. On the other hand, if the reaction proceeds through the charge transfer from the HOMO of donor to the LUMO of acceptor, retaining such a symmetry as is characterized by the definition "$2_s + 2_s$" by Woodward and Hoffmann, the ground state of the whole system should fall into an open shell in the regioselectivity-determining stage.

(46) D. C. Dittmer, G. C. Levy, and G. E. Kuhlmann, *J. Am. Chem. Soc.*, **89**, 2793 (1967).

(47) An alternative explanation has been given by Epiotis [footnote 17 in ref 9a]. It is interesting that the rotation in the acceptor molecule was observed in the case of the cis donor, but not in the case of the trans donor. The difference is also explicable on the basis of the three-centered geometry. This problem may be closely related to the orientation of the tailing center in the acceptor. The tail atom may occupy the same side as the substituent of the donor with a lone pair, due to nonbonded attraction between the tail atom and the lone pair similarly coming from the HOMO–LUMO interaction.[48] The preferential orientation of the tail atom is expected to occur toward the trans donor and to fix the configuration on the tail atom. On the other hand, the neighboring substituent, an alkyl group, in the cis donor may hinder the similar orientation due to the steric interference.

(48) S. Inagaki and K. Fukui, unpublished results. The authors are greatly indebted to Professor C. S. Foote.

2670

On the Donor–Acceptor Relationship in Cyclic Additions

Sir:

It is well established that the Diels–Alder reaction is favored by the presence of one or more electron-attracting groups in conjugation with the carbon–carbon double bond of dienophile. Alkyl and alkoxyl groups on diene also facilitate the reaction.[1] The donor-acceptor relation between diene and dienophile was studied by Houk,[2] Sustmann,[3] and Herndon[4] from an energetical viewpoint, applying the concept of orbital interaction.[5] Sustmann's classification of the additions into the "normal" and "inverse" Diels–Alder reactions was derived on the basis of relative strength of electron transfer from diene to dienophile and that from dienophile to diene.[3] In this communication, concentrating our attention to chemical bonds, we show that the orbital interaction between antisymmetric MO's can furnish cyclic additions with concerted character. Major interaction between symmetric MO's seems to imply the possibility of another reaction mechanism. The energy is an overall quantity integrated over all the space, while chemical bonds are local in nature. Hence, examination of bond formation in chemical reactions may supply us with information which has not been reached by the usual energy criteria.

Let us consider an interaction between a donor, say A, and an acceptor, say B. For the sake of the simplicity of discussion, we assume here that the electronic structure of the interacting system composed of A and B is described by the combination of two configurational wave functions,[6]

$$\Psi = C_0\Psi_0 + C_{i\to l}\Psi_{i\to l} \quad (1)$$

where Ψ_0 is the original configuration in which both of A and B interact with the retention of their closed-shell structures in an isolated state and $\Psi_{i\to l}$ stands for an electron transfer from the ith occupied MO, ϕ_i, of A to the lth unoccupied MO, ψ_l, of B. Then, the electron density is given by,[7,8]

$$\begin{aligned}\rho(1) &= \rho_{0,0}(1) + C_{i\to l}{}^2(\rho_{i\to l,i\to l}(1) - \rho_{0,0}(1)) \\ &\quad + 2C_0C_{i\to l}(\rho_{0,i\to l}(1) - S_{0,i\to l}\rho_{0,0}(1)) \\ &\cong \rho_{0,0}(1) + C_{i\to l}{}^2(\psi_l(1)^2 - \phi_i(1)^2) \\ &\quad + 2\sqrt{2}C_0C_{i\to l}(\psi_l(1)\phi_i(1) - \phi_i(1)^2 s_{il}) \end{aligned} \quad (2)$$

in which

$$\rho_{p,q}(1) = \int \Psi_p{}^*(1,2,\ldots,N)\Psi_q(1,2,\ldots,N) \times d\xi_1 d\tau_2\ldots d\tau_N \times N$$

$$S_{0,i\to l} = \int \Psi_0{}^*\Psi_{i\to l}d\tau,\ s_{il} = \int \phi_i\psi_l\, dv$$

It should be noted that donor MO ϕ_i loses electrons in two distinctive ways. A fraction of the electron density is directly transferred from ϕ_i to ψ_l (sum to $C_{i\to l}{}^2$) and another fraction is consumed to create the intermolecular bonds through the orbital overlap between these MO's (sum to $2\sqrt{2}C_0C_{i\to l}s_{il}$).[9]

When the MO ϕ_i donates its electrons into the intermolecular region through an electron-transfer interaction, the nuclei of A will follow the reorganization of the electron density, pulled by the electrostatic force originating from the foregoing electrons. At the same time, the nuclei of B will also have attraction from the electron density accumulated in the intermolecular region.[10] In case of the Diels–Alder reaction, one of the termini of diene α should be bonded tightly to one of the reaction sites of dienophile β, and the other terminal of diene γ should be linked to the other reaction site of dienophile δ. In order to have the fixed framework of a six-membered ring, the interaction between α and δ and that between γ and β should be antibonding (nonbonding) or only slightly bonding. Accordingly, the electrons donated from the MO ϕ_i must be accumulated on either wings of intermolecular region where new chemical bonds will appear, connecting the reaction sites of diene and dienophile. The overlap density between two non-

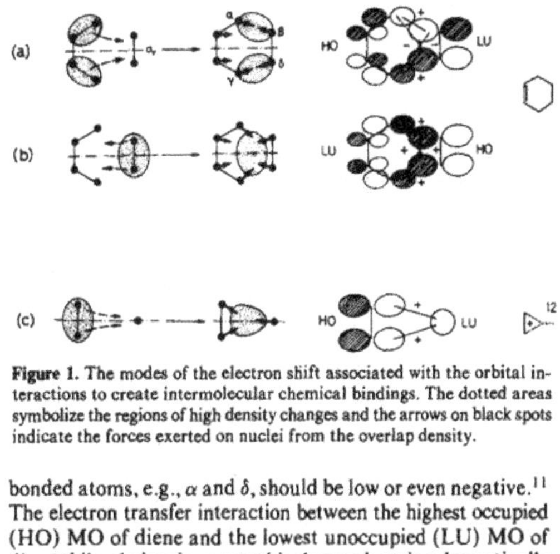

Figure 1. The modes of the electron shift associated with the orbital interactions to create intermolecular chemical bindings. The dotted areas symbolize the regions of high density changes and the arrows on black spots indicate the forces exerted on nuclei from the overlap density.

bonded atoms, e.g., α and δ, should be low or even negative.[11] The electron transfer interaction between the highest occupied (HO) MO of diene and the lowest unoccupied (LU) MO of dienophile obviously meets this demand, as is schematically illustrated in Figure 1a. The interaction of the HOMO of dienophile with the LUMO of diene results in an increase of electron density in the central part of the intermolecular region (Figure 1b), hence is not responsible for the formation of two apparently separated bonds between diene and dienophile (quasi)simultaneously. In case of the formation of a bridged cation, on the other hand, the incoming ion must be attracted from both of the carbons of an olefinic double bond. Thus, the electron transfer from the symmetric MO's should be predominant as shown in Figure 1c, to generate positive overlap density in the central region between the two species.

Our density analysis can locate a clearcut distinction of the consequence of the interaction between two antisymmetric MO's from that of the interaction between two symmetric MO's in cyclic additions. According to our present analysis, concerted formation of two bonds can be expected only if the dominant orbital interaction is the one between two antisymmetric MO's. Electron releasing groups on diene and electron withdrawing groups on dienophile promote the electron transfer from the HOMO of diene to the LUMO of dienophile, both being antisymmetric, in the Diels–Alder reaction. If the major interaction takes place between symmetric MO's, the cyclic addition will presumably be less concerted or experience complex formation before giving the final product. This proposition may supply us with a novel view on the mechanisms of cyclic additions and substituent effects. Further experimental study of the substituent effects on thermal cyclic additions of $4N$ and $4N + 2$ conjugated chains seems to be needed.

Acknowledgment. Support of this work by the Ministry of Education, Japan, is gratefully acknowledged (Grant-in-Aid 047068).

References and Notes

(1) See, for instance, E. S. Gould, "Mechanism and Structure in Organic Chemistry", Holt, Rinehart and Winston, New York, N.Y., 1959, pp 533–537.
(2) K. N. Houk, *Acc. Chem. Res.*, **11**, 361 (1975), and his papers cited therein.
(3) R. Sustmann, *Tetrahedron Lett.*, 2721 (1971).
(4) W. C. Herndon, *Chem. Rev.*, **72**, 157 (1972).
(5) (a) K. Fukui in "Molecular Orbitals in Chemistry, Physics, and Biology", P.-O. Löwdin and B. Pullman, Ed., Academic Press, New York, N.Y., 1964, p 513; (b) K. Fukui and H. Fujimoto in "Mechanisms of Molecular Migrations", B. S. Thyagarajan, Ed., Academic Press, New York, N.Y., 1969, pp 117–190; (c) R. B. Woodward and R. Hoffmann, *Angew. Chem., Int. Ed. Engl.*, **8**, 781 (1969); (d) M. J. S. Dewar, *ibid.*, **10**, 761 (1971); (e) L. Salem, *J. Am. Chem. Soc.*, **90**, 543, 553 (1968).
(6) H. Fujimoto and K. Fukui in "Chemical Reactivity and Reaction Paths", G. Klopman, Ed., Wiley-Interscience, New York, N.Y., 1974, pp 23–54.

[Reprinted from the Journal of the American Chemical Society, **98**, 2670 (1976).]

(7) H. Fujimoto, S. Yamabe, and K. Fukui, *Bull. Chem. Soc. Jpn.*, **44**, 2936 (1971).
(8) H. Fujimoto and R. Hoffmann, *J. Phys. Chem.*, **78**, 1874 (1974).
(9) The changes in the nuclear configurations of reactants associated with the donation and acceptance of electrons were discussed previously by the present authors. See K. Fukui and H. Fujimoto, *Bull. Chem. Soc. Jpn.*, **42**, 3399 (1969). Equation 2 indicates that donor MO loses electrons more rapidly than acceptor MO receives. Since the weakening of bonds between reaction sites and adjacent atoms in either reactant upon acceptance of electrons into unoccupied MO's and/or donation of electrons from occupied MO's gets smaller as the conjugation chain becomes longer, the larger conjugated system would preferably play the donor part in order to bring about harmonious displacements of nuclei in both of the reactants along a reaction path.
(10) This conclusion is derived by the application of the Hellmann–Feynman theorem to chemically interacting systems (H. Hellmann, "Einführung in Quantenchemie", Franz Deuticke 1937; R. P. Feynman, *Phys. Rev.*, **56**, 340 (1939)). See, also, H. Nakatsuji, *J. Am. Chem. Soc.*, **95**, 345 (1973).
(11) The term "overlap density" is used here to denote the part of the interference density (K. Ruedenberg, *Rev. Mod. Phys.*, **32**, 335 (1960)) which is responsible for the intermolecular bond formation.
(12) G. A. Olah in "Chemical Reactivity and Reaction Paths", G. Klopman, Ed., Wiley-Interscience, New York, N.Y., 1974, pp 253–273.

Hiroshi Fujimoto,* Satoshi Inagaki, Kenichi Fukui
Faculty of Engineering, Kyoto University
Kyoto 606, Japan
Received November 20, 1975

Orbital Mixing Rule[1]

Satoshi Inagaki, Hiroshi Fujimoto, and Kenichi Fukui*

Contribution from the Department of Hydrocarbon Chemistry, Faculty of Engineering, Kyoto University, Kyoto 606, Japan.
Received September 15, 1975

Abstract: The interaction among more than two molecular orbitals is considered. An orbital (ψ_{Ai}) of a system, say A, mixes into itself the other orbital (ψ_{Aj}) of A, which is originally orthogonal to ψ_{Ai}, through the interaction with ψ_{Bk} of the other system, B. The sign relation of ψ_{Ai}, ψ_{Aj}, and ψ_{Bk} in the perturbed orbital ψ_{Ai}' is definitely given by molecular orbital perturbation theory (orbital mixing rule). The rule gives rise to an important view, origin, and direction of nonequivalent orbital extension. The importance of the nonequivalency in frontier orbital extension is exemplified by electrophilic exo additions to norbornenes and related compounds, and by the syn–anti stereoselectivity in Diels–Alder reactions of 5-monosubstituted cyclopentadienes. It is also suggested that a vacant d orbital of sulfur plays a significant role in determining the stereochemical courses of the reactions where thiocarbonyl compounds and thioalkoxy derivatives of ethylene and acetylene accept nucleophilic reagents. A transannular cross σ-bond formation between proximal double bonds caused by electrophiles was explained by the rule. Furthermore, the strength and the direction of π-orbital polarization were predicted and used in interpreting the chemical behavior of π bonds.

The concept of orbital interaction has played a predominant role in the theoretical investigation of the mechanisms of chemical reactions and the electronic structures of molecules. Attention has been paid almost exclusively to interaction between a pair of orbitals. The interaction of an orbital with a second orbital necessarily involves others (orbital mixing), appreciably in some cases. The orbital mixing analyses have been employed by Libit and Hoffmann,[2a] Inagaki and Fukui,[1b] Fujimoto and Hoffmann,[2b] and Imamura and Hirano[2c] to elucidate different chemical phenomena. The purpose of this paper is to clarify, in our own way, the orbital mixing rule on the basis of perturbation theory, and to shed light on the chemical phenomena of which the reasoning remains uncertain or unchallenged.

Theoretical Background

Let us concern ourselves with two orthonormal sets of orbitals between which interaction is allowed. Orbitals of one set are represented by ψ_{Ai} and ψ_{Aj}, those of the other set being represented by ψ_{Bk}. The orbital energy is denoted by ϵ, with the subscript standing for the orbital. One obtains a theoretical formulation for mixing among nondegenerate orbitals in the second-order perturbed orbital as follows (see Appendix):

$$\psi_{Ai}' = \psi_{Ai} + \frac{h_{ik}h_{jk}}{(\epsilon_{Ai} - \epsilon_{Aj})(\epsilon_{Ai} - \epsilon_{Bk})}\psi_{Aj} + \frac{h_{ik}}{\epsilon_{Ai} - \epsilon_{Bk}}\psi_{Bk} \qquad (1)$$

in which the prime stands for the perturbed orbital.

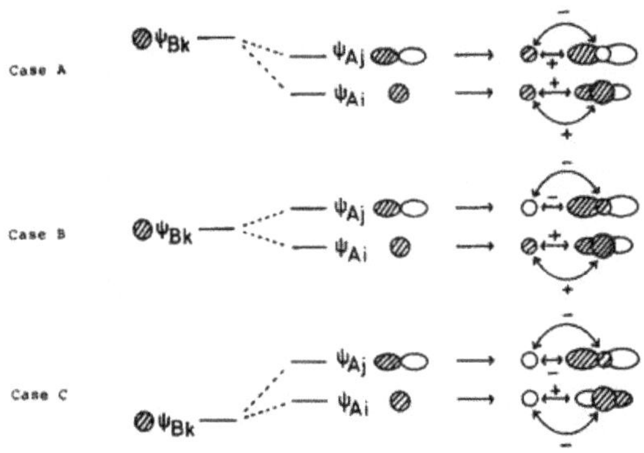

Figure 1. The sign relations upon mixing of three nondegenerate orbitals.

The sign relation among the coefficients of ψ_{Ai}, ψ_{Aj}, and ψ_{Bk} is found in eq 1 to be uniquely determined for a given system. The signs of the resonance integrals (h_{ab}'s) and the relationship of the orbital energies are important factors. One can assume without loss of generality that the signs of all basis orbitals involved, which are arbitrary in principle, are defined so that a positive overlap integral will show bonding character. This assumption is intended to avoid the confusion that negative overlap integrals imply bonding character. Then the resonance integrals (h_{ab}'s) are usually negative. Accordingly, the numerators in eq 1 have the minus sign in the first-order coefficient of ψ_{Bk} and the plus sign in the second-order coefficient of ψ_{Aj}. The signs of the denominators, $(\epsilon_{Ai} - \epsilon_{Bk})$ and $(\epsilon_{Ai} - \epsilon_{Aj})(\epsilon_{Ai} - \epsilon_{Bk})$, are readily derived from the relative energies of the orbitals. For instance, consider the modifications of the lowest lying orbital, say ψ_{Ai}, by mixing in ψ_{Aj} through the interactions of both with ψ_{Bk} ($\epsilon_{Ai} < \epsilon_{Aj}$ and ϵ_{Bk}). The coefficient of the zero-order orbital, ψ_{Ai}, is assumed to be +1. The sign of ψ_{Bk} is plus, since $h_{ik} < 0$ and $\epsilon_{Ai} - \epsilon_{Bk} < 0$. The coefficient of ψ_{Aj}, $h_{ik}h_{jk}/(\epsilon_{Ai} - \epsilon_{Aj})(\epsilon_{Ai} - \epsilon_{Bk})$, has the plus sign, since $h_{ik}h_{jk} > 0$ and the denominator is plus. In other words, ψ_{Aj} and ψ_{Bk} have the same signs. In a similar way, the sign relations in the cases of the other orbital energy relations are drawn from eq 1 and are visualized in Figure 1. Atomic orbitals are used to represent the sign relations. The arrows from the left to the right in Figure 1 connect the unperturbed orbitals, ψ_{Ai} and ψ_{Aj} (left), to the perturbed orbitals (right). The signs, + and −, denote in-phase and out-of-phase combinations, respectively. As a result, one can derive the orbital mixing rule: (1) the lowest orbital, ψ_{Ai} in case A or case B of Figure 1, is modified so that both combinations between ψ_{Ai} and ψ_{Bk} and ψ_{Bk} and ψ_{Aj} are in phase; (2) the highest orbital, ψ_{Aj} in case B or case C, mixes the others into itself so that both combinations between ψ_{Aj} and ψ_{Bk} and ψ_{Bk} and ψ_{Ai} are out of phase; (3) the orbital, ψ_{Aj} in case A or ψ_{Ai} in case C, mixes the others into itself in the opposite phase relations between ψ_{Ai} and ψ_{Bk} and ψ_{Bk} and ψ_{Aj}. The corresponding conclusions can be analogously drawn for the mixing-in of more than three-orbital systems and degenerate orbital systems.

Application and Discussion

The simple formula for orbital mixing can be used to elucidate the mechanisms of a variety of interesting chemical reactions. Most of the applications given here are concerned with intramolecular orbital mixing in the construction of the molecular orbitals of the whole molecules from the orbitals of the composite subsystems. One of the applications is to clarify the origin of the nonequivalent extension of orbitals under the influence of the conjugative entity breaking down the equivalent environments, and in turn to predict the direction. The electronic structures of plane-asymmetrically substituted olefins, predicted from the orbital mixing theory, are found to be consistent with the stereochemical behaviors within the framework of the frontier electron theory or the concept of orbital interaction.[3] Secondly, a rationale is provided for the difference in the regioselectivity of the reactions between oxygen and sulfur analogues. The discussions on the regioselectivity gave an interesting result, that is, theoretical support for the d-orbital participation of the sulfur atom. Thirdly, the direction of polarization of typical olefins by the substituents is discussed. Finally, a rationale for a transannular cross σ-bond formation between proximal double bonds caused by electrophiles is also discussed.

Origin and Direction of Nonequivalent Orbital Extension.[1,4] Exo stereoselection has been observed in the reactions of norbornene with a variety of electrophiles (eq. 2).[5-12] The ratio-

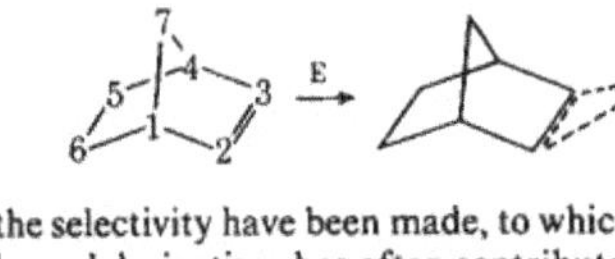

(2)

nales of the selectivity have been made, to which the solvolysis of 2-norbornyl derivatives has often contributed. Winstein et al.[6] stressed the electronic cause, characteristic of the bicyclo[2.2.1]hydrocarbon skeleton, or the C(1)–C(6) σ-bond participation stabilizing the transition state of the solvolytic reaction of the exo isomers (**1**). Brown[7] and Schleyer[8] attributed this selectivity to stereochemical features of norbornene, i.e., the steric interference by the endo hydrogens at C(5) and C(6) (**2**), and a torsional effect exerted between the bridgehead hydrogens and the neighboring olefinic hydrogens (**3**), re-

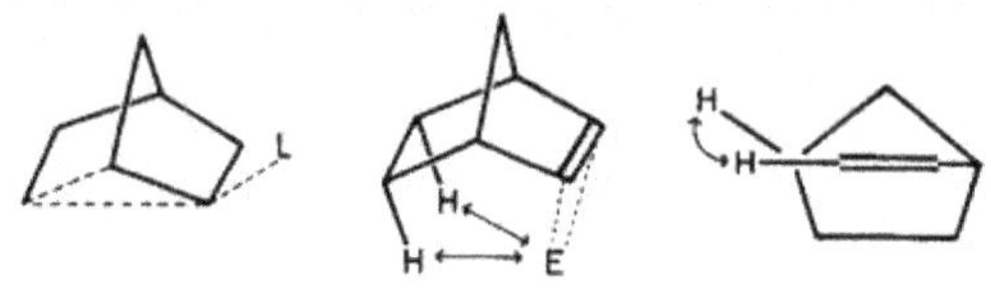

1 2 3

spectively. Brown's explanation is questionable. The steric crowding in the exo region, that is, the introduction of methyl groups at the 7 position, did not change the selectivity from exo to endo in the oxymercuration[9,10] or the hydrochlorination[11] of norbornenes. The following experimental results also throw some doubt on the predominant role of Schleyer's torsional effect: (1) the reactions of 1-methylnorbornene with unsymmetrical electrophiles such as mercuric salts[9] and formic acid[12] yielded the possible isomeric exo products in about equal amounts (the torsional effect should have favored the initial attack of the electrophiles on C(2) rather than on C(3)); (2) the hydrochlorinations of 1-methylnorbornene gave the exo products in the inverted ratios of one to the other under slightly different conditions.[12] In this section, we will propose another reason for these matters on the basis of the electronic structure of norbornene predicted by the orbital mixing rule.

According to the concept of orbital interaction,[3] the charge transfer interaction involving the highest occupied molecular orbital (HOMO) is the most important in characterizing the behavior toward electrophile. A detailed description of the HOMO is indispensable in scrutinizing electrophilic reaction mechanisms. Briefly, for the present purpose, an electrophile attacks norbornene in the direction of higher HOMO electron density.

The HOMO of norbornene extends to a nonequivalent degree in the exo and endo directions. This rehybridization originates from mixing of σ orbital(s) in the ethylenic π orbital under the influence of an conjugative entity. The unperturbed π orbital is the predominant component of the norbornene

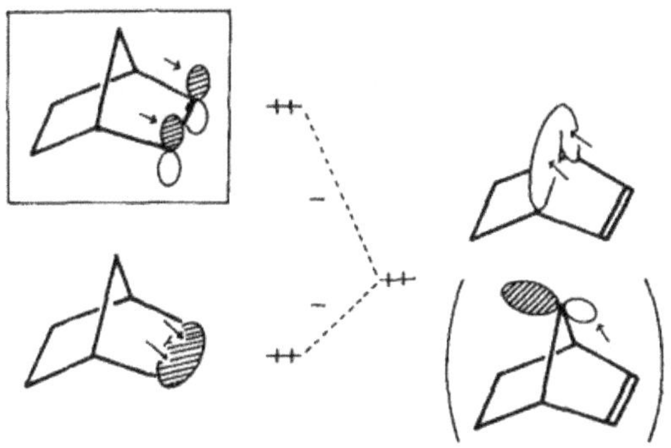

Figure 2. Origin and direction of nonequivalent HOMO extension of norbornene.

HOMO. The σ orbital mixed in order to cause the rehybridization is on the C(2)–C(3) bond. The question is whether the interaction of the π orbital occurs preferentially in the exo direction or in the endo direction. The extended Hückel[13] and CNDO/2[14] MO calculations on the assumed structure constructed by hybridizing norbornane[15] for the saturated part and norbornadiene[16] for the unsaturated part suggest two modes of significant interactions of the π orbital. One is a stronger hyperconjugation with the methano bridge than with the ethano bridge. The methano-bridge orbital is shown in Figure 2. The other important interaction is a back-side interaction with the anti C–H bond at the 7-carbon.[17] Both interactions occur at the exo side of the π orbital. We can now predict that the HOMO extends in the exo direction (see Figure 2). (In this and the following figures the rectangular frame surrounds the perturbed orbital of interest and the arrows attached to the orbital lobes show the interaction points.) The high-lying π orbital is mixed out of phase with the perturbing σ orbital(s) in the exo region and with the σ orbital on the unsaturated carbons in an out-of-phase relation to the exo σ orbital(s) (ψ_{Aj} in case B or case C of Figure 1). The prediction was confirmed by the calculated sign relation between the mixing π and σ orbitals: the s AO's of C(2) and C(3) and the pσ component of the C(2)–C(3) bond have the same signs as that of the exo lobes of the π orbital in the HOMO of norbornene. The mixing-in of the s and pσ orbitals gives rise to greater orbital overlapping or to more electron crowding in the exo region. These can be responsible for the exo electrophilic attack. In order to visualize the spatial extension of the HOMO, a contour map of the plane perpendicular to the C(1)C(2)-C(3)C(4) coplane was examined. A nonequivalent extension along the exo–endo direction is obviously discernible in Figure 3.

This explanation is similarly applicable to the exo stereoselectivity in free-radical reactions of 2-norbornyl[18,19] and 7-oxa-2-norbornyl,[19] and to the relative rate of base-catalyzed deuterium exchange of 2-norbornanones.[20]

Additional support for the direction of nonequivalent orbital extension and its relation to exo–endo stereoselectivity is available, although steric factors are possible and contribute to some degree. The π-HOMO of bicyclo[2.1.0]pentene is predicted to extend in the exo direction. The role of the methano bridge (and the anti C(7)–H bond) in norbornene is considered in this case to be replaced by that of the bridging σ bond of the bicyclopentene bent in the exo direction. The prediction is also confirmed by the contour map depicted in the plane perpendicular to the C(1)C(2)C(3)C(4) coplane (Figure 3). The electron density is calculated by the extended Hückel method.[13] The calculation was carried out on the molecular structure determined by microwave spectra.[21] The electronic

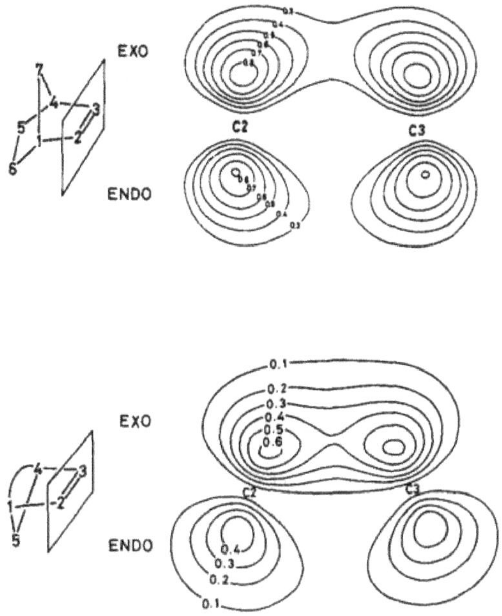

Figure 3. The HOMO electron density distributions of norbornene and bicyclo[2.1.0]pentene.

feature is consistent with the direction of attack of diimide,[22] which is an electrophilic reagent (eq 3).[23] The similar role of the bent bond was observed in electrophilic additions to bicyclo[3.1.0]hexene (eq 4).[24] The stereochemistry of the methanolysis products from 4-bicyclo[3.1.0]hexenyl trifluoroacetate

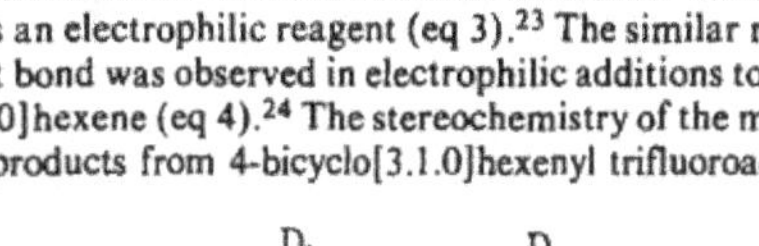

(3)

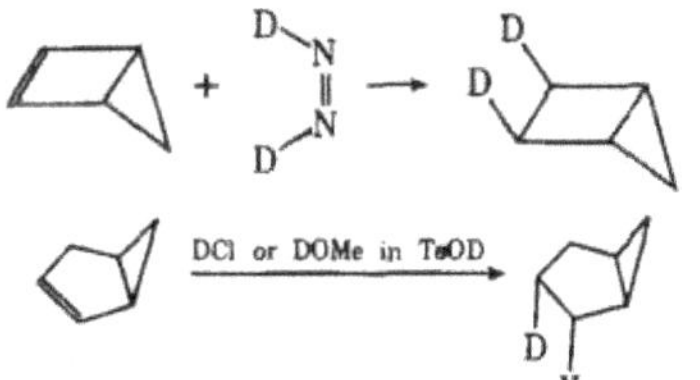

(4)

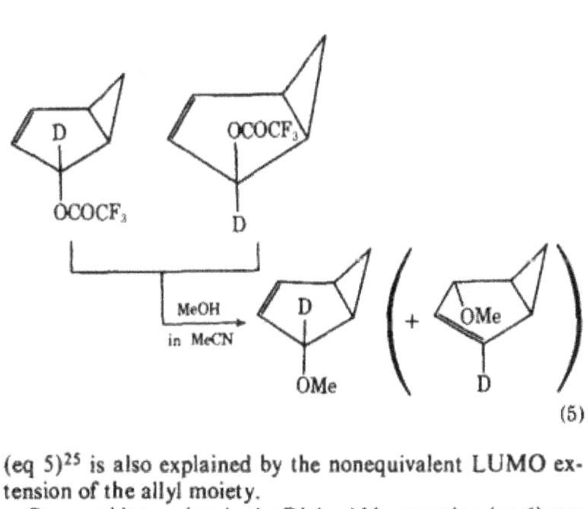

(5)

(eq 5)[25] is also explained by the nonequivalent LUMO extension of the allyl moiety.

Syn–anti isomerism in the Diels–Alder reaction (eq 6) can also be discussed in terms of nonequivalent extension of a frontier orbital. Recently, Williamson and his co-workers systematically investigated an aspect of syn–anti stereoselec-

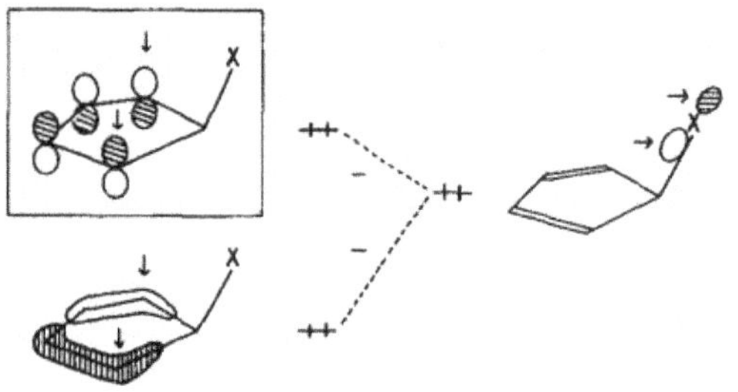

Figure 4. Origin and direction of nonequivalent HOMO extension of 5-substituted cyclopentadienes.

tivity in the cycloadditions of 1,2,3,4,5-pentachlorocyclopentadiene with a variety of dienophiles and demonstrated that the bond formation occurs preferentially at the sterically

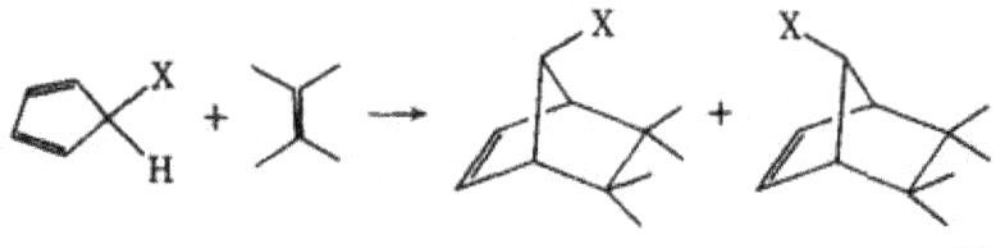

(6)

hindered side syn to the 5-chlorine atom.[26] The similar stereoselectivity was also observed in the reaction of ethylene with 5-acetoxycyclopentadiene.[27] On the other hand, the reaction of 5-methylcyclopentadiene with *N*-phenylmaleimide yields both *syn*- and *anti*-methylnorbornenes in about equal amounts.[28]

The dienes favoring the syn orientation bear lone-pair electrons on the 5-substituent. The nonbonding orbital breaks down the symmetry with respect to the molecular plane to a considerable degree. The π-HOMO of the dienes is perturbed only by the nonbonding orbital antisymmetric with respect to reflection in the mirror plane passing through C(5), since the π-HOMO is antisymmetric. The perturbation allows low-lying σ orbitals of the carbon skeleton to mix into the π-HOMO. According to the orbital mixing rule, the π-HOMO combining with the n-orbital out of phase mixes the σ-orbital in such a way that the n-orbital and the σ-orbital are out of phase (Figure 4). It is predicted in Figure 4 that the HOMO of the dienes with an antisymmetrical lone-pair orbital in the 5-substituent extends in the direction of the substituent. In the case of the methyl substituent, the perturbation of the π-HOMO is small, since the energy of the methyl C–H bond orbital lies too low to interact with the π-HOMO as efficiently as the n orbital.

The semiempirical SCF calculations[29] unquestionably show that the HOMO electron cloud of 5-chlorocyclopentadiene is seriously biased in the region syn to Cl, while in the HOMO of 5-methylcyclopentadiene, such deviation of the HOMO electron distribution is small (Figure 5).[30] The HOMO electron density contours were depicted in the plane vertical to the molecular plane. One can say, the syn attack of the electron-accepting dienophile is favored by the nonequivalent extension of the HOMO of dienes. This conclusion is consistent with the reluctance of styrene and propylene to syn attack on 1,2,3,4,5-pentachlorocyclopentadiene[26a] and with the increased probability of syn attack in the presence of aluminum chloride.[26b] These phenomena reflect the poor capacity of styrene and propylene to accept electrons and the enhanced electron affinity of the dienophiles by the complex formation with the Lewis acid, slighting and esteeming the nature of the HOMO of the diene, respectively.[31]

Theoretical Support for d-Orbital Participation of Sulfur. Participation of d orbitals of the second-row elements has been an enduring problem. No unequivocal evidence for d-orbital

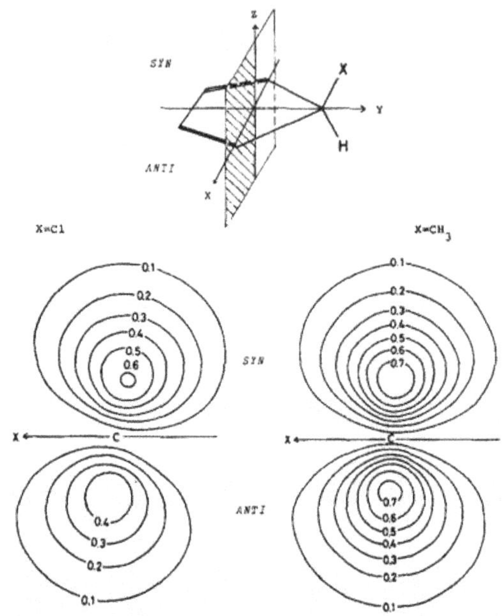

Figure 5. The HOMO electron density distributions of 5-chloro- and 5-methylcyclopentadiene.

participation has been offered thus far. The authors have met with this problem during the work on thermal 2 + 2 cycloaddition reactions and ene reactions.[32] In a previous paper[32] it was concluded that ene reactions, as well as 2 + 2 cycloadditions of electron-accepting heteronuclear unsaturated bonds with olefins, begin with the three-centered interaction involving two nucleophilic donor atoms and a more electrophilic atom of heteronuclear acceptors. The intermolecular arrangement is most stabilized by the interaction between the HOMO of the olefins and the LUMO of the acceptor. The conclusion is consistent with the observed orientation in ene reactions of carbonyl compounds; the bond formation with olefinic carbon takes place exclusively at the carbonyl carbon with the larger LUMO amplitude (eq 7a).[32] A thiocarbonyl compound, per-

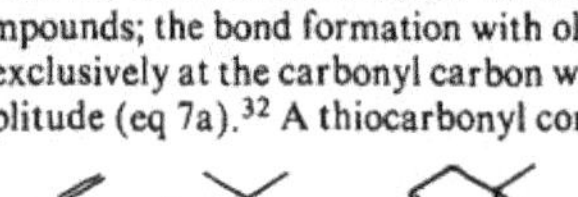

(7a)

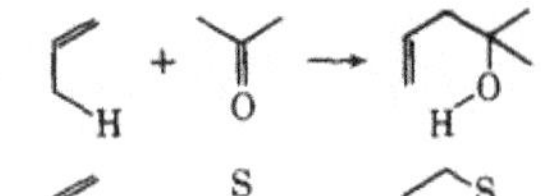

(7b)

$$Ph_2C{=}S + RMgX \longrightarrow\longrightarrow Ph_2CH_2SR \qquad (8a)$$

$$R_2C{=}C{=}S + RLi \longrightarrow\longrightarrow R_2C{=}CHSR \qquad (8b)$$

fluorothioacetone, was surprisingly reported to form a bond with olefins at the sulfur atom (eq 7b).[33] Thiocarbonyl groups have been reported to undergo nucleophilic attacks of organometallic compounds at the sulfur atom (8),[34] in contrast with the oxygen analogues.

Orbital energies of the p AO's of our present interest were estimated to decrease in the order of C, S, and O.[35,36] The π MO's of carbonyl and thiocarbonyl, obtained on the basis of the above order, have similar features in the orbital amplitude or larger AO coefficients at the carbon in the π LUMO and the heteroatom in the π HOMO. This suggests that nucleophilic attack occurs preferentially on the carbon of either group. This is inconsistent with the observed orientations. The contradiction between the theoretical prediction and the ex-

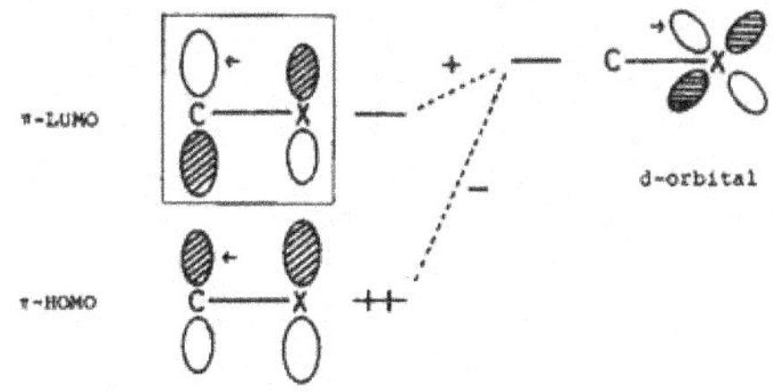

Figure 6. Mixing of a π HOMO into a π LUMO through sulfur d orbitals in thiocarbonyl, giving an AO coefficient ratio reverse to that in the LUMO of the carbonyl group.

perimental results disappears with d-orbital participation of sulfur. If the LUMO of thiocarbonyl is composed mainly of the sulfur d orbital, it is reasonable that nucleophiles should form a bond with the sulfur atom. However, this assumption may not be appropriate. We are interested in the property of the LUMO for the case where the d orbital mixes itself to the LUMO to a lesser, but appreciable degree. This situation is conveniently represented by the case in which the d orbital lies higher than the π LUMO (Figure 6). According to the orbital mixing rule, the π LUMO is combined with the d orbital in phase; the p-orbital component on the carbon and the d orbital may have the same signs at the interacting lobes. In turn, the π HOMO mixes itself into the π LUMO through interaction with the d orbital. The sign of the π HOMO is opposite to that of the d orbital at the interacting carbon p-orbital lobe. The sign relation between the π HOMO and the π LUMO suggests that the π LUMO may be modified by the d-orbital participation in order to have a larger coefficient at the sulfur of thiocarbonyl, since the π HOMO to be mixed into the π LUMO has the same signs at the sulfur atom and the opposite signs at the carbon.

Interestingly, the regioselection in 1,3-dipolar cycloadditions of diazomethane to a thiocarbonyl compound has been reported to change from one to the other under the influence of solvents.[37] In petroleum ether, ether, or benzene as solvents, the orientation is compatible with that of the aforementioned reactions (eq 7b, 8a, and 8b); the nucleophilic carbon of diazomethane is preferentially combined with the thiocarbonyl sulfur atom (eq 8c). In contrast, the reverse orientation is

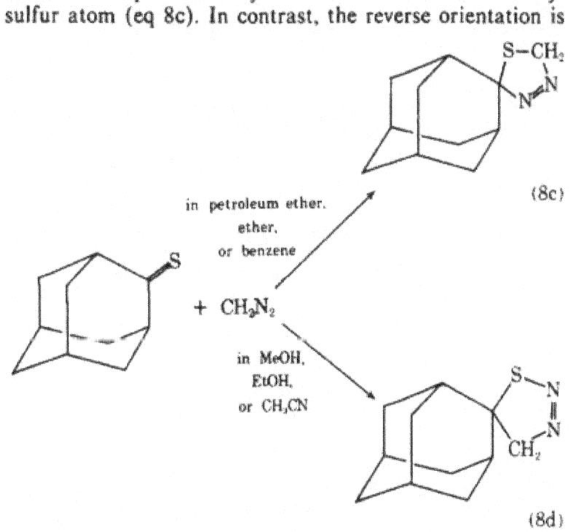

(8c)

(8d)

preferred in such solvents as ethanol, methanol, or acrylonitrile (eq 8d). It is believed that the stereochemical difference is due to the occupation of the reactive sulfur atom by active lone-pair electrons of ethanol, methanol, or acrylonitrile prior to the attack of the diazomethane.

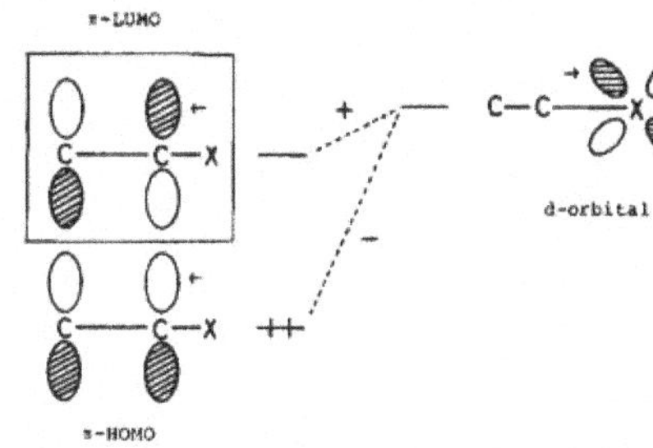

Figure 7. Mixing of a π HOMO into a π LUMO through sulfur d orbitals in thioalkoxy derivatives of ethylene and acetylene, which gives a large AO coefficient at the β carbon in the LUMO.

The vacant d-orbital effect of sulfur may be responsible for the opposite regioselections in the 1,3-dipolar cycloaddition reactions of alkoxy and thioalkoxy derivatives of acetylene with diazomethane (eq 9a and 9b).[38] It has theoretically been

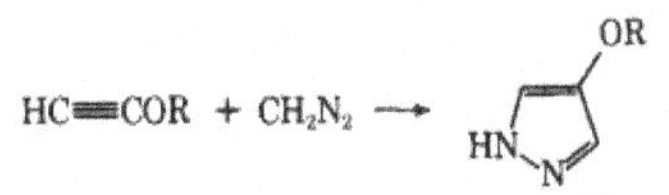 (9a)

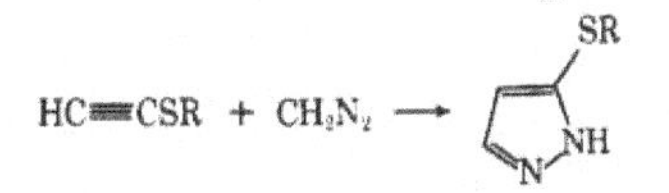 (9b)

proposed that the bond formation in 1,3-dipolar cycloadditions take place in such a way that the atom of the larger AO coefficient of the donor HOMO combines with that of the larger AO coefficient of the acceptor LUMO.[39] Diazomethane, the electron-donating partner in 1,3-dipolar cycloaddition reactions,[40] has larger HOMO amplitude at the carbon. The LUMO of alkoxyacetylene has a larger AO coefficient on the α carbon because of the influence of the oxygen lone-pair electrons (this is inferred from the orbital mixing rule). Accordingly, the regioselectivity with alkoxyacetylene is consistent with the present knowledge. The similar property of the LUMO of thioalkoxyacetylene can be concluded as long as only the effect of the sulfur lone-pair electrons on π-orbital polarization is taken into account. The perturbation of the π LUMO by a vacant d orbital, but not by the n orbital, gives rise to the LUMO of thioalkoxyacetylene compatible with the regioselectivity. The LUMO is constructed by first mixing the vacant d orbital in phase with the π LUMO and then by mixing the π HOMO out of phase with the d orbital (Figure 7). It is suggested here that the LUMO of thioalkoxyacetylene has the larger molecular orbital amplitudes at the β carbon. This explains the contrasted regioselectivity of the sulfur analogue.

Such a d-orbital participation is compatible with the β attack of the anionic hydrocarbon groups of organolithium compounds to vinyl sulfides (eq 10).[41]

$$PhSCH{=}CH_2 + RLi \longrightarrow PhSCHLi{-}CH_2R \quad (10a)$$

$$EtSCH{=}CH_2 + RLi \longrightarrow EtSCHLi{-}CH_2R \quad (10b)$$

Direction of Polarization and Chemical Properties of Ethylene Derivatives. Recently Libit and Hoffmann[2a] examined the substituent effect on π-electron distribution of olefins and aromatics. The polarization effect can be used to interpret the familiar facts that the addition of hydrogen halide to alkylethylenes and butadiene occurs according to Markownikoff's rule, while that to acrylic acid occurs in the anti-Markownikoff manner.[42] The HOMO electron distribution has a maximum at the β carbon of alkylethylenes, at the terminal carbon of butadiene, and at the α carbon of acrylic acid. The key factor

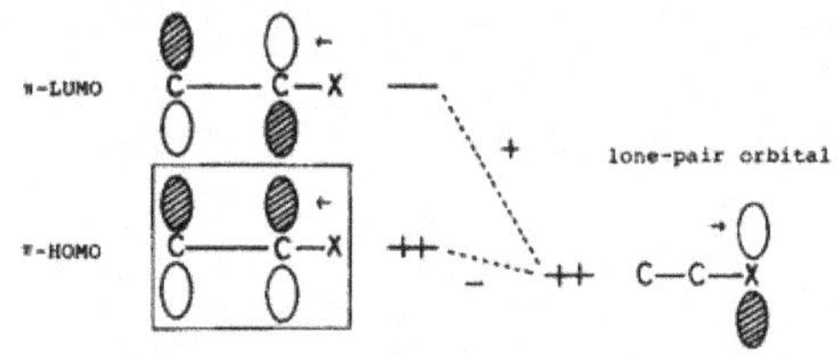

Figure 8. Mixing of a π-LUMO into a π-HOMO through a lone-pair orbital which gives a large AO coefficient at the β carbon in the HOMO

determining the direction of polarization is the relative energy of the perturbing orbital of the substituent group. The HOMO electron density of ethylenic compounds is concentrated on the β carbon when the perturbing orbital lies at equal (the π HOMO of the other ethylenic moiety in butadiene) or lower (C-CH_3 σ orbital in alkylethylene) energy levels. The case is reversed with the high-lying orbitals (the π LUMO of the carboxylic group in acrylic acid). The above discussions are based on the orbital mixing rule.

The polarization effect on the ethylenic π orbital is applicable to a recent interesting observation; see the pair of reactions in eq 11.[43] The difference in the course of the reaction is

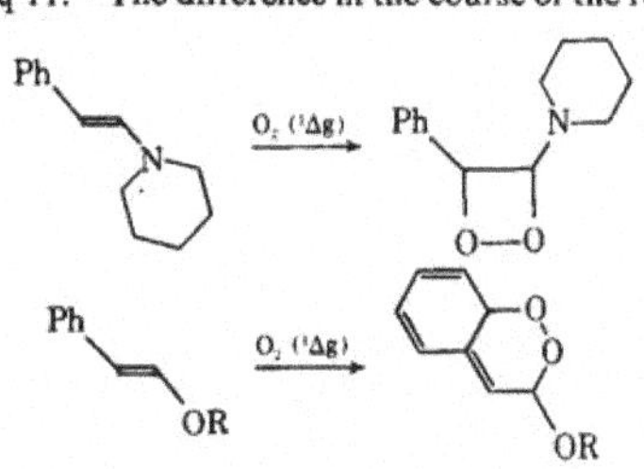

(11a)

(11b)

attributable to the strength of the polarization by the substituents. The lone-pair orbitals of nitrogen and oxygen presumably lie between the energy levels of the σ CH in propylene and the perturbing ethylenic π orbital in butadiene. Accordingly, the n orbitals are expected to concentrate the HOMO electron density on the β carbon, with respect to its own position, or on the α carbon of the styrene derivative of interest. However, the perturbation by the phenyl π orbitals leads to the opposite trend, concentrating it on the β carbon of styrene. The n orbital of nitrogen, higher in energy than that of oxygen, is likely to polarize the ethylenic π orbital appreciably more, overcoming the effect of the phenyl ring. The mechanism of the polarization by the n orbital is schematically represented in Figure 8. The degree of polarization is seen in the denominator of eq 1 to depend on the energy separation between the π HOMO and the n orbital.

The present authors have pointed out previously that the 2 + 2 cycloaddition reactions and ene reactions between donor and acceptor begin with a three-centered interaction among the nucleophilic atoms of the donors and an electrophilic atom of the acceptors.[32,44] The model holds for a slightly polar π bond of the donors. In the case of the amino substituent, the transient three-membered ring structure may collapse at an earlier stage of the reaction; the oxygen atom attacking the olefinic π bond tends preferentially to come to the α carbon of styrene with the higher HOMO electron density. It follows that the 4 + 2 cycloaddition reaction invoving the β carbon of styrene is disfavored in the enamine.

Transannular Cross-Bond Formation. A transannular σ-bond formation between proximal double bonds (eq 12) has recently been reported[45-48] to take place on attacks by an electrophile, and subsequently by a nucleophile. These reactions proceed by cross bonding.

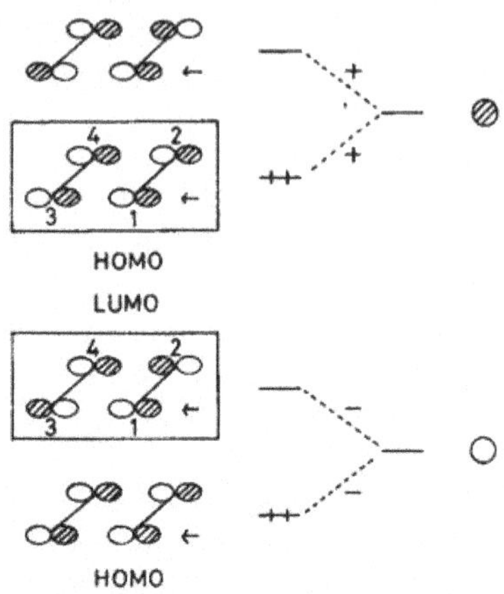

Figure 9. Mixing of a π HOMO and a π LUMO of a double-bond system into each other through the LUMO of an electrophile.

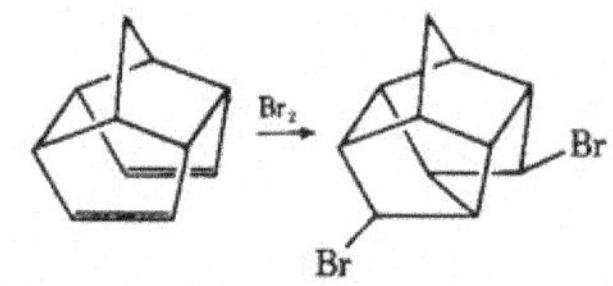

(12a)

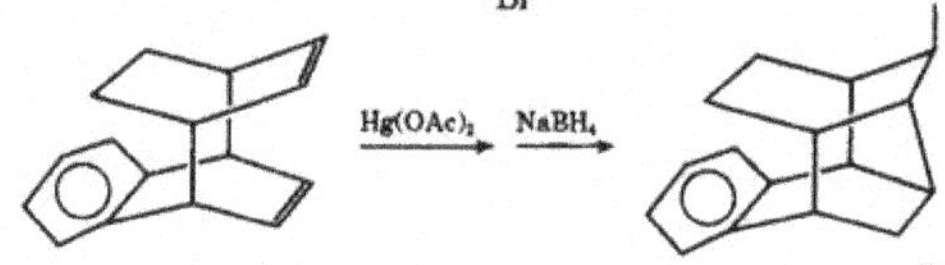

(12b)

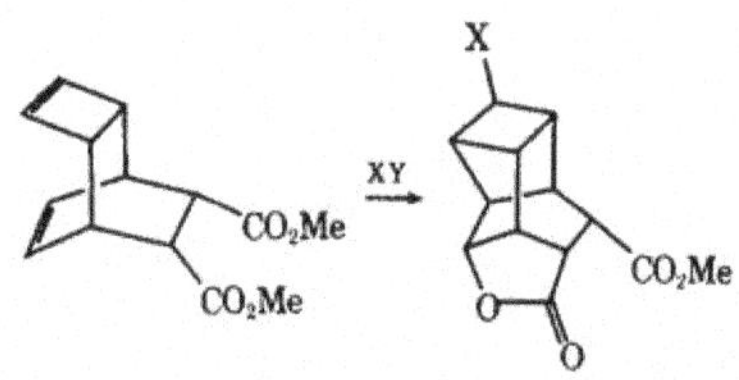

(12c)

An electronic factor of the cross linkage between the double bonds is given by the orbital mixing rule. The modifications of the HOMO and LUMO of the double-bond system interacting with the LUMO of electrophiles are depicted in Figure 9. The original HOMO of the double-bond system has in-phase relations between the p AO's on the carbons linked by the π bonds and out-of-phase relations between the double bonds. The HOMO electrons contribute to the repulsion between the double bonds. When an electrophile approaches a terminus of a double bond, say C(1), the LUMO of the double-bond system is mixed in the HOMO through the interaction with the LUMO of electrophiles. According to the orbital mixing rule, the signs of the HOMO and the LUMO coefficients at C(1) are the same (Figure 9). The modified HOMO has larger amplitudes at C(1) and C(4) and smaller amplitudes at C(2) and C(3) (Figure 9). As a result, the anti-bonding property in the HOMO decreases between C(2) and C(3) and increases between C(1) and C(4). This is partially the reason that cross linkage occurs. The modified LUMO of the double-bond system is similarly shown by the orbital mixing rule to have larger amplitudes at C(1) and C(4). It follows that the attack of a nucleophilic entity occurs preferentially at C(4) rather than at C(3). The change of the HOMO and the LUMO of the double-bond system by an attacking electrophile is clearly shown to favor the cross bonding. A related reaction accompanied by the ring opening of epoxide under the influence of acids (eq 13)[49] is similarly explained.

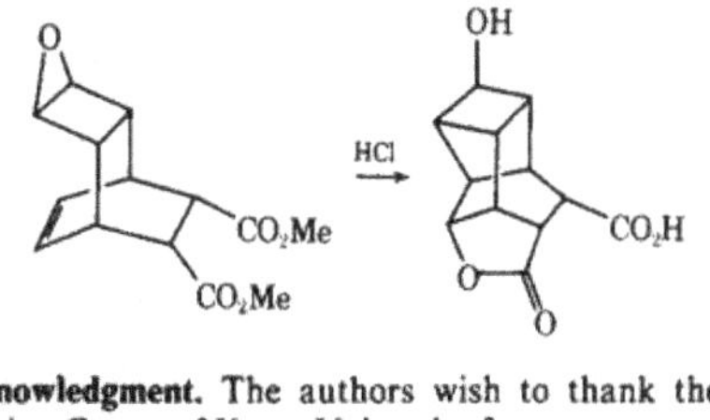

(13)

Acknowledgment. The authors wish to thank the Data Processing Center of Kyoto University for generous use of the FACOM 230-60 computer.

Appendix

Consider a molecule A subjected to the effect of an outer field due to an approaching molecule B. Let the MO's of A and B be ψ_{Ai} and ψ_{Bk} and their energies ϵ_{Ai} and ϵ_{Bk}. The one-electron Hamiltonian of the total system is h, and h_A' is the change of the one-electron Hamiltonian of molecule A due to the approach of molecule B, h_B' being the analogous quantity for molecule B. The following integrals are defined for constructing the perturbed secular determinant.[50]

$$s_{ik} = \int \psi_{Ai}\psi_{Bk}\, dv$$

$$h_{ik} = \int \psi_{Ai} h \psi_{Bk}\, dv$$

$$h_{ij}' = \int \psi_{Ai} h_A' \psi_{Aj}\, dv$$

$$h_{kl}' = \int \psi_{Bk} h_B' \psi_{Bl}\, dv$$

On solving the perturbed secular equation, the perturbed MO's and their energies of A are obtained as

$$\begin{aligned}
\psi_{Ai}' = & \left\{1 - \sum_k^B \frac{s_{ik}(h_{ik} - s_{ik}\epsilon_{Ai})}{\epsilon_{Ai} - \epsilon_{Bk}} - \frac{1}{2}\sum_k^B \frac{(h_{ik} - s_{ik}\epsilon_{Ai})^2}{(\epsilon_{Ai} - \epsilon_{Bk})^2} \right. \\
& \left. - \frac{1}{2}\sum_{j(\neq i)}^A \frac{h_{ij}'^2}{(\epsilon_{Ai} - \epsilon_{Aj})^2}\right\} \psi_{Ai} + \sum_{j(\neq i)}^A \left\{\frac{h_{ij}'}{\epsilon_{Ai} - \epsilon_{Aj}} \right. \\
& \left. + \sum_k^B \frac{(h_{ik} - s_{ik}\epsilon_{Ai})(h_{jk} - s_{jk}\epsilon_{Ai})}{(\epsilon_{Ai} - \epsilon_{Aj})(\epsilon_{Ai} - \epsilon_{Bk})} - \frac{h_{ij}'(h_{ii}' - h_{jj}')}{(\epsilon_{Ai} - \epsilon_{Aj})^2}\right\} \psi_{Aj} \\
& + \sum_k^B \left\{\frac{h_{ik} - s_{ik}\epsilon_{Ai}}{\epsilon_{Ai} - \epsilon_{Bk}} + \sum_{j(\neq i)}^A \frac{h_{ij}'(h_{jk} - s_{jk}\epsilon_{Ai})}{(\epsilon_{Ai} - \epsilon_{Aj})(\epsilon_{Ai} - \epsilon_{Bk})} \right. \\
& - \frac{s_{ik}h_{ii}'}{\epsilon_{Ai} - \epsilon_{Bk}} + \sum_{l(\neq k)}^B \frac{h_{kl}'(h_{il} - s_{il}\epsilon_{Ai})}{(\epsilon_{Ai} - \epsilon_{Bk})(\epsilon_{Ai} - \epsilon_{Bl})} \\
& \left. - \frac{(h_{ik} - s_{ik}\epsilon_{Ai})(h_{ii}' - h_{kk}')}{(\epsilon_{Ai} - \epsilon_{Bk})^2}\right\} \psi_{Bk} + 0(\Delta^3) \quad \text{(A-1)}
\end{aligned}$$

and

$$\begin{aligned}
\epsilon_{Ai}' = & \epsilon_{Ai} + h_{ii}' + \sum_k^B \frac{(h_{ik} - s_{ik}\epsilon_{Ai})^2}{\epsilon_{Ai} - \epsilon_{Bk}} + \sum_{j(\neq i)}^A \frac{h_{ij}'^2}{\epsilon_{Ai} - \epsilon_{Aj}} \\
& + 2\sum_{\substack{j<m \\ (j,m\neq i)}}^A \frac{h_{ij}'h_{im}'h_{jm}'}{(\epsilon_{Ai} - \epsilon_{Aj})(\epsilon_{Ai} - \epsilon_{Am})} \\
& + 2\sum_{j(\neq i)}^A \sum_k^B \frac{h_{ij}'(h_{ik} - s_{ik}\epsilon_{Ai})(h_{jk} - s_{jk}\epsilon_{Ai})}{(\epsilon_{Ai} - \epsilon_{Aj})(\epsilon_{Ai} - \epsilon_{Bk})} \\
& + 2\sum_{k<l}^B \frac{h_{kl}'(h_{ik} - s_{ik}\epsilon_{Ai})(h_{il} - s_{il}\epsilon_{Ai})}{(\epsilon_{Ai} - \epsilon_{Bk})(\epsilon_{Ai} - \epsilon_{Bl})} \\
& - 2\sum_k^B \frac{s_{ik}h_{ii}'(h_{ik} - s_{ik}\epsilon_{Ai})}{\epsilon_{Ai} - \epsilon_{Bk}} - \sum_{j(\neq i)}^A \frac{h_{ij}'^2(h_{ii}' - h_{jj}')}{(\epsilon_{Ai} - \epsilon_{Aj})^2} \\
& - \sum_k^B \frac{(h_{ii}' - h_{kk}')(h_{ik} - s_{ik}\epsilon_{Ai})^2}{(\epsilon_{Ai} - \epsilon_{Bk})^2} + 0(\Delta^4) \quad \text{(A-2)}
\end{aligned}$$

where Δ signifies the first-order quantities like s_{ik}, h_{ik}, h_{ij}', h_{kl}', and so on, and $0(\Delta^3)$ implies a small quantity of order Δ^3.[50]

An important consequence of these equations may be the rule of intramolecular orbital mixing by the orbitals of a different molecule, which is given by the coefficient of ψ_{Aj} in ψ_{Ai}'. This is written as

$$\frac{h_{ij}'}{\epsilon_{Ai} - \epsilon_{Aj}} + \sum_k^B \frac{(h_{ik} - s_{ik}\epsilon_{Ai})(h_{jk} - s_{jk}\epsilon_{Ai})}{(\epsilon_{Ai} - \epsilon_{Aj})(\epsilon_{Ai} - \epsilon_{Bk})} \quad \text{(A-3)}$$

if we neglect the remaining terms as less important than these two. The first term originates from the "static" effect of the field of molecule B upon the Hamiltonian for molecule A, while the second is caused by the "dynamic" effect of the orbitals of the second molecule.[2c] Both constitute the effect of polarization of molecule A.

Next, we take only the direct, "dynamic" orbital effect among three orbitals, ψ_{Ai}, ψ_{Aj}, and ψ_{Bk}, into account, and consider the phase relation in the perturbed orbital ψ_{Ai}' represented by eq 1 in the text, where the overlap integrals are neglected as small.

References and Notes

(1) (a) S. Inagaki, Ph.D. Dissertation, Kyoto University, 1975. (b) Preliminary communication on this subject was made: S. Inagaki and K. Fukui, *Chem. Lett.*, 509 (1974).

(2) (a) L. Libit and R. Hoffmann, *J. Am. Chem. Soc.*, **96**, 1370 (1974); (b) H. Fujimoto and R. Hoffmann, *J. Phys. Chem.*, **78**, 1874 (1974); (c) A. Imamura and T. Hirano, *J. Am. Chem. Soc.*, **97**, 4192 (1975).

(3) K. Fukui in "Molecular Orbitals in Chemistry, Physics, and Biology", P.-O. Löwdin and B. Pullman, Ed., Academic Press, New York, N.Y., 1964, p 513; *ibid.* in "Modern Quantum Chemistry", Vol. 1, O. Sinanoglu, Ed., Academic Press, New York, N.Y., 1965, p 49; K. Fukui and H. Fujimoto, *Mech. Mol. Migr.*, **2**, 118 (1969); K. Fukui, "Theory of Orientation and Stereoselection", Springer-Verlag, Heidelberg, 1970; *Acc. Chem. Res.*, **4**, 57 (1971).

(4) Other approaches to related problems appeared elsewhere: (a) K. Fukui, *Tetrahedron Lett.*, 2427 (1965); (b) K. Fukui and H. Fujimoto, *Bull. Chem. Soc. Jpn.*, **39**, 2116 (1966); (c) N. T. Anh, O. Eisenstein, J.-M. Lefour, and M.-E. Trân Huu Dâu, *J. Am. Chem. Soc.*, **95**, 6146 (1973); (d) J. Klein, *Tetrahedron Lett.*, 4307 (1973); *Tetrahedron*, **30**, 3349 (1974); (e) C. L. Liotta, *Tetrahedron Lett.*, 519, 523 (1975).

(5) For example, see R. C. Fahey, *Top. Stereochem.*, **3**, 237 (1968).

(6) S. Winstein and D. Trifan, *J. Am. Chem. Soc.*, **74**, 1147, 1154 (1952).

(7) H. C. Brown, *Chem. Br.*, **2**, 199 (1966).

(8) P. v. R. Schleyer, *J. Am. Chem. Soc.*, **89**, 701 (1967).

(9) T. T. Tidwell and T. G. Traylor, *J. Org. Chem.*, **33**, 2615 (1968).

(10) H. C. Brown, J. H. Kawakami, and S. Ikegami, *J. Am. Chem. Soc.*, **89**, 1525 (1967).

(11) H. C. Brown and K.-T. Liu, *J. Am. Chem. Soc.*, **89**, 3898, 3900 (1967).

(12) P. v. R. Schleyer, *J. Am. Chem. Soc.*, **89**, 3901 (1967).

(13) R. Hoffmann, *J. Chem. Phys.*, **39**, 1397 (1963).

(14) J. A. Pople and D. L. Beveridge, "Approximate Molecular Orbital Theory", McGraw-Hill, New York, N.Y., 1970.

(15) Y. Morino, K. Kuchitsu, and A. Yokozeki, *Bull. Chem. Soc. Jpn.*, **40**, 1552 (1967).

(16) J. F. Chiang, C. W. Wilcox, Jr., and S. H. Bauer, *J. Am. Chem. Soc.*, **90**, 3149 (1968).

(17) This interaction has not been referred to in the preliminary report.[1b]

(18) H. Fujimoto and K. Fukui, *Tetrahedron Lett.*, 5551 (1966), and references cited therein.

(19) T. Kawamura, T. Koyama, and T. Yonezawa, *J. Am. Chem. Soc.*, **92**, 7222 (1970); **95**, 3220 (1973).

(20) S. P. Jindal, S. S. Sohoni, and T. T. Tidwell, *Tetrahedron Lett.*, 779 (1971); S. P. Jindal and T. T. Tidwell, *ibid.*, 783 (1971).

(21) S. L. Hsu, A. H. Andrist, T. D. Gierke, R. C. Benson, W. H. Flygare, and J. E. Baldwin, *J. Am. Chem. Soc.*, **92**, 5250 (1970).

(22) P. G. Gassman, K. T. Mansfield, and T. J. Murphy, *J. Am. Chem. Soc.*, **91**, 1684 (1969).

(23) T. Takagi, *Kogyo Kagaku Zasshi*, **70**, 892 (1967).

(24) P. K. Freeman, M. F. Grostic, and F. A. Raymond, *J. Org. Chem.*, **30**, 771 (1965); P. K. Freeman, F. A. Raymond, and M. F. Grostic, *Ibid.*, **32**, 24 (1967).

(25) J. A. Berson and N. M. Hasty, *J. Am. Chem. Soc.*, **93**, 1549 (1971).

(26) (a) K. L. Williamson, Y.-F. Li Hsu, R. Lacko, and C. H. Youn, *J. Am. Chem. Soc.*, **91**, 6129 (1969); (b) K. L. Williamson and Y.-F. Li Hsu, *ibid.*, **92**, 7385 (1970).

(27) Dissertations (Harvard): P. Wilder, Jr. (1950), R. E. Vanelli (1950), and C. J. Norton (1955), quoted by S. Winstein, M. Shavatsky, C. J. Norton, and R. B. Woodward, *J. Am. Chem. Soc.*, **77**, 4183 (1955).

(28) S. McLean and P. Haynes, *Tetrahedron*, **21**, 2313 (1965).

(29) T. Yonezawa, K. Yamaguchi, and H. Kato, *Bull. Chem. Soc. Jpn.*, **40**, 536 (1967).

(30) Similar results are obtained by an extended Hückel MO calculation.[13]

(31) It is necessary to refer to the intermolecular nonbonded attraction recently proposed to determine the syn–anti stereoselection of the Diels–Alder reactions [N. T. Anh, *Tetrahedron*, **29**, 3227 (1973)]. He stressed the

charge transfer from the n orbital on the 5-substituent to the LUMO of the dienophile. However, this mode of nonbonded interaction may occur at the cost of more significant interaction between the π HOMO of the diene and the π LUMO of the dienophile indispensable to the cycloaddition. The HOMO of the cyclopentadiene with an n orbital at the 5 position is out-of-phase combination of the π HOMO and the n orbital. Accordingly, if the nonbonded attraction occurs effectively, or if the LUMO of dienophile is in-phase combined with the n-orbital component of the HOMO, the antibonding nature appears between the π HOMO component of diene and the LUMO of dienophile. This suggests that the nonbonded attraction of the n orbital with the reaction center of the dienophile disfavors the leading interaction in the Diels–Alder reactions.

(32) S. Inagaki, T. Minato, S. Yamabe, H. Fujimoto, and K. Fukui, *Tetrahedron*, **30**, 2165 (1974).

(33) W. J. Middelton, *J. Org. Chem.*, **30**, 1395 (1965); W. J. Middelton, E. G. Howard, and W. H. Shakey, *J. Am. Chem. Soc.*, **83**, 2589 (1961); see also J. A. Boerma, N. H. Nilsson, and A. Senning, *Tetrahedron*, **30**, 2735 (1974).

(34) (a) P. Beak and J. W. Worley, *J. Am. Chem. Soc.*, **92**, 4142 (1970); **94**, 597 (1972); (b) P. Beak, J. Yamamoto, and C. J. Upton, *J. Org. Chem.*, **40**, 3052 (1975); (c) M. Dagonneau, J.-F. Hemidy, D. Cornet, and J. Vialle, *Tetrahedron Lett.*, 3003 (1972); (d) A. Ohno, K. Nakamura, M. Uohama, and S. Oka, *Chem. Lett.*, 983 (1975); A. Ohno, K. Nakamura, M. Uohama, S. Oka, T. Yamabe, and S. Nagata, *Bull. Chem. Soc. Jpn.*, **48**, 3718 (1975); (e) E. Schaumann and W. Walter, *Chem. Ber.*, **107**, 3562 (1974). (f) Allylic Grignard reagents [M. Dagonneau and J. Vialle, *Tetrahedron*, **30**, 415 (1974)] and benzylic lithium reagents[34b] were reported to attack the thiocarbonyl carbon. A possible explanation is that the relative stability of allylic and benzylic anions may allow the organometallic reagents to dissociate and to coordinate the metal cation on the sulfur atom, prior to the attack of the carbanionic part on thiocarbonyl groups, while alkyl, vinyl, and phenyl organometallylic reagents may attack directly on thiocarbonyl groups dissociating into carbanion and metal cation. See ref 2b and 2c for the localization of the LUMO of cation-coordinated thiocarbonyl groups on the carbon caused by the electrostatic field. (g) One of the authors (S. I.) would like to express his appreciation to Dr. A. Ohno, Dr. T. Yamabe, Dr. S. Nagata, and Dr. K. Nakamura for their stimulating discussion and current information on thiocarbonyl chemistry.

(35) H. O. Pritchard and H. A. Skinner, *Chem. Rev.*, **55**, 745 (1955).

(36) Electronegativity of p AO's estimated by Klopman is consistent with this order: G. Klopman, *J. Am. Chem. Soc.*, **86**, 1463 (1964).

(37) A. P. Krapcho, M. P. Silvon, I. Goldberg, and E. G. E. Jahngen, Jr., *J. Org. Chem.*, **39**, 860 (1974).

(38) S. H. Groen and J. F. Arens, *Recl. Trav. Chim. Pays-Bas*, **80**, 879 (1961).

(39) (a) K. N. Houk, J. Sims, C. R. Watts, and L. J. Luskus, *J. Am. Chem. Soc.*, **95**, 7301 (1973); (b) T. Minato, S. Yamabe, S. Inagaki, H. Fujimoto, and K. Fukui, *Bull. Chem. Soc. Jpn.*, **47**, 1619 (1974).

(40) R. Sustmann, *Tetrahedron Lett.*, 2717 (1971).

(41) (a) F. G. Bordwell and G. D. Cooper, *J. Am. Chem. Soc.*, **73**, 5187 (1951); (b) P. D. Bartlett, S. Friedman, and M. Stiles, *ibid.*, **75**, 1771 (1953); (c) W. E. Parham, M. A. Kalnins, and D. R. Theissen, *J. Org. Chem.*, **27**, 2698 (1962); W. E. Parham and R. F. Motter, *J. Am. Chem. Soc.*, **81**, 2146 (1959).

(42) C. K. Ingold, "Structure and Mechanism in Organic Chemistry", Cornell University Press, New York, N.Y., 1969, p 944.

(43) (a) C. S. Foote, S. Mazur, P. A. Burns, and D. Lerdal, *J. Am. Chem. Soc.*, **95**, 586 (1973); (b) C. S. Foote and J. W.-P. Lin, *Tetrahedron Lett.*, 3267 (1968).

(44) S. Inagaki, S. Yamabe, H. Fujimoto, and K. Fukui, *Bull. Chem. Soc. Jpn.*, **45**, 3510 (1972); S. Inagaki and K. Fukui, *J. Am. Chem. Soc.*, **97**, 7480 (1975); see also S. Inagaki and K. Fukui, *Bull. Chem. Soc. Jpn.*, **46**, 2240 (1973).

(45) G. R. Underwood and B. Ramamoorthy, *Chem. Commun.*, 12 (1970); *Tetrahedron Lett.*, 4125 (1970).

(46) N. C. Yang and J. Libman, *J. Am. Chem. Soc.*, **94**, 9228 (1972); **95**, 4473 (1973). One of the authors (K.F.) acknowledges a private discussion on this problem with Professor Yang of Chicago University.

(47) T. Sasaki, K. Kanematsu, and A. Kondo, *J. Org. Chem.*, **39**, 2246 (1974).

(48) For the related reactions, see (a) I. Tabushi, K. Fujita, and R. Oda, *J. Org. Chem.*, **35**, 2376 (1970); (b) S. Uemura, A. Onoe, and M. Okano, *J. Chem. Soc., Chem. Commun.*, 210 (1975).

(49) T. Sasaki, K. Kanematsu, and A. Kondo, *J. Org. Chem.*, **40**, 1642 (1975).

(50) K. Fukui, C. Nagata, T. Yonezawa, H. Kato, and K. Morokuma, *J. Chem. Phys.*, **31**, 287 (1957); K. Fukui, K. Morokuma, T. Yonezawa, and C. Nagata, *Bull. Chem. Soc. Jpn.*, **33**, 963 (1960).

Orbital Interaction in Three Systems

Satoshi Inagaki, Hiroshi Fujimoto, and Kenichi Fukui*

Contribution from the Department of Hydrocarbon Chemistry, Faculty of Engineering, Kyoto University, Kyoto 606, Japan. *Received August 21, 1975*

Abstract: The three-system interaction theory previously developed for giving an orbital interaction rationale for the role of the catalysts is shown to cover various organic chemical reactions of both ground- and excited-state molecules. The mechanisms of thermal and photochemical multicycloadditions of layered compounds with olefins and the effects of protic acids and of substituents on symmetry-disfavored reactions are clarified, especially in connection with orbital phase relation. Bicyclic compounds with π bonds on each bridge are classified into three groups, i.e., electron-delocalizing, semielectron-delocalizing, and electron-localizing systems. The bridge flipping in such bicyclic compounds is also discussed.

Molecular orbital (MO) perturbation methods have played conspicuous roles in developing the theory of chemical reactions.[1-4] Further theoretical foundation and its application and extension have been made on the basis of the concept of configuration interaction.[5-10] Theories of organic chemical reactions thus far developed have been engaged in the two-system interaction, but a few have been engaged in the three-system interaction.[11] Recently the present authors have given a theoretical basis on which one can discuss mechanisms of metal-catalyzed reactions in a unified manner by taking a catalyst as one of the members of three interacting systems.[8] In the present paper the authors refine the procedure of deriving the orbital phase condition for stabilizing three systems, newly apply the theory to photochemical, thermal, and nonmetal catalyzed reactions, and discuss the stabilization of bicyclic compounds with π bonds on each bridge.

Theory

Suppose that an eigenfunction of interacting system Ψ_I is approximately given by a linear combination of various configuration functions Φ_K's constructed from the MO's of isolated subsystems, A, B, and C:

$$\Psi_I = \sum_K C_K \Phi_K$$

The eigenvalue E_I of Ψ_I is approximately derived by expanding the secular equation at $E = H_{II}$ where the main configuration in Ψ_I is supposed to be Φ_I.

$$E_I = H_{II} + \sum_K \frac{|H_{IK} - S_{IK}H_{II}|^2}{H_{II} - H_{KK}} + \sum_K \sum_L \frac{(H_{IK} - S_{IK}H_{II})(H_{KL} - S_{KL}H_{II})(H_{LI} - S_{LI}H_{II})}{(H_{II} - H_{KK})(H_{II} - H_{LL})} + \ldots \quad (1)^8$$

The integrals, H_{MN} and S_{MN}, are defined as

$$H_{MN} = \int \Phi_M^* H \Phi_N \, d\tau$$

$$S_{MN} = \int \Phi_M^* \Phi_N \, d\tau$$

in which H is the total Hamiltonian operator of the whole system. From the condition that the third-order energy term contributes to stabilization, i.e.,

$$S_{IK}S_{KL}S_{LI} > 0 \quad (2)^8$$

the present authors have drawn the orbital-phase relation for energetically favorable modes of three-system interactions.[8]

There are two ways of dividing three systems by their roles. They are two donors and one acceptor in one way (case I), and two acceptors and one donor in the other way (case II). In case I the orbital phase relation for stabilization has been found[8] to be in phase (+) between the HOMO of donor A and the LUMO of acceptor C, in phase between the HOMO of donor B and the LUMO of acceptor C, and out of phase (−) between the $(HOMO)_A$ and $(HOMO)_B$, according to the convention (Figure 1). In case II each relation should be in phase between $(LUMO)_A$ and $(HOMO)_C$, between $(LUMO)_B$ and $(HOMO)_C$, and between $(LUMO)_A$ and $(LUMO)_B$ as depicted in Figure 1.[8] The orbital phase restriction comes out when one takes into consideration the interaction between the important transferred configurations besides the interaction of the initial configuration with them. The transferred configurations are obtained from the initial configuration by shifting an electron from the HOMO of either donor to the LUMO of the acceptor in case I, or from the HOMO of the donor to the LUMO of either acceptor in case II. The direction of the electron shifts is shown by the arrow in the donor–acceptor diagram in Figure 1. The dashed line shows that the interaction between the transferred configurations involves the HOMO–HOMO interaction between the donors in case I or the LUMO–LUMO interaction between the acceptors in case II.

The orbital phase condition in the excited-state interaction

4694

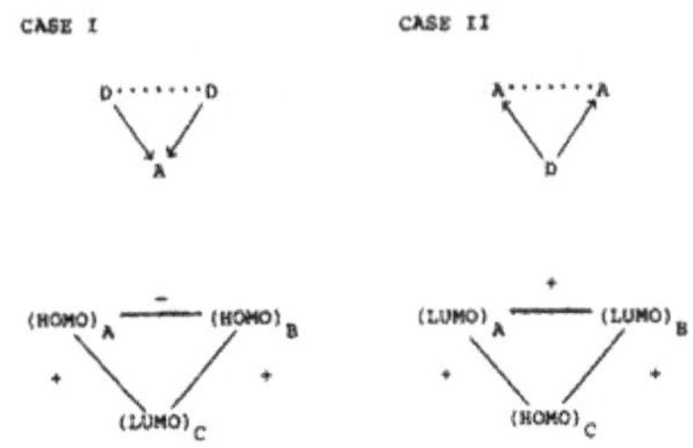

Figure 1. A conventional presentation of orbital-phase conditions for ground-state interactions.

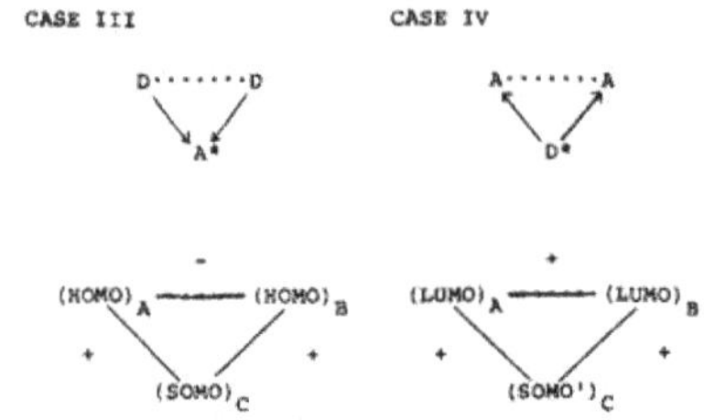

Figure 3. A conventional presentation of orbital phase conditions for excited-state interactions.

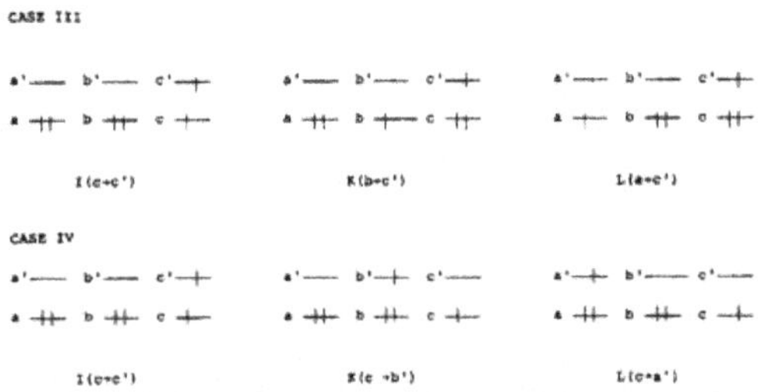

Figure 2. A simplified presentation of important configurations. Case III: excited molecule C is an acceptor. Case IV: excited molecule C is a donor.

is obtained similarly. Suppose that molecule C is excited electronically, two singly occupied MO's being denoted by SOMO and SOMO′ (lower and higher ones in energy, respectively). The excited system acts as an acceptor through the interaction of the SOMO with the HOMO of the other system (case III), or as a donor through the interaction of the SOMO′ with the LUMO of the other system (case IV). The configurations involved in the significant interaction are illustrated in Figure 2. The stabilization conditions in case III and case IV are given by

$$S_{c\to c',b\to c'}S_{b\to c',a\to c'}S_{a\to c',c\to c'} \simeq -s_{ab}s_{bc}s_{ca} > 0 \quad (3a)$$

$$S_{c\to c',c\to b'}S_{c\to b',c\to a'}S_{c\to a',c\to c'} \simeq s_{a'b'}s_{b'c'}s_{c'a'} > 0 \quad (3b)$$

These conditions are visualized in Figure 3.

Summarizing the orbital phase condition for stabilizing three systems in both ground and excited states, one can obtain a rule.

Rule I: (A) The orbitals between which a partial electron is transferred, i.e., HOMO and LUMO, HOMO and SOMO, and LUMO and SOMO′, should be combined in phase. (B) The HOMO's from which a partial electron is transferred to the LUMO or the SOMO of the third system should be combined out of phase. (C) The LUMO's to which a partial electron is transferred from the HOMO or the SOMO′ of the third system should be combined in phase.

We consider the acyclic interaction among three systems. There are four typical interaction patterns, i.e., D–D–A, A–A–D, D–A–D, and A–D–A. The term "acyclic" is used to mean the absence of the direct MO overlap between the terminal systems.

The effect of three-system interaction is less appreciable in the acyclic case. This is shown by estimating the magnitude of the third-order energy in terms of the MO overlap order on the assumption that $H_{MN} - S_{MN}H_{II} \propto S_{MN} = o(s_{AB}{}^{\lambda}s_{BC}{}^{\mu}s_{CA}{}^{\nu})$ where λ, μ, and ν stand for the minimum numbers of the electron shifts between A and B, between B and C, and between C and A required to obtain the electron configuration M from N.[7a,8,9] The third-order energy is estimated by the magnitude of $S_{IK}S_{KL}S_{LI}$ (see eq 1). In the cyclic interaction $S_{IK}S_{KL}S_{LI} = o(s_{AB}{}^1s_{BC}{}^1s_{CA}{}^1)$ where I is the initial configuration, K and L being the important transferred configurations. In the acyclic interaction of the A–B–C system, $S_{IK}S_{KL}S_{LI} = o(s_{AB}{}^2s_{BC}{}^2)$ since the overlap between a pair of the important configurations, approximated to be s_{CA} in the cyclic interaction, should be replaced by $s_{AB}s_{BC}$ because of $s_{CA} = 0$.[12] Accordingly, the third-order perturbation energy of interest is smaller in the acyclic interaction. As a result it may be said that the acyclic interaction of three systems is usually considered as two sets of the two-system interactions between the neighboring ones. The familiar consequences of the two-system interactions[1,7] can be applied to the acyclic interactions. The HOMO–LUMO interaction between the neighboring systems contributes most to electron delocalization or stabilization. The HOMO–HOMO interaction between them destabilizes the system through exchange repulsion. The LUMO–LUMO interaction has no effect since no electron is involved.

In a special case, however, there is a conspicuous effect peculiar to the three-system interaction even in the acyclic interaction. Suppose that the donor–acceptor relation between a pair of systems is so appreciable that the initial configuration and a transferred configuration (denoted by D) are nearly degenerate ($H_{II} \simeq H_{DD}$). Then one can obtain a new type of perturbation energy term[9]

$$\sum_K \frac{|H_{DK} - S_{DK}H_{DD}|^2}{H_{DD} - H_{KK}} \quad (4)$$

which has the same form as the second-order energy term in the nondegenerate case. This shows that both I and D contribute equivalently in the degenerate case. The perturbation energy term 4 is chemically of great importance.

The D–D–A and the A–A–D systems, in which the donor–acceptor relation between a D–A pair is significant, appreciably gain the additional stabilization (the term 4) through the HOMO–HOMO interaction between D's and the LUMO–LUMO interaction between A's, respectively, while the corresponding D–A–D and A–D–A systems do not. The important configurations, D and K, contained in the energy term 4, are visualized in Figure 4. In the D–D–A interaction pattern they are two transferred configurations where an electron is shifted from the HOMO of either of the D's to the LUMO of A. The configuration interaction involves the HOMO–HOMO interaction between D's. The key orbital interaction is depicted on the right of the important configurations in Figure 4. In the A–A–D interaction pattern the important configurations are the transferred ones where an electron is shifted from the HOMO of D to the LUMO of either of the A's. The configuration interaction involves the LUMO–LUMO interaction between A's. In the above cases the important configuration interactions are both reduced to the key orbital interaction between the neighboring systems. In D–A–D or A–D–A the important configuration interactions are, however, reduced to the orbital interactions between the

4695

terminal systems; the HOMO-HOMO interaction between the terminal D's in D-A-D and the LUMO-LUMO interaction between the terminal A's in A-D-A do not occur efficiently on account of the special separation. As a result, one may say the following with respect to the ground-state interaction:

Rule II: In the acyclic D-D-A or A-A-D systems where the donor-acceptor relation between the neighboring D-A pair is significant, the HOMO-HOMO interaction between D's or the LUMO-LUMO interaction between A's appreciably contributes to the stabilization.

The "abnormal" stabilization through the HOMO-HOMO interaction or through the LUMO-LUMO interaction is attributed to the SOMO-like property of the HOMO having released electrons or the SOMO′-like property of the LUMO having accepted electrons. Such stabilizing HOMO-HOMO and LUMO-LUMO interactions have also been recognized in the bimolecular interactions involving pseudoexcitation.[9,10]

The stabilization of the acyclic three-system interactions does not depend on the sign relation among the particular orbitals involved, although the orbital phase continuity between each pair of the neighboring systems should be satisfied if the cyclic conjugation occurs between the two. This is evident. There are no energy terms relating the signs of three orbitals significantly involved in the acyclic interaction. There is no orbital overlap between the terminal systems.

Rule II is here given in another way. The method rests on the procedure of sequential orbital interactions. In usual cases the HOMO of the D-A system is composed of an out-of-phase combination of the HOMO's of D and A, and the LUMO of the D-A system is composed of an in-phase combination of the LUMO's of D and A. In the D-D-A the interaction between the HOMO of the terminal D and the LUMO of the D-A is important. The interaction involves the HOMO-LUMO interaction between D's. In the A-A-D the interaction between the LUMO of the terminal A and the HOMO of the D-A is important. The interaction also involves the HOMO-LUMO interaction between A's. The HOMO-LUMO interaction between the adjacent systems is similarly found to be important in A-D-A and D-A-D. The situation changes when the donor-acceptor relation is significant between the neighboring D-A pair. In this case the HOMO and the LUMO of D-A are composed of an in-phase and an out-of-phase combination of the HOMO of D and the LUMO of A, respectively. The important interaction between the HOMO of the terminal D and the LUMO of D-A in D-D-A involves the HOMO-HOMO interaction between D's. The important interaction between the LUMO of the terminal A and the HOMO of D-A in A-A-D involves the LUMO-LUMO interaction between A's. The interaction between the HOMO of D and the LUMO of A-D in D-A-D and the interaction between the LUMO of A and the HOMO of D-A in A-D-A involve the HOMO-LUMO interaction between the neighboring D and A.

In the excited-state interactions the donor-acceptor relation between an excited molecule and a ground-state molecule is expected to be significant even in usual cases since the SOMO or an originally doubly occupied orbital demands electrons strongly, and since the SOMO′ or an originally unoccupied orbital readily releases electrons. The excited molecule is relatively strong donor (D*) or acceptor (A*). Therefore, there may also be a similar effect peculiar to the three-system interaction even in the acyclic interaction.

Rule III: In the acyclic D-D-A* or A-A-D* excited-state interactions the HOMO-HOMO(-SOMO) interaction or the LUMO-LUMO(-SOMO′) interaction contributes to stabilization and to bond formation between D's or A's.

Application and Exemplification

(I) Cyclic Interaction. The orbital phase condition derived

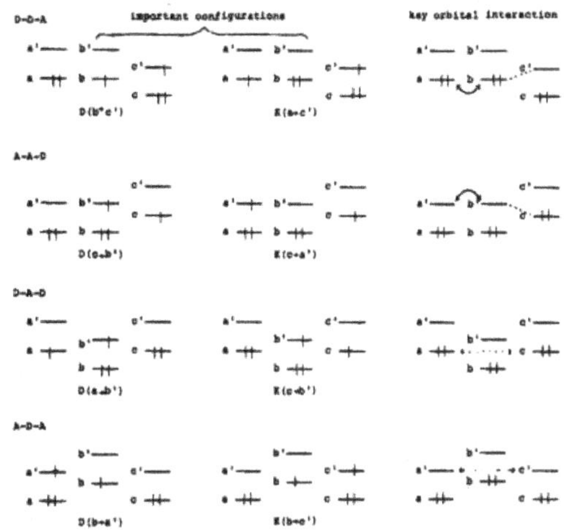

Figure 4. A simplified presentation of important configurations and key orbital interactions peculiar to acyclic three-system interaction with remarkable donor-acceptor relation.

from the third-order perturbation energy is compatible with the Woodward-Hoffmann rule.[13] The merit of the three-system interaction scheme developed here consists in allowing one to consider the interaction on the basis of each functional group, and to take into account the detailed aspect of chemical reactions through the donor-acceptor property of the interacting entities. One can find some suitable examples in the following reactions.

Homo-Diels-Alder Reactions.[13] Diels-Alder reactions have been interpreted as simultaneous bond-forming processes. The interactions between the HOMO of diene and the LUMO of dienophile and between the LUMO of diene and the HOMO of dienophile are favored by the symmetries of the particular orbitals. It seems more appropriate to interpret homo-Diels-Alder reactions in terms of the three-system interaction among two π bonds in diene and a π bond in dienophile since in all homo-dienes the double bonds do not interact with each other so strongly that the two π bonds are taken as a single functional group. In eq 5 is depicted one of the particular orbital sets, the

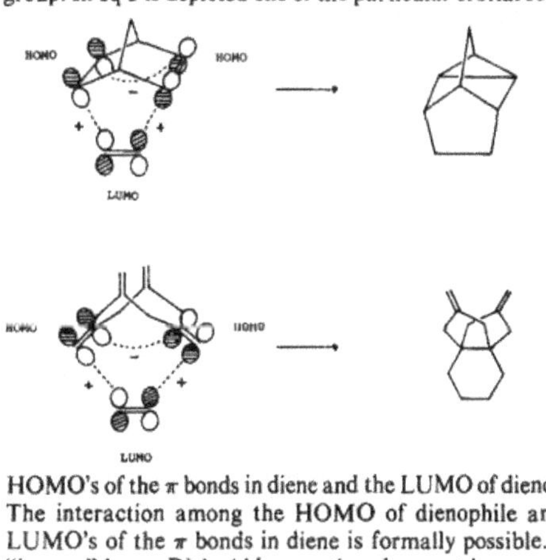

(5)

HOMO's of the π bonds in diene and the LUMO of dienophile. The interaction among the HOMO of dienophile and the LUMO's of the π bonds in diene is formally possible. Such "inverse" homo-Diels-Alder reactions have not been reported as yet.

The out-of-phase combination between the HOMO's does not stand for the repulsive force between the π bonds. Although the HOMO-HOMO interaction usually causes only exchange repulsion in the case of interaction between the two systems, the HOMO-HOMO interaction in the three systems contributes to stabilization under the influence of the third system. The stabilization may be attributed to a SOMO-like property of the original HOMO's resulting from the partial electron release to the LUMO of the third system.

Cope Rearrangements.[14,15] The merit of the three-system interaction theory is unquestionable when the orbital phase aspect of Cope rearrangement is considered in terms of orbital interaction. Two orbital interaction schemes were thus far required to understand the mechanism of Cope rearrangements in detail: one for the orbital interaction weakening the C-C σ bond to yield a pair of allyl fragments and the other for the interaction stabilizing the transition state.[1e] According to the three-system interaction theory, the nature, especially the orbital phase aspect, of Cope rearrangements is simply described by a single scheme of the interaction among the HOMO's of the two π bonds and the LUMO of the σ bond, or among the LUMO's of the π bonds and the HOMO of the σ bond. Only cyclic (HOMO, HOMO, LUMO) interaction is depicted here.

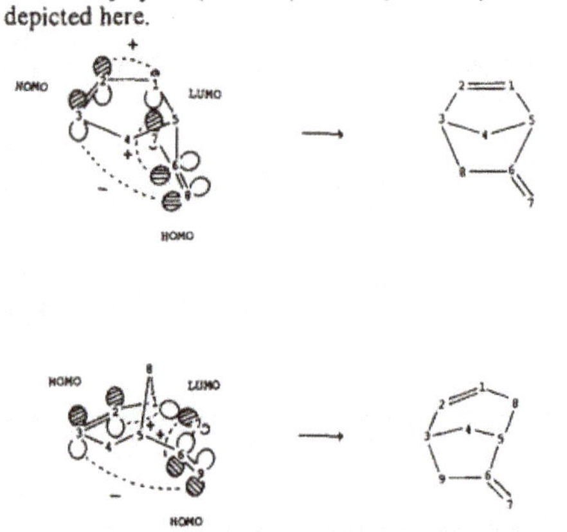

(6)

2,3-Sigmatropic Shifts.[13,17,18] The mechanism of 2,3-sigmatropic shifts is also better interpreted as an interaction among three functional groups. The HOMO of the lone pair or something like this, the LUMO of the π bond, and the LUMO of the σ bond may interact most significantly since the ability of the lone pair to donate electrons may be the most prominent. The cyclic (HOMO, LUMO, LUMO) interaction scheme depicted here implies that the rearrangements begin with the simultaneous charge transfers from the HOMO of the lone pair to the LUMO of the π bond (nucleophilic attack) and to the LUMO of the σ bond (weakening of the σ bond), accompanying the interaction between the transferred configurations or the LUMO-LUMO interaction between the π bond and the σ bond (formation of a new π bond) (eq 7).

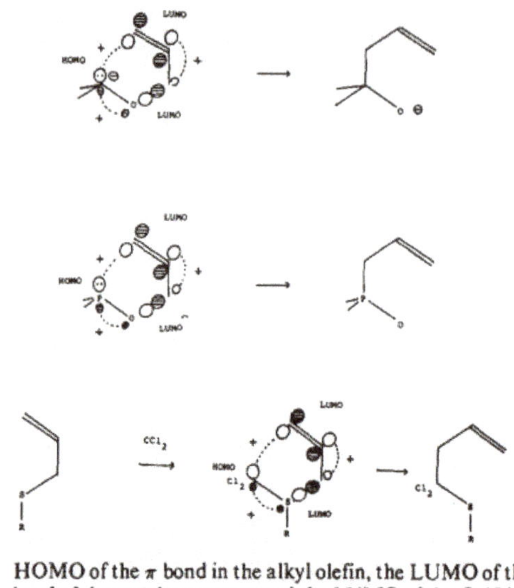

(7)

Ene and Retroene Reactions.[18,19] One feels inclined to draw from the Woodward-Hoffmann rule the concerted property of ene and retroene reactions since both reactions involve six electrons in the cyclic interactions. Indeed, the AO sign relation of the HOMO and the LUMO of alkyl olefins is suitable for simultaneous interaction with unsaturated species at either reacting site. However, the particular orbitals of alkyl olefins have too small an amplitude at the C-H site to cause an effective cyclic interaction. An intermediary mechanism between the concerted and the two-step mechanisms has been proposed; at an initial stage a cyclic three-centered interaction occurs among the p orbitals of the π bond of alkyl olefins and a p orbital of an electrophilic center of the reaction partner, and then the transient three-membered ring collapses with hydrogen transfer and the π bond shift.[20] The mechanism is here reinterpreted in terms of the three-orbital interaction among the HOMO of the π bond in the alkyl olefin, the LUMO of the π bond of the reaction partner, and the LUMO of the C-H bond of the olefin. The interaction scheme implies that the partial electron transfer from the π HOMO of the alkyl olefin to the LUMO of the attacking species is sequentially followed by "back donation" from the LUMO having accepted electrons to the LUMO of the C-H bond, or that the charge transfers from the HOMO of the π bond of the olefin intermolecularly to the LUMO of the reaction partner and intramolecularly to the LUMO of the C-H bond accompany the interaction between the LUMO's having accepted electrons. The intramolecularly transferred configuration is an excited configuration. The ground-state process where an excited configuration of one or both reactants is involved significantly was previously referred to as pseudoexcitation.[9] The reactions accompanying pseudoexcitation proceed by an intermediary mechanism between typical concerted and two-step mechanisms even if the high stereoselectivity happens to be observed.

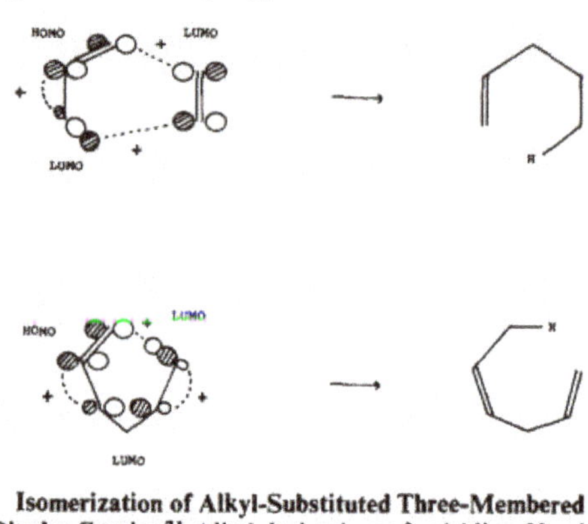

(8)

Isomerization of Alkyl-Substituted Three-Membered Ring Dipolar Species.[21] Alkyl derivatives of aziridine *N*-oxide or episulfoxide rearrange to the allylic compounds through an intramolecular abstraction of an alkyl hydrogen by the oxygen atom (Y) accompanying the C-X bond cleavage. A nucleophilic attack of the negatively charged oxygen atom on the C-H hydrogen is unquestionably one of the important interactions. From the prominent ability of the oxygen lone pair to donate electrons the cleaving C-X bond is expected to act as an acceptor toward the lone pair. The cyclic (HOMO, LUMO, LUMO) interaction is depicted here.

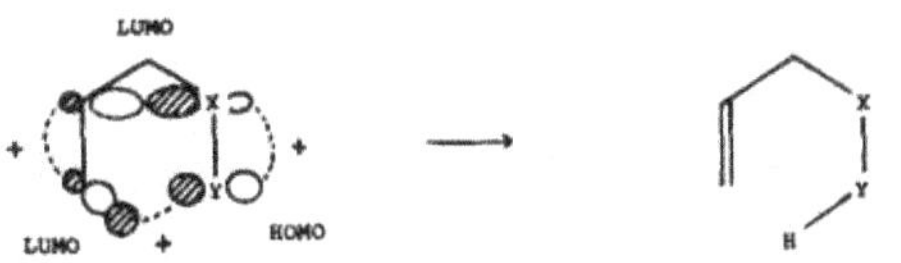

(9)

Degenerate Isomerization of Snoutene.[22] This is also a typical reaction of three bonds. The orbital interaction scheme is better described as a three-system interaction rather than any two-system interactions. Here the HOMO of the π bond and the LUMO's of the σ bonds are used in the illustration, although the quantitative evidence is not available for the preference over the cyclic interaction among the HOMO's of the σ bonds and the LUMO of the π bond.

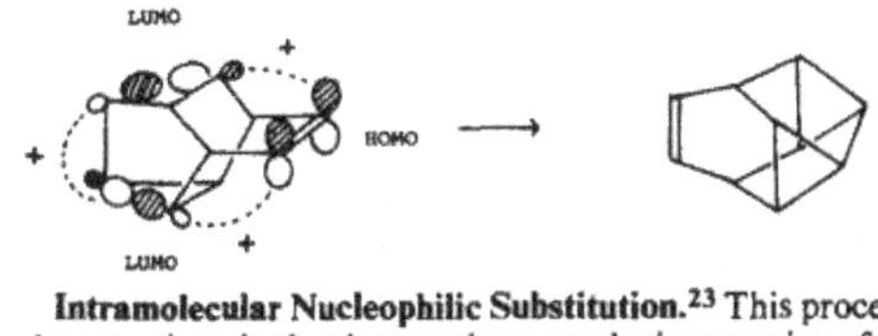

(10)

Intramolecular Nucleophilic Substitution.[23] This process is almost unique in that it sometimes results in retention of configuration about the α carbon atom. The simultaneity of the bond interchange is not so complete as that of typical concerted reactions, for example, Diels-Alder reactions. However, the reaction is considered to take place through a four-centered transition state. Here an orbital interaction scheme is depicted for the transition state. The C-O bond and the S-Cl bond are weakened. The slightly negatively charged chlorine is bound to the positively charged carbon. The key orbitals should be the HOMO of chloride ion, and the LUMO's of the carbonium cation and the SO_2 part. The intramolecular nucleophilic substitution reaction of chroloformate is explained similarly.

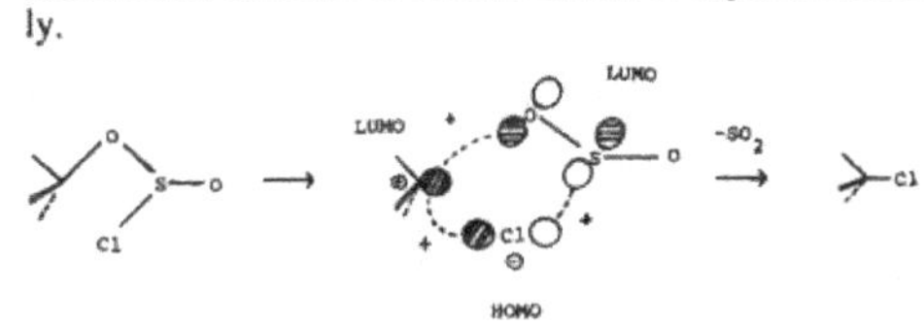

(11)

Insertion of Sulfur Dioxide into a Metal-Carbon Bond.[24] Interestingly, the insertion of SO_2 into the Fe-R bond of (η^5-ring)$Fe(CO)_2R$ results in inverting the configuration at the α carbon of alkyl R. The observed rate dependence on the nature of R suggests that the cleavage of the Fe-R bond is a heterolytic process caused by the back-side attack of SO_2 on the α carbon of R and that a contact ion pair forms. One reasonably imagines the process as follows; at first the interaction between the bonding orbital of the Fe-R bond and the LUMO of sulfur dioxide (with the largest amplitude at sulfur) weakens and lengthens the Fe-R bond with the charge separation between the iron and the carbon, and a sequential interaction among the HOMO of the anionic carbon, the LUMO of the cationic iron, and the LUMO of sulfur dioxide stabilizes the transition state. It is interesting that the orbital interaction scheme is quite similar to that of the intramolecular nucleophilic substitution.

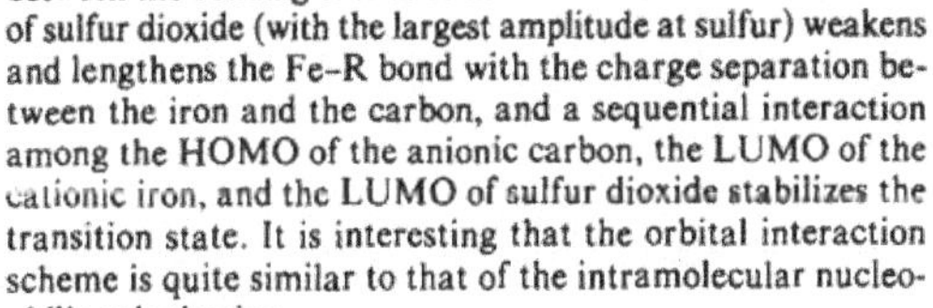

(12)

Di-π-methane Rearrangement.[25] The photochemical conversion of 1,4-dienes to vinylcyclopropanes, known as di-π-methane rearrangement, proceeds with stereospecificity. The configuration integrity about the π bond in the migrating moiety (C_4=C_5) is maintained; the substituents at C_5 cis and trans to the C_3-C_4 bond in the reactant are ones cis and trans to the C_2-C_4 bond in the product, respectively.[25b] With respect to ring closure between C_1 and C_3 there exists a strong preference for occurring at the side anti to the migrating π bond,[25c,d] with inversion of configuration at the methane carbon (C_3).[25e]

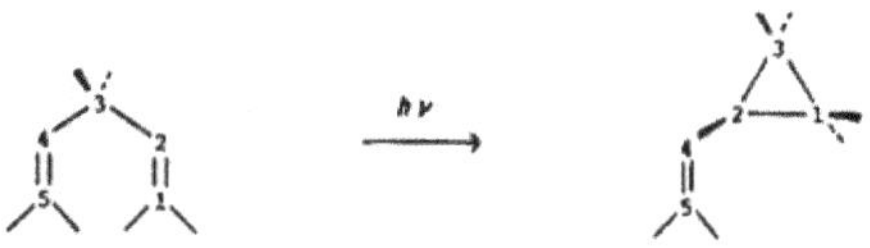

(13)

The observed stereochemistry of the products of the di-π-methane rearrangements is compatible with the two-system interaction scheme (**1**). The key parts are the π bond and the σ bond that are broken during the reaction. However, an inevitable question comes into our minds. Why does the C_4=C_5 π bond remain intact through the process? We believe that the π orbital seemingly not involved plays a significant role, as Zimmerman has noted. In fact, it has been demonstrated that neither the singlet nor the triplet of 1,1,5,5-tetraphenyl-3,3-dimethyl-1-pentene (monoene) rearranged in the di-π-methane-like fashion.[26]

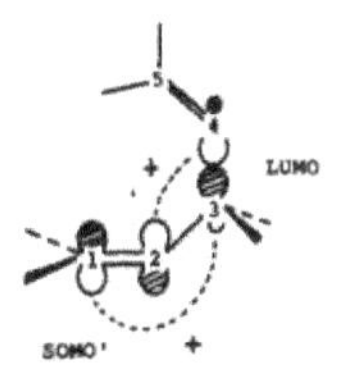

1

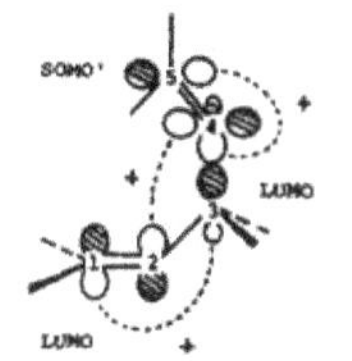

2

The interaction involving the C_4=C_5 π bond is best described by the three-system interaction scheme (**2**). An electron in the SOMO' of a π bond first delocalizes to the other π bond through the interaction with the LUMO between C_2 and C_4. The interaction causes folding of the structure of the di-π-methane. The geometrical change necessarily involves the bending of one or both of the C_2-C_3 and the C_3-C_4 σ bonds out of the incipient three-membered ring. The direction of the bending is both geometrically and from the viewpoint of orbital phase continuity suitable for the cyclic (SOMO', LUMO, LUMO) interaction involving the LUMO of the C_3-C_4 σ bond or the LUMO of the C_2-C_3 σ bond (the former case is depicted in **2**). The C_4 front-side lobe of the σ LUMO interacts with the C_5 exo lobe of the π SOMO' to accept electrons. The C_3 back-side lobe of the σ LUMO interacts with the C_1 exo lobe of the π LUMO.

As a result the three-orbital phase condition is completely compatible with the smooth change of the geometry required to account for the stereochemistry of the di-π-methane rearrangement. This conclusion is also obtained from the cyclic (HOMO, HOMO, SOMO) interaction.

In the above discussion the interaction between the π SOMO' and the π LUMO has been described as preceding the others. The separate discussion on the orbital interactions comes mainly from the convenience. It is likely that the di-π-methane rearrangement may occur through the sequential interactions, and that one interaction between a subsystem pair never begins after another is completed. Otherwise, one should have concluded a randomization of the stereochemistry.

The orbital interaction scheme for the di-π-methane rearrangement is applicable to similar rearrangements of dienones and bicyclo[3.2.1]octa-2,6-diene (eq 14).[27]

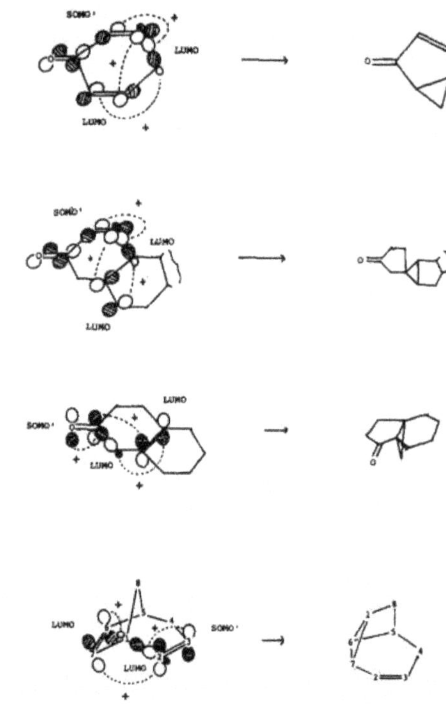

(14)

Stabilization of Bicyclic π-Conjugate Molecules. The interaction of π bonds in bicyclic compounds (**3**) is suitably treated on the basis of the three-system interaction theory. It will be shown that such bicyclic compounds may be classified into the three groups electron-localizing (**4**), semielectron-delocalizing (**5**), and electron-delocalizing (**6**) systems. They

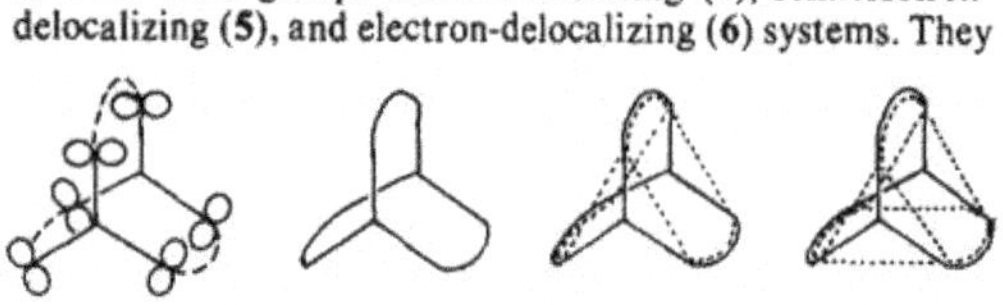

3 4 5 6

have been divided into the two classes, stabilized or destabilized systems, in the pioneering work of Goldstein and Hoffmann.[11a,b] A rationale for the bridge flipping will also be given by the present theory.

The bicyclic compounds in which each HOMO-LUMO interaction is symmetry disfavored between any π-system pair are classified as electron localizing (**4**). The systems in which the HOMO-LUMO interaction is symmetry favored between two pairs of π systems yield the other two classes. The compounds dissatisfying and satisfying the three-orbital phase condition (rule I) are referred to as semielectron-delocalizing (**5**) and electron-delocalizing (**6**) systems, respectively.

An important corollary, closely related to the bridge flipping, follows from rule II. Some semielectron-delocalizing and electron-delocalizing molecules can be distorted in favor of the HOMO-LUMO interaction at the expense of the other. Such geometrical change implies a shift of the interaction pattern from the cyclic (D, D, A) or (A, A, D) to the acyclic D-D-A or A-A-D. The HOMO-LUMO interactions between D's and between A's are disfavored by the orbital symmetry in **5** or **6**. Accordingly, no additional stabilization is usually expected to come from the interaction between D's or A's. The HOMO-HOMO interaction between D's and the LUMO-LUMO interaction between A's are favored by the symmetry property. If these interactions contribute to stabilization, the acyclic D-D-A and A-A-D are more likely to be stable relatively. This is the case with the bicyclic compounds where the donor-acceptor relation is significant between a D-A pair.

Corollary: The HOMO-HOMO(-LUMO) interaction between D's and the LUMO-LUMO(-HOMO) interaction between A's contribute, at least in part, to transforming the symmetrical structure of electron-delocalizing and semielectron-delocalizing compounds with a significant D-A pair into the distorted structure.

The electron-delocalizing 7-norbornadienyl cation **7** is known to adopt an unsymmetrical structure[28-31] represented by **8**.[32] This is explained in the corollary. The introduction of a second double bond in the norbornane framework causes a further rate enhancement for ionization at C_7 of 10^3 relative to the already anchimerically accelerated *anti*-7-norbornenyl derivative.[29,30] The degree of rate enhancement is smaller than that of the effect of the introduction of the first π bond (10^{11}).[29,33] The additional rate enhancement is here interpreted to come from the HOMO-HOMO interaction between the π bonds. If the HOMO-HOMO interaction does not contribute to stabilization, the solvolysis rate of 7-norbornadienyl derivatives must be comparable to that of *anti*-7-norbornenyl derivatives since in the distorted 7-norbornadienyl cation there is no significant interaction but one between the HOMO of a π bond and the LUMO of the ionizing center, common to *anti*-7-norbornenyl cation.

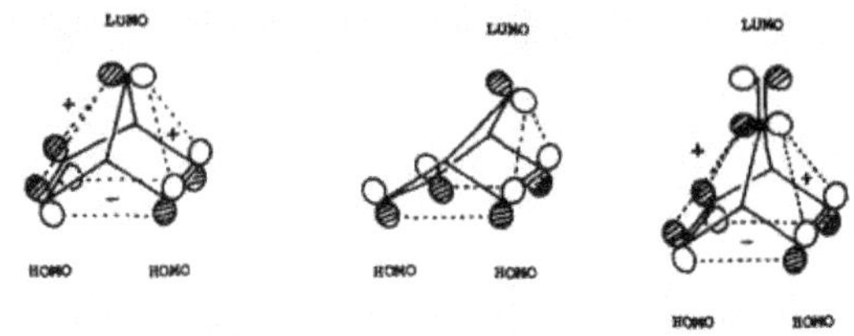

7 8 9

7-Methylenenorbornadiene (**9**) is also an electron-delocalizing system. Comparable interactions between the vinylene bridges and the C_7 p orbital are expected from the experimental results.[34] The donor-acceptor relation between any pairs of subsystems is too ambiguous to cause bridge flipping.[35] The observed dipolar (negative terminus at C_8)[34] supports the orbital interaction scheme.

A typical semielectron-delocalizing system is 4-bicyclo[3.2.2]nonatrienyl anion **10**. Although the two-system interactions between the HOMO of the allyl anion moiety with the LUMO's of the double bonds are favored by the symmetry property, **10** is not favored by the three-orbital phase condition.

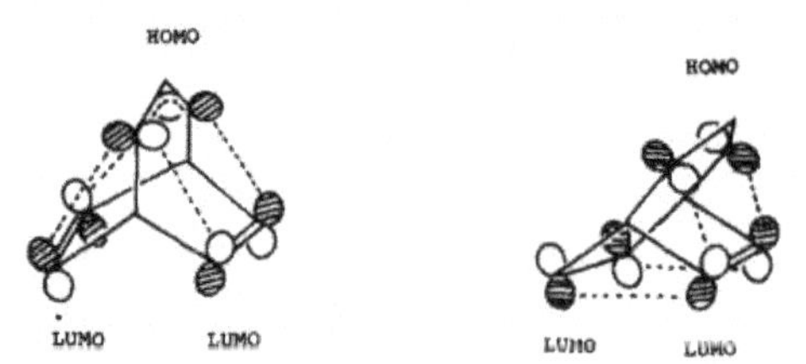

10 11

There is a possibility that the anion **10** undergoes such bridge flipping (**11**) as **7** since the donor–acceptor relation is significant between the allyl anion and the π bonds. If there are some factors preventing the bridge flipping, an important one may be an exchange repulsion between the symmetrical lowest lying π orbital of the allyl fragment and the HOMO of the double bond. Electron delocalization has been shown to occur to an appreciable degree.[36] No convincing evidence for a symmetrical or unsymmetrical equilibrium geometry is available now.

7-Norbornadienyl anion (**12**),[37] barrelene (**13**),[38] and bicyclo[3.2.2]nonatrienyl cation (**14**)[39] are electron-localizing

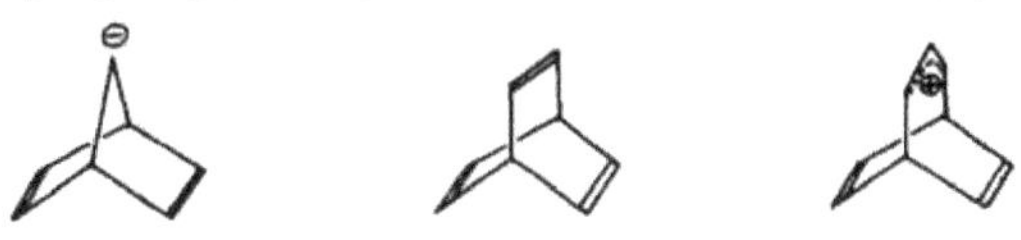

12 13 14

systems. These molecules have been classified by Goldstein and Hoffmann[11b] as the systems that are not stabilized.

Unfortunately, no evidence for or against the practical significance of the classification into the electron-delocalizing and the semielectron-delocalizing systems is available now. Experimental efforts on this problem are to be expected.

It would be meaningful to refer to an essential difference between such an aromatic compound as benzene and the electron-delocalizing bicyclic compounds. Electron delocalization in **6** occurs through the interaction among a set of three particular orbitals as depicted. On the other hand, there are many sets of electron-delocalizing orbital interactions in the aromatic hydrocarbons. For example, in a hypothetical compound, cyclohexatriene, there are six sets of three-orbital interactions.

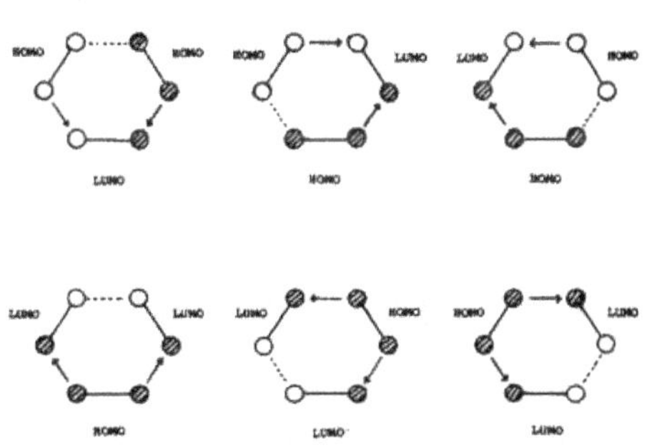

(15)

Thus, the degree of electron delocalization is remarkably different between the electron-delocalizing systems **6** and the aromatic compounds.

The present method is different from the Goldstein and Hoffmann's method[1a,b] of treating bicyclics in the procedure and in the conclusion. Their procedure begins by first permitting two of the systems to interact. The resulting composite orbitals are then allowed to interact further with orbitals of the third system. According to our findings one can draw the stabilization condition directly from the donor–acceptor property of each system and the symmetry property of the particular orbitals. In conclusion, the discrimination of the electron-delocalizing **5** from the semielectron-delocalizing **6** system is neglected and both are classified as the stabilized systems. It is in question whether the 7-norbornadienyl cation (**7**) and bicyclo[3.2.2]anion (**10**) are both stabilized (0,2,2) systems. In **7** the HOMO of the combined system of the two π bonds is favored by the orbital phase to interact with the LUMO of the cationic center at the 7 position, while in **10** the LUMO of the combined system of the two π bonds is not favored by the orbital phase to interact with the HOMO of the allyl anion fragment.

Quadricyclyl Ions. Quadricyclyl cation and anion are taken as being composed of three functional groups, two cyclopropane rings, and an ionic center. The important orbitals in the cation are obviously the HOMO's of the cyclopropane parts and the LUMO of the cation (**15**). Those orbitals in the anion

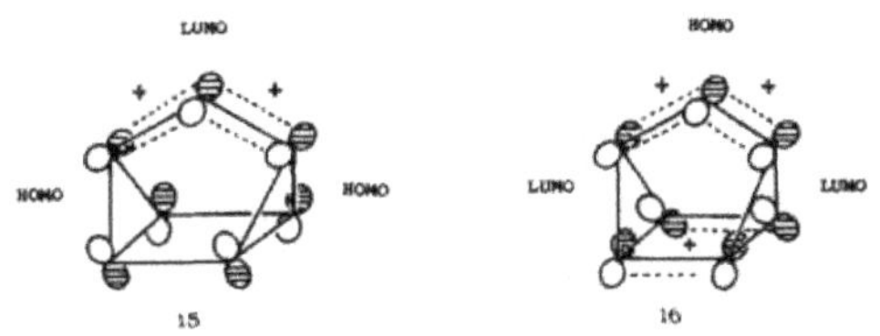

15 16

are the LUMO's of the cyclopropane parts and the HOMO of the anionic center (**16**). Both ions are stabilized by the HOMO–LUMO interactions between the ionic center and each cyclopropane ring. However, the cyclic (HOMO, HOMO, LUMO) three-system interaction in the cation does not meet the orbital phase requirement while the cyclic (HOMO, LUMO, LUMO) interaction in the anion is favored by the orbital phase requirement (rule I). These properties suggest that quadricyclyl cation (with two cyclopropane rings) may be more stabilized than nortricyclyl cation (with one cyclopropane ring) owing to the HOMO–LUMO interaction between each cyclopropane ring and the cation center, but that the destabilization from the cyclic three-system interaction partially diminishes the stabilization.

Loew and Wilcox[40] considered the orbital interaction between the two parts, cyclopropane and cyclopropylcarbinyl cation, and answered the problem that the second cyclopropane provides a mere 20-fold solvolysis rate increase whereas the strict additivity of stabilization energies would dictate 6×10^8-fold enhancement relative to the nortricyclyl system. Although no experimental data are now available for showing the relative stabilities of 7-norbornanyl, nortricyclyl, and quadricyclyl anion, one may say the stabilization by the second cyclopropane ring exceeds the stabilization by the first cyclopropane ring since the cyclic (HOMO, LUMO, LUMO) interaction is stabilizing in contrast to the cyclic (HOMO, HOMO, LUMO) interaction in the cation.

(II) Acyclic Interaction. The interaction in the A–B–C system where A and B, and B and C interact but A and C do not interact is approximately taken as the two independent A–B and B–C interactions in usual cases (see the theoretical section). The stabilization of such systems can be roughly discussed in terms of the HOMO–LUMO interactions between the neighboring systems. The HOMO–HOMO interaction between the D's in the D–D–A system and the LUMO–LUMO interaction between A's in the A–A–D system contribute to stabilization, appreciable as the donor–acceptor relation is significant between the D–A pair (rule II). This "abnormal" effect has been proposed to be an important electronic factor of bridge flipping of the bicyclic compound **3**. The stereochemical courses of some catalyzed "symmetry-forbidden" reactions such as Ag^+-catalyzed electrocyclic reactions and valence isomerization of strained cage compounds and 1,3-sigmatropic rearrangements were also elucidated similarly.[8] Here the important role of the HOMO–HOMO or the LUMO–LUMO interaction in the presence of the third entity interacting with one of them will be exemplified by the thermal and the photochemical reactions and by the nonmetal-catalyzed reactions.

Multicycloaddition Reactions. Interesting cycloaddition reactions have recently been observed in the thermal reactions of paracyclophadiynes with an electron-demanding olefin, tetracyanoethylene (TCNE),[41] and in the photochemical reactions of [10][9.10]anthracenopha-4,6-diyne with furan (X = O) or cyclopentadiene (X = CH_2).[42] The thermal reaction is understood in terms of the interaction among the HOMO of the benzene ring, the HOMO of the diyne part, and the LUMO of TCNE.[32] It should be noted that the bond formation between the benzene ring and the diyne part is not favored by the HOMO–LUMO symmetry properties. A highly electron-deficient TCNE attacking the diyne allows the HOMO–HOMO interaction between the diyne part and the benzene ring to contribute to the stabilization and the bond formation between them (rule II). The photochemical multicycloaddition reactions are also understood along the line of rule III in a similar manner. The key orbital interaction is one among the SOMO′ of the anthracene part, the LUMO of the diyne, and the LUMO of cyclopentadiene, or among the SOMO of the anthracene, the HOMO of the diyne, and the HOMO of cyclopentadiene.[32]

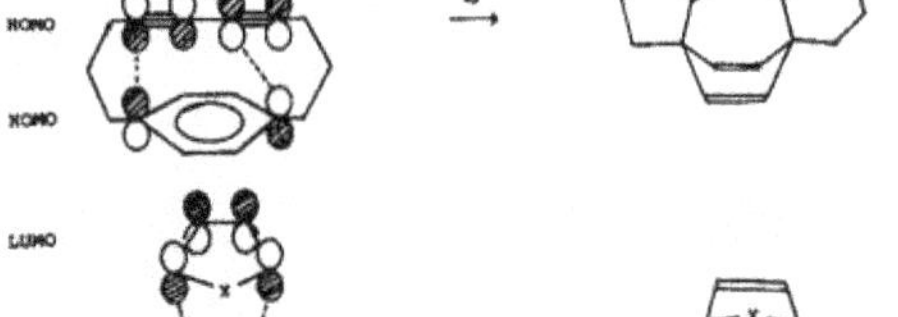

(16)

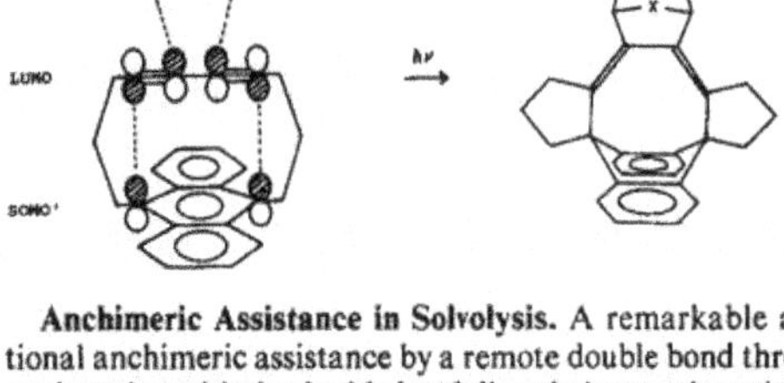

(17)

Anchimeric Assistance in Solvolysis. A remarkable additional anchimeric assistance by a remote double bond through conjugation with the double bond directly interacting with the ionizing center has been observed in the solvolysis of **17**.[43] The remote double bond makes **17** ca. 10^3 times more reactive than related *anti*-7-norbornenyl systems. This may also be ascribed to the stabilizing HOMO–HOMO interaction between the double bonds (rule II).[32]

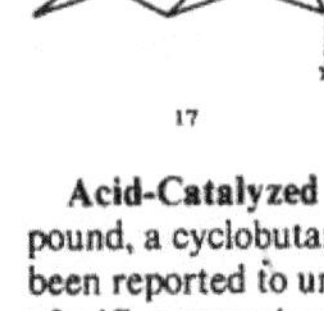

17

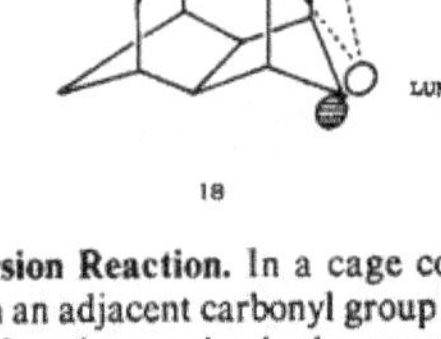

18

Acid-Catalyzed Cycloreversion Reaction. In a cage compound, a cyclobutane ring with an adjacent carbonyl group has been reported to undergo 2 + 2 cycloreversion in the presence of trifluoroacetic acid.[44] This symmetry-disfavored transformation can be completely explained by rule II. In this case, a protonated carbonyl group works as a strong acceptor[45] to allow the HOMO–HOMO interaction to be bonding between the σ bonds.[32]

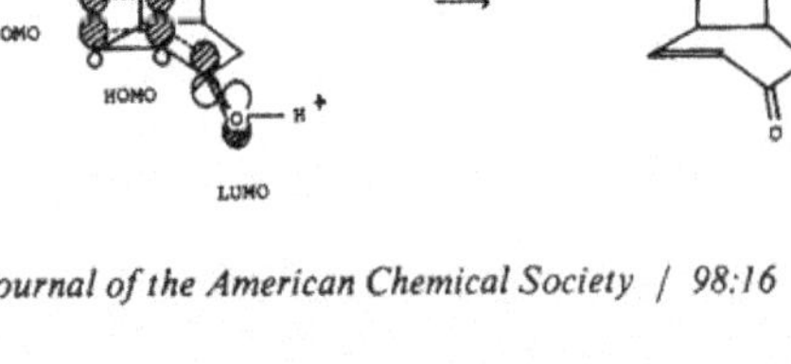

(18)

A Substituent Effect on Symmetry-Disfavored Reaction. Another interesting symmetry-disfavored reaction has been observed. The thermal 1,3-sigmatropic shift in an olefin with electron-demanding cyano groups takes place with 90% retention of configuration at the migrating carbon center.[46] Usual thermal 1,3-sigmatropic shifts are interpreted in terms of the cyclic HOMO–LUMO interaction between the double bond and the σ bond. The cyano substituents, although they do not take part in the cyclic interaction directly, may have influence through the interaction with a member of the cycle. One may expect according to rule II that the HOMO–HOMO interaction between the double bond and the σ bond is appreciably stabilizing under the influence of the interaction between the HOMO of the σ bond and the LUMO of the cyano groups.[32,47]

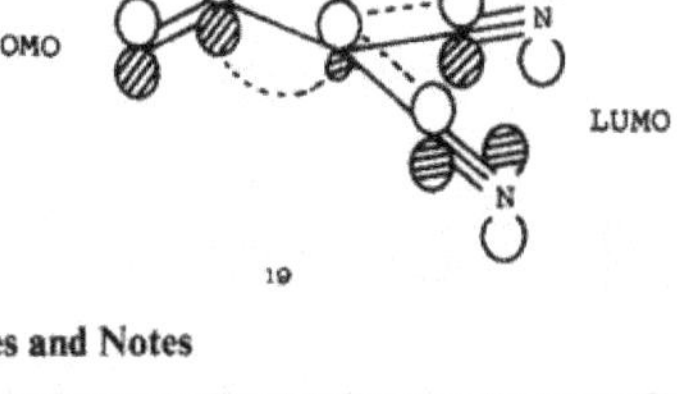

19

References and Notes

(1) (a) K. Fukui, "Molecular Orbitals in Chemistry, Physics, and Biology", P.-O. Löwdin and B. Pullman, Ed., Academic Press, New York, N.Y., 1964, p 513, and references cited therein; (b) K. Fukui, "Modern Quantum Chemistry", O. Sinanoglu, Ed., Vol. 1, Academic Press, New York, N.Y., 1965, p 49; (c) K. Fukui and H. Fujimoto, "Mechanisms of Molecular Migrations", Vol. 2, B. S. Thyagarajan, Ed., Interscience, New York, N.Y., 1969, p 118; (d) K. Fukui, "Theory of Orientation and Stereoselection", Springer-Verlag, Heidelberg, 1975; (e) K. Fukui, *Acc. Chem. Res.*, **4**, 57 (1971).

(2) G. Klopman, *J. Am. Chem. Soc.*, **90**, 223 (1968).

(3) L. Salem, *J. Am. Chem. Soc.*, **90**, 543, 553 (1968).

(4) (a) W. C. Herndon, *Chem. Rev.*, **72**, 157 (1972); (b) R. F. Hudson, *Angew. Chem.*, **85**, 63 (1973); (c) J. D. Bradley and G. C. Gerrans, *J. Chem. Educ.*, **50**, 463 (1973); (d) K. N. Houk, *Acc. Chem. Res.*, **8**, 361 (1975).

(5) (a) J. N. Murrell, M. Randić, and D. R. Williams, *Proc. R. Soc. London, Ser. A*, **284**, 566 (1965); (b) J. N. Murrell and G. Shaw, *J. Chem. Phys.*, **46**, 1768 (1967).

(6) R. Rein and M. Pollak, *J. Chem. Phys.*, **47**, 2039 (1967).

(7) (a) K. Fukui and H. Fujimoto, *Bull. Chem. Soc. Jpn.*, **41**, 1989 (1968); (b) K. Fukui, *Int. Congr. Pure Appl. Chem. 23rd*, **1**, 65 (1971); (c) H. Fujimoto and K. Fukui, *Adv. Quantum Chem.*, **6**, 177 (1972).

(8) K. Fukui and S. Inagaki, *J. Am. Chem. Soc.*, **97**, 4445 (1975).

(9) S. Inagaki, H. Fujimoto, and K. Fukui, *J. Am. Chem. Soc.*, **97**, 6108 (1975)

(10) N. D. Epiotis, *Angew. Chem., Int. Ed. Engl.*, **13**, 751 (1974), and his preceding papers. Some models employed in his discussion are unlikely.[9,20]

(11) (a) M. J. Goldstein, *J. Am. Chem. Soc.*, **89**, 6357 (1967); (b) M. J. Goldstein and R. Hoffmann, *Ibid.*, **93**, 6193 (1971); (c) H. Fujimoto, S. Kato, S. Yamabe, and K. Fukui, *Bull. Chem. Soc. Jpn.*, **46**, 1071 (1973); (d) F. Bernardi, N. D. Epiotis, and R. L. Yates, *J. Am. Chem. Soc.*, **97**, 1334 (1975).

(12) In the D–D–A system the overlap integral between the initial configuration and the transferred configuration where an electron is shifted from the terminal D to A should be reduced to the overlap integral between the HOMO of the terminal D and the LUMO of A. In the A–A–D system the overlap between the initial configuration and the transferred configuration where an electron is shifted from D to the terminal A should be reduced to the overlap between the HOMO of D and the LUMO of the terminal A. In the D–A–D system the overlap between the transferred configurations where an electron is shifted from each D to A should be reduced to the overlap between the HOMO's of D's. In the A–D–A system the overlap between the transferred configurations where an electron is shifted from D to each A should be reduced to the overlap integral between the LUMO's of A's. These MO overlaps between the nonadjacent systems are assumed to be zero in the acyclic interaction. Accordingly, the above overlap integrals between the configurations are approximated as the products of the two MO overlaps involving the MO of the intermediary system.

(13) R. B. Woodward and R. Hoffmann, *Angew. Chem., Int. Ed. Engl.*, **8**, 781 (1969), and references cited therein.

(14) D. Hasselmann, *Tetrahedron Lett.*, 3465 (1972); 3739 (1973).

(15) J. A. Berson and J. M. Janusz, *J. Am. Chem. Soc.*, **96**, 5939 (1974).

(16) V. Mark, "Mechanisms of Molecular Migrations", Vol. 2, B. S. Thyagarajan, Ed., Interscience, New York, N.Y., 1969, p 320.

(17) G. A. Russell and G. J. Mikol, "Mechanisms of Molecular Migrations", Vol. 1, B. S. Thyagarajan, Ed., Interscience, New York, N.Y., 1968, p 157.

(18) For example, see: H. M. R. Hoffmann, *Angew. Chem.*, **81**, 597 (1969).
(19) R. M. Roberts, R. G. Landolt, R. N. Greene, and E. W. Heyer, *J. Am. Chem. Soc.*, **89**, 1404 (1967), and references cited therein.
(20) S. Inagaki, S. Yamabe, H. Fujimoto, and K. Fukui, *Bull. Chem. Soc. Jpn.*, **45**, 3510 (1972); S. Inagaki, T. Minato, S. Yamabe, H. Fujimoto, and K. Fukui, *Tetrahedron*, **30**, 2165 (1974); S. Inagaki and K. Fukui, *J. Am. Chem. Soc.*, **97**, 7480 (1975).
(21) J. E. Baldwin, G. Höfle, and S. C. Choi, *J. Am. Chem. Soc.*, **93**, 2810 (1971); J. E. Baldwin, A. K. Bhatnager, S. C. Choi, and T. J. Shortridge, *ibid.*, **93**, 4082 (1971).
(22) L. A. Paquette and J. C. Stowell, *J. Am. Chem. Soc.*, **93**, 2459 (1971).
(23) E. S. Gould, "Mechanism and Structure in Organic Chemistry", Henry Holt, New York, N.Y., 1960, p 295.
(24) (a) G. M. Whitesides and D. J. Boschetto, *J. Am. Chem. Soc.*, **93**, 1529 (1971); (b) S. E. Jacobson and A. Wojcicki, ibid., **95**, 6962 (1973).
(25) (a) H. E. Zimmerman and G. L. Grunewald, *J. Am. Chem. Soc.*, **88**, 183 (1966); (b) H. E. Zimmerman and A. C. Pratt, *ibid.*, **92**, 6267 (1970); (c) H. E. Zimmerman and G. E. Samuelson, *ibid.*, **89**, 5971 (1967); **91**, 5307 (1969); H. E. Zimmerman, P. Hackett, D. F. Juers, and B. Schroder, *ibid.*, **89**, 5973 (1967); **93**, 3653 (1971); (d) P. S. Mariano and J. K. Ko, *ibid.*, **94**, 1766 (1972); (e) H. E. Zimmerman, J. D. Robbins, R. D. McKelvey, C. J. Samuel, and L. R. Sousa, *ibid.*, **96**, 4630 (1974). (f) For a review of much of the di-π-methane rearrangement literature, see S. S. Hixon, P. S. Mariano, and H. E. Zimmerman, *Chem. Rev.*, **73**, 531 (1973).
(26) H. E. Zimmerman and R. D. Little, *J. Am. Chem. Soc.*, **94**, 8256 (1972); **96**, 5143 (1974).
(27) Reference cited in ref 13 where the two-part $\pi_s + \pi_s$ interaction scheme is adopted.
(28) (a) P. R. Story and M. Saunders, *J. Am. Chem. Soc.*, **84**, 4876 (1962); (b) R. K. Lustgarten, M. Brookhart, and S. Winstein, *ibid.*, **89**, 6350 (1967); M. Brookhart, R. K. Lustgarten, and S. Winstein, *ibid.*, **89**, 6352, 6354 (1967); R. K. Lustgarten, M. Brookhart, and S. Winstein, *ibid.*, **90**, 7364 (1968).
(29) S. Winstein, *Q. Rev., Chem. Soc.*, **23**, 141 (1969).
(30) S. Winstein and C. Ordronneau, *J. Am. Chem. Soc.*, **82**, 2084 (1960).
(31) The barrier to the bridge flipping in 7-norbornadienyl cation is greater than 19.6 kcal/mol at 45 °C. According to the corollary the displacement of the vinylene hydrogens by electron-donating substituents such as alkoxy or amino groups, and of the C_7 hydrogen by an electron-accepting substituent such as carbonyl or cyano groups, is expected to make the equilibrium geometry more distorted and to make the barrier higher.
(32) Either sign of the particular MO's can be chosen in principle without changing the physical meaning, as described in the theoretical section.
(33) S. Winstein, M. Shatavsky, C. Norton, and R. B. Woodward, *J. Am. Chem. Soc.*, **77**, 4183 (1955); S. Winstein and M. Shatavsky, *ibid.*, **78**, 592 (1956); S. Winstein, A. H. Lewin, and K. C. Pande, *ibid.*, **85**, 2324 (1963).
(34) (a) R. W. Hoffmann, R. Schüttler, W. Schäfer, and A. Schweig, *Angew. Chem.*, **84**, 533 (1972); (b) L. A. Paquette and M. J. Broadhurst, *J. Org. Chem.*, **38**, 1893 (1973).
(35) It would be interesting to examine experimentally the bridge flipping in **9** with highly electron-demanding substituents such as cyano or carbonyl groups at the methylene carbon and with highly electron-releasing substituents such as amino or alkoxy groups at the vinylene carbons.
(36) (a) S. W. Staley and D. W. Reichard, *J. Am. Chem. Soc.*, **91**, 3998 (1969); (b) M. V. Moncur and J. B. Grutzner, *ibid.*, **95**, 6449 (1973); (c) M. J. Goldstein and S. Natowsky, *ibid.*, **95**, 6451 (1973).
(37) (a) G. Wittig and E. Hahn, *Angew. Chem.*, **72**, 781 (1960); (b) R. A. Finnegan and R. S. McNees, *J. Org. Chem.*, **29**, 3234 (1964); (c) A. Streitwieser, Jr., and R. A. Caldwell, *ibid.*, **27**, 3360 (1962); (d) G. Wittig and J. Otten, *Tetrahedron Lett.*, 601 (1963); G. Wittig and G. Klumpp, *ibid.*, 607 (1963).
(38) R. B. Turner, *J. Am. Chem. Soc.*, **86**, 3586 (1964); cf. R. B. Turner, P. Goebel, B. J. Mallon, W. E. Doering, J. F. Coburn, Jr., and M. Pomerantz, *ibid.*, **90**, 4315 (1968).
(39) (a) P. Ahlberg, D. L. Harris, and S. Winstein, *J. Am. Chem. Soc.*, **92**, 4454 (1970); (b) J. C. Barborak, J. Daub, D. M. Follweiler, and P. v. R. Schleyer, *ibid.*, **91**, 7760 (1969); J. C. Barborak and P. v. R. Schleyer, *ibid.*, **92**, 3184 (1970); (c) J. B. Grutzner and S. Winstein, *ibid.*, **92**, 3186 (1970).
(40) L. M. Loew and C. F. Wilcox, *J. Am. Chem. Soc.*, **97**, 2296 (1975), and references cited therein.
(41) T. Kaneda, T. Ogawa, and S. Misumi, *Tetrahedron Lett.*, 3373 (1973).
(42) T. Inoue, T. Kaneda, and S. Misumi, *Tetrahedron Lett.*, 2969 (1974).
(43) E. L. Allred and J. C. Hinshaw, *Tetrahedron Lett.*, 1293 (1968).
(44) (a) T. Iwakuma, O. Yonemitsu, N. Kanamaru, K. Kimura, and B. Witkop, *Angew. Chem.*, **85**, 85 (1973); (b) T. Iwakuma, K. Hirao, and O. Yonemitsu, *J. Am. Chem. Soc.*, **96**, 2570 (1974). (c) The transannular 2 + 2 cycloaddition reaction between a C═C bond and one of the C–C σ bonds of cyclopropane ring has been observed to occur presumably via acylation on the carbonyl oxygen attached to the cyclopropane ring [E. J. Corey and R. D. Balanson, *Tetrahedron Lett.*, 3153 (1973)]. This reaction is taken as being "catalyzed" by acyl cation instead of proton.
(45) π LUMO of a carbonyl compound, acrolein, has been shown by a SCF calculation to be appreciably (~5 eV) lowered through the protonation on a lone pair of the carbonyl oxygen [K. Fukui and H. Fujimoto, *Nippon Kagakusenì Koen-shu*, **29**, 27 (1972)].
(46) R. C. Cookson and J. E. Kemp, *Chem. Commun.*, 385 (1971).
(47) Another electronic cause of symmetry-disfavored 1,3-sigmatropic shift in antibisallylic systems has recently been proposed [S. Inagaki, T. Minato, H. Fujimoto, and K. Fukui, *Chem. Lett.*, 89 (1976)]. The multicyclic interaction effect should be essentially distinguished both from the Berson–Salem's "subjacent orbital" effect and from the three-system interaction substituent effect described here.

6 BULLETIN OF THE CHEMICAL SOCIETY OF JAPAN, VOL. 48 (1), 6—12 (1975) [Vol. 48, No. 1

An MO-Theoretical Treatment of the Cationic Ring-Opening Polymerisation. I. Ethylene Oxide

Gernot FRENKING, Hiroshi KATO,* and Kenichi FUKUI

Department of Hydrocarbon Chemistry, Faculty of Engineering, Kyoto University, Sakyo-ku, Kyoto 606

**College of General Education, Nagoya University, Chikusa-ku, Nagoya 464*

(Received July 26, 1974)

The stereochemical course of the acid-catalyzed ring-opening polymerisation of ethylene oxide (e.o.) has been investigated on the basis of the CNDO/2 method. By the calculation of the total energy of the interacting system of e.o. and protonated e.o. the most favored reaction path is indicated. The activation energy is found to consist only of the desolvation process. The effect of the deformation of the molecular geometry upon the net charge, the total energy, and the shape of particular MO's is discussed. It is concluded that the ring strain does not influence the rate. By means of configuration analysis, the dominance of the HOMO–LUMO interaction in the later stage of the reaction has been established. The initiation and propagation reactions follow the same reaction mechanism.

Extensive experimental studies in the field of ring-opening polymerisation (ROP) have been made.[1] In contrast to numerous papers reporting those results, however, there have been very few theoretical treatments of this field.[2] The experimental results concerning the reaction path of the acid-catalyzed ROP of cyclic ethers can be summarized as follows:[3] The initiator is first formed by a reversible protonation:

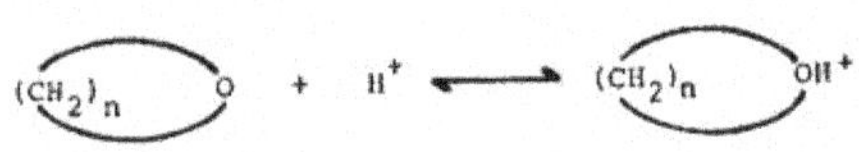

The initiation is considered to be the addition of a monomer to this protonated monomer. There is much evidence that the process of the ring-opening addition takes place as an S_N2-like backside attack. The propagation step is essentially constructed by a succession of similar processes at the terminal monomeric unit of a polymer chain:

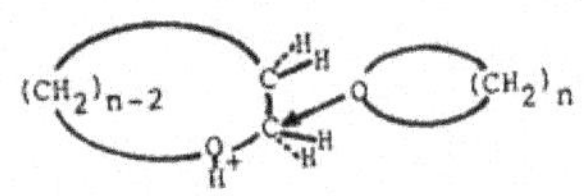

The backside attack is indicated by an inversion at the attacked carbon atom. However, still remains some doubt about the sequence of the bond interchange and the transition state. There has been discussed a "borderline S_N2-mechanism,"[4] where the old C–O bond is already partly broken at a rather early stage of the attack. Recent kinetic studies of the cationic ROP of some four-membered cyclic ethers show that the entropy of activation is the dominating factor in the reactivity.[5]

In the present and forthcoming papers, we want to report the results of our MO-theoretical treatment of various aspects of the reactivity and stereochemistry of heterocyclic-ring compounds in ROP. We start with ethylene oxide (e.o.), since it may serve as a simple model case for other, more complicated compounds.[6] Using the CNDO/2 method, we treat the isolated e.o. and its protonated compound, and also their interacting system. In order to determine the most favored geometries and reaction path, we examine the effect of changes in several geometrical parameters. Furthermore, utilizing the frontier theory[8] and configuration analysis,[9] the mechanisms of cationic ROP are investigated.

Calculations and Used Geometries

All our MO-calculations were based upon the CNDO/2 model of Pople *et al.*[7]

Figure 1(a) shows the energy levels of the MO's of e.o. and protonated e.o., together with the chosen geometry. Figure 1(b) shows the resulting net charges.

We calculated the total energy of the combined system of e.o. and protonated e.o. with different geometries. Figure 2 shows the geometries which gave the energy minimum at the fixed C–O′ distance for each of the frontside and backside attacks.[16] Further-

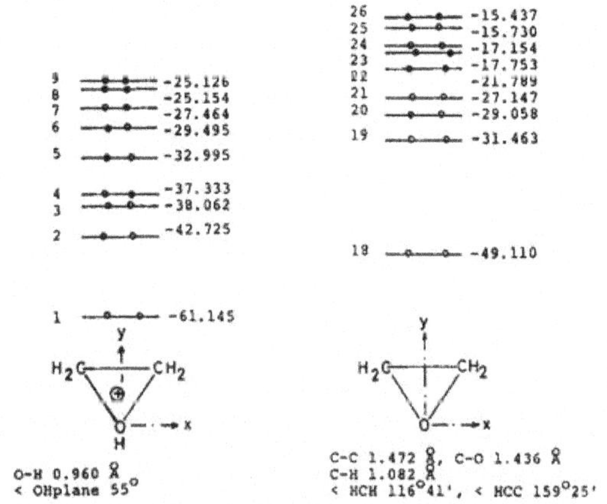

Fig. 1(a). MO energy levels and geometry of ethylene oxide and protonated ethylene oxide.

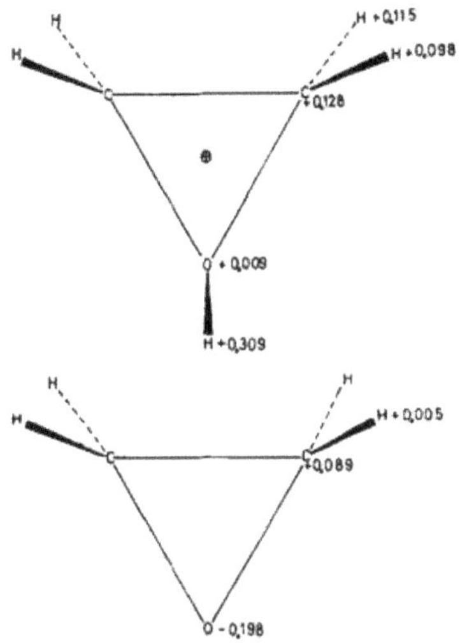

Fig. 1(b). Net charges of ethylene oxide and protonated ethylene oxide.

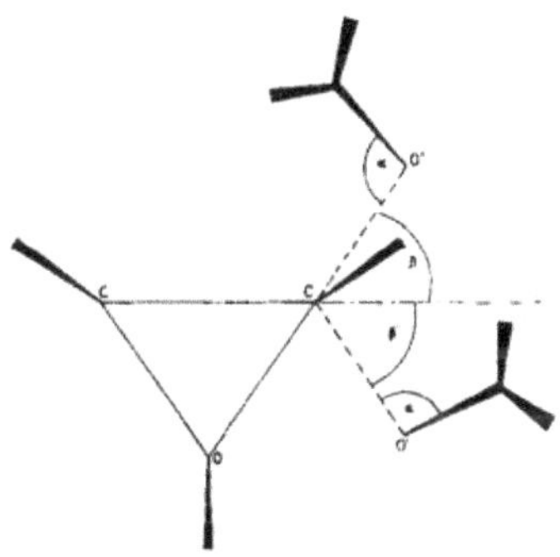

Fig. 2. Geometry of the molecules in the two attacking directions, the backside attack upper and the frontside attack lower.
$\alpha=110°$, $\beta=60°$, $\beta'=60°$.

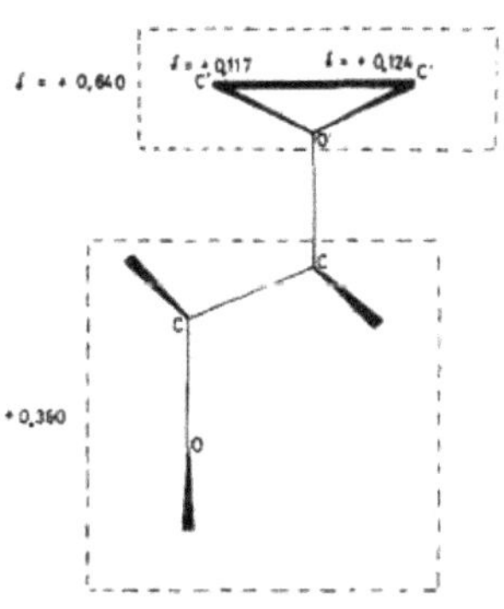

Fig. 3. Geometry of the final product of the initiation reaction and net charge distribution. The ring has the same geometry as e.o. (see Fig. 1).
O′–C 1.430 Å; C–C 1.540 Å; C–O 1.430 Å; C–H 1.090 Å; O–H 0.960 Å; <O′CC 110°; <HC plane =<OC plane 55°.

TABLE 1. VALUES OF TOTAL ENERGY, ENERGY BETWEEN THE ATTACKING OXYGEN (O′) AND THE ATTACKED CARBON (C), AND THE NET CHARGE TRANSFER WITH DIFFERENT GEOMETRIES

$r_{C-O'}$ [Å]	Stretching of r_{C-O} [Å]	Bending of C–H bond γ [degree]	E_{tot} [a.u.]	$E_{C-O'}$ [a.u.]	Transferred charge
Attack along the y-axis					
2.5	0	0	−71.8140	−0.013	0.002
$\beta=60°$ (backside attack)					
2.5	0	0	−71.8068	−0.012	0.005
		−30	−71.8100	−0.019	0.009
	0.2	0	−71.7529	−0.018	0.012
		−60	−71.7463	−0.032	0.019
2.0	0	0	−71.8114	−0.064	0.047
		−30	−71.8302	−0.104	0.061
		−60	−71.8017	−0.134	0.084
	0.2	0	−71.7709	−0.095	0.083
		−30	−71.8031	−0.140	0.095
		−60	−71.7962	−0.177	0.125
1.5	0	0	−71.7323	−0.341	0.246
		−30	−71.8262	−0.462	0.290
		−60	−71.8528	−0.548	0.352
	0.2	0	−71.7412	−0.434	0.354
		−30	−71.8445	−0.546	0.376
		−60	−71.8951	−0.632	0.424
$\beta'=60°$ (frontside attack)					
2.5	0	0	−71.8146	−0.019	0.009
	0.2	0	−71.7583	−0.027	0.012
		30	−71.7185	−0.026	0.014
2.0	0	0	−71.8245	−0.094	0.048
		30	−71.7973	−0.099	0.052
	0.2	0	−71.7880	−0.151	0.089
		30	−71.7561	−0.161	0.109
1.5	0	0	−71.7833	−0.493	0.250
		30	−71.7929	−0.538	0.285
	0.2	0	−71.8460	−0.645	0.408
		30	−71.8595	−0.697	0.451
final product of the initiation reaction					
			−71.9441	−0.890	0.640

more, the neutral e.o. has been placed in a planar and perpendicular position to the protonated e.o. plane. Among all of these calculations, the geometries adapted in Fig. 2 gave the lowest energy values with regard to the above-mentioned parameters (see Table 1).

In the next step, we examine the energy minimum as to the attacking direction with the two C–O′ distances of 2.0 and 2.5 Å. We also calculate the energy of the final product of this reaction of ROP. Its geometry is shown in Fig. 3, while the resulting energy is listed in Table 1.

Finally, we investigate the deformation of the ring of protonated e.o., since the geometry of the attacked protonated e.o. changes toward the geometry of the final product throughout the reaction. The considered deformations are indicated in Fig. 4, where the deformations of 1 and/or 4 are taken in the frontside attack course and those of 2, 3, and/or 4 for the backside attack.

The results are listed in Table 1. The table also

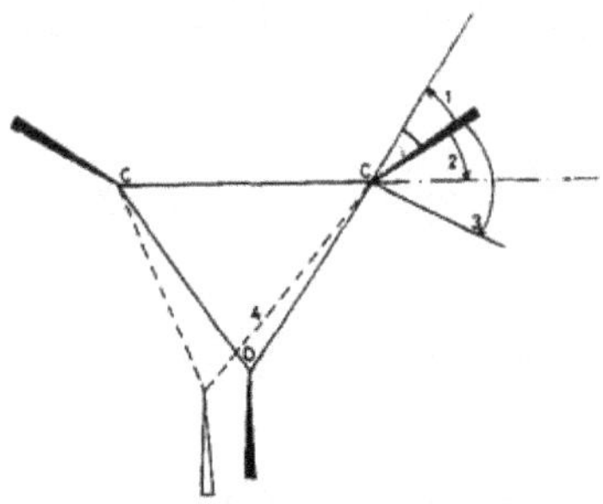

Fig. 4. Assumed deformations of the protonated ethylene oxide during the attack of the neutral molecule.
1: Hydrogens shifted up to $\gamma=30°$
2: Hydrogens shifted down to $\gamma=-30°$
3: Hydrogens shifted down to $\gamma=-60°$
4: One C–O bond length stretched by 0.2 Å.

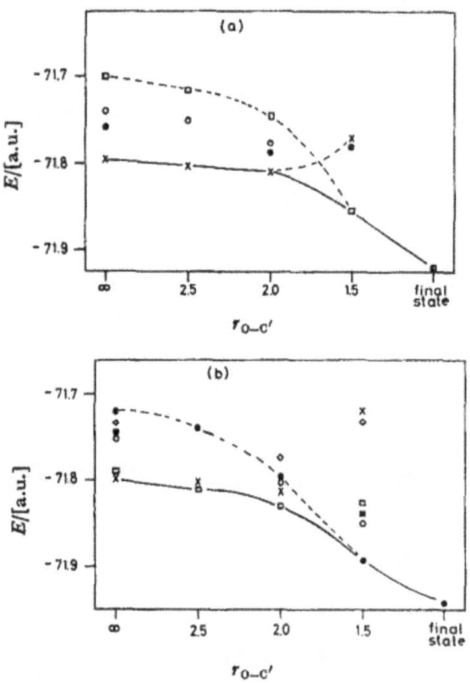

Fig. 5. The changes of energy levels along the two paths of attacking with the change of geometry indicated.
(a): Frontside attack
C–O bond length normal γ: 0° (×), γ: 30° (●)
C–O bond length +0.2 Å δ: 0° (○), δ: 30° (□)
(b): Backside attack
C–O bond length normal γ: 0° (×), γ: −30° (□), γ: −60° (○).
C–O bond length +0.2 Å γ: 0° (◇), γ: −30° (■), γ: −60° (●).

gives the values of the transferred charge from the neutral to the protonated compound and the values of the energy of the C–O′ bonding (E_{AB} in CNDO/2). Figures 5(a) and 5(b) show the total energies in relation to the C–O′ distance. The datum for $r=\infty$ is the simple sum of the values for the separated molecules.

Discussion

In spite of the use of many approximations, the results give interesting insights into the reaction path. Comparing the frontside and the backside attacks with the energy values shown in Fig. 5(a) and 5(b), it can be seen that there is not much difference with $r_{C-O'}$ values larger than 2.0 Å. After this stage, the backside attack is strongly favored from an energetic point of view.

It has been proposed in several papers by Fukui *et al.*[8] that "a majority of chemical reactions should take place at the position and in the direction of the maximum overlapping of the HOMO (Highest Occupied MO's) and the LUMO (Lowest Unoccupied MO's)." The HOMO of e.o., which is not shown here, is indeed mainly localized on the lone-pair oxygen p_z orbital (AO coefficient 0.88). There are two nearly degenerated LUMO's of the protonated e.o. (a' at -3.086 eV and a'' at -2.419 eV, with the point group C_s). They correspond to the a_1 and b_2 LUMO's of e.o. (point group C_{2v}). The consideration of both of them has previously been emphasized in the calculations of the anionic ROP of e.o.[10] The shapes of

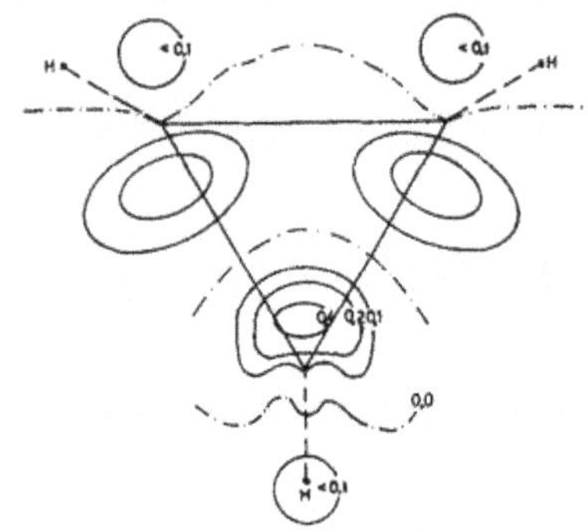

(a)

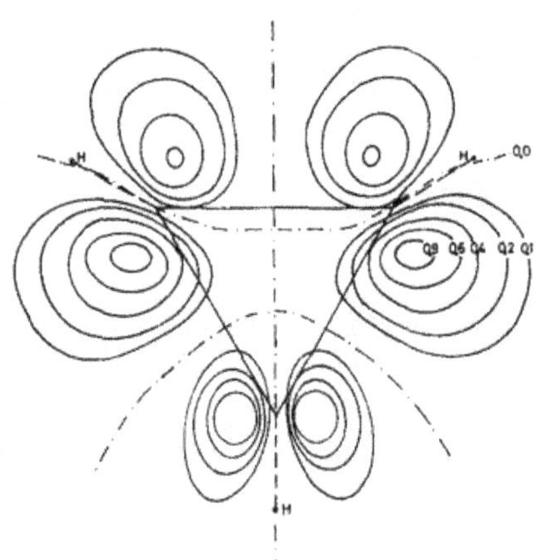

(b)

Fig. 6. The LUMO's of protonated e.o. The figures indicate the values of the real MO's in the molecular plane. The contour lines correspond to the density values indicated.
(a): a' LUMO (b): a'' LUMO

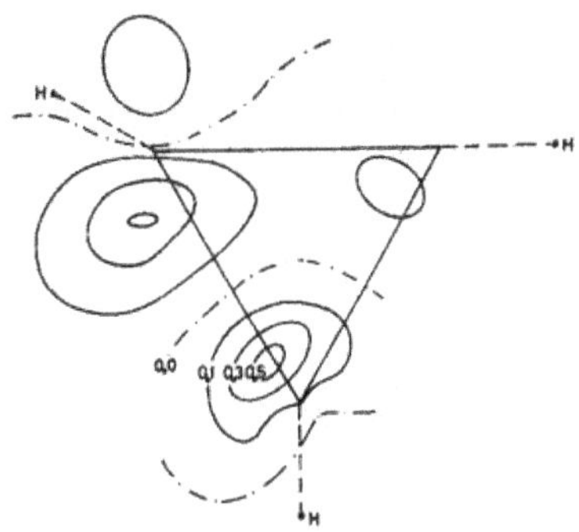

(a)

(a): *a*′ LUMO at normal C–O bond length, H-shift with $\gamma=-30°$.

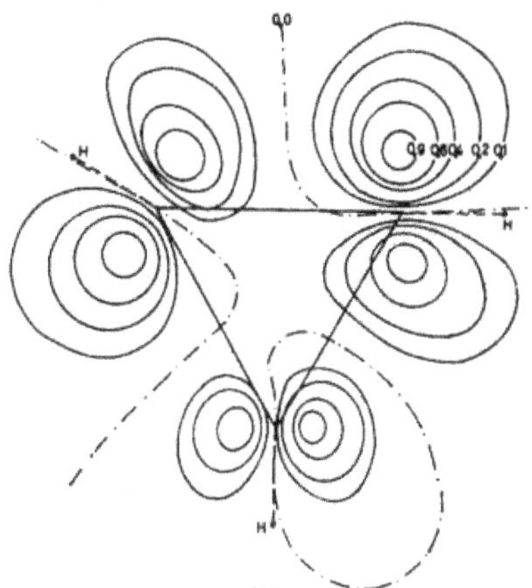

(b)

(b): *a*″ LUMO at normal C–O bond length, H-shift with $\gamma=-30°$.

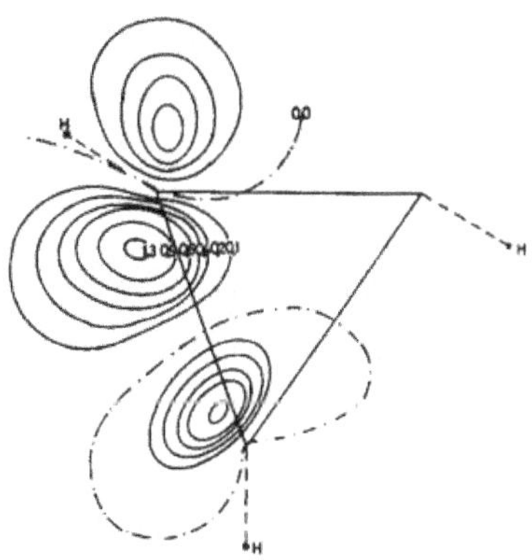

(c)

(c): *a*′ LUMO at C–O bond length +0.2 Å, H-shift with $\gamma=-60°$.

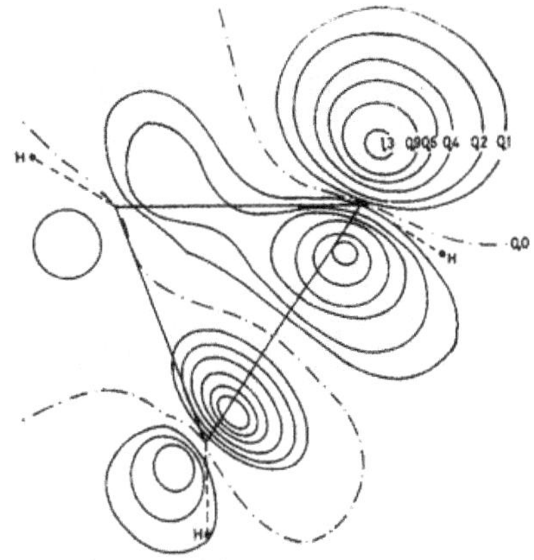

(d)

(d): *a*″ LUMO at C–O bond length +0.2 Å, H-shift with $\gamma=-60°$.

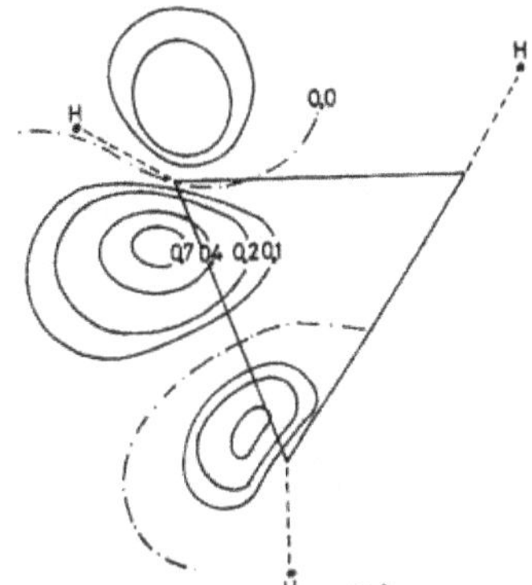

(e)

(e): *a*′ LUMO at C–O bond length +0.2 Å, H-shift with $\gamma=30°$.

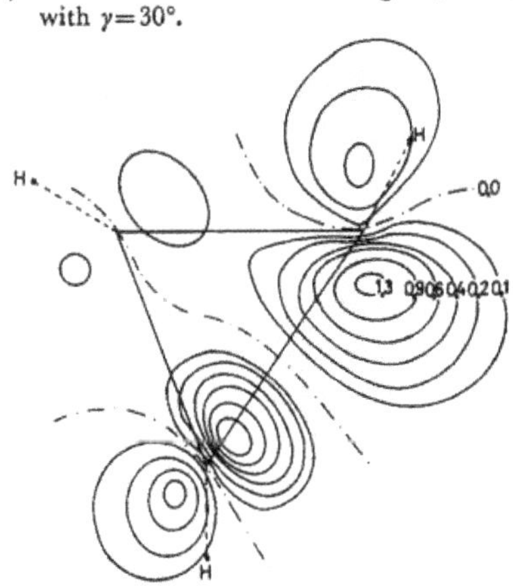

(f)

(f): *a*″ LUMO at C–O bond length ×0.2 Å, H-shift with $\gamma=30°$.

Fig. 7. Change of the two LUMO's under some deformations of the geometry of protonated e.o.

these LUMO's in the molecular plane are shown in Figs. 6(a) and 6(b). Both of them have a node between oxygen and carbon, but the *a*′ LUMO has its main extension in an antibonding region between oxygen and hydrogen. (In Fig. 6(a) the contour lines not on the molecular plane do not appear.) Therefore, it may be presumed that it is the *a*″ LUMO which plays the dominant role in the ring cleavage.

This LUMO mainly extends in the initial geometry at the frontside. However, it is noticed that the shapes and energies of these LUMO's may vary with the geometry. Figures 7(a)—(f) collect the shapes of the LUMO's with various geometries which give the local minima for the respective $r_{C-O'}$ values, as given in Table 1. With these geometries, the notations a' and a'' lose their exact meaning because of the loss of symmetry, but they are still kept for the sake of convenience. Figure 7(b) clearly indicates that the shift of the hydrogens, which is expected to occur at an earlier stage of the reaction, as is shown in Table 1 and Fig. 5, results in a reversal of the maximum extension of the a'' LUMO from the frontside to the backside direction, whereas the extension of the a' LUMO at the attacked carbon atom is reduced. This tendency is even more accelerated with the deformation of the ring frame and a further hydrogen shift, as is shown in Figs. 7(c) and 7(d). Figures 7(e) and 7(f) show a similar trend for the frontside reaction.

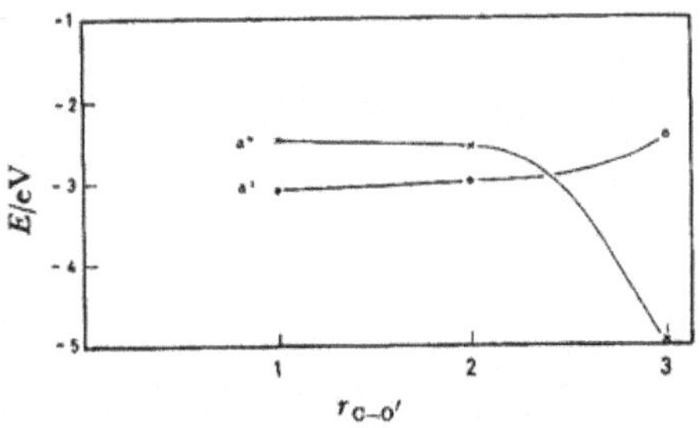

Fig. 8. Change of the energy level of the two LUMO's of protonated e.o. with some deformations which correspond to the reaction path of the backside attack (see Fig. 5(b)).
1: initial geometry, 2: C–O bond length normal, H-shift γ $-30°$, 3: C–O bond length $+0.2$ Å, H-shift γ $-60°$.

That it is the a'' LUMO and not the a' LUMO which is dominating in the HOMO–LUMO interaction can also be seen from the change in the energy level along with the change in the geometry, as is indicated for the backside attack in Fig. 8. The narrowing of the frontier orbitals has already been pointed out by Fukui *et al.*[8)]

Some additional calculations have been carried out for the backside attack. We calculated the composed system by placing the neutral e.o. at a $r_{C-O'}$ distance of 2.5 Å along the y-axis in a planar arrangement to the protonated compound, as the Coulomb attraction should be the dominating force in this case. The numerical results, given in Table 1, show a greater stabilisation than the corresponding value for the assumed backside path. Furthermore, we varied the geometry parameters with a $r_{C-O'}$ distance of 1.5 Å. The values for β (see Fig. 2) have been taken as 75 and 45°, and the hydrogen shift with $\gamma=-45°$ (see Fig. 4). However, all these calculations gave no further stabilisation compared to the values for $\beta=60°$ and $\gamma=-60°$ (see Table 1). As our results for the principal reaction mechanism are independent of the exact location, we did not make any further calculations concerning the exact reaction path.

Additional information is given by the results of the configuration analysis[9,11)] listed in Table 2, where some absolute values of coefficients for the ground state, monotransferred (delocalized), and intramolecular monoexcited (polarized) configurations are presented. The large value of the coefficient of the ground-state configuration at the early stage of the reaction indicates the dominance of the Coulomb interaction. However, the dramatically increasing value of the monotransferred configuration at $r_{C-O'}=1.5$ Å, especially that corresponding to the transfer from of e.o. the HOMO to the LUMO of protonated e.o.,[12)] shows the importance of this kind of interaction at the final stage of the reaction.

From Table 1 it can be seen that the values for $E_{C-O'}$ and the quantities of charge transfer show no remarkable parallelism with the total energy, nor is there any interdependence between them. Roughly speaking, both values increase with the lowering of the total energy. It is interesting that, with the distance $r_{C-O'}$ of 1.5 Å, the backside attack shows a significantly lower total energy, but smaller $E_{C-O'}$ and transferred charge, than the frontside attack.

It is worthwhile to examine the charge transfer in more detail. Table 3 shows the change in the net charge of each atom at the different reaction stages for the local minimum energy (see Fig. 5). The numbers represent the differences from the net charge calculated for the isolated molecules with the respective geometry, where the + sign refers to the decrease in the atomic electron density caused by the interaction, and the − sign, to its increase. In contradiction to the whole protonated e.o. molecule coming more negative (less positive), the attacked carbon atom, 2, be-

Table 2. Absolute values of the coefficients for the ground state, some monotransferred and monoexcited configurations along the backside and frontside paths
Only the one or two largest values are indicated

Attacking-direction	$r_{C-O'}$ [Å]	Ψ_0	Coefficients for monotransferred configurations	Coefficients for monoexcited configurations
Backside ($\beta=60°$)	2.5	0.999	$0.06(\Psi_{26\to11})$; $0.03(\Psi_{21\to11})$	$0.02(\Psi_{24\to27})$
	2.0	0.960	$0.17(\Psi_{36\to11})$; $0.08(\Psi_{21\to11})$	$0.03(\Psi_{6\to11})$; $0.03(\Psi_{23\to27})$
	1.5	0.690	$0.39(\Psi_{26\to10})$; $0.18(\Psi_{21\to10})$	$0.19(\Psi_{8\to10})$; $0.13(\Psi_{5\to10})$
Frontside ($\beta'=60°$)	2.5	0.992	$0.04(\Psi_{36\to11})$; $0.04(\Psi_{26\to10})$	$0.03(\Psi_{23\to27})$
	2.0	0.959	$0.13(\Psi_{36\to11})$; $0.07(\Psi_{26\to10})$	$0.04(\Psi_{23\to27})$
	1.5	0.598	$0.39(\Psi_{36\to10})$; $0.18(\Psi_{21\to10})$	$0.33(\Psi_{9\to10})$; $0.14(\Psi_{5\to10})$

TABLE 3. DIFFERENCE OF THE NET CHARGES AT EACH ATOM BETWEEN THE COMBINED SYSTEM AND ISOLATED MOLECULES WITH THE RESPECTIVE GEOMETRY. THE GEOMETRIES ARE THOSE ONES WHICH GAVE MINIMA IN THE ENERGY CALCULATIONS (see Table 1)

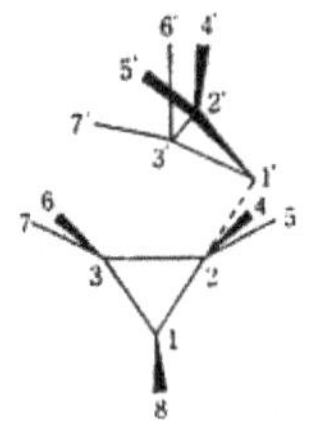

Atom	$r_{C-O'}=2.5Å$	$r_{C-O'}=2.0Å$	$r_{C-O'}=1.5Å$
1	−0.007	−0.029	−0.128
2	+0.011	+0.042	+0.084
3	−0.004	−0.021	−0.094
4	+0.003	−0.008	−0.064
5	+0.002	−0.010	−0.066
6	−0.004	−0.011	−0.046
7	−0.004	−0.010	−0.043
8	−0.003	−0.013	−0.073
1′	−0.027	−0.018	+0.132
2′	+0.003	+0.006	+0.021
3′	+0.003	+0.005	+0.020
4′	+0.030	+0.042	+0.079
5′	−0.014	−0.007	+0.046
6′	+0.030	+0.042	+0.079
7′	−0.015	−0.008	+0.045

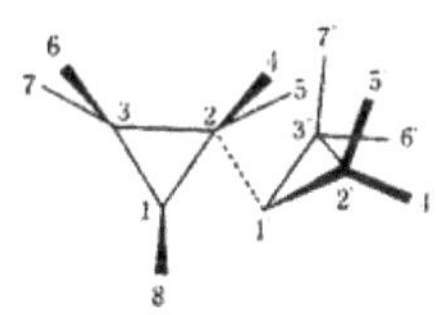

Atom	$r_{C-O'}=2.5Å$	$r_{C-O'}=2.0Å$	$r_{C-O'}=1.5Å$
1	+0.003	−0.014	−0.173
2	+0.014	+0.061	+0.184
3	−0.010	−0.037	−0.102
4	−0.006	−0.017	−0.084
5	−0.007	−0.019	−0.056
6	−0.004	−0.009	−0.035
7	−0.005	−0.011	−0.100
8	+0.005	−0.001	−0.086
1′	−0.058	−0.053	+0.160
2′	+0.009	+0.011	+0.016
3′	+0.001	0.0	+0.008
4′	+0.030	+0.039	+0.080
5′	−0.001	+0.007	+0.055
6′	+0.030	+0.038	+0.078
7′	−0.003	+0.006	+0.054

comes more positive. Similarly, the oxygen atom, 1′, of the neutral molecule is still at $r_{C-O'}=2.0$ Å, more negative than in the isolated molecule, before it becomes more positive at $r_{C-O'}=1.5$ Å.

The potential curve of the reaction path anticipated from Fig. 5 indicates the absence of an energy barrier. This can easily be understood in view of the important role played by the solvent in ionic reactions. It has been found that this is particularly important for the reactivity of cyclic ethers in ROP.[13] Polar solvents stabilizing the monomers much better bring about a remarkable decrease in the reactivity of the monomers in ROP, compared with less polar solvents. Our calculations indicate that the activation energy originates only from the desolvation process. Therefore, the location of the transition state along the reaction path of Fig. 5(a) remains rather uncertain, however, the facts that the deformation of the ring frame occurs rather late ($r_{C-O'}<2.0$ Å) and is accompanied by a rapid decrease in the total energy suggest that the transition state is placed somewhere earlier, with a geometry where the ring frame of the protonated e.o. is not yet deformed. If this is true, it follows that, in kinetically-controlled reactions where the activation energy determines the reaction rate, the ring strain is not influential on the rate. This has already been pointed out by Saegusa *et al.*[14] in reporting on their kinetic studies of 4-, 5-, and 7-membered cyclic ethers in cationic ROP. They found that the activation entropy is the most important factor in determining the reactivity.

There remains one problem. So far we have dealt only with the initiation step, but it has been stressed that the propagation step is quite different in its reaction rate, the latter being faster than the former.[15] The difference is as follows: in the propagation step, e.o. attacks an oxyethylated e.o. cation instead of a protonated one, and the length of the substituent chain is supposed to be not very important in connection with the reaction rate. Accordingly, we calculated the LUMO of the product of the initiation step. Figure 9 shows its shape in the ring plane. As the AO coefficients of this MO at atoms not placed in the ring plane are smaller than 0.03, it can be thought that the charges are located only in this plane. A comparison with the a'' LUMO of the protonated e.o. (Fig. 6(b)) shows a great similarity. In the case of alkylated e.o., however, the LUMO is not degenerated, being the lowest one not undergoing deformation due to the interaction. Also, the charge distributions (Fig. 3) of

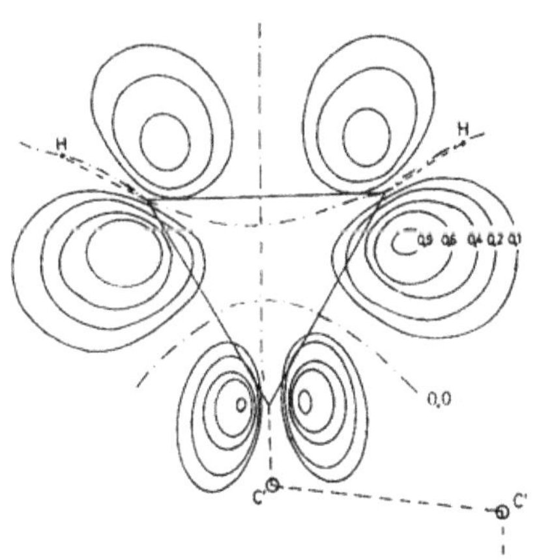

Fig. 9. The LUMO of the final product of the initiation reaction with the geometry of Fig. 3, indicating the values in the ring plane.

the atoms located at the ring are nearly the same as those of the protonated e.o. (Fig. 1(b)). From those facts, it can be expected that the mechanisms of the propagation reaction is not very different from the initiation.

The question is left why the rate of the propagation reaction is higher than the initiation reaction. The orbital energies of the LUMO's of the respective cations lead us to the opposite results. The LUMO of protonated e.o. (−3.086 eV) is lower than the LUMO of the oxyethylated e.o. (−1.629 eV). However, our results have shown that the bond breaking is more important than the bond making. The E_{AB} values for the C–O bond which is going to break during the reaction indicate a weaker bond for the oxyethylated e.o. (−0.816 eV) than for the protonated e.o. (−0.829 eV). This gives an explanation of the difference in the reaction rate.

From these calculations, the stereochemical course can be summarized as follows: the attack of the e.o. starts along the y-axis and is, in the beginning, governed by the Coulomb attraction only. The approach toward one carbon atom is accompanied by a shift of the connected hydrogens at a rather early stage of the reaction and produces a change in the *a''* LUMO extension towards the attacking oxygen. This in turn accelerates the reaction by an increase in the charge transfer resulting from better HOMO–LUMO interaction. The deformation of the ring frame at a later stage accelerates this process still more and causes remarkable decreases in the total energy and also the C_2–O_1 bonding energy and increases in the amount of charge transfer and the C–O′ bonding energy. In this way, the deformation of the protonated compound leads to a reaction path with a lower energy for the backside attack than for the frontside attack.

One of the authors (G.F.) wishes to express his gratitude to Professor Takeo Saegusa and Dr. Shiro Kobayashi for a valuable discussion; he is also indebted to Dr. Hiroshi Fujimoto, Mr. Satoshi Inagaki, Mr. Tsutomu Minato, and Mr. Shigeki Kato for making avalaible the computer programs.

All three authors wish to express their appreciation to the Data Processing Center of Kyoto University for the generous permission to use the FACOM 230-60 computer there.

Thank is given to the Japanese Minister of Education and the Deutscher Akademischer Austauschdienst for giving grant to one of the authors (G.F.).

References

1) a) R. E. Parker and N. S. Isaacs, *Chem. Rev.*, **59**, 737 (1959); b) Y. Ishii and S. Sakai, "Ring Opening Polymerisation," ed. by K. C. Frisch and S. L. Reegen, Marcel Dekker, (1969), pp. 13—109.

2) K. Fukui, H. Kato, T. Yonezawa, and S. Okamura, This Bulletin, **37**, 904 (1964).

3) See Ref. 1b), p. 16.

4) See Ref. 1a), p. 746.

5) T. Saegusa and S. Kobayashi, *Progr. Polym. Sci.*, **6**, 107 (1973).

6) a) J. Koskikallio and E. Whalley, *Trans. Faraday. Soc.*, **59**, 815 (1959); b) C. C. Price and R. Spector, *J. Amer. Chem. Soc.*, **88**, 4171 (1966).

7) J. A. Pople, D. P. Santry, and G. A. Segal, *J. Chem. Phys.*, **43**, 129 (1965).

8) See *e.g.* K. Fukui, *Fortschr. Chem. Forsch.*, **15**, 1 (1970) and literatures cited therein.

9) See *e.g.* H. Baba, S. Suzuki, and T. Takemura, *J. Chem. Phys.*, **50**, 2028 (1969); H. Fujimoto, S. Kato, S. Yamabe, and K. Fukui, *ibid.*, **60**, 572 (1974).

10) H. Fujimoto, M. Katata, S. Yamabe, and K. Fukui, This Bulletin, **45**, 1320 (1971); note the choice of a different principal axis in the determination of the symmetry group.

11) S. Yamabe, T. Minato, H. Fujimoto, and K. Fukui, *Theor. Chim. Acta*, **32**, 187 (1974).

12) The reader is reminded that the succession of the orbitals denoted for the initial geometry in Fig. 1(a) changes in part because of the change in geometry. Therefore, the *a''* LUMO of the protonated e.o., which plays the dominating part in the HOMO–LUMO interaction, is (in the succession of Fig. 1(a)) in the beginning the 11, but later the 10. See also Fig. 8.

13) See Ref. 5, p. 114, Table 1.

14) See Ref. 5, p. 145, Table 8.

15) See Ref. 5, p. 129.

16) The calculations where geometrical data are not listed have been carried out for the C–O′ distance of 1.5 Å and for the values for β=30, 60, 90°; β'=30, 60, and 90°, where β refers to the angle of backside attack, and β', to the frontside attack, as is indicated in Fig. 2.

June, 1977] BULLETIN OF THE CHEMICAL SOCIETY OF JAPAN, VOL. 50 (6), 1391—1398 (1977) 1391

A Theoretical Study on Biradicals. I. Theoretical Characteristics of Biradicals

Kenichi FUKUI and Kazuyoshi TANAKA

Department of Hydrocarbon Chemistry, Faculty of Engineering, Kyoto University, Sakyo-ku, Kyoto 606

(Received October 28, 1976)

An interpretation is given on the characteristics of biradicaloids, such as the bonding, polar, and biradical characters of singlet biradicals. Systematization of the reactivity of singlet and triplet biradicals is based on the principle that 1) deformation takes place in the direction of bonding in singlet biradicals, and 2) deformation or bond formation occurs to separate unpaired-electrons from each other in triplet biradicals. Combination of the two principles is applied to the theory of orientation and stereoselection in excited-state reactions.

The word "biradical" in a chemical sense makes us think of a species having an odd electron on each of two sites of a molecule. This conforms to the interpretation of a biradical in the triplet state. However, problems may arise in the case of a singlet biradical.

The conversion of biradical species along the path of ground- and excited-state reactions has been treated theoretically.[1-4] Salem *et al.* discussed the nature of a singlet biradical by taking ionic states into account and explained a number of experimental results in regard to various reaction intermediates in photochemistry.[5-10] The present paper gives some new material.

Stabilization of Singlet Biradical by a Correlated Motion of Electrons

Consider a pair of two independent normalized one-electron space functions a and b whose overlap integral is s. In the case of biradical problems these two orbitals can be made to represent the two "unpaired"-electron orbitals essentially localized at each radical site. The following three configurations are here taken into consideration to treat the problem as a two-electron system:

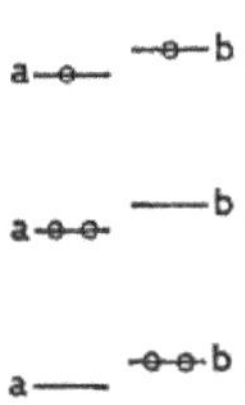

$$\left.\begin{aligned} \Phi_1(1,2) &= \frac{1}{\sqrt{2+2s^2}}\{a(1)b(2)+b(1)a(2)\} \\ \Phi_2(1,2) &= a(1)a(2) \\ \Phi_3(1,2) &= b(1)b(2) \end{aligned}\right\} \quad (1)$$

It is assumed that orbitals a and b are real and their signs are chosen in such a way that s becomes positive.

The electronic states of this two-electron system are then given by

$$\Psi = \frac{1}{\sqrt{N}}(C_1\Phi_1+C_2\Phi_2+C_3\Phi_3), \quad (2)$$

where C_1, C_2, and C_3 can be obtained by solving the eigenvalue problem of the Hamiltonian matrix, and $1/\sqrt{N}$ is the normalization factor. They are the usual energy-extremized wave functions in which the effect of correlation in electron motion is taken into account.

Direct consideration of the correlated motion of electrons might also be considered. The extent of correlation in orbital motion could roughly be represented by the averaged reciprocal interelectronic distance,

$$\overline{\frac{1}{r_{12}}} = \frac{2}{\nu(\nu-1)}\iint\rho(1,2|1,2)\frac{1}{r_{12}}dv(1)dv(2), \quad (3)$$

(ν: number of electrons)

where r_{12} is the distance of electrons 1 and 2, and

$$\rho(1',2'|1,2) = \frac{\nu(\nu-1)}{2}\int\Psi^*(r_1'\sigma_1, r_2'\sigma_2, r_3\sigma_3, \cdots) \times\Psi(r_1\sigma_1, r_2\sigma_2, r_3\sigma_3, \cdots)d\sigma_1 d\sigma_2 dr_3 d\sigma_3\cdots \quad (4)$$

is the second-order spinless density matrix. For the present two-electron problem, we have

$$\gamma \equiv \overline{\frac{1}{r_{12}}} = \iint\Psi^*(1,2)\frac{1}{r_{12}}\Psi(1,2)dv(1)dv(2). \quad (5)$$

The extremization of $\overline{1/r_{12}}$ is achieved by obtaining the stationary values satisfying

$$\delta\gamma = 0. \quad (6)$$

The variation with respect to C_1, C_2, and C_3 leads to the secular equation

$$\begin{vmatrix} \gamma_{11}-\gamma & \gamma_{12}-S_{12}\gamma & \gamma_{13}-S_{13}\gamma \\ \gamma_{21}-S_{21}\gamma & \gamma_{22}-\gamma & \gamma_{23}-S_{23}\gamma \\ \gamma_{31}-S_{31}\gamma & \gamma_{32}-S_{32}\gamma & \gamma_{33}-\gamma \end{vmatrix} = 0, \quad (7)$$

where

$$\gamma_{jk} = \iint\Phi_j^*(1,2)\frac{1}{r_{12}}\Phi_k(1,2)dv(1)dv(2)$$

and

$$S_{jk} = \iint\Phi_j^*(1,2)\Phi_k(1,2)dv(1)dv(2)$$

$(j, k=1, 2, \text{and } 3)$.

The coefficient of Φ_j in Eq. 2 can be obtained simultaneously, giving correlation-extremized wave functions $\{\Psi^{(i)}\}$ which are always mutually orthogonal. Both energy-extremization and correlation-extremization give wave functions with an essentially parallel trend in nonpolar species.[11] Thus in such favourable cases, we can use correlation-extremized wave functions as an approximate substitute of the usual energy-extremized functions. Such a consideration of the interaction of at least three configurations as mentioned above is essential in the theoretical interpretation of biradicals.

Bonding Character between Two Radical Sites

The bonding character between two radical sites each containing essentially one electron can be discussed by using Ψ of Eq. 2. A most reasonable scale of bonding character[12] might be the magnitude of accumulation of electron population in the intermediate region between two radical sites.[13-15]

The distribution of electron density is given by

$$\rho(1) = 2\int \Psi^*(1,2)\Psi(1,2)dv(2). \tag{8}$$

The density $\rho(1)$ is divided into three terms as

$$\rho(1) = A_{11}a(1)^2 + 2A_{12}a(1)b(1) + A_{22}b(1)^2. \tag{9}$$

The bonding strength can be determined by the second term of r.h.s. since the coefficient A_{12} of the cross term $a(1)b(1)$ contributes to the accumulation of electrons in the intermediate region.

It follows from Eq. 2 that

$$A_{12} = \frac{1}{N}\left\{\frac{s}{1+s^2}C_1^2 + \sqrt{\frac{2}{1+s^2}}C_1(C_2+C_3) + 2sC_2C_3\right\}, \tag{10}$$

where

$$N = C_1^2 + C_2^2 + C_3^2 + \frac{2\sqrt{2}\,s}{\sqrt{1+s^2}}C_1(C_2+C_3) + 2s^2C_2C_3.$$

In general, C_j may be complex. However, for the sake of simplicity, Eq. 9 is written for the real values of C_j. If we allow C_1, C_2, and C_3 of Eq. 2 to take *any* real values, the maximum and minimum values which A_{12} can take are $1/(1+s)$ and $-1/(1-s)$ corresponding to the wave functions

$$\left.\begin{aligned} \Psi^{(B)} &= \frac{1}{\sqrt{2}}\frac{1}{1+s}\left\{\sqrt{1+s^2}\Phi_1 + \frac{1}{\sqrt{2}}(\Phi_2+\Phi_3)\right\} \\ \text{and} \quad \Psi^{(AB)} &= \frac{1}{\sqrt{2}}\frac{1}{1-s}\left\{\sqrt{1+s^2}\Phi_1 - \frac{1}{\sqrt{2}}(\Phi_2+\Phi_3)\right\}, \end{aligned}\right\} \tag{11}$$

respectively. Thus, we see that the most bonding state ($\Psi^{(B)}$) or the most antibonding state ($\Psi^{(AB)}$) is obtained in the case of *moderate* mixing of three configurations Φ_1, Φ_2, and Φ_3. This implies that admixture of the ionic configuration Φ_2 or Φ_3 with the covalent configuration Φ_1, or the delocalization of each odd electron to the other radical site, serves as an effective origin of bonding character between two weakly interacting radical sites. The importance of such electron delocalization in the bond formation between molecules or radicals was early pointed out.[16-19] In particular, special importance of a cross term arising from two configurations, corresponding in the present case to the terms of C_1C_2 or C_1C_3 in A_{12}, was stressed.[20-22]

The bonding character of a singlet biradical is thus represented by sA_{12}. In the light of the relation

$$A_{11} + 2sA_{12} + A_{22} = 2,$$

we can also measure the less bonding character by the quantity $1/2(A_{11}+A_{22})$.

Polar Character of Two Radical Sites

The density $\rho(1)$ also provides information as to what extent the two radical sites are polar. Evidently, polarity parallels $|A_{11}-A_{22}|$. We tentatively define the polar character of a singlet biradical by means of

$$\Pi = \left|\frac{A_{11}-A_{22}}{A_{11}+A_{22}}\right|, \tag{12}$$

in which

$$\left.\begin{aligned} A_{11} &= \frac{1}{N}\left(\frac{1}{1+s^2}C_1^2 + \frac{2\sqrt{2}\,s}{\sqrt{1+s^2}}C_1C_2 + 2C_2^2\right), \\ A_{22} &= \frac{1}{N}\left(\frac{1}{1+s^2}C_1^2 + \frac{2\sqrt{2}\,s}{\sqrt{1+s^2}}C_1C_3 + 2C_3^2\right). \end{aligned}\right\} \tag{13}$$

These are the coefficients of the $a(1)^2$ and $b(1)^2$ terms in $\rho(1)$ of Eq. 9 which is derived from a wave function Ψ already *correlation-extremized* according to the procedure in previous section.

By this definition, we have

$$\Pi(\Phi_1) = 0, \quad \Pi(\Phi_2) = \Pi(\Phi_3) = 1,$$

provided that Φ_1, Φ_2, and Φ_3 are correlation-extremized states.

A change in electron distribution can be expected to arise causing bonding stabilization through a moderate mixing of ionic structures. In this connection the work of Wulfman and Kumei,[23] who first pointed out the high possibility of polarization in singlet excited states of alkenes, is of interest.

Biradical Character of Singlet Biradicals

Since the bonding character between two radical sites becomes maximum only in the case of moderate admixing of ionic configurations with a covalent configuration, the less ionic character can not be adopted for the measure of biradical character.

We must take into account the following requisites which qualify a biradical:

1) The bonding character between two radical ends should be small since a species with too large bonding character might be called a molecule, but not a biradical. The value $1/2(A_{11}+A_{22})$ should not be small.

2) The polar character should not be large since a species with a strong inequality in electron density at the radical ends should be called a zwitterion. Namely, $(1-\Pi)$ should not be small.

A conventional but reasonable definition of the biradical character $\mathfrak{B}$ may therefore be given by

$$\begin{aligned} \mathfrak{B} &= \frac{1}{2}(A_{11}+A_{22})(1-\Pi) = \frac{1}{2}\{A_{11}+A_{22}-|A_{11}-A_{22}|\} \\ &= (\text{the smaller of } A_{11} \text{ and } A_{22}). \end{aligned} \tag{14}$$

Thus, the maximum possible biradical character in a *bonding* state is unity $(=1/(1+s))$, and the maximum biradical character in an *antibonding* state is larger than unity $(=1/(1-s))$. We also have

$$\mathfrak{B}(\Phi_1) = \frac{1}{1+s^2}, \quad \mathfrak{B}(\Phi_2) = 0, \quad \mathfrak{B}(\Phi_3) = 0,$$

in which Φ_1, Φ_2, and Φ_3 are the functions of Eq. 1, assuming that they are correlation-extremized ones.

The discussion can be easily extended to the case of complex wave functions.

Nonbonding Biradicals

It may happen that each odd electron occupies each of two orbitals orthogonal to each other so that the overlap integral s in Eq. 9 disappears. Hence, the

$\Psi^{(i)}$			γ	
$\frac{1}{\sqrt{2}}\{a(1)a(2)+b(1)b(2)\}$	$^1A_1$	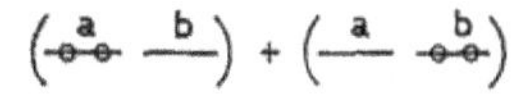	$(aa\|aa)+(ab\|ab)$	≈ 0.89120(a.u.)
$\frac{1}{\sqrt{2}}\{a(1)a(2)-b(1)b(2)\}$	$^1B_2$	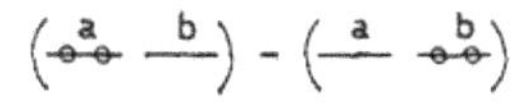	$(aa\|aa)-(ab\|ab)$	≈ 0.88969
$\frac{1}{\sqrt{2}}\{a(1)b(2)+b(1)a(2)\}$	$^1B_1$	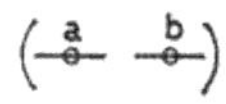	$(aa\|bb)+(ab\|ab)$	≈ 0.39420
$\left(\frac{1}{\sqrt{2}}\{a(1)b(2)-b(1)a(2)\}\right.$	$^3A_2$	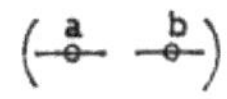	$(aa\|bb)-(ab\|ab)$	$\left.\approx 0.39268\right)$

maximal biradical character is unity.

An example is perpendicular ethylene in which two orthogonal radical orbitals a and b are $(2p_x)_A$ and $(2p_y)_B$, respectively, A and B denoting two carbon atoms. In this case, it follows from symmetry relations that

$$\gamma_{12}=\gamma_{13}=0,\quad S_{12}=S_{13}=0$$

in Eq. 7. Thus, $\Phi_1=1/\sqrt{2}\{a(1)b(2)+b(1)a(2)\}$ is already a correlation-extremized wave function with which Φ_2 or Φ_3 of different symmetry does not mix. From the definition of Eq. 14, the state $\Psi^{(1)}=\Phi_1$ has the $\mathfrak{B}$ value of unity and can be called "purely" biradical. The other two states ($\Psi^{(2)}=1/\sqrt{2}(\Phi_2+\Phi_3)$ and $\Psi^{(3)}=1/\sqrt{2}(\Phi_2-\Phi_3)$) also have the $\mathfrak{B}$ value of unity. Such a biradical species may be called a *nonbonding biradical.*

The level situation of three correlation-extremized states is given above, together with the value of γ. The result is qualitatively consistent with that of numerical calculations.[24]

It is to be noted that the level gap between $^1B_1$ and $^3A_2$ states is very small (*ca.* 1 kcal/mol) in this example. The small S–T separation is characteristic of nonbonding biradicals. From an energetic point of view, they are expected to interchange easily between singlet and triplet states.

Biradical with Cyclic Orbitals

Singlet Oxygen. Let two ($^1\pi_g$) orbitals of the oxygen molecule be

$$\phi_x=\frac{1}{\sqrt{2-2s}}\{(2p_x)_1-(2p_x)_2\},$$

$$\phi_y=\frac{1}{\sqrt{2-2s}}\{(2p_y)_1-(2p_y)_2\},$$

where O_1-O_2 axis is parallel to the z-axis and s is the overlap integral of $(2p_x)_1$ and $(2p_x)_2$ or that of $(2p_y)_1$ and $(2p_y)_2$. Three configurations are specified by

$$\Phi_1=\frac{1}{\sqrt{2}}\{\pi^+(1)\pi^-(2)+\pi^-(1)\pi^+(2)\},$$

$$\Phi_2=\pi^+(1)\pi^+(2),$$

$$\Phi_3=\pi^-(1)\pi^-(2),$$

where

$$\pi^+=\frac{1}{\sqrt{2}}(\phi_x+i\phi_y),$$

$$\pi^-=\frac{1}{\sqrt{2}}(\phi_x-i\phi_y).$$

Φ_1, Φ_2, and Φ_3 of different symmetries are correlation-extremized wave functions corresponding to the stationary values of γ.

$$\Psi^{(i)}\qquad\qquad\gamma$$

$$\Phi_1=\frac{1}{\sqrt{2}}\{\phi_x(1)\phi_x(2)+\phi_y(1)\phi_y(2)\}\qquad {}^1\Sigma_g^+\quad (xx|xx)+(xy|xy),$$

$$[\Phi_2,\Phi_3]=\left[\frac{1}{\sqrt{2}}\{\phi_x(1)\phi_x(2)-\phi_y(1)\phi_y(2)\},\ \frac{1}{\sqrt{2}}\{\phi_x(1)\phi_y(2)+\phi_y(1)\phi_x(2)\}\right]\qquad {}^1\Delta_g\quad \frac{1}{2}(xx|xx)+\frac{1}{2}(xx|yy),$$

$$\left(\frac{1}{\sqrt{2}}\{\phi_x(1)\phi_y(2)-\phi_y(1)\phi_x(2)\}\qquad {}^3\Sigma_g^-\quad (xx|yy)-(xy|xy)\right)$$

(($xy|xy$), *etc.*, denote ($\phi_x\phi_y|\phi_x\phi_y$), *etc.*).

The uppermost level corresponds to $^1\Sigma_g^+$ excited singlet state and the lower degenerate level to $^1\Delta_g$ state. The former stands for the configuration

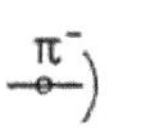

with $\mathfrak{B}=1$. The species can thus be called purely biradical.

In the latter state, the correlation-extremized wave functions consist of purely doubly-occupied configurations

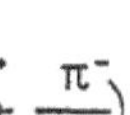 and 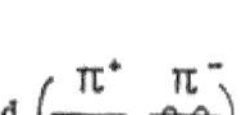

with $\mathfrak{B}=0$.

It should be noted that we have to employ the symmetry orbitals π^+ and π^- to construct wave functions with definite angular momenta.

Such cyclic orbitals, however, do not reflect the concept of radical site. The interpretation of $\mathfrak{B}$-values in this case should be made with certain reservation.

Cyclobutadiene (Planar Square). The orthogonal nonbonding orbitals can be taken as

$$\phi_1=\frac{1}{\sqrt{2-2s'}}\{(2p_z)_4-(2p_z)_2\},$$

$$\phi_2=\frac{1}{\sqrt{2-2s'}}\{(2p_z)_1-(2p_z)_3\},$$

where $(2p_z)_j$ ($j=1$, 2, 3, and 4) are the four $2p_z$ orbitals and s' is the overlap integral of $(2p_z)_1$ and $(2p_z)_3$. The three states are as follows.

$\Psi^{(i)}$		γ
$\frac{1}{\sqrt{2}}\{\phi_1(1)\phi_1(2)+\phi_2(1)\phi_2(2)\}$	${}^1A_{1g}$	$(11\|11)+(12\|12)$,
$\frac{1}{\sqrt{2}}\{\phi_1(1)\phi_1(2)-\phi_2(1)\phi_2(2)\}$	${}^1B_{1g}$	$(11\|11)-(12\|12)$,
$\frac{1}{\sqrt{2}}\{\phi_1(1)\phi_2(2)+\phi_2(1)\phi_1(2)\}$	${}^1B_{2g}$	$(11\|22)+(12\|12)$,
$\frac{1}{\sqrt{2}}\{\phi_1(1)\phi_2(2)-\phi_2(1)\phi_1(2)\}$	${}^3A_{2g}$	$(11\|22)-(12\|12)$,

($(11|22)$, *etc.*, denote $(\phi_1\phi_1|\phi_2\phi_2)$, *etc.*).

The level situation is consistent with the results of more elaborate calculations.[25)]

Singlet Planar Trimethylenemethane.[26)] The two orthogonal orbitals containing nonbonding electrons are written as

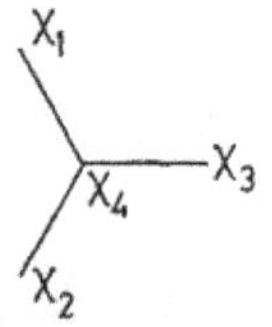

$$\phi_1=\sqrt{\frac{2}{3(1-s')}}\left\{\chi_1-\frac{1}{2}(\chi_2+\chi_3)\right\},$$

$$\phi_2=\frac{1}{\sqrt{2(1-s')}}(\chi_2-\chi_3),$$

where χ_j ($j=1$, 2, 3, and 4) are the four $2p_z$ orbitals, and s' is the overlap integral of χ_1 and χ_2. The resulting three states are

$${}^1A_1' \quad \frac{1}{\sqrt{2}}\{\phi_1(1)\phi_1(2)+\phi_2(1)\phi_2(2)\},$$

$${}^1E' \left\{\begin{array}{l}\frac{1}{\sqrt{2}}\{\phi_1(1)\phi_2(2)+\phi_2(1)\phi_1(2)\},\\ \frac{1}{\sqrt{2}}\{\phi_1(1)\phi_1(2)-\phi_2(1)\phi_2(2)\},\end{array}\right.$$

$$\left({}^3A_2' \quad \frac{1}{\sqrt{2}}\{\phi_1(1)\phi_2(2)-\phi_2(1)\phi_1(2)\}\right).$$

The density of ${}^1E'$ ground state becomes

$$\rho(1)=\frac{2}{3(1-s')}[\{\chi_1(1)^2+\chi_2(1)^2+\chi_3(1)^2\}$$
$$-\{\chi_1(1)\chi_2(1)+\chi_1(1)\chi_3(1)+\chi_2(1)\chi_3(1)\}]$$

showing that this is an antibonding biradical with $\mathfrak{B}=1/(1-s')$. The antibonding character might primarily cause the nonpolar geometry of singlet ground-state trimethylenemethane.[27)]

General Characterization of Biradicals

The correlation-extremization approach developed in previous section is convenient for grasping a qualitative feature, but it can not be used in polar species where the potential field is partial to either one of the two radical sites. In such cases, the usual energy-extremization procedure, *viz.*, the configuration interaction (CI) approach, must be taken.

We use a CI wave function of n electron system $\Psi(1, 2,\cdots n)$ to construct the density distribution $\rho(1)$, which is decomposed in such a way as

$$\rho(1)=A_{11}a(1)^2+2A_{12}a(1)b(1)+A_{22}b(1)^2$$
$$+2\sum_j\{B_{aj}a(1)+B_{bj}b(1)\}\chi_j(1)$$
$$+\sum_{j,k}B_{jk}\chi_j(1)\chi_k(1), \qquad (15)$$

The set of orbitals, a, b, and χ_j, in principle atomic hybrids or usual atomic orbitals, is obtained by a relevant transformation of the set of valence-shell atomic orbitals in such a way that orbitals a and b are appropriately associated with essentially located radical sites.

The bonding character and polar character can be represented by sA_{12} and Π of Eq. 12, respectively. We define a new concept, the delocalized character of a biradical, $\mathscr{D}$, by

$$\mathscr{D}=2-(A_{11}+2sA_{12}+A_{22}). \qquad (16)$$

The biradical character of Eq. 14 can then be modified as

$$\mathfrak{B}=\frac{A_{11}+A_{22}}{2-\mathscr{D}}(1-\Pi)$$
$$=\frac{2}{2-\mathscr{D}}(\text{the smaller of } A_{11} \text{ and } A_{22}). \qquad (17)$$

Bonding Deformation of Singlet Biradicals

In the following examples, where the polar character Π can be put equal to zero, the biradical character $\mathfrak{B}$ depends only upon the overlapping between radical sites. Thus singlet biradicals stabilize with increasing overlapping between radical sites, which is in accordance with the direction of decreasing biradical character $\mathfrak{B}$. In contrast, the triplet species tend to decrease overlapping to stabilize.[29,30)]

A molecular deformation in singlet biradicals to raise overlapping may even take place when stabilization overcomes destabilization, if it arises, due to the deformation. It may happen that a triplet biradical with minimal overlapping shifts to the singlet state through intersystem crossing to cause stabilization by bonding deformation. Let us discuss the bonding deformation in the direction of decreasing $\mathfrak{B}$ with regard to several examples of singlet biradicals.

The dependence of energy on the torsional angle θ in perpendicular ethylene discussed in previous section is shown below.[24)] The lowest singlet has a deformation-instability in addition to the triplet-instability near $\theta=\pi/2$. The ${}^3A_2$ ground state of perpendicular geometry once formed may tend towards the planar form through the intersystem crossing to produce ${}^1A_g$ ground-state ethylene.

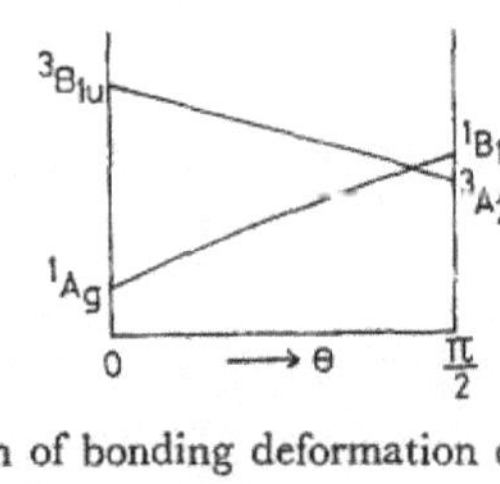

The direction of bonding deformation of perpendicular

species is simply represented by the following orbital phase scheme.[28)]

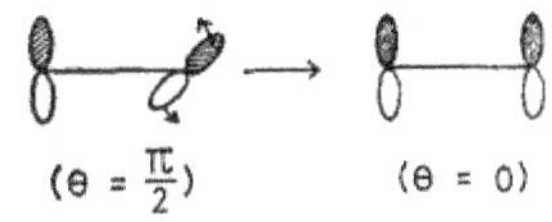

Similarly, the direction of deformation of square cyclobutadiene is given by the direction of in-phase overlapping of two odd-electron orbitals.

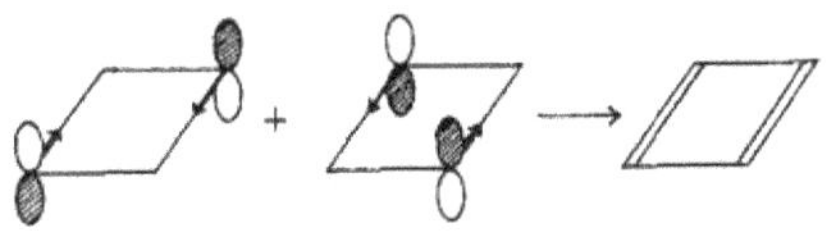

Also planar octagonal cyclooctatetraene will deform in the following way.

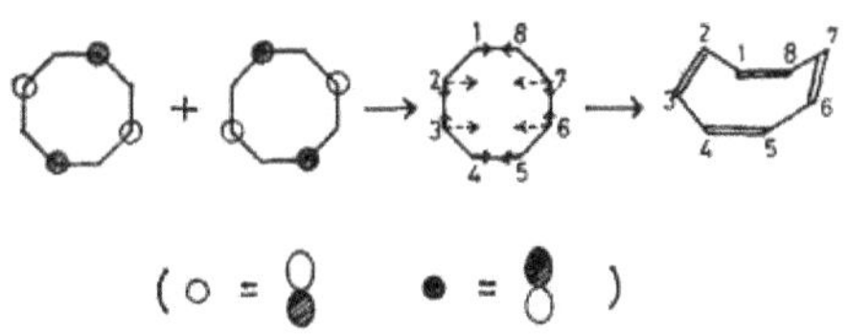

The bonding deformation consideration is employed to predict the geometry of stable intermediates in singlet radical reactions.

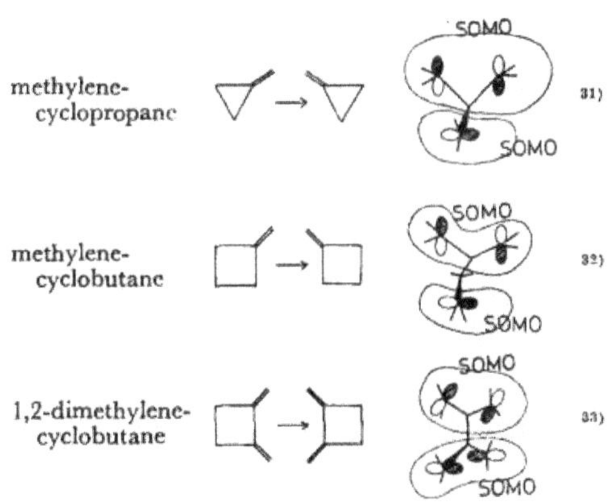

SOMO denotes a singly occupied molecular orbital.

The bond alternation in carbon $2p\pi$ chains can be explained by the following. The highest occupied (HO) and the lowest unoccupied (LU) MO's of a chain of equally distanced four p orbitals result approximately from the combination of the following two orbitals *a* and *b*.

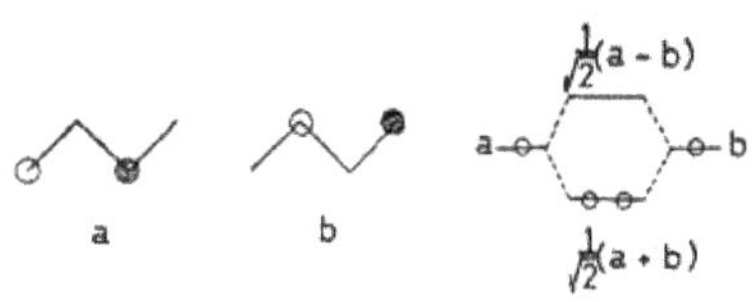

In this fictitious singlet biradical, the direction of deformation is also caused by the bonding overlapping of *a* and *b* to form a butadiene molecule, shown by

The lower occupied π orbital has hardly any connection with this deformation owing to its less nodal property.

Similarly, the difference in bond lengths in hexatriene, fulvene, naphthalene, *etc.*, is explained by considering the next HOMO's which contribute to the deformation less than HOMO because of their less nodal character.

An interesting application of the bonding deformation approach may be the prediction of the reaction path in excited states. In the π-π^* excited singlet state of hexatriene, the highest half-occupied orbital (a') and the lowest unoccupied orbital (b') are represented by

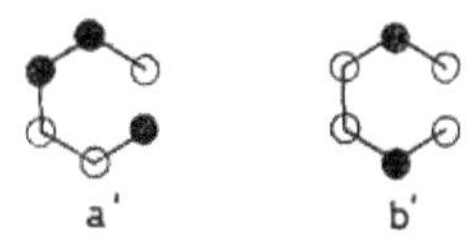

In order to consider bonding stabilization of the electron in orbital a', let us construct in the place of a' and b' the following quasiorthogonal orbitals.

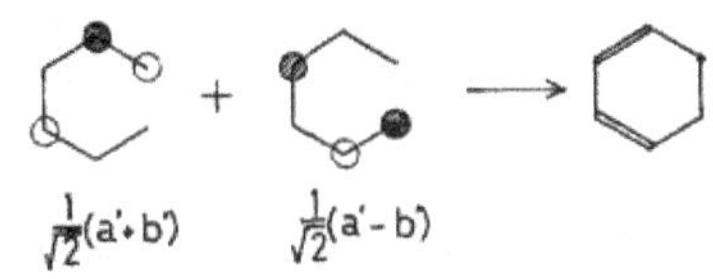

Combinations of these orbitals lead to orbitals a' and b', the former becoming occupied by a single electron to contribute to the bonding stabilization, and the direction of ring closure, if it occurs simultaneously, should be conrotatory. The conclusion is unchanged in the excited state with doubly occupied a'. In contrast, in the excited butadiene, the direction of hypothetical 1,4-bonding would be disrotatory.

The lower singly occupied orbitals would contribute less by a similar discussion in terms of their less nodal property.

The simple approach is useful since it can be applied to a rough estimation of the direction of bonding in higher excited singlet states of a known electron configuration. The configuration might have several SOMO's. The general procedure is as follows:

i) By a linear combination of the HOMO (singly or doubly occupied) and the LUMO, say a' and b', we construct two MO's, a and b, which are mutually orthogonal and spatially separated from each other to construct a fictitious singlet biradical species.

ii) The direction of maximal overlapping of a and b to decrease the biradical character $\mathfrak{B}$, obtained by the usual procedure in the orbital interaction approach,[28)] leads to a favourable reaction path.

In some cases, an excited state of the species before deformation may correlate to the ground state of the species after deformation, as in the following,

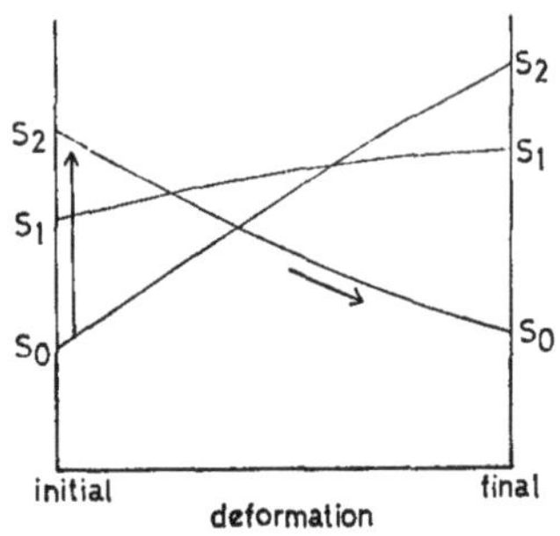

An example is the hypothetical disrotatory cyclization of S_2 excited butadiene to S_0 cyclobutene in the Longuet-Higgins-Abrahamson state correlation diagram.[34] The direction of such processes can be discussed by the method mentioned above.

Unpaired-Electron Isolating Deformation or Bonding in Triplet Biradicals

Once a triplet state is formed photochemically or thermally, a deformation or bond formation may be liable to take place to acquire stabilization in such a way that each of the two unpaired-electrons enters each of two orthogonal orbitals, or they become separated from each other by the newly formed bond as far away as possible.[9,30] In this connection Michl's conception of "loose" and "tight" biradicaloids is useful.[30c] The following examples are given for illustration.

Consider a π-π^* triplet hexatriene molecule. Twisting of the carbon chain would certainly diminish the overlapping of the two unpaired-electron orbitals. Two possibilities exist in regard to the twisting.

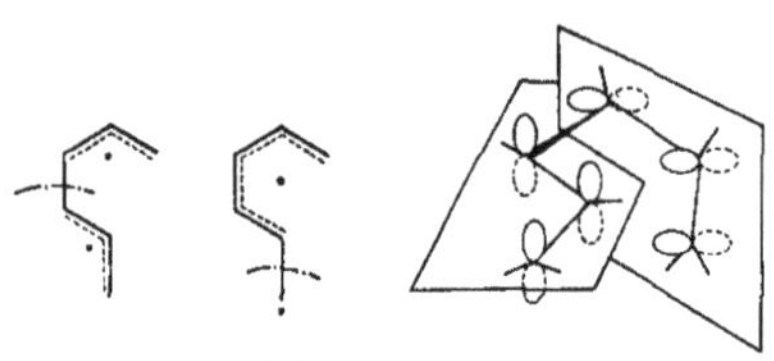

(I) (II) (III)

The sum of conjugation stabilization in the two separated parts in (I) is greater than in (II). The resulting geometry is shown in (III). Courtot and Salaün[35a] showed that sensitized photoreactions of hexatrienes preferentially cause the *cis-trans* isomerization of the *central* double bond, and that the singlet reaction causes isomerization of the terminal double bonds. The results are consistent with the theoretical prediction by Baird and West[35b] that the twisting of inner bonds should be preferred in the lowest triplet excited state of polyenes. In the singlet reaction a structure essentially like

may be prevalent on account of the bonding nature of singlet biradicals, which will favour the terminal bond isomerization.

Geometries in which the planes associated with two unpaired-electron orbitals are perpendicular to each other might be expected more widely.[35c]

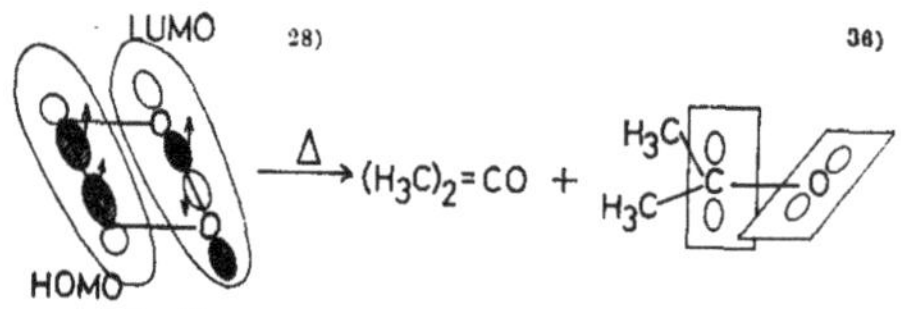

1,2-dioxetane (triplet)

(Similar reactions are considered in chemiluminescent decomposition of 1,2-dioxetanone derivatives.[37])

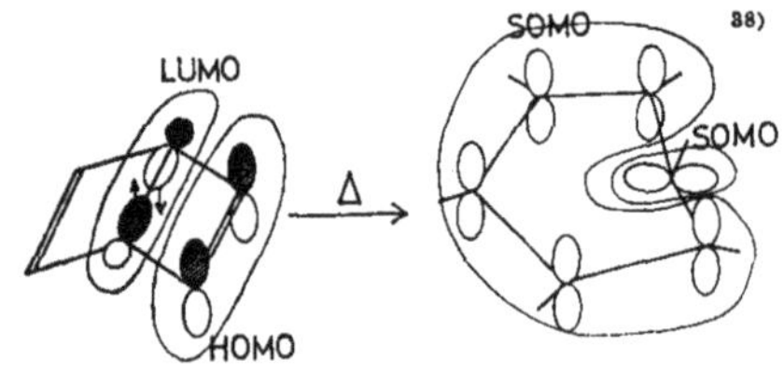

Dewar benzene (triplet)

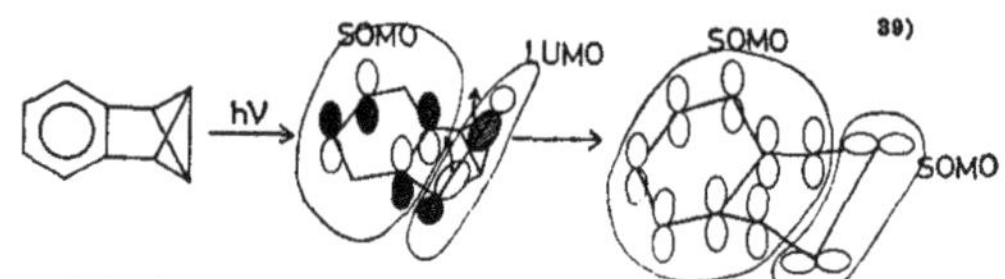

naphthvalene (S_1) (T_1)

The above are examples of the separation of unpaired-electron orbitals by deformation. The isolation of unpaired-electrons also takes place by the formation of separating bonds.

40)

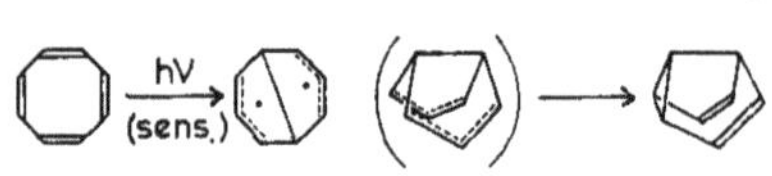

41)

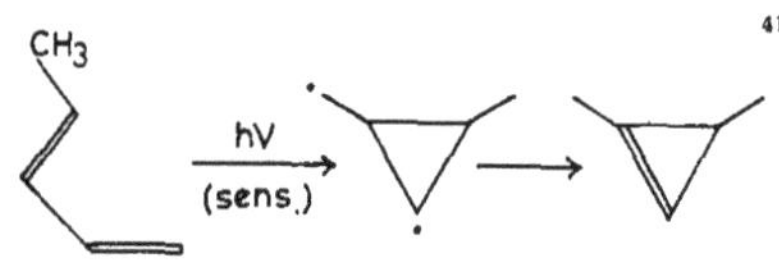

In the singlet-state reaction, a similar reaction takes place. In this case a certain concertedness may be mixed in the mechanism.[28]

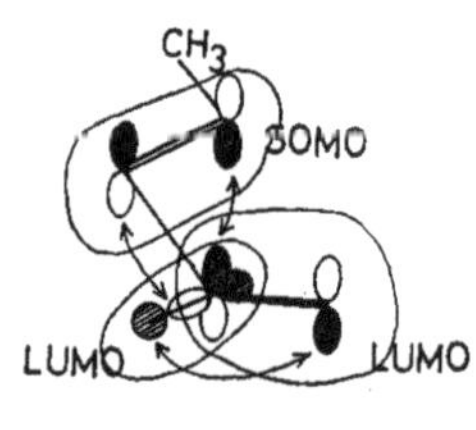

It appears that combination of the principle of unpaired-electron isolation in the triplet state with that of bonding deformation in the singlet state brings about a plausible mechanism occasionally prevalent in photochemical processes in which biradicaloid intermediates play an important role.

Suppose that the lowest triplet state (T_1) is produced, in some cases from higher excited singlet states (S_n) *via* S_1, through radiationless transitions (triplet biradical formation by a nonsensitized photoprocess). A possible deformation path might be composed of stages (I) and (II) as shown in Fig. 1.

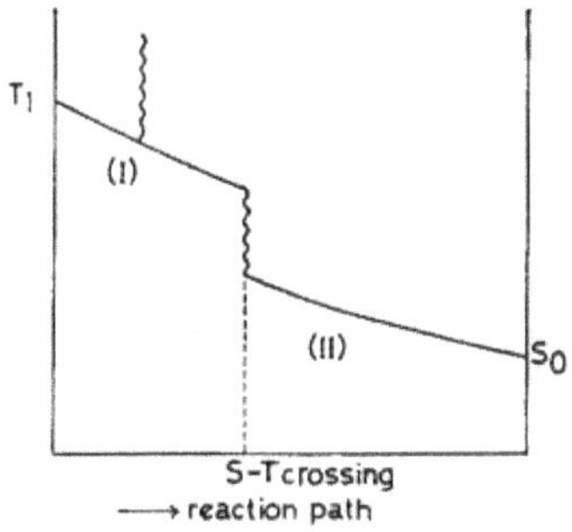

Fig. 1. A possible reaction path. (S_0: singlet ground state; T_1: first triplet state; ～～～: radiationless transition)

Stage (I) involves separation of unpaired-electron orbitals by orthogonally twisting deformation or bond formation, occuring in the first triplet state. Stage (II) is that of bonding deformation or bond formation in the singlet ground state, S_0.

By considering intermediate T_1, stereoselective phenomena can be explained by the augmenting chance to undergo further conversion due to the long lifetime of the triplet state.

This mechanism of the intermediation of triplet state is supported by Kushick and Rice[42] as regards the effectuation of the S-T transition by dynamical coupling of the torsional motion around the C_1–C_2 bond in the *cis-trans* photoisomerization of butadiene.

Let us consider excited-state reactions of benzene, six π MO's of which are given below.

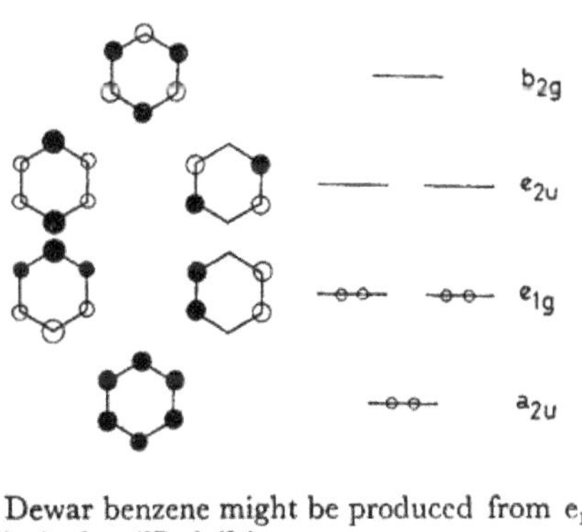

Thus, Dewar benzene might be produced from $e_{1g}\to e_{2u}$ excited singlet ($^1B_{1u}$),[43a)]

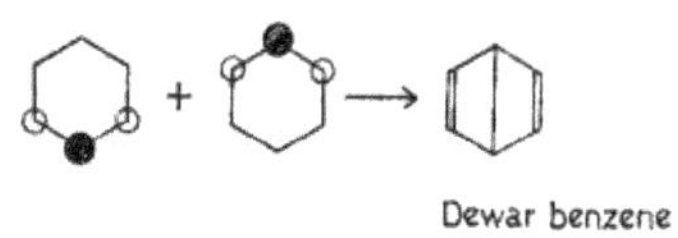

and the formation of benzvalene, prismane, and Dewar benzene would be favoured by triplet intermediation.

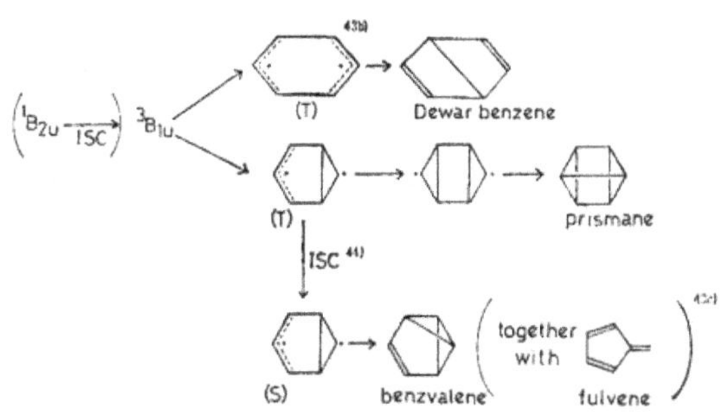

The mechanism involved might control the direction of reaction, giving a new principle of orientation or stereoselection in excited-state reactions.

This work was partly supported by Grants-in-Aid for Scientific Research from the Ministry of Education of Japan (No. 047068 and No. 139012). The authors express their appreciation to Prof. Hiroshi Kato, Nagoya University, for valuable discussions. One of the authors (K. F.) wished to thank Prof. P. Courtot for the interesting photochemical problem which stimulated him to carry out the present work.

References

1) R. B. Woodward and R. Hoffmann, *Angew. Chem. Int. Ed. Engl.*, **8**, 781 (1969).

2) J. A. Berson and L. Salem, *J. Am. Chem. Soc.*, **94**, 8917 (1972).

3) M. J. S. Dewar, S. Kirschner, and H. W. Kollmar, *J. Am. Chem. Soc.*, **96**, 7579 (1974).

4) D. R. Roberts, *J. Chem. Soc., Chem. Commun.*, **1974**, 683.

5) L. Salem and C. Rowland, *Angew. Chem. Int. Ed. Engl.*, **11**, 92 (1972); L. Salem, *Pure Appl. Chem.*, **33**, 317 (1973); L. Salem, W. G. Dauben, and N. J. Turro, *J. Chim. Phys.*, **70**, 694 (1973); L. Salem, *J. Am. Chem. Soc.*, **96**, 3486 (1974).

6) L. Salem, C. Leforestier, G. Segal, and R. Wetmore, *J. Am. Chem. Soc.*, **97**, 479 (1975).

7) V. Bonačić-Koutecký, P. Bruckmann, P. Hiberty, J. Koutecký, C. Leforestier, and L. Salem, *Angew. Chem.*, **87**, 599 (1975); *Angew. Chem. Int. Ed. Engl.*, **14**, 575 (1975).

8) W. G. Dauben, L. Salem, and N. J. Turro, *Acc. Chem. Res.*, **8**, 41 (1975).

9) L. Salem and P. Bruckmann, *Nature*, **258**, 526 (1975).

10) L. Salem, *Science*, **191**, 822 (1976).

11) The word "extremized" is used here only in the category of the present two-electron, two-orbital approximation. The parallel trend mentioned is evident for the systems with equivalent radical centres (see text). However, it is not necessarily guaranteed for general cases.

12) K. Ruedenberg, *Rev. Mod. Phys.*, **34**, 326 (1962).

13) H. Fujimoto, S. Yamabe, and K. Fukui, *Bull. Chem. Soc. Jpn.*, **44**, 2936 (1971).

14) K. Fukui, XXIII rd Internl. Congress of Pure and Applied Chemistry, Vol. 1, Butterworths, London (1971), p. 65.
15) H. Fujimoto, S. Yamabe, and K. Fukui, *Tetrahedron Lett.*, **1971**, 443.
16) S. Yamabe, T. Minato, H. Fujimoto, and K. Fukui, *Theoret. Chim. Acta*, **32**, 187 (1974).
17) S. Yamabe, S. Kato, H. Fujimoto, and K. Fukui, *Bull. Chem. Soc. Jpn.*, **46**, 3619 (1973).
18) H. Fujimoto, S. Yamabe, T. Minato, and K. Fukui, *J. Am. Chem. Soc.*, **94**, 9205 (1972).
19) K. Fukui, S. Kato, and H. Fujimoto, *J. Am. Chem. Soc.*, **97**, 1 (1975).
20) H. Fujimoto and K. Fukui, *Adv. Quantum Chem.*, **6**, 177 (1972).
21) H. Fujimoto and K. Fukui, in "Chemical Reactivity and Reaction Paths," ed by G. Klopman, John Wiley & Sons, Inc., New York (1974), p. 23.
22) S. Yamabe, S. Kato, H. Fujimoto, and K. Fukui, *Bull. Chem. Soc. Jpn.*, **48**, 1 (1975).
23) C. E. Wulfman and S. Kumei, *Science*, **172**, 1061 (1971).
24) R. J. Buenker, S. D. Peyerimhoff, and H. L. Hsu, *Chem. Phys. Lett.*, **11**, 65 (1971).
25) *E.g.* D. P. Craig, *Proc. R. Soc. London, Ser A*, **202**, 498 (1950).
26) cf. W. T. Borden, *J. Am. Chem. Soc.*, **98**, 2695 (1976).
27) M. J. S. Dewar and J. S. Wasson, *J. Am. Chem. Soc.*, **93**, 3081 (1971); D. R. Yarkony and H. F. Schaefer III, *ibid.*, **96**, 3754 (1974); W. J. Hehre, L. Salem, and M. R. Wilcott, *ibid.*, **96**, 4328 (1974).
28) K. Fukui, "Theory of Orientation and Stereoselection," Springer-Verlag, Berlin (1975).
29) See p. 81 of Ref. 28.
30) (a) S. Kita and K. Fukui, *Bull. Chem. Soc. Jpn.*, **42**, 66 (1969); (b) K. Fukui, *Acc. Chem. Res.*, **4**, 57 (1971); (c) J. Michl, *J. Am. Chem. Soc.*, **93**, 523 (1971); *Mol. Photochem.*, **4**, 257 (1972); "Chemical Reactivity and Reaction Paths," ed by G. Klopman, John Wiley & Sons, Inc., New York (1974), p. 301; (d) H. E. Zimmermann and G. A. Epling, *J. Am. Chem. Soc.*, **94**, 3647 (1972).
31) E. F. Ullman, *J. Am. Chem. Soc.*, **82**, 505 (1960); J. J. Gajewski, *J. Am. Chem. Soc.*, **93**, 4450 (1971); W. von E. Doering and H. D. Roth, *Tetrahedron*, **26**, 2825 (1970); W. von E. Doering and L. Birladeanu, *ibid.*, **29**, 499 (1973).
32) W. von E. Doering and J. C. Gilbert, *Tetrahedron Suppl.*, **7**, 397 (1966).
33) J. J. Gajewski and C. N. Shih, *J. Am. Chem. Soc.*, **89**, 4532 (1967); W. von E. Doering and W. R. Dolbier, Jr., *ibid.*, **89**, 4534 (1967).
34) H. C. Longuet-Higgins and E. W. Abrahamson, *J. Am. Chem. Soc.*, **87**, 2045 (1965).
35) (a) P. Courtot and J. Y. Salaün, *J. Chem. Soc., Chem. Commun.*, **1976**, 124; (b) N. C. Baird and R. M. West, *J. Am. Chem. Soc.*, **93**, 4427 (1971); (c) W. G. Dauben and J. S. Ritscher, *ibid.*, **92**, 2925 (1970).
36) N. J. Turro and P. Lechtken, *J. Am. Chem. Soc.*, **95**, 264 (1973); N. J. Turro, H. C. Steinmetzer, and A. Yetka, *ibid.*, **95**, 6468 (1973); N. J. Turro and P. Lechtken, *Pure Appl. Chem.*, **33**, 363 (1973); M. J. S. Dewar and S. Kirschner, *J. Am. Chem. Soc.*, **96**, 7578 (1974); N. J. Turro, P. Lechtken, N. E. Schore, G. Schuster, H. C. Steinmetzer, and A. Yetka, *Acc. Chem. Res.*, **7**, 97 (1974).
37) T. Goto and Y. Kishi, *Angew. Chem. Int. Ed. Engl.*, **7**, 407 (1968); W. D. McElroy, H. H. Seliger, and E. H. White, *Photochem. Photobiol.*, **10**, 153 (1969); E. H. White, J. D. Miano, C. J. Watkins, and E. J. Breaux, *Angew. Chem.*, **86**, 292 (1974).
38) P. Lechtken, R. Breslow, A. H. Schmidt, and N. J. Turro, *J. Am. Chem. Soc.*, **95**, 3025 (1973).
39) N. J. Turro, P. Lechtken, A. Lyons, R. R. Hautala, E. Carnahan, and T. J. Katz, *J. Am. Chem. Soc.*, **95**, 2035 (1973).
40) H. E. Zimmermann and H. Iwamura, *J. Am. Chem. Soc.*, **90**, 4763 (1968).
41) S. Boué and R. Srinivasan, *J. Am. Chem. Soc.*, **92**, 3226 (1970).
42) J. N. Kushick and S. A. Rice, *J. Chem. Phys.*, **64**, 1612 (1976).
43) (a) I. Haller., *J. Chem. Phys.*, **47**, 1117 (1967); (b) A. A. Gwaiz, M. A. El-Sayed, and D. S. Tinti, *Chem. Phys. Lett.*, **9**, 454 (1971); (c) H. R. Ward and J. S. Wishnok, *J. Am. Chem. Soc.*, **90**, 5353 (1968).
44) The possibility of the triplet intermediation in non-sensitized photolysis should be further examined. The sudden polarization of singlet biradicals[23] could be combined with the intersystem crossing $T_1 \to S_1$. The molecular geometry having minimum energy in the excited state is one with which the state T_1 has minimum energy. Such T_1 geometry would play some role in excited-state reactions (*e.g.*, see K. Yamaguchi, T. Fueno, and H. Fukutome, *Chem. Phys. Lett.*, **22**, 466 (1973)).

[January, 1978] BULLETIN OF THE CHEMICAL SOCIETY OF JAPAN, VOL. 51 (1), 1—10 (1978) 1

A Molecular Orbital Calculation of Chemically Interacting Systems. Interaction between Two Radicals

Tsutomu MINATO, Shinichi YAMABE,* Hiroshi FUJIMOTO, and Kenichi FUKUI

Faculty of Engineering, Kyoto University, Sakyo-ku, Kyoto 606

**Department of Chemistry, Nara University of Education, Takabatake-cho, Nara 630*

(Received April 29, 1977)

The reaction path of radical-radical recombination and disproportionation has been studied by means of a perturbational method with a clear orbital interaction concept, and the mechanism of these reactions has been elucidated theoretically. It has been confirmed that these termination reactions do not have a common transition state. The recombination is found to take the path which gives the maximum interaction between the singly occupied molecular orbitals (SOMO's) of two radical species, whereas the disproportionation is shown to take the route which gives the significant charge transfer interaction from the particular doubly occupied (DO) MO of one radical to the SOMO of the other radical. The DOMO localized at the C–H bond donates electrons to the SOMO to cause the hydrogen abstraction. In the process of disproportionation, one radical which abstracts a hydrogen atom acts as an electron-acceptor and the other radical with the hydrogen atom to be abstracted as an electron-donor. The difference between the mechanism of the recombination and that of the disproportionation is clarified in terms of the mode of the orbital interaction.

When organic free radicals are generated in a reaction, the unpaired-electron species eventually undergo a termination reaction, yielding stable molecules with closed-shell electronic structures through two reaction pathways. One of them is the recombination in Eq. 1 and the other is the disproportionation with the abstraction of a hydrogen atom β to the radical center in Eq. 2.[1)]

$$RCH_2CH_2\cdot + R'CH_2CH_2\cdot \xrightarrow{k_c} RCH_2CH_2CH_2CH_2R' \quad (1)$$

$$RCH_2CH_2\cdot + R'CH_2CH_2\cdot \xrightarrow{k_d} RCH_2CH_3 + R'CH{=}CH_2 \quad (2)$$

The knowledge of termination modes together with their relative rates is very important in predicting and interpreting the radical chain reactions and their products. For example, a change from recombination to disproportionation can effect a twofold variation in polymer length.

Early pictures of radical-radical reactions assumed a common transition state for the two processes of Eqs. 1 and 2, involving the excited dimer which could either be rearranged to give the disproportionation products or be quenched to give the recombination product.[2)] This, however, is ruled out experimentally by a number of observations.[3)] Bradley and Rabinovitch suggested an alternative interpretation of the mechanisms of the recombination and disproportionation of alkyl radicals.[4)] They rejected the description of the activated complex for the disproportionation as an excited dimer and suggested that different configurations of the composite system may take part involving couplings of the radical species through three-centered hydrogen atom bonds already suggested for the recombination.[5)] Consequently, the terms, "head-to-head" and "head-to-tail", used to describe the configuration of the activated complexes are discarded in the Bradley-Rabinovitch model.

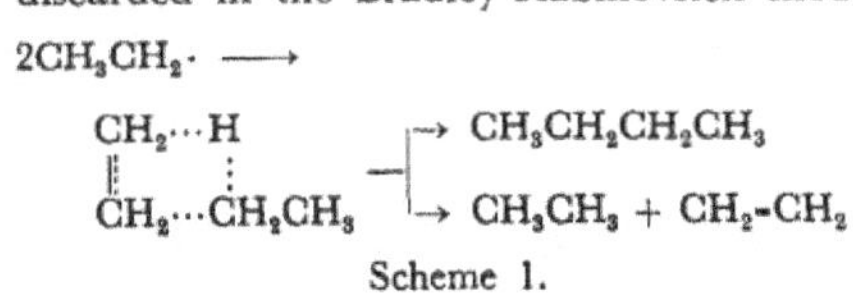

Scheme 1.

Benson put forward the argument that both reactions, though quite different from each other, involve loose transition states with a significant ionic character.[3a)]

$$2CH_3\cdot \longrightarrow \overset{-}{C}H_3 : \overset{+}{C}H_3 \text{ or } \overset{+1/2}{CH_2}\text{-}H\cdots\overset{-1/2}{CH_3} \longrightarrow CH_3CH_3$$

Scheme 2a.

$$2CH_3CH_2\cdot \longrightarrow CH_3\overset{-}{C}H_2{:}H\text{-}CH_2\overset{+}{C}H_2 \text{ or } CH_3\overset{+}{C}H_2{:}\overset{-}{H}\text{-}CH_2CH_2 \longrightarrow CH_3CH_3 + CH_2{=}CH_2$$

Scheme 2b.

Although it is accepted that the transition states of two radical-radical termination reactions differ, there are two incompatible ways of speculation for these processes as shown in Schemes 1, 2a, and 2b. Since the reaction paths of the recombination and disproportionation are still undetermined, it seems useful to study radical destruction processes theoretically.

Recently, the ΔW method which deals with the radical-radical reacting system in terms of the molecular interaction was developed by the present authors and was applied to the recombination of two methyl radicals.[6)] As the result of calculation, the interaction between both the singly occupied molecular orbitals (SOMO's) and the deformation of the methyl radical were found to be of importance for progress of the reaction. The result also demonstrates the applicability of the ΔW method further to larger reacting systems. In the present work, we have made a comparative study on the recombination and disproportionation by the use of the ΔW method. The purpose of this work is to shed light on the disputing point of the mechanism of two reaction pathways. Furthermore, the general aspect of radical-radical reaction is explained conveniently in terms of the shape of MO's of several isolated radical species, since the value of the interaction energy, ΔW, is completely dependent on the overlap of MO's and gap of orbital-energy levels.

Method of Calculation

A system composed of an ethyl radical (ER) and a methyl radical (MR) is selected as a reaction model in the ΔW method. In this reaction system the recombination gives propane and the disproportionation produces methane and ethylene.

In the ΔW method, the wave function (Ψ) of a chemically-interacting system between ER and MR is represented in the configuration interaction (CI) procedure:

$$\begin{aligned}\Psi = C_0\Psi_0 &+ \sum_{i=1}^{6}\sum_{l=5}^{7} C_{i\to l}\Psi_{i\to l} + \sum_{k=1}^{3}\sum_{j=8}^{13} C_{k\to j}\Psi_{k\to j} \\ &+ \sum_{i=1}^{6} C_{i\to o'}\Psi_{i\to o'} + \sum_{k=1}^{3} C_{k\to o}\Psi_{k\to o} + \sum_{l=5}^{7} C_{o\to l}\Psi_{o\to l} \\ &+ \sum_{j=8}^{13} C_{o'\to j}\Psi_{o'\to j} + C_{o\to o'}\Psi_{o\to o'} + C_{o'\to o}\Psi_{o'\to o}. \qquad (3)\end{aligned}$$

In order to describe MO's of ER and MR in their isolated states, a semiempirical all-valence-electron SCF-UHF method including overlap integrals is employed.[7] The way of estimating semiempirical parameters to evaluated MO integrals is the same as that used in a previous paper.[8] Without the lowest occupied ls-like MO, ER has six doubly occupied (DO) MO's and MR has three DOMO's. In Eq. 3, Ψ_0 denotes the adiabatically-interacting configuration in which neither electron transfer nor electron excitation takes place as is shown in Fig. 1. The letters "i" and "j" indicate a DOMO and an unoccupied MO, respectively, of ER, and "k" and "l" a DOMO and an unoccupied MO, respectively, of MR. "o" is the SOMO of the former radical, and "o'" that of the latter. $\Psi_{i\to l}$ denotes a configuration of one-electron shift from the i-th DOMO to the l-th unoccupied MO. C_0, $C_{i\to l}$, $C_{k\to j}$, $\cdots$, are the CI coefficients obtained by solving the usual CI secular equation. Our attention is focused on the reacting system with the singlet spin state, because in the system with the triplet spin state a radical-radical reaction is far from occurring. In this calculation, no configurations with intramolecular excitations are taken into account in Eq. 3, since it requires much more CPU time to compute the molecular integrals appearing in the CI matrix elements of the above configurations. Also, two reactants are neutral and consequently the excited configurations do not contribute as much as the charge-transferred ones to the stabilization of the reacting systems.[6]

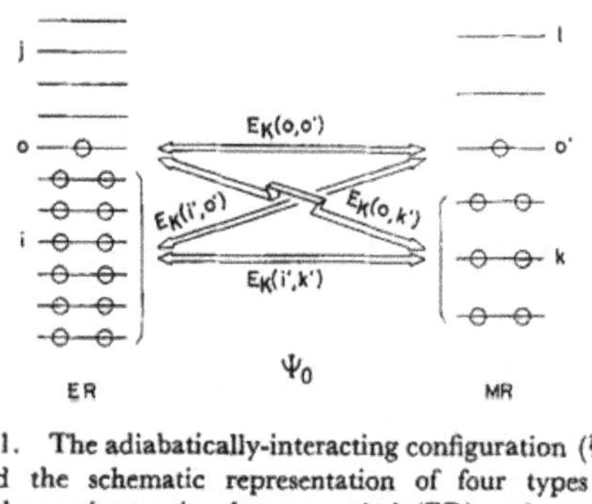

Fig. 1. The adiabatically-interacting configuration (Ψ_0) and the schematic representation of four types of exchange interaction between ethyl (ER) and methyl (MR) radicals. The lowest DOMO of two radicals, *i.e.*, the 1s-type one is omitted in this figure.

When the CI secular equation is expanded perturbationally, the total stabilization energy (ΔW) due to the CI mentioned above can be partitioned into the following three terms, Coulomb (E_Q), exchange (E_K), and delocalization (D).

$$\Delta W = E_Q + E_K - D. \qquad (4)$$

E_Q is the classical electronic interaction energy between ER and MR. In both the singlet spin state and the triplet spin state, E_K is found to be the repulsive term (>0) in the system of two radicals of the usual organic-chemical size. E_K can be partitioned approximately into four terms.[6]

$$E_K = E_K(i',k') + E_K(i',o') + E_K(o,k') + E_K(o,o'). \qquad (5)$$

In Fig. 1, four components of E_K are shown schematically. D is the stabilization energy due to the mixing-in of the charge transferred configurations with the adiabatically-interacting configuration. D can be represented by the sum of eight components.[6]

$$\begin{aligned}D = D(i\to l) &+ D(k\to j) + D(i\to o') + D(k\to o) + D(o\to l) \\ &+ D(o'\to j) + D(o\to o') + D(o'\to o). \qquad (6)\end{aligned}$$

In order to visualize the change of electron density along the reaction progress, a contour map of the difference electron density is given. The difference density, $\Delta\rho(1|1)_{E_K}$, of the exchange interaction is partitioned into four terms, each of which corresponds to that of the energy decomposition in Eq. 5.

$$\begin{aligned}\Delta\rho(1|1)_{E_K} = \Delta\rho(1|1)_{E_K(i',k')} &+ \Delta\rho(1|1)_{E_K(i',o')} \\ +\Delta\rho(1|1)_{E_K(o,k')} &+ \Delta\rho(1|1)_{E_K(o,o')}. \qquad (7)\end{aligned}$$

$\Delta\rho(1|1)_{E_K(i',k')}$, for instance, indicates the difference electron density caused by the exchange interaction between both DOMO's. The difference electron density including the exchange and charge transfer (CT) interactions, $\Delta\rho(1|1)$, is also given.

Figure 2 illustrates the geometry of assumed reaction models. The geometry of the isolated ER comes from Pople's *ab initio* calculation,[9] and that of the isolated MR is referred to Herzberg's book.[10] The length of all C–H bonds and the C_1–C_2 bond of two isolated systems is kept frozen throughout the calculation. Three reaction models (I, II, and III) are shown in Fig. 2. They are distinguished by the position of MR. Models I and II represent a recombination in which MR attacks the radical center (C_2) of ER, MR approaching ER in the *cis*-conformation in I and in the *trans*-conformation in II. Model III represents a disproportionation in which MR attacks a hydrogen atom (H_3) of ER. In this calculation seven geometrical parameters are introduced to indicate various reaction processes. β is the angle between the methylene plane (H–C_2–H) and C_1–C_2 bond axis of ER in Models I, II, and III. R denotes the distance between C_M of MR and C_2 of ER, and α is the angle between $C_M\cdots C_2$ line and the C–H bond of MR in I and II. θ and θ'

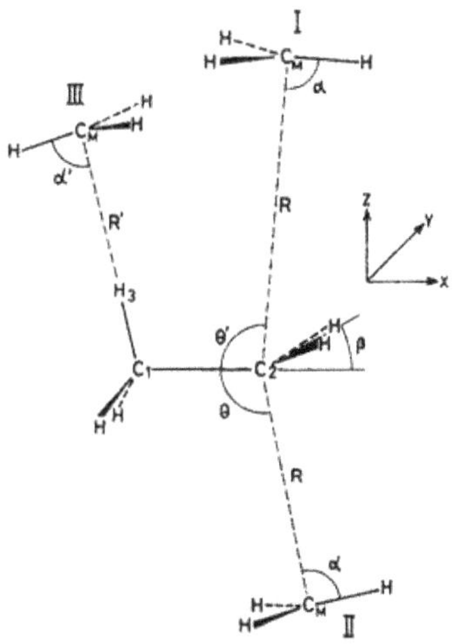

Fig. 2. Three assumed reaction models employed for the calculation of the interaction energy.

stand for the angle between $C_M \cdots C_2$ line and C_1–C_2 bond axis in Models I and II. R' denotes the distance between C_M of MR and H_3 of ER, and α' the angle between $C_M \cdots H_3$ line and the C–H bond of MR in Model III.

Results of Calculation

The interaction energy (ΔW) of Models I and II is given in Table 1. The ΔW of Model I (R=2.7 Å, θ'=100°) is compared with that of Model II (R=2.7 Å, θ=100°). We see that Model II is more favorable for the recombination. The superiority or inferiority of the recombination path is attributable to the two terms of exchange energy, $E_K(i', k')$ and $E_K(i', o')$, (see Fig. 1) and the components originating from o-o' interaction, $E_K(o, o')$, $D(o \to o')$, and $D(o' \to o)$. Let us examine the possibility of the three-centered bond exhibited in Scheme 1. If the termination reaction proceeds along Scheme 1, the approach of MR to the C_1–H_3 bond is expected to reduce the ΔW value. In order to see the effect of this geometrical change on each energy, the angle θ' is varied tentatively from 100 to 90° in Model I. This is found to bring about a larger ΔW value, *i.e.*, destabilization of the system (Table 1). In view of the two ΔW's (θ'=100, 90°) of Model I, it is realized that this destabilization comes primarily from the increase of the exchange energy, E_K, and slightly from the decrease of the delocalization energy, D. The analysis of four components of E_K shows that its enlargement is due to the increase of both $E_K(i', k')$ and $E_K(i', o')$ and the decrease of the stabilizing term, $E_K(o, o')$. $D(o \to o')$ and $D(o' \to o)$ which are the stabilization energies are reduced by that variation of the geometry. The increase of $E_K(i', k')$ and $E_K(i', o')$ is due to the approach of the electronic cloud of MR to that of the C_1–H_3 bond of ER, whereas the decrease of three stabilizing terms, $E_K(o, o')$, $D(o \to o')$, and $D(o' \to o)$ is explained as follows. The SOMO of ER has a node between C_1 and C_2. As MR approaches the C_1–H_3 bond of ER, the out-of-phase overlapping of the SOMO of MR with the lobe of the SOMO of ER localized at C_1 and H_3 becomes larger. The overlap between both the SOMO's becomes small and the stabilization due to $E_K(o, o')$, $D(o \to o')$, and $D(o' \to o)$ is reduced by θ'=100→90°.

TABLE 1. CALCULATED INTERACTION ENERGIES FOR THE RECOMBINATION AND DISPROPORTIONATION OF ETHYL AND METHYL RADICALS

Geometrical parameters, R, R', α, α', β, θ, and θ' are those shown in Fig. 2.

	Model I		Model II		Model III	
R or R' [Å]	2.7	2.7	2.7	2.4	2.4	2.0
α or α' [°]	90	90	90	105	90	105
β [°]	0	0	0	43.2	0	0
θ or θ'[°]	100	90	100	105	—	—
E_Q [eV]	0.020	0.022	0.022	0.069	−0.012	−0.033
$E_K(i', k')$	0.459	0.666	0.394	0.479	0.180	0.452
$E_K(i', o')$	0.190	0.227	0.174	0.338	0.120	0.428
$E_K(o, k')$	0.072	0.057	0.087	0.229	0.010	0.023
$E_K(o, o')$	−0.162	−0.141	−0.178	−0.622	−0.014	−0.047
E_K [eV]	0.559	0.809	0.477	0.424	0.296	0.856
$D(i \to l)$	0.003	0.007	0.003	0.011	0.001	0.001
$D(k \to j)$	0.006	0.011	0.004	0.011	0.016	0.030
$D(i \to o')$	0.072	0.089	0.063	0.108	0.063	0.233
$D(k \to o)$	0.033	0.030	0.036	0.084	0.001	0.003
$D(o \to l)$	0.003	0.002	0.003	0.002	0.000	0.000
$D(o' \to j)$	0.003	0.003	0.003	0.002	0.008	0.019
$D(o \to o')$	0.103	0.087	0.114	0.377	0.016	0.049
$D(o' \to o)$	0.110	0.100	0.117	0.372	0.004	0.012
D [eV]	0.333	0.329	0.343	0.967	0.109	0.347
ΔW [eV]	0.246	0.502	0.156	−0.474	0.175	0.476

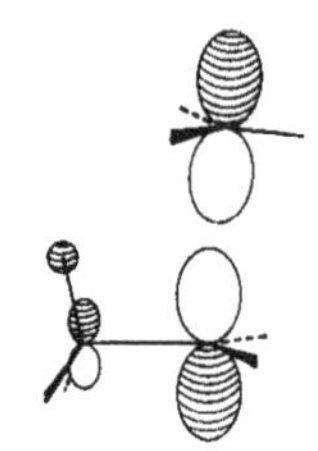

the SOMO of ER

The results show that the process of the radical destruction is not described by Scheme 1 and the recombination takes the route in which MR approaches the radical center, C_2, of ER in the *trans*-conformation to the C_1–H_3 bond of a methyl group of ER.

In Model II (R=2.4 Å, θ=105°), the structural deformation of two radicals is taken into account, *i.e.*, α=90→105° and β=0→43.2°. This deformation operates to reduce the E_K and to increase the absolute value of the stabilization terms, $E_K(o, o')$, $D(o\to o')$, and $D(o'\to o)$, making the ΔW negative (Table 1). Thus, the need of the structural deformation of the radical species to promote the reaction is explicable in this MO scheme. The deformation associates the change of the extent of the hybridization ($sp^2\to sp^3$) of two radical centers, resulting in the growth of the o-o' MO interaction which contributes overwhelmingly to the stabilization of the system. As for Model II (R=2.4 Å), the CI coefficients of Eq. 3 are obtained.

$$\Psi = 0.918\Psi_0 - 0.123\Psi_{o\to o'} - 0.116\Psi_{o'\to o} + \cdots. \quad (8)$$

This equation is an alternative expression of the importance of $D(o\to o')$ and $D(o'\to o)$ to promote the reaction. From the equation the transition state of the recombination hardly seems to have ionic character due to the almost equal weight of two competing ($o\to o'$ and $o'\to o$) CT interactions.

The difference density in the reaction process was investigated with respect to Model II (R=2.4 Å, θ=105°). The density map helps us to visualize the movement of electronic cloud through some types of interactions between ER and MR. Figure 3-a shows the electron redistribution by $\Delta\rho(1|1)_{E_K(i',k')}$ caused by mutual overlap between both the DOMO's. The electron redistribution gives negative contour lines amid the $C_2\cdots C_M$ region, corresponding to the repulsive $E_K(i',k')$. The electron density maps of $\Delta\rho(1|1)_{E_K(i',o')}$ and $\Delta\rho(1|1)_{E_K(o,k')}$ are shown in Figs. 3-b and 3-c. Two terms of the exchange interaction, (i', o') and (o, k'), are seen to reduce the electron density along the $C_2\cdots C_M$ line, reflecting the repulsive character of $E_K(i', o')$ and $E_K(o, k')$. The last component of $\Delta\rho(1|1)_{E_K}$, $\Delta\rho(1|1)_{E_K(o,o')}$, is shown in Fig. 3-d. Since the $E_K(o, o')$ is a considerable stabilization energy (Table 1), $\Delta\rho(1|1)_{E_K(o,o')}$ is expected to give the large bonding density in the middle region between ER and MR. The expectation is confirmed by the central huge build-up of the electron density in this figure. The change of the electron distribution, $\Delta\rho(1|1)$, due to the exchange and CT interactions is shown in Fig. 3-e. It is seen that the addition of CT interaction to the exchange interaction accumulates the electron density significantly in the $C_2\cdots C_M$ region. The electron transfer from ER to MR and the compensating one from MR to ER take place equivalently, the transition state having no ionic character. This result is in line with that pointed out in Eq. 8. The trend of these electron redistributions is similar to that of the recombination of two methyl radicals.[6] From Eq. 8 and Fig. 3, the overlap of two SOMO's, the common origin of $E_K(o, o')$, $D(o\to o')$, and $D(o'\to o)$, is shown to contribute remarkably to the new bond formation.

Calculation of the ΔW was carried out with respect to Model III. The results are also given in Table 1. It should be noted that the stabilization components, $E_K(o, o')$, $D(o\to o')$, and $D(o'\to o)$, are much smaller (absolute value) in this model than in the two recombination models I and II. The difference is attributable to the fact that the SOMO of ER is localized to a greater extent at the C_2 atom, and the overlap between both two SOMO's in Model III is small. Of the many components of ΔW in Model III, it is $D(i\to o')$ that gives the greatest stabilization energy to the disproportionation path. Comparing the two geometries of R'=2.4 Å and R'=2.0 Å of Model III, we find that D of R'=2.0 Å is larger than that of R'=2.4 Å, which is mainly due to the increase of its one component, $D(i'\to o)$, along the decrease of R'. The result shows that the $i\to o'$ type CT interaction becomes more important with the progress of reaction. In spite of the increase of D, ΔW of R'=2.0 Å is larger that that of R'=2.4 Å (Table 1). The global destabilization of Model III (R'=2.0 Å) is attributable to the increase of the repulsive term, E_K. In this calculation, the extension of the C_1–H_3 bond of ER in the course of disproportionation is not taken into account. In the realistic reaction model with the elongated C_1–H_3 bond, however, E_K would become small because of the separation of MR from ethylene fragment of ER and accordingly its ΔW would be much smaller than the ΔW of Model III which has the constant C_1–H_3 bond length regardless of R'.

In order to find the charge-transferred configuration which contributes most to the stabilization of the reacting system, the CI coefficients are obtained with respect to Model III (R'=2.0 Å).

$$\Psi = 0.955\Psi_0 - 0.120\Psi_{i_4\to o'} + 0.054\Psi_{o\to o'} + 0.035\Psi_{i_5\to o'} + 0.034\Psi_{i_2\to o'} + 0.024\Psi_{o'\to o} + \cdots. \quad (9)$$

Terms, i_2, i_4, and i_5, denote the DOMO's of ER and suffix attached to "i" stands for the numbering order counted from the lowest DOMO (see Fig. 1). Equation 9 shows that the CT interaction from i_4 (a') to o' (a_1) is the most important.[11] a' denotes the MO of ER within the X-Z plane belonging to the C_s point group and a_1 the MO of MR along the $C_M\cdots H_3$ line belonging to the C_{3v} point group. The i_4 is localized substantially at the C_1–H_3 bond and is the highest occupied (HO) MO of the C–H bond cleaved by disproportionation.

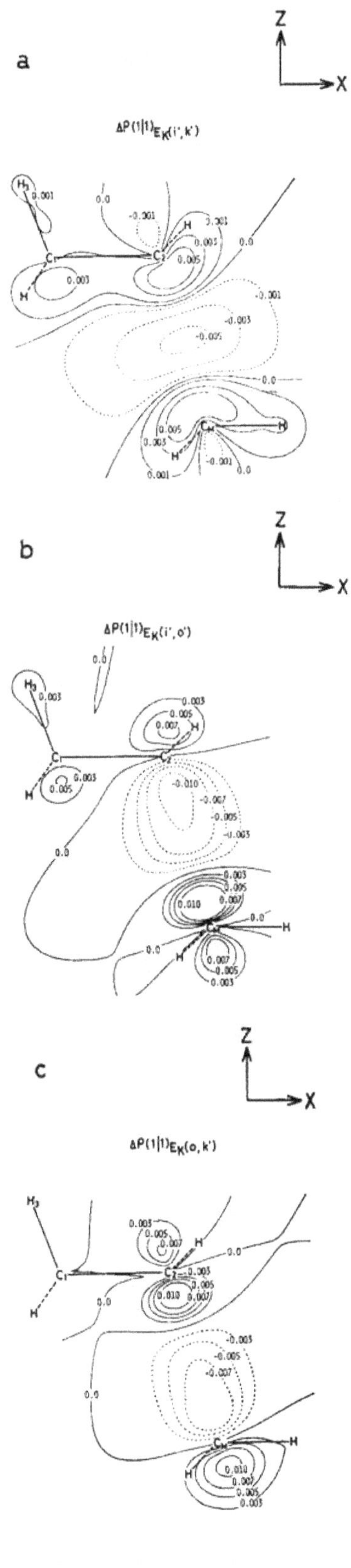

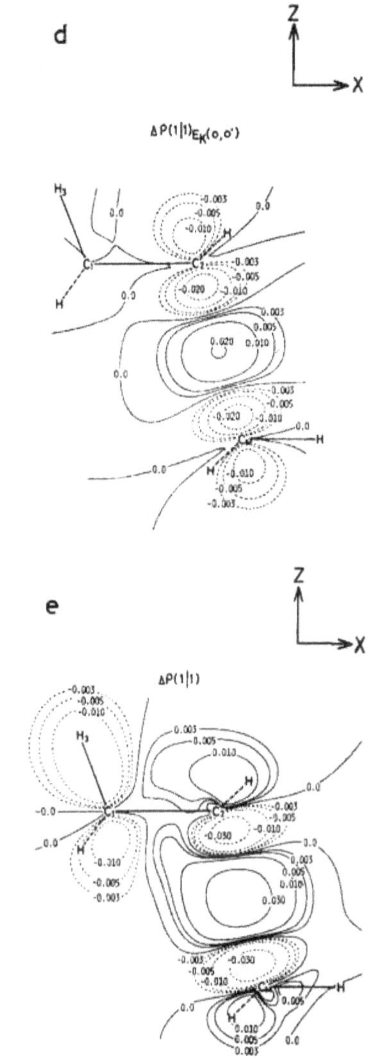

Fig. 3. The difference electron density maps of the recombination between ER and MR in Model II. The number attached to each contour line is in e/Å^3.

a The difference density map of $\Delta\rho(1|1)_{E_K(i',k')}$.
b The difference density map of $\Delta\rho(1|1)_{E_K(i',o')}$.
c The difference density map of $\Delta\rho(1|1)_{E_K(o,k')}$.
d The difference density map of $\Delta\rho(1|1)_{E_K(o,o')}$.
e The difference density map of $\Delta\rho(1|1)$.

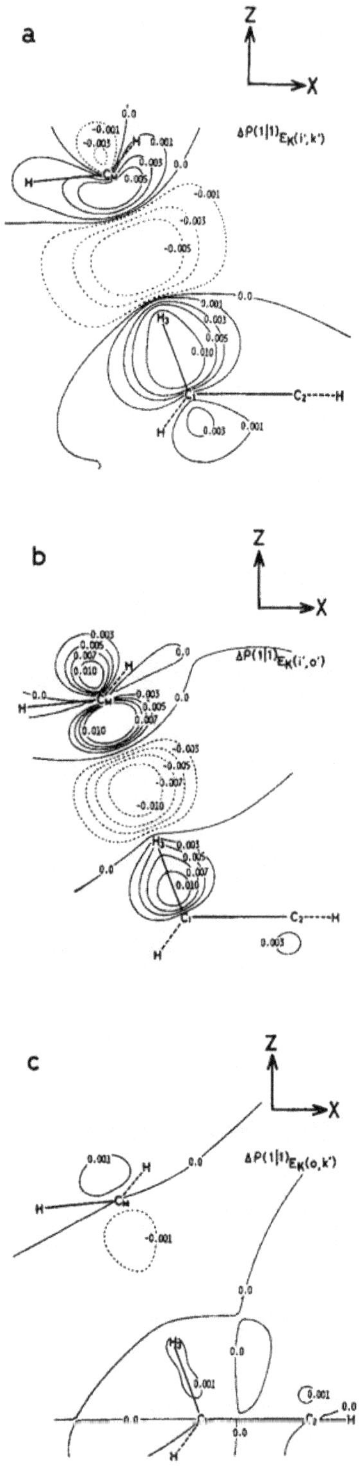

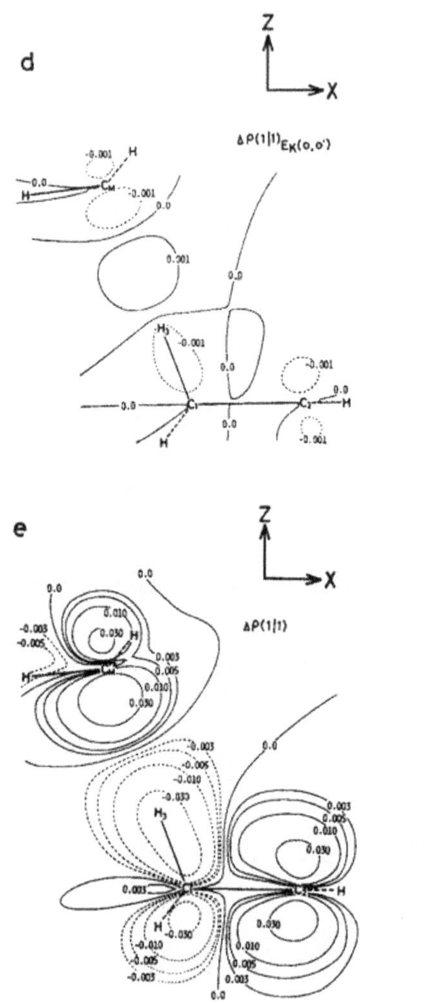

Fig. 4. The difference electron density maps of the disproportionation between ER and MR in Model III. The number attached to each contour line is also in e/Å^3.

a The difference density map of $\Delta\rho(1|1)_{E_K(i',k')}$.
b The difference density map of $\Delta\rho(1|1)_{E_K(i',o')}$.
c The difference density map of $\Delta\rho(1|1)_{E_K(o,k')}$.
d The difference density map of $\Delta\rho(1|1)_{E_K(o,o')}$.
e The difference density map of $\Delta\rho(1|1)$.

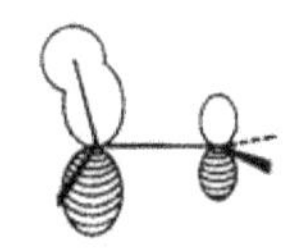

the i_4 MO of ER

The MO of this type is hereafter called the "C–H HOMO." In Model III representing disproportionation, this particular CT interaction plays an important role to weaken the C_1–H_3 bond through the loss of electron density of the i_4, forming a new $C_M \cdots H_3$ bond. From Eq. 9, MR is found to be an electron acceptor and ER to be an electron donor during the course of reaction. This donor-acceptor relation was also observed in a previous work dealing with the hydrogen abstraction reaction of a methane by a methyl radical.[12)]

The electron redistribution in disproportionation is evaluated in the same way as in the recombination. Four components of $\Delta\rho(1|1)_{E_K}$ in Eq. 7 are shown separately in Fig. 4. They exhibit the same trend of electron movement as is observed in Fig. 3 of the recombination. $\Delta\rho(1|1)_{E_K(i',k')}$ (Fig. 4-a) show the large amount of the decreased density in the $C_M \cdots H_3$ region. The electron density, running away from this region, is piled up around C_M and the C_1–H_3 bond. The electron density around C_M resembles that of the recombination (Fig. 3-a). The density redistribution, $\Delta\rho(1|1)_{E_K(i',o')}$ (Fig. 4-b) is found to give the decreased density in the $C_M \cdots H_3$ line. Since the overlap between the SOMO of ER and any DOMO of MR is small, the change of the electron arrangement brought about by $\Delta\rho(1|1)_{E_K(o,k')}$ is expected to be minor (Fig. 4-c). The three difference densities, $\Delta\rho(1|1)_{E_K(i',k')}$, $\Delta\rho(1|1)_{E_K(i',o')}$, and $\rho\Delta(1|1)_{E_K(o,k')}$, corresponding to three repulsive energy terms of E_K in Eq. 4 do not contribute to the formation of a new C_M–H_3 bond. The contour map of the electron density of $\Delta\rho(1|1)_{E_K(o,o')}$ is shown in Fig. 4-d. Since the overlap between both SOMO's is smaller in Model III than that in Model II, the density redistribution through $\Delta\rho(1|1)_{E_K(o,o')}$ is not so noticeable in the former model. Figure 4-e shows $\Delta\rho(1|1)$, representing the electron redistribution due to the exchange and CT interactions. The considerable accumulation of the electron density around C_M of MR and C_2 of ER is observed. The diminution of the electron density appears at the C_1–H_3 bond of ER. The electron redistribution shows that the exchange and CT interactions, especially the $i_4 \rightarrow o'$ CT interaction, make electrons of ER flow into MR and that the C_1–H_3 bond is weakened by the loss of bonding electrons. The accumulated electron density around C_M operates to pull the H_3 atom, resulting in the formation of the new C_M–H_3 bond.

The change of atomic charges in Model III (R'= 2.0 Å) is shown in Table 2. We see that the exchange and CT interactions make the charge on the C_M atom of MR more negative and that on the H_3 atom of ER more positive. This indicates that the C_M atom of MR has a negative charge and the H_3 atom of ER has a positive charge at the transition state of disproportionation. The tendency is almost similar to that suggested in the case of the disproportionation of alkoxy radicals. The result showing that the transition state of disproportionation has the ionic character is in agreement with Benson's scheme (Scheme 2b), while that of the recombination has no ionic character as shown with respect to Model II.

Table 2. Change of net charges by the exchange and CT interactions in Model III
A more negative value indicates a larger electron population.

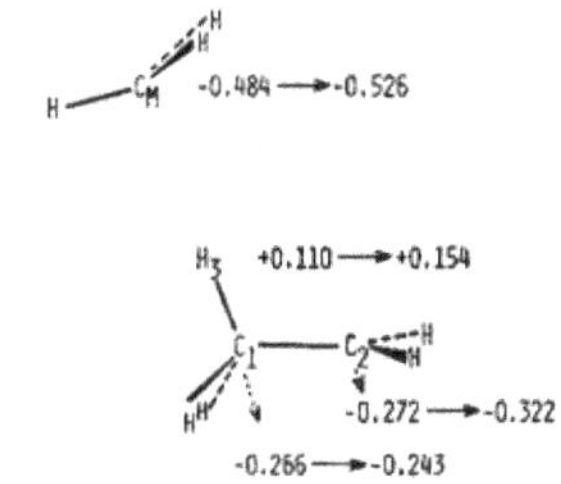

No formation of the π bond between C_1 and C_2 of ER is observed in Fig. 4-e. The absence of the bonding component which is the origin of the π bond in ethylene is attributable to the neglect of the excited configurations in the wave function Ψ of Eq. 3. The process of the π bond formation can be predicted qualitatively, even if the polarization effect is not taken into account explicitly. The orbitals which contribute to the π bond formation are considered to be the i_4 and o, since all residual MO's of ER have no large p_z atomic orbital (AO) component, that is, a π bond character. A schematic representation of the π bond formation by the orbital mixing of i_4 with o through o' is illustrated in Fig. 5.[13)] The interaction of o' with o and i_4 gives the ψ_1, ψ_2, and ψ_3 MO's. In a DOMO, ψ_2, o, and o' overlap in phase, whereas i_4 and o' overlap

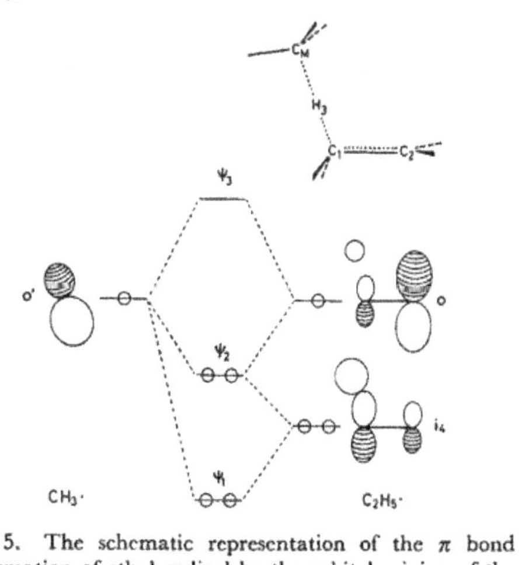

Fig. 5. The schematic representation of the π bond formation of ethyl radical by the orbital mixing of the i_4 and o through the o'. The signs of two MO's (i_4 and o) are chosen so that both of them may have positive overlap integrals with the o'. $\psi_2 = o - i_4 + o'$.

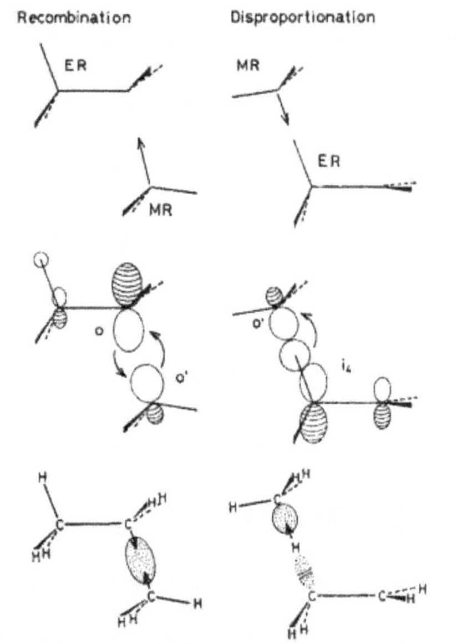

Fig. 6. Schematic representation of the reaction paths of recombination and disproportionation.

out of phase. Consequently in the ψ_2 orbital the interaction between the lobe at the C_1 atom of i_4 and the lobe at the C_2 atom of o contributes to the π bond formation. Thus, the π bond formation of ER is explained by the orbital mixing of i_4 with o through o'.

Let us summarize our results obtained so far. The reaction pathway of recombination differs a great deal from that of disproportionation, the reactions having no common transition state. This is illustrated schematically in Fig. 6. The SOMO-SOMO interaction is of importance for the progress of the reaction in recombination, whereas the CT interaction from the C–H HOMO (i_4) to the SOMO of one radical directed toward the hydrogen atom of the other is of importance in disproportionation. In the disproportionation, one of radical species which abstracts a hydrogen atom is an electron acceptor and the other radical species which has a hydrogen atom to be abstracted is an electron donor. Consequently the transition state of the disproportionation is expected to have an ionic character. In the recombination, however, this tendency is not observed.

Comparison with Experimental Data

Since the interaction energy, ΔW, is entirely controlled by the extent of the overlap of MO's obtained independently in two radical species, its value can be estimated roughly by analyzing the shape of these MO's. Instead of actual calculation of ΔW, the reactivity of several radical species is conveniently discussed in terms of the shape of some particular MO's. In this section, the theoretical prediction given by such a simple analysis is compared with available experimental data as regards recombination and disproportionation.

Recombination. The simplest way in which odd-electron species are transformed into stable compounds is through the chemical bonding between the radical centers of two such species. However, in a few cases, the recombination takes an unusual path. The dimerization of cyanomethyl radical leads to a ketenimine as well as to the "normal" head-to-head dimer.[14] The former product is an unstable intermediate which rearranges quantitatively to the latter.

$$2R_2\dot{C}\text{—CN} \longrightarrow R_2\underset{\text{CN}}{\underset{|}{C}}\text{—N=C=CR}_2 + R_2\underset{\text{CN}}{\underset{|}{C}}\text{——}\underset{\text{CN}}{\underset{|}{CR_2}}$$

In order to study the reactivity of this radical, the SOMO of cyanomethyl radical is obtained by the extended Hückel (EH) MO method.[15] The shape of the SOMO is as follows.

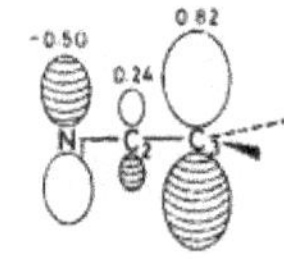

The numbers attached to each lobe denote the coefficients of the AO's of the SOMO. Judging from the shape of the SOMO of cyanomethyl radical, two recombination paths are possible. The first is the formation of the C_3–C_3 bond, giving the latter stable product, the C_3 atom having the largest lobe of the SOMO. The second is the linkage of the N_1 of one radical with the C_3 of the other, giving an abnormal product (ketenimine). The second mode of recombination is probably due to the following. Since N_1 is charged more negatively than C_3, reflecting the difference of electronegativity, the Coulomb long-range force represented by E_Q of Eq. 4 becomes largely attractive by the approach of the $C_3\cdots N_1$. Thus, the second path of recombination is more favorable than the first in the initial stage of reaction. The abnormal recombination product is formed when the carbon radical center of one radical attacks the nitrogen atom of the other.

Disproportionation. *Isotopic Effect*: The isotopic effect is expected in the disproportionation path shown in Fig. 6. The substitution from H to D on the carbon atom adjacent to the radical center reduces the disproportionation rate constant, k_d, whereas in the recombination path k_c does not seem to be affected so much by this substitution. In Eq. 11, isotopic substitution on the carbon atom adjacent to the radical center has a large effect on the disproportionation, but a small effect on the recombination in Eq. 10 and the disproportionation in Eq. 12. Consequently, the isotopic substitution is expected to reduce k_d/k_c. It is experimentally evident that in the gas, liquid, and solid phases there is a significant kinetic isotopic effect by which the substitution from H to D reduces k_d/k_c.[18] Gibian and Corley measured the k_d/k_c value of $C_6H_5CHCH_3$ and $C_6H_5CHCD_3$ in the benzene solution, and found the $[k_d/k_c(\text{H})]/[k_d/k_c(\text{D})]$ value to be 1.87.[16] Boddy and Steacie[17] found the k_d/k_c value of $C_2D_5\cdot$ to be 0.0985 in the gase phase, and that of $C_2H_5\cdot$ 0.13, leading to the ratio $[k_d/k_c(\mathrm{H})]/[k_d/k_c(\mathrm{D})]=1.32$. The data support the reaction paths of recombination and disproportionation assumed throughout the present work.

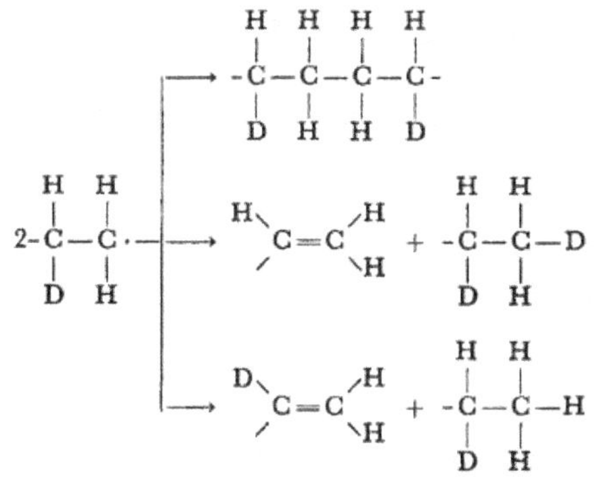

(10)

(11)

(12)

Substitution Effect: The CT interaction from the C–H HOMO of one radical to the SOMO of the other radical is found to be of importance to cause disproportionation. In our calculation of ΔW, the former radical is ER and the latter MR. The result together with $\Delta\rho(1|1)$ of Fig. 4-e leads to the prediction that the disproportionation occurs more easily when the radical receiving an abstracted hydrogen (R) has an electron-accepting group or the radical releasing a hydrogen (R′) has an electron-donating group. Let us examine this substitution effect in more detail. For the sake of convenience, the ratio of the rate constants, k_d/k_c, is denoted by $\Delta(\mathrm{R, R'})$. The number of the β-hydrogens of R′ is defined as βH. The large value of $\Delta(\mathrm{R, R'})/\beta\mathrm{H}$ indicates the ease of the occurrence of disproportionation.

Table 3 gives the $\Delta(\mathrm{R, R'})$ of some small alkyl radical reactions.[18] The orbital energies of the C–H HOMO of R′ are calculated by the EHMO method. The elevation of the level of the C–H HOMO corresponds to the higher reactivity of the electron-donating radical. The C–H HOMO level of the ethyl radical is found to be raised by the electron donating group. In the disproportionation between the methyl radical and the substituted ethyl radical, the $\Delta(\mathrm{R, R'})/\beta\mathrm{H}$ increases as the level of the C–H HOMO of the latter radical rises. This tendency is also observed when the ethyl radical reacts with the substituted ethyl radical (Table 3). This is also understandable from the C–H HOMO level. The inclusion of the electron acceptor group, Cl, to the R′ makes the $\Delta(\mathrm{CCl_3, C_2H_4Cl})/\beta\mathrm{H}$ smaller than the $\Delta(\mathrm{CCl_3, C_2H_5})/\beta\mathrm{H}$. Although the recombination is assumed to be uninfluenced by the substitution effect, the trend of the above data suggests that the disproportionation rate constant is affected by the energy gap between the SOMO of the radical receiving the hydrogen atom and the C–H HOMO of the radical releasing the hydrogen atom when the height of the SOMO is kept frozen.

Table 3. Ratio of rate constants with the substitution effect

R	R′	$\Delta(\mathrm{R, R'})$	$\Delta(\mathrm{R, R'})/\beta\mathrm{H}$	HOMO level[a] (eV)
$CH_3\cdot$	$CH_3CH_2\cdot$	0.039	0.012	−15.18
	$CH_3CH_2CH_2\cdot$	0.058	0.029	−14.00
	$(CH_3)_2CHCH_2\cdot$	—	—	−13.38
$C_2H_5\cdot$	$CH_3CH_2\cdot$	0.065	0.022	−15.18
	$CH_3CH_2CH_2\cdot$	0.066	0.033	−14.00
	$(CH_3)_2CHCH_2\cdot$	0.04	0.04	−13.38
$CCl_3\cdot$	$CH_3CH_2\cdot$	0.22	0.073	−15.18
	$CH_2ClCH_2\cdot$	0.14	0.070	−15.23

a) The calculated orbital energy of the C-H HOMO (see Text) is indicated.

Solvent Effect: The density map, $\Delta\rho(1|1)$, (Fig. 4-e) shows that the transition state of the disproportionation is polarized in spite of the reaction of two neutral radicals, but not that of recombination. This leads to the expectation that the disproportionation is accelerated in polar solvents, but not recombination. Stefani measured the ratio of the rate constants, k_d/k_c, for the termination reaction of ethyl radicals in various sol-

Table 4. Disproportionation-recombination of ethyl radicals as a function of solvents[a)]

	Scavenger	k_d/k_c	δ[b)]	Ω	Z[c)]	E_T[c)]
Gas phase	None	0.14				
Isooctane	None	0.144±0.003	48		60.1	
Isooctane	Styrene	0.145±0.007	48		60.1	
Limonene	None	0.158±0.005	56			
Ethylbenzene	None	0.156±0.001	59			
m-Xylene	None	0.165±0.003	61			
Toluene	None	0.167±0.003	62			33.9
2-Butanol	None	0.168±0.003	68			
2-Propanol	Styrene	0.178±0.001	70		76.3	48.6
1-Propanol	Styrene	0.181±0.003	72		78.3	50.7
Aniline	Styrene	0.195±0.001	75			44.3
Acetonitrile	None	0.200±0.006	83	0.692	71.3	46.0
Acetonitrile	Styrene	0.208±0.006	83	0.692	71.3	46.0
Ethylene glycol	None	0.241±0.003	97		85.1	56.3

a) From Ref. 18. b) Estimated from Fig. 1 of Ref. 18. c) *Cf.*, M. Seno and T. Arai, "Yukikagakuhanno Ni Okeru Yobaikoka," ed by T. Asahara, Sangyotosho, Tokyo (1970).

vents.[18] He found that $\log(k_d/k_c)$ is a linear function of $\sqrt{\delta}$ where δ is the Hildebrand solubility parameter, expressing the internal pressure of the solvent.[19] Herbrandson and Neufeld proposed[20] that δ correlated with various polarity parameters such as Y,[21] Ω,[22] Z,[23] and E_T.[24] The values of k_d/k_c, δ, and these four polarity parameters are given in Table 4. The larger the values of the parameters, the larger the solvent polarity. The transition state of disproportionation is expected to have a polar character to be stabilized by the solvent. This is in line with our proposal of the different ionic characters of recombination and disproportionation.

Alkoxy radicals are known to give very high fractions of disproportionation in comparison with alkyl radicals.[1a] It is suggested that this phenomenon is attributable to the increased polar conditions by an electronegative oxygen atom at the transition state.[25]

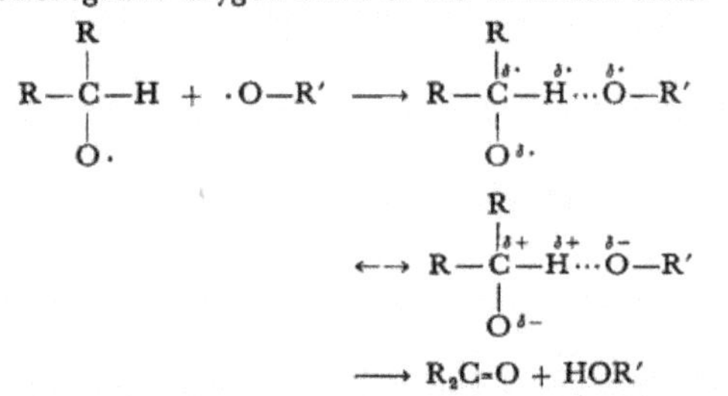

The suggestion that the transition state of the disproportionation of alkoxy radicals has an ionic character supports our view of the transition state of the disproportionation of alkyl radicals.

Conclusion

Recombination, one of two termination reactions, takes a reaction path differing from that of the other termination reaction, disproportionation. These two reactions are shown to have no common transition state. The recombination path is the route by which two radicals approach in such a direction that the overlap between both SOMO's may become maximum. It is suggested that the transition state of recombination is not much polarized. The disproportionation path is the route by which one radical receiving a hydrogen atom attacks the β hydrogen atom of the other radical. The CT interaction from the C–H HOMO of one radical to the SOMO of the other abstracting a hydrogen atom plays an important role to cause disproportionation. The radical receiving a hydrogen atom is found to be an electron acceptor and the radical with the C–H bond broken in the disproportionation an electron donor. The transition state of disproportionation is shown to have a polar character.

We would like to express our gratitude to the staff of the Data Processing Center, Kyoto University, for permission to use the FACOM 230-75 computer. The MO calculation was carried out with financial support of the Ministry of Education (Grant-in-Aid 139012).

References

1) (a) M. J. Gibian and R. C. Corley, *Chem. Rev.*, **73**, 441 (1973); (b) J. A. Kerr, "Free Radicals," Vol. 1, ed by J. K. Kochi, John Wiley & Sons, New York, N. Y. (1973).

2) J. A. Kerr and A. F. Trotman-Dickenson, *Prog. React. Kinet.*, **1**, 107 (1961).

3) (a) S. W. Benson, *Adv. Photochem.*, **2**, 1 (1964); (b) H. A. Gillis, *Can. J. Chem.*, **49**, 2861 (1971).

4) J. N. Bradley and B. S. Rabinovitch, *J. Chem. Phys.*, **36**, 3498 (1962).

5) O. K. Rice, *J. Phys. Chem.*, **65**, 1588 (1961).

6) S. Yamabe, T. Minato, H. Fujimoto, and K. Fukui, unpublished result (1977).

7) T. Yonezawa, K. Yamaguchi, and H. Kato, *Bull. Chem. Soc. Jpn.*, **40**, 536 (1967).

8) K. Fukui, H. Fujimoto, and S. Yamabe, *J. Phys. Chem.*, **76**, 232 (1972).

9) W. A. Lathan, W. J. Hehre, and J. A. Pople, *J. Am. Chem. Soc.*, **93**, 808 (1971).

10) G. Herzberg, "Molecular Spectra and Structure," D. Van Nostrand Co., New York (1966).

11) It should be noted that when a radical receiving a hydrogen atom has a high-lying SOMO, or a radical giving a hydrogen atom has the lowest unoccupied (LU) MO with a very low energy localized at the C–H bond, the CT interaction from the SOMO to the LUMO could be more important than that from the particular DOMO to the SOMO.

12) H. Fujimoto, S. Yamabe, T. Minato, and K. Fukui, *J. Am. Chem. Soc.*, **94**, 9205 (1972).

13) S. Inagaki, H. Fujimoto, and K. Fukui, *J. Am. Chem. Soc.*, **98**, 4054 (1976).

14) (a) C. G. Hammond, O. D. Trapp, R. T. Keys, and D. L. Neff, *J. Am. Chem. Soc.*, **81**, 4878 (1959); (b) G. Ayrey, K. L. Evans, and D. J. D. Wong, *Eur. Polym. J.*, **9**, 1347 (1973).

15) R. Hoffmann, *J. Chem. Phys.*, **39**, 1397 (1963).

16) M. J. Gibian and R. C. Corley, *J. Am. Chem. Soc.*, **94**, 4178 (1972).

17) P. J. Boddy and E. W. Steacie, *Can. J. Chem.*, **38**, 1576 (1960).

18) A. P. Stefani, *J. Am. Chem. Soc.*, **90**, 1694 (1968).

19) J. H. Hildebrand and R. L. Scott, "Regular Solutions," Prentice-Hall, Inc., Englewood Cliffs, N. J. (1962).

20) H. F. Herbrandson and F. R. Neufeld, *J. Org. Chem.*, **31**, 1140 (1966).

21) A. H. Fainberg and S. Winstein, *J. Am. Chem. Soc.*, **78**, 2770 (1956).

22) J. A. Berson, Z. Hamlet, and W. A. Mueller, *J. Am. Chem. Soc.*, **84**, 297 (1962).

23) E. M. Kosower, *J. Am. Chem. Soc.*, **80**, 3253 (1958).

24) K. Dimroth, C. Reichardt, T. Siepmann, and F. Bohlmann, *Ann. Chem.*, **661**, 1 (1963).

25) D. F. Dever and J. G. Calvert, *J. Am. Chem. Soc.*, **84**, 1362 (1962).

[Reprinted from the Journal of Physical Chemistry, 81, 727 (1977).]

A Molecular Orbital Theoretical Study on $(SN)_2$ Molecules at the Initial Stage of Polymerization to $(SN)_x$

Tokio Yamabe, Kazuyoshi Tanaka, Kenichi Fukui,*

Department of Hydrocarbon Chemistry, Faculty of Engineering, Kyoto University, Sakyo-ku, Kyoto 606, Japan

and Hiroshi Kato

College of General Education, Nagoya University, Chikusa-ku, Nagoya 464, Japan *(Received May 24, 1976; Revised Manuscript Received December 28, 1976)*

Concerning the initial stage of solid state polymerization of $(SN)_x$, semiempirical INDO-type ASMO-SCF calculations are performed for the precursor S_2N_2, and several "deformed" structures of $(SN)_2$, the dimeric unit $(SN)_4$, and the trimeric unit $(SN)_6$. According to the results of calculations, the triplet biradical nature emerges in appropriately "deformed" structures of $(SN)_2$. This agrees well with the experimental results wherein paramagnetism (g = 2.005) is observed at the initial stage of polymerization. $(SN)_4$ and $(SN)_6$, however, show no triplet nature probably corresponding to the fact that the system gradually becomes diamagnetic as the polymerization proceeds. The mechanism of polymerization at the initial stage is discussed based on our calculated results.

Introduction

There has recently been considerable interest in the covalent polymer, polymeric sulfur nitride, $(SN)_x$, since the discovery that it is a low-dimensional metallic conductor.[1,2] The theoretical treatment for $(SN)_x$ has also been carried out to reveal the metallic character of the $(SN)_x$ in its band structure.[3] Recently, MacDiarmid et al. have reported that the crystalline monomeric S_2N_2 polymerizes thermally in the solid state to form the $(SN)_x$ polymer.[4] They have also determined the structure of $(SN)_x$ as well as S_2N_2 shown in Figure 1A and 1B by x-ray diffraction. The initially colorless tabular monoclinic crystal of S_2N_2 turns intense blue-black and becomes paramagnetic giving a weak free radical signal at g = 2.005. This substance then changes to golden-colored, diamagnetic crystals of the $(SN)_x$ polymer.

For the mechanism of the solid state polymerization, they have also proposed, particularly at the first step, the thermal opening (widening) of one of the S–N bonds in S_2N_2 to form a biradical species. This radical can then link up with another S_2N_2 molecule in the crystal to give the partly polymerized, paramagnetic $(SN)_2$ of blue-black color, and finally the golden $(SN)_x$ polymer. Moreover, Baughman et al. have studied the polymerization including

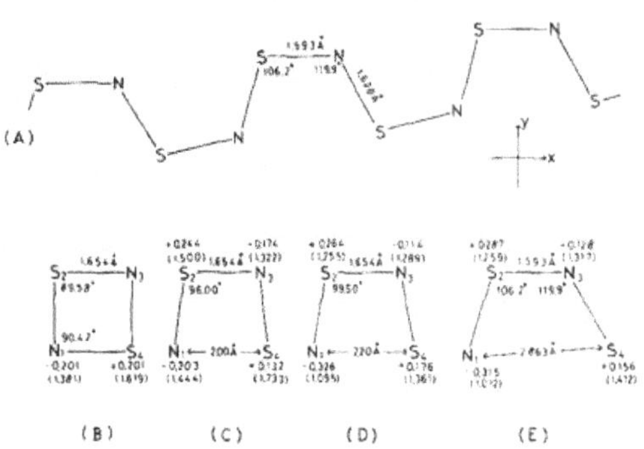

Figure 1. Geometries of the polymer $(SN)_x$ (A), the precursor S_2N_2 (B), and several deformed $(SN)_2$ molecules (C, D, and E) with atomic net charges and π electron densities (bracketed values). For (D) and (E), those of the triplet states are shown.

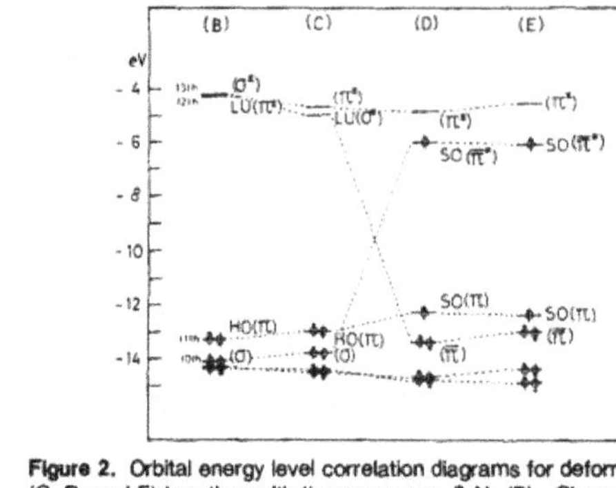

Figure 2. Orbital energy level correlation diagrams for deformed $(SN)_2$ (C, D, and E) together with the precursor, S_2N_2 (B). Changes of the orbital energy levels below the eighth MO are not essential and omitted here.

TABLE I: Transition Energies and CI Improvements for the Stability of the Triplet States of the Deformed $(SN)_2$ (in eV)

Structure	Δ^3E_{π,π^*} (11th MO → 13th MO)	Δ^3E_{π,σ^*} (11th MO → 12th MO)	Δ^3E_{π,σ^*} (after CI)
B	1.922[a]	2.514[b]	2.867
C	1.401	1.223	1.627
D[c]	1.060	−0.693- ($\Delta^3E_{\pi,\bar{\pi}^*}$)	−0.068- ($\Delta^3E_{\pi,\bar{\pi}^*}$)
E[c]	1.876	−0.544- ($\Delta^3E_{\pi,\bar{\pi}^*}$)	−0.153- ($\Delta^3E_{\pi,\bar{\pi}^*}$)

[a] In this case, 11th MO → 12th MO. [b] In this case, 11th MO → 13th MO. [c] In these structures Δ^3E_{π,σ^*} is replaced by $\Delta^3E_{\pi,\bar{\pi}^*}$ because of the orbital crossing occurring in the deformation from C to D (see text).

defect structure crystallographically.[5]

In the present paper, we try to elucidate the polymerization reaction, particularly at the initial stage, and to confirm the mechanism proposed above by MO calculations. At first, we calculate the electronic states of the precursor S_2N_2, and then, of the several "deformed" structures of $(SN)_2$ as shown in Figure 1 in order to investigate the thermal initiation reaction. The calculations are performed with the use of the semiempirical INDO-type ASMO–SCF method for valence electrons including sulfur 3d orbitals,[6a] and the stability of the triplet state (open shell) is examined by the sign of the transition energy from the lowest singlet state (closed shell).[6b] This method has given fairly reasonable results, especially for the transition energies[6c-e] because of good parametrization for Coulomb repulsion integrals.[7] Examined further are the electronic states of the dimeric unit $(SN)_4$ and the trimeric unit $(SN)_6$.

Electronic State of S_2N_2

The molecular structure of S_2N_2 employed here is shown in Figure 1B, which has been determined by MacDiarmid et al.[4a] This is square-planar (D_{2h}) with 90° of the S–N–S angle[8] and almost equal bond lengths, 1.654 Å.[9] The calculation of the electronic state of S_2N_2 has been carried out by using the SCF–Xα scattered wave method, which gives a rather highly polar S–N bond ($S^{0.48+}$–$N^{0.48-}$).[10] This is probably because the method overestimates the contribution from the ionic structure.

The atomic net charges and π electron densities calculated here are also shown in Figure 1. This molecule has six π electrons: two π electrons from each sulfur atom and one π electron from each nitrogen atom. In consequence of the planar D_{2h} structure, the occupied orbitals are separated into σ and π orbitals. In the S–N bond, the charge densities of σ electrons are 4.180 and 3.820 on S and N, respectively, whereas those of π electrons are 1.619 and 1.381, respectively. This means that the charge transfer of π electrons from S to N (0.381) exceeds that of σ electrons from N to S (0.180). Consequently, the net charges on S and N are +0.201 and −0.201, respectively, as is usual for S–N bonds in other molecules[11] reflecting the fact that sulfur is less electronegative than nitrogen. As a matter of course, the highest occupied molecular orbital (HOMO) is of typical π MO, following another π and two σ MOs close together as shown in Figure 2. It should be noticed here that there are two extraordinary low lying unoccupied MOs with negative eigenvalues and large S–N antibonding characters, i.e., π^* and σ^* being very close each other. Such low lying σ^* MO may often induce bond cleavages or drastic rearrangement among MOs.

Electronic States of "Deformed" $(SN)_2$

According to the mechanism of the polymerization reaction proposed by MacDiarmid et al., S_2N_2 is considered to be deformed thermally (near 0 °C) as the first step to open one S–N bond, giving biradical character. In this section, we investigate how the biradical character appears according to the deformation. The geometries of $(SN)_2$ chosen here are shown in Figure 1C–E. In Figure 1C and 1D, the bond lengths of the sides opposite to the opened S–N bond are taken as being equal to the S–N bond length in the precursor, S_2N_2 (Figure 1B), since shortening of them is unfavorable from the viewpoint of nuclear repulsion as found in this calculation. Figure 1E is employed from the fragment of the $(SN)_x$ polymer in Figure 1A. Sulfur 3d orbitals are requisite for the SCF convergence in the calculations for the open forms even though their contribution is not significant. As previously mentioned, we devoted our attention to these calculations in order to investigate the possibility of the triplet biradical by examining the transition energy from the lowest singlet state to the triplet state in each structure.

In the cases of the precursor (Figure 1B) and the slightly deformed structure (Figure 1C), the ground states are, not unexpectedly, singlet, although the triplet π–π^* transition energy (Δ^3E_{π,π^*}) and the triplet π–σ^* transition energy (Δ^3E_{π,σ^*}) of the latter are positively small as shown in Table I. Next, we tried the case of a little more deformed structure (Figure 1D). Δ^3E_{π,π^*} remains positive and almost the same value, while $\Delta^3E_{\pi,\bar{\pi}^*}$ is found to be negative. Hereafter, we define the signs π and $\bar{\pi}$ to represent the

TABLE II: α Spin Densities in $(SN)_2$ of Structures D and E[a]

Atom	s	p_x	p_y	$\bar{\pi}$ spin density	p_z	π spin density
			D			
N_1	0.000	0.272	0.000	0.273	0.187	0.187
N_3	0.008	0.081	0.036	0.125	0.145	0.145
S_2	0.001	0.069	0.005	0.092	0.036	0.082
S_4	0.000	0.469	0.026	0.510	0.572	0.586
			E			
N_1	0.000	0.214	0.000	0.215	0.168	0.168
N_3	0.009	0.067	0.098	0.175	0.171	0.171
S_2	0.000	0.046	0.004	0.052	0.008	0.060
S_4	0.000	0.364	0.172	0.559	0.584	0.601

[a] AO spin densities of 3d orbitals are of small values and hence omitted here. The values of $\bar{\pi}$ and π spin densities, however, contain the 3d contributions.

$\bar{\pi}$-type MO perpendicular to the molecular plane and that in the molecular plane, respectively. The above result means that the ground state of such a structure is a triplet formed from two singly occupied (SO) MOs, i.e., π SOMO and $\bar{\pi}$* SOMO, as shown in Figure 2. The same result was obtained for structure E (Figure 1). Furthermore, these results were improved by the configuration interaction (CI) method including one electron excitations for the triplet states and pairing excitations for the singlet states within all ranges from (HO − 3)MO to (LU + 3)MO. The situations are essentially not altered after CI as shown in Table I. Thus, it is probable that the triplet state is energetically favorable for the ground state of such deformed structures rather than the singlet state. The atomic net charges and π electron densities of each structure are also shown in Figure 1, where the values for Figure 1D and 1E are those of the triplet states. We assume here that the open-shell MOs in the triplet state are occupied with two more electrons of α spins. The α spin densities of the π MO and the $\bar{\pi}$* MO are shown in Table II. As a whole, one cannot find a large change of the atomic net charges even in the triplet states of Figure 1D and 1E. This implies that the transition from the singlet state to the triplet state is not due to intramolecular charge transfer. On the other hand, such a transition would be expected to cause a drastic change of the number of π electrons from 6 to 5, as one of the π electrons is transferred from the π MO (11th) perpendicular to the molecular plane to the $\bar{\pi}$* MO (12th) in the plane in the biradicals of the structures D and E (Figure 2). The large spin densities of both the π electrons and the $\bar{\pi}$ electrons are mainly on S_4 and on N_1. The $\bar{\pi}$ radical electron may be more reactive than π radical electron because the orbital energy of $\bar{\pi}$* SOMO is much higher than that of π SOMO. Interestingly, the dominant component of such $\bar{\pi}$* SOMO are p_x and as such are consistent with the direction of the polymerization as shown in Figure 1A.

To understand such singlet–triplet transition in detail, we give correlation diagrams of the orbital energy levels for each structure as shown in Figure 2 and pictures of some of these MOs of both Figure 2C and 2D are given in Figure 3. The 10th MO and the 12th MO in Figure 3C are of bonding (pσ) and antibonding (pσ*) character, respectively, to both the N_1–S_4 and the S_2–N_3 bonds, and are of antibonding (p$\bar{\pi}$*) and bonding (p$\bar{\pi}$) character, respectively, to both the N_1–S_2 and the N_3–S_4 bonds. Thus, if the p$\bar{\pi}$ bond should be stronger than the pσ bond by the elongation of the pσ bond, there may be a possibility of level crossing between the 10th MO and the 12th MO. In fact, when the structure of $(SN)_2$ changes from Figure 1C to 1D, the occurrence of the level crossing between the

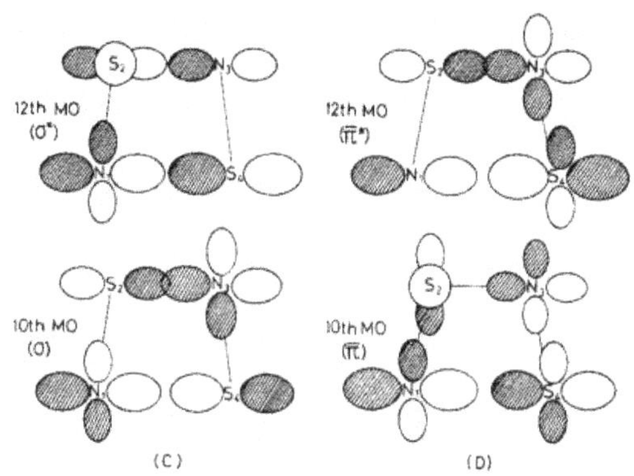

Figure 3. Pictures of 10th and 12th MOs of $(SN)_2$ having structures (C) and (D), respectively. Only the major AO components are depicted.

TABLE III: Values of Fock Matrix Elements Influencing 10th Orbital Energies of C and D (in eV)

r AO	s AO	(C) $C^{10}_r C^{10}_s \langle r\|F\|s\rangle$[a]	(D) $C^{10}_r C^{10}_s \langle r\|F\|s\rangle$
N_1 $2p_x$	S_2 $3p_x$	0.373	−0.147
N_1 $2p_x$	S_4 $3p_x$	−1.461	−0.556
N_3 $2p_x$	S_4 $3p_x$	0.372	−0.439
N_3 $2p_y$	S_4 $3p_x$	−0.088	0.133

[a] See the text about the notations C^{10}_r and C^{10}_s.

12th MO (σ*) and the 10th MO (σ) is seen as shown in Figure 2. Namely, the 10th MO and the 12th MO in Figure 3C are transformed to give bonding (p$\bar{\pi}$) and antibonding (p$\bar{\pi}$*) character in Figure 3D, respectively, mainly with respect to the N_3–S_4 bond, and somewhat less with respect to the N_1–S_2 bond. These details are clarified by examining the values of the main Fock matrix elements, $C^{10}_r C^{10}_s \langle r|F|s\rangle$, where r and s are the component atomic orbitals (AOs) of the 10th MO, and C is the AO coefficient. It is seen from Table III that, in Figure 3C, the $N_1(2p_x)$–$S_4(3p_x)$ matrix element contributes to stabilization for the orbital energy, and that the $N_1(2p_x)$–$S_2(3p_x)$ and $N_3(2p_x)$–$S_4(3p_x)$ matrix elements contribute to destabilization. However, in Figure 3D, the $N_1(2p_x)$–$S_4(3p_x)$ element leads to less stabilization and the $N_3(2p_x)$–$S_4(3p_x)$ to stabilization. Since the orbital level crossing in going from Figure 3C to 3D is caused mainly by the changing of the sign of the $3p_x$ orbital on S_4 and the larger contribution of the 3s orbital on S_2, it would not be so difficult to bring about such a triplet transition, if the widening of the N_1–S_4 bond takes place easily.

In fact, according to the IR spectrum of the solid S_2N_2,[12] the strong bands at 476.2 and 785 cm^{-1} are assigned to the B_{2u} and B_{3u} modes, respectively, which correspond to a widening of one of the S–N bonds (and hence shortening of the opposite S–N bond). These extraordinary low frequencies imply the shallow potential of the S–N bond for such vibrational modes and therefore the widening may easily induced, even thermally, because, e.g., 476.2 cm^{-1} is equal to 1.36 kcal/mol, which is the same order of RT at room temperature (ca. 0.6 kcal/mol). This consideration indicates that the polymerization may be accelerated by applying the appropriate IR radiation.

Thus it is expected that, at the point where the level crossing occurs, the N_1–S_4 pσ bond vanishes and the N_3–S_4 p$\bar{\pi}$ bond prevails, and that the triplet state emerges in the course of the deformation probably due to vibrational motion. Furthermore, we have attempted to estimate the mixing of the triplet configuration with the singlet closed-shell configuration for Figure 2C, and the mixing of the singlet (original closed-shell) configuration with the triplet (biradical) configuration for Figure 2D and 2E through spin–orbit coupling. Considering the perturbation correction,[13] the state function, required at present, ${}^{1(3)}\Phi_i$, is given by

$$ {}^{1(3)}\Phi_i = {}^{1(3)}\Phi_i{}^0 + \sum_k \frac{\langle {}^{1(3)}\Phi_i{}^0 | H_{SO} | {}^{3(1)}\Phi_k{}^0 \rangle}{\Delta {}^3E_{i,k}} {}^{3(1)}\Phi_k{}^0 $$

where ${}^{1(3)}\Phi_i{}^0$ and ${}^{3(1)}\Phi_k{}^0$ are the unperturbed ith singlet (triplet) ASMO and the unperturbed kth triplet (singlet) ASMO, respectively. H_{SO} is usual spin–orbit coupling operator, and $\Delta^3E_{i,k}$ is the energy difference. In the case of Figure 2C, ${}^1\Phi_i{}^0$ refers to the closed-shell configuration and ${}^3\Phi_k{}^0$ to the $\pi \rightarrow \sigma^*$ (11th MO → 12th MO) triplet configuration, whereas in Figure 2D and E, ${}^3\Phi_i{}^0$ refers to the $\pi \rightarrow \bar{\pi}^*$ (11th MO → 12th MO) triplet configuration and ${}^1\Phi_k{}^0$ to the original closed-shell configuration. These ${}^{3(1)}\Phi_k{}^0$ states seem to contribute mainly to the second term in the above correction rather than to other configurations, because the energy differences Δ^3E_{π,σ^*} ($\Delta^3E_{\pi,\bar{\pi}^*}$) are considerably smaller. The matrix element in the numerator is further reduced to $\langle \Psi_\pi{}^{11} | H_{SO} | \Psi_{\sigma^*}{}^{12} \rangle$ ($\langle \Psi_\pi{}^{11} | H_{SO} | \Psi_{\bar{\pi}}{}^{*12} \rangle$), where $\Psi_\pi{}^{11}$ and $\Psi_{\sigma^*}{}^{12}$ ($\Psi_{\bar{\pi}^*}{}^{12}$) denote the 11th π MO and the 12th σ^* ($\bar{\pi}^*$) MO, respectively. Taking only one-center terms into account as to the matrix element and employing the spin–orbit coupling constants ζ of 74.2 and 363 cm^{-1} for nitrogen and sulfur, respectively,[14] we obtain

$$ {}^1\Phi_i = {}^1\Phi_i{}^0 + 0.009\,{}^3\Phi_k{}^0 \quad \text{for Figure 2C} $$
$$ {}^3\Phi_i = {}^3\Phi_i{}^0 - 0.019\,{}^1\Phi_k{}^0 \quad \text{for Figure 2D} $$

and

$$ {}^3\Phi_i = {}^3\Phi_i{}^0 - 0.006\,{}^1\Phi_k{}^0 \quad \text{for Figure 2E} $$

The orders of the mixing coefficients are considerably larger than those in usual organic hetero compounds[15] on account of the heavy atom (sulfur) effect. Moreover, the 11th π MO and the 12th σ^* ($\bar{\pi}^*$) MO are advantageous, considering the character of the orbital angular momentum operator involved in H_{SO}. Thus, in $(SN)_2$, the degree of mixing of the singlet state is essentially remarkable. In particular, the biradical resulting from the π–$\bar{\pi}^*$ triplet transition would easily be formed from the singlet state. It should be stressed here that the energy difference between the triplet and the singlet state is small (approximately 1 kcal/mol) and hence it would not be unreasonable if the population of the singlet state, even if it were energetically higher, should increase with an increase of temperature according to the Boltzmann distribution. In this respect, it is interesting to note that the ESR spectrum of blue-black, partly polymerized, $(SN)_2$ decreases in intensity with increase in the temperature at which it is measured, e.g., −100 °C vs. 25 °C.[16]

Electronic States of $(SN)_4$ and $(SN)_6$

In order to understand the polymerization steps further, we calculated the electronic states of linear $(SN)_4$ and $(SN)_6$. The geometries employed are those from appropriate fragments of the polymer in Figure 1A. The calculated atomic net charges and π electron densities are shown in Figure 4 and the orbital energy diagrams for these molecules are shown in Figure 5. It is seen that π electron densities are distributed less on N_1 and S_8 in $(SN)_4$, and on N_1 and S_{12} in $(SN)_6$, and that $(SN)_4$ and $(SN)_6$ have a total of 10 and 16 π electrons, respectively. Furthermore the ground states of these molecules are found to be singlet

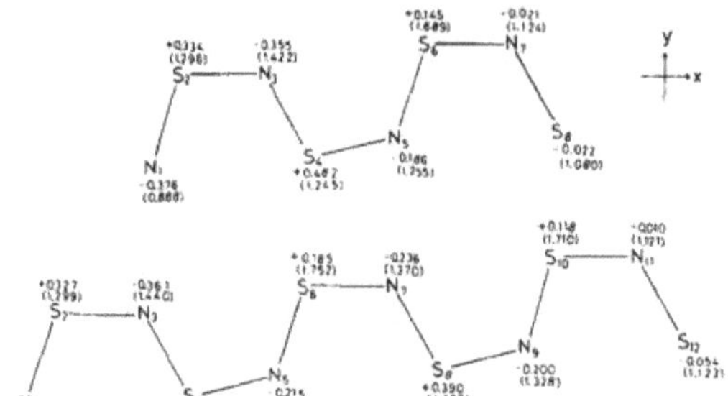

Figure 4. The calculated results for atomic net charges and π electron densities (bracketed values) of $(SN)_4$ and $(SN)_6$ molecules. Note that the total number of π electrons are 10 and 16, respectively, for these fragments.

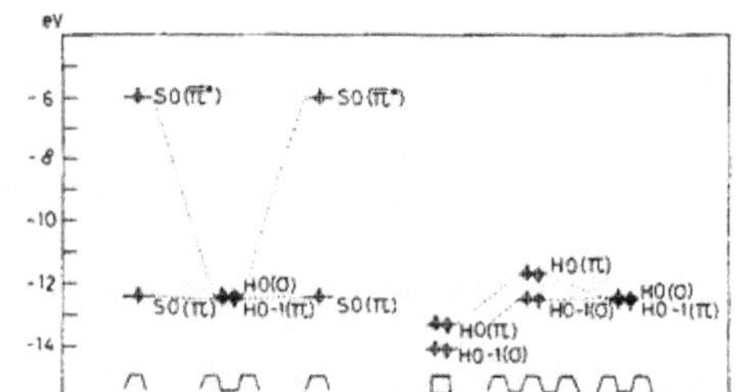

Figure 5. Orbital energy diagrams of S_2N_2, $(SN)_2$, $(SN)_4$, and $(SN)_6$.

rather than triplet states. The HOMO (σ) in $(SN)_4$ is considerably stabilized by the SOMO ($\bar{\pi}^*$)–SOMO ($\bar{\pi}^*$) interaction between two $(SN)_2$ biradicals. The (HO − 1)MO (π) is almost unchanged from the original π MO in structure E (Figure 5) and these two are nearly degenerate. In $(SN)_6$, however, the HOMO (π) is somewhat separated from the (HO − 1)MO (σ). These results and their HOMOs and (HO − 1)MOs suggest that $(SN)_4$ is derived from two $(SN)_2$ molecules in the triplet state ($5\pi + 5\pi$), and that $(SN)_6$ is derived from $(SN)_4$ and square S_2N_2 in the singlet state ($10\pi + 6\pi$).

Thus from the results of these calculations, the paramagnetism is expected to vanish as the polymerization proceeds, which is in agreement with the experimental results.

Conclusion

We have examined the electronic states of the square S_2N_2 molecule and several deformed $(SN)_2$ molecules, and of $(SN)_4$ and $(SN)_6$ molecules.

In the deformed $(SN)_2$, the triplet biradical nature begins to emerge as one of the S–N bonds opens. The appearance of the triplet state seems to correspond to the interaction of the opened S–N pσ bond with the adjacent S–N p$\bar{\pi}$ bond in the molecular plane.

From a simple estimation of the spin–orbit coupling in the deformed $(SN)_2$ molecules, there appears to be considerable mixing of the singlet state with the triplet state or of the triplet state with the singlet state. The opening of the S–N bond might be expected to occur easily, i.e., thermally, in view of the IR spectroscopic data of the appropriate vibrational modes.

$(SN)_4$ and $(SN)_6$, however, do not show triplet ground states. $(SN)_4$ is believed to result through the dimerization of two $(SN)_2$ biradicals, whereas $(SN)_6$ may be regarded as being formed by the addition of singlet $(SN)_4$ to singlet S_2N_2. Analogous processes would then lead to higher polymers.

Acknowledgment. We are grateful to the Data Processing Center of Kyoto University for its generous permission to use the FACOM 230-75 Computer. We owe thanks to Professor Alan G. MacDiarmid and to his outstanding co-workers for informing us about their valuable experimental result, and to the former for improving our English before the submission. We are also grateful to Dr. Hideyuki Konishi for his kind help in performing the MO calculations. This work was supported by a Grant-in-Aid for Scientific Research from the Ministry of Education of Japan (No. 065101).

References and Notes

(1) (a) V. V. Walatka, Jr., M. M. Labes, and J. H. Perlstein, *Phys. Rev. Lett.*, **31**, 1139 (1973); (b) C. Hsu and M. M. Labes, *J. Chem. Phys.*, **61**, 4640 (1974).

(2) Some recent references are (a) R. L. Greene, P. M. Grant, and G. B. Street, *Phys. Rev. Lett.*, **34**, 89 (1975); (b) A. A. Bright, M. J. Cohen, A. F. Garito, A. J. Heeger, C. M. Mikulski, P. J. Russo, and A. G. MacDiarmid, *ibid.*, **34**, 206 (1975); (c) R. L. Greene, G. B. Street, and L. J. Suter, *ibid.*, **34**, 577 (1975); (d) W. D. Gill, R. L. Greene, G. B. Street, and W. A. Little, *ibid.*, **35**, 1732 (1975); (e) P. M. Grant, R. L. Greene, and G. B. Street, *ibid.*, **35**, 1743 (1975); (f) L. Ley, *ibid.*, **35**, 1796 (1975); (g) P. Mengel, P. M. Grant, W. E. Rudge, B. H. Schlechtman, and D. W. Rice, *ibid.*, **35**, 1803 (1975); (h) L. Pintschovius, H. P. Geserich, and W. Möller, *Solid State Commun.*, **17**, 477 (1975); (i) C. H. Chen, J. Silcox, A. F. Garito, A. J. Heeger, and A. G. MacDiarmid., *Phys. Rev. Lett.*, **36**, 525 (1976).

(3) (a) D. E. Parry and J. M. Thomas, *J. Phys. C*, **8**, L45 (1975); (b) W. E. Rudge and P. M. Grant, *Phys. Rev. Lett.*, **35**, 1799 (1975); (c) W. I. Friesen, A. J. Berlinsky, B. Bergersen, L. Weiler, and T. M. Rice, *J. Phys. C*, **8**, 3549 (1975); (d) V. T. Rajan and L. M. Falicov, *Phys. Rev.*, **B12**, 1240 (1975); (e) A. Zunger, *J. Chem. Phys.*, **63**, 4854 (1975); (f) H. Kamimura, A. M. Glazer, A. J. Grant, Y. Natsume, M. Schreiber, and A. D. Yoffe, *J. Phys. C*, **9**, 291 (1976); (g) A. A. Bright and P. Soven, *Solid State Commun.*, **18**, 317 (1976).

(4) A. G. MacDiarmid, C. M. Mikulski, P. J. Russo, M. S. Saran, A. F. Garito, and A. J. Heeger, *J. Chem. Soc., Chem. Commun.*, 476 (1975); (b) C. M. Mikulski, P. J. Russo, M. S. Saran, A. G. MacDiarmid, A. F. Garito, and A. J. Heeger, *J. Am. Chem. Soc.*, **97**, 6358 (1975).

(5) R. H. Baughman, R. R. Chance, and M. J. Cohen, *J. Chem. Phys.*, **64**, 1869 (1976).

(6) (a) T. Yonezawa, H. Konishi, and H. Kato, *Bull. Chem. Soc. Jpn.*, **42**, 933 (1969); (b) C. C. J. Roothaan, *Rev. Mod. Phys.*, **23**, 69 (1951); (c) H. Konishi, H. Kato, and T. Yonezawa, *Theor. Chim. Acta (Berl.)*, **19**, 71 (1970); (d) H. Yamabe, H. Kato, and T. Yonezawa, *Bull. Chem. Soc. Jpn.*, **44**, 22 (1971); (e) H. Yamabe, H. Kato, and T. Yonezawa, *Ibid.*, **44**, 611 (1971).

(7) The two-center Coulomb repulsion integrals are calculated by the Ohno approximation (K. Ohno, *Theor. Chim. Acta (Berl.)*, **2**, 219 (1964)), and the one-center exchange integrals are evaluated by the Slater–Condon parameters estimated by Hinze and Jaffé (J. Hinze and H. H. Jaffé, *J. Chem. Phys.*, **38**, 1834 (1963)).

(8) 90.42 and 89.58° for the S–N–S and the N–S–N angles, respectively.

(9) 1.651 and 1.657 Å alternatively.

(10) D. R. Salahub and R. P. Messmer, *J. Chem. Phys.*, **64**, 2039 (1976).

(11) D. B. Adams, A. J. Banister, D. T. Clark, and D. Kilcast, *Int. J. Sulfur Chem.*, **1**, 143 (1971).

(12) J. Bragin and M. V. Evans, *J. Chem. Phys.*, **51**, 268 (1969).

(13) For example, see D. S. McClure, *J. Chem. Phys.*, **20**, 682 (1952).

(14) M. Blume and R. E. Watson, *Proc. R. Soc. (London), Ser. A*, **271**, 565 (1963).

(15) For example, see T. Yonezawa, H. Katô, and H. Kato, *Theor. Chim. Acta (Berl.)*, **13**, 125 (1969).

(16) Unpublished observations, A. G. MacDiarmid and M. S. Saran.

Orbital Interaction in the Dimerization of Disulfur Dinitride to Tetrasulfur Tetranitride

Sir: In recent years, disulfur dinitride, S_2N_2, and tetrasulfur tetranitride, S_4N_4, shown in Figure 1, have been of much interest in that they are the precursors of polymeric sulfur nitride, $(SN)_x$,[1] which is a low-dimensional metallic conductor and even becomes a superconductor at 0.3 K.[2] On the other hand, it has been also reported that S_2N_2 rapidly dimerizes, in certain organic solvents with a trace of alkali, to S_4N_4,[3] which is known to have a geometry with coplanar N atoms from X-ray diffraction analysis.[4] It is an interesting problem to explain the process of the formation of S_4N_4 from S_2N_2 by considering orbital interactions.[5]

From the atomic arrangement of S_4N_4 and the principle of maximum overlap of the highest occupied (HO) MO and the lowest unoccupied (LU) MO between two S_2N_2 molecules, two kinds of S_2N_2 stacking interaction models A and B are considered as shown in Figure 2. They are of essentially different stacking configurations when S_2N_2 approaches each other. Several spectroscopic and MO theoretical studies have been performed as to the assignment of orbital patterns in S_2N_2.[6,7] The HOMO and the LUMO of S_2N_2 employed here are the same as those obtained previously by the present authors with the use of the ASMO-SCF method.[8] The shape of the HOMO is consistent with that analyzed by photoelectron spectroscopic study[6a] and that of the LUMO is the same with those obtained by others with various MO calculations.[7b-f]

The HOMO-LUMO interactions of S_1-N_3 and N_2-S_4 are "bonding" in model A, and thereby two new $p\sigma$ type bonds are ready to be formed between them, accordingly as the π bonds of S_1-N_2 and N_3-S_4 are weakened. On the contrary, interactions of N_1-S_3 and S_2-N_4 are "antibonding" and, therefore, do not contribute to form new bonds. The intermediate-range interaction of S_2-S_3, which is favorable for bonding, should be taken into account as well even though the interaction is weak, since the π AO of S stretches considerably, and the overlap between S_2 and S_3 is not then negligible.

On the other hand, the stacking interaction in model B merely favors the S_2-S_4 bond formation. This type of orbital interaction is disadvantageous in bringing about the subsequent process of dimerization; model B should be discarded. It should be noted that the stacking in model B coincides with that in the S_2N_2 crystal along the *b* axis.[1c]

One can explain the process of the formation of S_4N_4 with coplanar N atoms with the use of model A as follows: the main bonds to be formed are S_1-N_3 and N_2-S_4 $p\sigma$ bonds while the previous S_1-N_2 and N_3-S_4 cleave simultaneously with the formation of new bonds as shown in Figure 2A. In the framework of this dimerization model, the reaction proceeds quite easily only through bond interchange without large dislocation among nuclei and, therefore, without going over a large potential barrier. A slight deformation then takes place until the most stable geometry in Figure 1 is accomplished along the S_2-S_3 bond which plays the role of a fulcrum as illustrated in Figure 3. The most stable geometry is then expected to be completed when the weak S_1-S_4 bond is formed. The existence of these transannular S-S bonds has also been suggested by the MO calculations for the S_4N_4 molecule itself.[7b,7c,9]

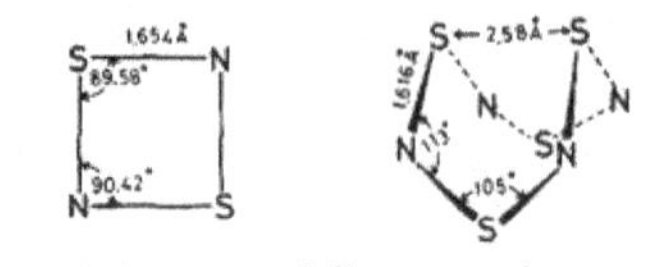

Figure 1. Geometries of S_2N_2[1a,1b] and S_4N_4.[4]

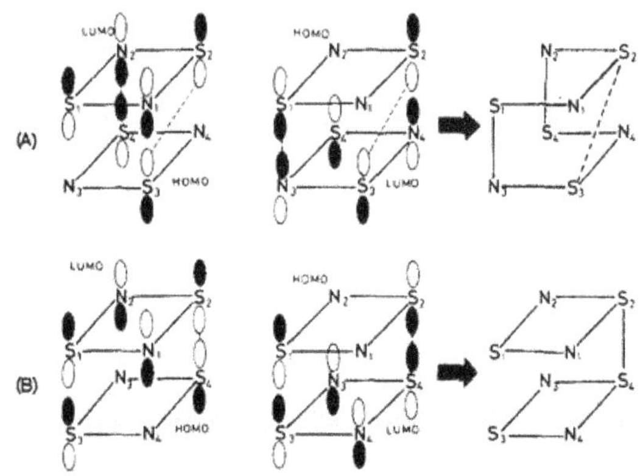

Figure 2. Two stacking interaction models A and B to form S_4N_4 by the dimerization of S_2N_2. The absolute value of AO coefficient on S in the HOMO is 0.707, and those on S and N in the LUMO are 0.412 and 0.511, respectively. The contributions from 3d AO's of S to those MO's are small and neglected here.

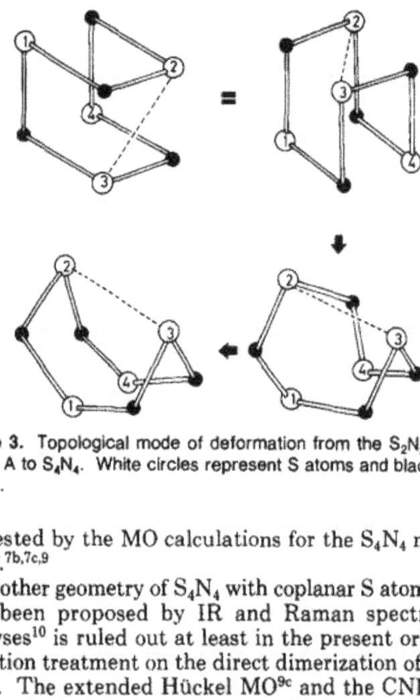

Figure 3. Topological mode of deformation from the S_2N_2 stacking model A to S_4N_4. White circles represent S atoms and black ones N atoms.

Another geometry of S_4N_4 with coplanar S atoms which had been proposed by IR and Raman spectroscopic analyses[10] is ruled out at least in the present orbital interaction treatment on the direct dimerization of S_2N_2 to S_4N_4. The extended Hückel MO[9c] and the CNDO/BW MO[9g] calculations have also supported a geometry with coplanar N atoms from the viewpoint of the total energy of S_4N_4. However, there still may be a possibility for the interconversion of S_4N_4 between the geometries with coplanar N atoms and with coplanar S atoms making use of some appropriate vibrational modes each other. In this

[Reprinted from the Journal of Physical Chemistry, 82, 1453 (1978)]

1454 The Journal of Physical Chemistry, Vol. 82, No. 12, 1978

point, further investigation on these geometries taking also vibrational analysis into account would be desirable.

Acknowledgment. This work was partly supported by Grants-in-Aid for Scientific Research from the Ministry of Education of Japan (No. 065101 and No. 255315).

References and Notes

(1) (a) A. G. MacDiarmid, C. M. Mikulski, P. J. Russo, M. S. Saran, A. F. Garito, and A. J. Heeger, *J. Chem. Soc., Chem. Commun.*, 476 (1975); (b) C. M. Mikulski, P. J. Russo, M. S. Saran, A. G. MacDiarmid, A. F. Garito, and A. J. Heeger, *J. Am. Chem. Soc.*, **97**, 6358 (1975); (c) M. J. Cohen, A. F. Garito, A. J. Heeger, A. G. MacDiarmid, C. M. Mikulski, M. S. Saran, and J. Kleppinger, *ibid.*, **98**, 3844 (1976).

(2) R. L. Greene, G. B. Street, and L. J. Suter, *Phys. Rev. Lett.*, **35**, 1799 (1975).

(3) H. J. Emeléus, *Endeavour*, **32**, 76 (1973).

(4) B. D. Sharma and J. Donohue, *Acta Crystallogr.*, **16**, 891 (1963).

(5) K. Fukui, "Theory of Orientation and Stereoselection", Springer-Verlag, Berlin, 1975.

(6) (a) D. C. Frost, M. R. LeGeyt, N. L. Paddock, and N. P. C. Westwood, *J. Chem. Soc., Chem. Commun.*, 217 (1977); (b) J. Sharma, D. S. Downs, Z. Iqbal, and F. J. Owens, *J. Chem. Phys.*, **67**, 3045 (1977).

(7) (a) M. P. S. Collins and B. J. Duke, *J. Chem. Soc., Chem. Commun.*, 701 (1976); (b) D. R. Salahub and R. P. Messmer, *J. Chem. Phys.*, **64**, 2039 (1976); (c) W. R. Salaneck, J. W-p Lin, A. Paton, C. B. Duke, and G. P. Ceasar, *Phys. Rev. B*, **13**, 4517 (1976); (d) M. Kertész, S. Suhai, A. Ažman, D. Kocjan, and A. I. Kiss, *Chem. Phys. Lett.*, **44**, 53 (1976); (e) J. A. Jafri, M. D. Newton, T. A. Pakkanen, and J. L. Whitten, *J. Chem. Phys.*, **66**, 5167 (1977); (f) I. P. Batra, S. Ciraci, and W. E. Rudge, *Phys. Rev. B*, **15**, 5858 (1977); (g) W. Y. Ching, J. G. Harrison, and C. C. Lin, *ibid.*, **15**, 5975 (1977); (h) P. W. Deutsch and L. A. Curtiss, *Chem. Phys. Lett.*, **51**, 125 (1977).

(8) T. Yamabe, K. Tanaka, K. Fukui, and H. Kato, *J. Phys. Chem.*, **81**, 727 (1977).

(9) (a) D. Chapman and T. C. Waddington, *Trans. Faraday Soc.*, **58**, 1679 (1962); (b) F. S. Braterman, *J. Chem. Soc. A*, 2297 (1965); (c) A. G. Turner and F. S. Mortimer, *Inorg. Chem.*, **5**, 906 (1966); (d) J. B. Mason, *J. Chem. Soc. A*, 1567 (1969); (e) R. Gleiter, *ibid.*, 3174 (1970); (f) P. Cassoux, J. F. Labarre, O. Glemser, and W. Koch, *J. Mol. Struct.*, **13**, 405 (1972); (g) M. S. Gopinathan and M. A. Whitehead, *Can. J. Chem.*, **53**, 1343 (1974).

(10) E. R. Lippincott and M. C. Tobin, *J. Chem. Phys.*, **21**, 1559 (1953).

Department of Hydrocarbon Chemistry
Faculty of Engineering
Kyoto University
Kyoto 606, Japan

Kazuyoshi Tanaka
Tokio Yamabe
Akinori Noda
Kenichi Fukui*

College of General Education
Nagoya University
Nagoya 464, Japan

Hiroshi Kato

Received January 16, 1978

[Reprinted from the Journal of Physical Chemistry 82, 272 (1978).]

Electronic Behavior of Amorphous Chalcogenide Models

Akitomo Tachibana, Tokio Yamabe, Mariko Miyake, Kazuyoshi Tanaka, Hiroshi Kato,[†] and Kenichi Fukui*

Department of Hydrocarbon Chemistry, Faculty of Engineering, Kyoto University, Kyoto 606, Japan and College of General Education, Nagoya University, Chikusa-ku, Nagoya 464, Japan *(Received September 12, 1977)*

Molecular analogue models, As_4S_6 and As_4S_4 (units), are examined for the determination of localized paramagnetic states in amorphous As_2S_3 glass. As has been suggested by Ovshinsky et al. and Duke et al., the As_4S_6 unit is the smallest molecular unit that is able to be present in the As_2S_3 glass satisfying the stoichiometry and the electronic structure of valence states. It is shown in the present paper that the As_4S_6 unit acts as a reasonable electron-trapping center having a function analogous to that of the cavity-center model for the study of solvated electrons. The distribution of trapped unpaired electron by As_4S_6 accounts well for the anionic ESR center localized on As atoms observed in the glass. Since the presence of the As_4S_4 unit in As_2S_3 glass is implied from the recent measurement of Raman spectra, the electronic behaviors of As_4S_6 and As_4S_4 are compared and discussed.

Introduction

Amorphous chalcogenides have recently received much attention in particular in consequence of Ovshinsky and his co-workers, especially on the applications to switching, memory, and imaging devices.[1] The prominent functions of amorphous chalcogenide materials are considered as largely concerned with the short-range structure and hence the local nature of active electrons rather than the long-range order.[2] In the present paper attention is focused on the electronic structure of amorphous As_2S_3 which has recently stimulated great interest.[3] The dominant geometrical structure of the As_2S_3 glass has been considered as a 12-membered ring structure of two-dimensional layered crystalline As_2S_3 (orpiment).[4-6] Recently, however, De Neufville, Moss, and Ovshinsky suggested from x-ray diffraction data that the amorphous As_2S_3 is a van der Waals molecular glass composed of As_4S_6 molecules which are present in the vapor.[7,8] From observations of photoemission and ultraviolet absorption spectra as well as from the theoretical study of CNDO/S MO calculations, Duke et al. showed that there is a considerable similarity between the electronic structures of the glass (As_2S_3) and gaseous molecules (As_4S_6).[8] It is implied therefore that amorphous As_2S_3 contains a considerable amount of As_4S_6 units in the glass matrix.[8] In other words, in general, a short-range structure essentially like that of the isolated molecular species, for example, the As_4S_6 unit in the As_2S_3 glass, may play an important role in interpreting the characteristic electronic properties of amorphous chalcogenides. Such a short-range structure of nonrigid, inconstant nature may tentatively be referred to as a "casket" in the present paper.

Recently Bishop, Strom, and Taylor[9] have found, after low-temperature ultraviolet irradiation (2.41 eV) of As_2S_3 glass, optically induced localized paramagnetic states in the forbidden gap associated with induced broad midgap optical absorption. Subsequently it has become revealed that for As-containing chalcogenide glasses such as As_2S_3 and As_2Se_3 the ESR spectrum is composed of two signals:[10,11] (i) a relatively narrow resonance of width about 75 G, and (ii) a broader resonance with shoulder, of width about 700 G for As_2S_3 glass, which is superimposed on the first resonance. The narrow resonance has been shown to be due to the unpaired electrons localized on chalcogen atoms probably as cationic radicals,[9-11] while the broad shoulder has been attributed to second anionic centers localized on ^{75}As atoms because of the large hyperfine interaction.[10-12] It is of interest that the intensities of both the induced ESR and optical absorption grow at nearly the same rate as irradiation proceeds. Conversely irradiation with photon energies corresponding to the induced absorption band, which extends from as low as about 0.8 eV to the original band edge of about 2.0 eV, bleaches the induced absorption and ESR signal simultaneously. The similarity of the growth and bleaching rates for both the ESR and optical absorption demonstrates that these effects are related to localized centers closely connected with each other.[11] The optically induced states are unique to glasses because for ordered As_2S_3 layered crystals the photoinduced phenomena are not observed.[11] Thus the localized electronic states in the gap of amorphous chalcogenides have been of major interest hitherto.[13,14]

In the present paper it is attempted to show by semiempirical MO calculations that the As_4S_6 unit can be a reasonable candidate for the molecular species (analogue) of the "casket" acting as an electron-trapping center in the As_2S_3 glass. In the glass the photoinduced hole centers on chalcogen atoms may be stable against recombination because of the presence of electron-trapping centers, As_4S_6 units as will be demonstrated, but may not be stable in the ordered crystal where there are no such caskets. In this model the energy of detachment from the casket corresponds to the photoinduced midgap absorption. Moreover, the ESR hyperfine coupling constant is calculated to be consistent with the experimental value as shown in the text.

The proposed presence of caskets acting as electron-trapping centers is reasonable also from observations such as (i) holes are mobile carriers in amorphous As_2S_3 films whereas electrons are quite immobile;[15] (ii) photoemitted electrons suffer considerable hindrance during their passing through As_2S_3 films.[16]

It should be noted here that recent measurements of Raman spectra[17] suggest, in spite of nonstoichiometry, the existence of As_4S_4 units in amorphous As_2S_3 film. However, it has been shown from MO calculation of the DOVS (density of valence states)[18] that existence of a high density of As_4S_4 units is questionable.[8] In addition to this, according to the present calculated results, the As_4S_4 unit is likely to undergo cleavage of As–As bonds via electron capture and therefore, if present in the As_2S_3 glass, this species may not be a reasonable candidate for the casket acting as stable electron-trapping center.

[†] College of General Education, Nagoya University.
* Address correspondence to this author at Kyoto University.

0022-3654/78/2082-0272$01.00/0

The Journal of Physical Chemistry, Vol. 82, No. 3, 1978 **273**

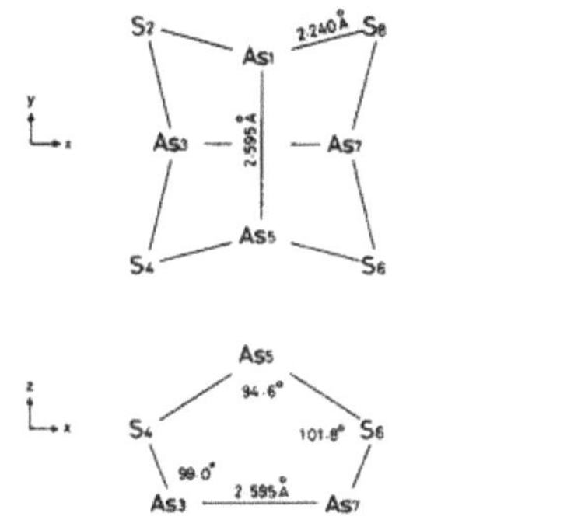

Figure 1. Molecular geometry of As_4S_4 which belongs to point group D_{2d}.

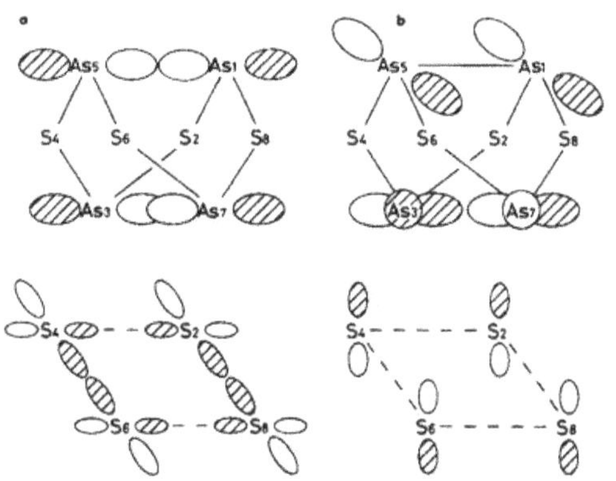

Figure 2. (a) HOMO and (b) LUMO patterns of deformed As_4S_4. The deformation is assumed symmetric with As–As bond lengths each stretched ±0.05 Å, all the four sulfurs being clamped as described in the text.

TABLE I: Total Energies of $As_4S_4^-$ and $As_4S_6^-$ (in eV)[a]

	Normal/deformed $As_4S_4^-$	Normal/deformed $As_4S_6^-$
E	-0.980/-1.110	-2.638/-2.632

[a] The energies of the neutral species of normal geometry are set equal to zero as standards.

TABLE II: Changes of E_{AB} by Electron Attachment and Deformation

	Normal As_4S_4	Deformed $As_4S_4^-$	Normal As_4S_6	Deformed $As_4S_6^-$
As1-As5	−10.513	−11.913	−0.035	+0.757
As3-As7	−10.513	−8.104	−0.035	+1.498

TABLE III: Changes of Bond Orders and AO Densities by Electron Attachment and Deformation (in eV)

	Normal As_4S_4/ deformed $As_4S_4^-$			Normal As_4S_6/ deformed $As_4S_6^-$	
	(s)—As1—(p_x)			(s)—As1—(p_y)	
As1(s)	1.749/ 1.730		(s)	1.628/ 1.714	
(p_x)	0.000/ 0.000	0.667/ 0.903	(p_y)	0.393/ 0.283	1.119/ 1.265
As5(s)	0.058/ 0.067	0.000/ 0.000	(s)	0.025/ −0.064	−0.014/ 0.093
(p_x)	0.000/ 0.000	0.090/ 0.350	(p_y)	0.014/ −0.093	0.120/ 0.238
	(s)—As3—(p_x)			(s)—As3—(p_y)	
As3(s)	1.749/ 1.774		(s)	1.628/ 1.635	
(p_x)	−0.128/ −0.080	1.029/ 1.100	(p_y)	0.000/ 0.000	0.685/ 0.716
As7(s)	0.058/ 0.031	0.292/ 0.237	(s)	0.025/ 0.023	0.000/ 0.000
(p_x)	−0.292/ −0.237	−0.799/ −0.734	(p_y)	0.000/ 0.000	0.026/ 0.039

We have carried out semiempirical ASMO calculations for all valence electrons (INDO type).[19] This method is different from CNDO-type calculations of Pople et al.[20] in that it includes the one- and two-center electron repulsion integrals evaluated semiempirically,[21] giving reasonable results, especially for transition energies.[19,22] For anion radicals the UHF treatment has been adopted. Spin contamination is excluded by Löwdin's projection operator method[23] simplified by Amos and Snyder.[24]

Electronic Structures of As_4S_4 and $As_4S_4^-$

At first, the electronic states of As_4S_4 and the anion radical $As_4S_4^-$ are calculated. The molecular geometry[25] is shown in Figure 1. The valence bands of As_4S_4 (realgar) were studied by x-ray photoemission spectroscopy and interpreted utilizing the CNDO/S MO method by Salaneck et al.[8a,18] Neglecting the electron correlation as well as the rearrangement, the orbital energies of the individual MO's were fitted fairly well to the observed ionization potentials of the solid state within an additive constant. The upper part of the valence band is, as pointed out by Salaneck et al.,[18] composed almost entirely of p electrons, and, what is more important, HOMO consists of As–As σ bonds as shown in Figure 2a. This bond is weak since the As–As bond length, about 2.6 Å, is longer than the corresponding covalent bond length, about 2.4 Å.[18] The weakness of the As–As bond may relate to the experimental evidence that As_4S_4 (realgar) changes very easily to As_2S_3 (orpiment), especially on exposure to light,[25a,26] which may relate to the origin of catastrophic changes observed in amorphous chalcogenide materials.[1]

Interestingly, it is found that the LUMO (one of the doubly degenerate LUMO's) of As_4S_4 consists of one $\bar{\pi}$ MO for the As1–As5 bond and one σ^* MO for the As3–As7 bond as shown in Figure 2b, where $\bar{\pi}$ denotes the π-type MO parallel to the four-S plane. Through electron capture into this LUMO (the orbital energy is −0.942 eV) the $\bar{\pi}$ MO stabilizes the As1–As5 bond whereas the σ^* MO destabilizes the As3–As7 bond and this may initiate the deformation of the As1–As5 bond shortening, and of the As3–As7 bond elongation. The other LUMO causes the same deformation in the opposite direction. This point is worth discussing in more detail.

The $As_4S_4^-$ anion is more stable in the deformed structure than in that of D_{2d} symmetry; the total energies are summarized in Table I. The deformation is assumed symmetric with As–As bond lengths each stretched ±0.05 Å, all the four sulfurs being clamped. The vertical electron affinity is calculated to be 0.980 eV and the anion is further stabilized by 0.130 eV along the deformation, while the neutral As_4S_4 molecule is stable against the deformation.[27] The distribution of the atomic net charges, As (+0.456) and S (−0.456), supports the large electron affinity on the As atoms. The stabilization energy of the $As_4S_4^-$ anion along the deformation can be analyzed by examining the diatomic contribution E_{AB}[28,20] to the total energy as summarized in Table II. After the deformation takes place, the As1–As5 bond is stabilized, from −10.513 to −11.913 eV, and the As3–As7 bond is destabilized, from −10.513 to −8.104 eV, compared with the neutral molecule of D_{2d} symmetry. This tendency is confirmed by exam-

274 The Journal of Physical Chemistry, Vol. 82, No. 3, 1978

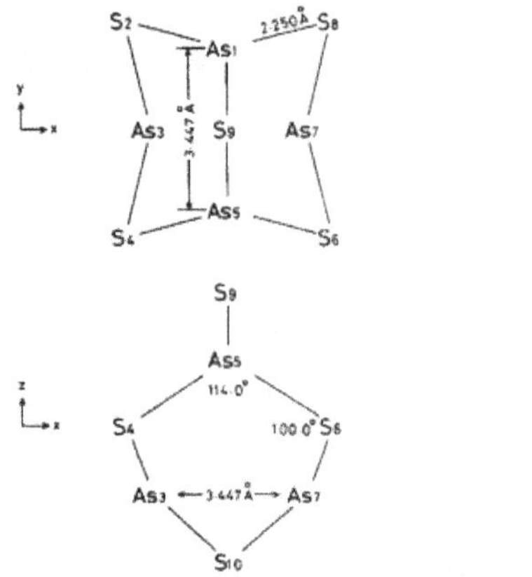

Figure 3. Molecular geometry of As_4S_6 which belongs to point group T_d.

ining the changes of the electron densities and bond orders which are summarized in Table III. For a neutral As_4S_4 molecule with D_{2d} symmetry, the As–As $\bar{\pi}$ bond order is 0.090 and the As–As σ bond order is -0.799, which indicates that the As–As bond has a σ bond nature associated with a small amount of $\bar{\pi}$ bond nature. Once the deformation takes place, the As1–As5 $\bar{\pi}$ bond order changes from 0.090 to 0.350 and the AO density changes from 0.667 to 0.903 per atom. This is a consequence of electron trapping into the $\bar{\pi}$ MO, which makes the $\bar{\pi}$ bond stronger. On the other hand, the small amount of electrons which flows into the As3–As7 σ^* MO causes a weakening of the bond; the σ bond order increases from -0.799 to -0.734. Moreover, the decrease of s-electron density in As1 and As5, from 1.749 to 1.730 per atom, intensifies the $\bar{\pi}$ bond nature of the As1–As5 bond, whereas the increase of s-electron density in As3 and As7, from 1.749 to 1.774 per atom, supports the increase of stabilization for the radical cleavage of the As3–As7 bond.

It may be demonstrated therefore that the $As_4S_4^-$ anion is not suitable for interpreting the stable electron centers localized on As atoms which are produced in the As_2S_3 glass by irradiation. Although excess electrons are easily captured by the As_4S_4 unit (further calculations show that more than one electron is not accepted), the electron is apt to be transferred away as the unit deforms. If there is room for As_4S_4 to deform freely in the As_2S_3 glass matrix the optimum structure of the anion would be attained, but the possibility is not examined in the present paper.

Electronic Structures of As_4S_6 and $As_4S_6^-$

The As_4S_6 molecular geometry which belongs to point group T_d[26] is depicted in Figure 3. One can convert As_4S_4 to As_4S_6 simply by inserting a sulfur atom between each As–As pair. However the electronic structures of As_4S_6 become quite different from those of As_4S_4.

First, as pointed out by Duke et al.,[8] the DOVS of As_4S_6 is similar to that of experimental measurements for As_2S_3 glass, whereas that of As_4S_4[18] is not.[8]

Secondly, As_4S_6 is stable against electron attachment; the total energies are summarized in Table I in order to show the stability of As_4S_6 against electron attachment and deformation. The deformation is the same as for As_4S_4. The calculated vertical electron affinity is 2.638 eV, which is 1.658 eV larger than that of As_4S_4, and the anion as well as the neutral species are stable against deformation.[27] In this case both the low orbital energy, -2.433 eV, of the LUMO and the distribution of the atomic net charges, As

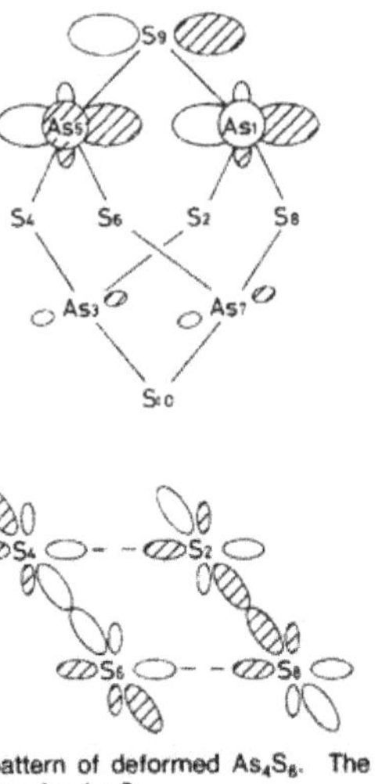

Figure 4. LUMO pattern of deformed As_4S_6. The deformation is assumed the same as for As_4S_4.

(+0.666) and S (-0.444), indicate that the positively charged As atoms attract excess electrons more easily than that for As_4S_4. The results of E_{AB} analysis are summarized in Table II. The essential difference from As_4S_4 is that in this case the destabilization of the shorter As1–As5 bond occurs, from -0.035 to +0.757 eV, as electrons enter through the shortened As1–As5 bond.[29] The mechanism of repression reflects the nodal properties of the LUMO of the neutral As_4S_6. As As_4S_6 deforms one of the triply degenerate LUMO's undergoes drastic change in nodal property compared with the LUMO of As_4S_4 as illustrated in Figure 4 (cf. Figure 2b). It is to be noted that the LUMO on the shorter As1–As5 bond is an antibonding σ^* type and on the longer As3–As7 bond is, though slightly, a bonding $\bar{\pi}$ type; this nature of the LUMO of As_4S_6 is totally distinguishable from that of As_4S_4. The electron trapped by this LUMO represses the deformation by making As1 and As5 repel each other, and by binding As3 and As7. This result is confirmed by examining the changes of the bond orders and AO densities which are summarized in Table III. After the deformation the bond order of the As1–As5 bond increases from 0.120 to 0.238, and the AO density from 1.119 to 1.265 per atom. This means that excess electrons go into the σ^* MO on the As1–As5 bond, which represses the shortening of this bond. On the other hand the $\bar{\pi}$ bond order of the longer As3–As7 bond increases from 0.026 to 0.039, which acts as resistor against the elongation. As a result, the deformation is repressed.

Consequently, the stability against electron attachment is verified for the As_4S_6 unit, which is not expected in the As_4S_4 unit. In this case the energy of detachment of the attached electron through the midgap excitation is estimated to be[30] $h\nu_{detach} = 2.220$ eV. However, this is too large to account for the optically induced midgap absorption in As_2S_3 glass which extends from as low as about 0.8 eV to the original band edge of about 2.0 eV.[11] This difficulty will be solved by adopting another model for electron capture as shown in the next section.

Cavity-Center Model

On account of the T_d symmetry, there are three sets of equivalent As–As pairs for As_4S_6 molecules; As1–As5/As3–As7, As5–As7/As1–As3, and As7–As1/As3–As5. For each set the same stabilizing mechanism exists, and hence

TABLE IV: 1s-Like Cavity-Center Models

$U_{1s,1s}$	ζ_{1s}	ΔE^a	ϵ_{LUMO}	$h\nu^b$	E_{AB} 1s–As	E_{AB} 1s–S
	0.3	−3.653	−3.073	3.056	−1.903	−0.009
0.000	0.4	−3.785	−3.602	2.503	−1.976	−0.014
	0.5	−3.469	−3.916	2.144	−1.819	−0.020

^{a} Energy differences from the model without 1s center. ^{b} Minimum singlet excitation (HOMO → LUMO) energies.

excess electrons may be stable in the As_4S_6 unit with the induced totally symmetric motion. The induced orbital for excess electrons may be represented by an almost spherically symmetric orbital concentric with the As_4S_6 unit. Generally speaking, it is to be recognized that the dynamic electron–casket interaction of attractive nature is now described by a static orbital. This is analogous to the cavity-center model for the study of solvated electrons, which has succeeded in describing the nature of the capturing center, especially the magnetic property of the trapped electron.[31-33] To represent the induced orbital fitted for the subsequent UHF treatment a 1s part is conventionally added to the linear combination of degenerate LUMO's. In the present case the semiempirical parameters chosen for the 1s-like part and the calculated results are summarized in Table IV. The ionization potential for the added 1s-like part is taken to be the orbital energy, −2.433 eV, of the LUMO of As_4S_6 without the 1s center. This choice corresponds to that of −3.0 eV for $-(I_{1s} + A_{1s})/2$ of the INDO scheme as chosen by Ishimaru et al.[33] For the value of the local core matrix element $U_{1s,1s}$, the most appropriate one of $U_{1s,1s} = 0.0$ eV is chosen after Ishimaru et al.,[33] which ensures that the stabilization energy for excess electrons is solely due to the energy of interaction with the parent As_4S_6 unit. Variations in terms of orbital exponents indicate that $\zeta_{1s} = 0.4$ is the optimum value. This selection of a diffuse orbital for excess electrons is consistent with those results of Jortner[32] and Ishimaru et al.[33] Thus the LUMO of the As_4S_6 unit becomes nearly totally symmetric which is localized mainly on the 1s-like part with the AO coefficient 0.631, forming a sphere-quantized MO.[34] The As_4S_6 unit itself has a calculated minimum singlet excitation energy of 2.503 eV for the HOMO → LUMO transition (2.709 eV without 1s-like center), which does not contradict the characteristic low absorption band edge of about 2.0 eV of amorphous As_2S_3 that does not appear in the crystalline phase.[35] Now, from E_{AB} analysis (see Table IV) the bonding character of the 1s-like center with each As atom is stronger than with each S atom, which suggests that the trapped electron will attract the nearest As atoms ($R_{1s\text{-}As} = 2.111$ Å, $R_{1s\text{-}S} = 2.665$ Å) and will retain the stability of the T_d symmetric structure of the As_4S_6 casket. Indeed the vertical electron affinity of the As_4S_6 unit now becomes 3.681 eV which is 1.043 eV greater than the former result (without 1s-like center), and the anion is stable against the deformation as shown in Table V. The detachment energy of trapped electrons is estimated to be 0.996 eV from the results of closed shell calculation,[30] and 1.083 eV from the results of UHF calculation,[36] $h\nu_{detach} = 0.996{\sim}1.083$ eV. This corresponds to the optically induced midgap absorption in As_2S_3 glass which extends from about 0.8 eV to the band edge of about 2.0 eV.[11]

It should be noted here that from further calculations the energies of multiple anions are obtained and summarized as follows. $As_4S_6^-$ is −3.681 eV (−2.638 eV); $As_4S_6^{2-}$ is −3.712 eV (−0.539 eV); $As_4S_6^{3-}$ is +1.291 eV (+4.434 eV), relative to the corresponding neutral species. The values in parentheses represent energies without the 1s-like center. The double anion of As_4S_6 presumably exists and

TABLE V: Total Energies of $As_4S_4^-$ and $As_4S_6^-$ by the 1s-Like Cavity-Center Models (in eV)^a

	Normal/deformed $As_4S_4^-$	Normal/deformed $As_4S_6^-$
E	−1.210/−1.330	−3.681/−2.837

^{a} The energies of the neutral species of normal geometry are set equal to zero as standards.

turns out to be particularly more stable than the single anion provided that the 1s-cavity center model is adopted. It may be tentatively concluded therefore that the dynamic electron–casket interaction can mediate an effective attractive interaction between a pair of electrons which overcomes direct Coulomb repulsion, and then it produces a localized two-electron state which lies lower than the one-electron state.

Now we calculate the hyperfine coupling constant for the second component of the optically induced ESR spectrum, that is, the broad resonance with shoulder of width about 700 G which is superimposed on the first sharp resonance. The resonance magnetic field H and the resonance frequency $h\nu$ are related as

$$H = h\nu/g\mu_B \pm A/2$$

where g is the magnitude of electronic spin gyromagnetic tensor, μ_B is the Bohr magneton, and A is the magnitude of the hyperfine tensor. For the paramagnetic centers localized on ^{75}As atoms the second term is predominant because of the large hyperfine interaction. If the orientation of the paramagnetic site is represented by θ and ϕ, Euler angles of rotation, g and A are given by[11]

$$g^2 = g_1^2 \sin^2\phi \sin^2\theta + g_2^2 \cos^2\phi \sin^2\theta + g_3^2 \cos^2\theta$$

$$g^2A^2 = g_1^2A_1^2 \sin^2\phi \sin^2\theta + g_2^2A_2^2 \cos^2\phi \sin^2\theta + g_3^2A_3^2 \cos^2\theta$$

where g_i and A_i are the components of the corresponding tensors. Bishop et al.[11] considered various cases that allow the unpaired spin to be delocalized on more than one equivalent As atoms. In the present As_4S_6 casket model, the number of equivalent As atoms is four, which was not considered by Bishop et al.[11] If one follows the procedure of analytical fit adopted by Bishop et al. to the experimental ESR spectrum, one obtains $g_1 = 2.00$, $g_2 = 2.02$, $g_3 = 2.07$ (±0.02), and $A_\perp = 7.125$ G, $A_\parallel = 50.0$ G (±50 G), for the broad resonance with shoulder of width 700 G. On the other hand, from the calculated spin density one obtains hyperfine coupling constants which are almost isotropic, $|A_\perp| = |A_\parallel| = 30.141$ G (see Table VI). In this case the atomic s- and p-hyperfine coupling constants for As are chosen as 4500 and 119 G, respectively.[11,37] The agreement with the experimental result is good within experimental error. If we do not adopt the 1s-like cavity-center model the hyperfine coupling constants are estimated to be $|A_\perp| = |A_\parallel| = 215.044$ G, which are several times larger than the experimental result.[38,39]

The cavity center model is similarly examined with respect to the As_4S_4 unit. Also in this case the LUMO localizes mainly on the added 1s-like center with the

276 The Journal of Physical Chemistry, Vol. 82, No. 3, 1978

TABLE VI: Spin Densities and Hyperfine Coupling Constants[a]

	Present calculations (before annihilation)	
(1) Spin Densities		
Added 1s	0.381	(0.383)
As 4s	−0.007	(−0.021)
$4p_x$	0.032	(0.044)
$4p_y$	0.032	(0.044)
$4p_z$	0.032	(0.044)
S9 3s	0.002	(0.000)
$3p_x$	−0.005	(−0.014)
$3p_y$	−0.005	(−0.014)
$3p_z$	0.051	(0.058)

(2) Hyperfine Coupling Constants

	A, G
With 1s center	$\lvert A_\perp \rvert = \lvert A_\parallel \rvert = 30.141$
Without 1s center	$\lvert A_\perp \rvert = \lvert A_\parallel \rvert = 215.044$
Exptl[b]	$A_\perp = 7.125$; $A_\parallel = 50.0$
	$(\sim A_\perp = 14.25$; $A_\parallel = 100.0)$

[a] Parameters for 1s-like cavity-center model are $U_{1s,1s} = 0.000$ eV and $\zeta_{1s} = 0.4$. [b] Analytical fit of the data from ref 11. The experimental errors are ±50 G.

optimum ζ_{1s} value of 0.5. However, when an electron is attached (further calculations show that more than one electron cannot be attached) to the normal and deformed As_4S_4 molecule the SOMO (singly occupied MO) does not localize on the 1s-like center and the anion is not stable against the deformation as shown in Table V. So, it is hard again to assume that there exists a stable $As_4S_4^-$ anion in As_2S_3 glass to be observed by ESR measurement.

Thus the casket model is of great advantage in studying the local electronic structure of glassy states and hence the local topological arrangements of atoms in the glass. Furthermore the casket model is supposed to have potential utility in the interpretation of various transport, optical absorption, photoconductivity, and photoluminescence experiments carried out on the amorphous chalcogenide materials.

Conclusion

The electronic structures of As_4S_4, $As_4S_4^-$, As_4S_6, and $As_4S_6^-$ have been examined.

It has been shown that the As_4S_6 unit is stable, whereas the As_4S_4 unit is unstable against electron attachment. The stability of the $As_4S_6^-$ anion is also elucidated by the cavity-center model. Not only the low excitation energy of the parent As_4S_6 unit as a constituent of As_2S_3 glass, but also the optically induced ESR spectrum and midgap absorption in the glass have been well accounted for by this model.[29]

Thus it might be concluded that the As_4S_6 unit is a reasonable candidate for the molecular analogue of the "casket" acting as a stable electron-trapping center in the As_2S_3 glass.

Acknowledgment. The authors are grateful to the Data Processing Center of Kyoto University for generous use of the FACOM M190 Computer. A. T. thanks Mr. H. Kobayashi for his kind information on semiempirical parameters for the As atom. This work was supported by a Grant-in-Aid for Scientific Research from the Ministry of Education of Japan.

References and Notes

(1) (a) S. R. Ovshinsky, *J. Non-Cryst. Solids*, **2**, 99 (1970); (b) S. R. Ovshinsky and K. Sapru, "Proceedings of the 5th International Conference on Amorphous and Liquid Semiconductors", J. Stuke and W. Brenig, Ed., Taylor and Francis, Garmisch-Partenkirchen, 1974, p 447; (c) S. R. Ovshinsky and H. Fritzsche, *IEEE Trans. Electron Devices*, **ED-20**, 91 (1973); (d) S. R. Ovshinsky, "4th International Congress for Reprography and Information", Hanover, Germany, 1975.

(2) D. Adler, *Sci. Am.*, **236**, 36 (1977).

(3) (a) A. J. Apling, "Electronic and Structual Properties of Amorphous Semiconductors", P. G. Lecomber and J. Mort, Ed., Academic Press, London, 1973, p 243; (b) A. L. Renninger and B. L. Averbach, *Phys. Rev. B*, **8**, 1507 (1973); (c) A. J. Leadbetter and A. J. Apling, *J. Non-Cryst. Solids*, **15**, 250 (1974); (d) G. Lucovsky and R. M. Martin, *ibid.*, **8–10**, 185 (1972).

(4) I. Chen, *Phys. Rev. B*, **8**, 1440 (1973).

(5) T. Shimizu and N. Ishii, *Phys. Status Solidi b*, **74**, K39 (1976).

(6) S. G. Bishop and N. J. Schevchik, *Phys. Rev. B*, **12**, 1567 (1975).

(7) J. P. De Neufville, S. C. Moss, and S. R. Ovshinsky, *J. Non-Cryst. Solids*, **13**, 191 (1973).

(8) (a) C. B. Duke, W. R. Salaneck, A. Paton, K. S. Liang, N. O. Lipari, and R. Zallen, "Structure and Excitations of Amorphous Solids", G. Lucovsky and F. Galeener, Ed., AIP, New York, 1976, p 23; (b) W. R. Salaneck and R. Zallen, *Nuovo Cimento B*, **38**, 248 (1977).

(9) S. G. Bishop, U. Strom, and P. C. Taylor, *Phys. Rev. Lett.*, **34**, 1346 (1975).

(10) S. G. Bishop, U. Strom, and P. C. Taylor, *Phys. Rev. Lett.*, **36**, 543 (1976).

(11) S. G. Bishop, U. Strom, and P. C. Taylor, *Phys. Rev. B*, **15**, 2278 (1977).

(12) S. G. Bishop, U. Strom, and P. C. Taylor, *Solid State Commun.*, **18**, 573 (1976).

(13) (a) R. A. Street and N. F. Mott, *Phys. Rev. Lett.*, **35**, 1293 (1975); (b) E. N. Economou, K. L. Ngai, and T. L. Reinecke, *ibid.*, **39**, 157 (1977); (c) M. Kastner, D. Adler, and H. Fritzsche, *ibid.*, **37**, 1504 (1976).

(14) S. R. Ovshinsky, *Phys. Rev. Lett.*, **36**, 1469 (1976).

(15) S. W. Ing, Jr., J. H. Neyhart, and F. Schmidlin, *J. Appl. Phys.*, **42**, 696 (1971).

(16) W. R. Salaneck and R. Zallen, *Solid State Commun.*, **20**, 793 (1976).

(17) S. A. Solin and G. N. Papatheodorou, *Phys. Rev. B*, **15**, 2084 (1977).

(18) W. R. Salaneck, K. S. Liang, A. Paton, and N. O. Lipari, *Phys. Rev. B*, **12**, 725 (1975).

(19) (a) T. Yonezawa, H. Konishi, and H. Kato, *Bull. Chem. Soc. Jpn.*, **42**, 933 (1969); (b) H. Konishi, H. Kato, and T. Yonezawa, *Theor. Chim. Acta (Berl.)*, **19**, 71 (1970); (c) H. Yamabe, H. Kato, and T. Yonezawa, *Bull. Chem. Soc. Jpn.*, **44**, 22, 611 (1971).

(20) J. A. Pople, D. P. Santry, and G. A. Segal, *J. Chem. Phys.*, **43**, S129, S136 (1965).

(21) The two-center Coulomb repulsion integrals are estimated by the Ohno approximation (K. Ohno, *Theor. Chim. Acta (Berl.)*, **2**, 219 (1964)), and the one-center exchange integrals are evaluated by Slater–Condon parameters calculated by Hinze and Jaffé (S. Hinze and H. H. Jaffé, *J. Chem. Phys.*, **38**, 1834 (1963)).

(22) (a) T. Yamabe, K. Tanaka, K. Fukui, and H. Kato, *J. Phys. Chem.*, **81**, 727 (1977); (b) T. Yamabe, K. Tanaka, and K. Fukui, *Ann. N.Y. Acad. Sci.*, in press.

(23) P. -O. Löwdin, *Phys. Rev.*, **97**, 1509 (1955).

(24) A. T. Amos and L. C. Snyder, *J. Chem. Phys.*, **41**, 1773 (1964).

(25) (a) T. Ito, N. Morimoto, and R. Sadanaga, *Acta Crystallogr.*, **5**, 775 (1952); (b) P. Goldstein and A. Paton, *Acta Crystallogr., Sect. B*, **30**, 915 (1974).

(26) C. -S. Lu and J. Donohue, *J. Am. Chem. Soc.*, **66**, 818 (1944).

(27) The Jahn–Teller effect may appear. The deformation should occur obeying that of normal vibrational modes, however, the precise analysis of this point was omitted.

(28) (a) M. S. Gordon, *J. Am. Chem. Soc.*, **91**, 3122 (1969); (b) S. E. Ehrenson and S. Seltzer, *Theor. Chim. Acta*, **20**, 17 (1971).

(29) The distances As1–As5 and As3–As7 are very long, 2.595 Å, which is too long to be a bond distance. However, the pairs As–As are called "bonds" here for the sake of comparison to those of As_4S_4.

(30) Based on closed-shell calculations, the detachment energy is estimated to be $h\nu_{detach} = \epsilon_{LU+1} - \epsilon_{LU}$.

(31) For a review article on solvated electrons, see N. R. Kestner, "Electron-Solvent and Anion-Solvent Interactions", L. Kevan and B. C. Webster, Ed., Elsevier, Amsterdam, 1976, p 1.

(32) J. Jortner, *Radiat. Res. Suppl.*, **4**, 24, (1964).

(33) (a) S. Ishimaru, T. Yamabe, K. Fukui, and H. Kato, *Chem. Phys. Lett.*, **17**, 264 (1972); (b) S. Ishimaru, H. Kato, T. Yamabe, and K. Fukui, *J. Phys. Chem.*, **77**, 1450 (1973).

(34) This is the difference from the solvated electron model adopted by Ishimaru et al. in ref 33 where the LUMO is localized predominantly on the added 1s orbital with the AO coefficient almost unity. In the present model the added 1s orbital supports the formation of a sphere-quantized LUMO through the interaction with the vacant orbitals on the constituent atoms in order to represent the induced orbital by excess electron–casket interaction. This effect may be obtained by extensive CI calculations instead, but the possibility is not examined in the present paper.

(35) R. Zallen, R. E. Drews, R. L. Emerald, and M. L. Slade, *Phys. Rev. Lett.*, **26**, 1564 (1971).

(36) For localized electrons, Ishimaru et al. have given in ref 33 the expression of excitation energy for $m \rightarrow n$ transition as $h\nu \simeq \epsilon_n^\beta - \epsilon_m^\alpha - J_{mn}^\alpha$.

(37) C. M. Hurd and P. Coodin, *Phys. Chem. Solids*, **20**, 523 (1966).

(38) In ref 11 the width of the shoulder is suggested to extend as large as 1400 G. In this case the result of the analytical fit gives the experimental hyperfine coupling constants as $A_{\perp}$ = 14.25 G and $A_{\parallel}$ = 100 G (±50 G), and for the interpretation of the experimental result either model will do, with or without the 1s-like cavity center in the As_4S_6 casket.

(39) There is no decisive evidence to determine which model is convincing, with or without the 1s-cavity center in the As_4S_6 casket. If negative spin density ρ_s is found on the As atoms in the As_4S_6 casket, the 1s-cavity center model is adopted. However there are no available NMR data as far as the authors know. Moreover the influence of surrounding medium will be important in the glass matrix; presumably, for example, the layered structure of crystalline As_2S_3 may hold the major part of the origin of anisotropy of the glass, which is not examined in the present paper.

[Reprinted from the Journal of Physical Chemistry, 83, 767 (1979).]

Orbital Interaction in the Cycloaddition Reactions of Tetrasulfur Tetranitride, Tetraarsenic Tetrasulfide, and Tetraarsenic Tetraselenide with Olefins

Sir: Tetrasulfur tetranitride, S_4N_4, has been of gradually increasing importance as a starting material of sulfur-nitrogen chemistry.[1] Among other reasons, it has recently drawn much attention as the precursor of polymeric sulfur nitride, $(SN)_x$,[2] which is a low-dimensional metallic conductor, even becoming a superconductor at 0.3 K.[3] Meanwhile, the reaction of S_4N_4 with ordinary organic molecules is also a matter of continued interest.[4] Several examples of cycloaddition reactions of S_4N_4 with olefins have been originally reported by Becke-Goehring and Schläfer.[5] They have naively supposed the products 1 from

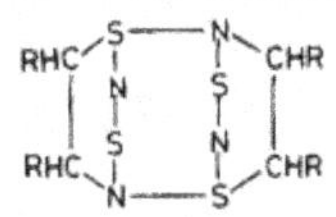

1

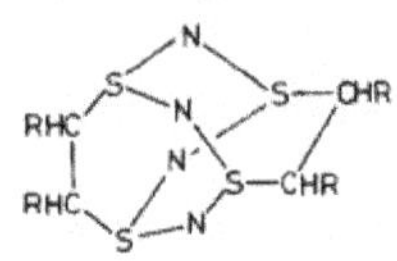

2

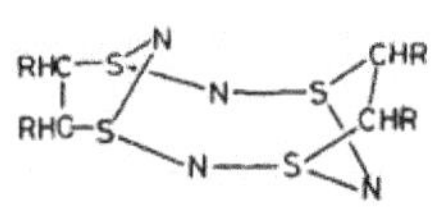

3

an analogy with the usual Diels–Alder reaction, the S_4N_4 acting as a diene. On the other hand, Gleiter has proposed using the MO concept, structure 2 which would be ob-

768 *The Journal of Physical Chemistry, Vol. 83, No. 6, 1979* Communications to the Editor

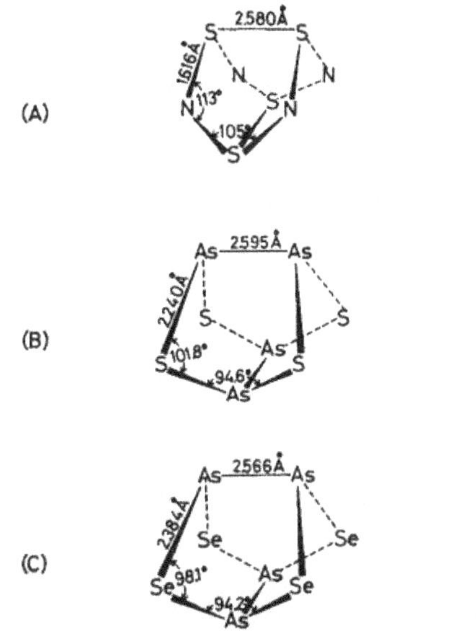

Figure 1. View of the molecular structures of S_4N_4 (A), As_4S_4 (B), and As_4Se_4 (C).

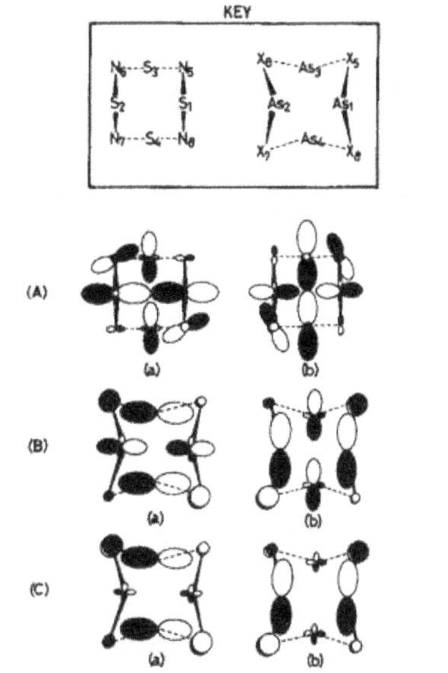

Figure 2. Patterns for the LUMO's of S_4N_4 (A), As_4S_4 (B), and As_4Se_4 (C). X designates S or Se in As_4S_4 or As_4Se_4, respectively.

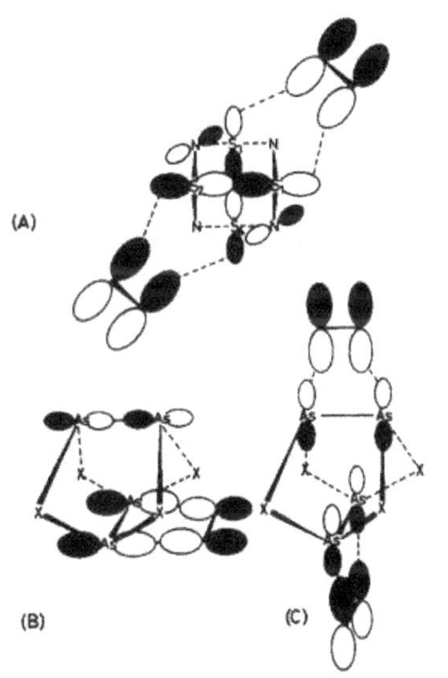

Figure 3. Orbital interactions of olefin HOMO with S_4N_4 LUMO (A), As_4X_4 LUMO (B), and with As_4X_4 (LU + 1) MO (C).

tained by the attack of two olefins on transannular S–S bonds.[6] Recent crystallographic analyses concerned with the bis adduct of S_4N_4 with norbornadienes using X-ray diffraction[7] or with *trans*-cyclooctenes using a NMR technique,[8] however, have definitely shown a different structure, 3. In this case, the C=C bonds of two olefins add across two S–N–S units to give two five-membered C–S–N–S–C rings retaining the eight-membered $(SN)_4$ ring.

Here we attempt to reexamine the orientation resulting from the cycloaddition of S_4N_4 with olefins based on a knowledge of the frontier orbital patterns[9] of S_4N_4 with the structure in Figure 1A.[10] It is also of interest to study the reactions of analogous compounds such as tetraarsenic tetrasulfide, As_4S_4 (realgar), and tetraarsenic tetraselenide, As_4Se_4, which have similar geometries to S_4N_4 as shown in Figure 1B and 1C,[11] although there have been no experimental reports concerning cycloaddition reactions of these compounds.

The MO calculations are performed with a semiempirical INDO-type ASMO-SCF method.[12] According to the calculation, the lowest unoccupied (LU) MO's of S_4N_4, As_4S_4, and As_4Se_4 are all doubly degenerate and their levels are negative (−2.689, −0.942, and −1.871 eV, respectively). The patterns of the LUMO's are shown in Figure 2. These energetically low-lying LUMO's imply electron deficient properties for these compounds as the present authors have pointed out.[13] Therefore, the interaction between the highest occupied (HO) MO of olefin and the LUMO's of S_4N_4, As_4S_4, or As_4Se_4 is dominant. Although there are two degenerate LUMO's in these species, one of them, e.g., LUMO (a) in Figure 2, will become slightly lower than the other, LUMO (b), according to the initial charge transfer from the approaching olefins.

For S_4N_4, considering the phase and spatial distribution of the LUMO of S–S pσ^* type, one of the olefins will attack concertedly S_1 and S_3, and the other S_2 and S_4 as illustrated in Figure 3A, and vice versa for the other LUMO (b), giving structure 3 observed in crystallographic analyses. Another structure of two five-membered C–N–S–N–C rings proposed earlier on the basis of NMR and other spectroscopic methods[14] is thus ruled out from the present orbital interaction treatment, as has been rejected in recent experimental works.[7,8]

On the other hand, As_4X_4 (X = S or Se), an olefin will attack either of two LUMO's of As–As p$\bar{\pi}$ types, affording a four-membered C–As–As–C ring as shown in Figure 3B, where $\bar{\pi}$ denotes π-type MO parallel to the four S (or Se) plane. The spatial distribution of the LUMO, however, may not be necessarily effective for cycloaddition compared with that of the (LU + 1) MO of As–As pπ type, the level of which is also negative (−0.152 eV for X = S and −1.251 eV for X = Se). Therefore, there will be another possibility that two olefins attack the (LU + 1) MO to yield two four-membered C–As–As–C rings as in Figure 3C. Anyway, for the reactions of As_4X_4 with olefins, the

The Journal of Physical Chemistry, Vol. 83, No. 6, 1979 769

structure >As–C–C–As<, which is different from the case for S_4N_4, is conjectured. It would be of interest to examine the orientation of olefins in suitable reaction systems.

Acknowledgment. We are grateful to the Data Processing Center of Kyoto University for its permission to use the FACOM M190 computer. This work was supported by a Grant-in-Aid for Scientific Research from the Ministry of Education of Japan (No. 255315).

References and Notes

(1) H. J. Emeléus, *Endeavour*, **32**, 76 (1973).
(2) (a) A. G. MacDiarmid, C. M. Mikulski, P. J. Russo, M. S. Saran, A. F. Garito, and A. J. Heeger, *J. Chem. Soc., Chem. Commun.*, 476 (1975); (b) C. M. Mikulski, P. J. Russo, M. S. Saran, A. G. MacDiarmid, A. F. Garito, and A. J. Heeger, *J. Am. Chem. Soc.*, **97**, 6358 (1975); (c) M. J. Cohen, A. F. Garito, A. J. Heeger, A G. MacDiarmid, C. M. Mikulski, M. S. Saran, and J. Kleppinger, *ibid.*, **98**, 3844 (1976).
(3) R. L. Greene, G. B. Street, and L. J. Suter, *Phys. Rev. Lett.*, **35**, 1799 (1975).
(4) For example, see (a) M. Goehring, *Q. Rev. Chem. Soc.*, **10**, 437 (1956); (b) H. W. Roesky, "The Sulfur-Nitrogen Bond," in "Sulfur in Organic and Inorganic Chemistry", A Senning, Ed., Vol. 1, Marcel Dekker, New York, 1971, Chapter 3.
(5) M. Becke-Goehring and D. Schläfer, *Z. Anorg. Chem.*, **356**, 234 (1968).
(6) R. Gleiter, *J. Chem. Soc. A*, 3174 (1970).
(7) A. M. Griffin and G. M. Sheldrick, *Acta Crystallogr., Sect. B*, **31**, 895 (1975).
(8) W. L. Mock and I. Mehrotra, *J. Chem. Soc., Chem. Commun.*, 123 (1976).
(9) As to the importance of the frontier orbitals in chemical reactions, see (a) K. Fukui, *Acc. Chem. Res.*, **4**, 57 (1971); (b) K. Fukui, "Theory of Orientation and Stereoselection", Springer-Verlag, West Berlin, 1975.
(10) B. D. Sharma and J. Donohue, *Acta Crystallogr.*, **16**, 891 (1963).
(11) (a) T. Ito, N. Morimoto, and R. Sadanaga, *Acta Crystallogr.*, **5**, 775 (1952); (b) P. Goldstein and A. Paton, *Acta Crystallogr., Sect. B*, **30**, 915 (1974).
(12) T. Yonezawa, H. Konishi, and H. Kato, *Bull. Chem. Soc. Jpn.*, **42**, 933 (1969).
(13) K. Tanaka, T. Yamabe, A. Tachibana, H. Kato, and K. Fukui, *J. Phys. Chem.*, **82**, 2121 (1978).
(14) M. R. Brinkman and C. W. Allen, *J. Am Chem. Soc.*, **94**, 1550 (1972).

Department of Hydrocarbon Chemistry
Faculty of Engineering
Kyoto University
Kyoto 606, Japan

Tokio Yamabe
Kazuyoshi Tanaka
Akitomo Tachibana
Kenichi Fukui*

College of General Education
Nagoya University
Nagoya 464, Japan

Hiroshi Kato

Received June 23, 1978; Revised Manuscript Received October 10, 1978

Part IV

Reaction Coordinate

In discussing reaction mechanisms chemists assume rather intuitively a reaction path, starting from the initial stage to reach the final product stage. Bonds are formed transiently between some pairs of atoms and some bonds are weakened at the transition state. Changes in hybridization states may occur at the reaction sites. If one could trace by computations such changes in the structure of a reacting system along the reaction path, reaction mechanisms would be discussed more in detail.

It was rather strange that the theory which defined the center line of the reaction path on the potential energy surface had not been provided until the concept of *"intrinsic reaction coordinate"* was proposed in 1970 (paper no. 38). A general equation to define the line of force was given and the theory was applied to several systems (papers no. 40, no. 41 and no. 42). We called the method of determining automatically the structural change along the reaction path as *"reaction ergodography"*. It is pleasant to find these days that the IRC is calculated easily by using softwares such as *Gaussian*, HONDO, GAMESS and MOPAC that are very familiar to chemists.

[Reprinted from the Journal of Physical Chemistry, 74, 4161 (1970).]

A Formulation of the Reaction Coordinate

by Kenichi Fukui[1]

Department of Chemistry, Illinois Institute of Technology, Chicago, Illinois 60616 *(Received June 17, 1970)*

Since the discovery of orbital symmetry control[2,3] in determining the favorable path of chemical reactions, the concept of the reaction coordinate has been used in the molecular orbital theory of chemical reactivity more frequently than before. The symmetry arguments of

(1) To whom correspondence should be addressed at Department of Hydrocarbon Chemistry, Kyoto University, Japan.

molecular orbitals (MO), in particular of the HO (highest occupied) MO and the LU (lowest unoccupied) MO, have been made in relation to the favorable path of chemical reactions,[4-8] and the nature of HOMO–LUMO interaction has been understood in connection with the change of molecular geometry along the reaction coordinate.[9,10] The conservation of orbital symmetry along the reaction coordinate has been discussed in connection with the stereoselection rule.[3,11]

In this connection, it is desirable to formulate the "reaction coordinate." Although in actual sense this can obviously never be a simple concept, a conventional definition would be helpful. We tentatively define the reaction coordinate as a curve passing through the initial and the transition points and orthogonal to energy equipotential contour surface. A reaction coordinate, defined as such, may be called "intrinsic reaction coordinate."

Let the number of nuclei of the reacting system be N. There are $3N - 6$ independent internuclear distances which are enough to determine the potential energy W. These variables are denoted by $x_1x_2 \text{----} x_m$ $(n = 3N - 6)$. An equipotential surface can be given by a hypersurface in the n-dimensional $x_1x_2 \text{----} x_n$ space as $c = W(x_1x_2 \text{----} x_n)$ with respect to a given electronic state in which c is a constant value of potential energy. The points corresponding to the initial, final, transition, and intermediate states must satisfy the condition

$$\frac{\partial W}{\partial x_1} = \frac{\partial W}{\partial x_2} = \text{----} = \frac{\partial W}{\partial x_n} = 0 \qquad (1)$$

These points together are referred to as "equilibrium points." The sets x_i $(i = 1, 2, \text{----}\ n)$ representing these points are, respectively, denoted by $\xi_i^{(\text{init})}$, $\xi_i^{(\text{fin})}$, $\xi_i^{(\text{trans})}$, and $\xi_i^{(\text{interm})}$.[12] The reaction coordinate as defined above is given by a solution of the following simultaneous differential equations which passes through the points $\xi_i^{(\text{init})}$, $\xi_i^{(\text{trans})}$, and $\xi_i^{(\text{fin})}$.

$$\mathrm{d}x_1 \Big/ \frac{\partial W}{\partial x_1} = \mathrm{d}x_2 \Big/ \frac{\partial W}{\partial x_2} = \text{----} = \mathrm{d}x_n \Big/ \frac{\partial W}{\partial x_n} \qquad (2)$$

The derivation of eq 2 is obvious. The tangential plane of the surface $c = W(x_1x_2 \text{----} x_n)$ at a point x_i is given by $\sum_i(\partial W/\partial x_i)(X_i - x_i) = 0$ where X_i are the current coordinates. If we represent a reaction coordinate as $x_1 = T_1(t)$, $x_2 = T_2(t)$, $\text{----}\ x_n = T_n(t)$ using a parameter t, the tangent of this curve at the point x_i will be $(X_1 - x_1)/(\mathrm{d}T_1/\mathrm{d}t) = (X_2 - x_2)/(\mathrm{d}T_2/\mathrm{d}t) = \text{----} = (X_n - x_n)/(\mathrm{d}T_n/\mathrm{d}t)$, and the condition that this line should be perpendicular to the above-mentioned tangential plane is nothing but eq 2.

Since the reaction coordinate is orthogonal to the tangential plane of the equipotential surface, the potential energy value on this tangential plane has always an extremum[13] at the point crossing the reaction coordinate.

Equation 2 is worthwhile since this will connect the chemical concept of reaction path to numerical computation in combination with the recent growth of possibility of obtaining the values of W for a reacting system with a tolerable reliability. But this equation is also useful in getting some qualitative information about the nature of reaction coordinate.

Equation 2 represents an extremely simple fact that the direction of displacement conforms to the direction of "force," namely the vector $(-\partial W/\partial x_1, -\partial W/\partial x_2 \text{----} -\partial W/\partial x_n)$. In this manner, at a nonequilibrium point, the driving force for a configuration change acts in the direction of extremum potential energy change. The coordinate, ρ, in the direction of configuration change at any point is thus simply given by

$$\frac{\partial W}{\partial \rho} = \text{extremum} \qquad (3)$$

If we designate the direction cosines of a coordinate as $l_1l_2 \text{----}\ l_n$ and regard the potential energy gradient in this direction as a function of l_j eq 3 leads to the relation

$$\frac{1}{l_1}\frac{\partial W}{\partial x_1} = \frac{1}{l_2}\frac{\partial W}{\partial x_2} = \text{----} = \frac{1}{l_n}\frac{\partial W}{\partial x_n} \qquad (3')$$

determining the l_j's for the reaction coordinate ρ at any point.

Any nonequilibrium point on a reaction coordinate has *one* direction satisfying eq 3. Along the reaction coordinates the curvature of potential energy increase has to be *maximum* with respect to the variation of direction; this is a requisite for the occurrence of a chemical reaction with an energy barrier to be overcome.

A particular situation may arise at an equilibrium point ξ_i at which all $\partial W/\partial x_i$ vanish. At a point close to ξ_i, say $\xi_i + \Delta x_i$, however, we have

(2) K. Fukui, "Molecular Orbitals in Chemistry, Physics, and Biology," P.-O. Löwdin and B. Pullman, Ed., Academic Press, New York, N. Y., 1964, p 513.

(3) R. B. Woodward and R. Hoffmann, *J. Amer. Chem. Soc.*, **87**, 395 (1965), and their papers cited in *Angew. Chem. Int. Ed. Engl.*, **8**, 781 (1969).

(4) K. Fukui, *Bull. Chem. Soc. Jap.*, **39**, 498 (1966).

(5) K. Fukui and H. Fujimoto, *ibid.*, **39**, 2116 (1966).

(6) K. Fukui and H. Fujimoto, "Mechanisms of Molecular Migrations," Vol. 2, B. S. Thyagarajan, Ed., Interscience, New York, N. Y., 1969, p 117.

(7) L. Salem, *Chem. Brit.*, 449 (1969); *Chem. Phys. Lett.*, **3**, 99 (1969).

(8) L. Salem and J. S. Wright, *J. Amer. Chem. Soc.*, **91**, 5947 (1969).

(9) K. Fukui and H. Fujimoto, *Bull. Chem. Soc. Jap.*, **41**, 1989 (1968).

(10) K. Fukui and H. Fukimoto, *ibid.*, **42**, 3399 (1969).

(11) R. G. Pearson, *Theor. Chim. Acta*, **16**, 107 (1970).

(12) Here a stable initial state refers to the ground-state reaction. In an excited state, the possibility of involvement of a chemical relaxation which may start from a nonequilibrium point cannot entirely be rejected.

(13) For this path to have a definite physical meaning of reaction coordinate, this extremum should be an absolute minimum.

$$W/\partial x_i \sim \sum_j (\partial^2 W/\partial x_i \partial x_j)_{x_k=\xi_k} \Delta x_j$$

In order for this point to lie on one reaction coordinate, it follows from eq 2 that

$$\frac{\Delta x_1}{\sum_j \left(\frac{\partial^2 W}{\partial x_1 \partial x_j}\right)_{x_k=\xi_k} \Delta x_j} = \frac{\Delta x_2}{\sum_j \left(\frac{\partial^2 W}{\partial x_2 \partial x_j}\right)_{x_k=\xi_k} \Delta x_j} = \text{-----} = \frac{\Delta x_n}{\sum_j \left(\frac{\partial^2 W}{\partial x_n \partial x_j}\right)_{x_k=\xi_k} \Delta x_j} = \text{const.} = \frac{1}{\kappa}$$

which immediately leads to the secular equation

$$\begin{vmatrix} b_{11} - \kappa & b_{12} & \cdots & b_{1n} \\ b_{21} & b_{22} - \kappa & \cdots & b_{2n} \\ \cdots & \cdots & \cdots & \cdots \\ b_{n1} & b_{n2} & \cdots & b_{nn} - \kappa \end{vmatrix} = 0$$

$$\left(b_{ij} = b_{ji} = \left(\frac{\partial^2 W}{\partial x_i \, \partial x_j}\right)_{x_k=\xi_k}\right)$$

giving Δx_i by which we can determine the direction of displacement. Accordingly, in this case we have instead of eq 3

$$\frac{\partial^2 W}{\partial \rho^2} = \text{extremum} \tag{4}$$

as the condition to be satisfied by a reaction coordinate ρ.[14] This means, similar to eq 3′, that the direction cosines l_j of the reaction coordinate ρ are specified as satisfying the relation

$$\frac{1}{l_1}\frac{\partial}{\partial x_1}\sum_j l_j \frac{\partial W}{\partial x_j} = \frac{1}{l_2}\frac{\partial}{\partial x_2}\sum_j l_j \frac{\partial W}{\partial x_j} = \text{-----} = \frac{1}{l_n}\frac{\partial}{\partial x_n}\sum_j l_j \frac{\partial W}{\partial x_j} \tag{4'}$$

Once the system starts to displace along one of these reaction coordinates, the route is controlled by eq 2 up to a transition-state or another equilibrium point.

At an equilibrium point, we have in general n directions which satisfy eq 4. At a stable equilibrium point (*e.g.*, corresponding to $\xi_i^{(\text{init})}$, $\xi_i^{(\text{fin})}$, or $\xi_i^{(\text{interm})}$, $\partial^2 W/\partial\rho^2$ is positive in the direction of all of n such directions. At a transition-state point $\partial^2 W/\partial\rho^2$ is negative in direction of the reaction coordinate and positive in the others.

Lastly, it is interesting to consider the symmetry of the reaction coordinate[3,11] by the use of eq 2. Equation 2 prescribes completely the locus of the motion of nuclei along the intrinsic reaction coordinate. Equation 2 shows that the direction of displacement of nuclear positions is in conformity with the direction of gradient of the potential energy of the entire reacting system. In other words, the reaction coordinate always belongs to the totally symmetric representation of the point group to which the symmetry of the nuclear arrangement belongs. Therefore, it is self-evident that the symmetry, if any exists in the system, is conserved in the deformation along the reaction coordinate. In this way, it results that each reaction coordinate has a definite symmetry of its own.

The relation between the symmetry possessed by a system and the symmetry of the MO's associated with that system is a matter of common knowledge. According to the group-theoretical requirement, the molecular orbital symmetry is necessarily conserved along one reaction coordinate throughout, provided the symmetry of the reaction coordinate is conserved throughout. In this way, the *principle of conservation of orbital symmetry* is derived from eq 2. This in fact made possible the dexterous use of the correlation diagrams by Woodward and Hoffmann.[3]

But it should be noted that the orbital symmetry is conserved in relation to *every* possible reaction coordinate which obeys the control of eq 2, no matter whether it is of high energy or low energy. Therefore, the principle of conservation of orbital symmetry itself cannot tell us the reason why a particular reaction path proves to be favorable and another unfavorable on the correlation diagram. In order to explore this reason, we have to look into the nature of the change in potential energy of the system along the reaction coordinate.

Bader[15,16] and Pearson[17] showed in regard to many examples that at an equilibrium point the value of $\partial^2 W/\partial\rho^2$ is seriously reduced if the reaction coordinate ρ belongs to the same irreducible representation as the direct product of the representations of HOMO and LUMO of the entire reacting system. The mechanism of self-accelerating reduction of the interaction energy according to the proceeding of reaction along the reaction coordinate has been clarified[2,4-10] in relation to the HOMO–LUMO interaction between the reactants. In the case of two-molecule reactions, the interrelation between these two approaches is self-evident. In intramolecular reactions, too, the correspondence is a matter of approximation if we use the partition technique[6] in the reacting molecule. The localized nature[10] of usual chemical reactions almost permits us in many cases to consider an intramolecular reaction as a reaction between two parts of one molecule.

Acknowledgment. This work was carried out during the author's stay at Illinois Institute of Technology as a National Science Foundation Senior Foreign Scientist Fellow. The author thanks Dr. S. I. Miller of IIT and Dr. H. Fujimoto of Kyoto University for helpful comments.

(14) In some cases, the secular equation factors, and Δx_i's separate into several noncombining groups.

(15) R. F. W. Bader, *Mol. Phys.*, **3**, 137 (1960).
(16) R. F. W. Bader, *Can. J. Chem.*, **40**, 1164 (1962).
(17) R. G. Pearson, *J. Amer. Chem. Soc.*, **91**, 1252, 4947 (1969).

THE CHARGE AND SPIN TRANSFERS IN CHEMICAL REACTION PATHS

KENICHI FUKUI

Dept. of Hydrocarbon Chemistry, Kyoto University, Kyoto, Japan

Contents:

1. Introduction

The role of the interaction of particular molecular orbitals (MO) in chemical reactions in general has long been recognized as an essential factor determining the orientation (Fukui *et al.*, 1952, 1954a; Fukui, 1964, 1965a, 1970a, b, 1971b; Klopman and Hudson, 1967; Klopman, 1968; Salem, 1968, 1969a, b). The interaction of particular MO's also relates to the stereoselection rule and is believed to control the stereochemical path (Fukui, 1964, 1965b, 1971a; Woodward and Hoffmann, 1965, 1969a, b; Fukui and Fujimoto, 1969a; Anh, 1970), and the mechanism of such particular MO controls has qualitatively been discussed in relation to the nature of chemical reactions (Fukui *et al.*, 1954b; Fukui and Fujimoto, 1968, 1969b; Fujimoto *et al.*, 1971b; Fujimoto and Fukui, 1972).

The particular MO's which play such an essential part in chemical reactions are the highest occupied MO (or sometimes high-lying occupied MO's) (HOMO), and the lowest unoccupied MO (or sometimes low-lying unoccupied MO's) (LUMO). In donor molecules, the behavior of HOMO is more important than that of LUMO, while in acceptor molecules this is reversed. In open-shell molecules, i.e. in radicals and excited-state molecules, the above-stated circumstances are valid in the unrestricted Hartree–Fock sense. In the restricted Hartree–Fock sense, the singly occupied MO's (SOMO) play this part. The overlapping interaction between HOMO of one specie and LUMO of another one, hereafter simply referred to as HOMO–LUMO interaction, has an essential significance in chemical reactions between two closed-shell molecules. Similarly, HOMO–SOMO, SOMO–LUMO, and SOMO–SOMO interactions, in the restricted Hartree–Fock sense, are the other types of particular orbital interaction.

These circumstances are essentially general, irrespective of whether the molecule is saturated or unsaturated, aliphatic or aromatic, and whether the intermolecular reaction is unicentric or multicentric. The criterion of particular orbital interaction can be applied also to intramolecular reactions in which two parts in the molecule essential to the reaction are regarded as if they were two molecules.

R. Daudel and B. Pullman (eds.), The World of Quantum Chemistry, 113–141.

The purpose of the present paper is to give a more detailed, quantitative description for the mode of particular orbital interactions in chemical processes. Stress is laid on the role and significance of charge and spin transfers in the molecular deformation and the bond interchange occurring in mutually reacting chemical species. It is also expected that such a treatment serves to give an illuminating interpretation for the selection of favorable paths in actual chemical reactions.

2. Description of Mutually Interacting Systems

Consider most generally a system composed of N nuclei and a definite number of electrons. We are able to calculate the adiabatic potential of the system with *fixed* nuclei in the frame of Born–Oppenheimer approximation. The location of a nucleus α in usual three-dimensional Cartesian space is denoted by X_α, Y_α. Z_α, and the distance of two nuclei α and β by $R_{\alpha\beta}$. The adiabatic potential W is given by a function of $(3N-6)$ independent $R_{\alpha\beta}$'s. As these $R_{\alpha\beta}$'s, we may choose for instance R_{12}, R_{13}, R_{23}, $R_{i\mu}$ ($i=1, 2, 3$; $\mu=4, 5, \ldots N$) by specifying any three nuclei which are not collinear. We write these ($3N$-6) particular pairs of nuclei as ($\alpha\beta$).

Since we consider here a nonequilibrium system with fixed nuclei in general, each nucleus is subject to a force in the direction of grad W. If we permit an *infinitely slow* motion from the initial fixed nuclear configuration, the displacement of each nucleus will be given by

$$\cdots = \frac{M_\alpha \mathrm{d}X_\alpha}{\partial W/\partial X_\alpha} = \frac{M_\alpha \mathrm{d}Y_\alpha}{\partial W/\partial Y_\alpha} = \frac{M_\alpha \mathrm{d}Z_\alpha}{\partial W/\partial Z_\alpha} = \cdots, \quad (\alpha = 1, 2, \ldots N) \tag{1}$$

in which M_α is the mass of nucleus α, and $\mathrm{d}X_\alpha$, $\mathrm{d}Y_\alpha$, $\mathrm{d}Z_\alpha$ are the components of displacement vector of nucleus α,d $\mathbf{R}_\alpha$. This displacement obviously satisfies the conditions

$$\begin{aligned} &\sum_\alpha M_\alpha \mathrm{d}X_\alpha = \sum_\alpha M_\alpha \mathrm{d}Y_\alpha = \sum_\alpha M_\alpha \mathrm{d}Z_\alpha = 0 \\ &\sum_\alpha M_\alpha (Y_\alpha \mathrm{d}Z_\alpha - Z_\alpha \mathrm{d}Y_\alpha) = \sum_\alpha M_\alpha (Z_\alpha \mathrm{d}X_\alpha - X_\alpha \mathrm{d}Z_\alpha) = \\ &\qquad = \sum_\alpha M_\alpha (X_\alpha \mathrm{d}Y_\alpha - Y_\alpha \mathrm{d}X_\alpha) = 0, \end{aligned} \tag{2}$$

since no force acts to translate the center of gravity nor to rotate the system around it.

The locus of Equation (1) is given by solving this equation regarded as a differential equation. If we consider such a locus passing through a transition state, this becomes an '*intrinsic*' reaction coordinate (Fukui, 1970c). An *equilibrium point* of function W may be defined as satisfying the condition

$$\partial W/\partial R_{\alpha\beta} = 0 \quad (\text{with respect to all } (\alpha\beta)). \tag{3}$$

Accordingly, a reaction intermediate, an initial state or final state of a reacting system, and a transition state correspond to equilibrium points. The initial and final states of a reacting system which is composed of two independent species are represented by those points of function W which correspond to the infinite mutual separation of these species. At an equilibrium point Equation (1) is modified to be

$$\cdots = \frac{M_\alpha \mathrm{d}X_\alpha}{\sum_\beta \left(\mathrm{grad}_\beta \frac{\partial W}{\partial X_\alpha} \cdot \mathrm{d}\mathbf{R}_\beta\right)} = \frac{M_\alpha \mathrm{d}Y_\alpha}{\sum_\beta \left(\mathrm{grad}_\beta \frac{\partial W}{\partial Y_\alpha} \cdot \mathrm{d}\mathbf{R}_\beta\right)} = \frac{M_\alpha \mathrm{d}Z_\alpha}{\sum_\beta \left(\mathrm{grad}_\beta \frac{\partial W}{\partial Z_\alpha} \cdot \mathrm{d}\mathbf{R}_\beta\right)} = \cdots, \quad (4)$$

in which grad_β signifies the vector $(\partial/\partial X_\beta, \partial/\partial Y_\beta, \partial/\partial Z_\beta)$. Equation (4) leads to a usual secular equation of $3N$ dimensions to give $(3N-6)$ directions of displacement which are nothing but the normal coordinates. The other six zero-roots of the secular equation correspond to the freedom of translation and rotation as a whole.

In order to treat Equation (1) further we have to know the change of adiabatic potential with nuclear displacement. By the use of the Hellmann–Feynman theorem we have

$$\frac{\partial W}{\partial X_\alpha} = \frac{\partial V_{nn}}{\partial X_\alpha} + \int \varrho(1) \frac{\partial V_n(1)}{\partial X_\alpha} \mathrm{d}v(1), \quad \text{etc.,} \quad (5)$$

where V_{nn} is the nuclear–nuclear repulsion, $V_n(1)$ is the nuclear attraction of electron 1, $\int \mathrm{d}v(1)$ means the integration over the space coordinates of electron 1, and $\varrho(1)$ is the first-order density given by

$$\varrho(1) = n_e \int \mathrm{d}\sigma(1) \int \Psi^*(1, 2, 3, \ldots)\, \Psi(1, 2, 3, \ldots)\, \mathrm{d}\tau(2)\, \mathrm{d}\tau(3) \ldots, \quad (6)$$

where n_e is the total number of electrons, Ψ is the electronic wave function of the total system, $\int \mathrm{d}\sigma(1)$ implies the integration with respect to the spin coordinate of electron 1, and $\int \mathrm{d}\tau(2)$ signifies the integration over space and spin coordinates of electron 2.

At an equilibrium point the displacement of nuclei can be described in terms of normal coordinates, $Q^{(1)}, Q^{(2)}, \ldots Q^{(3N-6)}$. The adiabatic potential W is expressed in the form

$$W = W_0 + \tfrac{1}{2} \sum_{i=1}^{3N-6} \kappa^{(i)} Q^{(i)2} \quad (7)$$

in the vicinity of the equilibrium point. The coefficient $\kappa^{(i)}$ represents $\partial^2 W/\partial Q^{(i)2}$ and is generally positive with respect to $(3N-6)$ normal coordinates in stable equilibrium points. At a transition state one of $\kappa^{(i)}$'s is negative, and the corresponding direction is that of reaction coordinate. At initial and final states of intermolecular reactions in which two species are infinitely separate from each other, the six $\kappa^{(i)}$'s corresponding to one mutual approach, and five mutual rotations including torsion with respect to the axis connecting the two centers of gravity of each species are all infinitesimal. The others are those of normal vibrations localizing at each species. In such systems the intrinsic reaction coordinate at infinite separation is obviously the normal coordinate of mutual approach.

The $\kappa^{(i)}$ can be a measure of the favorableness of normal coordinate $Q^{(i)}$ as a reaction coordinate. The reaction starts from a given initial state most favorably along the normal coordinate of minimum κ. In intermolecular reactions the infinitesimal κ's never serve as the measure of favorableness, and the comparison of stability of the combined system at finite separation of the components becomes significant.

The second-order perturbation treatment of the total system at an equilibrium point in which the nuclear displacement is regarded as a perturbation provides a convenient expression of κ (Bader, 1962), i.e.

$$\kappa = \frac{\partial^2 V_{nn}}{\partial Q^2} + \int \varrho_{00}(1) \frac{\partial^2 V_n(1)}{\partial Q^2} \, dv(1) + 2 \sum_k{}' \frac{\left| \int \varrho_{0k}(1) \frac{\partial V_n(1)}{\partial Q} \, dv(1) \right|^2}{W_0 - W_k}. \tag{8}$$

In this expression, the symbols are the same as employed in Equation (5) except that subscript 0 and k signify the ground and kth excited electronic states at the equilibrium point. The transition density $\varrho_{0k}(1)$ is defined by

$$\varrho_{0k}(1) = n_e \int d\sigma(1) \int \Phi_0^*(1, 2, 3 \ldots) \, \Phi_k(1, 2, 3 \ldots) \, d\tau(2) \, d\tau(3) \ldots. \tag{9}$$

The usual symbol $\sum_k'$ implies the summation over the excited states k (excepting 0). The electron distribution for the perturbed system, i.e. for a point on the normal coordinate Q corresponding to a nuclear configuration near the equilibrium point is given by

$$\varrho(1) = \varrho_{00}(1) + 2Q \sum_k{}' \varrho_{0k}(1) \frac{\int \varrho_{k0}(2) \frac{\partial V_n(2)}{\partial Q} \, dv(2)}{W_0 - W_k}, \tag{10}$$

where Q is the nuclear displacement from the equilibrium point along the normal coordinate Q (Bader, 1962). This expression plays a significant role in the discussion of favorableness of the reaction coordinate in relation to actual occurrence (see Section 7).

3. Wave Function Suitable for Configuration and Population Analyses

We present the electronic wave function, Ψ, for the combined system consisting of two reacting species, A and B, with a nuclear configuration fixed at one point on the reaction coordinate. Each of the species A and B in a mutually interacting condition may have in general a geometry which is different from that of the equilibrium state, i.e. the most stable one possessed in the free condition. The total system is supposed to be composed of such a nonequilibrium species of generally *deformed* geometry. The condition that the total system lies on a reaction coordinate is never an essential one, but we can treat a more general case similarly. The procedure of constructing the wave function for the total system is chosen so as to be suitable for the present purpose to elucidate the mechanism of chemical reaction in terms of particular orbital interaction. Two methods are introduced here.

A. CI WAVE FUNCTION

In this procedure the wave function Ψ is represented by a linear combination of many configuration functions, Φ_M's, i.e. a configuration interaction (CI) technique is adopted. Each Φ_M is built from the MO's calculated by an appropriate Fock–Roothaan process with respect to each of the species A and B which exists separately with the *same* geometry as in the combined system (Fukui and Fujimoto, 1968). We call hereafter these species as 'isolated' species A and B. Never confound this with a species in the free, equilibrium condition. Each normalized configuration function Φ_M is antisymmetrized in a determinantal form and is constructed so as to represent a spin eigenstate. In this way, we have

$$\Psi = \sum_M C_M \Phi_M . \tag{11}$$

We designate the configuration function corresponding to the electron configuration which is the same as in each of the isolated species A and B by Φ_0 especially. The coefficient C_M in Equation (1) is determined so as to minimize the total energy

$$W = \int \Psi^* H \Psi \, \mathrm{d}\tau \bigg/ \int \Psi^* \Psi \, \mathrm{d}\tau \tag{12}$$

by solving the secular equation

$$\begin{vmatrix} H_{00} - W & H_{01} - S_{01}W \ldots \\ H_{10} - S_{10}W & H_{11} - W \quad \ldots \\ \ldots\ldots\ldots\ldots & \ldots\ldots\ldots\ldots \end{vmatrix} = 0, \tag{13}$$

in which H is the Hamiltonian operator of the combined system and

$$\begin{aligned} H_{MN} &= \int \Phi_M^* H \Phi_N \, \mathrm{d}\tau \\ S_{MN} &= \int \Phi_M^* \Phi_N \, \mathrm{d}\tau . \end{aligned} \tag{14}$$

The wave function, energy, and the coefficients may be specified by the electronic state to which they belong in such a way as $\Psi^{(n)}$, $W^{(n)}$, and $C_M^{(n)}$.

In order to explain how a configuration function, Φ_M, is constructed, we take as an example the case of two closed-shell species. We denote an occupied MO and an unoccupied MO of isolated species A by a_i and a_j, respectively, and those of species B by b_k and b_l, respectively. 'Unoccupied' MO's are defined here as corresponding to the nonrealistic solutions of the Fock equations for the ground state of an isolated species.

Consider an electron configuration in which one electron originally occupying the MO a_i jumps into the MO a_j. The function Φ_M corresponding to this configuration is denoted by $\Phi_{i\to j}$. In a similar manner $\Phi_{i\to l}$ stands for the configuration in which one electron in a_i of species A is transferred to one of the unoccupied MO's of species

B, b_l. The configuration $\Phi_{i\to j}$ may be referred to as a *monoexcited* one, and $\Phi_{i\to l}$ a *monotransferred* one. In the same way we can consider a *diexcited* configuration, say $\Phi_{i'\to j'}^{i\to j}$, a *ditransferred* one, say $\Phi_{i'\to l'}^{i\to l}$, and a *monoexcited monotransferred* one, say $\Phi_{i'\to j'}^{i\to j}$, and so forth. It is a matter of course that the accuracy of the wave function Ψ increases with more configurations taken into account. The function Φ_M may be specified by indicating the spin eigenvalue as ${}^1\Phi_M$ (singlet function), ${}^3\Phi_M$ (triplet function), and so on. The original configuration, Φ_0, may be called *zero*-configuration.

In this way, for instance ${}^1\Phi_{i\to l}$ is easily written as

$$ {}^1\Phi_{i\to l} = 2^{-1/2}\{\mathcal{N}\mathcal{A}[a_1\bar{a}_1a_2\bar{a}_2\ldots\, a_{i-1}\bar{a}_{i-1}a_ia_{i+1}\bar{a}_{i+1}\ldots\, a_n\bar{a}_nb_1\bar{b}_1\ldots\, b_m\bar{b}_m\bar{b}_l] - \mathcal{N}\mathcal{A}[a_1\bar{a}_1a_2\bar{a}_2\ldots\, a_{i-1}\bar{a}_{i-1}\bar{a}_ia_{i+1}\bar{a}_{i+1}\ldots\, a_n\bar{a}_nb_1\bar{b}_1\ldots\, b_m\bar{b}_mb_l]\}\,, \quad (15) $$

in which m and n are the numbers of doubly occupied MO's of isolated species A an B, respectively, a_i and $\bar{a}_i$ represent the spin orbitals with spin function of α and β, respectively, and $\mathcal{N}\mathcal{A}$ implies the normalization-antisymmetrization operator.

The coefficient C_M in Equation (1) represents the degree of participation of electron configuration Φ_M in the wave function for the mutually interacting system. The difference in electron distribution between the isolated species and the interacting system is caused by the contribution of these configurations which represent local excitations or mutual transfers of electrons, since each Φ_M has an electron distribution generally different from Φ_0.

The total energy of the combined system is interpreted in terms of the MO's of isolated species A and B. The picture adopted here concerns the rearrangement of electrons in the deformed nonequilibrium species on the occasion of mutual interaction. This energy of rearrangement may be referred to as 'interaction energy', which can be attributed to the mixing of Φ_M and the accompanying electron 'migration' in and between the species A and B. The merit of such a treatment is to enable an analysis of the origin of interaction energy in terms of MO's of each reactant molecule.

We assume here the normal case of chemical reactions between two ground-state species in which H_{00} is lower than any of H_{MM}'s.

The treatment mentioned above leads to the following form of the ground-state wave function:

$$ \Psi = C_0\Phi_0 + \left\{ \sum_M^{\text{monoex.}} + \sum_M^{\text{monotr.}} + \sum_M^{\text{diex.}} + \sum_M^{\text{ditr.}} + \sum_M^{\substack{\text{monoex.}\\ \text{monotr.}}} + \ldots \right\} C_M\Phi_M \ldots, \quad (16) $$

in which -ex. and -tr. stand for -excited and -transferred. The interaction energy is defined by

$$ \Delta W = W - (W_{A0} + W_{B0}), \quad (17) $$

where W is the lowest energy of the system composed of mutually interacting species A and B, W_{A0} and W_{B0} being the lowest energy of the isolated species A and B. In order to catch the essential feature of the interaction, it is often convenient to adopt an adequate procedure of approximation. For instance, the energy constituents can be recognized by means of the following approximate expression:

$$\Delta W \sim \{H_{00} - (W_{A0} + W_{B0})\} - \left(\sum_{M}^{\text{monoex.}} + \sum_{M}^{\text{monotr.}} \right) \frac{|H_{0M} - S_{0M}H_{00}|^2}{H_{MM} - H_{00}}, \tag{18}$$

in which multiexcited and multitransferred terms and higher perturbation terms are neglected. The contribution of each electron configuration is separately estimated by this equation. The term of $\{H_{00} - (W_{A0} + W_{B0})\}$ contains the effect of interaction without change in electron configuration and is divided into usual *Coulomb* and *exchange* terms. The second term refers to the influence of mixing of different configurations, and the terms corresponding to the excited configurations may be named *polarization* term and those corresponding to the transferred configurations can be called *delocalization* term.

In conformity with this approximation, the wave function of Equation (16) can be approximated as

$$\Psi \sim C_0\Phi_0 + \left(\sum_{M}^{\text{monoex.}} + \sum_{M}^{\text{monotr.}} \right) C_M\Phi_M, \tag{19}$$

where

$$C_0 \sim 1 - \left(\sum_{M}^{\text{monoex.}} + \sum_{M}^{\text{monotr.}} \right) \frac{S_{M0}(H_{0M} - S_{0M}H_{00})}{H_{00} - H_{MM}},$$
$$C_M \sim \frac{H_{M0} - S_{M0}H_{00}}{H_{00} - H_{MM}}. \tag{20}$$

Since the magnitude of the numerator of C_M parallels the overlap integral S_{M0}, the mixing of a configuration into the original one is proportional to the overlapping between these and inversely proportional to the level separation. This relation is important with regard to the mechanism of chemical interaction.

B. LCMO WAVE FUNCTION

A similar object is accomplished by another method of constructing the wave function. We calculate the MO's of the combined system A+B, ϕ_p's, as well as the MO's of each isolated species A and B. We proceed with the closed-shell case in which the concept of double occupation is employed, for simplicity. Each MO, ϕ_p, is expressed in the following form:

$$\phi_p = \sum_{i}^{\text{occ}} C_i^{(p)}a_i + \sum_{j}^{\text{uno}} C_j^{(p)}a_j + \sum_{k}^{\text{occ}} C_k^{(p)}b_k + \sum_{l}^{\text{uno}} C_l^{(p)}b_l, \quad (p = 1, 2, \ldots \nu) \tag{21}$$

where $\sum^{\text{occ}}$ and $\sum^{\text{uno}}$ respectively denote summation over all the occupied and the unoccupied MO's (Fujimoto *et al.*, 1974), and ν is the total number of occupied and unoccupied MO's of A and B. It is supposed here that all MO's are composed of linear combinations of independent atom-centered functions, usually atomic orbitals (AO). Since the number of AO's adopted agrees with that of MO's in each of systems A, B, and A+B, the coefficients in the right-hand side of Equation (21) can easily be determined by comparing the AO coefficients of both sides of the equation. The ground-state wave function using the MO's of Equation (21) leads to the following density function:

$$\begin{aligned}\varrho(1) &= \sum_{p}^{\text{occ}} \phi_p^*(1)\,\phi_p(1), \\ &= \varrho^{(\text{I})}(1) + \varrho^{(\text{II})}(1) + \varrho^{(\text{III})}(1) + \varrho^{(\text{IV})}(1),\end{aligned} \tag{22}$$

in which $\varrho^{(\text{I})}$ contains the terms of the type $a_i^*(1)a_i(1)$, $a_j^*(1)a_j(1)$, $b_k^*(1)b_k(1)$, and $b_l^*(1)b_l(1)$ and reflects the intramolecular change in bonding situation; the term $\varrho^{(\text{II})}$ consists of the terms $a_i^*(1)a_{i'}(1)$, $a_i^*(1)a_j(1)$, etc.; $\varrho^{(\text{III})}$ is a similar term concerning the MO's of B; $\varrho^{(\text{II})}$ and $\varrho^{(\text{III})}$ are the origin of intramolecular rearrangement of electrons in each of species A and B and vanish upon integration with respect to the space coordinates; and the last term $\varrho^{(\text{IV})}$ is composed of the products of an MO of A and an MO of B, and is principally accountable for the formation of new bonds between A and B. The overlapping of the occupied MO's of A and the occupied MO's of B gives an antibonding effect for the intermolecular bond formation through exchange repulsion. The electron density accumulating in the intermolecular region necessary for the formation of new bonds chiefly results from the occupied-unoccupied orbital overlapping between A and B. Of course, in contrast with the remarkable contributions of these orbital overlappings, the unoccupied–unoccupied orbital overlapping between two species is much less significant.

As a result of interaction, each of the initially doubly occupied MO's of isolated species A and B turns to have an occupation number less than 2 in general, and each of the initially unoccupied MO's comes to have a finite occupation number. These changes can be discussed in terms of each MO of the isolated species A and B.

The wave function of the combined system composed of MO's of Equation (21) is decomposed into the products of MO's of constituent species A and B. Comparison with the result of similar decomposition of the right-hand side of Equation (16) easily gives the coefficients C_0 and C_M's of Equation (16) in terms of $C_i^{(p)}$, etc. of Equation (21). This is the procedure of configuration analysis (Baba *et al.*, 1969), and we can thus discuss the degree of contribution of each electron configuration.

The essential feature of the methods, described above under A and B, is that these methods enable configuration and population analyses in terms of electrostatic, exchange, polarization, and delocalization effects separately. Each effect can further be divided into the contributions of each MO, if necessary. The role of particular orbital interactions can be elucidated in this way. Also the chemically obscure concept of 'donation and back-donation' can easily be made quantitative.

4. Transfer of Charge and Spin Through Orbital Interactions

A. SEMIEMPIRICAL CALCULATION BY CI APPROACH

It is worthwhile to show how the methods mentioned in the previous section work in elucidating the mechanism of chemical interaction on molecular, or even 'orbital' basis. Several examples in which a semiempirical all-valence-electron SCF (self-consistent-field) MO calculation is employed have already been reported (Fukui, 1971b; Fukui *et al.*, 1972; Fujimoto *et al.*, 1972c). The reactions treated there are as follows:

$$Cl^- + CH_3Cl \rightarrow ClCH_3 + Cl^-$$

(Fukui, 1971b; Fukui *et al.*, 1972; Fujimoto *et al.*, 1971a);

$$H^- + \underset{\diagdown\ \diagup \atop O}{CH_2{-}CH_2} \rightarrow CH_3 - CH_2 - O^-$$

(Fukui, 1971b; Fujimoto *et al.*, 1972c);

$$HCl + C_2H_4 \rightarrow \begin{array}{c} CH_2 = CH_2 \\ \vdots \\ H \\ | \\ Cl \end{array}$$

(Fukui, 1971b).

Also the dimerization of methylenes to ethylene (Fujimoto *et al.*, 1972a), the addition of singlet methylene to ethylene to form cyclopropane (Fujimoto *et al.*, 1972b), the hydrogen abstraction from methane by methyl radical (Fujimoto *et al.*, 1972d), and the addition of methyl radical to ethylene (Fuljimoto *et al.*, 1972d) have been dealt with by similar calculations. Calculations of these examples are carried out by Method A (CI wave function) mentioned in the previous section. A brief summary is given here.

(1) *Bimolecular Nucleophilic Substitution of Methyl Chloride by Chloride Ion*

The preference of inversion path to retention path is shown to result from a difference in stabilization energy. The stability of inversion path is attributed mainly to the large delocalization energy, which is understood to originate from the effective overlapping of the lone-pair orbital of Cl^- with LUMO of CH_3Cl extending along the C—Cl bond on the back of the C atom. This consists of a favorable HOMO-LUMO interaction. Population analysis along this reaction path can show that a considerable amount of electrons is transferred from Cl^- to the antibonding LUMO localizing on the C—Cl bond of CH_3Cl, resulting in the weakening of original C—Cl bond and an accumulation of electrons in the intermolecular region to form a new C—Cl bond.

(2) *Ring Opening of Ethylene Oxide by Hydride Anion*

The two lowest lying unoccupied MO's of ethylene oxide have large amplitude on the back of C atoms. The unsymmetrical approach of H^- along the path of *trans*-opening is shown to be more favorable than *cis*-opening. This way of approach is explained by the shape of two LUMO's, the overlapping of which with the hydride anion orbital brings about the delocalization stabilization. The deformation of C_2H_4O molecule due to the unsymmetrical approach of H^- causes a growth of LUMO amplitude near the reaction center which makes the charge transfer easier and easier.

(3) *Addition of Hydrogen Chloride to Ethylene*

The LUMO of hydrogen chloride has a large amplitude at hydrogen atom, extending to the direction opposite to the chlorine atom, and has a node between Cl—H bond. The HOMO of ethylene is π bonding orbital. The symmetrical approach favors the HOMO–LUMO overlapping. The configuration Φ_0 in Equation (16), which represents the adiabatic interaction without mixing of transferred or excited configurations, produces a density distribution in which electrons are expelled from the intermolecular region by exchange repulsion. Calculation shows that the mixing of transferred configurations, chiefly of HOMO–LUMO interaction, gathers electrons in the region between two molecules, which contribute to form a new bond.

(4) *Dimerization of Singlet Methylenes to Ethylene*

Singlet methylene has a C_{2v} bent structure, HOMO being the lone-pair orbital extending in the direction of the molecular plane and LUMO the π-orbital. The HOMO–LUMO criterion is very useful for guessing the favorable path. A rough inference at once leads to the path indicated in Figure 1a, in which HOMO$^{(1)}$ of $CH_2^{(1)}$ overlaps with LUMO$^{(2)}$ of $CH_2^{(2)}$ along the line of approach to form a σ-bond of ethylene, while HOMO$^{(2)}$ parallels LUMO$^{(1)}$ to form a π-bond. However, a further detailed investigation suggests a little modification of the model (a), since LUMO$^{(1)}$–HOMO$^{(2)}$ bonding would more effectively favor the stabilization in σ-type one rather than π-type. A model expected as the result would be between (a) and (b) of Figure 1. An

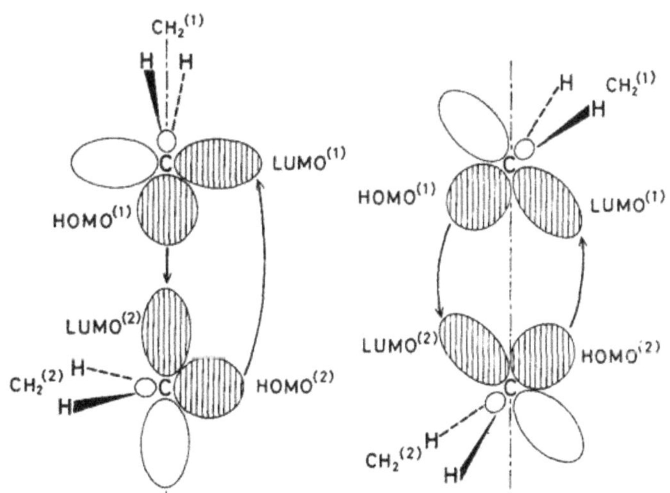

Fig. 1. Unsymmetrical (a) and symmetrical (b) approach of two methylenes favorable to double HOMO–LUMO overlappings.

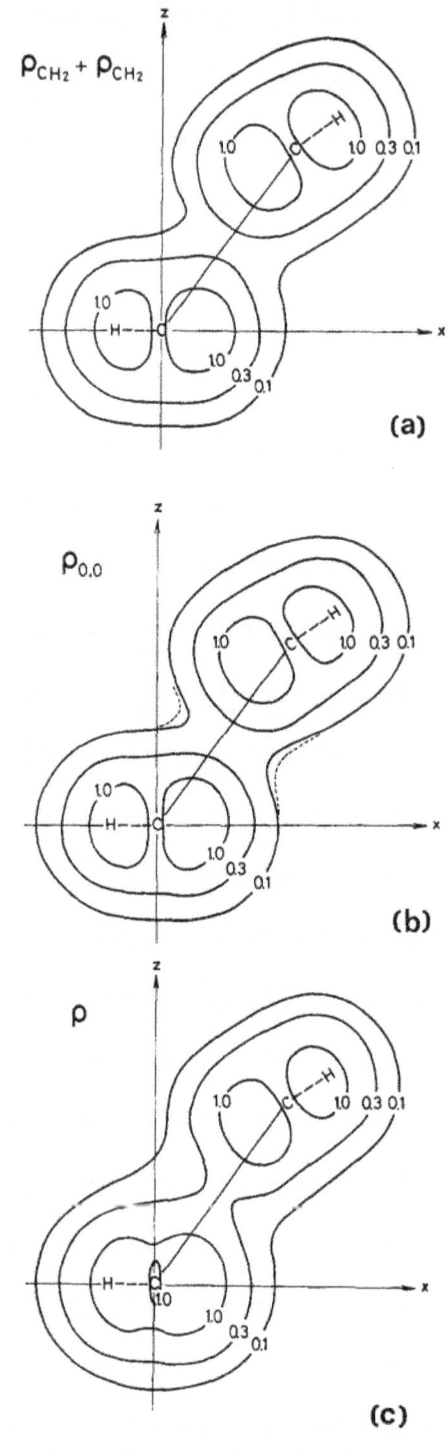

Fig. 2. Changes of electron density distribution in the dimerization of methylenes. (a) The sum of electron densities of two methylenes with no interaction. (b) The electron density of the system of two adiabatically interacting methylenes. (c) The electron density of the system of two interacting methylenes with the mixing of transferred configurations.

extended Hückel calculation (Hoffmann *et al.*, 1970) indicates that the deviation of each angle between CH_2-plane and the C—C axis from the model (a) to the direction of the model (b) is 6° at the C—C distance of 3.0 Å. This model shows that the contribution to the large delocalization energy mainly comes from the HOMO–LUMO interaction.

The changes in electron densities are shown in Figure 2. The simple sum of electron distributions of two methylenes with no interaction at all is (a), and (b) indicates the electron density of two methylenes which interact with each other adiabatically, i.e., without mixing of charge-transfer or excited configurations. It is seen in (b) that the exchange repulsion excludes electrons out of the intermolecular region. It is the effect of mixing of charge-transfer configurations that the electron density in that region grows to contibute to the bond formation.

(5) *Addition of Singlet Methylene to Ethylene*

Similar investigation is made with regard to the addition of singlet methylene to ethylene to produce cyclopropane. The HOMO–LUMO overlapping is effectuated by the mode of interaction shown in Figure 3, which is essentially the same as obtained by an extended Hückel calculation (Hoffmann, 1968). No explanation will be needed with regard to Figure 3. The calculation with respect to this model by the method described in Section 3 gives the following result. The two transferred configurations (from the HOMO of etylene to the LUMO of methylene, and from the HOMO of methylene to the LUMO of ethylene) have a dominant contribution to the wave function. The electron density maps of Figure 4 show the important role of the HOMO–LUMO interaction in the genesis of newly forming bonds. The exchange repulsion expels electrons from the intermolecular region and the mixing of charge-transfer states calls electrons back in this region. But almost the same maps of (c) and (d) imply that the contribution of the HOMO–LUMO interaction is decisive also in this case.

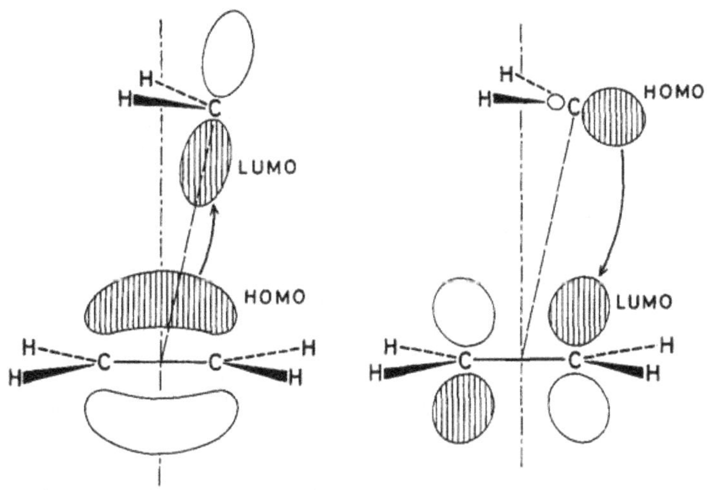

Fig. 3. The mode of HOMO–LUMO overlapping in the interaction between methylene and ethylene.

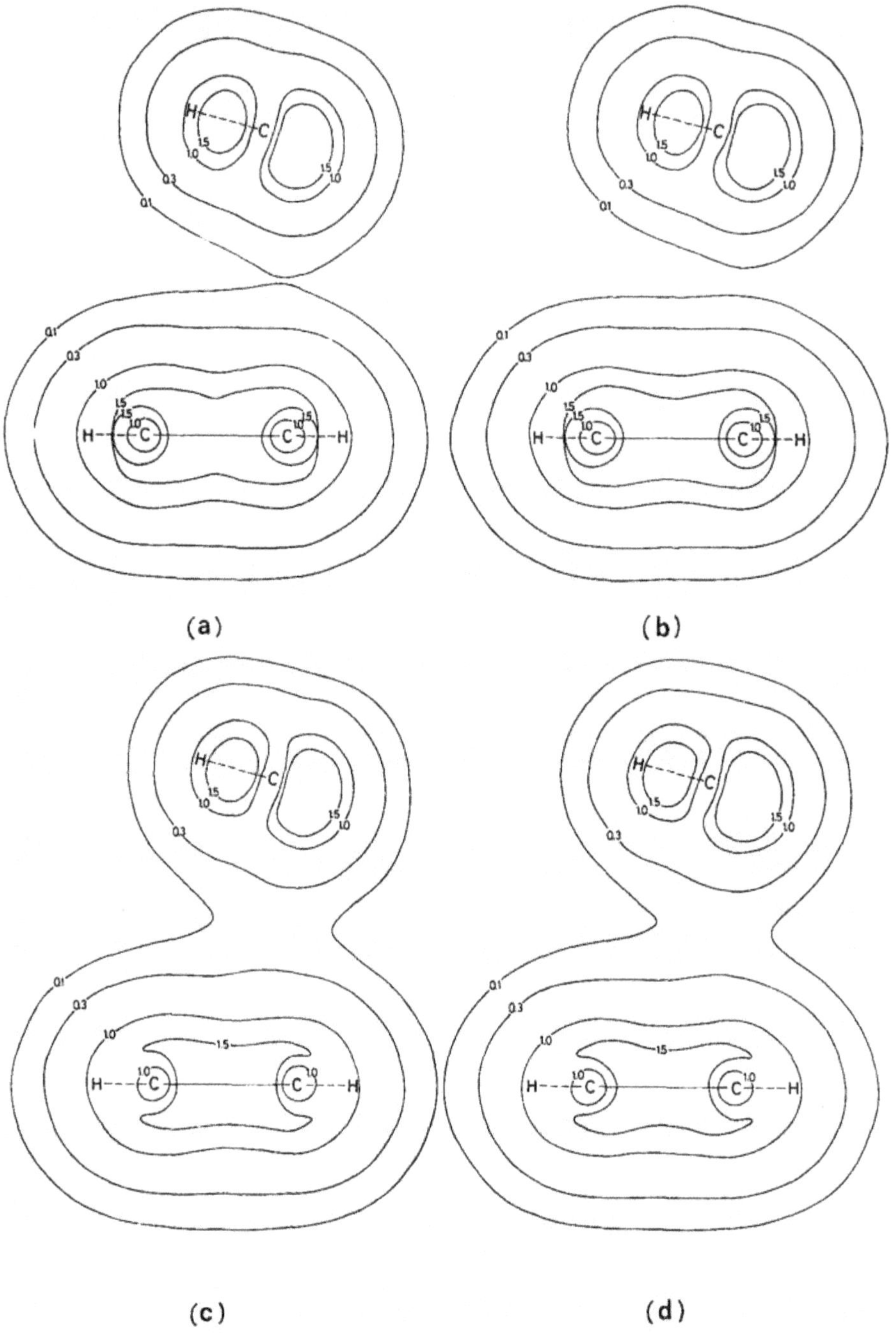

Fig. 4. Electron density maps for the system ($CH_2 + C_2H_4$). (a) The simple sum of electron densities of CH_2 and C_2H_4. (b) Adiabatic interaction without mixing of charge-transfer configurations. (c) Interaction with mixing of charge-transfer configurations. (d) With the consideration of the HOMO–LUMO interaction only.

(6) *Hydrogen Abstraction of Methane by Methyl*

The model used for the treatment of this reaction is shown in Figure 5. The unpaired electron of methyl radical is assumed to have α spin without loss of generality. The unrestricted open-shell calculation is made to obtain electron density distribution. Figure 6 indicates the density maps with regard to spins α and β, respectively.

As in the case of interactions of two closed-shell molecules, the electron exchange causes expulsion of electrons from the intermolecular region also in the system composed of a radical and a closed-shell, as is seen by the comparison of Figures 6a and b. The difference of maps (b) and (c) indicates that the electron density in the intermolecular region is supplied by the charge-transfer interaction. The most important contribution is the β-spin transfer from CH_4-HOMO to the LU-β-MO of methyl radical, and that of the α-spin transfer from HO-α-MO of methyl to CH_4-LUMO is the next.

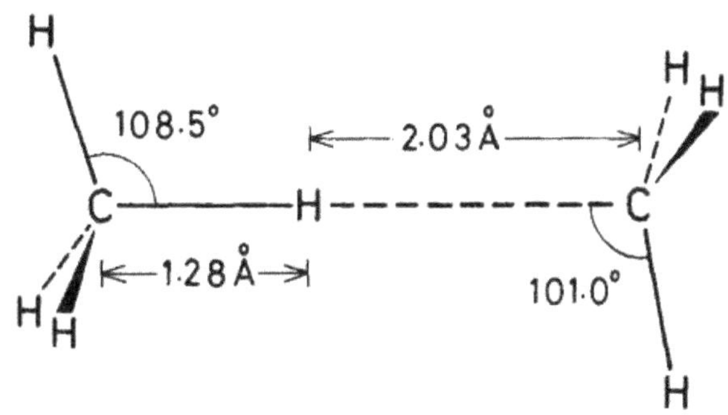

Fig. 5. Model for the interaction of CH_4 and CH_3.

The increased amount of α-spin in the region between methane and methyl is almost cancelled by the exchange repulsion as is illustrated in (a-1) and (b-1), and the maps of (c-1) and (a-1) are eventually not much different. It is hence β-spin electrons in this case that contribute to the formation of a new C–H bond on account of the trend of pairing with α-spin electrons. In this fashion, the exchange mechanism of spin migration is really remarkable, but for the problem of chemical reactions the charge-transfer mechanism has a more decisive significance. The polarization mechanism is less important than these two.

The donor–acceptor relation in regard to α- and β-spins may depend on the polarity of the radical and on the combination of radical and molecule. Population analysis in Figure 7 shows that the methyl radical behaves as an electrophile and attracts β-spin electrons from methane molecule.

(7) *Addition of Methyl to Ethylene*

A similar calculation is applied to the addition of methyl radical to ethylene molecule. An asymmetrical model is used (Hoyland, 1971). The charge-transfer interaction is

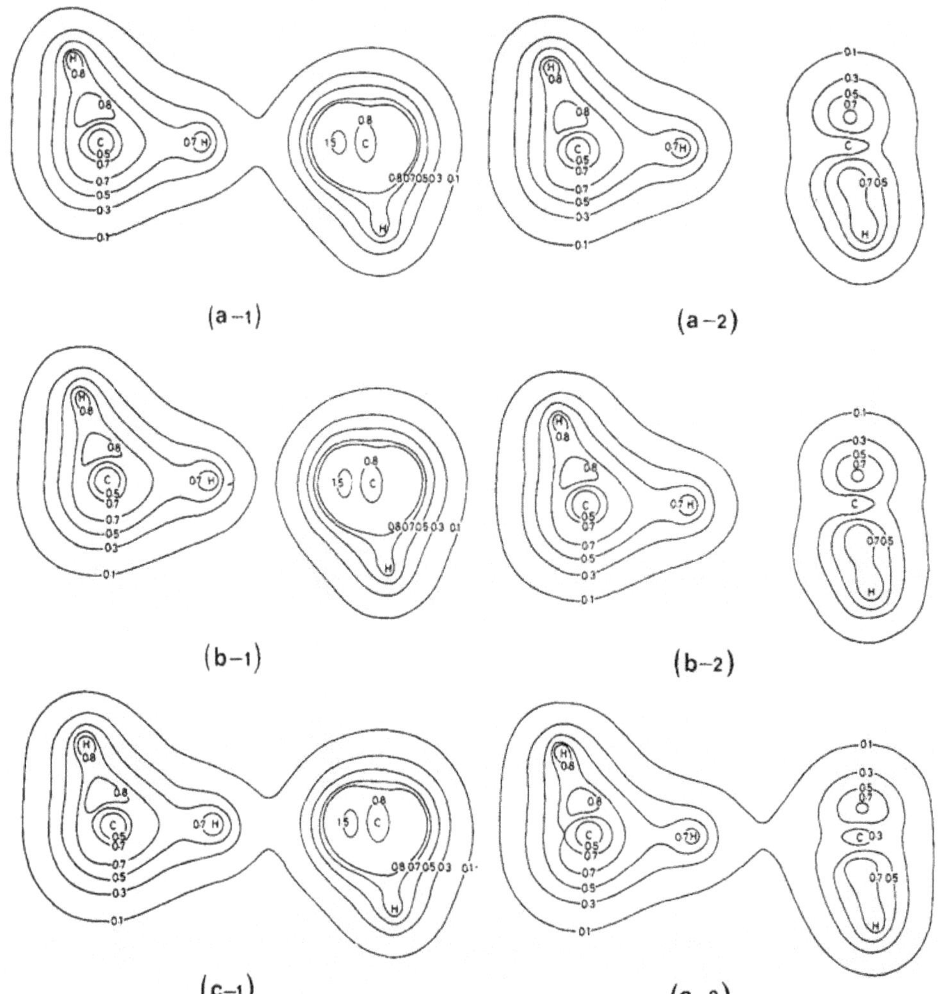

Fig. 6. The spin distributions of the system composed of methane and methyl. (a) The simple sum of electron densities of CH_4 and CH_3 without interaction. (b) The electron density of the system ($CH_4 + CH_3$) with adiabatic interaction. (c) The electron density with the mixing of charge-transfer taken into account. -1 and -2 refer to α and β spins.

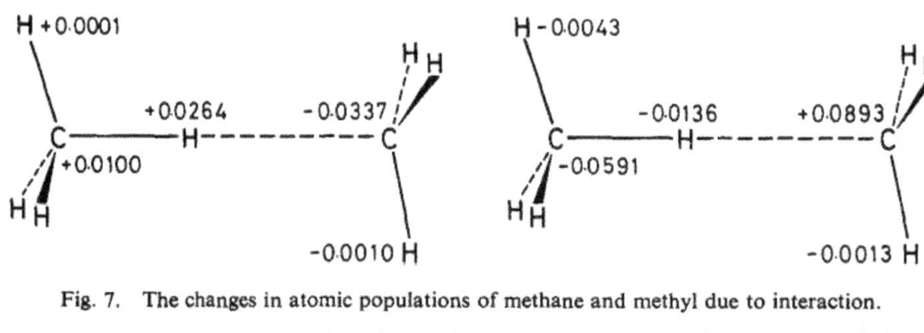

Fig. 7. The changes in atomic populations of methane and methyl due to interaction.

shown to be essential also in this case, and the α-spin transfer from the HO-α-MO of CH_3 to LUMO of ethylene exhibits the largest contribution and the β-spin transfer from HOMO of ethylene to the LU-β-MO of CH_3 is the next. The former causes an increase of electron density in the intermolecular region but this is wiped off by exchange interaction. The net accumulation of electrons between methyl carbon and one of ethylene carbons is effectuated by the latter β-spin transfer. The population analysis gives a net positive charge on methyl radical. This time methyl radical behaves as a nucleophile in the course of interaction.

B. AB INITIO CALCULATION BY LCMO APPROACH

In addition to the results of semiempirical calculation described above, an *ab initio* MO calculation has been applied to similar problems for the present purpose. The examples to be discussed here are diborane (B_2H_6) as a dimeric form of BH_3 (Yamabe *et al.*, 1973a), the NH_3-HF system (Yamabe *et al.*, 1973b), borazane as a complex of NH_3 and BH_3 (Fujimoto *et al.*, 1974), and borine carbonyl as a complex of BH_3 and CO (Kato *et al.*, 1974). All of these calculations have recourse to method B (LCMO wave function) described in the previous section. The minimal basis set of Slater-type orbitals, each of which is expanded in three Gaussian-type orbitals, is used for the *ab initio* calculation.

(1) B_2H_6 *Molecule as a Dimer of* BH_3

The electronic structure of diborane, supposed to be of bridged D_{2h} geometry, is a suitable example to be discussed conveniently by the present method. Two BH_3 molecules are deformed so as to make up the D_{2h} dimer. The nature of the bridge bonds in B_2H_6 is elucidated in terms of the MO's of such deformed BH_3. The HOMO of one of these deformed BH_3's and the LUMO of the other are indicated in Figures 8a and 8b, respectively. The energy of HOMO is elevated and the LUMO is stabilized in comparison with undeformed BH_3, as well as the orbital lobes grow so as to make the HOMO–LUMO interaction easier (cf. Section 5). We can clearly see how the deformation of molecules favors the HOMO–LUMO interaction by overlapping. Reflecting this circumstance, the configuration analysis gives an outstanding contribution of the configurations corresponding to the $(\mathrm{HOMO})_\mathrm{I} \rightarrow (\mathrm{LUMO})_\mathrm{II}$ and $(\mathrm{HOMO})_\mathrm{II} \rightarrow (\mathrm{LUMO})_\mathrm{I}$ transfers, and also of the ditransferred configuration corresponding to the *mutual* HOMO → LUMO transfers. The concept of donation and back-donation is thus made distinct.

The electron density appearing in the intermolecular region results largely from the mixing of transferred configurations, among which the two HOMO–LUMO transfers are conspicuous. It is obvious that this *double* HOMO–LUMO overlapping becomes the main origin of bridge bonding structure of B_2H_6. Such *double* single-center interaction illustrated in Figure 9 should not be confounded with the two-center interaction which frequently causes the stereoselectivity in the concerted interaction (see Section 5). In this case, the exchange repulsion is also shown to expel electrons from intermolecular area to result in exerting an antibonding effect.

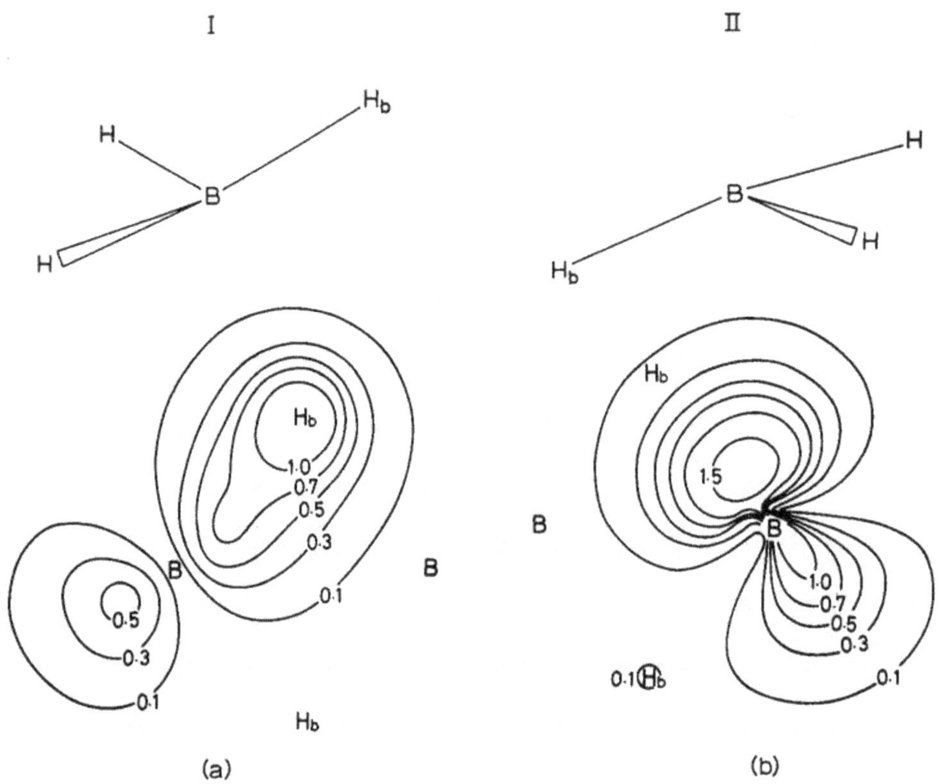

Fig. 8. The density maps of HOMO and LUMO of deformed BH_3. (a). The HOMO of $(BH_3)_I$. (b). The LUMO of $(BH_3)_{II}$.

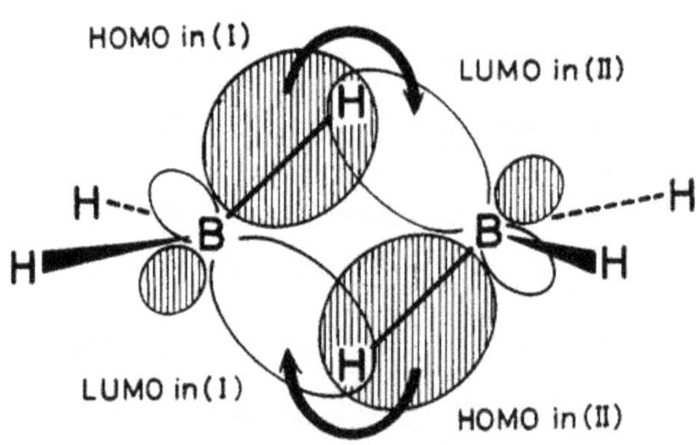

Fig. 9. The double HOMO–LUMO overlapping in B_2H_6 ((I) and (II) imply $(BH_3)_I$ and $(BH_3)_{II}$ of Figure 8, respectively).

(2) *NH_3-HF complex*

The nature of the hydrogen bond in NH_3-HF complex is also a good example. A linear C_{3v} model is employed. The electron density in the intermolecular region is shown to originate chiefly from the electron transfers from occupied NH_3 MO's to unoccupied HF MO's, among which the transfer from HOMO of NH_3 to LUMO of HF contributes 96.31% of the total contribution of transferred configurations. The back donation from HF to NH_3 is small in this case.

(3) *Borazane as a Complex of NH_3 and BH_3*

The staggered C_{3v} geometry is assumed for the model of the NH_3-BH_3 complex. With the use of a geometry shown in Figure 10, the total amount of electrons transferred from NH_3 to BH_3 is calculated as 0.4370. The decrease in occupation number

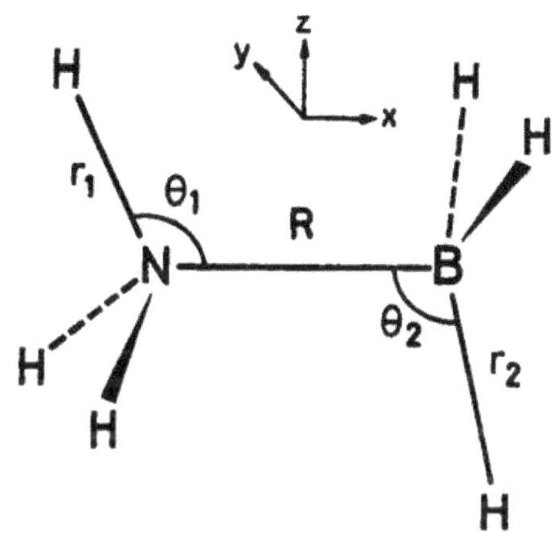

Fig. 10. The model of the geometry of NH_3—BH_3 complex.
$R = 1.561$ Å; $r_1 = 1.011$ Å; $r_2 = 1.222$ Å; $\theta_1 = 116.0°$; $\theta_2 = 105.4°$.

of NH_3-HOMO ($3a_1$) is 0.4394, while the increase of BH_3-LUMO ($3a_1$) is 0.4759. The overlap population of the B—N bond comes largely from that between axially extending $2p$ orbitals and $2s$ orbitals of N and B, arising from the interaction between occupied MO's of NH_3 and unoccupied MO's of BH_3, reflecting the relation that the directions of extension of NH_3-HOMO and BH_3-LUMO are both axial and opposite to the bonds with hydrogen atoms.

The configurational expansion of the wave function of the combined system, represented by Equation (16), is as follows:

$$\Psi = 0.60755\,\Phi_0 + 0.36786\,\Phi_{HO\to LU} + 0.08742\,\Phi_{HO\to LU}^{HO\to LU} + \cdots, \tag{23}$$

where HO and LU signify the NH_3-HOMO and the BH_3-LUMO, and, hence, $\Phi_{HO\to LU}^{HO\to LU}$ denotes the ditransferred configuration in which two electrons shift from the NH_3-HOMO to the BH_3-LUMO. The coefficient of this ditransferred configuration is

TABLE I

The constituents of atom bond population of N—B in NH_3BH_3

	ϱ	π
$(NH_3 . BH_3 \mid NH_3 . BH_3)$	−0.0971	−0.0242
$(NH_3 . BH_3 \mid NH_3^+ . BH_3^-)$	+0.3723	−0.0145
$(NH_3 . BH_3 \mid NH_3^- . BH_3^+)$	+0.0004	+0.0142
$(NH_3^+ . BH_3^- \mid NH_3^+ . BH_3^-)$	+0.0227	−0.0089
$(NH_3^- . BH_3^+ \mid NH_3^- . BH_3^+)$	−0.0009	−0.0002
$(NH_3 . BH_3 \mid NH_3^* . BH_3)$	−0.0003	−0.0033
$(NH_3 . BH_3 \mid NH_3 . BH_3^*)$	+0.0530	−0.0002
$(NH_3^* . BH_3 \mid NH_3^* . BH_3)$	−0.0021	−0.0004
$(NH_3 . BH_3^* \mid NH_3 . BH_3^*)$	−0.0070	−0.0005

larger than any of monotransferred configurations except HO → LU and also any of the monoexcited configurations.

Table I shows the contributions of each 'state' to the atom bond population of N—B in NH_3BH_3. The notation $(NH_3.BH_3 \mid NH_3^+.BH_3^-)$ represents the contribution from the overlap densities

$$\varrho_{0K}(1) = \int \Phi_0^*(1, 2, 3, \ldots)\, \Phi_K(1, 2, 3, \ldots)\, d\tau(2)\, d\tau(3) \ldots,$$

in which Φ_K represents one of the configurations corresponding to one electron transfer from NH_3 to BH_3. The others imply similar meaning, an asterisk denoting excited state. Figure 11 indicates the difference density, $\varrho_{NH_3BH_3}(1) - (\varrho_{NH_3}(1) + \varrho_{BH_3}(1))$, which clearly illustrates the effect of interference between two interacting species. All of the facts pointed out above represent the great disparity in donor–acceptor relationship between NH_3 and BH_3.

(4) *Borine Carbonyl as a Complex of BH_3 and CO*

The geometry of borine carbonyl is assumed here as a C_{3v} structure, i.e.

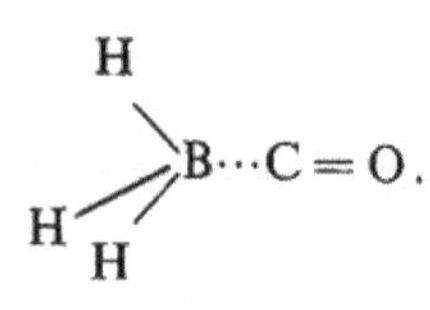

The total wave function is expanded into the following form:

$$\begin{aligned}\Psi(BH_3{-}CO) = {} & 0.5475\, \Psi(BH_3.CO) + 0.2977\, \Psi(BH_3^-.CO^+) + \\ & + 0.1470\, \Psi(BH_3^+.CO^-) + 0.0560\, \Psi(BH_3.CO^*) + \\ & + 0.0705\, \Psi(BH_3^*.CO) + 0.0602\, \Psi(BH_3^{2-}.CO^{2+}) + \\ & + 0.0243\, \Psi(BH_3^{2+}.CO^{2-}) + 0.0683\, \Psi(BH_3^{\mp}.CO^{\pm}) + \\ & + 0.0074\, \Psi(BH_3^*.CO^*) + \cdots . \end{aligned} \tag{24}$$

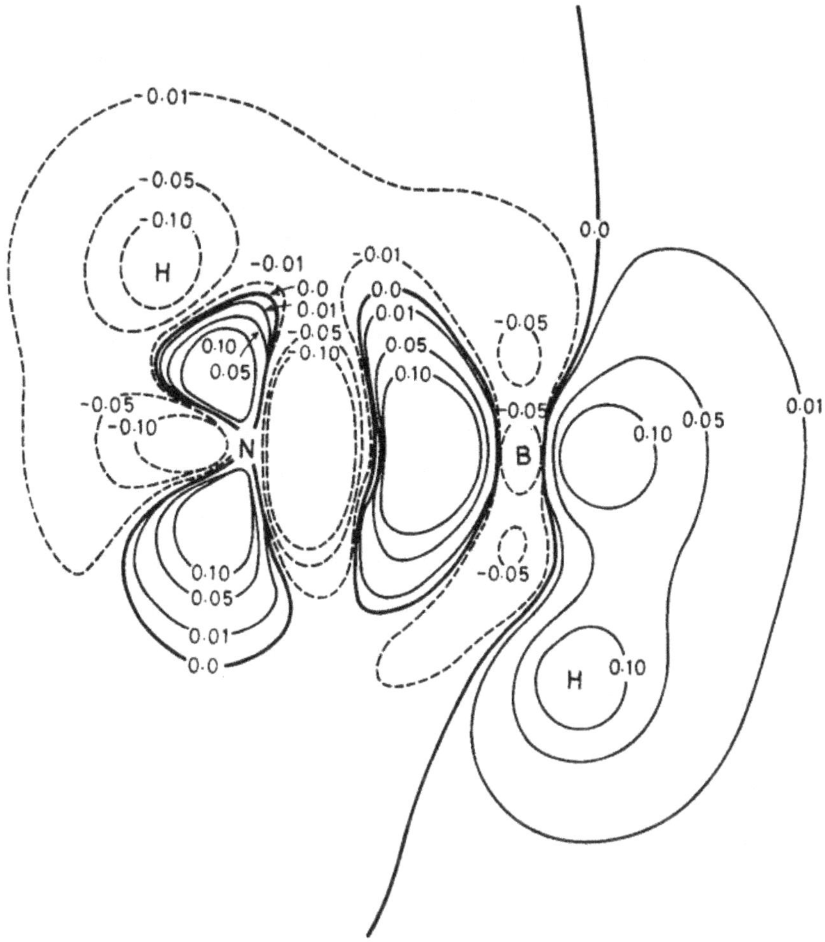

Fig. 11. The difference density, $\varrho_{NH_3BH_3}(1) - (\varrho_{NH_3}(1) + \varrho_{BH_3}(1))$.

The square of coefficient of each 'state function' is the sum of the squares of coefficients of all possible configurations. The charge-transfer interaction plays a distinct role in the formation of this complex. The contribution of the HOMO–LUMO interaction is of discriminative importance also in this case. The constitution of the B—C 'bond' is analysed in terms of each state. The B—C atom bond population chiefly consists of the σ-component of ($BH_3^- . CO^+$) state and, next, the π-component of ($BH_3^+ . CO^+$) state. These components come from the overlapping of these respective states with the zero-configuration of Equation (16) in the expression of the electron density function; i.e., the density terms of such a type as

$$\sum_{M}^{\text{monotr.}} 2C_0 C_M \varrho_{0M}(1), \tag{25}$$

in which $\varrho_{0M}(1)$ implies the overlap density like the form of Equation (19) between

TABLE II

The constituents of atom bond population of B—C in BH_3CO

	ϱ	π
(BH_3.CO | BH_3.CO)	−0.0954	−0.0007
(BH_3.CO | BH_3^-.CO^+)	+0.2521	−0.0052
(BH_3.CO | BH_3^+.CO^-)	−0.0124	+0.0314
(BH_3^-.CO^+ | BH_3^-.CO^+)	+0.0280	−0.0023
(BH_3^+.CO^- | BH_3^+.CO^-)	−0.0069	−0.0003
(BH_3.CO | BH_3.CO^*)	−0.0021	−0.0007
(BH_3.CO | BH_3^*.CO)	+0.0441	−0.0006
(BH_3.CO^* | BH_3.CO^*)	−0.0007	−0.0001
(BH_3^*.CO | BH_3^*.CO)	−0.0060	−0.0001

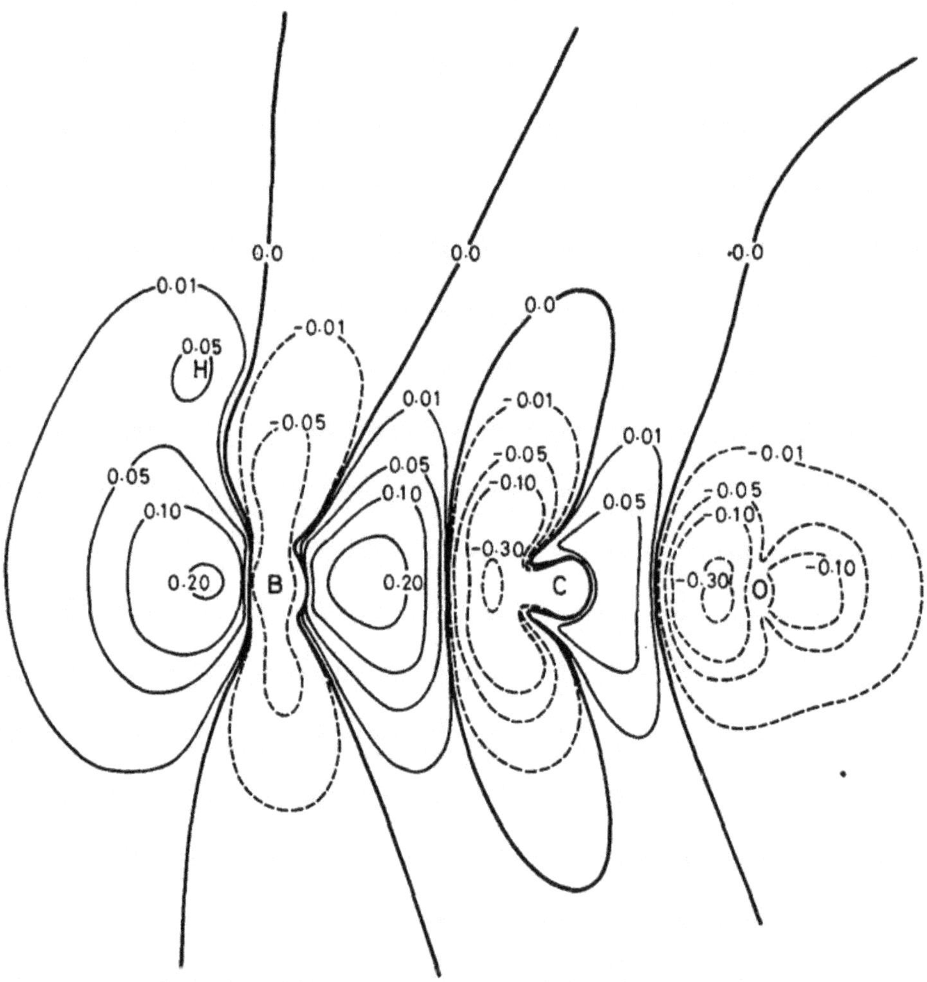

Fig. 12. The difference density, $\varrho_{BH_3CO}(1) - (\varrho_{BH_3}(1) + \varrho_{CO}(1))$.

configurations 0 and M. The σ- and π-components are distinguished by means of the type of overlapping of each pair of constituent atomic orbitals.

Table II indicates the constituents of the B—C bond in borine carbonyl, in which $(BH_3.CO \mid BH_3^-.CO^+)$ signifies the contribution to the atom bond population from the overlap densities $\varrho_{KL}(1)$ where configurations K and L belong to the state functions $\Psi(BH_3.CO)$ and $\Psi(BH_3^-.CO^+)$, respectively.

Figure 12 shows the difference density, $\varrho_{BH_3CO}(1)-(\varrho_{BH_3}(1)+\varrho_{CO}(1))$.

5. Favorableness of a Reaction Coordinate

The conspicuous role played by the particular orbital interaction in charge and spin transfers is obvious in view of the various examples mentioned in the last section, and should be closely related to the condition for actual occurrence of a chemical reaction and further, to the selection of favorable path of chemical reactions.

Consider a chemical reaction occurring between molecules A and B. The 'intrinsic' reaction coordinate mentioned in Section 2 begins from the infinite separation of A and B. But, practically, we are forced to discuss the most stable mutual arrangement of A and B at large, but finite, separation. This can be accomplished by using Equations (18) and (19). As the various results given in the previous section clearly show, the dominant contribution to the stabilization originates from monotransferred configurations, particularly from HOMO-to-LUMO transfers. The most contributing terms among each summation of the right-hand side of Equations (18) and (19) are

$$\sim -\,|H_{0,HO\to LU} - S_{0,HO\to LU}H_{00}|^2/(H_{HO\to LU,HO\to LU} - H_{00}) \tag{26}$$

and

$$C_{HO\to LU} \sim (H_{HO\to LU,\,0} - S_{HO\to LU,\,0}H_{00})/(H_{00} - H_{HO\to LU,\,HO\to LU}), \tag{27}$$

respectively. Roughly speaking, these terms parallel with the HOMO–LUMO overlapping and are inversely proportional to the HOMO–LUMO level separation between reactant molecules. Hence, the consideration of HOMO–LUMO overlapping can be a criterion of the stable spatial arrangement of A and B at large separation. At a long distance, the far-reaching effect of charge distribution is believed to dominate the orientation in charged or polar molecules. But even in such cases, the consequence of HOMO–LUMO criteria almost parallels with that of electrostatic consideration.

It happens occasionally that the species A and B form a loose intermediate and the reaction starts from such a complex. The formation of such a complex can be discussed in terms of HOMO–LUMO overlapping, similarly as mentioned before. When the intrinsic reaction coordinate begins from this equilibrium point, this will start along one of normal coordinates of this complex (cf. Section 2). The favorableness of this normal coordinate is characterized by κ of Equation (8). Evidently, the smaller the value of κ is, the more favorable is that normal coordinate. The reduction of the value of κ has recourse to the negative third term of the right-hand side of Equation (8) (Bader, 1962). Very roughly, the wave function of the ground state can be approximated by the zero-configuration function and that of an excited state k by each corresponding configuration function, say K: i.e.

$$\Psi_0 \sim \Phi_0 \quad \text{and} \quad \Psi_k \sim \Phi_K,$$

provided that the interaction is small, so that the transition density ϱ_{0k} is also approximated by

$$\varrho_{0k}(1) \sim \varrho_{0K}(1),$$

in which $\varrho_{0K}(1)$ is overlap density between configurations 0 and K. Obviously, the main contribution to the third term of the right-hand side of Equation (8) will come from the term in which Φ_K is $\Phi_{HO\to LU}$, since we are discussing the complex favored with effective HOMO–LUMO overlapping. The normal coordinate corresponding to the reaction coordinate must be related to the direction of closer mutual approach. But the overlap density $\varrho_{0,HO\to LU}$ has appreciable values in the intramolecular region, since

$$\varrho_{0,HO\to LU}(1) \sim c\{a_{HO}(1)\,b_{LU}(1) + 0(s_2)\}, \tag{28}$$

where c is a numerical constant, a_{HO} and b_{LU} is HOMO of A and LUMO of B, respectively, and $0(s_2)$ denotes small terms of the square of overlap integrals between the MO's of A and B. Accordingly, the nuclear motion to this intermolecular region will cause stabilization. Such a circumstance reduces the value of κ and contributes to the favorableness of this reaction coordinate, so that in such cases the overall reaction will be facilitated by this complex formation.

Once the reaction coordinate departs from an equilibrium point, the energy change, ΔW, due to the small nuclear displacement, is given by Equation (5), i.e. as

$$\Delta W = \Delta V_{nn} + \int \varrho(1)\,\Delta V_n(1)\,\mathrm{d}v(1), \tag{29}$$

where ΔV_{nn} is the change of nuclear–nuclear repulsion due to this nuclear displacement and the $\Delta V_n(1)$ is the electron–nuclear attraction for electron 1. The favorableness of a reaction coordinate depends on the degree of the reduction of unstabilization due to ΔV_{nn} by the negative contribution of the second term of the right-hand side of Equation (29) in case of mutual approach of two reactant molecules along the reaction coordinate. If the electron density function, $\varrho(1)$, shows an accumulation in the intermolecular area in the ground state of the combined system, the approach of nuclei to this area will give a *negative* value of the integral. In this connection, it seems useful to consider a simple case in which only the zero-configuration and one HO → LU transfer configuration, denoted simply by subscript K, are mixed. Ground states and excited states are specified by superscripts (g) and (e), respectively. The approximate wave functions are represented by

$$\Psi^{(g)} \simeq \left\{1 - \frac{S_{K0}(S_{0K}H_{00} - H_{0K})}{H_{KK} - H_{00}}\right\}\Phi_0 + \frac{S_{K0}H_{00} - H_{K0}}{H_{KK} - H_{00}}\,\Phi_K,$$

$$\Psi^{(e)} \simeq -\frac{S_{OK}H_{KK} - H_{OK}}{H_{KK} - H_{OO}}\Phi_0 + \left\{1 - \frac{S_{OK}(S_{KO}H_{KK} - H_{KO})}{H_{KK} - H_{OO}}\right\}\Phi_K, \tag{30}$$

in the approximation of Equation (19). The density distribution is written as

$$\varrho(1) = F_{OO}(1) + F_{KK}(1) + F_{OK}(1), \tag{31}$$

in which F_{OO}, etc. stand for the contribution from $\Phi_0\Phi_0$, etc., and it follows that

$$\begin{aligned} F_{OK}^{(g)}(1) &\sim \frac{S_{OK}(S_{KO}H_{OO} - H_{KO})}{H_{KK} - H_{OO}} \cdot \frac{\varrho_{OK}(1)}{S_{OK}}, \\ F_{OK}^{(e)}(1) &\sim -\frac{S_{KO}(S_{OK}H_{KK} - H_{OK})}{H_{KK} - H_{OO}} \cdot \frac{\varrho_{KO}(1)}{S_{KO}}. \end{aligned} \tag{32}$$

The function of the second factor of the right-hand side obviously has positive values in the intermolecular region, and the coefficient has positive definite sign in the ground-state and negative sign in the excited state in usual charge-transfer interactions. Hence, the ground-state charge-transfer interaction will contribute to stabilization in the nuclear motion of mutual approach of the reactants. If the circumstances favor the HOMO–LUMO interaction to make this effect strong enough for overcoming the effect of exchange repulsion which expels electrons out of the intermolecular region, the electron density will become positive in this region to aid the new bond formation. It has been shown that all of the various reactions cited in Section 4 satisfy this condition for actual occurrence.

As has been shown in Equation (10), the difference of electron density, $\Delta\varrho(1)$, and the difference of electron–nuclear attraction, ΔV_n, are connected by

$$\Delta\varrho(1) = 2\sum_k{}' \varrho_{Ok}(1)\frac{\int \varrho_{kO}(2)\,\Delta V_n(2)\,\mathrm{d}v(2)}{W_0 - W_k}. \tag{33}$$

The notation of $\Delta V_n(2)$ signifies the quantity referred to the coordinates of electron 2. This equation is applicable to any small nuclear displacement. If it is permitted that only one important term in the summation is taken, this equation becomes

$$\Delta\varrho(1) \sim 2\varrho_{O,\,HO\to LU}(1)\frac{\int \varrho_{HO\to LU,\,O}(2)\,\Delta V_n(2)\,\mathrm{d}v(2)}{H_{OO} - H_{HO\to LU,\,HO\to LU}}, \tag{34}$$

where the energies W_0 and W_k are approximated by H_{OO} and $H_{HO\to LU,HO\to LU}$, respectively. Since the overlap density $\varrho_{HO\to LU,\,O}$ has appreciable values only in the intermolecular region as has been mentioned above, the nuclear displacement to this region gives a negative value of the numerator of the right-hand side of Equation (34) in the intermolecular region. Accordingly, the mutual approach of two molecules A and B, keeping the arrangement of HOMO–LUMO overlapping, causes an accumulation of electrons in the intermediary region between two molecules. This is the genesis of a chemical bond between two reactants.

The electron accumulation in the intermolecular region originates largely in the charge-transfer from HOMO of one molecule A to LUMO of the other B, or vice versa. But both the electron escape from a bonding MO and the electron acceptance in an antibonding MO will cause the weakening of bonds. In this fashion, the formation of the new bonds is accomplished at the sacrifice of the bond fission. This is the bond exchange in chemical reactions and the cause of molecular deformation. As the chemical interaction becomes stronger in favored cases, the following three principles work cooperatively, giving rise to an accelerating effect for the promotion of reaction (Fukui and Fujimoyo, 1969b; Fujimoto and Fukui, 1972):

(1) *Positional Parallelism between the Charge-Transfer and the Bond Interchange*

At an early stage, the interaction begins between the positions where HOMO of one reactant and LUMO of the other overlap effectively. But in a majority of organic compounds, saturated or unsaturated, the position of large amplitude of HOMO is at the same time the position where the bond order with the other part of molecule is large and, hence, the position liable to undergo loosening of the bonds with the other part of the molecule in case of electron runaway from HOMO. Similarly, the position of large LUMO extension is also easily subject to weakening of the neighboring bonds in case of electron acceptance in LUMO (Fukui and Fujimoto, 1969b). This relation holds also in deformed molecules at a point on the reaction coordinate, since such a parallel relation is somewhat of mathematical nature, being almost valid in any nuclear arrangement (Fukui and Fujimoto, 1969b).

(2) *Successive Narrowing of the HOMO-LUMO Level Separation*

Since a free molecule takes a nuclear arrangement of the least energy, the deformation due to bond exchange gives rise to unstabilization, which is caused by the elevation of levels of occupied MO's. The MO which dominantly contributes to this unstabilization must be HOMO, since in this MO the actual bond loosening takes place at the bonds where the partial bond order is large, resulting from the positional parallelism between the reaction and the bond weakening. In consequence, the HOMO level rises conspicuously. Similarly, unoccupied orbitals fall according to the molecular deformation in general, among which LUMO goes down most remarkably. Many instances are available (Fukui and Fujimoto, 1969b). As the reaction proceeds, the HOMO–LUMO separation between the two reactants becomes smaller. The charge-transfer interaction takes place more easily when the HOMO–LUMO level difference becomes smaller. This is clearly understood by Equation (20).

(3) *Gradual Increase of the Localizing Character of HOMO and LUMO at the Reaction Center*

The molecular deformation accompanied by the proceeding of reaction gives rise to a growth of HOMO or LUMO amplitude near the reaction center. This relation applies in a wide range of compounds, and is understood by the following consideration. The conjugation between the reaction site and the remainder decreases successively with proceeding of the reaction. At the limit of vanishing conjugation, an MO localizing at the reaction center appears. In this way, the MO is confined within a narrow region in the vicinity of the reaction site. Such a tendency is valid even not in this extreme case. The HOMO and the LUMO amplitudes increase at the reaction center, so that the HOMO–LUMO transfer interaction grows, since the numerator of C_M in Equation (20) is roughly proportional to the product of HOMO and LUMO amplitudes of both reactants.

As has been mentioned, the charge-transfer interaction takes place between the positions of large HOMO and LUMO extension. Such positions are at the same time those which are liable to loose the neighboring bonds. The bond interchange causes molecular deformation, giving rise to the narrowing of HOMO–LUMO separation and the growth of HOMO and LUMO extension at the reaction center. Consequently, the charge-transfer and the bond exchange are promoted. In this way, initially very weak interaction is magnified to be a strong one, so that the reaction is accelerated. This is the mechanism of favored occurrence of a reaction.

In complicated polyatomic molecules, many occupied and unoccupied levels exist other than HOMO and LUMO. The difference of energy level can not discriminate HOMO and LUMO only. The reason why the particular MO interaction has a decisive significance in many reactions will first be clear with the explanation presented above.

The particular role of HOMO and LUMO in chemical reactions is thus evident. But the level of each MO is influenced by many factors. Also an energetically highest occupied MO happens to be unable to overlap with LUMO of the other molecule e.g. on account of symmetry relation in order to cause a given reaction. In such cases, the reaction will generally be not smooth, but when another occupied MO satisfying the necessary condition lies in the vicinity of the HOMO level, it is possible that this MO will play the part of HOMO. The same can be said with respect to LUMO. In these cases, the particular MO's can be *high-lying* occupied MO's and *low-lying* unoccupied MO's.

All the discussions made so far in this section are referred to the singele-center interaction in which only one bond is newly forming between two reactants A and B. The interaction of three species has been treated simply (Fujimoto *et al.*, 1973). In the reaction in which two bonds are simultaneously formed, the circumstances are more complicated. Here, it is recalled to mind that the double single-center reactions mentioned in Section 4B(1) should be distinguished from the 'concerted' two-center reactions, which is the object of immediate discussion. The electron distribution in the intermolecular region is almost dominated by the HOMO–LUMO interaction, as has been explained before. But it is evident that, if the HOMO–LUMO overlappings in the vicinity of these two centers of reaction are comparable to each other and have different signs, all the effects of HOMO–LUMO interaction contributing to the favorableness of the reaction coordinate hitherto discussed are cancelled, and other repulsive effects will be controlling. Therefore, the condition of actual occurrence of a concerted two- (or, more generally, many-) center reaction is that the overlap density has the same sign in the centers of bond formation. This conclusion agrees with the result already pointed out (Fukui, 1964).

The arguments so far made can essentially be applied also to the *intramolecular* reaction. In this case, the HOMO–LUMO criteria are employed with respect to the two parts of the molecule appropriately selected by the consideration of the essential relation to that reaction (Fukui, 1971a).

6. Theoretical Tracing of Reaction Paths

The locus of the intrinsic reaction coordinate is given by the differential equation, Equation (1), in which the denominators are replaced by Equation (5):

$$\cdots = \frac{M_\alpha \, \mathrm{d}X_\alpha}{\dfrac{\partial V_{nn}}{\partial X_\alpha} + \displaystyle\int \varrho(1) \frac{\partial V_n(1)}{\partial X_\alpha} \, \mathrm{d}v(1)} = \frac{M_\alpha \, \mathrm{d}Y_\alpha}{\dfrac{\partial V_{nn}}{\partial Y_\alpha} + \displaystyle\int \varrho(1) \frac{\partial V_n(1)}{\partial Y_\alpha} \, \mathrm{d}v(1)} = \frac{M_\alpha \, \mathrm{d}Z_\alpha}{\dfrac{\partial V_{nn}}{\partial Z_\alpha} + \displaystyle\int \varrho(1) \frac{\partial V_n(1)}{\partial Z_\alpha} \, \mathrm{d}v(1)} = \cdots . \tag{35}$$

If the nuclear configuration of a point on the reaction coordinate is given, the direction of displacement of each nucleus in the usual three-dimensional space is obtained so as to determine the 'next' point. Repetition of such procedure results in a locus. The determination of positions of nuclei of this starting point is always possible by finding the geometry of minimum energy, by varying the geometry of the total system, with a given, definite distance between the centers of gravity of two reactant molecules. An appropriate electronic device might be able to visualize even a quasi continuous motion of nuclei along the intrinsic reaction coordinate. Such a theoretical tracing of reaction path may be called 'reaction ergodography', and realization of this would be of practical significance from the standpoint of instructive puposes. How to overcome the difficulties expected from the accumulation of error due to successive plots and the essential error accompanying the Hellmann–Feynman usage of nonaccurate wave functions will be the point.

7. Conclusion

Nonempiricism having been introduced recently into chemistry, which is originally an 'empirical' science where division was the means of control, seems to serve in enabling generalization and synthesis of chemistry and chemical concept. Now, for instance, tolerable methods of nonempirical calculation for chemically interacting systems are available to seek for a unique common criterion for judging the favorableness of chemical interactions, whether or not interior intramolecular reactions of various types, heterolytic or homolytic, singlecentered or multicentered, and also molecule or molecular complex formations, and irrespective of the kind of the molecules interacting, inorganic or organic, saturated or unsaturated, and radical or excited molecules. However, it must be premised that such a criterion should be employable for the 'chemical' design by chemists. In this sense, the concept of particular orbital interaction between molecular species is expected to be qualified as a candidate for such a criterion, since this is based on the presumably immutable chemical concept of 'molecule', and is probably suitable for the understanding of usual chemists who can hardly get rid of the usage of isolated molecule concept.

Acknowledgements

The author thanks Mr Shinichi Yamabe, Mr Shigeki Kato, Mr Tsutomu Minato, and Mrs Chizuko Tanaka for their cooperation.

References

Anh, N. T.: 1970, *Les Règles de Woodward-Hoffmann*, Edscience, Paris.
Baba, H., Suzuki, S., and Takemura, T.: 1969, *J. Chem. Phys.* **50**, 2078.
Bader, R. F. W.: 1962, *Can. J. Chem.* **40**, 1164.
Fujimoto, H. and Fukui, K.: 1972, *Adv. Quant. Chem.* **6**, 177.
Fujimoto, H., Yamabe, S., and Fukui, K.: 1971a, *Tetrahedron Letters*, pp. 439, 443.
Fujimoto, H., Yamabe, S., and Fukui, K.: 1971b, *Bull. Chem. Soc. Jap.* **44**, 2936.
Fujimoto, H., Yamabe, S., and Fukui, K.: 1972a, *Bull. Chem. Soc. Jap.* **45**, 1566.
Fujimoto, H., Yamabe, S., and Fukui, K.: 1972b, *Bull. Chem. Soc. Jap.* **45**, 2424.
Fujimoto, H., Katata, M., Yamabe, S., and Fukui, K.: 1972c, *Bull. Chem. Soc. Jap.* **45**, 1320.
Fujimoto, H., Kato, S., Yamabe, S., and Fukui, K.: 1973, *Bull. Chem. Soc. Jap.* **46**, 1071.
Fujimoto, H., Kato, S., Yamabe, S., and Fukui, K.: 1974, *J. Chem. Phys.* (in press); compare also the references cited therein.
Fujimoto, H., Yamabe, S., Minato, T., and Fukui, K.: 1972d, *J. Am. Chem. Soc.* **94**, 9205.
Fukui, K.: 1964, in *Molecular Orbitals in Chemistry, Physics, and Biology* (ed. by P.-O. Löwdin and B. Pullman), Academic Press, New York, p. 513.
Fukui, K.: 1965a, in Modern Quantum Chemistry, Istanbul Lectures (ed. by O. Sinanoğlu), Academic Press, New York, Part I, p. 49.
Fukui, K.: 1965b, *Tetrahedron Letters*, p. 2009.
Fukui, K.: 1970a, in *Sigma Molecular Orbital Theory* (ed. by O. Sinanoğlu and K. B. Wiberg), Yale University Press, New Haven, Conn., U.S.A., p. 121.
Fukui, K.: 1970b, *Fortschr. Chem. Forsch.* **15**, 1.
Fukui, K.: 1970c, *J. Phys. Chem.* **74**, 4161.
Fukui, K.: 1971a, *Accounts Chem. Res.* **4**, 57.
Fukui, K.: 1971b, in *XXIIIrd Int. Congr. Pure Appl. Chem. (IUPAC)*, Vol. 1, Butterworths, London, p. 65.
Fukui, K. and Fujimoto, H.: 1968, *Bull. Chem. Soc. Japan* **41**, 1989; and compare the references cited therein.
Fukui, K. and Fujimoto, H.: 1969a, in *Mechanisms of Molecular Migrations* (ed. by B. S. Thyagarajan), Vol. 2, Wiley-Interscience, New York, p. 117.
Fukui, K. and Fujimoto, H.: 1969b, *Bull. Chem. Soc. Jap.* **42**, 3399.
Fukui, K., Yonezawa, T., and Shingu, H.: 1952, *J. Chem. Phys.* **20**, 722.
Fukui, K., Yonezawa, T., Nagata, C., and Shingu, H.: 1954a, *J. Chem. Phys.* **22**, 1433.
Fukui, K., Yonezawa, T., and Nagata, C.: 1954b, *Bull. Chem. Soc. Jap.* **27**, 423.

Fukui, K., Fujimoto, H., and Yamabe, S.: 1972, *J. Phys. Chem.* **76**, 232.
Hoffmann, R.: 1968, *J. Am. Chem. Soc.* **90**, 1475.
Hoffmann, R., Gleiter, R., and Mallory, F. B.: 1970, *J. Am. Chem. Soc.* **92**, 1460.
Hoyland, J. R.: 1971, *Theor. Chim. Acta* **22**, 229.
Kato, S., Fumimoto, H., Yamabe, S., and Fukui, K.: 1974, *J. Am. Chem. Soc.*, in press.
Klopman, G.: 1968, *J. Am. Chem. Soc.* **90**, 223.
Klopman, G. and Hudson, R. F.: 1967, *Theor. Chim. Acta* **8**, 165.
Salem, L.: 1968, *J. Am. Chem. Soc.* **90**, 543, 553.
Salem, L.: 1969a, *Chem. Phys. Letters* **3**, 99.
Salem, L.: 1969b, *Chem. Brit.* **5**, 449.
Woodward, R. B. and Hoffmann, R.: 1965, *J. Am. Chem. Soc.* **87**, 395, 2511.
Woodward, R. B. and Hoffmann, R.: 1969a, *Angew. Chem.* **81**, 797.
Woodward, R. B. and Hoffmann, R.: 1969b, *The Conservation of Orbital Symmetry*, Academic Press, New York.
Yamabe, S., Minato, T., Fujimoto, H., and Fukui, K.: 1973a, *Theor. Chim. Acta*, in press.
Yamabe, S., Kato, S., Fujimoto, H., and Fukui, K.: 1973b, *Bull. Chem. Soc. Jap.* **46**, 3619.

Constituent Analysis of the Potential Gradient Along a Reaction Coordinate. Method and an Application to CH_4 + T Reaction

Kenichi Fukui,* Shigeki Kato, and Hiroshi Fujimoto

Contribution from the Department of Hydrocarbon Chemistry, Kyoto University, Sakyo-ku, Kyoto, Japan. *Received June 18, 1974*

Abstract: On the basis of a formulation of the chemical reaction coordinate previously proposed the nuclear displacement on the reaction coordinate is discussed in terms of the potential energy gradient which is determined by the aid of the Hellmann-Feynman theorem from the electron density distribution and the nuclear arrangement of the total reacting system. The electron density is partitioned into several physically significant terms including exchange, delocalization, and polarization densities by the use of configuration interaction procedure, so that the potential gradient is also divided into the corresponding terms to discuss the origin of the favorableness or unfavorableness of a reaction path. An application is made to the methane-tritium reaction by the employment of an *ab initio* unrestricted Hartree-Fock molecular orbital method. Theoretical information is obtained with regard to submolecular mechanisms of the reaction.

The molecular orbital (MO) theory has been utilized in multifarious fashions to interpret the process of chemical reactions from an early stage of its development.[1] Various theories of orientation[2a] and stereoselection[2] have been proposed essentially in the frame of the activation energy criterion.[1,2] A progress has been made in that the partitioning of interaction energy,[3] for instance into Coulomb, exchange, delocalization, and polarization energies,[3b] assisted us in understanding the nature of phenomena occurring on the reaction path. An MO theoretical reasoning was given to the concept of a "favorable" path. The important role of the electron delocalization through the interaction of particular MO's (*i.e.*, the highest occupied (HO) MO, the lowest unoccupied (LU) MO, and the singly occupied (SO) MO) has been recognized.[3,4]

The description of the chemical process employed in these studies has mostly involved some assumptions concerning the reaction path, *i.e.*, the change in molecular geometry of each reactant along the reaction path. But this would be known by obtaining theoretically the complete surface for the potential energy of the reacting system. Such calculations were performed by the use of semiempirical and *ab initio* MO procedures.[5] These studies afforded useful information with respect to the reaction path, activation energy, or height of other barriers, geometry of the transition state, and so on. The only difficulty in this method lies in the drawing of a multidimensional potential energy surface for rather complicated systems.

In view of these circumstances it is desirable to obtain directly the information concerning the process along the reaction path. Bader[6a] gave the force acting along the reaction coordinate and discussed the change of electron density distribution and molecular geometry. Bader and Chandra,[6b] with the use of very accurate wave functions, calculated the interatomic binding force in the hydrogen formation reaction by means of the Hellmann-Feynman theorem.[7] This theorem was applied by Nakatsuji[8] to the discussion of molecular geometry and to the chemical interaction by the use of a chemically intuitive partitioning of forces into atomic dipole, exchange, and gross charge forces. Wang and Karplus[9a] obtained the trajectory of the path of methane formation from CH_2 and H_2 by calculating directly the forces, without the use of the Hellmann-Feynman force, given from the CNDO[10] potential surface. McIver and Komornicki[9b] presented a method to search for the transition state geometry by calculating the potential gradients.

The advantage of the physical simplicity and visuality in such force calculations will be multiplied by the combination with the theoretical formulation of reaction coordinate.[11,4c] An information on the reaction path would be obtained without resort to the potential surface calculation.

The present paper treats the nuclear displacement along the reaction coordinate, analyzing constituents of the potential energy gradient in terms of various modes of configuration interaction between reactant molecules. This is for the purpose of elucidating what is the most important factor for determining the favorable reaction path. An application is made to a simple system; *i.e.*, the substitution of a hydrogen atom of methane by tritium.[12]

$$CH_4 + T \longrightarrow CH_3T + H$$

Description of Reaction Coordinate

In order to describe a reacting system, the nuclei are

2

commonly treated as material particles which follow classical mechanics. The trajectory is given as a solution of a set of Lagrange's equations of motion

$$\frac{d}{dt}\frac{\partial L}{\partial \dot{\xi}_i} - \frac{\partial L}{\partial \xi_i} = 0 \quad (i = 1, 2, \ldots, f) \qquad (1)$$

where L is the Lagrangian, ξ_i and $\dot{\xi}_i$ are generalized coordinate and velocity, respectively, and f is the number of internal degrees of freedom. The Lagrangian is given as

$$L = (1/2)\sum_{i,j}^{f} a_{ij}(\xi)\dot{\xi}_i\dot{\xi}_j - W(\xi) \qquad (2)$$

in which W is the potential energy function and ξ stands for the set of ξ_i's. The coefficients a_{ij} in the quadratic form are easily obtained from the corresponding formula expressed in the Cartesian coordinates by eliminating the translations and the rotations of the whole system.

To solve eq 1, one must determine the initial condition. But the initial condition is associated with the statistical distribution of molecular states, and the solutions obtained, *i.e.*, trajectories, depend on the choice of initial conditions.[9] If we permit an infinitely slow motion of nuclei, one can get a unique trajectory. This definition of trajectory corresponds to the reaction coordinate which is imagined in the discussion of chemical reactions based on the enthalpy criterion. This was defined and named as the "intrinsic" reaction coordinate by one of the present authors.[11]

Direction of Nuclear Displacement. Let us consider one nonequilibrium point on the reaction coordinate. By solving eq 1 under the condition mentioned above, one can get the infinitesimal displacements of coordinates as the solution of the following simultaneous equations[13]

$$\frac{\sum_{j=1}^{f} a_{ij}(\xi)\Delta\xi_j}{\partial W/\partial \xi_i} = \text{const} \quad (i = 1, 2, \ldots, f) \qquad (3)$$

where $\Delta\xi_i$ is the value of displacement of coordinate ξ_i. Furthermore, if we write the matrix (a_{ij}) as $\mathbf{A}$ and put $\mathbf{B} = \mathbf{A}^{-1}$, eq 3 is transformed to

$$\frac{\Delta\xi_i}{\sum_{j=1}^{f} b_{ij}(\xi)\frac{\partial W}{\partial \xi_j}} = \text{const} \quad (i = 1, 2, \ldots, f) \qquad (4)$$

where b_{ij} is the matrix element to $\mathbf{B}$. Thus, when one $\Delta\xi_i$ is determined, all other $\Delta\xi_j$'s $(j \neq i)$ are determined by eq 4 to give the direction of nuclear displacement.

Next, at an equilibrium point which corresponds to the initial, final, transition, or an intermediate state, the energy gradients $\partial W/\partial \xi_i$ vanish, and we can obtain $\Delta\xi_i$'s by simply putting

$$\frac{\partial W}{\partial \xi_i} \sim \sum_{j=1}^{f} \frac{\partial^2 W}{\partial \xi_i \partial \xi_j}\Delta\xi_j \quad (i = 1, 2, \ldots, f) \qquad (5)$$

in eq 4. Thus, it is shown that the reaction coordinate starts from one of the normal coordinates at an equilibrium point, provided that they are nondegenerate.

Intermolecular Interaction Mode. In order to calculate the energy gradient, we use the Hellmann-Feynman theorem

$$\frac{\partial W}{\partial \xi_i} = \langle \Psi | \frac{\partial H}{\partial \xi_i} | \Psi \rangle = \int \rho(1)\frac{\partial V_{ne}^{(1)}}{\partial \xi_i}d\mathbf{r}(1) + \frac{\partial V_{nn}}{\partial \xi_i} \qquad (6)$$

where Ψ is the electronic wave function, H is the Hamiltonian, V_{ne} and V_{nn} are the electron-nuclear and the nuclear-nuclear potentials in H, respectively, and $\mathbf{r}$ implies the space variables. The diagonal element of the first-order density matrix is given as

$$\rho(1) = n_e \int \Psi^*\Psi d\mathbf{s}_1 d\tau_2 \ldots d\tau_{n_e} \qquad (7)$$

in which n_e is the number of electrons of the system, and s and τ are the spin and the spin-space variables. It is well known that the Hellmann-Feynman relation holds only for the exact wave function or "stable" wave function like the exact Hartree-Fock approximation.[14]

In the treatment of a mutually interacting system, the wave function is conveniently approximated by a combination of configuration functions, corresponding to charge-transferred, locally excited, and various other configurations, which are represented in terms of the MO's of each reactant molecule.[15,16] This configuration interaction (CI) representation gives a chemically graspable picture concerning the nature of the intermolecular interaction and chemical reactions. The perturbation method[15] can be used for the system of weaker interactions, but for stronger ones the configuration analysis,[16] that is, the method using the Hartree-Fock wave function expanded in terms of such configuration functions, gave a more reliable result.[17]

In these considerations, the configuration analysis technique is employed also in the present treatment. Since the details of configuration analysis have been described previously,[17a,b] we present here the outline pertinent to the purpose to treat the $(CH_4 + T)$ system. To calculate the wave function of open-shell system, we employ the unrestricted Hartree-Fock (UHF) method.[18] Performing the configuration analysis, we can represent the wave function of the reacting system, Ψ, in the following CI form

$$\Psi = C_0\Psi_0 + \sum_{l_\alpha}^{uno} C_{i_\alpha \to l_\alpha}\psi_{i_\alpha \to l_\alpha} + \sum_{k_\beta}^{occ} C_{k_\beta \to i_\beta}\psi_{k_\beta \to i_\beta} + \sum_{k_\alpha}^{occ}\sum_{l_\alpha}^{uno} C_{k_\alpha \to l_\alpha}\psi_{k_\alpha \to l_\alpha} + \sum_{k_\beta}^{occ}\sum_{l_\beta}^{uno} C_{k_\beta \to l_\beta}\psi_{k_\beta \to l_\beta} + \ldots \qquad (8)$$

where 0 indicates the zero configuration, that is the original one, i_α and i_β are the occupied and unoccupied orbitals of tritium, and $k_{\alpha,\beta}$ and $l_{\alpha,\beta}$ mean an occupied and an unoccupied MO of methane, respectively. The symbols Σ^{occ} and Σ^{uno} imply the summation over occupied and unoccupied MO's, respectively, and $i_\alpha \to l_\alpha$, for instance, stands for the one α-electron transfer from the ith MO to the lth MO. The coefficients of eq 8, C_0, $C_{i_\alpha \to l_\alpha}$, etc., are given in the same manner as in our previous papers[17] with some modifications characteristic to the UHF case.

Partitioning of Density Matrix. Using the CI representation of the wave function of the reacting system, eq 8, we construct the first-order density matrix which is directly connected with the energy gradient through the Hellmann-Feynman theorem. The electron density of the reacting system, $\rho(CH_4 + T)$, is first divided into two parts

$$\rho(CH_4 + T) = \{\rho(CH_4) + \rho(T)\} + \Delta\rho^I \qquad (9)$$

where $\rho(CH_4)$ and $\rho(T)$ are the densities of CH_4 are T in their isolated states, respectively. The term $\Delta\rho^I$ is the difference density which is caused by the effect of interference between two interacting species and is termed the interference density.[19] Integration over all space provides

$$\int \rho(CH_4)d\mathbf{r} = 10, \quad \int \rho(T)d\mathbf{r} = 1 \qquad (10)$$

$$\int \Delta\rho^I d\mathbf{r} = 0 \qquad (11)$$

The interference density is further decomposed into the exchange, delocalization, polarization density, and so on.

3

$$\Delta\rho^{I} = \rho_K + \rho(CH_4T\,|\,CH_4^-T^+) + \rho(CH_4^-T^+\,|\,CH_4^-T^+) + \rho(CH_4T\,|\,CH_4^+T^-) + \rho(CH_4^+T^-\,|\,CH_4^+T^-) + \rho(CH_4T\,|\,CH_4{}^*T) + \rho(CH_4{}^*T\,|\,CH_4{}^*T) + \ldots \quad (12)$$

where the notation $CH_4^-T^+$ implies the state of one-electron transfer from T to CH_4, and $CH_4{}^*$ stands for the monoexcited states of CH_4. The definitions of specialized interference densities are as follows

exchange density

$$\rho_K = \rho_{00} - \{\rho(CH_4) + \rho(T)\} \quad (13)$$

delocalization density

cross terms

$$\rho(CH_4T\,|\,CH_4^-T^+) = 2\sum_{l_\alpha}^{uno} C_0 C_{i_\alpha - l_\alpha}(\rho^\alpha{}_{0,i_\alpha - l_\alpha} - S^\alpha{}_{0,i_\alpha - l_\alpha}\rho^\alpha{}_{00})$$

$$\rho(CH_4T\,|\,CH_4^+T^-) = 2\sum_{k_\beta}^{occ} C_0 C_{k_\beta - i_\beta}(\rho^\beta{}_{0,k_\beta - i_\beta} - S_{0,k_\beta - i_\beta}\rho^\beta{}_{00}) \quad (14)$$

pure terms

$$\rho(CH_4^-T^+\,|\,CH_4^-T^+) = \sum_{l_\alpha}^{uno}\sum_{l_\alpha'}^{uno} C_{i_\alpha - l_\alpha}C_{i_\alpha - l_\alpha'}(\rho^\alpha{}_{i_\alpha - l_\alpha, i_\alpha - l_\alpha'} - S^\alpha{}_{i_\alpha - l_\alpha, i_\alpha - l_\alpha'}\rho_{00})$$

$$\rho(CH_4^+T^-\,|\,CH_4^+T^-) = \sum_{k_\beta}^{occ}\sum_{k_\beta'}^{occ} C_{k_\beta - i_\beta}C_{k_\beta' - i_\beta}(\rho^\beta{}_{k_\beta - i_\beta, k_\beta' - i_\beta} - S^\beta{}_{k_\beta - i_\beta, k_\beta' - i_\beta}\rho_{00}) \quad (15)$$

polarization density

cross term

$$\rho(CH_4T\,|\,CH_4{}^*T) = 2\left\{\sum_{k_\alpha}^{occ}\sum_{l_\alpha}^{uno} C_0 C_{k_\alpha - l_\alpha}(\rho^\alpha{}_{0,k_\alpha - l_\alpha} - S_{0,k_\alpha - l_\alpha}\rho^\alpha{}_{00}) + \sum_{k_\beta}^{occ}\sum_{l_\beta}^{uno} C_0 C_{k_\beta - l_\beta}(\rho^\beta{}_{0,k_\beta - l_\beta} - S_{0,k_\beta - l_\beta}\rho^\beta{}_{00})\right\} \quad (16)$$

pure term

$$\rho(CH_4{}^*T\,|\,CH_4{}^*T) = \sum_{k_\alpha, k_\alpha'}^{occ}\sum_{l_\alpha, l_\alpha'}^{uno} C_{k_\alpha - l_\alpha}C_{k_\alpha' - l_\alpha'}(\rho^\alpha{}_{k_\alpha - l_\alpha, k_\alpha' - l_\alpha'} - S^\alpha{}_{k_\alpha - l_\alpha, k_\alpha' - l_\alpha'}\rho_{00}) + \sum_{k_\beta, k_\beta'}^{occ}\sum_{l_\beta, l_\beta'}^{uno} C_{k_\beta - l_\beta}C_{k_\beta' - l_\beta'} \times (\rho^\beta{}_{k_\beta - l_\beta, k_\beta' - l_\beta'} - S_{k_\beta - l_\beta, k_\beta' - l_\beta'}\rho^\beta{}_{00}) \quad (17)$$

where ρ^α and ρ^β are the α-spin and β-spin components of density, respectively. The other kinds of terms of smaller contributions are neglected since the terms taken above can always cover large fractions of the total density (see Table III).

In eq 13–17[20]

$$\rho^\alpha{}_{00} + \rho^\beta{}_{00} = n_e\int \Psi_0{}^*\Psi_0 ds_1 d\tau_2 \ldots d\tau_{n_e} \quad (18)$$

$$\rho^\alpha{}_{0,i_\alpha - l_\alpha} + \rho^\beta{}_{0,i_\alpha - l_\alpha} = n_e\int \Psi_0{}^*\Psi_{i_\alpha - l_\alpha} ds_1 d\tau_2 \ldots d\tau_{n_e} \quad (19)$$

$$\rho^\alpha{}_{i_\alpha - l_\alpha, i_\alpha - l_\alpha'} + \rho^\beta{}_{i_\alpha - l_\alpha, i_\alpha - l_\alpha'} = n_e\int \Psi_{i_\alpha - l_\alpha}\Psi_{i_\alpha - l_\alpha'} ds_1 d\tau_2 \ldots d\tau_{n_e} \quad (20)$$

and

$$S_{0,i_\alpha - l_\alpha} = \int \Psi_0{}^*\Psi_{i_\alpha - l_\alpha} d\tau \quad (21)$$

The physical meaning of these interference densities is easily understood. Also each density vanishes upon integration over the whole space.

Using the Hellmann-Feynman theorem, we can divide the energy gradient in terms of the contribution of each electron configuration. The quasi-classical contribution, $\rho(CH_4) + \rho(T)$, is partitioned into two parts: the intramolecular and the intermolecular Coulombic parts. The nuclear part of the Hellmann–Feynman force is also divided into the intramolecular and the intermolecular contributions. The interference density is decomposed into various contributions according to the density partitioning. The formulation of these partitioned energy gradients is made in a straightforward fashion and is not given.

Result of Calculation

Some MO calculations (CNDO and extended Hückel) have been carried out on the potential energy surface for the $CH_4 + H$ system and the results have been compared with respect to several different types of reaction, *i.e.*, abstraction, substitution, and so forth.[21] Recently *ab initio* calculations for this system were performed by Morokuma and Davis[22a] (RHF) and Ehrenson and Newton[22b] (UHF), and the potential energy surface was obtained to get reliable values of the transition state geometry and energy differences between these types of reactions. These studies suggest that the most favorable type of reaction is abstraction. Experimental results show that the normal reaction is abstraction and in high-energy conditions the substitution occurs additionally.

Notwithstanding these results, the present calculation is made with respect to substitution, since this is suitable for examining the utility of the method proposed. As the substitution, the inversion model is adopted here because the retention path requires much more energy than the former,[22a,b] although experimentally the occurrence of both inversion and retention is reported.[22c]

We first calculated the potential energy surface for this model by the UHF-MO method with the minimal basis set of Slater type orbitals (STO). The integrals are calculated by expanding STO's with three-term Gaussian functions (STO-3G).[23,24] The orbital exponents of STO's are 5.67 for C(1S), 1.72 for C(2S) and C(2P), and 1.18 for H(1S). The reacting system is assumed to maintain C_{3v} symmetry. Parameters used to describe the system are shown in Figure 1. The nonreactive C–H distances, R_{CH_2}, R_{CH_3}, and R_{CH_4}, are fixed to be 1.097 Å throughout, which is the optimized value for the isolated methane molecule with the basis set used here. The two parameters, $\theta_{H_1CH_2}$ and R_{CH_1}, are optimized with respect to various C–T distances, R_{C-T}. In Figure 2a, the minimum energy curve for R_{CT} is plotted. The calculated barrier height is 71.2 kcal/mol, which is comparable with the results of Morokuma and Davis.[22a,25] We show in Figure 2b the correlation of three parameters along the minimum energy reaction path. The geometry of the saddle point is calculated to be $R_{CT} = R_{CH_1} = 1.33$ Å, $\theta_{H_1CH_2} = 90°$.

In order to calculate the direction of nuclear displacement, we must determine the internal coordinates. Although the number of internal degrees of freedom for the six-atomic system is 12 as indicated in Table I, we can reduce it to 4 (ξ_1, ξ_2, ξ_3, ξ_4) by the symmetry property of the reacting system. These four coordinates are not orthogonal to each other but are suitable for representing the nuclear rearrangements in this reaction model. If we assume that

4

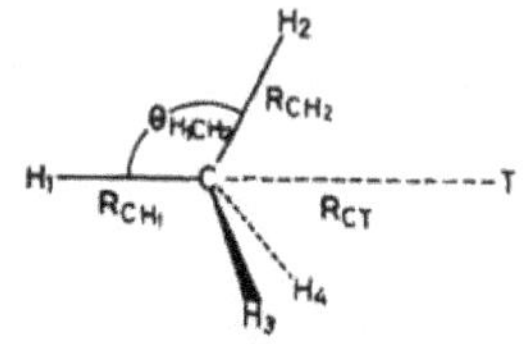

Figure 1. Model and parameters used for the reacting system.

Table I. Twelve Internal Coordinates

$$\xi_1 = R_{CT}$$
$$\xi_2 = \frac{1}{\sqrt{6}}(\theta_{H_1CH_2} + \theta_{H_1CH_3} + \theta_{H_1CH_4} - \theta_{H_2CH_3} - \theta_{H_3CH_4} - \theta_{H_4CH_2})$$
$$\xi_3 = R_{CH_1}$$
$$\xi_4 = \frac{1}{\sqrt{3}}(R_{CH_2} + R_{CH_3} + R_{CH_4})$$
$$\xi_5 = \frac{1}{\sqrt{6}}(2\theta_{TCH_2} - \theta_{TCH_3} - \theta_{TCH_4})$$
$$\xi_6 = \frac{1}{\sqrt{2}}(\theta_{TCH_3} - \theta_{TCH_4})$$
$$\xi_7 = \frac{1}{\sqrt{6}}(2\theta_{H_1CH_2} - \theta_{H_1CH_3} - \theta_{H_1CH_4})$$
$$\xi_8 = \frac{1}{\sqrt{6}}(2\theta_{H_2CH_3} - \theta_{H_3CH_4} - \theta_{H_4CH_2})$$
$$\xi_9 = \frac{1}{\sqrt{2}}(\theta_{H_1CH_3} - \theta_{H_1CH_4})$$
$$\xi_{10} = \frac{1}{\sqrt{2}}(\theta_{H_3CH_4} - \theta_{H_4CH_2})$$
$$\xi_{11} = \frac{1}{\sqrt{6}}(2R_{CH_2} - R_{CH_3} - R_{CH_4})$$
$$\xi_{12} = \frac{1}{\sqrt{2}}(R_{CH_3} - R_{CH_4})$$

tritium approaches the C atom by Δx, *i.e.*, $\Delta\xi_1 = \Delta x$ (<0), we can determine the displacements of the other three coordinates by eq 4. We calculated them at one point on the reaction coordinate where $\xi_1 = 1.6$ Å. The result is given in Table II. The relative nuclear displacements calculated from the potential energy relations of Figure 2 give results of $\Delta\xi_2/\Delta\xi_1 \sim 0.1$ and $\Delta\xi_3/\Delta\xi_1 \sim -0.3$ at $\xi_1 = 1.6$ Å, which are in good agreement with those values of Table II. This seems to imply that the Hellmann–Feynman theorem is sufficiently applicable to the present system which contains no lone-pair electrons and consists of several small atoms located around the central large atom. The direction of nuclear displacement climbing the potential barrier is opposite to that of the coupled force, $\Sigma_{j=1}{}^{12} b_{ij}(-\partial W/\partial\xi_j)$, and corresponds to the steepest ascending path on the potential energy surface. In general, the reacting system which involves an activation energy must ascend the potential energy barrier opposing the "forces." Actually, the energy which is needed to overcome the barrier is introduced from the kinetic energy of nuclei initially furnished.

Table II. Direction of Nuclear Displacement at $\xi_1 = 1.6$ Å[a]

	ξ_1	ξ_2	ξ_3	ξ_4
$\partial W/\Delta\xi_i$	0.04577	0.00597	−0.00788	0.00040
$\Sigma_{j=1}{}^{12} b_{ij}\frac{\partial W}{\partial \xi_j}$	0.05028	0.00479	−0.01240	0.00244
$\Delta\xi_i$	Δx	$0.0953\Delta x$	$-0.246\Delta x$	$0.0485\Delta x$

[a] The values in this table are given in atomic units.

It may be of interest to discuss the energy change caused by the infinitesimal nuclear displacement. On the reaction coordinate, S, both W and all ξ_1, ξ_2, ξ_3, and ξ_4 are functions of a single variable S; *i.e.*,

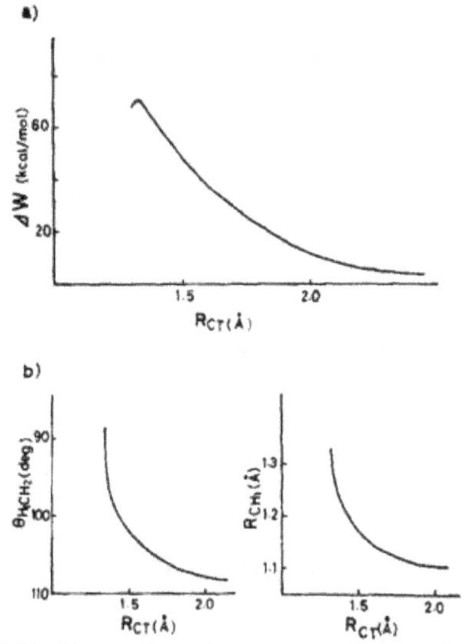

Figure 2. (a) Change of potential energy along the reaction path. (b) Correlations of three parameters.

$$W = W(S),\ \xi_1 = \xi_1(S),\ \xi_2 = \xi_2(S),\ \xi_3 = \xi_3(S),\ \xi_4 = \xi_4(S) \quad (22)$$

Hence, the small energy change on the reaction coordinate is written as

$$\Delta W = \frac{dW}{dS}\Delta S = \sum_{i=1}^{4}\frac{\partial W}{\partial \xi_i}\frac{d\xi_i}{dS}\Delta S = \sum_{i=1}^{4}\frac{\partial W}{\partial \xi_i}\Delta\xi_i \quad (23)$$

At the point $\xi_1 = 1.6$ Å, ΔW is calculated to be $0.0483\Delta x$ au ($30.3\Delta x$ kcal/mol) and its components are 28.7 Δx for ξ_4 in kcal/mol unit. The energy change associated with the tritium approach onto the carbon atom is largest at this point.

Origin of the Potential Energy Gradient

We have seen the acclivity of potential energy surface which the reacting system has to ascend to reach the transition state. We now analyze the slope of potential surface along the reaction coordinate in terms of the mode of intermolecular interaction and clarify the essential factors for raising and lowering the gradient.

Electronic Structure of the Reacting System. Before analyzing the potential energy gradient, we will give a brief discussion on the electronic structure of the reacting system. If we apply the configuration analysis, the CI representation of the wave function of the system contains all possible configurations in principle. Then we truncate the wave function to include the zero (original) and all one-electron transferred and all one-electron excited configurations. We collect contributions from the configuration functions of the same character in one term and express it in a more chemically graspable form. For example, the configuration wave function of one-electron transfer from tritium to methane is given as

$$\psi(\mathrm{CH_4^-T^+}) = \sum_{l_\alpha}^{\mathrm{uno}} C_{l_\alpha^- l_\alpha}\psi_{l_\alpha^- l_\alpha} \Big/ \left(\sum_{l_\alpha}^{\mathrm{uno}}\sum_{l_\alpha'}^{\mathrm{uno}} C_{l_\alpha^- l_\alpha} C_{l_\alpha'^- l_\alpha'} S_{l_\alpha^- l_\alpha,\, l_\alpha'^- l_\alpha'}\right) \quad (24)$$

5

In Table III, the coefficients of various electronic config-

Table III. Coefficients of Various Configuration Wave Functions

	2.0	1.8	1.6	1.4	1.35
$\psi(CH_4T)$	0.9855	0.9709	0.9376	0.8336	0.7028
$\psi(CH_4^-T^+)$	0.0316	0.0520	0.0892	0.1651	0.2153
$\psi(CH_4^+T^-)$	0.0569	0.0799	0.1132	0.1721	0.2222
$\psi(CH_4^*T)$	0.0141	0.0287	0.0612	0.1458	0.2341
Total wt	99.97	99.88	99.37	95.44	86.62

urations and the total weight of the limited number of CI wave functions taken there are shown for several values of ξ_1. These configurations cover the total weight of more than 99 except near the transition state. At an early stage of the reaction, the contributions from the transferred and the excited configurations are small (the former is the delocalization term and the latter is the polarization term). As the reaction proceeds, however, their contributions increase and the cross terms with the zero configuration become important. In particular the α-electron transfer from tritium to $4a_1$[26] unoccupied orbitals of the methane system and the β-electron transfer from $3a_1$ occupied orbital to tritium play dominant roles in the increase of the contribution of transferred configurations, which is shown in Figure 3. The MO's of deformed methane (C_{3v}) are also indicated in Figure 4. It is seen that the nodal property of unoccupied $4a_1$ orbital serves to weaken the C–H bond to be broken.

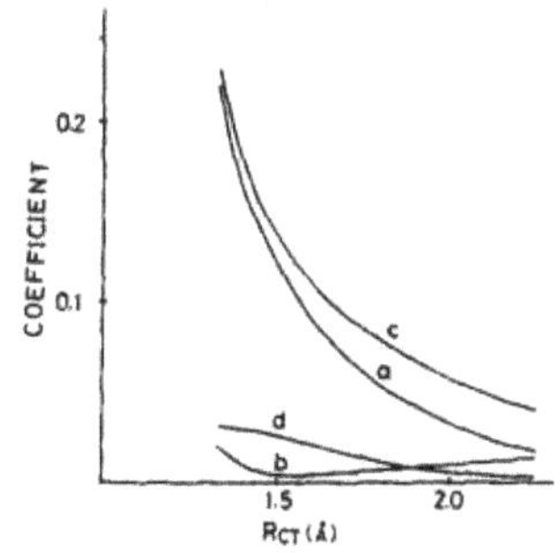

Figure 3. Change of coefficients of several transferred configurations along the reaction path: (a) from tritium 1s to $4a_1$ of C_{3v} methane; (b) from tritium 1s to $5a_1$ of C_{3v} methane; (c) from $3a_1$ of C_{3v} methane to tritium 1s; (d) from $2a_1$ of C_{3v} methane to tritium 1s.

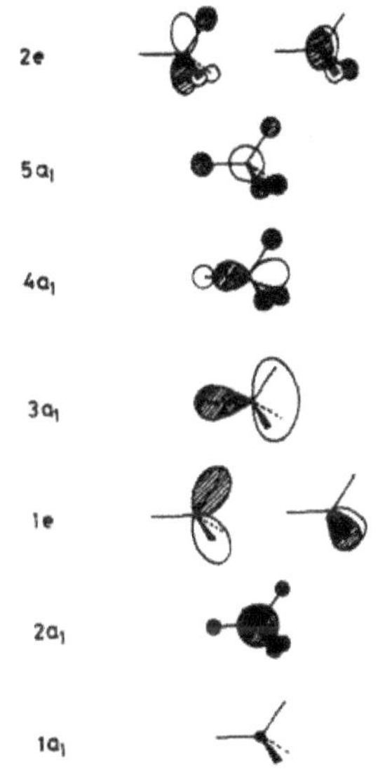

Figure 4. Schematic representation of MO's of C_{3v} methane.

Energy Gradient. As shown in eq 23, the energy gradient along the reaction coordinate, S, can be divided into the terms which correspond to each internal coordinate. Therefore, we can examine them independently in relation to the electronic structure of the reacting system. In this system the sign of displacement is negative for $\Delta\xi_1$ and $\Delta\xi_2$ and positive for $\Delta\xi_3$, and hence the terms which contribute to making $(\partial W/\partial\xi_i)\Delta\xi_i$ negative are considered to lower the energy gradient and to make the reaction easier.

The first matter to be considered is the freedom of tritium approach to the carbon atom. We give the components of $-\partial W/\partial\xi_i$ at three points on the reaction coordinate, ξ_1 = 1.8, 1.6, and 1.4 Å in Table IV. It is shown in Table IV that

Table IV. Constituents of Energy Gradient with Respect to ξ_1

	1.8	1.6	1.4
Intramolecular	0.00047	0.00052	0.00060
Coulombic	0.00301	0.00556	0.00825
Exchange	0.02835	0.05616	0.09261
$(CH_4T\vert CH_4^-T^+)$	−0.00208	−0.00654	−0.01905
$(CH_4^-T^+\vert CH_4^-T^+)$	−0.00017	−0.00093	−0.00494
$(CH_4T\vert CH_4^+T^-)$	−0.00467	−0.00842	−0.01419
$(CH_4^+T^-\vert CH_4^+T^-)$	0.00023	0.00079	0.00275
$(CH_4T\vert CH_4^*T)$	0.00082	0.00342	0.01027
$(CH_4^*T\vert CH_4^*T)$	−0.00001	−0.00005	−0.00066

the exchange term increases mainly the gradient, while the cross term of transferred configurations (of both directions of electron transfer) with the zero configuration contributes dominantly to lower it. In a neutral system such as CH_4 + T, as expected, the Coulombic interaction is not important. The cross term coming from excited configurations with the zero one acts as repulsive. The effects of the pure term of both the delocalized and the polarized structures are small in comparison with the cross terms. In order to clarify the different roles between the exchange term and the cross terms coming from the transferred and the zero configurations, the electron population of the reacting system may be partitioned according to the density partitioning mentioned above. The partitioned population at ξ_1 = 1.6 Å is shown in Table V. The exchange term yields a strong antibonding population in the carbon–tritium region, indicating the repulsion between the occupied orbitals. On the other hand, the cross terms create bonding components, which are mainly caused by the overlapping of the occupied orbitals of one subsystem with the unoccupied orbitals of the other. Consequently, the contribution of α-electron transfer is canceled by the exchange repulsion and the β-electron transfer plays a significant role for the new bond formation.[15c]

Of interest next is the origin of Walden inversion. This is involved in the energy gradient with respect to ξ_2. The components of $-\partial W/\partial\xi_2$ are shown in Table VI. The exchange repulsion term extremely reduces the energy slope, which is caused by the strong antibonding densities between tritium and nonreactive hydrogen atoms. The contribution of the cross term of polarized configurations with the zero one causes the change of hybridization on the carbon atom and acts to promote the methyl inversion next to the exchange term. The exchange interaction is the intermolecular origin and the polarization is the intramolecular one for bringing about the Walden inversion. On the contrary, the deformed methane is forced to get loose to its equilibrium geometry. Hence the intramolecular contribution raises the energy gradient. The cross term coming from the configurations of electron-transfer from methane to tritium with the zero configuration creates the bonding components between tri-

6

Table V. Change of Electron Population due to the Interaction between CH_4 and T

	T	C	H_1	H_2
	(i) Atom Population			
$\rho(CH_4) + \rho(T)$	1.0	6.16273	0.97623	0.95368
ρ_K	0	−0.00800	+0.00685	+0.00038
$\rho(CH_4T \vert CH_4^-T^+)$	−0.01120	−0.00094	+0.00819	+0.00132
$\rho(CH_4^-T^+ \vert CH_4^-T^+)$	−0.00796	+0.00501	+0.00256	+0.00013
$\rho(CH_4T \vert CH_4^+T^-)$	+0.03241	−0.01319	−0.00994	−0.00309
$\rho(CH_4^+T^- \vert CH_4^+T^-)$	+0.01282	−0.00646	−0.00401	−0.00078
$\rho(CH_4T \vert CH_4^*T)$	0	−0.04349	+0.04255	+0.00031
$\rho(CH_4^*T \vert CH_4^*T)$	0	−0.00008	+0.00007	+0.00001

	C–T	$T–H_2$
(ii) Atom Bond Population Intermolecular		
$\rho(CH_4) + \rho(T)$	0	0
ρ_K	−0.20437	−0.04772
$\rho(CH_4T \vert CH_4^-T^+)$	+0.06519	−0.01391
$\rho(CH_4^-T^+ \vert CH_4^-T^+)$	+0.00163	+0.00038
$\rho(CH_4T \vert CH_4^+T^-)$	+0.04137	+0.00932
$\rho(CH_4^+T^- \vert CH_4^+T^-)$	−0.00085	−0.00021
$\rho(CH_4T \vert CH_4^*T)$	+0.00242	+0.00041
$\rho(CH_4^*T \vert CH_4^*T)$	+0.00009	+0.00004

	$C–H_1$	$C–H_2$	$H_1–H_2$	$H_2–H_3$
	Intramolecular			
$\rho(CH_4) + \rho(T)$	0.72697	0.77348	−0.05624	−0.04036
ρ_K	−0.00467	0.02365	−0.00426	0.00378
$\rho(CH_4T \vert CH_4^-T^+)$	0.00995	0.00339	−0.00447	0.00229
$\rho(CH_4^-T \vert CH_4^-T^+)$	−0.00509	−0.00166	−0.00114	0.00010
$\rho(CH_4T \vert CH_4^+T^-)$	−0.00174	−0.00715	0.00299	−0.00148
$\rho(CH_4^+T^- \vert CH_4^+T^-)$	−0.00249	−0.00055	0.00054	−0.00017
$\rho(CH_4T \vert CH_4^*T)$	−0.01738	0.00413	−0.00016	0.00014
$\rho(CH_4^*T \vert CH_4^*T)$	−0.00518	−0.00063	−0.00011	−0.00008

Table VI. Constituents of Energy Gradient with Respect to ξ_2 (au)

	1.8	1.6	1.4
Intramolecular	0.01353	0.02540	0.04707
Coulombic	−0.00119	−0.00241	−0.00412
Exchange	−0.01448	−0.02027	−0.03101
$(CH_4T \vert CH_4^-T^+)$	−0.00020	−0.00061	−0.00146
$(CH_4^-T^+ \vert CH_4^-T^+)$	−0.00027	−0.00094	−0.00313
$(CH_4T \vert CH_4^+T^-)$	0.00271	0.00413	0.00438
$(CH_4^+T^- \vert CH_4^+T^-)$	−0.00013	−0.00028	−0.00058
$(CH_4T \vert CH_4^*T)$	−0.00365	−0.00651	−0.00865
$(CH_4^*T \vert CH_4^*T)$	−0.00004	−0.00029	−0.00215

tium and nonreactive hydrogens, as shown in Table V, and is obstructive of the inversion.

Lastly, we examine the leaving of the hydrogen atom, H_1. The components of $\partial W/\partial \xi_3$ are shown in Table VII. The intramolecular charge distribution raises the gradient. The crossed product of the polarized configuration (the most contributing configuration is, of course, $3a_1 \rightarrow 4a_1$ (see Figure 4)) with the zero configuration plays the most important role to weaken the carbon–hydrogen bond. It is clear from Table V that this term decreases the carbon–hydrogen-1 overlap population most distinctly. The importance of the $3a_1 \rightarrow 4a_1$ excitation is easily seen from the characteristic feature of a radical reaction.

Table VII. Constituents of Energy Gradient with Respect to ξ_3 (au)

	1.8	1.6	1.4
Intramolecular	0.02717	0.04603	0.07275
Coulombic	0.00040	0.00056	0.00071
Exchange	0.00259	0.00419	0.00422
$(CH_4T \vert CH_4^-T^+)$	−0.00073	−0.00114	−0.00148
$(CH_4^-T^+ \vert CH_4^-T^+)$	−0.00013	−0.00032	−0.00085
$(CH_4T \vert CH_4^+T^-)$	−0.00203	−0.00354	−0.00588
$(CH_4^+T^- \vert CH_4^+T^-)$	−0.00110	−0.00204	−0.00435
$(CH_4T \vert CH_4^*T)$	−0.01583	−0.02838	−0.04976
$(CH_4^*T \vert CH_4^*T)$	−0.00021	−0.00091	−0.00458

Notice should be taken here of the use of the Hellmann-Feynman theorem which yields a chemically graspable picture in the constituent analysis of energy gradient. As is well known, however, the condition to satisfy this theorem is very rigid. It is expected that any calculation with a limited basis set causes an appreciable error attributable to the nonsatisfaction of this condition. Nevertheless, we believe that the essential results obtained in the present paper will not be altered by the use of more rigorous calculations which are employable in principle with no difficulty except a larger consumption of computer time.

Acknowledgment. Authors express their appreciation to the Data Processing Center, Kyoto University, for generous permission to use the FACOM 230-75 computer.

References and Notes

(1) For instance, see A. Streitwieser, Jr., "Molecular Orbital Theory for Organic Chemists," Wiley, New York, N.Y., 1961.

(2) For instance, see (a) K. Fukui, "Theory of Orientation and Stereoselection," Springer-Verlag, Berlin, 1970; (b) R. B. Woodward and R. Hoffmann, "The Conservation of Orbital Symmetry," Academic Press, New York, N.Y., 1969, and the papers cited therein.

(3) (a) J. N. Murrell, M. Randic, and D. R. Williams, *Proc. Roy. Soc., Ser. A*, **284**, 566 (1965); (b) K. Fukui and H. Fujimoto, *Bull. Chem. Soc. Jap.*, **41**, 1989 (1968); (c) L. Salem, *J. Amer. Chem. Soc.*, **90**, 543, 553 (1968); A. Devaquet and L. Salem, *ibid.*, **91**, 3793 (1969); A. Devaquet, *Mol. Phys.*, **18**, 233 (1970); (d) G. Klopman, *J. Amer. Chem. Soc.*, **90**, 223 (1968); (e) R. Sustmann and G. Binsch, *Mol. Phys.*, **20**, 1, 9 (1971).

(4) (a) K. Fukui and H. Fujimoto, *Bull. Chem. Soc. Jap.*, **42**, 3399 (1969); (b) K. Fukui, "XXIII rd International Congress on Pure and Applied Chemistry," Vol. 1, Butterworth, London, 1971, p 65, and references cited therein; (c) K. Fukui, "Proceeding of the First International Conference on Quantum Chemistry," Reidel, 1974, p 113, and references cited therein.

(5) See, for instance (a) R. Hoffmann, R. Gleiter, and F. B. Mallory, *J. Amer. Chem. Soc.*, **92**, 1460 (1970); (b) K. Hsu, R. J. Buenker, and S. D. Peyerimhoff, *ibid.*, **94**, 5639 (1972); (c) J. A. Horsley, Y. Jean, L. Moser, L. Salem, R. M. Stevens, and J. S. Wright, *ibid.*, **94**, 279 (1972); (d) M. J. S. Dewar and M. C. Kohn, *ibid.*, **94**, 2704 (1972); (e) A. C. Trindle and D. D. Shillady, *ibid.*, **95**, 703 (1973); (f) P. Merlet, S. D. Peyerimhoff, R. J. Buenker, and S. Shih, *ibid.*, **96**, 959 (1974).

(6) (a) R. F. W. Bader, *Can. J. Chem.*, **40**, 1164 (1962); (b) R. F. W. Bader and A. K. Chandra, *ibid.*, **46**, 955 (1968).

(7) (a) H. Hellmann, "Einführung in die Quantenchemie," Deuticke, Leipzig, 1937; (b) R. P. Feynman, *Phys. Rev.*, **56**, 340 (1939); (c) for the applications and extensions of this theorem, see, *e.g.*, B. M. Deb, *Rev. Mod. Phys.*, **45**, 22 (1973).

(8) (a) H. Nakatsuji, *J. Amer. Chem. Soc.*, **95**, 345, 354, 2084 (1973); (b) H. Nakatsuji, *ibid.*, **96**, 24, 30 (1974).

(9) (a) I. S. Y. Wang and M. Karplus, *J. Amer. Chem. Soc.*, **95**, 8160 (1973); (b) J. W. McIver, Jr., and A. Komornicki, *ibid.*, **94**, 2625 (1972).

(10) J. A. Pople, D. P. Santry, and G. A. Segal, *J. Chem. Phys.*, **43**, S 129 (1965); J. A. Pople and G. A. Segal, *ibid.*, **43**, S 136 (1965).

(11) K. Fukui, *J. Phys. Chem.*, **74**, 4161 (1970).

(12) (a) R. Wolfgang, *Progr. React. Kinet.*, **3**, 97 (1965); R. Wolfgang, *Accounts Chem. Res.*, **2**, 248 (1969); **3**, 48 (1970); (b) R. W. Walker, *J. Chem. Soc. A*, 2391 (1968); (c) M. J. Kurylo and R. B. Timmons, *J. Chem. Phys.*, **50**, 5133 (1969); (d) C. C. Chou and F. S. Rowland, *ibid.*, **50**, 2763, 5133 (1969).

(13) This general eq 3 can be written[4c] in the form

$$\cdots = \frac{M_\alpha \Delta X_\alpha}{\partial W/\partial X_\alpha} = \frac{M_\alpha \Delta Y_\alpha}{\partial W/\partial Y_\alpha} = \frac{M_\alpha \Delta Z_\alpha}{\partial W/\partial Z_\alpha} = \cdots$$
$$(\alpha = 1, 2, \ldots)$$

by the use of usual three-dimensional Cartesian coordinates X_α, Y_α, and Z_α for nucleus α of mass M_α, in which the conditions of no translation nor rotation of the center of gravity

$$\sum_\alpha M_\alpha \Delta X_\alpha = \sum_\alpha M_\alpha \Delta Y_\alpha = \sum_\alpha M_\alpha \Delta Z_\alpha = 0$$

$$\sum_\alpha M_\alpha (Y_\alpha \Delta Z_\alpha - Z_\alpha \Delta Y_\alpha) = \sum_\alpha M_\alpha (Z_\alpha \Delta X_\alpha - X_\alpha \Delta Z_\alpha) = \sum_\alpha M_\alpha (X_\alpha \Delta Y_\alpha - Y_\alpha \Delta X_\alpha) = 0$$

7

are automatically satisfied since W is a function of internuclear distances only. The original equation given in ref 11 is obtained by the substitution

$$x_1 = M_1^{1/2}X_1,\ x_2 = M_1^{1/2}Y_1,\ x_3 = M_1^{1/2}Z_1,$$
$$x_4 = M_2^{1/2}X_2,\ x_5 = M_2^{1/2}Y_2,\ x_6 = M_2^{1/2}Z_2,\ \ldots$$

and this should have explicitly been specified there.

(14) (a) A. C. Hurley, "Molecular Orbitals in Chemistry, Physics, and Biology," P.-O. Löwdin and B. Pullman, Ed., Academic Press, New York, N.Y., 1964, p 161; (b) G. G. Hall, *Phil. Mag.*, **6**, 249 (1961).

(15) (a) H. Fujimoto, S. Yamabe, and K. Fukui, *Bull. Chem. Soc. Jap.*, **44**, 2936 (1971); **45**, 1566, 2424 (1972); (b) K. Fukui, H. Fujimoto, and S. Yamabe, *J. Phys. Chem.*, **76**, 232 (1972); (c) H. Fujimoto, S. Yamabe, T. Minato, and K. Fukui, *J. Amer. Chem. Soc.*, **94**, 9205 (1972).

(16) H. Baba, S. Suzuki, and T. Takemura, *J. Chem. Phys.*, **50**, 2078 (1969).

(17) (a) H. Fujimoto, S. Kato, S. Yamabe, and K. Fukui, *J. Chem. Phys.*, **60**, 572 (1974); (b) S. Kato, H. Fujimoto, S. Yamabe, and K. Fukui, *J. Amer. Chem. Soc.*, **96**, 2024 (1974); (c) S. Yamabe, S. Kato, H. Fujimoto, and K. Fukui, *Theor. Chim. Acta*, **30**, 327 (1973); (d) S. Yamabe, T. Minato, H. Fujimoto, and K. Fukui, *Theor. Chim. Acta*, **32**, 187 (1974).

(18) J. A. Pople and R. K. Nesbet, *J. Chem. Phys.*, **22**, 571 (1954).

(19) K. Ruedenberg, *Rev. Mod. Phys.*, **34**, 326 (1962).

(20) These hold the following relations in the UHF treatment:

$$\rho^{\beta}_{0,l_\alpha-l_\alpha} = S^{\alpha}_{0,l_\alpha-l_\alpha}\rho_{00},\ \rho^{\beta}_{l_\alpha-l_\alpha,l_{\alpha'}-l_{\alpha'}} = S_{l_\alpha-l_\alpha,l_{\alpha'}-l_{\alpha'}}\rho^{\beta}_{00},$$
$$\rho^{\alpha}_{0,l_\beta-l_\beta} = S^{\alpha}_{0,l_\beta-l_\beta}\rho_{00},$$

and

$$\rho^{\alpha}_{l_\beta-l_\beta,l_\beta'-l_\beta'} = S^{\alpha}_{l_\beta-l_\beta,l_\beta'-l_\beta'}\rho_{00}$$

(21) (a) J. J. Kaufmann, J. J. Harkins, and W. S. Koski, *Int. J. Quantum Chem., Symp.*, **1**, 261 (1967); (b) R. E. Oeston, Jr., and S. Ehrenson, *Chem. Phys. Lett.*, **9**, 351 (1971).

(22) (a) K. Morokuma and R. E. Davis, *J. Amer. Chem. Soc.*, **94**, 1060 (1972); (b) S. Ehrenson and M. D. Newton, *Chem. Phys. Lett.*, **13**, 24 (1972); (c) a referee kindly pointed out the experimental result of Professor Bunker.

(23) (a) K. O-ohata, H. Taketa, and S. Huzinaga, *J. Phys. Soc. Jap.*, **21**, 2306 (1966); H. Taketa, S. Huzinaga, and K. O-ohata, *ibid.*, **21**, 2313 (1966); (b) W. J. Hehre, R. F. Stewart, and J. A. Pople, *J. Chem. Phys.*, **51**, 2651 (1969).

(24) In the present calculation the expansion coefficients and the Gaussian orbital exponents are due to R. F. Stewart, *J. Chem. Phys.*, **52**, 431 (1970).

(25) For minimal basis set RHF–SCF calculation, 82.6 kcal/mol. For double ζ-SCF, 63.7 kcal/mol, and 41.7 kcal/mol for double ζ and CI calculation.

(26) The terminology of MO's of the methane part follows C_{3v} symmetry.

[Reprinted from the Journal of the American Chemical Society, 98, 6395 (1976).]

Reaction Ergodography. Methane–Tritium Reaction

Sir:

Detailed description of chemical reactions has long been one of the goals of theoretical chemistry. Since chemical reactions are usually attended with much complexity in the modes of nuclear rearrangements corresponding to the change of the reacting system from initial to final, the concept of the reaction coordinate has been conveniently utilized to abstract the essential feature for the rationalization of chemical reactions. Many theoretical attempts have been made to determine the reaction coordinate or the minimum energy path of various types of chemical reactions with the aid of semiempirical and ab initio molecular orbital procedures.[1] The only difficulty in this direction of research is the multidimensionality of the potential-energy surfaces. In this connection, a proposal was made previously about the theoretical formulation of the reaction coordinate and it is named the intrinsic reaction coordinate.[2] Such a theoretical formulation of the reaction coordinate enables a theoretical tracing of the change of geometry of the reacting system along the reaction path, which was called reaction ergodography.[3] Also an analysis was made on the potential-energy gradient on the reaction coordinate to discuss the favorableness or the unfavorableness of a given reaction path in relation to the mode of interaction between two reacting species.[4]

Another utility of the reaction coordinate is demonstrated in the description of the reactive collision process. Marcus[5] defined the natural collision coordinate in the three-body reactive scattering. In order to treat the dynamics of a polyatomic system, a more generalized definition of the reaction coordinate is required. The intrinsic reaction coordinate is suitable for such a purpose.

In this communication, we illustrate the intrinsic reaction coordinate of the methane–tritium reaction system and its isotopic analogy.[6] Both the abstraction,

$$CH_4 + T \rightarrow CH_3 + HT \quad \text{(I)}$$

$$CD_4 + T \rightarrow CD_3 + DT \quad \text{(II)}$$

and the substitution,

$$CH_4 + T \rightarrow CH_3T + H \quad \text{(III)}$$

$$CD_4 + T \rightarrow CD_3T + D \quad \text{(IV)}$$

were examined.

Since the formulation of the intrinsic reaction coordinate was given previously,[2-4] we give a brief description here. Consider a reacting system composed of N atoms, the reaction coordinate is defined in terms of the gradient of the potential-energy function, $W(\xi_1, \xi_2, \ldots, \xi_f)$, in which ξ_i is the independent internal coordinate such as the bond length or the bond angle, and f is the number of the internal degrees of freedom. At a nonequilibrium point on the reaction coordinate, the direction of the displacement vector along the reaction coordinate is defined as

$$\frac{\sum_{j=1}^{f} g_{ij}(\xi)\mathrm{d}\xi_j}{\partial W/\partial \xi_i} = \text{constant} \quad (i = 1, 2, \ldots, f) \qquad (1)$$

and

$$g_{ij}(\xi) = \sum_{k=1}^{3N} \frac{\partial x_k}{\partial \xi_i}\frac{\partial x_k}{\partial \xi_j}$$

where x_k is the mass-weighted Cartesian coordinate which represents the position of an atom in the configuration space. Equation 1 can be rewritten with the mass-weighted Cartesian coordinates as follows,

6396

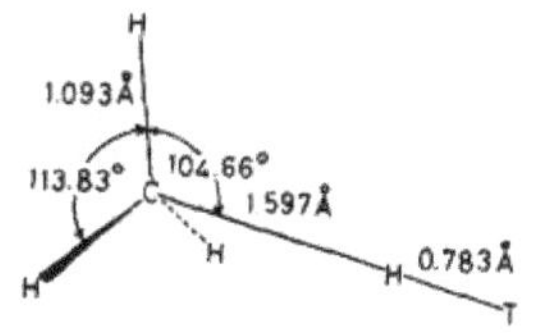

Figure 1. The geometry of the transition state for the abstraction mode.

$$\frac{dx_i}{dS} = \frac{\partial W/\partial x_i}{dW/dS} \quad (i = 1, 2, \ldots\ldots\ldots, 3N) \tag{2}$$

and

$$\frac{dW}{dS} = \pm\sqrt{\sum_{k=1}^{3N}\left(\frac{\partial W}{\partial x_k}\right)^2} \tag{3}$$

In eq 2, S means the length measured along the reaction coordinate. The sign of eq 3 is adopted to be plus for the ascending path and to be minus for the descending path, respectively. The Eckart condition[7] to fix the translation and the rotation about the center of gravity is automatically satisfied in eq 2. At an equilibrium point on the reaction coordinate, the gradient of the potential energy vanishes and the displacement vector becomes one of the normal coordinates at that point. When we solved the simultaneous differential equation, eq 2, by setting the initial condition such that the solution conforms to one of the normal coordinates of an equilibrium point, the nuclear configuration along the reaction coordinate is given.

For the potential-energy function of CH_4 + T and CD_4 + T systems, we adopted the analytical function proposed by Raff.[8] He adjusted the results of the ab initio and the INDO[9] calculations of the potential-energy surface to a simple analytic function and performed a classical trajectory calculation of these systems.

We first calculated the reaction coordinate of the abstraction mode, I and II. The geometry of the transition state was obtained by means of Powell's function minimization procedure,[10] where the mean square of the gradient vector was made to be zero. The normal coordinates were given as the eigenvectors of the secular equation,

$$\det\left(\frac{\partial^2 W}{\partial x_i \partial x_j} - \kappa\delta_{ij}\right) = 0$$

The masses of atoms used here are 12.0 for C, 1.008 for H, 2.014 for D, and 3.016 for T, respectively. The displacement vector on the reaction coordinate at the transition state is the eigenvector whose eigenvalue is uniquely negative. The geometry of the transition state is shown in Figure 1. We solved eq 2 by imposing the displacement vector at the transition state as the initial condition. As to the integration, we employed the variable step-size Runge–Kutta–Gill method.[11] Figure 2 (top) illustrates the nuclear configuration and the displacement vector at several points on the reaction coordinate of reaction mode I. The magnitude of vectors is scaled to the Cartesian coordinate representation. The change of the potential energy for modes I and II is also shown in Figure 2 (bottom). The displacement vector along the reaction coordinate varies gradually and converges well at the initial and the final states. At the initial stage of reaction ($S < -2.0$), the component of the reaction coordinate is the relative translational mode between the centers of gravity of CH_4 and T. With the approach of T to CH_4, the CH bond lengthens and the methyl group tends to be slightly planar ($-2.0 < S < -0.8$). At $S = -0.73$, the tritium reverses its direction and the stripping of the hydrogen atom is accelerated. After crossing over the transition state, the displacement vector is a combination of the leaving mode of H_2 and the bending mode of the methyl group to the planar form, and becomes the relative translation between the centers of gravity of CH_3 and H_2 at the final state. As seen in

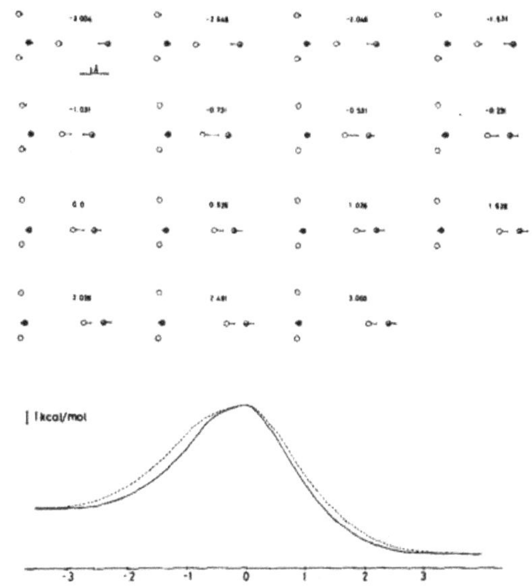

Figure 2. The change of nuclear configuration, displacement vector, and energy along the reaction coordinate of the abstraction mode. The scale of the reaction coordinate is $\sqrt{M_n}$ au, where M_n is the ratio of the atomic mass unit with the mass of electron. The location of atom is represented by: ●, C; ⊖, T; and ○, H. The dashed line is for reaction II, while the solid line is for reaction I.

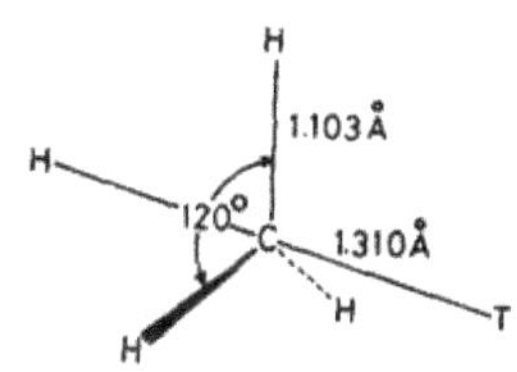

Figure 3. The geometry of transition state for the substitution mode.

Figure 2 (bottom), the potential barrier along the reaction coordinate is widened by the deuteration.

Next, the calculation was performed on the substitution mode, III and IV. Since the transition state corresponds to a cusp point of the potential surface with this approximation, we minimized the energy within the framework of the D_{3h} point group in the nuclear configuration space to calculate the geometry of the transition state. The optimization was carried out by Powell's procedure. In Figure 3, the nuclear configuration of the transition state is given. A set of nuclear configurations and the displacement vectors for mode III are shown in Figure 4 (top) and the comparison is made on the potential energies of III and IV in Figure 4 (bottom). Figure 4 (top) indicates a typical mode of the methyl inversion and the bond interchange in the substitution reaction mechanism. It should be noted that the difference of the masses of tritium and hydrogen atoms brings the difference of the magnitude of the displacement vectors. The effect of deuteration is more remarkable than in the case of abstraction mode. The width of the potential barrier is much affected by the deuteration in the region between the transition state and the final state, while it does not change remarkably in the region between the initial and the transition states. This is easily understood from the fact that, at the former region, the displacement is mainly constituted by the leaving mode of the deuterium and the inversion of the CD_3 group in reaction IV so that the ratio of the masses of deuterium and hydrogen reflects directly. This isotope effect

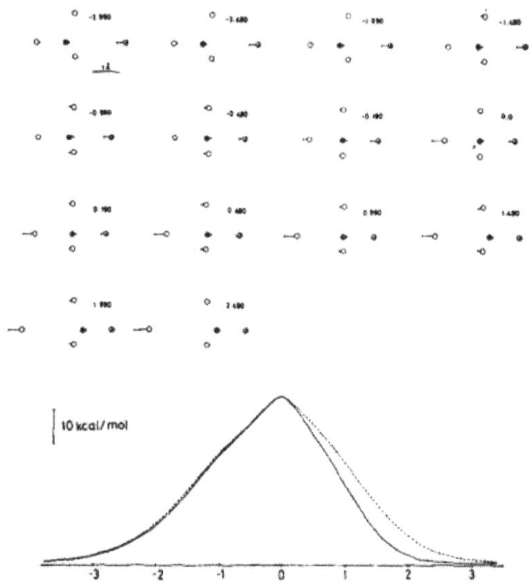

Figure 4. The change of nuclear configuration, displacement vector, and energy along the reaction coordinate of the substitution mode. The solid line is for reaction III and the dashed line is for reaction IV. Others are the same as in Figure 2.

on the width of potential barrier becomes important when the tunneling effect is discussed. The tunneling is expected not to be negligible near the top of the potential barrier even in the polyatomic reaction.

Many calculations of the trajectories of chemical reactions have been performed and afforded valuable information with respect to dynamic features of reactions.[12] Wang and Karplus[13] applied the trajectory method to the quantum mechanical surface. The trajectory, nevertheless, depends on the initial condition chosen. The intrinsic reaction coordinate is independent of the initial condition and is uniquely determined for a given reaction path on a given potential surface.

Applications of the intrinsic reaction coordinate to the dynamic problems of polyatomic reacting systems also have some advantages. The reaction coordinate automatically includes the reduced mass of the reacting system, and we can easily construct the equation of motion or the Hamiltonian of the system with considering the coordinate system perpendicular to the reaction coordinate. As shown in the present calculation, the effect of isotopic nuclei is directly realized in the width of the potential barrier. An application to the dynamics of polyatomic systems will be presented in the near future.

Acknowledgment. The authors express their appreciation to the Data Processing Center, Kyoto University, for the generous permission to use the Facom 230-75 computer. This work was supported by a Grant-in-Aid for Scientific Research for the Ministry of Education of Japan (No. 047068).

References and Notes

(1) For instance, see (a) R. Hoffman, R. Gleiter, and F. B. Mallory, *J. Am. Chem. Soc.*, **92**, 1460 (1970); (b) M. J. S. Dewer and M. C. Kohn, *ibid.*, **94**, 2704 (1972); (c) P. Marlet, S. D. Peyerimhoff, R. J. Buenker, and S. Shin, *ibid.*, **96**, 959 (1974); (d) K. Morokuma and R. E. Davis, *ibid.*, **94**, 1060 (1972).
(2) K. Fukui, *J. Phys. Chem.*, **74**, 4161 (1970).
(3) K. Fukui in "The World of Quantum Chemistry", R. Daudel and B. Pullman, Ed., D. Reidel Publishing Co., Dordrecht, Holland, 1974, p 113.
(4) K. Fukui, S. Kato, and H. Fujimoto, *J. Am. Chem. Soc.*, **97**, 1 (1975).
(5) R. A. Marcus, *J. Chem. Phys.*, **49**, 2610, 2617 (1968).
(6) (a) R. Wolfgang, *Prog. React. Kinet.*, **3**, 97 (1965); R. Wolfgang, *Acc. Chem. Res.*, **2**, 248 (1969); **3**, 48 (1970); (b) R. W. Walker, *J. Chem. Soc. A*, 2391 (1968); (c) M. J. Kurylo and R. B. Timmons, *J. Chem. Phys.*, **50**, 5113 (1969); (d) C. C. Chou and F. S. Rowland, *ibid.*, **50**, 2763, 5133 (1969).
(7) C. Eckart, *Phys. Rev.*, **47**, 552 (1935).
(8) L. M. Raff, *J. Chem. Phys.*, **60**, 2220 (1974).
(9) J. A. Pople and D. L. Beveridge, "Approximate Molecular Orbital Theory", McGraw-Hill, New York, N.Y., 1970.
(10) M. J. D. Powell, *Compt. J.*, **7**, 155 (1964).
(11) S. Gill, *Proc. Cambridge Philos. Soc.*, **47**, 96 (1951).
(12) For instance, see (a) M. Karplus, R. N. Porter, and R. D. Sharma, *J. Chem. Phys.*, **43**, 3259 (1965); (b) D. L. Bunker and M. D. Pattengill, *ibid.*, **52**, 3041 (1970); T. Valencich and D. L. Bunker, *ibid.*, **61**, 4161 (1974).
(13) I. S. Y. Wang and M. Karplus, *J. Am. Chem. Soc.*, **95**, 8060 (1973).

Shigeki Kato, Kenichi Fukui*
Faculty of Engineering, Kyoto University
Sakyo-ku, Kyoto, Japan
Received April 13, 1976

684

A Theoretical Treatment on the Behavior of the Hydrogen-Bonded Proton in Malonaldehyde

Shigeki Kato, Hiroshi Kato, and Kenichi Fukui*

Contribution from the Department of Hydrocarbon Chemistry, Kyoto University, Kyoto, Japan. *Received May 3, 1976*

Abstract: The quantum mechanical behavior of the hydrogen-bonded proton is treated by means of the intrinsic reaction coordinate previously proposed. The reaction coordinate is obtained by the aid of the gradient of potential energy determined by CNDO/2 procedure. The effective motion along the reaction coordinate is calculated considering all vibrational motions perpendicular to the reaction coordinate. The statistical feature of vibrational state of malonaldehyde is discussed on the basis of the canonical ensemble in a heat bath of a given temperature, and the density matrix is calculated in order to examine the proton distribution on the reaction coordinate and the rate of proton transfer with relation to the temperature. An ellipse-like orbit of proton motion is suggested.

The proton moves through the hydrogen bonds in many chemical and biological systems. Both theoretical and experimental chemists have been much interested in such a problem for a long time.[1] On theoretical ground, the detailed description of the proton transfer or exchange in hydrogen bonding systems seems to include two separated steps. The first step is to calculate the potential energy surface for the proton motion. For this purpose, semiempirical and ab initio molecular orbital (MO) calculations have been carried out for a variety of systems including intra- and intermolecular hydrogen bondings.[2] These studies have given encouraging results in predicting the proton potentials as well as the stable geometries, hydrogen bond energies, and spectroscopic properties. The origin of hydrogen bond formation has also been explained by analyzing the binding energy or the charge distribution.[3]

The second problem is to solve the equation of motion for protons on a given potential surface and to evaluate the equilibrium constant or the rate of proton transfer after statistical averaging. Löwdin[4] reviewed the quantum mechanical treatments of proton tunneling in a double well potential of the hydrogen bonding in nucleotide base pairs and discussed them in connection with the biological duplication. Although further developments in this direction of research have been achieved,[5] the greater part of these works has been based on the model or experimentally adjusted empirical potentials.

In order to describe such a process, one has to define the reaction coordinate, the curve leading from the bottom valley of initial state over the transition state to the valley of final state on the potential energy surface, inherent to the system concerned.

In the present paper, we proposed a method to treat the proton exchange in the intramolecular hydrogen bonding system by means of "intrinsic" reaction coordinate[6] and apply it to the enol tautmer of malonaldehyde, which is the simplest 1,3-dione compound. The reaction coordinate is defined by the use of the gradient of the quantum mechanical hypersurface of reacting species and is here obtained by the semiempirical CNDO/2 MO method.[7]

Some MO theoretical studies of malonaldehyde have been presented.[8-10] The position of the proton has been given much attention in these works. Schuster[8] first calculated the potential surface of the proton by the CNDO/2 method. His results predicted that the energy barrier for the proton transfer is quite low (0.5 kcal/mol). Recently ab initio MO procedure has been applied by two groups. Karlström et al.[9] concluded that the stable configuration of this compound is asymmetric and the barrier height is calculated to be 11.5 kcal/mol. Isaacson and Morokuma[10] employed semiempirical (CNDO/2 and INDO) and ab initio calculations on this molecule and proposed the condition for a neutral hydrogen bonding system to have a symmetric configuration. They suggested the possibility of malonaldehyde having a symmetric hydrogen-bonding proton on the basis of their hypothesis.

After our work had been submitted for publication, the microwave study by Rowe, Duerst, and Wilson[11] was presented. The existence of a symmetric double minimum potential with a relatively low barrier has been concluded from their experiments.

The behavior of proton in malonaldehyde is characterized by the rapid movement from one well to another in the thermal energy range. Therefore, the determination of the position of proton is a statistical or stochastic problem. In view of these circumstances, it seems to be interesting to interpret the statistical feature of proton exchange in malonaldehyde.

The purpose of the present paper is to calculate the quantum mechanical state of malonaldehyde on the basis of the CNDO/2 potential surface. The distribution of proton on the reaction coordinate and the rate of transformation between two conformers have also been calculated and discussed with relation to the temperature.

Theoretical Description of Quantum Mechanical State

In this section, we give a brief description of the quantum mechanical model for the motion of the reacting system.

Reaction Coordinate. The reaction coordinate is defined in the framework of the Born-Oppenheimer potential

$$H_{el}(r,\mathcal{R})\Phi_{el}(r,\mathcal{R}) = W(\mathcal{R})\Phi_{el}(r,\mathcal{R}) \tag{1}$$

where H_{el} is the electronic hamiltonian, Φ_{el} is the electronic wave function, W is the Born-Oppenheimer energy, and r and $\mathcal{R}$ are the electronic and the nuclear position vectors, respectively.

Let us consider a reacting system including N atoms. At a nonequilibrium point on the reaction coordinate, the direction of displacement of nuclear configuration along the reaction coordinate is given by the use of potential energy gradients as follows:[6]

$$\ldots = \frac{M_\alpha\, dX_\alpha}{\partial W/\partial X_\alpha} = \frac{M_\alpha\, dY_\alpha}{\partial W/\partial Y_\alpha} = \frac{M_\alpha\, dZ_\alpha}{\partial W/\partial Z_\alpha} = \ldots \quad (\alpha = 1, 2, \ldots, N) \tag{2}$$

where X_α, Y_α, and Z_α are the Cartesian coordinates of nucleus of mass M_α. The Eckart condition to fix the translation and rotation of the center of gravity[12] is automatically satisfied in eq 2:[6]

$$\sum_\alpha M_\alpha\, dX_\alpha = \sum_\alpha M_\alpha\, dY_\alpha = \sum_\alpha M_\alpha\, dZ_\alpha = 0$$

and

[Reprinted from the Journal of the American Chemical Society, 99, 684 (1977).]

$$\sum_\alpha M_\alpha(X_\alpha\,dY_\alpha - Y_\alpha\,dX_\alpha) = \sum_\alpha M_\alpha(Y_\alpha\,dZ_\alpha - Z_\alpha\,dY_\alpha) = \sum_\alpha M_\alpha(Z_\alpha\,dX_\alpha - X_\alpha\,dZ_\alpha) = 0 \quad (3)$$

Here we adopt the mass-weighted Cartesian coordinates, $x_1, x_2, \ldots, x_{3N}$, so that eq 2 is rewritten as

$$\frac{dx_1}{\partial W/\partial x_1} = \frac{dx_2}{\partial W/\partial x_2} = \ldots = \frac{dx_{3N}}{\partial W/\partial x_{3N}} \quad (2a)$$

These equations determine an infinite number of curves in the $3N$-dimensional space spanned by these coordinates. If we designate the length measured along one such curve as S, we have

$$\frac{dx_i}{dS} = \frac{\partial W/\partial x_i}{dW/dS} \; (i = 1, 2, \ldots, 3N) \quad (4)$$

where

$$\frac{dW}{dS} = \pm\left(\sum_{i=1}^{3N}\left(\frac{\partial W}{\partial x_i}\right)^2\right)^{1/2} \quad (5)$$

The sign of eq 5 is plus for an ascending path and minus for a descending path. Such a curve which passes three equilibrium points, the initial system, the transition state, and the final system of a given reaction path, is termed an "intrinsic" reaction coordinate of that reaction.[6]

At an equilibrium point, the displacement vector in the direction of the reaction coordinate is given by one of the eigenvectors derived from the following secular equation:

$$\det\left(\frac{\partial^2 W}{\partial x_i \partial x_j} - \delta_{ij}\kappa\right) = 0 \quad (6)$$
$$(i, j = 1, 2, \ldots, 3N; \kappa = \text{const})$$

Among $3N$ eigenvectors, six with zero eigenvalue correspond to the translation of the center of gravity and the rotation about the center of gravity. The nuclear configuration along the reaction coordinate is obtained when we solve the simultaneous equation, eq 4, by setting the initial condition such that the solution conforms to one of the normal coordinates of an equilibrium configuration.

In order to describe the internal motion of the reacting system, we have to determine the $3N - 6$ internal coordinates. Since six redundant coordinates can be eliminated by the use of eq 3, we can define the local coordinate system $\{\xi_1(S), \xi_2(S), \ldots, \xi_f(S)\}$ at any point on the reaction coordinate, which is connected with the components of the infinitesimal displacement vector

$$dx_i = \sum_{j=1}^{f} \frac{\partial x_i}{\partial \xi_j} d\xi_j$$

where $f = 3N - 6$ is the internal degree of freedom. When we set $d\xi_1$ to the displacement vector along the reaction coordinate, the remaining coordinates ξ_i $(i = 2, 3, \ldots, f)$ satisfy the relation

$$\frac{\partial W}{\partial \xi_i} = \sum_{j=1}^{3N} \frac{\partial W}{\partial x_j}\frac{\partial x_j}{\partial \xi_i} = \frac{dW}{dS}\sum_{j=1}^{3N}\frac{dx_j}{dS}\frac{\partial x_j}{\partial \xi_i} = \frac{dW}{dS}\delta_{1i} \quad (7)$$

Then we can get the coordinate system $\{\xi_1, \xi_2, \ldots, \xi_f\}$ which is orthogonal to the reaction coordinate and satisfies eq 7.

Schrödinger Equation. The general form of the quantum mechanical hamiltonian can be written by the use of the coordinate system defined in the previous section as:[13,14]

$$H = -\tfrac{1}{2} g^{1/4}\left\{\frac{\partial}{\partial \xi_1} g_{11} g^{-1/2} \frac{\partial}{\partial \xi_1} + \sum_{i=2}^{f}\left(\frac{\partial}{\partial \xi_1} g_{1i} g^{-1/2}\frac{\partial}{\partial \xi_i} + \frac{\partial}{\partial \xi_i} g_{i1} g^{-1/2}\frac{\partial}{\partial \xi_1}\right) + \sum_{i,j=2}^{f}\frac{\partial}{\partial \xi_i} g_{ij} g^{-1/2}\frac{\partial}{\partial \xi_j}\right\} g^{1/4} + W(\xi_1, \xi_2, \ldots, \xi_f) \quad (8)$$

with

$$g = \det(g_{ij})$$

and

$$(g_{ij}) = (\bar{g}_{ij})^{-1}$$

The matrix $(\bar{g}_{ij})$ is the ordinary kinetic tensor with its elements

$$\bar{g}_{ij} = \sum_{k=1}^{3N} \frac{\partial x_k}{\partial \xi_i}\frac{\partial x_k}{\partial \xi_j} \quad (9)$$
$$(i, j = 1, 2, 3, \ldots, f)$$

From the definition of the reaction coordinate, the kinetic tensor is diagonal on the reaction coordinate. When we expand them by a power series of the vibrational coordinates, ξ_i $(i = 2, 3, \ldots, f)$,

$$\bar{g}_{11} = 1 + \sum_{i=2}^{f}\left(\frac{\partial \bar{g}_{11}}{\partial \xi_i}\right)_0 \xi_i$$

$$\bar{g}_{1i} = \bar{g}_{i1} = \sum_{j=2}^{f}\left(\frac{\partial \bar{g}_{1i}}{\partial \xi_j}\right)_0 \xi_j$$

and

$$\bar{g}_{ij} = \delta_{ij}$$

the kinetic energy term of eq 8 is approximated to the first order of the expansion:

$$T \simeq -\tfrac{1}{2}\left[\left\{1 - \sum_{i=2}^{f}\left(\frac{\partial \bar{g}_{11}}{\partial \xi_i}\right)_0 \xi_i\right\}\frac{\partial^2}{\partial \xi_1^2} - \sum_{i=2}^{f}\left(\frac{\partial \bar{g}_{1i}}{\partial \xi_i}\right)_0 \frac{\partial}{\partial \xi_1} - \sum_{i,j=2}^{f}\left(\frac{\partial \bar{g}_{1i}}{\partial \xi_j}\right)_0 \xi_j\left(\frac{\partial^2}{\partial \xi_1 \partial \xi_i} + \frac{\partial^2}{\partial \xi_i \partial \xi_1}\right) + \sum_{i=2}^{f}\frac{\partial^2}{\partial \xi_i^2}\right] \quad (10)$$

The potential part of the hamiltonian is divided into two parts

$$W(\xi_1; \xi_2, \xi_3, \ldots, \xi_f) = W_1(\xi_1) + W_2(\xi_1; \boldsymbol{\xi}) \quad (11)$$

and

$$W_2(\xi_1; 0) = 0$$

where $\boldsymbol{\xi} = (\xi_2, \xi_3, \ldots, \xi_f)$. The first term of eq 11 means the change of potential energy along the reaction coordinate and the second one is that of vibrational coordinates except for the reaction coordinate. This term is approximated as the sum of the contributions of each vibrational coordinate

$$W_2(\xi_1; \boldsymbol{\xi}) \simeq \sum_{i=2}^{f} W_2{}^i(\xi_1; \xi_i) \quad (12)$$

We now define the zeroth order vibrational hamiltonian, which corresponds to the motion perpendicular to the reaction coordinate

$$H_V = -\frac{1}{2}\sum_{i=2}^{f}\left\{\frac{\partial^2}{\partial \xi_i^2} + W_2{}^i(\xi_1; \xi_i)\right\} \quad (13)$$

The eigenfunction is expressed as the product of respective vibrational wave function,

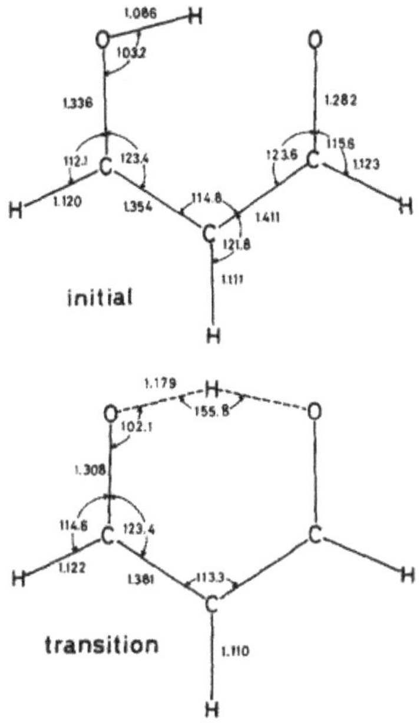

Figure 1. Optimized geometries of the initial and transition states. Bond length and angle are given in Å and deg, respectively.

$$H_V\Psi_K = E_K{}^V(\xi_1)\psi_K \tag{14}$$

and

$$\psi_K = \prod_{i=2}^{f} \chi_{K_i} \tag{15}$$

In eq 14 and 15, the suffix K represents the combination of the quantum number of respective modes. The wave function χ_{K_i} satisfies the relation

$$\left\{-\frac{1}{2}\frac{\partial^2}{\partial\xi_i{}^2} + W_2{}^i(\xi_1;\xi_i)\right\}\chi_{K_i} = \epsilon_{K_i}(\xi_1)\chi_{K_i} \tag{16}$$

and

$$E_K{}^V(\xi_1) = \sum_{i=2}^{2} \epsilon_{K_i}(\xi_1) \tag{17}$$

The wave function of the reacting system can be expanded by the vibrational wave function 15:

$$\Psi(\xi_1;\xi) = \sum_K u_K(\xi_1)\psi_K(\xi_1;\xi) \tag{18}$$

Using eq 9, 13, 14, and 18, the Schrödinger equation

$$H\Psi(\xi_1;\xi) = E\Psi(\xi_1;\xi)$$

can be written by the following coupled equation:

$$\begin{aligned}\left\{-\frac{1}{2}\frac{\partial^2}{\partial\xi_1{}^2} + W_1(\xi_1) + E_K{}^V(\xi_1) - E\right\}u_K(\xi_1) &= \frac{2}{\partial W/\partial\xi_1}\sum_{i=2}^{f}\frac{\partial^2 W}{\partial\xi_i\partial\xi_1}\langle\chi_{K_i}|\xi_i|\chi_{K_i-1}\rangle \\ &\quad\times\left(-\frac{1}{2}\frac{\partial^2}{\partial\xi_1{}^2}\right)u_{K-1(i)}(\xi_1) \\ &\quad+\frac{2}{\partial W/\partial\xi_1}\sum_{i=2}^{f}\frac{\partial^2 W}{\partial\xi_i\partial\xi_1}\langle\chi_{K_i}|\xi_i|\chi_{K_i+1}\rangle \\ &\quad\times\left(-\frac{1}{2}\frac{\partial^2}{\partial\xi_1{}^2}\right)u_{K+1(i)}(\xi_1) + \text{higher order terms}\end{aligned} \tag{19}$$

where $K - 1(i)$ and $K + 1(i)$ imply to up and down one quantum number of the ith vibrational mode. The derivation of eq 19 is straightforward and is not shown here.

The vibrational adiabatic approximation is imposed by the rhs of eq 19 to be zero. The conditions where the nonadiabatic correction is important are (1) the motion along the reaction coordinate has high kinetic energy and (2) the curvature of the reaction coordinate is large, which are easily expected from eq 19. When we treat the statistical ensemble of the proton motion in the thermal region, the nonadiabatic correction can be assumed to be negligible or cancelled. Thus eq 19 is approximated as

$$\left\{-\frac{1}{2}\frac{\partial^2}{\partial\xi_1{}^2} + W_1(\xi_1) + E_K{}^V(\xi_1) - E\right\}u_K(\xi_1) = 0$$

Effective Potential for Proton Transfer

At first, we calculated the reaction coordinate for the proton exchange in malonaldehyde within the CNDO/2 approximation.

The reaction coordinate is assumed to have B_2 symmetry of the C_s point group. The geometry of the equilibrium points on the reaction coordinate was optimized by the use of the variable metric procedure of McIver and Komornicki.[15] The derivative of the potential energy was obtained by the CNDO/2 version of the SCF formalism for a system with doubly occupied orbitals[16]

$$\begin{aligned}\frac{\partial W}{\partial x_k} = \sum_{B\neq A}&\left\{Z_AZ_B\frac{\partial R_{AB}{}^{-1}}{\partial x_k} - (P_{AA}Z_B + P_{BB}Z_A - P_{AA}P_{BB})\frac{\partial\gamma_{AB}}{\partial x_k}\right\} + \sum_{B\neq A}\sum_{\mu}^{A}\sum_{\nu}^{B} \\ &\times\left(2P_{\mu\nu}\frac{\partial\beta_{\mu\nu}}{\partial x_k} - \tfrac{1}{2}P_{\mu\nu}{}^2\frac{\partial\gamma_{AB}}{\partial x_k}\right)\end{aligned} \tag{20}$$

where the coordinate x_k belongs to atom A. The notation of eq 20 is the same as in the original CNDO/2 formulation of Pople et al.[7] The geometries of the initial and transition state are shown in Figure 1. The CNDO/2 optimization of geometries within C_s symmetry framework predicts that the asymmetric configuration is stable and the barrier height is 1.04 kcal/mol. The difference in the results of Schuster or Isaacson and Morokuma seems to originate from the optimization of the backbone structure. The normal coordinates of the equilibrium points were calculated as the eigenvectors of eq 6, where the second derivative of the potential energy is calculated by the numerical differentiation of the potential gradient. In eq 6, the mass of the H atom has a value of 1836.12 au. The C and O atoms are 12 and 16 times the mass of H, respectively. Figure 2 shows the relative magnitude and the direction of the displacement vectors along the reaction coordinate at the initial, transition, and final states. The reaction coordinate was calculated by the approximate form of eq 4.

$$\frac{\Delta x_i}{\Delta S} = \frac{\partial W/\partial x_i}{\mathrm{d}W/\mathrm{d}S} \tag{21}$$

Equation 21 was solved with the step size $0.05(M_H)^{1/2}$ au, where M_H is the mass of the H atom. The energy change caused by one step is about 0.1 kcal/mol. The potential energy along the reaction coordinate, $W_1(\xi_1)$, is shown in Figure 3. The initial state is located at $-0.63(M_H)^{1/2}$ au, when the point of the transition state is chosen to be zero.

The second step of the calculation is to determine the potentials for the vibrational coordinates which are perpendicular to the reaction coordinate. We calculated the normal coordinates at eight points on the reaction coordinate from $\xi_1 = -\infty$ to 0. Among 20 vibrational coordinates, the number of in-plane modes is 14 and the remainder is out-of-plane modes. It is very

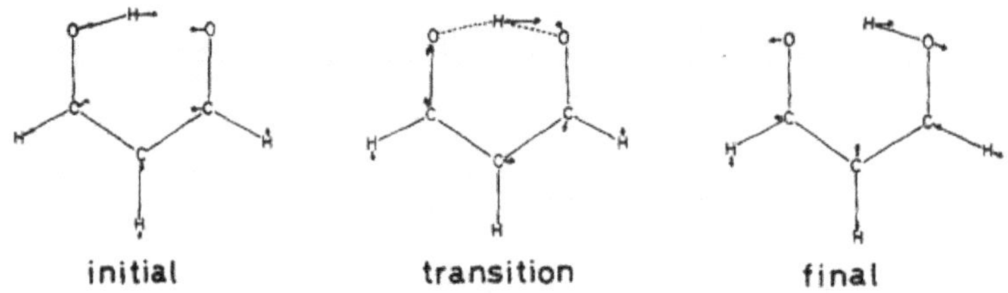

Figure 2. Displacement vectors along the reaction coordinate at three equilibrium points.

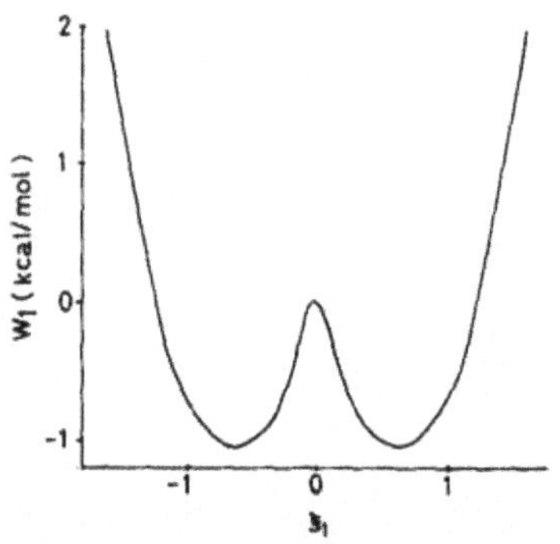

Figure 3. Potential energy profile along the reaction coordinate. The scale of reaction coordinate is $(M_H)^{1/2}$ au.

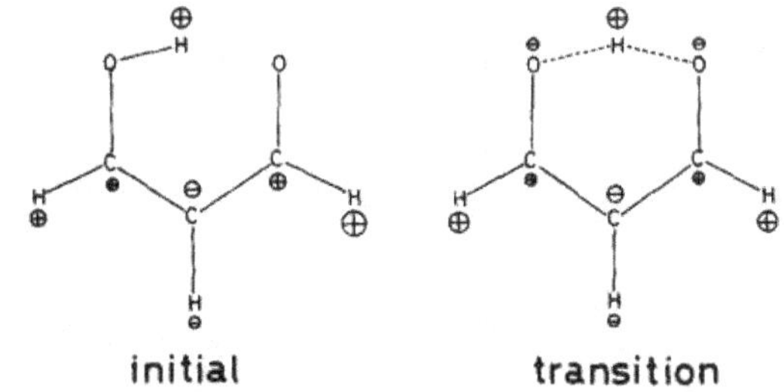

Figure 4. Displacement along one out-of-plane vibrational coordinate with a negative force constant.

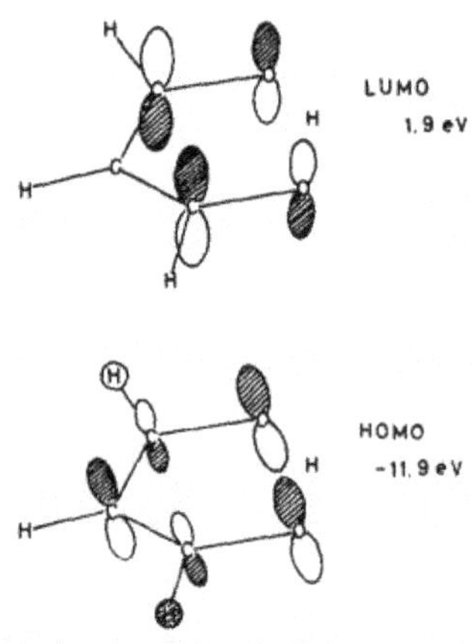

Figure 5. Schematic representation of nodal properties of HOMO and LUMO and their energies.

interesting for us that there is one out-of-plane vibrational mode with a negative force constant. The component of the displacement vector corresponding to this vibrational mode is given in Figure 4. Although malonaldehyde has been predicted to be a planar molecule in previous studies, the result of CNDO/2 calculation forecasts that the stable geometry of this molecule is nonplanar. The reason for this negative force constant is expected by the consideration of the symmetry properties of the frontier orbitals in malonaldehyde. The second order perturbation expression of force constant is given as[17]

$$\frac{\partial^2 W}{\partial \xi_i{}^2} = \left\langle 0 \left| \frac{\partial^2 W}{\partial \xi_i{}^2} \right| 0 \right\rangle + \sum_I \frac{\left| \left\langle 0 \left| \frac{\partial V}{\partial \xi_i} \right| I \right\rangle \right|^2}{E_0 - E_I} \tag{22}$$

where V is the sum of nuclei-electron and nuclear repulsion potentials in the electronic hamiltonian, the bra-ket is the electronic wave function, and the suffixes 0 and I imply the ground and excited states, respectively. As Bader[18] suggested, if (1) the direct product of the irreducible representation of the highest occupied (HO) MO and the lowest unoccupied (LU) MO coincides with the irreducible representation of the vibrational coordinate and (2) the transition density of these MO's is localized to the direction of the displacement vector, the second term of eq 21 plays a dominant role to reduce the force constant. Figure 5 shows the schematic representation of HOMO and LUMO at the transition state, where the above-mentioned conditions are well satisfied. This vibrational coordinate is termed as ξ_2 and the remaining coordinates as ξ_i ($i = 3, 4, \ldots, 21$). For 19 vibrational coordinates having the positive force constants, the potentials are approximated by the harmonic form

$$W_2{}^i(\xi_1;\xi_i) = {}^1\!/_2 K_i(\xi_1)\xi_i{}^2 \tag{23}$$

In Table I, the vibrational frequencies and assignment of them at the transition state are given. Although there is, to our knowledge, no available experimental data of vibrational spectra of malonaldehyde, the enol acetylacetone, the dimethyl derivative of malonaldehyde, has been extensively studied by means of an IR and a Raman spectroscope in both gas and liquid phases.[19] The normal coordinate analysis has alo been performed by Ogoshi and Nakamoto.[20] As seen in Table I, the CNDO/2 calculation of the vibrational frequencies somewhat overestimated on the whole, comparing with the experimental values of acetylacetone. The stretching mode shows the most serious discrepancy, which is 1.6 times the experimental values. Since the vibrational frequencies depend on the reaction coordinate, they are represented as

$$\omega_i(\xi_1) = \omega_i(0) + \Delta\omega_i(\xi_1) \tag{24}$$

where $\omega_i(0)$ is the frequency at $\xi_1 = 0$ and $\Delta\omega_i(\xi_1)$ denotes the difference from $\omega_i(0)$. The change of vibrational frequencies with respect to the reaction coordinate, $\Delta\omega_i(\xi_1)$, is shown in Figure 6. The O–H stretching mode couples with the reaction coordinate most distinctly and changes its frequency in the process of the proton transfer.

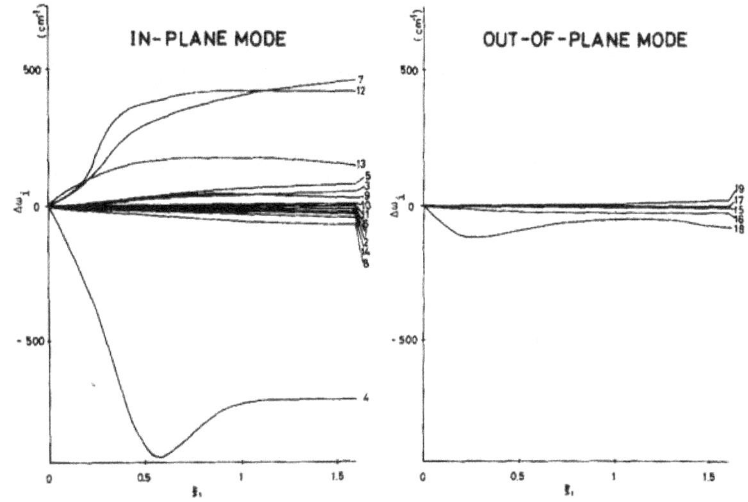

Figure 6. Change of vibrational frequencies with respect to the reaction coordinate. The number of vibrational modes is the same as in Table I. The scale of the reaction coordinate is the same as in Figure 3.

Table I. Normal Coordinates at the Transition State

No.	Freq, cm^{-1}	Symmetry	Nature of vib
		(a) In-Plane Mode	
1	4684	A_1	C-H stretching
2	4538	A_1	C-H stretching
3	4536	B_2	C-H stretching
4	3094	B_2	O-H stretching
5	2647	A_1	C-O stretching
6	2217	B_2	C-O stretching
7	2039	A_1	O-H-O bending
8	1897	A_1	C-C stretching
9	1468	B_2	C-H bending
10	1424	A_1	C-H bending
11	1280	A_1	C-H bending
12	973	A_1	Ring deformation
13	672	A_1	Ring deformation
14	572	B_2	Ring deformation
		(b) Out-of-Plane Mode	
15	1221	A_2	C-H bending
16	1186	B_1	C-H bending
17	944	B_1	O-H-O bending + C-H bending
18	791	B_1	O-H-O bending + C-C-C bending
19	358	A_2	Ring deformation

For the vibrational mode with negative force constant, ξ_2, the potential is approximated by the sixth order polynomial. The wave function is expressed by a linear combination of 20 harmonic functions

$$\chi_2(\xi_2) = \sum_{n=0}^{19} C_n N_n H_n(\alpha\xi_2) \exp(-\tfrac{1}{2}\alpha\xi_2{}^2) \tag{25}$$

where N_n is the normalization constant and $H_n(\alpha\xi_2)$ is the Hermite polynomial. The coefficients, C_n, are determined by the variational procedure. The scaling factor, α, is optimized by the minimization of the sum of root mean square deviation of eigenvalues[21]

$$\Delta E = \sum_{m=0}^{M} (\langle m|H|m\rangle^2 - \langle m|H^2|m\rangle)^{1/2} \tag{26}$$

where H is the vibrational hamiltonian for ξ_2 and M is chosen to be 1. The mean square error ΔE is calculated to be within 1% of the difference between the eigenvalues of the ground state and the first excited one for all values of ξ_1. In Figure 7, some lower eigenvalues of vibration ξ_2 are shown. The ground state energy is slightly below the top of the barrier at the transition state and becomes higher with distance from this point.

We can thus determine the effective potentials for the motion on the reaction coordinate

$$V_K{}^{\text{eff}}(\xi_1) = W_1(\xi_1) + E_K{}^{\text{V}}(\xi_1)$$

which is specified by the combination of the quantum number of vibrations perpendicular to the reaction coordinate

$$K = (n_2, n_3, \ldots, n_{21})$$

Figure 8 shows the effective potential for the zero-point vibration, where all vibrational quantum numbers are zero. The energy barrier height for the proton transfer reaction is lowered to 0.6 kcal/mol, and the position of the potential minimum becomes closer to the transition state, which is mainly caused by the contibution of the out-of-plane ξ_2 vibrational mode.

Statistical State of Malonaldehyde

Density Matrix. The analytical form of the effective potential along the reaction coordinate is expressed by the combination of the eighth order polynomial and one Gaussian function

$$V_K{}^{\text{eff}}(\xi_1) = \sum_{i=0}^{8} a_{K,i}\xi_1{}^i + A_K \exp(-B_K\xi_1{}^2) \tag{27}$$

The parameters $a_{K,i}$, A_K, and B_K were determined by the least-squares method. The variational procedure, with 20 harmonic functions, was used to obtain the eigenvalues of them as in the case of $W_2{}^2(\xi_1)$. The matrix elements are easily derived by the method of Chan and Stelman.[22] The scaling factor was determined by the minimization of eq 26. The optimization was carried out to the zero-point potential in which all vibrations have zero quantum number. The same scaling factor was taken for all the effective potentials throughout. Some eigenvalues of the potential $V_0{}^{\text{eff}}(\xi_1)$ are shown in Figure 8. The ground state energy is slightly below the top of the barrier. The spacing of the energy spectrum differs from the case of harmonic potential. The remaining potentials have on the whole

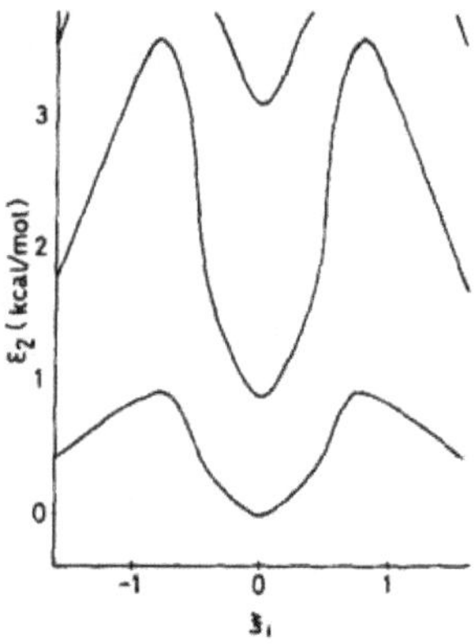

Figure 7. Change of eigenvalues for the ξ_2 vibrational mode with respect to the reaction coordinate. The scale of the reaction coordinate is the same as in Figure 3.

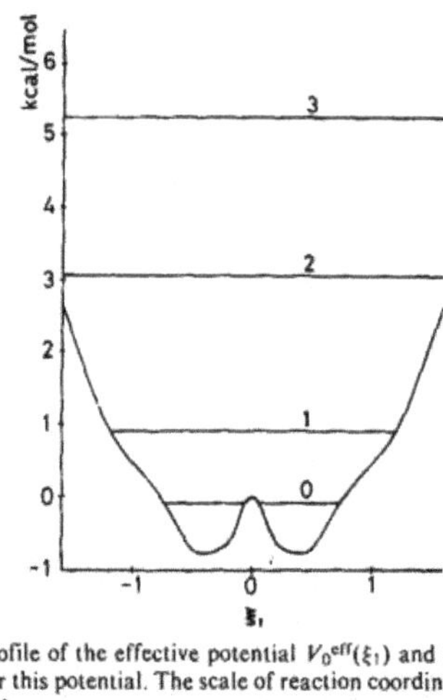

Figure 8. Profile of the effective potential $V_0^{eff}(\xi_1)$ and some lower eigenvalues for this potential. The scale of reaction coordinate is the same as in Figure 3.

a similar form of energy the level interval as $V_0^{eff}(\xi_1)$. Some of them have the ground state energy level highly bounded to the wells.

In order to describe the quantum statistical state of malonaldehyde, we have assumed the canonical ensemble in the heat bath of a given temperature. The vibrational state of the molecule is represented by the density matrix. The density matrix of the steady state is diagonal and given as

$$\rho(\xi_1,\xi) = \sum_K \sum_{m \in K} \exp(-E_{Km}/\kappa T) \times \Psi_{Km}^*(\xi_1;\xi)\Psi_{Km}(\xi_1;\xi)/Z(T) \quad (28)$$

and

$$Z(T) = \sum_K \sum_{m \in K} \exp(-E_{Km}/\kappa T)$$

where E_{Km} is the mth eigenvalues of the Kth effective potential, κ is the Boltzmann constant, T is the absolute temperature, and $Z(T)$ is the vibrational partitioning function. When we integrate eq 28 by the vibrational coordinates, ξ, we can define the density matrix on the reaction coordinate

$$\rho(\xi_1) = \int \rho(\xi_1;\xi)\,d\xi$$

$$= \sum_K \sum_{m \in K} \exp(-E_{Km}/\kappa T) u_{Km}^*(\xi_1) u_{Km}(\xi_1)/Z(T) \quad (29)$$

This density matrix implies the projection of the molecular conformer on the reaction coordinate. The projected density, $\rho(\xi_1)$, was calculated for various temperatures from 0 to 500 K, where the summation of eq 29 was truncated when the condition

$$\exp[-(E_{Km} - E_{00})/1000\kappa] < 10^{-4}$$

is satisfied. The calculation includes about 70 effective potentials. Figure 9 illustrates the projected density of malonaldehyde, $\rho(\xi_1)$, at 0, 300, and 500 K, respectively. As is seen in Figure 9, there exists a hollow at the middle point of the reaction coordinate, and, with the increase of temperature, it becomes deep and the density distribution is biased to the side region. This effect is easily realized by the populations of energy levels of the motion on the reaction coordinate. In Table II, the occupancies of respective energy levels at 100, 300, and 500 K are listed.[23] As the temperature increases, the contributions of the excited states become large and the first excited state plays the important role in the change of density distribution.

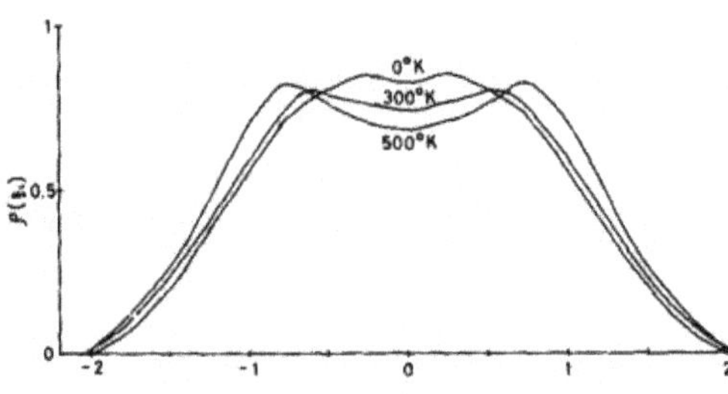

Figure 9. Density projected on the reaction coordinate at 0, 300, and 500 K. The scale of the reaction coordinate is the same as in Figure 3.

Proton Transfer Rate. In order to obtain the proton transfer rate, we have separated the conformation of malonaldehyde into two parts, which is realized by dividing the effective potential, $V_K^{eff}(\xi_1)$, at the transition state. The phenomenological expression for the proton transfer rate is given as in the conventional form:

$$\frac{dn_A(t)}{dt} = -k_{AB}(t)n_A(t) + k_{BA}(t)n_B(t) \quad (30)$$

here $n_A(t)$ and $n_B(t)$ are the population of wells A and B, $k_{AB}(t)$ is the rate from A to B, and $k_{BA}(t)$ is the rate of the reverse direction, respectively. Since the potential is symmetric, $k_{AB}(t)$ is equal to $k_{BA}(t)$. The mean rate is defined from eq 30:

$$\bar{k} = \lim_{t \to \infty} \frac{\int_0^t k_{AB}(t')\,dt'}{t}$$

$$= -\lim_{t \to \infty} \frac{\ln\,[2n_A(t)/n_A(0) - 1]}{t} \quad (31)$$

When we define the projection operator

$$P_A = \sum_{i_A} P_{i_A} = \sum_{i_A} \phi_{i_A}\phi_{i_A}^* \quad (32)$$

the population of the well A, $n_A(t)$, is given by

$$n_A(t) = T_r(\rho(\xi_1;\xi;t), P_A) \quad (33)$$

Table II. Population of Energy Levels for the Motion on the Reaction Coordinate

Level No.	100 K	300 K	500 K
0	0.9928	0.8464	0.7289
1	0.0072	0.1495	0.2404
2	0.0000	0.0040	0.0270
3	0.0000	0.0001	0.0036
4	0.0000	0.0000	0.0001

where ϕ_{i_A} is the ith eigenfunction of the well A and $\rho(\xi_1;\xi;t)$ is the density matrix when the time is t. The eigenfunction ϕ_{i_A} is variationally determined by imposing the boundary condition that there is no amplitude in region B. The wave function is approximated by

$$\phi_{i_A}(\xi_1) = \theta(\xi_1) \times \left\{\sum_{m=0}^{19} C_{i_A}{}^m N_m H_m(\alpha\xi_1)\exp(-\tfrac{1}{2}\alpha\xi_1{}^2)\right\} \quad (34)$$

where the normalization constant, N_m, is given as

$$N_m = \left(\frac{\alpha}{(\pi)^{1/2}\,2^{m-1}m!}\right)^{1/2}$$

and the function $\theta(\xi_1)$ is a step function satisfying the condition:

$$\theta(\xi_1) = \begin{cases} 1 & (\xi_1 < 0) \\ 0 & (\xi_1 \geq 0) \end{cases} \quad (35)$$

The same scaling factor as in the parent double-well potential is used to calculate $\phi_{i_A}(\xi_1)$. The density matrix of the initial time is expressed by the use of the statistical operator

$$\rho(\xi_1;\xi;0) = \sum_{i_A} \phi_{i_A}\, n_{i_A}(0)\phi_{i_A}{}^* \quad (36)$$

and the initial population is given as

$$n_{i_A}(0) = T_r(\rho(\xi_1;\xi), P_{i_A}) \quad (37)$$

where $\rho(\xi_1;\xi)$ is the density matrix of steady state. The time development of the density matrix is obtained by the use of the evolution operator

$$U(t) = \exp(-iHt)$$

Then the population of the region A at the time t is calculated by

$$n_A(t) = \sum_K \left[\sum_{i,j\in K} \exp\{i(E_{Ki} - E_{Kj})t\} \times \sum_{k_A,l_A} n_{k_A}(0) S_{i,k_A}{}^K S_{j,k_A}{}^K S_{i,l_A}{}^K S_{j,l_A}{}^K\right] \quad (38)$$

where $S_{i,k_A}{}^K$ is the overlap integral

$$S_{i,k_A}{}^K = \int u_{Ki}(\xi_1)\phi_{i_A}{}^K(\xi_1)\,d\xi_1$$

In Figure 10, the time evolution of the population $n_A(t)$ is shown, in which some different features compared with the classical analogue are seen. A steady region exists where $n_A(t)$ is about 0.75, though the classical propagation is characterized by the exponential decay. The steady state is considered to originate from the interference between the wave propagating from A to B and the wave of opposite direction.

The rate of the proton transfer was calculated from eq 31. The rate is almost constant over the whole temperature range and is calculated to be $9.77 \times 10^{13}\ s^{-1}$. The activation energy of this process is nearly zero. The time interval of proton transfer from one well to another is 1.02×10^{-12} s.

The calculated result of proton transfer time implies that the movement of the hydrogen-bonded proton is very quick with the order of a low frequent molecular vibration.

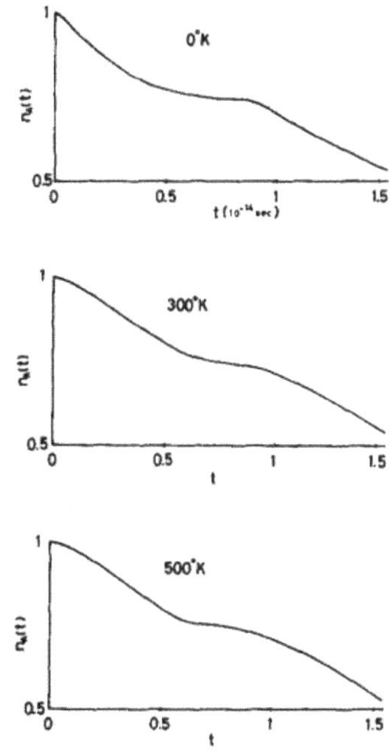

Figure 10. Time development of the population in the region A at 0, 300, and 500 K.

Discussion

Until now we have treated the dynamic behavior of the hydrogen-bonded proton of malonaldehyde by means of the "intrinsic" reaction coordinate. All vibrational motion perpendicular to the proton transfer coordinate (reaction coordinate) was taken into account. These vibrational motions were seen to affect the motion on the reaction coordinate in the case when the potential barrier is very low as in the present system compared with the usual chemical reaction where the activation barrier is moderately high.

The calculation of the vibrational state of malonaldehyde shows that the rapid transformation occurs between the two asymmetric conformers through the low potential barrier. This result is consistent with the microwave prediction of proton motion by Rowe, Duerst, and Wilson.[11] In some previous MO calculations,[8-10] the position of proton in this molecule has received much attention and different conclusions have been derived. The proton is a quantum mechanical particle and has the zero-point energy even at 0 K, so the position of the proton is a probability problem. In our statistical calculation, the distribution of the hydrogen-bonded proton on the reaction coordinate varies with the increase of temperature, and the mean position of the proton is symmetric in the low-temperature range and becomes oriented asymmetrically at the high temperature.

Our CNDO/2 calculation of the potential energy surface of malonaldehyde suggested the existence of one out-of-plane vibrational motion with a negative force constant. Then the energy minimum path for the proton has an ellipse like profile and its motion is considered to contain the orbital angular momentum. When we assumed that the vibrational frequency is proportional to the energy gap between the ground state and the first excited one, the frequencies of the motion on the mode

ξ_2 and the reaction coordinate are the same order (see Figures 7 and 8), which further convinces one of the elliptical orbital of the hydrogen-bonded proton of malonaldehyde.

Of course, the result in this paper fairly depends on the accuracy of the calculated potential energy surface. It is well known that the CNDO/2 method reproduces the bond length, bond angle, and hydrogen bond energy well. The overestimate of the C–H stretching frequencies does not affect the results, since the change of frequencies along the reaction coordinate is sufficiently small. The complete geometry optimization and the normal coordinate calculation are of advantage in the present work. It should be noted that an arbitrary displacement of out-of-plane coordinate does not lower the potential energy,[8] whereas only one of six vibrational coordinates has a negative force constant on the reaction coordinate.

Compared with the microwave spectra, the calculated potential barrier along the reaction coordinate may be underestimated. Nevertheless, we believe that the essential result obtained in the present paper will not be altered by the use of a more reliable potential surface.

The method proposed in the present paper is also applicable to any type of unimolecular reaction. The dynamic nature or the rate of unimolecular reaction is under development at the present. These works are in progress and will be presented in the near future, so this paper will be regarded as a preliminary of this series.

Acknowledgment. This work was supported by a Grant-in-Aid for Scientific Research from the Ministry of Education of Japan (No. 047068). Authors express their appreciation to the Data Processing Center of Kyoto University for generous permission to use the Facom 230-75 computer.

References and Notes

(1) R. P. Bell, "The Proton in Chemistry", Methuen and Company Ltd., London, 1959; S. Bratoz, *Adv. Quantum Chem.*, **3**, 209 (1967); G. C. Pimentel and A. L. McClellan, *Annu. Rev. Phys. Chem.*, **72**, 347 (1971).

(2) See review, P. A. Kollman and L. C. Allen, *Chem. Rev.*, **72**, 283 (1972).

(3) S. Yamabe, S. Kato, H. Fujimoto, and K. Fukui, *Bull. Chem. Soc. Jpn.*, **48**, 1 (1975); K. Morokuma, *J. Chem. Phys.*, **55**, 1236 (1971); S. Iwata and K. Morokuma, *J. Am. Chem. Soc.*, **95**, 7563 (1973).

(4) P.-O. Löwdin, *Adv. Quantum Chem.*, **2**, 213 (1965); see also, P.-O. Löwdin, *Rev. Mod. Phys.*, **35**, 724 (1963); *Biopolym. Symp.*, **1**, 161 (1964).

(5) E. A. Psehenichov and N. D. Sokolov, *Int. J. Quantum Chem.*, **1**, 885 (1967); J. Brickmann and H. Zimmermann, *J. Chem. Phys.*, **50**, 1608 (1969).

(6) K. Fukui, *J. Phys. Chem.*, **74**, 4161 (1970); K. Fukui, S. Kato, and H. Fujimoto, *J. Am. Chem. Soc.*, **97**, 1 (1975).

(7) J. A. Pople and D. L. Beveridge, "Approximate Molecular Orbital Theory", McGraw-Hill, New York, N.Y., 1970.

(8) P. Schuster, *Chem. Phys. Lett.*, **3**, 433 (1969).

(9) G. Karlström, H. Wennerström, B. Jönsson, S. Forsén, J. Almlöf, and B. Roos, *J. Am. Chem. Soc.*, **97**, 4188 (1975).

(10) A. D. Isaacson and K. Morokuma, *J. Am. Chem. Soc.*, **97**, 4453 (1975).

(11) W. F. Rowe, Jr., R. W. Duerst, and E. B. Wilson, *J. Am. Chem. Soc.*, **98**, 4021 (1976).

(12) C. Eckart, *Phys. Rev.*, **47**, 552 (1935).

(13) B. Podolsky, *Phys. Rev.*, **32**, 812 (1928).

(14) G. L. Hofacker, *Z. Naturforsh., A*, **18**, 607 (1963); J. Brickmann, *ibid.*, **28**, 1759 (1973); J. Brickmann, *J. Chem. Phys.*, **62**, 1086 (1975).

(15) J. W. McIver and A. Komornicki, *Chem. Phys. Lett.*, **10**, 303 (1971).

(16) J. Gerrat and I. M. Mills, *J. Chem. Phys.*, **49**, 1719 (1969); J. Pancir, *Theor. Chim. Acta*, **29**, 21 (1973).

(17) Although this formula is true for the exact wave function of the hamiltonian, it is valid for a qualitative discussion based on an approximate wave function.

(18) R. F. Bader, *Can. J. Chem.*, **40**, 1164 (1962).

(19) R. S. Rosmussen, D. D. Tunnicliff, and R. P. Brattain, *J. Am. Chem. Soc.*, **75**, 1068 (1947); G. G. Smith, *ibid.*, **75**, 1134 (1954); S. Bratoz, D. Hadzi, and G. Rossy, *Trans. Faraday Soc.*, **52**, 464 (1956); R. Mecke and E. Funke, *Z. Elektrochem.*, **60**, 1124 (1956); H. Musso and H. Junge, *Chem. Ber.*, **101**, 801 (1968).

(20) H. Ogoshi and K. Nakamoto, *J. Chem. Phys.*, **45**, 3113 (1966).

(21) J. Goodisman, *J. Chem. Phys.*, **47**, 5247 (1967).

(22) S. I. Chan and D. Stelman, *J. Chem. Phys.*, **39**, 545 (1963).

(23) The occupancy of the energy level n is defined by

$$\sum_{K} \exp(-E_{Kn}/\kappa T) \Big/ \sum_{K} \sum_{m \in K} \exp(-E_{Km}/\kappa T)$$

Theoret. Chim. Acta (Berl.) 51, 189–206 (1979)

THEORETICA CHIMICA ACTA

Intrinsic Dynamism in Chemically Reacting Systems

Akitomo Tachibana and Kenichi Fukui

Department of Hydrocarbon Chemistry, Faculty of Engineering, Kyoto University, Kyoto, Japan

The geometrical cell structure of the chemically reacting system is discussed. The boundary of the cell uniquely defines the dividing surface between the initial reactant side and the final product side. Introducing the concept of the intrinsic reaction time (IRT) and the accumulation time (AT) of reaction along the meta-IRC (intrinsic reaction coordinate), the intrinsic dynamism in the cell is discussed. Then, the stable limit theorems with respect to the intrinsic nature of the normal vibrations are derived. The theory is elucidated by using a model potential surface.

Key words: Intrinsic dynamism – Intrinsic reaction coordinate (IRC) approach – Chemically reacting systems, cell structure of ~ – Stable limit theorems

1. Introduction

Since the establishment of Eyring's absolute rate theory [1], the concept of reaction coordinates has been a guiding principle in studying the complicated modes of chemical reactions. The fundamental assumptions of this theory consist of E1) the presence of the activated complex (or transition state) B in the course of the chemical reaction A → C:

$$A \rightarrow B \rightarrow C, \tag{1.1}$$

and E2) the chemical equilibrium between A and B. Usually, the location of the activated complex has been ascribed to the transition point, that is the saddle, on the adiabatic potential energy surface. In this connection, recent quantum mechanical calculations have succeeded in developing the efficient algorithms of searching for the transition point on the surface [2–4].

Recently, Miller [5–7] has devoted himself to the quantum mechanical reformulation of the absolute rate theory. In this case, the nuclear configuration of the

0040–5744/79/0051/0189/$03.60

activated complex itself is not the major subject; but the hypersurface S appears on which the flux of the trajectories in the phase space takes the minimal value [8, 5–7]. Then, the surface S defines the dividing surface by which the initial reactant side A and the final product side C are discriminated. Also, the classical rate formula of the absolute rate theory has been interpreted quantum mechanically by replacing the partition function of the activated complex with the flux integral on the dividing surface S [5–7]. But it should be noted that the flux integral itself is dependent on the total energy of the reacting system and hence, it is not guaranteed that the only one dividing surface S is determined for the given reaction (1.1). Also, the relation between the geometrical property of S and that of the adiabatic potential energy surface has not been clarified; for example, the transition point B is not necessarily located on the dividing surface S.

On the other hand, Hofacker and his collaborators [9–14] have directed their attention to the geometrical property of the reaction path and its relationship with the chemical dynamism in terms of the local quantum mechanics along the reaction path. Particularly, they have defined an unbranched reaction path and discussed the vibrational (non-)adiabaticity along the reaction path [9, 10]; the essential assumptions of their treatments are summarized as follows: H1) the reaction path is a curve of minimum Gaussian curvature, and H2) the first derivative of the adiabatic potential vanishes on the surface perpendicular to this path. The quantum mechanical developments of their theory have been based on the perturbational treatments of Podolsky's Hamiltonian [15, 16], where the following simplifications are adopted: H3) the kinetic energy operator is separable; that is, the sum of squares of individual linear momentum operators, and H4) harmonic oscillator approximation for the bound degrees of freedom perpendicular to the reaction path. The reaction path characterized by these simplifications has also been referred to as the dynamic reaction path [14]; in this connection, Marcus has utilized the concept of the dynamic reaction path for three-body reactions [14, 17–20]. Note, however, that there are some difficulties in the treatments described above. First, there has been no mathematical proof for the presence of the unbranched reaction path satisfying the criterions H1) and H2). Moreover, since the first derivative of the adiabatic potential energy completely vanishes at any equilibrium point, one cannot find the direction of the reaction path at the equilibrium point. Second, there is an unsatisfactory balance in the perturbational order inherent to the treatment of the dynamic reaction path: the simplification H3) is obtained by the zeroth-order perturbation theory of the kinetic energy operator in terms of the deviation of coordinates from the reaction path, while the simplification H4) is obtained by the second-order perturbation theory of the potential energy operator [9, 10, 12–14]. Although the first-order correction to the kinetic energy operator has been partly taken into account [9, 10, 12–14], the zeroth-order kinetic energy operator is supposed to be insufficient for the quantum mechanical description of the chemical dynamism in the region of finite volume.

Now, Fukui has introduced a concept of intrinsic reaction coordinate (IRC) as the reaction path from the initial A to the final C basin (or valley) by way of the transition point B on the adiabatic potential energy surface, and given the differential equation of the IRC [21]. Also, Fukui *et al.* [22–25] and Morokuma *et al.* [26, 27] have developed the reaction ergodography in terms of the IRC. The IRC possesses the outstanding properties that F1) this is the intrinsic solution of Lagrange's equation of motion [23] and F2) this converges to one of the stable normal coordinates of positive force constants at the stable equilibrium point and to the unstable normal coordinate of negative force constant at the transition point [21]. Likewise, the general solution of the IRC equation allows the unique definition of the reaction coordinate for the reaction starting from any non-equilibrium point on the potential energy surface: this solution has been referred to as the meta-IRC, and the IRC has been given as the particular case of the meta-IRC [28]. In this way, the utility of the IRC approach over the usual qualitative and convenient concept of reaction coordinates has been corroborated [21–28].

In the present paper, we discuss the dynamism of chemically reacting systems in terms of the IRC approach. Particularly, it is shown that the chemically reacting system has a cell structure as if the system were an organism. We can then tread the cell of A along the IRC, pass through the intercell boundary at the transition point B, and finally enter the cell of C along the IRC. The intercell boundary is uniquely determined in terms of the geometry of the adiabatic potential energy surface. Hence, the intercell boundary allows the unique and natural definition of the dividing surface S separating the initial reactant side A and the final product side C. Moreover, it should be noted that the solution of the meta-IRC satisfies the criterion H2) of the reaction path defined by Hofacker *et al.* [28]. But the meta-IRC cannot be always identified with the reaction path defined by Hofacker *et al.*, because the criterion H1) for the latter concerns the second-order derivative of the adiabatic potential energy surface U and the IRC equation concerns only the first derivative of U. Furthermore, it will be shown that in the IRC approach, the perturbation-free kinetic energy operator is obtained by the sum of squares of individual local momentum operators. This property may prove useful for one to overcome the difficulty inherent to the approach of the dynamic reaction path [14], as well as to improve the accuracy inherent to the approximate treatment of the IRC approach [25] and other approaches [17–20, 29–34] of the same kind.

The intrinsic dynamism of the chemically reacting system is formulated using a pair of time-like parameters: the intrinsic reaction time (IRT) and the accumulation time (AT) of reaction. The IRT is defined uniformly all over the configuration space as if it were the realistic time, while the AT is defined within the cell or the boundary excluding the equilibrium points. In particular, the IRC equation is simplified, as is shown in (3.19), if the intrinsic dynamism is represented with respect to the AT. Note that this simplified equation (3.19) is similar to the well-known differential equation of the dynamical system [35]. Different from the usual dynamical system described in the Euclidean space [35], 1) the intrinsic dynamism is formulated in the Riemannian space. Moreover, 2) from the definition (3.18), the AT diverges to infinity in the limit at the equilibrium point. The limiting property of the AT may represent the accumulating behavior of the meta-IRC at the equilibrium point; this limiting property of the AT is analogous to that of the time-like parameter used in the usual dynamical system [35]. However, it should be noted that the limiting property of the AT can be deduced logically from its definition, while the analogous property of the time-like parameter in the usual dynamical system is not provided with the *a priori* reasoning itself. As far as the authors are aware, there is no detailed theory that deals with the particular case of the intrinsic dynamism having the characteristics 1) and 2). Also shown are the stable limit theorems of the IRC approach, which describe an intrinsic nature of the normal vibrations at the stable equilibrium point.

2. Cell Structure in Chemically Reacting Systems

Let us consider the chemical reaction (1.1) where the reaction proceeds from the reactant side A through the transition point B to the product side C. In the case of isomerization reaction, the structure of the reacting system is illustrated in Fig. 1 using the reaction pattern [28]: A and C correspond to the minima of the potential energy surface and B corresponds to the saddle. The isomerization reaction is described by the idealized locus of nuclei, namely the IRC, as is illustrated in Fig. 1; note that we consider in this simple reaction scheme that there are no other minima on the potential energy surface. Then, any non-equilibrium point on the path AB of the IRC corresponds to the nuclear configuration on the reactant side; conversely, any non-equilibrium point on the path BC of the IRC corresponds to the nuclear configuration on the product side. Hence, the discrimination of the nuclear configuration according to the reactant side and the product side is very clear on the IRC. However, in the actual dynamical processes of the chemical reaction, the system may take other nuclear configurations bypassing the transition point B. Then, the definition for the vague notion of the reactant "side" and the product "side" should be firmly established.

It should be noted here that the vague entity of the "atomic region" in a molecule has become firmly established on a solid theoretical foundation by Bader *et al.* [36–43]; a certain atomic region (or atomic group region) in a molecule is circumscribed in terms of the distribution of the electron density [36–40], and the regional quantum mechanics has been formulated [41–43].

Now, in the chemically reacting system, we shall utilize the geometrical property of the adiabatic potential energy surface in order to obtain the discernment of the reactant side and the product side. In this case, in terms of the mathematical treatments of abstract gradient vector fields, a manifold is found in the neighborhood of the stable equilibrium point of the potential and is often referred to as a basin [35, 44]. Since the sophisticated mathematical treatments are forbidding, we shall conveniently interpret for the chemically reacting system the concept of the purely mathematical object, namely the basin, as the cell, and analyze the intrinsic dynamism of the chemical reaction (1.1) in terms of the cell structure of the system. In our previous paper [28], we have shown that any reaction starting from a non-equilibrium point in the neighborhood of the stable equilibrium point, say A, finally reaches A by following the unique meta-IRC. So we have the cell of the stable equilibrium point A as follows: the cell of A, or the Cell (A) to be short, is a set of the non-equilibrium points on the meta-IRC along which the direction of the gradient force field $F = -\text{grad}\ U$ does not vanish and finally converges to the stable equilibrium point A. Thus, for any stable equilibrium point A_i, we have the unique Cell (A_i). Within the cell, the progress of any intrinsic reaction is described by the meta-IRC, being accompanied by the vibrational degrees of freedom on the equi-potential surface which is perpendicular to the meta-IRC [28]; in terms of the contravariant character of coordinates, the promoting mode of the intrinsic chemical reaction itself is measured by the normal meta-IRC [28]. Then, as a natural extension of the notion of the transition point, the intercell boundary ((A_i, A_j)) between a pair of the stable equilibrium points A_i and A_j can be defined by the intersection of the boundaries as

$$((A_i, A_j)) = \partial\,\text{Cell}\,(A_i) \cap \partial\,\text{Cell}\,(A_j) \tag{2.1}$$

where ∂ Cell (A_i) denotes the boundary of the Cell (A_i). The IRC which connects A_i and A_j has the unique accumulation point in the intercell boundary ((A_i, A_j)) from either cell, Cell (A_i) or Cell (A_j). This point is nothing but the transition point. In other words, any *idealized* flux of particles which leave A_i for A_j along the meta-IRC becomes a minimum at the intercell boundary ((A_i, A_j)). In this connection, the intercell boundary may serve the minimum-of-flux criterion for the definition of the dividing surface S [8, 5] in terms of the *intrinsic dynamism* of a reaction.

Let us obtain the quantum mechanical expression of the rate constant $k_{C\leftarrow A}$ for the reaction (1.1). The general formula has been shown by Miller as follows [5, 6]:

$$k_{C\leftarrow A}(T) = Q_A^{-1}\,\text{tr}\,[e^{-\beta H}\,\delta(f(\boldsymbol{q}))(\partial f(\boldsymbol{q})/\partial \boldsymbol{q})\cdot \boldsymbol{p}m^{-1}\mathscr{P}], \tag{2.4a}$$

where $\beta = (kT)^{-1}$, $(\boldsymbol{p}, \boldsymbol{q})$ are the vectors of the nuclear momenta and coordinates of the system, Q_A is the partition function per unit volume of the reactant, H is the total Hamiltonian of the system, $\mathscr{P}$ projects onto all the states that have evolved in the infinite past from the reactant, and $f(\boldsymbol{q})$ is a function which defines the dividing surface S via

$$f(\boldsymbol{q}) = 0; \quad \text{on S.} \tag{2.4b}$$

Here, the dividing surface S is given by the intercell boundary:

$$S = ((A, C)) \tag{2.4c}$$

(of course, we consider now the simplified example of the isomerization reaction (see Fig. 1), where the transition point B is isolated on the adiabatic potential energy surface and there are no intermediates in the course of the reaction). From the definition, the intercell boundary is energy-independent. Moreover, the IRC crosses the intercell boundary at right angles and the critical promoting mode of negative force constant at the transition point is exactly isolated from the stable normal modes of positive force constants lying on the intercell boundary [28].

From the 2-dimensional cross-sectional viewpoint, the cell and the intercell boundary are directly visualized using the pattern of reaction [28] as in Fig. 2.

3. Intrinsic Dynamism in the Cell

3.1. *The Intrinsic Reaction Time* (IRT) *and the Accumulation Time* (AT) *of Reaction*

Suppose the n generalized coordinates q^i ($i = 1, \ldots, n$) which describe the variation of the nuclear geometry of the chemically reacting system: the mass-weighted Cartesian coordinates x^t ($t = 1, \ldots, 3N$), N being the number of the nuclei, are then described by

$$x^t = x^t(q^1, \ldots, q^n) \quad (t = 1, \ldots, 3N). \tag{3.1}$$

Using the multidimensional vector [28]

$$\boldsymbol{x} = (x^1, \ldots, x^{3N}), \tag{3.2}$$

the infinitesimal displacement vector of the chemical reaction is given by

$$d\boldsymbol{x} = \boldsymbol{e}_i \, dq^i \tag{3.3}$$

(Einstein's summation convention is adopted in this paper) where

$$\boldsymbol{e}_i = \partial \boldsymbol{x}/\partial q^i \quad (i = 1, \ldots, n). \tag{3.4}$$

The covariant components are found to be

$$d\boldsymbol{x} = \boldsymbol{e}^i \, dq_i. \tag{3.5}$$

The relationship with the contravariant components are represented by

$$dq_i = a_{ij} \, dq^j \quad (i = 1, \ldots, n) \tag{3.6}$$

and hence

$$\boldsymbol{e}_i = a_{ij} \boldsymbol{e}^j \quad (i = 1, \ldots, n), \tag{3.7}$$

where

$$a_{ij} = \boldsymbol{e}_i \cdot \boldsymbol{e}_j = \sum_{t=1}^{3N} (\partial x^t/\partial q^i)(\partial x^t/\partial q^j) \quad (i, j = 1, \ldots, n). \tag{3.8}$$

Likewise, the gradient vector field of U is defined in terms of the generalized coordinate system by

$$\text{grad}\, U = \boldsymbol{e}^i \, \partial U/\partial q^i = \boldsymbol{e}_i \, \partial U/\partial q_i. \tag{3.9}$$

Note that all these vectors are considered to be the vectors in the n-dimensional Riemannian space R_n whose metric tensor is a_{ij} (3.8) [28]. Using the above, the IRC equation [28] is now rewritten as

$$dq_i/ds = (\partial U/\partial q^i)/(dU/ds) \quad (i = 1, \ldots, n), \tag{3.10}$$

where ds is the infinitesimal line element of the meta-IRC, which is represented by the first fundamental form of R_n as

$$ds^2 = a_{ij} \, dq^i \, dq^j, \tag{3.11}$$

and where dU/ds is given by

$$dU/ds = (\partial U/\partial q^i)(dq^i/ds). \tag{3.12}$$

On the other hand, let the intrinsic rate of the chemical reaction along the meta-IRC be $c(x)$ and the intrinsic reaction time (IRT) be t' (the IRT is a measure of the idealized locus of motion along the meta-IRC, which is not identified with the realistic time t), then the infinitesimal vector of the chemical reaction is $c\, dt'$:

$$dx = c\, dt' \tag{3.13}$$

and we have

$$dq_i = c_i\, dt' \quad (i = 1, \ldots, n), \tag{3.14}$$

where

$$c_i = e_i \cdot c \quad (i = 1, \ldots, n). \tag{3.15}$$

Thus, we obtain from (3.10) and (3.14)

$$dq_i/dt' = c_i = (\partial U/\partial q^i)(ds/dt')/(dU/ds) \quad (i = 1, \ldots, n) \tag{3.16}$$

and the locus of the meta-IRC has the form

$$q_i(t') = q_i(t'; x(t'_0)) \quad (i = 1, \ldots, n) \tag{3.17a}$$

$$q^i(t') = q^i(t'; x(t'_0)) \quad (i = 1, \ldots, n), \tag{3.17b}$$

where t'_0 is the initial value of the IRT. This form defines the point transformation using the IRT as the parameter, which forms a one-parameter continuous group of transformations [45]. This group is referred to as G_1.

Now, let us introduce a new time-like parameter τ of G_1 by

$$\tau = \int^{t'} [(ds/dt')/(dU/ds)]\, dt' = \int^{s(t')} ds/(dU/ds), \tag{3.18}$$

which will be referred to as the accumulation time (AT) of reaction. Clearly, the AT is singular at any equilibrium point, where $dU/ds = 0$; hence, we can use the AT as the continuous parameter of G_1 only within the cell or the boundary excluding the equilibrium points. Using the AT, the fundamental equation of G_1 is simplified as

$$dq_i(\tau)/d\tau = \partial U/\partial q^i(\tau) \quad (i = 1, \ldots, n) \tag{3.19}$$

and then, we have the locus of the meta-IRC in the form

$$q_i(\tau) = e^{X(\tau-\tau_0)} q_i(\tau_0) \quad (i = 1, \ldots, n) \tag{3.20a}$$

or

$$q^i(\tau) = e^{X(\tau-\tau_0)} q^i(\tau_0) \quad (i = 1, \ldots, n), \tag{3.20b}$$

where τ_0 is the initial value of the AT, and where X is the generator of G_1 [45]:

$$X = (\partial U/\partial q^i)\, \partial/\partial q_i = (\partial U/\partial q_i)\, \partial/\partial q^i. \tag{3.21}$$

Now, it should be remembered that the IRT may change uniformly along the meta-IRC, while the AT directly depends on the reaction path, which is not uniform in general. In particular, at the stable equilibrium point P_{eq}, we have the limiting value

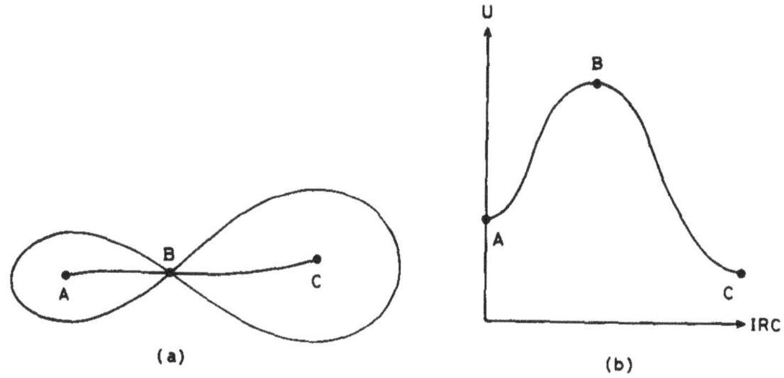

Fig. 1. (a) Pattern of isomerization reaction [28]: A (reactant) → B (transition point) → C (product). (b) Potential barrier along the IRC

of the AT as $\tau = -\infty$ irrespective of the direction of the meta-IRC which accumulates at the P_{eq}. At the transition point, however, we have $\tau = +\infty$ along the IRC and $\tau = -\infty$ along the meta-IRC which belongs to the intercell boundary. This is easily realized by examining the model reaction system undergoing the isomerization reaction (1.1) as is illustrated in Figs. 1, 3. Moreover, at any equilibrium point P_{eq}, the intrinsic rate $\mathbf{c}$ of the reaction is given with respect to the IRT by

$$dx/dt' = \mathbf{c} = \mathbf{v}\, ds/dt'; \quad \text{at } P_{eq}, \tag{3.22}$$

where $\mathbf{v}$ is the vector of the normal vibration [28]. On the other hand, in terms of the AT, the "rate" becomes zero at the P_{eq} as is clear from (3.19):

$$dx/d\tau = 0; \quad \text{at } P_{eq}. \tag{3.23}$$

The heterogeneous behavior of (3.23) is characteristic of the intrinsic reaction dynamism with respect to the AT.

It should be remarked here that the limiting property of the AT at the equilibrium point will play an important role when we discuss the stable limit theorems in Sect. 5.

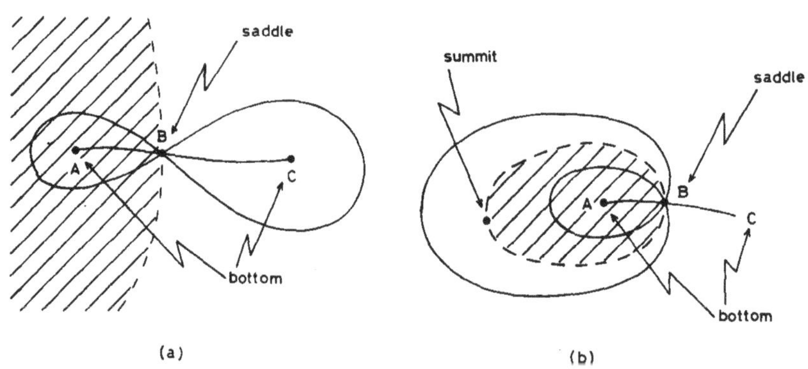

Fig. 2. Cell (A) (shaded region), Cell (C) (plain region), and intercell boundary ((A, C)) (dashed curve) in terms of the patterns of (a) isomerization reaction, and (b) fragmentation reaction [28]

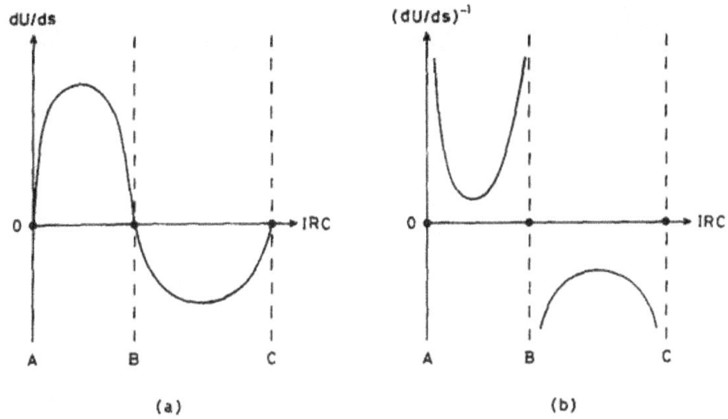

Fig. 3. Path-dependent values of (a) dU/ds, and (b) $(dU/ds)^{-1}$ along the IRC given in Fig. 1

3.2. *The Accumulation Time of Reaction and the Meta*-IRC

In this subsection, we investigate the relationship between the AT, the solution of the meta-IRC, and the adiabatic potential U.

Performing a coordinate transformation, we have the solution of the meta-IRC as [28]

$$dq'_n \neq 0, \qquad dq'_j = 0 \quad (j = 1, \ldots, n-1), \tag{3.24}$$

$$\partial U/\partial q'^n \neq 0, \qquad \partial U/\partial q'^j = 0 \quad (j = 1, \ldots, n-1). \tag{3.25}$$

In other words, the displacement vector along the meta-IRC has the one-dimensional form [28]

$$dx = e'^n \, dq'_n. \tag{3.26}$$

Furthermore, let the succeeding coordinate transformation be

$$q''_n = \int^{q'_n} dq'_n/(\partial U/\partial q'^n), \qquad q''_j = q'_j \quad (j = 1, \ldots, n-1). \tag{3.27}$$

In this q''-coordinate system, we have

$$\partial U/\partial q''^n = 1, \qquad \partial U/\partial q''^j = 0 \quad (j = 1, \ldots, n-1) \tag{3.28}$$

This shows that the q''^n is equivalent to the potential U itself:

$$q''^n = U + \text{const} \tag{3.29}$$

where const is an additive constant.

Now, differentiating

$$U = \text{const}$$

we have

$$0 = dq^i \, \partial U/\partial q^i.$$

This shows that a displacement vector $dx = e_i\, dq^i$ lying on the U = const surface is orthogonal to the gradient vector field grad $U = e^i\, \partial U/\partial q^i$ of U. Therefore, we can choose a coordinate system q''' by making the $n - 1$ coordinates span the equi-potential surface and hence be orthogonal to the q''^n-curve:

$$q'''^n = q''^n = U + \text{const} \quad , \quad q'''^k\,(k = 1, \ldots, n-1) \tag{3.30}$$

span the equi-potential surface.

In this coordinate system, we have

$$a'''_{kn} = a'''^{kn} = 0 \quad (k = 1, \ldots, n-1) \tag{3.31}$$

and the first fundamental form of R_n becomes

$$ds^2 = a'''_{kl}\, dq'''^k\, dq'''^l + a'''_{nn}(dq'''^n)^2 \quad (k, l = 1, \ldots, n-1). \tag{3.32}$$

This shows that the q'''^n-curve is nothing but the normal meta-IRC [28] (see Fig. 4). Then, we have

$$\partial U/\partial q'''^n = 1, \qquad \partial U/\partial q'''^k = 0 \quad (k = 1, \ldots, n-1). \tag{3.33}$$

In this case, the generator X becomes

$$X = \partial/\partial q'''_n \tag{3.34}$$

and the finite equations of the group G_1 are found to be

$$\begin{aligned} q'''_n(\tau) &= q'''_n(\tau_0) + \tau - \tau_0, \\ q'''_k(\tau) &= q'''_k(\tau_0) \quad (k = 1, \ldots, n-1) \end{aligned} \tag{3.35}$$

or, the infinitesimal components are

$$dq'''_n = d\tau, \qquad dq'''_k = 0 \quad (k = 1, \ldots, n-1). \tag{3.36}$$

The meta-IRC is now represented by the q'''_n-curve and is equivalent to the AT itself; indeed, (3.26) is written as

$$dx = e'''^n\, d\tau, \tag{3.37}$$

where e'''^n is the tangent vector of the meta-IRC. Also, using the relationship for the normal meta-IRC [28], we have

$$dq'''^n = a'''^{nn}\, dq'''_n = a'''^{nn}\, d\tau, \qquad dq'''^k = 0 \quad (k = 1, \ldots, n-1). \tag{3.38}$$

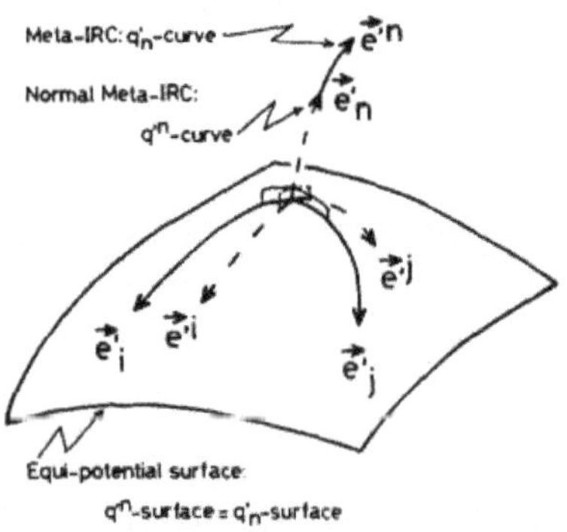

Fig. 4. Geometry of the reaction path represented in terms of the normal coordinate system

On the other hand, we have from (3.30)

$$dq''^{n} = dU. \tag{3.39}$$

Hence, we obtain the functional relationship of U with τ as

$$dU = a''^{nn}\, d\tau. \tag{3.40}$$

4. Quantum Mechanical Representation of the Local Intrinsic Dynamism

In [28], we have introduced the local coordinates Δu^i ($i = 1, \ldots, n$) in order to describe the local structure of the chemically reacting system and besides, as the more convenient local coordinates, the local normal coordinates Δy^α ($\alpha = 1, \ldots, n$) or the reduced local normal coordinates $\Delta \bar{y}'^\alpha$ ($\alpha = 1, \ldots, n$); remark that $(a'_{nn})^{1/2}\Delta u'^n$ in [28] is now conveniently written as $\Delta \bar{y}'^n$ in terms of the normal coordinate system in which the q'^n-coordinate is the normal meta-IRC (see Fig. 4). By considering the Δy^α or $\Delta \bar{y}'^\alpha$ ($\alpha = 1, \ldots, n$) as the independent variables, the quantum mechanical Hamiltonian H of the system will be locally given by

$$H(x + \Delta x) = (1/2) \sum_{\alpha=1}^{n} p_\alpha^2 + U(x + \Delta x) \tag{4.1a}$$

$$= (1/2) \sum_{\alpha=1}^{n} \bar{p}_\alpha'^2 + U(x + \Delta x) \tag{4.1b}$$

where

$$p_\alpha = -i\hbar\, \partial/\partial \Delta y^\alpha \tag{4.2a}$$

or

$$\bar{p}'_\alpha = -i\hbar\, \partial/\partial \Delta \bar{y}'^\alpha. \tag{4.2b}$$

It should be noted that, unlike the global Hamiltonian obtained by Podolsky [15], these local Hamiltonians (4.1a) or (4.1b) have *separable kinetic energy operators*. Then, the adiabatic potential is also represented in the alternative forms as follows.

A) Δy-coordinate system. In this case, we have

$$U(x + \Delta x) = \sum_{r=0}^{\infty} (1/r!) \sum_{\alpha_1, \cdots, \alpha_r = 1}^{n} \Delta y^{\alpha_1} \cdots \Delta y^{\alpha_r} D^{(y)}_{\alpha_1 \cdots \alpha_r} U \tag{4.3a}$$

where the ($r = 0$)-term is nothing but $U(x)$ and the others are given by

$$D^{(y)}_{\alpha_1 \cdots \alpha_r} = v^{i_1}_{(\alpha_1)} \cdots v^{i_r}_{(\alpha_r)} D_{i_1 \cdots i_r}. \tag{4.3b}$$

In this expression, $D_{i_1 \cdots i_r}$ is the differential operator, satisfying

$$D_{i_1 \cdots i_r i_{r+1}} = \partial D_{i_1 \cdots i_r}/\partial q^{i_{r+1}} - \sum_{p=1}^{r} \Gamma^k_{i_p i_{r+1}} D_{i_1 \cdots i_{p-1} k i_{p+1} \cdots i_r}, \tag{4.4}$$

where Γ^k_{ij} is the Christoffel symbol of the second kind [28]. For example, we have [28]

$$D_i = \partial/\partial q^i, \tag{4.5a}$$

$$D_{ij} = \partial^2/\partial q^i\,\partial q^j - \Gamma^k_{ij}\,\partial/\partial q^k. \tag{4.5b}$$

The explicit form of (4.3) has been given in [28] up to the second order of the expansion.

B) $\Delta\bar{y}'$-coordinate system. In this case, we have

$$U(x + \Delta x) = \sum_{r=0}^{\infty} (1/r!) \sum_{\alpha_1,\cdots,\alpha_r=1}^{n} \Delta\bar{y}'^{\alpha_1} \cdots \Delta\bar{y}'^{\alpha_r} D^{(\bar{g}')}_{\alpha_1\cdots\alpha_r} U, \tag{4.6a}$$

$$D^{(\bar{g}')}_{\alpha_1\cdots\alpha_r} = \bar{v}'^{i_1}_{(\alpha_1)} \cdots \bar{v}'^{i_r}_{(\alpha_1)} D'_{i_1\cdots i_r}, \tag{4.6b}$$

where the reduced vector components $\bar{v}'^{i}_{(\alpha)}$ ($\alpha, i = 1, \ldots, n-1$) are given by diagonalizing the Hessian matrix in the equi-potential surface [28] and the others are now defined by

$$\bar{v}'^{n}_{(\alpha)} = \bar{v}'_{(\alpha)n} = 0 \qquad (\alpha = 1, \ldots, n-1), \tag{4.6c}$$

$$\bar{v}'^{i}_{(n)} = (a'_{nn})^{-1/2}\delta_{ni} \quad (i = 1, \ldots, n), \tag{4.6d}$$

$$\bar{v}'_{(n)i} = (a'_{nn})^{1/2}\delta_{ni} \quad (i = 1, \ldots, n). \tag{4.6e}$$

Neglecting the higher-order terms, we have

$$U(x + \Delta x) = U(x) + \Delta\bar{y}'^n D^{(\bar{g}')}_n U + (1/2) \sum_{\alpha=1}^{n} \bar{\mu}_\alpha (\Delta\bar{y}'^\alpha)^2 + \sum_{\alpha=1}^{n-1} \Delta\bar{y}'^n \Delta\bar{y}'^\alpha D^{(\bar{g}')}_{n\alpha} U + O((\Delta\bar{y}')^3). \tag{4.7}$$

In this case, although the off-diagonal parts of the second-order terms cannot be eliminated, we have only one linear term of $\Delta\bar{y}'^n$ in the expansion (4.7). Thus, the Hamiltonian in terms of the $\Delta\bar{y}'$-coordinate system is more appropriate for the description of 1) the one-dimensional promotion of the chemical reaction along the meta-IRC, and 2) the residual vibrations confined in the equi-potential surface which is perpendicular to the meta-IRC. If the requirement of the vibrational adiabaticity is fulfilled, then the two representations of the Hamiltonian, (4.1a) and (4.1b), of course become identical. As a whole, it should be emphasized that the perturbation-free separable kinetic energy operator is more tractable than Podolsky's Hamiltonian with respect to the description of the local quantum mechanics of the chemically reacting system.

5. Stable Limit Theorems

In this section, the limiting behavior of the meta-IRC at an equilibrium point P_{eq} is discussed.

From (3.10), we have by virtue of l'Hospital's theorem

$$dq_i/ds = (\partial^2 U/\partial q^i\,\partial q^j)(dq^j/ds)/(d^2U/ds^2); \quad \text{at } P_{eq} \tag{5.1}$$

and therefore

$$(\partial^2 U/\partial q^i \partial q^j - (d^2U/ds^2)a_{ij})\, dq^j/ds = 0; \quad \text{at } P_{eq}. \tag{5.2}$$

This shows that the motion along the meta-IRC should become one of the normal vibrations at the P_{eq}; then, the force constant is given by d^2U/ds^2.

Since the stable equilibrium point is the center of the cell where the idealized reaction may start or finish, it will be of primary importance to investigate which normal vibrational mode the meta-IRC should converge to, or which normal vibration the reaction may make useful for the initializing motion of the reaction.

We show that the weakest normal mode has the unique nature compared with the stronger normal modes in terms of the boundary condition with respect to the intrinsic dynamism at the stable equilibrium point.

In the present treatment, the characteristic limiting value of the AT plays an important role: τ become $-\infty$ if the meta-IRC converges to the stable mode of positive force constant, while τ becomes $+\infty$ if the meta-IRC converges to the unstable normal mode of negative force constant. This heterogeneous character of the AT has been clearly distinguished from the uniform character of the IRT in the text (see the discussions following (3.18)). Provided that we were working in an Euclidean space, the analogous situation is well known: in the usual textbook approach [35, 46] and recently in the study of the orthogonal trajectories of the electron density by Collard and Hall [47], one may find the infinite character of the AT-like parameter, though without showing the *ab initio* reasoning of the character itself.

Now, in terms of the local normal coordinates Δy^α ($\alpha = 1, \ldots, n$), the IRC equation is reduced from (3.19) to

$$d\Delta y^\alpha/d\tau = \partial U(x + \Delta x)/\partial \Delta y^\alpha \quad (\alpha = 1, \ldots, n), \tag{5.3}$$

where the r.h.s. of this equation is evaluated by (4.3). In the neighborhood of the equilibrium point, the local IRC equation (5.3) can then be approximated to be

$$d\Delta y^\alpha/d\tau \sim \mu_\alpha \Delta y^\alpha \quad (\alpha = 1, \ldots, n). \tag{5.4}$$

Neglecting the higher order contributions of unharmonicity, the solution becomes

$$\Delta y^\alpha \sim c_\alpha \exp(\mu_\alpha \tau) \quad (\alpha = 1, \ldots, n), \tag{5.5}$$

where c_α ($\alpha = 1, \ldots, n$) are the integral constants which govern the boundary condition of the meta-IRC at the equilibrium point. Then, we obtain

$$d\Delta y^\alpha/ds \sim c_\alpha \mu_\alpha \exp(\mu_\alpha \tau) \Big/ \left[\sum_{\beta=1}^{n} c_\beta^2 \mu_\beta^2 \exp(2\mu_\beta \tau) \right]^{1/2} \quad (\alpha = 1, \ldots, n). \tag{5.6}$$

Note that at the *stable* equilibrium point, τ becomes $-\infty$, and then, by omitting the orientation of the vector from our consideration, we have

$$d\Delta y^\alpha/ds \sim (c_\alpha \mu_\alpha / c_{\alpha=\min} \mu_{\alpha=\min}) \exp[(\mu_\alpha - \mu_{\alpha=\min})\tau] \quad (\alpha = 1, \ldots, n), \tag{5.7}$$

where $\alpha = \min$ designates *the normal mode of minimal force constant whose contribution is non-zero*. It follows that

$$d\Delta y^{\alpha}/ds = 1; \quad \text{if } \alpha = \min$$
$$= 0; \quad \text{otherwise.} \tag{5.8}$$

Hence, we have

$$dq^{i}/ds = v^{i}_{(\alpha=\min)}, \quad \text{i.e.,} \quad dx/ds = v_{(\alpha=\min)}; \quad \text{at } \tau = -\infty. \tag{5.9}$$

This shows that any idealized locus of the chemical reaction proceeding along the meta-IRC should converge to *the weakest normal mode that is available* at the stable equilibrium point. Using this important property, we shall prove the theorem:

Theorem 1. It is *most probable* for any intrinsic chemical reaction promoting along the meta-IRC to converge to *the weakest normal mode* at the stable equilibrium point.

This is referred to as the stable limit theorem 1.

Proof. Any intrinsic chemical reaction in the Cell (P_{eq}) is represented by a point in the space spanned by c_{α} ($\alpha = 1, \ldots, n$) which specify the boundary condition at the stable equilibrium point P_{eq}: this space is referred to as V. Suppose the force constants at the P_{eq} be given in the sequence of $\mu_i > \mu_j$ if $i < j$:

$$\mu_1 > \mu_2 > \cdots > \mu_n > 0 .$$

Then, it is found that the meta-IRC should converge to the weakest normal mode n unless $c_n = 0$. In other words, the meta-IRC cannot converge to the weakest normal mode n if the representative point of the meta-IRC belongs to the ($c_n = 0$)-hypersurface in V. Clearly, the ($c_n = 0$)-hypersurface has *measure zero* in V; therefore, we can find a meta-IRC which converges to the weakest normal mode n *almost everywhere* in V. This proves the theorem.

Moreover, using the property that the meta-IRC is perpendicular to the equi-potential surface in the cell, we have

Theorem 2. It is *most probable* that the normal modes on the equi-potential surface along the meta-IRC should converge to *the non-weakest normal modes* at the stable equilibrium point.

This is referred to as the stable limit theorem 2. The proof is trivial.

To summarize, we conclude that almost all the intrinsic chemical reactions, in their idealized loci, start by using the weakest normal vibration at the stable equilibrium point, proceed along the IRC accompanying the stronger vibrations in the equi-potential surface (in the initial stage of the reaction), and reach the transition point to give the critical normal vibration of negative force constant; after overriding the transition point, the reaction along the IRC is most likely to converge to the weakest normal mode of the product.

Using a model potential surface, the applicability of the stable limit theorems will be clearly shown in the next section.

Furthermore, at any saddle, we have two kinds of normal modes: the ($\tau = -\infty$)-modes and the ($\tau = +\infty$)-modes corresponding to the positive force constants and the negative force constants, respectively (at the summit of the potential surface, we have only the ($\tau = +\infty$)-modes). If the converged normal mode belongs to the former set of ($\tau = -\infty$)-modes, then we have (5.9) in the same way. On the other hand, if the converged normal mode belongs to the latter set of ($\tau = +\infty$)-modes, then we have a quite different result:

$$dq^i/ds = v^i_{(\alpha=\max)}, \quad \text{i.e.,} \quad dx/ds = \boldsymbol{v}_{(\alpha=\max)}; \quad \text{at } \tau = +\infty, \tag{5.10}$$

where $\alpha = \max$ designated the normal mode of *maximal* (remark that this mode is the *weakest* in the ($\tau = +\infty$)-modes) force constant whose contribution is non-zero. Note that at any unstable equilibrium point, these two kinds of mixed characters, (5.9) and (5.10), will be useful for the analysis of the intrinsic dynamism according as the meta-IRC lies on the intercell boundary or not; the mixed characters themselves may be arranged to be referred to as the unstable limit theorem.

In the study of chemical reactions that are selected by the excitation of a single normal mode at a stable equilibrium point [48], the implications of the stable limit theorems may play an important role. Also, then, the effect of the conservation of symmetry may be significant, particularly for the distribution of the initially activated vibrational modes [21, 49, 50]. Moreover, the detailed information with respect to the geometry of the adiabatic potential energy surface itself will be useful when we consider the nonadiabatic reactions where the state degeneracy gives the significant effect in the geometry of the surface. Finally, it should be noted that the force-theoretical arguments [51, 52] are helpful for the IRC approach; the electronic properties of the chemically reacting system are intimately related to the variation of the gradient field of the adiabatic potential energy surface.

6. Example

Here, we give a model of the isomerization reaction (1.1) and elucidate some of the results of the present theory.

The shape of the model potential is depicted in Fig. 6 of [28]. The solution of the meta-IRC is represented by [28]

$$\begin{aligned} x_1 &= 0, \tfrac{1}{2}, 1 \\ x_2 &= c \times [|x_1| \times |x_1 - 1| \times |x_1 - 1/2|^{-2}]^{a^2/b^2}; \quad a > b, \end{aligned} \tag{6.1}$$

where x_1 and x_2 are the generalized coordinates in an Euclidean coordinate space, and where c is a constant.

We shall examine the behavior of the reduced normal vibrations on the equi-potential surface along the meta-IRC in Cell (A); in Cell (C), we have the mirror image of the situation in Cell (A). The eigenvector and the eigenvalue of the Hessian matrix on the equi-potential surface are obtained in Cell (A) as follows:

$$\bar{\boldsymbol{v}}_{(1)} = (f/(f^2 + g^2)^{1/2}, \quad -g/(f^2 + g^2)^{1/2}), \tag{6.2}$$

$$\bar{\mu}_1 = [(12/a^2)(x_1 - \alpha_+)(x_1 - \alpha_-)f^2 + (2/b^2)g^2]/(f^2 + g^2), \tag{6.3}$$

where $\alpha_\pm$ are the certain constants [28], and where

$$f = (2/b^2)c/(2/a^2)x_1(x_1 - 1)(2x_1 - 1), \tag{6.4}$$

$$g = c/x_2. \tag{6.5}$$

Note that the orientation of the vector is omitted from our consideration. Now, at the center A of Cell (A), we have (see Fig. 5a)

$$\bar{v}_{(1)} = (0, 1); \quad \text{at A}, \tag{6.6}$$

$$\bar{\mu}_1 = 2/b^2; \quad \text{at A}. \tag{6.7}$$

This is the hardest normal mode at A, and hence this confirms the stable limit theorem 2. Interestingly, this normal mode is invariably transferred in a parallel way along the IRC (see Fig. 5b):

$$\bar{v}_{(1)} = (0, 1); \quad x_2 = 0, \tag{6.8}$$

$$\bar{\mu}_1 = 2/b^2; \quad x_2 = 0. \tag{6.9}$$

On the other hand, the isolated solution of the meta-IRC that converges smoothly to the hardest normal mode at A is

$$x_1 = 0. \tag{6.10}$$

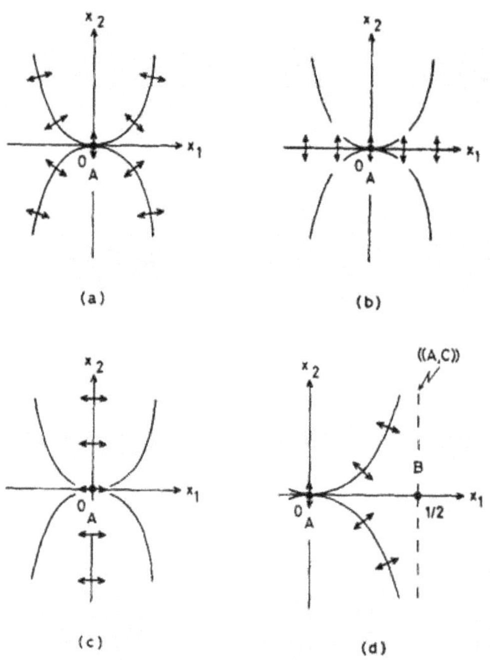

Fig. 5. Vectors ↔ of the reduced normal vibrations on the equi-potential surface along (**b**) the IRC, (**c**) the isolated meta-IRC which is rarely found on account of the stable limit theorem 2, and (**a, d**) all the other meta-IRC's in the Cell (A). Some of the representative meta-IRC's are drawn by solid curves and lines. Also, (**d**) shows the behavior of the reduced normal vibration in the limit at the intercell boundary ((A, C))

Along this meta-IRC, in order to make the value of x_2 finite, the constant c should formally diverge to infinite. Then we have (see Fig. 5c)

$$\bar{v}_{(1)} = (1, 0); \quad x_1 = 0, \tag{6.11}$$

$$\bar{\mu}_1 = 2/a^2; \quad x_1 = 0. \tag{6.12}$$

This is of course the weakest normal mode at A. Moreover, it may be worthwhile to examine the behavior of the reduced normal vibration in the limit at the intercell boundary ((A, C)). Along the IRC, we have (6.8) and (6.9), while the others satisfy (see Fig. 5d)

$$\bar{v}_{(1)} \to (1, 0); \quad x_1 \to 1/2, \quad x_2 \neq 0, \tag{6.13}$$

$$\bar{\mu}_1 \to -1/a^2; \quad x_1 \to 1/2, \quad x_2 \neq 0. \tag{6.14}$$

This is nothing but the critical normal mode of negative force constant at the transition point B [28]. This mode may act as the escaping mode of the intrinsic reactive flux that oozes from the Cell (A) to the adjacent Cell (C) through the intercell boundary ((A, C)).

Acknowledgement. This work was supported by a Grant-in-Aid for Scientific Research from the Ministry of Education of Japan.

References

1. Glasstone, S., Laidler, K., Eyring, H.: The theory of rate processes. New York: McGraw-Hill 1941
2. McIver, Jr., J. W., Komornicki, A.: Chem. Phys. Letters **10**, 303 (1971); J. Am. Chem. Soc. **94**, 2625 (1972)
3. Halgren, T. A., Pepperberg, I. M., Lipscomb, W. N.: J. Am. Chem. Soc. **97**, 1248 (1975)
4. Peterson, M. R., Csizmadia, I. G., in: Application of MO theory in organic chemistry, Csizmadia, I. G., ed. Amsterdam: Elsevier 1977
5. Miller, W. H.: J. Chem. Phys. **61**, 1823 (1974)
6. Miller, W. H.: Accounts Chem. Res. **9**, 306 (1976)
7. Miller, W. H.: J. Chem. Phys. **65**, 2216 (1976)
8. Keck, J. C.: Advan. Chem. Phys. **13**, 85 (1967)
9. Fischer, S. F., Hofacker, G. L., Seiler, R.: J. Chem. Phys. **51**, 3951 (1969)
10. Hofacker, G. L., Levine, R. D.: Chem. Phys. Letters **9**, 617 (1971)
11. Hofacker, G. L., Levine, R. D.: Chem. Phys. Letters **15**, 165 (1972)
12. Hofacker, G. L., Rösch, N.: Ber. Bunsenges. Physik. Chem. **77**, 661 (1973)
13. Hofacker, G. L., Michel, K. W.: Ber. Bunsenges. Physik. Chem. **78**, 174 (1974)
14. Hofacker, G. L., Levine, R. D.: Chem. Phys. Letters **33**, 404 (1975)
15. Podolsky, B.: Phys. Rev. **32**, 812 (1928)
16. Hofacker, G. L.: Z. Naturforsch. **18a**, 607 (1963)
17. Marcus, R. A.: J. Chem. Phys. **45**, 4493 (1966)
18. Marcus, R. A.: J. Chem. Phys. **45**, 4500 (1966)
19. Marcus, R. A.: J. Chem. Phys. **49**, 2610 (1968)
20. Marcus, R. A.: J. Chem. Phys. **49**, 2617 (1968)
21. Fukui, K.: J. Phys. Chem. **74**, 4161 (1970)
22. Fukui, K., in: The world of quantum chemistry, Daudel, R., Pullman, B. eds. Dordrecht, Holland: D. Reidel Publ. Co. 1974
23. Fukui, K., Kato, S., Fujimoto, H.: J. Am. Chem. Soc. **97**, 1 (1975)

24. Kato, S., Fukui, K.: J. Am. Chem. Soc. **98**, 6395 (1976)
25. Kato, S., Kato, H., Fukui, K.: J. Am. Chem. Soc. **99**, 684 (1977)
26. Ishida, K., Morokuma, K., Komornicki, A.: J. Chem. Phys. **66**, 2153 (1977)
27. Joshi, B. D., Morokuma, K.: J. Chem. Phys. **67**, 4880 (1977)
28. Tachibana, A., Fukui, K.: Theoret. Chim. Acta (Berl.) **49**, 321 (1978)
29. Brickmann, J.: Z. Naturforsch. **28a**, 1759 (1973)
30. Brickmann, J.: Intern. J. Quantum Chem. **S7**, 427 (1973)
31. Russegger, P., Brickmann, J.: Chem. Phys. **62**, 1086 (1975)
32. Russegger, P., Brickmann, J.: Chem. Phys. Letters **30**, 276 (1975)
33. Brickmann, J.: Z. Naturforsch. **30a**, 1730 (1975)
34. Russegger, P., Brickmann, J.: J. Chem. Phys. **66**, 1 (1977)
35. Hirsch, M. W., Smale, S.: Differential equations, dynamical systems, and linear algebra. New York: Academic Press 1974
36. Bader, R. F. W., Beddall, P. M.: J. Chem. Phys. **56**, 3320 (1972)
37. Bader, R. F. W., Beddall, P. M., Peslak, Jr., J.: J. Chem. Phys. **58**, 557 (1973)
38. Bader, R. F. W., Runtz, G. R.: Mol. Phys. **30**, 117 (1975)
39. Bader, R. F. W.: Accounts Chem. Res. **8**, 34 (1975)
40. Runtz, G. R., Bader, R. F. W., Messer, R. R.: Can. J. Chem. **55**, 3040 (1977)
41. Srebrenik, S., Bader, R. F. W.: J. Chem. Phys. **63**, 3945 (1975)
42. Srebrenik, S., Bader, R. F. W., Dang, T. T. N.: J. Chem. Phys. **68**, 3667 (1978)
43. Bader, R. F. W., Srebrenik, S., Dang, T. T. N.: J. Chem. Phys. **68**, 3680 (1978)
44. Thom, R.: Structural stability and morphogenesis. London: Benjamin 1975
45. Eisenhart, L. P.: Riemannian geometry. Princeton: Princeton Univ. Press 1963
46. Pontrjagin, L.: Ordinary differential equations. New York: Addison-Wesley 1962
47. Collard, K., Hall, G. G.: Intern. J. Quantum Chem. **12**, 623 (1977)
48. Ambartsumian, R. V., Letokhov, V. S., in: Chemical and biochemical applications of lasers, vol. 3, Moore, C. B. ed. New York: Academic Press 1977
49. Bader, R. F. W.: Mol. Phys. **3**, 137 (1960)
50. Bader, R. F. W.: Can. J. Chem. **40**, 1164 (1962)
51. Deb, B. M.: Rev. Mod. Phys. **45**, 22 (1973)
52. Nakatsuji, H.: J. Am. Chem. Soc. **96**, 24; **96**, 30 (1974)

Received August 29, 1978

THEORETICA CHIMICA ACTA
© by Springer-Verlag 1979

Theoret. Chim. Acta (Berl.) 51, 275—296 (1979)

Intrinsic Field Theory of Chemical Reactions*

Akitomo Tachibana and Kenichi Fukui

Department of Hydrocarbon Chemistry, Faculty of Engineering, Kyoto University, Kyoto, Japan

An intrinsic principle of least action is presented for the intrinsic dynamism of chemical reactions. Then, as the stationary trajectory, a meta-IRC (intrinsic reaction coordinate) draws a geodesic curve in a rigged Riemannian space. This establishes a geodesic law for the intrinsic dynamism. Moreover, a diagrammatic perturbation theory is formulated for the intrinsic dynamism, and a dynamical interaction between a chemically reacting system and a background system is investigated. Then, the structural stability of the system is discussed using a new concept of the dynamical potential field. An example is given in order to elucidate the present theory.

Key words: Intrinsic field theory – Intrinsic reaction coordinate (IRC) approach – Intrinsic principle of least action – Geodesic law – Structural stability – Dynamical potential field

1. Introduction

The ultimate goal of modern theoretical chemistry is supposed to be the establishment and examination of the foundation of the principle of chemical reactions. Quantum chemistry provides not only a theoretical framework for this activity, but it also provides a scheme of calculation to make a quantitative assessment of the theory. In particular, recent developments in computational quantum chemistry have acquired an outstanding progress in the computation of the adiabatic potential hypersurfaces of chemically reacting systems, and the development of molecular beam experiments has added direction and impetus. The determination of the hypersurface itself is one of the most difficult computational problems, but there

* Dedicated to Prof. Hermann Hartmann on the occasion of his 65th birthday.

0040-5744/79/0051/0275/$04.40

still remains the problem of how the chemically reacting system evolves through it. In this connection, a significant advance came with the introduction of the frontier orbital theory [1] and the Woodward–Hoffmann rules [2] with respect to stereo-selection and orientation of chemical reactions.

In the present paper, however, we shall direct our attention at another approach to chemical reactions. This is the IRC approach presented by Fukui [3] and developed by Fukui *et al.* [4–7] and Morokuma *et al.* [8, 9]; with a potential surface established, we seek the IRC through it, and arrive at the reaction rate constant by a suitable computational technique. Since the differential equation postulated by Fukui for the definition of the IRC is based on the gradient field of the potential [3], study on the geometrical properties of the chemically reacting system is central to the theoretical foundation of the IRC approach. Recently, from the detailed geometrical study of the adiabatic potential hypersurface, the intrinsic dynamism inherent to the IRC approach has been developed [10, 11]. In particular, by introducing a concept of meta-IRC, it has been clarified that a stable equilibrium point, that is the point of an equilibrium structure of the system, is realized as an accumulation point of the meta-IRC's which tie in its neighborhood a set of non-equilibrium points of the potential. This leads to a natural concept of cell structure of the chemically reacting system, where an intercell boundary appears by which an initial reactant cell and a final product cell are discriminated [11]. The IRC is then given by a particular meta-IRC which smoothly connects the terminal cells of the chemical reaction through the transition point located on the intercell boundary [11]. Stable limit theorems with respect to the intrinsic nature of the normal vibrations at the equilibrium point are also characteristic of the intrinsic dynamism [11].

On the other hand, as is well known, the fundamental principles of mechanics are formulated in terms of variational principles. The most famous one is the principle of least action presented by Maupertuis, who originally attempted to obtain for the corpuscular theory of light a theorem analogous to Fermat's principle of least time [12]. Maupertuis' principle of least action provides a real dynamical trajectory with a stationary character: for holonomic conservative system, an action integral evaluated along a real trajectory has the stationary value against deformation of the integral path with respect to a virtual trajectory adjacent to the real trajectory, with the time being correlated to the coordinates in such a way as to satisfy the conservation law of energy [12]. Maupertuis' principle of least action can be reformulated into Hamilton's principle, where, for a virtual trajectory, the time has the same terminal values as the real trajectory [12].

Likewise, from a geometrodynamical viewpoint, a locus of a trajectory satisfies a geodesic law: a trajectory draws a geodesic curve in a rigged Riemannian space. For example, Einstein's geodesic variational principle for the world line is the most famous; in the relativistic theory of gravitation, the world line draws a geodesic curve in a rigged Riemannian space whose metric is determined by the distribution of the source of gravitation [13]. Also, in the non-relativistic limit, Jacobi's geodesic principle is available; note that this is a reformulation of Maupertuis' principle of least action [12]. In this connection, Einstein's geodesic variational principle for the world line can be considered to be a reformulation of a relativistic principle of least action [13].

In the first part of the present paper, we attempt to establish two kinds of fundamental variational principles for the intrinsic dynamism of chemical reactions. One is the *intrinsic principle of least action*, (2.5), and the other is the *geodesic variational principle*, (2.13). The former "dynamical" variational principle states that the "extended" action integral as defined by (2.2) has a stationary value for any part of an actual meta-IRC in the cell, as compared with neighboring paths which have the same terminal points and the same accumulation time (AT) τ [11] as the actual meta-IRC. The latter "geometrical" variational principle states that a meta-IRC is a geodesic curve in a rigged Riemannian space: this space is hereafter referred to as R_n^*. The relationship between these "dynamical" and "geometrical" variational principles is also shown, as in (2.20).

On the other hand, everything in the universe is interconnected and correlated with everything else. If we single out a chemically reacting system, we have divided the universe into two parts. All we have not singled out will be called the environment or the background. The background dependence of a chemically reacting system may be measured as a distortion of the rigged Riemannian space R_n^*, because a meta-IRC should draw a geodesic curve in R_n^* even under the influence of the background effect. (A particular perturbation may destroy the geodesic law, but such a case is not treated in the present paper.) Since the metric tensor g_{ij} of R_n^* is determined by the adiabatic potential U and the metric tensor a_{ij} of the primordial Riemannian space R_n, as is shown in (2.11) and (2.12), it may be of primary importance how to correlate the background effect into the perturbation of U and a_{ij} in terms of the intrinsic dynamism.

Thus, in the second part of the present paper, we shall develop a perturbation theory of the intrinsic dynamism of chemical reactions, which allows the perturbational treatment of U and a_{ij}. Then, a diagrammatic perturbation series is found for the perturbed solution of meta-IRC. Using this perturbation theory, a mechanism of dynamical energy transfer between a chemically reacting system and a background system will be discussed in terms of a new concept of the dynamical potential field, where the distortion of the metric tensor a_{ij} plays an important role. The dynamical potential field is found to be intimately related with the stability criterion of a chemically reacting system.

2. Variational Principles

2.1. *Intrinsic Principle of Least Action*

Let the "extended" action A with respect to a time-like parameter θ be defined by

$$A = \int 2T(q^i, dq^i/d\theta)\, d\theta \tag{2.1}$$

with

$$2T = a_{ij}(dq^i/d\theta)(dq^j/d\theta), \qquad (2.2)$$

where $q^i(i = 1, \ldots, n)$ are the n generalized coordinates in the configuration Riemannian space R_n which describes the nuclear geometry of the system as a function of θ, T is the "extended" kinetic energy in terms of the "extended" velocities $dq^i/d\theta$ $(i = 1, \ldots, n)$, and a_{ij} is the metric tensor of R_n. The extended action A can be interpreted as the classical "action" if we multiply an appropriate scaling factor, if necessary.

As is well known, Maupertuis' principle of least action states that, along a real trajectory, the action is stationary with respect to virtual variation of the trajectory [12]:

$$\delta A = 0 \text{ along non-relativistic trajectory with } \theta = t, \qquad (2.3)$$

where t is the realistic time. Likewise, Einstein's geodesic variational principle for the world line can be put into a form of relativistic principle of least action [13]:

$$\delta A = 0 \text{ along the world line with } \theta = s_w, \qquad (2.4)$$

where s_w is the length of the world line chosen as a parameter of the world line.

Now, in the case of the intrinsic dynamism with respect to the AT τ, we can show that an "intrinsic" principle of least action holds as

$$\delta A = 0 \text{ along the meta-IRC with } \theta = \tau. \qquad (2.5)$$

More precisely, the extended action A has a stationary value for any part of a meta-IRC in the cell, as compared with neighboring paths between the same termini for which the AT is correlated to the coordinates in such a way as to have same terminal values.

Proof of the intrinsic principle of least action. First, we shall prove that the extended kinetic energy T relates with the AT-dependent behavior of the adiabatic potential U along a meta-IRC in the form:

$$2T = dU/d\tau \text{ along the meta-IRC.} \qquad (2.6)$$

Proof. The AT-dependence of U is represented by

$$dU/d\tau = (\partial U/\partial q^i)(dq^i/d\tau). \qquad (2.7)$$

Using the IRC equation, of the form (3.19) in [11], the first derivative of U is given by

$$\partial U/\partial q^i = a_{ij}\, dq^j/d\tau. \qquad (2.8)$$

Substituting (2.8) into (2.7), we have (2.6). Second, we shall prove that a meta-IRC satisfies the Euler equation of the variational problem (2.5) as

$$(d/d\tau)[\partial(2T)/\partial(dq^i/d\tau)] - \partial(2T)/\partial q^i = 0 \text{ along the meta-IRC.} \qquad (2.9)$$

Proof. Using (2.6) and (2.7), we have $2T = (\partial U/\partial q^i)(dq^i/d\tau)$, and then we obtain (2.9) immediately:

$$(d/d\tau)[\partial(2T)/\partial(dq^i/d\tau)] = (d/d\tau)[\partial U/\partial q^i]$$
$$= (\partial^2 U/\partial q^i\,\partial q^j)(dq^j/d\tau)$$
$$= \partial(2T)/\partial q^i.$$

As a whole, the intrinsic principle of least action is proved.

Moreover, we have obtained from (2.1) and (2.6) the stationary value of the extended action A along meta-IRC; this is the increment of the potential energy from the initial terminus to the final terminus:

$$A = \Delta U \text{ along the meta-IRC.} \tag{2.10}$$

2.2. Geodesic Variational Principle

Here, we shall show that a meta-IRC is a geodesic curve in the rigged Riemannian space R_n^* whose metric tensor is given by

$$g_{ij} = \Delta_1 U a_{ij} \tag{2.11}$$

where $\Delta_1 U$ is Beltrami's differential parameter of the first kind with respect to the adiabatic potential U:

$$\Delta_1 U = a^{ij}(\partial U/\partial q^i)(\partial U/\partial q^j). \tag{2.12}$$

Then, a meta-IRC satisfies the geodesic variational principle:

$$\delta \int ds_g = 0 \text{ along the meta-IRC,} \tag{2.13}$$

where the line element ds_g of the meta-IRC is represented by using the first fundamental form of R_n^* as

$$ds_g^2 = g_{ij}\,dq^i\,dq^j. \tag{2.14}$$

Note that the relationship with the primordial configuration Riemannian space R_n is found to be

$$ds_g^2 = \Delta_1 U\,ds^2, \tag{2.15}$$

where ds^2 is the first fundamental form of R_n:

$$ds^2 = a_{ij}\,dq^i\,dq^j. \tag{2.16}$$

The conspicuous property of the meta-IRC as a stationary curve is illustrated in Fig. 1.

Proof of the geodesic variational principle. First, we shall prove that Beltrami's differential parameter of the first kind (2.12) is given along a meta-IRC by

$$\Delta_1 U = (dU/ds)^2 \text{ along the meta-IRC.} \tag{2.17}$$

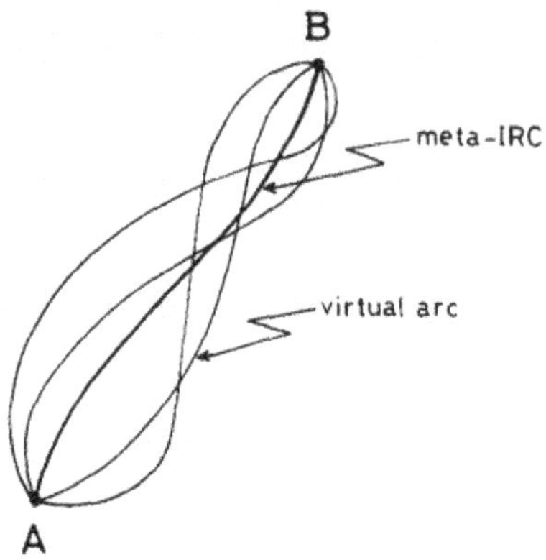

Fig. 1. Meta-IRC draws a geodesic curve ($A \to B$) in the rigged Riemannian space R_n^*

Proof. Using the IRC equation, of the form (3.10) in [11], we have

$$\partial U/\partial q^i = (dU/ds)a_{ij}(dq^j/ds). \tag{2.18}$$

Substituting (2.18) into (2.12) and using (2.16), we have (2.17). Second, we shall prove

$$ds_g^2 = (2T\, d\tau)^2 \text{ along the meta-IRC.} \tag{2.19}$$

Proof. Substituting (2.17) into (2.15) and using (2.6), we have (2.19). This proves that the variational principles (2.5) and (2.13) are identical along the meta-IRC in the cell; and therefore, the meta-IRC satisfies the geodesic variational principle, (2.13), as well as the intrinsic principle of least action, (2.5).

It should be noted that this geometrical variational principle, (2.13), is not merely a reformulation of the dynamical principle, (2.5). The two variational principles are presented independently. But, interestingly, the absolute values of the two kinds of functionals become the same along a meta-IRC in the cell:

$$|A| = \int ds_g = |\Delta U|. \tag{2.20}$$

Note that the intrinsic principle of least action and the geodesic variational principle are valid in the interior of a cell, where the stationary curve, that is a meta-IRC, is given as a smooth function of AT. Of course, it is possible to apply these variational principles for the intrinsic dynamism confined in the boundary of a cell excluding the equilibrium points. Also, by an appropriate limiting technique of the integration, it may be possible to extend the variational principles in order to allow the equilibrium points to be the termini of the integral. However, in order to establish the stationary nature of an IRC, which is nothing but a particular meta-IRC, a subtle problem arises; since an IRC crosses an intercell boundary, from a reactant cell to a product cell, a difficulty about the variational problem through the boundary of a cell should be overcome. Here, it should be remembered that an IRC has an accumulation point, that is a transition point, on the intercell boundary from either cell [11]. Hence, a remedy for this kind of difficulty may be given by separating the integrand into two parts: from a reactant cell to a transition point, and from a transition point to a product cell. Then, the variational principles may be applicable in due course.

A clear illustration of the applicability of the geodesic variational principle to the IRC will be shown in Sect. 4 using a model potential surface.

Thus, the stationary nature of the intrinsic dynamism of chemical reactions is found to be governed by the variational principles. One may say that this establishes a teleological character of the intrinsic dynamism.

3. Perturbation Theory

In this section, we shall develop a perturbation theory of the intrinsic dynamism.

3.1. General Theory

3.1.1. IRC Equation

The IRC equation has been given in a compact form as [11]

$$dq^i(\tau)/d\tau = v^i(\tau) \quad (i = 1, \ldots, n) \tag{3.1}$$

where v^i $(i = 1, \ldots, n)$ are the contravariant components of the gradient field of the adiabatic potential U:

$$v^i = a^{ij}\, \partial U/\partial q^j. \tag{3.2}$$

The general solution of the IRC equation, that is the meta-IRC, constitutes a one-parameter continuous group G_1. Using the generator X of G_1, namely

$$X = v^i\, \partial/\partial q^i, \tag{3.3}$$

the locus of the meta-IRC can be represented by [11]

$$q^i(\tau) = R(\tau, \tau_0)q^i(\tau_0), \qquad q_i(\tau) = R(\tau, \tau_0)q_i(\tau_0) \tag{3.4}$$

with

$$R(\tau, \tau_0) = e^{X(\tau - \tau_0)}. \tag{3.5}$$

Here, R is an operator which governs the time (AT)-evolution of the reacting system along the meta-IRC: this operator is now referred to as the intrinsic reaction operator.

Now, it should be noted that the IRC equation can be reduced from (3.1) to an alternative form using the generator X of G_1 as

$$dq^i(\tau)/d\tau = Xq^i(\tau), \qquad dq_i(\tau)/d\tau = Xq_i(\tau). \tag{3.6}$$

The analogy of this form (3.6) with the form of the well-known time-dependent Schrödinger equation is remarkable:

$$d|\Psi(t)\rangle/d(-it/\hbar) = H|\Psi(t)\rangle, \tag{3.7}$$

where $\Psi(t)$ is the time-dependent wave vector and H is the Hamiltonian. Then, the following correspondence is found between the IRC equation (3.6) and the time-dependent Schrödinger equation (3.7):

$$\tau \leftrightarrow -it/\hbar \tag{3.8}$$

$$q^i(\tau) \text{ or } q_i(\tau) \leftrightarrow |\Psi(t)\rangle \tag{3.9}$$

$$X \leftrightarrow H. \tag{3.10}$$

Likewise, we obtain the following correspondence between the intrinsic reaction operator and the quantum mechanical evolution operator U:

$$R(\tau, \tau_0) \leftrightarrow U(t, t_0). \tag{3.11}$$

It follows that the intrinsic reaction operator satisfies the basic equations, analogous to those with respect to the quantum mechanical evolution operator, as follows: 1) the differential equation,

$$dR(\tau, \tau_0)/d\tau = XR(\tau, \tau_0), \tag{3.12}$$

and 2) the integral equation,

$$R(\tau, \tau_0) = 1 + \int_{\tau_0}^{\tau} XR(\tau', \tau_0)\, d\tau'. \tag{3.13}$$

Now, let us consider a perturbation of a meta-IRC. Suppose a perturbation of the gradient field in the form

$$v^i = v^{(0)i} + \tilde{v}^i, \tag{3.14}$$

where $v^{(0)i}$ $(i = 1, \ldots, n)$ are the unperturbed components and $\tilde{v}^i$ $(i = 1, \ldots, n)$ are the components of the perturbation. Then, the generator X is represented by

$$X = X^{(0)} + \tilde{X}, \tag{3.15}$$

where

$$X^{(0)} = v^{(0)i}\, \partial/\partial q^i, \tag{3.16}$$

$$\tilde{X} = \tilde{v}^i\, \partial/\partial q^i. \tag{3.17}$$

We assume that the unperturbed solution of the intrinsic reaction operator is available:

$$R^{(0)}(\tau, \tau_0) = e^{X^{(0)}(\tau - \tau_0)} \tag{3.18}$$

and that the unperturbed solution is obtained as

$$\begin{aligned} q^{(0)i}(\tau) &= R^{(0)}(\tau, \tau_0) q^i(\tau_0), \\ q_i^{(0)}(\tau) &= R^{(0)}(\tau, \tau_0) q_i(\tau_0). \end{aligned} \tag{3.19}$$

Then, let R be given by

$$R = R^{(0)}\tilde{R}, \tag{3.20}$$

where $\tilde{R}$ is the intermediate reaction operator. The intermediate reaction operator satisfies the following differential equation:

$$d\tilde{R}/d\tau = \tilde{\mathfrak{X}}_I\tilde{R}, \tag{3.21}$$

where $\tilde{\mathfrak{X}}_I$ is the intermediate generator defined by

$$\tilde{\mathfrak{X}}_I = R^{(0)-1}\tilde{\mathfrak{X}}R^{(0)}. \tag{3.22}$$

Also, the intermediate reaction operator satisfies the integral equation

$$\tilde{R}(\tau, \tau_0) = 1 + \int_{\tau_0}^{\tau} \tilde{\mathfrak{X}}_I(\tau')\tilde{R}(\tau', \tau_0)\, d\tau' \tag{3.23}$$

and the solution may be given by an iterative procedure as

$$\tilde{R}(\tau, \tau_0) = 1 + \sum_{m=1}^{\infty} \tilde{R}^{(m)}(\tau, \tau_0), \tag{3.24a}$$

where

$$\tilde{R}^{(m)}(\tau, \tau_0) = \int_{\tau>\tau_m>\cdots>\tau_1>\tau_0} d\tau_m \ldots d\tau_1\, \tilde{\mathfrak{X}}_I(\tau_m)\cdots\tilde{\mathfrak{X}}_I(\tau_1). \tag{3.24b}$$

Using (3.20), (3.22), (3.24), and Dyson's chronological operator P [14], the intrinsic reaction operator is found to be

$$R(\tau, \tau_0) = R^{(0)}(\tau, \tau_0) P \exp\left[\int_{\tau_0}^{\tau} d\tau' \tilde{\mathfrak{X}}_I(\tau')\right] \tag{3.25a}$$

or

$$R(\tau, \tau_0) = R^{(0)}(\tau, \tau_0) + \sum_{m=1}^{\infty} R^{(m)}(\tau, \tau_0), \tag{3.25b}$$

where

$$R^{(m)}(\tau, \tau_0) = \int_{\tau>\tau_m>\cdots>\tau_1>\tau_0} d\tau_m \ldots d\tau_1\, R^{(0)}(\tau, \tau_m)\tilde{\mathfrak{X}}(\tau_m)$$
$$\times \cdots \times R^{(0)}(\tau_2, \tau_1)\tilde{\mathfrak{X}}(\tau_1)R^{(0)}(\tau_1, \tau_0). \tag{3.25c}$$

Each term in this expansion (3.25b) describes the individual contribution from the perturbation of the gradient field as is shown in Fig. 2. For example, the zeroth-order diagram found in Fig. 2a describes the unperturbed path, and the first-order diagram found in Fig. 2b describes the sequence of 1) the unperturbed reaction path from τ_0 to τ_1, 2) the application of the perturbation at τ_1, and 3) the unperturbed reaction path from τ_1 to τ under the change of direction caused by the perturbation. If we evaluate the $m + 1$ terms $R^{(0)}, R^{(1)}, \ldots, R^{(m)}$ in (3.25b), then the meta-IRC is represented in terms of the mth degree of accuracy with respect to the perturbation.

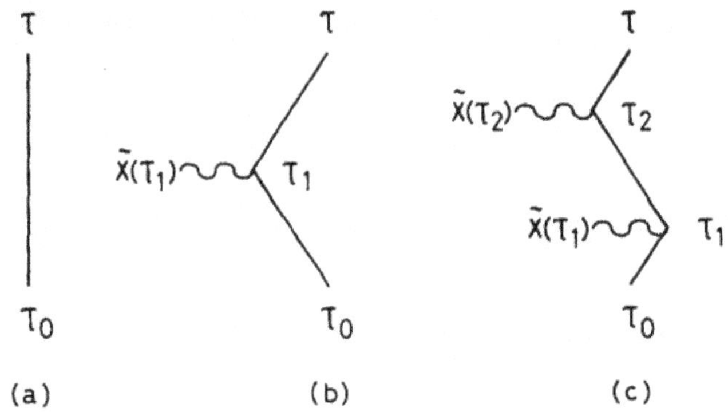

Fig. 2. Diagram of a perturbed meta-IRC: a) zeroth-order, b) first-order, c) second-order diagram. Solid line shows an unperturbed path, and wavy line shows a perturbation caused by a perturbation generator $\tilde{X}$ at each AT

Furthermore, it may be more tractable to obtain first, the unperturbed solution (3.19), and second, the deviation from the unperturbed solution using an alteration operator A in the form:

$$q^i(\tau) = A(\tau, \tau_0)q^{(0)i}(\tau), \qquad q_i(\tau) = A(\tau, \tau_0)q_i^{(0)}(\tau). \tag{3.26}$$

In this case, the alteration operator is represented by

$$A = RR^{(0)-1} \tag{3.27}$$

or

$$A = R^{(0)}\tilde{R}R^{(0)-1}. \tag{3.28}$$

For small value of $\Delta\tau \equiv \tau - \tau_0$, the series expansion of the alteration operator with respect to $\Delta\tau$ may be useful:

$$A(\tau, \tau_0) = \sum_{m=0}^{\infty} (\Delta\tau^m/m!)f_m, \tag{3.29a}$$

where

$$f_m = \sum_{r=0}^{m} (-1)^r_m C_r[X^{(0)} + \tilde{X}]^{m-r}[X^{(0)}]^r. \tag{3.29b}$$

For example, we have

$$f_0 = 1, \tag{3.29c}$$

$$f_1 = \tilde{X}, \tag{3.29d}$$

$$f_2 = X^{(0)}\tilde{X} - \tilde{X}X^{(0)} + \tilde{X}^2. \tag{3.29e}$$

Provided that the operators $X^{(0)}$ and $\tilde{X}$ commute, then we have

$$f_m = \tilde{X}^m \tag{3.30}$$

and the alteration operator is given in a simple form

$$A(\tau, \tau_0) = e^{\mathfrak{X}(\tau-\tau_0)}. \tag{3.31}$$

3.1.2. Hessian Matrix

Using a Dirac bracket representation, the inner product in R_n [10] is conveniently represented by

$$\langle e^i \mid e_j \rangle = \delta^i_j, \qquad \langle e_i \mid e_j \rangle = a_{ij}. \tag{3.32}$$

Then, introducing a Hessian operator $\mathscr{H}$ in the form,

$$\mathscr{H} = |e^i\rangle H_{ij} \langle e^j|, \tag{3.33}$$

the eigenvalue equation of the normal vibrations [10] can be represented by

$$\mathscr{H}|v_{(\alpha)}\rangle = \mu_\alpha |v_{(\alpha)}\rangle \qquad (\alpha = 1, \ldots, n) \tag{3.34}$$

with

$$\langle v_{(\alpha)}|\mathscr{H}|v_{(\beta)}\rangle = \mu_\alpha \delta_{\alpha\beta}, \tag{3.35}$$

$$\langle v_{(\alpha)} \mid v_{(\beta)}\rangle = \delta_{\alpha\beta}, \tag{3.36}$$

where

$$|v_{(\alpha)}\rangle = v^i_{(\alpha)} |e_i\rangle. \tag{3.37}$$

The formal manipulation of the perturbational treatment with respect to the normal vibrations is analogous to that of the quantum mechanical eigenvalue problem of the Hamiltonian H, with H replaced by $\mathscr{H}$.

These perturbational treatments of the meta-IRC and the Hessian matrix are general, and therefore, the perturbation of 1) the metric tensor a_{ij} of the Riemannian space R_n, as well as 2) the potential function U can be treated on an equal theoretical footing.

We shall apply the present perturbation theory for a particular problem in Sect. 3.2.

3.2. Application to a New Concept of the Dynamical Potential Field

3.2.1. Dynamical Perturbation

Consider a chemically reacting system A composed of N_A nuclei and a background system M composed of N_M nuclei as is schematically illustrated in Fig. 3. Then, the Lagrangian of the total system A + M is given by

$$L = (1/2)\dot{x}(A)^2 + (1/2)\dot{x}(M)^2 - U_A - U_M - U_{AM}, \tag{3.38}$$

where $U_{A,M}$ is the potential function of each system A, M, and U_{AM} is the interaction potential function between A and M (such as intermolecular potential).

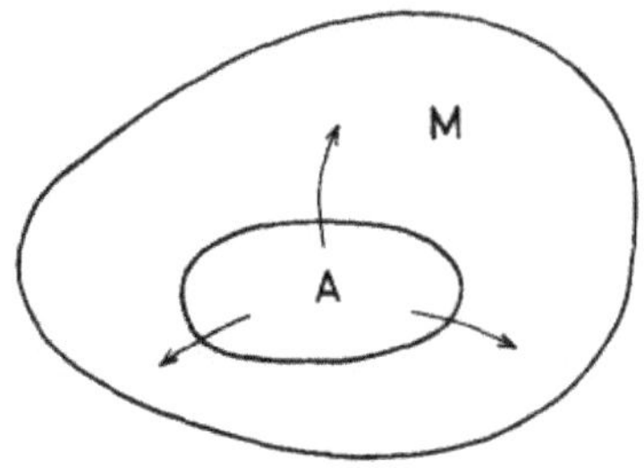

Fig. 3. Chemically reacting system A and initially frozen background system M. An arrow indicates dynamical energy flow

The configuration vectors are represented in the $3(N_{\rm A} + N_{\rm M})$-dimensional Euclidean space as

$$\boldsymbol{x} = (\boldsymbol{x}({\rm A}), \boldsymbol{x}({\rm M})) \tag{3.39}$$

with

$$\boldsymbol{x}({\rm A}) = (x^1, \ldots, x^{3N_{\rm A}}), \tag{3.40}$$

$$\boldsymbol{x}({\rm M}) = (x^{3N_{\rm A}+1}, \ldots, x^{3N_{\rm A}+3N_{\rm M}}). \tag{3.41}$$

Of course, the configuration space of each of the individual systems A and M itself constitutes the $3N_{\rm A}$- or $3N_{\rm M}$-dimensional Euclidean space, respectively. The line element ds in each space is given by the first fundamental form as

$$ds({\rm A})^2 = \sum_{t=1}^{3N_{\rm A}} (dx^t)^2, \tag{3.42}$$

$$ds({\rm M})^2 = \sum_{t=3N_{\rm A}+1}^{3N_{\rm A}+3N_{\rm M}} (dx^t)^2. \tag{3.43}$$

The kinetic energy is then given by

$$(1/2)\dot{\boldsymbol{x}}({\rm A})^2 = (1/2)(ds({\rm A})/dt)^2, \tag{3.44}$$

$$(1/2)\dot{\boldsymbol{x}}({\rm M})^2 = (1/2)(ds({\rm M})/dt)^2. \tag{3.45}$$

Now, provided that the interaction $U_{\rm AM}$ is very weak,

$$U_{\rm AM} = \text{weak}, \tag{3.46}$$

then we may consider that there exists only the isolated system A, and that the presence of the background system M may be neglected. In this case, suppose that the system A is described by the n generalized coordinates q^i ($i = 1, \ldots, n$) then the first fundamental form (3.42) is written as

$$ds({\rm A})^2 = a_{ij}^{(0)}\, dq^i\, dq^j \tag{3.47}$$

with

$$a_{ij}^{(0)} = \sum_{t=1}^{0N_{\rm A}} (\partial x^t/\partial q^i)(\partial x^t/\partial q^j). \tag{3.48}$$

Note that the generalized coordinate system constitutes the n-dimensional Riemannian space $R_n^{(0)}$ whose metric tensor is given by (3.48). As a whole, the Lagrangian of the total system can be approximated to be

$$L \simeq L^{(0)} \equiv (1/2)a_{ij}^{(0)}\dot{q}^i\dot{q}^j - U^{(0)}, \tag{3.49}$$

where

$$U^{(0)} = U_A. \tag{3.50}$$

This is the starting point to the succeeding treatment.

We shall treat the perturbation theory of $L^{(0)}$ in a particular case. The treatment proceeds in three stages as the accuracy of the treatment improves. In the third stage of the treatment, a new concept of the dynamical potential field will appear.

I) *First stage.* Let us consider the case where the system A is selectively activated under the condition (3.46), and where the background system M is held fixed as if the zero-point vibrational motions were frozen. This case may occur, for example, when a) a molecular system A in a large molecular aggregate (such as a molecular crystal) is activated by a laser beam, or b) in a molecular system, a set of normal modes A are selectively activated by a laser beam (in this case, the separation of A with the background system M may be performed with respect to the normal coordinates rather than the Cartesian coordinates), and c) a chemical reaction in a system A is initiated under a weak interaction with the solvent M. Then, the Lagrangian of the total system may be represented by that of the isolated system A under the perturbation of a field Φ:

$$L = L^{(0)} - \Phi. \tag{3.51}$$

In this representation, the perturbation field is given by

$$\Phi = U_{AM} + \Phi_M, \tag{3.52}$$

where Φ_M is brought about from the Lagrangian of the background system M:

$$\Phi_M = U_M - (1/2)\dot{\boldsymbol{x}}(M)^2. \tag{3.53}$$

Under the frozen condition for the dynamical motion of M, namely,

$$\boldsymbol{x}(M) = \text{constant vector}, \tag{3.54}$$

the field Φ_M is merely a constant:

$$\Phi_M(\text{frozen}) = U_M = \text{const}. \tag{3.55}$$

Hence, the Lagrangian of the total system can be represented by

$$L = L^{[I]} \equiv (1/2)a_{ij}^{(0)}\dot{q}^i\dot{q}^j - U^{[I]}, \tag{3.56}$$

where $U^{[I]}$ is the potential field of the system A under the static perturbation U_{AM}:

$$U^{[I]} = U_A + U_{AM} + \text{const}. \tag{3.57}$$

Apparently, in this stage, the perturbation theory of Sect. 3.1 is applicable.

II) *Second stage.* Let us consider the second stage of approximation, where the dynamical motion of the background system M is almost fixed frozen but a slight configurational change may take place under the influence of the system A. In other words, the background system M is quasi-frozen and the time-evolution is brought about through the dynamics of A. This stage may correspond to a situation where a part of M which is most closely connected with A through the perturbation U_{AM} becomes slowly activated. In this stage, as the motion of the quasi-frozen configuration of M is slow, the kinetic energy part of the field Φ_M may be neglected:

$$\Phi_M(\text{quasi-frozen}) = U_M = \text{a function of } q^i. \tag{3.58}$$

Hence, the Lagrangian of the total system may be represented by

$$L = L^{[II]} \equiv (1/2)a_{ij}^{(0)}\dot{q}^i\dot{q}^j - U^{[II]}, \tag{3.59}$$

where $U^{[II]}$ is the potential field of the system A under the static perturbation $U_{AM} + U_M$:

$$U^{[II]} = U_A + U_{AM} + U_M = \text{a function of } q^i. \tag{3.60}$$

Also, at this stage, the perturbation theory of Sect. 3.1 is applicable without difficulty.

III) *Third stage.* Now, we shall consider the dynamical motion of the background system M. In this case, the dynamical part of the field Φ_M should be explicitly taken into account:

$$\Phi_M(\text{dynamical}) = U_M - (1/2)\dot{x}(M)^2. \tag{3.61}$$

It should be noted that the time-evolution of the system M is assumed to be brought about only through that of A; therefore, the dynamical part of Φ_M is given from (3.45) by

$$(1/2)\dot{x}(M; x(A))^2 = (1/2)(\partial s(M)/\partial q^i)(\partial s(M)/\partial q^j)\dot{q}^i\dot{q}^j, \tag{3.62}$$

where q^i ($i = 1, \ldots, n$) are the generalized coordinates of A. Thus, the Lagrangian of the total system may be represented by

$$L = L^{[III]} \equiv (1/2)a_{ij}\dot{q}^i\dot{q}^j - U^{[III]}, \tag{3.63}$$

where a_{ij} and $U^{[III]}$ are given by

$$a_{ij} = a_{ij}^{(0)} + (\partial s(M)/\partial q^i)(\partial s(M)/\partial q^j), \tag{3.64}$$

$$U^{[III]} = U^{[II]} = U_A + U_{AM} + U_M. \tag{3.65}$$

Here, the $s(M) = s(M; q^i(A))$ in (3.62) and (3.64) will be referred to as the dynamical potential field. We have obtained now the Lagrangian in the n-dimensional Riemannian space R_n whose metric tensor is given in the modified form (3.64). Since the coordinate system is mass-weighted, the modification of the metric tensor implies that the dynamical perturbation results in renormalization of nuclear masses of the system A. In other words, under the dynamical perturbation of this kind, the "dressed-particle" having the effective mass appears (note that this kind of perturbation is of purely dynamical nature, and hence, may in some cases compensate that of purely static nature such as U_{AM} and U_{M}).

To conclude, we may say that the dynamical motion of the system A disturbs the background field Φ, which finally affects the dynamism of A, yielding the dressed-particles.

In this connection, it is well-known, in the perturbation theory of the effective mass of an electron, that the interaction of electron with phonon affects the effective mass of the electron [15]; we have shown, in a closed form, a new type of the concept of effective mass that is of purely dynamical origin.

Since the dynamical potential is given by the length of the dynamical motion of the multidimensional vector $\boldsymbol{x}$(M), it is positive definite, and we can estimate the warming efficiency of the background system M in terms of the gradient of the dynamical potential.

In this stage, also, the perturbation theory of Sect. 3.1 will be applicable. Some of the results peculiar to the present stage of the dynamical perturbation will be discussed in detail in Sect. 3.2.2.

3.2.2. Application of the Perturbation Theory

I, II) First and second stages. In these stages, the perturbation field is of purely static origin. If we write

$$U = U^{(0)} + U^{(1)}, \tag{3.66}$$

where $U^{(1)}$ is the perturbation, then we have, for the solution of the alteration operator (3.29),

$$f_1 = a^{(0)ij}(\partial U^{(1)}/\partial q^j)\,\partial/\partial q^i, \tag{3.67}$$

$$\begin{aligned} f_2 = {} & a^{(0)ij}(\partial U^{(0)}/\partial q^j)(D^{(0)}_{ik}U^{(1)})a^{(0)kl}\,\partial/\partial q^l \\ & - a^{(0)ij}(\partial U^{(1)}/\partial q^j)(D^{(0)}_{ik}U^{(0)})a^{(0)kl}\,\partial/\partial q^l \\ & + a^{(0)ij}(\partial U^{(1)}/\partial q^j)(D^{(0)}_{ik}U^{(1)})a^{(0)kl}\,\partial/\partial q^l \\ & + a^{(0)ij}(\partial U^{(1)}/\partial q^j)a^{(0)kl}(\partial U^{(1)}/\partial q^l)D^{(0)}_{ik}, \end{aligned} \tag{3.68}$$

where $D^{(0)}_{ij}$ is the second-order differential operator given in $R^{(0)}_n$ [11]:

$$D^{(0)}_{ij} = \partial^2/\partial q^i\,\partial q^j - \Gamma^{(0)k}_{ij}\,\partial/\partial q^k. \tag{3.69}$$

Note that the first-order correction of f_1 exactly vanishes on the surface perpendicular to the gradient field of $U^{(1)}$ in $R^{(0)}_n$. For the normal vibrations, we have the Hessian operator in the form

$$\mathscr{H} = \mathscr{H}^{(0)} + \mathscr{H}^{(1)}, \tag{3.70}$$

where

$$\mathscr{H}^{(0)} = |e^{(0)i}\rangle H^{(0)}_{ij}\langle e^{(0)j}|, \tag{3.71}$$

$$\mathscr{H}^{(1)} = |e^{(0)i}\rangle H^{(1)}_{ij}\langle e^{(0)j}|, \tag{3.72}$$

with

$$H_{ij}^{(0)} = D_{ij}^{(0)} U^{(0)}, \tag{3.73}$$

$$H_{ij}^{(1)} = D_{ij}^{(0)} U^{(1)}. \tag{3.74}$$

Then, following the usual procedure of the perturbed eigenvalue problem, we can easily obtain the perturbed solution of the eigenvalue equation for the Hessian operator in due course. Using the zeroth-order solutions,

$$\mathscr{H}^{(0)}|v_{(\alpha)}^{(0)}\rangle = \mu_\alpha^{(0)}|v_{(\alpha)}^{(0)}\rangle, \tag{3.75}$$

$$\langle v_{(\alpha)}^{(0)} \mid v_{(\beta)}^{(0)}\rangle = \delta_{\alpha\beta}, \tag{3.76}$$

we have the Rayleigh–Schrödinger type perturbation series as

$$v_{(\alpha)} = v_{(\alpha)}^{(0)} + v_{(\alpha)}^{(1)} + \cdots, \tag{3.77}$$

$$\mu_\alpha = \mu_\alpha^{(0)} + \mu_\alpha^{(1)} + \cdots. \tag{3.78}$$

For example, we obtain

$$v_{(\alpha)}^{(1)} = \sum_{\beta(\neq\alpha)} (\mu_\alpha^{(0)} - \mu_\beta^{(0)})^{-1} \cdot v_{(\beta)}^{(0)} \cdot \langle v_{(\beta)}^{(0)}|\mathscr{H}^{(1)}|v_{(\alpha)}^{(0)}\rangle, \tag{3.79}$$

$$\mu_\alpha^{(1)} = \langle v_{(\alpha)}^{(0)}|\mathscr{H}^{(1)}|v_{(\alpha)}^{(0)}\rangle. \tag{3.80}$$

III) Third stage. In this stage, the dynamical potential field disturbs the metric of the configuration Riemannian space itself. The basis vectors of R_n may then be given by

$$e_i = e_i^{(0)} + \boldsymbol{i}\partial s(\mathrm{M})/\partial q^i \quad (i = 1, \ldots, n) \tag{3.81}$$

where $\boldsymbol{i}$ is a unit vector satisfying

$$\langle e_i^{(0)} \mid \boldsymbol{i}\rangle = \langle \boldsymbol{i} \mid e_i^{(0)}\rangle = 0, \qquad \langle \boldsymbol{i} \mid \boldsymbol{i}\rangle = 1. \tag{3.82}$$

Also, the contravariant components of the basis vectors and the metric tensor are given by

$$e^i = e^{(0)i} - [1 + \Delta_1^{(0)} s(\mathrm{M})]^{-1} a^{(0)ij} (\partial s(\mathrm{M})/\partial q^j) \times (e^{(0)k}\, \partial s(\mathrm{M})/\partial q^k - \boldsymbol{i}), \tag{3.83}$$

$$a^{ij} = a^{(0)ij} - [1 + \Delta_1^{(0)} s(\mathrm{M})]^{-1} \times (a^{(0)ik}\, \partial s(M)/\partial q^k)(a^{(0)jl}\, \partial s(\mathrm{M})/\partial q^l), \tag{3.84}$$

where

$$\Delta_1^{(0)} s(\mathrm{M}) = a^{(0)ij} (\partial s(\mathrm{M})/\partial q^i)(\partial s(\mathrm{M})/\partial q^j) \tag{3.85}$$

is Beltrami's differential parameter of the first kind with respect to $s(\mathrm{M})$ obtained in $R_n^{(0)}$. Likewise, the Christoffel symbols are obtained as

$$[ij, l] = [ij, l]^{(0)} + (\partial^2 s(\mathrm{M})/\partial q^i\, \partial q^j)(\partial s(\mathrm{M})/\partial q^l), \tag{3.86}$$

$$\Gamma_{ij}^k = \Gamma_{ij}^{(0)k} + [1 + \Delta_1^{(0)} s(M)]^{-1} (a^{(0)kl}\, \partial s(\mathrm{M})/\partial q^l) D_{ij}^{(0)} s(\mathrm{M}). \tag{3.87}$$

Using the above, we shall analyze first the locus of the deformed meta-IRC, and second, the perturbed normal vibrations. Finally, we shall examine a mechanism of energy transfer between A and M in terms of the dynamical potential field.

First, let $\tilde{a}^{ij}$ be given by

$$\tilde{a}^{ij} \equiv a^{ij} - a^{(0)ij}, \tag{3.88}$$

then, for the solution of the alteration operator (3.29), we have

$$f_1 = (\tilde{a}^{ij}\, \partial U/\partial q^j)\, \partial/\partial q^i, \tag{3.89}$$

$$\begin{aligned} f_2 = {} & (a^{(0)ij}\, \partial \tilde{a}^{kl}/\partial q^j)(\partial U/\partial q^i)(\partial U/\partial q^l)\, \partial/\partial q^k \\ & + a^{(0)ij}\tilde{a}^{kl}(D^{(0)}_{jl}U)[(\partial U/\partial q^i)\, \partial/\partial q^k - (\partial U/\partial q^k)\, \partial/\partial q^i] \\ & + a^{(0)ij}\tilde{a}^{kl}(\partial U/\partial q^j)\Gamma^{(0)m}_{il}[(\partial U/\partial q^m)\, \partial/\partial q^k + (\partial U/\partial q^k)\, \partial/\partial q^m] \\ & + \tilde{a}^{ij}(\partial \tilde{a}^{kl}/\partial q^j)(\partial U/\partial q^i)(\partial U/\partial q^l)\, \partial/\partial q^k \\ & + \tilde{a}^{ij}\tilde{a}^{kl}(\partial U/\partial q^j)[(D^{(0)}_{il}U)\, \partial/\partial q^k + (\partial U/\partial q^k)D^{(0)}_{il}] \\ & + \tilde{a}^{ij}\tilde{a}^{kl}(\partial U/\partial q^j)\Gamma^{(0)m}_{il}[(\partial U/\partial q^m)\, \partial/\partial q^k + (\partial U/\partial q^k)\, \partial/\partial q^m]. \end{aligned} \tag{3.90}$$

For example, we obtain the perturbed solution of the meta-IRC up to the first order in $\Delta\tau$ as

$$\begin{aligned} q^i(\tau) = {} & q^i(\tau_0) - \Delta\tau[1 + \Delta^{(0)}_1 s(\mathrm{M})]^{-1}\langle \mathrm{grad}^{(0)} s(\mathrm{M}) \mid \mathrm{grad}^{(0)} U\rangle \\ & \times \langle \mathrm{grad}^{(0)} s(\mathrm{M}) \mid e^{(0)i}\rangle + 0(\Delta\tau^2), \end{aligned} \tag{3.91}$$

where $\mathrm{grad}^{(0)}$ is the first-order differential operator given in $R^{(0)}_n$:

$$\mathrm{grad}^{(0)} = e^{(0)i}\, \partial/\partial q^i. \tag{3.92}$$

The first-order deviation of the meta-IRC vanishes when 1) the gradient field of the dynamical potential $s(\mathrm{M})$ is orthogonal to that of U in $R^{(0)}$, or, for the ith component, 2) the gradient field of the dynamical potential $s(\mathrm{M})$ lies entirely on the q^i-surface in $R^{(0)}_n$.

Second, let us analyze the dynamical background effect to the normal vibrations. In the present stage, the perturbed Hessian operator is represented by the infinite series expansion:

$$\mathscr{H} = \mathscr{H}^{(0)} + \mathscr{H}^{(1)} + \mathscr{H}^{(2)} + \cdots \tag{3.93}$$

where the perturbation parameter is that for the dynamical potential $s(\mathrm{M})$. For example, we have

$$\mathscr{H}^{(0)} = |e^{(0)i}\rangle H_{ij}^{(0)}\langle e^{(0)j}|, \tag{3.94}$$

$$\mathscr{H}^{(1)} = |e^{(0)i}\rangle H_{ij}^{(0)}\langle ia^{(0)jk}\,\partial s(\mathrm{M})/\partial q^k| + |ia^{(0)ik}\,\partial s(\mathrm{M})/\partial q^k\rangle H_{ij}^{(0)}\langle e^{(0)j}|, \tag{3.95}$$

$$\begin{aligned}\mathscr{H}^{(2)} = &- |e^{(0)i}\rangle H_{ij}^{(0)}\langle a^{(0)jk}(\partial s(\mathrm{M})/\partial q^k)(\partial s(\mathrm{M})/\partial q^l)e^{(0)l}| \\ &- |a^{(0)ik}(\partial s(\mathrm{M})/\partial q^k)(\partial s(\mathrm{M})/\partial q^l)e^{(0)l}\rangle H_{ij}^{(0)}\langle e^{(0)j}| \\ &- |e^{(0)i}\rangle \Gamma_{ij}^{(2)k}\,\partial U/\partial q^k\langle e^{(0)j}| \\ &+ |ia^{(0)ik}\,\partial s(\mathrm{M})/\partial q^k\rangle H_{ij}^{(0)}\langle ia^{(0)jk}\,\partial s(\mathrm{M})/\partial q^k|,\end{aligned} \tag{3.96a}$$

with

$$\Gamma_{ij}^{(2)k} = a^{(0)kl}(\partial s(\mathrm{M})/\partial q^l)D_{ij}^{(0)}s(\mathrm{M}). \tag{3.96b}$$

In this case, on account of the orthogonality relationship (3.82), all the first-order quantities are zero: [the intermediate normalization of the eigenvector is adopted: $\langle v_{(\alpha)}^{(0)} \mid v_{(\alpha)}^{(m)}\rangle = 0$ $(m = 1, 2, \ldots)$]

$$v_{(\alpha)}^{(1)i} = 0, \tag{3.97}$$

$$\mu_\alpha^{(1)} = 0. \tag{3.98}$$

The second-order quantities are obtained as

$$v_{(\alpha)}^{(2)i} = \sum_{\beta(\neq\alpha)} (\mu_\beta^{(0)} - \mu_\alpha^{(0)})^{-1} v_{(\beta)}^{(0)i} c_{\beta\alpha} \tag{3.99a}$$

with

$$c_{\beta\alpha} = \mu_\alpha^{(0)}\langle v_{(\alpha)}^{(0)} \mid \mathrm{grad}^{(0)}\, s(\mathrm{M})\rangle\langle v_\beta^{(0)} \mid \mathrm{grad}^{(0)}\, s(\mathrm{M})\rangle + \langle \mathrm{grad}^{(0)}\, U \mid \mathrm{grad}^{(0)}\, s(\mathrm{M})\rangle \sum_\gamma \omega_\gamma \langle v_{(\alpha)}^{(0)} \mid w_{(\gamma)}\rangle\langle v_{(\beta)}^{(0)} \mid w_{(\gamma)}\rangle, \tag{3.99b}$$

and

$$\mu_\alpha^{(2)} = -\mu_\alpha^{(0)} \mid \langle v_{(\alpha)}^{(0)} \mid \mathrm{grad}^{(0)}\, s(\mathrm{M})\rangle|^2 - \langle \mathrm{grad}^{(0)} U \mid \mathrm{grad}^{(0)}\, s(\mathrm{M})\rangle \sum_\gamma \omega_\gamma |\langle v_{(\alpha)}^{(0)} \mid w_{(\gamma)}\rangle|^2. \tag{3.100}$$

In these formulas, the contributions from the second derivatives of the dynamical potential $s(\mathrm{M})$ are simplified by diagonalizing the second-order terms as follows:

$$(D_{ij}^{(0)}s(\mathrm{M}) - \omega_\gamma a_{ij}^{(0)})w_{(\gamma)}^j = 0 \quad (\gamma = 1, \ldots, n), \tag{3.101a}$$

$$a_{ij}^{(0)} w_{(\gamma)}^i w_{(\delta)}^j = \delta_{\gamma\delta} \quad (\gamma, \delta = 1, \ldots, n) \tag{3.101b}$$

with

$$w_{(\gamma)} = e_i^{(0)} w_{(\gamma)}^i \quad (\gamma = 1, \ldots, n). \tag{3.101c}$$

Now, let us examine a mechanism of energy transfer between A and M. In order to clarify the nature of the dynamical potential field $s(\mathrm{M})$ on the energy transfer, we shall classify two typical cases such as (*A*): $\mathrm{grad}^{(0)}\, s(\mathrm{M})$ is parallel to the equi-potential surface, and (*B*): $\mathrm{grad}^{(0)}\, s(\mathrm{M})$ is perpendicular to the equi-potential surface. Other cases may have a mixed nature of (*A*) and (*B*).

Case (*A*). In this case, vibrational motion on the equi-potential surface directly induces the variation of $s(\mathrm{M})$, and then the kinetic energy of the background system M will increase. A normal mode which has maximal overlap with $\mathrm{grad}^{(0)}$ $s(\mathrm{M})$ has a leading contribution to this kind of vibrational energy transfer.

Now, remember that a normal mode α confined on the equi-potential surface is characterized by the vibrational quantum number n_α. Since the dynamical interaction with M is weak so as to be treated by perturbation theory, the quantum number n_α may be conserved during the dynamical process of energy transfer; that is, the vibrational quantum number n_α may be a good quantum number. Hence, in order to account for the energy outflow from A to M, the unit of quantum of the vibrational mode α should be reduced. In other words, softening of phonon should occur for the normal mode α. Clearly, the result of the perturbation theory deduced from (3.100) represents this mechanism of energy transfer in terms of the softening of phonon, that is the weakening of the normal mode α:

$$\mu_\alpha^{(2)} = -\mu_\alpha^{(0)} |\langle v_{(\alpha)}^{(0)} \mid \mathrm{grad}^{(0)}\, s(\mathrm{M})\rangle|^2. \tag{3.102}$$

Case (*B*). In this case, promotion of chemical reaction along the meta-IRC produces kinetic energy of M through the coupling with $\mathrm{grad}^{(0)}\, s(\mathrm{M})$. This kind of translational energy transfer is further classified according as (*B*-1): $\mathrm{grad}^{(0)}\, s(\mathrm{M})$ is parallel to $\mathrm{grad}^{(0)}\, U$, and (*B*-2): $\mathrm{grad}^{(0)}\, s(\mathrm{M})$ is anti-parallel to $\mathrm{grad}^{(0)}\, U$. Additionally, for the vibrational mode α on the equi-potential surface, we have

$$\mu_\alpha^{(2)} = -\langle \mathrm{grad}^{(0)}\, U \mid \mathrm{grad}^{(0)}\, s(\mathrm{M})\rangle \sum_\gamma \omega_\gamma |\langle v_{(\alpha)}^{(0)} \mid w_{(\gamma)}\rangle|^2. \tag{3.103}$$

Note that the factor $\langle \mathrm{grad}^{(0)}\, U \mid \mathrm{grad}^{(0)}\, s(\mathrm{M})\rangle$ in (3.103) changes its sign accordingly, which effect is significant when we consider an indirect scheme of vibrational energy transfer that may occur on the equi-potential surface.

Case (*B*-1). In this case, as U increases along the meta-IRC, the dynamical potential $s(\mathrm{M})$ increases in the same direction. In other words, as the system A destabilizes, the increase of the net energy transfer from A to M is brought about. This mechanism of energy flow results in a lowering of the total energy of the system A, and finally inhibits the destabilizing motion of A. Conversely, as U decreases, the $s(\mathrm{M})$ decreases in the same direction. This mechanism shows that the approach of the system A to a stable equilibrium structure is supported by the positive energy supply from the background system M; finally, the system A reproduces its equilibrium structure. As a whole, the structural stability of the system A is guaranteed.

This is confirmed by another approach of analysis on the mechanism of energy transfer that may occur on the equi-potential surface. Since the gradient field

$\text{grad}^{(0)} s(M)$ vanishes on the equi-potential surface, the direct mode of energy transfer in terms of the dynamical potential field $s(M)$ is not present. But an indirect scheme of vibrational energy transfer is present, being governed by the vibration-like mode γ of $s(M)$, as defined by (3.101), through the coupling with the vibrational mode α of the system A. Clearly, the mode of positive ω_γ governs the energy transfer from A to M, while the mode of negative ω_γ governs the energy transfer from M back to A in the reverse direction. Under the conservation law of vibrational quantum number with respect to the normal mode α, the allowable coupling scheme should be

$$\omega_\gamma > 0 \rightarrow \text{softening of normal mode } \alpha$$

$$\omega_\gamma < 0 \rightarrow \text{hardening of normal mode } \alpha. \tag{3.104}$$

These postulates are confirmed by the formula (3.103). Case (*B*-2). In this case, as U increases, the $s(M)$ decreases. This mechanism shows that the destabilizing motion of A is facilitated by the energy flow from M. Conversely, as U decreases, the $s(M)$ increases. This mechanism implies that the dynamics of A at the stable equilibrium structure becomes seriously disturbed by the dynamical perturbation with M, which may finally become so serious that the perturbational treatment may not be adequate. As a whole, the structural stability of the system A is not guaranteed.

Likewise, the vibration-like coupling of $s(M)$ with the normal mode of A on the equi-potential surface has completely opposite effect, as compared with (3.104), because of the negative contribution of the factor $\langle \text{grad}^{(0)} U \mid \text{grad}^{(0)} s(M) \rangle$ in (3.103):

$$\omega_\gamma > 0 \rightarrow \text{hardening of normal mode } \alpha$$

$$\omega_\gamma > 0 \rightarrow \text{softening of normal mode } \alpha. \tag{3.105}$$

This kind of heterogeneous character of the coupling scheme may result in the growth of unbalanced distribution of the vibrational energy on the equi-potential surface. This, again, indicates the structural instability of A.

Note that, at an equilibrium point of A, we have the same situation as in the Case (*A*). However, since the dynamical event is somewhat of a global nature, the critical mode of dynamical energy transfer should be analyzed from a global coupling scheme as described above.

To summarize, a mechanism of dynamical energy transfer from a chemically reacting system A to a background system M is discussed. We have clarified the typical patterns of energy transfer for a structurally stable system and for a structurally unstable system; the former case is found in Case (*A*) and Case (*B*-1), and the latter case is found in Case (*B*-2). A structurally stable system allows an energy exchange with a background system and the structure of the system itself remains stable, while a structurally unstable system allows an amplification of the interaction mode with a background system and may finally undergo a fragmentation.

For a structurally stable system, 1) an energy outflow occurs in the Case (*A*) through a vibrational mode that lies in the direction of $\mathrm{grad}^{(0)} s(\mathrm{M})$, and in the Case (*B*-1) through the promoting mode of chemical reaction and a vibrational mode that couples with a vibration-like mode of positive ω_γ, and 2) an energy inflow occurs in the Case (*B*-1) through the promoting mode of chemical reaction and a vibrational mode that couples with a vibration-like mode of negative ω_γ. This classification is exact if the requirement of the vibrational adiabaticity is fulfilled; for a general case where the vibrational coupling with the promoting mode of chemical reaction is present, the off-diagonal part of the Hessian matrix between the reduced normal modes on the equi-potential surface and the promoting mode lying along meta-IRC should be taken into account [10].

In the present paper we have laid great emphasis on chemically reacting systems. However, the potential applicability for the system having the Lagrangian of the form (3.38) and (3.51) is supposed to be universal. The intrinsic nature of the dynamical process of the system can be analyzed in terms of the intrinsic dynamism. The quantum mechanical formulation is also available [11].

4. Example

In this section, we shall elucidate the applicability of the fundamental principle of the intrinsic field theory of chemical reactions, that is the geodesic variational principle to an IRC, by using a model potential surface.

The shape of the model potential is illustrated in Fig. 6 of [10]. This model shows a typical pattern of isomerization reaction $A \to B \to C$, A and C being the minima and B the transition point. The IRC appropriate for the isomerization reaction is depicted in Fig. 4.

Suppose a variational parameter ε, being connected with a virtual trajectory as

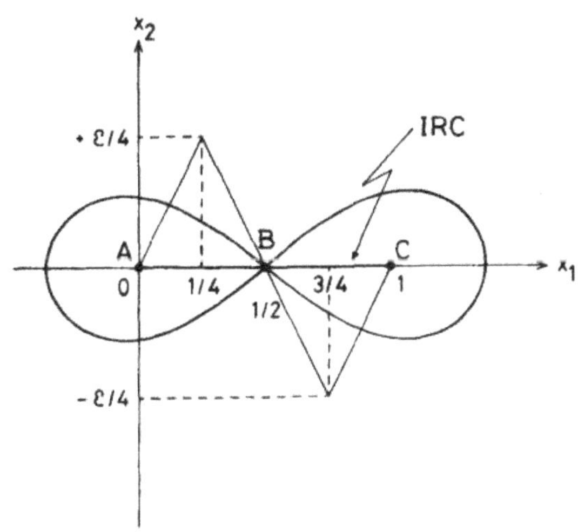

Fig. 4. Pattern of an isomerization reaction model $A \to B \to C$ [10]. Folded line from A through B to C shows a virtual trajectory used in (4.1) of the text

shown in Fig. 4. Then, we have the length of the trajectory in the rigged Riemannian space R_n^* as

$$\int_A^C ds_g \equiv \int_A^B ds_g + \int_B^C ds_g$$
$$= |\Delta U(A \to B)| + |\Delta U(B \to C)| + h\varepsilon^2 + 0(\varepsilon^3) \quad (4.1)$$

with

$$|\Delta U(A \to B)| = |\Delta U(B \to C)| = 1/16a^2, \quad (4.2)$$

$$h = (a^2/b^4) \ln (3\sqrt{3}/4) + 1/16a^2 \quad (>0), \quad (4.3)$$

where a and b are certain constants [10]. Clearly, the IRC corresponding to the locus of $\varepsilon = 0$ satisfies the geodesic variational principle (2.13), yielding the minimal value of the functional.

Acknowledgement. This work was supported by a Grant-in-Aid for Scientific Research from the Ministry of Education of Japan.

References

1. Fukui, K.: Theory of orientation and stereoselection. Berlin: Springer-Verlag 1970; Fleming, I.: Frontier orbitals and organic chemical reactions. London: Wiley 1976
2. Woodward, R. B., Hoffmann, R.: The conservation of orbital symmetry. New York: Academic Press 1969
3. Fukui, K.: J. Phys. Chem. **74**, 4161 (1970)
4. Fukui, K., in: The world of quantum chemistry, Daudel, R., Pullman, B. Eds. Holland: D. Reidel Publ. Co. 1974
5. Fukui, K., Kato, S., Fujimoto, H.: J. Am. Chem. Soc. **97**, 1 (1975)
6. Kato, S., Fukui, K.: J. Am. Chem. Soc. **98**, 6395 (1976)
7. Kato, S., Kato, H., Fukui, K.: J. Am. Chem. Soc. **99**, 684 (1977)
8. Ishida, K., Morokuma, K., Komornicki, A.: J. Chem. Phys. **66**, 2153 (1977)
9. Joshi, B. D., Morokuma, K.: J. Chem. Phys. **67**, 4880 (1977)
10. Tachibana, A., Fukui, K.: Theoret. Chim. Acta (Berl.) **49**, 32 (1978)
11. Tachibana, A., Fukui, K.: Theoret. Chim. Acta (Berl.) (in press) (1979)
12. Whittaker, E. T.: A treatise on the analytical dynamics of particles and rigid bodies. Cambridge: Cambridge Univ. Press 1937
13. Møller, C.: The theory of relativity. London: Oxford Univ. Press 1952
14. Dyson, F. J.: Phys. Rev. **75**, 486 (1949)
15. Kittel, C.: Quantum theory of solids. New York: Wiley 1963

Received October 11, 1978

Variational Principles in a Chemical Reaction

KENICHI FUKUI
Department of Hydrocarbon Chemistry, Kyoto University, Kyoto, Japan

Abstract

Analogous to other dynamical phenomena, a chemical reaction is controlled by variational principles. It is shown that the intrinsic reaction coordinate (IRC) formalism is essential to derive these results. Hamilton's principle and the geodesic principle are explained in terms of the IRC approach. In this connection the use of differential geometrical methods is explained.

1. Introduction

It is well known that a majority of fundamental laws in physics are formulated as a variational equation. Recognition of physical laws in terms of a variational principle has its origin in Fermat's principle of least time, and Maupertuis' principle of least action (1747). Lagrange's function L plays an essential role in the variational principle in dynamics. The actual path for a conservative system is given by the variational equation

$$\delta \int L\, dt = 0$$

for which Euler's differential equation becomes the equation of motion. This is what is called Hamilton's principle.

The purpose here is to present a variational principle for chemical reactions which specifies the actual path as having a stationary value of a certain quantity as compared with the other given paths.

The first step is to formulate the actual path of reaction. This has been carried out unequivocally by Fukui [1, 2], in the form of simultaneous differential equations, in the configuration space in which a point determines the relative nuclear configuration. The coordinates of this space is called "internal coordinates."

The potential energy function, W, for a chemically reacting system is obtained on the basis of the Born–Oppenheimer approximation as a function of internal coordinates. Hence, the value of potential energy is associated with each point of the configuration space.

To discuss the motion of chemically reacting systems in the configuration space, it is necessary to separate the internal motion from others, for instance, the transition and the rotation as a whole. This can be carried out by Wilson's method [3] even in the nonrigid, reacting system [4]. The space in which the internal motion separated from others in this fashion is properly discussed is no more Euclidean but Riemannian in general. Therefore, some mathematical preparations are needed. For this purpose, Synge's early work [5] is quite helpful.

International Journal of Quantum Chemistry: Quantum Chemistry Symposium 15, 633–642 (1981)

CCC 0161-3642/81/010633-10$01.00

2. Synge's Geometry of Dynamics

More than a half century ago Synge [5] introduced tensor calculus into dynamical problems [6]. The present problem also serves as a suitable example for displaying the utility of differential geometry [7]. For this purpose a n-dimensional configuration space with generalized coordinates q^i ($i = 1, 2, \ldots n$) is considered. The line element ds of this space is defined as

$$ds^2 = 2T\,dt^2 = a_{ij}dq^i dq^j \tag{1}$$

The superscript i in the generalized coordinates q^i corresponds to the traditional index for a *contravariant* vector, dq^i, and subscript i and j in a_{ij} imply indices of a *covariant* tensor, in accord with the usual convention of differential geometry. Einstein's convenient notation that the repeated index in a product implies summation with respect to that index from 1 to n, e.g.,

$$a_{ij}dq^j = \sum_{j=1}^{n} a_{ij}dq^j$$

is also used.

From Eq. (1) $(a_{ij}dq^i dq^j)^{1/2}$ represents the distance of two adjacent points of this space. The line element represented by this equation corresponds to the ordinary geometrical line element for a particle of unit mass moving in this space.

Since a_{ij} in Eq. (1) is obtained from the kinetic energy expression for the $3N$-dimensional Cartesian space, N being the number of atoms, by a reduction of six variables corresponding to the translation and the rotation as a whole, the space of Eq. (1) is generally non-Euclidean, and hence differential geometry occupies the whole stage. Here, it is also noted that the internal Riemannian space is positive definite.

3. The IRC Formalism

Consider a point in the configuration space. Suppose that the reacting system is in the nuclear configuration corresponding to that point and starts to move with infinitely small velocity. In what direction will the reacting system will move? This quasistatic infinitesimal displacement of nuclear configuration was called "intrinsic motion," and the locus resulting from a succession of these infinitesimal motions was given by a solution curve of the differential equation [8, 9]

$$\frac{a_{1j}dq^j}{\dfrac{\partial W}{\partial q^1}} = \frac{a_{2j}dq^j}{\dfrac{\partial W}{\partial q^2}} = \cdots = \frac{a_{nj}dq^j}{\dfrac{\partial W}{\partial q^n}} \qquad (n = 3N - 6) \tag{2}$$

in which $a_{ij} = a_{ji}$ is a function of $q^1, q^2, \ldots q^n$, and each denominator is the gradient of the potential energy function. This implies that such a locus is nothing but a line of force. Also the solution curves of Eq. (2) are the orthogonal trajectories of the family of equipotential surfaces $W(q^1, q^2, \ldots q^n) = \text{const.}$

The solution of Eq. (2) is given by

$$f_i(q^1,q^2,\ldots q^n) = C_i \qquad (i = 1, 2, \ldots n-1)$$

The solution curve is determined by assigning a set of $(n-1)$ parameters C_i.

Equation (2) has an infinite number of solution curves, all of which start from an equilibrium point to run toward a $(n-1)$-dimensional surface surrounding that equilibrium point. The whole assembly of these solution curves which start from an equilibrium point form a cell [10, 11]. The equilibrium point is called a cell center. The whole configuration space is divided into a definite number of cells. Each cell boundary has a transition-state point to connect that cell to another.

The solution curve of Eq. (2) that starts from a cell center to arrive the transition-state point is called "intrinsic reaction coordinate (IRC)." It is reasonable to regard that this is the "center line" of the path of chemical reactions.

The IRC formalism was used to discuss the geometry change along the reaction path [8, 12–14], the shape of potential brrier [12], the detailed reaction mechanism based on the geometry thus determined [8], the rate calculation along the IRC [15], and so forth.

The cell structure is characteristic of a given reacting system. The cell concept is essential in the wave-mechanical interpretation of chemical reaction between two equilibrium points.

A. *Differential Geometry of IRC Formalism*

One of the most important advantages of using differential geometrical methods exists in that the relation expressed by a tensor equation is valid independent of the coordinate transformation. For instance, the scalar equation

$$\delta(dW/ds) = 0$$

in which variation is carried out with respect to dq^i, gives the direction of IRC. This relation holds for any generalized coordinate, and implies that the IRC is the steepest descent (or ascent) path in the space of Eq. (1).

If the right-hand side of Eq. (2) is put equal to $d\tau$, eq. (2) is written as

$$a_{ij}\frac{dq^j}{d\tau} = \frac{\partial W}{\partial q^i} \qquad (i = 1, 2, \ldots n) \tag{3}$$

The parameter τ is a monotonously increasing, time-like parameter associated with the direction of length of each solution curve within one cell or within one subspace of intercell boundary. Along one solution curve, τ varies from $-\infty$ to $+\infty$. In the vicinity of an equilibrium point, τ varies infinitely.

The solution of Eq. (3) is represented by

$$q^i = q^i(\tau, c_i) \qquad (i = 1, 2, \ldots n;\ c_i\text{: a constant})$$

with mediation of the parameter τ. The relation between τ and the length of the solution curve, denoted by s, is given by $d\tau = ds/(dW/ds)$ since along a solution curve it follows immediately that

$$\frac{dW}{d\tau} = \frac{\partial W}{\partial q^i}\frac{dq^i}{d\tau} = a_{ij}\frac{dq^i}{d\tau}\frac{dq^j}{d\tau} = \left(\frac{ds}{d\tau}\right)^2$$

If we use contravariant tensor a^{ij}, which is obtained from a_{ij} by the well-known relation [7]

$$\mathbf{a}^{ij} = |a_{kl}|^{-1}\frac{\partial |a_{kl}|}{\partial a_{ij}} \qquad (a^{ij} = a^{ji})$$

$|a_{kl}|$ denoting the determinant of matrix $(\mathbf{a}_{ij})$, we get in place of Eq. (3)

$$\frac{dq^i}{d\tau} = a^{ij}\frac{\partial W}{\partial q^j} \tag{4}$$

since $a^{ik}a_{kj} = \delta_{ij}$. This is another useful formulation of the reaction path.

B. *Curvature of* IRC

The curvature κ of a curve $q^i(s)$ $(i = 1, 2, \ldots n)$ in the space of Eq. (1) is defined [5, 10] as

$$\kappa^2 = a_{ij}\kappa^i\kappa^j$$

where the vector κ^i has the form of

$$\kappa^i = \frac{d^2q^i}{ds^2} + \Gamma^i_{jk}\frac{dq^j}{ds}\frac{dq^k}{ds}$$

in which Γ^i_{jk} is the Christoffel symbol of the second kind [7].

As was mentioned in Sec. 2, a wave-mechanical calculation of the rate of chemical reactions was made practicable by the IRC approach [15]. In that case, to make the problem one dimensional, it was essentially necessary to use an "adiabaticity" assumption between the "translational" motion along the IRC and the vibrational motions in perpendicular directions. As is easily understood, the vibrational adiabaticity becomes unassured when the curvature of IRC is large [15]. In this way, the curvature of IRC in the configuration space of Eq. (1) possesses a close relation to the energy transfer in the motion along the reaction path.

C. *Normal Coordinates at a Nonequilibrium Point*

The normal coordinates of a semirigid molecule are usually obtained by diagonalizing the Hessian matrix—$\partial^2W/\partial q^i\partial q^j$ in which q^i's are internal coordinates—at an equilibrium point where $\partial W/\partial q^i = 0$. But at a nonequilibrium point the Hessian matrix should be defined as [10]

$$H_{ij} = \frac{\partial^2 W}{\partial q^i \partial q^j} - \Gamma^k_{ij}\frac{\partial W}{\partial q^k} \qquad (i = 1, 2, \ldots n) \tag{5}$$

so as to secure the invariance under coordinate transformations. In geometrical terminologies, H_{ij} is the covariant derivative of covariant vector $\partial W/\partial q^i$. The normal vibrations are obtained by diagonalizing the Hessian matrix thus defined [10].

4. Variational Principles of Chemical Reaction

By way of the IRC concept, the chemical reaction path can be derived from some variational principles just as in other dynamical phenomena.

A. Synge's Geodesic Principle

In contrast with the space of Eq. (1) mentioned in the preceding section, Synge [5] considered another interesting space, in which the line element is defined as

$$ds^2 = 2(E - W)T\,dt^2 = (E - W)a_{ij}dq^i dq^j \tag{6}$$

where E is the total energy of any dynamical system which is assumed *conservative.* This line element was named "action line element," while the line element of Eq. (1) was called "kinematical line element" [5]. It is well known that among the motions of E = const. the actual path is determined by the *principle of least action*

$$\delta \int 2T\,dt = 0$$

But in this space the relation $2T\,dt = \sqrt{2}ds$ holds along the actual path. Therefore, the actual motion is ruled by a variational equation

$$\delta \int ds = 0$$

if ds of Eq. (6) is applied here. As is well known [7], the Euler equation for this variation problem is nothing but the geodesic equation

$$\frac{d^2 g^i}{ds^2} + \Gamma^i_{jk}\frac{dq^j}{ds}\frac{dq^k}{ds} = 0 \qquad (i = 1, 2, \ldots n) \tag{7}$$

in which ds is given by Eq. (6). Hence, in this special space *the actual motion occurs along the geodesic line.*

Synge's geodesic principle is also easily ascertained directly by the use of geodesic equation, Eq. (7). The proof is as follows. The Lagrange equation of motion for the present case is given by [7]

$$a_{ij}\frac{d^2 g^j}{dt^2} + [kl, i]\frac{dq^k}{dt}\frac{dq^1}{dt} = -\frac{\partial W}{\partial q^i} \qquad (i = 1, 2, \ldots n) \tag{8}$$

in which $[kl, i]$ is the Christoffel symbol of the first kind [7]. It can be shown by a simple manipulation that the direction given by Eq. (8) satisfies Eq. (7) if we employ the rlation of Eq. (6) in Eq. (7).

B. Chemodynamical Hamilton's Principle

Let us again get back to the space of Eq. (1). We noticed in Sec. 2.A that the parameter τ is of a time-like nature. Actually, we can realize this analogy in the following argument for the principle of least action [16, 17].

As a matter of fact, instead of the time used in the traditional principle of least action mentioned in Sec. 3.A, the parameter τ is used in intrinsic dynamics to derive an analogous principle. For this purpose, "extended Lagrangian" was firstly defined as

$$L_\tau = T_\tau + \tfrac{1}{2}\Delta_1 W \tag{9}$$

in which

$$T_\tau = \tfrac{1}{2}\, a_{ij} \frac{dq^i}{d\tau}\frac{dq^j}{d\tau}$$

$$\Delta_1 W = a^{ij} \frac{\partial W}{\partial q^i}\frac{\partial W}{\partial q^j}$$

and T_τ is the "kinetic energy" term and $\Delta_1 W$ is Beltrami's differential parameter of the first order with respect to W [7].

If we use Eq. (4) it follows immediately that

$$2T_\tau = (ds/d\tau)^2 = (dW/d\tau) = (\partial W/\partial q^i)\,(dq^i/d\tau) = a^{ij}(\partial W/\partial q^i)(\partial W/\partial q^j) = \Delta_1 W$$

which is valid along a solution curve. Therefore, we have the relation

$$\int_{\tau_0}^{\tau_1} 2T_\tau\, d\tau = \int_{W(\tau_0)}^{W(\tau_1)} dW = W(\tau_1) - W(\tau_0)$$

along a solution curve. Accordingly, along two actual paths starting from the cell center at the "time" $\tau_0 = -\infty$ arriving an equipotential surface at the same "time" τ_1, this integral takes the same value [16].

It can be shown in this connection, as will be mentioned below, that the IRC is derived from "Hamilton's principle." Namely, the integral $\int_{\tau_0}^{\tau_1} L_\tau\, d\tau$ is defined with respect to any path $q^i = q^i(\tau)$ $(i = 1, 2, \ldots n)$ which passes two given points lying on one solution curve. Then, the path of actual motion results from the variation equation

$$\delta \int_{\tau_0}^{\tau_1} L_\tau\, d\tau = 0 \tag{10}$$

with the constraint that $q^i(\tau_0)$ and $q^i(\tau_1)$ are fixed for every path [17]. That is, along the actual path the integral $\int_{\tau_0}^{\tau_1} L_\tau\, d\tau$ has a stationary value [16]. In order to convince skeptics of this fact, it will be effective to imagine a virtual path $q^i = q^i(\tau,\epsilon)$ $(i = 1, 2, \ldots n)$ in which ϵ is a small parameter and $q^i(\tau,0)$ is assumed to satisfy the IRC equation. After a little lengthy, but common, calculation, it can be shown that

$$\int_{\tau_0}^{\tau_1} \left[\frac{\partial}{\partial \epsilon}(L_\tau)_{q^i = q^i(\tau,\epsilon)}\right]_{\epsilon=0} d\tau = 0$$

if we note that the relations (see Fig. 1)

$q^i(\tau_0,\epsilon) = q^i(\tau_0,0)$, $q^i(\tau_1,\epsilon) = q^i(\tau_1,0)$ and $(dq^i/d\tau)_{\epsilon=0} = (a^{ij}\partial W/\partial q^j)_{\epsilon=0}$

Since on a solution curve T_τ is equal to $\frac{1}{2}\Delta_1 W$, Eq. (10) can be substituted by the "principle of least action"

$$\delta \int_{\tau 0}^{\tau 1} 2T_\tau d\tau = 0$$

for the paths satisfying the relation

$$T_\tau - \tfrac{1}{2}\Delta_1 W = 0$$

Furthermore, the integral of Eq. (10) actually takes a minimum value along the solution curve [17]. This can be proved by the following logic: the integral

$$\int_{\tau 0}^{\tau 1} \left[\frac{\partial^2}{\partial \epsilon^2}(L_\tau)_{q^i=q^i(\tau,\epsilon)}\right]_{\epsilon=0} d\tau$$

is shown equal to

$$\int_{\tau 0}^{\tau 1} \left[\frac{\partial}{\partial \epsilon}\left\{\left(a_{ij}\frac{dq^j}{d\tau} - \frac{\partial W}{\partial q^i}\right)\right\}_{q^1=q^1(\tau,\epsilon)}\right]_{\epsilon=0} \left[\frac{\partial}{\partial \epsilon}\left\{\left(\frac{dq^i}{d\tau} - a^{ik}\frac{\partial W}{\partial q^k}\right)\right\}_{q^1=q^1(\tau,\epsilon)}\right]_{\epsilon=0} d\tau$$

The integrand has a form of $A_i A^i$, i.e., the magnitude of a contravariant vector A^i. But the metric of this space is positive definite, and accordingly, the integral is positive definite.

C. Chemodynamical Geodesic Principle

A more interesting variational problem is brought about by considering a Riemannian space in which the actual path becomes geodesics. Such a Riemannian space is constructed by taking the metric as [17]

$$ds^2 = (\Delta_1 W)a_{ij}dq^i dq^j \tag{11}$$

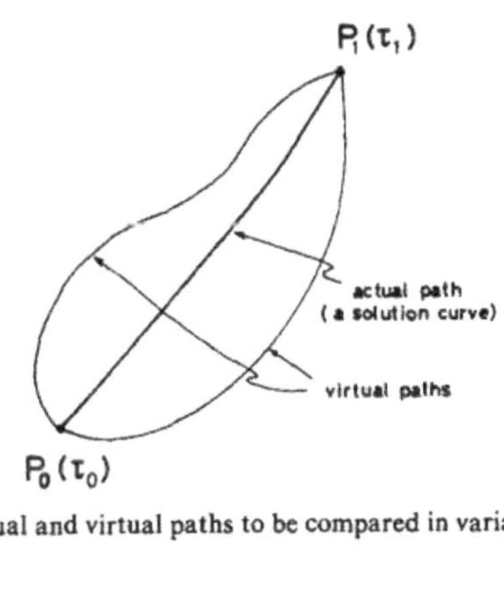

Figure 1. Actual and virtual paths to be compared in variational problems.

as corresponding to Eq. (6). In such a non-Euclidean space the line element on a solution curve becomes equal to the potential energy difference, since along a solution curve we have (see Sec. 3.B)

$$ds/d\tau = [(2T_\tau)(\Delta_1 W)]^{1/2} = 2T_\tau = dW/d\tau$$

Hence, the length of actual paths starting from the cell center getting to an equipotential surface is all equal in this special space.

Now, it is shown that the IRC equation is derived from the variational equation

$$\delta \int_{P_0}^{P_1} ds = 0 \tag{12}$$

where P_0 and P_1 lie on one solution curve. Eq. (12) is replaced by

$$\delta \int_{\tau_0}^{\tau_1} (ds/d\tau)\, d\tau = 0$$

with $q^i(\tau_0)$ and $q^i(\tau_1)$ fixed at points P_0 and P_1 just like in Sec. 3.B. If the variation of q^i is considered similarly, this equation is substituted by

$$\int_{\tau_0}^{\tau_1} \left[\frac{\partial}{\partial \epsilon}(ds/d\tau)_{q^i=q^i(\tau,\epsilon)}\right]_{\epsilon=0} d\tau$$

$$= \int_{\tau_0}^{\tau_1} \left[\left\{(\Delta_1 W)\frac{\partial}{\partial \epsilon}(T_\tau)_{q^i=q^i(\tau,\epsilon)} + T_\tau \frac{\partial}{\partial \epsilon}(\Delta_1 W)_{q^i=q^i(\tau,\epsilon)}\right\}\Big/(2T_\tau)\right]_{\epsilon=0} d\tau$$

$$= \int_{\tau_0}^{\tau_1} \left[\frac{\partial}{\partial \epsilon}(L_\tau)_{q^i=q^i(\tau,\epsilon)}\right]_{\epsilon=0} d\tau = 0$$

and this is true as was already mentioned in Sec. 3.B. An alternative direct proof is also possible, by showing that the curve given by the IRC equation satisfies the geodesic equation, Eq. (7), for this space.

This result implies that, in this interesting space, the actual path is the *shortest* path from a stable equilibrium point to the transition-state point, and the length of the IRC is equal to the activation energy [17].

Synge [5] established the condition that the orthogonal trajectories of a family of surfaces $F(q^1, q^2, \ldots q^n) =$ const. should be geodesics of the space $ds^2 = g_{ij}dq^idq^j$ is

$$(g^{ij}F_{ki}F_j)/F_k = (g^{ij}F_{li}F_j)/F_l \tag{13}$$

in which $F_k = \partial F/\partial q^k$ and $F_{ij} = \partial F_i/\partial q^j - \Gamma^k_{ij}F_k$. It can easily be shown that in our IRC problem we have

$$g^{ij}H_{ki}\partial W/\partial q^j = 0 \qquad \text{(independent of } k\text{)}$$

if we take $g_{ij} = (\Delta_1 W)a_{ij}$ according to Eq. (11). This equation is included in the condition of Eq. (13) when F is put equal to W, so that we can see Synge's condition herein satisfied.

The results mentioned in Secs. 3.A–C, are more understandable by the following analogy between traditional and intrinsic dynamics [17], which is indicated in Table I.

TABLE I

Variational principles	Traditional dynamics	Intrinsic dynamics
Hamilton's principle	$\delta \int L\, dt = 0$ $L = T - W$	$\delta \int L_\tau\, d\tau = 0$ $L_\tau = T_\tau + (1/2)\Delta_1 W$
Principle of least action	$\delta \int 2\, T\, dt = 0$ for paths with $T + W = E$	$\delta \int 2\, T_\tau\, d\tau = 0$ for paths with $T_\tau - (1/2)\Delta_1 W = 0$
Geodesic principle	$\delta \int ds = 0$ in the space of $ds^2 = (E - W)\, 2T\, dt^2$	$\delta \int ds = 0$ in the space of $ds^2 = (\Delta_1 W)\, 2T_\tau d\tau^2$

Bibliography

[1] K. Fukui, J. Phys. Chem. **74,** 4167 (1970).

[2] K. Fukui, *The World of Quantum Chemistry.* Proceedings of the First International Congress of Quantum Chemistry, Menton, 1973, R. Daudel and B. Pullman, Eds. (Reidel, Dordrecht, 1974).

[3] E. B. Wilson, Jr., J. Chem. Phys. **7,** 1047 (1939).

[4] K. Fukui, A. Tachibana, and K. Yamashita, Int. J. Quantum Chem., to appear.

[5] J. L. Synge, Phil. Trans. **A226,** 31 (1926).

[6] E. T. Whittaker, *A Treatise on the Analytical Dynamics of Particles and Rigid Bodies* (Cambridge University Press, Cambridge, 1937).

[7] See any text book on differential geometry, such as D. Laugwitz, *Differential and Riemannian Geometry* (Academic, New York, 1965); L. P. Eisenhart, *Riemannian Geometry* (Princeton U.P., Princeton, 1963).

[8] K. Fukui, S. Kato, and H. Fujimoto, J. Am. Chem. Soc. **97,** 1 (1972).

[9] K. Fukui, Recl. Trav. Chim. Pays-Bas **98,** 75 (1979).

[10] A. Tachibana and K. Fukui, Theoret. Chim. Acta (Berlin) **49,** 321 (1978).

[11] A. Tachibana and K. Fukui, Theoret. Chim. Acta (Berlin) **51,** 189 (1979).

[12] S. Kato and K. Fukui, J. Am. Chem. Soc. **98,** 6395 (1976).

[13] K. Ishida, K. Morokuma, and A. Komornicki, J. Chem. Phys. **66,** 2153 (1977).

[14] B. D. Joshi and K. Morokuma, J. Chem. Phys. **67,** 4880 (1977).

[15] S. Kato, H. Kato, and K. Fukui, J. Am. Chem. Soc. **99,** 684 (1977).

[16] A. Tachibana and K. Fukui, Theoret. Chim. Acta (Berlin) **51,** 275 (1979).
[17] A. Tachibana and K. Fukui, Theoret. Chim. Acta (Berlin) **57,** 81 (1980).

Received April 20, 1981

Reprinted from Accounts of Chemical Research, 1981, 14, 363–368.

The Path of Chemical Reactions — The IRC Approach

KENICHI FUKUI

Department of Hydrocarbon Chemistry, Kyoto University, Kyoto, Japan

Received February 23, 1981 (Revised Manuscript Received August 27, 1981)

The absolute rate theory of Eyring,[1] which provided a highly useful theoretical tool for studying chemical reactions, is essentially associated with the concept of "transition state", a unique state on the "potential energy surface" of the reacting system. To obtain the absolute rate theoretically involves the following steps: (i) calculating the potential energy function in terms of the atomic positions; (ii) obtaining the transition state and reactant and product points; (iii) unequivocally determining the reaction path connecting the relevant stable points and the transition state point; and (iv) calculating the absolute rate of the nuclear rearrangement along the reaction path by appropriate quantum statistical mechanical approaches.

Modern ab initio quantum-chemical method are capable of calculating the total energy of a molecular system with a fixed nuclear arrangement—adiabatic potential, V—based on the Born–Oppenheimer approximation.[2] We represent the nuclear configuration of a chemically reacting system composed of N nuclei by a point in $3N$-dimensional configuration space with space-fixed $3N$ Cartesian coordinates, X_α, Y_α, Z_α (α = 1, 2, ... N). The potential energy function, V, can be represented in terms of X_α, Y_α, Z_α, and we can obtain the equilibrium points as satisfying

$$\frac{\partial V}{\partial X_\alpha} = \frac{\partial V}{\partial Y_\alpha} = \frac{\partial V}{\partial Z_\alpha} = 0 \quad (\alpha = 1, 2, \ldots N)$$

Practical methods of calculating the energy gradient and of locating the equilibrium points have been developed.[3]

On the basis of results already established for steps i and ii, the present Account aims to discuss steps iii and iv. Molecules react with each other while vibrating about an equilibrium configuration, rotating as a whole around the center of mass, and exchanging energies in all degrees of freedom. The molecular changes in nuclear configuration are so varied that it is difficult to imagine a simple classical picture of the motion. On the other hand, the chemical reaction path is a concept which has a definite foundation from empirical observations. The problem of how to connect this empirical concept of a reaction path to a model relevant to theoretical treatment is a matter of deep concern in chemistry. A number of different approaches have been developed for defining the reaction path.[4-7] Here, the discussion is focused on our "intrinsic reaction coordinate" or "IRC" approach.

The IRC approach is based on the classical equations of motion, written as

$$\frac{\mathrm{d}}{\mathrm{d}t}(M_\alpha \dot{X}_\alpha) = -\frac{\partial V}{\partial X_\alpha} \quad \text{etc.}$$

Consider movement of the nuclei from a given point with an *infinitesimal* velocity. The velocity of nuclei is given by

$$M_\alpha \dot{X}_\alpha = -\frac{\partial V}{\partial X_\alpha} t + \text{constant, etc.}$$

for a small time interval. Since, by assumption, the nuclei start to move with an infinitesimal velocity, i.e.,

Kenichi Fukui is Professor of Chemistry in the Faculty of Engineering of Kyoto University, Kyoto, Japan. He was born in Nara, the ancient imperial capital of Japan, and received his bachelor's and doctor's degrees from Kyoto University. Professor Fukui's research interests primarily involve the theory of chemical reactions. He is chiefly responsible for developing "frontier orbital theory" which is now widely used to treat orientation and stereoselection in chemical reactions. He is a Foreign Associate of the National Academy of Sciences and a Member of the International Academy of Quantum Molecular Science. He was awarded the 1981 Nobel Prize in Chemistry.

(1) (a) S. Glasstone, K. Laidler, and H. Eyring, "The Theory of Rate Processes", McGraw-Hill, New York, 1941; (b) also see, e.g., K. J. Laidler and A. Tweedale, *Adv. Chem. Phys.*, **21**, 113 (1971); L. Hofacker, *Int. J. Quantum Chem.* **35**, 33 (1969).

(2) (a) D. R. Yarkony, W. J. Hunt, and H. F. Schaefer, *Mol. Phys.*, **26**, 941 (1973), and references cited therein; (b) B. Liu, *J. Chem. Phys.*, **58**, 1924 (1973); (c) H. F. Schaefer in "Atom-Molecule Collision Theory: A Guide for the Experimentalists", R. B. Bernstein, Ed., Plenum, New York, 1979, p 45.

(3) (a) J. W. McIver, Jr., and A. Komornicki, *J. Am. Chem. Soc.*, **94**, 2625 (1972); (b) A. Komornicki, K. Ishida, K. Morokuma, R. Ditchfield, and M. Conrad, *Chem. Phys. Lett.*, **45**, 595 (1977).

(4) (a) R. A. Marcus, *J. Chem. Phys.*, **45**, 4493, 4500 (1966); **49**, 2610, 2617 (1968); (b) J. C. Light, *Adv. Chem. Phys.*, **19**, 1 (1971); (c) P. Russegger and J. Brickmann, *J. Chem. Phys.*, **66**, 1 (1977); P. Russegger, *Chem. Phys.*, **22**, 41 (1977); **34**, 329 (1978); **41**, 299 (1979).

(5) (a) R. A. Marcus, *J. Chem. Phys.*, **45**, 2138, 2630 (1966); (b) K. Morokuma and M. Karplus, *ibid.*, **55**, 63 (1971); (c) D. L. Bunker, *Methods Comput. Phys.*, **10**, 287 (1971); (d) R. N. Porter, *Annu. Rev. Phys. Chem.*, **25**, 317 (1974); (e) W. H. Millers, *J. Chem. Phys.*, 61, 1823 (1974); (f) R. B. Walker and J. C. Light, *Annu. Rev. Phys. Chem.*, **31**, 401 (1980).

(6) R. F. Nalewajski, *Int. J. Quant. Chem.*, **12**, 87 (1978).

(7) (a) J. N. Murrell, *Struct. Bonding (Berlin)* **32**, 93 (1977); (b) J. F. McNutt and R. E. Wyatt, *J. Chem. Phys.*, **70**, 5307 (1979); (c) K. Müller, *Angew. Chem., Int. Ed. Engl.*, **19**, 1 (1980); K. Müller and L. D. Brown, *Theor. Chim. Acta*, **53**, 75 (1979); (d) M. V. Basilevsky and V. M. Ryaboy, *Chem. Phys.*, **50**, 231 (1980); (e) P. G. Mezey, *Theor. Chim. Acta*, **54**, 95 (1980); **58**, 309 (1981); (f) K. Jug, *ibid.*, **54**, 263 (1980); (g) M. Sana, G. Reckinger, and G. Leroy, *ibid.*, **58**, 145 (1981).

$\dot{X}_\alpha = 0$ at $t = 0$, the integration constant should be zero. This incremental change introduces the concept of "intrinsic motion".

In this way, we obtain the simultaneous equations[8]

$$\cdots = \frac{M_\alpha dX_\alpha}{\frac{\partial V}{\partial X_\alpha}} = \frac{M_\alpha dY_\alpha}{\frac{\partial V}{\partial Y_\alpha}} = \frac{M_\alpha dZ_\alpha}{\frac{\partial V}{\partial Z_\alpha}} = \cdots \quad (1)$$

If we adopt the "mass-weighted" cartesian coordinates x_i ($i = 1, 2, \ldots 3N$) in which

$$M_\alpha^{1/2} X_\alpha = x_{3\alpha-2}, \quad M_\alpha^{1/2} Y_\alpha = x_{3\alpha-1}, \quad M_\alpha^{1/2} Z_\alpha = x_{3\alpha}$$

eq 1 becomes

$$\frac{dx_1}{\frac{\partial V}{\partial x_1}} = \frac{dx_2}{\frac{\partial V}{\partial x_2}} = \cdots = \frac{dz_{3N}}{\frac{\partial V}{\partial x_{3N}}} \quad (2)$$

with the kinetic energy expression

$$T = \frac{1}{2} \sum_{i=1}^{3N} \dot{x}_i^2$$

The basic idea of this formalism was proposed in 1970,[9] and the solution of this equation which reaches a transition-state point from a stable equilibrium point was named as "intrinsic reaction coordinate (IRC)".[10] Equation 1 as well as eq 2 may be called the "IRC equation". These are regarded as the fundamental equations which determine the "center line" of the reaction path region, since the IRC represents the vibrationless–rotationless motion path of the reacting system. The IRC equation has an infinite number of solutions in addition to the IRC. These solutions are generally called "meta-IRC's".[11]

Nature of the IRC

We designate the infinitesimal distance of the mass-weighted cartesian configuration space as ds. If we denote each side of eq 2 as $d\tau$, we have the relation

$$ds^2 = \sum_{i=1}^{3N} dx_i^2 = \sum_{i=1}^{3N} \left(\frac{\partial V}{\partial X_i} \right)^2 d\tau^2$$

which is valid on a meta-IRC. Also it should be noticed that on a meta-IRC the useful relation

$$d\tau = \frac{ds}{\frac{dV}{ds}} \quad (3)$$

holds with respect to the parameter τ, since on a meta-IRC we have

$$dV = \sum_{i=1}^{3N} \frac{\partial V}{\partial x_i} dx_i = \sum_{i=1}^{3N} \left(\frac{\partial V}{\partial x_i} \right)^2 d\tau$$

Therefore, we can express the IRC equation in another compact form as

$$\frac{dx_i}{ds} = \frac{\frac{\partial V}{\partial x_i}}{\frac{dV}{ds}} \quad (i = 1, 2, \cdots 3N) \quad (4)$$

If the space is provided with the metric $ds^2 = 2T dt^2$, the IRC is then shown to be the path for which the gradient of the potential energy, dV/ds, is an extremum.[9,12] In this sense, the IRC descending from the transition-state point may be called the "steepest descent path". It is evident from eq 3 that the symmetry, if any exist in the geometry of the reacting system, is conserved along the IRC, so that the orbital symmetry is also conserved along the IRC.[9,13]

Reaction Ergodography

Equation 1 or 2 comprises a set of differential equations. As stated before, they have an infinite number of solution curves. The particular solution which corresponds to the IRC is easily obtained if we start from the transition-state point and successively plot the neighboring points by a numerical gradient calculation along the direction given by the IRC equation. The following examples have been considered in several papers[14–19] in which the change in geometry of the reacting molecules along the IRC was given:

$$CH_4 + T \rightarrow CH_3T + H \quad (i^{14})$$

$$CD_4 + T \rightarrow CD_3T + D \quad (ii^{15})$$

$$CH_4 + T \rightarrow CH_3 + HT \quad (iii^{15})$$

$$CD_4 + T \rightarrow CD_3 + DT \quad (iv^{15})$$

$$HNC \rightarrow HCN \quad (v^{16})$$

$$CH_4 + H^- \rightarrow H^- + CH_4 \quad (vi^{17})$$

$$CHFO \rightarrow CO + HF \quad (vii^{18})$$

$$C_2H_5F \rightarrow C_2H_4 + HF \quad (viii^{19})$$

The procedure of plotting the IRC for a chemical reaction is called "reaction ergodography".[15]

The IRC calculation determines the energy as well as the geometry of the reacting system. The energy change along the IRC gives the potential barrier shape. In this case the width of the barrier provides information about the isotope effect.[15] For instance, the effect of deuteration in the methane–tritium reactions was studied.[15] In the abstraction mode ((iii) and (iv)) the barrier broadening was remarkable *before* reaching the

(8) K. Fukui, "The World of Quantum Chemistry. Proceedings of the First International Congress of Quantum Chemistry, Menton, 1973", R. Daudel and B. Pullman, Eds., D. Reidel, Dordrecht, Holland, 1974, p 113.

(9) K. Fukui, *J. Phys. Chem.*, **74**, 4161 (1970); also see ref 13 in ref 14.

(10) Why the term "intrinsic" was used here was to introduce the concept of "intrinsic motion"—a quasistatic nuclear displacement. The term "intrinsic path" was suggested by Mezey[7e] instead of "intrinsic coordinate". This may be reasonable since an IRC is a mathematical curve, but this curve has been obtained through a number of too simplified assumptions to be named "path" of a chemical reaction.

(11) A. Tachibana and K. Fukui, *Theor. Chim. Acta*, **49**, 321 (1978).

(12) (a) K. Fukui, A. Tachibana, and K. Yamashita, *Int. J. Quantum Chem.*, in press; (b) K. Fukui, *ibid.*, in press, and many papers cited therein.

(13) K. Fukui, *Recl. Trav. Chim. Pays-Bas*, **98**, 75 (1979).

(14) K. Fukui, S. Kato, and H. Fujimoto, *J. Am. Chem. Soc.*, **97**, 1 (1975).

(15) S. Kato and K. Fukui, *J. Am. Chem. Soc.*, **98**, 6395 (1976).

(16) K. Ishida, K. Morokuma, and A. Komornicki, *J. Chem. Phys.*, **66**, 2153 (1977).

(17) B. D. Joshi and K. Morokuma, *J. Chem. Phys.*, **67**, 4880 (1977).

(18) K. Morokuma, S. Kato, and K. Hirao, *J. Chem. Phys.*, **72**, 6800 (1980).

(19) S. Kato and K. Morokuma, *J. Chem. Phys.*, **73**, 3900 (1980).

transition state, whereas for the substitution mode ((i) and (ii)) it occurred, even more remarkably than the abstraction case, *after* the transition state. These results suggest that the mass effect mainly relates to the C–H stretching mode. The effect in the abstraction is reduced by that on the H–T bond to be formed, whereas in the substitution case the effect in the CH_3 inversion adds.

Once the geometry along the IRC has been determined, we can discuss the force acting on atoms to displace them.[14] The potential energy gradient is partitioned by dividing the electron density into several terms, including exchange, delocalization, and polarization terms.[20] By such treatments the origin of favorableness and unfavorableness of a given reaction path is discussed in terms of orbital interactions.[21] In the case of methane–tritium reaction i, for instance, the importance of $3a_1 \rightarrow 4a_1$ local excitation configuration in methane was quantitatively pointed out.[14]

If desired, we can combine the IRC approach with the considerations of the interaction of particular orbitals[8,21]—the interaction of the highest occupied molecular orbital (HOMO) and the lowest unoccupied molecular orbital (LUMO)—to discuss the nature of reaction paths and the detailed mechanism of that reaction. With each geometry along the reaction path, the importance of a particular orbital interaction can be identified.[22-25] A recent study to visualize the phase pattern of one single "interaction frontier orbital"[26] or "interactive hybrid MO"[27] for each reactant molecule helps to elucidate the mechanism of a chemical reaction. An animated picture displaying the essential features of a chemical reaction would be a powerful heuristic device.

The IRC Equation with Internal Coordinates

Although the use of $3N$ Cartesian coordinates encounters essentially no significant difficulty in the reaction ergodography, it is ovbiously not suitable for discussing the potential energy surface, since these coordinates include translational and rotational motion. Thus, it is worthwhile to reduce the number of variables to $3N - 6$ by eliminating the translation and the rotation of the reacting system as a whole.

Consider any set of internal coordinates which is sufficient to uniquely describe the nuclear arrangment of the reacting system. Denote these by $q_1, q_2, \ldots q_n$ ($n = 3n - 6$), which are expressible in terms of $3N$ Cartesian coordinates X_α, Y_α, Z_α. Then, the following relations are obtained.

$$dq_i = \sum_{\alpha=1}^{N}\left(\frac{\partial q_i}{\partial X_\alpha}dX_\alpha + \frac{\partial q_i}{\partial Y_\alpha}dY_\alpha + \frac{\partial q_i}{\partial Z_\alpha}dZ_\alpha\right) \quad (i = 1, 2, \ldots n) \tag{5}$$

We add the following six relations after Eckart:[28]

$$\sum_\alpha M_\alpha dX_\alpha = \sum_\alpha M_\alpha dY_\alpha = \sum_\alpha M_\alpha dZ_\alpha = 0$$

$$\sum_\alpha M_\alpha(Y_\alpha dZ_\alpha - Z_\alpha dY_\alpha) = \sum_\alpha M_\alpha(Z_\alpha dX_\alpha - X_\alpha dZ_\alpha) = \sum_\alpha M_\alpha(X_\alpha dY_\alpha - Y_\alpha dX_\alpha) = 0 \tag{6}$$

which assures a zero total momentum and zero total angular momentum motion. Note that the increments given by eq 1 satisfy the relation of eq 6.[8] Then, we get $3N$ linear simultaneous equations with respect to $3N$ unknowns dX_α, dY_α, dZ_α, from which we obtain

$$dX_\alpha = \sum_{i=1}^{n}\frac{\partial X_\alpha}{\partial q_i}dq_i, \quad dY_\alpha = \sum_{i=1}^{n}\frac{\partial Y_\alpha}{\partial q_i}dq_i, \quad dZ_\alpha = \sum_{i=1}^{n}\frac{\partial Z_\alpha}{\partial q_i}dq_i \tag{7}$$

If we put these relations into the kinetic energy formula, we obtain

$$T = \tfrac{1}{2}\sum_{i,j=1}^{n} a_{ij}\dot{q}_i\dot{q}_j \tag{8}$$

in which

$$a_{ij} = \sum_{\alpha=1}^{N} M_\alpha\left(\frac{\partial X_\alpha}{\partial q_i}\frac{\partial X_\alpha}{\partial q_j} + \frac{\partial Y_\alpha}{\partial q_i}\frac{\partial Y_\alpha}{\partial q_j} + \frac{\partial Z_\alpha}{\partial q_i}\frac{\partial Z_\alpha}{\partial q_j}\right)$$

can be represented in tems of *internal* coordinates q_i's.[12] In actuality the direct calculation of the elements a_{ij} is never an easy task. But the inverse of the matrix (a_{ij}), denoted by $(a^{-1}{}_{ij})$, can easily be obtained from

$$a^{-1}{}_{ij} = \sum_{\alpha=1}^{N} M_\alpha^{-1}\left(\frac{\partial q_i}{\partial X_\alpha}\frac{\partial q_j}{\partial X_\alpha} + \frac{\partial q_i}{\partial Y_\alpha}\frac{\partial q_j}{\partial Y_\alpha} + \frac{\partial q_i}{\partial Z_\alpha}\frac{\partial q_j}{\partial Z_\alpha}\right)$$

The proof that this is an element of the inverse matrix of (a_{ij}) was established by Wilson.[29] Therefore, a_{ij} can be obtained by calculating the inverse matrix of $(a^{-1}{}_{ij})$.

We can follow the same procedure which was used to derive eq 1 to derive the IRC equation with internal motion,

$$\frac{\sum_{j=1}^{n} a_{1j}dq_j}{\dfrac{\partial V}{\partial q_1}} = \frac{\sum_{j=1}^{n} a_{2j}dq_j}{\dfrac{\partial V}{\partial q_2}} = \ldots = \frac{\sum_{j=1}^{n} a_{nj}dq_j}{\dfrac{\partial V}{\partial q_n}} \tag{9}$$

or, alternatively, if written in terms of $a^{-1}{}_{ij}$

$$\frac{dq_1}{\sum_{j=1}^{n} a^{-1}{}_{1j}\dfrac{\partial V}{\partial q_j}} = \frac{dq_2}{\sum_{j=1}^{n} a^{-1}{}_{2j}\dfrac{\partial V}{\partial q_j}} = \ldots = \frac{dq_n}{\sum_{j=1}^{n} a^{-1}{}_{nj}\dfrac{\partial V}{\partial q_j}} \tag{10}$$

It is easy to prove that the trace of geometry of the

(20) H. Fujimoto and K. Fukui, *Israel J. Chem.*, **19**, 317 (1980).
(21) K. Fukui, "Theory of Orientation and Stereoselection", Springer, Berlin, 1970, 1975; *Acc. Chem. Res.*, **4**, 57 (1971).
(22) K. Fukui, "23rd International Congress of Pure and Applied Chemistry", Vol. 1, Butterworths, London, 1971, p 65.
(23) H. Fujimoto, S. Yamabe, and K. Fukui, *Bull. Chem. Soc. Jpn.*, **45**, 1556, 2424 (1972); H. Fujimoto, S. Yamabe, T. Minato, and K. Fukui, *J. Am. Chem. Soc.*, **94**, 9205 (1972); S. Kato, H. Fujimoto, S. Yamabe, and K. Fukui, *ibid.*, **96**, 2024 (1974); S. Yamabe, T. Minato, H. Fujimoto, and K. Fukui, *Theor. Chim. Acta*, **32**, 187 (1974); H. Fujimoto, S. Kato, S. Yamabe, and K. Fukui, *J. Chem. Phys.*, **60**, 572 (1974).
(24) H. Fujimoto and K. Fukui in "Chemical Reactivity and Reaction Paths", G. Klopman, Ed., Wiley, New York, 1974, p 23.
(25) T. Minato, S. Yamabe, H. Fujimoto, and K. Fukui, *Bull. Chem. Soc. Jpn.*, **51**, 1 (1978); S. Yamabe, T. Minato, H. Fujimoto, and K. Fukui, *ibid.*, **52**, 3243 (1979).
(26) K. Fukui, N. Koga, and H. Fujimoto, *J. Am. Chem. Soc.*, **103**, 196 (1981).
(27) H. Fujimoto, N. Koga, M. Endo, and K. Fukui, *Tetrahedron Lett.* **22**, 1263 (1981); *ibid.*, in press; H. Fujimoto, N. Koga, and K. Fukui, *J. Am. Chem. Soc.*, in press.

(28) C. Eckart, *Phys. Rev.*, **47**, 552 (1935).
(29) E. B. Wilson, Jr., *J. Chem. Phys.*, **7**, 1047 (1939); E. B. Wilson, Jr., J. C. Decius, and P. C. Cross, "Molecular Vibrations. The Theory of Infrared and Raman Vibrational Spectra", McGraw-Hill, New York, 1955.

reacting molecules given by eq 9 agrees with that obtained by eq 1.

For the purpose of deriving various analytical relations, one of the most convenient sets of internal coordinates is the following. If we denote the internuclear distance between two nuclei α and β as $R_{\alpha\beta}$, we can take as the variables $q_1, q_2, \ldots q_n$.

$$q_1 = R_{23}, \quad q_2 = R_{31}, \quad q_3 = R_{12}$$

$$q_{3r+m} = R_{m,3+r} \ (r = 1, 2, \ldots N-3;\ m = 1, 2, 3)$$

This set of coordinates is, for brevity, referred to as "internal F coordinates".[12b] If we adopt the internal F-coordinate system, all elements of the matrix $(a^{-1}{}_{ij})$ are conveniently and explicitly expressed in terms of $q_1, q_2, \ldots q_n$.[30]

Perhaps the simplest illustration of the internal IRC is for the collinear triatomic reaction. Here, the IRC equation is

$$\mathrm{d}q_1/\mathrm{d}q_2 = \frac{(M_2^{-1} + M_3^{-1})\partial V/\partial q_1 - M_2^{-1}\partial V/\partial q_2}{-M_2^{-1}\partial V/\partial q_1 + (M_1^{-1} + M_2^{-1})\partial V/\partial q_2}$$

with the nuclear arrangment

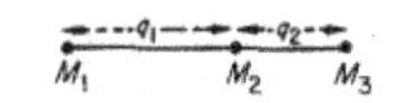

This corresponds to a kinetic energy expression with constant a_{ij}'s

$$T = 2^{-1}(M_1 + M_2 + M_3)^{-1}\{M_3(M_1 + M_2)\dot{q}_1^2 + 2M_1M_3\dot{q}_1\dot{q}_2 + M_1(M_2 + M_3)\dot{q}_2^2\}$$

Since this quadratic form is converted to a diagonal one by a linear transformation of q_1 and q_2, the IRC can be discussed in cartesian space by the use of skewed coordinate axes taken by Eyring.[1a]

Complete Classical Hamiltonian of a Reacting System

Recently, the kinetic energy formula for large amplitude motions in nonrigid molecules has attracted the attention of theoreticians. Several papers have been published which discuss the Hamiltonian appropriate for these problems.[31-33]

One of the least rigid molecules is the chemically reacting system regarded as a supermolecule. An *exact* kinetic energy formula pertinent to this system is represented by[12a,32]

$$T = (1/2)\sum_{i,j=1}^{3N-6} a_{ij}\dot{q}_i\dot{q}_j + (M/2)(\dot{X}^2 + \dot{Y}^2 + \dot{Z}_2) + (1/2)(I_1\sigma_1^2 + I_2\sigma_2^2 + I_3\sigma_3^2) + (1/2)(I_1^{-1}m_1^2 + I_2^{-1}m_2^2 + I_3^{-1}m_3^2) + (\sigma_1 m_1 + \sigma_2 m_2 + \sigma_3 m_3) \quad (11)$$

in which the first term on the right-hand side is the internal motion within the conditions of eq 6, the second is the translational energy of the center of mass, M being the total mass, X, Y, Z are the space-fixed cartesian coordinates of the center of mass, and the remaining terms originate from nonzero total angular momentum. The third term is the usual rotational energy, σ_r (r = 1, 2, 3) being the components of angular velocity about the instantaneous principal axes of inertia with instantaneous principal moments of inertia I_r (r = 1, 2, 3). The angular velocity components σ_r are given by the Eulerian angles θ, ϕ, ψ[34] and their time derivatives as

$$\sigma_1 = (\sin\psi)\dot{\theta} - (\sin\theta\cos\psi)\dot{\phi}$$

$$\sigma_2 = (\cos\psi)\dot{\theta} + (\sin\theta\sin\psi)\dot{\phi}$$

$$\sigma_3 = \dot{\psi} + (\cos\theta)\dot{\phi}$$

The fourth term arises from the internal rotational motion of each nucleus around the center of mass with respect to the principal axes of inertia. The m_r (r = 1, 2, 3) are given by

$$m_1 = \sum_{\alpha=1}^{N} M_\alpha(\xi_{2\alpha}\dot{\xi}_{3\alpha} - \xi_{3\alpha}\dot{\xi}_{2\alpha}), \quad \text{etc.}$$

in which $\xi_{1\alpha}$, $\xi_{2\alpha}$, $\xi_{3\alpha}$ are the Cartesian coordinates of nucleus α whose axes are coincident with the three principal axes of inertia. The I_r as well as $\xi_{r\alpha}$ are obtained as functions of the internal coordinates q_i by use of the equations

$$q_i' = q_i'(q_1, q_2, \ldots q_{3N-6}) \ (i = 1, 2, \ldots 3N-6)$$

$$(q_i')^2 = (\xi_{1\alpha} - \xi_{1\beta})^2 + (\xi_{2\alpha} - \xi_{2\beta})^2 + (\xi_{3\alpha} - \xi_{3\beta})^2$$

$$\sum_\alpha M_\alpha\xi_{1\alpha} = \sum_\alpha M_\alpha\xi_{2\alpha} = \sum_\alpha M_\alpha\xi_{3\alpha} = 0$$

$$\sum_\alpha M_\alpha\xi_{2\alpha}\xi_{3\alpha} = \sum_\alpha M_\alpha\xi_{3\alpha}\xi_{1\alpha} = \sum_\alpha M_\alpha\xi_{1\alpha}\xi_{2\alpha} = 0$$

in which q_i denote any set of internal coordinates and q_i' are the internal F coordinates. The last term of the right-hand side of eq 11 stands for the interaction between the rotation "as a whole" and the internal (vibrational) rotation of each nucleus.

The utility of eq 11 is that it is a general expression of kinetic energy for a reacting system and is represented in terms of $3N$ coordinates X, Y, Z, θ, ϕ, ψ, q_i (i = 1, 2, ... $3N-6$) and their time derivatives. Accordingly, we can easily set up Lagranges's equations of motion to discuss the classical dynamics. Also the momenta conjugate to these coordinates, and the classical Hamiltonian, may be obtained by the usual procedures[12,32] which leads to the quantum-mechanical operator form.[35]

The Cell Structure of the Configuration Space

One of the most important features of the IRC approach with internal coordinates comes about in connection with the cell structure of the internal configuration space.[36] The whole space is divided into cells by each set of meta-IRC's, each containing one stable equilibrium point normally. Each meta-IRC belonging to a cell starts from the equilibrium point and runs toward the cell boundary. The intercell boundary is composed of $(n-1)$-dimensional subspaces, each containing normally one transition-state point which connects that cell to another. The stable equilibrium point of a cell is called the "cell center". When we discuss

(30) See, however, the remarks mentioned in Appendix 1 of ref 12a.
(31) R. Meyer and Hs. H. Günthard, *J. Chem. Phys.*, 49, 1510 (1968).
(32) H. M. Pickett, *J. Chem. Phys.*, 56, 1715 (1972).
(33) (a) P. R. Bunker, *J. Chem. Phys.*, 47, 718 (1967); 48, 2832 (1968); J. T. Hougen, P. R. Bunker, and J. W. C. Jons, *J. Mol. Spectros.*, 34, 136 (1970); (b) A. Sayvetz, *J. Chem. Phys.*, 7, 383 (1939); J. K. G. Watson, *Mol. Phys.*, 15, 479 (1968); 19, 465 (1970); C. R. Quade, *J. Chem. Phys.*, 65, 700 (1976); (c) D. C. Moule and Ch. V. S. Ramachandra Rao, *J. Mol. Spectros.*, 45, 120 (1973); J. C. D. Brand and Ch. V. S. Ramachandra Rao, *ibid.*, 61, 360 (1976); P. G. Mezey and V. S. Ramachandra Rao, *J. Chem. Phys.*, 72, 121 (1980); (d) F. Jørgensen, *Int. J. Quantum Chem.*, 14, 55 (1978).
(34) E. T. Whittaker, "A Treatise on the Analytical Dynamics of Particles and Rigid Bodies", Cambridge University Press, Cambridge, 1937.
(35) B. Podolsky, *Phys. Rev.*, 32, 812 (1928).
(36) A. Tachibana and K. Fukui, *Theor. Chim. Acta*, 51, 189 (1979).

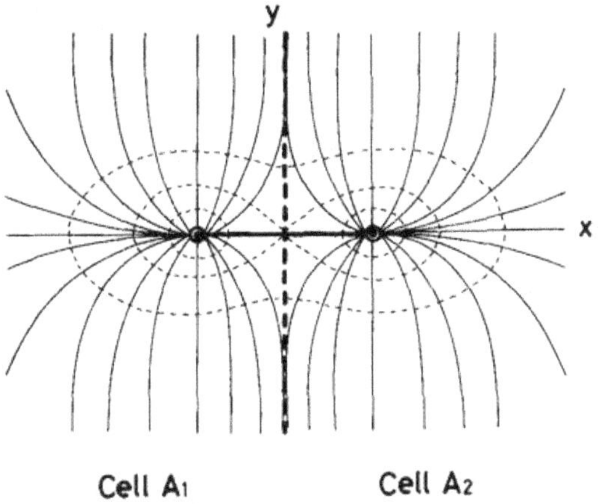

Figure 1. The cell structure indicated by a simple model. (⊙) Cell center; (×) transition-state point; (- - -) equipotential surface (line; (—) meta-IRC: A heavy dashed line represents the intercell boundary and a heavy straight line the IRC. $y = \text{const.}[|x^2 - 1|^{1/2}/|x|]^{b^2/2a^2}$ $(b^2 > 2a^2)$.

a chemical reaction between two points corresponding to initial and final states, the cell containing the initial state is named the initial cell and that containing the final state, the final cell.

The characteristics of the cell structure are most tangibly understood by adopting the following simple model:[11]

$$ds^2 = dx^2 + dy^2 = 2Tdt^2$$
$$V = a^2(1 - x^2)^2 + b^2y^2$$

It is seen in Figure 1 that meta-IRC's are perpendicular to equipotential surfaces. This is a general feature of IRC. It is possible that on the cell boundary surrounding one cell there may exist more than one transition state point. An illustrative model is

$$ds^2 = dr^2 + r^2d\phi^2 = 2Tdt^2$$

$$V = (r - a)^2 + a^2[\sin^2 (n/2)]\phi \quad (r, \phi\text{:polar coordinates})$$

The result is given in Figure 2.

The cell structure is characteristic of a given reacting system. The cell concept is essential for a wave-mechanical interpretation of a chemical reaction between two neighboring cells. As might be anticipated, it has been shown that techniques from differential geometry[37] are effective in developing the IRC approach.[11,36,38-40] Several interesting mathematical results have been obtained concerning the IRC.[11,12,36,38-40]

Absolute Rate of Chemical Reactions

The cell structure of configuration space is essentially important for the wave-mechanical evaluation of the absolute rate of chemical reactions. The chemical reaction is regarded as a large-amplitude nuclear motion in the internal configuration space. The amplitude may in many cases be infinitely large. Here, however, we imagine the atomic rearrangement only between finite cells, corresponding to intramolecular reactions.

(37) J. L. Synge, *Philos. Trans. R. Soc. London, Ser. A*, **226**, 31 (1926); also see any textbook available on differential geometry.
(38) A. Tachibana and K. Fukui, *Theor. Chim. Acta*, **51**, 275 (1979).
(39) A. Tachibana and K. Fukui, *Theor. Chim. Acta*, **57**, 81 (1980).
(40) A. Tachibana, *Theor. Chim. Acta*, **58**, 301 (1981).

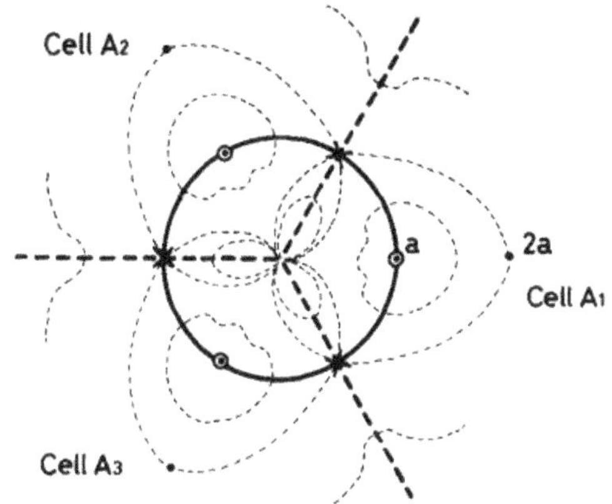

Figure 2. A model for the two-dimensional n-cell space. (The case $n = 3$ is indicated.) See Figure 1 for identification of symbols.

Consider one of the cells as the initial cell. Let the Hamiltonian operator for the internal motion of the reacting system be $\mathcal{H}$. As has been mentioned before, we can assume that the classical expression of Hamiltonian of the form

$$\mathcal{H} = \tfrac{1}{2}\sum_{i,j} a^{-1}{}_{ij}P_iP_j + V$$

is known, in which $a^{-1}{}_{ij}$ is a function of coordinates q_i $(i = 1, 2, \ldots n)$ and P_i is the momentum which conjugates to q_i. Then, the quantum-mechanical expression of Hamiltonian operator given by Podolsky[36]

$$\mathcal{H} = \tfrac{1}{2}\sum_{i,j} a^{-1/4}\left(-i\hbar\frac{\partial}{\partial q_i}\right)a^{-1}{}_{ij}a^{1/2}\left(-i\hbar\frac{\partial}{\partial q_j}\right)a^{-1/4} + V(q_1, q_2, \ldots q_n)$$

is applicable, where $a = \det(a_{ij})$. If the Hamiltonian is applicable to more than one cell region, we call it the "global" Hamiltonian. In this expression, $\mathcal{H}$ is taken so that the integral

$$\int \ldots \int \Psi^*(q_1, q_2, \ldots q_n)\Psi(q_1, q_2, \ldots q_n)\, dq_1\, dq_2 \ldots dq_n$$

in which $\Psi(q_1, q_2, \ldots q_n)$ is a global wavefunction, becomes unity. If the global wavefunction were known, a full description of dynamics of nuclear rearrangement from one cell to another would be easy in the framework of the general wave-mechanical treatment for time-dependent processes.[41] The method consists of[12] (a) obtaining the cell Hamiltonian and its eigenfunctions with respect to the initial cell which is surrounded by a hypothetical wall of infinitely high potential, (b) constructing the initial density function for the initial cell by use of the cell wavefunctions obtained in (a) and the initial statistical distribution, (c) obtaining the time development of the cell wavefunctions obtained in (b) in terms of global wavefunctions, and finally (d) calculating the time-dependent population of the initial cell from the time-dependent density function. However, owing to the great deal of computational difficulties involved in the procedure, no actual calculation has yet been done. Certainly, a recourse to some

(41) E.g., see D. N. Zubarev, "Nonequilibrium Statistical Thermodynamics", Consultants Bureau, New York, 1974.

368

physically reasonable simplification is needed.

One way of making the rate calculation practicable is to reduce the problem to one dimension. The first such attempt was made in the absolute rate calculation of proton migration in the enol-form malonaldehyde.[42] We can imagine a picture of a large-amplitude nuclear motion in which the reacting system proceeds along the IRC with normal vibrational motions in the plane perpendicular to the direction of IRC. We assume in this approximation a "vibrational adiabaticity"[43] with regard to the large amplitude "translational" motion along the IRC.

The result[42] for malonaldehyde proton migration did not agree with experiment, although the adiabatic approach illustrated the one-dimensional reduction. The rate constant obtained ($\sim 10^{14}$ s^{-1}) was 10^2–10^3 times larger than that estimated from experimental studies ($\sim 10^{11}$ s^{-1}). Obviously, this result was partially caused by the small potential barrier, less than 1 kcal/mol, adopted in the calculation. The numerical result could have been no doubt improved by taking the barrier height to be 11 kcal/mol which was recently obtained by theoretical calculations.[44] Such a presumption is actually ascertained by a simpler tunneling calculation.[12a]

The Hamiltonian valid for a narrow region along the IRC is called the IRC Hamiltonian. The effect of vibrational *non*adiabaticity[45] can be taken into account in the IRC Hamiltonian. A calculation of the tunneling rate of the reaction HNC → HCN was carried out[46a] by the sue of the IRC Hamiltonian of Miller and coworkers,[46b] obtaining the interesting result that the rate constant value $\sim 10^5$ s^{-1} was achieved about 8 kcal/mol below the classical threshold.

Concluding Remarks

The IRC approach in chemodynamical studies has just begun to take shape. The present stage of the theory is mainly to establish a framework. With this in mind the theory has been formulated to be as broadly applicable as possible. At this stage, it may be tolerable to sacrifice pragmatism because of the shortage of efficient computing procedures. It is expected that in the near future, however, actual calculations will become feasible using a new generation of computers. In anticipation of these advances, the theory should be armed with mathematical foundations.

In particular, the *global* character of configuration space should be clarified by the use of global coordinates. The solution of the multidimensional Schrödinger equations, the rate problem in general nonadiabatic cases, the intercell tunneling rate involving infinite cells, the treatment of the nonzero angular momentum case using the general kinetic energy formula of eq 11, and IRC approach for excited state reactions, and so forth will be explored by mathematical methods.

It is a pleasure to acknowledge the reviewers of the manuscript and the editors for their suggestions and comments.

(42) S. Kato, H. Kato, and K. Fukui, *J. Am. Chem. Soc.*, **99**, 684 (1977).

(43) R. A. Marcus, *J. Chem. Phys.*, **43**, 1598 (1966); **46**, 959 (1967); M. Quack and J. Troe, *Ber. Bunsen ges. Phys. Chem.*, **78**, 240 (1974); **79**, 170 (1975).

(44) G. Karlström, H. Wennerström, B. Jönsson, and S. Forsén, *J. Am. Chem. Soc.*, **97**, 4188 (1975); E. M. Fluder and J. R. de la Vega, *ibid.*, **100**, 5265 (1978); J. Catalán, M. Yáñez, and J. I. Fernández-Alonso, *ibid.*, **100**, 6917 (1978).

(45) R. A. Marcus and M. E. Coltrin, *J. Chem. Phys.*, **67**, 2609 (1977); R. A. Marcus, *J. Phys. Chem.*, **83**, 204 (1979).

(46) (a) S. K. Gray, W. H. Miller, Y. Yamaguchi, and H. F. Schaefer, III, *J. Chem. Phys.*, **73**, 2733 (1980); (b) W. H. Miller, N. C. Handy, and J. E. Adams, *ibid.*, **72**, 99 (1980).

Part V

A New Development

The frontier orbital theory was proposed originally to discuss chemical reactivities of conjugated molecules. The number of the constituent AO's was not large in those systems. The HOMO was separated well from other occupied MO's and the LUMO was separated well from other unoccupied MO's in orbital energy. Thus, the frontier orbitals were chosen automatically by looking at their orbital energies. In these days, chemists apply the MO theory to various molecular systems, large in size and complicated in structure. Naturally, those systems have many MO's, placed close to each other in orbital energy. In addition, each MO tends to be delocalized over many AO's. It is sometimes difficult to find out the real frontier MO's.

As has been shown in the preceding parts of this volume, we have made many attempts to clarify the reasons why the frontier MO's play crucial roles in chemical reactions. We should notice that chemical interactions are local by nature. Chemists have interpreted and predicted their experimental results in terms of some useful notions, such as bonds, functional groups, substituent groups and so on. In other words, chemists pay attention to some specific parts of reactant molecules to characterize their reactivity trend. In order to represent local characteristics of interactions between large species, we introduced a new concept of orbital interactions called *"interaction frontier orbitals"* (paper no. 47) or *"interacting hybrid molecular orbitals"* (paper no. 48), depending on the types of orbital interactions involved.

We could show that molecules that were different in size and structure but had the same functional unit gave the orbitals very similar in shape to each other for a given type of reaction. Those orbitals were found to be localized beautifully on the reaction sites to tell us what part of molecules participated actively in interaction with a reagent. On the other hand, we could show that a molecule could provide orbitals of different patterns for different types of reactions. Multifarious activities of a molecule and a similarity that a variety molecules exhibited in a reaction could thus be represented by the interaction orbitals. Now, attempts are made to connect the results of MO calculations with familiar notions in chemistry, such as functional groups, substituent groups, acid and base, chemical hardness and softness and electronegativity of an atom or a group in a molecule. The old electronic theory appears to have been replaced by the orbital theory.

196 *J. Am. Chem. Soc.* **1981**, *103*, 196–197

Interaction Frontier Orbitals

Kenichi Fukui, Nobuaki Koga, and Hiroshi Fujimoto*

Faculty of Engineering
Kyoto University, Kyoto 606, Japan

Received September 15, 1980

Electron delocalization is known to be a major driving force of chemical reactions, yielding accumulation of the electron density in the intermolecular region between the reaction sites.[1,2] Frontier orbitals often play dominant roles,[3,4] but other molecular orbitals (MO's) also show subtle effects,[2,5,6] especially in large molecular systems. We report here that a pair of transformations of the canonical MO's of the fragments in a composite reacting system give rise to presumably the most succinct description of the delocalization interaction, including the contributions of all the relevant MO's.

Let the interaction between the occupied MO ϕ_i ($i = 1, 2, \ldots, s$) of the donor fragment and the unoccupied MO ψ_l ($l = 1, 2, \ldots, t$) of the acceptor fragment be denoted by D_{il}.[7] The "delocalization energy matrix" **D** comprising the elements D_{il} is rectangular, in general, of order $s \times t$, and, hence, one may construct the product $\mathbf{D}^\dagger\mathbf{D}$. This matrix is Hermitian and, therefore, can be diagonalized by a unitary transformation $\mathbf{V} = (V_{ij})$, giving the nonnegative eigenvalues $\Lambda = (\lambda_i)$:

$$\mathbf{D}^\dagger\mathbf{D}\mathbf{V} = \mathbf{V}\Lambda \qquad (1)$$

Assuming here that $s \le t$ and that λ_i's are not zero for $i = 1, 2, \ldots, s$,[8,9] the following two sets of transformations convert the canonical MO's into a set of paired molecular hybrids of the fragments:

$$\phi_i' = \lambda_i^{-1/2}\sum_{j=1}^{s}\sum_{k=1}^{t} D_{jk}V_{ki}\phi_j \qquad (i = 1, 2, \ldots, s) \qquad (2)$$

and

$$\psi_l' = \sum_{k=1}^{t} V_{kl}\psi_k \qquad (l = 1, 2, \ldots, t) \qquad (3)$$

Unlike the corresponding orbitals by Amos and Hall,[10] one has to repeat the transformations by evaluating the matrix elements D_{il}' for the new set of orbitals to complete the true "pairing" of orbitals.[11] Upon convergence one gets $D_{ii}' = -\lambda_i'^{1/2}$ and $\sum_{i=1}^{s}\sum_{l=1}^{t} D_{il} = \sum_{i=1}^{s} D_{ii}'$.[12] That is, the hybrid orbital ϕ_i' interacts only with the counterpart ψ_i' and vice versa and the interactions between the canonical MO's, $s \times t$ in number, are reduced to the interactions of, at the most, the s pairs of hybrids, s being the smaller of the number of the occupied MO's and the number of the unoccupied MO's participating in the interaction.

The square root of the eigenvalue λ_i' represents the strength of the interaction between ϕ_i' and ψ_i'. The largest one grows steadily during the iterations and the corresponding hybrid gains amplitude in the region of the interaction in each fragment.[13] Incidentally, one may carry out similar transformations with regard to the combination of the unoccupied MO's of the donor and the occupied MO's of the acceptor. The new orbitals ϕ' and ψ' obtained each constitute an orthonormal set, retaining the electronic structures of the fragments unchanged.

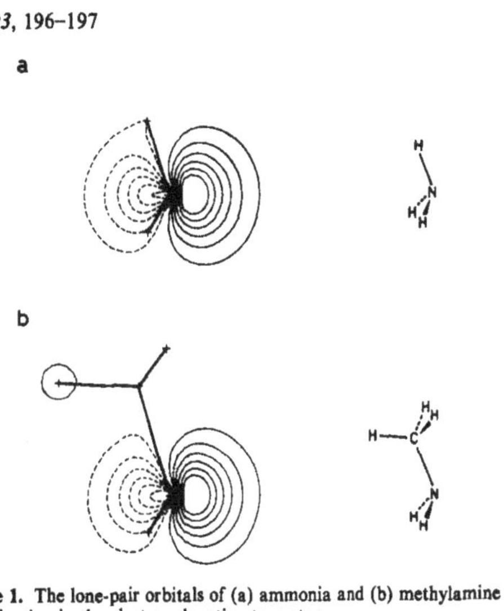

Figure 1. The lone-pair orbitals of (a) ammonia and (b) methylamine participating in the electron donation to proton.

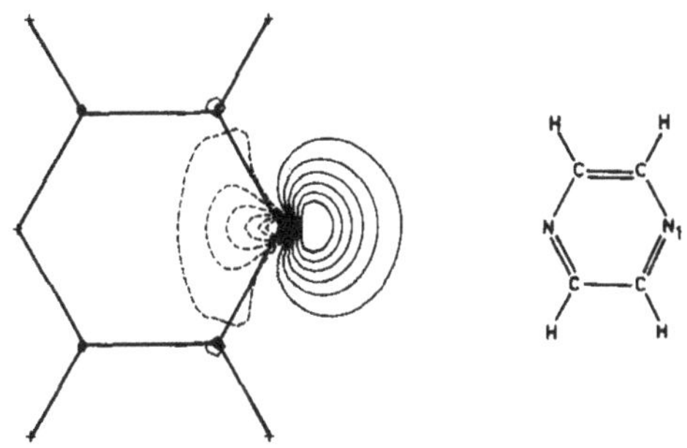

Figure 2. Lone-pair orbital of pyrazine taking part in the electron donation to proton.

The following ab initio STO-6G MO calculations provide examples to illustrate the foregoing argument.[14] By placing a proton, for simplicity, on the C_3 axis of ammonia 1.03 Å away from the nitrogen atom, we got a single hybrid taking part in the delocalization interaction, as shown in Figure 1.[15] Of the remaining four occupied orbitals, one constitutes the inner shell and the other three are localized in the N–H bond regions and have zero matrix elements in **D** with the proton 1s orbital. The hybrid is shown to be of stronger p character than the lone-pair hybrid calculated by the Edmiston–Ruedenberg procedure.[16] These results suggest that the lone-pair orbital having the large s component stabilizes the molecule in an isolated state, but it has to enhance the p nature in order to create the new N–H bond in ammonium ion. The lone-pair hybrid of methylamine calculated similarly is seen to have almost the same shape as that of ammonia.

Figure 2 shows the lone-pair orbital of pyrazine, participating exclusively in the delocalization interaction with a proton located 1.03 Å away from the N_1 atom on the C_2 axis. It is given approximately by a 1:1 mixture of the $6a_1$ (the highest occupied) MO and the $5b_{2u}$ (the next highest occupied) MO with concom-

(1) (a) Fukui, K.; Fujimoto, H. *Bull. Chem. Soc. Jpn.* **1968**, *41*, 1989. (b) Fujimoto, H.; Yamabe, S.; Fukui, K. *Ibid.* **1971**, *44*, 2936.
(2) Fujimoto, H.; Sugiyama, T. *J. Am. Chem. Soc.* **1977**, *99*, 15.
(3) Fukui, K. In "Molecular Orbitals in Chemistry, Physics, and Biology"; Lödwin, P.-O.; Pullman, B., Eds.; Academic Press: New York, 1964; p 513.
(4) Houk, K. N. *Acc. Chem. Res.* **1975**, *8*, 361.
(5) Fujimoto, H.; Inagaki, S. *J. Am. Chem. Soc.* **1977**, *99*, 7424.
(6) Fujimoto, H.; Osamura, Y.; Minato, T. *J. Am. Chem. Soc.* **1978**, *100*, 2959.
(7) Fujimoto, H.; Kosugi, N. *Bull. Chem. Soc. Jpn.* **1977**, *50*, 2209.
(8) If $s \ge t$, the transformations given by eq 2 and 3 are interchanged.
(9) The treatment of zero eigenvalues is trivial.
(10) Amos, A. T.; Hall, G. G. *Proc. R. Soc. London, Ser. A* **1961**, *263*, 483.
(11) Lödwin, P.-O. *J. Appl. Phys. Suppl.* **1962**, *33*, 251.
(12) Prime is used to distinguish the molecular hybrids from the canonical MO's.
(13) In BH_3–NH_3, for example, two pairs of hybrids with the a_1 symmetry take part in the electron donation from NH_3 to BH_3. The ratio of $\lambda'^{1/2}$ was found to be about $1{:}10^{-4}$.
(14) Hehre, W. J.; Lathan, W. A.; Ditchfield, R.; Newton, M. D.; Pople, J. A. GAUSSIAN 70, Program No. 236, QCPE, Indiana University, Bloomington, IN.
(15) The hybrid determined in this manner depends more or less on the kind of reagent and on the geometry of mutual approach.
(16) Edmiston, C.; Ruedenberg, K. *Rev. Mod. Phys.* **1963**, *35*, 457.

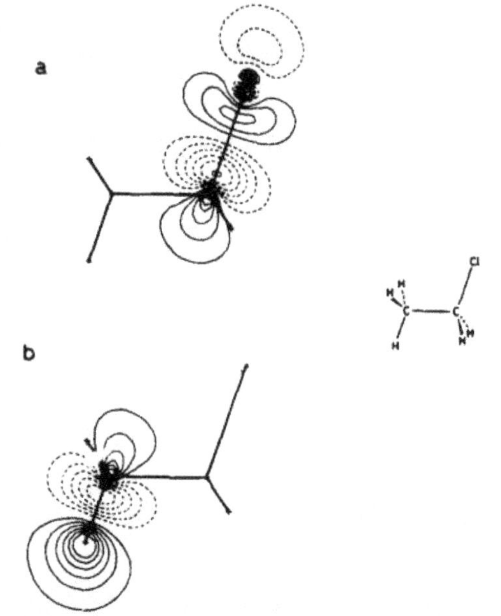

Figure 3. Unoccupied molecular hybrids of ethyl chloride participating in the delocalization interactions with hydride ion. (a) S_N2 and (b) E2.

itant mixing in of the lower-lying occupied MO's. The in-phase combination of these delocalized MO's at N_1 causes the resultant orbital to be localized almost perfectly on the site under attack. The orbital is of stronger s character relative to the lone-pair hybrid of ammonia.

Molecular hybrids of ethyl chloride are presented in Figure 3. They were calculated by placing a hydride ion on the C_α–Cl bond axis (in the S_N2 type model) and the C_β–H bond axis (in the E2 type model) at a distance which is 1.5 times the normal C–H and H–H bond lengths, away from C_α and H_β, respectively. The remaining six unoccupied orbitals have greatest amplitudes in other bond regions and do not contribute to the delocalization interaction in each case. The results of calculation suggest that the inversion at C_α in the S_N2 reaction and the breaking of the C_α–Cl bond in the E2 reaction should be ascribed to other sorts of interactions, probably to the exchange and polarization interactions.[5,6]

It was shown above that the one-electron orbitals participating virtually in the delocalization interaction are localized surprisingly well in the region of initial attacks. The stronger the localization is, the greater the delocalization stabilization would be. The present method will be of use to clarify the outcome of the delocalization interaction, particularly in the interactions of large systems with small reagents. A pair of hybrids, having the amplified lobes around the frontier of interaction, are determined with regard to each symmetry element of the composite reacting system.

Acknowledgment. This work was supported by the Ministry of Education, Japan, under Grant-in-Aid for Scientific Research (No. 435045, 465266).

7452 J. Am. Chem. Soc. 1981, 103, 7452–7457

A Coupled Fragment Molecular Orbital Method for Interacting Systems

Hiroshi Fujimoto,* Nobuaki Koga, and Kenichi Fukui

Contribution from the Faculty of Engineering, Kyoto University, Kyoto 606, Japan.
Received December 1, 1980

Abstract: A succinct orbital description of molecular interactions is presented. This covers automatically all the orbital interactions and orbital mixings in perturbation theory and visualizes their consequences in terms of hybrid molecular orbitals which are localized in the region of interaction. These hybrid orbitals are determined within the framework of the pairing theorem so as to offer a clearer view of bond-forming processes along a reaction coordinate. Of these paired hybrid orbitals, one or a few pairs play the dominant role and they thereby define the domains of reagent and reactant that participate actively in the interaction. These hybrid molecular orbitals are suggested to give the basis of frontier orbitals.

Almost 30 years have passed since the first paper on the frontier electron theory came out.[1] By utilizing the mathematical simplicity of the primitive Hückel molecular orbital (MO) method, reactivities of polyenes and condensed aromatic systems were discussed extensively.[2] The extended Hückel,[3] CNDO,[4] and other semiempirical MO methods appearing in 1960's magnified the applicability of the theory to sizable molecules with three-dimensional frameworks.[5] The stereoselection rules[6] seem also to be explainable by means of interactions between frontier orbitals,[7] though several alternative ways of interpretation surely exist.[6,8–10]

The basic frontier orbital concept appears to have grown to an efficient tactic to understand the results of semiempirical and ab initio MO calculations, being called frequently orbital interactions and orbital mixings.[11–18] The inclusion of other MOs than the highest occupied MO and/or the lowest unoccupied MO is clearly of significance for a detailed study of the geometric changes and electron reorganization in chemically interacting systems along a reaction coordinate.[16–18] Perturbation theory has most often been utilized to elucidate how the MOs of fragments interact or intermix through interactions.[11–15] The expansion of composite MOs in terms of the fragment MOs has led, on the other hand, to a powerful method of analyzing weak and moderately strong interactions by means of electron configurations of the reagent and reactant.[19,20] It is obvious that the frontier electron theory is related closely in this respect to the theory of charge transfer developed by Mulliken.[21]

Very recently, we have proposed a novel concept of "interaction frontier orbitals".[22] By taking not only the frontier MOs but also all the relevant MOs properly into account, we intended to condense the entire consequence of delocalization interaction into a pair or, at most, a few pairs of hybrid orbitals of fragments which are localized in the region of interaction. An application to some systems demonstrated the utility of the method in elucidating the role of electron delocalization in chemical interactions.[22] In this paper, we show an extensive generalization of the method to cover the significant effects of the exchange and polarization interactions in chemical reactions. Reactive domains of molecules are illustrated.

Procedure of Calculations

To begin with we may mention orbital interactions between the fragments A and B in a composite system A–B. The MOs of the system A–B can be expanded as usual in linear combinations of the MOs of fragments:

$$\Phi_m = \sum_i^{A} c_{m,i}\phi_i + \sum_k^{B} c_{m,k}\psi_k \qquad (1)$$

It is easy to calculate exactly the coefficients $c_{m,i}$ and $c_{m,k}$ by the use of expansion theorems or by solving linear simultaneous equations.[20]

From eq 1 we define an *intermolecular interaction matrix* **P** for some appropriate one-electron operator $\mathbf{p}(1)$;

$$\mathbf{P} = (P_{i,k}) \qquad (2)$$

$$i = 1, 2, \cdots, s;\ k = 1, 2, \cdots, t$$

where s and t signify the number of relevant MOs of A and of B, respectively. In the case of the system A–B having a closed-shell electronic structure the ground state of which is described by a single determinant wave function, we get:

$$P_{i,k} = 4\sum_m^{occ} c_{m,i}c_{m,k}\int \phi_i(1)\mathbf{p}(1)\psi_k(1)\,dv_1 \qquad (3)$$

The modifications associated with an employment of multiconfiguration wave function and applications to open-shell systems are straightforward.

The matrix **P** is rectangular, in general, of order $s \times t$. All the occupied and unoccupied MOs of fragments are included in

(1) Fukui, K.; Yonezawa, T.; Shingu, H. *J. Chem. Phys.* **1952**, *20*, 722–5.
(2) Fukui, K. In "Molecular Orbitals in Chemistry, Physics, and Biology", Löwdin, P.-O., Pullman, B., Eds.; Academic Press: New York, 1964; pp 513–37.
(3) (a) Hoffmann, R. *J. Chem. Phys.* **1963**, *39*, 1397–1412. (b) Hoffmann, R.; Lipscomb, W. N. *Ibid.* **1962**, *32*, 2179–89; **1962**, *37*, 2872–83.
(4) Pople, J. A.; Beveridge, D. L. "Approximate Molecular Orbital Theory"; McGraw-Hill: New York, 1970.
(5) Fujimoto, H.; Fukui, K. In "Chemical Reactivity and Reaction Paths", Klopman, G. Ed.; Wiley-Interscience: New York, 1974; pp 23–54.
(6) Woodward, R. B.; Hoffmann, R. "The Conservation of Orbital Symmetry"; Academic Press: New York, 1969.
(7) Fukui, K. "Theory of Orientation and Stereoselection"; Springer Verlag: Berlin, 1974.
(8) Longuet-Higgins, H. C.; Abrahamson, E. W. *J. Am. Chem. Soc.* **1965**, *87*, 2045–6.
(9) Zimmerman, H. E. *J. Am. Chem. Soc.* **1966**, *88*, 1564–5; 1566–7.
(10) Dewar, M. J. S. *Tetrahedron Suppl.* **1966**, *8*, 75–92.
(11) Herndon, W. C. *Chem. Rev.* **1972**, *72*, 157–79.
(12) Hudson, R. F. *Angew. Chem.* **1973**, *85*, 63–84.
(13) Houk, K. N. *Acc. Chem. Res.* **1975**, *8*, 361–9.
(14) Libit, L.; Hoffmann, R. *J. Am. Chem. Soc.* **1974**, *96*, 1370–83.
(15) Inagaki, S.; Fujimoto, H.; Fukui, K. *J. Am. Chem. Soc.* **1976**, *98*, 4054–61.
(16) Fujimoto, H.; Sugiyama, T. *J. Am. Chem. Soc.* **1977**, *99*, 15–22.
(17) Fujimoto, H.; Inagaki, S. *J. Am. Chem. Soc.* **1977**, *99*, 7424–32.
(18) Fujimoto, H.; Osamura, Y.; Minato, T. *J. Am. Chem. Soc.* **1978**, *100*, 2954–9.
(19) Baba, H.; Suzuki, S.; Takemura, T. *J. Chem. Phys.* **1969**, *50*, 2078–86.
(20) Fujimoto, H.; Kato, S.; Yamabe, S.; Fukui, K. *J. Chem. Phys.* **1974**, *60*, 572–8.
(21) Mulliken, R. S. *J. Am. Chem. Soc.* **1952**, *74*, 811–24.
(22) Fukui, K.; Koga, N.; Fujimoto, H. *J. Am. Chem. Soc.* **1981**, *103*, 196–7.

order to represent all the orbital effects, e.g., the exchange, delocalization, and polarization interactions. The product $\mathbf{P}^\dagger\mathbf{P}$ is an Hermitian matrix and gives nonnegative eigenvalues upon diagonalization;

$$\mathbf{P}^\dagger\mathbf{P}\mathbf{U} = \mathbf{U}\Gamma \tag{4}$$

where:

$$\mathbf{U} = (U_{j,l}) \tag{5}$$

$$\Gamma = (\gamma_i) \tag{6}$$

Assuming tentatively that $s \leq t$ and that the eigenvalues γ_i are not zero for $i = 1, 2, \cdots, s$,[23] the canonical MOs of A and those of B are converted into separate orthonormal sets of hybrid MOs by the following pairwise orbital transformations;

$$\phi'_i = \gamma_i^{-1/2} \sum_{j=1}^{s} \sum_{l=1}^{t} P_{j,l} U_{l,i} \phi_j \tag{7}$$

$$i = 1, 2, \cdots, s$$

and:

$$\psi'_k = \sum_{l=1}^{t} U_{l,k} \psi_l \tag{8}$$

$$k = 1, 2, \cdots, t$$

Note that replacement of the overlap integral matrix **S** for the matrix **P** reduces these transformations to the corresponding orbital formalism by Amos and Hall.[24] In the present case, we have to iterate the whole procedure by evaluating the matrix elements $P'_{i,k}$ for the new sets of orbitals ϕ'_i and ψ'_k to make the transformations self-consistent.[25,26]

Upon convergence the orbitals ϕ' of the fragment A and the orbitals ψ' of the fragment B yield a unique set of orbital couples in which an orbital ϕ'_i interacts only with its counterpart ψ'_i and vice versa.[27] We get, therefore:

$$\sum_{i=1}^{s} \sum_{k=1}^{t} P_{i,k} = \sum_{i=1}^{s} P'_{i,i} \tag{9}$$

The $s \times t$ orbital interactions can thus be reduced to the interactions of paired hybrid MOs of fragments, s in number. Note here that s is the smaller of the numbers of MOs in the fragments, A and B. The orbitals ϕ' are constructed only from the canonical MOs of the fragment A and the orbitals ψ' consist only of the canonical MOs of the fragment B, but they are no longer constrained by the local symmetry of fragments. They are adapted in pairs for the symmetry of the composite system.

By tracing the procedure of orbital transformations mentioned above, one may recognize that a pair of orbitals, say ϕ'_f and ψ'_f, corresponding to the largest eigenvalue γ_f, should have the greatest amplitude in the region of mutual interaction in A and in B, respectively, with regard to each symmetry element of the composite system, provided that the one-electron operator **p**(1) was chosen adequately to represent the interaction. Other orbitals are localized naturally in other regions to determine the electronic structure of the fragments in interaction and, therefore, they will not participate significantly in the intermolecular interaction. The *interactive hybrid MOs* ϕ'_f and ψ'_f do not have to be either occupied or unoccupied. They are something in-between, depending on the type of interaction.

(23) The treatment of zero eigenvalues is trivial.

(24) Amos, A. T.; Hall, G. G. *Proc. R. Soc. London, Ser. A* **1961**, *263*, 483–93. See, also: King, H. F.; Stanton, R. E.; Kim, H.; Parr, R. G. *J. Chem. Phys.* **1967**, *47*, 1936–41.

(25) The matrix element $P_{i,k}$ is not directly proportional to the overlap integral $S_{i,k}$, in general.

(26) Prime is used to distinguish the hybrid MOs from the canonical MOs.

(27) The remaining $(t - s)$ orbitals of B find no counterpart in A and, therefore, do not contribute to the intermolecular interaction.

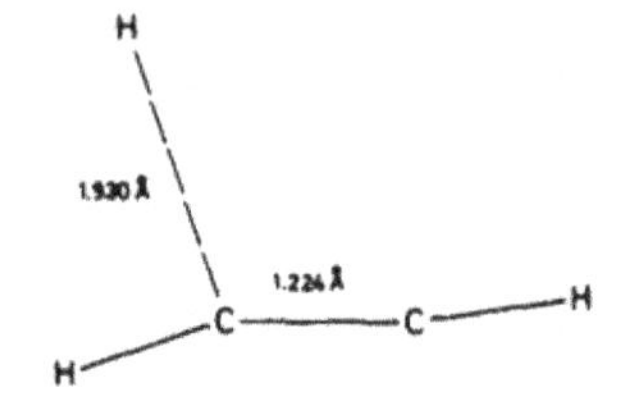

Figure 1. A sketch of reaction model for addition of hydrogen atom to acetylene.

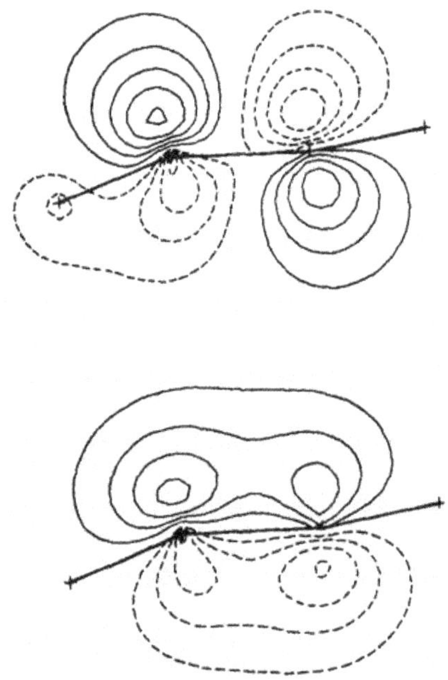

Figure 2. Interaction frontier orbitals of acetylene to interact with a hydrogen atom in the α spin part (upper) and in the β spin part (lower).

It is a general extension of the early frontier electron theory to try to locate the particular orbital(s) which shall play a dominant role in reactions. It is obvious that the frontier orbitals, as well as the interaction frontier orbitals,[22] are nothing but simplified versions of the interactive hybrid MOs, determined for minor sets of canonical MOs. When the matrix **P** spans only the occupied subset of donor orbitals and the unoccupied subset of acceptor orbitals, the orbital transformations yield the *interaction frontier orbitals*.[28] A combination of the highest occupied MO of donor and the lowest unoccupied MO of acceptor gives rise at once to the frontier MOs. In order to clarify the relation between several types of interaction MOs, we have begun our discussion above by representing the intermolecular part of the density matrix in terms of the canonical MOs of fragments. An important aspect of the interactive hybrid MOs is that they can also be derived directly from the atomic orbital density matrix of the composite system without reference to the fragment canonical MOs.[29]

Results of Calculation

Addition of hydrogen atom to acetylene provides a simple but suitable example to illustrate the foregoing argument. The reaction model shown in Figure 1 was taken from an extensive calculation by Nagase and Kern.[30] We used an ab initio UHF SCF MO

(28) Interaction frontier orbitals were defined more rigorously in ref 22 with regard to the Hamiltonian operator of the composite interacting system. Hybrid MOs depend more or less on the choice of matrix **P**.

(29) Atomic orbitals should once be converted to some orthonormal set of orbitals within each fragment.

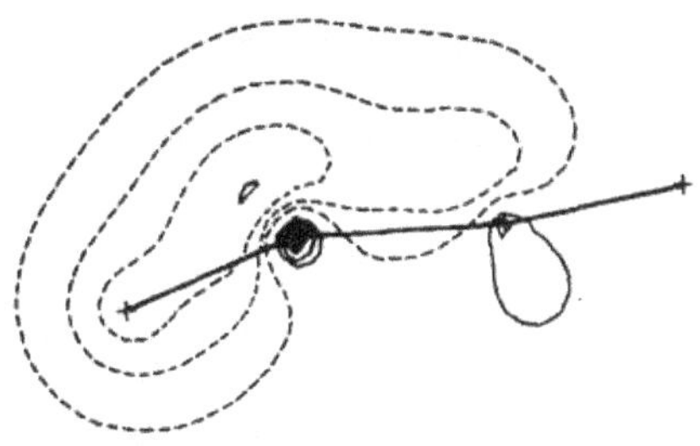

Figure 3. Hybrid MO of acetylene taking part in overlap repulsion.

method with the STO-6G basis set to calculate the electron density between the two species.[31] Without loss of generality, we assigned the odd electron with the spin α.

In Figure 2 we show the interaction frontier orbitals. Here, we took tentatively the intermolecular overlap populations[32] between the fragments as the elements of **P**.[33] Since a hydrogen atom possesses a single orbital, a single hybrid MO participates in the delocalization interaction in each spin part of acetylene. In the α spin part, electron delocalization takes place from hydrogen atom to acetylene and, therefore, the interaction frontier orbital is obtained with regard to the matrix **P** defined for the unoccupied subset of canonical MOs. The resultant orbital is given by an admixture of unoccupied MOs. In the β spin part, on the contrary, the delocalization interaction takes place between the occupied MOs of acetylene and the orbital of the hydrogen radical. The interaction frontier orbital of acetylene for β spin is defined, therefore, with respect to the occupied subset of canonical MOs of the acetylene fragment. Though the unoccupied and occupied π-type MOs of the deformed acetylene molecule are the predominant constituent of the interaction frontier orbital for the α spin part and that for the β spin part, respectively, contributions of σ-type MOs have also been taken obviously into the hybrid MO in each spin part by this procedure.

An overlap repulsion arises in the α spin part due to the interactions between the occupied MOs of the acetylene fragment and the singly occupied MO of the attacking radical.[34] By carrying out the orbital transformation within the occupied MOs of acetylene, we obtained a hybrid MO shown in Figure 3. It is seen clearly that not only the site under attack but also the bonds connecting the reaction site and the adjacent atoms take part in the overlap repulsion. Rehybridization at the reaction center to bend the molecule should reduce effectively the repulsive interaction.[18]

Figure 4 shows the interactive hybrid MOs of acetylene for the α and β spin parts. All the occupied and unoccupied canonical MOs participate in these orbitals to determine the regions of bonding and antibonding interactions with the incoming hydrogen atom. Because of a strong competition between the exchange and delocalization interactions as demonstrated by the orbitals shown in Figures 2 and 3, the interactive hybrid MO for the α spin part does not show signficant amplitude in the region of interaction.[35] The interaction between the hybrid MO and the hydrogen 1s orbital is antibonding. In the β spin part, on the other hand, the interactive hybrid MO is localized well around the site under attack to give a bonding interaction with the attacking radical. The hybrid MO contains explicitly, unlike the interaction frontier orbital given in Figure 2, the important effect of polarization to specify the domain of acetylene molecule which takes part actively

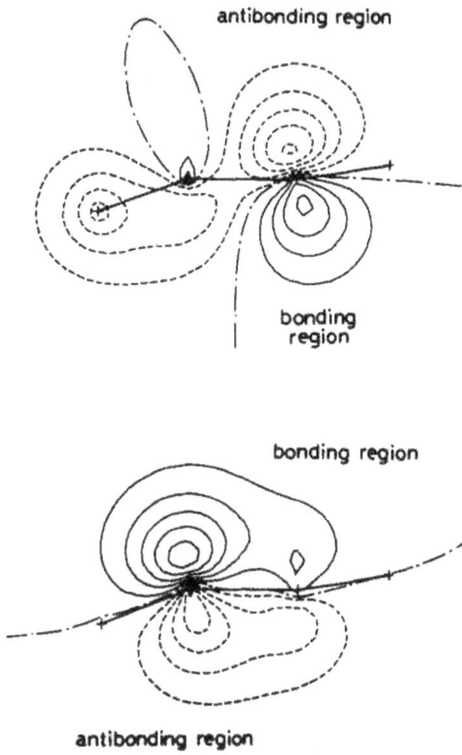

Figure 4. Interactive hybrid MOs of acetylene to interact with a hydrogen atom in the α spin part (upper) and in the β spin part (lower).

Table I. Coefficients of Canonical MOs in Interactive Hybrid MOs of Acetylene

	α spin			β spin	
MO	all MOs[a]	occupied subset[b]	unoccupied subset[c]	all MOs[a]	occupied subset[c]
σ'_1	−0.0343	−0.0437		−0.0051	−0.0060
σ'_2	−0.0501	−0.0669		−0.0105	−0.0112
σ'_3	−0.4889	−0.6771		0.0304	0.0320
σ'_4	−0.3426	−0.4708		0.0654	0.0720
σ'_5	−0.2010	−0.2774		0.0383	0.0426
π	0	0		0	0
π'	−0.3537	−0.4881		0.9538	0.9959
π'^*	0.6876		0.9991	0.2886	
π^*	0		0	0	
σ'_6	0.0383		0.0406	−0.0108	
σ'_7	−0.0019		−0.0024	0.0128	
σ'_8	0.0085		0.0141	0.0004	

[a] The hybrid MOs shown in Figure 4. [b] The hybrid MO shown in Figure 3. [c] The hybrid MOs shown in Figure 2.

in the interaction. It seems to be more favorable for a carbon–carbon π bond to donate electron density in part via unoccupied MOs, particularly the π^* MO, than to donate all the electron density directly from the π MO in order to create a new single bond between a carbon atom and an attacking reagent.

The coefficients of canonical MOs in the hybrid MOs are presented in Table I. The canonical MOs of the deformed acetylene molecule having maximal amplitudes in the plane of hydrogen attack were denoted by σ' and π' to retain approximately a correlation with the σ and π MOs of linear acetylene. The signs of these canonical MOs were chosen so as to give positive overlap integrals between the fragment MOs. The π and π^* MOs which have the direction of extension perpendicular to the plane do not

(30) Nagase, S.; Kern, C. W. *J. Am. Chem. Soc.* **1979**, *101*, 2544–49.
(31) Hehre, W. J.; Lathan, W. A.; Ditchfield, R.; Newton, M. D.; Pople, J. A. *QCPE No. 236*.
(32) Mulliken, R. S. *J. Chem. Phys.* **1955**, *23*, 1833–40.
(33) Fujimoto, H.; Koga, N.; Endo, M.; Fukui, K. *Tetrahedron Lett.* **1981**, *22*, 1263–6.
(34) The exchange interaction in homolytic reactions was studied in detail previously. See: Fujimoto, H.; Yamabe, S.; Minato, T.; Fukui, K. *J. Am. Chem. Soc.* **1972**, *94*, 9205–10.
(35) Note here that the interactive hybrid MO for the α spin part is not given by the 1:1 mixture of the orbitals shown in Figures 2 and 3.

 J. Am. Chem. Soc., Vol. 103, No. 25, 1981 7455

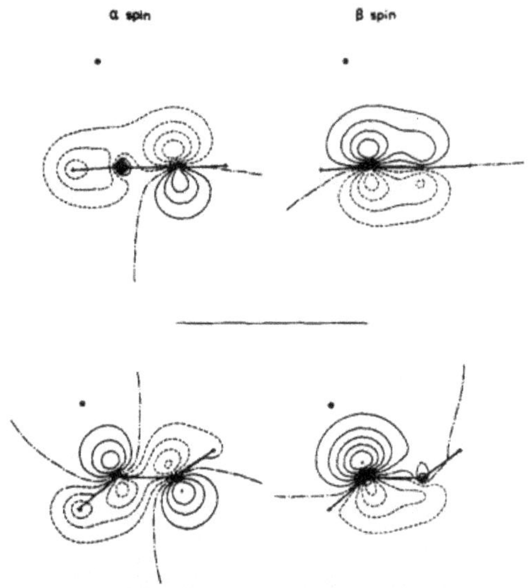

Figure 5. Interactive hybrid MOs of acetylene in the acetylene–hydrogen system located before (upper) and after (lower) the transition state. Position of the attacking hydrogen atom is marked by a dot.

interact with the 1s orbital of the attacking hydrogen radical by symmetry.

The lowest unoccupied π'^* MO is found to play an exclusively important role among the unoccupied MOs of the α spin part, while the highest occupied π' MO is seen to participate dominantly in the interaction among the occupied MOs of the β spin part, demonstrating a characteristic of reactions of small conjugated molecules. The occupied π' MO is shown to take part in the interaction in an antibonding manner in the α spin part, indicating the presence of a strong exchange repulsion. The unoccupied π'^* MO contributes to the bonding interaction from the beginning of reaction in the β spin part to construct a localized interaction orbital. A basic aspect of homolytic reactions is represented by these hybrid MOs.

This calculation suggests that addition of electrophiles to unsaturated bonds is facile, whereas addition of nucleophiles is unfavorable in accordance with a general trend observed experimentally. One may suppose, therefore, that an attacking radical should play the electron-acceptor part in homolytic additions. This is unlikely, however. Replacement of an electron-releasing group for the hydrogen which is attached to the carbon under attack, for instance, may strengthen the bonding interaction in the β spin part, but will lead concurrently to a strengthening of the overlap repulsion and a weakening of the delocalization interaction in the α spin part. The substituent does not seem to favor the addition. Introduction of an electron-withdrawing substituent will suppress the antibonding interaction and promote the bonding interaction in the α spin part, though the bonding interaction in the β spin part is weakened to some extent. It is suggested, therefore, that a π bond would play preferably an electron-acceptor part in homolytic additions, in agreement with a broad generalization of observed trends by Hefter, Hecht, and Hammond.[36] An application of the present method to addition of proton to propene has shown that a selective attack to the terminal carbon is interpreted in terms of a more efficient localization of bonding region in the vicinity of the interaction center.[37]

Figure 5 presents the interactive hybrid MOs for two models of the acetylene–hydrogen atom system, one being located before and the other being located after the transition state illustrated in Figure 1.[30] By comparing the hybrid MOs with those given in Figure 4, we can trace an evolution of reactive domains with the progress of reaction in terms of the hybrid MOs. The occupied π' MO begins to participate in the bonding interaction in the α spin part and the unoccupied π'^* MO plays an increasingly important role in the β spin part at a later stage of reaction.

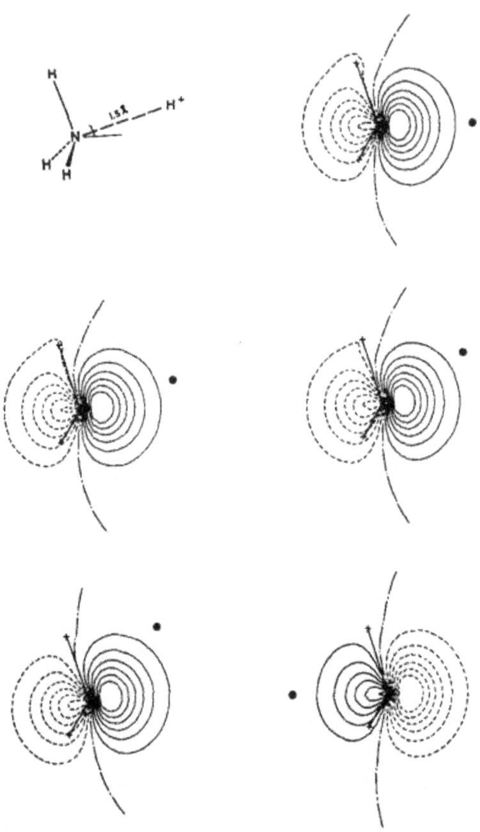

Figure 6. Interactive hybrid MO of ammonia in the ammonia–proton system. Location of proton is marked by a dot.

Figure 6 shows the interactive hybrid MOs of ammonia calculated by placing a proton at several different positions. Since no saddle point was found within an STO-6G level study, the $N-H^+$ distance was chosen tentatively at 1.5 Å. The orbitals look very similar to the interaction frontier orbital obtained in our previous calculation.[38] The polarization interaction is not important in this case and the highest occupied canonical MO dominates the interaction with proton. The interactive hybrid MO, as well as the interaction frontier orbital, of ammonia was found to be of stronger p character than the localized molecular orbital obtained by the use of the Edmiston–Ruedenberg procedure.[39] In an isolated state of ammonia, the nitrogen 2s orbital may constitute a large portion of the lone pair to stabilize the molecule. When a proton attacks, the lone-pair orbital should rearrange itself to strengthen the ability for electron donation. The 2p component should be enhanced in the lone-pair orbital. The localized hybrid orbitals generated by the Edmiston–Ruedenberg procedure and by our method appear to each symbolize the static and dynamic aspect of lone-pair orbitals, respectively. Incidentally, the lone-pair orbital of ammonia which interacts with a proton approaching

(36) Hefter, H. J.; Hecht, T. A.; Hammond, G. S. *J. Am. Chem. Soc.* **1972**, *94*, 2793–7. See, also ref 34.

(37) Fujimoto, H.; Koga, N., unpublished result.

(38) See Figure 1 in ref 22. The $N-H^+$ distance was taken somewhat longer in the present calculation.

(39) Edmiston, C.; Ruedenberg, K. *Rev. Mod. Phys.* **1963**, *35*, 457–465.

7456 J. Am. Chem. Soc., Vol. 103, No. 25, 1981

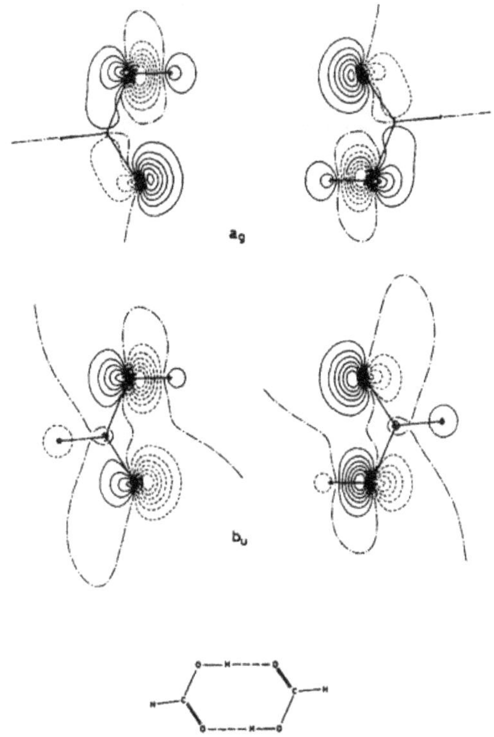

Figure 7. Two pairs of interactive hybrid MOs in formic acid dimer. Contours are shown for separated monomers.

from the backside shows a weaker p nature relative to others, suggesting that such an interaction is unfavorable.

It was shown in Figure 6 that, like the bond orbitals in stable molecules,[40,41] the hybrid MO of ammonia could hardly follow the motion of proton. When a proton was placed in a position 45° away from the symmetry axis, for example, the hybrid MO was calculated to decline only about 10°. The hybrid MOs taking part in interactions seem also to be governed rigorously by the structure of fragments, notwithstanding that these orbitals are derived only by looking at the intermolecular part of the density matrix. It is obvious that the structural constraint arises indirectly by projecting the geometry and orientation of a fragment in space on the other fragment. Thus, by virtue of the delicately arranged valence-active electron density between fragments, whole molecular frameworks participate actively in the determination of reaction pathways, even if interactions are local by nature.

In the acetylene–hydrogen atom system presented above, the interactive hybrid MO of acetylene tends to point nicely at the incoming hydrogen both in the α spin part and in the β spin part, as demonstrated in Figure 5. This signifies that the addition is a typical nonpolar reaction in which the formation of a new covalent bond is the principal driving force of the process. When the interactive hybrid MOs would deviate in the direction of its extension from the path of reagent approach, some other factors than orbital effects, most probably Coulombic attractions or repulsions, should come to play a significant role. The maximum overlap of the hybrid MOs seems to provide the simplest criterion to interpret the pathway of nonpolar reactions.

Both in the acetylene–hydrogen atom system and in the ammonia–proton system, the attacking reagent possesses a single

(40) Nakatsuji, H. *J. Am. Chem. Soc.* **1974**, *96*, 30–37.
(41) Chipman, D. M.; Palke, W. E.; Kirtman, B. *J. Am. Chem. Soc.* **1980**, *102*, 3373–83.

Table II. Contributions of Canonical MOs to Interactive Hybrid MOs of Formic Acid Molecule

	MO	weight %	
		a_g	b_u
occupied	σ_4	0.00	5.90
	σ_5	6.92	0.52
	σ_6	0.37	13.85
	σ_7	15.96	0.31
	σ_8	18.65	0.06
	σ_9	3.40	30.23
	σ_{10}	21.95	21.11
unoccupied	σ_{11}	23.31	20.06
	σ_{12}	8.62	6.83
	σ_{13}	0.22	0.42
	σ_{14}	0.59	0.70

orbital. Accordingly, all the orbital effects are condensed in a single hybrid MO of the reactant in these cases. In the majority of chemical reactions, the growth of a new bond between the particular sites of reagent and reactant is the main driving force. A single pair of hybrid MOs will suffice for describing virtually all orbital interactions, even when the attacking reagent has several orbitals. In highly symmetrical multicentric processes, however, several pairs of fragment MOs are supposed to participate in interactions. In Figure 7, we show, as an example, the interactive hybrid MOs for dimeric formic acid.

We obtain two pairs of interactive hybrid MOs which contribute to the hydrogen bonding. The fragment MOs obtained are seen clearly to be adapted in pairs for the symmetry of the composite system. The overlap population between the hybrid MOs, $P'_{f,f}$, was calculated to be small but positive, being 0.0474 for the a_g pair and 0.0367 for the b_u pair. The sum of occupancies in the pair of MOs is 2.728 and 2.949, indicating the existence of a pretty strong overlap repulsion which has somehow been counterbalanced by the delocalization, polarization, and other interactions.[42] This result is in line with our analysis of the interaction energy by the use of an ab initio perturbation scheme.[43] We obtained −0.0232, 0.0367, −0.0343, and −0.0014 au for the Coulomb, exchange, delocalization, and polarization interactions, respectively.[44] Constituents of the hybrid MOs are presented in Table II in which σ denotes the canonical MOs of monomeric formic acid having maximal amplitudes in the molecular plane. Incidentally, no bonding pair of hybrid MOs was found in the case of cyclic water dimer. The electrostatic interaction between water molecules was shown previously to be the most important component of the bonding force.[45]

Discussion

We may examine next some aspects of the interactive hybrid MOs. As it has thoroughly been discussed in our previous papers,[5,20] the ground state of a composite reacting system is presented by a combination of various electron configurations of the fragments, A and B:

$$\Psi = \sum_p C_p \Psi_p \tag{10}$$

Electron configurations that we should note here are the original one Ψ_0, electron-transferred ones $\Psi_{i\to l}$ $(\Psi_{k\to j})$ and $\Psi_{i\to l,j\to l}$, and locally excited ones $\Psi_{i\to j}$ $(\Psi_{k\to l})$, where i and k denote the occupied MOs of A and B, respectively, and j and l signify the unoccupied MOs of A and B, respectively. In the original configuration, the fragments retain their ground-state configurations that they possess

(42) The interaction between a doubly occupied orbital and a singly occupied orbital is mostly destabilizing. See ref 34 and, also: Bernardi, F.; Epiotis, N. D.; Cherry, W.; Schlegel, H. B.; Whangbo, M.-H.; Wolfe, S. *J. Am. Chem. Soc.* **1976**, *98*, 469–78.
(43) Fujimoto, H.; Kosugi, N. *Bull. Chem. Soc. Jpn.* **1977**, *50*, 2209–14.
(44) The delocalization interaction was found to be weak in a semiempirical perturbation calculation by Nagase and Fueno. See: Nagase, S.; Fueno, T. *Theor. Chim. Acta* **1974**, *35*, 217–30.
(45) Umeyama, H.; Morokuma, K. *J. Am. Chem. Soc.* **1977**, *99*, 1316–32.

Table III. Overlap Populations between the Hydrogen 1s Orbital and the Canonical MOs of Acetylene Molecule

acetylene MO	α spin	β spin
σ'_1	−0.0001	−0.0000
σ'_2	−0.0002	−0.0001
σ'_3	−0.0166	0.0008
σ'_4	−0.0092	0.0013
σ'_5	−0.0030	0.0004
π	0	0
π'	−0.0186	0.0381
π'^*	0.0313	0.0093
π^*	0	0
σ'_6	0.0001	−0.0001
σ'_7	−0.0000	0.0000
σ'_8	0.0000	0.0000

in an isolated state. The shift of an electron and that of a pair of electrons are distinguished from each other by the suffix $i \rightarrow l$ and the suffix $i \rightarrow l, i \rightarrow l$, respectively. Diexcited configurations $\Psi_{i \rightarrow j, i \rightarrow j}$ ($\Psi_{k \rightarrow l, k \rightarrow l}$) arise mainly from the intramolecular part and, therefore, scarcely take part in the intermolecular interaction.[43]

Besides the well-known effect of shifting the electronic charge from one fragment to the other fragment,[46] a mixing in of electron-transferred configurations possesses another important consequence, i.e., accumulation of the valence-active density in the intermolecular region.[16] For instance, the interaction between the configurations Ψ_0 and $\Psi_{i \rightarrow l}$ gives rise to the overlap density term $\phi_i(1)\psi_l(1)$. A part of electron population donated from the occupied MO ϕ_i is retained in the region between A and B through the overlap of ϕ_i with the unoccupied MO ψ_l. Similarly, the ditransferred configuration $\Psi_{i \rightarrow l, i \rightarrow l}$ yields the density term ϕ_i-$(1)\psi_l(1)$ by means of the interaction with the monotransferred configuration $\Psi_{i \rightarrow l}$.

The locally excited configurations contribute primarily to electron reorganization within the fragments induced by several causes.[17] They also yield overlap density terms by mixing with electron-transferred configurations. We obtain, for example, $\phi_i(1)\psi_k(1)$ from the interaction between $\Psi_{k \rightarrow l}$ and $\Psi_{i \rightarrow l}$ and $\phi_j(1)\psi_l(1)$ from the interaction between $\Psi_{i \rightarrow j}$ and $\Psi_{i \rightarrow l}$. That is, the occupied MO ψ_k of the acceptor fragment participates in the generation of the intermolecular overlap density through the interaction with the electron hole ϕ_i and the unoccupied MO ϕ_j of the donor fragment does so by promoting an electron from the occupied MO ϕ_i in the course of interactions. Thus, the intermolecular interaction matrix **P** involves the contributions of all the important electron configurations and, therefore, the interactive hybrid MOs represent important orbital effects. The hybrid MOs are given accordingly by linear combinations of the occupied and unoccupied (and, also, singly occupied) canonical MOs in each fragment with respect to each symmetry element of the composite interacting system.

(46) Mulliken, R. S.; Person, W. B. "Molecular Complexes"; Wiley: New York, 1969.

Conclusion

The hybrid molecular orbitals obtained here have been found to be localized in the frontier of mutual interaction. They may deserve, therefore, to be called the original of the frontier orbitals. Then, several possibilities of modification arise, the highest occupied MO or the lowest unoccupied MO being regarded as the simplest prototype.[1] Once again, we may stress here the significant assistance of other MOs throughout the course of chemical reactions. An important consequence of mixings between the frontier MO and extrafrontier MOs is the construction of reactive domain in a particular region of the reactant. The growth of new covalent bonds begins between the reaction sites.

The hybrid MO description of interactions will offer hopefully a clearer insight into the bond-forming processes by visualizing, in line with our chemical intuition, effects of various factors that will possibly have a great influence on reactivities of molecules and selectivities of reactions. Especially, analyses of interactions between a large molecular system and a small reagent will be much simplified, since they are represented in terms of several pairs of fragment hybrid MOs. Finally, we may notice that the present approach seems to provide a concrete application of the principle of maximum overlap by Pauling[47,48] and the overlap and orientation principle by Mulliken[49] to practical problems.

Acknowledgment. This work was supported by the Ministry of Education, Japan, under Grant-in-Aid for Scientific Research (465266, 521314).

Appendix

Table III shows the overlap populations between the 1s orbital of the attacking hydrogen radical and the canonical MOs of acetylene molecule with respect to a reaction model sketched in Figure 1. According to eq 4 we constructed a matrix of order 12 × 12 which was diagonalized to yield a set of new orbitals for each spin part. Then, the MOs of the composite interacting system were expanded in terms of the hydrogen 1s orbital and the new orbitals of acetylene and, thus, a new overlap population matrix was obtained in each spin part. The whole procedure of orbital transformations was repeated till a self-consistent result was generated. We got one positive eigenvalue and eleven zero eigenvalues in this case in the final matrix product. Twenty-two cycles and ten cycles of iterative calculations for the α spin part and for the β spin part, respectively, paying attention to the signs of interim orbitals to facilitate the convergence, reached our tentative self-consistency limit, $|P'_{l,k}| < 1 \times 10^{-5}$ ($l \neq k$). Total computing time used to calculate the interactive hybrid MOs for both spins was 2.1 s on a FACOM M200 at the Data Processing Center, Kyoto University. The calculation of hybrid MOs for subsets of canonical MOs was found to be much faster.

(47) Pauling, L. C. "The Nature of the Chemical Bond"; Cornell University Press: Ithaca, N.Y., 1960.
(48) Weinhold, F.; Brunck, T. K. *J. Am. Chem. Soc.* **1976**, *98*, 3745–9.
(49) Mulliken, R. S. *Recl. Trav. Chim. Acta* **1956**, *75*, 845–52.

REMINISCENCES
—— MY FORTY YEARS' STUDY OF CHEMICAL REACTIONS

Kenichi Fukui*
Kyoto Institute of Technology
Matsugasaki-Hashigamicho, Sakyo-ku
Kyoto 606, Japan

ABSTRACT A scientific biography of K. Fukui as one of the inquirers into the "chemical reaction" is presented in the form of a narrative story. The emphasis is placed on the circumstances of developments and applications of the frontier orbital concept and the intrinsic reaction coordinate approach.

INTRODUCTION

Good Morning, Ladies and Gentlemen.

I feel greatly privileged to talk at this symposium. To do so is also a great pleasure, of course, and my pleasure would be much greater, if I were here still as an active researcher. Actually, I retired from Kyoto University in April, 1982 due to the age-limit regulation, and am now an administrator at a national university in Kyoto. Such an age-limit system is very common in national universities in Japan, and is uniformly applied to everyone without exception. My present job is really pleasant and enjoyable, but you can easily imagine that this new job is not convenient for research. I have no laboratory, no staff, no more young students working with me, and have very little time to spend for research. I cannot see my scientific journals without going to my former Department which is a few kilometers away from my present office. So, it is difficult for me to keep pace with you all in reading new scientific papers, and in commenting or discussing new topics. That is the reason why this morning I am talking essentially about past things, and I hope you will forgive my intention in doing so.

---

*Also belongs to : Institute for Fundamental Chemistry, 15 Morimoto-cho, Shimogamo, Sakyo-ku, Kyoto 606, Japan

V. H. Smith, Jr. et al. (eds.), Applied Quantum Chemistry, 1–25.

EXPERIMENTAL STUDIES, 1940-45

My research life started with experimental studies in 1940 when I was a third year student at Kyoto University. The subject of my studies was the reaction of paraffin hydrocarbons with antimony pentachloride. A paraffin molecule having tertiary carbon atoms reacted with $SbCl_5$ to form an orange-red precipitate, which after awhile converted itself to tarry matter after vigorous evolution of HCl gas, and if ethyl ether was added to that, it turned into a beautiful deep blue solution. On the other hand such a distinctive color reaction was not seen in paraffins with no tertiary carbon atoms, and the former behavior was attributed to the initial formation of a carbonium ion. The remarkable difference in reactivity between a tertiary carbon and a secondary or primary carbon atom attracted my deep interest. Incidentally, such reactivity problems of paraffin hydrocarbons were treated theoretically more than twenty years later by myself with my colleagues.

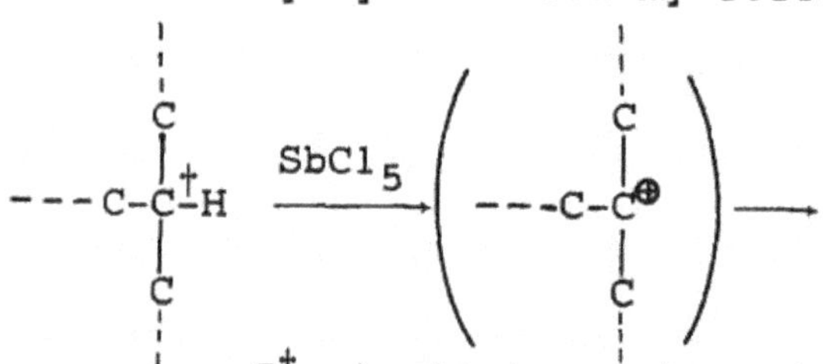

$C^{\dagger}$: tertiary carbon atom

My second experimental work on hydrocarbon chemistry was the isomerization of n-butene to isobutene, which was carried out in 1942-1943. I learned that the migration of methyl group was facilitated by the electron deficiency in the double bond caused by the action of heterogeneous acidic catalysts. This study went well and brought me a prize in 1944 when I was 25 years old.

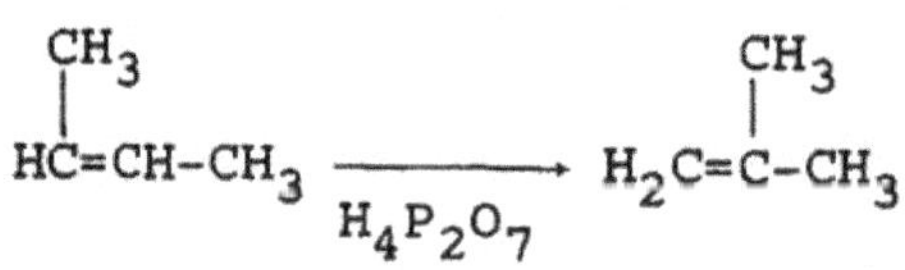

Again incidentally, this problem appeared before my eyes later as an example of the reaction generally called "[1,2]-sigmatropic shift" in the terminology of the Woodward-Hoffmann rule [1].

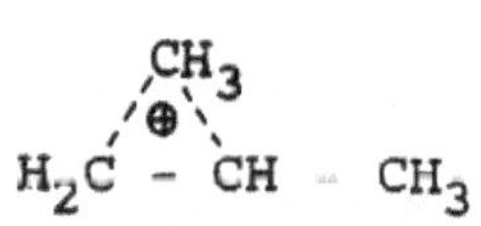

a two-electron cycle

BIRTH OF THE FRONTIER ORBITAL CONCEPT

The fact that I had been studying hydrocarbon chemistry as my specialty had a close connection with the birth of my frontier orbital concept. The reason was that the then prevailing "electronic theory" was not a suitable means for explaining the reactivity of hydrocarbons where the electric charge distribution in the molecule is almost uniformly zero. Such a situation drew my attention particularly to aromatic hydrocarbons in which an interesting "orientation" problem had already been well known in organic chemistry.

I calculated simply the Hückel molecular orbital, coefficients of the highest occupied molecular orbital (MO) of several aromatic hydrocarbons. The figures represented well the reactive positions in electrophilic aromatic substitutions. This particular orbital was named as the "frontier orbital". This paper appeared in the Journal of Chemical Physics in 1952 [2], the same year Prof. Mulliken's important paper on the theory of charge-transfer complexation appeared in the Journal of American Chemical Society [3]. This Mulliken paper was illuminating to me in interpreting my result as an electron delocalization between molecules [4].

Encouraged by this interesting result, an extension to the nucleophilic reaction was attempted. In this case the lowest unoccupied MO played the role of the frontier orbital. In radical substitutions both of these orbitals cooperated as the frontier orbitals [5]. A recently calculated contour diagram of the highest occupied MO of naphthalene is indicated in Figure 1.

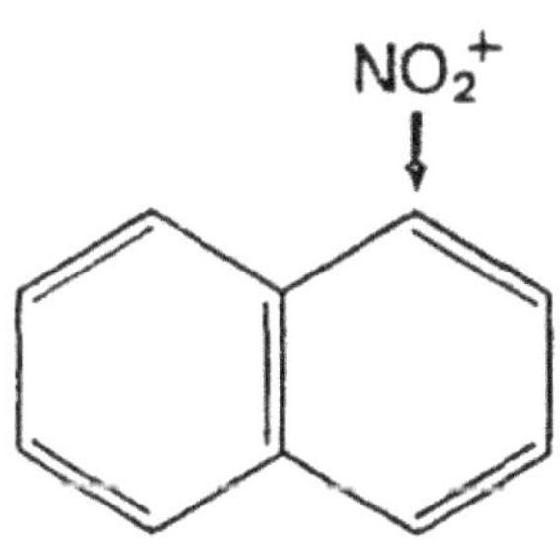

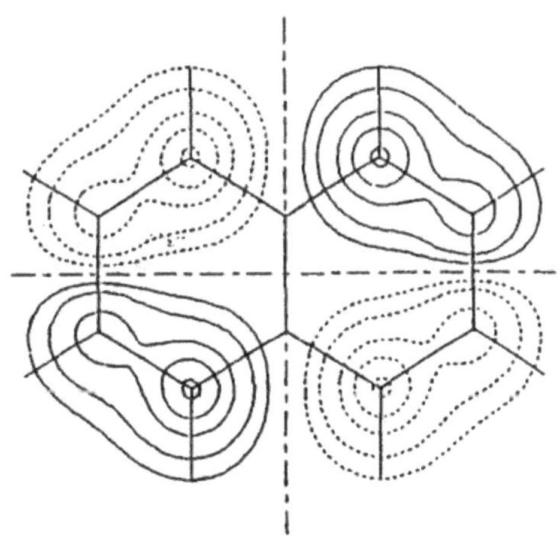

Orientation in the nitration of naphthalene.

Figure 1. The highest occupied MO of naphthalene.

The result of such a generalization was published at the Nikko Symposium of the International Conference on Theoretical Physics held in Tokyo in 1953.

The problem already mentioned was the selection of positions in a molecule in a reaction. To compare the reactivity of different molecules, the use of some theoretical indices which are calculated with respect to each position of each molecule may be convenient. I derived an index called "superdelocalizability" for aromatic substitutions in 1954 [4]. This index resulted as a natural consequence of the frontier orbital scheme. At the time when this index was proposed, there existed various theoretical indices, each based on different models. All of these indices explained the reactivity of condensed aromatic hydrocarbons almost equally well. Among these indices I found an interesting mathematical interrelation that was valid for a large fraction of aromatic hydrocarbons. These indices were all expressed in integral form in terms of the integrand of superdelocalizability [6].

Interrelation of various reactivity indices:

$$S_r = \frac{(-\beta)}{\pi}\int_{-\infty}^{\infty} G_r(y)\,dy \tag{1}$$

$$\left.\begin{aligned}
F_r &= \frac{-1}{\pi(-\beta)}\int_{-\infty}^{\infty}\{1-y^2G_r(y)\}\,dy + N_{max}\\
L_r &= \frac{1}{\pi}\int_{-\infty}^{\infty}\ln\{y^2G_r(y)\}^{-1}\,dy\\
H_r &= \frac{1}{\pi}\int_{-\infty}^{\infty}\ln\{1+y^2G_r(y)\}\,dy\\
\pi_{rr} &= \frac{(-\beta)}{\pi}\int_{-\infty}^{\infty}y^2\{G_r(y)\}^2\,dy\\
l_r &= \frac{2}{(-\beta)}\{G_r(0)\}^{-1/2} = 2(\bar{S}_r)^{-1/2}.
\end{aligned}\right\} \tag{2}$$

S_r : superdelocalizability
F_r : free valence
L_r : localization energy
H_r : hyperconjugation energy
π_{rr} : self-polarizability
l_r : Dewar's reactivity number

(See K. Fukui et al., *J. Chem. Phys.* 26, 831 (1957))

You may easily expect a parallelism in interpreting the reactivity of a majority of aromatic hydrocarbons by the use of these indices.

The density distribution of the lowest unoccupied MO of halogenated paraffins is indicated in Figure 2. The figures represent well the experimental behavior towards elimination [7]. Such a propensity was hardly expected in other reactivity theories.

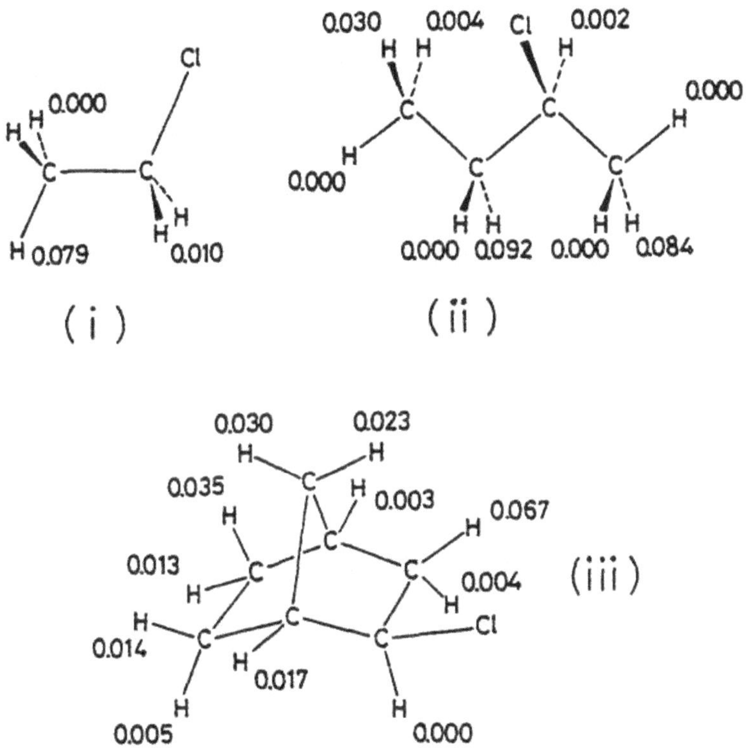

Figure 2. The lowest unoccupied MO of halogenated paraffins.
(i) ethyl chloride
(ii) 2-chlorobutane
(iii) 1-*exo*-chloronorbornane

HOMO-LUMO ELECTRON DELOCALIZATION

You may easily guess from the already-mentioned results that my model for a chemical reaction is based on the mutual delocalization of electrons between two kinds of frontier orbitals, as shown in Figure 3. These two frontier orbitals can be conveniently distinguished by calling them in an abbreviated form as HOMO and LUMO. I do not know who first began to call them so. Probably it was Roald Hoffmann. At least I can say that this naming became common only after his proposal of the Woodward-Hoffmann rule [1]. It was really the publication of this rule that made my theory known to the majority of the chemical community.

In this manner, the frontier orbital scheme extended its availability to compounds beyond the so-called π-electron compounds. In this connection, the relation between the frontier orbital theory and the "electronic theory" may be interesting. The electronic theory existed at that time as a most powerful means of explaining the delicate difference in reactivity of various complicated organic molecules. It is easy to realize the parallel relation between the frontier orbital approach and the electronic theoretical approach, because an electron-rich position in a molecule is usually also the place where the HOMO amplitude is large, and an electron-deficient position is generally also the place where the LUMO amplitude is large. Accordingly, a parallel relation between the frontier orbital approach and the electronic theory was in a general sense obvious.

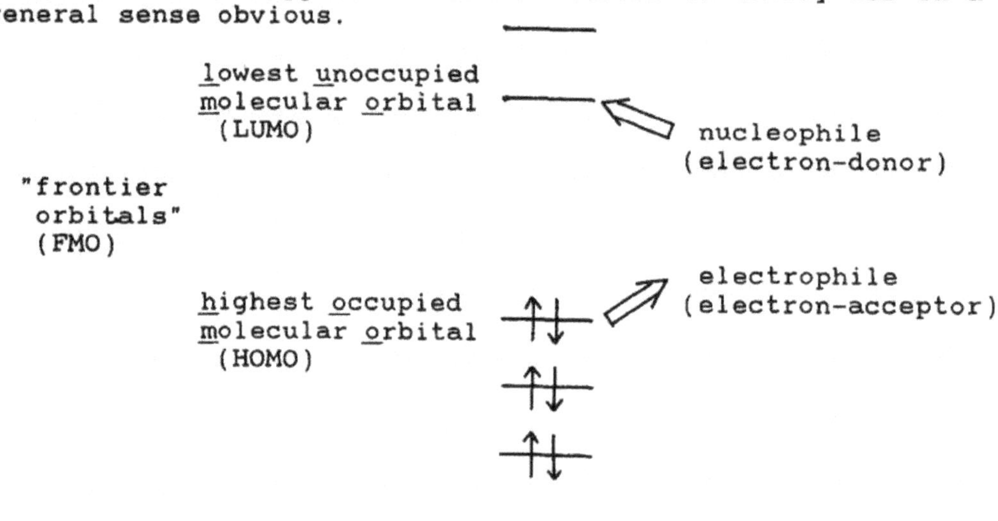

Figure 3. Delocalization model for chemical reactions. (see K. Fukui, *"Theory of Orientation and Stereoselection"*, Springer-Verlag, Berlin (1970 and 1975)).

A more remarkable relation of the electronic theory to the frontier orbital scheme was pointed out concerning the Hammett rule by Henri-Rousseau and Texier [8], as shown in Figure 4. The Hammett rule was an extremely convenient method of discussing the reactivity of substituted hydrocarbons in terms of so-called Hammett constants. The meaning of Hammett constants, particularly that of the sign of the reaction constant, became clear, and these constants were made theoretically calculable to a certain approximation by the aid of a perturbation treatment.

$$\log\frac{k_j}{k_0} = \rho\sigma_j.$$

ρ : reaction constant
σ_j: substituent const. (the value of σ_j is parallel to the electron attracting character of the substituent)

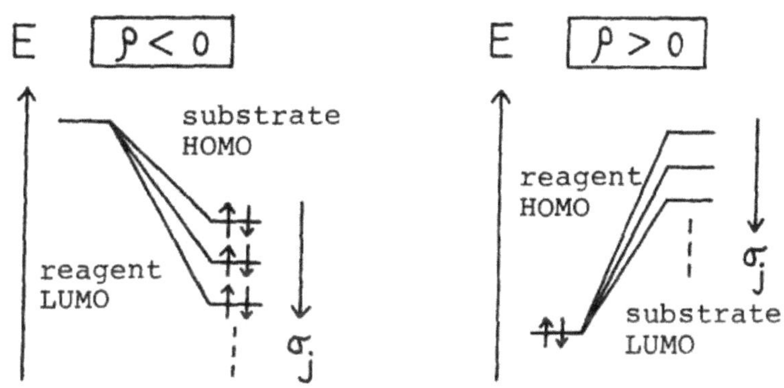

The case of electrophilic reagents and nucleophilic substrates

The HOMO energies of substrates are lowered by increasing the σ_j values of the substituents, the interaction energy decreases so that the rate decreases.

The case of nucleophilic reagents and electrophilic substrates

The LUMO energies of substrates are lowered by increasing the σ_j values of the substituents, the interaction energy increases so that the rate increases.

Figure 4. The Hammett rule and the frontier orbital approach. (See O. Henri-Rousseau and F. Texier, *J. Chem. Educ.* 55, 437 (1978)).

One of the best examples of mutual HOMO-LUMO delocalization interaciton may be that of the Diels-Alder diene synthesis, as shown in Figure 5. The orbital phase situation of the diene HOMO and the dienophile LUMO, and that of the dienophile HOMO and diene LUMO, are both favorable for a bond exchange of a more or less concerted nature between these molecules. That is to say, if electrons go out from the diene HOMO, the bonds that are bonding in that MO will be weakened while the bond that is antibonding will be strengthened. Obviously, these changes contribute to the bond exchange necessary for the occurrence of the reaction. The same thing can be said when electrons enter into the diene LUMO. A similar discussion is possible with respect to the dienophile molecule. At the same time, it is easily seen that these electron delocalizations would place electrons into the reaction sites to form the two new bonds.

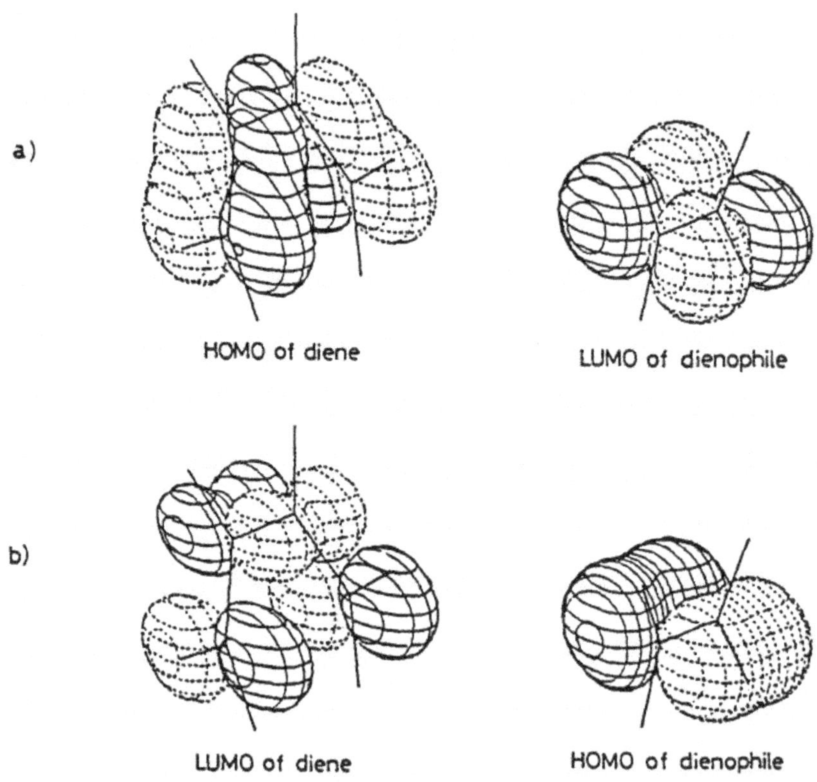

Figure 5. A model reaction for the diene synthesis.
a) Diene HOMO-dienophile LUMO interaction
b) Dienophile HOMO-diene LUMO interaction

The essence of this paper appeared in a monograph published in 1964, edited by Löwdin and Pullman as a tribute to Prof. Mulliken [9]. But that paper had not been publicly known until Roald Hoffmann kindly cited it in his very widely read paper on the Woodward-Hoffmann rule [1].

Later, similar HOMO-LUMO interaction schemes were applied by organic chemists, particularly by Houk, to the discussion of a wide variety of cyclic reactions —— in particular, the problem of regioselection in substituted diene syntheses, 1,3-dipolar additions, and so on [10]. The results of various applications were presented in two monographs written by authors from the United Kingdom [11].

The preceding discussions were essentially associated with the intermolecular cyclic HOMO-LUMO interaction. A similar HOMO-LUMO interaction scheme is also conceived in intramolecular interactions. One of its applications can be made to the problem of aromaticity, as shown in Figure 6. The fact that anthracene is less aromatic than phenanthrene is well understood by this scheme. In phenanthrene, the HOMO-LUMO phase relation is favorable with respect to any virtual division of the π-electron network, while in anthracene such a beautiful relation does not exist.

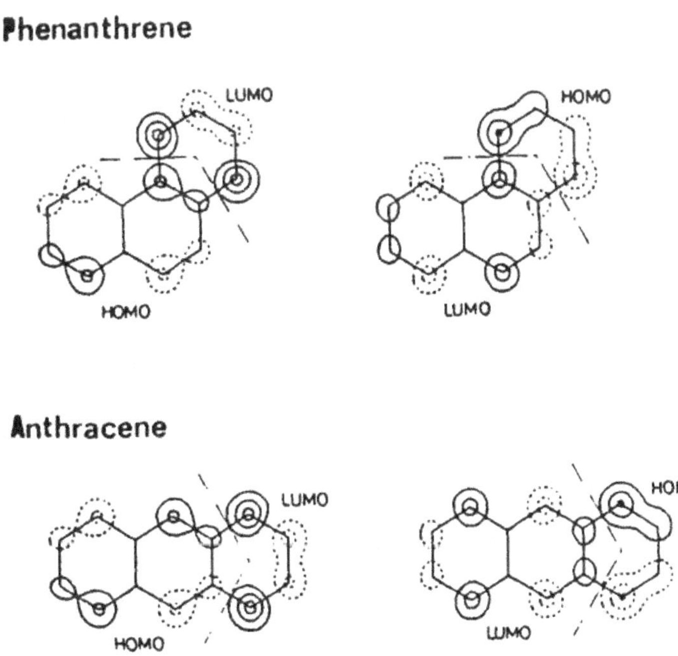

(-·-·- section of virtual division)

Figure 6. The HOMO-LUMO phase relation in virtual division of phenanthrene and anthracene.

INTRINSIC REACTION COORDINATE

In 1970, I proposed the concept of "intrinsic reaction coordinate" [12]. This name is often abbreviated as IRC. This is to define the "center line" of the reaction path on the potential energy surface, as shown in Figure 7. In a classical mechanical sense, the reaction path is represented by a zigzag course dependent on the initial condition. It was my intention to make my reactivity theory more quantitative by defining the reaction path which is not affected by the initial condition.

The following story may be interesting to you. When I submitted this paper to the Journal of Physical Chemistry, the letter from the referee read, although I do not remember the sentence exactly, "This paper is not very original, nevertheless it is publishable ...". The referee's judgement was quite pertinent, since my paper was too simple to be appreciated as "original". However, this concept was later developed in many directions which will be discussed below, and so it was really publishable. Thus, the referee's judgement was quite correct. So, I have been admiring this anonymous referee. Since this present symposium is a large assembly of theoretical chemists a major portion of which are Americans, it is highly probable that we have that referee here as one of participants of this symposium.

If you use a simple model potential function, you can easily find the reaction coordinate and formulate it. But on actual potential surfaces, some care must be taken.

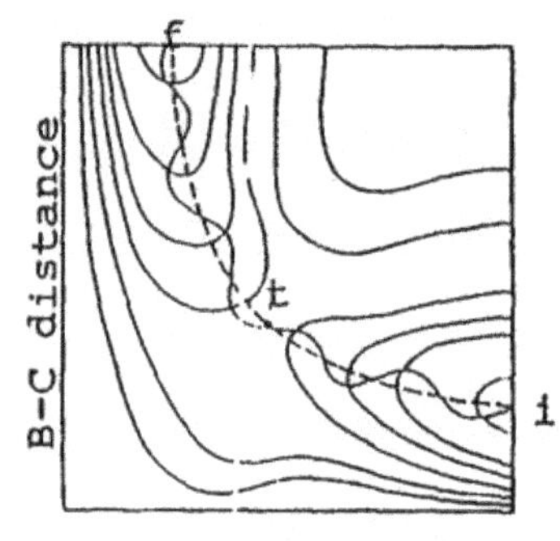

--- : IRC
i : initial point
t : transition-state point
f : final point

Figure 7. A schematic diagram of the reaction coordinate of the reaction AB+C→A+BC.

I defined the IRC as the locus of infinitely slow change of configuration of reacting molecular systems governed by classical dynamical laws. So, the energetics are quantum-mechanical, but the dynamics are essentially classical. In due course the shift to quantum-mechanics is possible by the aid of known procedure.

For the purpose of obtaining the IRC itself, it is convenient and sufficient to use Cartesian coordinates. To discuss the rate of reaction in the frame of the IRC formalism, it is more convenient to use an appropriate internal coordinate system to obtain the **generalized** IRC equation, it should be written in a form which is invariant with coordinate transformation by the aid of usual differential-geometric conventions.

$$\frac{a_{1j}dq^j}{\frac{\partial W}{\partial q^1}} = \frac{a_{2j}dq^j}{\frac{\partial W}{\partial q^2}} = \cdots\cdots = \frac{a_{nj}dq^j}{\frac{\partial W}{\partial q^n}}. \qquad (3)$$

$(n=3N-6)$

q^j : internal coordinate $(j=1,2,\cdots n)$
ds : line element, $ds^2 = a_{ij}dq^i dq^j$
W : energy

The IRC is defined as the solution curve which starts from the initial stable point and reaches the final stable point passing through the transition-state point on the potential energy surface.

It is easily shown that the generalized IRC equation is derived from the following variation equation.

$$\delta\left(\frac{dW}{ds}\right) = 0. \qquad (4)$$

This implies that the IRC already defined is nothing but the **steepest descent path** from the transition-state point to a basin.

Figure 8 indicates a simple schematic model for a two-dimensional configuration space which is divided into two cells by the assemblies of solution curves of the IRC equation.

The mode of division of the configuration space into cells is more clearly shown by another example in Figure 9. An n-cell problem is derived from a simple form of a potential function . Each cell contains one stable point separated by walls from neighboring cells. The walls contain transition-state points within them, through each of which two different stable points are connected by IRC's.

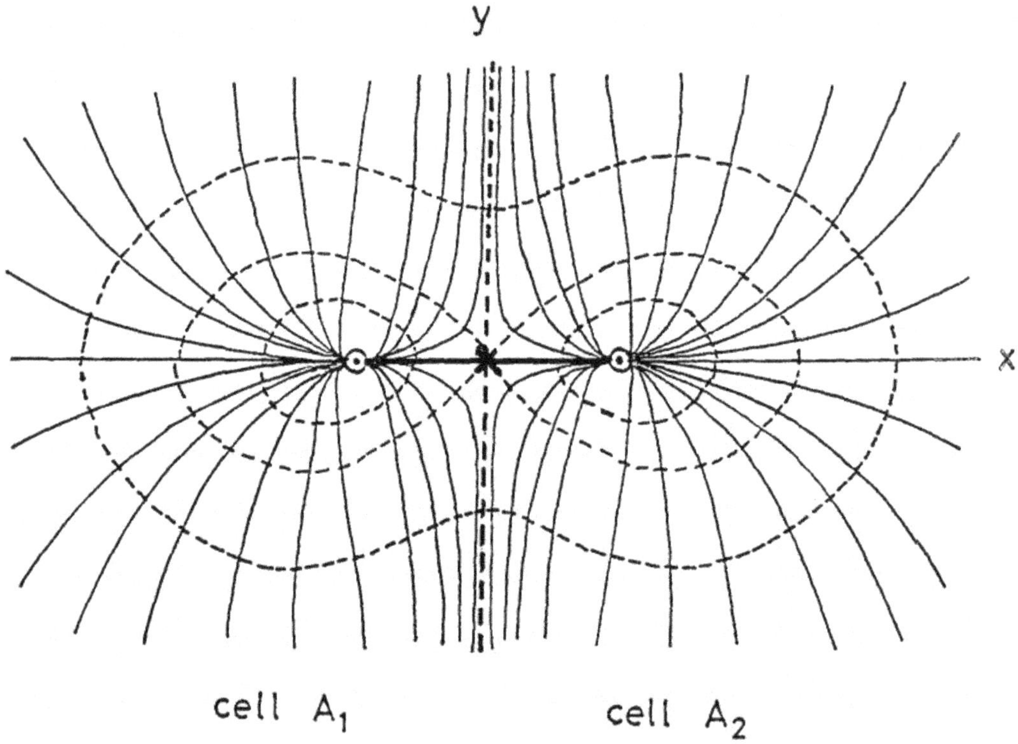

solution curves:

$$y=\text{const.}\times\left[\frac{|x^2-1|^{\frac{1}{2}}}{|x|}\right]^{b^2/2a^2} \qquad (b^2>2a^2)$$

⊙ : cell center ---: equipotential surface
X : transition-state point ---: intercell boundary
▬ : IRC ——: solution curve

$$ds^2 = dx^2+dy^2$$
$$W = a^2(1-x^2)^2+b^2y^2$$

x,y : Cartesian coordinates

Figure 8. The cell structure produced by a simple model potential function.

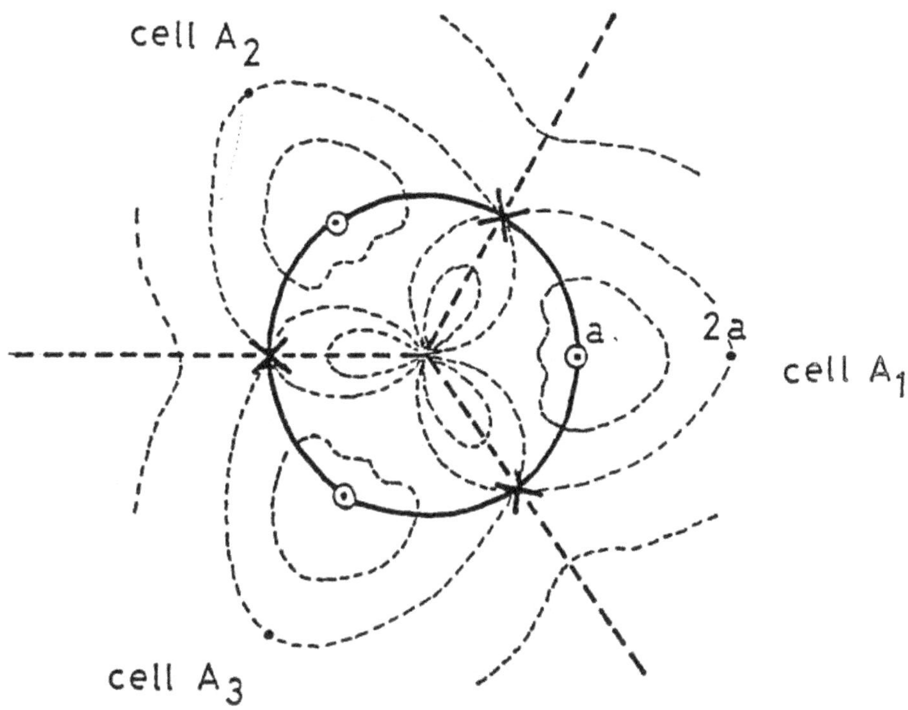

⊙ : cell center ---: equipotential line
X : transition-state point ▬▬▬: intercell boundary ("separatrix")
▬: IRC

$$ds^2 = dr^2 + r^2 d\theta^2$$

$$W = (r-a)^2 + a^2 \sin^2(\tfrac{3}{2}\phi)$$

r, ϕ : polar coordinates

Figure 9. A model for the two-dimensional n-cell space. (The case n=3 is indicated)

In one of the simplest approaches to a kinetic study, you consider the energy of vibrations in the direction perpendicular to the IRC as providing an "effective" potential for the "translational" motion along the IRC. Such an approach is analogous to the Born-Oppenheimer approximation in which the electronic energy serves to provide an effective potential for the vibrational motion.

More generally, you can look at an arbitrary reacting system consisting of several atoms of any kind, say

$$C_iH_jO_kN_l \cdots\cdot, \qquad i+j+k+l+ \cdots = N.$$

If you know the potential energy function calculated quantum-mechanically in terms of (3N-6) internal coordinates, you will obtain all of the equilibrium points, and the whole configuration space is divided into a definite number of cells by the assemblies of the solution curves of the IRC equation, as indicated in Figure 10.

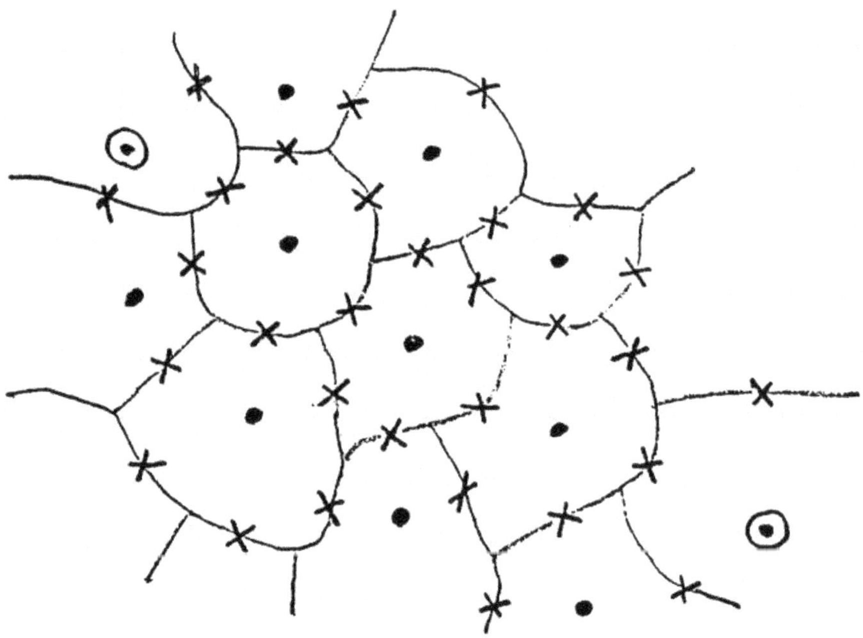

⊃ : cell
× : transition-state point
• : stable equilibrium point
⊙ : {initial reactants point / final products point}

Figure 10. A labyrinth of an arbitrary reacting system.

If you allow me to tell you a very speculative story, the chemical reaction can be treated theoretically as the process of wave-mechanical transfer of a wave packet, initially confined to the initial cell corresponding to the reactants, to the final cell corresponding to the products. This model is helpful in obtaining a most general theoretical picture of a chemical reaction involving the tunnelling process and multistep processes. In this connection, it is noteworthy that Paul Mezey [13] is vigorously developing a theory of global behavior of the reaction space —— "reaction topology", and an interesting "directed graphs" theory is being developed by Oktay Sinanoglu [14].

TABLE I. VARIOUS VARIATIONAL FORMULAE FOR IRC PATH

L: Lagrangian, T: kinetic energy, W: potential energy

L_τ: extended Lagrangian, T_τ: $\frac{1}{2}a_{ij}\frac{dq^i}{d\tau}\frac{dq^j}{d\tau}$

τ: time parameter defined by $a_{ij}\frac{dq^j}{d\tau} = \frac{\partial W}{\partial q^i}$

$\Delta_1 W$: Beltrami's differential parameter of the first order with respect to W

Variational principles	Traditional dynamics	Intrinsic dynamics
Hamilton's principle	$\delta\int L dt = 0$ $L = T - W$	$\delta\int L_\tau d\tau = 0$ $L_\tau = T_\tau + (1/2)\Delta_1 W$
Principle of least action	$\delta\int 2T\,dt = 0$ for paths with $T + W = E$	$\delta\int 2T_\tau d\tau = 0$ for paths with $T_\tau - (1/2)\Delta_1 W = 0$
Geodesic principle	$\delta\int ds = 0$ in the space of $ds^2 = (E-W)\,2T\,dt^2$	$\delta\int ds = 0$ in the space of $ds^2 = (\Delta_1 W)\,2T_\tau\,d\tau^2$

(See K. Fukui, *Intern. J. Quant. Chem., Quant. Chem. Symp.* 15, 633 (1981)).

Shamefully I confess that no appreciable numerical result has yet been obtained by myself except for a simplest treatment for the hydrogen migration in the enol-form of malonaldehyde [15]. Fortunately, however, new alternative, much more effective methods were developed by American theorists, particularly by the groups of Miller [16], and Truhlar [17]. I am anticipating the progress of these approaches instead of developing my own.

The solution curve of the IRC equation satisfies several variational equations [18]. These are indicated in TABLE I. You can see that the path of chemical reaction also fulfils variational principles just as an ordinary dynamical path does.

One of the important applications of the IRC approach is to draw various sorts of <u>correlation diagrams</u> along the IRC. The quantities to be plotted may be the total energy of the reacting system, the electronic energy levels, the MO energies, the forces acting on each atom, the vibrational modes and frequencies, the curvature of the IRC, and any other quantities calculable at each point on the IRC.

As an example, I will show you a vibrational correlation diagram. Figure 11 shows the frequencies of CH_2O along the isomerization path to HC̈-OH (hydroxycarbene) [19].

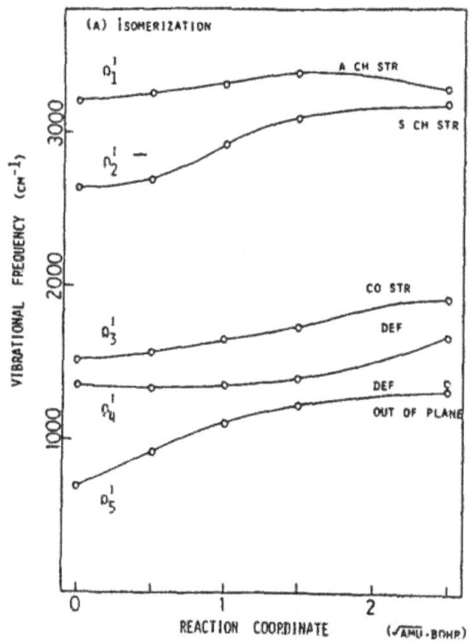

Figure 11. The variation of vibrational frequencies of normal coordinates perpendicular to IRC for the isomerization of CH_2O. (Reproduced by permission from *Chem. Phys. Letters*, 84, 123 (1981))

A diagram like this will be useful in relation to the study of laser-controlled mode-selective chemical reactions. You can see the antis. def. mode of 1377 cm^{-1} of CH_2O is solely left unconnected in the diagram. This indicates that the IRC, going down from the transition state (RC=0), converges asymptotically to this mode at the initial equilibrium point. The mode Q_4 also interacts with the IRC, but in this case the mode coupling is small as is seen in a moderate change of frequency.

Figure 12 is the diagram for the decomposition of CH_2O into H_2 and CO [19]. In this case, too, the IRC is connected to antis. def. mode of 1377 cm^{-1}. But along the decomposition path the mode coupling is strong. In particular, the sym. CH-str. mode of 3192 cm^{-1} plays an important role in the coupling with the product channel. Accordingly, it is expected that the excitation of this mode will probably be effective to the selective occurrence of decomposition if other subsidiary conditions are favorably satisfied.

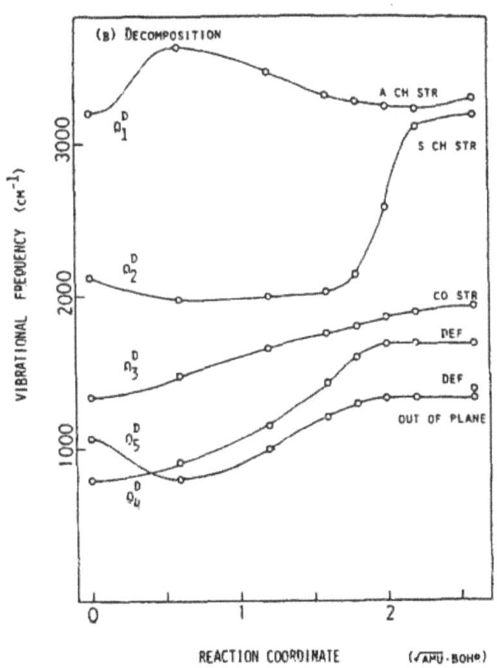

Figure 12. The variation of vibrational frequencies of normal coordinates perpendicular to IRC for the decomposition of CH_2O. (Reproduced by permission from *Chem. Phys. Letters*, **84**, 123 (1981))

Figure 13 shows an application to the decomposition of FCH_2OH to CH_2O and HF [20]. It will be interesting to discuss this reaction in relation to the problem of chemical laser systems. You see the OH-str. mode of the starting molecule is connected to the HF-str. mode of the product system. This mode interacts with the IRC, changing its frequency drastically. As a consequence, the IRC is intensely bent immediately before and after the transition state. An excitation of the OH-str. mode must therefore have a close connection with the excitation of the HF-str. mode of the product.

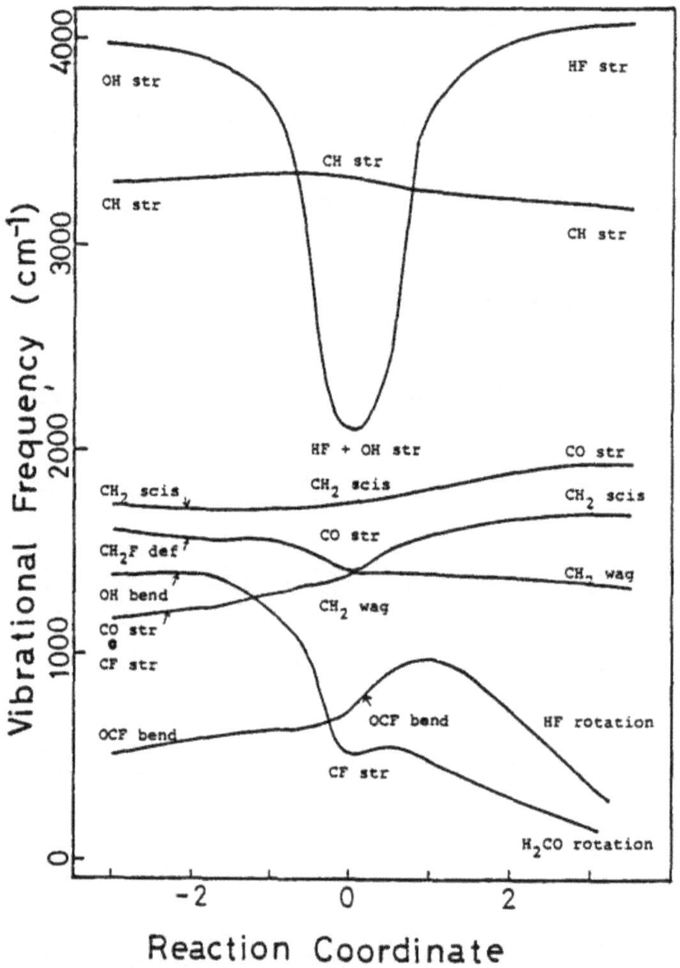

Figure 13. Vibrational frequency correlation diagram of the HF elimination of FCH_2OH. (Reproduced by permission from *Theoret. Chim. Acta (Berl.)* 60, 523 (1982))

Chemical HF laser emission was observed from the reaction of CH_3F with the oxygen atom in the ^3P and the ^1D state. This implies that the lifetime of the insertion-elimination intermediate is too short to be energetically randomized before decomposition. This sort of discussion with such a vibrational correlation diagram will therefore be useful in the study of chemical laser systems.

You can see from the previous diagrams that the IRC curvature is important in discussing the intrasystem energy transfer. If you want to calculate the curvature, you may use the following formula [18].

$$\left.\begin{aligned} \kappa^2 &= a_{ij}\kappa^i\kappa^j \\ \kappa^i &= \frac{d^2q^i}{ds^2} + \Gamma^i_{jk}\frac{dq^j}{ds}\frac{dq^k}{ds}. \end{aligned}\right\} \quad (5)$$

κ : curvature of the curve $q^i(s)$, $i=1,2,\cdots n$, in the space of $ds^2=a_{ij}dq^idq^j$

Γ^i_{jk} : the Christoffel symbol of the second kind

It may happen sometimes that one needs to discuss the effect of motion of the reacting system <u>**as a whole**</u> upon the IRC. The <u>**extended**</u> IRC equation is as follows [21]:

$$a_{ij}dq^j/\left[\frac{\partial W}{\partial q^i} + \frac{1}{2}\sum_{r=1}^{3}\frac{\partial I_r^{-1}}{\partial q^i}(\omega_0-\Omega_0)_r^2 - \sum_{(r,s,t)}\mu_{ri}I_r^{-1}(I_s^{-1}-I_t^{-1})(\omega_0-\Omega_0)_s(\omega_0-\Omega_0)_t\right] =(\text{independent of } i). \quad (6)$$

ω_0 : total angular momentum

Ω_0 : angular momentum of the center of mass

(See K. Fukui et al., *Intern. J. Quant. Chem., Quant. Chem. Symp.* 15, 621 (1981)).

If you put the velocity and the angular momentum of the whole system equal to zero, you obtain the previous general form of IRC equation.

INTERACTION FRONTIER ORBITAL

If you combine the classical frontier orbital concept to the IRC approach, you will reach to the frontier orbitals of mutually interacting molecules. That is to say, the **interaction frontier orbital**, often conveniently abbreviated as IFO [22]. This can be obtained by orthogonalizing the delocalization energy with the use of pairing technique analogous to the Amos-Hall corresponding orbital formalism [23]. The IFO is sometimes useful for visualizing the mode of formation and breaking of chemical bonds in a reaction. You can do this by obtaining the succession of IFO's along the IRC. That is the **IFO correlation diagram**. Several examples of IFO's are shown in Figures 14-16.

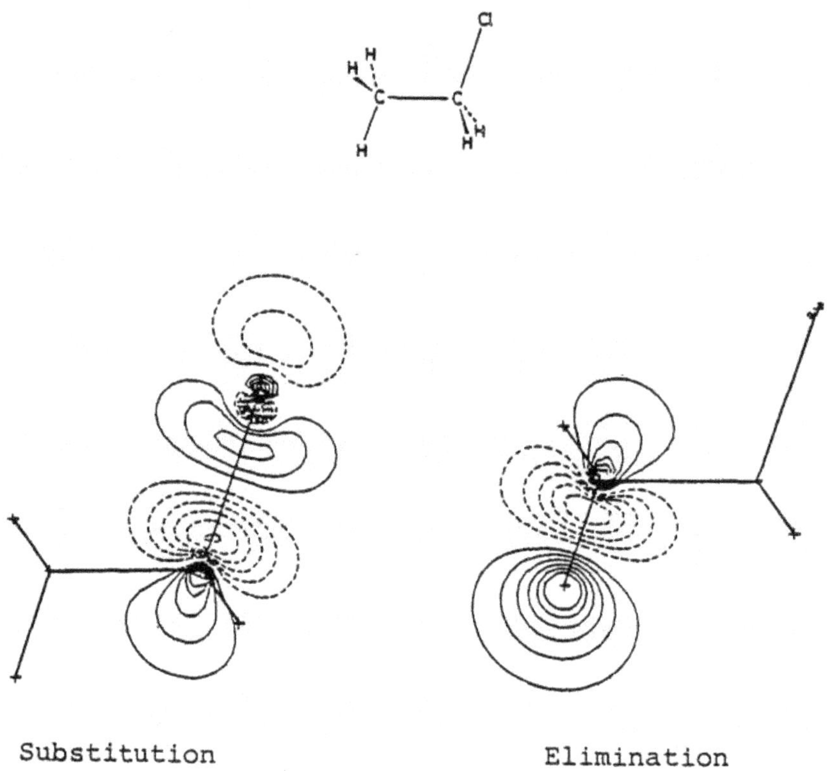

Substitution Elimination

Figure 14. The IFO's of ethyl chloride in electron-accepting interaction with hydride anion [22]. (Reproduced by permission from *J. Am. Chem. Soc.* 103, 196 1981))

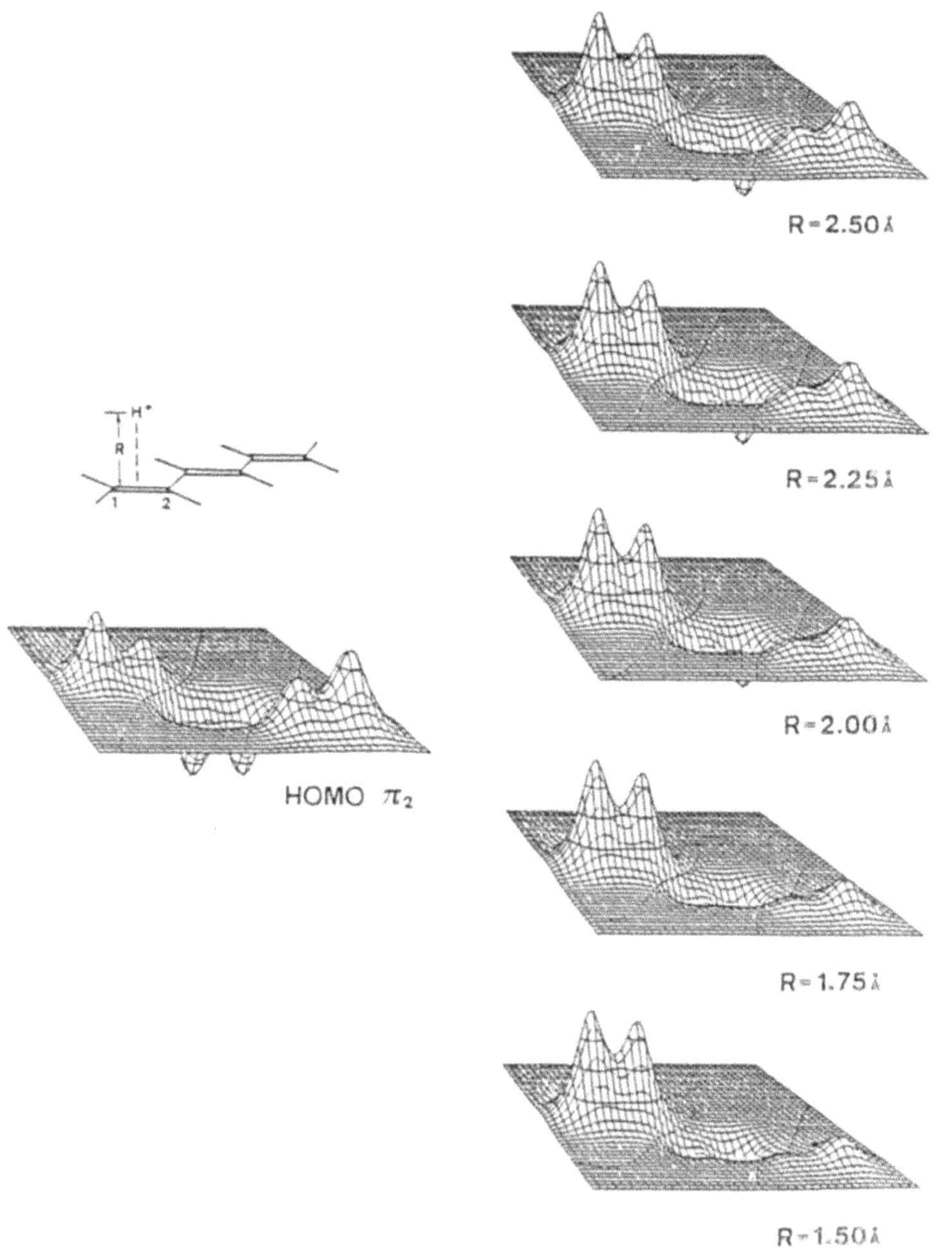

Figure 15. The IFO of hexatriene in delocalizing interaction with proton [24]. (Reproduced by permission from *J. Phys. Chem.* **88**, 3539 (1984))

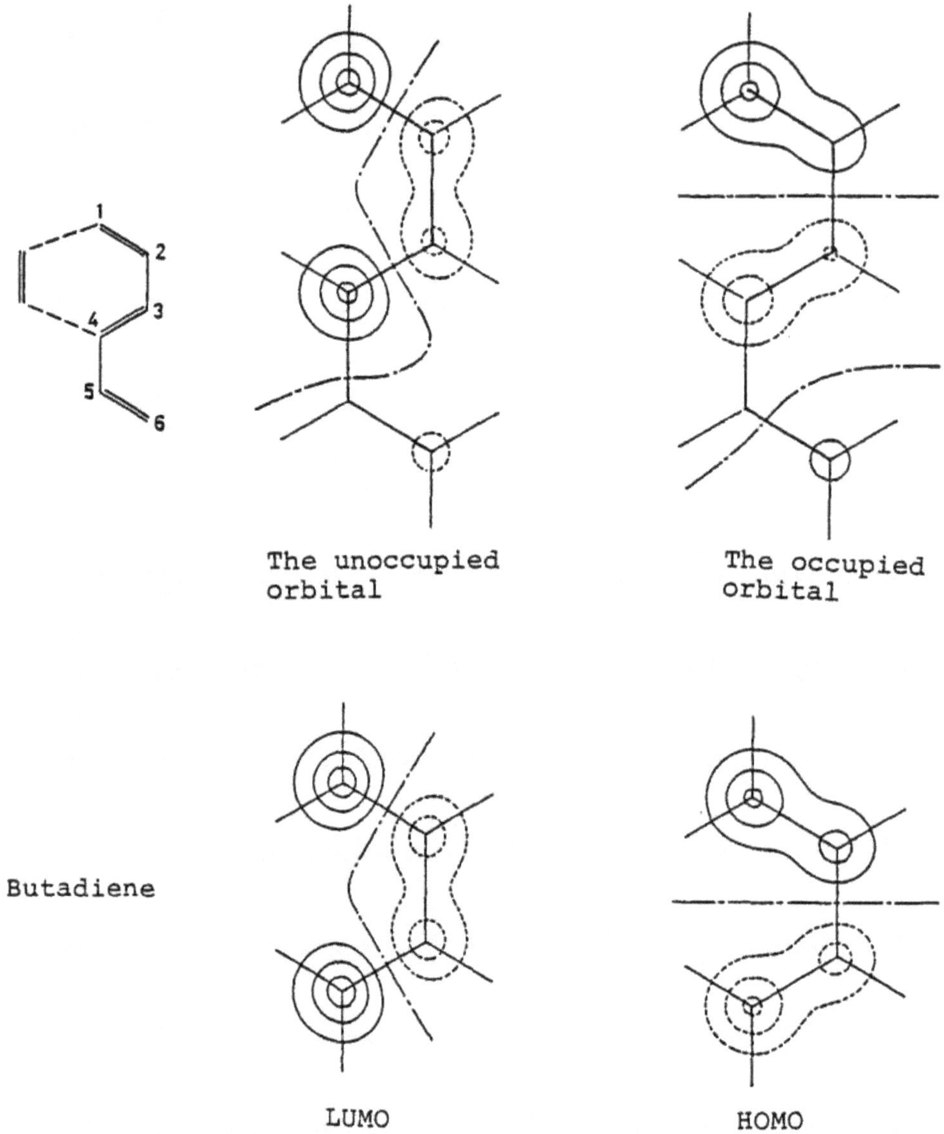

Figure 16. The IFO's of hexatriene in delocalizing interaction with ethylene, closely resembling the HOMO and LUMO of butadiene [24]. (Reproduced by permission from *J. Phys. Chem.* 88, 3539 (1984))

You have seen that the direction of progress of a chemical reaction is determined by the IRC formalism. In this connection, it may be interesting for you to know, as shown in Figure 17, that the nuclear vibration in this direction serves to absorb electrons from the region of negative HOMO-LUMO overlapping and pour them into the region of positive HOMO-LUMO overlapping. I owe this formalism to Prof. Bader [25]. You can imagine that suitably large values of the vibrational quantum number n will make this effect appreciable. In this way, you can see that the molecular vibration plays a crucial role in the electron-transfer or -conveying process. The physics and chemistry of organic semiconductors or conductors will be an

---

The Bader type formula [25]:

$$\overline{\Delta\rho}(1) \sim 2c^2\psi_{HO}(1)\psi_{LU}(1)\int x\phi_n^2(x)dx \times \int \psi_{HO}(2)\psi_{LU}(2)\frac{\partial V_{Ne}(2)}{\partial x}dv(2)/(W_0-W_{HO\to LU}).$$

$\Delta\rho$: the charge of electron density
x: nuclear displacement in the direction of IRC mode

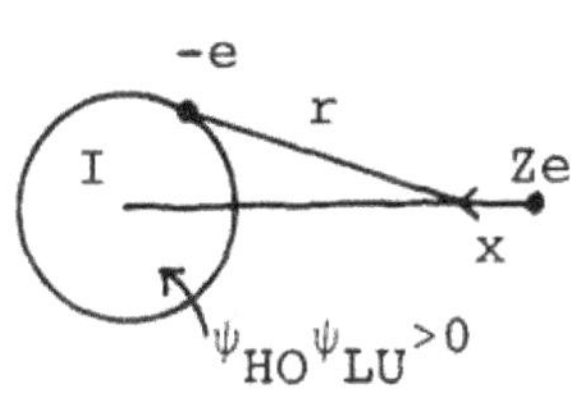

region I : the region where HOMO and LUMO effectively overlap*

$\overline{\Delta\rho}(1)>0$ in I if the symmetry of $\psi_{HO}\psi_{LU}$ conforms to the symmetry of V_{Ne}.

THE NUCLEAR VIBRATION IN THE DIRECTION OF IRC SERVES TO ABSORB ELECTRONS FROM THE REGION OF **NEGATIVE** OVERLAPPING OF HOMO AND LUMO AND POUR THEM INTO THE REGION OF **POSITIVE** OVERLAPPING.

Suitably large values of vibrational quantum number n cause this effect large!

$\phi_n(x)$: nth vibrational wavefunction
c : a constant
* The signs of HOMO and LUMO are taken so that $\psi_{HO}\psi_{LU}>0$ in I.

Figure 17. The change of electron density due to the vibrational motion along the IRC.

interesting area for discussing effects of this sort. A few papers were published along this line [26].

In the frame of his density functional study, very recently, Prof. Bob Parr derived a beautiful relationship between the chemical potential of the assembly of electrons in the molecule and the frontier orbital distribution [27]. This provides a thermodynamical rationalization of the frontier orbital approach.

REFERENCES

1. R.B. Woodward and R. Hoffmann, *Angew. Chem.* **81**, 797 (1969).
2. K. Fukui, T. Yonezawa, and H. Shingu, *J. Chem. Phys.* 20, 722 (1952).
3. R.S. Mulliken, *J. Am. Chem. Soc.* **74**, 811 (1952).
4. K. Fukui, T. Yonezawa, and C. Nagata, *Bull. Chem. Soc. Japan* 27, 423 (1954).
5. K. Fukui, T. Yonezawa, C. Nagata, and H. Shingu, *J. Chem. Phys.* 22, 1433 (1954).
6. K. Fukui, T. Yonezawa, and C. Nagata, *J. Chem. Phys.* 26, 831 (1957).
7. K. Fukui, "*Theory of Orientation and Stereoselection*", Springer-Verlag, Berlin (1970 and 1975).
8. O. Henri-Rousseau and F. Texier, *J. Chem. Education* 55, 437 (1978).
9. K. Fukui in "*Molecular Orbitals in Chemistry, Physics and Biology*", P.-O. Löwdin and B. Pullman, eds., Academic Press, New York (1969), p.513.
10. K.N. Houk, *Acc. Chem. Res.* **8**, 361 (1975) and references cited therein.
11. I. Fleming, "*Frontier Orbitals and Organic Reactions*", John Wiley, New York (1976); T.L. Gilchrist and R.C. Storr, "*Organic Reactions and Orbital Symmetry*", Cambridge Univ. Press, 2nd ed., London (1979).
12. K. Fukui, *J. Phys. Chem.* **74**, 4161 (1970).
13. P. Mezey, Proceedings of the Nobel Laureate Symp. on *Applied Quant. Chem.*, Dec. 1984, Honolulu, Hawaii, and references cited therein.
14. A. Fernandez and O. Sinanoglu, *Theoret. Chim. Acta (Berl.)* 65, 179 (1984).
15. S. Kato, H. Kato and K. Fukui, *J. Am. Chem. Soc.* **99**, 684 (1977).
16. W.H. Miller, N.C. Handy, and J.E. Adams, *J. Chem. Phys.* 72, 99 (1980); W.H. Miller, *J. Phys. Chem.* 87, 3811 (1983).
17. D.G. Truhlar and A.D. Isaacson, *J. Chem. Phys.* 77, 3516 (1982); D.G. Truhlar, W.L. Hase, and J.T. Hynes, *J. Phys. Chem.* 87, 2664 (1983).
18. A. Tachibana and K. Fukui, *Theoret. Chim. Acta (Berl.)* 57, 81 (1980); K. Fukui, *Intern. J. Quant. Chem., Quant. Chem. Symp.* **15**, 633 (1981).

19. K. Yamashita, T. Yamabe, and K. Fukui, *Chem. Phys. Letters* **84**, 123 (1981).
20. K. Yamashita, T. Yamabe, and K. Fukui, *Theoret. Chim. Acta (Berl.)* **60**, 523 (1982).
21. K. Fukui, A. Tachibana, and K. Yamashita, *Intern. J. Quant. Chem., Quant. Chem. Symp.* **15**, 621 (1981).
22. K. Fukui, N. Koga, and H. Fujimoto, *J. Am. Chem. Soc.* **103**, 196 (1981).
23. A.T. Amos and G.G. Hall, *Proc. Roy. Soc. London*, Ser. **A 1961**, 263, 483.
24. H. Fujimoto, N. Koga, and I. Hataue, *J. Phys. Chem.* **88**, 3539 (1984).
25. R.F.W. Bader, *Can. J. Chem.* **40**, 1164 (1962).
26. A. Tachibana, T. Yamabe, K. Hori, and Y. Asai, *Chem. Phys. Letters* **106**, 36 (1984); A. Tachibana, K. Hori, and T. Yamabe, *ibid.* **112**, 279 (1984); A. Tachibana, K. Hori, Y. Asai, and T. Yamabe, *J. Chem. Phys.* **80**, 6170 (1984).
27. J.P. Perdew, R.G. Parr, M. Levy, and J.L. Balduz, Jr., *Phys. Rev. Lett.* **49**, 1691 (1982); R.G. Parr and W. Yang, *J. Am. Chem. Soc.* **106**, 4049 (1984).

Part VI

Other Topics

[Reprinted from the Journal of Physical Chemistry, 78, 148 (1974).]

On the Behavior of the Excess Electron in Methane

Shingo Ishimaru, Tokio Yamabe, Kenichi Fukui,*

Department of Hydrocarbon Chemistry, Kyoto University, Kyoto, Japan

and Hiroshi Kato

Department of General Education, Nagoya University, Nagoya, Japan *(Received July 24, 1973)*

The electronic state of the trapped electron in methane is calculated by the use of the INDO-UHF method and the minimum lifetime of the electron is also estimated from the above result. Consequently, it is concluded that (1) the traps in methane do not play a substantial role in the motion of the excess electron, (2) the absorption spectra of the excess electron in methane will be difficult to obtain or quite broad, if detected, and (3) the present molecular orbital treatment is expected to lead to reasonable results for other hydrocarbons in which the trapped electron is more stable than in methane.

Introduction

There are many theoretical investigations[1-7] which confirm the existence of trapped electrons in cavities in polar solvents. For example, in previous papers,[7] the present authors have shown with the aid of the semiempirical Hartree-Fock method that, using the cavity model, it is possible to interpret not only the thermal and optical but also the magnetic properties of the solvated electron.

On the other hand, several authors[8-11] suggest that electrons in liquid hydrocarbons are also trapped in cavities which are formed by the fortuitous aggregation of molecules. From the standpoint of electron trapping by cavities in such nonpolar solvents, the following may be easily conjectured: the electron-trap potential in hydrocarbons, unlike the cases of polar solvents, will be generally weak at attraction and mainly caused by the repulsion between the excess electron and the electron cloud of molecules. Hence the potential will not always be enough to create the stable, bound state of the excess electron. If the electron cannot be bound, it will stay quasistationarily for the trap time τ until removed to the outside of the trap thermally or by tunnelling. In case τ is so small that the above collective effect of molecules plays only a trivial role, the electron will be "quasifree" and its property may be understood by considering the transport phenomena of an electron with thermal velocity among molecules.[12-16]

Thus it is probable that one of the factors which dominate the behavior of the excess electron in hydrocarbon will be the electronic state of the electron in the cavity. Therefore, we investigate in the present paper the electronic state of the trapped electron in a hydrocarbon, selecting the simplest case which is that of methane, with the use of the hypothetical cavity model and the Hartree-Fock method as before. The lifetime of the trapped electron is also discussed based on that result.

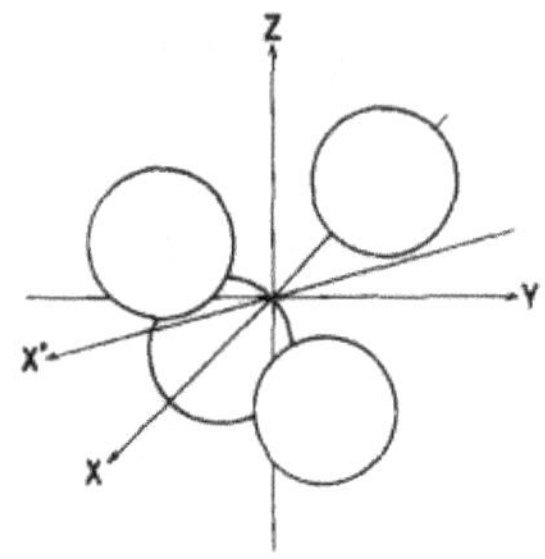

Figure 1. Four methane molecules in a tetrahedral array. X' axis lies on $X-Y$ plane and intersects X axis with a 45° angle.

TABLE I: The Total Energies of the Trap with an Excess Electron in the Ground State[a]

R, Å	$E(e_M^-)$ $\zeta = 0.2$	$\zeta = 0.3$	$\zeta = 0.4$
1.5	−1049.638 (−1049.618)	−1042.425	−1044.027 (−1044.025)
2.0	−1076.756	−1078.202	−1079.836
2.5	−1073.804 (−1073.803)	−1075.209	−1076.440 (−1076.439)

[a] In parentheses, the values after annihilation are cited, only when they are varied by the manipulation.

TABLE II: The Calculated Results for the Trapped Electron in Methane[a]

ζ, au^{-1}	ΔE, eV	ϵ_{HO}^{α}, eV	$h\nu$, eV	ρ_H
0.2	−0.254	−2.094	4.441	0.007
0.3	0.578	0.030	0.252 (0.253)	0.017 (0.016)
0.4	0.234	0.357	−4.316 (−4.312)	0.025 (0.023)

[a] In parentheses, the values after annihilation are cited, only when they are varied by the manipulation.

Model and Method

The trap is assumed to be composed of four methane molecules coordinating tetrahedrally with respect to the carbon atoms of each toward the center (Figure 1). The distance from the cavity center to the carbon atom R is set at 1.5, 2.0, and 2.5 Å, the second of which is a rough estimation based on the structure of solid methane.[17]

The calculation is performed by the unrestricted Hartree-Fock method[18] with INDO approximation.[19] Then, in the same way as the cases of polar solvents,[7] the additional two groups of hydrogen-like atomic orbitals (AO) located at the cavity center are used in addition to valence AO's of solvent molecules because as was shown in our previous paper,[7] such a treatment makes it very easy to describe the state of an excess electron, if it is localized in the cavity. For the sake of simplicity of calculation, moreover, the ground and excited states of an excess electron are obtained separately with the additional 1s AO for the former state and the 2s and three 2p AO's for the latter state, respectively (method 2 in ref 7). The orthogonalization condition between the above two states will be approximately satisfied, if an excess electron is localized in the additional AO's. Parameters concerned with the additional two groups of AO's are analogous to those used in ref 7; the orbital exponents, ζ, are set at 0.2, 0.3, and 0.4 au^{-1}, all the bonding parameters are set at −4.5 eV, and the Mulliken-type electronegativities are set so that core integrals for the ground and excited states of an excess electron become nearly 0.0 and 4.0 eV. This last value is about 0.5 eV higher than that of method 3 in ref 7, according to which the excitation energies obtained were most reasonable.

Besides, because the UHF wave function is not a pure spin multiplet,[20,21] we examine the effect of spin contamination with the aid of the single annihilation method of Amos and Snyder,[22] although the resulting correction will be small when an excess electron is localized in the additional AO's.

Results and Discussion

At first, we will discuss the electronic state of the trapped electron. Table I shows the total energies of the trap with an excess electron in the ground state, $E(e_M^-)$, for three values of R; in parentheses, the values after annihilation are cited only when they are varied[23] by this manipulation. The result seems to indicate the anion state is most stable when R = 2.0 Å. Hence, in the following discussion, only the case of R = 2.0 Å is treated. In Table II, ΔE[24] represents the stabilization energy given by the difference between the energy of the anion radical group and the neutral group which is calculated using the same basis set as that for the open shell system, ϵ_{HO}^{α} is the energy of the highest occupied molecular orbital (with α spin), $h\nu$ is the energy of the first excitation, and ρ_H is the inside proton spin density. The values after annihilation are cited in parentheses only when they are varied by the manipulation. Since the excess electron in liquid methane has been found (from the mobility data[25,26]) to be quasifree, the positive ΔE for ζ = 0.3 and 0.4, meaning that the excess electron in the cavity cannot be stabilized, will be reasonable.[27] In these cases, ϵ_{HO}^{α} is also positive and small. It is to be noted that the highest occupied molecular orbitals are dominated by the additional 1s AO[28] like those for polar solvents;[7] *i.e.*, the coefficients of this AO before annihilation are 0.964 (ζ = 0.2), 0.947 (ζ = 0.3), and 0.933 (ζ = 0.4), respectively. Concerning the excitation energy and the proton spin density, there has been no observation because of the high mobility in this liquid,[25,26] but $h\nu$ in methane will presumably be smaller than 0.6 eV since the maximum absorption probably occurs at wavelengths $\lambda \gtrsim$ 2000 nm in liquid propane.[29] Moreover, accepting the tetrahedral model, ρ_H would be expected to be of the order of 0.01 in terms of the values of the epr line width ΔH in other hydrocarbons at 77°K[30-32] and the formula by Kip, *et al.*[33] According to the present calculation, the excess electron in the lowest excited state is localized in 2p MO and the value of $h\nu$ at ζ = 0.3 seems to be most appropriate. Besides, it is noticeable that all of $|\rho_H|$ are of the same order as those of other hydrocarbons and that ρ_H is positive contrary to the result for polar solvents.[7] The latter result is ascribed to the relatively large contribution of the spin delocalization term, which was negligibly small in the cases of polar solvents.[7] It should be remarked that, like the cases of polar solvents,[34] ρ_H is mainly determined by the pair-wise interaction between a methane molecule and a localized electron, because according to the calculation of this system using the valence AO's of a methane molecule and the 1s AO (ζ = 0.3 and its center is 2.0 Å distant from the carbon atom), ρ_H after annihilation is 0.016 and coincides with the corresponding

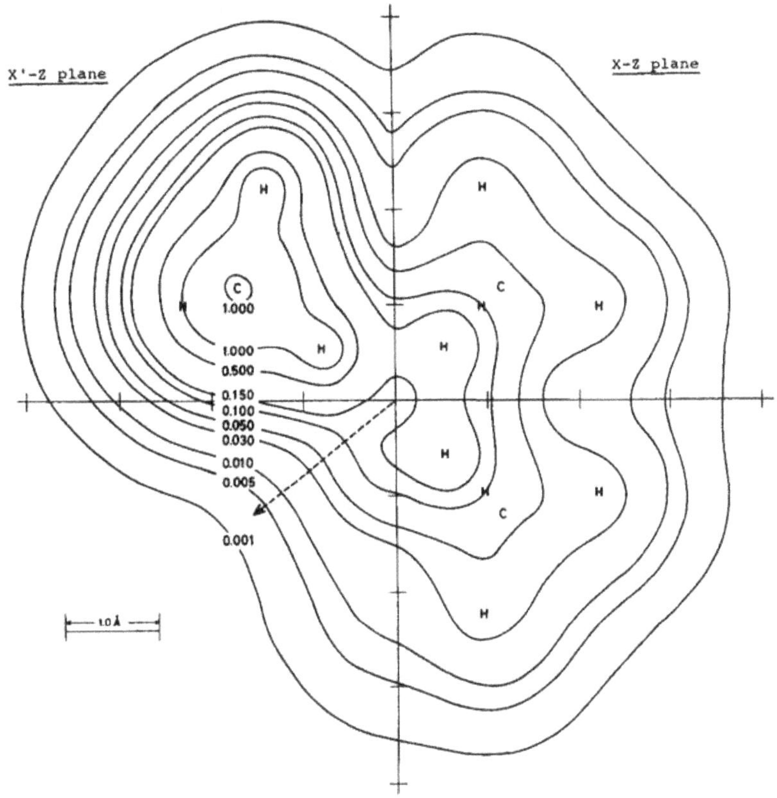

Figure 2. The electron distribution of methane molecules (ζ = 0.3). The atoms are projected on two planes.

value in Table II. It is probable that, in the present case of nonpolar solvent, the influence of molecules outside the first coordination shell will be not greater than that of adjacent molecules in this shell and that therefore the former influence will be also insignificant for ρ_H.

As is easily expected from the result that an excess electron in both the ground and excited states is localized in the additional AO's, the effect of spin contamination is negligibly small on the whole. In fact, the values of $\langle S^2 \rangle$ are 0.750 both before and after annihilation except in the case of the excited state with ζ = 0.4, even in which case the value is 0.752.

Summarizing the above results, the most appropriate ζ seems to be 0.3. In Table III, the calculated charge densities and the dipole moments, none of which changes by annihilation, are shown. From this result, it is supposed that at the first stage when only methane molecules aggregate instantaneously with the approximate vibrational period of 10^{-13} sec, the distortion of each molecule is small (the dipole moment D = 0.039 D outward from the cavity), secondly, that if an excess electron comes into the system and is localized in the cavity, these molecules are further polarized to have D = 0.406 D in each one toward the cavity center in a time less than 10^{-15} sec (electronic polarization time), and finally, that thereby the electron itself is stabilized[35] compared with its initial state in the cavity. Figure 2 shows the electron distribution of four methane molecules obtained without the additional 1s AO; the atoms are projected on X-Z and X'-Z planes. In the cavity, there is a place surrounded by the electron cloud of methane and a very shallow depression of the distribution exists near the center (compare this with our previous result for water[7]). It may be natural to consider that the excess electron will most probably go out along the direction combining the carbon atom of a methane molecule and the cavity center, which is shown with a broken line in Figure 2. In Figure 3, the potential felt by an electron moving along this direction, which is approximated by ΔE (ζ = 0.3) as in our previous work, is shown by a solid line; on the horizontal axis, the distance from the cavity center is indicated. It should be stressed that, even in this direction, there is a peak with a height of 1.13 eV at a distance of 1.65 Å from the center though ΔE is always positive. Hence an electron which has come into this cavity will be expected to stay for a time and to go outward thermally or by tunnelling. In the following, assuming electron tunnelling, we will estimate the trap lifetime of the electron by the use of the above potential which gives the minimum value of the lifetime.

For convenience, we approximate the curve of ΔE by the function[36]

$$V(x) = a - v \cosh^2 \mu [\tanh |(x - b) - \mu| + \tanh \mu]^2 \quad (1)$$

where x is the coordinate in the horizontal axis and a, b, v, and μ are parameters determined as follows: (1) $V(x)$ has the maximum value, a, at $x = b$; hence a = 1.13 and

TABLE III: The Charge Densities and the Dipole Moments of Methane[a]

Charge densities[b]				Dipole moment of each molecule, D[c]	Induced dipole moment, D[c]
q_H	q_C	$q_{H'}$	q_{1s}		
0.817 (0.996)	4.184 (3.977)	0.991 (1.009)	1.111 (...)	+0.406[d] (−0.039)	+0.445

[a] These results are not varied by annihilation. [b] The subscripts H, C, H′, and 1s indicate the inside hydrogen, the carbon, the outside hydrogen, and the additional 1s AO (ζ = 0.3), respectively. Values in parentheses are those for the neutral state calculated without the additional 1s AO. [c] The positive sign indicates the direction toward the cavity center and vice versa. [d] In each of the molecules, the sum of the net charges is not zero, but the absolute value of the positive charge is very small ($\lesssim$0.03) and so the dipole moment can be approximately estimated.

b = 1.65. (2) At $x \to +\infty$, $V(x)$ gives the asymptotic value, $a - \upsilon \exp(2\mu)$, which is zero in the present case. (3) At $x \to -\infty$, $V(x)$ gives the asymptotic value, $a - \upsilon \exp(-2\mu)$, which is put equal to ΔE at x = 0, *i.e.*, 0.578.[37] (4) From (2) and (3), υ = 0.790 and μ = 0.179. The concrete form of $V(x)$ thus obtained is shown in Figure 3 with a broken line (the part for $x < 0$ is abbreviated). The coincidence in $x \lesssim 2.0$ Å is good, but in the outer region there is some discrepancy between two curves. However, such a discrepancy may be permissible, because the excess electron will diffuse in this region and hence the real electron-trap potential will be lower than the curve of ΔE. The Schrödinger equation for an electron moving in the field of $V(x)$ is exactly solved,[36] giving the wave function which represents a transmitted beam at $x = +\infty$ and an incident and reflected beam at $x = -\infty$ under the condition that the kinetic energy of the electron is positive for all x. The resulting transmissivity is[36]

$$T = \frac{2 \sinh (\pi k_+) \sinh (\pi k_-)}{\cosh [\pi(k_+ + k_-)] + \cosh (\pi \beta)} \quad (2)$$

where

$$k_+ = \sqrt{(2m/\hbar^2)[\epsilon - (a - \upsilon \exp(2\mu))]}$$
$$k_- = \sqrt{(2m/\hbar^2)[\epsilon - (a - \upsilon \exp(-2\mu))]} \quad (3)$$
$$\beta = \sqrt{(8m/\hbar^2)\, \upsilon \cosh^2 \mu - 1}$$

in which ϵ is the energy of the electron, m is the mass of an electron, and $\hbar = h/2\pi$ (h is Planck's constant). Then the trap lifetime τ is roughly estimated by[38]

$$\tau \sim \frac{r}{T(p/m)} \quad (4)$$

where r is the coordinate of the inside turning point with regard to the curve of ΔE and p/m corresponds to the velocity of the electron. The result is listed in Table IV. The transmissivity is near unity for the electron with the kinetic energy of 0.1 ~ 0.5 eV and is 0.328 even for the electron with the small kinetic energy of 0.01 eV which is of the order of kT (k is Boltzmann's constant and T is the absolute temperature); accordingly, the lifetimes are quite short (10^{-15} to 10^{-16} sec). Therefore, the incident electron has probably no sooner polarized the methane molecules than it goes outward from the cavity. It is to be noted that as already mentioned the lifetimes obtained are the minimum values because in other directions the repulsive potential will be higher and thicker and so the transmissivity will decrease.

Thus, the following may be concluded from the above result. (1) The traps formed fortuitously do not play a substantial role in the motion of the excess election in methane. That is, the excess electron in methane will be quasifree as pointed out by several authors.[25,26,39] (2) The absorption spectra of the excess electron in methane will be difficult to obtain or quite broad, if detected, as easily expected from the uncertainty principle, $\Delta\epsilon \gtrsim \hbar/\tau \sim 0.7$ eV. (3) In view of the calculation of the electronic state of the trapped electron in methane, it is probable that the present molecular orbital treatment will lead to reasonable results for other hydrocarbons in which the trapped electron is more stable than in methane.

Further theoretical investigations seem to be needed in order to clarify the nature of the trapped electron in hydrocarbons.

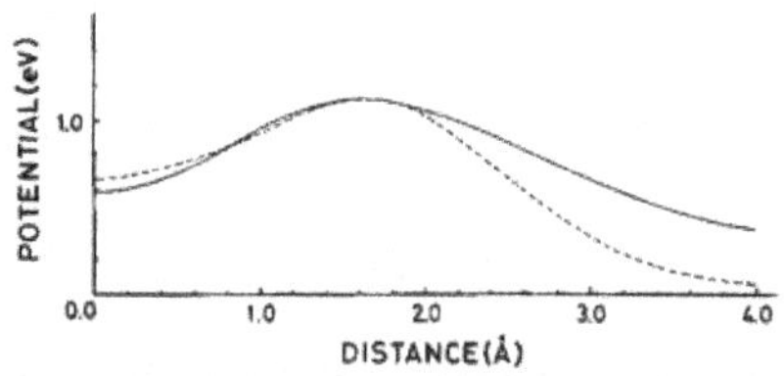

Figure 3. The shape of the trap potential (ζ = 0.3) in the direction of the broken line in Figure 2. The solid line is the potential approximated by ΔE and the broken line is the result of eq 1.

TABLE IV: The Calculated Properties of the Excess Electron in Methane

Kinetic energy, eV	Velocity $\times 10^8$ cm/sec	Transmissivity, T	Lifetime $\times 10^{-15}$ sec
0.010	0.059	0.328	0.723
0.100	0.188	0.704	0.348
0.200	0.265	0.812	0.342
0.300	0.325	0.864	0.303
0.400	0.375	0.894	0.316
0.500	0.419	0.915	0.344

Acknowledgment. The authors wish to thank Professor K. Fueki for his valuable suggestions. It is a pleasure to express our gratitude to the Data Processing Center, Kyoto University, for its permission to use the Facom 230-60 computer. S. Ishimaru gratefully acknowledges the financial support of the Japanese Society for the Promotion of Science.

References and Notes

(1) J. Jortner, *J. Chem. Phys.*, **27**, 823 (1957); **30**, 839 (1959).
(2) M. Natori and T. Watanabe, *J. Phys. Soc. Jap.*, **21**, 1573 (1966); M. Natori, *ibid.*, **24**, 913 (1968).
(3) K. Iguchi, *J. Chem. Phys.*, **48**, 1735 (1968).
(4) D. A. Copeland, N. R. Kestner, and J. Jortner, *J. Chem. Phys.*, **53**, 1189 (1970).
(5) K. Fueki, Da-Fei Feng, and L. Kevan, *J. Phys. Chem.*, **74**, 1976 (1970); K. Fueki, Da-Fei Feng, L. Kevan, and R. E. Christoffersen, *ibid.*, **75**, 2297 (1971).
(6) G. Howat and B. C. Webster, *J. Phys. Chem.*, **76**, 3714 (1972).
(7) S. Ishimaru, H. Kato, T. Yamabe, and K. Fukui, *Chem. Phys. Lett.*, **17**, 264 (1972); *J. Phys. Chem.*, **77**, 1450 (1973).
(8) G. R. Freeman and J. M. Fayadh, *J. Chem. Phys.*, **43**, 86 (1965).
(9) J. M. Buchanan and F. Williams, *J. Chem. Phys.*, **44**, 4377 (1966).
(10) W. F. Schmidt and A. O. Allen, *J. Chem. Phys.*, **52**, 4788 (1970).
(11) R. M. Minday, L. D. Schmidt, and H. T. Davis, **54**, 3112 (1971); *J. Phys. Chem.*, **76**, 442 (1972).

(12) J. Jortner, N. R. Kestner, S. A. Rice, and M. H. Cohen, *J. Chem. Phys.*, **43**, 2614 (1965).
(13) K. Hiroike, N. R. Kestner, S. A. Rice, and J. Jortner, *J. Chem. Phys.*, **43**, 2625 (1965).
(14) J. Lekner, *Phys. Rev.*, **158**, 130 (1967).
(15) M. H. Cohen and J. Lekner, *Phys. Rev.*, **158**, 305 (1967).
(16) B. E. Springett, J. Jortner, and M. H. Cohen, *J. Chem. Phys.*, **48**, 2720 (1968).
(17) H. M. James and T. A. Keenan, *J. Chem. Phys.*, **31**, 12 (1959).
(18) J. A. Pople and R. K. Nesbet, *J. Chem. Soc.*, **22**, 571 (1954).
(19) J. A. Pople, D. L. Beveridge, and P. A. Dobosh, *J. Chem. Phys.*, **47**, 2026 (1967).
(20) P.-O. Löwdin, *Advan. Chem. Phys.*, **2**, 207 (1959).
(21) K. F. Berggren and R. F. Wood, *Phys. Rev.*, **130**, 198 (1963).
(22) T. Amos and L. C. Snyder, *J. Chem. Phys.*, **41**, 1773 (1964).
(23) To three decimal places.
(24) This refers to the "corrected stabilization energy" in ref 7.
(25) P. G. Fuochi and G. R. Freeman, *J. Chem. Phys.*, **56**, 2333 (1972).
(26) W. F. Schmidt and G. Bakale, *Chem. Phys. Lett.*, **17**, 617 (1972).
(27) However, it is to be noticed that the destabilization may be partly attributed to the overestimated charge density of the hydrogen atom of methane in the INDO calculation; for example, according to the *ab initio* calculation, the charge density of the hydrogen atom of isolated methane is 0.867 with a C–H bond length of 1.094 Å (see W. E. Palke and W. N. Lipscomb, *J. Amer. Chem. Soc.*, **88**, 2384 (1966)); nevertheless, the INDO calculation with a C–H bond length of 1.090 Å gives 1.009 for the value. Therefore, ΔE was also recalculated by correcting the electronegativity of the hydrogen atom so that it might give a reasonable charge density for this atom, 0.867. ΔE consequently obtained is -1.868 eV, which seems to be too low and to have errors caused by the collapse of the consistency of parametrization. Hence we will maintain further discussions, keeping in mind that ΔE will probably be slightly lowered by more accurate calculations.
(28) Such a molecular orbital is called 1s MO.
(29) H. A. Gillis, N. V. Klassen, G. G. Teather, and K. H. Lokan, *Chem. Phys. Lett.*, **10**, 481 (1971).
(30) A. Ekstrom and J. E. Willard, *J. Phys. Chem.*, **72**, 4599 (1968).
(31) K. Tsuji, H. Yoshida, and K. Hayashi, *J. Chem. Phys.*, **46**, 810 (1967).
(32) K. Tsuji and T. Williams, *J. Amer. Chem. Soc.*, **89**, 1526 (1967).
(33) A. F. Kip, C. Kittel, R. A. Levy, and A. M. Portis, *Phys. Rev.*, **91**, 1066 (1953).
(34) S. Ishimaru, H. Kato, H. Tomita, T. Yamabe, and K. Fukui, *Chem. Phys. Lett.*, in press.
(35) According to the results that $|\rho_H| \ll 1$ and that q_{1s} is near unity, the correlation energy may be considered to be small.
(36) P. H. Morse and H. Feshbach, "Methods of Theoretical Physics," McGraw-Hill, New York, N. Y., 1953, p 1659.
(37) Hence the transmissivity for the electron with the energy near ΔE at $X = 0$ will slightly be overestimated.
(38) D. Bohm, "Quantum Theory," Prentice-Hall, New York, N. Y., 1951, Chapter 12.
(39) H. Fueki, *Can. J. Chem.*, **50**, 3379 (1972).

Theoret. Chim. Acta (Berl.) 39, 103—110 (1975)

Calculations of the Pseudopotential for the Excess Electron in Water and Methane

Shingo Ishimaru*, Kenichi Fukui

Department of Hydrocarbon Chemistry, Faculty of Engineering, Kyoto University, Sakyo-ku, Kyoto, Japan

Hiroshi Kato

Department of General Education, Nagoya University, Chikusa-ku, Nagoya, Japan

Received March 26, 1975/April 22, 1975

The potentials for excess electrons in cavities of water and methane are analyzed with the use of the pseudopotential theory. The results are consistent with the previous discussions; the excess electron in water can probably be trapped in the cavity and that in methane will be quasifree. In the case of methane, the effect of the molecular coordination on the potential is further discussed by varying the cavity radius.

Key words: Pseudopotentials, calculations of ~ – Excess electrons in water and methane

1. Introduction

In previous papers [1, 2], the behavior of excess electrons in polar and nonpolar media was investigated by using semiempirical and nonempirical Hartree-Fock methods. In those works, the potentials for excess electrons in cavities of water and methane were obtained by taking the stabilization energy of an excess electron as an approximate potential; the former shows that the excess electron in water will be trapped by the potential well in the cavity and the latter shows that the excess electron in methane will be quasifree.

In the present paper, we will analyze the potentials for excess electrons in cavities of water and methane with the use of the pseudopotential theory [3–8] and try to throw more light on the electron-trap interaction.

2. Potentials

Firstly, it is assumed that molecules surrounding the cavity do not interact and that the potential V for an excess electron is the sum of potentials due to these independent molecules:

$$V=\sum_n v^n \tag{1}$$

where v^n denotes the potential of nth molecule and the summation runs over all surrounding molecules. Secondly, it is assumed that no distortion of molecules is

* Present address: Institut fur physikalische Chemie, Freie Universität Berlin, 1 Berlin 33, Thielallee 63–67.

induced by the interaction with an excess electron; that is, the discussion will be carried out in the framework of the first-order perturbation theory.

Then, according to the pseudopotential theory [3–8], v^n is approximated by a pseudopotential v^n_{ps} of nth molecule[1]:

$$v^n = v^n_{ps} = v^n_{\mathrm{HF}} + v^n_R \tag{2}$$

$$V = V_{ps} \equiv \sum_n v^n_{ps} \qquad V_{\mathrm{HF}} \equiv \sum_n v^n_{\mathrm{HF}} \qquad V_R \equiv \sum_n v^n_R$$

where, in case the molecule has a closed-shell structure with occupied molecular orbitals ψ_i, the Hartree-Fock potential v^n_{HF} is given by the nuclear attraction potential (v^n_{nuc}) and the Coulomb ($\mathscr{J}^n_i$) and exchange ($\mathscr{K}^n_i$) operators as follows:

$$v^n_{\mathrm{HF}} = v^n_{\mathrm{nuc}} + \sum_i (2\mathscr{J}^n_i - \mathscr{K}^n_i) \tag{3}$$

in which the summation runs over all occupied orbitals, and where v^n_R, a repulsion potential based on the Pauli principle, is defined by [5]:

$$v^n_R|\phi_e\rangle = \sum_i |\psi^n_i\rangle\langle\psi^n_i|\mathscr{G}|\phi_e\rangle \tag{4}$$

in which ϕ_e is the smooth pseudowave function of an excess electron and $\mathscr{G}$ is an arbitrary operator. A useful form of $\mathscr{G}$ seems to be $-v^n_{\mathrm{HF}}$ [5, 6]. Taking a plane wave as ϕ_e and making its wave vector negligibly small[2] [6], we can rewrite v^n_{ps} as:

$$v^n_{ps} = v^n_{\mathrm{HF}} - \sum_i |\psi^n_i\rangle\langle\psi^n_i|v^n_{\mathrm{HF}}\rangle\,. \tag{5}$$

Since the exchange potential will usually small compared with the Coulomb potential, we will neglect it in the calculation of v^n_{HF} (this approximation will be examined in Sect. 4). Then, v^n_{HF} becomes the electrostatic potential[3] [9–12]. On the other hand, since v^n_R will be of short range, so far as v^n_R near the cavity center is discussed, it may be permitted to neglect the contribution from the atoms except the inside hydrogen atoms, that is, to replace the surrounding molecules by only the inside hydrogen atoms with the electron density in their original molecules. This simplification makes it quite easy to calculate v^n_R.

3. Traps and SCF Wave Functions

In both cases of water and methane, the trap is assumed to be formed by four molecules coordinating one of OH or CH bonds toward the cavity center respectively (Fig. 1). The distance from the oxygen atom to the center is set at 2.9200 Å according to the X-ray diffraction experiments [13, 14] and the outside hydrogen atoms are placed at the nearest points to the X-Y plane [1]. The distance from the carbon atom to the center is set at 2.0[4] and 2.5 Å, the former of which

[1] Following the second assumption, the effect of the core polarization [6] is not included.

[2] In case a localized state of an excess electron is considered taking 1s atomic orbital as ϕ_e [1], Eq. (5) is given by making its orbital exponent negligibly small.

[3] $-v^n_{\mathrm{HF}}$ is frequently used for the studies on the sites of protonation and reactions with electrophilic reagents. For example, see Ref. [12].

[4] The value of 2.0 Å seems to give the most stable anion state; see Ref. [2].

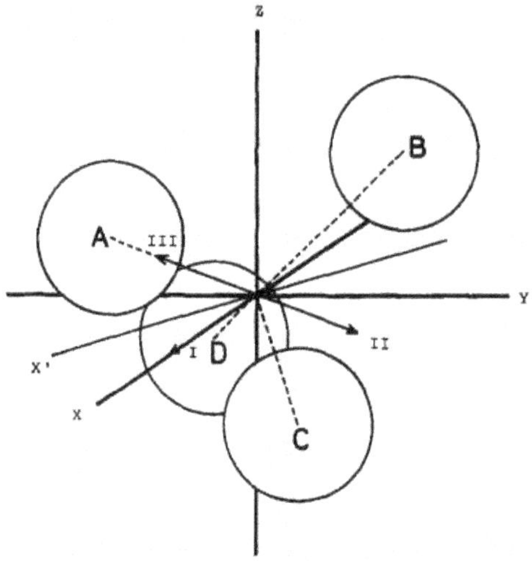

Fig. 1. Four molecules, A, B, C, and D, in a tetrahedral array; one of OH or CH bonds of each molecule points toward the cavity center. X' axis lies on X-Y plane and intersects X axis with a 45° angle. In the case of water, two molecules, A and B, are on X'-Z plane and four outside hydrogen atoms are placed at the nearest points to X-Y plane. In the case of methane, two CH bonds of molecule A (or B) are on X'-Z plane. I indicates the direction along X axis and II and III indicate two opposite directions along the line combining the oxygen (or carbon) atom and the cavity center

is a rough estimation based on the structure of solid methane [15], and the outside hydrogen atoms are placed so that this molecular aggregation may have Td-symmetry[5].

SCF wave functions of water and methane molecules are obtained by using a minimal basis set composed of the Slater-type orbitals with the standard exponents [16]; each orbital is given by a least-squares representation [17] of a sum of three Gaussian-type orbitals.

4. Results and Discussion

4.1. Electrostatic Potentials

In Figs. 2 and 3, the electrostatic potentials in cavities of water and methane are drawn in a.u.; the atoms are projected on X-Z and X'-Z planes. In both cases, there is a flat area near the center and it is clearly shown that the electrostatic potential near a molecule is not much affected by other molecules[6]. The values in the methane cavity are rather lower than those of water, which is mainly brought about by the smallness of the radius of the methane cavity. However, the potential in methane is steeper in direction II than that in water.

[5] In Ref. [2], the whole system is not accurately in Td-symmetry.

[6] For example, compare the present result for the water cavity with the electrostatic potential (for a proton) of an isolated water molecule in Ref. [18].

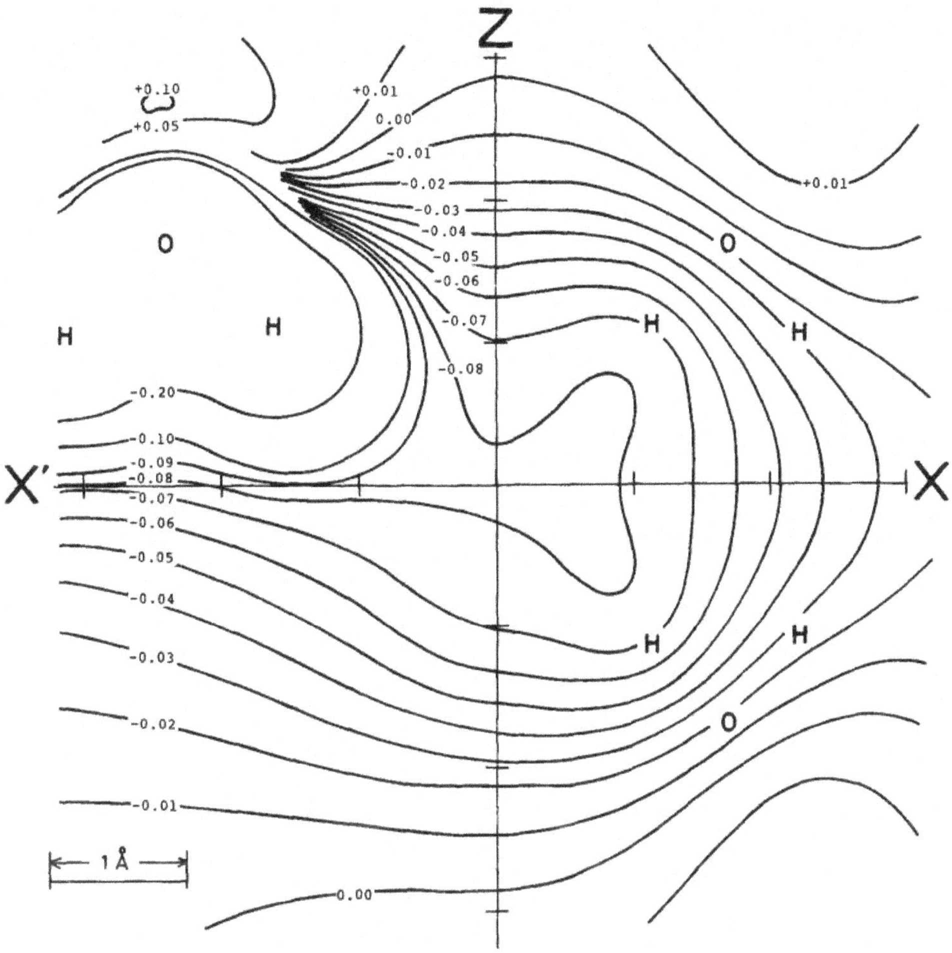

Fig. 2. The electrostatic potential in the case of water (in a.u.)

4.2. The Effect of V_R

By using the approximations described in Sect. 2, v_R^n is given by:

$$v_R^n = (4 - \tfrac{88}{27} P_H^n)\zeta_H^n \exp(-\zeta_H^n r_H^n) \quad \text{(in a.u.)} \tag{6}$$

in the framework of the minimal basis set; P_H^n is the hydrogen atom population of n-th molecule (0.8219 in water molecule and 0.9400 in methane molecule in the present calculation), and r_H^n is the distance (in a.u.) between an excess electron and the nucleus of the hydrogen atom with 1*s* AO (its orbital exponent $\zeta_H^n = 1.24$ a.u.$^{-1}$).

The results are displayed in Figs. 4–7 where (a) and (b) indicate the curves of V_{HF} and V_{ps}, respectively. In Figs. 5–7, the curves in the right- and left-hand sides show the results in directions II and III, respectively and the broken line shows V_{ps} additionally including the exchange potential of only the inside hydrogen

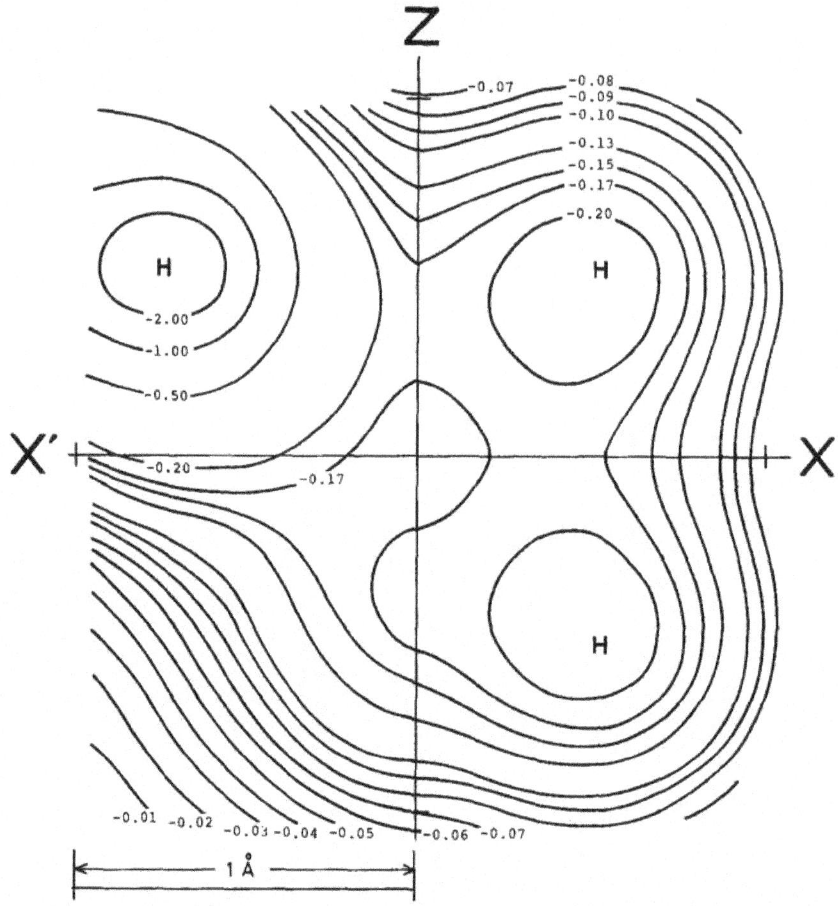

Fig. 3. The electrostatic potential in the case of methane (in a.u.)

atoms. Figs. 4 (direction I) and 5 (directions II and III) are the results of water. Figure 4, which is of course symmetric concerning the cavity center, corresponds to Fig. 6 in a previous paper [1] (therein the value of the potential was all negative and was about -1.0 eV at $X=0$). Curve (a) in Fig. 4 is much elevated by the effect of V_R and, as a result, curve (b) goes positive at $X>0.85$ Å. In the cavity, however, there is surely a potential well (at $X=0$, $V_{ps}=-0.0146$ a.u.$=-0.397$ eV) which resembles the previous one in its shape. In Fig. 5, V_{ps} in direction II becomes rather gentle owing to the rapid decrease of V_R. Though this potential depression will play an indispensable role in the electron trapping, the depth, -0.397 eV, seems to be insufficient to overcome the kinetic energy effect of an excess electron[7] by itself. However, in case the core polarization of the nearest neighboring molecules and the long-range polarization are taken into account [19], the electron

[7] The ground state of an excess electron in polar solvents is often described by using 1s atomic orbital with the orbital exponent near 0.3 a.u. [1]; for the value of 0.3, the kinetic energy of the electron is about 1.2 eV.

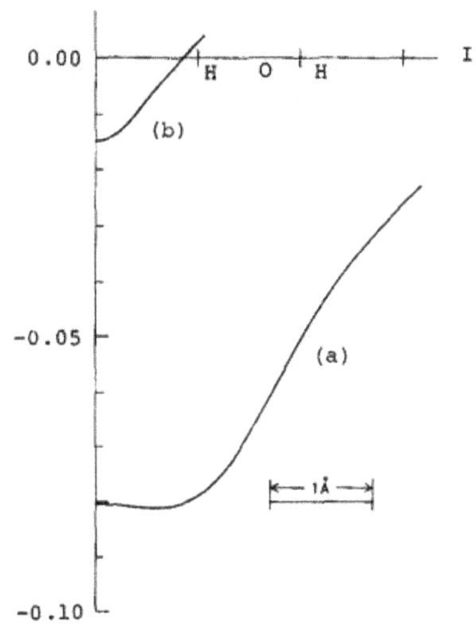

Fig. 4

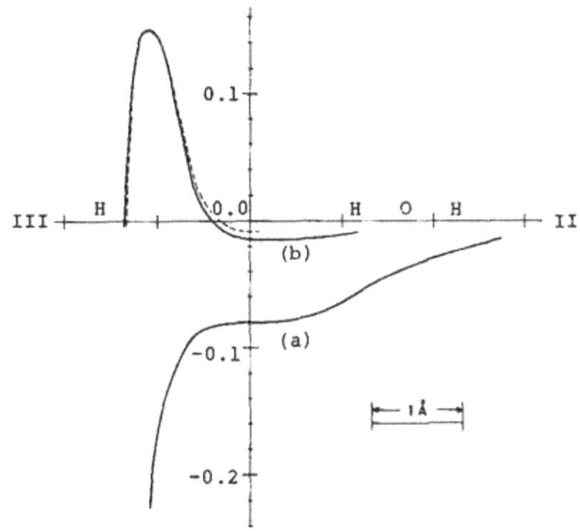

Fig. 5

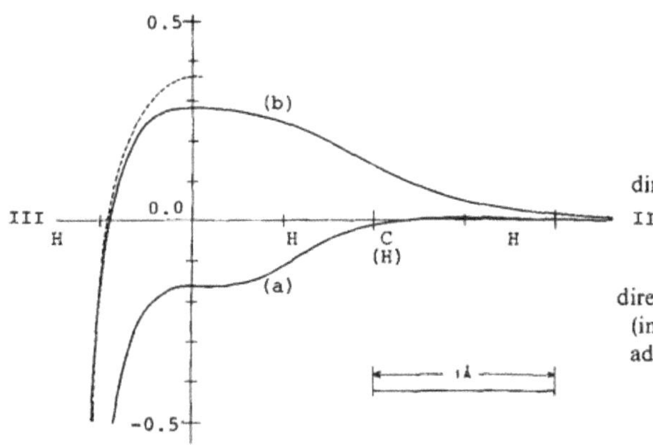

Fig. 6

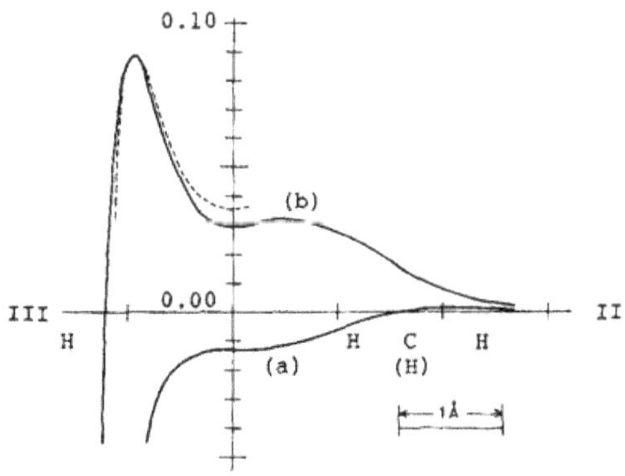

Fig. 7

Fig. 4
The curves of V_{HF} (a) and V_{ps} (b) in direction I in the water cavity (in a.u.)

Fig. 5
The curves of V_{HF} (a) and V_{ps} (b) in directions II and III in the water cavity (in a.u.). The broken line indicates V_{ps} additionally including the effect of the exchange potential of the inside hydrogen atoms

Fig. 6
The curves of V_{HF} (a) and V_{ps} (b) in directions II and III in the methane cavity (in a.u.). The distance between the carbon atom and the center is set at 2.0 Å. The broken line indicates V_{ps} additionally including the effect of the exchange potential of the inside hydrogen atoms

Fig. 7
The curves of V_{HF} (a) and V_{ps} (b) in directions II and III in the methane cavity (in a.u.). The distance between the carbon atom and the center is set at 2.5 Å. The broken line indicates V_{ps} additionally including the effect of the exchange potential of the inside hydrogen atoms

will be bound in the cavity. As previously mentioned [1], the electron moving along direction III encounters at first a potential barrier (with a height of 0.177 a.u. = 4.82 eV in the present calculation) caused by the large repulsion of V_R near the hydrogen atom. The exchange potential is repulsive in the cavity but its effect seems to be trivial.

Fig. 6 shows the results of methane; the distance from the carbon atom to the cavity center is set at 2.0 Å. In this case, the cavity radius is rather small compared with that of water, which causes (1) the deep depression of the electrostatic potential near the center [curve (a)] though it is repulsive in the region $1.16 < r < 2.49$ Å along direction II (r: the distance from the cavity center) and (2) the large repulsion of V_R which makes V_{ps} positive everywhere in the cavity [curve (b)]. It should be noticed that, if V_R and the exchange potential are accurately estimated, V_{ps} in direction III will once ascend toward the outside except in the regions around nuclei. The obvious discrepancy between curve (b) in the right-hand side and the previous curve [2] suggests us that, when the cavity radius is so small, it is indispensable to take into account of the induced polarization, the charge transfer, and so on[8] (the exchange potential seems to be small according to the result of the broken line). In any case, it is sure that, unlike the case of water, an excess electron feels the repulsion potential in the intermolecular region and is quasifree. It will then be worthwhile to clarify the electron-methane molecule potential after an excess electron goes outward from the cavity in liquid methane. Here, we will present the result of the potential in the cavity with a slightly larger radius – the distance from the carbon atom to the cavity center is 2.5 Å – as a datum to consider such a situation. In Fig. 7, it is found that the values of curve (b), in which the exchange potential is relatively small like in Fig. 6, decrease to about a tenth of those in Fig. 6 in the intermolecular region, though V_{ps} is still positive, and that curve (b) has a shallow depression near the center; according to a previous calculation [2], however, such a depression would play only a trivial role in the motion of an excess electron. Furthermore, there appears a peak with a height of 0.098 a.u. = 2.7 eV in curve (b) in direction III. The distance from the carbon atom to the peak is about 1.5 Å and is not noticeably different from the scattering length [6] 1.02 Å estimated approximately by Fueki [21]. These results disclose that the expansion of the cavity by only 0.5 Å brings the feature of the potential of an isolated methane molecule into relief in the total potential. From the above discussions, it is anticipated that an excess electron in liquid methane drifts repeating the scattering by a molecule and the transient stay in the fortuitous molecular aggregation [22–25]. The circumstance would be similar in other saturated hydrocarbons.

Thus, by using the pseudopotential theory, the potential in the cavity has been analyzed and the important role of the repulsion potential due to molecular electrons in the vicinity of a molecule has been pointed out. As a matter of course, the short-range part (except the polarization part) in the isotropic model potentials presented by other authors [26, 27] should be regarded as the orientation-averaged result of the anisotropic potential like the present one. The above discussion will help to consider the dynamical motion of electrons in liquids and solids [28].

[8] Concerning the relation between the pseudowave function and the charge transfer and polarization effects, a discussion is carried out in Ref. [20].

Acknowledgment. We wish to thank Mr. K. Ishida for his help in making the program for the calculation of the electrostatic potential. We also wish to thank Professor J. Jortner for his comments. It is a pleasure to express our gratitude to the Data Processing Center, Kyoto University, for its permission to use the FACOM 230-75 computer.

References

1. Ishimaru,S., Kato,H., Yamabe,T., Fukui,K.: J. Phys. Chem. **77**, 1450 (1973)
2. Ishimaru,S., Yamabe,T., Fukui,K., Kato,H.: J. Phys. Chem. **78**, 148 (1974)
3. Phillips,J.C., Kleinman,L.: Phys. Rev. **116**, 287 (1959)
4. Cohen,M.H., Heine,V.: Phys. Rev. **122**, 1821 (1961)
5. Austin,B.J., Heine,V., Sham,L.J.: Phys. Rev. **127**, 276 (1962)
6. Jortner,J., Kestner,N.R., Rice,S.A., Cohen,M.H.: J. Chem. Phys. **43**, 2614 (1965); New developments in quantum chemistry, Istanbul Lectures, Part II, Sinanoğlu, O. (Ed.), p. 129. New York: Academic Press 1965
7. Weeks,J.D., Hazi,A., Rice,S.A.: Advan. Chem. Phys. **16**, 283 (1969)
8. Chang,T.C., Habitz,P., Pittel,B., Schwarz,W.H.E.: Theoret. Chim. Acta (Berl.) **34**, 263 (1974). Some recent articles concerning the pseudopotential theory are summarized in this paper
9. Truhlar,D.G., Rice,J.K.: J. Chem. Phys. **52**, 4480 (1970)
10. Trajmar,S., Truhlar,D.G., Rice,J.K.: J. Chem. Phys. **52**, 4502 (1970)
11. Truhlar,D.G., Van-Catledge,F.A., Dunning,T.H.: J. Chem. Phys. **57**, 4788 (1972)
12. Giessner-Prettre,C., Pullman,A.: Theoret. Chim. Acta (Berl.) **33**, 91 (1974). Some recent articles using the electrostatic potential for a proton are summarized in this paper
13. Morgan,J., Warren,B.E.: J. Chem. Phys. **6**, 666 (1938)
14. Brady,G.W., Romanov,W.J.: J. Chem. Phys. **32**, 306 (1960)
15. James,H.M., Keenan,T.A.: J. Chem. Phys. **31**, 12 (1959)
16. Hehre,W.J., Stewart,R.F., Pople,J.A.: J. Chem. Phys. **51**, 2657 (1969)
17. Stewart,R.F.: J. Chem. Phys. **52**, 431 (1970)
18. Giessner-Prettre,C., Pullman,A.: Theoret. Chim. Acta (Berl.) **25**, 83 (1972)
19. Newton,M.D.: J. Chem. Phys. **58**, 5833 (1973)
20. Kato,H., Ishimaru,S.: Bull. Chem. Soc. Japan **46**, 3336 (1973)
21. Fueki,K.: Can. J. Chem. **50**, 3379 (1972)
22. Freeman,G.R., Fayadh,J.M.: J. Chem. Phys. **43**, 86 (1965)
23. Buchanan,J.M., Williams,F.: J. Chem. Phys. **44**, 4377
24. Schmidt,W.F., Allen,A.O.: J. Chem. Phys. **52**, 4788 (1970)
25. Minday,R.M., Schmidt,L.D., Davis,H.T.: J. Chem. Phys. **54**, 3112 (1971); J. Phys. Chem. **76**, 442 (1972)
26. Copeland,D.A., Kestner,N.R., Jortner,J.: J. Chem. Phys. **53**, 1189 (1970)
27. Fueki,K., Feng,D.-F., Kevan,L.: J. Am. Chem. Soc. **95**, 1398 (1973)
28. Tachiya,M.: J. Chem. Phys. **60**, 2275 (1974)

Prof. Dr. K. Fukui
Department of Hydrocarbon Chemistry
Faculty of Engineering
Kyoto University
Sakyo-ku, Kyoto, Japan

J. Phys. B: Atom. Molec. Phys., Vol. 10, No. 16, 1977. Printed in Great Britain. © 1977

Theory of decaying states based on a method of coupled equations: Kapur–Peierls and Siegert resonant states and the 'extended' Hellmann–Feynman theorem

Akitomo Tachibana, Tokio Yamabe and Kenichi Fukui

Department of Hydrocarbon Chemistry, Faculty of Engineering, Kyoto University, Kyoto, Japan

Received 2 February 1977, in final form 26 April 1977

Abstract. This paper gives a theoretical treatment of long-living decaying states in resonance phenomena from the viewpoint of decaying bound states. The effective Hamiltonian for the Kapur–Peierls resonant state is constructed in this theory by distinct potentials or operators. This theory is based on a method of coupled equations similar to the famous Faddeev equation, and hence we need no projection operators to construct the effective Hamiltonian. We shall disclose and develop the 'extended' Hellmann–Feynman theorem applied to the resonant states. Particularly, the energy gradient of the Kapur–Peierls eigenvalue $\omega_n(E)$ is found in a compact form as

$$\partial\omega_n(E)/\partial E = \langle\phi_n|[\partial t/\partial E]|\phi_n\rangle$$

where ϕ_n is the original bound state which decays and t is the reaction operator. Furthermore, the difference between the Siegert resonant-state energy and the particular solution of the Kapur–Peierls eigenvalue problem at the resonant-state energy is estimated. We shall also elucidate the present theory in a simple example of tunnelling problems.

1. Introduction

There are a number of studies dealing with resonances of quantum systems and usually the resonances are described by scattering theory (see, for example, Blatt and Weiskopf 1952, Newton 1960, 1966, De Alfaro and Regge 1965). From the scattering viewpoint, the starting wavefunction ϕ_A is the solution of the asymptotic Hamiltonian H_A. The system has a continuous energy spectrum but certain energies are singled out as resonances caused by the scattering interaction $V_B = H - H_A$. This description, however, seems to lose sight of the most striking characteristic of a narrow resonance which is always associated with a long-living decaying state, a state which physically resembles a true discrete eigenstate (Goldberger and Watson 1964, More 1971a). More and Gerjuoy (More 1971b, More and Gerjuoy 1973) have demonstrated that a resonance wavefunction is effective for describing such a narrow resonance and discussed some of the properties, particularly the connection between the resonance wavefunction and the Kapur–Peierls (KP) wavefunction (Kapur and Peierls 1938, Peierls 1948). In these papers, the resonance wavefunctions have been defined by the residues of the Green's function at its complex poles. The Green's function itself has been regarded as the resolvent of an effective Hamiltonian, which has the KP eigenfunctions and eigenvalues. The starting effective Hamiltonian used by More and Gerjuoy (1973), however, has not been written in terms of the actual Hamiltonians or potentials and hence it may be of limited use for further theoretical extensions or applications. Of course, the properties of the resonance wavefunction itself have been studied by many authors; for instance, Frey and Thiele (1968) have studied the time development and Huestis (1975) has investigated the superposition principle of the resonance wavefunctions.

Another approach suitable for describing such a decaying state is the bound-state viewpoint (Miller 1966). Here one introduces an unperturbed Hamiltonian H_B which is assumed to have discrete eigenstates $\{\phi_n\}$. From the decaying bound-state viewpoint, one of these, ϕ_n, is taken as the initial state of the system. The remainder of the Hamiltonian $V_A = H - H_B$ causes the time dependence in the occupation of the state ϕ_n. For a clear illustration of the present situation, let us consider the problem of Stark broadening; an external uniform electrostatic field F applied to bound electrons in atoms and molecules induces not only energy shifts but also broadenings of energy levels. In this case, V_A corresponds to the field applied to the unperturbed system H_B, and finally a loosely bound electron is taken away, as in the case of field emission from metals. Then, in the final channel, one takes an unperturbed Hamiltonian H_A as the sum of a free-electron Hamiltonian and that of the residual ion (core), the interaction V_B being that of the final free electron with the core. For this kind of problem, the decaying bound-state viewpoint seems to be particularly effective (Yamabe *et al* 1977a, b, Tachibana *et al* 1977), in contrast with the usual opinion of scattering theory that the bound-state viewpoint is neither mathematically nor physically elegant (More 1971b).

On the other hand, recent theoretical treatments of resonant states (Taylor *et al* 1966, Nicolaides 1972 and references cited therein) have been widely developed on the basis of Feshbach's PQ projection-operator formalism (Feshbach 1962). The equivalent general formalism has been found by Fano (1961), albeit in a different fashion, and the relation between the works of Fano and Feshbach has been clarified by O'Malley and Geltman (1965). The formalism usually requires the diagonalisation of the total Hamiltonian H in P and Q spaces, which are mutually orthogonal: $PQ = QP = 0$. The off-diagonal matrices (PHQ and QHP) are then treated as the perturbation. The formalism has been used effectively to obtain the physical resonance energy by solving the Schrödinger equation in the truncated Hilbert space (Q space) (O'Malley and Geltman 1965, Lipsky and Russek 1966). However, in actual problems, particularly when the nature of V_A and V_B are significantly different from each other, we often encounter the case where there is an adequate (quasi-) projection operator P' possessing the same character as P but not always orthogonal to Q (Temkin *et al* 1972, Bhatia and Temkin 1976). In this case, the basic relationship for Feshbach's formalism, $PQ = QP = 0$, is no longer valid. Thus, it may be worthwhile to formulate the theory of the long-living decaying state from the bound-state viewpoint without using the projection operators.

The main subjects of this paper are (i) to give an explicit form for a long-living decaying state from the bound-state viewpoint by using two sets of eigenfunctions of the distinct Hamiltonians H_A and H_B and (ii) to discuss some of the properties. The two Hamiltonians H_A and H_B can be given independently by dividing the space into two distinct regions. In the first of those regions, localised round the atom or molecule, the unperturbed Hamiltonian is H_B, and in the second, the asymptotic region, it is H_A. The KP wavefunctions $\{\xi_n(E)\}$ and the complex eigenvalues $\{\omega_n(E)\}$ are then obtained by solving a newly defined KP eigenvalue problem. The intimate relationships with the original bound-state wavefunction and its real energy are clearly exhibited. The complex energy of the Siegert resonance (Siegert 1939) can also be introduced. The KP eigenvalue $\omega_n(E)$ contains E, the energy of the system, itself as a parameter and then we shall apply the Hellmann–Feynman theorem (Hellmann 1937, Feynman 1939), leading to an 'extended Hellmann–Feynman theorem'. We then investigate the variational method for the presented treatment. In the last section the theory will be elucidated in a simple example of the Stark-broadening problem. The applicability of the present treatment seems to be limited but autoionisation, predissociation and the general tunnelling problems are potential applications.

2. Coupled-equations theory

As mentioned previously, the total Hamiltonian can be written in two ways:

$$H = H_B + V_A = H_A + V_B. \tag{2.1}$$

H_B governs the bound region while H_A belongs to the unbound region of the resonant state. Here the eigenfunctions of H_A and H_B are given by

$$(E - H_A)|\phi_A(E)\rangle = 0 \qquad \langle\phi_A(E)|\phi_A(E')\rangle = \delta(E - E') \tag{2.2}$$

$$(E_n^{(0)} - H_B)|\phi_n\rangle = 0 \qquad \langle\phi_n|\phi_m\rangle = \delta_{nm} \tag{2.3}$$

where we disregard the degeneracy of the energy spectra. Then there exist bound states ϕ_n of H_B and the continuum $\phi_A(E)$ of H_A for the energy region $E \sim E_n^{(0)}$. Due to the interaction V_A the bound states ϕ_n become metastable and finally decay into the continuum $\phi_A(E)$. We admit, now, an overlap between ϕ_n and $\phi_A(E)$, which may be reasonable from the consideration that the initial bound state and the asymptotic continuum may overlap in the intermediate region. Moreover, in the present treatment, we assume that the total wavefunction Ψ can be written as the sum of two functions Φ_A and Φ_B projected as

$$\Psi = \Phi_A + \Phi_B \tag{2.4}$$

$$\Psi \rightarrow \Phi_B \qquad \text{as } V_A \rightarrow 0 \tag{2.5a}$$

$$\Psi \rightarrow \Phi_A \qquad \text{as } V_B \rightarrow 0. \tag{2.5b}$$

These limiting properties show that Φ_A represents the unbound and Φ_B the bound region of the decaying state, respectively. The Schrödinger equation

$$H\Psi = E\Psi$$

can then be split into two coupled equations (see equation (4.8) of Hahn (1968a)):

$$(E - H_A)\Phi_A = V_A\Phi_B \tag{2.6}$$

$$(E - H_B)\Phi_B = V_B\Phi_A \tag{2.7}$$

as may be seen by adding $-V_B\Phi_A$ to both sides of (2.6), and $-V_A\Phi_B$ to both sides of (2.7), and then adding the resulting equations. Note that we need no projection operators in order to distinguish H_A from H_B and Φ_A from Φ_B. These are the basic coupled equations in our treatment. The coupled-equations theory is potentially applicable to a wide class of physical and chemical problems, particularly, as has been pointed out by Krüger and Levin (1976, 1977), in the bound-state problems of a many-electron system, where the use of different basis sets has been shown to yield remarkably good results. We shall try to apply the coupled-equations theory to the problem of decaying states.

Now, imposing the initial condition of the incident wave of H_A, one can solve equation (2.6) to give

$$\Phi_A = \phi_A^{(+)} + G_A^+ V_A \Phi_B \tag{2.8}$$

where $\phi_A^{(\pm)}$ specify the outgoing (+) and the incoming (−) wave solutions of H_A, and where G_A^+ is the Green's operator of H_A, defined for real E as

$$G_A^+(E) = \mathrm{P}(E - H_A)^{-1} - \mathrm{i}\pi\delta(E - H_A) \tag{2.9}$$

(P means the principal part). Substituting equation (2.8) into equation (2.7), we have the inhomogeneous equation for Φ_B:

$$[E - H_B^+(E)]\Phi_B = V_B\phi_A^{(+)} \tag{2.10}$$

$$H_B^+(E) = H_B + V(E). \tag{2.11}$$

Here we obtain $H_B^+(E)$ as the effective Hamiltonian for the KP resonant states, which contains the non-symmetric energy-dependent effective potential $V(E)$ given for real E as

$$V(E) = V_B G_A^+(E) V_A = V_B \mathrm{P}(E - H_A)^{-1} V_A - \mathrm{i}\pi V_B \delta(E - H_A) V_A. \tag{2.12}$$

Thus the effective potential $V(E)$ is written in terms of the known functions $\{\phi_A\}$ and the actual potentials V_A and V_B. It is clear that $V(E)$ contains an imaginary part which is responsible for the decay process. Interestingly, the imaginary part can be expressed in the simple closed form $-\mathrm{i}\pi V_B\delta(E - H_A)V_A$. This will make it easy to survey and figure out the properties of decaying states in a direct way. Note that the effective Hamiltonian $H_B^+(E)$ is non-Hermitian because of $V(E)$. Therefore the eigenvalues are complex; for a fixed energy E, $H_B^+(E)$ provides the KP complex eigenvalues $\{\omega_n(E)\}$ associated with the KP eigenfunctions $\{\xi_n(E)\}$. The KP eigenvalue $\omega_n(E)$ is written as

$$\omega_n(E) = \epsilon_n(E) - \tfrac{1}{2}\mathrm{i}\gamma_n(E) \tag{2.13}$$

where $\epsilon_n(E)$ and $\gamma_n(E)$ are real if E is real, corresponding to the resonance energy and the width of the KP resonant state, respectively. The new KP eigenvalue equations are

$$[\omega_n(E) - H_B^+(E)]|\xi_n(E)\rangle = 0 \tag{2.14}$$

$$\langle\eta_n(E)|[\omega_n(E) - H_B^+(E)] = 0. \tag{2.15}$$

Here $\{\eta_n(E)\}$ are the dual states of the KP resonant states satisfying the bi-orthogonal relations (Morse and Feshbach 1953)

$$\langle\eta_n(E)|\xi_m(E)\rangle = N_n(E)\delta_{nm} \tag{2.16}$$

where $\{N_n(E)\}$ are the normalisation constants.

Now the KP complex eigenvalue depends on the energy of the system as a kind of parameter. Therefore, it may be worthwhile to examine the Hellmann–Feynman theorem (Hellmann 1937, Feynman 1939, Epstein *et al* 1967) for such a metastable state. The Hellmann–Feynman theorem has been applied to bound states in order to study the parameter dependence of the eigenstates. It will also be possible to apply the Hellmann–Feynman theorem to the KP resonant states because they are eigenstates of the effective Hamiltonian. In this case, an 'extended' integral Hellmann–Feynman theorem has the following form for the energy E as well as another parameter v:

$$[\omega_n(E, v) - \omega_n(E, v')]\langle\eta_n(E, v)|\xi_n(E, v')\rangle = \langle\eta_n(E, v)|H_B^+(E, v) - H_B^+(E, v')|\xi_n(E, v')\rangle . \tag{2.17}$$

Here v can be chosen, for instance, as the nuclear coordinate for the molecular resonant state or the field strength F for the Stark-broadening problem. In the limit $v' \rightarrow v$, we have the 'extended' differential Hellmann–Feynman theorem:

$$\partial\omega_n(E, v)/\partial v = \langle\eta_n(E, v)|[\partial H_B^+(E, v)/\partial v]|\xi_n(E, v)\rangle/N_n(E, v) \tag{2.18}$$

$$= \langle\eta_n|[\partial H_B/\partial v]|\xi_n\rangle/N_n + \langle\eta_n|[\partial V(E)/\partial v]|\xi_n\rangle/N_n . \tag{2.18a}$$

It is interesting to compare this theorem with the usual one pertaining to bound states. The first term is the familiar one where only the bound-state Hamiltonian is differentiated. The second term, due to the effective potential $V(E)$ defined by equation (2.11), is characteristic of the resonant state. Furthermore, supposing the order parameter of $V_{A,B}$ as $\lambda_{A,B}$, and assuming that V_A and V_B are independent of each other, we have

$$\partial\omega_n(E)/\partial\lambda_B = \langle\eta_n|V_B + V_B G_A^+(E)V_A|\xi_n\rangle/\lambda_B N_n \tag{2.19}$$

$$\partial\omega_n(E)/\partial\lambda_A = \langle\eta_n|V_B G_A^+(E)V_A + V_B G_A^+(E)V_A G_A^+(E)V_A|\xi_n\rangle/\lambda_A N_n . \tag{2.20}$$

It is interesting that the higher-order scattering processes introduced by G_A^+ become sequentially important for the estimates of these quantities. Subsequently for a different E, which is also characteristic of the present problem, we have

$$[\omega_n(E) - \omega_n(E')]\langle\eta_n(E)|\xi_n(E')\rangle = \langle\eta_n(E)|V_B[G_A^+(E) - G_A^+(E')]V_A|\xi_n(E')\rangle \tag{2.21}$$

and hence

$$\partial\omega_n(E)/\partial E = \langle\eta_n(E)|V_B[\partial G_A^+(E)/\partial E]V_A|\xi_n(E)\rangle/N_n(E) \tag{2.22}$$

$$= -\langle\eta_n(E)|V_B[G_A^+(E)]^2 V_A|\xi_n(E)\rangle/N_n(E). \tag{2.22a}$$

Another tractable expression for $\partial\omega_n(E)/\partial E$ will be found later from equation (3.12). These theorems may be useful for various problems of physical and chemical phenomena having metastable nature.

Now let us discuss the relationship with the Siegert resonance (Siegert 1939). The Siegert complex energy has been defined by the pole of the effective Hamiltonian (More and Gerjuoy 1973). In our treatment, the Siegert energy will be introduced by the complex pole of the resonant scattering amplitude. By using the KP bi-orthogonal complete set, the transition matrix $\mathscr{T}_{\alpha\to\beta}$, giving the transition amplitude from a channel α to a channel β, is obtained as

$$\mathscr{T}_{\alpha\to\beta} = \mathscr{T}^{(0)}_{\alpha\to\beta} + \mathscr{T}^{(r)}_{\alpha\to\beta} \tag{2.23}$$

$$\mathscr{T}^{(r)}_{\alpha\to\beta} = \sum_n \langle\phi^{(-)}_{A\beta}|V_A|\xi_n\rangle\langle\eta_n|V_B|\phi^{(+)}_{A\alpha}\rangle/(E-\omega_n)N_n \tag{2.24}$$

where $\mathscr{T}^{(0)}_{\alpha\to\beta}$ is the direct scattering amplitude and $\mathscr{T}^{(r)}_{\alpha\to\beta}$ is the resonant amplitude. Accordingly, the pole of the resonant amplitude is given by the root of the equation

$$E = \omega_n(E). \tag{2.25}$$

Then we define the Siegert complex energy by the solution $E = W_n$ of this equation

$$W_n = E_n - \tfrac{1}{2}\mathrm{i}\Gamma_n \tag{2.26}$$

where E_n is the resonance energy and Γ_n is the width. At the same time, the Siegert resonant state is also defined by the KP eigenfunction $\xi_n(W_n)$ evaluated at $E = W_n$. Clearly, under the condition of no incident wave for Ψ, the function $\Phi_B(W_n)$ evaluated at $E = W_n$ is nothing but the Siegert wavefunction. In this case, the total wavefunction is written simply enough in the form

$$\Psi(W_n) = [1 + G^{+}_{A}(W_n)V_A]\xi_n(W_n). \tag{2.27}$$

We can easily estimate the differences between $\epsilon_n(E_n)$ and E_n, or $\gamma_n(E_n)$ and Γ_n as follows:

$$\epsilon_n(E_n) - E_n = \tfrac{1}{8}[\mathrm{d}^2\epsilon_n(E_n)/\mathrm{d}E_n^2]\Gamma_n^2 + \tfrac{1}{4}[\mathrm{d}\gamma_n(E_n)/\mathrm{d}E_n]\Gamma_n + \mathrm{O}(\Gamma_n^4) \tag{2.28}$$

$$\gamma_n(E_n)/\Gamma_n = 1 - \mathrm{d}\epsilon_n(E_n)/\mathrm{d}E_n + \mathrm{O}(\Gamma_n^2). \tag{2.29}$$

These equations imply that the KP eigenvalue is not exactly consistent with the Siegert energy even when the energy E of the system is equal to the resonance energy E_n. Usually $\epsilon_n(E_n)$ and $\gamma_n(E_n)$ are used as the equivalents (see e.g. Miller 1966) or the approximants (see e.g. O'Malley and Geltman 1965) to E_n and Γ_n respectively, and critical attention is not being paid to these differences. It may be interesting and worthwhile to estimate these differences. The Hellmann–Feynman theorems, equations (2.22) or (3.12), immediately yield estimates of the energy dependence of the KP eigenvalue.

3. Relationships with bound states

We have shown in equation (2.26) that the energy is 'quantised', $E = W_n$, on the complex energy plane. It seems then that there is a close relationship between the Siegert resonant state $\xi_n(W_n)$ and the original bound state ϕ_n. Here, instead of finding the direct relationship between them, we shall try to clarify the relationship between the KP resonant state and the bound state. The relationship will be described by the wave operators W_ξ and W_η as follows:

$$|\xi_n\rangle = W_\xi|\phi_n\rangle \tag{3.1}$$

$$\langle\eta_n| = \langle\phi_n|W_\eta \tag{3.2}$$

where we suppose the intermediate normalisations

$$\langle\phi_n|\xi_n\rangle = \langle\phi_n|W_\xi|\phi_n\rangle = 1 \tag{3.3}$$

$$\langle\eta_n|\phi_n\rangle = \langle\phi_n|W_\eta|\phi_n\rangle = 1. \tag{3.4}$$

Substituting equations (3.1) and (3.2) into equations (2.14) and (2.15) respectively, we have

$$W_\xi = 1 + T(\omega_n)V \tag{3.5}$$

$$W_\eta = 1 + VT(\omega_n). \tag{3.6}$$

Here, V is the effective potential defined by equation (2.12) and T is the reduced resolvent of H_B^+,

$$T(\lambda) = \frac{p_n}{\lambda - H_\mathrm{B}^+} \tag{3.7}$$

where p_n is the projection operator defined by

$$p_n = 1 - |\phi_n\rangle\langle\phi_n|. \tag{3.8}$$

From equations (2.14) and (2.15), one obtains

$$\omega_n = E_n^{(0)} + \langle\phi_n|V|\phi_n\rangle + \langle\phi_n|VT(\omega_n)V|\phi_n\rangle \tag{3.9}$$

$$= E_n^{(0)} + \langle\phi_n|t(\omega_n)|\phi_n\rangle \tag{3.9a}$$

where t is the reaction operator given by

$$t = VW_\xi = W_\eta V = V + VTV. \tag{3.10}$$

These are the exact functional relationships which connect the KP resonant state with the original bound state. It should be noted here that, owing to the non-symmetric nature of the effective potential, it is necessary to introduce the additional wave operator W_η for the dual state η_n; this is one of the essential differences compared to the usual wave-operator formalism for real eigenvalue problems of Hermitian operators (Löwdin 1965). Likewise, one obtains

$$N_n = 1 + \langle\phi_n|VT^2V|\phi_n\rangle. \tag{3.11}$$

Moreover, the Hellmann–Feynman theorem concerning the energy dependence of the KP eigenvalue can be written by using the closed form of $\omega_n(E)$ from equation (3.9) as

$$\partial\omega_n/\partial E = \langle\phi_n|[\partial t/\partial E]|\phi_n\rangle. \tag{3.12}$$

This expression will prove effective if we know the energy dependence of the reaction operator t, even though in some approximate form.

These formal expressions can be transformed in terms of the eigenfunctions $\{\phi_n\}$ of H_B. First, on expanding Φ_B by $\{\phi_n\}$,

$$\Phi_B = \sum_n b_n \phi_n \tag{3.13}$$

we have from equation (2.10)

$$b_n(E - E_n^{(0)} - \langle\phi_n|V|\phi_n\rangle) - \sum_{m(\neq n)} b_m \langle\phi_n|V|\phi_m\rangle = \langle\phi_n|V_B|\phi_A^{(+)}\rangle. \tag{3.14}$$

The coefficients $\{b_n(E)\}$ are then determined as functions of E by solving this set of inhomogeneous linear equations. Furthermore, the KP eigenfunctions and eigenvalues will be obtained by solving the secular equation of the homogeneous part (see equation (3.25)). It may be useful now to develop the variational principle for the KP resonant states in the variational manifold spanned by $\{\phi_n\}$. Some variational principles have been developed by Bardsley *et al* (1966) and by Miller (1966) to obtain complex energies. In our treatment, the KP resonant state can be written from equations (3.1) and (3.5) as

$$|\xi_n\rangle = |\phi_n\rangle + TV|\phi_n\rangle \equiv |\phi_n\rangle + |\theta_n\rangle \tag{3.15}$$

and similarly the dual state as

$$\langle\eta_n| = \langle\phi_n| + \langle\phi_n|VT \equiv \langle\phi_n| + \langle\pi_n|. \tag{3.16}$$

By using these functions, θ_n and π_n, the KP eigenvalue ω_n is represented in the alternative forms

$$\omega_n = E_n^{(0)} + \langle\phi_n|V|\phi_n\rangle + \langle\phi_n|V|\theta_n\rangle \tag{3.17}$$

$$= E_n^{(0)} + \langle\phi_n|V|\phi_n\rangle + \langle\pi_n|V|\phi_n\rangle. \tag{3.18}$$

We now look for a variational principle for $\theta_n(E)$ and $\pi_n(E)$ and try to obtain $\omega_n(E)$ for arbitrary energy E. The essential difficulty of our problem is to seek the variational principle for the non-Hermitian or non-symmetric operator $H_B^+(E)$. In this case, the inhomogeneous equations satisfied by $\theta_n(E)$ and $\pi_n(E)$ are

$$[H_B^+(E) - \omega_n(E)]|\theta_n(E)\rangle + V(E)|\phi_n\rangle = 0 \tag{3.19}$$

$$\langle\pi_n(E)|[H_B^+(E) - \omega_n(E)] + \langle\phi_n|V(E) = 0 \tag{3.20}$$

where the necessary conditions are

$$\langle\phi_n|\theta_n\rangle = \langle\pi_n|\phi_n\rangle = 0. \tag{3.21}$$

As θ_n and π_n are extremised independently, we find the functional

$$\Delta\omega_n[\tilde{\theta}_n, \tilde{\pi}_n] = \langle\tilde{\pi}_n|p_n(H_B^+ - \omega_n)p_n|\tilde{\theta}_n\rangle + \langle\tilde{\pi}_n|p_nV|\phi_n\rangle + \langle\phi_n|Vp_n|\tilde{\theta}_n\rangle \tag{3.22}$$

where $\tilde{\theta}_n$ and $\tilde{\pi}_n$ are the corresponding trial functions. The KP eigenvalue ω_n can now be given by

$$\omega_n = E_n^{(0)} + \langle\phi_n|V|\phi_n\rangle + \Delta\omega_n[\theta_n, \pi_n]. \tag{3.23}$$

This procedure is similar to the one used by Miller (1966) for real resonance energy E_n. We note that this is equivalent to the direct variation of the functional

$$\omega_n[\tilde{\xi}_n, \tilde{\eta}_n] = \langle \tilde{\eta}_n | H_B^+ | \tilde{\xi}_n \rangle / \langle \tilde{\eta}_n | \tilde{\xi}_n \rangle \tag{3.24}$$

and solve the resultant secular equation

$$\det|(\omega_n - E_n^{(0)})\delta_{nm} - V_{nm}| = 0 \tag{3.25}$$

where the expansion coefficients for the trial functions $\tilde{\xi}_n$ and $\tilde{\eta}_n$ are extremised independently. The Siegert energy W_n can also be obtained by solving $W_n = \omega_n(W_n)$ iteratively.

Furthermore, the direct application of perturbation theory to the formal expressions may in some cases be meaningful. Particularly, in the case of isolated resonances, the higher-order scattering processes may be neglected, and the level shift Δ_n and the width γ_n may be approximated from equation (3.9) as

$$\omega_n \sim E_n^{(0)} + \langle \phi_n | V | \phi_n \rangle = E_n^{(0)} + \Delta_n^{(1)} - \tfrac{1}{2}\mathrm{i}\gamma_n^{(1)} \tag{3.26}$$

where

$$\Delta_n^{(1)}(E) = \langle \phi_n | V_B \mathrm{P}(E - H_A)^{-1} V_A | \phi_n \rangle \tag{3.27}$$

$$\gamma_n^{(1)}(E) = 2\pi \langle \phi_n | V_B \delta(E - H_A) V_A | \phi_n \rangle . \tag{3.28}$$

These quantities are energy dependent. The non-symmetric components of the integrands result from the overlaps between the bound state ϕ_n and the continuum of H_A. By using equations (2.2) and (2.3), these are separated in two ways:

$$\begin{aligned}\Delta_n^{(1)}(E) &= \langle \phi_n | V_A | \phi_n \rangle + \langle \phi_n | V_A \mathrm{P}(E - H_A)^{-1} V_A | \phi_n \rangle \\ &\quad - (E - E_n^{(0)}) \langle \phi_n | \mathrm{P}(E - H_A)^{-1} V_A | \phi_n \rangle \qquad (3.29a)\\ &= -\langle \phi_n | V_B | \phi_n \rangle + \langle \phi_n | V_B \mathrm{P}(E - H_A)^{-1} V_B | \phi_n \rangle \\ &\quad + (E - E_n^{(0)}) \langle \phi_n | V_B \mathrm{P}(E - H_A)^{-1} | \phi_n \rangle \qquad (3.29b)\end{aligned}$$

$$\begin{aligned}\gamma_n^{(1)}(E) &= 2\pi \langle \phi_n | V_A \delta(E - H_A) V_A | \phi_n \rangle - (E - E_n^{(0)}) 2\pi \langle \phi_n | \delta(E - H_A) V_A | \phi_n \rangle \qquad (3.30a)\\ &= 2\pi \langle \phi_n | V_B \delta(E - H_A) V_B | \phi_n \rangle + (E - E_n^{(0)}) 2\pi \langle \phi_n | V_B \delta(E - H_A) | \phi_n \rangle . \qquad (3.30b)\end{aligned}$$

The non-symmetric contributions disappear when $E = E_n^{(0)}$ at the origin of the resonance, and then the level shift is obtained up to second order in V_A as a kind of perturbation energy and the width is given by a 'golden-rule' type formula:

$$\Delta_n^{(1)}(E_n^{(0)}) = \langle \phi_n | V_A | \phi_n \rangle + \langle \phi_n | V_A \mathrm{P}(E_n^{(0)} - H_A)^{-1} V_A | \phi_n \rangle \tag{3.31}$$

$$\gamma_n^{(1)}(E_n^{(0)}) = 2\pi \langle \phi_n | V_A \delta(E_n^{(0)} - H_A) V_A | \phi_n \rangle . \tag{3.32}$$

Hence we can obtain reasonable results by using the explicit forms of the operators.

We have thus developed the theory of resonant states, clarifying the similarity with the usual bound states and retaining the physically distinct image by using the actual Hamiltonians or operators. Specifically, at the limit $V_A \to 0$, we immediately obtain

$$\lim_{V_A \to 0} \xi_n, \eta_n = \phi_n \qquad \text{and} \qquad \lim_{V_A \to 0} \omega_n = E_n^{(0)} . \tag{3.33}$$

These relationships hold for arbitrary energy E and hence, for the Siegert resonance, similar results will hold. As a corollary, by using equations (2.27) and (2.25), we find

$$\lim_{V_A \to 0} \Psi(W_n) = \phi_n \qquad \text{and} \qquad \lim_{V_A \to 0} W_n = E_n^{(0)}. \tag{3.34}$$

These relationships say that the resonant states (and also the very state of the total system) should become the true bound states when the operator V_A, introducing the metastable nature, vanishes.

4. An example

Here we shall elucidate the presented theory in a simple example, the Stark-broadening problem, by using the following one-dimensional potentials:

$$V_A = -Fx \tag{4.1}$$

$$V_B = -V_0\delta(x) \tag{4.2}$$

where F is the strength of the external electrostatic field. This model has been used to study the numerical methods for finding the effect of the electric field on bound states (Lukes and Somaratna 1969, Moyer 1973). Here, we intend to obtain the KP eigenvalue which can be used to examine the relationships developed in the text. The partitioning of the Hamiltonian is as follows:

$$H = -(\hbar^2/2m)\,d^2/dx^2 - V_0\delta(x) - Fx = H_B + V_A = H_A + V_B \tag{4.3}$$

where

$$H_B = -(\hbar^2/2m)\,d^2/dx^2 - V_0\delta(x) \tag{4.4}$$

$$H_A = -(\hbar^2/2m)\,d^2/dx^2 - Fx. \tag{4.5}$$

H_B has only one bound state ϕ_0 (Morse and Feshbach 1953);

$$\phi_0(x) = \alpha^{1/2}\exp(-\alpha|x|) \tag{4.6}$$

$$E_0^{(0)} = -\hbar^2\alpha^2/2m \qquad \alpha = mV_0/\hbar^2 \tag{4.7}$$

and this state becomes metastable by application of V_A. On the other hand, the solutions of H_A are Airy functions Ai and Bi (Antosiewicz 1965, Aspnes 1966), and the Green's function is given explicitly by (Lukes and Somaratna 1969, Whitcombe 1971)

$$G_A^+(x, x'; E) = \frac{1}{i\hbar}\left(\frac{m}{2\pi i\hbar}\right)^{1/2}\int_0^\infty dt\, t^{-1/2}\exp\left[\frac{i}{\hbar}\left(\frac{m(x-x')^2}{2t} + Et + \frac{F(x+x')t}{2} - \frac{F^2t^3}{24m}\right)\right] \tag{4.8a}$$

$$= -\left(\frac{2mF}{\hbar^2}\right)^{2/3}\frac{\pi}{F}\begin{cases}[\mathrm{Bi}(z') + \mathrm{i\,Ai}(z')]\,\mathrm{Ai}(z) & x \leqslant x' \\ \mathrm{Ai}(z')\,[\mathrm{Bi}(z) + \mathrm{i\,Ai}(z)] & x \geqslant x'\end{cases} \tag{4.8b}$$

and the associated functions

$$\langle x|P(E - H_A)^{-1}|x'\rangle = -\left(\frac{2mF}{\hbar^2}\right)^{2/3} \frac{\pi}{F} \begin{cases} \mathrm{Bi}(z')\mathrm{Ai}(z) & x \leqslant x' \\ \mathrm{Ai}(z')\mathrm{Bi}(z) & x \geqslant x' \end{cases} \tag{4.8c}$$

$$\langle x|\delta(E - H_A)|x'\rangle = \left(\frac{2mF}{\hbar^2}\right)^{2/3} \frac{1}{F} \mathrm{Ai}(z')\mathrm{Ai}(z) \tag{4.8d}$$

where

$$z = (2mF/\hbar^2)^{1/3}(-x - E/F). \tag{4.8e}$$

Note that the Green's function in equation (4.8*b*) is an entire function of E (Moyer 1973), so that the associated functions given by equations (4.8*c*,*d*) can be defined unambiguously in the complex E plane. The energy-dependent KP eigenvalue is now obtained by using equations (3.26)–(3.30). The present treatment is identical to that of the variational principle, equation (3.24), with the use of the trial functions $\tilde{\xi}_n = \tilde{\eta}_n = \phi_n$. When the external field F is small, the level shift $\Delta_0^{(1)}(E)$ and the width $\gamma_0^{(1)}(E)$ are small and for $E \sim E_0^{(0)}$ are as follows:

$$\Delta_0^{(1)}(E) \sim (\hbar^2\alpha^2/m)d(1 + d)^{-1} - (\hbar^2\alpha^2/2m)d(1 + \tfrac{1}{2}d)(1 + d)^{-1/2} \times \{\exp[y^3(d^2 + \tfrac{2}{3}d^3)] + \exp[-y^3(d^2 + \tfrac{2}{3}d^3)]\} \tag{4.9a}$$

$$\gamma_0^{(1)}(E) \sim (\hbar^2\alpha^2/m)(1 + d)^{-1}\exp[-\tfrac{4}{3}y^3(1 + d)^3] + (\hbar^2\alpha^2/2m)d(1 + \tfrac{1}{2}d)(1 + d)^{-1/2} \times (1 + 2\pi^{1/2}y^{3/2})\exp[-\tfrac{4}{3}y^3(1 + 3d + \tfrac{9}{4}d^2 + \tfrac{1}{2}d^3)] \tag{4.9b}$$

where the F^2 term in $\Delta_0^{(1)}(E)$ is omitted here and for negative energy $E = -\hbar^2\kappa^2/2m$ we define the variables

$$\kappa = \alpha(1 + d) \tag{4.9c}$$

$$y = (\hbar^2/2mF)^{1/3}\alpha. \tag{4.9d}$$

The KP eigenvalue obtained by equations (4.9*a*,*b*) is shown in figures 1 and 2. The energy-dependent level shift $\Delta_0^{(1)}(E)$ changes smoothly around $E \sim E_0^{(0)}$. This indicates that, as far as the real energy E is considered, the variational calculation of the KP eigenvalue will yield a good result near the resonance energy. On the other hand, the width $\gamma_0^{(1)}(E)$ shows a rather sharp change. This suggests that the decay rate of the KP resonant state may change critically with the variation of the propagating energy of the particle in the asymptotic region. The F dependence of the KP eigenvalue is also shown in figures 1 and 2; the lower-energy side $E < E_0^{(0)}$ becomes gradually stabilised as F decreases, while the other side $E > E_0^{(0)}$ seems to be less stabilised. On the other hand, the decay rate diminishes uniformly.

Substituting $E = E_0^{(0)}$ into equations (3.31) and (3.32), one obtains the Siegert complex energy $W_0 = E_0 - \tfrac{1}{2}\mathrm{i}\Gamma_0$ at the particular value of the KP solution as follows:

$$E_0 \sim E_0^{(0)} + \Delta_0^{(1)}(E_0^{(0)}) = -E_0^{(0)} + 2E_0^{(0)}\pi\alpha(4\hbar^2/mF)^{1/3}\mathrm{Ai}(z^{(0)})\mathrm{Bi}(z^{(0)}) \tag{4.10}$$

$$\Gamma_0 \sim \gamma_0^{(1)}(E_0^{(0)}) = -4E_0^{(0)}\pi\alpha(4\hbar^2/mF)^{1/3}\mathrm{Ai}^2(z^{(0)}) \tag{4.11}$$

where $z^{(0)} = (\hbar^2/2mF)^{2/3}\alpha^2$. More refined results will be found by using equations (2.28) and (2.29) with the 'extended' Hellmann–Feynman theorems from equations (2.22) or

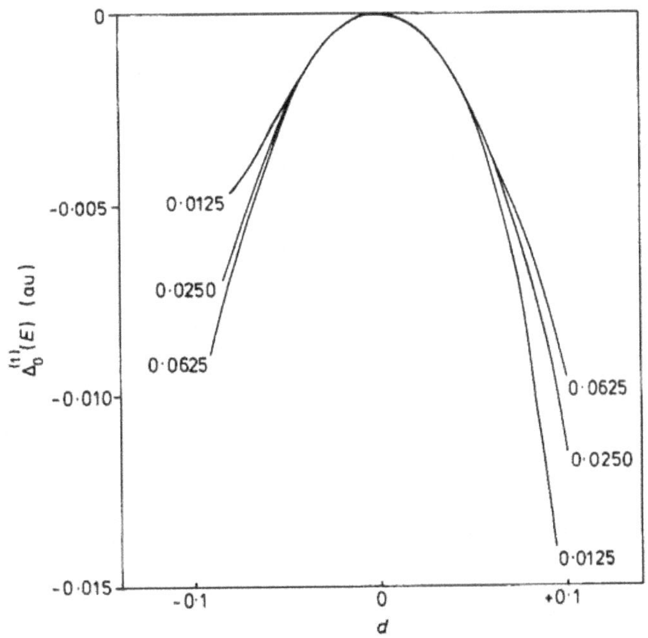

Figure 1. Level shift $\Delta_0^{(1)}(E)$ of the Kapur–Peierls eigenvalue with field strength $F = 0{\cdot}0125$, 0·0250 and 0·0625 (au). d is a dimensionless parameter defined by equation (4.9c): $d \gtrless 0$ corresponds to $E \lessgtr E_0^{(0)}$. α is chosen to be 1 au.

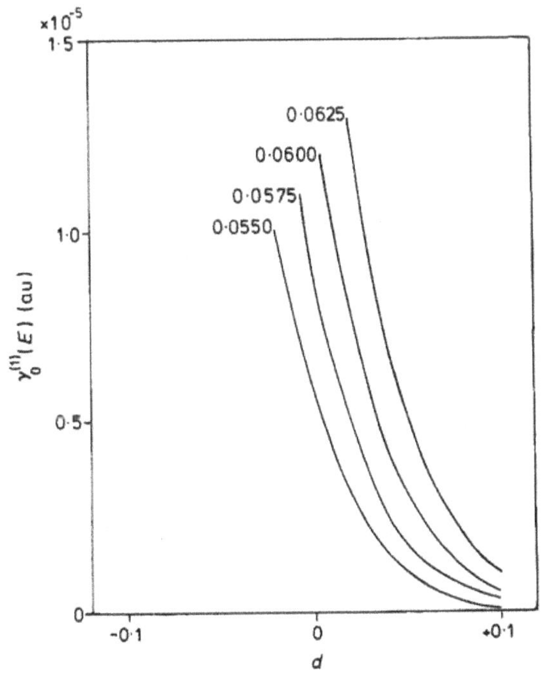

Figure 2. Width $\gamma_0^{(1)}(E)$ of the Kapur–Peierls eigenvalue with field strength $F = 0{\cdot}0550$, 0·0575, 0·0600 and 0·0625 (au). d is as defined in figure 1. α is chosen to be 1 au.

(3.12). In the weak-field limit, i.e. for large $z^{(0)}$, these are obtained in the asymptotic form as follows:

$$E_0 \sim E_0^{(0)} + \Delta_0^{(1)}(E_0^{(0)}) = E_0^{(0)} + 2E_0^{(0)} \sum_{n=1}^{\infty} \left(\frac{(6n-1)!!}{n!96^n\alpha^{6n}}\right)\left(\frac{2mF}{\hbar^2}\right)^{2n} \tag{4.12}$$

$$\Gamma_0 \sim \gamma_0^{(1)}(E_0^{(0)}) \sim (\hbar^2\alpha^2/m)\exp(-2\hbar^2\alpha^2/3mF). \tag{4.13}$$

Moyer (1973) has also shown the existence of the complex pole W_0 in this model problem and estimated it as $W_0 - E_0^{(0)} = O(F^2)$. In the present treatment, we have been able to find the higher-order F dependence of the resonance energy E_0 and the leading term of the width Γ_0. The exponential factor of Γ_0 is characteristic for the Stark problem.

Furthermore, at the state of approximation in equation (4.9*a*), we have

$$d\epsilon_0(E_0^{(0)})/dE_0^{(0)} \sim 0$$

$$d^2\epsilon_0(E_0^{(0)})/dE_0^{(0)2} \sim 1/E_0^{(0)} < 0.$$

Likewise, the imaginary part becomes

$$d\gamma_0^{(1)}(E_0^{(0)})/dE_0^{(0)} \sim (2\alpha/F)\Gamma_0 > 0.$$

Substituting these expressions into equations (2.28) and (2.29), we have the following results:

$$\epsilon_0(E_0) - E_0 \sim (\alpha/2F)\Gamma_0^2 + O(\Gamma_0^4) \tag{4.14}$$

$$\gamma_0(E_0)/\Gamma_0 \sim 1 + O(\Gamma_0^2). \tag{4.15}$$

It follows that the lifetime of the KP resonant state grows in the lower-energy side $E < E_0^{(0)}$ rather than in the other side $E > E_0^{(0)}$, and particularly at the resonance energy it coincides, with reasonable accuracy, with that of the Siegert resonant state.

Acknowledgment

This work was supported by a Grant in Aid (#139012) for Scientific Research from the Ministry of Education of Japan.

References

Antosiewicz H A 1965 *Handbook of Mathematical Functions* ed M Abramowitz and I A Stegun (Washington, DC: National Bureau of Standards)

Aspnes D E 1966 *Phys. Rev.* **147** 554–66

Bardsley J N, Herzenberg A and Mandl F 1966 *Proc. Phys. Soc.* **89** 305–19

Bhatia A K and Temkin A 1976 *Phys. Rev.* A **13** 2322–4

Blatt J M and Weiskopf V F 1952 *Theoretical Nuclear Physics* (New York: Wiley)

De Alfaro V and Regge T 1965 *Potential Scattering* (Amsterdam: North-Holland)

Epstein S T, Hurley A C, Wyatt R F and Parr R G 1967 *J. Chem. Phys.* **47** 1275–86

Fano U 1961 *Phys. Rev.* **124** 1866–78

Feshbach H 1962 *Ann. Phys., NY* **19** 287–313

Feynman R P 1939 *Phys. Rev.* **56** 340–3

Frey R B and Thiele E 1968 *J. Chem. Phys.* **48** 3240–5

Goldberger M L and Watson K M 1964 *Collision Theory* (New York: Wiley)
Hellmann H 1937 *Einführung in die Quantenchemie* (Leipzig and Vienna: Deuticke)
Huestis D L 1975 *J. Math. Phys.* **16** 2148–52
Kapur P and Peierls R 1938 *Proc. R. Soc.* A **166** 277–95
Krüger H and Levin F S 1976 *Phys. Lett.* **65B** 109–12
——1977 *Chem. Phys. Lett.* **46** 95–8
Lipsky L and Russek A 1966 *Phys. Rev.* **142** 59–71
Löwdin P O 1965 *Perturbation Theory and its Applications in Quantum Mechanics* ed C H Wilcox (New York: Wiley)
Lukes T and Somaratna K T S 1969 *J. Phys. C: Solid St. Phys.* **2** 586–92
Miller W H 1966 *Phys. Rev.* **152** 70–3
More R M 1971a *Phys. Rev.* A **3** 1217–20
——1971b *Phys. Rev.* A **4** 1782–90
More R M and Gerjuoy E 1973 *Phys. Rev.* A **7** 1288–303
Morse P M and Feshbach H 1953 *Methods of Theoretical Physics* (New York: McGraw-Hill)
Moyer C A 1973 *J. Phys. C: Solid St. Phys.* **6** 1461–6
Newton R G 1960 *J. Math. Phys.* **1** 319–47
——1966 *Scattering Theory of Waves and Particles* (New York: McGraw-Hill)
Nicolaides C A 1972 *Phys. Rev.* A **6** 2078–92
O'Malley T F and Geltman S 1965 *Phys. Rev.* **137** A1344–52
Peierls R 1948 *Proc. Camb. Phil. Soc.* **44** 242–50
Siegert A J F 1939 *Phys. Rev.* **56** 750–2
Tachibana A, Yamabe T and Fukui K 1977 to be published
Taylor H S, Nazaroff G V and Golebiewski A 1966 *J. Chem. Phys.* **45** 2872–88
Temkin A, Bhatia A K and Bardsley J N 1972 *Phys. Rev.* A **5** 1663–71
Whitcombe A R 1971 *J. Phys. C: Solid St. Phys.* **4** 1–2
Yamabe T, Tachibana A and Silverstone H J 1977a *J. Phys. B: Atom. Molec. Phys.* **10** 2083–100
——1977b *Phys. Rev.* A to be published

www.ingramcontent.com/pod-product-compliance
Lightning Source LLC
LaVergne TN
LVHW070906300926
849162LV00012B/1864
9789810222413